Human Factors In Computing Systems

Chi '95 Conference Proceedings

CHI '95
MOSAIC OF CREATIVITY

CONFERENCE ON
HUMAN FACTORS IN
COMPUTING SYSTEMS
MAY 7-11 1995
DENVER COLORADO USA

D0144505

Editors:
Irvin R. Katz
Robert Mack
Linn Marks

Mary Beth Rosson
Jakob Nielsen

Ordering Information

The Association for Computing Machinery, Inc.
1515 Broadway, 17th Floor
New York, NY 10036

Sample Citation Information: SIGCHI '95 (Denver, Colorado, May 7-11, 1995). In *Human Factors in Computing System* Proceedings, Annual Conference Series, 1995, ACM SIGCHI, pp. xx-yy.

ORDERING INFORMATION

Orders from nonmembers of ACM placed within the United States should be directed to:

Addison-Wesley Publishing Company Order Department
Jacob Way
Reading, MA 01867
Tel: 1-800-447-2226

Addison-Wesley will pay postage and handling on orders accompanied by check. Credit card orders may be placed by mail or by calling the Addison-Wesley Order Department at the number above. Followup inquiries should be directed to the Customer Service Department at the same number. Please include the Addison-Wesley ISBN number with your order:
  A-W Softcover Proceedings
  ISBN:0-201-84705-1

Orders from nonmembers of ACM placed from outside the United States should be addressed as noted below.

Europe/Middle East
Addison-Wesley Publishing Group
Concertgebouwplein 25
1071 LM Amsterdam
The Netherlands
  Tel: +31 20 6717296
  Fax: +31 20 6645334

Germany/Australia/Switzerland
Addison-Wesley Verlag Deutschland
GmbH
Hildachstraβe 15d
Wachsbleiche 7-12
53111 Bonn
Germany
  Tel: +49 228 98 515 0
  Fax: +49 228 98 515 99

United Kingdom/Africa
Addison-Wesley Publishers Ltd.
Finchampstead Road
Wokingham, Berkshire RG11 2NZ
United Kingdom
  Tel: +44 734 794000
  Fax: +44 734 794035

Asia
Addison-Wesley Singapore Pte. Ltd.
15 Beach Road
#05-02/09/10 Beach Centre
Singapore 0718
  Tel: +65 339 7503
  Fax: +65 339 9709

Japan
Addison-Wesley Publishers Japan Ltd.
Nichibo Building
1-2-2 Sarugakucho
Chiyoda-ku, Tokyo 101
Japan
  Tel: +81 33 2914581
  Fax: +81 33 2914592

Australia/New Zealand
Addison-Wesley Publishers Pty.Ltd.
6 Byfield Street
North Ryde, N.S.W. 2113
Australia
  Tel: +61 2 878 5411
  Fax: +61 2 878 5830

Latin America
Addison Wesley Iberoamericana S.A.
Boulevard de las Cataratas #3
Colonia Jardines del Pedregal
Delegacion Alvaro Obregon
01900 Mexico D.F.
  Tel: +52 5 660 2695
  Fax: +52 5 660 4930

Canada
Addison-Wesley Publishing (Canada) Ltd.
26 Prince Andrew Place
Don Mills, Ontario M3C 2T8 Canada
  Tel: 416-447-5101
  Fax: 416-443-0948

Orders from ACM Members:

A limited number of copies are available at the ACM member discount. Send order with payment in U.S. dollars to:

ACM Order Department
P.O. Box 12114
Church Street Station
New York, NY 10257

OR, for information on accepted European currencies and exchange rates, contact:

ACM European Service Center
Avenue Marcel Thiry 204
1200 Brussels
Belgium
  Tel: +32 2 774 9602
  Fax: +32 2 774 9690
  Email: acm_europe@acm.org

Credit card orders from U.S. and Canada:
  1-800-342-6626
Credit card orders may also be placed by mail.

Credit card orders from the New York metropolitan area and outside the U.S.:
  +1 212-626-0500

Single-copy orders placed by fax:
  +1 212-944-1318

Electronic mail inquiries may be directed to acmhelp@acm.org.

Please include your ACM member number and the ACM order number with your order.

ACM Order Number: 608950

ACM ISBN: 0-89791 694-8

From the ACM/SIGCHI Chair and Executive Vice-Chair

On behalf of the Executive Committee of ACM's Special Interest Group on Computer-Human Interaction [ACM/SIGCHI], welcome to CHI '95, SIGCHI's annual conference on Human-Computer Interaction (HCI). Each CHI conference offers a snapshot of the HCI field as it grows and evolves and, collectively, the Proceedings of these conferences provide a history of the growth and evolution of HCI. The nature of these Proceedings has also evolved and a version of the CHI '95 Proceedings was made available on CD-ROM to conference participants. We are grateful to the CHI '95 committee for taking this step and look forward to using what they learn this year in evolving Proceedings in future years.

While the CHI conference is likely the most visible activity of SIGCHI, we also:

- Support other conferences in Computer-Supported Cooperative Work, Multimedia, and User Interface Software and Technology.

- Support workshops on special topics, such as the First International Workshop on Intelligent User Interfaces and the upcoming Designing Interactive Systems workshop.

- Distribute the highly respected quarterly SIGCHI Bulletin to members and to others.

- Supported the "birth" and continue to support the evolution of two new publications - *Transactions on Computer-Human Interaction* (*TOCHI*), a member of the ACM Transactions series; and *interactions*, ACM's first magazine devoted to HCI practitioners.

- Provide support to a number of activities proposed by our members that provide benefit to both SIGCHI and the field of HCI.

Why do people join SIGCHI? SIGCHI is a worldwide community of individuals who share an interest in one or more of the diverse and evolving facets of HCI. This conference and other SIGCHI activities provide forums for these people to meet, both formally and informally, to exchange views, share experiences, and learn. We invite you to join and stay involved with this vibrant community. There is a SIGCHI membership application at the back of this volume. We also invite you to join with other volunteers in providing SIGCHI's leadership.

On behalf of ACM/SIGCHI, we thank and congratulate the conference committee and all the volunteers who make this conference possible for a job well done.

Welcome to CHI '95.

Jim Miller
ACM/SIGCHI Chair

Michael E. Atwood
ACM/SIGCHI Executive Vice Chair

From the Conference Co-Chairs

We are happy to welcome you to Denver, Colorado, for CHI '95. The field of human-computer interaction (HCI) has matured. A decade ago the field consisted entirely of new faces from other disciplines. Now companies employ HCI professionals, there are HCI consulting firms, journals and magazines are devoted to HCI issues, and universities offer degree programs in HCI. There are 'senior' and 'junior' members of the field. What does this mean for our field and our conference?

One result of the maturity of the field is enhanced creativity and impact. With time, people in the field of HCI have been able to absorb the techniques, philosophy, and practice of user-centered design and development. Once these skills are taken for granted, people can put them to work. Instead of using creativity largely to shape the field of HCI, much of our creative energy can now go into shaping the world of computing environments—a world in which more and more people live. The new Design Briefings track at CHI '95 shows the impact of HCI on real design teams and real computing environments.

Another result of the maturity of the field is increased diversity. The exposure of more types of professionals and academics to HCI means that it is attaining breadth. Different people understand HCI in different ways, bringing different perspectives to its theory and practice. This enriches the field and adds to the mosaic. Participants in the conference show an extraordinary range of professions. An examination of the application domains covered in the technical material reveals the scope of the field.

Will the field of HCI become too diverse, too big, too mature? As a result, will our view of the mosaic be limited to a very small piece? CHI conference participants and others interested in HCI face this challenge every day. This conference was designed to meet this challenge by exposing participants to the breadth of the field while still enabling them to focus on their specific areas.

As Co-Chairs, we were privileged to add more pieces to the mosaic. We are interested to see where you will take the field in the future.

Teresa Roberts and Scott Robertson
CHI '95 Conference Co-Chairs

Soyez les bienvenus

Velkommen

ようこそ

Welkom

Bien Venido

歡迎

dobro nam došli

vítáme Vás

환영

ברוכים הבאים

Bine aţi venit

Benvenuto

歡迎 Добро пожаловать

Välkommen

Üdvözölve

Ongi ettori

dobro nam došli

Benvenuto

ようこそ

Üdvözölve

Bien Venido

Ongi ettori

vítáme Vás

Welcome

Welkom

Velkommen

dobro nam došli

Willkommen

환영

Bine aţi venit!

CHI '95 is sponsored by ACM's Special Interest Group on Computer-Human Interaction (ACM/SIGCHI).

Cooperating Societies

ACM Special Interest Group
on Computer Graphics (ACM/SIGGRAPH)

ACM Special Interest Group
on Computers and the Physically Handicapped
(ACM/SIGCAPH)

ACM Special Interest Group on Office
Information Systems (ACM/SIGOIS)

Association for Artificial Intelligence of Russia
(RAAI)

Association Française des Sciences et
Technologies de l'information et des Systèmes
(AFCET)

Association pour la Recherche Cognitive
(ARC)

Austrian Computer Society (OCG)

British HCI Group (BCS-HI)

Cognitive Science Society

The Division of Applied Experimental
and Engineering Psychologists of the
American Psychological Association
(Div. 21 of APA)

Dutch Computer Society (NGI)

European Association of
Cognitive Ergonomics (EACE)

Gesellschaft für Informatik,
Fachgruppe Software-Ergonomic (GI)

Human Factors and
Ergonomics Society (HFES)

Human Factors and Ergonomics
Society, European Chapter
(HFES, Euro. Chapter)

Human Interface (SICE-HI Japan)

IEEE Computer Society

Italian Association for Artificial
Intelligence (AIIA)

Software Psychology Society
(Potomac Chapter)

Conference Co-Chairs
Teresa Roberts
U S WEST Advanced Technologies (USA)

Scott Robertson
U S WEST Advanced Technologies (USA)

CMC Liaison
John 'Scooter' Morris
Genentech, Inc. (USA)

ACM Liaison
Diane Darrow, CAE
ACM (USA)

Technical Program Co-Chairs
Clayton Lewis
University of Colorado (USA)

Peter Polson
University of Colorado (USA)

Demonstrations
Catherine Wolf
IBM T.J. Watson Research Center (USA)

Design Briefings
Jakob Nielsen
SunSoft, Inc. (USA)

Doctoral Consortium
Catherine Marshall
Collaborative Technologies (USA)

Interactive Experience
Eric Bergman
SunSoft, Inc. (USA)

Paul Reed
AT&T Bell Laboratories (USA)

Organization Overviews
Robert Rist
University of Technology, Sydney (AUS)

Jean Scholtz
Intel Corp. (USA)

Panels
S. Joy Mountford
Interval Research Corporation (USA)

Kurt Schmucker
Apple Computer, Inc. (USA)

Research Symposium
Janni Nielsen
Copenhagen Business School (DK)

Cathleen Wharton
U S WEST Advanced Technologies (USA)

Papers
Mary Beth Rosson
*Virginia Polytechnic Institute &
State University (USA)*

Short Papers and Posters
Stacey Ashlund
Lotus Development Corp. (USA)

Jonathan Grudin
University of California, Irvine (USA)

Muneo Kitajima
*National Institute of Bioscience and
Human Technology, Japan (J)*

Social Action Posters
Pamela Burke
AT&T Bell Laboratories (USA)

Michael Muller
U S WEST Advanced Technologies (USA)

Tutorials
Ruven Brooks
Schlumberger Austin Research (USA)

Larry Diamond
Schlumberger Austin Research (USA)

Videos
Angela Lucas
Logica Cambridge, Ltd. (UK)

Tracy Roberts
Bell Northern Research, Ltd. (CDN)

Workshops and SIGs
George Engelbeck
U S WEST Advanced Technologies (USA)

Wayne Gray
George Mason University (USA)

**Accompanying Persons
and Childcare**
Kate Ehrlich
Lotus Development Corp. (USA)

Carrie Rudman
U S WEST Advanced Technologies (USA)

Audio-Visual Support
Mark Wilkes
IBM Corp. (USA)

Computing Support
Chris Esposito
Boeing Computer Sciences (USA)

Exhibits
Michael E. Atwood
NYNEX Science & Technology, Inc. (USA)

Jay Elkerton
NYNEX Science & Technology, Inc. (USA)

Hypermedia Information Access
Keith Instone
Bowling Green State University (USA)

Industry Liaisons
Ian McClelland
Philips Corporate Design (NL)

Jean Scholtz
Intel Corp. (USA)

Local Arrangements
Barbara Diekmann
U S WEST Advanced Technologies (USA)

Merchandising
Rhona Charron
*Ontario Institute for Studies
in Education (CDN)*

Publications
Irvin R. Katz
Educational Testing Service (USA)

Robert Mack
IBM T.J. Watson Research Center (USA)

Linn Marks
IBM T.J. Watson Research Center (USA)

Publicity
Blanche Cohen
Teknikos (USA)

Registration
Steve Anderson
*Lawrence Livermore National
Laboratory (USA)*

Chris Ghinazzi
*Lawrence Livermore National
Laboratory (USA)*

Signage
Robert Fein
U S WEST Advanced Technologies (USA)

Melissa Schofield
Seattle University (USA)

Special Needs and Access
John Bennett
Independent Consultant (USA)

Student Volunteers
Chantal Kerssens
Universiteit van Amsterdam (NL)

Astrid Kerssens
Universiteit van Amsterdam (NL)

Steven Pemberton
*CWI: Dutch National Centre for
Research in Mathematics and
Computer Science (NL)*

Surveys and Evaluations
Lorraine Borman *(USA)*

Technical Support
Alan Edwards
Unisys (USA)

Treasurer
Thea Turner
NYNEX Science & Technology, Inc. (USA)

Regional Coordinators

Eastern Europe
Peter Brusilovsky
*International Centre for Scientific
and Technical Information (Russia)*

Western Europe
Françoise Détienne
INRIA (F)

North and South America
Susan McDaniel,
University of Michigan (USA)

Pacific Rim
Hirotada Ueda
Hitachi, Ltd. (J)

Professional Staff

Conference Management
Paul Henning
Conference & Logistics Consultants (USA)

Conference Administrator
Deborah Compere
Conference & Logistics Consultants (USA)

Public Relations
Rosemary Wick Stevens
Ace Public Relations (USA)

Publications Coordinator
Rick Gondella
Conference & Logistics Consultants (USA)

Design Firm
Energy Energy Design (USA)

Process Advisor
Carol Klyver
Foundations of Excellence (USA)

Papers Committees

Associate Papers Chairs
Ronald Baecker
University of Toronto (CDN)

Meera M. Blattner
*University of California, Davis, and
Lawrence Livermore National Laboratory
(USA)*

Sara A. Bly
Consultant (USA)

Joëlle Coutaz
*Laboratoire de Génie Informatique
(IMAG) (F)*

Alan Dix
University of Huddersfield (UK)

Marc Eisenstadt
The Open University (UK)

Ellen P. Francik
Pacific Bell (USA)

George W. Furnas
Bellcore (USA)

Michael D. Good
Xtensory Inc. (USA)

Wayne D. Gray
George Mason University (USA)

David R. Hill
University of Calgary (CDN)

H. Rex Hartson
*Virginia Polytechnic Institute &
State University (USA)*

Karen Holtzblatt
InContext Enterprises, Inc. (USA)

Robin Jeffries
SunSoft, Inc. (USA)

Jock Mackinlay
Xerox PARC (USA)

Michael J. Muller
U S WEST Advanced Technologies (USA)

Kumiyo Nakakoji
*University of Colorado at Boulder,
Software Research Associates, Inc.,
and Nara Advanced Institue of
Science and Technology (USA)*

Peter Pirolli
Xerox PARC (USA)

Gitta Salomon
IDEO Product Development (USA)

Lynn Streeter
U S West Advanced Technologies (USA)

Bradley T. Vander Zanden
University of Tennessee (USA)

Colin Ware
University of New Brunswick (CDN)

Papers Review Committee
Mark Abel
Intel Corp. (USA)

Beth Adelson
Rutgers University (USA)

Klaus H. Ahlers
*European Computer-Industry Research
Centre (ECRC) (D)*

Bengt Ahlström
Royal Institute of Technology, Stockholm (S)

Robert B. Allen
Bellcore (USA)

Carl Martin Allwood
University of Göteborg (S)

Mark W. Altom
AT&T Bell Laboratories (USA)

James L. Alty
*Loughborough University of Technology
(UK)*

Richard I. Anderson
Usability Adventures Consulting (USA)

Mark Apperley
University of Waikato (NZ)

Udo Arend
SAP AG (D)

Stacey Ashlund
Lotus Development Corp. (USA)

Liam J. Bannon
University of Limerick (IE)

Mathilde M. Bekker
Delft University of Technology (NL)

Matt Belge
Vision & Logic (USA)

Rachel Bellamy
Apple Computer, Inc. (USA)

Lucy Berlin
Taligent, Inc. (USA)

Nigel Bevan
National Physical Laboratory (UK)

Randolph G. Bias
IBM Corporation (USA)

Andre Bisseret
INRIA (F)

Brad Blumenthal
University of Illinois at Chicago (USA)

Susanne Bodker
Aarhus University (DK)

Heinz-Dieter Böecker
*German National Research Center for
Computer Science and Information
Technology (D)*

Kellogg Booth
University of British Columbia (CDN)

Dr. P. A. Booth
University of Salford (UK)

Susan Bovair
Georgia Institute of Technology (USA)

Guy Boy
EURISCO (F)

Jerome Broekhuijsen
*WordPerfect, The Novell Applications
Group (USA)*

John Brooke
Digital Equipment Corp. Ltd. (UK)

Ruven Brooks
Schlumberger Austin Research (USA)

Maddy D. Brouwer-Janse
Philips Research Laboratories - IPO (NL)

Marc H. Brown
*Digital Equipment Corp. Systems Research
Center (USA)*

Hans Brunner
U S WEST Advanced Technologies (USA)

Peter Brusilovsky
ICSTI (Russia)

Mary Beth Butler
Lotus Development Corp. (USA)

Stuart Card
Xerox PARC (USA)

John M. Carroll
*Virginia Polytechnic Institute &
State University (USA)*

Richard Catrambone
Georgia Institute of Technology (USA)

Joan M. Cherry
University of Toronto (CDN)

Gilbert Cockton
University of Glasgow (UK)

Ellis S. Cohen
Open Software Foundation (USA)

Bill Curtis
TeraQuest Metrics & SEI-CMU (USA)

Catalina Danis
IBM T.J. Watson Research Center (USA)

Mary Carol Day
AT&T Bell Laboratories (USA)

Tom Dayton
Bellcore (USA)

Maurizio De Cecco
Ircam (F)

Françoise Détienne
INRIA (F)

Keith R. Dickerson
BT/RACE Industrial Consortium (BLG)

Stephanie Doane
University of Illinois (USA)

Miwako Doi
Toshiba Corporation (J)

Sarah Douglas
University of Oregon (USA)

Paul Dourish
Rank Xerox EuroPARC (UK)

Susan M. Dray
Dray & Associates (USA)

Susan T. Dumais
Bellcore (USA)

Wolfgang Dzida
*GMD-System Design Technology
Institute (D)*

Jonathan V Earthy
Lloyd's Register (UK)

Alan Edwards
Unisys (USA)

Roger Ehrich
*Virginia Polytechnic Institute &
State University (USA)*

Kate Ehrlich
Lotus Development Corp. (USA)

Michael Eisenberg
University of Colorado (USA)

Jay Elkerton
NYNEX Science & Technology, Inc. (USA)

Thomas Erickson
Apple Computer, Inc. (USA)

Steven K. Feiner
Columbia University (USA)

Steve Fickas
University of Oregon (USA)

Rob Fish
Bellcore (USA)

James D. Foley
Georgia Institute of Technology (USA)

Stephen Freeman
*Rank Xerox Research Centre,
Grenoble France (F)*

Sharon R. Garber
3M (USA)

Bill Gaver
Royal College of Art (UK)

Don Gentner
SunSoft, Inc. (USA)

Andreas Girgensohn
NYNEX Science & Technology Inc. (USA)

Robert J. Glushko
Passage Systems, Inc. (USA)

Louis M. Gomez
Northwestern University (USA)

Peter Gorny
University of Oldenburg (D)

Saul Greenberg
University of Calgary (CDN)

Raymonde Guindon
Hewlett-Packard (USA)

Nils-Erik Gustafsson
ELLEMTEL Telecom Systems Labs (S)

Volker Haarslev
University of Hamburg (D)

Simon R. Hakiel
IBM UK Laboratories Ltd. (UK)

Nick Hammond
University of York (UK)

Mary L. Hardzinski
Ameritech (USA)

Jean-Paul Haton
University of Nancy I and CRIN/INRIA (F)

Anker Helms Jørgensen
Copenhagen University (DK)

Scott Henninger
University of Nebraska-Lincoln (USA)

James D. Herbsleb
Software Engineering Institute (USA)

David R. Hill
University of Calgary (CND)

Ralph D. Hill
Bellcore (USA)

William C. Hill
Bellcore (USA)

Ken Hinckley
University of Virginia (USA)

Hans-Jüergen Hoffmann
University at Darmstadt (D)

James D. Hollan
University of New Mexico (USA)

Erik Hollnagel
Human Reliability Associates Ltd. (UK)

H. Ulrich Hoppe
GMD-IPSI (D)

Stephanie L. Houde
Apple Computer, Inc. (USA)

Andrew Howes
University of Cardiff (UK)

Scott Hudson
Georgia Institute of Technology (USA)

Keith Instone
Bowling Green State University (USA)

Hiroshi Ishii
NTT Human Interface Laboratories (J)

Hiroo Iwata
University of Tsukuba (J)

Robert J.K. Jacob
Tufts University (USA)

Pertti Jarvinen
University of Tampere (SF)

Bonnie E. John
Carnegie Mellon University (USA)

Jeff Johnson
Sun/First Person (USA)

Clare-Marie Karat
IBM T.J. Watson Research Center (USA)

John Karat
IBM T. J. Watson Research Center (USA)

Demetrios Karis
GTE Labs (USA)

Solange Karsenty
DEC Paris Research Laboratory (F)

Irvin R. Katz
Educational Testing Service (USA)

Dr. Paul Kearney
Sharp Laboratories of Europe (UK)

Reinhard Keil-Slawik
*Heinz Nixdorf Institute,
University of Paderborn (D)*

Rudolf K. Keller
Université de Montréal (CDN)

Wendy A. Kellogg
IBM T. J. Watson Research Center (USA)

Susan Kirschenbaum
*Naval Undersea Warfare Center Division
Newport (USA)*

Muneo Kitajima
*National Institute of Bioscience and
Human-Technology (J)*

Alfred Kobsa
University of Konstanz (D)

Marja-Riitta Koivunen
Helsinki University of Technology (SF)

Joseph Konstan
University of Minnesota (USA)

David Kurlander
Microsoft Research (USA)

Masaaki Kurosu
Hitachi Ltd. (J)

Gordon Kurtenbach
Alias Research (CDN)

Hideaki Kuzuoka
University of Tsukuba (J)

Alison Lee
NYNEX Science & Technology, Inc. (USA)

John J. Leggett
Texas A&M University (USA)

Andreas C. Lemke
Alcatel SEL (D)

Mark Linton
Silicon Graphics, Inc. (USA)

Gerald L. Lohse
*The Wharton School of the
University of Pennsylvania (USA)*

John Long
University College London (UK)

Jonas Löwgren
Linköping University (S)

Arnold M. Lund
Ameritech (USA)

Gene Lynch
Tektronix, Inc. (USA)

Robert L. Mack
IBM T.J. Watson Research Center (USA)

Wendy E. Mackay
Rank Xerox EuroPARC (F)

Allan MacLean
*Rank Xerox Research Centre,
Cambridge (UK)*

Gary Marchionini
University of Maryland (USA)

Aaron Marcus
Aaron Marcus and Associates, Inc (USA)

Linn Marks
IBM T.J. Watson Research Center (USA)

Hans Marmolin
UI Design AB (S)

Kevin A. Mayo
Science Applications International Corporation (SAIC) (USA)

Jean McKendree
University of York (UK)

Cliff McKnight
Loughborough University of Technology (UK)

William W. McMillan
Eastern Michigan University (USA)

Jim Miller
Apple Computer, Inc. (USA)

Naomi Miyake
Chukyo University (J)

Tom Moher
University of Illinois at Chicago (UK)

Johanna Moore
University of Pittsburgh (USA)

Thomas P. Moran
Xerox PARC (USA)

John 'Scooter' Morris
Genentech, Inc. (USA)

Jane N. Mosier
MITRE Corp. (USA)

Bob Mulligan
AT&T Bell Laboratories (USA)

Alice M. Mulvehill
MITRE Corp. (USA)

Brad A. Myers
Carnegie Mellon University (USA)

Manfred Nagl
Aachen University of Technology (D)

Yasushi Nakauchi
National Defense Academy (J)

Frieder Nake
Universität Bremen (D)

Christine Neuwirth
Carnegie Mellon University (USA)

William Newman
Rank Xerox Research Centre (UK)

Janni Nielsen
Copenhagen Business School (DK)

Erik L. Nilsen
Lewis & Clark College (USA)

Haruhiko Nishiyama
Keio University (J)

Lorraine F. Normore
Chemical Abstracts Service (USA)

Mark Notess
Hewlett-Packard Company (USA)

David G. Novick
Oregon Graduate Institute (USA)

Else Nygren
Uppsala University, Center for Human-Computer Studies (S)

Horst Oberquelle
Hamburg University (D)

Dan R. Olsen Jr.
Brigham Young University (USA)

Gary M. Olson
University of Michigan (USA)

Judith Olson
University of Michigan (USA)

Tim Oren
CompuServe, Inc. (USA)

Sharon Oviatt
Oregon Graduate Institute of Science & Engineering (USA)

John F. Patterson
Lotus Development Corp. (USA)

Richard W. Pew
BBN Inc. (USA)

Ken Pier
Xerox PARC (USA)

Steven E. Poltrock
Boeing Computer Services (USA)

Karen Rafnel
Intel Corp. (USA)

Robert J. Remington
Lockheed Missiles & Space Company (USA)

Douglas Riecken
AT&T Bell Laboratories (USA)

John Rieman
University of Colorado (USA)

Robert S. Rist
University of Technology, Sydney (AUS)

George G. Robertson
Xerox PARC (USA)

Robert W. Root
Bellcore (USA)

Steven Roth
Carnegie Mellon University (USA)

Peter L. Rowley
The Ontario Institute for Studies in Education (CDN)

Richard Rubinstein
Cognitive Construction Co. (USA)

Carrie Rudman
U S WEST Advanced Technologies (USA)

Alexander I. Rudnicky
Carnegie Mellon University (USA)

Pamela Samuelson
University of Pittsburgh (USA)

Angela Sasse
University College London (UK)

Mark Schlager
SRI International (USA)

Franz Schmalhofer
DFKI (D)

Chris Schmandt
MIT Media Lab (USA)

Jean Scholtz
Intel Corp. (USA)

Andrew Sears
DePaul University (USA)

Chris Shaw
University of Alberta (CDN)

Gurminder Singh
National University of Singapore (SGP)

Gerda Smets
Delft University of Technology (NL)

Randall Smith
Sun Microsystems Laboratories (USA)

Elliot Soloway
University of Michigan (USA)

James F. Sorce
GTE Laboratories (USA)

James C. Spohrer
Apple Computer, Inc. (USA)

Jared M. Spool
User Interface Engineering (USA)

Loretta Staples
U dot I, Inc. (USA)

Karl E. Steiner
The University Of Illinois at Chicago (USA)

Paulien F. Strijland
Apple Computer, Inc. (USA)

Piyawadee 'Noi' Sukaviriya
Georgia Institute of Technology (USA)

Kent Sullivan
Microsoft Corporation (USA)

Alistair Sutcliffe
City University (UK)

Pedro Szekely
USC/Information Sciences Institute (USA)

Gerd Szwillus
Universitaet - GH - Paderborn (D)

Haruo Takemura
Nara Institute of Science & Technology (J)

John C. Tang
SunSoft, Inc (USA)

Masayuki Tani
Hitachi, Ltd. (J)

Michael J. Tauber
University of Paderborn (D)

Loren Terveen
AT&T Bell Laboratories (USA)

Linda Tetzlaff
IBM T.J. Watson Resesarch Center (USA)

John C. Thomas
NYNEX Science & Technology, Inc. (USA)

Thea Turner
NYNEX Science & Technology, Inc. (USA)

Claus Unger
University of Hagen (D)

Charles van der Mast
Delft University of Technology (NL)

Hans van der Meij
University of Twente (NL)

Gerrit C. van der Veer
Vrije Universiteit (NL)

Floris L. van Nes
*Institute for Perception Research/IPO -
Philips Research Laboratories (NL)*

Bill Verplank
Interval Research Corporation (USA)

Kim Vicente
University of Toronto (CDN)

Robert A. Virzi
GTE Laboratories, Inc. (USA)

Willemien Visser
INRIA (F)

Yvonne Waern
Linköping University (S)

Clive P. Warren
*British Aerospace (Operations) Ltd.,
Sowerby Research Center (UK)*

John A. Waterworth
University of Umea (S)

Gerhard Weber
University of Stuttgart (D)

Alan Wexelblat
MIT Media Lab (USA)

Steve Whittaker
Lotus Development Corp. (USA)

Mark Wilkes
IBM Corp. (USA)

Beverly Williges
*Virginia Polytechnic Institute &
State University (USA)*

Robert Williges
*Virginia Polytechnic Institute &
State University (USA)*

Russel Winder
University College London (UK)

Terry Winograd
Stanford University (USA)

Ian H. Witten
University of Waikato (NZ)

Dennis Wixon
Digital Equipment Corp. (USA)

Peter Wright
University of York (UK)

Nicole Yankelovich
Sun Microsystems Laboratories (USA)

Richard M Young
MRC Applied Psychology Unit (UK)

Jürgen Ziegler
Fraunhofer Institute IAO (D)

Design Briefings Review Committee

Karen Bedard
SunSoft, Inc. (USA)

Kathryn Best
Resonant Reality, Inc. (USA)

Patricia A. Billinsley
The Merritt Group (USA)

Stephen A. Brewster
VTT Information Technology (SF)

Brenda J. Burkhart
Bellcore (USA)

Allison Druin
NYU Media Research Lab (USA)

Tom Erickson
Apple Computer, Inc. (USA)

Michael Harris
*AT&T Human Interface Technology
Center (USA)*

Austin Henderson
Apple Computer, Inc. (USA)

Shifteh Karimi
Apple Computer, Inc. (USA)

Masaaki Kurosu
Hitachi Ltd. (J)

David Leffler
General Magic (USA)

Daniel T. Ling
Microsoft Corp. (USA)

Arnold Lund
Ameritech (USA)

Stephen M. Madigan
Microsoft Corp. (USA)

Mark Malamud
Microsoft Corp. (USA)

Michael McDevitt
Microsoft Corp. (USA)

Naomi Miyake
Chukyo University (J)

Michael F. Mohageg
Silicon Graphics, Inc. (USA)

Rolf Molich
Kommunedata (DK)

Osamu Morikawa
*National Institute of Bioscience
and Human-Technology (J)*

Kevin Mullet
Macromedia, Inc. (USA)

Shogo Nishida
Mitsubishi Electric Corporation (J)

Terje Norderhaug
Media Design inProgress (USA)

Kimberly O'Brien
SunSoft, Inc. (USA)

Kara Pernice
Lotus Development Corp. (USA)

Dick Rijken
Utrecht School of the Arts (NL)

Daniel Rosenberg
Oracle Corporation (USA)

Hugh Rubin
Microsoft Corp. (USA)

Keiichi Sato
Kyoto Institute of Technology (J)

Sara Sazegari
Taligent, Inc. (USA)

Lauren Schwartz
Microsoft Consulting Services (USA)

Eviatar Shafrir
Hewlett-Packard Corp. (USA)

Tim Shea
Marcam Corp. (USA)

Tim Skelly
Microsoft Corp. (USA)

Harald Stegavik
SINTEF (NO)

Desiree Sy
Information Design Solutions (CDN)

Talin
The Dreamers Guild, Inc. (USA)

Carlos Teixiera
CCG/ZGDV (PT)

Akifumi Tokosumi
Tokyo Institute of Technology (J)

Harry Vertelney
*Pacific Telesis Electronic
Publishing Services (USA)*

Maria G. Wadlow
Transarc Corporation (USA)

Annette Wagner
SunSoft, Inc. (USA)

Brad Weed
Microsoft Corp. (USA)

Ellen White
Bellcore (USA)

Chauncey E. Wilson
Human Factors International, Inc. (USA)

Terry Winograd
Stanford University (USA)

Irene Wong
Apple Computer, Inc. (USA)

Kristina Hooper Woolsey
Apple Computer, Inc. (USA)

Acknowledgements
CHI '95 acknowledges the work of Peter Foltz of the University of Pittsburgh and Adrienne Lee of New Mexico State University for production of the Keyword Index.

Photographs courtesy of The Denver Metro Convention and Visitors Bureau.

Special Thanks

The CHI '95 conference would not be possible without the efforts of the conference committee volunteers. Thanks to all these individuals and the following institutions whose support has made their participation possible:

Apple Computer, Inc.

AT&T

Bell Northern Research

Boeing

Bowling Green State University

Collaborative Technologies

Copenhagen Business School

CWI

Educational Testing Service

Genentech, Inc.

George Mason University

IBM

Intel Corporation

Interval Research Corporation

Lawrence Livermore National Laboratories

Logica Cambridge, Ltd.

National Institute of Bioscience and
    Human Technology, Japan

NYNEX

Ontario Institute for Studies in Education

Philips

Schlumberger

SunSoft, Inc.

Teknikos

U S WEST

Unisys

Universiteit van Amsterdam

University of Colorado

University of Technology, Sydney

Virginia Polytechnic Institute & State University

CHI '95 acknowledges the conference corporate sponsors for their generous support:

ENERGY ENERGY DESIGN

NYNEX

SunSoft
A Sun Microsystems, Inc. Business

XEROX

Like the conference itself, the CHI '95 Conference Proceedings and the CHI '95 Conference Companion form a mosaic of many topics and many modes of presentation. They present state-of-the-art work in human-computer interaction (HCI) from around the world. Each of the many elements of the technical program, showcased in these volumes, is described briefly on this page.

The CHI '95 Conference Proceedings contains the Technical Papers. Over the years, papers in the CHI Conference Proceedings have been an important archival resource for the field, in part because they are rigorously refereed. The CHI '95 Proceedings also contains Design Briefings, a new feature of the conference program which may take its place beside the papers as an important resource for the CHI community.

The CHI '95 Conference Companion contains the Short Papers and summaries of all the other parts of the technical program. The Companion volume helps participants recall the sessions they attended and provides a point of reference for others seeking information and opportunities related to their interests in HCI.

Clayton Lewis
Peter Polson
CHI '95 Technical Program Co-Chairs

In the Conference Proceedings:

Design Briefings
Design Briefings present notable user interface designs. Special emphasis is placed on conceptual issues embodied in the designs and on the design and evaluation methods used during development. Design Briefings are rigorously reviewed by an international committee. The full texts are published. The Design Briefings portion of the Table of Contents begins on page viii.

Papers
Papers are thorough descriptions of innovative and important work in the research and practice of HCI. Papers are rigorously reviewed by an international committee of experts from academics and industry. The full texts are published. The Papers portion of the Table of Contents begins on page iii.

In the Conference Companion:

Demonstrations
Many systems are demonstrated at the CHI conference. Summaries of the Demonstrations briefly explain the important aspects of these systems.

Doctoral Consortium
The Doctoral Consortium is a meeting of Ph.D. candidates who have completed their dissertation proposals in HCI or a closely related field. It is the future of HCI. Participants' thesis summaries are published.

Interactive Experience
This event is an ongoing, hands-on, exploratory activity that features the latest in unusual concepts and creations for interaction between humans and computers. Summaries describe the installations at this event.

Interactive Posters
Interactive Posters are visual presentations on display throughout the conference. They are reviewed by a committee and summaries of their content are published.

Organization Overviews
These summaries describe the work of leading organizations in the HCI field, including product development and research centers.

Panels
Panels offer discussion and controversy on emerging ideas and views in HCI. Panel summaries include a list of the panelists and their positions.

Plenaries
The Plenary addresses are invited presentations by prominent members of the field. The addresses open and close the conference, setting a theme and offering a challenge to people interested in HCI.

Short Papers
Late-breaking work and emerging concepts are described in short papers. A review committee carefully selects exciting Short Papers at the closest possible date to the conference.

Special Interest Groups
Special Interest Groups (SIGs) are gatherings that enable individuals sharing a common interest to meet informally for discussion. SIGs are among the most spontaneous conference events, combining the technical with the social.

Tutorials
Tutorials are short courses that provide opportunities to develop new skills and knowledge through extended interaction with expert instructors. The tutorial summaries contain the instructors' names and descriptions of the courses.

Videos
Summaries of the material in the Video Program provide descriptions of these highly visual and auditory presentations.

Workshops
Workshops provide an extended forum for small groups to exchange ideas on a specific topic of common interest. Workshop participants are selected based on position statements. The Research Symposium is a two-day meeting of HCI researchers who discuss a variety of HCI-related research ideas.

Display Navigation by an Expert Programmer: A Preliminary Model of Memory

Erik M. Altmann

Computer Science
Carnegie Mellon University
412-268-5728
altmann@cs.cmu.edu

Jill H. Larkin

Psychology
Carnegie Mellon University
412-268-3785
jhl@cs.cmu.edu

Bonnie E. John

Computer Science, Psychology,
and HCI Institute
Carnegie Mellon University
412-268-7182
bej@cs.cmu.edu

ABSTRACT

Skilled programmers, working on natural tasks, navigate large information displays with apparent ease. We present a computational cognitive model suggesting how this navigation may be achieved. We trace the model on two related episodes of behavior. In the first, the user acquires information from the display. In the second, she recalls something about the first display and scrolls back to it. The episodes are separated by time and by intervening displays, suggesting that her navigation is mediated by long-term memory, as well as working memory and the display. In the first episode, the model automatically learns to recognize what it sees on the display. In the second episode, a chain of recollections, cued initially by the new display, leads the model to imagine what it might have seen earlier. The knowledge from the first episode recognizes this image, leading the model to scroll in search of the real thing. This model is a step in developing a psychology of skilled programmers working on their own tasks.

KEYWORDS: Psychology of programming, user models, expert programmers, display navigation, program comprehension, memory, learning, Soar.

INTRODUCTION

Skilled programmers in natural task environments navigate large volumes of information, including code and streams of output from the programs they use. As we discuss, this navigation ability requires a programmer to encode (learn) knowledge specific to the situation. For example, a programmer would learn that some important information once appeared on the computer screen but is now hidden.

We observed a skilled programmer stepping interactively through a program trying to comprehend it in detail before changing it. Here we examine two related episodes of behavior from this session. In the first, she attends to a feature on the display. In the second, she is reminded of this feature. Though it has long since disappeared from view, she finds it readily, by scrolling the buffer she is working in.

Behavior of this kind raises the following questions: What kinds of memory for the display do programmers encode? How does a programmer gain access to such memories, and how do they lead to navigation? We discuss a cognitive computational model with knowledge and mechanisms that could account for such behavior.

This work is a step in developing a psychology of skilled programmers working on their own tasks. In particular, we address the encoding of situation knowledge and its use in navigation. Ultimately, a detailed understanding of such mechanisms could speak to the design of environments and languages for efficient expert use.

Task and Data

The programming session we observed was part of an ongoing effort to create a natural-language comprehension program using Soar, a production-based cognitive-modeling architecture [12]. The task the programmer set herself for this session was to comprehend and modify a part of this program originally written by a colleague.

The programmer displayed and edited code files, and ran her program interpretively in a GNU Emacs process buffer. During most of the session she stepped through the program slowly, issuing commands to display program state. The interpreter output, including long traces, appeared at the bottom of the process buffer, with Emacs automatically scrolling the buffer when more room was needed to display the output.

We recorded a think-aloud protocol, and captured the display and the user's gestures on video. We instrumented Emacs to record a timed keystroke protocol, the contents of the process buffer, and code files examined and edited.

Global Observations

To return to previously-displayed information, the programmer used *scrolling* and *string searching*, with scrolling predominant. We define a *scrolling sequence* as consecutive, same-direction scrolling commands. The user issued 26 such sequences in the 81-minute session, or roughly one sequence every 3 minutes. In total she scrolled 2482 lines of text through her 60-line window, or roughly 41 pages. Figure 1 shows the distribution by sequence length. Most sequences (14) covered only one page, and the longest sequence covered only 6 pages. This pattern has the skew one would expect, with no extremely long sequences biasing the overall count.

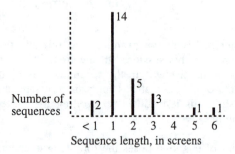

Figure 1: Scrolling sequences, by sequence length.

Scrolling relied heavily on long-term memory. In two-thirds of the sequences (17 out of 26), the target information had not been on the display in the past minute. There is little chance in these cases that working memory or any visual scratchpads could account for the user's memory of what she was looking for.

String searching was used only three times. One instance succeeded, with roughly 2 pages between start position and target. The two other instances failed, with the string not found. Both times the user then tried scrolling; both scrolling sequences also failed, one after 3 pages and the other after 6 pages. While the very limited use of methods may seem surprising, it is consistent with a finding that experienced users use small subsets of the commands available to them in an editor, ignoring even important cursor-movement commands [14].

These data suggest that scrolling in particular and navigation in general are common during real programming. Understanding the underlying mental processes could have practical importance for the design of navigation support.

An Example of Navigation

From our data we chose two related episodes, introduced in Figure 2. The display areas examined by the user are on the left, the user's utterances are on the right, and the time course for the utterances runs down the center. The task-specific references in the utterances will be defined later, as needed.

In E1, the user is looking at a *chunk*, or production. She expresses a lack of knowledge about its *tests*, or conditions (t18). She examines them, then notes an *operator test* (t21).

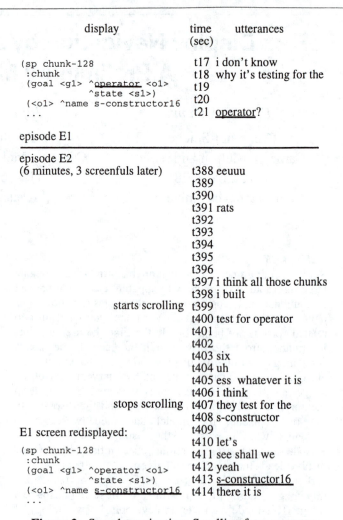

Figure 2: Sample navigation: Scrolling from memory.

Later, in E2, she recalls something about an operator test (t400). This prompts her to scroll back to the screen from E1.

The two episodes are 6 minutes apart, in which time the user issued 2 other scrolling sequences and 28 commands to the Soar interpreter. The output from these commands added 3 new pages to the bottom of the process buffer. Thus there is enough separation, in terms of time, tasks, and display changes, to indicate that the user's knowledge of the E1 display is encoded in her long-term memory.

The goal of this paper, and of the model we describe next, is to explain how this memory was encoded, what it represents, and how it was activated.

A COGNITIVE COMPUTATIONAL MODEL

Our model has three components that combine to produce an explanation of the behavior in the example above: knowledge, an encoding mechanism provided by an underlying cognitive architecture, and mechanisms for memory retrieval. Knowledge is the primary ingredient in most performance models of human behavior in HCI. The encoding mechanism automatically encodes information about the programming session into long-term memory, accounting for the memory implied by our user's behavior. Retrieval mechanisms on top of the architecture produce the effortful retrieval process that accounts for her navigation.

Knowledge

Our model has access to three distinct kinds of knowledge: *expert*, *external*, and *situation*. Expert knowledge is what we would expect a skilled programmer to bring to a programming task. This comprises knowledge about the particular program to be modified, including specific data structures and functions; knowledge about the implementation language, including its central concepts and idioms; and knowledge of computer science fundamentals, like data structures and algorithms. Such knowledge is typically found in expert systems and other symbolic AI programs. Expertise also includes knowledge of the programming environment, including procedures for navigation. This is the kind of expertise represented in the operators, methods, and selection rules of GOMS models [3].

Our model has expert knowledge about objects and their relations. Figure 3 shows the hierarchy. Some of this knowledge is specific to the user's *program*, and some of it is general to the *language* the program is implemented in (Soar). The model also has expertise about scrollable buffers; specifically, that if something was once on the screen but isn't now, scrolling through previous pages will bring it back into view.

External knowledge rests in the display, coordinating with internal knowledge to extend a problem-solver's effective memory [4, 5, 9, 10]. It has an immediately-available component, which is visible. It also has a hidden component, which can be made accessible by navigation. Computational models that address the display as external knowledge often treat only the immediately-available component(e.g., [1, 5, 8]).

Situation knowledge describes a particular session with the interpreter. For example, our user's navigation behavior shows that she *knows* she has seen something earlier in the session, and also that she has some idea of what objects she tried to comprehend earlier in the session. Situation knowledge is an important driver of navigation through her

process buffer, because navigation is often in pursuit of something she has seen before. Since this information only arises in the course of a session, for it to influence behavior it must be learned on the fly. Although HCI has investigated the acquisition of expert procedural knowledge (e.g., [6, 15, 16, 18]), this encoding and retrieval of situation knowledge has not yet appeared in many user modeling efforts.

Encoding Situation Knowledge

Our programmer seems to put little effort into encoding situation knowledge. It could be that she somehow keeps all such knowledge in working memory for a long time while juggling other tasks, but this is not cognitively plausible. An alternative is to posit a limited working memory, dictating that the programmer must externalize situation knowledge onto paper or the display [4, 5, 17]. However, there is no evidence that our programmer externalizes knowledge. The remaining alternative is that some process encodes situation knowledge into long-term memory (LTM) automatically, without any effort that would surface in a verbal protocol.

To get this kind of automatic encoding in our model, we have adopted Soar [12] as the underlying cognitive architecture. This gives us integrated learning and performance. It also affords the opportunity to connect to other relevant Soar models; such connections have led to improved and integrated coverage of complex user data [11].

As it functions in our model, Soar proceeds by trying to comprehend program objects, one at a time. For brevity, we refer to the object being comprehended as the *goal*. Soar tries to comprehend the goal by generating knowledge, either by retrieving knowledge from its own LTM, or by consulting external memory, or by some combination. This generated knowledge accumulates in working memory (WM), an ephemeral store created anew for each goal.

As a side-effect of generating new knowledge, Soar encodes a new rule in LTM, to avoid having to generate this knowledge again. This rule transforms existing knowledge (its conditions) into new knowledge (its actions). The situation rule can only activate when the existing knowledge is present in WM. Situation rules can be sensitive to both display features and goals. Display conditions arise when the model looks to the display for new knowledge. Goal conditions arise because the object being comprehended typically helps determine what new knowledge is relevant.

Situation rules depend on highly-specific WM contents. To make use of these rules requires mechanisms, on top of the architecture, that try to activate rules by intelligently generating WM contents.

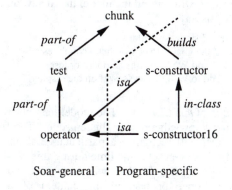

Figure 3: Objects and relations in the user's program.

Retrieving Situation Knowledge

With Soar as the underlying architecture, the model automatically learns rules that recognize specific situations. But how can we get the model to *recall* something later when the goal or the display has changed? Just as with people, recall is more difficult than recognition, and our model must use a more complex mechanism.

The model has to search LTM intelligently, or it will never activate anything. We refer to this guided search as *probing*. At a high level, probing is similar to Rist's cue-based memory search [17]. It also corresponds closely to the mechanistic aspects of the Model Human Processor's long-term memory [3].

Probing generates two kinds of cues: *semantic* and *image*. These correspond to the two kinds of conditions in situation rules: goals and display features. Semantic cues are objects that the model may have tried to comprehend during the programming session, and so might know something about. Image cues are objects that the model may have seen on the display. Probing generates these cues by drawing upon expert knowledge about what cues may be useful.

In principle, probing can activate any of the model's knowledge, both expert and situation. The extent to which it does in practice depends on its success at generating cues. With respect to Figure 3, the more objects and relations the model has access to, the better it will be able to search its LTM. Thus as a general problem-solving mechanism, probing has the important property that it can transform additional expertise into better performance.

The components of our model are knowledge (expert, external, and situation), automatic encoding (to learn situation knowledge), and probing (to activate knowledge). They combine to produce the behavior of our expert programmer, as we describe in the next section.

TRACING THE MODEL ON OUR EXAMPLE

Below we discuss our model in detail as it emulates the behavior in our navigation example. The first three columns of Figure 4 replicate the display, time course, and utterances from Figure 2. The right column is new, and shows a trace of the model's rules that follow the programmer's behavior. The dashed lines make connections where display features and utterances are explicit evidence for the model's knowledge.

Episode E1: Learning Situation Knowledge

The programmer's high-level task for this session is to modify the program so it will learn slightly different natural-language comprehension rules than it currently learns. Therefore, she wants to understand what the program currently learns (the chunks it builds) before she modifies the code. Just before E1, she printed out the chunk shown in the upper left of Figure 4, to try to comprehend it. During E1 she phrases a question about "it", specifically about what "it's testing" (t17-18). She pauses, searching for tests on the display, then determines that the chunk tests an "operator" (t21).

Setting a goal. The model, reflecting what the user has attended to at this point, has a chunk represented in WM. From expert knowledge about Soar, the model knows that chunks are important to comprehend. It also knows that chunks have tests, and that tests have an important functional role within a chunk. It therefore sets a goal to understand *chunk tests*.

> Expert rule (Soar knowledge):
> 1.1 if WM says that a chunk exists, but
>   says nothing about its tests,
>   set a goal to comprehend chunk tests.

Learning situation knowledge about the goal. The first thing the model does when it sets a new goal is encode into LTM that it has indeed set that goal during this session. This is a mechanism that supports probing. If the model sets the same goal in the future, the rule will fire as a hint to the model that it might already know something about this goal, from a previous time.

> **Learned situation rule**:
> 1.2 if the goal is to comprehend tests,
>   add to WM that this goal was set before.

Attending to the display. Trying to comprehend chunk tests, the model looks at the display. Whether a chunk contains an operator test is critical to understanding its high-level functionality. Operator tests are therefore important to look for when trying to comprehend a chunk's tests.

For the model, the string ^operator, underlined on the upper left of Figure 4, is the *beacon* [2, 17, 19] for the existence of an operator test. No matter what other details of a specific operator appear in a chunk test, the chunk will contain the beacon if and only if the chunk tests an operator. This beacon knowledge represents expertise about the language.

> Expert rule (Soar knowledge):
> 1.3 if the goal is to comprehend chunk tests,
>   and we see an operator test,
>   add to WM what that we see an operator test.

Learning situation knowledge about the display. The operator test is new knowledge about the chunk's tests. The model captures this knowledge by creating a new rule. This rule will fire whenever the goal is to comprehend a chunk and when working memory contains an operator test. The rule firing will remind the model that it saw an operator test on the display.

> **Learned situation rule:**
> 1.4 if the goal is to comprehend chunk tests,
>   and WM contains an operator test,
>   add to WM that we saw it on the display.

Implications of this Learning. The model learned two pieces of situation knowledge during this episode. The first (1.2) is a rule that will recognize a particular goal and note in WM that it has been set some time during this session. The second (1.4) is a rule that will recognize when WM contains an operator test, in the context of a goal to

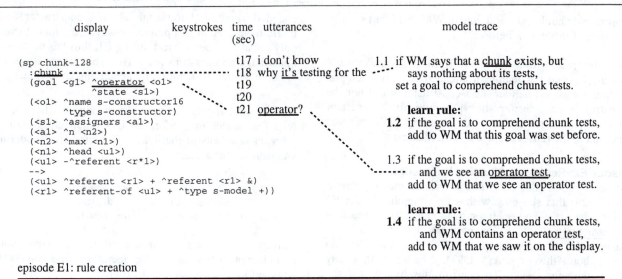

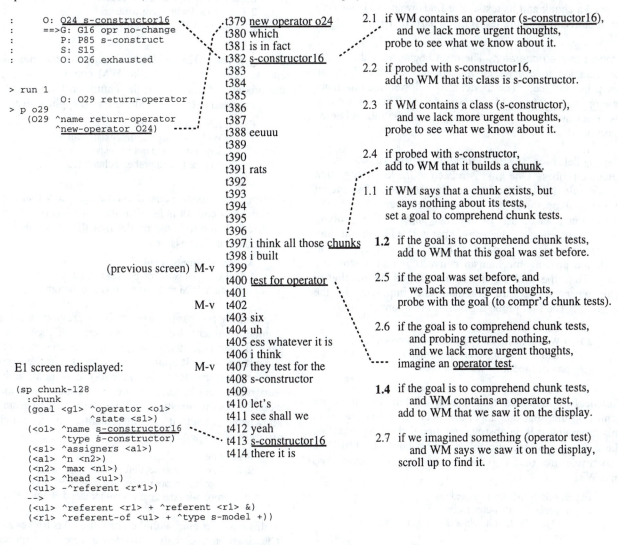

Figure 4: Learning and activating situation rules for scrolling.

Dashed lines make connections where display features and utterances are explicit evidence for the model's knowledge. Boldface indicates learned rules.

comprehend chunk tests, and note in WM that it has seen an operator test in a chunk before.

What the model did not learn from this episode was anything beyond the feature it attended to. For example, the model encoded neither the s-constructor16 test nor the s-constructor test, despite their proximity to the operator test. This narrowly-focussed encoding will become important in the retrieval episode, described next.

Episode E2: Retrieving Situation Knowledge

Just prior to E2, the programmer printed out a program object (o29) that she now wishes to comprehend. As E2 begins, she sees new-operator o24. She then attends to s-constructor16 (t382), following the shared identifier o24. Pausing, she says "eeuuu rats" (t388) and articulates an insight about "those chunks" (t397). She thinks they may contain operator tests. To confirm this hypothesis, she scrolls back (t399-407) to the display from E1, which showed a chunk and its tests. She finds what she is looking for, identifying a chunk test for s-constructor16.

Probing with an operator. The trace begins after the model has attended to everything immediately connected to the just-printed object. The model also followed the link from new-operator o24 to o24 s-constructor16. There is nothing else relevant to look at on the display, so the model begins to probe.

The model uses several kinds of expert knowledge to generate probes. One kind is related to the knowledge that guided rule 1.3 to attend to an operator test. The model knows that operators are both ubiquitous and one of the main functional units in the language. The probing mechanism guesses that any operator will inform any context in which it appears. If any operator appears in WM, the probing mechanism converts it into a semantic cue. Thus the probing mechanism draws on familiarity with what objects are common and important in the language. The first probe it generates is s-constructor16.

> Probe rule:
> 2.1 if WM contains an operator (s-constructor16),
>   and we lack more urgent thoughts,
>   probe to see what we know about it.

Retrieval in response to a probe. While operators are central to the Soar language, the s-constructor is central to this program. It is an abstract functional class, representing the collection of specific tokens like s-constructor16. An expert rule fires putting information about the abstract class into WM.

> Expert rule (program knowledge):
> 2.2 if probed with s-constructor16,
>   add to WM that its class is s-constructor.

In general, any number of rules like this can fire in parallel in response to a probe. Thus different pieces of information about s-constructor16 could appear at this point in the trace. For example, had the model attended to the name of the operator in E1 (which was s-constructor16), it would have

encoded situation knowledge about s-constructor16 that might have fired at this point. However, the model learned only about the operator test during E1, thus has no situation knowledge about s-constructor16. This means it must try again to trigger its memory, by generating another cue.

Probing with a class. Expert knowledge of the language tells the model that information about a class could be relevant to tokens of that class. Therefore the model uses s-constructor as a cue.

> Probe rule:
> 2.3 if WM contains a class (s-constructor),
>   and we lack more urgent thoughts,
>   probe to see what we know about it.

This probe retrieves an association between s-constructor and chunks. All s-constructor operators build chunks; this is expert knowledge about the program.

> Expert rule (program knowledge):
> 2.4 if probed with s-constructor,
>   add to WM that it builds a chunk.

Setting a goal. The model has recalled something about a chunk. This creates a similar WM context as in E1, after the model had attended to the chunk on the display. The same knowledge that applied there also applies here.

> Expert rule (Soar knowledge):
> 1.1 if WM says that a chunk exists, but
>   says nothing about its tests,
>   set a goal to comprehend chunk tests.

Situation knowledge applied to probing. Now that a goal to comprehend chunks is in WM, the situation rule learned in E1 fires. It reminds the model that this goal has been set previously in this session.

> **Activated situation rule:**
> **1.2** if the goal is to comprehend chunk tests,
>   add to WM that this goal was set before.

The probing mechanism knows that a previously-set goal is a good candidate for a semantic cue. This is the most general heuristic that the probing mechanism has for generating cues. It applies automatically, in the sense that the model learns to recognize every new goal, and every recognized goal becomes a probe. Probing with the goal when it was previously set allows the model to accumulate knowledge about it in increments, in a pattern of progressive deepening [7, 13].

> Probe rule:
> 2.5 if the goal was set before, and
>   we lack more urgent thoughts,
>   probe with the goal (to comprehend chunk tests).

In this case, probing with the current goal retrieves nothing. The situation knowledge encoded in rule 1.4 requires both that an operator test be in WM and that the goal be to comprehend chunk tests. This points to the crux of the problem with retrieving situation knowledge. The automatic encoding provided by the underlying architecture produces knowledge that is so specific to the situation that it cannot be evoked with just a partial match to that situation. In this case, both the chunk-tests goal and the operator test must be present in WM for rule 1.4 to fire.

The model can tell that the probe failed, because no additional knowledge is retrieved. However, it still has the information that the current goal was selected earlier; the model must have learned *something* then. The question is how to evoke it.

In E1, with the same goal as now, the model attended to an operator test on the display (rule 1.3), because expert knowledge about the language said this was a useful thing to do. The probing mechanism now uses the same knowledge here, to generate an operator test as an image cue. In this way the probing mechanism is essentially *imagining* beacons.

> Probe rule:
> 2.6 if the goal was to comprehend chunk tests,
>   and probing returned nothing,
>   and we lack more urgent thoughts,
>   imagine an operator test.

Situation knowledge drives navigation. After the previous probe, WM contains an operator test. Rule 1.4 now fires, and the model retrieves the fact that it saw an operator test on the display. The protocol contains explicit evidence that the programmer is thinking about an operator test (t400).

> Activated situation rule:
> 1.4 if the goal is to comprehend chunk tests,
>   and WM contains an operator test,
>   add to WM that we saw it on the display.

Expert knowledge about the programming environment says that in a process buffer, all user input and process output occurs at the bottom. To find a hidden feature, there is nowhere to look but up, so the model scrolls up through previous pages.

> Expert rule (programming environment):
> 2.7 if we imagined something (operator test)
>   and WM says we saw it on the display,
>   scroll up to find it.

Implications of this Retrieval. The programmer retrieves information about a previous situation. This raises a question: Why does she need to navigate at all? Couldn't she simply search internally to evoke the right memory?

The model suggests an explanation: she didn't encode all the details of the E1 display. The model encoded only that it saw an operator test. Even though s-constructor16 was on the screen in E1, and is on the screen during E2, and is even explicitly referred to by the programmer during E2, there is no situation knowledge that encodes s-constructor16 as a chunk test. This is consistent with the E1 protocol, in which the user says nothing about s-constructor16 as she examines the chunk's tests. Thus the model suggests that her need to scroll to the E1 display arises because she did not encode the information that would have answered her hypothesis about chunk tests (t397-408).

Similarly, the user refers to the class s-constructor during E2. For the model, the presence of s-constructor in WM is again not enough to retrieve any explicit memories from E1, because it was not encoded in any situation knowledge. With the chunk redisplayed, however, the model can use its knowledge of the language to follow the common symbol <o1> from the ^operator beacon to its name (s-constructor16) and its knowledge of the program to connect this to s-constructor. Thus the model suggests that she did not encode s-constructor, but that by looking once again at the E1 display she can confirm that "those chunks .. test for the s-constructor" (t397-408).

DISCUSSION

We have presented a model of how an experienced programmer may manage the large amount of internal and external memory needed for advanced programming. The model demonstrates that what we have termed situation knowledge mediates behavior that expertise, external memory, and working memory cannot explain by themselves.

Specifically, the model provides hypotheses about how programmers navigate to find hidden information.

• Situation knowledge, sensitive to both goal and display, is encoded in LTM automatically as a byproduct of problem solving. In our example, the model learns to recognize goals it set before and display features to which it attended.

• Activating such memories requires recreating a WM context like the one in which they were encoded. When a memory was encoded in a situation other than the current, this requires search to find the right context. This approach to memory management, with most of the effort for retrieval occurring at retrieval time, has been used in other models that learn and use recognition knowledge in situated tasks [1, 6, 16, 18].

• Deliberate search of LTM (probing) is a fallback problem-solving strategy. Probing uses semantic cues to recreate goal conditions and image cues to recreate display conditions. Together these trigger memories of previous situations, which may lead to navigation.

This work is a step in developing a psychology of natural programming. While our model is preliminary, our data are rich and come from a skilled programmer working in her own environment on her own difficult task. They therefore have more ecological validity than data from experiments where the environment is restricted or unfamiliar, or where the task is set by the experimenter and may be unnaturally easy. Understanding the cognition of programming, at the detailed level of a computational model, could ultimately increase the effectiveness of programmers through improved design of languages, programming environments, and training.

ACKNOWLEDGMENTS

This work was supported in part by the Wright Laboratory, Aeronautical Systems Center, Air Force Materiel Command, USAF, and ARPA under grant number F33615-93-1-1330, in part by the Office of Naval Research, Cognitive Science Program, Contract Number N00014-89-J-1975N158, and in part by the Advanced Research Projects Agency, DoD, monitored by the Office of Naval Research under contract N00014-93-1-0934. The views and conclusions contained in this document are those of the authors and should not be interpreted as representing the official policies or endorsements, either expressed or implied, of Wright Laboratory, the Advanced Research Projects Agency, the Office of Naval Research, or the U S Government. We would like to thank Francesmary Modugno, Herbert A Simon, and the CHI reviewers for extensive and helpful comments on this paper.

REFERENCES

1. Bauer, M and John, B E. Modeling time-constrained learning in a highly interactive task. *Human Factors in Computing Systems: Proc. CHI '95.* ACM, New York, 1995.

2. Brooks, R E. Towards a theory of the comprehension of computer programs. *Internat. J. Man-Machine Studies 18* (1983), 543-554.

3. Card, S K, Moran, T P, and Newell, A. *The Psychology of Human-Computer Interaction.* Erlbaum, Hillsdale NJ, 1983.

4. Davies, S P. Externalising information during coding activities: Effects of expertise, environment, and task. *Emp. Studies of Programmers: 5th workshop.* Ablex, Norwood NJ, 1993. 42-61.

5. Green, T R G, Bellamy, R K E, Parker, J M. Parsing and gnisrap: A model of device use. *Emp. Studies of Programmers: 2nd workshop.* Ablex, Norwood NJ, 1987. 132-146.

6. Howes, A. A model of the acquisition of menu knowledge by exploration. *Human Factors in Computing Systems: Proc. CHI '94.* ACM, New York, 1994. 445-451.

7. Kant, E, and Newell A. Problem solving techniques for the design of algorithms. *Information Processing & Management 20*, 1-2 (1984), 97-118.

8. Kitajima, M, and Polson, P G. A computational model of skilled use of a graphical user interface. *Human Factors in Computing Systems: Proc. CHI '92.* ACM, New York, 1992. 241-249.

9. Larkin, J H. Display-based problem solving. In *Complex Information Processing: The impact of Herbert A Simon*, D Klahr and K Kotovsky, Eds. Erlbaum, Hillsdale NJ, 1989. 319-341.

10. Larkin, J H. and Simon, H A. Why a diagram is (sometimes) worth ten thousand words. *Cognitive Science 11* (1987), 65-99.

11. Nelson, G, Lehman, J F, and John, B E. Integrating cognitive capabilities in a real-time task. *Proc. 16th Annual Conf. Cognitive Science Society.* 1994.

12. Newell, A. *Unified Theories of Cognition.* Harvard University Press, Cambridge, 1990.

13. Newell, A and Simon, H A. *Human Problem Solving.* Prentice-Hall, Englewood Cliffs NJ, 1972.

14. Payne, S J. Display-based action at the user interface. *Internat. J. Man-Machine Studies 35* (1991), 275-289.

15. Polson, P G and Lewis, C H. Theory-based design for easily learned interfaces. *Human-Computer Interaction 5* (1990), 191-220.

16. Rieman, J, Lewis, C, Young, R M, and Polson, P G. "Why is a raven like a writing desk"? Lessons in interface consistency and analogical reasoning from two cognitive architectures. *Human Factors in Computing Systems: Proc. CHI '94.* ACM, New York, 1994. 438-444.

17. Rist, R S. Program structure and design. To appear in *Cognitive Science.*

18. Vera, A H, Lewis, R L, and Lerch, F J. Situated decision-making and recognition-based learning: Applying symbolic theories to interactive tasks. *Proc. 15th Annual Conf. Cognitive Science Society.* Erlbaum, Hillsdale NJ, 1993. 84-95.

19. Wiedenbeck, S. The initial stage of comprehension. *Internat. J. Man-Machine Studies 35* (1991), 517-540.

Predictive Engineering Models Using the EPIC Architecture for a High-Performance Task

David E. Kieras
Artificial Intelligence Laboratory
Electrical Engineering & Computer
Science Department
University of Michigan
1101 Beal Avenue
Ann Arbor, Michigan 48109-2110
(313) 763-6739
kieras@eecs.umich.edu

Scott D. Wood
Artificial Intelligence Laboratory
Electrical Engineering & Computer
Science Department
University of Michigan
1101 Beal Avenue
Ann Arbor, Michigan 48109-2110
(313) 763-6448
swood@eecs.umich.edu

David E. Meyer
Department of Psychology
University of Michigan
330 Packard Road
Ann Arbor, Michigan 48104
(313) 763-1477
David_E._Meyer@um.cc.umich.edu

ABSTRACT

Engineering models of human performance permit some aspects of usability of interface designs to be predicted from an analysis of the task, and thus can replace to some extent expensive user testing data. Human performance in telephone operator tasks was successfully predicted using engineering models constructed in the EPIC (**E**xecutive **P**rocess-**I**nteractive **C**ontrol) architecture for human information-processing, which is especially suited for modeling multimodal, complex tasks. Several models were constructed on an *a priori* basis to represent different hypotheses about how users coordinate their activities to produce rapid task performance. All of the models predicted the total task time with useful accuracy, and clarified some important properties of the task.

INTRODUCTION

Engineering models for human performance permit some aspects of user interface designs to be evaluated analytically for usability, without consuming resources for empirical user testing, by making usability predictions based on an analysis of the user's task in conjunction with principles and parameters of human performance [4, 7]. This paper reports results on a new class of engineering models for a type of high-performance task, namely the telephone operator tasks studied by Gray, John, and Atwood [5]. By "high performance" we mean that the task is time-stressed; the total execution time must be minimized, and the user of the workstation (the telephone operator) is well-practiced. These tasks are scientifically interesting because they are multimodal, involving speech reception and production as

well as the usual visual display and keystrokes, and also because they are *active system* tasks [7] in that the user must respond to events produced by the external environment, unlike *passive system* text editing, which is basically paced by the user. As pointed out by John and Kieras [7], engineering models for active system tasks are currently under-developed. Finally, predicting performance in such tasks can be economically important; a detailed information-processing analysis of telephone operator tasks, the Gray, John, and Atwood CPM-GOMS models [5], were of considerable economic value in this domain where a second's reduction in average task completion time represents considerable financial savings.

Background on CPM-GOMS

Since the CPM-GOMS methodology and its most noteworthy application [5] is the precursor to the present work, some background is important to make the contribution of the present work clear (see also [7] for a general discussion of GOMS methodologies). CPM-GOMS is based on the Model Human Processor (MHP) [4], which is a proposal for how human information processing is performed by a set of perceptual and motor processors surrounding a cognitive processor; these processors operate in parallel with each other. During performance of a task, the human engages in perceptual, cognitive, and motor activities; but since these activities can overlap each other in time, the total time to execute the task is far less than the total of the times for the individual activities. Predicting the time required to execute the task thus requires determining which individual perceptual, cognitive, and motor activities are overlapped.

In the CPM-GOMS methodology, the analyst constructs a schedule chart (PERT chart) to represent the temporal dependencies between the various sequential and parallel activities. Once this network of activities is constructed, the predicted execution time between the very first and the very last activity is the total of the times on the *critical path* through the network, which is the longest duration pathway along the dependencies between the task start and completion. The critical path can then be examined to determine which activities actually determine the time required to complete the task.

However, the practical problem with CPM-GOMS methodology is that constructing the schedule charts required to analyze an interface design is quite labor-intensive. The analysis is performed on a set of benchmark task scenarios, or task instances. For each task instance and interface design, the interface analyst must choose the particular hypothetical pattern of perceptual, cognitive, and motor activities, and construct the schedule chart that shows which MHP processors are active in what order, and which processor actions depend on which other actions. Of course, the analyst may be able to reuse large portions of the schedule charts if the alternative designs or tasks involve only small variations that can be represented just by rearrangements of portions of the schedule charts (as in the Ernestine models, [5]). But due to the work involved, the CPM-GOMS method is recommended for predicting execution time only when there is a small number of benchmark tasks to be analyzed (see [7]).

Generative Models of Interface Procedures

This paper presents a new family of engineering models based on the EPIC (**E**xecutive **P**rocess-**I**nteractive **C**ontrol) human information processing architecture developed by Kieras and Meyer [9, 12], and the earlier Cognitive Complexity Theory (CCT) production-system analysis of human-computer interaction [3, 10]. EPIC is similar to the Model Human Processor (MHP) [4], but EPIC incorporates many recent theoretical and empirical results about human performance in the form of a computer simulation modeling software framework. Using EPIC, a *generative*[1] model can be constructed that represents the general procedures required to perform a complex multimodal task as a set of production rules. When the model is supplied with the external stimuli corresponding to a specific task instance, it will then execute the procedures in whatever specific way the task instance requires, thus simulating a human performing the task, and *generating* the predicted actions and their time course.

If a generative model based on EPIC can be applied to predicting execution time in a high-performance task, it should be considerably more efficient than the CPM-GOMS approach. Preliminary work with an EPIC model of the telephone operator tasks [13] was encouraging, showing

---

[1] The term *generative* is used analogously to its sense in formal linguistics. The syntax of a language can be represented compactly by a generative grammar, a set of rules for *generating* all of the grammatical sentences in the language.

fairly good accuracy in predicting task and event times for a small set of task instances. However, this preliminary model was constructed in a "scientific" mode, in which the model was developed iteratively to provide a good fit to a single protocol, and was then validated against two other protocols. But for a engineering model to be most useful, it should be usefully accurate in an *a priori* mode, requiring little or no "tuning" based on empirical task observation.

Thus the work reported here investigated the extent to which usefully accurate predictions could be made with predictive EPIC models that are based on *a priori* task analysis and principles of construction.

THE EPIC ARCHITECTURE

EPIC was designed to explicitly couple the basic information processing and perceptual-motor mechanisms represented in the MHP with a cognitive analysis of procedural skill, namely that represented by production-system models such as CCT [3], ACT [1], and SOAR [11]. Thus, EPIC has a production-rule cognitive processor surrounded by perceptual-motor peripherals; applying EPIC to a task situation requires specifying *both* the production-rule programming for the cognitive processor, and also the relevant perceptual and motor processing parameters. EPIC computational task models are generative in that the production rules supply general procedural knowledge of the task, and thus when EPIC interacts with a simulated task environment, the model generates the specific sequence of serial and parallel activities required to perform specific tasks. The model is driven by a task instance description that consists only of the sequence and timing of events external to the user, such as which characters appear at what location on the screen at what time, possibly in response to actions performed by the user. Thus the task analysis reflected in the model is general to a class of tasks, rather than reflecting specific task scenarios.

Figure 1 shows the overall structure of processors and memories in the EPIC architecture. At this level, EPIC is rather conventional, and closely resembles the MHP. However, there are some important new concepts in the EPIC architecture; this brief presentation will highlight some key properties of EPIC that both distinguish it from the MHP and are important for the work reported here. More details can be found in [9, 12]. It is important to note that EPIC was used "as is" for the modeling work reported here; the details and parameters of the architecture had been developed in other task domains and modeling projects.

As shown in Figure 1, there is a conventional flow of information from sense organs, through perceptual processors, to a cognitive processor (consisting of a production rule interpreter and a working memory), and finally to motor processors that control effector organs.

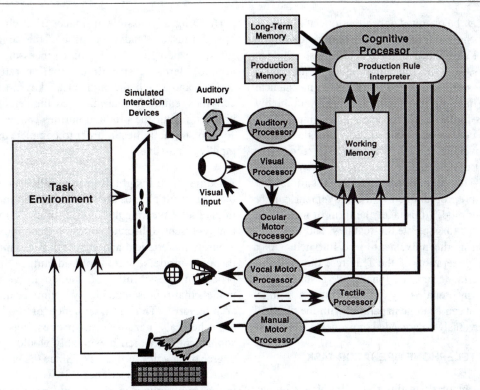

Figure 1. Overall structure of the EPIC architecture showing information flow paths as solid lines, mechanical control or connections as dotted lines. The processors run independently and in parallel; task performance is simulated by having the EPIC model interact with a simulated task environment.

There are separate perceptual processors with distinct processing time characteristics, and separate motor processors for vocal, manual, and oculomotor (eye) movements. There are feedback pathways from the motor processors, as well as tactile feedback from the effectors, which are important in coordinating multiple tasks. The declarative/procedural knowledge distinction of the "ACT-class" cognitive architectures [1] is represented in the form of separate permanent memories for production rules and declarative information. Working memory (WM) contains all of the temporary information tested for and manipulated by the production rules, including control information such as task goals and sequencing information, and also conventional working memory items, such as representations of sensory inputs.

A single stimulus input to a perceptual processor can produce multiple outputs to be deposited in WM at different times. The first output is a representation that a perceptual event has been detected, followed later by a representation that describes the recognized event. The perceptual processors in EPIC are "pipelines," in that an input produces an output at a certain later time, independently of what particular time it arrives.

The cognitive processor is programmed in terms of production rules, and so in order to model a task, we must supply a set of production rules that specify what actions in what situations must be performed to do the task. We are using the Parsimonious Production System (PPS) interpreter, which is especially suited to task modeling work, as in the CCT models [3]. One important feature of PPS is that control information such as the current goals is simply another type of WM item, and so can be manipulated by rule actions. The cognitive processor accepts input only at the beginning of each cycle, and produces output at the end of the cycle, whose mean duration we estimate at 50 ms. A critical difference with the MHP and many other production system architectures is that on each cognitive processor cycle, any number of rules can fire and execute their actions; this parallelism is a fundamental feature of PPS. Thus, unlike the MHP, the EPIC cognitive processor is not constrained to be doing only one thing at time. Rather, multiple processing threads can be represented simply as sets of rules that happen to run simultaneously.

The EPIC motor processors are much more elaborate than those in the MHP. Certain results motivate our assumptions that the motor processors operate independently, but the hands are bottlenecked through a single manual processor, and so normally can be operated only either one at a time, or synchronized with each other. Current research on movement control suggests that movements are specified in terms of features, and the time to produce a movement depends on its feature structure as well as its mechanical properties. We have represented this property in highly simplified models for the motor processors. The input to the motor processors consist of a symbolic name for the desired movement, or movement feature. The processor recodes the symbol into a set of movement features, and then initiates the movement. An important empirical result is that effectors can be preprogrammed if the movement can be anticipated. In our model, this takes the form of instructing the motor processor to generate the features, and then at a later time instructing the movement to be initiated. As a result of the pre-generation of the features, the resulting movement will be made sooner. Finally, we assume that a motor processor can prepare only one movement at a time, but this preparation can be done in parallel with the physical execution of a previously commanded movement.

MODELING THE TELEPHONE OPERATOR TASK
Task Summary
Briefly, the tasks analyzed in this report involve a human operator who sits at a computer-based workstation and assists customers to complete telephone calls. The specific class of tasks analyzed were ones in which the customer dials "0" followed by the destination telephone number, but then needs to supply orally a billing number to the operator (hereafter termed the *user*). The task begins when the workstation beeps to announce the arrival of a call, and then the workstation displays a variety of items on the screen about the call characteristics. The user must greet the customer with one of two greetings depending on whether the customer is calling from a pay phone or a private phone.

The major activity in the task is to use the screen information and the customer speech to determine which keys to press to specify the billing class of the call and then enter the billing number into the workstation, which then checks the number for validity. After getting the billing information from the customer, the user says "thank you." When the workstation validates the number, the user presses the POSITION RELEASE key to allow the call to proceed and signal readiness to handle the next call.

Many of the task activities can be overlapped; for example, the user typically starts pressing keys while the customer is still speaking, and can overlap much of his or her own speech with such activity and while waiting for the workstation to respond.

A Set of *A-Priori* Models
To simulate the user's performance, EPIC was "programmed" with a set of production rules capable of performing all possible instances of a class of telephone operator tasks. Under direction of the cognitive processor rules, the perceptual and motor processors move the eyes around, perceive stimuli on the operator's workstation screen, and reach for and strike keys. The time these activities require is determined by the perceptual and motor processors, but the production rules can arrange to overlap some of the activities in order to complete the entire task as rapidly as possible.

An immediate insight from the EPIC architecture is that there are multiple possibilities for performing task activities in parallel. Accordingly, a series of models was constructed that represented discrete points on a continuum starting with a purely hierarchical and sequential description of the task, through models that took advantage of the parallel processing possibilities of the cognitive architecture, to models that represented highly optimized utilizations of the architecture. Thus the sequence of models represent a hypothetical increase in processing efficiency and sophistication, which presumably would be related to the degree of practice in the task. Since the users producing the data were highly experienced, it was expected that one of the more optimized models would provide the best account of their performance.

The first model, termed the *Hierarchical Motor-Sequential* model, was based on a straightforward GOMS model for the task that followed the NGOMSL notation [8] for describing task procedures as a hierarchical set of methods consisting of sequential executed actions. Figure 2 shows the hierarchy of Goals and Methods for the GOMS model for the task. The production rules implemented this GOMS model in a style similar to the CCT templates described in [3]. In particular, each GOMS method entailed executing a pair of "housekeeping" productions corresponding to the entry and return from a submethod (see [3]), and a separate production rule for each basic perceptual or motor operator step in the method. The production rule for each step always waited for any motor action to be completed before it would fire to instruct the next motor action, or to invoke a submethod. Likewise, if an action was taken to acquire perceptual information, such as an eye movement, the next rule to fire always waited until the perceptual information was available. This model had a total of 50 production rules; one rule for each step in each method plus the additional "housekeeping" rules for each method.

Although it had strictly sequential methods, the Hierarchical Motor-Sequential model overlaps some of the task activities, for example, typing the billing number can begin while the customer is still speaking digits. Using a "pipeline" approach similar to John's [6] model of transcription typing, as each digit arrives in Working Memory, the cognitive processor sends the corresponding

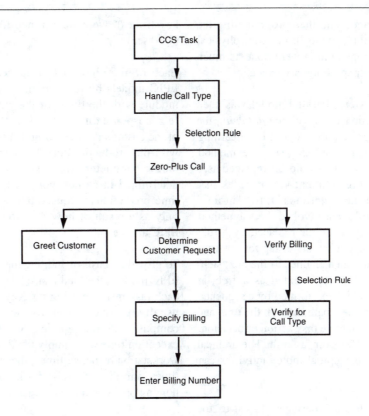

Figure 2. The hierarchy of goals and methods in the GOMS model for the telephone operator task. Connections labeled as selection rules indicate possible additional subgoals.

key press command to the manual motor processor as soon as it finishes the previous keystroke.

The second model, the *Hierarchical Motor-Parallel* model, assumed that the user could take advantage of the motor processor's ability to prepare the next movement while a movement is currently underway, and so physical execution of the next movement can be initiated as soon as the current movement is complete. The production rules from the previous model were simply modified so that they no longer waited for actions to be completed. Rather, each rule that instructed a motor processor merely waited for the relevant motor processor to be ready to prepare a new movement, and if the next rule did not use that processor, it did not have to wait for it to finish. Thus, activities involving different processors could be performed in parallel, and preparations for the next movement could be made in parallel with the execution of a movement. As a result, many purely cognitive activities, such as the rules performing method housekeeping, could then execute while perceptual-motor actions were taking place.

The third model, the *Hierarchical Prepared Motor-Parallel* model, assumed that the user would anticipate the eye or hand movements by instructing the motor processor to prepare movements in advance, as soon as it was ready to

accept movement instructions, and as early as logically possible. This advance preparation results in substantial time savings (typically 100-250 ms) when the movement is actually to be made. Note that EPIC's motor processors do not impose a time penalty for a movement preparation that is subsequently not used or is overwritten by a different movement instruction. Thus it is possible to speed up performance if the likely next keystroke can be predicted.

This model was constructed by adding additional production rules to the Motor-Parallel model to send the preparation instructions to the motor processors at the right time. Such preparation was possible only for movements that could be assumed to be constant at that point in the task; for example, typing a digit of the billing number could not be prepared in advance, since the billing number would vary from task to task. In contrast, pressing the billing category key could be prepared far in advance, given that the task structure makes it reasonable to assume that this key is probably the next one to be hit.

A fourth model, the *Hierarchical Premove/Prepared Motor-Parallel* model went further, by actually making the movements in advance if logically possible, and then preparing for any subsequent motion. Thus certain keystrokes could be anticipated by moving the hand to the location of the key in advance, and then programming the actual keystroke movement. Thus both the physical movement and the motor programming were done as much in advance as possible, further speeding task execution.

The original Hierarchical Motor-Parallel model was then modified in a different direction, one involving *flattening* the methods. According to principles proposed in learning theories such as ACT [2] and SOAR [11], the method housekeeping and other such rules would be replaced as a result of practice by a more efficient set of rules that effectively turn "subroutine" methods into "in-line" methods. For example, a rule that invoked the submethod for entering a billing number would be replaced by a rule that simply performed the first substantive step for entering the billing number, and which then chained to the next step. The resulting rule set could be represented as a tree, in which each class of task would be performed by a sequence of rule firings along a single linear path through the tree, and each rule performs some substantive task action or decision, with no housekeeping rules. However, as in the Hierarchical Motor-Parallel models, the perceptual-motor activities can overlap substantially.

The Flattened Method models are perhaps closest to the CPM-GOMS models for the telephone operator tasks [5], in that the methods consist simply of sequences of operators, with no hierarchical submethod structure (see [7, 8] for more discussion of this distinction).

The rule set for the fifth model, the *Flattened Motor-Parallel* model was constructed by modifying the Hierarchical Motor-Parallel model to concatenate the steps of separate methods, with selection rules being replaced by simple conditional tests on each branch. A sixth model, the *Premove/Prepared Flattened Motor-Parallel* model, incorporated the same advance movement and preparation as the Premove/Prepared Hierarchical Motor-Parallel model. Because the minimum number of activities are on the critical path, this model produces the fastest execution times.

COMPARISON OF THE MODELS TO DATA
Observed and Predicted Times

The basic question is how well the *a priori* constructed models predict actual task performance data. Using videotaped task performances collected, but not analyzed, during the Gray, John, and Atwood [5] Project Ernestine, we selected task instances covered by the models, and in which the operator made no substantial overt errors in performance, and the customer provided the relevant task information smoothly, without discussion with the operator. A set of four task instances for each of two users were selected. The video and audio recordings of the selected task instances were digitized at full frame rate, and the times of individual events (display changes, words of speech, and keystrokes), were determined to the nearest video frame (1/30 sec).

Each of the eight task instances was simulated with the EPIC models by programming the environment simulation module with the times of the externally-determined events (e.g., response time of the workstation, timing of each word of the customer's speech), and then running the EPIC system with the production rules for each model. All perceptual-motor parameters were kept fixed at values previously determined in earlier work [12, 13]. Thus the execution time predictions produced by the different models differed only as a result of how the production rules controlled the EPIC architecture.

To provide a basis of judging the relative contribution of the EPIC models, the total task execution time was predicted for each task instance using the Keystroke-Level Model, which usually produces usefully accurate results in ordinary computer interface applications [4, 7]. The predicted task execution time was simply the total of the observed relevant workstation response times, the customer and user speaking times, and the total time for keystrokes (280 ms each) and homing operators (400 ms each).

Results

Total Task Execution Times. Predicting the total task execution time is the key contribution of engineering models for this type of task. The normal definition of this time is the duration between the initial call arrival signal tone and the last keystroke, the POSITION RELEASE key. According to the workstation training materials, a certain screen event is the proper signal for hitting this key, and this was assumed in the GOMS analysis underlying the models. However, according to our informants, the POSITION RELEASE key is not very constrained by the task structure; in fact, it can be pressed at a variety of times, even well in advance of the screen event, and indeed the timing of this keystroke was quite unstable in the observed data. Accordingly, the total task execution time was calculated as the time to press the penultimate key, the START key, which is struck immediately after the last digit of the billing number is entered. The predicted task execution times for each model were compared to these observed task execution times.

All of the models, even the Keystroke-Level Model, accounted for a statistically significant 83% or more of the variance in the task execution times. This is due to the fact that the major determinant of the task execution time is the length of the billing number supplied by the customer, and all the models predict that the execution time will be longer as the length of the billing number increases. However, the goal of good engineering models is to supply predicted values of usability metrics that are not merely correlated with the empirically measured values, but are actually similar in numerical value. Figure 3 shows the *average absolute error* in prediction, expressed as a percentage of average observed value. The dotted line shows 10% error, a common rule of thumb for a useful level of prediction accuracy. The average absolute error of prediction ranges from 7% for the relatively simple Hierarchical Motor-Parallel model, to 14% for the worst-fitting EPIC model, to 28% for the Keystroke-Level Model.

All of the EPIC models appear to be usefully accurate in predicting total task execution time because they all represent to some extent how the task activities can be overlapped with each other (e.g., the billing number can be keyed in while the customer is still speaking the digits), so they all do a reasonable job of predicting overall task execution time. In contrast, the Keystroke-Level Model is much less accurate because it does not overlap any activities.

The surprise is that the highly optimized models did not fit the data as well as the simple Hierarchical Motor-Parallel model, which is only moderately efficient. This suggests that while users take advantage of the parallel preparation and execution capabilities of their motor processors to speed up their performance, they make little use of pre-positioning the eyes and hands in advance. Also, there is a hint that the flattened method models are "too fast" compared to the hierarchical models, but the difference is not large. Unfortunately, in these tasks, the Hierarchical model rules for submethod "calls" and "returns" tend to be overlapped with perceptual motor processing or external events, and so do not contribute to task performance time. A different type of task may be required to clearly distinguish these two families of models.

Individual Event Times. The scientific accuracy of the models can be tested more thoroughly by examining the predicted and observed timing of individual events such as keystrokes. These times were predicted very poorly by some models, and only moderately well by the best-fitting model, the Hierarchical Motor-Parallel model. Most of these events consist of typing the digits of the billing numbers. Detailed examination shows that in the observed task instances, the customer speaks the digits at a rate typically slower than the model (and apparently the actual users) can make the corresponding keystrokes, but the exact timing in the model is very sensitive to the delays in the situation, both those resulting from the workstation design and those due to speech recognition. Further work to characterize the details of the individual event timing is in progress.

One important implication of the detailed results is that apparently the rate at which the customer speaks the digits, not the rate at which the user can type, is probably the major bottleneck in the task execution time. A second implication is that perhaps the reason why the users appear to be following a task strategy that is only moderately efficient is that the task is so limited by the customer's speaking rate that there is no need for the greater efficiency of the more highly optimized models.

CONCLUSIONS

Some EPIC models for a high-performance task were constructed using *a priori* task analysis, construction principles, and parameter values, and these models were able to predict total task execution times with an accuracy

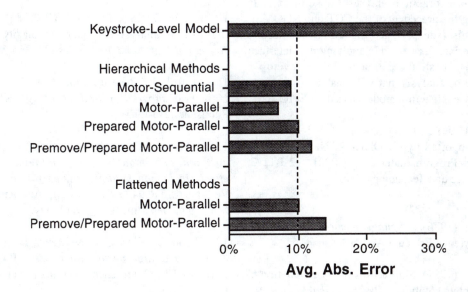

Figure 3. Average absolute error of prediction of total task execution time for each model.

high enough to be useful as engineering models for interface design. The detailed properties of the models suggest that the required level of optimization on the part of the user in these tasks may not be very high, although the users are highly practiced and execution speed is important. These results show the potential for EPIC to provide a framework for engineering models in complex, high-performance domains in which the user's performance time depends on the overlapped activity of separate processing capabilities.

The effort required to construct EPIC models seems to be considerably less than that for CPM-GOMS. In both approaches, the analyst must make many decisions about the details of task execution, such as when eye movements are necessary, but for EPIC models, these decisions are made only once for the general task procedures, rather than possibly multiple times in each specific benchmark task instance. Constructing the present models was relatively easy; the initial GOMS model was routine [7] once the information on the actual task procedures became available. Building the production-rule models was a matter of applying templates, both existing [3], or readily standardizable. Finally, the EPIC architecture itself was fixed and required no development for this analysis. In return for the rather modest construction effort, the resulting EPIC model can generate predicted execution times for all possible task instances within the scope of the GOMS model. Thus EPIC models would appear to be very efficient engineering models for high performance tasks.

At this point, EPIC is definitely a research system, and certainly is not ready for routine use by most interface designers. However, note that in some situations, such as the Ernestine project [5], the economics of the interface evaluation problem can make even a novel and demanding analysis approach a practical and useful solution. In addition, following the precedent of the CCT and NGOMSL engineering models (see [7]), as the EPIC architecture stabilizes and experience is gained in applying it to interface analysis problems, it should be possible to develop a simplified method of analysis that will enable designers to conveniently apply engineering models based on EPIC.

ACKNOWLEDGEMENT

This work was supported by the Office of Naval Research Cognitive Sciences Program under grant N00014-92-J-1173, and NYNEX Science and Technology, Inc.

REFERENCES

1. Anderson, J.R. (1976). *Language, memory, and thought*. Hillsdale, NJ: Lawrence Erlbaum Associates.

2. Anderson, J. R. (1987). Skill acquisition: Compilation of weak-method problem solutions. *Psychological Review*, 94, 192-210.

3. Bovair, S., Kieras, D. E., & Polson, P. G. (1990). The acquisition and performance of text editing skill: A cognitive complexity analysis. *Human-Computer Interaction*, 5, 1-48.

4. Card, S. K., Moran, T. P., & Newell, A. (1983). *The psychology of human-computer interaction*. Hillsdale, NJ: Lawrence Erlbaum Associates.

5. Gray, W. D., John, B. E., & Atwood, M. E. (1993) Project Ernestine: A validation of GOMS for prediction and explanation of real-world task performance. *Human-Computer Interaction*, 8, 3, pp. 237-209.

6. John, B. E. (1988) *Contributions to engineering models of human-computer interaction*. Doctoral dissertation, Carnegie Mellon University.

7. John, B. E. & Kieras, D. E. (1994) *The GOMS family of analysis techniques: Tools for design and evaluation.* Carnegie Mellon University School of Computer Science Technical Report No. CMU-CS-94-181. Also appears as the Human-Computer Interaction Institute Technical Report No. CMU-HCII-94-106.

8. Kieras, D. E. (1988). Towards a practical GOMS model methodology for user interface design. In M. Helander (Ed.), *Handbook of Human–Computer Interaction* (pp. 135–158). Amsterdam: North–Holland Elsevier.

9. Kieras, D.E., & Meyer, D.E. (1994). *The EPIC architecture for modeling human information-processing: A brief introduction.* (EPIC Tech. Rep. No. 1, TR-94/ONR-EPIC-1). Ann Arbor, University of Michigan, Department of Electrical Engineering and Computer Science.

10. Kieras, D. E., & Polson, P. G. (1985). An approach to the formal analysis of user complexity. *International Journal of Man-Machine Studies*, **22**, 365-394.

11. Laird, J., Rosenbloom, P, & Newell, A. (1986) *Universal subgoaling and chunking*. Kluwer Academic Publishers: Boston.

12. Meyer, D.E., & Kieras, D.E. (1994). *EPIC computational models of psychological refractory-period effects in human multiple-task performance.* (EPIC Tech. Rep. No. 2, TR-94/ONR-EPIC-2). Ann Arbor, University of Michigan, Department of Psychology.

13. Wood, S., Kieras, D., & Meyer, D. (1994). *An EPIC model for a high-performance task.* Poster presented at CHI'94 ACM Conference on Human Factors in Computing, Boston, April 25-28.

Modeling Time-constrained Learning in a Highly Interactive Task

Malcolm I. Bauer
Department of Computer Science
and HCI Institute
Carnegie Mellon University
Pittsburgh, PA 15213
(412) 268-7002
malcolm@cs.cmu.edu

Bonnie E. John
Departments of Computer Science and Psychology
and HCI Institute
Carnegie Mellon University
Pittsburgh, PA 15213
(412) 268-7182
bej@cs.cmu.edu

ABSTRACT

We investigate whether a memory-based learning procedure can explain the development of expertise within the time-constraints of a fast-paced highly interactive task. Our computational cognitive model begins with novice-like knowledge of a domain, and through experience converges on behavior that matches a pre-existing GOMS model of expert human performance. The model coordinates perception, comprehension, strategic planning, learning, memory, and motor action to respond to the time demands of the task while continually improving its performance. Because the model was constructed within the Soar architecture, it is able to make predictions of learning and performance time.

KEYWORDS: learning, GOMS, Soar, cognitive models.

INTRODUCTION

Issues of learnability are critical in almost any HCI design. Pinpointing the difficult subtasks of a domain, predicting the time it takes to master a system, and discovering the variability in learning of users, are all important factors in designing and evaluating HCI systems. Many real-time systems require that novices perform adequately while learning about a system and improving their performance. Tasks can place stringent time constraints on the learning and performance strategies people use (e.g. in air traffic control, financial trading, video games, etc.). In response to these phenomena, our research investigates the interaction between learning and performance in highly interactive real-time systems. We will describe a model that starts with little knowledge of a domain, uses a memory-based reasoning strategy to learn about the domain and converges on a recognition strategy that matches an existing GOMS model of the domain, while working under the strict time-constraints demanded by the task. It is our hope that as we better understand the mechanisms of learning in real-time systems, it will be possible to develop a full-fledged extension to GOMS that will allow HCI designers to make design decisions based upon learnability predictions. We view the current work as a contribution to this effort.

GOMS

Predicting Performance. GOMS models have successfully described expert performance in many domains. Some researchers [1] have suggested that GOMS-style analyses cannot model behavior in real-time, dynamic, interactive tasks because the time demands of these tasks prohibit the planning implied by the goal decomposition and method selection of GOMS. However, John and colleagues have demonstrated the utility of using GOMS in highly interactive domains such as video games [9,10] and telephone toll assistance services [6].

Predicting Learning Time. GOMS has also been used to predict the time associated with learning new methods and selection rules [e.g., 3,5]. These predictions assume that the user knows how to execute the primitive operators that comprise the methods, e.g., point with a mouse or type a command [8]. The learning time is then for assimilating the knowledge necessary to combine these operators into methods and evoke these methods in the appropriate circumstances. GOMS does not provide a model of learning mechanisms, rather, it provides a process for estimating learning time from a knowledge analysis of the task.

Modeling Learning Under Real-Time Constraints. The question we wish to answer is whether it is possible to model the learning that leads up to expert performance in a highly-interactive time-constrained task. Such a model would illustrate mechanisms for learning and strategies people might adopt to allow them to learn while performing in real-time and eventually converge on expert behavior that can be described by a GOMS model. To address this question we have modeled learning in the video game SUPER MARIO BROS. 3®. (SUPER MARIO BROS. 3®, Nintendo, and the characters discussed in this paper are trademarks of Nintendo of America Inc.) This domain had the advantages that there already exists an accurate GOMS model of expert performance [9,10], and it provides a rich highly interactive environment with many stringent time constraints.

GOMS and Soar

John and Vera [9] implemented their GOMS model within Soar [12], a production-rule-based cognitive architecture designed to model many aspects of human behavior including learning, memory, reasoning, problem-solving, perception and action. As a cognitive architecture, Soar offers many advantages for studying learning and performance in real-time. Soar can be used to construct runnable GOMS models of expert performance. It also includes an inherent learning mechanism that offers an established framework for moving beyond GOMS to learning. Most importantly for this research, Soar makes time predictions grounded in basic psychological studies of perception, cognition, and action [11, 12, 16]. This is important because it allows us to run our cognitive model in simulated time to explore the time-constraints placed on the model by the domain.

THE TASK DOMAIN

SUPER MARIO BROS. 3® is a video game in which a user maneuvers a screen character (Mario) through a two dimensional simulated world in a fixed period of time while avoiding dangers, gaining points by collecting treasures or killing enemies, and acquiring new powers. Commands are sent to Mario through a hand-held controller. A 4-way direction button, manipulated with the left thumb, determines which way Mario will move (left, right, and sometimes up or down) while two buttons controlled with the right thumb command Mario to jump, increase speed, or pick up items. The world is inhabited by several types of creatures who attempt to kill Mario by running into him or shooting him. Treasures are collected from blocks labeled with question marks ("question blocks" or QBs) by hitting them from particular directions, usually from underneath. Some real-time tasks that must be accomplished for successful play include judging trajectories to jump on top of objects while moving towards them, avoiding enemies as they approach, and completing a course of action (e.g. collecting a treasure) in the time before an enemy attacks.

GOMS Models of the Task

John and Vera [9, 10] modeled an expert playing world 1, level 1 of SUPER MARIO BROS. 3®. They constructed two GOMS analyses. One of the analyses was at the *functional-level* where operators correspond to activities such as searching a block or attacking an enemy (and are called function-level operators or FLOs). The FLOs can employ different methods for carrying out their slated tasks. For example, the attack-enemy FLO can employ the *stomp-on* method in which Mario jumps on top of an enemy to kill it, or the *hit-with-tail* method where, when equipped with a tail, Mario can kill the enemy by hitting it with his tail. The second GOMS analysis was at the *keystroke-level*, where operators correspond to finger motions on the game's control panel (and are called keystroke-level operators or KLOs). The functional-level analysis in John and Vera's GOMS model predicted close to 100% of the expert performance, while the keystroke level analysis predicted about 60%.

DESCRIPTION OF THE COGNITIVE MODEL
Time and Space

We are specifically interested in how learning can occur under time-constraints, so our model includes an explicit representation of time, both in our simulation of the game world, and in our model of the person playing the game. We wrote the game simulation as a set of production rules in Soar that manipulate objects in a special area in working memory allocated to the "external world." This external world consists of a simulated Nintendo controller, a large

data structure containing all the entities in the game world and their coordinate positions, and a simulated screen display that acts as a window onto that data structure. Production rules that simulate commands to the controller alter the coordinates of entities in the world and when Mario moves, the coordinates of the simulated screen.

To model time and space, we took measurements of aspects of the video game and used them to set several parameters in the simulation. Because Soar makes the prediction that a cognitive operator takes a fixed period of time of roughly 50ms [7, 11, 16], we chose 50ms as the grain size for the simulation. From the actual game, we measured the position of each object in the external world and its height and width. By taking repeated measurements of the game play, we determined that on a 19 inch television screen, Mario moves horizontally at roughly 3.46 in/sec at his "walking" pace." This implies that Mario can move 0.173 inches during each 50ms operator. (There is actually a very slight period of acceleration and deceleration, but this has been ignored for simplicity). We took similar measurements for Mario's vertical movement. Accordingly, in our simulation, every time a new operator is selected the display screen is updated by 50ms and Mario's coordinates are changed appropriately. In this way, Mario moves through the our simulated world at the same rate as in the actual game.

Our cognitive model perceives the current state of the screen in the external world and constructs or updates its internal representation of that screen in a single 50 ms time slice. The internal model does not contain coordinate positions but uses a qualitative representation involving relations such as "above", "adjacent", "underneath", etc. The model reasons about the entities in its internal model to decide its next action, and can put intentions to press or release keys in working memory. Motor commands fire 70 ms [4] (of simulated time) after the appearance of these intentions in working memory and cause the appropriate keys to be pressed or released on the simulated controller.

With this sort of grounding of the simulation in the real-time aspects of the game, and the grounding of the model in Soar's time predictions of cognitive processes, we were able to make detailed predictions of learning and performance under the time-constraints provided by the game.

Initial Knowledge

The cognitive model is of a person who has read the instructions for the game but has not actually played the game. It knows that to win it will be necessary to search QBs for treasure, attack enemies, avoid being killed, etc. It knows the mappings from actions to keys, for example, it knows that pressing the A-button causes Mario to jump, and pressing the right-direction-key causes Mario to move right. The model also has the ability to recognize the objects on the screen, so for example, if it sees an enemy, it knows it is an enemy and if it sees a QB, it knows it is something that contains a treasure. In addition, it knows methods for accomplishing goals, for example, to kill an enemy it knows to jump on its back (the *stomp-on* method). The model starts with no knowledge of the layout of the world, nor of exactly how fast Mario can move in the world and how far he can jump, in other words things that cannot be acquired by reading the manual.

This initial model differs from the John and Vera GOMS model [9, 10] in that, since this model explicitly does not know the dynamics of the game, it does not know *when* to execute keystrokes in service of a goal. For example, to kill an enemy it knows that it must use the stomp-on method and hit the A-key to jump on its back, but not when to hit that key to time the jump just right (as nine-year-olds say, it doesn't have "Nintendo fingers").

Learning

We focus our discussion of the model's learning procedure on one scenario, the threat of an enemy at the beginning of the game. To successfully kill the enemy, as Mario and the enemy approach one another, the model must cause Mario to jump at the appropriate distance away from the enemy to land on top of the it.

The model learns when to propose and perform actions. The model learns from experience by constructing success or failure cases upon the completion of a goal. It retrieves these cases in similar situations and uses them to construct new strategies for when to propose the operator. In the attack-enemy scenario above, the model must learn when to jump in order to land on top of the enemy and kill it, i.e. it must learn how far away it must be to propose the jump-on operator which issues the motor command to press the "A" key. This learning strategy emerges from Soar's inherent impasse-driven learning mechanism.

We are interested in determining if a memory-based learning procedure could operate effectively in the context of a fast-paced video game. Is there be enough time to comprehend the screen, recall previous instances and decide upon a new course of action within the time constraints of the game? It is quite possible that this kind of cognition is far too complex for someone to use while interacting with the game effectively. We present several runs of the model in the next section to address this question.

PERFORMANCE OF THE COGNITIVE MODEL

The graphs in Figures 1a through 1e depict the sequence of operators applied over time in a successive runs of the model. Time is represented on the horizontal axis in 50 ms intervals, the time of one operator application. The vertical axis lists possible operators, each of which will be described as they arise in the runs. The picture accompanying each graph displays the game behavior for the particular run.

Run 1: No Prior Experience (Figure 1a)

As displayed in the operator sequence graph of the first run, the model initially comprehends the screen. This application of the comprehend operator constructs the model's internal representation of the entities displayed on the screen, including representations of Mario, the 2 QBs, and the enemy. Because the enemy is seen as a threat, the model selects the attack-enemy operator next, and chooses the stomp-on method as a way to attack the enemy. At this

point the model hits an impasse because it does not know when to jump to successfully land on top of the enemy. To resolve this impasse, the model imagines possible outcomes of jumping (under-jump, over-jump, and success) in a manner similar to [2] and places nearby where it might have jumped from before. Memories of previous attempts that happen to match one of these possible outcomes appear in working memory i.e. they are retrieved as a function of the Soar architecture, not as an act of an operator [12, 15]. However, because this is the first attempt at jumping on an enemy, no previous cases appear in working memory. Given that it has no knowledge of when to jump, the model falls back on its default of jumping immediately at the perceived distance away from the enemy. Three strategy-learning operators generate this proposal strategy (using a method similar to the process described in [13]). Once the model has generated the strategy, it applies it: the next operator selected is jump-on, which issues the intention to press the A-key and the right-direction-key. As seen in the operator-sequence graph, motor commands cause the keys to be pressed 70 ms later (at point 1). For the next 0.7 seconds the model repeatedly applies the comprehend operator to monitor the jump. The last comprehend operator detects that Mario has under-jumped. Two memory operators then encode this information about the failure in a production (capturing the knowledge of the user thinking, "*that* sure didn't work!"). The learning-inhibition operator suppresses the use of the failed strategy (capturing the knowledge of the user thinking "Don't want to do *that* again!").

The comprehend operator that follows tells the model that the enemy is still a threat, and so the model's knowledge about attacking enemies is evoked again (note the identical shapes of the peaks at the beginning of the graph and at the end). The model is so busy thinking about attacking the enemy that it selects jump-on before comprehending that Mario has been killed. Thus, at this point in this first run, the model is not being immediately responsive to changes in the external world. The outward behavior in the game is displayed in the picture accompanying the operator sequence graph of the first run: Mario jumps too soon, lands in front of the enemy, attempts to jump again, but is killed.

The model predicts that this entire cognitive sequence takes place in roughly 1.6 seconds. In that time, the model comprehends the screen, chooses and implements a course of action, watches the effects of that action, and constructs a memory of the episode. The dynamics of the actual game provide sufficient time for such behavior to occur because if the user does not move Mario, it takes the enemy roughly six seconds to reach and kill him.

Run 2: Using Prior Experience (Figure 1b)

There are some notable differences in the operator-sequence graph between Run 2 and Run 1. In the second run, after comprehending the screen and deciding to use the stomp-on method to attack the enemy, the model imagines the possible outcomes of the jump and the failure associated with the first run appears in working memory. Since the previous jump failed because Mario under-jumped, the model reasons (using the memory reasoning operator) and decides it

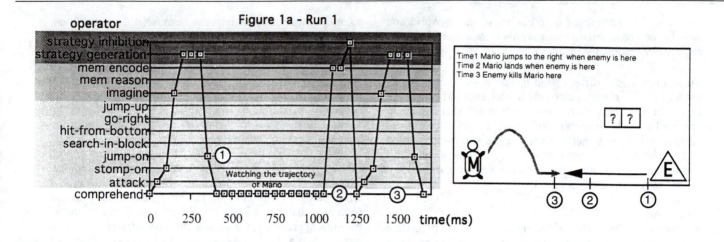

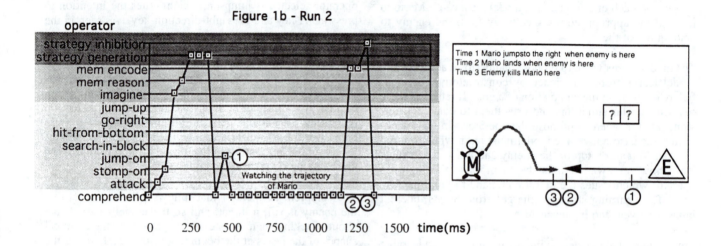

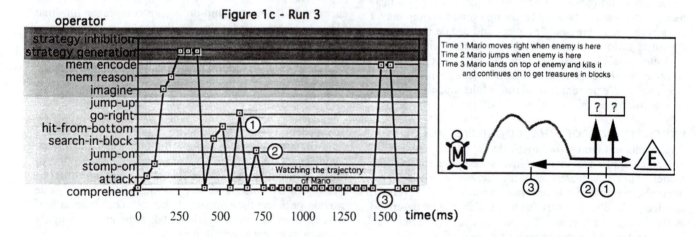

Figure 1a-c. Operator sequence graphs and pictures of the model's behavior in the first three of five successive runs. Shading in around the operator names indicates different types of operators: comprehend operators are the lightest. operators using GOMS-like knowledge of the task domain are medium-light, operators that deal with episodic memory creation or retrieval are medium-dark, and strategy operators are dark gray (Figure 1 continued on next page).

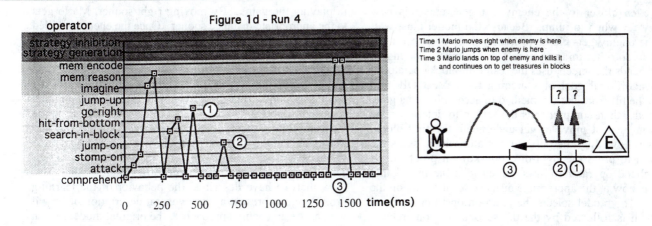

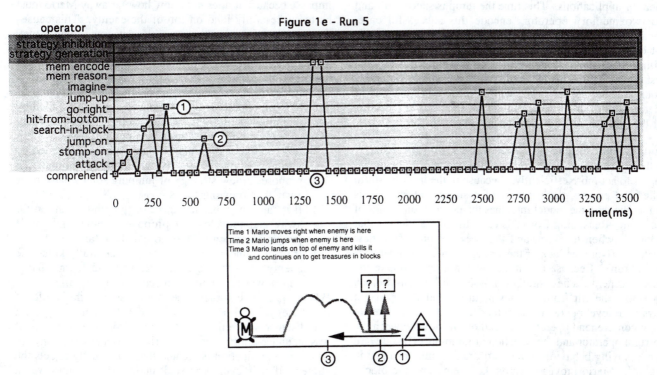

Figure 1d-e. Operator sequence graphs and pictures of the model's behavior in the last two of five successive runs. Shading in around the operator names indicates different types of operators: comprehend operators are the lightest, operators using GOMS-like knowledge of the task domain are medium-light, operators that deal with episodic memory creation or retrieval are medium-dark, and strategy operators are dark gray.

must jump when Mario is closer to the enemy. This new strategy is generated and as a result, the jump-on operator is not proposed immediately. Instead, a comprehend operator is applied after the learning of this strategy to watch the enemy approach. When the enemy is closer than it was in Run 1, the model selects the jump-on operator, pressing the A-key (at point 1). Once again, after 0.7 seconds of comprehension, the jump results in an under-jump failure, and the model constructs a second failure memory and inhibits this second failed strategy. Note that unlike the first run, the second attack sequence does not occur in the second run. Because the jump causes Mario to land closer to

the enemy, he is killed before the model even has a chance to think about attacking again.

Run 3: First Success (Figure 1c)
In the operator-sequence graph for the third run, the attack sequence at the beginning is the same as before: the model comprehends the screen and selects the attack-enemy and stomp-on operators. Upon hitting the impasse associated with not knowing when to jump, the model imagines the possible outcomes of the jump and this triggers the memories of the previous two failures. From these two cases the model reasons that it must cause Mario to jump

when even closer to the enemy. It generates this new strategy for when to jump. Because the model must wait until the enemy gets closer to Mario, there is time for the model to do other things before Mario jumps. The model comprehends the screen, uses the search-in-block operator to set an additional goal of searching the closest QB, and selects the hit-from-bottom method to accomplish this goal. The model then comprehends the screen to determine the direction it must move to get underneath the first block. This is followed by the go-right operator which issues the intention to press the right-direction-key (at point 2). A comprehend operator follows, which tells the model that Mario is now at the appropriate distance for jumping on the enemy. The model selects the jump-on operator (at point 2), which is followed by the 0.7 seconds of comprehend operator applications. This time the jump is successful, and the two memory operators encode this successful case. Because the jump was successful, this jumping strategy is not inhibited, unlike the past two runs. After successfully killing the enemy, Mario is still moving right towards the first block and the model continues to pursue the search-in-block goal set earlier. (We will describe this sequence of operators only in the final run, as the sequence is the same for the 3rd, 4th, and 5th runs, except shifted in time).

Run 4: Learning to Know It Knows (Figure 1d)

In the operator-sequence graph for the fourth run, the model begins to act more efficiently. Because in the last run Mario killed the enemy, the memory of a successful episode is triggered when the model imagines the possible outcomes of the jump. Reasoning from this case, the models learns that it knows when to jump, and the does not propose a new strategy. (Note the lack of the strategy generation operators in this run). Because the model does not need to take the time to generate a new strategy, it issues the search-in-block operator, the hit-from-bottom method, and the go-right keystroke-level operator earlier than in run 3. Note there are three comprehend operators instead of one now between the go-right operator and the selection of jump-on. The outward game playing behavior in the fourth run is similar to that in the 3rd. Mario moves towards the enemy and the blocks somewhat sooner than in the 3rd run, jumps, successfully kills the enemy and searches the blocks.

Run 5: Expert Performance (Figure 1e)

The fifth run is characterized by pure recognition-based proposals of the all operators. There is no imagining and reasoning from previous cases or strategy generation and inhibition. After comprehending the screen, the model selects attack-enemy operator and stomp-on. Because the model learned that in knows when to jump in Run 4, it goes directly on to comprehend the screen, select search-in-block and hit-from-bottom and issue the go-right operator (at point 1). When Mario gets to the appropriate distance from the enemy, the jump-on operator is selected. Once again the jump (at point 2) is followed by 0.7 seconds of comprehend operator applications, but note that Mario kills the enemy even earlier than last time. This occurs because the go-right operator is issued earlier than in run 4, so that although Mario is at the same distance from the enemy when the model proposes the jump-on operator, less time has passed

in playing the game. By moving right sooner, Mario gets to the appropriate distance sooner. Once the enemy is killed the model continues to comprehend the screen as Mario moves towards the first block. When it is under the first block, the model selects the jump-up operator and then comprehends the screen until Mario successfully hits the block and gains a treasure. Even before landing, the model selects the operators to search in the second block, using the hit-from-bottom method, and the go-right operator, so that Mario hits the ground running to the next block.

The General Learning Procedure

Now that we have described the behavior of the learning procedure, we present a more general description of how it works. As an enemy approaches, the original model hits an impasse because it does not know how far away Mario must be to successfully land on top of the enemy. This causes the model to imagine the possible outcomes, which may trigger memories of previous episodes. If no previous episodes are triggered, by default it proposes a strategy of jumping immediately. If previous cases are triggered, it uses the following rules to decide on a new place to jump:

1. If it can remember only instances in which it failed by jumping from too far away, it proposes a new strategy of waiting until it is closer than previous times.
2. If it can remember only instances in which it failed by jumping when the enemy was too close already, it proposes a new strategy of jumping somewhat farther away than previous times.
3. If it can remember instances of jumping when too far away and too close, it proposes a new strategy of jumping between the too close and too far cases.
4. If it remembers a case where it successfully killed the enemy, it uses this same strategy and learns that it knows when to jump in the current situation.

When a jump is complete, the model assesses the result. If the strategy failed, a memory is created that jumping from that distance was either too far away from the enemy (if it was an under-jump) or too close (if it was an over-jump). In addition, a production is created that specifically rejects the strategy if it is ever proposed again. If the jump was a success, a memory is created that jumping from that distance was a success and the strategy is not inhibited.

Overall, the model uses a simple episodic memory to store the results of previous jump-attempts in Soar's long term production memory, then reasons about these memories to generate and inhibit candidate jumping strategies until it experiences a successful jump. Interestingly, Soar's inherent learning mechanism requires this model to "imagine" actions and particular outcomes in order to learn jumping strategies. That is, to learn a new jumping strategy, the model must create a temporary state that matches the state of the external world in which it would like to jump, and then apply a simulated jump-on operator (that does not issue the actual motor commands to press the A button.). This mechanism is similar to that of the Soar and ACT-R models in [13]. Similarly, in order to remember previous actual outcomes of jumping, the model must hypothesize possible current outcomes. Memories are retrieved only if they match the imagined contexts. In this sense, the memory and learning systems have the property of *encoding specificity* proposed by Thomson and Tulving [14]. Note that we are not suggesting that the model predicts that people are consciously aware of this imagining process in the way that they are aware of lengthy daydreams. Indeed, the model predicts these processes occur over an extremely short period of time ranging from 50-150 ms.

DISCUSSION
From Novice to Expert Behavior
The five runs described above display an important trend. In the first two runs the model spends time retrieving memories, learning strategies, and inhibiting strategies, and it performs poorly -- Mario dies in both of these runs. In the third run, the jump is successful and no strategy inhibition occurs. In the fourth run, there is no strategy generation and inhibition, but episodic memories must still be triggered and reasoned about to recall the correct strategy. Finally in the fifth run, actions are initiated based upon the recognition of features of the current external world rather than reasoning about hypothetical or previous states of affairs. In other words, the runs show a trend from novice performance with little knowledge of the domain through intermediate stages in which learning and memory play an important role, to smooth expert performance characterized by efficient recognition-initiated action.

By the fifth run, we are left with a model that matches the functional-level operators (FLOs) selected by the GOMS model of John and colleagues [9, 10] without any memory or strategy operators preceding action. The model also displays the same sequence of keystroke-level operators (KLOs) as the GOMS model. The relationship between the FLOs and the KLOs is much like that of operators in CPM-GOMS models [e.g. 6, 8], where operators that serve different higher-level goals are interspersed. That is, the FLO setting the goal to attack-enemy is followed by a FLO to search-in-block and its KLO to go-right, before the appropriate time for the KLO to jump-on (the enemy) occurs. Comprehension operators (which do not appear explicitly in the previous GOMS models) are frequent in the fifth run, allowing the model to be highly responsive to changes in the external world.

Because memory plays such an important role in this learning procedure, the model predicts that even if someone got lucky and killed the enemy on the first try, it would take two more trials to reach expert performance. In the initial run, the model would have to generate the correct strategy. In the next run, the model would have to reason from the successful case to determine that it knows when to jump. Finally in the third run, no learning or memory processes would be involved to cause Mario to jump at the right time.

Most importantly, all runs in the model are carried out within the time constraints of the task. The cycle of perceiving, comprehending, intending and acting is fast enough for the model to be responsive to changes in the display. By coordinating its use of learning, memory, and comprehension, the model can learn new strategies even while playing this fast-paced game.

Emergent Allocation of Resources
All strategic learning that occurs in the model is geared toward discovering when to propose KLOs. In the runs we described, the only new rules learned by the model involve when to propose the jump-on KLO, i.e., when to press the direction-key and the A-key. Yet this learning had a significant effect on what additional FLOs were chosen. As the model's ability to decide when to propose jump-on improved, it freed up the model's cognitive resources to consider and apply other FLOs, in this case search-in-block and its associated method, hit-from-bottom. It also freed up resources so that the comprehend operator could be applied more often and, thus, the model could be more responsive to changes in the external world. This aspect of the behavior of the model is an important characteristic of performance and learning in time-constrained tasks: improving one area of performance can have a significant effect on other areas because it frees up resources that can be devoted elsewhere.

Other Possible Learning Strategies
We have demonstrated how a model with a memory-based learning procedure can perform a fast-paced highly-interactive task and still devote significant cognitive resources to developing expertise. We do not claim that this is the only model that can satisfy the time-constraints of performance and converge on a known GOMS model of expertise in the domain. Our model is really an existence proof that this class of models is potentially useful in explaining human behavior. In this section we will outline several other strategies that are possible.

We mentioned above that all strategic learning occurs at the KLO level in the current model. Knowledge acquired through learning is encoded in rules that propose the jump-on operator at the correct distance. This model issues the attack-enemy FLO whenever an enemy is on the screen but only presses the keys when the enemy is close enough. Another approach is to learn to attack an enemy only when it is strategically useful. A procedure similar to the one we described for learning when to issue a KLO could be used to learn in what situations to issue the attack-enemy FLO. In other words, the same memory-based learning procedure could apply at the FLO level as well as the KLO level.

Another kind of learning not addressed by this model is learning what to do next when there are several relevant but potentially conflicting FLOs and KLOs. This model allows several FLOs to be active simultaneously, in other words, it allows for multiple task goals to be present in working memory (as is common with the CPM-GOMS version of GOMS, [6, 8]). Alternative methods of goal management are possible, in fact, John and Vera [9] briefly described some preliminary work on learning selection rules to chose between simultaneously-proposed FLOs.

Lastly, strategic learning can occur by reflecting on longer sequences of actions taken in the world. The memories used in our model consist of very simple cause and effect episodes (e.g. "jumping from far away caused Mario to under-jump the enemy") and are reasoned about on the fly. People are certainly capable of remembering and reasoning about more complex schemata, especially if they have time to reflect. Such a reflection strategy would require a model to perform more memory-related processes while playing the game in order to store away the events for future reflection. Between rounds the model could reflect on the sequences of events that took place, and propose new actions off-line when there is more time to devote to reasoning and strategy generation.

The profusion of learning strategies just discussed may raise questions as to the value of a model that seems to have a single learning strategy. Certainly, in this paper we have described our model as having only a single memory-based learning strategy. However, single-strategy learning is merely a property of this particular instantiation of our Soar model. Just as an individual human being can display different strategies under different circumstances, the Soar architecture allows many strategies to exist in a single model, each evoked either deliberately through the application of knowledge, or as an emergent behavior in response to specific task demands. Thus, this model stands as an existence proof that symbolic cognitive architectures, like Soar can reasonably learn and improve performance, even in a highly-interactive task with severe real-time constraints. It also stands as a basis for adding other strategies to model the full range of human learning, a necessary step along the way to a predictive tool for learnability in real-time system design.

ACKNOWLEDGMENTS

This work was supported by the Office of Naval Research, Cognitive Science Program, Contract Number N00014-89-J-1975N158. The views and conclusions contained in this document are those of the authors and should not be interpreted as representing the official policies, either expressed or implied, of the Office of Naval Research or the U. S. Government. We would like to thank Erik Altmann and Craig Miller for their extensive and helpful discussions about programming in Soar.

REFERENCES

1. Agre, P. E. & Chapman, D. (1987) Pengi: An implementation of a theory of activity. In *Proceedings of the Sixth National Conference on Artificial Intelligence*, 268-272. Menlo Park, CA: American Association for Artificial Intelligence.

2. Altmann, E.M., Larkin, J.H., & John, B.E. (1995). Display navigation by an expert programmer: A preliminary model of memory. In Proceedings of CHI, 1995 (Denver, Colorado) ACM, New York.

3. Bovair, S., Kieras, D. E., & Polson, P. G. (1990). The acquisition and performance of text editing skill: A cognitive complexity analysis. *Human-Computer Interaction, 5*, 1-48.

4. Card, S. K., Moran, T. P., & Newell, A. (1983) *The psychology of human-computer interaction.* Lawrence Erlbaum, Associates, Hillsdale, NJ.

5. Gong, R. (1993). *Validating and refining the GOMS model methodology for software user interface design and evaluation.* Ph.D. dissertation, University of Michigan, 1993.

6. Gray, W. D., John, B.E., & Atwood, M. E. (1993) Project Ernestine: validating a GOMS analysis for predicting and explaining real-world task performance. *Human-computer interaction.*

7. John, B. E. (1988) *Contributions to engineering models of human-computer interaction.* Doctoral dissertation, Carnegie Mellon University.

8. John, B. E. & Kieras, D. E. (1994) *The GOMS family of analysis techniques: Tools for design and evaluation.* Carnegie Mellon University School of Computer Science Technical Report No. CMU-CS-94-181. Also appears as the Human-Computer Interaction Institute Technical Report No. CMU-HCII-94-106.

9. John, B. E. & Vera, A. H. (1992) A GOMS analysis for a graphic, machine-paced, highly interactive task. In Proceedings of CHI, 1992 (Monterey, California, May 3-May 7, 1992) ACM, New York, 1992. pp. 251-258.

10. John, B. E., Vera, A. H., & Newell, A. (1994). Towards real-time GOMS.: a model of expert behavior in a highly interactive task. *Behavior and information technology*, 13, 255-267.

11. Nelson, G., Lehman, J. F., & John, B. E. (1994) Integrating cognitive capabilities in a real-time task. *Proceedings of the 16th annual conference of the cognitive science society,* Atlanta, Georgia.

12. Newell, A. (1990) *Unified theories of cognition.* Cambridge, Mass: Harvard University Press.

13. Rieman, J., Lewis, C., Young, R. M., & Polson, P. G. (1994) "Why is a raven like a writing desk?" Lessons in interface consistency and analogical reasoning from two cognitive architectures. In proceedings of *CHI 1994,* (Boston, MA April 24-28) ACM, New York, pp. 438-444.

14. Thomson, D. M., & Tulving, E. (1970) Associative encoding and retrieval: Weak and strong cues. *Journal of experimental psychology*, 86, pp. 255-262.

15. Vera, A. H., Lewis, R.L., & Lerch, F.J. (1993) Situated decision-making and recognition-based learning: Applying symbolic theories to interactive tasks. *Proceedings of the 15th annual conference of the cognitive science society,* Boulder, Colorado.

16. Wiesmeyer, M. D. (1992). *An operator-based model of human covert visual attention.* Ph.D. thesis, University of Michigan.

KIDSIM: END USER PROGRAMMING
OF SIMULATIONS

Allen Cypher and David Canfield Smith

Advanced Technology Group
Apple Computer, Inc.
Cupertino, CA 95014
cypher@apple.com, dsmith@apple.com

ABSTRACT

KidSim is an environment that allows children to create their own simulations. They create their own characters, and they create rules that specify how the characters are to behave and interact. KidSim is programmed by demonstration, so that users do not need to learn a conventional programming language or scripting language. Informal user studies have shown that children are able to create simulations in KidSim with a minimum of instruction, and that KidSim stimulates their imagination.

KEYWORDS: end user programming, simulations, programming by demonstration, graphical rewrite rules, production systems, programming by example, user programming.

INTRODUCTION

KidSim is an environment that allows children to create their own simulations. In schools today, students express their ideas through drawings, written descriptions, and perhaps video and multimedia. Simulations open up an additional dimension, since they enable students to express ideas about the way objects behave and interact.

The educational philosophy of constructivism states that children learn best when they actively create. So we wanted to provide a tool that would allow children to construct simulations themselves. In order to create a system that would be engaging and appealing to children, we decided that it would be unacceptable to require them to use a conventional programming language or scripting language. Therefore, one of the major aims of our project was to find a simple, natural way for children to program simulations. We wanted to provide a set of capabilities that was sufficiently rich and expressive that children could build an interesting and diverse set of simulations, but we were willing to trade off computational power in order to maintain ease of use.

Based on a variety of informal user tests, we settled on a design that allows children to create simple rules of behavior by demonstration. To express more abstract behaviors, such as becoming tired over time or only eating when one is hungry, we devised interfaces that use a mixture of demonstration and direct manipulation. The resulting system, with its hybrid of interaction techniques, has satisfied our goals of both ease of use and expressiveness.

I. THE COMPONENTS OF KIDSIM

Using KidSim, children create worlds that contain a variety of characters, which have rules, appearances, and properties. A rule defines what a character is to do in a particular situation. Appearances allow the character to change its visual appearance, and properties are used to maintain information about the character.

When the simulation is running, KidSim continually advances its *clock*, and on each tick of the clock, each character on the board is given a chance to act according to its rules. Figure 1 shows a typical KidSim screen.

A. Rules

Each character has a list of rules. When a character is given its turn to act, KidSim searches down the list for the first rule that is applicable to the current state of the board. This rule is then executed.

Creating a Rule

Simple rules in KidSim are created by demonstration. That is, the user performs by hand each of the actions that the rule is to perform. KidSim then generalizes these actions to create a rule that will automatically perform those actions whenever it is executed.

The simplest rules to create in KidSim are rules which cause a character to move. For instance, Figure 2 shows how to create a rule which causes a character to move one square to the right.

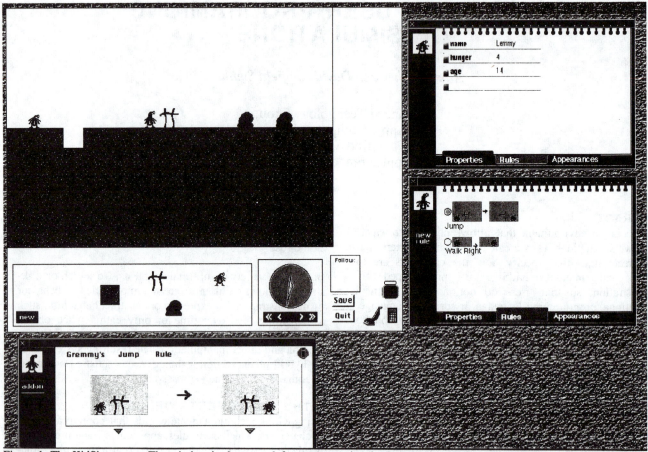

Figure 1. The KidSim screen. The window in the upper left corner contains the Board, with the Copy Box below it. On the right, the user has opened notebooks for the two gremlins on the Board. The upper notebook is open to the Properties page, and the lower is open to the Rules page. At the bottom of the screen is the Rule Editor, which is displaying the "Jump" rule.

When the user clicks on a character's "New Rule" button, the entire board darkens, except for a *spotlight* around the character. The user reshapes this spotlight to specify the context for the new rule. For the example in Figure 2a, the spotlight has been enlarged to include the square to the right of the character.

Next, the user demonstrates what the rule should do by moving the objects in the spotlight. In Figure 2b, the user moves the character into the square on the right, and then clicks on the done button.

The visual representation for a rule shows a picture of the "before" state of the rule on the left, a picture of the "after" state of the rule on the right, and an arrow between the pictures. The rule in Figure 2 can be read as "If there is an empty space to the right of me, move into it".

Since rules are tested on every tick of the clock, this one rule is sufficient to cause the character to move across the board when the simulation is running.

B. Appearances
A character can have a variety of appearances. By creating rules which change a character's appearance, the user can an-

imate the character's motion, change the direction it is facing, change its facial expression, and so on.

Figure 3 shows a rule for jumping over fences. Note that the *actions box* on the right side of the rule shows a description of each action that will take place when the rule is executed. To create this rule, the user moves the character to the square above the fence, drags the "jumping" appearance into the *current appearance box* on the appearance page of the character's notebook, moves the character to the square to the right of the fence, and then moves the "normal" appearance into the character's current appearance box.

C. Properties
The features of KidSim described so far make it easy for children to create rules that cause their characters to move in different ways, depending on the objects around them. Since this is the essence of many simulations, we designed an interface which was specially tailored to this capability. However, simulations of this sort are limited in that the only way a character can be different from one time to the next is by its location on the board, and by its appearance. But interesting simulations often depend on the fact that characters change as the simulation progresses. They may get old, or sick, or stronger, or smarter. In order to allow

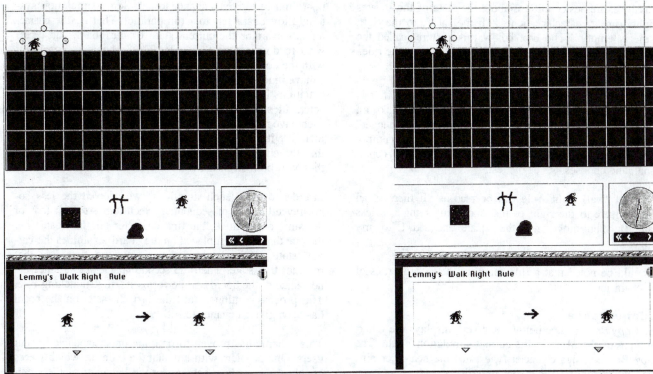

Figure 2. Creating a "Move Right" rule. The user shapes the spotlight to include the square to the right of the character, and then drags the character into that square. The rule, displayed at the bottom, shows the initial and final states for the rule.

for internal state that is not bound to the visual appearance of the character, KidSim characters have *properties*.

There are a few system-defined properties, such as *name* and *location*, but users can also create their own properties. For instance, a user could create the properties "age", "hunger", and "skills" for a character. Properties serve the role of instance variables in object-oriented programming.

Modifying properties

Figure 4 shows a modified "walk right" rule, where walking makes the character become hungry. The user creates the "hunger" property by typing "hunger" and "0" into the empty property box at the bottom of the list of properties.

To add on to the "walk right" rule, the user clicks on the "Add On" button for that rule. This executes the actions that are already part of the rule, and then lets the user demonstrate further actions. In this case, the user clicks on the "0" in the "hunger" property, and changes it to "1". KidSim generalizes this demonstration to mean "Add 1 to my hunger". If this is not the intended interpretation of the demonstration, the user can edit the description to, for instance, "Put 1 into my hunger". KidSim includes a calculator so that the user can create more complicated rules.

Testing properties

Besides modifying property values, it is important that rules be able to test properties. The picture on the left side

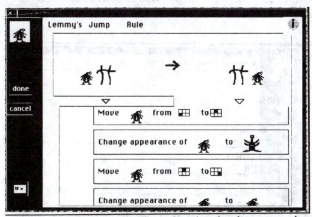

Figure 3. The "Jump" rule. The actions box has been opened to show each of the actions performed by the rule.

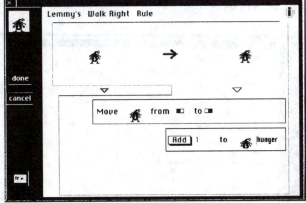

Figure 4. The "Walk Right" rule, modified to increment the character's hunger property.

of a rule expresses conditions about the state of the squares neighboring a character, such as "If the square to the right of me is empty". The box below this picture (called the *conditions box*) is used to add further conditions to the rule.

Figure 5 shows a further modification of the "walk right" rule, which will prevent the character from walking if it is too hungry. To create this conditional, the user clicks on the character in the left part of the rule, drags the "hunger" property from the menu that appears, drops it into an empty conditional, types in "10" on the right side of the conditional, and chooses "<" as the test.

The final "walk right" rule can be read as "If there is an empty square to the right of me, and if my hunger is less than 10, then move into the square and add 1 to my hunger".

It should be noted that a rule can refer to the properties of *any* character.

D. Inheritance
The *Copy Box*, located below the board, displays the different types of characters in the current simulation world. The Copy Box contains character *types*, and the board contains character *instances*. When the user drags a character type from the Copy Box onto the board, an instance of that type of character is created on the board. The user can create new types of characters by clicking on the *New* button in the Copy Box.

Each type of character has its own properties, appearances, and rules. Every instance of that type of character will have the same properties, appearances, and rules. When the user adds a property, appearance, or rule to any instance of a character, all characters of that type also acquire that property, appearance, or rule. However, each character instance has its own *value* for a property. So, for example, if gremlins have an "age" property, each individual gremlin on the board may have a different age. Furthermore, each character can have its own drawing for each appearance. Thus, individual gremlins need not all look the same.

E. Rule Execution
On each tick of the simulation clock, every character on the

board must be given a chance to act. Some simulation systems allow characters to act in parallel. That is, the characters all examine the current state of the board, and decide what to do based on that state. This scheme is consistent with the common sense notion of how independent entities behave in the real world. However, it can lead to confusing situations in a simulation. For instance, suppose two characters are standing on either side of an empty chair, and they each have rules stating "If I am next to an empty chair, sit in it." With parallel execution, both characters will notice that the chair is empty, and so when the simulation runs, they will both sit in the chair.

In order to avoid such situations, we adopted the less sophisticated model of sequential execution. At each tick of the simulation clock, the first character on the board (i.e. the one that was first placed on the board) examines the current state of the board and performs the first rule that matches this state. Then the second character examines the new state of the board and performs its first matching rule. This process continues until the last character on the board has been given a chance to act.

This scheme has proven to be quite understandable by our users. One problem with it is that the order in which pieces are placed on the board can affect the behavior of the simulation. Rather than trying to counter this side-effect, we simply provide user feedback to make the situation comprehensible: when the user steps the clock one tick at a time, KidSim highlights each character in turn as it is given a chance to act.

II. USER STUDIES
A. Early Studies
When we started to design KidSim, we conducted several informal user tests with fifth-grade children. In order to test whether graphical rewrite rules — rules which show a "before" picture on the left, and an "after" picture on the right — were a viable approach, we posed problems to the students, such as creating a character which could walk to the right and jump over obstacles. We asked them to draw before-after pictures on "Post-It" notes, and we then had them act out the roles of characters in the simulation, following the rules they had written. We were encouraged that the students were able to write rules in this format, and that they

Figure 5. Creating a conditional. The user clicks on the character to bring up the list of properties and drags the hunger property into the empty conditional box. The user then types in the value "10", and chooses the "<" test from a popup menu.

quickly understood how to test the rules against the current state of the "world". Furthermore, children were able to understand rules that had been written by other children.

In another informal study, four students created a video prototype of the KidSim board, in the style of Vertelney [14], using magnets under a piece of Plexiglas to move their characters. This gave us confidence that our use of a clock, with characters taking turns to act, was appropriate.

Even more informally, three fifth-grade teachers were early (and patient) users of the first working versions of our program. Even when the program was crashing quite regularly, they were able to construct a scenario where it started to rain, and characters went for shelter in a nearby house. Once the prototypes became somewhat more stable, we tested KidSim regularly on the students in their classes.

These early tests with teachers and students were invaluable for finding problems in specific user interface objects and interaction techniques. For instance, we originally used double-clicks on a character to bring up the notebook for the character, and single-clicks on buttons to create new rules. This inconsistency was confusing for all of our users, so we eliminated all double-clicks from our design. Some other examples: We found that a trash can was inappropriate for disposing characters, that allowing users to erase individual items in a conditional expression was excessive and unnecessary, and that adding the text "and if" at the beginning of each conditional expression made rules easier to understand.

These early tests also pointed out some more fundamental problems. Users would sometimes expect that there was only a fixed set of names that they could use for properties, as if the properties were predetermined by the system. We surmised that part of the problem was that the system properties, like "name", were listed together with the user properties. Although system properties are distinguished by displaying their names in blue, this distinction was evidently too subtle. Also, it seems that the visual appearance of properties is too heavy-weight, and that it implies that the computer is doing something special with them. We are therefore switching to a design that uses Boxer-style boxes [4], and that displays system properties in a separate area.

An encouraging result of our studies is that girls seem to enjoy using KidSim just as much as boys. We want to design a system that does not have a gender bias, and we are interested in conducting further studies to better understand which features of an interactive environment are particularly appealing to girls, and which are particularly unappealing.

B. The UK Study
Once we had created a working version of KidSim, with all of the basic features functioning as we intended, an informal user study of KidSim was conducted at the Centre for Research in Development, Instruction and Training at the University of Nottingham by Prof. David Gilmore. The study involved 56 children between the ages of 8 and 14.

Their exposure to KidSim varied between 1 hour and 8 hours, in multiple sessions, with the children working predominantly in groups of 2 or 3. Most of the children claimed some computer experience, though generally not with Apple Macintosh computers. Minimal instruction in KidSim was provided, consisting of approximately 10 minutes with an introductory worksheet. Generally two researchers, who had about 2 weeks experience with KidSim, were present in the room at each session. A session consisted of 4 groups of students.

The sessions were quite open-ended. Initially, the students were given some ideas of rules to write, such as "move a creature rightwards along the ground". All sessions were video-taped and audio-taped.

Most all children found the rule-writing interface easy to use and were able to generate multiple rules for their characters. Furthermore, the system provoked their imaginations, and the children invented goals for themselves and created their own characters and situations. They created a soccer game, PacMan, a maze traversal game, ninja turtles, and an aquarium.

This study showed that children can create rules, and that they can read an individual rule. However, it was not clear from this study whether they can understand sets of rules, and how multiple rules and characters interact.

C. Design Changes
Our main design goal was to produce a tool that children would be able to use to create their own simulations. Every step in the design involved tradeoffs between making KidSim powerful enough that children would find it expressive and engaging, and making it simple enough that they would not find it frustrating or confusing. Our periodic user testing has helped us to see where our initial design choices erred in one direction or the other.

Simple Inheritance
In the original version of KidSim (see [12]), it was possible to create arbitrarily deep hierarchies of character types. For instance, one could create clown fish, which are a type of fish, which are a type of animal, which is a type of object. We wanted users to create new rules and properties by adding them to a particular character on the game board, but we were not satisfied with any of our schemes for determining how to propagate new rules and properties up the inheritance hierarchy. Furthermore, the deep hierarchy led to potentially confusing situations, since users could instantiate abstract superclasses. Thus, it was possible to have an object on the game board which was an instance of animal, while other objects were instances of clown fish and sharks.

As a result of these difficulties, we changed to a simple inheritance scheme that admits only a single level of character types. This mechanism is certainly less powerful. For example, to add a "swimming" rule to all clown fish, sharks, and whales, users must put a copy of that rule in each of these three character types.

Independent of the "deep hierarchy" issue, we have been interested in allowing children to create new categories at any time. After creating a rule to jump over fences, for instance, a user might want to use the same rule to jump over rocks. We therefore added a new feature to KidSim, called *Jars* (see Figure 6). The user can create a new jar, name it "Obstacles", and put Fences and Rocks in the jar. When the user clicks on a fence in a rule (see Figure 7), a popup menu lists "Fence 27", "Fence", "Object", and "Obstacles". The user can select any one as the desired interpretation of what that object is to represent in the rule. Thus Jars can compensate somewhat for the absence of a deep hierarchy.

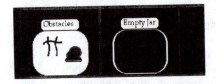

Figure 6. An "Obstacle" Jar, containing Fences and Rocks.

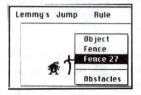

Figure 7. Selecting the desired generality for an object in a rule.

Characters larger than a square

Our current implementation assumes that every character fits into a single square on the board. This simplifying assumption makes it much easier to specify rules in terms of the squares neighboring a character. Although we were quite content with this simple approach, our users were not. They frequently want to create worlds where some characters are much larger than others. This means that our initial design decision resulted in a tool that was not sufficiently expressive. For instance, one user wanted to create a large horse that could carry several riders, and found it very unsatisfying to have to draw the horse in a single square. Therefore, our user studies have convinced us to modify KidSim to allow characters larger than a square, even though this will complicate the rule system.

III. KIDSIM AND END USER PROGRAMMING

Rule creation in KidSim uses of mix of Programming by Demonstration (PBD) and Direct Manipulation techniques. PBD is used during rule creation to show how a character moves, to change appearances, to create and delete characters (by dragging a character from the Copy Box into the spotlight, or by vacuuming a character), and to change the values of a property. Direct manipulation is used during rule creation to specify the context (by reshaping the spotlight), to generalize a character (by choosing the instance, the type, "Object", or a jar from a popup menu), to specify conditional tests (e.g. "my hunger < 10"), and to change the values of a property.

One of the reasons why programming is difficult is that programs are abstract. The strength of Programming by Demonstration comes from the fact that it allows the user to create an abstract rule by demonstrating what that rule should do in a specific situation.

KidSim rules are designed to bridge the gap between the specific and the abstract by conflating the specific example with the abstract rule. That is, the graphical representation of a rule has been designed so that the user can view this single representation in two different ways: 1) during rule creation, the rule should serve as a representation of the actions that the user has just performed, and 2) during rule application, the rule should serve as a representation of the general situation in which the rule is to apply, and of the general actions that are to occur when the rule is applied. For instance, while the rule in Figure 3 is being created, it should seem to the user that the fence in the rule refers to the fence to the right of the gremlin on the board. Later, when this rule is applied to a different gremlin confronting a different fence, it should seem natural to the user that the fence in the rule now refers to this completely different piece of fence.

As the user creates a rule, KidSim immediately abstracts the rule in a variety of ways, and the rule representation has been designed with the intent that none of these abstractions will confuse the user.

• **Instance abstraction.** For every character *instance* in the spotlight, the rule substitutes the character *type*. At creation time, since the appearance of the character type is likely to be quite similar to the appearance of the character instance on the board, this substitution should not create too large of a gap for the user to understand. And at application time, the appearance of the character instance on the board should be sufficiently close to the current appearance of the matching character in the rule that the user is able to understand that the characters do indeed match.

• **Motion abstraction.** When the user moves a character in the spotlight, the rule interprets this as a move *relative* to the starting location of the matching character. Users in fact expect this generalization, and have never expected the alternative interpretation that the move is to apply to the character instance with which the rule was originally created, or that the move is to this absolute location on the board.

• **Property abstraction.** When the user drags a property into the conditional box (on the left side of the rule) or the action box (on the right side of the rule), the rule generalizes this to refer to the property value, at the time of execution, of the matching character. At application time, users should not expect the original value from the time of rule creation to be used.

• **Property change abstraction.** When the user changes the value of a character's property, the rule displays a generalization that will produce the same change in the future (such as incrementing the value by 1). At creation time, users should not be surprised by the corresponding expression that appears in the rule.

Our user studies have shown that our rule representation is generally successful in serving its dual roles. However, the studies showed the last "should" mentioned above (for property changes) to be incorrect. In the original version of KidSim, rules to change the values of a property could only be created by demonstration. And some users were indeed confused during rule creation when, upon changing the value of a property, a statement of the form "Add 1 to my hunger" appeared in the rule.

In response to our user tests, we added a feature so that users could create the property-change expression by hand: the user selects "Put, Add, or Subtract" from a popup menu, types a number in the space to the right, and then drops in a property, to create a statement such as "Add 1 to my hunger". When we showed this new interface to one user, she found it understandable and was able to use it. We then explained how the same statement could be created by demonstration. She was now able to understand immediately this procedure which had previously confused her, and went so far as to state that it was much easier than creating the expression by hand. She now uses the demonstrational approach exclusively. We are planning to add a "recording agent" animation to KidSim, which will appear on the screen, watch the user change a property value, and then explicitly add the appropriate property-change expression to the rule. Perhaps with this additional feedback, the demonstrational approach will be easier for first-time users to understand.

Computer Literacy

Alan Kay has argued that there are powerful, yet difficult, concepts in programming, and that it is important that end user programming does not give up on teaching these concepts, in favor of making things easy [8]. Our intent is to remove the unimportant difficulties (like syntax) that unnecessarily impede students, so that they can proceed to work with the important concepts (like abstraction).

The main value of KidSim lies in an easy-to-use interface that allows children to create a fairly rich set of rules of behavior. Although KidSim is not a complete programming language — for instance, its control structure is limited to a fixed sequence for testing rules — it nonetheless embodies some important concepts in computer programming.

One of the advantages of KidSim's approach to programming is that it has eliminated many of the syntactic and semantic problems that confront users of conventional programming languages and scripting languages. Users need not worry about whether statements must end with a semicolon, or whether the word "the" is needed in an expression.

By working with KidSim, users can learn about the step-by-step execution of commands, about simple inheritance of properties and rules, about variables (i.e. properties), and about abstraction in rules.

IV. RELATED WORK

KidSim draws on four traditions in computer programming and human-computer interaction: production systems, graphical rewrite rules, programming by demonstration, and simulations. Executing the first matching rule in a list of test-action rules comes from production systems [3]. Graphical rewrite rules are used as a programming language in BITPICT [6], and are used to program simulations in Tableau [7], ChemTrains [1], and AgentSheets [11]. Mondrian [10] uses programming by demonstration to create rules represented by before-after pictures, similar to the storyboard representation of Chimera's macros [9]. The mixed icons and text used to represent program steps was used in Shoptalk [2].

KidSim is a successor to Playground [5]. Playground had the same goal of enabling children to create simulations. The most important positive thing that we learned from Playground was that its basic model of allowing children to create their own characters, and to attach rules to characters, was powerful and engaging. The most important negative thing that we learned was that scripting languages are too hard for children, even though Playground characters had a nice structure for storing programs, and the system provided a structure editor for creating syntactically correct statements. We also learned that characters should be allowed to directly manipulate other characters, since Playground did not allow this, and users found it too restrictive.

The system closest to KidSim is AgentSheets [11]. AgentSheets is a general-purpose simulation environment, which experienced developers can use to create a variety of domain-specific applications. That is, AgentSheets itself is not intended for end users, but it provides a set of tools that programmers can use to produce a great variety of systems tailored to specific end users. Notably, Repenning produced a domain-specific application for Turing machines that uses graphical rewrite rules, with end users creating Turing machine programs by demonstrating the steps in the program [11, pp. 80 - 82].

LiveWorld [13] is similar to KidSim in the style of the simulations that its users can create. Users make objects that move around on a board and interact with nearby objects. However, all programming in LiveWorld is done in Lisp, and therefore it is a much more powerful environment. Also, LiveWorld characters can contain sensors, which are a general and very effective means for specifying regions and objects of interest in a rule. LiveWorld also employs a powerful inheritance scheme.

Pinball Construction Set and SimCity are games that allow users to create interesting simulations, but users are limited to the pre-programmed behavior built into the objects that come with the application. KidSim is more akin to Rocky's Boots, which lets users create simple programs from a small set of primitives.

V. CONCLUSION
Future Directions

There are several features that we are planning to add to KidSim. We would like rules to be able to produce sounds, and to be able to test for sounds. We plan to introduce a feature to rules for referring to objects that are not spatially close to the character. We want to add buttons and switches, so that users can interact with a simulation as it is running, as in video games. And we are adding subroutines, so that rules can be grouped together. Subroutines should help to manage the complexity of having large numbers of rules, and also provide some additional control over the order in which rules are tested.

We would like to conduct further user tests to see whether children are able to use properties effectively, and whether they can understand large sets of rules. We would also like to determine whether KidSim is suitable for younger children.

Summary

The KidSim environment enables children to create their own simulations, and it encourages them to explore dynamic interactions between objects. Through programming by demonstration, children are able to create rules without using a conventional programming language. We have described how a process of prototyping and iterative design with informal user studies has helped us to build a system which is easy for end users to use, and which provides them with an expressive medium.

Acknowledgments

Warm and special thanks go to the teachers who motivated us, struggled with our early prototypes, and helped design KidSim: Betty Jo Allen-Conn, Julaine Salem, and Phyllis Lewcock.

We would like to thank Alan Kay for many important ideas. His respect for teachers, for the art of teaching, and for the importance of powerful ideas has informed our work.

Thanks to Jim Spohrer, David Maulsby, Edwin Bos, Kurt Schmucker, Stephanie Houde, Jeff Bradshaw and Rachel Bellamy for many valuable discussions.

We appreciate the following people for advocating important features: Alex Repenning for sequential execution, Enio Ohmaye for jars, and Peter Jensen for simple inheritance.

Many thanks to the programmers who have assisted us: Peter Jensen for the port to Prograph, Rodrigo Madanes for appearances, Dave Vronay for the drawing editor, David Maulsby for jars, Edwin Bos for subroutines, and Don Tilman and Ramón Felciano.

For his strikingly creative animations and graphics, we thank Mark Loughridge.

Thanks to the SK8 team. KidSim wouldn't exist without this wonderful environment: Ruben Kleiman (architecture), Adam Chipkin (scripting), Hernan Epelman-Wang (graphics), Brian Roddy (interface), and Alan Peterson (support).

For management and support, we thank Dana Schockmel, Kurt Schmucker, Jim Spohrer, Mark Miller, and Rick LeFaivre.

For the UK user test, we thank David Gilmore, Karen Pheasey, and Jean and Geoff Underwood. Their study was funded by the UK Economic and Social Research Council and NATO.

And finally, thanks to all of the students who tested KidSim.

REFERENCES

1. Bell, B. & Lewis, C. ChemTrains: A Language for Creating Behaving Pictures. In *1993 IEEE Workshop on Visual Languages*. Bergen, Norway, 1993, pp. 188-195.

2. Cohen, P. Synergistic Use of Direct Manipulation and Natural Language. In *Proceeding of CHI '89*. ACM, New York, 1989, pp. 227 - 234..

3. Davis, R. & King, J. An overview of production systems. Rep. STAN-CS-75-524, Computer Science Dept., Stanford Univ., Stanford, CA, 1975.

4. diSessa, A. A principled design for an integrated computational environment. In *Human-Computer Interaction*, 1, 1985, pp. 1 - 47.

5. Fenton, J. & Beck, K. Playground: An object-oriented simulation system with agent rules for children of all ages. In *Proceedings of OOPSLA '89*. ACM, New York, 1989, pp. 123 -137.

6. Furnas, G. New graphical reasoning models for understanding graphical interfaces. In *Proceeding of CHI '91*. ACM, New York, 1991, pp. 71 - 78.

7. Kay, A. Tableau, unpublished manuscript.

8. Kay, A. Foreword to Cypher, A. (Ed.), *Watch What I Do: Programming by Demonstration*, MIT Press, Cambridge, MA, 1993.

9. Kurlander, D. & Feiner, S. A History-Based Macro by Example System, In Cypher, A. (Ed.), *Watch What I Do: Programming by Demonstration*, MIT Press, Cambridge, MA, 1993, pp. 323 - 340.

10. Lieberman, H. Mondrian: A Teachable Graphical Editor, In Cypher, A. (Ed.), *Watch What I Do: Programming by Demonstration*, MIT Press, Cambridge, MA, 1993, pp. 341 - 360.

11. Repenning, A. AgentSheets: A tool for building domain-oriented dynamic, visual environments. Ph.D. dissertation, Dept. of Computer Science, University of Colorado at Boulder, 1993.

12. Smith, D.C., Cypher, A. & Spohrer, J. KidSim: Programming Agents Without a Programming Language. In *Communications of the ACM*, 37(7), July 1994, pp. 54 - 67.

13. Travers, M. Recursive Interfaces for Reactive Objects, In *Proceeding of CHI '94*. ACM, Boston, 1994, pp. 379 - 385.

14. Vertelney, L. Using Video to Prototype User Interfaces. In *SIGCHI Bulletin*, 21(2), October 1989, pp. 57 - 61.

Building Geometry-based Widgets by Example

Dan R. Olsen, Jr., Brett Ahlstrom, Douglas Kohlert
Computer Science Department, Brigham Young University
Provo, UT 84602
{olsen,ahlst,dkohl}@issl.cs.byu.edu

ABSTRACT

Algorithms are presented for creating new widgets by example. The basic model is one of an editable picture which can be mapped to control information. The mappings are learned from examples. The set of possible maps is readily extensible.

KEYWORDS: Widgets, Demonstrational Interfaces, Tool kit Builder, User Interface Software

INTRODUCTION

This paper describes a tool called LEMMING (Learning Easy Maps to Make Interesting New Gadgets) which allows new geometry based widgets to be created by example. A major portion of most direct manipulation user interfaces consist of interactive widgets of various sorts. These small fragments of interactive behavior are composed together to form complex interfaces for a variety of tasks. In almost all cases the model for these widgets is event based. That is, a majority of the interactive functionality is based on the handling of input events.

There are, however, some widgets which are geometrically defined. A prime example being the scroll bar. Although the normal scroll bar does have event behavior its primary model is the geometric manipulation of the thumb. The position of the thumb is tied to the value being controlled by the widget. Dragging the thumb changes the value. Changing the value, moves the thumb. There are, however, a variety of other possibilities like those shown in Figure 1. One manipulates speed / direction. The other controls pan and zoom.

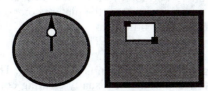

Figure 1 - Geometric widgets

Current event-based tool kits provide little help for creating such geometrically defined widgets. Programmers must, by hand, work out the geometry and interactive behavior.

This paper describes a system for creating new geometry-based widgets by example.

Geometry-based Widgets by Demonstration

The model that we are using for a widget is that it consists of a presentation part and a control part. The presentation part can be interactively changed by a user. These changes must then be mapped to corresponding changes to the control part. The application code can also change control data which the widget must map to changes in the presentation part and then must update the widget's display. The approach then is to provide general purpose mechanisms for users to change the presentation and then automatically learn the mapping between the presentation and the control.

To illustrate how new widgets are created, let us take the speed and direction widget shown in Figure 1. Figure 2 shows three examples of the widget along with the appropriate values controlled by the widget.

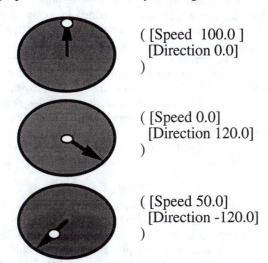

Figure 2 - Three widget examples

The designer would draw these three examples along with specifying the control values for each example. The widget system then learns the relationship between the arrow and the Direction value and between the oval position and the Speed value. Once the relationships have been learned, a new widget can be generated which has Speed and Direction as its control values and uses a highly restricted version of the draw editor as its user interface. This restricted draw editor can only drag objects that are not frozen and edit text that is not a constant.

Overview of the Widget Learning Technique

For purposes of our discussion the drawing of the widget will be called the presentation data and the information on the right of Figure 2 will be referred to as the control data. The widget creation tool must provide editors for both presentation data and control data. Internally both presentation and control data are represented in the same way.

A standard drawing program is not sufficient to serve as the presentation editor. The reason for this is that the drawing must be constrained in how it can be changed. The circle that forms the dial base must be fixed so that it cannot be moved. The arrow must be fixed so that only its head can move and that only along the perimeter of the circle. The small circle must be fixed to move only along the shaft of the arrow. We have designed just such a draw editor which allows users to readily define constrained movements without complex equations or geometric constructions. This technique based on parametric coordinate spaces is essential to our widget tool but is beyond the scope of this paper. This technique will only be describe in general terms.

For our example, the arrow's position is defined in the parametric coordinate system of the large circle. The parametric coordinate system of a circle is polar, which means that the arrow's points are defined in terms of an angle and a distance from the center. The small circle's position is defined in terms of the parametric coordinates of the arrow. The arrow's parametric coordinates are a Cartesian system with one axis along the arrow and one axis perpendicular to the arrow. Our draw editor provides a natural interface for defining these relationships that is not significantly more difficult than normal draw editors. In fact other than the general nature of the coordinates (polar, Cartesian or curved) of each object, widget designers are completely unaware of what the actual coordinate values are.

Using the editors for presentation and control data the designer provides the system with snapshots of presentation / control pairs such as those shown in Figure 2. The algorithm first analyzes the presentation and control examples independently to determine what has changed. This identifies the variable information that must ultimately be linked between the two models to form the working widget. In our speed/direction widget the presentation changes in one of the parametric coordinates of the arrow head and in one of the parametric coordinates of the small circle's position along the arrow. The control data changes in the values of Speed and Direction.

Because the arrow is defined in the parametric coordinates of the dial (which are polar) rather than in Cartesian coordinates, there is only one variable (radial angle) when the arrow head is moved around the perimeter of the circle, rather than the two which would be required for Cartesian-based drawing. Because the small circle is defined in the parametric coordinates of the arrow, it also changes in only one variable (along the shaft) regardless of the fact that the arrow's position changes radically between the three examples.

Once the changing components of both sets of data have been identified, the algorithm then learns mapping relationships between the presentation variables and the control variables. In our example there is a linear relationship between the arrow head position and the Direction control and there is a linear relationship between the small circle's position along the arrow and the Speed control. The algorithm also automatically learns the two coefficients for each linear relationship.

Our map learning algorithm is general enough to work in Cartesian space but there are serious problems in learning the maps and in constraining the drawing. The pan and zoom widget in Figure 1 or a simple scroll bar could easily be learned using a Cartesian rather than parametric drawing tool. To learn the dial in Cartesian space would require learning sinusoidal relationships between the (X,Y) position of the arrow head and the Direction control. General sinusoidal relationships have four coefficients which must be learned. This greatly complicates the learning algorithm. The relationship between the oval position and Speed in Cartesian space is unlearnable for practical purposes, as will be discussed later.

Once we have identified the variable parts of each side and the mapping relationships that connect those parts, we have learned the heart of our widget. To create a runtime version of the widget we combine a stripped down version of the presentation editor with the mapping information.

Other Widget Building Techniques

As discussed earlier, traditional widget building techniques such as Motif [1], Visual C++[2] or MacApp [3] are based on the handling of events and leave all geometric issues to the programmer. Peridot [4] provides mechanisms for creating the constraints among the picture components by example but provides little help in connecting the visual components to application data. The Peridot technique is in some sense related to the way in which we learn mapping relationships but this connection is very distant. ThingLab [5] provides mechanisms for visually creating constraints but the designer is required to visually program the actual constraints and to work out the geometric mathematics before building the constraints.

Juno [6] is much closer to the type of tool that we have built. Juno provided a visual mechanism for defining constraints on drawings. The designer would not need to understand or solve the underlying equations but would need understand compass and straightedge style geometric constructions. The set of constraints that Juno can handle is fixed and not easily extended.

Juno also only provided constraints between picture elements. As a drawing tool rather than a widget design tool, it had no mechanism to tie pictures to control data. The Juno approach was extended to widget design in GITS [7]. Additional constraints were added to tie pictures to data and an efficient solver was added to overcome the numerical slowness of Juno. The complexities of compass and straight-edge were still present, however, and the set of constraints was even more rigidly limited than in Juno.

Our goal was to provide a widget builder which would not require the designer to understand the underlying mathematics or geometry. It should be extensible in that new kinds of relationships could be added to the algorithm and it could be applied to a wide variety of presentations. This is in contrast with Wolber's DEMO [8] system which only supports linear relationships. In this paper we will focus on presentations which are 2D drawings in parametric coordinates but the algorithm is in no way limited to that domain.

Overview

In discussion the implementation of our widget design system we will first discuss the NIC data model and pattern language on which the algorithms are based. The concepts embodied in this data model and its pattern language are key to the widget learning algorithm. This data model provides the extensibility in the kinds of presentations which can be used to learn new widgets. This concepts and algorithms can be readily adapted to other data models without difficulty. We next discuss the class of MapObjects which are the heart of the mechanism for relating presentation variables to control variables. This class provides extensibility in the kinds of relationships that can be learned. This discussion is followed by the complete algorithm for learning the mapping between the presentation and the control. Lastly these techniques are assembled together in a working tool for designing widgets by example.

NIC INTRODUCTION

LEMMING is implemented in NIC (Nucleus for Interactive Computing) which is being developed as a comprehensive system for providing user interfaces to information. This introduction only covers those portions of NIC which are essential to LEMMING. For a more complete discussion of NIC the reader is referred elsewhere [9]. The central concept of NIC is that, like Lisp, everything in the system is represented in a single uniform data model.

Widgets, drawings, scripting language and user information are all represented in the same data model. Also defined in this model is a pattern language based loosely on unification which provides for general manipulation of data. These two concepts are the heart of the map learning algorithm for creating widgets by example.

Data Model

As with Lisp, the NIC data model provides a set of primitive values. These include long integers, floats, characters, symbols (as in Lisp) and an empty value. Instead of Lisp lists, NIC defines an abstract class Object which is defined in C++. Every object in NIC has an attribute part and an array part. In the attribute part, component values are referenced by symbolic name.

As in Lisp, NIC defines a textual notation for its objects. For example, Figure 3 shows an object with the attributes A and Name and the values 1,George and 2 in its array part. This example is an object of class :NewCls. Attributes are given with their names and values in square brackets. Array values are simply listed in order without names. Note that the attribute Name has another object for its value.

```
(:NewCls
        [A 74]
        [Name ( "Fred" "Allen" ) ]
        1 George 2
)
```

Figure 3 - Sample NIC object

There is a type Value which can hold any primitive or object. Most objects store Values as their components which provides a data modeling capability which is as flexible as Lisp.

The Object class does not define an implementation for objects. It only defines the standard protocol for setting and getting attribute values and indexed values. This puts a very flexible facade over very efficient implementations. For example the Image class presents itself with all of its pixels in its array part. In reality the pixels are stored in X image format for efficient display. The built in push button object is a special NIC class where the attributes Label and Action have special semantics. The point, however, is that given the Object class interface, all of these kinds of objects can be manipulated in terms of attributes and array indices without knowing how they are actually implemented.

Our speed / direction widget uses parametric drawing for its presentation. These drawings are defined as a set of special classes that know about drawing and editing. We can have a very simple NIC view of a Line object such as
        (:Line (10 20) (500 20) [Color Black]).
This view of a line describes the essential properties of the line including its color and endpoint coordinates. The line objects in our draw editor are somewhat more complex than this, but not in any ways that are important to the widget learning algorithm. The Line class, however, in addition to is external appearance as a NIC object contains a variety of functionality to draw itself, perform mouse hit detection and other interactive chores. For our purposes the NIC facade provides us with a uniform model the essential data

values. The map learning algorithm is based on this view of the data and is essential to the extendibility of the algorithm.

Pattern Language
In order to learn a map between the presentation data and the control data the algorithm needs to learn a general description of each of the data objects. For this we use NIC's pattern language. Only a brief introduction is given here. More extensive documentation is elsewhere [10]. An object can itself be a pattern. More general patterns can be created using don't care symbols (_). For example, the pattern

        ([Name (_ "Jones")]
            [Age 24]   )

will match any object which has Jones as the second element of the Name attribute and has 24 as its Age attribute. Attributes which are don't cares are omitted as well as any trailing don't cares in the array part.

Variables can also be added to patterns. A variable is a symbol beginning with an exclamation point (!). For example the pattern

        ([Name (!FN !LN)] [Age 24))

will match any object which has a Name attribute with an object having two array elements and has an Age attribute containing 24. If this pattern is matched against the object

        ([Name (**Dan Jones**)]
            [Age 24] [Weight 190])

the match will succeed and the symbol Dan will be assigned to the variable !FN and Jones will be assigned to the variable !LN. This is very similar to what might happen if a Prolog clause were matched against a data object. The generalization of objects into patterns and the assignment of values to variables as part of the match operation are central to our map learning algorithm.

There is an additional pattern operation called modify. It works like matching except that the object is changed by the variables rather than assigning to the variables. For example if !FN = Fred and !LN = Smith and the example pattern is used to modify the preceding data, the resulting object would be

        ([Name (**Fred Smith**)]
            [Age 24] [Weight 190])

Match assigns object values to variables. Modify assigns variable values into objects. The two operations are converses of each other.

Maps With Match and Modify
The combination of the match and modify operations on patterns provide the basis for a mapping between two objects. Take for example the objects in Figure 4.

P: (:Line (12 35) (**14** 2) [Color Black])
C: ([Location **14**])

Figure 4 - objects to be mapped.

P is the presentation object and C is the control object. What we would like is to link the X coordinate of the line's second point to the Location attribute of the control data. The algorithm can do this with the two patterns shown in Figure 5.

P: (_ (!V))
C: ([Location !V])

Figure 5 - Identity mapping patterns.

Suppose that the Line object, P, changes its X coordinate to 20. The system can then perform the presentation to control mapping by matching pattern P against the new object P, which will assign the 20 into !V. It can then modify object C with pattern C which will change the Location attribute to 20. If it was the Location attribute that changed the process would be reversed to perform the control to presentation mapping. This form of identity mapping between objects is too restrictive for our needs but it is the foundation for the more complete mapping algorithm.

Generalizing Examples into Patterns
When designers demonstrate new widgets they provide a set of snapshots of what is wanted. The first step of our algorithm is to identify what varies among the examples. Consider, for example, the three snapshots shown in Figure 6.

P1: (:Line(2 5) (**6 6**)[Color Black])
C1: ([Width **14**] [Height **12**])

P2: (:Line(2 5) (**7 10**)[Color Black])
C2: ([Width **16**] [Height **20**])

P3: (:Line(2 5) (**4 12**)[Color Black])
C3: ([Width **10**] [Height **24**])

Figure 6 - Examples to learn a map

In this case the mapping between the X coordinate of the line's second end point and the Width attribute in the control object is a linear mapping rather than an identity mapping. The same is true of the Y coordinate and the Height attribute. By comparing P1, P2 and P3 and also comparing C1, C2 and C3 the system can generalize the snapshot objects into the patterns Pp and Cp shown in Figure 7.

```
Pp: (:Line (2 5) ( !P1 !P2 ) [Color Black] )
Cp: ( [Width !C1] [Height !C2])
```

Figure 7 - Patterns generalized from examples

Since the mappings only require the variable parts of patterns the system can prune away all of the constant portions of these patterns to produce the patterns in Figure 8. The reader is referred to the pattern language documentation for details of this pruning algorithm. The pruning step is not required but significantly improves run-time performance on complex objects by eliminating non-essential pattern information.

```
Pp: ( _ (!P1 !P2) )
Cp: ( [ Width !C1] [Height !C2])
```

Figure 8 - Pruned patterns

The remainder of the map learning problem is to learn the linear mapping between !P1 and !C1 and between !P2 and !C2. It is also a problem to learn that !P1 is related to !C1 and not to !C2. A careful inspection of the examples will show that such a linear map cannot exist but the algorithm must learn this automatically.

MAP FUNCTION OBJECTS

There are a large number of ways in which variables in the presentation pattern might be related to variables in the control pattern. Our examples have shown the identity maps and linear maps. These maps must be two-way in the sense that if either side of the map changes the other side must be recalculated and modified to maintain the consistency. Examples of such mapping functions are found in Figure 9.

```
Identity: P = C
Linear: P = a C + b
Log: P = a log(C) + b
TextToFloat: P = FloatToStr( C )
```

Figure 9 - Example mapping functions

Each of these maps is invertable in the sense that P can be calculated from C and C from P. What we need, however, is not a fixed set of such functions but rather an open architecture for any such mappings.

The Abstract MapObject Class

In order to create an open architecture we define an abstract C++ class called MapObject. This class has three methods which are important to our discussion. These are:

float LearnMap(PresentSamples, ControlSamples, Accuracy)

Value MapToPresent (Value)
Value MapToControl (Value)

To create a new map of a particular type between two variables !P and !C, a new map object is created. Lists of sample values for !P and !C are passed to the method LearnMap along with a floating point number which indicates how accurate the samples must be in order to the map to be successfully recognized. For geometric maps this accuracy must be quite loose, since designers are not very accurate in the inputs that they provide by drawing. Based on these two lists of sample data, the MapObject will attempt to discover a map which is consistent with the data. Any information about the map that is learned is stored as attributes on the MapObject. An example of this would be the coefficients for a linear map.

The result of LearnMap is a confidence that a map of the desired type was actually learned. If, for example a LinearMap object is given samples which are not linearly related, LearnMap will return 0.0. If all samples are collinear it returns 1.0. If the samples are close to linear then a value between 0.0 and 1.0 is returned depending on how close they really are. The computation and meaning of the confidence depends on the particular class.

The method MapToPresent is invoked whenever the control variable is changed. This will compute the presentation value from the control value using the information learned by the map object. The method MapToControl will compute the control value from a modified presentation value.

If we take the snapshots in Figure 6 and the patterns in Figure 8 we would have the following lists of variable values:

!P1: (6, 7, 4)	!P2: (6, 10, 12)
!C1: (14, 16, 10)	!C2: (12, 20, 24).

If we use an IdentityMapObj to compare !P1 and !C1 the LearnMap method will return 0.0 because 6 is not remotely close to 14. If we use a LinearMapObj to compare !P2 and !C1 it will also return 0.0 because there are no coefficients a and b which will map values of !C1 to corresponding values of !P2. If the LinearMapObj is used to compare !P1 to !C1 it would return 1.0 and would set its attributes to be:

(:LinearMapObj [Present !P1] [Control !C1][a 0.5][b -1.0])

Based on the MapObject abstract class a wide variety of maps can be implemented as subclasses of MapObject. The map learning algorithm can use any MapObject without knowing the exact map being learned.

Number of examples required

An important question in demonstrational techniques is the number of examples that a designer is required to create. In most cases this can be determined by the number of unknowns in the mapping function plus one. If, as in the linear map, there are two coefficients, any two pairs of real numbers will produce a linear mapping. The LinearMap object requires at least three samples. Any two of the samples define a line with the third sample confirming that a line really is the appropriate choice. Any additional samples improve the estimate of the accuracy of the map. One of the reasons that sinusoidal maps are not used is that the equation "P = a sin(b C + d) + e" has four coefficients. This would mean that its map object would require 5 samples to confirm that a sinusoidal map is consistent with the data. This is generally too many for users to tolerate or to enter accurately.

A Complete Map Model

Based on the MapObjects we now can define a complete interactive map as consisting of
- a presentation pattern
- a control pattern
- a list of map objects which ties
  them together.

Given the data in Figure 6 the algorithm can produce the map shown in Figure 10.

```
Pp: ( _ (!P1 !P2) )
Cp: ( [ Width !C1] [Height !C2])
Maps:
     (:LinearMap [Present !P1][Control !C1][a 0.5][b -1.0] )
     (:LinearMap [Present !P2] [Control !C2][a 0.5][b 0.0] )
```

Figure 10 - Complete data mapping

The mapping algorithm then is

```
if the object P changes:
     Match pattern P against object P
     Run all maps using MapToControl
     Modify object C with pattern C
```

If object C is changed, then the reverse process is performed. This gives us a general model for computing mapping constraints between two arbitrary data objects.

A MAP LEARNING ALGORITHM

With the MapObject and the facilities of the pattern language we can now define the complete map learning algorithm. The inputs to the algorithm are:
- a list of (Present, Control) data pairs
- a list of MapObjects (one for each class to be used) in highest to lowest priority order.

Discovering Patterns

The algorithm takes all of the Present data values and generalizes them into a pattern using don't cares wherever the values differ. It then replaces all don't cares in the pattern with variables beginning with "!P", and then prune out all constant information. The same is done for the list of Control data values using variables beginning with "!C".

Discovering Maps

In order to discover the maps between variables in the Present and Control patterns the system must extract sample data. For each snapshot we match the Present pattern against the Present data and the Control pattern against the Control data. Using the examples in Figure 6 and the patterns in Figure 10 the system produces the objects shown in Figure 11. Note that variables are used as attributes. This is the variable storage model used by both match and modify.

```
1: ([!P1 6] [!P2 6] [!C1 14] [!C2 12])
2: ([!P1 7] [!P2 10] [!C1 16] [!C2 20])
3: ([!P1 4] [!P2 12] [!C1 10] [!C2 24])
```

Figure 11 - Variable values for snapshots

The map learning algorithm then is as follows:

```
For each presentation variable PV
     { For each control variable CV
          { LearnedMap = empty
          For each map object MO in ascending order
               { CMO = copy of MO
               if CMO.LearnMap( Listfor(PV),
                    Listfor(CV), accuracy)
                    > threshold then
               LearnedMap = CMO
               }
          if LearnedMap ≠ empty then
               add LearnedMap to list of maps
          }
}
```

This algorithm compares all variables on the presentation side to all variables on the control side using all possible mappings. The fact that this algorithm is $O(N^2M)$, where N is the number of variables and M is the number of maps, is not a problem. Because a widget with 20 variables would require at least 40 examples to learn, the limiting factor on this algorithm is designer patience rather than algorithm performance. It is rarely the case that a widget requires more than 4 or 5 variables.

GEOMETRIC WIDGET BUILDER

A cartoon of the interface for building geometric widgets is shown in Figure 12. Using the two editors, the designer creates snapshots and then presses the "Learn Map" button to invoke the learning algorithm. The interface provides feedback to indicate the success or failure of the process. Errors occur when some variables end up without maps attached or when presentation variables end up mapped to more than one control variable. This style of feedback is acceptable for small examples. In more complex widgets the designer is informed that there is a problem but is left with little help in determining what the problem is.

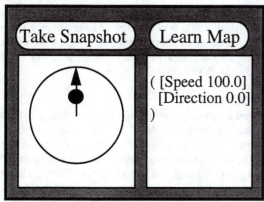

Figure 12 - Geometry Widget Builder

What is needed is an indication of what the learned maps actually are. The problem is finding mechanisms to convert the abstract information in the presentation pattern back into something visual that the designer can understand. As yet we have not solved this problem.

Run Time

The output of the geometry builder is a special widget class constructed as a stripped down version of the draw editor. Only the constrained dragging facilities remain. When a user drags an unconstrained part of the object, the Present-to-control mapping is run and when the control values are changed the reverse is performed. The generated widgets can then be installed in NIC's interface builder and used by interface designers as part of new interfaces. The generated widgets behave programmatically in the same manner as the built-in widgets in C++. The NIC interface system cannot really tell the difference.

MULTI-WAY MAPS

The problem of adequate feedback has been discussed earlier. There are two additional problems with multi-way mapping relationships which need to be addressed. They are multiple maps on the same variable and mapping objects with more than two variables.

Multiple maps on a variable

To illustrate this problem consider scroll bar in shown in Figure 13. It has a single control value Cur.

([Cur 20])

Figure 13 - Scroll bar widget

Three examples where the white rectangle is moved while the Cur attribute is changed will demonstrate this widget. The problem is that in the presentation, there are two variables (the left and right X coordinates of the rectangle) which both map to the same control variable. Let us suppose that the end user interactively drags only the left edge. A match will be performed which extracts the new value for the left edge and the old, unchanged, value of the

right edge. If the left edge map is run first it will change Cur to the new value. If the right edge map is then run, it will change Cur back to the old value.

This problem is overcome by checking presentation variables for change before running the maps. Since the right edge variable does not change, the system will not run that map. When the presentation to control mapping has been run the system then runs all of the control to presentation mappings. This will cause the right edge to change to correspond to the new value of Cur. Although we allow multiple maps on a control variable we do not allow multiple maps to apply to the same presentation variable since this would not make any sense in our widget design environment.

Mapping objects with more than two variables

The scroll bar is also an example of a more significant problem with our map learning technique. In most cases a scroll bar does not operate between two fixed bounds but rather between variable bounds. For example Cur might operate between two other attributes Min and Max. If the Max attribute changes (such as when new lines are added to a text document) the mapping of the current location to the scroll bar position also changes. This would require a mapping equation of the form

$$P = a\left(\frac{C - \min}{\max - \min}\right) + b$$

The first problem is that our simple $O(N^3)$ algorithm for discovering maps is not adequate because the equation defines a four-way relationship among variables rather than a two-way relationship. A simple minded approach to this problem is $O(N^5)$ which for 5 variables is $O(3125)$ which is doable but significantly more expensive. A second and more serious problem is that C, min and max can be bound in any permutation to Cur, Min or Max and still produce a valid set of coefficients a and b. Exchanging the map bindings of Min and Max will have no impact on the widget user. Exchanging the map bindings of Cur and Max will mean that if P is interactively changed, Max is changed rather than Cur. Solving this problem will involve more sophisticated analysis of the examples than the current system can provide. We pose it as a problem without a current solution.

LARGE EXAMPLE

Figure 14 shows a rather complex widget being constructed by example using the widget design interface. This widget is intend to be a 3D scrolling widget. The X dimension is controlled either by the slider across the front of the base or the position of the vertical post. The Y dimension is controlled by the slider on the vertical post and the Z dimension is controlled either by the slider along the right edge of the base or the position of the vertical post.

The drawing was created in our parametric drawing editor. The X and Z sliders have been defined in the coordinates of the parallelogram that forms the base. The vertical post is also defined relative to the base. The parametric coordinates of the base parallelogram follow the edges of the parallelogram rather than normal screen coordinates. This allows the Z slider to be naturally constrained to follow the edge rather than being free moving. The post's position is specified in the skewed coordinates of the base which correspond to those of the X and Z sliders. The Y slider is defined in relative to the post so that when the post moves the slider goes with it and the slider is constrained to only move vertically along the post.

The mapping of this drawing to the X, Y and Z controls requires 7 examples. The first three examples hold Y and Z constant while demonstrating three values of X by moving the X slider and post appropriately. The next two examples hold X and Y constant while demonstrating two additional examples for Z. The last two hold X and Z constant while demonstrating two additional values for Y. Creating these examples is not a real problem for a designer since the only trick is to handle each control independently. This chunking of examples by the user will make larger sets of examples manageable by designers.

SUMMARY

The widget by example algorithm presented here is defined exclusively in terms of NIC objects. The widgets and widget builder can be constructed from any presentation model, not just 2D drawings. All that is required is a NIC representation of the model and an editor. The map learning algorithm will learn maps between any two NIC objects and is readily extensible in the kinds of mappings used.

Although cartoon drawings are used in this paper instead of screen shots (for space reasons) this mechanism has been fully implemented on X windows with the generated widgets integrated into NIC's interactive tool kit.

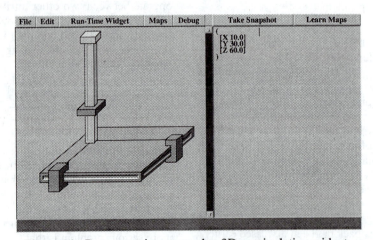

Figure 14 - Demonstrating a complex 3D manipulation widget

REFERENCES

1. McMinds, D.L., *Mastering OSF/Motif Widgets.* 1992, Canada: Hewlett-Packard Company.

2. Kruglinski, D.J., *Inside Visual C++.* 1993, Redmond, WA: Microsoft Press.

3. Wilson, D.A., L.S. Rosenstein, and D. Shafer, *Programming with MacApp.* 1990, Canada: .

4. Myers, B.A. and W. Buxton, *Creating Highly Interactive and Graphical User Interfaces by Demonstration.* Computer Graphics, 1986. **20**(4): p. 249-258.

5. Borning, A. *Defining Constraints Graphically.* in *CHI '86.* 1986. Boston: ACM.

6. Nelson, G., *Juno, a Constraint-based Graphics System.* Computer Graphics, 1985. **19**(3): p. 235-243.

7. Olsen, D.R. and K. Allan. *Creating Interactive Techniques by Symbolically Solving Geometric Constraints.* in *Third Annual Symposium on User Interface Software and Technology.* 1990. Snowbird, Utah: ACM.

8. Wolber, D. and G. Fisher. *A Demonstrational Technique for Developing Interfaces with Dynamically Created Objects.* in *UIST '91.* 1991. Hilton Head, South Carolina: ACM.

9. Olsen, D.R., *NIC: Documentation.* 1994, WWW URL: ftp://issl.cs.byu.edu/docs/NIC/home.html:

10. Olsen, D.R., *NIC: Pattern Language.* 1994, WWW URL: ftp://issl.cs.byu.edu/docs/NIC/NICPat.intro.html

Interactive Sketching for the Early Stages of User Interface Design

James A. Landay and Brad A. Myers
Computer Science Department
Carnegie Mellon University
5000 Forbes Ave.
Pittsburgh, PA 15213, USA
Tel: 1-412-268-3608
E-mail: landay@cs.cmu.edu
WWW Home Page: http://www.cs.cmu.edu:8001/Web/People/landay/home.html

ABSTRACT

Current interactive user interface construction tools are often more of a hindrance than a benefit during the early stages of user interface design. These tools take too much time to use and force designers to specify more of the design details than they wish at this early stage. Most interface designers, especially those who have a background in graphic design, prefer to sketch early interface ideas on paper or on a whiteboard. We are developing an interactive tool called SILK that allows designers to quickly sketch an interface using an electronic pad and stylus. SILK preserves the important properties of pencil and paper: a rough drawing can be produced *very quickly* and the medium is *very flexible*. However, unlike a paper sketch, this electronic sketch is *interactive* and can easily be *modified*. In addition, our system allows designers to examine, annotate, and edit a complete history of the design. When the designer is satisfied with this early prototype, SILK can *transform* the sketch into a complete, operational interface in a specified look-and-feel. This transformation is guided by the designer. By supporting the early phases of the interface design life cycle, our tool should both ease the development of user interface prototypes and reduce the time needed to create a final interface. This paper describes our prototype and provides design ideas for a production-level system.

KEYWORDS: User interfaces, design, sketching, gesture recognition, interaction techniques, programming-by-demonstration, pen-based computing, Garnet, SILK

INTRODUCTION

When professional designers first start thinking about a visual interface, they often sketch rough pictures of the screen layouts. In fact, everyone who designs user interfaces seems to do this, whether or not they come from a graphic

design background. Their initial goal is to work on the overall layout and structure of the components, rather than to refine the detailed look-and-feel. Designers, who may also feel more comfortable sketching than using traditional palette-based interface construction tools, use sketches to quickly consider various interface ideas.

Additionally, research indicates that designers should *not* use current interactive tools in the early stages of development since this places too much focus on design details like color and alignment rather than on the major interface design issues, such as structure and behavior [23]. What designers need are computerized tools that allow them to sketch rough design ideas quickly [22].

We are developing an interactive tool called SILK, which stands for <u>S</u>ketching <u>I</u>nterfaces <u>L</u>ike <u>K</u>razy, that allows designers to quickly sketch an interface using an electronic stylus. SILK then retains the "sketchy" look of the components. The system facilitates rapid prototyping of interface ideas through the use of common gestures in sketch creation and editing. Unlike a paper sketch, the electronic sketch allows the designer or test subjects to try out the sketch before it becomes a finished interface. At each stage of the process the interface can be tested by manipulating it with the mouse, keyboard, or stylus. Figure 1 illustrates a simple sketched interface. The interface has a scrollbar and a window for the scrolling data. It also has several buttons at the bottom, a palette of tools at the right, and four pulldown menus at the top.

Traditional user interface construction tools are often difficult to use and interfere with the designer's creativity. Our goal is to make SILK's user interface as unintrusive as pencil and paper. In addition to providing the ability to rapidly capture user interface ideas, SILK will allow a designer to edit the sketch using simple gestures. Furthermore, SILK's history mechanisms will allow designers to reuse portions of old designs and quickly bring up different versions of the same interface design for testing or comparison. Changes and written annotations made to a design over the course of a project can also be reviewed. Thus, unlike paper sketches, SILK sketches can evolve without forcing the designer to continually start over with a blank slate.

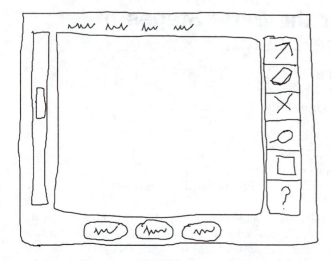

Figure 1: Sketched application interface created with SILK.

Unlike most existing tools, SILK will support the entire design cycle — from developing the initial creative design to developing the prototype, testing the prototype, and implementing the final interface. Our tool will provide the efficiency of sketching on paper with the ability to turn the sketches into real user interfaces for actual systems without re-implementation or programming. To some extent, SILK will be able to replace prototyping tools (*e.g.*, HyperCard, Director, and Visual Basic) and user interface builders (*e.g.*, the NeXT Interface Builder) for designing, constructing, and testing user interfaces (see Figure 2). At this time we have built a prototype of SILK that implements only a subset of the features described in this paper.

This paper describes how SILK functions and how it can be used effectively by user interface designers. The first section gives an overview of the problems associated with current tools and techniques. In the second section, we discuss the advantages of electronic sketching for user interface design. To ensure that the system would work well for its intended users, we took an informal survey of professional user interface designers to determine the techniques they now use for interface design. The results of the survey and a discussion of how these results were used in the design of SILK are presented in the next two sections. In the fifth section, we describe the sketch recognition algorithm. Finally, we summarize the related work and the status of SILK to date.

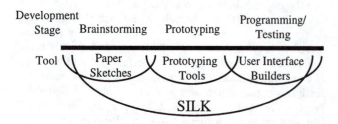

Figure 2: SILK can be used during all stages of user interface design, construction, and testing.

DRAWBACKS OF CURRENT DESIGN METHODS

User interface designers have become key members of software development groups. Designers often use sketching and other "low-fidelity techniques" [19] to generate early interface designs. *Low-fidelity techniques* involve creating mock-ups using sketches, scissors, glue, and post-it notes. Designers use mock-ups to quickly try out design ideas. Later they may use prototyping tools or user interface builders, or they may hand off the design to a programmer.

Prototyping tools allow non-programmers to write simple application mock-ups in a fraction of the time required using traditional programming techniques. *User interface builders,* the most common type of user interface construction tools, have become invaluable in the creation of both commercial and in-house computer applications. They allow the designer to create the look of a user interface by simply dragging widgets from a palette and positioning them on the screen. This facilitates the creation of the widget-based parts of the application user interface with little low-level programming, which allows the engineering team to concentrate on the application-specific portions of the product. Unfortunately, prototyping tools, user interface builders, and low-fidelity techniques have several drawbacks when used in the early stages of interface design.

Interface Tools Constrain Design

Traditional user interface tools force designers to bridge the gap between how they think about a design and the detailed specification they must create to allow the tool to reflect a specialization of that design. Much of the design and problem-solving literature discusses drawing rough sketches of design ideas and solutions [2], yet most user interface tools require the designer to specify more of the design than a rough sketch allows.

For example, the designer may decide that the interface requires a palette of tools, but she is not yet sure which tools to specify. Using SILK, a *thumbnail sketch* can easily be drawn with some rough illustrations to represent the tools (see Figure 1). This is in contrast to commercial interface tools that require the designer to specify unimportant details such as the size, color, finished icons, and location of the palette. This over-specification can be tedious and may also lead to a loss of spontaneity during the design process. Thus, the designer may be forced to abandon computerized tools until much later in the design process or forced to change design techniques in a way that is not conducive to early creative design.

One of the important lessons from the interface design literature is the value of iterative design; that is, creating a design prototype, evaluating the prototype, and then repeating the process several times [6]. Iterative design techniques seem to be more valuable as the number of iterations made during a project becomes larger. It is important to iterate quickly in the early part of the design process because that is when radically different ideas can and should be generated and examined. This is another area in which current tools fail during the early design process. The ability to turn out new designs quickly is hampered by the requirement for detailed designs. For example, in one test the interface sketched using SILK in Figure 1 could be created in just 70 seconds (sketched on paper it took 53 seconds), but to produce it with a traditional user interface builder (see Figure 3) took 329 seconds, which is nearly five times longer. In addition, the interface builder time does not include adding real icons to the tool palette due to the excessive time required to design or acquire them.

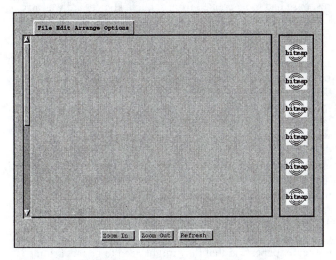

Figure 3: The sketched application from Figure 1 created with a traditional user interface builder.

HyperCard and Macromedia's Director are two of the most popular prototyping tools used by designers. Though useful in the prototyping stages, both tools come up short when used either in the early design stages or for producing production-quality interfaces. HyperCard's "programming" metaphor is based on the sequencing of different cards. HyperCard shares many of the drawbacks of traditional user interface builders: it requires designers to specify more design detail than is desired and often it must be extended with a programming language (HyperTalk) when the card metaphor is not powerful enough. In addition, HyperCard cannot be used for most commercial-quality applications due to its poor performance, which usually forces the development team to reimplement the user interface with a different tool.

Director was designed primarily as a media integration tool. Its strength is the ability to combine video, animation, audio, pictures, and text. This ability, along with its powerful scripting language, Lingo, has made it the choice of multimedia designers. These strengths, however, lead to its weaknesses when used as a general interface design tool. It is very hard to master the many intricate effects that Director allows. In addition, it lacks support for creating standard user interface widgets (i.e., scrollbars, menus, and buttons) and specifying their behavior in a straightforward manner. Finally, its full-powered programming language is inappropriate for non-programmers. This is also the major drawback to using Visual Basic, which is becoming increasingly popular for interface prototyping due to its complete widget set and third-party support.

Drawbacks of Sketching on Paper

Brainstorming is a process that moves quickly between radically different design ideas. Sketches allow a designer to quickly preserve thoughts and design details before they are forgotten. The disadvantage of making these sketches on paper is that they are hard to modify as the design evolves. The designer must frequently redraw the common features that the design retains. One way to avoid this repetition is to use translucent layers [24, 7]. Another solution is to use an erasable whiteboard. Both of these approaches are clumsy at best. In order to be effective, translucent layers require forethought on the part of the designer in terms of commonality and layout of components. Whiteboards make it hard to scale and move compound objects, and they do not allow the designer to delete elements from a list easily. None of these solutions help with the next step when a manual translation to a computerized format is required, either with a user interface builder or by having programmers create an interface from a low-level toolkit. This translation may need to be repeated several times if the design changes.

Another problem with relying too heavily on paper sketches for user interface design is the lack of support for "design memory" [9]. The sketches may be annotated, but a designer cannot easily search these annotations in the future to find out why a particular design decision was made. Practicing designers have found that the annotations of design sketches serve as a diary of the design process, which are often more valuable to the client than the sketches themselves [2]. Sketches made on paper are also difficult to store, organize, search, and reuse.

One of the biggest drawbacks to using paper sketches is the lack of interaction possible between the paper-based design and a user, which may be one of the designers at this stage. In order to actually see what the interaction might be like, a designer needs to "play computer" and manipulate several sketches in response to a user's verbal or gestural actions. This technique is often used in low-fidelity prototyping [19].

Designers need tools that give them the freedom to sketch rough design ideas quickly, the ability to test the designs by interacting with them, and the flexibility to fill in the design details as choices are made.

ADVANTAGES OF ELECTRONIC SKETCHING

Electronic sketches have most of the advantages described above for paper sketches: they allow designers to quickly record design ideas in a tangible form. In addition, they do not require the designer to specify details that may not yet be known or important. Electronic sketches also have the advantages normally associated with computer-based tools: they are easy to edit, store, duplicate, modify, and search. Thus a computer-based tool can make the "design memory" embedded in the annotations even more valuable.

The other advantages of electronic sketching pertain to the background of user interface designers and the types of comments sketches tend to garner during design evaluations. A large number of user interface designers, and particularly the intended users of SILK, have a background in graphic design or art. These users have a strong sketching background and our survey (see the next section) shows they often prefer to sketch out user interface ideas. An electronic stylus is similar enough to pencil and paper that most designers should be able to effectively use our tool with little training.

Anecdotal evidence shows that a sketchy interface is much more useful in early design reviews than a more finished-looking interface. Wong [23] found that rough electronic sketches kept her team from talking about unimportant low-level details, while finished-looking interfaces caused them to talk more about the "look" rather than interaction issues. The belief that colleagues give more useful feedback when evaluating interfaces with a sketchy look is commonly held in the design community. Designers working with other low-fidelity prototyping techniques offer similar recommendations [19]. In the field of graphic design, Black's user study found that "the finished appearance of screen-produced drafts shifts attention from fundamental structural issues" [1].

SURVEY OF PROFESSIONAL DESIGNERS

In order to focus on how best to support user interface design, we surveyed sixteen professional designers to find out what tools and techniques they used in all stages of user interface design. We also asked them what they liked and disliked about paper sketching, and what they liked and disliked about electronic tools. We also asked the designers to send us sketches that they had made early in the design cycle of a user interface. Using this information, we designed SILK to support the types of elements designers currently sketch when designing interfaces.

The designers we surveyed have an average of over six years experience designing user interfaces. They work for companies from around the world that focus on areas such as desktop applications, multimedia software, telephony, and computer hardware manufacturing. In addition, like our intended users, they all have an art or graphic design background.

Almost all of the designers surveyed (94%) use sketches and storyboards during the early stages of user interface design. Some reported that they illustrate sequences of system responses and annotate the sketches as they are drawn. The designers said that user interface tools, such as HyperCard, would waste their time during this phase. They said that a drawing and an explanation could be presented to management and tested with users before building a prototype. One designer stated that in the early stages of design "iteration is critical and must happen as rapidly as possible — as much as two or three times a day." The designer said that user interface builders always slowed the design process, "especially when labels and menu item specifics are not critical." Most of the designers also cited their familiarity with paper as a graphic designer. The pencil and paper "interface" was described as intuitive and natural.

Almost all of the designers surveyed (88%) use either HyperCard, Director, or Visual Basic during the prototyping stage of interface design. Some also use high-powered user interface builders. The designers reported that Director was only useful for "movie-like" prototyping, *i.e.*, as a tool to illustrate the functionality of the user interface without the interaction. In addition, the designers disliked Director because it lacked a widget set, the designs could not be used again in the final product, and every control and system response had to be created from scratch. HyperCard was also cited for its lack of some necessary user interface components.

In contrast, the designers complimented the user interface builders on their complete widget sets and the fact that the designs could be used in the final product. The difficulty of learning to use these tools, especially those with scripting languages, was considered a drawback. Also, the designers wanted the ability to draw arbitrary graphics and some tools did not allow this. In fact, most of the designers expressed an interest in being able to design controls with custom looks. Twelve of the sixteen (75%) reported that 20% or more of their time was spent designing this type of widget.

The designers reacted favorably to a short description we gave them about SILK. Some were concerned that it was not really paper and that they might need to get accustomed to it. The designers felt our system would allow quick implementation of design ideas and it would also help bring the sketched and electronic versions of a design closer together. In addition, the designers were happy with the ability to quickly iterate on a design and to eventually use that design in the final product. All but two expressed a willingness to try such a system.

DESIGNING INTERFACES WITH SILK

SILK blends the advantages of both sketching and traditional user interface builders, yet it avoids many of the limitations of these approaches. SILK enables the designer to move quickly through several iterations of a design by using gestures to edit and redraw portions of the sketch. Our system tries to recognize user interface widgets and other interface elements *as they are drawn* by the designer. Although the recognition takes place as the sketch is made, it is unintrusive and users will only be made aware of the recognition results if they choose to exercise the widgets. As soon as a widget has been recognized, it can be exercised. For example, the "elevator" of the sketched scrollbar in Figure 1 can be dragged up and down.

Next, the designer must specify the behavior *among* the interface elements in the sketch. For example, SILK knows how a button operates, but it cannot know what interface action should occur when a user presses the button. Some of this can be inferred either by the type of the element or with by-demonstration techniques [4, 16], but much of it may need to be specified using a visual language we are designing or even a scripting language for very complex custom behaviors. Our prototype does not yet support the specification of behaviors.

When the designer is happy with the interface, SILK will replace the sketches with real widgets and graphical objects; these can take on a specified look-and-feel of a standard graphical user interface, such as Motif, Windows, or Macintosh. The transformation process is mostly automated, but it requires some guidance by the designer to finalize the details of the interface (*e.g.*, textual labels, colors, *etc.*) At this point, programmers can add callbacks and constraints that include the application-specific code to complete the application. Figure 3 illustrates what the finished version of the interface illustrated in Figure 1 *might* look like had it been transformed by SILK (although SILK would have retained the sketched palette icons from Figure 1.)

Feedback

Widgets recognized by the system appear on the screen in a different color to give the designer feedback about the inference process. In addition, the type of the widget last inferred is displayed in a status area. Although both of these mechanisms are unobtrusive, the feedback can be disabled to allow the designer to sketch ideas quickly without any distractions.

The designer can help the system make the proper inference when either the system has made the wrong choice, no choice, or the widget that was drawn is unknown to the system. In the first case the designer might use the "cycle" gesture (see Figure 4) to ask the system to try its next best choice. Alternatively, the designer can choose from a list of possible widget types. If SILK made no inference on the widget in question, the designer might use the grouping gesture to force the system to reconsider its inference and focus on the components that have been grouped together. Finally, if the designer draws a widget or graphical object that SILK does not recognize, the designer can group the relevant components and then specify a name for the widget. This will allow the system to recognize the widget in the future.

Editing Sketches

One of the advantages of interactive sketches over paper sketches is the ability to quickly edit them. When the user holds down the button on the side of the stylus, SILK interprets strokes as editing gestures instead of gestures for creating new objects. These gestures are sent to a different classifier than the one used for recognizing widget components. The power of gestures comes from the ability to specify, with a single mark, a set of objects, an operation, and other parameters [3]. For example, deleting or moving a section of the drawing is as simple as making a single stroke with the stylus.

SILK supports gestures for cycling among inferences, deleting, moving, copying, and grouping basic components or widgets. The grouping gesture acts as a "hint" in the search for sequence and nearness relationships. Examples of these gestures are illustrated in Figure 4. As we test our system with more interface designers we expect to add gestures for other common operations.

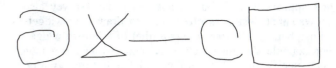

Figure 4: Gestures for cycling, deleting, moving, copying, and grouping.

Design History Support

One of the important features of SILK is its strong support for design history. A designer will be able to save designs or portions of designs for later use or review. Multiple designs can be displayed at the same time in order to perform a side-by-side comparison of their features or to copy portions of one design into a new design. SILK can also display several designs in a miniaturized format so that the designer can quickly search through previously saved designs visually rather than purely by name.

Another important history mechanism is SILK's support for annotations. The system will allow the designer to annotate a design by either typing or sketching on an *annotation layer*. This layer can be displayed or hidden with the click of a button in the SILK control panel. In addition, the annotations that were made using the keyboard can be searched later using a simple search dialog box. SILK will also support multiple layers, allowing different members of the design team to create personal annotations.

Specifying Behavior

In addition to editing and creating new objects in sketch mode, SILK also supports run mode and behavior mode. Run mode, which can be turned on from the SILK control panel, allows the designer to test the sketched interface. For example, as soon as SILK recognizes the scrollbar shown in Figure 1, the designer can switch to run mode and operate the scrollbar by dragging the "elevator" up and down. The buttons in Figure 1 can be selected with the stylus or mouse and they will highlight while the button is held down.

Easing the specification of the interface layout and structure solves much of the design problem, but a design is not complete until the behavior has also been specified. Unfortunately, the behavior of individual widgets is insufficient to test a working interface. Behavior mode will be used to specify the dynamic behavior between widgets and the basic behavior of new widgets or application-specific objects drawn in sketch mode.

We have identified two basic levels of behavior that the system must be able to handle. Sequencing between screens, usually in conjunction with hand drawn storyboards, is a behavior that has been shown to be a powerful tool for designers making concept sketches for early visualization [2]. The success of HyperCard has demonstrated that a significant amount of behavior can be constructed from sequencing screens upon button presses. For example, the designer may wish to specify that a dialog box appears when a button is pressed. Our survey also showed that designers often want to draw new widgets whose behavior is *analogous* to that of a known widget. For example, designers frequently need to draw a new icon and specify that it should act like a button. Finally, designers occasionally need to sketch a widget that has an entirely new behavior. Our experience is that this is not very common and thus we do not plan on supporting the definition of entirely new behaviors.

We are investigating several alternative ways to specify these behaviors. Programming-by-demonstration (PBD) is a technique in which one specifies a program by directly operating the user interface. In sketch mode we specify the layout and structure of the interface as described above, while in behavior mode we could demonstrate possible end-user actions and then specify how the layout and structure should change in response. A similar technique is used to describe new interface behaviors in the Marquise system [17].

A critical problem with PBD techniques is the lack of a static representation that can be later edited. Marquise and Smallstar [8] use a textual language (a formal programming language in the latter case) to give the user feedback about the system's inferences. In addition, scripts in these languages can then be edited by the user to change the "program". This solution is not acceptable considering that the intended users of SILK are user interface designers who generally do not have programming experience.

We may be able to solve this problem by combining PBD techniques with visual languages as in the Pursuit [13] and Chimera [11] systems. We are especially interested in using a visual notation that is made directly on the interface whose behavior is being described. Marks or symbols layered on top of the interface are used for feedback indicating graphical constraints in Briar [5] and Rockit [10]. In Rockit, the marks kept the user informed of the current inference of the system.

Using a notation of marks that are made directly on the sketch is beneficial for several reasons. One of the most important reasons is we can now use the same visual language for both the specification of the behavior and the editable representation indicating which behavior has been inferred or specified by the designer. In addition, these sketchy marks might be similar to the types of notations that one might make on a whiteboard or paper when designing an interface. For example, sequencing might be expressed by drawing arrows from buttons to windows or dialog boxes that appear when the button is pressed. Like the annotation layer described earlier, the layer that contains these behavioral marks can be turned on and off.

Another technique used to specify a behavior is to select it from a list of known behaviors and attach it to a drawn element. This seems well-suited for specifying analogous behaviors. This technique can be very limiting if the default behaviors do not include what a designer wishes to specify. We intend to survey many commercial applications and

designers to see if there is a reasonably small number of required behaviors so that they can be presented in list form. The success of the Garnet Interactor model indicates that a small list may be sufficient [14].

The visual language and PBD approaches provide specification methods that are similar to the way the interface is used. The list approach, however, is easy to use and may be quite successful for common behaviors. We expect to combine these techniques and then conduct user testing to refine the interface.

RECOGNIZING WIDGETS

Allowing designers to sketch on the computer, rather than on paper, has many advantages as we have already described. Several of these advantages cannot be realized without software support for recognizing the interface widgets in the sketch. Having a system that recognizes the drawn widgets gives the designer a tool that can be used for designing, testing, and eventually producing a final application interface. SILK's recognition engine identifies individual user interface components as they are drawn, rather than after an entire sketch has been completed. This way the designer can test the interface at any point without waiting for the entire sketch to be recognized. Working within the limited domain of common 2-d interface widgets (*e.g.*, scrollbars, buttons, pulldown menus, *etc.*) facilitates the recognition process. This is in contrast to the much harder problems faced by systems that try to perform generalized sketch recognition or beautification [18]. Our sketch recognition algorithm uses a rule system that contains basic knowledge of the structure and make-up of user interfaces to infer which widgets are included in the sketch.

Recognizing Widget Components

The recognition engine uses Rubine's gesture recognition algorithm [20] to identify the basic components that make up an interface widget. These basic components are then combined to make more complex widgets. For example, the scrollbar in Figure 1 was created by sketching a tall, thin rectangle and then a small rectangle (though the order in which they were sketched does not matter). Each of the basic components of a widget are trained by example using the Agate gesture training tool [12].

The algorithm currently limits our system to single-stroke gestures for the basic components. This means that the designer drawing the scrollbar in Figure 1 must use a single stroke of the pen for each of the rectangles that comprise the scrollbar. We intend to develop a better algorithm so that the designer can use multiple-strokes to draw the basic components.

Rubine's algorithm uses statistical pattern recognition techniques to train classifiers and recognize gestures. These techniques are used to create a classifier based on the features extracted from several examples. In order to classify a given input gesture, the algorithm computes the distinguishing features for the gesture and returns the best match with the learned gesture classes.

Composing Components

In order to recognize interface widgets, our algorithm must combine the results from the classification of the single-stroke gestures that make up the basic components. As each component is sketched and classified it is passed to an algorithm that looks for the following relationships:

1) Does the new component *contain* or is it *contained by* another component?

2) Is the new component *near* another component?

3) Is the new component in a *sequence* of components of the same type?

The first relationship is the most important for classifying widgets. We have noticed that many of the common user interface widgets can be expressed by containment relationships between more basic components. For example, the scrollbar in Figure 1 is a tall, skinny rectangle that contains a smaller rectangle. The second relationship allows the algorithm to recognize widgets such as check boxes, which usually consist of a box with text next to it. The final relationship allows for groupings of related components that make up a set of widgets (*e.g.*, a set of radio buttons.)

After identifying the basic relationships between the new component and the other components in the sketch, the algorithm passes the new component and the identified relationships to a rule system that uses basic knowledge of the structure and make-up of user interfaces to infer which widget was intended. Each of the rules that matches the new component and relationships assigns a confidence value that indicates how close the match is. The algorithm then takes the match with the highest confidence value and assigns the component to a new aggregate object that represents a widget. If none of the rules match, the system assumes that there is not yet enough detail to recognize the widget.

Adding new components to the sketch can cause the system to revise previously made widget identifications. This will only occur if the new component causes the rule system to identify a different widget as more likely than its previous inference. Similarly, deleting components of a widget can cause a new classification of the rest of the sketch.

Each of the widgets that SILK recognizes has corresponding Garnet objects that use the Garnet Interactor mechanism [14] to support interaction and feedback. When SILK identifies a widget, it attaches the sketched components that compose it to an instance of an interactor object that implements the required interaction.

STATUS

We currently have a prototype of SILK running under Common Lisp on both UNIX workstations and an Apple Macintosh with a Wacom tablet attached. The prototype is implemented using Garnet [15]. The prototype supports recognition and operation of several standard widgets. In addition, the system can transform SILK's representation of the interface to an interface with a Motif look-and-feel. The system currently only recognizes a few ways of drawing each widget. Using the sketches sent to us by designers, we

plan to extend the rule system to recognize more alternatives. SILK does not yet allow the specification of behavior between the widgets (*i.e.*, the sketchy scrollbar can be scrolled but it cannot yet be attached to the window containing data to scroll.) In addition, annotations are the only implemented history mechanism.

We are currently adding support to recognize more widgets and application-specific graphics. In addition, we are looking at ways to support multiple stroke recognizers. We plan to have design students use SILK in a user interface design course to see how it performs in practice. We have also begun designing a formal study to compare the types of problems found when performing an evaluation on both sketchy and finished-looking interfaces.

RELATED WORK

Wong's work on scanning in hand-drawn interfaces was the major impetus for starting our work in this area [23]. Our work differs in that we give designers a *tool* that allows them to create both the look and behavior of these interfaces directly with the computer. In addition, we will try to show that Wong's anecdotal evidence is supported in practice by comparing the types of comments made and the problems found when performing an interface evaluation on both sketchy and finished-looking interfaces.

Much of the work related to our system is found in the field of design tools for architects. For example, Strothotte reports that architects often sketch over printouts produced by CAD tools before showing works in progress to clients [21]. This seems to lend further evidence to the assertion that a different level of feedback is obtained from a sketchy drawing. In fact, Strothotte has produced a system that can render precise architectural drawings in a sketchy look.

Another important architectural tool allows architects to sketch their designs on an electronic pad similar to the one we are using [7]. Like SILK, this tool attempts to recognize the common graphic elements in the application domain — architectural drawings. Our tool differs in that it allows the specification and testing of the *behavior* of the drawing, whereas the architectural drawing is fairly static.

CONCLUSIONS

We envision a future in which most of the user interface code will be generated by user interface designers using tools like SILK rather than by programmers writing the code. We have designed our tool only after surveying the intended users of the system. These designers have reported that current user interface construction tools are a hindrance during the early stages of interface design; we have seen this both in our survey and in the literature. Our interactive tool will overcome these problems by allowing designers to quickly sketch an interface using an electronic stylus. Unlike a paper sketch, an electronic sketch will allow the designer or test subjects to interact with the sketch before it becomes a finalized interface. We believe that an interactive sketching tool that supports the entire interface design cycle will enable designers to produce better quality interfaces in a shorter amount of time than with current tools.

ACKNOWLEDGMENTS
The authors would like to thank Dan Boyarski, David Kosbie, and Francesmary Modugno for their helpful comments on this work. We would also like to thank the designers who responded to our survey. Finally, we would like to thank Stacey Ashlund, Elizabeth Dietz, and Dale James for help with technical writing.

This research was sponsored by NCCOSC under Contract No. N66001-94-C-6037, ARPA Order No. B326. The views and conclusions contained in this document are those of the authors and should not be interpreted as representing the official policies, either expressed or implied, of NCCOSC or the U.S. Government.

REFERENCES
1. Black, A. Visible planning on paper and on screen: The impact of working medium on decision-making by novice graphic designers. *Behaviour & Information Technology 9*, 4 (1990), 283–296.

2. Boyarski, D. and Buchanan, R. Computers and communication design: Exploring the rhetoric of HCI. *Interactions 1*, 2 (April 1994), 24–35.

3. Buxton, W. There's more to interaction than meets the eye: Some issues in manual input. In *User Centered Systems Design: New Perspectives on Human-Computer Interaction,* Norman, D.A. and Draper, S.W., Lawrence Erlbaum Associates, Hillsdale, N.J., 1986, pp. 319–337.

4. Cypher, A. *Watch What I Do: Programming by Demonstration,* MIT Press, Cambridge, MA (1993).

5. Gleicher, M. and Witkin, A. Drawing with constraints. *The Visual Computer 11*, 1 (1995), To appear.

6. Gould, J.D. and Lewis, C. Designing for usability: Key principles and what designers think. *Communcations of the ACM 28*, 3 (March 1985), 300–311.

7. Gross, M.D. Recognizing and interpreting diagrams in design. In *Proceedings of the ACM Conference on Advanced Visual Interfaces '94,* Bari, Italy, June 1994.

8. Halbert, D.C. *Programming by Example*, Ph.D. dissertation, Computer Science Division, EECS Department, University of California, Berkeley, CA, 1984.

9. Herbsleb, J.D. and Kuwana, E. Preserving knowledge in design projects: What designers need to know. In *Proceedings of INTERCHI '93: Human Factors in Computing Systems,* Amsterdam, The Netherlands, April 1993, pp. 7–14.

10. Karsenty, S., Landay, J.A., and Weikart, C. Inferring graphical constraints with Rockit. In *HCI '92 Conference on People and Computers VII,* British Computer Society, September 1992, pp. 137–153.

11. Kurlander, D. *Graphical Editing by Example*, Ph.D. dissertation, Department of Computer Science, Columbia University, July 1993.

12. Landay, J.A. and Myers, B.A. Extending an existing user interface toolkit to support gesture recognition. In *Adjunct Proceedings of INTERCHI '93: Human Factors in Computing Systems,* Amsterdam, The Netherlands, April 1993, pp. 91–92.

13. Modugno, F. and Myers, B.A. Graphical representation and feedback in a PBD system. In *Watch What I Do: Programming by Demonstration.* MIT Press, Cypher, A., Ch. 20, pp. 415–422, Cambridge, MA, 1993.

14. Myers, B.A. A new model for handling input. *ACM Transactions on Information Systems 8*, 3 (July 1990), 289–320.

15. Myers, B.A., Giuse, D., Dannenberg, R.B., Vander Zanden, B., Kosbie, D., Pervin, E., Mickish, A., and Marchal, P. Garnet: Comprehensive support for graphical, highly-interactive user interfaces. *IEEE Computer 23*, 11 (November 1990), 71–85.

16. Myers, B.A. Demonstrational Interfaces: A step beyond direct manipulation. *IEEE Computer 25*, 8 (August 1992), 61–73.

17. Myers, B.A., McDaniel, R.G., and Kosbie, D.S. Marquise: Creating complete user interfaces by demonstration. In *Proceedings of INTERCHI '93: Human Factors in Computing Systems,* Amsterdam, The Netherlands, April 1993, pp. 293–300.

18. Pavlidis, T. and Van Wyk, C.J. An automatic beautifier for drawings and illustrations. *Computer Graphics 19*, 3 (July 1985), 225–234, ACM SIGGRAPH '85 Conference Proceedings.

19. Rettig, M. Prototyping for tiny fingers. *Communications of the ACM 37*, 4 (April 1994), 21–27.

20. Rubine, D. Specifying gestures by example. *Computer Graphics 25*, 3 (July 1991), 329–337, ACM SIGGRAPH '91 Conference Proceedings.

21. Strothotte, T., Preim, B., Raab, A., Schumann, J., and Forsey, D.R. How to render frames and influence people. In *Proceedings of Eurographics '94,* Oslo, Norway, September 1994, pp. 455–466.

22. Wagner, A. Prototyping: A day in the life of an interface designer. In *The Art of Human-Computer Interface Design.* Addison-Wesley, Laurel, B., pp. 79–84, Reading, MA, 1990.

23. Wong, Y.Y. Rough and ready prototypes: Lessons from graphic design. In *Short Talks Proceedings of CHI '92: Human Factors in Computing Systems,* Monterey, CA, May 1992, pp. 83–84.

24. Wong, Y.Y. Layer Tool: Support for progressive design. In *Adjunct Proceedings of INTERCHI '93: Human Factors in Computing Systems,* Amsterdam, The Netherlands, April 1993, pp. 127–128.

Information Foraging in Information Access Environments

Peter Pirolli
Xerox Palo Alto Research Center
3333 Coyote Hill Road
Palo Alto, CA 94304
pirolli@parc.xerox.com

Stuart Card
Xerox Palo Alto Research Center
3333 Coyote Hill Road
Palo Alto, CA 94304
card@parc.xerox.com

ABSTRACT

Information foraging theory is an approach to the analysis of human activities involving information access technologies. The theory derives from optimal foraging theory in biology and anthropology, which analyzes the adaptive value of food-foraging strategies. Information foraging theory analyzes trade-offs in the value of information gained against the costs of performing activity in human-computer interaction tasks. The theory is illustrated by application to information-seeking tasks involving a Scatter/Gather interface, which presents users with a navigable, automatically computed, overview of the contents of a document collection arranged as a cluster hierarchy.

KEYWORDS: Information foraging theory, information access.

INTRODUCTION

Recent years have witnessed an explosive increase in public interest in information access and communication technologies. Along with this burgeoning excitement, the rapid growth of electronically available information sources, such as those available on the Internet, has further exacerbated the need for effective and efficient tools for information workers and consumers. For researchers and developers in human-computer interaction, this increases the need for models and analysis techniques that allow us to determine the value added by particular information access, manipulation, and presentation techniques, and to reveal design elements that may yield further enhancements.

To motivate previous designs of interactive information systems [6], we appealed to mechanisms of cognitive science [5], and to general principles of information science [17]. We have argued that in an information-rich world, the real design problem to be solved is not so much how to collect more information, but rather, how to optimize the user's time, and we have deployed these principles in an attempt to increase relevant information gained per unit time expended. But for task analysis, design exploitation, and evaluation of information systems, a more developed theory is needed.

In this paper, we lay out the framework for an approach we call *information foraging theory*. This approach considers the adaptiveness of human-system designs in the context of the information ecologies in which tasks are performed. Typically, this involves understanding the variations in activity afforded by some space of human-system design parameters, and understanding how these variations trade-off the value of information gained against the costs of performing the activity. While complementary with information processing approaches to computer interaction theory, such as those in the GOMS family [5, 14], information foraging theory emphasizes a larger time-scale of behavior, the cost structure of external information-bearing environments, and human adaptation.

Consider the time-scales of activity outlined by Newell [13]. The sorts of information-seeking and sensemaking activities of interest to us span from the middle of the *cognitive band* of activity (~100 ms - 10 s), across the *rational band* (minutes to hours), and perhaps into the *social band* (days to months). Typically, information-processing models of cognition have addressed behavior at the cognitive band, and elementary cognitive mechanisms and processes (e.g., such as those summarized in the Model Human Processor, [5]) play a large part in shaping observed behavior at that grain size. As the time scale of activity increases, "there will be a shift towards characterizing a system...without regard to the way in which the internal processing accomplishes the linking of action to goals" (Newell [13], p 150). Such explanations assume that behavior is governed by *rational* principles and largely shaped by the constraints and affordances of the task environment. Rather than assuming classical normative rationality, one may assume that the rationale for behavior is its adaptive fit to its external ecology [4]. This is the essence of an *ecological stance* (Neisser, as cited in Bechtel [4]) towards cognition. Whereas information-processing models, such as GOMS, provide mechanistic accounts of *how* cognition operates, ecological models address *why* it operates that way, given the ecological context in which it occurs. This kind of integrated explanatory framework has been promoted by Marr [12] and, more recently, Anderson [1, 2] in cognitive science.

INFORMATION FORAGING TASKS

Information foraging refers to activities associated with assessing, seeking, and handling information sources. Such search will be adaptive to the extent that it makes optimal use of knowledge about expected information value and expected costs of accessing and extracting the relevant information.

We use the term "foraging" both to conjure up the metaphor of organisms browsing for sustenance and to indicate a connection to the more technical *optimal foraging theory* found in biology and anthropology [21, 22].

Animals adapt their behavior and their structure through evolution to survive and reproduce to their circumstance. Essentially animals adapt, among other reasons, to increase their rate of energy intake. To do this they evolve different methods: a wolf hunts ("forages") for prey, but a spider builds a web and allows the prey to come to it. Humans seeking information also adopt different strategies, sometimes with striking parallels to those of animal foragers. The wolf-prey strategy bears some resemblance to classic information retrieval, and the spider-web strategy is like information filtering. Human hunter-foragers have been observed to hunt in groups when the variance of finding food is high. They accept a lower expected mean in order to minimize the probability of several days without food. Similarly, we have observed, in the field, professional market analysts who had developed an ethic of cross-referring information, essentially information-foraging in groups, so as to reduce the probability of missing important literature.

Optimal foraging theory is a theory that has been developed within biology for understanding the opportunities and forces of adaptation. We believe elements of this theory can help in understanding existing human adaptations for gaining and making sense out of information. It can also help in task analysis for understanding how to create new interactive information system designs.

Optimality models in general include the following three major components.

- *Decision assumptions* that identify which of the problems faced by an agent are to be analyzed. This might involve decisions about whether to pursue a given type of information item upon encountering it, the time to spend processing a collection of information items, the choice of moves to make in navigation, the choice of strategy under uncertainty, or degree of resource sharing.

- *Currency assumptions*, which identify how choices are to be evaluated. The general assumption in ecological analyses is that some feature x will exist over other features, if x satisfies some existence criteria. Existence criteria have two parts (a) a *currency*, and (b) a *choice principle*. Typically, optimal foraging models in anthropology and biology assume energy as a currency. Information foraging theory will assume information value as currency. Choice principles include maximization, minimization, and stability.

- *Constraint assumptions*, which limit and define the relationships among decision and currency variables. These will include constraints that arise out of the task structure, interface technology, and the abilities and knowledge of a user population.

We assume that information foraging is usually a task that is embedded in the context of some other task and its value and cost structure is consequently defined in relation to the embedding task and often changes dynamically over time [3, 18]. The *value* of information [16] and the *relevance* of specific sources [18] are not intrinsic properties of information-bearing representations (e.g., documents) but can only be assessed in relation to the embedding task environment.

Usually, the embedding task is some ill-structured problem for which additional knowledge is needed in order to better define goals, available courses of action, heuristics, and so on [15, 20]. Such tasks might include such things as choosing a good graduate school, developing a financial plan for retirement, developing a successful business strategy, or writing an acceptable scientific paper. The structure of processing and the ultimate solution are, in large part, a reflection of the particular knowledge used to structure the problem space. Consequently, the value of the external information may often reside in the improvements to the outcomes of the embedding task.

The use of optimality models should not be taken as a hypothesis that human behavior is classically rational, with perfect information and infinite computational resources. A more successful hypothesis about humans is that they exhibit *bounded rationality* or make choices based on *satisficing* [19]. However, satisficing can often be characterized as localized optimization (e.g., hill-climbing) with resource bounds and imperfect information as included constraints [23]. Optimality models do *not* imply that animals or information foragers will necessarily develop so as to embrace the simple optimum. Rather, they describe the possibilities of a niche, a possible advantageous adaptation if not blocked by other forces (for example, the consequences of another adaptation). For us, these models help fill in what Anderson [1] calls the Rational Level theory of information access.

OVERVIEW OF THE EXAMPLES

We present several examples of foraging analyses to illustrate some of the range of problems and insights that may be addressed. Our coverage has to be limited, so we use three relatively concrete and detailed analyses from a particular system.

In the context of a an analysis of the Scatter/Gather document browser [9] we introduce two simple models, the *information patch model*, and the *information diet model*, borrowed rather directly from optimal foraging theory. These "conventional" models derive from Holling's disc equation [22], which states that rate of currency intake, R, is the ratio of net amount of currency gained (energy in the case of biological systems; information value in our case), U, divided by the total amount of time spent searching, T_S, and exploiting, T_h,

$$R = \frac{U}{T_s + T_h}. \qquad (1)$$

The net gain, U, is the overall currency intake (the gross amount foraged), U_f, less the currency expended in foraging, C_f or $U = U_f - C_f$. It can be shown that broadly applicable stochastic assumptions approximate Holling's disc equation.

If we assume that information workers or consumers encounter information items as a linear function of time, the total number of items encountered while searching can be represented as λT_S, where λ is the rate of encounter with items per unit time. The average rate of currency intake will be $\bar{u} = \dfrac{U_f}{\lambda T_S}$ and the average cost of handling items will be $\bar{h} = \dfrac{T_h}{\lambda T_S}$, and so $U_f = \bar{u}\lambda T_S$ and $T_h = \bar{h}\lambda T_s$. If s is the search cost per unit time then the total cost of search will be sT_S. Substituting these into Equation 1, we have

$$R = \frac{\bar{u}\lambda T_S - sT_S}{T_S + \bar{h}\lambda T_S}$$

$$= \frac{\lambda\bar{u} - s}{1 + \lambda\bar{h}} \qquad (2)$$

The information patch model and the information diet model are formulated as variants of Equation 2. We discuss the analytic optimal solutions to these models in the context of the illustrations.

We also develop a more comprehensive dynamic model that incorporates the information patch model and information diet model as subcomponents. Using dynamic programming we illustrate how one may determine the optimal human-system strategies using dynamic programming techniques.

EXAMPLE 1: FORAGING IN INFORMATION PATCHES

A user's encounters with valuable or relevant information will typically have a clumpy structure over space and time. Information items are often grouped into collections such as libraries, databases, and wire services. The biological analogy is that an organism's ecology may have a variety of food patches of differing characteristics and the organism must decide how to best allocate its foraging time. Models of this situation are called *patch models* [7, 10, 22]. We discuss an information patch model in the context of the Scatter/Gather text database browser.

Scatter/Gather Interface

Figure 1 presents a typical view of the Scatter/Gather interface.[1] The emphasis in this browsing paradigm is to present users with a kind of automatically computed overview of the contents of a document collection, and to provide a method for navigating through this summary at different levels of granularity. This is achieved by organizing the collection into a *cluster hierarchy*. Conceptually, a collection may be clustered, for instance, into $B = 10$ groups of related documents. Each cluster is represented by a separate area as in Figure 1. For each cluster, the user is presented with typical words that occur in the text contents of documents in a cluster, as well as the titles of the three most prototypical documents. The user may *gather* some subset of those clusters, by pointing and selecting buttons above each cluster, and then ask the system to *scatter* that subcollection into another B subgroups, by selecting a Scatter/Gather button at the top of the display in Figure 1. The clustering is based on a form of inter-document similarity computation based on representations of text contents. Scatter/Gather browsing and clustering employs methods that can occur in constant interaction-time [8].

Patch Analysis

We can view Scatter/Gather clusters as *information patches*. Foraging in a cluster-patch corresponds to selecting a cluster, displaying the document titles belonging to the cluster in a scrollable window, scanning/scrolling through the listed titles, and for each title deciding if it is relevant or not to the query at hand. If the title is judged relevant, then it is handled by selecting, cutting, and pasting it to a query record window. Relevant and irrelevant documents will be randomly intermixed in the display window.

This simple loop of activity can be characterized by a cumulative gain function $g_i(t)$ that indicates how much information value is acquired over time t in cluster-patches of type i. In our empirical studies, we used specific task instructions that indicated that the information value was simply the number of relevant documents collected. The proportion of relevant documents in a cluster is the *precision*, P, of that cluster

$$P = \frac{N_R}{N_T}, \qquad (3)$$

where N_R is the number of relevant documents and N_T is the total number of documents in the cluster. The rate of encounter, λ_P with relevant items while scanning/scrolling through a list is

$$\lambda_P = \frac{P}{t_s}, \qquad (4)$$

where t_s is the time it takes to scan and judge the relevance of a title. If the time to handle a relevant item were $t_h = 0$, then we would have

$$g_i(t) = t\lambda_P. \qquad (5)$$

However, when handling costs are non-negligible,

---

[1]This interface was developed by Marti Hearst at Xerox PARC.

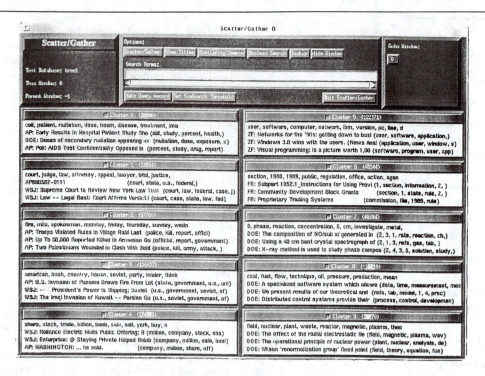

Figure 1. The Scatter/Gather interface for navigation through large document collections

$$g_i(t) = t\left(\frac{\lambda_p}{1 + \lambda_p t_h}\right)$$

$$= \frac{N_R t}{N_T t_s + N_R t_h} \tag{6}$$

In this case, $g_i(t)$ is just a linear function of t. This is illustrated qualitatively by function $g_A(t)$ in Figure 2. Next we illustrate how a foraging model reveals the improvements that would be made by introducing an interface method that displayed cluster titles ordered by the similarity of document contents to a user's query rather than in the unordered presentation captured by Equation 6. This method is currently unavailable in Scatter/Gather, but certainly possible and advisable given our analysis.[2] This foraging model illustrates explicitly how the overall rate of information gain would change, as well as how the users' foraging time in each cluster would be reduced.

If the document titles in a cluster were ordered by their probability of relevance to a query, then the gain function would shift qualitatively to a diminishing returns curve such as $g_B(t)$ in Figure 2. That is, there would be higher number of relevant documents encountered while scanning the initial parts of the display of document titles, but this encounter rate would diminish as one continues scanning. Simple patch-foraging models [22] identify the overall maximum rate of information and the optimal time at which to stop foraging in a patch. In these traditional patch models, each patch type i

with gain function $g_i(t_i)$, is additionally characterized by the rate at which such patch-gain functions will be encountered during prolonged foraging, λ_i. For instance, over a sufficiently long run of work with the Scatter/Gather interface one might be able to classify encountered clusters by their expected gain functions, g_i, and to additionally associate with each cluster type i the rate, λ_i, at which one expects to encounter clusters with gain functions g_i.

The optimization problem is to determine the optimal vector of patch residence times $(t_1, t_2, ..., t_n)$ that maximizes the currency intake rate R.

$$R(t_1, t_2, \ldots, t_n) = \frac{\sum_{i=1}^{n} \lambda_i g_i(t_i)}{1 + \sum_{i=1}^{n} \lambda_i t_i}. \tag{7}$$

Charnov's marginal value theorem [7] states that the long-term rate of currency intake is maximized by choosing patch residence times so that the marginal value (instantaneous rate) of the currency gain at the time of leaving each patch equals the long-term average rate across all patches in the ecology.[3]   More technically, the full vector of rate maximizing t_i's, $(\hat{t}_1, \hat{t}_2, \ldots, \hat{t}_n)$, must fulfill the condition specified by

---

[2]The current Scatter/Gather does have a similarity search operator that works over the entire collection but cannot be restricted to selected clusters.

[3]We are assuming that it is long-term average net gain that is being maximized. See Hames [10] for other optimization policies.

$$g'_1(\hat{t}_1) = R(\hat{t}_1, \hat{t}_2, \ldots, \hat{t}_n)$$

$$g'_2(\hat{t}_2) = R(\hat{t}_1, \hat{t}_2, \ldots, \hat{t}_n)$$

$$\vdots \qquad\qquad (8)$$

$$g'_n(\hat{t}_n) = R(\hat{t}_1, \hat{t}_2, \ldots, \hat{t}_n)$$

Charnov [7] shows that this theorem implies that residence times decrease as overall rate of gains from the information ecology improve and the optimal patch residence time occurs when the instantaneous gain rate equals the overall rate of gain from all patches that are foraged.

Figure 2 shows the graphical representation of Charnov's marginal value theorem that appears in many discussions of optimal foraging theory. Figure 2 captures the basic relations for the situation in which there is just one kind of patch-gain function. Travel time between patches (the inverse of patch encounter rate) is plotted on the abscissa, starting at the origin and moving to the left. To determine the overall maximum rate of gain R_B, one draws a line tangent to the gain function $g_B(t)$ and passing through $1/\lambda$ to the left of the origin. The point of tangency also provides the optimal maximum foraging time $\hat{t}_B$. Reasoning graphically with Figure 2, one should see that if one further improves the method of ordering of relevant documents presented to a user, then one changes the gain function $g_B(t)$ to $g_C(t)$, and that (keeping the tangent line anchored at $1/\lambda$) the rate of return improves to R_C, and the optimal foraging time in a cluster-patch decreases to $\hat{t}_C$. Given sufficient information about the relevant quantities and functions, one could make precise predictions about the impact of interface changes on optimal user performance. However, using the graphical reasoning just illustrated one may also make useful qualitative predictions.

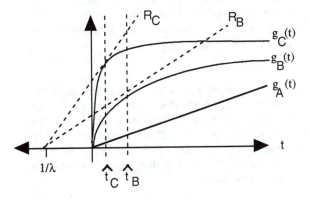

Figure 2. Optimal foraging in information cluster-patches. $g_A(t)$ is the current information-gain function, and $g_B(t)$ and $g_C(t)$ illustrate the effects of ordering by relevancy.

We suspect that in most tasks the rate of gaining information relevant to some embedding task from external collections diminishes (or is finite) with time. This occurs, for instance, in situations in which there is some redundancy among relevant documents. Consequently, the patch model analysis ought to have broad applicability.

EXAMPLE 2: INFORMATION DIET

The problem of selecting which clusters to pursue can be viewed as analogous to the problem faced by a predator who must select the optimal mix of prey to pursue upon encounter. Models of such choices are often called *prey selection* or *diet selection* models in optimal foraging theory. In our case, we have a problem of determining the *optimal information diet*.

Discussion of the analytic formulation and solution of prey selection models may be found in Stephen and Krebs [22], but here we focus on three general results: (a) ranking by profitability, (b) the zero-one rule, and (c) the independence of inclusion from encounter rate. Assume that the items being selected into the diet can be categorized as to their *profitability*, which is the expected net gain divided by the amount of time it takes to handle the item. In our case, the cluster profitabilities will be proportional to λ_p. To select the optimal diet, the items should be ranked by their profitability. Starting with the most profitable items, include lower-ranked items in the diet only if their profitability is greater than or equal to the average rate of gain of all the higher ranked items. Including items with profitability lower than the average rate of gain of the all the higher ranked items can only decrease the average. This is an instance of the *principle of lost opportunity*: by handling lower-ranked items in the diet one loses the opportunity to go after higher-ranked items. The zero-one result simply states that the optimal diet will be one in which items of a given profitability level are chosen all-or-none. The independence of inclusion from encounter rate result states that the inclusion of lower-ranked items in a diet is solely dependent on their profitability and not upon the rate at which they are encountered. To use an everyday analogy: if reading junk mail is categorized as a too-low profitability item (because there are better things to pursue), then the decision to ignore junk mail when it is encountered should be made regardless of the amount or rate of junk mail received.

EXAMPLE 3: A DYNAMIC FORAGING MODEL

Information patch models and information diet models are examples of *static* optimization models that do not take into account changes in human-system state that occur over time. Furthermore, the analysis above focused on localized subcomponents of the overall interaction with Scatter/Gather. It ignored the fact that users navigate through the cluster hierarchy presumably making choices about the best subclusters of documents to pursue. We now illustrate a more complete analysis characterized as a *dynamic* optimization model.

Tasks

For our model development we analyzed a version of Scatter/Gather that operates on the NIST Tipster corpus. This contains over one million documents collected from the Wall Street Journal, the AP newswire, Department of Energy technical abstracts, the Federal Register, and computer articles published by Ziff-Davis. This is a test corpus that is extensively used by the information retrieval community. One reason for using Tipster is that there are sets of information retrieval tasks (queries) that have been defined with associated listings of relevant and non-relevant Tipster documents, as judged by experts. If not an objective measure of task success, this at least provides us with a common standard to compare performance against.

Subjects, Measurements, and Procedures

To get measures of basic costs of using the Scatter/Gather interface we asked four subjects to retrieve as many relevant documents as possible for as many TREC queries as possible in one hour. Subjects were Xerox PARC researchers (including the two authors) whose experience with Scatter/Gather-Tipster were 96 hrs, 20 hrs, 10 hrs, and 2 hrs The purpose of this study was to obtain measurements of time costs associated with various Scatter/Gather browsing operations. Video tapes of subjects were parsed into the relevant events and time-coded to the nearest second. In addition, we conducted separate experiments on system response times. The relevant time costs are presented in Table 1.

Table 1: Empirical costs of Scatter/Gather use.

Average time on a query	$T = 720$ s
Scanning a cluster and judging relevance	$t_{sc} = 3$ s
Adding a cluster to gather list	$t_{gc} = 5$ s
Scanning a document title and judging relevance	$t_s = 1$ s
Selecting, cutting, and pasting title to record window	$t_h = 5$ s
System time to scatter new clusters	$t_{cl} = 23$ s
System time to display titles in a cluster	$t_d = 20$ s

Dynamic Programming

Dynamic programming has also been used to develop optimal foraging models [11]. Assuming a state variable X, one may define an evaluation function $\phi(x)$ that specifies the value of a state at some task deadline time T. In our task, this is the number of relevant documents collected by the end of the task. A state-change operator, $\delta_i(x_t) = x_{t+1}$, that produces a new state x_{t+1} from a current state x_t, can be additionally characterized by the time, $C_i(t)$, it takes to perform the operator. The final fitness value $F(x,T)$—in our case number of documents one can expect to collect by the end of the task—for any state x at any time $0 \leq t \leq T$, can be defined recursively as

$$F(x,t) = \begin{cases} \phi(x), & \text{if } t = T \\ \max_i[F(\delta_i(x), t + C_i(x))], & \text{if } t < T \end{cases} \quad (9)$$

That is, assuming that the optimal choices (or *optimal policy*) will be followed from states at $t + 1$, make the optimal choice at time t.

Dynamic Programming Model for Scatter/Gather

For Scatter/Gather, we developed a multidimensional, approximate representation of the state variable X. We define a set of accessor functions that return the relevant components of a state variable x:

$$\begin{cases} N(x) = & \text{No. documents in current cluster,} \\ R(x) = & \text{No. relevant in current cluster,} \\ f_D(x) = & \text{No. foraged in current cluster,} \\ f_C(x) = & \text{No. subclusters foraged,} \\ Q(x) = & \text{No. documents collected so far,} \\ H(x) = & \text{Clustering history.} \end{cases} \quad (10)$$

where "clustering history" represents the path by which the user has navigated through the cluster hierarchy by iteratively gathering then scattering clusters.

A reasonably average initial state x_0 would be

$$\begin{cases} N(x_0) = 10^6 \\ R(x_0) = 10^3 \\ f_D(x_0) = 0 \\ f_C(x_0) = 0 \\ Q(x_0) = 0 \end{cases} \quad (11)$$

Next, we define a set of operators, $\delta_k(x)$, $k = 0, 1, ..., K$, that specify the possible changes in human-system interaction state, and an associated set of time-cost functions, $C_k(x)$, that specify how long it takes to perform the operations.

The set of activity involved in displaying a cluster, scanning and judging the displayed titles, and collecting the relevant ones can be captured by an operator $\delta_0(x)$. When applied to a state x_n, it yields a new state x_{n+1}

$$x_{n+1} = \delta_0(x_n). \quad (12)$$

At a given time, t, with $T - t$ time remaining until the deadline is hit, the number of relevant documents collected from the displayed cluster titles will be either (a) all the relevant cluster titles, if there is sufficient remaining time to go through the whole cluster, or (b) as many relevant documents as can be collected in the remaining time, as characterized by Equation 6:

$$Q(x_{n+1})$$
$$= \begin{cases} Q(x_n) + R(x_n), & \text{if } C_Q(x_n) < (T - t) \\ Q(x_n) + g_A(T - t), & \text{otherwise} \end{cases} \quad (13)$$

The associated time-cost function is

$$C_Q(x_n) = \min \begin{cases} N(x_n)t_s + R(x_n)t_h \\ T - t \end{cases} \quad (14)$$

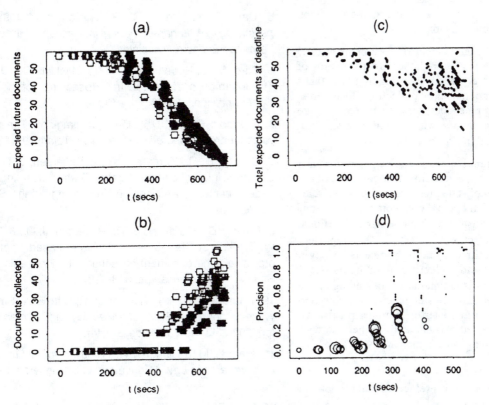

Figure 3. Expected gains and optimal operator choices in Scatter/Gather navigation.

Unless the time cost of the operator takes the human-system interaction past the deadline time, T, interaction returns to the cluster that is superordinate in the cluster hierarchy, and the number of foraged documents, remaining relevant documents, and foraged subclusters are updated appropriately, as is the history.

It remains to specify the operators for navigating through the Scatter/Gather cluster hierarchy. At each cluster in the hierarchy, there is a set of 10 subclusters that may be gathered and then rescattered. Of the $2^9 - 2$ possible gatherings at each node in the cluster hierarchy, the analysis of information diet suggests that the optimum can be found by considering only nine possibilities. The clusters can be ranked by their profitabilities λ_p and we can consider gatherings of the top 1, top 2, ... top 9 clusters to find the optimal "cluster diet" (in the following, for practical reasons, we only considered the top 3).

Operators for choosing the highest-precision $i = 1, 2, ...9$ clusters ca be defined by operators δ_i,

$$y_{n+1} = \delta_i(y_n). \tag{15}$$

If a cluster is divided into B subclusters (e.g., $B = 10$, in our case), then the number of documents collected together by gathering i subclusters together is, in the average case,

$$N(y_{n+1}) = \frac{N(y_n)i}{B}. \tag{16}$$

The number of relevant documents that will be found in the newly gathered cluster will be

$$R(y_{n+1}) = D_i(y_n) - f_D(y_n), \tag{17}$$

where $D_i(y_n)$ is the Cumulative Cluster Relevancy Distribution for a gathering of i subclusters, and $f_D(y_n)$ is the number of relevant documents that have already been foraged through. $D_i(y_n)$ assumes that the relevant documents in the gathered clusters are distributed exponentially when the clusters are ranked from highest to lowest in terms of their precisions, λ_p, bounded by the maximum number of documents that can be allotted to a cluster. We used the exponential exp(-.8 * cluster rank) in our calculations.

The cost of the gathering operation is

$$C_i(y_n) = Bt_{sc} + it_{gc} + t_{cl} \tag{18}$$

Figure 3 presents a summary of the expected value and the optimal operator-choice policies for states in the state-space of Scatter/Gather navigation. Each point in Figure 3 represents a state reachable from the initial state specified in (11) by the iterative application of the Scatter/Gather operators specified above: display-scan-select relevant documents from a cluster (δ_0), scatter-gather one (δ_1), two (δ_2), or three clusters (δ_3). For instance, the leftmost point in Figures 3a-3c plots the initial state and four of the state-points to right of that point were generated by application of the four operators (each state-point in Figures 3a-3c is similarly associated with four points to the right, up to the deadline time).

Figure 3a presents the number of documents that are optimally expected to yet be gained before the task deadline arrives, $E^*[g_i(T - t)]$, for each Scatter/Gather state. Each state is plotted as a box, whose degree of fill is proportional to the precision, λ_p, of clusters at that state. $E^*[g_i(T - t)]$

decreases as available time decreases and as the precision decreases, as suggested by Equation 6.

Figure 3b presents, $Q(x)$, the number of documents already collected at each state of Scatter/Gather, as a function of time. State precisions are plotted as in Figure 3a. It takes about half the task time to begin collecting documents. Figure 3c plots (ignoring precisions) the sum of $Q(x)$ (Figure 3b) and $E*[g_i(T - t)]$ (Figure 3a), which is the fitness, $F(x, t)$ of each state x, as defined in Equation 9.

Figure 3d shows the optimal operator-choice policy for each state encountered up to the decision to begin to collect documents, as a function of state precision and time. Circles represent states from which optimal operators are δ_1. δ_2, and δ_3 for scatter-gathering clusters, with circle size proportional to the optimal number of clusters selected. Dots represent states for which the optimal operator choice is to begin displaying, scanning, and selecting documents (δ_0). Scatter-gathering clusters is optimal at the beginning of the task when time available and cluster precision is low and displaying-selecting is optimal later or at higher precisions. The optimal number selected clusters tends to increase with overall state precision, and precision increases with time.

CONCLUSION

Stephen's and Krebs begin the preface of their book on *Foraging Theory*, thus:

> This book analyzes feeding behavior in the way an engineer might study a new piece of machinery. An engineer might ask, among other questions, about the machine's purpose: is it for measuring time, wind speed, or income tax? This is a worthwhile question for the engineer, because machines are built with a purpose in mind, and any description of a machine should refer to its purpose. . . .
> Biologists also ask questions about purpose, about what things are for in contrast to the engineer, the biologist thinks of design or purpose as the product of natural selection, rather than as the product of a conscious creator. (Stephens & Krebs, [22], p. ix)

In this paper, we have tried to make a start at reversing the above analogy—exploiting theory developed in the service of behavioral ecology to analyze information ecologies and the design of interactive information systems. It is hoped that this line of research will give us a new set of tools to aid in the design of these systems.

REFERENCES

1. Anderson, J. R. (1990). *The adaptive character of thought* . Hillsdale, NJ: Lawrence Erlbaum Associates.

2. Anderson, J. R. (1993). *Rules of the mind*. Hillsdale, NJ: Lawrence Erlbaum Associates.

3. Bates, M. J. (1989). The design of browsing and berrypicking techniques for the online search interface. *Online Review, 13*, 407-424.

4. Bechtel, W. (1985). Realism, instrumentalism, and the intentional stance. *Cognitive Science, 9*, 473-497.

5. Card, S. K., Moran, T. P., & Newell, A. (1983). *The psychology of human-computer interaction* . Hillsdale, NJ: Lawrence Erlbaum Associates.

6. Card, S. K., Robertson, G. G., & Mackinlay, J. D. (1991). The Information Visualizer. In Proceedings of the *CHI '92 Conference* (pp. 181-188), New Orleans.

7. Charnov, E. L. (1976). Optimal foraging: The marginal value theorem. *Theoretical Population Biology, 9*, 129-136.

8. Cutting, D. R., Karger, D. R., & Pedersen, J. O. (1993). Constant interaction-time Scatter/Gather browsing of very large document collections. In Proceedings of the *SIGIR '93 Conference*.

9. Cutting, D. R., Karger, D. R., Pedersen, J. O., & Tukey, J. W. (1992). Scatter/gather: A cluster-based approach to browsing large document collections. In Proceedings of the *SIGIR '92 Conference* (pp. 318-329).

10. Hames, R. (1992). Time allocation. In E. A. Smith & B. Winterhalder (Ed.), *Evolutionary ecology and human behavior* (pp. 203-235). New York: de Gruyter.

11. Mangel, M. & Clark, C. W. (1988). *Dynamic modeling in behavioral ecology* . Princeton, NJ: Princeton University Press.

12. Marr, D. (1982). *Vision* . San Francisco: W.H. Freedman.

13. Newell, A. (1990). *Unified theories of cognition*. Cambridge, MA: Harvard University Press.

14. Olson, J. R. & Olson, G. M. (1990). The growth of cognitive modeling in human-computer interaction since GOMS. *Human-Computer Interaction, 5*, 221-265.

15. Reitman, W. R. (1965). *Cognition and thought: An information-processing approach* . New York: Wiley.

16. Repo, A. J. (1986). The value of information: Approaches in economics, accounting, and management science. *Journal of the American Society for Information Science, 40*, 68-85.

17. Resnikoff, H. L. (1989). *The illusion of reality* . New York: Springer-Verlag.

18. Schamber, L., Eisenberg, M. B., & Nilan, M. S. (1990). A re-examination of relevance: Towards a dynamic situational definition. *Information Processing and Management, 26*, 755-776.

19. Simon, H. A. (1955). A behavioral model of rational choice. *Quarterly Journal of Economics, 69*, 99-118.

20. Simon, H. A. (1973). The structure of ill-structured problems. *Artificial Intelligence, 4*, 181-204.

21. Smith, E. A. & Winterhalder, B. (1992). *Evolutionary ecology and human behavior* . New York: de Gruyter.

22. Stephens, D. W. & Krebs, J. R. (1986). *Foraging theory* . Princeton, NJ: Princeton University Press.

23. Stigler, G. J. (1961). The economics of information. *The Journal of Political Economy, 69*, 213-225.

TileBars: Visualization of Term Distribution Information in Full Text Information Access

Marti A. Hearst
Xerox Palo Alto Research Center
3333 Coyote Hill Rd, Palo Alto, CA 94304
(415) 812-4742; hearst@parc.xerox.com

ABSTRACT

The field of information retrieval has traditionally focused on textbases consisting of titles and abstracts. As a consequence, many underlying assumptions must be altered for retrieval from full-length text collections. This paper argues for making use of text structure when retrieving from full text documents, and presents a visualization paradigm, called Tile-Bars, that demonstrates the usefulness of explicit term distribution information in Boolean-type queries. TileBars simultaneously and compactly indicate relative document length, query term frequency, and query term distribution. The patterns in a column of TileBars can be quickly scanned and deciphered, aiding users in making judgments about the potential relevance of the retrieved documents.

KEYWORDS: Information retrieval, Full-length text, Visualization.

INTRODUCTION

Information access systems have traditionally focused on retrieval of documents consisting of titles and abstracts. As a consequence, the underlying assumptions of such systems are not necessarily appropriate for full text documents, which are becoming available online in ever-increasing quantities. Context and structure should play an important role in information access from full text document collections. A critical structural aspect of a full-length text is the pattern of distributions of the terms that comprise it. When a system retrieves a document in response to a query, it is important to indicate not only how strong the match is (e.g., how many terms from the query are present in the document), but also how frequent each term is, how each term is distributed in the text and where the terms overlap within the document. This information is especially important in long texts, since it is less clear how the terms in the query contribute to the ranking of a long text than a short abstract. The need for this kind

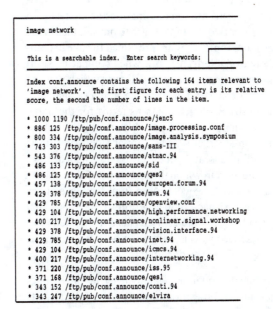

```
image network

This is a searchable index.  Enter search keywords:  [      ]

Index conf.announce contains the following 164 items relevant to
'image network'.  The first figure for each entry is its relative
score, the second the number of lines in the item.

* 1000 1190 /ftp/pub/conf.announce/jenc5
*  886  125 /ftp/pub/conf.announce/image.processing.conf
*  800  334 /ftp/pub/conf.announce/image.analysis.symposium
*  743  303 /ftp/pub/conf.announce/sans-III
*  543  376 /ftp/pub/conf.announce/atnac.94
*  486  133 /ftp/pub/conf.announce/sid
*  486  125 /ftp/pub/conf.announce/qes2
*  457  138 /ftp/pub/conf.announce/europen.forum.94
*  429  378 /ftp/pub/conf.announce/mva.94
*  429  785 /ftp/pub/conf.announce/openview.conf
*  429  104 /ftp/pub/conf.announce/high.performance.networking
*  400  217 /ftp/pub/conf.announce/nonlinear.signal.workshop
*  429  378 /ftp/pub/conf.announce/vision.interface.94
*  429  785 /ftp/pub/conf.announce/inet.94
*  429  104 /ftp/pub/conf.announce/icmcs.94
*  400  217 /ftp/pub/conf.announce/internetworking.94
*  371  220 /ftp/pub/conf.announce/iss.95
*  371  168 /ftp/pub/conf.announce/qes1
*  343  152 /ftp/pub/conf.announce/conti.94
*  343  247 /ftp/pub/conf.announce/elvira
```

Figure 1: A sketch of the results of a WAIS search on *image* and *network* on a dataset of conference announcements.

of distributional information has not been emphasized in the past, perhaps in part because researchers had not focused on long texts.

To address these issues, I introduce a new display paradigm called *TileBars* which allows users to simultaneously view the relative length of the retrieved documents, the relative frequency of the query terms, and their distributional properties with respect to the document and each other. TileBars seem to be a useful analytical tool for understanding the results of Boolean-type queries, and preliminary work indicates they are useful for determining document relevance when applied to sample queries from a standard full-text test collection. This approach to visualization of the role of the query terms within the retrieved documents may also help explain why standard information retrieval measures succeed or fail for a given query.

BACKGROUND: STANDARD INFORMATION RETRIEVAL

The purpose of information retrieval is to help users effectively access large collections of objects with the goal of satisfying the users' stated information needs [6].[1] The most common approaches to text retrieval are Boolean term specification and similarity search. I use the term "similarity search" as an umbrella term covering the vector space model [26], probabilistic models [5], [12] and any other approach which attempts to find the documents that are most similar to a query or to one another based solely or primarily on the terms they contain.

Similarity search, in effect, ranks documents according to how close, in a multidimensional term space, combinations of the documents' terms are to combinations of the terms in the query. The closer two documents are to one another in the term space, the more topics they are presumed to have in common. This is a reasonable framework when comparing short documents, since the goal is often to discover which pairs of documents are most alike. For example, a query against a set of medical abstracts which contains terms for the name of a disease, its symptoms, and possible treatments is best matched against an abstract with as similar a constitution as possible. In similarity search, the best overall matches are not necessarily the ones in which the largest percentage of the query terms are found, however. For example, given a query of T terms, the vector space model permits a document that contains only a subset S of the query terms to be ranked relatively high if these terms occur infrequently in the corpus as a whole but frequently in the document.

In Boolean retrieval a query is stated in terms of disjunctions, conjunctions, and negations among sets of documents that contain particular words and phrases. Documents are retrieved whose contents satisfy the conditions of the Boolean statement. The users can have more control over what terms actually appear in the retrieved documents than they do with similarity search. In its basic form, Boolean search does not produce a ranking order, although ranking criteria as used in similarity search are often applied to the results of the Boolean search [11].

The Problem with Ranking

There is great concern in the information retrieval literature about how to rank the results of Boolean and similarity searches. I contend that this concern is misplaced. Once a manageable subset of the thousands of available documents has been found, then the issue becomes a matter of providing the user with information that is informative and compact enough that it can be interpreted swiftly.[2] As discussed in the next subsection, there are many different ways in which a long text can be "similar" to the query that issued it, and so

a system should supply the user with a way to understand the relationship between the retrieved documents and the query.

Furthermore, the standard approach to document ranking is opaque; users are unable to see what role their query terms played in the ranking of the retrieved documents. An ordered list of titles and probabilities is under-informative. The link between the query terms, the similarity comparison, and the contents of the texts in the dataset is too underspecified to assume that a single indicator of relevance can be assigned.

Instead, the representation of the retrieval results should present as many attributes of the texts and their relationship to the queries as possible, and present the information in a compact, coherent and accurate manner. Accurate in this case means a true reflection of the relationship between the query and the documents.

Consider for example what happens when one performs a keyword search using WAIS [18]. If the search completes, it results in a list of document titles and relevance rankings. The rankings are based on the query terms in some capacity, but it is unclear what role the terms play or what the reasons behind the rankings are. The length of the document is indicated by a number, which although interpretable, is not easily read from the display. Figure 1 represents the results of a search on *image* and *network* on a database of conference announcements. The user cannot determine to what extent either term is discussed in the document or what role the terms play with respect to one another. If the user prefers a dense discussion of images and would be happy with only a tangential reference to networking, there is no way to express this preference.

Attempts to place this kind of expressiveness into keyword based system are usually flawed in that the users find it difficult to guess how to weight the terms. If the guess is off by a little they may miss documents that might be relevant, especially because the role the weights play in the computation is far from transparent. Furthermore, the user may be willing to look at documents that are not extremely focused on one term, so long as the references to the other terms are more than passing ones. Finally, the specification of such information is complicated and time-consuming.

The Importance of Document Structure

A problem with applying similarity search to full-length text documents is that the structure of full text is quite different from that of abstracts. Abstracts are compact and information-dense. Most of the (uncommon) terms in an abstract are salient for retrieval purposes because they act as placeholders for multiple occurrences of those terms in the original text, and because generally these terms pertain to the most important topics in the text. Consequently, if the text is of any sizeable length, it will contain many subtopic discussions that are never mentioned in its abstract, if one exists. On the other hand, an expository text may be viewed as a sequence of subtopics set against a "backdrop" of one or

---

[1] This paper will focus on collections of textual information only, although other media types apply as well.

[2] As further evidence for this viewpoint, Noreault et al. [23] performed an experiment on bibliographic records in which they tried every combination of 37 weighting formulas working in conjunction with 64 combining formulas on Boolean queries. They found that the choice of scheme made almost no difference: the best combinations got about 20% better than random ordering, and no one scheme stood out above the rest. These results imply that small changes to weighting formulas don't have much of an effect.

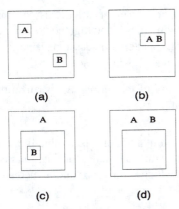

(a)   (b)

(c)   (d)

Figure 2: Possible relationships between two terms in a full text.   (a) The distribution is disjoint, (b) co-occurring locally, (c) term A is discussed globally throughout the text, B is only discussed locally, (d) both A and B are discussed globally throughout the text.

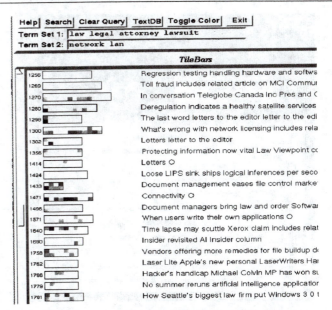

Figure 3: The TileBar display paradigm. Rectangles correspond to documents, squares correspond to text segments, the darkness of a square indicates the frequency of terms in the segment from the corresponding Term Set. Titles and the initial words of a document appear next to its TileBar.

two main topics. A long text is often comprised of many different subtopics which may be related to one another and to the backdrop in many different ways. The main topics of a text are discussed in its abstract, if one exists, but subtopics usually are not mentioned. Therefore, instead of querying against the entire content of a document, a user should be able to issue a query about a coherent subpart, or subtopic, of a full-length document, and that subtopic should be specifiable with respect to the document's main topic(s).

Figure 2 illustrates some of the possible distributional relationships between two terms in the main topic/subtopic framework. An information access system should be aware of each of the possible relationships and make judgments as to relevance based in part on this information. Thus a document with a main topic of "cold fusion" and a subtopic of "funding" would be recognizable even if the two terms do not overlap perfectly. The reverse situation would be recognized as well: documents with a main topic of "funding policies" with subtopics on "cold fusion" should exhibit similar characteristics.

The idea of the main topic/subtopic dichotomy can be generalized as follows: different distributions of term occurrences have different semantics; that is, they imply different things about the role of the terms in the text. The possible distribution relations that can hold between two sets of terms, and predictions about the usefulness of each distribution type, are enumerated and explained in [14].

TextTiling: Automatic Discovery of Document Structure

To determine the kind of document structure described above, I have developed an algorithm, called *TextTiling*, that partitions expository texts into multi-paragraph segments that reflect their subtopic structure [15]. (Since the segments are adjacent and non-overlapping, they are called TextTiles.) The algorithm detects subtopic boundaries by analyzing the term

repetition patterns within the text. The main idea is that terms that describe a subtopic will co-occur locally, and a switch to a new subtopic will be signalled by the ending of co-occurrence of one set of terms and the beginning of the co-occurrence of a different set of terms. In texts in which this assumption is valid, the central problem is determining where one set of terms ends and the next begins. The algorithm is domain-independent, and is fully implemented. The results of TextTiling are difficult to evaluate; comparisons to human judgments show the results are imperfect, as is often the case in fuzzy natural language processing tasks, but serviceable for their application to the task described below.

TILEBARS

This section presents one solution to the problems described in the previous subsections. The approach is synthesized in reaction to three hypotheses:

- Long texts differ from abstracts and short texts in that, along with term frequency, term distribution information is important for determining relevance.

- The relationship between the retrieved documents and the terms of the query should be presented to the user in a compact, coherent, and accurate manner (as opposed to the single-point of information provided by a ranking).

- Passage-based retrieval should be set up to provide the user with the context in which the passage was retrieved, both within the document, and with respect to

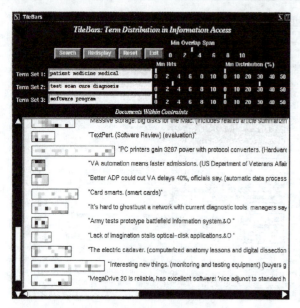

Figure 4: TileBar search on (*patient medicine medical* AND *test scan cure diagnosis* AND *software program*) with some distribution constraints.

the query.

Figure 3 shows an example of a new representational paradigm, called TileBars, which provides a compact and informative iconic representation of the documents' contents with respect to the query terms. TileBars allow users to make informed decisions about not only which documents to view, but also which passages of those documents, based on the distributional behavior of the query terms in the documents. As mentioned above, the goal is to simultaneously indicate:

(1) The relative length of the document,
(2) The frequency of the term sets in the document, and
(3) The distribution of the term sets with respect to the document and to each other.

Each large rectangle indicates a document, and each square within the document represents a TextTile. The darker the tile, the more frequent the term (white indicates 0, black indicates 8 or more instances, the frequencies of all the terms within a term set are added together). Since the bars for each set of query terms are lined up one next to the other, this produces a representation that simultaneously and compactly indicates relative document length, query term frequency, and query term distribution. The representation exploits the natural pattern-recognition capabilities of the human perceptual system [21]; the patterns in a column of TileBars can be quickly scanned and deciphered.

Term overlap and term distribution are both easy to compute and can be displayed in a manner in which both attributes together create easily recognized patterns. For example, overall darkness indicates a text in which both term sets are discussed

in detail. When both term sets are discussed simultaneously, their corresponding tiles blend together to cause a prominent block to appear. Scattered discussions have lightly colored tiles and large areas of white space.

TileBars make use of the following visualization properties (extracted from [27]):

- A variation in position, size, value [gray scale saturation], or texture is ordered [ordinal] that is, it imposes an order which is universal and immediately perceptible. [3]

- If shading is used, make sure differences in shading line up with the values being represented. The lightest ("unfilled") regions represent "less", and darkest ("most filled") regions represent "more". [20]

- Because they do have a natural visual hierarchy, varying shades of gray show varying quantities better than color. [29]

Note that the stacking of the terms in the query-specification portion of the document is reflected in the stacking of the tiling information in the TileBar: the top row indicates the frequencies of terms from Term Set 1 and the bottom row corresponds to Term Set 2. Thus the issue of how to specify the keyterms becomes a matter of what information to request in the interface. There is an implicit OR among the terms within a term set and an implicit AND between the term sets. Retrieved documents must have at least K hits from each term set, where K is an adjustable parameter.

TileBars allow users to be aware of what part of the document they are about to view before viewing it. To see what the document is about overall, they can simply mouse-click on the part of the representation that symbolizes the beginning of the document. Alternatively, they may go directly to a segment in the middle of the text in which terms from both term sets overlap, knowing in advance how far down in the document the passage occurs.

The TileBar representation allows for grouping by distribution pattern. Each pattern type occupies its own window in the display and users can indicate preferences by virtue of which windows they use. Thus there is no single correct ranking strategy: in some cases the user might want documents in which the terms overlap throughout; in other cases isolated passages might be appropriate. A variation of the interface organizes the retrieval results according to the distribution pattern type.

Networks and the Law
Figure 3 shows some of the TileBars produced for the query on the term sets (*law legal attorney lawsuit*) AND (*network lan*) on the ZIFF collection [13]. (ZIFF is comprised mainly of commercial computer news.) In response to this query

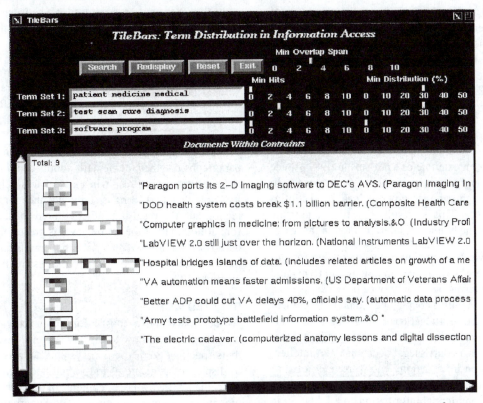

Figure 5: TileBar search on (*patient medicine medical* AND *test scan cure diagnosis* AND *software program*) with stricter distribution constaints.

one might expect documents about computer networks used in law firms, lawsuits involving illegal use of networks, and patent battles among network vendors. Since retrieval is on a collection of commercial computer texts, most instances of the word *network* will refer to the computer network sense, with exceptions for neural networks and perhaps some references to computer science theory and telephone systems. Since *legal* is an adjective, it can be used as a modifier in a variety of situations, but a strong showing of hits in its term set should indicate a legitimate legal discussion.

In the figure, the results have not been sorted in any manner other than document ID number. It is instructive to compare what the bars imply about the content of the texts with what actually appears in the texts. Document 1433 stands out because it appears to discuss both term sets in some detail. Documents 1300 and 1471 are also prominent because of a strong showing of the network term set. Document 1758 also has well-distributed instances of both term sets, although with less frequency than in document 1433. Legal terms have a strong distributional showing in 1640, 1766, 1781 as well. There are also several documents with very few occurrences of either term, although in some cases terms are more locally concentrated than in others. Most of the other documents look uninteresting due to their lack of overlap or infrequency of term occurrences.

Looking now at the actual documents we can determine the

accuracy of the inferences drawn from the TileBars. Clicking on the first tile of document 1433 brings up a window containing the contents of the document, centered on the first tile. The search terms are highlighted with two different colors, distinguished by term set membership, and the tile boundaries are indicated by ruled lines and tile numbers. The document describes in detail the use of a network within a legal office.

Looking at document 1300, the intersection between the term sets can be viewed directly by clicking on the appropriate tile. From the TileBar we know in advance that the tile to be shown appears about three quarters of the way through the document. Clicking here reveals a discussion of legal ramifications of licensing software when distributing it over the network. Document 1471 has only the barest instance of legal terms and so it is not expected to contain a discussion of interest – most likely a passing reference to an application. Indeed, the term is used as part of a hypothetical question in an advice column describing how to configure LANs. Note that a document like this would have been ranked highly by a mechanism that only takes into account term frequency.

The remaining documents with strong distributions of legal terms, 1758, 1640, 1766, 1781, discuss a documentation management system on a networked PC system in a legal office, a lawsuit between software providers, computer crime, and another discussion of a law firm using a new networked software system, respectively. Only the latter has overlap with networking terms. Interestingly, the solitary mention of networking at the end of 1766 lists it as a computer crime problem to be worried about in the near future. This is an example of the suggestive nature of the positional information inherent in the representation.

Finally, looking at the seemingly isolated discussion of document 1298 we see a letter-to-the-editor about the lack of liability and property law in the area of computer networking. This letter is one of several letters-to-the-editor; hence its isolated nature. This is an example of a perhaps useful instance of isolated, but strongly overlapping, term occurrences. In this example, one might wonder why one legal term continues on into the next tile. This is a result of the tiling algorithm being slightly off in the boundary determination in this case.

As mentioned above, the remaining documents appear uninteresting since there is little overlap among the terms and within each tile the terms occur only once or twice. We can confirm this suspicion with a couple of examples. Document 1270 has one instance of a legal term; it is a passing reference to the former profession of an interview subject. Document 1356 discusses a court's legal decision about intellectual property rights on information. Tile 3 provides a list of ways to protect confidential information, one item of which is to avoid storing confidential information on a LAN. So in this case the reference to networks is only in passing.

Note that the conjunction of information about how much of each term set is present with how much the hits from each term set overlap provide indicate different kinds of information, which cannot be discerned from a ranking.

Computer-aided Medical Diagnosis

Figures 4 and 5 show the results of a query on three term sets in a version of the interface that allows the user to restrict which documents are displayed according to several constraints: minimum number of hits for each term set, minimum distribution (the percentage of tiles containing at least one hit), and minimum adjacent overlap span. In this example the user is interested in documents that discuss computer-aided techniques for medical diagnosis, and the query is a conjunction of three term sets: (*patient medicine medical*) AND (*test scan cure diagnosis*) AND (*software program*). In Figure 4 the user has indicated that the document must contain a substantive discussion of the diagnosis terms, and that overlap among all three term sets must occur at least once within the span of three adjacent tiles. Note that this looser restriction yields some documents about computer-aided diagnosis with only passing references to medicine, which may indeed meet the user's information need. In figure 5, the user has emphasized the importance of the medical terms as well by specifying that displayed documents must have hits in at least 30% of their tiles. Judging from the titles displayed, this restriction was indeed useful in isolating documents of interest. Placing such constraints may cause relevant documents to be discarded, but an interface like this allows the user some

control over the ever-present tradeoff between showing only relevant documents and showing all relevant documents.

Implementation Notes

The current implementation of the information access method underlying the TileBar display makes use of ≈ 132,000 documents ZIFF portion of the TREC/TIPSTER corpus [13]. The interface uses the Tcl/TK X11-based toolkit [24] and the search engine uses TDB [8], implemented in Common Lisp. The use of TextTiles is not critical to the implementation; paragraphs or other segmentation units could be substituted, although this could result in units of less helpful granularity. Note that TextTiling is run in advance for the entire collection and the resulting indices stored for later use; therefore although the time for retrieval is greater than for a standard Boolean full-text query, it is not significantly so. Performance issues for indexing with passages are discussed in, for example, [22].

RELATED WORK

As mentioned above, most information access systems have not grappled with how to display retrieval results from long texts specifically. Hypertext systems address issues related to display of contents of individual documents but are less concerned with display of contents of a large number of documents in response to a query. The Superbook system [10] shows where the hits from a query are in terms of the structure of a single, large, hierarchically structured document, but does not handle multiple documents simultaneously, nor does it show the terms of a multi-term query separately, nor does it display the frequencies graphically.

In general, document content information is difficult to display using existing graphical interface techniques because textual information does not conform to the expectations of sophisticated display paradigms, such as the techniques seen in the Information Visualizer [25]. These techniques either require the input to be structured (e.g., hierarchical, for the Cone Tree) or scalar along at least one dimension (e.g., for the Perspective Wall). The aspects of a document that satisfy these criteria (e.g., a timeline of document creation dates) do not illuminate the actual content of the documents.

Another graphical interface is that of Value Bars [4], which display relative attribute size for a set of attributes. The example in [4] shows a window listing a file directory's contents and vertical Value Bars alongside the window's scrollbar. Each horizontal slice of a Value Bar represents the size or the age of a listed file, although the attributes of the Value Bars do not align directly with window's contents nor with one another, thus precluding the perception of overlap among the displayed item's attributes. One could imagine using Value Bars for display of retrieval results by replacing the filenames with titles of retrieved documents and having the attributes correspond to the number of hits for term sets. However, the display would still not indicate term overlap or term distribution. Similar remarks apply to the Read Wear interface [17].

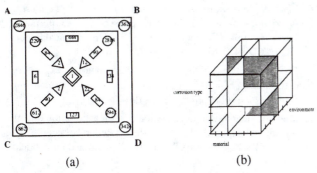

Figure 6: Sketches of (a) the InfoCrystal (b) the Cube of Contents.

Turning now to information retrieval systems, the simplest approach to displaying retrieval results is, of course, to list the titles or first lines of the retrieved documents and their ranks, and many systems do this. Existing systems that do more can be characterized as performing one of two functions: (1) displaying the retrieved documents according to their overall similarity to a query or other retrieved documents, and/or (2) displaying the retrieved documents in terms of keywords or attributes pre-selected by the user. Neither of these approaches address the issues of term distribution, frequency, and overlap that TileBars do. For reasons argued above, systems of type (1) are problematic, especially with respect to full-text collections.

Systems of type (2) show the relation of the contents of texts to user-selected attributes; these include VIBE [19], the InfoCrystal [28], the Cube of Contents [2], and the system of Aboud et al. [1]. These systems require users to select the classifications around which the display is organized. The goal of VIBE [19] is to display the contents of the entire document collection in a meaningful way. The InfoCrystal [28] is a sophisticated interface which allows visualization of all possible relations among N user-specified "concepts" (or Boolean keywords). The InfoCrystal displays, in a clever extension of the Venn-diagram paradigm, the number of documents retrieved that have each possible subset of the N concepts. Figure 6(a) shows a sketch of what the InfoCrystal might display as the result of a query against four keywords or Boolean phrases, labeled A, B, C, and D. The diamond in the center indicates that one document was discovered that contains all four keywords. The triangle marked with "12" indicates that twelve documents were found containing attributes A, B, and D, and so on. The Information Crystal does not indicate information about the distribution or frequency of occurrence of the query terms within the document. Thus it is perhaps more appropriate for titles and abstracts than for full text. The Cube of Contents [2] helps the user build a query by selecting values for up to three mutually exclusive attributes (Figure 6(b)). This assumes a text pre-labeled with relevant information and an understanding of domain-dependent structural information for the document set. Again, frequency and distribution information could not be indicated easily in this framework.

CONCLUSIONS AND FUTURE WORK

I have introduced a new display device, called TileBars, that vializes explicit term distribution information in a full text information access system. The representation simultaneously and compactly indicates relative document length, query term frequency, and query term distribution. The patterns in a column of TileBars can be quickly scanned and deciphered, aiding users in making fast judgments about the potential relevance of the retrieved documents. TileBars can be sorted or filtered according to their distribution patterns and term frequencies, aiding the users' evaluation task still more. An indepth description of an example helped show the semantic affects of various term distribution patterns. The TileBar representation should extend easily to representing media types other than text.

In the future user studies should be run to determine how users interpret the meaning of the term distributions and how they may be used in relevance feedback. It may be useful to determine in what situations the users' expectations are not met, in hopes of identifying what additional information will help prevent misconceptions. Another kind of evaluation is currently underway [16], exploring the effects of term distribution in the TREC/TIPSTER test collection [13] on individual queries. Associated with the documents in the TIPSTER collection are a set of queries and human-assigned relevance judgments. In the past two years there has been a spate of research on passage retrieval in this collection, but the results are mixed and difficult to interpret. The main trend seems to be that some combination of scores from the full document with scores from the highest scoring passage or segment yields a small improvement over the baseline of using the full document alone. The work reported in [16] attempts to determine how term distribution and overlap affects retrieval results in this task, and in the process provides an argument for the use of a TileBar-like display. Preliminary results indicate that scores can be improved by taking individual term distribution preferences for individual queries into account.

Information access mechanisms should not be thought of as retrieval in isolation. Cutting et al. [9] advocate a text access paradigm that "weaves together interface, presentation and search in a mutually reinforcing fashion"; this viewpoint is adopted here as well. For example, the user might send the contents of the a TileBar window to an interface like Scatter/Gather [7] which can cluster the document subset, and display their main topics. The user could then select a subset of the clusters to be sent back to the TileBar session. This kind of integration will be attempted in future work.

Acknowledgements

This paper has benefited from the comments of Jan Pedersen and six anonymous reviewers. I would also like to thank Robert Wilensky for supporting this line of research and Marc Teitelbaum for help in an earlier implementation.

REFERENCES

1. M. Aboud, C. Chrisment, R. Razouk, and F. Sedes. Querying a hypertext information retrieval system by the use of classification. *Information Processing and Management*, 29(3):387–396, 1993.

2. H. C. Arents and W. F. L. Bogaerts. Concept-based retrieval of hypermedia information – from term indexing to semantic hyperindexing. *Information Processing and Management*, 29(3):373–386, 1993.

3. Jacques Bertin. *Semiology of Graphics*. The University of Wisconsin Press, Madison, WI, 1983. Translated by William J. Berg.

4. Richard Chimera. Value bars: An information visualization and navigation tool for multi-attribute listings. In *Proceedings of the ACM SIGCHI Conference on Human Factors in Computing Systems*, pages 293–294, May 1992.

5. William S. Cooper, Fredric C. Gey, and Aitoa Chen. Probabilistic retrieval in the TIPSTER collections: An application of staged logistic regression. In Donna Harman, editor, *Proceedings of the Second Text Retrieval Conference TREC-2*, pages 57–66. National Institute of standard and Technology Special Publication 500-215, 1994.

6. W. Bruce Croft and Howard R. Turtle. Text retrieval and inference. In Paul S. Jacobs, editor, *Text-Based Intelligent Systems: Current Research and Practice in Information Extraction and Retrieval*, pages 127–156. Lawrence Erlbaum Associates, 1992.

7. Douglass R. Cutting, David Karger, and Jan Pedersen. Constant interaction-time Scatter/Gather browsing of very large document collections. In *Proceedings of the 16th Annual International ACM/SIGIR Conference*, pages 126–135, Pittsburgh, PA, 1993.

8. Douglass R. Cutting, Jan O. Pedersen, and Per-Kristian Halvorsen. An object-oriented architecture for text retrieval. In *Conference Proceedings of RIAO'91, Intelligent Text and Image Handling, Barcelona, Spain*, pages 285–298, April 1991. Also available as Xerox PARC technical report SSL-90-83.

9. Douglass R. Cutting, Jan O. Pedersen, Per-Kristian Halvorsen, and Meg Withgott. Information theater versus information refinery. In Paul S. Jacobs, editor, *AAAI Spring Symposium on Text-based Intelligent Systems*, 1990.

10. Dennis E. Egan, Joel R. Remde, Louis M. Gomez, Thomas K. Landauer, Jennifer Eberhardt, and Carol C. Lochbaum. Formative design evaluation of superbook. *Transaction on Information Systems*, 7(1), 1989.

11. Edward A. Fox and Matthew B. Koll. Practical enhanced Boolean retrieval: Experiences with the SMART and SIRE systems. *Information Processing and Management*, 24(3), 1988.

12. Norbert Fuhr and Chris Buckley. Optimizing document indexing and search term weighting based on probabilistic models. In Donna Harman, editor, *The First Text Retrieval Conference (TREC-1)*, pages 89–100. NIST Special Publication 500-207, 1993.

13. Donna Harman. Overview of the first Text REtrieval Conference. In *Proceedings of the 16th Annual International ACM/SIGIR Conference*, pages 36–48, Pittsburgh, PA, 1993.

14. Marti A. Hearst. *Context and Structure in Automated Full-Text Information Access*. PhD thesis, University of California at Berkeley, 1994. (Computer Science Division Technical Report UCB/CSD-94/836).

15. Marti A. Hearst. Multi-paragraph segmentation of expository text. In *Proceedings of the 32nd Meeting of the Association for Computational Linguistics*, June 1994.

16. Marti A. Hearst. An investigation of term distribution effects on individual queries. Technical Report Report Number ISTL-QCA-1994-12-06, Xerox PARC, 1995. Submitted for publication.

17. William C. Hill, James D. Hollan, Dave Wroblewski, and Tim McCandless. Edit wear and read wear. In *Proceedings of the ACM SIGCHI Conference on Human Factors in Computing Systems*, pages 3–9, May 1992.

18. Brewster Kahle and Art Medlar. An information system for corporate users: Wide area information servers. Technical Report TMC199, Thinking Machines Corporation, April 1991.

19. Robert R. Korfhage. To see or not to see – is that the query? In *Proceedings of the 14th Annual International ACM/SIGIR Conference*, pages 134–141, Chicago, 1991.

20. S. Kosslyn, S. Pinker, W. Simcox, and L. Parkin. *Understanding Charts and Graphs: A Project in Applied Cognitive Science*. National Institute of Education, 1983. ED 1.310/2:238687.

21. Jock Mackinlay. *Automatic Design of Graphical Presentations*. PhD thesis, Stanford University, 1986. Technical Report Stan-CS-86-1038.

22. Alistair Moffat, Ron Sacks-Davis, Ross Wilkinson, and Justin Zobel. Retrieval of partial documents. In Donna Harman, editor, *Proceedings of the Second Text Retrieval Conference TREC-2*, pages 181–190. National Institute of standard and Technology Special Publication 500-215, 1994.

23. Terry Noreault, Michael McGill, and Matthew B. Koll. A performance evaluation of similarity measures, document term weighting schemes and representations in a Boolean environment. In R. N. Oddy, S. E. Robertson, C. J. van Rijsbergen, and P. W. Williams, editors, *Information Retrieval Research*, pages 57–76. Butterworths, London, 1981.

24. John Ousterhout. An X11 toolkit based on the Tcl language. In *Proceedings of the Winter 1991 USENIX Conference*, pages 105–115, Dallas, TX, 1991.

25. George C. Robertson, Stuart K. Card, and Jock D. MacKinlay. Information visualization using 3D interactive animation. *Communications of the ACM*, 36(4):56–71, 1993.

26. Gerard Salton. *Automatic text processing: the transformation, analysis, and retrieval of information by computer*. Addison-Wesley, Reading, MA, 1988.

27. Hikmet Senay and Eve Ignatius. Rules and principles of scientific data visualization. Technical Report GWU-IIST-90-13, Institute for Information Science and Technology, The George Washington University, 1990.

28. Anselm Spoerri. InfoCrystal: A visual tool for information retrieval & management. In *Proceedings of Information Knowledge and Management '93*, Washington, D.C., Nov 1993.

29. Edward Tufte. *The Visual Display of Quantitative Information*. Graphics Press, Chelshire, CT, 1983.

An Organic User Interface For
Searching Citation Links

Jock D. Mackinlay, Ramana Rao and Stuart K. Card
Xerox Palo Alto Research Center
3333 Coyote Hill Rd; Palo Alto, CA 94304
{mackinlay, rao, card}@parc.xerox.com

ABSTRACT

This paper describes Butterfly, an Information Visualizer application for accessing DIALOG's Science Citation databases across the Internet. Network information often involves slow access that conflicts with the use of highly-interactive information visualization. Butterfly addresses this problem, integrating search, browsing, and access management via four techniques: 1) visualization supports the assimilation of retrieved information and integrates search and browsing activity, 2) automatically-created "link-generating" queries assemble bibliographic records that contain reference information into citation graphs, 3) asynchronous query processes explore the resulting graphs for the user, and 4) process controllers allow the user to manage these processes. We use our positive experience with the Butterfly implementation to propose a general information access approach, called *Organic User Interfaces for Information Access*, in which a virtual landscape grows under user control as information is accessed automatically.

KEYWORDS: information visualization, search, browsing, access management, information retrieval, organic user interfaces, data fusion, hypertext, citation graphs

INTRODUCTION

The proposed national information infrastructure includes the dream that people will have easy access to many large information databases across the Internet. Such increased access to information is likely to be useful, but the prospect of large amounts of information remotely linked together around the globe poses new challenges for the design of user interfaces for information access. Internet users have difficulty discovering information sources, have trouble learning how to search them, have suffered from slow network information access, and have generally complained of being "lost in cyberspace."

Two major approaches for network information access have emerged: search and browsing. Search is exemplified by the DIALOG system, which permits over 400 databases at a single site to be searched with a uniform command-line user inter-

face [5]. Browsing is exemplified by the Mosaic user interface and World-Wide Web protocol that permit the user to click on hypertext links and jump from document to document across multiple sites [2]. Although both systems have significantly advanced the accessibility of networked information, they also have limitations. DIALOG's query language is complex and difficult to learn. It works very well for professional librarians, who can spend the time to develop proficiency, but it can be a substantial barrier for casual users. Also, the nature of its scrolling teletype-style output makes it difficult to use the results of a previous search to formulate a new query. On the other hand, Mosaic provides easy browsing of the Web but its sequential access makes it difficult for users to orient themselves and see patterns in larger parts of the information space. Mosaic users frequently complain of having trouble knowing where to go and losing previously-visited nodes.

Recently, 3D interactive animation and "focus-plus-context" techniques have been developed for information visualization of large information structures [10]. However, the techniques have typically been applied to single information sets loaded into local memory. Extending them to work on information repositories across the Internet requires that we address the problem of coupling a high-bandwidth, local, interactive user interface with a slow-response, distant information repository (the "Fast UI/Slow Repository" problem). The user interface operates at the level of tenths of a second, whereas search results typically return in tens of seconds.

Remote access also gives rise to the more general possibility of interacting with multiple information repositories simultaneously (the "Fast UI/Slow Multiple Repository" problem). Multiple access can compensate the user for slow network access, but it poses additional challenges for the design of information access user interfaces. In particular, simultaneously accessing multiple repositories increases the difficulty of *access management*, which is the task of choosing where and when to search or browse.

In this paper, we address the more general Fast UI/Slow Multiple Repository problem. We also address a related problem, the graceful integration of search, browsing, and access management, which are all impacted by slow network access. We describe a system, Butterfly, for simultaneously exploring multiple DIALOG bibliographic databases across the Internet using 3D interactive animation techniques that we had previously developed for various Fast UI/Fast Repository applications [10]. Our goals are to allow richer, more rapid

Annotations	Typescript
select source	`?b 434`
	`<16 lines of accounting information removed>`
search on author	`?s au=card sk`
search S1 had 7 results	`    S1        7   AU=CARD SK`
type S1 format 3 result 1	`?t 1/3/1`
	`1/3/1`
	`DIALOG(R)File 434:Scisearch(R)`
	`(c) 1994 Inst for Sci Info. All rts. reserv.`
	`12204937   Genuine Article#: KU797   No. Reference...`
	`Title: INFORMATION VISUALIZATION USING 3D INTERAC...`
first author	`Author(s): ROBERTSON GG; CARD SK; MACKINLAY JD`
	`Corporate Source: XEROX CORP,PALO ALTO RES CTR,33...`
	`    ALTO//CA/94304`
year, vol, page	`Journal: COMMUNICATIONS OF THE ACM, 1993, V36, N4...`
	`ISSN: 0001-0782`
	`Language: ENGLISH   Document Type: ARTICLE`
search for citers	`?s cr=robertson gg, 1993, v36, p56, ?`
search S2 had 1 result	`    S2        1   CR=ROBERTSON GG, 1993, V36, P56, ?`

Figure 1: This annotated typescript from a DIALOG session shows a search of the Science Citation Database for articles that include S. K. Card as an author. Typescripts like this do not particularly show the structure of a search.

information assimilation than a conventional turn-taking interface (e.g., Mosaic or DIALOG), to avoid user overhead, to enable faster DIALOG investigations, to enable more complex investigations, and to reduce training time.

The key technique used by Butterfly is to embed access activity organically within an information visualization that supports the integrated management of search and browsing. In particular, Butterfly creates a virtual environment that grows under user control as asynchronous query processes link bibliographic records to form citation graphs. Our positive experience with the Butterfly implementation suggests that this approach may be quite useful in general for network information access. We conclude with a description of this approach, which we call *Organic User Interfaces for Information Access*.

CHALLENGES OF NETWORK INFORMATION ACCESS

Although search and browsing have received more attention than access management, the size and complexity of the Internet make access management an important and challenging task that deserves more attention. In particular, our studies show that people typically engage in a cyclic, dynamic process to discover where they should look in a database for information [11]. Network access makes this process more difficult to manage. Each access may involve substantial delays, either from communication overhead or the sheer size of the databases to be searched. Moreover, networked databases sometimes charge access fees that must be factored into access management decisions. Finally, the assimilation of results from search and browsing are not well supported, especially across different repositories.

Consider the standard method of interaction with DIALOG.

Figure 1 is an excerpt of a DIALOG search of the Science Citation Index in which the user retrieves a bibliographic item and then queries the database about articles that cite that item (called *citers*). In addition to the query language developed for professional librarians, this typescript illustrates two major challenges that DIALOG interaction has for casual users: 1) the textual stream of results is difficult to interpret, and 2) results are not easily incorporated into new queries for related information. The DIALOG system partially mitigates these problems by using labels such as `S1` for search results. These labels can be used to retrieve detailed information in a rich set of formats designed to reduce access costs. They can also be used to establish scopes for subsequent queries. However, these result labels are not particularly mnemonic and are somewhat difficult to find in the typescript. Even harder to find in the typescript is result information, such as an article's title and authors, that is typically used to formulate queries for related information. Furthermore, this information must be re-entered by the user. Local librarians have even been observed using a pencil to write down DIALOG output that is to be used as subsequent input.

A recent advance over traditional search-oriented information access applications such as DIALOG are graphical user interfaces for information visualization [1, 6, 10]. Such interfaces make access management easier because queries become user interface actions that occur immediately and are revisable. The user can rapidly explore alternative queries. The graphical display makes these explorations memorable. However, real-time visualization involves a tight coupling of the user interface to the information, which is expensive to achieve for networked databases. Users either need to invest in high-bandwidth network connections or enough disk storage so that they can retrieve the entire database locally for high-bandwidth access during visualization.

Browsing-oriented applications, such as Mosaic, can easily deal with the size and complexity of the network because they access information incrementally. The hypertext research community has developed a variety of techniques to help users manage such access activity [3, 4, 12]. These techniques include bookmarks so that the user can easily return to an interesting place, maps of the network surrounding the current node, and visual cues describing the browsing history. However, even with these techniques, browsing large databases can be laborious and complex, particularly when the network adds communication delays to every access. Ideally, the computer should shoulder some of the burden of processing the large collections while the user retains the responsibility of managing the access activity. One of the goals of the research described in this paper was to retain the naturalness of the browsing approach while incorporating the power of queries for processing databases.

THE BUTTERFLY APPLICATION

Butterfly is an Information Visualizer (IV) application for accessing three DIALOG databases, the Science Citation Index, the Social-Science Citation Index, and the IEEE Inspec database [5, 10]. These databases describe a large number of scholarly articles. For example, the 1977 Science Citation database contained approximately five hundred thousand articles that were the source of seven million reference links to three million articles [7].

Butterfly is based on four key ideas:

1. Visualizations of references and citers: We started with the idea to visualize scholarly articles as user interface objects with two wings, one wing for listing an article's references and the other wing for listing the article's citers. We called these objects "butterflies" for the obvious reason. Our goal was to use 3D interactive animation to support rapid browsing among butterflies by treating the wings as links from one butterfly object to related butterflies. We also designed various visual cues described below for the butterfly wings to help the user identify interesting areas to explore and otherwise manage their access.

2. Link-generating queries: DIALOG is a search-oriented application for accessing database records and does not provide much support for links among database records. The Science Citations Index databases, in particular, are collections of bibliographic records that describe references with strings of the form: first author, year, volume, page number, and an abbreviation of the book or journal title. We use this reference information to automatically create "link-generating" queries that link an article's record to the corresponding records for the article's references and citers.

3. Asynchronous query processes: Given link-generating queries, users can use the Butterfly visualization to browse the citation graphs implicit in the Science Citations Index databases by clicking on the butterfly wings. However, the size of the citation database makes such browsing laborious.

In particular, looking at articles listed on the wings of a butterfly is a repetitive process of manual navigation. Furthermore, a four second delay is inherent in the execution of every DIALOG command.

We address this problem with an idea as old as multi-processing window systems: Butterfly uses asynchronous query processes to access the databases. We have identified a number of advantages for using asynchronous processing for information access [9], including that the user does not have to wait for queries to complete. The interesting innovation in the Butterfly design is that it automatically creates query processes for the user to avoid explicit user management overhead. This innovation requires a policy for creating such processes because the branching factor of the citation graph would rapidly exhaust all machine resources if all branches were expanded simultaneously. Furthermore, indiscriminate expansion would be expensive given that DIALOG charges access fees. Thus, Butterfly uses a conservative policy: query process are only created automatically for the butterfly that is in the user's current focus of attention. The resulting behavior is very intuitive and lightweight. When the user pauses to view an interesting butterfly, asynchronous query processes are automatically created to gather additional information about that item. When the user moves on, these processes are automatically terminated.

4. Embedded process control: Sometimes, however, attention is not a sufficient heuristic for controlling query processes. For example, a user may just want to focus on the references of a survey article and stop processes from retrieving the citer information. Therefore, we represent the asynchronous query processes in the Butterfly visualization to give fine control. Thus, when the task requires it, the user can explicitly create and terminate query processes to direct them toward desirable information. This embedding is discussed in the next section.

Butterfly User Interface Components

This section describes the components of the Butterfly visualization shown in the screen snapshot in Figure 2 and Mackinlay Color Plate 1. The design of the Butterfly visualization combines query and browsing elements. The upper part of the snapshot focuses on queries and the lower part focuses on browsing. Users typically start with queries to find articles in topic areas of interest and then browse reference and citation links to find related articles. The ordering of the following Butterfly user interface components indicates a rough flow of user activity:

1. Sources. Three buttons at the top generate forms for entering queries to three DIALOG database sources: the Science Citation Index (SCI+), the Social-Science Citation Index (Soc-SCI), and an IEEE database (Inspec), which was included in this prototype because it includes computer science conference articles as well as the journal articles found in the other two databases. The colors of these three database buttons are used throughout the interface to indicate the source of information. The final button is used to read files in Refer format, a common bibliographic interchange standard.

2. Result pyramids. A pyramid of objects with the top of the pyramid facing toward the user is used to visualize the results of database queries. The pyramid in the snapshot is

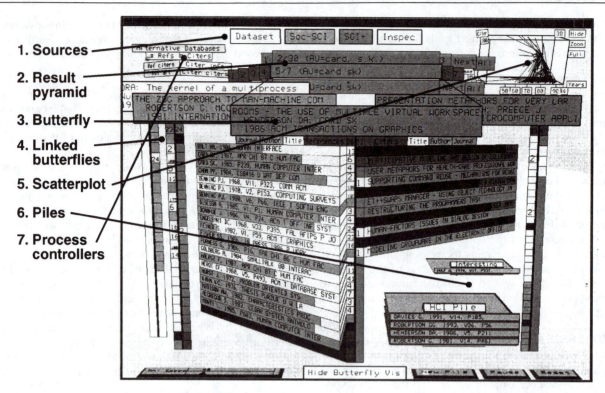

1. **Sources**
2. **Result pyramid**
3. **Butterfly**
4. **Linked butterflies**
5. **Scatterplot**
6. **Piles**
7. **Process controllers**

Figure 2: A snapshot of the Butterfly visualizer. The upper part focuses on search and the lower part focuses on browsing. We have grouped the user interface objects to indicate a rough flow of user activity. These groups are discussed in the text. Mackinlay Color Plate 1 is a color version of this snapshot.

partially covered by butterfly objects that have folded wings. A key issue for query visualization is that query results can have unpredictable size, which can make access management difficult because of the cost (both time and money) of retrieving large results. Butterfly addresses this problem by disclosing results progressively. Each query result is visualized as a horizontal layer in the pyramid colored to indicate the source database. The snapshot shows three queries each looking for articles authored by `card sk` in one of the three databases. The query result currently at the top of the pyramid is result 5 from the Science Citation database. This object is colored to indicate its source database. The colors of objects in the rest of the pyramid have been darkened to indicate that they are not the top object. New results are retrieved by clicking on the `Next` button. The `All` button is used to create an asynchronous query process that sequentially retrieves the items of a query result. The advantage of a pyramid shape is that it uses less screen space and reinforces the retrieval ordering of the results.

3. A butterfly. The current query result is described with a butterfly, which is in the center of the snapshot. The head of the butterfly lists the title, author, year, and journal of the article. The rest of a butterfly consists of a neck, body, and two wings. The wings of the butterfly list the article's references on the left and the article's citers found in the Science Citation database on the right. The items listed on a wing are called veins. Since an article can have many references and citers, the wings are limited to 22 veins to ensure that the text is legible. The body segments show the total number of items, and fan lines connect the visible segments to the veins. The neck contains a variety

of buttons for changing the views on the wings and two buttons for controlling the creation and termination of asynchronous query processes to retrieve information about the references and citers. A user can also click on a reference or citer vein to execute a link-generating query that creates a butterfly for the corresponding reference or citer. The relationship between such butterflies is shown by placing them next to each other as described in Group 4 below.

Color is used on the butterfly veins to provide the user with a rich collection of information for access management. Citations form complex graph structures that provide many paths for visiting an article. Sometimes the user will have already visited a citer or reference by some other path than the current article's butterfly. In this case, the vein color is darkened. References or citers visited directly from the current article's butterfly are shown with purple bars. The length of the bar indicates the order in which a given wing's articles have been visited, with the most-recently visited being longest. Yellow bars indicate the number of citers known to the system for each article. Articles with many citers are often a good place to browse. The goal here is to graphically encode information about about the unexplored parts of the citation graph so that the user can manage access effectively. The number of known citers is also printed in the corresponding segment on the body of the butterfly where it can be always be read, including when the wings are folded.

4. Linked butterflies. To the left and right of the current butterfly are butterflies with folded wings that have been explored by the user by following link-generating queries from the current butterfly. Each article in this chain of butterflies references the article to the left and is cited by the article to the right.

5. Scatterplot. In the upper right corner of the snapshot is a 3D scatterplot of the retrieved information. The points represent articles plotted against time, the alphabetic sort of the last name of the first author of an article, and the number of known citers for an article. Edges show the citation relationships among articles. Buttons provide a variety of viewing controls. Unlike butterfly chains, which show a few articles close to the current butterfly, the scatterplot provides a complete view of the citation graph that has been explored by the user. The scatterplot clearly shows the complexity of citation graphs and provides a contextual view for the user's activity. The current article and its reference and citer links are shown in red.

6. Piles. The objects below the butterflies are piles of articles that the user has selected. Piles allow the user to remember specific articles and to group related articles. An article is put in a pile by clicking on the head of the current butterfly, which produces a text widget with the corresponding bibliographic information in Refer file format. This text widget can be placed in a pile or stored to disk, where the information can be exchanged with other bibliographic applications.

7. Process controller. In the upper left corner is a stylized miniature of a butterfly for controlling the automatic creation of asynchronous query processes. Simple on/off switches control the automatic creation of different types of processes, such as whether an article is described in alternative databases, how many references or citers it has, and what the title, author, and journal are for its citers. A pale-green color indicates that the corresponding asynchronous query process may be started. A click turns the switch pale-red and terminates the corresponding process, which may be sequentially visiting the veins of a wing.

Butterfly Animation

Computers and people often operate at different time constants. For example, user interface transitions can often occur instantaneously, which can require cognitive effort as the user assimilates what has changed. Previous research on Information Visualizer developed an animation loop architecture for buffering the user from such mismatches in time constants. We have found that animated transitions let the human perceptual system track user interface changes with much less cognitive effort. For example, new results are brought to the top of the query pyramid with an animation that lets the user track the change perceptually.

An innovative aspect of the Butterfly visualizer is its use of the animation loop to buffer the user from long-running asynchronous query processes. Fundamentally, the animation loop and the asynchronous query process combine to give the Butterfly visualization an organic feel. Mackinlay Color Plate 2 shows the automatic growth of a butterfly as query processes retrieve information about an article that has just been put at the top of the query result pyramid. Pale green is used in the interface to indicate where query processes are active. Initially, nothing is known about the article except that it exists, which automatically starts a query process to retrieve its database record. While this query process is running, the user sees the

pale green butterfly head that is shown in the left panel of Plate 2 and is free to examine and/or manipulate any other part of the visualization. When the process completes, the retrieved database record contains enough information to grow the article's butterfly except for the right wing, which describes the article's citers. This partial butterfly is shown in the middle panel of Plate 2. It includes a pale green citer label to indicate that another query process has automatically started to query the database about the article's citers. The right panel in Plate 2 includes the right wing of the butterfly and two active query processes. A citer-wing process is retrieving bibliographic information for each citer and a reference-wing process is retrieving the number of citers (shown with yellow bars) for each reference.

Implementation Details

The Butterfly visualizer uses the object-oriented architecture of the Information Visualizer [10]. For example, most components in Figure 2 have top-level objects that control the positions of child objects. An earlier paper describes the architecture we use for accessing multiple databases and handling query results [9]. The implementation of query processes uses well-known computer science techniques for multiprocess programming, including process locks, critical sections, and a transaction-style modification of user interface objects.

Animation mitigates the apparent complexity of the Butterfly, making it easier to learn. Animated transitions often require precise adjustment of multiple objects. For example, a click on the query pyramid causes that clicked object to grow and replace the topmost object, which must shrink and move away. We added a new animation control mechanism to the IV architecture to accomplish this adjustment. This mechanism uses timer objects, which monitor elapsed real-time to adjust a visual value through a range. In the Butterfly, we use timer objects that adjust through the range zero to one, which makes it easy to adjust an arbitrary set of object properties through arbitrary ranges. The timer's value represents the percentage of growth, and one minus the timer value is the percentage of shrinkage. Typically, we use a single timer object for an animated transition, which ensures all objects adjust in unison.

Usage Studies

Butterfly has been successfully used for a number of complex searches of the Science Citation and Inspec databases. For example, Figure 3 describes a search for information visualization articles that resulted in over 40 relevant articles. This search, which was done by an alpha user who had no previous experience with the system, shows that Butterfly searches can have complex control structure and can proceed deeply into the search space.

The Butterfly visualizer generates and parses DIALOG typescripts that would normally have to be processed directly by the user. Clearly, users cannot generate and absorb typescripts at the same rate without the support of Butterfly's visual interface or similar techniques (although expert DIALOG users might deploy additional advanced DIALOG commands not supported by Butterfly). Furthermore, our alpha users managed useful and fairly efficient searches with only modest training (< 30 minutes). In contrast, text-based access to DIALOG typically involves 6 hours of formal training, but explains

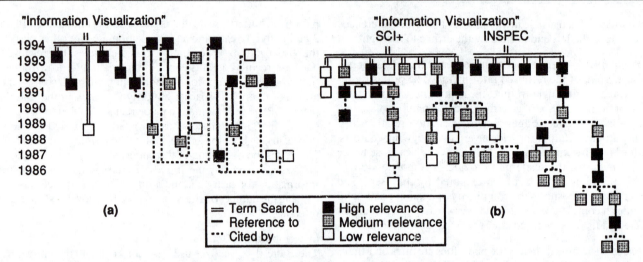

Figure 3: Two protocol fragments of a complex search for articles about information visualization. Diagram (a) shows that Butterfly searches can have complex control structure. Branching lines indicate that the user returned to a previous search state to explore another part of the database. Diagram (b) shows that Butterfly searches can proceed deeply into the search space.

more functionality than provided by Butterfly. A focus group of professional librarians (who were expert DIALOG users) generally thought the system would help them in their work.

Figure 3(a) shows the analysis of a protocol fragment from an alpha user creating a bibliography on the topic "information visualization". The diagram shows retrieved articles plotted by year, with user activity generally proceeding from left to right. Double lines show articles generated by a keyword request, dotted lines lead to articles found through reverse citation, and solid lines lead to articles found through references in a given article. The figure shows how the system made it possible to go forward and backward in time following chains of links forward and backward. In particular, the figure shows two cases of complex user control structure, where citing links led to reference links and then the user resumed from a previous state of the search.

Figure 3(b) suppresses time and plots a larger fragment of the protocol so as to show the depth and structure of citation chaining. The figure documents the use of chained combinations of forward and reverse citations up to 9 levels deep. By contrast, one of our professional librarians reported that most DIALOG users consider reference chains of length 3 to be extraordinarily deep [8]. The figure further shows many cases of falling back to previous parts of a chain and pursuing other linkages. This protocol clearly suggests that users are aided in doing complex searches by improving the speed of movement through the retrieved literary space, by allowing asynchronous query processes to carry out in parallel much of the overhead work of the searches, and by providing a manipulable visualization for the control structure.

DISCUSSION

The Butterfly application provides a high-bandwidth easy-to-learn user interface for accessing multiple slow repositories by organically embedding access activity, including search, browsing, and access management, within an information visualization that allows richer, more rapid information assim-

ilation than Dialog's typescript interface. Automatically formulated link-generating queries reduce the number of queries that must be formulated by the user, and asynchronous query processes, including automatic creation, reduce the overhead associated with accessing networked databases.

The Butterfly application was developed as part of the Information Visualizer (IV) project [10]. Although its major contribution to the project was exploring the use of visualization techniques for accessing multiple slow network repositories, it also continued our exploration of interactive animation and 3D graphics. Interactive animation, in particular, improves the understandability of the Butterfly design. 3D increases the density of the workspace, with the placement of tools, such as the piles and scatterplot, in the distance where they can be seen without taking up much scarce screen space. However, the folded butterfly wings and pyramid of results are mostly compensating for 3D fonts that must be larger than corresponding 2D fonts. It should be possible to develop 2D versions of these components.

ORGANIC USER INTERFACES

Our positive experience with the Butterfly implementation suggests this approach might be generally effective for building applications for accessing networked information. We believe these "Organic User Interfaces for Information Access" will include four components, which generalize the four key ideas underlying the Butterfly application:

1. Information landscapes: As butterflies fly through the forest, they help to construct *information landscapes*, which are virtual environments that hold retrieved information. These "workspaces" will be familiar places that persist over time as users organize retrieved information using tools like Butterfly's piles and scatterplot. Visualization techniques, such as animation and graphical presentation, will support rapid movement across these landscapes, which will include graphical indicators, similar to the yellow bars on the butterfly wings, that will indicate promising areas to explore in the "terra incognita" of information not yet retrieved.

2. Growth sites: Although databases typically store information in discrete chunks, information rarely exists in isolation. For example, the citation relationships that are automatically provided by Butterfly's link-generating queries are not the only links possible among scholarly articles. People often want to link bibliographic records by their shared authors, journals, subject matter, etc. These links form *growth sites* in the information landscape, where existing information can be used to automatically construct queries for related information.

3. Growth agents: As in nature, information landscapes will have a multitude of potential growth sites that can easily consume all human and computational resources. The requirement for human resources can be partially reduced with *growth agents*, asynchronous processes that grow an information landscape. A key issue is how these agents are deployed. Butterfly uses a conservative policy for automatic deployment, which is to grow where the user attention is focused.

4. Growth controllers: Finally, information landscapes will include mechanisms, such as the butterfly miniature, for managing growth. The *growth controllers* will let users direct the activity of growth agents, much like a gardener shapes a plant to maximize fruit production.

We believe the notion of an organic user interfaces for information access will expand to include a multiplicity of information landscapes for individuals, groups, and society. These landscapes will be used as workspaces and channels of communication.

Acknowledgments: Access to the DIALOG Information Service was part of a collaborative research agreement with Xerox PARC. George Robertson developed the Pile implementation and the IV architecture. Mark Stefix developed the Timer objects. Polle Zellweger suggested substantial improvements to the paper.

REFERENCES

1. Christopher Ahlberg and Ben Shneiderman. Visual information seeking: Tight coupling of dynamic query filters with starfield displays. In *Proceedings of ACM CHI'94 Conference on Human Factors in Computing Systems*, pages 313–317. ACM Press, 1994.

2. Tim Berners-Lee, Robert Cailliau, Jean-Francois Groff, and Bernd Pollermann. World-wide web: The information universe. *Electronic Networking:Research, Applications and Policy*, 1(2), 1992.

3. Mark Bernstein. The bookmark and the compass: Orientation tools for hypertext users. *ACM SIGOIS Bulletin*, 9(4):34–45, 1988.

4. P. J. Brown. Do we need maps to navigate round hypertext documents? *Electronic Publishing*, 2(2):91–100, 1989.

5. *DIALOG Pocket Guide 1994.* Dialog Information Services, 3460 Hillview Avenue, P.O. Box 10010, Palo Alto, CA 94303, 1994.

6. S. G. Eick. Graphically displaying text. *Journal of Computational and Graphical Statistics*, 3(2):127–142, 1994.

7. Eugene Garfield. *Citation Indexing.* ISI Press, 1975.

8. Giuliana Lavendel. Personal communication. 1994.

9. Ramana Rao, Daniel M. Russell, and Jock D. Mackinlay. System components for embedded information retrieval from multiple disparate information sources. In *Proceedings of the ACM Symposium on User Interface Software and Technology*, pages 23–33. ACM Press, Nov 1993.

10. George G. Robertson, Stuart K. Card, and Jock D. Mackinlay. Information visualization using 3d interactive animation. *Communications of the ACM*, 36(4), 1993.

11. Dan M. Russell, Mark J. Stefik, Peter Pirolli, and Stuart K. Card. The cost structure of sensemaking. In *Proceedings of InterCHI '93*, pages 269–276. ACM, April 1993.

12. Kenneth Utting and Nicole Yankelovich. Context and orientation in hypermedia networks. *ACM TOIS*, 7(1):58–84, 1989.

END-USER TRAINING: AN EMPIRICAL STUDY COMPARING ON-LINE PRACTICE METHODS

Susan Wiedenbeck, Patti L. Zila, and Daniel S. McConnell

Computer Science and Engineering Department
University of Nebraska, Lincoln, NE 68588, U.S.A.
phone: (402) 472-5006
susan@cse.unl.edu

ABSTRACT

An empirical study was carried out comparing three kinds of hands-on practice in training users of a software package: exercises, guided-exploration, and a combination of exercises and guided-exploration. Moderate to high experience computer users were trained. Subjects who were trained with exercises or the combined approach did significantly better in both time and errors than those trained using guided-exploration. There were no significant differences between the exercise and the combined approach groups. Thus, it appears that the better performance of these groups can be attributed to the exercise component of their practice.

KEYWORDS: training, practice methods, exercises, guided-exploration, minimal manual, end-users, tutorials.

INTRODUCTION

We are interested in refining the understanding of the training needs of users of software packages. Since the mid-1980s, Carroll and his associates have documented problems which users face in computer training [5, 6, 7]. Their work indicates that learners of software packages experience numerous problems in the initial learning of software systems, including voluminous materials, lack of focus on real users' tasks, absence of error recovery information, and misleading analogies drawn from non-computer experience. Carroll and his colleagues have proposed a solution which addresses many of the training problems listed above, which is referred to as minimalist training [4, 7]. Minimalist training has been tested with positive results, as outlined below. However, there is a need for further studies which focus on individual parts of their larger strategy and attempt to determine in detail the role that they play. Here we investigate the role of different kinds of hands-on computer practice within the paradigm of minimalist training. We evaluate three kinds of practice which differ in the amount of structure provided to the learner.

PREVIOUS RESEARCH AND RESEARCH QUESTIONS

The minimalist training model has the following characteristics [7]: focus on real tasks, reduction in the verbiage of training materials, and support for error recovery and recognition. These features are meant to maintain high motivation, promote active learning, and make the environment "safe" for learners in the sense that they can try features out without fear of becoming hopelessly mired in errors.

Carroll, et al.'s experimental studies [6, 7] support the effectiveness of the minimalist training approach compared to commercial tutorial materials. A recent replication by Lazonder and van der Meij [12] supported their findings. Black, Carroll, and McGuigan [2] tried to isolate some of the many dimensions on which the minimal manual differed from traditional tutorial manuals and found two significant features: the shorter length in itself and the incompleteness of the materials, which encouraged inferencing by the trainee. Gong and Elkerton [11] showed that learning improved when subjects used a briefer manual with procedural instructions. However, in their study the effect of error recovery information was uncertain. On the other hand, both Olfman [14] and Davis and Bostrom [10] compared an exploration training approach to a more traditional instruction-based approach and found no significant differences.

An aspect of training that is closely related to the concept of minimalist training is the kind of hands-on practice provided in the training materials. Minimalist training advocates argue that people prefer active learning or learning by doing rather than by reading a manual. If that is the case, then hands-on practice may be essential for both learning and motivation in training. Carroll et al. [7] did not explicitly study practice methods, but it is clear that hands-on practice is essential to minimalist training. Charney and her colleagues [8, 9] studied practice methods and found that methods involving problem-solving practice were superior to merely reading or typing worked-out examples.

Among methods involving problem-solving, there are two major categories: exercise and exploration. Exercises leave the learner to decide on a solution strategy and the method for implementing it, and thus appear to be much less mechanical than typing the keystrokes of a worked-out example. Exercises provided by the manual designers can be structured to cover the full range of functions, illustrate important concepts, and correct misconceptions [9]. They may be particularly good for assuring coverage of the basic functionality of a system.

Exploration leaves the choice of the practice itself, as well as the solution strategy and method of implementing it, to the discretion of the learner. Allowing learners to set their own goals has been seen as a key to maintaining high learner motivation [7]. Another possible advantage of exploration is that it may facilitate meaningful learning. Assimilation theory [1] distinguishes between meaningful and rote learning. In meaningful learning the learner actively manipulates new information to connect it conceptually to prior knowledge, leading to a deep understanding of the new information and the ability to apply it in novel ways. Rote learning is memorization with little concern for its connection to prior knowledge. While rote learning may lead to success in tasks which require the repetition of known information, it is not likely to lead to success in novel tasks. Exploration-based training may promote meaningful learning by encouraging the learner to set goals which go beyond the simple procedures in the training manual. However, a possible problem with exploration training is that practice devised by learners may not be optimal for covering the capabilities and difficulties of the software.

In a previous empirical comparisons of exercise and exploration practice methods, Sebrechts and Marsh [16], studying UNIX novices, found that the exercise group performed better than the exploration group. It should be noted that their exploration practice was completely open-ended and occurred when the subjects had finished reading the manual. For half the subjects the manual was not even available during the practice.

As one part of a larger study, we carried out an empirical comparison of three kinds of hands-on practice: exercise, guided-exploration, and the combination of exercises plus guided-exploration. Unlike, Carroll et al. [6, 7], the content of our training manuals was the same across all conditions, and only the practice instructions differed. The exercise practice posed specific problems for the learner to solve. The guided-exploration practice asked learners to set their own goals and create their own problems to meet the goals. However, in doing so they were guided by suggestions and questions, which focused their attention on a set of possible goals appropriate to the software. We use the term guided-exploration to distinguish this kind of exploration from completely open-ended exploration. The combined practice consisted of an exercise followed by guided-exploration instructions. Below are the research questions which we addressed:

— **Does guided-exploration lead to more successful training outcomes than exercises?** Charney and Reder [8] did not include exploration practice in their study. Sebrechts and Marsh [16] included completely open-ended exploration practice and found that exercises led to better performance than exploration. We hypothesized that a more focused exploration-based practice would improve the performance of exploration trainees. As a result, we designed our guided-exploration practice to be a little less open-ended. We asked learners to practice at specific points as they worked through their manual and asked some focusing questions about the topic under discussion at that point. This was intended to guide them to choose appropriate goals, while still leaving them to choose the specific goals and the amount of practice. We hoped that giving a bit more guidance in the exploration would allow us to detect the advantages of exploration if they exist.

— **Is the combination of exercises followed by guided-exploration better than guided-exploration alone?** It has been suggested [16] that a possible reason for poor outcomes of exploration-based practice is that the learner does not know enough about the software to devise adequate hands-on practice. A remedy is to first give learners practice with exercises, then have them explore further. Our combined practice was designed to allow us to judge whether guided-exploration is more effective when it follows the more structured practice embodied in exercises.

— **Does goal-setting during practice appear to aid in learning software?** Carroll et al. [6, 7] see goal-setting by learners as fundamental in computer training. The essential difference between exercises and guided-exploration is that learners set their own goals in guided-exploration. The comparison of guided-exploration to exercises will give us insights about the role of goal-setting.

— **Which practice method aids meaningful learning?** Past research [3, 10, 13, 17] has used near and far-transfer tasks to operationalize the difference between rote and meaningful learning. Near-transfer tasks are very similar to tasks taught in the training. Far-transfer tasks require the learner to go beyond what was explicitly taught. This could involve combining several operations in a novel way or using tools and operations not described in the training, but suggested by analogy to others that were taught. Singley and Anderson [18] discuss procedural-to-procedural transfer as the transfer that occurs when productions acquired in training apply directly to a transfer task. We see our near-transfer tasks as being essentially of this type. Singley and Anderson [18] define declarative-to-procedural transfer as occurring when declarative knowledge acquired during training aids the acquisition of productions during transfer. We would classify our far-transfer as largely declarative-to-procedural. If exercises aid in the acquisition of productions for basic operations during training, they may aid near-transfer. Exploration practice, on the other hand, may aid far-transfer by encouraging learners to create problems that go beyond the training materials, leading to discovery of new concepts or novel combinations of simple concepts. A training method combining exercises and exploration could facilitate both near and far-transfer. In their study, Charney and Reder [8] evaluated the learning of easy and difficult commands, but apparently neither they nor Sebrechts and Marsh [16] tested subjects with tasks which required using commands in a novel way.

METHODOLOGY

Subjects
A total of 51 subjects participated in the experiment. Subjects were volunteers recruited among upper-level undergraduate and graduate students and were randomly assigned to the three training conditions. Seventeen subjects served in each training condition.

The subjects were recruited to represent moderate to high general computer experience. This group was chosen because in today's business environment most information workers routinely use one or more software packages in their work. Thus, learners of a package usually are experienced with other software. We wanted to represent this reality in our study. The mean age of the subjects was 28 years. Eighty percent were male and 20 percent female; however, a preliminary analysis of the data using an Analysis of Covariance showed that sex was not a significant covariate of performance. The subjects came from a wide variety of fields, but the majority were from technical areas, such as business and engineering. Many of the subjects had programming experience, ranging from introductory programming courses to an undergraduate degree in computer science. All of the subjects had considerable experience with applications programs on personal computers. The most widespread experience was with word processors and spreadsheets, followed by graphics and communications programs.

Materials
In this experiment subjects were trained on the Hypercard™ software program. Hypercard is a hybrid program which contains an integrated set of text and graphics tools along with an end-user programming language, Hypertalk™, which can be used to create advanced applications. Training was given on the basic text and graphics features of Hypercard. The Hypertalk programming language was not taught.

A self-study training manual was created for the experiment. Text-based learning materials were chosen because of evidence of better retention and transfer in text-based materials as opposed to alternate media [15]. Like Carroll et al.'s manual [7], our manual was brief, consisting of approximately twelve pages of text. It had a single introductory paragraph which motivated the use of Hypercard by mentioning several typical applications: address book, calendar, etc. Descriptions of commands and procedures were grouped into topic areas named to match the tasks of new users. Information was included to help the reader coordinate reading with what was appearing on the screen. The user was left to infer procedures by analogy as much as possible. Explicit error recognition and recovery information was included in the manual.

At various points in the manual practice opportunities were given. In all there were 14 practice opportunties. Three different versions of each practice were developed: exercise, guided-exploration, and combined. These three practice types were embedded in the manual to create three different versions of the manual with identical text, but differing in the statement of the hands-on practice.

Exercises gave specific tasks for the learner to carry out. They told what object to operate on and what to do to the object. Exercises were stated with the intention that it be easy for learners to evaluate their success. Guided-exploration asked the learner to create tasks for themselves in order to try out the procedures and commands detailed in each section of the manual. Thus, the learner was encouraged to practice the same procedures as the exercise subjects, but the specific object on which to work was not specified, nor was the specific result desired (where the object should be, what it should look like), nor how much to practice. In the guided-exploration practice instructions, focusing questions were posed to draw the learner's attention to important aspects of the interaction. Combined practice consisted of an exercise followed by a guided-exploration instruction. Figure 1 gives examples of the three kinds of practice instructions.

A set of evaluation tasks was also developed to use in measuring the subjects' performance after training. There were fourteen evaluation tasks in total. Nine of the tasks were near-transfer tasks, consisting of commands and procedures which were covered in the training manual. An example of a near-transfer task is the following:

> "On Card 2 create a field called Name.
> The field should be able to hold 2 lines
> of text and should be approximately 4"
> wide. Position this field towards the
> top and in the center of Card 2."

This was classified as near-transfer because creating fields, naming them, and physically manipulating

This practice was given following the manual section that discussed using the background. The manual section on the background consisted of two pages containing an explanation of the background, procedural instructions for using it, and related error recognition and recovery information. This section occurred near the end of the training, when the subjects had already created a stack of cards and put information on them.

Exercise

Go to the background. Create a button on the background and call it "Next". Move it to the bottom of the screen. Leave the background and return to the card. The button should now appear on every card in the stack.

Guided-exploration

Think of information that is needed in your stack of cards. Can you think of any information which each card would have in common? With this in mind, place the common information on the background, including buttons, fields, and graphics. When you feel confident that you can successfully work in the background, move on to the next section.

Exercise + guided-exploration

Go to the background. Create a button on the background and call it "Next". Move it to the bottom of the screen. Leave the background and return to the card. The button should now appear on every card in the stack.

On your own
Think of other information that is needed in your stack of cards. Can you think of any information which each card would have in common? With this in mind, place the common information on the background, including buttons, fields, and graphics. When you feel confident that you can successfully work in the background, move on to the next section.

Figure 1: Examples of three kinds of practice

them were topics described in the manual. Exercise and combined condition subjects were given exercises to practice most of these elements, although never in a form identical to the evaluation task. Since guided-exploration subjects chose their own practice, it is not certain which elements they practiced, but they did read the same descriptive text as the exercise and combined condition subjects.

In addition, there were five far-transfer tasks which required the subject to go beyond procedures explicitly described in the manual. Far-transfer was operationally defined as one of three things: using a tool that had not been taught in the manual, doing an operation taught in the manual in some different context from the original, or combining a series of separate operations in some novel way to achieve a goal. An example of a far-transfer task is the following:

> "Place a copy of the portrait on Card 1
> onto Card 2. Place the copy to the left
> of the field Name."

In the manual the subjects were taught how to copy text. They were also taught how to select and move regular-shaped graphic objects, such as rectangles. However, the manual did not give instructions about copying an object such as the portrait mentioned in this exercise. The subject first had to recognize that the rather elaborate portrait was a graphic object, just like a rectangle, then conclude that it could be moved by a copy/paste operation as is text, then find a way to select an irregular-shaped graphic. Thus, this task was far-transfer because the subject had to infer how to do it by combining what they knew about copying text with their knowledge of manipulating graphics. In the task list the near and far-transfer tasks were intermixed.

Procedure

Subjects were run individually. The average time to complete a session was 2 hours, but some subjects took up to 2 1/2 hours depending on the amount of time they spent practicing and doing the evaluation tasks. First, the subject completed a questionnaire detailing his or her computer experience. Then the subject worked through the manual independently, carrying out the hands-on practice when instructed by the manual. Subjects were asked to work through the manual from beginning to end but were allowed to go back to earlier sections whenever they wished. Feedback was not given on the subject's work in order to simulate a real self-study environment where normally the only feedback is from the computer. Subjects were given a maximum of 90 minutes for training with the manual. If they finished working through the manual before the training period was over, they were allowed to continue their training, if they wished, in one of two ways, depending on their experimental condition. Guided-exploration and combined subjects were allowed to continue exploring

Hypercard on their own, reading and practicing on-line as they wished. Exercise subjects, on the other hand, were restricted to prevent unstructured exploration of the system. They were allowed to reread the manual, but the only practice they were allowed was to repeat exercises given in the manual. An experimenter was always present to monitor that subjects followed the instructions for their practice condition. The experimenter also kept detailed notes about the subjects' specific activities and their duration during the training phase. The training was stopped either after 90 minutes or when subjects indicated that they were finished. The total time spent on hands-on practice during training was recorded.

The evaluation phase of the experiment was conducted after a break, during which subjects filled out several questionnaires not reported on here. Subjects were given the set of 14 evaluation tasks to carry out. The manual was not available for use, but subjects were given a one page summary of the menu items and procedures taught in the manual to aid their memories. The tasks were timed and the subjects' work was saved. The evaluation phase lasted a maximum of 55 minutes, but subjects could stop sooner if they were finished or could make no further progress.

RESULTS

One subject completed a first pass through the manual but did not have time to go back to earlier sections for rereading and further practice. All other subjects had time to return to earlier sections of the manual. All but two subjects requested to go on to the evaluation tasks before the 90 minute training period was up. Thus, it appears that subjects had sufficient training time. Table 1 summarizes the means and standard deviations of the dependent variables, training time, near transfer time, far transfer time, near transfer correctness, and far transfer correctness for each practice condition. The following

paragraphs describe the results of the multivariate and univariate analyses.

A MANOVA including all the dependent variables was run. The MANOVA showed that the effect of practice type was significant ($F(10, 90) = 5.62$, $p < .05$), showing an overall difference in performance among the practice types. After the significant MANOVA, ANOVAs were run for the individual time and correctness variables.

Training time consisted of the total time that subjects spent in hands-on practice during the training period, excluding the time spent reading the manual, but including time for looking up information in the manual during practice. Near-transfer time was the sum of the times spent doing the 9 near-transfer tasks. Far-transfer time was the sum of times for the 5 far-transfer tasks. The ANOVA for training time showed that there was a main effect of practice type ($F(2,48) = 7.82$, $p < .05$). Newman-Keul's test for specific differences was run and showed that the exercise condition was faster than the guided-exploration or the combined condition ($p < .05$), but the guided-exploration and the combined condition did not differ significantly from each other. The advantage of the exercise condition was about 12 minutes over the combined condition and 17 minutes over guided-exploration. The ANOVA for near-transfer time revealed that there was a main effect of practice type ($F(2,48) = 11.11$, $p < .05$). Newman-Keul's test indicated that the guided-exploration subjects were significantly slower than the exercises or combined subjects on near-transfer tasks ($p < .05$). The ANOVA for far-transfer time also showed a significant difference based on practice type ($F(2,48) = 7.88$, $p < .05$). The exercise and combined conditions were significantly faster than guided-exploration (Newman-Keul's test, $p < .05$). The exercise and combined conditions did not differ significantly from each other.

Variable	Exercise	Guided Exploration	Combined
Training time (minutes)	31.25 (11.71)	48.11 (11.43)	43.44 (15.05)
Near-transfer time (minutes)	10.24 (3.45)	17.35 (5.92)	10.44 (5.31)
Far-transfer time (minutes)	7.28 (3.56)	11.26 (4.29)	6.46 (3.40)
Near-transfer correctness (maximum = 27)	24.71 (1.65)	18.59 (4.69)	22.65 (4.57)
Far-transfer correctness (maximum = 15)	14.24 (1.20)	7.82 (3.80)	12.18 (3.19)

Table 1. Mean and standard deviation (in parentheses) of dependent variables by practice type

The evaluation tasks were graded for correctness on a scale of 0 to 3 as follows:

     3 = task completely correct
     2 = task mostly correct (over 50% correct)
     1 = task mostly incorrect (50% or less correct)
     0 = task completely incorrect or not attempted.

The work of the subjects was graded independently by two judges using a set of detailed grading criteria developed in advance. The score assigned to each task was the average of the grades of the two judges. The inter-rater reliability was .97. For the analysis, the sum of scores for all near-transfer tasks was used as the subject's near-transfer correctness score, and the sum of the scores for all far-transfer tasks was used as the far-transfer correctness score. The ANOVA for near-transfer correctness indicated a main effect of practice type ($F(2,48) = 11.08$, $p < .05$). Newman-Keul's test showed that the exercise and combined conditions had higher correctness on near-transfer tasks than the guided-exploration condition ($p < .05$). However, the exercise and combined groups did not differ significantly from each other. The ANOVA for far-transfer correctness also showed a significant difference based on practice type ($F(2,48) = 22.68$, $p < .05$). Newman-Keul's test revealed that the guided-exploration group performed more poorly than the other groups ($p < .05$), but exercise and combined groups did not differ from each other.

DISCUSSION

The results regarding the time spent on hands-on practice during training reflect differences among the three practice methods. The combined method took longer than exercises because there was more for the learner to do. We can reasonably assume that the extra time represents time that learners spent on exploration. Since the combined group spent almost 40 percent more time, the exploration component of their practice was substantial. The exploration group also spent significantly more time on practice than did the exercise group. There are several possible explanations of their additional time. Part of the time can probably be attributed to goal-setting, since these subjects were responsible for choosing their own practice. Another explanation could be that guided-exploration subjects tried out more functions or more advanced functions. However, if this were the case, we would have expected guided-exploration subjects to perform better than exercise subjects on far-transfer tasks, something which we did not see. A third possibility is that exploration subjects made more errors in training and spent more time in error recovery. Unfortunately, we do not have the data to evaluate this explanation. It should be noted that our result that training was slower for the guided-exploration group does not contradict the results of Lazonder and van der Meij [12]. They found faster overall training time for subjects using a minimal exploration manual, but they were comparing a minimal manual to a much longer standard tutorial

manual, and they were focusing on the total training time, not just on the difference in time spent doing hands-on practice.

Our evaluation task results showed that exercise and combined practice tended to be superior to guided-exploration practice. Sebrechts and Marsh [16] had found that performance was poorer with completely open-ended exploration practice than with exercises. As a result, we made our guided-exploration moderately more structured to aid the learner in setting appropriate goals. In particular, we asked them to practice at given points in the manual, and we posed questions to them to focus their goal-setting and help them notice certain critical aspects of the interaction. Even so, exploration practice still led to poorer results. Since exercise subjects trained faster than guided-exploration subjects, the training time does not explain these results.

Performance in the combined condition was very similar to the exercise condition. This implies that the exercise practice was the essential factor leading to successful hands-on training. We had speculated that combined practice might be optimal because the exercise would give subjects a basic understanding of each concept, and further exploration would then allow them to expand their understanding. However, this expectation was not supported. We had also speculated that practice methods with an exploration component would lead to success on far-transfer tasks because they would encourage meaningful learning. Instead, we found that for far-transfer performance of subjects in the exercise condition equaled that of subjects in the combined condition and exceeded that of subjects in the guided-exploration condition. From this it appears that exercise practice allows the trainee to work with the material sufficiently to later apply it in novel tasks. Guided-exploration practice, with its goal-setting component, was not an aid to far-transfer.

Why did guided-exploration practice lead to unexpectedly poor results? Other researchers [9, 16] have speculated that novice learners have difficulty creating adequate practice on their own. Based on our results, we believe that the inability to formulate adequate practice may also apply to more experienced learners. It seems most likely to occur when learners are working with software packages dissimilar to those they already know. We believe that this was the case in our experiment. While most of the subjects had experience with both text processing and graphics software, they did not have experience with text and graphics in the integrated combination of Hypercard. We dealt with this in the exercise and combined conditions by providing a set of exercises that integrated text and graphics to build a very rudimentary address book-type of application. The guided-exploration practice instructions continually asked subjects to think of realistic information that they could group using the features of Hypercard, but this apparently was not successful in encouraging subjects to create a realistic application. In fact, our observations recorded during training showed that most exploration subjects (13 out of 17) created a series of practice problems that were discrete and unrelated to one another. Thus, they may not have gained a sense of how Hypercard could actually be used. This may have hurt their performance on far-transfer tasks, in which they needed to put separate pieces of knowledge together in novel combinations. Our observations during training also suggested that guided-exploration subjects did not practice basic skills as thoroughly as the exercise and combined subjects. They tended to skip practicing some functions described in the manual and to minimally practice others. This may have hurt them on near-transfer tasks. For example, the exercise given for putting text in a field required the subject to type the text, erase part of it, change the font, and change the style. Guided-exploration subjects usually did not practice all of these basic skills, even though they were all described in the manual. Also, guided-exploration subjects tended to be attracted to and spend proportionately more time on creating and manipulating graphics than fields or buttons. This is a poor distribution of effort in terms of Hypercard functionality. Buttons and fields are the basic elements used in information storage and manipulation, while graphics affect the look of a Hypercard application but not its functionality. Exercise and combined subjects tended to practice the basics because the set of exercises stressed them. Thus, the goal-setting of the guided-exploration subjects did not aid learning.

CONCLUSION

Based on these results, we recommend the inclusion of well-designed exercises in training materials. We see two main advantages to exercise practice. First, exercises can assure that the learner is exposed to the software's most important functionality. Second, a set of exercises can suggest typical uses for which the software is well-suited, thus helping users to see how it could be used to advantage in their own work. Of course, these advantages can only be realized in exercises which are carefully designed and tested. The requirement of careful design and testing applies not just to individual exercises but to the exercise set as a whole. A series of exercises must be well-integrated if it is to effectively suggest appropriate uses of the software.

We do not see our results as a challenge to the minimalist training model. In fact, our experiment was firmly situated in the minimalist tradition by our minimal manual which promoted active, hands-on learning and inferencing on the part of the user. Charney and Reder's results [8] supported the importance of problem-solving practice. This work suggests that it is important that the problem-solving practice be well-conceived. Even relatively high experience learners, such as ours, may experience difficulties in devising useful practice on their own,

and this seems particularly likely when the software domain is unfamiliar. Carroll et al. [6, 7] argue that it is important for learners to set their own goals in order to maintain motivation during training. We agree that doing real work is highly motivating. However, many times learners do not begin with real work to accomplish which can only be done using the new software. Rather they begin with questions about how the software can benefit them in their own work. In this situation, the user's goal is less immediate and concrete. The user wants to learn what the software offers and how to carry out typical operations. For these learners, the structuring inherent in a well-conceived set of exercises appears to be optimum. It should be noted that exercises in a self-study manual are only suggestions. Users whose goal is to carry out a specific piece of work should still be able to use a minimal style manual to advantage for that purpose.

ACKNOWLEDGMENTS.

This work was supported by NSF grant CDA-9200230 and by funds from the University of Nebraska. It was completed while the first author was a visitor at the INRIA. The authors would like to thank Jack Kant for his work in developing the training materials.

REFERENCES

1. Ausubel, D. P. Educational psychology: A cognitive view. Holt, Reinhart and Winston, NY, 1968.

2. Black, J.B., Carroll, J.M., and McGuigan, S.M. What Kind of Minimal Instruction Manual Is the Most Effective, in Proc. CHI+GI 1987 Human Factors in Computing Systems and Graphics Interface (Toronto, April 5-9, 1987), ACM Press, pp. 159-162.

3. Borgman, C.L. The user's mental model of an information retrieval system: an experiment on a prototype on-line catalog. International Journal of Man-Machine Studies, 24 (1986), 47-64.

4. Carroll, J.M. The Nurnberg Funnel: Designing Minimalist Instruction for Practical Computer Skill. MIT Press, Cambridge, MA, 1990.

5. Carroll, J.M. and Mack, R.L. Learning to Use a Word Processor: By Doing, By Thinking, and By Knowing. In Human Factors in Computer Systems, J.C. Thomas and M.L. Schneider, Eds., Ablex, Norwood, NJ, pp. 13-51.

6. Carroll, J.M., Mack, R.L., Lewis, C.L., Grischkowski, N.L., and Robertson, S.R. Exploring exploring a word processor. Human-Computer Interaction, 1, 3 (1985), 283-307.

7. Carroll, J.R., Smith-Kerker, P.L., Ford, J.R., and Mazur-Rimetz, S.A. The minimal manual. Human-Computer Interaction, 3, 2 (1987), 123-153.

8. Charney, D.H. and Reder, L.M. Designing interactive tutorials for computer users. Human-Computer Interaction, 2 (1986), 297-317.

9. Charney, D.H., Reder, L.M., and Wells, G.W. Studies of elaboration of instructional texts. In Effective Documentation: What Have We Learned From Research, S. Doheny-Farina, Ed., MIT Press, Cambridge, MA, 1988, pp. 47-72.

10. Davis, S.A. and Bostrom, R.P. Training end users: an experimental investigation of the roles of the computer interface and training methods. MIS Quarterly, 17 (1993), 61-85.

11. Gong, R. and Elkerton, J. Designing Minimal Documentation Using a GOMS Model: A Usability Evaluation of an Engineering Approach, in Proc. CHI'90 (Seattle, April 1-5, 1990), ACM Press, pp. 99-106.

12. Lazonder, A.W. and van der Meij, H. The minimal manual: is less really more? International Journal of Man-Machine Studies, 39 (1993), 729-752.

13. Mayer, R.E. The psychology of how novices learn computer programming. Computing Surveys, 13 (1981), 121-141

14. Olfman, L.A. A Comparison of Applications-Based and Construct-Based Training Methods for DSS Generator Software, unpublished doctoral dissertation, Indiana University, Bloomington, IN, 1987.

15. Palmiter, S. and Elkerton, J. An Evaluation of Animated Demonstrations for Learning Computer-Based Tasks, in Proc. CHI'91 (New Orleans, April 27-May 2, 1991), ACM Press, pp. 257-263.

16. Sebrechts, M.M. and Marsh, R.L. Components of computer skill acquisition; some reservations about mental models and discovery learning. In Designing and Using Human-Computer Interfaces and Knowledge-Based Systems, G. Salvendy and M. J. Smith, Eds., Elsevier Science Publishers B. V., Amsterdam, 1989, pp. 168-173.

17. Sein, M.K. and Bostrom, R.P. Individual differences and conceptual models in training novice users. Human-Computer Interaction, 4 (1989), 197-229.

18. Singley, M.K. and Anderson, J.R. The Transfer of Cognitive Skill. Harvard University Press, Cambridge, MA, 1989.

A Comparison of Still, Animated, or Nonillustrated On-Line Help with Written or Spoken Instructions in a Graphical User Interface

Susan M. Harrison

Department of Computer Science
University of Wisconsin-Eau Claire
Eau Claire, WI 54702
(715) 836-5381
harrison@uwec.edu

ABSTRACT

Current forms of on-line help do not adequately reflect the graphical and dynamic nature of modern graphical user interfaces. Many of today's software applications provide text-based on-line help to assist users in performing a specific task. This report describes a study in which 176 undergraduates received on-line help instructions for completing seven computer-based tasks. Instructions were provided in either written or spoken form with or without still graphic or animated visuals. Results consistently revealed that visuals, either still graphic or animated, in the on-line help instructions enabled the users to significantly perform more tasks in less time and with fewer errors than did users who did not have visuals accompanying the on-line help instructions. Although users receiving spoken instructions were faster and more accurate for the initial set of tasks than were users receiving written instructions, the majority of subjects preferred written instructions over spoken instructions. The results of this study suggest additional empirically-based guidelines to designers for the development of effective on-line help.

KEYWORDS: Graphical user interfaces, on-line help, visuals, user interface components.

INTRODUCTION

With the advancement of microcomputer technology and the use of graphical user interfaces, the number of features available within software applications has rapidly increased. Word 5.0a [22] for example, is a word processing package for the Macintosh that presents the user with a choice of 97 features in the opening screen of the text-based on-line help system. The Excel 4.0 spreadsheet package [23] lists 22 features in the first level of the text-based help directory, and a second listing appears containing 111 features for the spreadsheet's worksheet.

Unfortunately, advances in the design of on-line help systems have not matched recent advances in the design of interface technology. Researchers have observed that subjects in need of assistance often do not utilize the on-line help available [9,20,32]. Subjects cited "poor quality of the help information" as a common reason for avoiding on-line help. Researchers have found that the common textual form of on-line help tends to hinder rather than help the user in the performance of procedural tasks [9,10,15,32]. The visual and dynamic nature of modern graphical user interfaces is not captured by the traditional form of on-line help [33,37].

The on-line help of this study presented instructions for completing procedures involving the creation, editing, and deletion of five types of objects within a HyperCard environment. HyperCard is an integrated object-oriented programming system for the development of documents with linked segments of information of text, sound, and graphics [25]. The visual illustrations presented within the on-line help of this study included still graphics and animated visuals in the form of a computer-generated demonstration using animation.

The relative effectiveness of each design of the on-line help within this study was evaluated by the speed and accuracy in which the user performed each procedure. As suggested in performance measurement guidelines by Bovair and Kieras [8], dependent measures of completion rates, elapsed time, and error rates were used to evaluate the effectiveness of written or spoken instructions with or without still graphic or animated visuals in enhancing the user's performance of computer-based procedures within a graphical user interface environment.

BACKGROUND

Designers of on-line help presenting procedural instructions "should concentrate on ensuring that the procedure construction goes smoothly" [8,p.227]. Procedure construction is part of the process of acquiring a procedural skill—a process that researchers recognize as an aggregate of three phases. In the first phase relevant facts for performing the skill are encoded and committed to memory in a declarative form [3]. The learner focuses on understanding the skill by often verbally rehearsing the information as the skill is performed. Strategies for executing the procedure are formulated, tested, and executed in a consciously controlled way [31] during this phase, resulting in slow and error-prone performance [1]. During the second phase, execution errors are gradually detected and eliminated as the learner practices the skill [24]. The execution of the procedure becomes automated and faster through extensive additional practice as the set of individual steps are collapsed into one large unit during the third phase [2].

The amount of practice the user invests in learning a procedural skill is the user's choice, not the researcher's or the software designer's. This study investigated the effectiveness of design elements in on-line help without the variable of practice. By excluding the use of practice, this study was able to focus on the design effects of on-line help rather than on practice effects.

PREVIOUS RESEARCH

Only a few studies reported in the literature employed the use of on-line help for computer-based procedures and only three of those studies [27,28,37] investigated the use of graphics and animation in on-line help. These studies incorporated the use of practice in addition to the on-line help as a means of aiding the user in acquiring the procedural skills.

Shneiderman [33] contends that use of a graphical demonstration is the most direct way for novices to learn the basic functionality and steps necessary to perform procedures within the graphical user interface environment. A few researchers [26,28,37] have investigated the potential use of demonstrations within on-line help for presenting procedural information. Such demonstrations of real-time instantiations of computer-based procedures have been termed "animated demonstrations" [28]. Like still graphics, animated demonstrations immediately show what objects are used within the procedures. This approach was shown to be partially effective [27,28]. Segmented animated demonstrations, presenting one step of a multi-step task at a time, may aid the user in developing discrete steps of the procedure. Unless the animation is segmented, the user may not be able to identify and encode specific aspects of the procedure [8,26]. No studies were found, however, that investigated the use of segmented animated demonstrations.

Overall, the research with on-line help indicates that the current use of text-based on-line help is generally ineffective [10,11,12,15,18]. Researchers agree, however, that the effectiveness of on-line help may be improved if on-line help was properly developed using empirically based guidelines. The empirical evidence from research investigating the use of instructions, audio, visuals, and animation in the design of on-line help suggest that on-line help for procedural tasks may be more effective when:

- Explanations providing goal structure or organizational information are presented at the beginning of the procedural task [13];

- Procedural instructions are divided into small steps with each step labeled and presented in the order they are to be executed [8];
- Information within each step is explicitly stated, and the most important information is presented first [14];
- Visuals accurately depict the procedural step since users often prefer to follow the visual examples rather than the instructions [19] and visuals tend to help eliminate orientation errors [36];
- Visuals are accompanied by some form of written or spoken instructions in order to cue the user to the important aspects of the visuals [35];
- Spoken instructions begin simultaneously with or slightly after the visuals are presented [6,21];
- Animation is segmented to focus the user's attention on specific parts of the animated displays [29,30];
- Instructions are segmented to reinforce the concept of chunk or steps for completing the procedure [8]; and
- The opportunity to perform the procedure is delayed until all steps of the procedure have been illustrated [7].

By following these guidelines for the design of the on-line help, the user's mental representation of the procedure should be more accurately constructed, the referential links to the physical objects should be better connected, and the individual segmented steps of the procedure should be better organized within memory thus providing a proper foundation for the successful performance of the procedure.

EMPIRICALLY-BASED ON-LINE HELP DESIGN

The design of the elements in the on-line help adhered as close as possible to the empirically-based guidelines listed. The order and the language of the instructions, in both written and spoken forms, were based on the guidelines presented. Still graphic or animated visuals were added to the instructions. The spoken instructions accompanying the still graphics or animation were designed to begin immediately after the first visual appeared on the screen or as soon as the animation sequence started. The step-by-step procedural information was presented in a segmented form. In-depth information pertaining to the studies associated with the guidelines listed is found in [17].

The combinations of written or spoken instructions with or without still graphics or animated visuals in a segmented form made this study one of the first experiments of its kind to investigate the effectiveness of on-line help on the performance of computer-based procedures within a graphical user interface environment. This study was designed to investigate the relative effectiveness between:
- the use of on-line help with and without visual illustrations;
- the use of on-line help with still graphics and with animated visuals; and
- the use of on-line help with written instructions and with spoken instructions.

METHODOLOGY

The experimental context of this study was a HyperCard emulator with on-line help in which subjects received instructions for completing seven procedural tasks. This study used a 2 x 3 randomized factorial design with Verbal Instruction and Visual Illustration as the two treatment factors. The Verbal Instruction factor was the presentation of instructions either in spoken or written form. The Visual Illustration factor presented either no illustrations, still graphic illustrations, or animated graphics.

Participants

Participants for this study were drawn from an available pool of 215 undergraduate students enrolled in the Science Curriculum and Instruction course within the School of Education and in the Introduction to Computers and Their Uses course within the School of Arts and Science at a university of approximately 10,000 students. To be eligible to participate in the study, each student was required: (a) to demonstrate a basic ability in the use of a Macintosh computer, keyboard, mouse, and pull-down menus; (b) to have no previous experience in the authoring techniques of HyperCard [5]; (c) to have completed the HyperCard Tour 2.0v2 tutorial [4]; and (d) to have demonstrated an understanding of the basic computer and HyperCard concepts. These requirements reduced the potential sample size to 198. Equipment failures, language barriers, and lack of attendance during the experiment resulted in 176 participants being randomly assigned to one of the six treatment groups.

Materials

All materials used in this study, including the computer program, were designed and developed by the experimenter in consultation with experts in the areas of survey construction and simulation programming. A set of eight HyperCard authoring tasks were used in this study. An example of a task explanation, specific task description, and task steps for one of the tasks is shown in Figure 1.

Explanation: An icon (a small picture) can be added to a button to make the purpose of a button more meaningful to the user. If a non-meaningful icon already exists on the button, the icon of the button should be changed.

Task: Change the icon on the button from the Man icon to the Phone icon.

Steps presented one at a time:
1. Choose the Button tool from the Tools menu.
2. Double-click on the button to be modified.
3. Click on the Icon button.
4. Click once on the scroll down arrow to locate the icon image for the button.
5. Click once on the icon image.
6. Click once the Ok button.
7. Choose the Browse tool from the Tools menu when you are finished.

Figure 1. Details of Change Button Icon Task.

HyperCard Emulator Design. A HyperCard emulator was developed in SuperCard [34] to avoid potential difficulties involved in programming and controlling menu responses and capturing timing information and mouse locations while using the actual HyperCard application. The

HyperCard emulator included the screen images, menu options, and dialog boxes normally seen during the process of performing the HyperCard procedures used as tasks in the study. The basic layout of the screen shown in Figure 2 was divided into four functional areas [16].

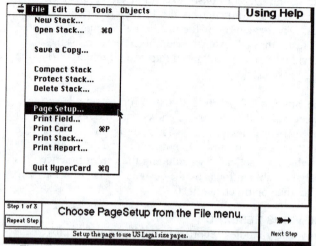

Figure 2: Screen Design with Four Functional Areas.

The large main functional area, the work space, covered three-fourths of the upper portion of the screen. The work space was used to display the simulated HyperCard menu bar and all still or animated graphics. Subjects performed all tasks in this area using the simulated menu bar and related dialog boxes. Welcome Information and Task Explanations covered this area when such information was displayed.

Information and options relating to the current step of a task being explained or performed were placed in the second functional area positioned at the bottom of the screen. To the left of this area, the number of the current step and the option to repeat a step were located. The main portion of the informational area contained the current step instruction. An abbreviated version of the task requirement was shown below the current step instruction.

Supplemental information concerning the current phase of the study was displayed in the third area located in the upper right corner of the screen. When the Abort option appeared after four incorrect attempts were made to complete a step, it also appeared in this area. The fourth area located at the bottom right of the screen contained the current navigational option. Options, displayed one at a time, included advancing to the next step, beginning the task test, continuing, beginning Help, and exiting.

Color was added to the screen design to help divide the functional areas. The dark areas shown on the screen were displayed as a medium to light blue. The background of any informational field was a light yellow. The colors for the screen design were specifically selected as a result of the user evaluations. The final version of the emulator, complete with all sounds for the spoken instructions and converted to a stand-alone application consumed 1,287K on a high-density diskette.

Treatment Designs. All steps for a given task were presented consecutively in all treatments. To maintain a sense of realism, a simulated HyperCard menu bar was visible across the top of the screen for all treatment groups. Subjects using spoken forms of the Verbal Help or seeing animated forms of the Visual Help were allowed to replay the current step of the task. Immediately after the Help presentation for the task, subjects were required to complete the task once without additional access to help. The completion of the task was part of the Immediate Exercise set, and the screens that were accessed were the same for all treatment groups.

The Written, Animated Treatment, *WA*, animated the cursor movement required to perform the instruction. This movement provided subjects with a visual cue to the important point of the instruction. Options on pull-down menus were highlighted as the cursor moved down through the choices. Figure 3 demonstrates the way the cursor moved across the screen and activated a pull-down menu on

the help screen for the *WA* treatment. Although four screens are used in Figure 3, all cursor movement in the treatment occurred on the same screen.

A still graphic that corresponded only to the written step instruction was shown in the work area of the screen for the Written, Still Graphic Treatment, *WS*. The graphic illustrated the way the screen looked as that step of the procedure was completed, screen (d) of Figure 3.

Subjects assigned to the Written, Nonillustrated Treatment, *WN*, were provided written step instructions, one at a time, in the same location at the bottom of the screen as illustrated in screen (a) of Figure 3. As in other written treatments, the task requirement was displayed below the written instruction and the step indicator was located to the left of the written instruction. Graphic examples were not provided for the *WN* treatment.

Subjects in all spoken treatments were presented each step instruction through a set of headphones. Spoken instructions were identical to the written instructions presented to written treatment groups.

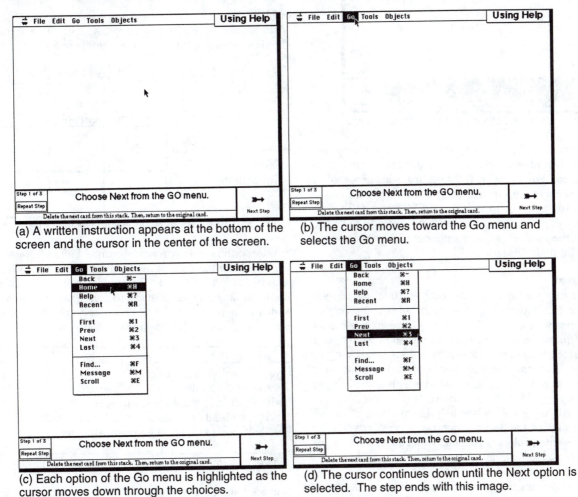

(a) A written instruction appears at the bottom of the screen and the cursor in the center of the screen.

(b) The cursor moves toward the Go menu and selects the Go menu.

(c) Each option of the Go menu is highlighted as the cursor moves down through the choices.

(d) The cursor continues down until the Next option is selected. The step ends with this image.

Figure 3: Four Screens Demonstrating an Example of Written Animated Help Screens.

No written instructions appeared on the screen in the Spoken, Nonillustrated Treatment, *SN*, treatment. The area normally used for written instructions was covered with a medium blue color. The task requirement, however, was shown at the bottom of the screen Graphic examples were not provided for the *SN* treatment. The Repeat Step option, located below the step indicator, was available to allow the subjects to replay the instruction.

Both the Spoken, Still Graphic Treatment, *SS*, shown in Figure 4 and the Spoken, Animated Treatment, *SA*, provided visuals similar to those described for the *WS* and *WA* treatments, respectively. Like the SN treatment, step instructions were verbally presented through a set of headphones and no written instructions appeared on the screen.

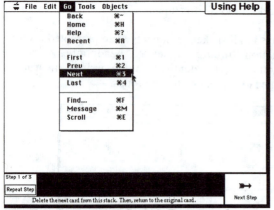

Figure 4: Spoken, Still Graphic Treatment Screen.

Questionnaire. Seven sets of mirror images of statements assessing the subjects' reactions to the help instructions were created. For example, if the original statement said "I prefer to receive help instructions in a written rather than spoken form," the mirror image said "I prefer to receive help instructions in a spoken rather than written form." The questionnaire used a six-point scale ranging from Strongly Agree to Strongly Disagree.

Procedures

Eight groups of subjects, composed of 20 to 25 people, completed the study using Macintosh LC II computers in the university computer laboratory. The experimenter arranged the subjects in the room so that subjects assigned nonillustrated help could not see the screens displaying of illustrated help and subjects assigned still graphic help could not see the screens displaying animated help. Subjects were guided through the steps of completing one practice task using the on-line help instructions. After all subjects had tried the task and questions were answered by the experimenter, subjects continued with the seven tasks of the Immediate Exercise set.

The following steps were repeated for each task in the set.
• On the computer screen, the usefulness of the task was explained and the specific task requirement for completing the task was stated.

• A written or spoken help instruction for completing the first step of the task was presented by the computer. At the subject's request, each additional step was given, one at a time, until all steps of the task were presented.
• After the last step was presented and when the subject was ready to continue, a message reminding the subject of the task requirement appeared on the screen.
• Subjects attempted to complete the entire procedure at their own pace. After four incorrect attempts, the option to abort appeared until the subject chose to abort or until the correct action was selected.

After the series of tasks were completed, subjects were asked to complete the questionnaire about help instructions.

Dependent Measures

Audit trails, commonly used in analyzing learner-computer interactions [38], were included in the HyperCard Emulator to record completion rate, elapsed time, and error rate. Values for each category were directly collected or calculated by the HyperCard Emulator for each step of a task and for each task of the set. Responses to the questionnaire were entered into a data file and verified by the experimenter.

RESULTS

Data were analyzed for 176 subjects randomly assigned to the six treatment groups as shown in Figure 5.

		Visual Illustration		
		Non-illustrated	Still Graphic	Animated
Verbal Instruction	Written	29	29	28
	Spoken	29	29	32

Figure 5: Two by Three Randomized Design of the Study.

A priori orthogonal contrasts, shown in Figure 6, were used for analyzing the data of the six treatment cells. Contrast *Visual (1)* compared treatment groups receiving nonillustrated and illustrated (still graphic and animated) forms of help. *Visual (2)* compared treatment groups receiving still graphic and animated forms of help.

Effect	df
verbal: written vs. spoken	1
visual (1): nonillustrated vs. illustrated	1
visual (2): still graphic vs. animated	1
verbal x visual (1): nonillustrated vs. illustrated	1
verbal x visual (2): still graphic vs. animated	1
	5

Figure 6: A Priori Orthogonal Contrasts Used in Analysis.

Although multiple contrasts were used in the analysis, only the comparison of the nonillustrated and illustrated groups revealed any significant differences in the study.

Effectiveness was measured in terms of the dependent variables for completion rate and error rate. Efficiency was measured in terms of the dependent variable for the time the subject used in attempting to complete a task or set of tasks.

Completion Rates for Exercise Sets

Completion rates were inspected in two ways: the percent of each treatment group completing all seven procedural tasks of the set and the average percent of tasks completed within each treatment group.

Percent of treatment group completing all tasks. A log-linear comparison of the cell proportions across treatment groups for subjects who completed all seven tasks of the Immediate Exercise set with subjects who did not complete all seven tasks revealed a significant difference for the nonillustrated versus illustrated contrast (coef. = .19, 95% CI = .08 to .30). An examination of cell frequencies showed that only 22 of the 58 subjects or 37.9% of the nonillustrated groups (*WN*, *SN*) completed the entire Immediate Exercise set. With the illustrated (*WS*, *SS*, *WA*, *SA*) groups, 78 of the 118 subjects or 66.1% completed the entire Immediate Exercise set.

Average percent of tasks completed. The number of tasks completed out of the seven possible tasks of the exercise set was converted to a percentage for each subject. A significant difference for the nonillustrated versus illustrated contrast was detected ($F(1, 170) = 12.99$, $p < .001$). Cell means for the percent of tasks completed within the Immediate Exercise set by treatment group reveal that the nonillustrated groups (*WN*, *SN*) have a combined mean completion rate of 83.5% ($n = 58$, $SD = 18.69$). The illustrated groups (*WS*, *WA*, *SS*, *WA*) have a combined mean completion rate of 92.5% ($n = 118$, $SD = 13.82$).

Error Rates for Exercise Sets

Since subjects were allowed to abort the performance of a procedural task on any step of the procedure, calculations for error rates were based on errors per step attempted. When error rates were evaluated, a significant difference was found for the nonillustrated versus illustrated contrast ($F(1, 170) = 22.48$, $p < .001$) when an α level of .01 was used. An evaluation of cell means shows the combined nonillustrated (*WN*, *SN*) conditions to have a higher error rate ($M = 1.83$, $SD = 1.33$, $n = 58$) than the error rate ($M = .95$, $SD = 1.08$, $n = 118$) of the combined illustrated conditions (*WS*, *WA*, *SS*, *WA*).

Elapsed Time for Exercise Sets

Because subjects did not always complete all of the steps of a given procedural task within the exercise set, elapsed times were measured based on the number of seconds a subject spent in attempting to complete a step of a procedure. To calculate elapsed time per step, the total amount of time elapsed while the subject attempted to complete all task steps was divided by the total number of steps the subject attempted to complete. Results of the MANOVA test for the illustration contrast and an

inspection of the cell means indicated that the significant differences occurred ($F(1, 170) = 28.99$, $p < .001$) with nonillustrated groups having a mean elapsed time of 13.80 seconds and illustrated groups having a mean of 9.58 seconds.

DISCUSSION

Nonillustrated versus Illustrated

The findings of this study add support to the notion that the presence of a visual, either a still graphic or an animation, aids the subjects in acquiring accurate representations of the steps involved in the procedure. At least two possible explanations exist for why subjects receiving illustrated on-line help exhibited faster and more accurate initial performances of the procedural tasks than did the other subjects.

First, the visuals contained pictures of the objects in the graphical user interface that were to be used in performing the procedure. Phrases such as Edit Menu, the OK button, and Field tool, scroll down arrow, and Shared Text box, mentioned in the instructions, could have been internally mapped to the actual object. Second, the visuals provided an example of the system response to the user input so the subject would not be surprised by the screen image displayed. Subjects in the illustrated groups were perhaps more prepared for the next step of the procedure since they would have known what image would appear as a result of their action. The presentation of the illustrations during the on-line help may have allowed subjects to visually rehearse and plan the actions necessary for the completion of the procedure. It is likely that subjects in the nonillustrated groups were unable to be prepared since they lacked the ability to mentally picture the screen image associated with the action's response.

Still Graphic versus Animated Illustrations

An unexpected result of this study was the lack of a significant difference between the subjects' performance in the still-graphic conditions and the animated conditions. However, the data suggested a slight advantage for animation over still graphics as expected. One possible explanation for the lack of difference is that the task performed in this study did not require the coding of motion information which would have been provided by the animation. In this study, the main difference between the information provided by the still graphic condition and the information provided by the animated condition was the movement of the cursor. It is likely that subjects inferred the movement of the cursor by mentally noting the change in cursor location from the graphic image illustrating the first step to the graphic image illustrating the second step. If the procedural tasks had involved the movement of an actual object such as the resizing of a button or the changing of a picture's location, it is possible that significant differences would have been found between the still-graphic and animated conditions.

A second possible explanation for the similar results between the still-graphics condition and the animated condition relates to the use of segmentation. Individual task steps were emphasized by each image presented in the still-graphics condition. If the animated presentation had been presented in a continuous form, like the animated demonstrations in the Palmiter [26] study, it may have been difficult for the subjects to identify the individual procedural steps. However, since the animation provided in this study was segmented to emphasize each step of the procedural task, subjects in both the still-graphics and animated conditions probably were equally able to identify each procedural step. The segmentation of the animated illustration may be one reason why the animated conditions consistently performed better than the other conditions across time and type of tasks. Although the better performance was nonsignificant when compared to the other groups, results showing that the animated groups tended to maintain their high performance across time are noteworthy since those results contradict the findings of the Palmiter [26] studies.

Implications of the Findings

Caution should be exercised in generalizing from this population to a larger population of college students and from the procedural tasks included in this study to other computer-based procedures. The robustness of the findings of this study, however, provides a strong basis for suggesting that instructions for computer-based procedures presented via on-line help should be accompanied by some form of visual illustration. Designers of current on-line help facilities in the graphical user interface environment should be encouraged to consider changing from the standard from of text-based on-line help to a form of on-line help that incorporates graphics.

Recommendations for Future Research

Based on the results of this study, the review of the literature, and the general observations made by the experimenter during this study, two areas of research for investigating the effectiveness of on-line help are recommended. Since tasks in this study did not involve the aspect of motion beyond the movement of the cursor, no significant advantages were revealed for the use of animation. Research needs to be conducted using a set of computer-based procedures involving the movement of different objects on the screen to help determine the effectiveness of animation in on-line help.

When the on-line help information in this study was presented, the work space on the subject's monitor was completely filled. Since the size of the typical computer monitor is rapidly increasing as technology advances, different conclusions about the effectiveness of on-line help may have been reached if the on-line help information remained visible on the screen as the subject completed the procedure. A study could be conducted to compare how the presence or absence of the on-line help during task performance affects the completion of a computer-based procedure.

The results from this study can be summarized in one statement: *The use of visuals within on-line help instructions for computer-based procedures enabled adult subjects to perform more procedural steps in less time and with fewer errors than subjects who received no visuals within on-line help instructions.* Insights gained from this investigation should contribute to the current knowledge base used by designers to guide their development of on-line help and ultimately help the end-user.

REFERENCES

1. Ackerman, P. L. (1989). Individual differences and skill acquisition. In P. L. Ackerman, R. J. Sternberg, & R. Glaser (Eds.), *Learning and individual differences: Advances in theory and research* (pp. 165-217). New York: W. H. Freeman.

2. Anderson, J. R. (1982). Acquisition of cognitive skill. *Psychological Review*, 89(4), 369-406.

3. Anderson, J. R. (1990). *Cognitive psychology and its implications* (3rd ed.). New York: W. H. Freeman.

4. Apple Computer (1990b). *HyperCard Tour 2.0* [Macintosh program]. Cupertino, CA: Apple Computer.

5. Apple Computer (1991). *HyperCard 2.0* [Macintosh program]. Cupertino, CA: Apple Computer.

6. Baggett, P. (1984). Role of temporal overlap of visual and auditory material in forming dual media associations. *Journal of Educational Psychology*, 76(3), 408-417.

7. Baggett, P. (1988). The role of practice in videodisc-based procedural instructions. *IEEE Transactions on Systems, Man, and Cybernetics*, 18(4), 487-496.

8. Bovair, S., & Kieras, D. E. (1991). Toward a model of acquiring procedures for text. In R. Barr, M. L. Kamil, P. B. Mosenthal, & P. D. Pearson (Eds.), *Handbook of reading research: Vol. II* (pp. 206-229). White Plains, NY: Longman.

9. Carroll, J. M., & Mazur, S. A. (1986). LisaLearning. *IEEE Computer*, (November), 35-49.

10. Cohill, A. M., & Williges, R. C. (1985). Retrieval of HELP information for novice users of interactive computer systems. *Human Factors*, 27(3), 335-343.

11. Czaja, S. J., Hammond, K., Blascovich, J. J., & Swede, H. (1986). Learning to use a word-processing system as a function of training strategy. *Behaviour and Information Technology*, 5(3), 203-216.

12. Czaja, S. J., Hammond, K., Blascovich, J. J., & Swede, H. (1989). Age related differences in learning to use a text-editing system. *Behaviour and Information Technology*, 8(4), 309-319.

13. Dixon, P. (1987a). The processing of organizational and component step information in written directions. *Journal of Memory and Language, 26*(1), 24-35.

14. Dixon, P., Faries, J., & Gabrys, G. (1988). The role of explicit action statements in understanding and using written directions. *Journal of Memory and Language, 27*(6), 649-667.

15. Dunsmore, H. E. (1980). Designing an interactive facility for non-programmers. In *Proceedings of the 1980 Annual Conference of the Association for Computing Machinery*, (pp. 475-483). Nashville, TN: Association for Computing Machinery.

16. Grabinger, R. S. (1993). Computer screen designs: Viewer judgments. *Educational Technology Research and Development, 41*(2), 35-73.

17. Harrison, S. (1994). Still, animated, or nonillustrated on-line help with written or spoken instructions for performance of computer-based procedures (Doctoral dissertation, University of Minnesota, 1994).

18. Hicks Jr., J. O., Hicks, S. A., & Sen, T. K. (1991). Learning spreadsheets: Human instruction vs. computer-based instruction. *Behaviour and Information Technology, 10*(6), 491-500.

19. LeFevre, J.-A., & Dixon, P. (1986). Do written instructions need examples? *Cognition and Instruction, 3*(1), 1-30.

20. Mack, R. L., Lewis, C. H., & Carroll, J. M. (1983). Learning to use word processors: Problems and prospects. *ACM Transactions on Office Information Systems, 1*(3), 254-271.

21. Mayer, R. E., & Anderson, R. B. (1991). Animations need narrations: An experimental test of a dual-coding hypothesis. *Journal of Educational Psychology, 83*(4), 484-490.

22. Microsoft Corporation (1992a). *Excel 4.0* [Macintosh program]. Redmond, WA: Microsoft.

23. Microsoft Corporation (1992b). *Microsoft Word 5.0a* [Macintosh program]. Redmond, WA: Microsoft.

24. Neves, D. M., & Anderson, J. R. (1981). Knowledge compilation: Mechanisms for the automatization of cognitive skills. In J. R. Anderson (Ed.), *Cognitive skills and their acquisition* (pp. 57-84). Hillsdale, NJ: Lawrence Erlbaum Associates.

25. Nielsen, J., Frehr, I., & Nymand, H. O. (1991). The learnability of HyperCard as an object-oriented programming system. *Behaviour and Information Technology, 10*(2), 111-120.

26. Palmiter, S. (1991). Animated demonstrations for learning procedural, computer-based tasks (Doctoral dissertation, University of Michigan, 1991). *Dissertation Abstracts International, 52,* 03B.

27. Palmiter, S., & Elkerton, J. (1991). An evaluation of animated demonstrations for learning computer-based tasks. In S. P. Robertson, G. M. Olson, & J. S. Olson (Eds.), *Proceedings of CHI '91 Human Factors in Computing Systems*, (pp. 257-263). New Orleans, LA: Association for Computing Machinery.

28. Palmiter, S., Elkerton, J., & Baggett, P. (1991). Animated demonstrations vs. written instructions for learning procedural tasks: A preliminary investigation. *International Journal of Man-Machine Studies, 34,* 687-701.

29. Rieber, L. P. (1989). The effects of computer animated elaboration strategies and practice on factual and application learning in an elementary science lesson. *Journal of Educational Computing Research, 5*(4), 431-444.

30. Rieber, L. P. (1991a). Animation, incidental learning, and continuing motivation. *Journal of Educational Psychology, 83*(3), 318-328.

31. Schneider, W., & Shiffrin, R. M. (1977). Controlled and automatic human information processing: I. Detection, search, and attention. *Psychological Review, 84*(1), 1-66.

32. Sellen, A., & Nicol, A. (1990). Building user-centered on-line help. In B. Laurel (Ed.), *The art of human-computer interface design* (pp. 143-153). Reading, MA: Addison-Wesley.

33. Shneiderman, B. (1983). Direct manipulation: A step beyond programming languages. *IEEE Computer, 16*(8), 57-69.

34. Silicon Beach Software (1991). *Aldus Super Card 1.6* [Macintosh program]. San Diego: Aldus.

35. Spangenberg, R. W. (1973). The motion variable in procedural learning. *Audio Visual Communication Review, 21*(4), 419-436.

36. Stone, D. E., & Glock, M. D. (1981). How do young adults read directions with and without pictures? *Journal of Educational Psychology, 73*(3), 419-426.

37. Sukaviriya, P., Isaacs, E., & Bharat, K. (1992). Multimedia help: A prototype and an experiment. In P. Bauersfeld, J. Bennett, & G. Lynch (Eds.), *Proceedings of CHI'92 Conference on Human Factors in Computing Systems*, (pp. 433-434). Monterey, CA: Association for Computing Machinery.

38. Williams, M. D., & Dodge, B. J. (1993). Tracking and analyzing learner-computer interaction. In M. R. Simonson & K. L. Abu-Omar (Eds.), *Proceedings of the 1993 Annual Convention of the Association for Educational Communications and Technology*, (pp. 1115-1129). New Orleans, LA: Iowa State University.

Dynamic Generation of Follow up Question Menus: Facilitating Interactive Natural Language Dialogues

Vibhu O. Mittal   Johanna D. Moore

Department of Computer Science &

Learning Research & Development Center.

University of Pittsburgh

Pittsburgh, PA 15260

e-mail: {*mittal,jmoore*}@*cs.pitt.edu*

ABSTRACT

Most complex systems provide some form of help facilities. However, typically, such help facilities do not allow users to ask follow up questions or request further elaborations when they are not satisfied with the systems' initial offering. One approach to alleviating this problem is to present the user with a menu of possible follow up questions at every point. Limiting follow up information requests to choices in a menu has many advantages, but there are also a number of issues that must be dealt with in designing such a system. To dynamically generate useful embedded menus, the system must be able to, among other things, determine the context of the request, represent and reason about the explanations presented to the user, and limit the number of choices presented in the menu. This paper discusses such issues in the context of a patient education system that generates a natural language description in which the text is directly manipulable – clicking on portions of the text causes the system to generate menus that can be used to request elaborations and further information.

Keywords: hyper-media, natural language, intelligent systems, user interface components, usability engineering

Introduction

Help facilities play an important role in user acceptance of many systems. An important characteristic of a help facility is the ability to allow the user to follow up on previous requests by asking for further elaborations or posing related questions in the context of the original request. Studies have shown that in human-human interactions, users typically follow up requests for information with further questions [6, 13, 17, 19]. This illustrates that initial explanations are seldom sufficient and unlikely to satisfy users in real life. This also underlines the need for computer based help systems to offer facilities for asking follow up questions, since the same explanation is unlikely to be satisfactory to the variety of different users that

may use the system. The ability to follow up on information requests becomes even more critical in applications such as medical informatics where misunderstanding could result in serious, unintended consequences.

Unfortunately, handling follow up questions in the form of unrestricted natural language queries is not yet possible. Using a restricted sub-language of English (or some other natural language) as a query language is also problematic, because users find it difficult to remember a circumscribed set of words and phrases they can use for expressing their questions [16]. Interference from other synonymous, natural ways of phrasing questions makes it difficult for users to recall the restricted set of allowable inputs [1], and thus users find restricted natural languages difficult to learn and frustrating to use [9]. Moreover, restricted query languages require that users be able to pose well-formulated follow up questions. Users who are confused may not know exactly what question they wish to ask – an analysis of student-TA help sessions found that confused students often responded to explanations given by the TA with expressions such as "huh?" [13]. However, in many cases, users can still pinpoint the source of the problem more precisely than this, while still being unable to formulate a question in the system's query language.

One method of alleviating this problem is to present users with a direct manipulation interface that allows them to point to the portions of system generated information that they do not understand or would like clarified. In response to the user's pointing action, the system then provides a menu of questions that may be asked about the highlighted text. By allowing users to point to the text they do not understand, many of the difficult referential problems in understanding free natural language input can be avoided. However, for such an interface to be feasible, the system must be able to understand what the user is pointing at, i.e., the system must understand its own explanations.

In this paper, we describe how the use of a plan based model of text generation can support this type of interface. The discourse structure of the explanation constructed by the planner records the reasoning behind the planning process, which can then be used by the system to understand both the context of the utterance the user is pointing at, as well as the

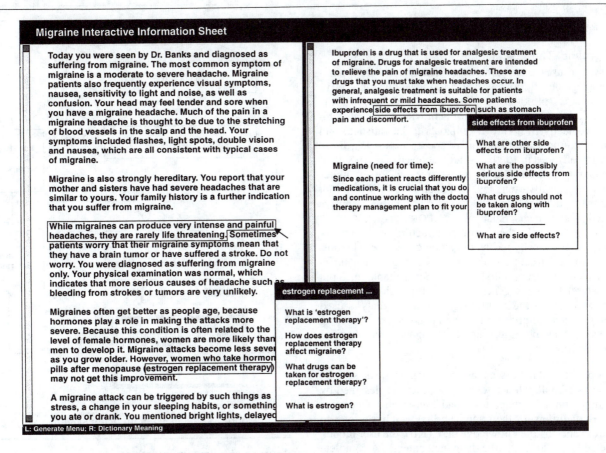

Figure 1: A snapshot of the system with some of the dynamically generated menus.

reasons for the generation of the utterance in the first place.

We discuss the issues in designing a system to dynamically generate embedded menus of follow up questions, and illustrate them in the context of our patient education system for migraine [2, 3]. The system generates descriptions about migraine headaches in English: these contain the doctor's diagnosis, recommendations and prescriptions, along with a description of migraine headaches, their causes and their implications for the patient. The user (patient) can query the system on any aspect of the explanation by pointing at the relevant portion of the text and clicking the mouse. As the user moves the mouse over the text, the system highlights portions of the text that have further information associated with them. Clicking on a mouse-sensitive text span causes the system to generate a menu that can be used to request answers to follow up queries and further information. Figure 1 shows a snapshot of the system. The figure shows the cursor (which is the arrow) being moved over the first sentence in the third paragraph, which is highlighted by the system. Two of the menus generated by clicking on other utterances are also shown in the figure.

In designing our system, we considered ethnographic analyses of doctor-patient interactions, as well as interviews with patients, to determine their information needs at different times and in different contexts. Based on these analyses, lists of questions that the patients had asked, or would have liked to have asked were developed. The discourse structures

for the explanations generated by the system were analyzed to determine the communicative goals which in ideal cases would have fulfilled the information needs expressed by the patients. Based on these analyses, we developed the heuristics used by the system to generate such questions dynamically. Our system builds upon previous research in both the CHI and the AI communities: empirical models of human interaction with menus – concerned with issues such as the number and aggregation of menu items, the order of their presentation, menu titles, etc., e.g., [15, 21, 23], as well as work on natural language generation – using the discourse structure to facilitate follow up questions, etc., e.g., [6, 13], were used as a basis in the design of this framework.

It is important to keep in mind that although the interface described here bears a resemblance to a hypertext-style interface, the system is *not* a hypertext system in the traditional sense, i.e., it is not organized as a collection of canned pieces of text interconnected by typed links. The explanations are dynamically generated by the system in response to a user's question or the system's need to communicate with the user. A hypertext system would have all of the text prepared *a priori*, and the user would have to browse through it to find the information required. Furthermore, in a hypertext system, all of the things that can be pointed to must be worked out in advance. It is easy to imagine that users may have questions about items in the texts that were not envisioned, and hence not provided for, by the hypertext designers. In our system, what can be pointed at is determined dynamically, and the links are not worked out in advance.

Background: The Text Planner

The system uses a text planner to generate coherent natural language descriptions: given a top level communicative goal (such as (DESCRIBE (CONCEPT MIGRAINE))),[1] the system finds operators capable of achieving this goal. Operators typically post further subgoals to be satisfied, and planning continues until primitive speech acts – i.e., those directly realizable in English – are achieved. The result of the planning process is a discourse tree, where the nodes represent goals at various levels of abstraction with the root being the initial goal, and the leaves representing primitive realization statements (such as (INFORM (ANALGESIC IBUPROFEN))). In the discourse structure, goals are related to other goals by means of certain coherence relations. This discourse structure is then passed to a grammar interface, which converts it into a set of inputs suitable for input to a natural language generation system, such as FUF [7]. Information about the relationships between goals is used by the grammar component to generate connectives and cue phrases, such as 'because,' 'however,' and so on in the final surface form.

Plan operators can be seen as small schemas (or scripts) which describe how to achieve a goal; they are designed by studying natural language texts and transcripts. Operators include conditions for their applicability, which can refer to resources such as the knowledge base (KB), the user model, or the context (including the dialogue context). A complete description of the generation system is beyond the scope of this paper – see [13, 14] for more details. Figure 2 shows a block diagram of the complete system. Figure 3 shows part of the discourse structure generated by the system for the description in Figure 1.

Generating Follow Up Questions

There are three main issues that must be addressed in designing a system to generate menus of follow up questions: (i) determining the types of text spans that can be queried by the user, (ii) identifying the various sources of information that can be used to generate possible questions about the highlighted text, and (iii) identifying factors that can be used to prune the set of candidate questions generated in (ii) so that the resulting set of questions can be presented to the user as a menu. This section discusses these three issues in the context of our migraine system.

Note that in our framework, utterances are generated from primitive speech acts, which are specified as follows:

(<speech act> <proposition>)

A <speech act> can of different types: INFORM, ASK, RECOMMEND or COMMAND; a <proposition> can be either a simple one, such as '(IS-A ANALGESIC-DRUG IBUPROFEN),' or a more complex one involving an action, such as, '(GOAL <goal> (STEP <step>))' where the goal could be to 'treat migraine' and the step could be 'prescribe medication'.

---

[1]The actual syntactic form used to represent the goals and speech acts in the implemented system is more complex; for the sake of clarity, we have shown simplified versions in this paper.

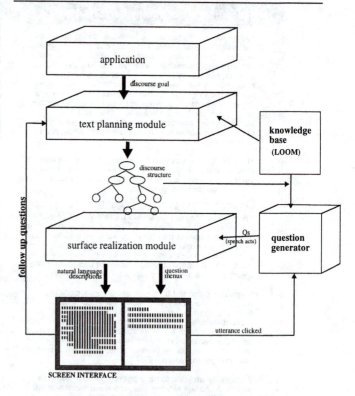

Figure 2: A block diagram of the system.

Sources of follow up questions

Since the system must highlight items that the user can ask questions about (when the mouse is moved over the items), it is necessary to determine *a priori* the types of text from which the system can generate follow up questions. Our system allows three types of text constituents to be highlighted, i.e., there are three sources of follow up questions (i) the entire clause, (ii) the individual constituent clauses of a complex clause, or (iii) the individual objects and entities that the clause refers to.

Selecting noun phrases: In the simplest possible case, the user can select a noun phrase, (a concept in the knowledge base) and the system generates the candidate question:

What is a <concept>?

thus generating questions such as *"What is cafergot?"* or *"What is estrogen replacement therapy?"*

Selecting simple clauses: When the user highlights a simple clause, the follow up questions generated by the system contain the questions generated for each of the noun phrases in the clause, as well as those resulting from the overall clause itself. Since simple clauses arise from single speech acts in the discourse structure, the questions about the overall clause depend upon the type of the underlying speech act: INFORM, ASK, RECOMMEND or COMMAND. The system reasons about two types of questions that can be asked about each of these speech acts: why-questions and how-questions. Since the interpretation of these questions depends upon both the type of the proposition, as well as the type of speech act, not all types of questions can be generated in all cases. We consider

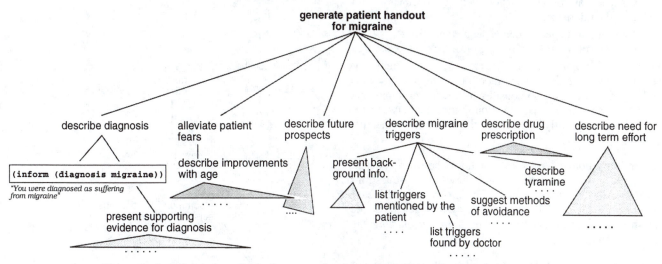

Figure 3: A skeletal view of the discourse structure generated by the text planner.

each case in turn below.

The speech act INFORM: The INFORM speech act results in an assertion being generated in the surface form. Both how- and why-questions can be generated in some cases. However, when the speech act is of the form:

```
(INFORM (IS-A <concept1> <concept2>))
(INFORM (HAS <concept1> <attribute>))
```

the system can generate candidate menu items by forming only a why-question from the speech act; a sensible how-question cannot be formed. For example, if the user highlights the utterance "Ibuprofen is a drug," or "Ibuprofen has analgesic properties", the system can form reasonable why-questions, i.e., *"Why is ibuprofen a drug"* or *"Why does ibuprofen have analgesic properties?"*, but not reasonable how-questions.

However, in cases where the proposition to the speech act involves an action, the system generates both how- and why-questions. Speech acts of the form

```
(INFORM (ACTION <action>))
```

result in the generation of assertions such as: *"I'm trying to do <action>"*. The why-question generated from this form is *"Why are you trying to do <action>?"*, while the how-question corresponds to *"How is this <action> done?"* If the <action> is being performed as part of a larger goal (the speech act form is (INFORM (GOAL <goal> STEP <action>), the system also generates: *"How does this <action> help achieve the <goal>?"*

The speech acts RECOMMEND and COMMAND: Two other types of speech acts are RECOMMEND and COMMAND. As their names indicate, these speech acts result in the generation of surface level forms that recommend or command the user to undertake some action. In both of these cases, the user can ask both a why- and a how-question. When a user asks "Why?" in response to a recommendation or a command, the user is asking to be persuaded that the action being recommended (or commanded) should be done. Thus,

speech acts of the form

```
(RECOMMEND (ACTION <action>))
(COMMAND (ACTION <action>))
```

result in questions being generated such as *"Why should this <action> be done?"*.

The how-question in response to a RECOMMEND or COMMAND speech act can be interpreted as requesting a method for performing the action being recommended (or commanded). Thus, the resulting question generated is: *"How would I accomplish this <action>?"*. For instance, if the user happened to point at 'Take cafergot once every day', the system would generate the how- and why-questions as follows:

Why should I take cafergot once every day?
How should I take cafergot once every day?

The speech act ASK: The fourth type of speech act – ASK – causes the surface realization component to phrase the proposition in the form of a question to the user. Questions can be highlighted with the mouse just as any other clause the system produces. The user may wish to understand why the system needs to know the answer to its question in order to perform its task, i.e., she may wish to ask "Why are you asking me this question?". For instance, the user could point at the question 'Do you have migraines on weekends more frequently than on weekdays?', and the system would generate the question *"Why are you asking me whether I have migraines more frequently on weekends than on weekdays?"*

Note that it is not possible to form a meaningful how-question for text produced by an ASK speech act. It does not make sense for the user to ask *"How are you asking me <question>?"*

Generating Menu Entries for Complex Clauses In the most complex case, the user can highlight an entire complex clause, i.e., two (or more) clauses related by a coherence relation. Complex clauses are generated as a result of realizing multiple adjacent speech acts in the discourse structure. The coherence relations are typically used by the surface realization component to generate appropriate cue phrases in the text, such as 'because', 'however', etc. Consider, for instance, the utterance:

(1) Migraines often get milder as people age, **since**
(2) hormones play a role in making the attacks more severe.

If the user highlights the complete utterance, the system generates follow up questions from both the simple clauses (1) and (2). It must also also consider follow up questions that arise because of the causal relation that exists between (1) and (2), here explicitly indicated by the use of the connective 'since'.

Depending upon the relation that holds between the text spans, the system can generate either how- or why-questions regarding the complex clause. For instance, in the case above, with the causal relation, the system generates the question: *"How do hormones play a role in making the attacks more severe?"*

In the case of the MEANS relation, which indicates that one of the entities is enabled by the other, the system can generate why-questions, as in the following case: *"Why are you trying to reduce the frequency of migraine attacks by suggesting that I avoid sleeping late on weekends?"* In other words, the user is asking the system for justification for the choice of a particular method (avoiding sleeping late on weekends) to achieve the goal in question (reducing the frequency of migraine attacks) as opposed to trying some other strategy.

Quite often, the questions generated from a complex clause are the same ones generated from either of its constituent clauses alone. The system prunes duplicate questions from its list while eliminating other irrelevant questions, as discussed later in this section.

Generating additional questions

In addition to the heuristics discussed above, the system also generates other questions based on either the previous interaction with the user, or the information in the knowledge base (KB). For instance, if the system has already presented a description of two sibling concepts (two different concepts with the same parent concept), and the user points to one of them, the system will generate (in addition to the questions described above) the candidate question: *"What is the difference between <current-concept> and <sibling-concept>?"*. Thus, if the terms 'migraine' and 'tension headaches' have been mentioned, and the user points at 'migraine', the system generates the two questions: *"What is migraine?"*, as well as, *"What is the difference between a migraine and a tension headache?"*

In addition to reasoning about the previous discourse, the system also reasons about the relationships in the KB to find other candidate questions that the user may wish to ask. We use a structured inheritance network knowledge representation language called Loom [12]. This allows the system to easily determine parent-child relationships between concepts in the KB. Consider the fragment of the discourse structure shown in Figure 4 about the medications prescribed by the doctor:

Motrin can help relieve the pain in most patients. Common

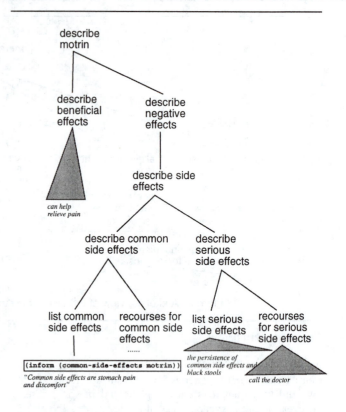

Figure 4: Discourse structure for describing a drug.

side effects are stomach pain and discomfort. If you can take food without vomiting, this may help you tolerate Motrin better. Call Dr. Rivers if these effects persist, or if you experience black stools.

If the user highlights the phrase 'Common side effects' the system generates not just the candidate questions *"What is a common side effect?"* and *"What is a side effect?"*, but also generates the following two questions:

- *What are other side effects of Motrin?*
- *What side effects of Motrin are serious enough to call the doctor?*

These two questions are generated by reasoning about the concept hierarchy in the KB: the goal being achieved is to list the common-side-effects (a relation in the knowledge base) of Motrin (the speech act is (inform (common-side-effects motrin))). Since there are two other sibling relations defined in the knowledge base under side-effects – other-side-effects and serious-side-effects – the system generates questions about Motrin for those relations as well.

Pruning the set of candidate questions

Once the system has generated candidate questions based on the type of speech act, the propositions in the speech act, as well as the discourse structure, the set of possible questions is pruned to remove questions that (i) have already been asked by the user, (ii) have already been answered, or (iii) may not be relevant based on the discourse context or the user model. This pruning is important because one of the main problems in hypermedia systems is that users tend to get lost in the network; doing so may even cause them to forget what they were seeking in the first place [4, 8]. (Presenting a user with a follow up question that the user thinks has already been answered may cause the user to think that she's even more confused than she actually is.)

Pruning previously asked questions is possible since the the discourse structure records all utterances, including the questions, in addition to the responses generated. Thus questions such as *"What is migraine?"*, which get generated in a large number of cases in our application, are not presented repeatedly to the user once they have been answered by the system.

In the second case, the system attempts to reason about complex clauses to prune those questions generated by one of the constituent clauses that are answered by other constituent clause(s). For instance, if the user were to highlight the utterance

(i) 'I am trying to reduce the frequency of your migraine attacks by
(ii) prescribing Elavil'

the system reasons about the discourse structure underlying this utterance. The two speech acts that give rise to (i) and (ii) are related via the relation MEANS (this generates the connective 'by' in the case above). Since the relation MEANS relates utterances about goals to utterances about methods (the general form is '<goal> BY <method>'), the system does not generate the how-question about the goal (*"How are you trying to reduce the frequency of migraine attacks?"*) because the answer has already been presented to the user.

However, the system does not attempt to prune *all* questions whose answers have been been generated previously. This is because the user may not have realized that an answer given in another context applies to the current situation as well. For instance, if the answer to a question *"Why should some drugs be taken after meals?"* occurred in the context of explaining how to take Motrin, and later, the user highlights the sentence 'Take the medicine after meals' in the context of a discourse plan explaining how to take ibuprofen, the question would be generated *again* in the new context. Even though the goal (RECOMMEND (TAKE MEDICINE (AFTER FOOD))) appears in both the discourse plans, the user may not remember the reasoning in the previous case. Therefore, in such cases, the system will generate a why-question for the recommendation 'Take it after food' in cases of other drugs as well.

The third factor used by the system in pruning questions is the user model [10, 18]. In our system, a skeletal model of the user (containing the previous medical history of the user) is available to the system. The information contained in this model is used to prune the candidate set of questions as well. For instance, in the case of a female user, highlighting the noun phrase 'level of female hormones' results in the question *"How do birth control pills affect the levels of female hormones"* being generated by the system since the knowledge base indicates a relationship between hormones and birth control medication. However, the system prunes this question for male patients.

Issues in Generating Embedded Menus

Applications such as our patient education system, in which users are presented with large amounts of text that they may wish to query, seem ideally suited for generating embedded menus [11, 15, 21, 23]. However, there are a number of other issues that play a role in designing effective menu systems. Issues such as the menu title, the number and ordering of menu items, their phrasing, and so on all impact on their effectiveness [15, 21]. Since our efforts have been concentrated on generating the follow up questions dynamically, some of these issues, such as the phrasing, and the number of menu items, for instance, are not as central to this framework as they are in the case of static, pre-determined menus. In our framework, the phrasing of the generated menu items is automatically consistent with the phrasing of the highlighted text, since they are both generated from the same underlying representation. In addition, all the menu items are phrased as questions that begin with the keywords 'what', 'why' or 'how'. These two factors result in menus that are consistent with guidelines developed empirically, e.g., [21, 22].

In the case of menu titles, it is less clear what would constitute an effective solution in our case. The range of questions that the system can generate is quite large, and the entities being queried can range from single words to entire sentences. Based on the fact that the title is also used as a marker to indicate what the menu is about [21, 24], our system uses the highlighted text (or the initial part of it) being queried as the menu title. Preliminary results from our evaluation show that the users seem to find this acceptable.

Our framework is well suited for dynamically determining the ordering and aggregation of menu items appropriately. Studies on ordering menu items have shown that a fixed ordering scheme is better than one in which menu items are changed randomly, because people tend to learn with practice [5, 24]. However, in our case, each menu generated is different from the previous ones. Thus, ordering of the menu items in our system is not constrained by previous presentations, but is determined dynamically, based on the utterance being queried, the previous questions asked, and the constituent parts of the utterance (clauses, as well as noun phrases), in decreasing order of importance.

Each utterance gives rise to some questions that are more closely related to the utterance than other questions that may be generated from its constituents, e.g., the question *"What is a symptom?"* is more germane to 'symptom' than *"What are common symptoms of migraine?"*. Of course, the reverse is true, if the user were to highlight the phrase 'common symptoms of migraine'. The system orders the questions based on this notion of closeness. The degree of closeness is computed by determining the distance of the different speech acts in the discourse structure (that give rise to questions) from the speech act whose utterance was highlighted by the user. This helps ensure that the user can quickly identify the central question. The aggregation of menu items is based on the utterances that the questions are generated from. For instance, if the user were to highlight the phrase 'estrogen replacement therapy,' the system would group questions about the phrase separately from questions about the constituent parts of the phrase. This results in different groupings, with one group consisting of questions such as *"What is estrogen replacement therapy?"* and *"Why does estrogen replacement therapy affect migraines?"* and another containing questions based on the constituent parts, such as the question *"What is estrogen?"*. Preliminary results from our evaluation suggest that users favor this aggregation scheme as compared to aggregating based on the types of questions (what-, why- and how-questions).

System Evaluation

As mentioned previously, our current system is designed in the framework of a patient education system for migraine headaches. It is implemented in Common Lisp and CLIM and can generate a total of about 180 different questions in this domain. The average number of menu items generated is 4. In a typical description generated for a migraine patient (about 3 screens), there are approximately 50 pieces of text that are mouse sensitive and result in menus being generated. All the menus are one level deep.

Our system is currently undergoing normative evaluation. We have had approximately 40 people use the dynamic menu facility to generate feedback on the ease of finding the desired follow up questions. Based on these results, the heuristics for generating questions were further refined so that the system generated all possible questions in that context while suppressing questions perceived as irrelevant. This evaluation also served to help us validate our models of titling the menus, as well as ordering and aggregating menu items.

The usability and utility of the migraine system has also been evaluated in two preliminary studies with actual patients suffering from migraine. In the first study, three patients used the system in the context of an actual visit with a neurologist. In the second study, thirteen persons with headache and one or more symptoms of migraine interacted with the interactive system without seeing the neurologist. In both of these studies, the patients were observed using the system, and were also interviewed afterwards regarding their session with the system. While we recognize that this is an evaluation of patients perceptions, and not a study of outcomes, we nevertheless believe that the results are helpful and encouraging. Table 1 shows an excerpt of patients' assessment of the system:

Questions	Answer Category	
	Yes	No
Did you like using the program?	16 (100%)	0 (0%)
Did all of the information presented make sense?	13 (81%)	3 (19%)
Did you feel comfortable about using a computer to get this kind of information?	16 (100%)	0 (0%)
Was the computer itself easy to use?	14 (88%)	2 (12%)
Did the program tell you anything you did not already know?	15 (94%)	1 (6%)
Do you think this information will help you manage your headaches better?	9 (56%)	7 (44%)
Did you learn anything that you would not have asked your doctor?	12 (75%)	4 (25%)

Table 1: Preliminary responses to the system.

In general, the average amount of time spent on the system by each patient was 46 minutes, with the minimum being 14 minutes and the maximum being 160 minutes. The average number of utterances clicked on by the patients was 11, with the minimum being 1 and the maximum being 29. The average number of questions asked was 16, ranging from a low of 3 to a high of 45.

Given that some of the patients in our evaluation had never used a mouse-based interface before, it was gratifying that none of them had any problems using this menu based interface.[2] The follow up interview also revealed that the users were satisfied with the range of choices that were generated by the system. They did not indicate any questions that they would have liked to have asked of the system, but were unable to do so. The consensus was that the system was easy to use and the interface helped them find relevant information easily. Twelve of the patients also stated that the questions in the menus had helped them "learn things that they would not have asked their doctor." These results suggest that our interface has been helpful to patients, *not just in presenting queries to the system, but also in exploring additional, related information that the system possesses about migraine.* It is clear that a more extensive and controlled evaluation is necessary before the actual benefits of such an interface can be determined. We are in the process of designing such an evaluation. In future, we plan to extend the range of application domains and evaluate its coverage and effectiveness in allowing the users to request further information or clarify ambiguities or misunderstandings.

Conclusions and Future Work

This paper has described our approach to dynamically generating menus of follow up questions in explanatory or help systems. The ability to handle follow up requests in context is essential in many applications, and can become crucial in situations such as the patient education system described here. Our approach avoids the problem of natural language understanding and instead adopts a pragmatic (and possibly more useful) approach of generating choices for the user to select from.

Our initial evaluations of the system reveal that users seem comfortable with the interface as a means of asking follow up questions. Our test subjects stated that they were satisfied with the range of questions generated by the system and that the interface helped them find relevant information. Initial surveys suggest that such an interface is helpful to patients, not just in presenting queries to the system, but also in exploring additional, related information that the system possesses about the domain. This is an important issue, and we are in the process of designing a more extensive and controlled evaluation of the interface.

Acknowledgements

This work was partially supported by grant number R01 LM05299 from the National Library of Medicine, National Institutes of Health. Its contents are solely the responsibility

---

[2]Our only hitch arose from the fact that the mouse on our system is a 3-button mouse on which the middle and right buttons have been disabled. Once this was clarified to the users, they had no problems.

of the author and do not necessarily represent the official views of the National Library of Medicine. The ethnographic analyses were carried out by Dr. D. E. Forsythe and Ms. M. Brostoff. We would also like to acknowledge all the other members of the migraine project: G. Banks, N. Bee, B. Buchanan, G. Carenini, S. Margolis and S. Ohlsson.

REFERENCES

1. BLACK, J. B., AND MORAN, T. P. Learning and remembering command names. In *Proceedings of the Conference on Human Factors in Computing* (Gaithersberg, MD, 1982), pp. 8–11.

2. BUCHANAN, B. G., MOORE, J., FORSYTHE, D., BANKS, G., AND OHLSSON, S. Involving patients in health care: Using medical informatics for explanation in the clinical setting. In *Proceedings of the Symposium on Computer Applications in Medical Care* (Washington, D. C., 1992).

3. BUCHANAN, B. G., MOORE, J., FORSYTHE, D., GIUSEPPE, C., BANKS, G., AND OHLSSON, S. Using medical informatics for explanation in the clinical setting. Tech. Rep. Number CS-93-16, Department of Computer Science, University of Pittsburgh, (To appear in *Artificial Intelligence in Medicine*).

4. CARANDO, P. Shadow: Fusing hypertext with AI. *IEEE Expert 4*, 4 (1989), 65–78.

5. CARD, S. K. User perceptual mechanisms in the search of computer command menus. In *Proceedings of CHI'82: Human Factors in Computing Systems* (1982), pp. 190–196.

6. CAWSEY, A. *Explanation and interaction: the computer generation of explanatory dialogues*. MIT Press, Cambridge, MA, 1992.

7. ELHADAD, M., AND ROBIN, J. Controlling content realization with functional unification grammars. In *Proceedings of the Sixth International Workshop on Natural Language Generation*, R. Dale, E. Hovy, D. Rosner, and O. Stock, Eds. Springer Verlag, 1992.

8. HALASZ, F. G. Reflections on NoteCards: Seven issues for the next generation of hypermedia systems. *Communications of the Association for Computing Machinery 31*, 7 (1988), 836–870.

9. KELLY, M. J., AND CHAMPANIS, A. Limited vocabulary natural language dialogue. *International Journal of Man-Machine Studies 9* (1977), 479–501.

10. KOBSA, A., AND WAHLSTER, W. User Models in Dialog Systems. In *User Models in Dialog Systems*, A. Kobsa and W. Wahlster, Eds. Springer-Verlag, Symbolic Computation Series, Berlin, 1989, pp. 4–34.

11. KOVED, L., AND SHNEIDERMAN, B. Embedded menus: Selecting items in context. *Communications of the ACM 29* (1986), 312–318.

12. MACGREGOR, R. A Deductive Pattern Matcher. In *Proceedings of the 1988 Conference on Artificial Intelligence* (St Paul, Mn, Aug. 1988), American Association of Artificial Intelligence.

13. MOORE, J. D. *Participating in Explanatory Dialogues: Interpreting and Responding to Questions in Context*. MIT Press, Cambridge, MA, 1994.

14. MOORE, J. D., AND PARIS, C. L. Planning text for advisory dialogues: Capturing intentional annd rhetorical information. *Computational Linguistics 19*, 4 (Dec. 1993), 651–694.

15. NORMAN, K. L. *The psychology of menu selection: designing cognitive control ot the human/computer interface*. Ablex Publishers, Norwood, NJ, 1991.

16. PETRICK, S. R. On natural language based computer systems. *IBM Journal of Research and Development 20* (1976), 314–325.

17. POLLACK, M., HIRSCHBERG, J., AND WEBBER, B. User participation in the reasoning processes of expert systems. In *Proceedings of the AAAI* (Pittsburgh, Pa, 1982), American Association of Artificial Intelligence.

18. RASKUTTI, B., AND ZUKERMAN, I. Generation and selection of likely interpretation during plan recognition in task-oriented consultation systems. *User Modeling and User-Adapted Interaction 1*, 4 (1991), 323–353.

19. RINGLE, M. H., AND BRUCE, B. C. Conversation Failure. In *Knowledge Representation and Natural Language Processing*, W. G. Lehnert and M. H. Ringle, Eds. Lawrence Erlbaum Associates, Hillsdale, New Jersey, 1981, pp. 203–221.

20. SAVAGE, R. E., HABINEK, J. K., AND BARNHART, T. W. The design, simulation and evaluation of a menu driven user interface. In *Proceedings of the Conference on Human Factors in Computer Systems* (1982), pp. 36–40.

21. SHNEIDERMAN, B. Designing menu selection systems. *Journal of the American Society For Information Science 37*, 2 (1986), 57–70.

22. SHNEIDERMAN, B. *Designing the user interface: Strategies for effective human computer interaction*. Addison-Wesley, Reading, MA, 1987.

23. SHNEIDERMAN, B., AND KEARSLEY, G. *Hypertext hands-on!* Addison-Wesley, Reading, MA., 1989.

24. TEITELBAUM, R. C., AND GRANDA, R. The effects of positional constancy on searching menus for information. In *Proceedings of CHI'83: Human Factors in Computing Systems* (Baltimore, MD, 1983), ACM, pp. 150–153.

A Generic Platform
for Addressing the Multimodal Challenge

Laurence Nigay, Joëlle Coutaz
Laboratoire de Génie Informatique (LGI-IMAG)
BP 53, 38041 Grenoble Cedex 9, France
Tel: +33 76-51-44-40 +33 76-51-48-54
E-mail: Laurence.Nigay@imag.fr Joelle.Coutaz@imag.fr

ABSTRACT

Multimodal interactive systems support multiple interaction techniques such as the synergistic use of speech and direct manipulation. The flexibility they offer results in an increased complexity that current software tools do not address appropriately. One of the emerging technical problems in multimodal interaction is concerned with the fusion of information produced through distinct interaction techniques. In this article, we present a generic fusion engine that can be embedded in a multi-agent architecture modelling technique. We demonstrate the fruitful symbiosis of our fusion mechanism with PAC-Amodeus, our agent-based conceptual model, and illustrate the applicability of the approach with the implementation of an effective interactive system: MATIS, a Multimodal Airline Travel Information System.

KEYWORDS: Multimodal interactive systems, software design, software architecture, I/O devices, interaction languages, data fusion.

INTRODUCTION

One new challenge for Human Computer Interaction (HCI) is to extend the sensory-motor capabilities of computer systems to better match the natural communication means of human beings. Towards this goal, multimodal interfaces are being developed to support multiple interaction techniques such as the synergistic use of speech and gesture. The power and versatility of multimodal interfaces result in an increased complexity that current design methods and tools do not address appropriately. As observed by B. Myers, "user interface design and implementation are inherently difficult tasks"[11]. Myers's assertion is even more relevant when considering the constraints imposed by the recent technological push. In particular, multimodal interaction requires [3]:

- the fusion of different types of data originating from distinct interaction techniques as exemplified by the "put that there" paradigm,

- the management of multiple processes including support for synchronization and race conditions between distinct interaction techniques.

Thus, multimodal interfaces make necessary the development of software tools that satisfy new requirements. Such tools are currently few and limited in scope. Either they address a very specific technical problem such as media synchronization [9], or they are dedicated to very specific modalities. For example, the Artkit toolkit is designed to support direct manipulation augmented with gesture only [7].

In this article, we propose a software architecture model, PAC-Amodeus, together with a generic fusion mechanism for designing and implementing multimodal interaction. The PAC-Amodeus model along with the fusion engine form a reusable global platform applicable to the software design and implementation of multimodal interactive systems.

The structure of the paper is as follow: first, we clarify the notion of interaction technique using the concepts of interaction language and physical device. We then present the principles of our software architecture model, PAC-Amodeus, and show how interaction languages and devices operate within the components of the architecture. Going one step further in the implementation process, we populate PAC-Amodeus with the presentation of our generic fusion mechanism. We conclude with an example that illustrates how PAC-Amodeus and the fusion engine function together. This example is based on MATIS whose main features are presented in the next section.

AN ILLUSTRATIVE EXAMPLE: MATIS

MATIS (Multimodal Airline Travel Information System) allows a user to retrieve information about flight schedules using speech, direct manipulation, keyboard and mouse, or a combination of these techniques [13]. Speech input is processed by Sphinx, a continuous speaker independent recognition engine developed at Carnegie Mellon University [10]. As a unique feature, MATIS supports both individual and synergistic use of multiple input modalities [13]. For example, using one single modality, the user can say "show me the USAir flights from Boston to Denver" or can fill in a form using the keyboard. When exploiting synergy, the user can also combine speech and gesture as in "show me the USAir flights from Boston to this city" along with the selection of "Denver" with the mouse. MATIS does not impose any dominant modality: all of the modalities have the same power of expression for specifying a request and the user can freely switch between them. The system is also able to support multithreading: a MATIS user can disengage from a partially formulated request, start a new one, and later in the interaction process, return to the pending request.

PHYSICAL DEVICES AND INTERACTION LANGUAGES

A *physical device* is an artefact of the system that acquires (input device) or delivers (output device) information. Examples of devices in MATIS include the keyboard, mouse, microphone and screen.

An *interaction language* defines a set of well-formed expressions (i.e., a conventional assembly of symbols) that convey meaning. The generation of a symbol, or a set of symbols, results from actions on physical devices. In MATIS, examples of interaction languages include pseudo-natural language and direct manipulation.

We define an *interaction technique* as the coupling of a physical device d with an interaction language L: <d, L>.

Interaction techniques supported by MATIS include speech, written natural language, graphic input and output:

- speech input is described as the couple <microphone, pseudo natural language NL>, where NL is defined by a specific grammar,
- written natural language input is defined as <keyboard, pseudo natural language NL> (a MATIS user can also type in NL sentences in a dedicated windows),
- graphic input is described in terms of <mouse, direct manipulation>, and
- graphic output corresponds to the couple <screen, tables>. (Flight schedules returned by MATIS are always presented in a tabular format.)

Physical devices and interaction languages are resources and knowledge that the system and the user must share to accomplish a task successfully. They cover "the articulatory and semantic distances" expressed in Norman's theory [16].

Adopting Hemjslev's terminology [6], the physical device determines the substance (i.e., the non analyzed raw material) of an expression whereas the interaction language denotes the form or structure of the expression.

In [15], we demonstrate the adequation of the notions of physical device and interaction language for classifying and deriving usability properties for multimodal interaction. In this article, we adopt a complementary perspective and examine the relevance of these notions for software design.

SOFTWARE DESIGN

One important issue in software design is the definition of software architectures that support specific quality factors such as portability and modifiability. PAC-Amodeus is a conceptual model useful for devising architectures driven by user-centered properties including multithreading and multimodality. PAC-Amodeus blends together the principles of both Arch [18] and PAC [1]. Arch and its companion, the "slinky" metamodel, provide the appropriate hooks for performing engineering tradeoffs such as identifying the appropriate level of abstraction for portability, making semantic repair or distributing semantics across the components of the architecture [4]. In particular, the five component structure of Arch includes two adapters, the Interface with the Functional Core and the Presentation Techniques Component, that allow the software designer to insulate the key element of the user interface (i.e., the Dialogue Controller) from the variations of the functional core and of the implementation tools (e.g., the X window environment). The Arch model however, does not provide any guidance about the decomposition of the Dialogue Controller nor does it indicate how salient features in new interaction techniques (such as parallelism, fusion and fission of information [3]) can be supported within the architecture. PAC, on the other hand, stresses the recursive decomposition of the user interface in terms of agents, but does not pay attention to engineering issues. PAC-Amodeus gathers the best of the two worlds. Figure 1a shows the resulting model.

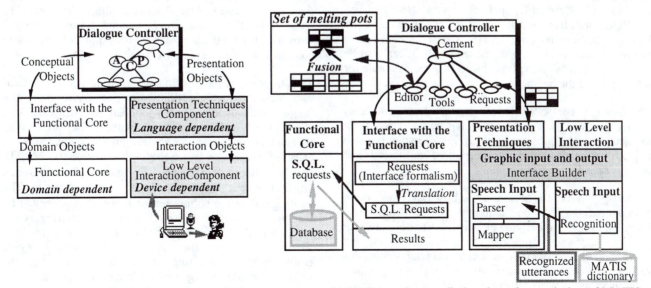

Figure 1: (a) The PAC-Amodeus software components. (b) PAC-Amodeus applied to the software design of MATIS.

A more detailed description of PAC-Amodeus can be found in [14]. Succinctly, the five components of the arch defines the levels of abstraction appropriate for performing engineering tradeoffs such as setting the boundaries between the levels of abstraction. We offer the notions of physical device and interaction language as criteria for setting these boundaries. For example, the designer may decide that the Low Level Interaction Component is device dependent. At a higher level of abstraction, the Presentation Techniques Component is device independent but language dependent. At the top of the Arch, the Dialogue Controller is both language and device independent.

PAC-Amodeus refines the Dialogue Controller into a set of cooperative agents that capture parallelism and information processing (e.g., data fusion) at multiple levels of abstraction. In turn, an agent is modelled as a three facet structure:

- the Presentation facet is in direct contact with the Presentation Techniques Component of the arch. It can be used to implement extensions on top of the Presentation Techniques Component;

- the Abstraction facet is in direct contact with the Interface with the Functional Core;

- the Control facet manages the links and constraints between its two surrounding facets (i.e., the Presentation and the Abstraction facets) as well as its relationships with other agents. As in ALV [8], the Control facet provides the hook for expressing constraints between different perspectives of the same concept.

In combining the Arch principles with PAC, one obtains an "engineerable" model that supports properties inherited from the agent paradigm. Figure 1b illustrates the application of PAC-Amodeus to the software design of MATIS. The Functional Core hosts the database of American cities, airline companies, flight numbers, departure and arrival times, etc. SQL requests are required to access information stored in the database. The Interface with the Functional Core (IFC) operates as a translator between the SQL formalism and the data structures used in the Dialogue Controller. In MATIS, the IFC serves as a communication bridge. As discussed in [2], it can also be used to restructure conceptual objects in a form suitable for the purpose of the interaction.

The Dialogue controller (DC) is organized as a two-level hierarchy of agents. This hierarchy has been devised using the heuristic rules presented in [15]. For example, because requests can be elaborated in an interleaved way, there is one agent per pending request.

At the other end of the spectrum, the Low Level Interaction Component (LLIC) is instantiated as two components inherited from the underlying platform: (1) The NeXTSTEP event handler and graphics machine, and (2) the Sphinx speech recognizer which produces character strings for recognized spoken utterances. Mouse-key events, graphics primitives, and Sphinx character strings are the interaction objects exchanged with the Presentation Techniques Component (PTC).

In turn, the Presentation Techniques Component (PTC) is split into two main parts: the graphics objects (used for both input and output) and the NL parser (used for input only). Graphics objects result from the code generation performed by Interface Builder. The Sphinx parser analyzes strings received from the LLIC using a grammar that defines the NL interaction language. As discussed above, the PTC is no longer dependent on devices, but processes information using knowledge about interaction languages.

Having presented the overall structure of PAC-Amodeus, we need now to address the problem of data fusion. As discussed in [14], fusion occurs at every level of the arch components. For example, within the LLIC, typing the option key along with another key is combined into one single event. In this article, we are concerned with data fusion that occurs within the Dialogue Controller.

THE FUSION MECHANISM

Within the Dialogue Controller, data fusion is performed at a high level of abstraction (i.e., at the command or task level) by PAC agents. As shown in Figure 1b, every PAC agent has access to a fusion engine through its Control facet. This shared service can be viewed either as a reusable technical solution (i.e., a skeleton) or as a third dimension of the architectural model.

Fusion is performed on the presentation objects received from the PTC. These objects obey to a uniform format: the melting pot. As shown in Figures 1b and 2, a melting pot is a 2-D structure. On the vertical axis, the "structural parts" model the composition of the task objects that the Dialogue Controller is able to handle. For example, request slots such as destination and time departure, are the structural parts of the task objects that the Dialogue Controller handles for MATIS. Events generated by user's actions are abstracted through the LLIC and PTC and mapped onto the structural parts of the melting pots. In addition, LLIC events are time-stamped. An event mapped with the structural parts of a melting pot defines a new column along the temporal axis.

The structural decomposition of a melting pot is described in a declarative way outside the engine. By so doing, the fusion mechanism is domain independent: structures that rely on the domain are not "code-wired". They are used as parameters for the fusion engine. Figure 2 illustrates the effect of a fusion on two melting pots: at time t_i, a MATIS user has uttered the sentence "Flights from Boston to this city" while selecting "Denver" with the mouse at t_{i+1}. The melting pot on the bottom left of Figure 2 is generated by the mouse selection action. The speech act triggers the creation of the bottom right melting pot: the slot "from" is filled in with the value "Boston". The fusion engine combines the two melting pots into a new one where the departure and destination locations are both specified.

The criteria for triggering fusion are threefold: the complementarity of melting pots, time, and context. When triggered, the engine attempts three types of fusion in the following order: microtemporal fusion, macrotemporal fusion, and contextual fusion.

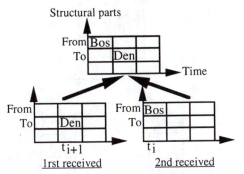

Figure 2: Fusion of two melting pots.

• *Microtemporal fusion* is used to combine related informational units produced in parallel or in a pseudo-parallel manner. It is performed when the structural parts of input melting pots are complementary and when these melting pots are close in time: their time interval overlaps. Figure 3 shows one possible configuration of temporal relationships between two melting pots m_i and $m_{i'}$ candidates for microtemporal fusion.

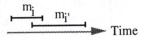

Figure 3: Two melting pots candidates for microtemporal fusion due to the intersection of their time intervals.

• *Macrotemporal fusion* is used to combine related informational units produced sequentially or possibly in parallel by the user but processed sequentially by the system, or even delayed by the system due to the lack of processing resources (e.g., processing speech input requires more computing resources than interpreting mouse clicks). Macrotemporal fusion is performed when the structural parts of input melting pots are complementary and when the time intervals of these melting pots do not overlap but belong to the same temporal window. Figure 4 illustrates the temporal constraints between two melting pots candidates for macrotemporal fusion.

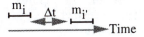

Figure 4: Two melting pots candidates for macrotemporal fusion.

• *Contextual fusion* is used to combine related informational units produced without attention for temporal constraints. For example, a MATIS user may

specify the destination, then give a call, and resume the task a couple of minutes later. Contextual fusion is driven by the current active context. In MATIS, the current context corresponds to the current active request. Contextual fusion combines a new input melting pot m with the melting pots M of the current context if the content of m is complementary with one of the melting pots M. (Melting pots in M are ordered according to their recency.)

Having presented the driving principles of the fusion mechanism, we now focus on the technical details.

INSIDE THE FUSION MECHANISM
Our fusion algorithm has been implemented in C and embedded in a PAC-Amodeus architecture. We first introduce the metrics associated with each melting pot, then describe the three types of fusion in detail. Finally, we present the management of the set of melting pots and their transfer within the hierarchy of PAC agents.

Metrics for a Melting Pot
Figure 5 portrays the metrics that describe a melting pot m_i:

$m_i=(p1, p2,..., pj,..., pn)$: m_i is comprised of n structures $p1, p2, ...pn$.

$info_{ij}$: piece of information stored in the structural part pj of m_i.

$Tinfo_{ij}$: time-stamp of $info_{ij}$.

$Tmax_i$: time-stamp of the most recent piece of information stored in m_i.

$Tmin_i$: time-stamp of the oldest piece of information stored in m_i.

$Temp\_win_i$: duration of the temporal window for m_i.

Δt: Remaining life span for m_i.

A melting pot encapsulates a set of structural parts p_1, $p_2,...p_n$. The content of a structural part is a piece of information that is time-stamped. Time stamps are defined by the LLIC when processing user's events. The engine computes the temporal boundaries ($Tmax$ and $Tmin$) of a melting pot from the time stamps of its pieces of information.

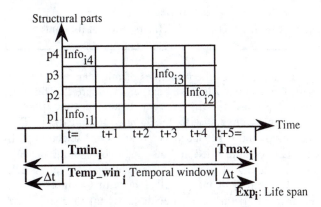

Figure 5: Metrics used to define a melting pot m_i.

So for $m_i=(p_1, p_2,...p_n)$, $Tmax_i=Max(Tinfo_{ij})$ and $Tmin_i=Min(Tinfo_{ij})$.

The temporal window of a melting pot defines the temporal proximity (+/- Δt) of two adjacent melting pots: for $mi=(p1, p2,...p_n)$, $Temp\_win_i=[Tmin_i-\Delta t, Tmax_i+\Delta t]$. Temporal windows are used to trigger macrotemporal fusion.

The last metrics used to manage a melting pot is the notion of life span, Exp_i: $Exp_i=Tmax_i+\Delta t=Max(Tinfo_{ij})+\Delta t$. This notion is useful for removing a melting pot from the set of candidates for fusion.

The Mechanism
The fusion mechanism is driven by a set of rules.

Rule 1 deals with *microtemporal fusion*. Because priority is given to parallelism at the user's level, microtemporal fusion is first attempted on the arrival of a new melting pot from the Presentation Techniques Component. Since it models a user's action at instant t', this melting pot is composed of one column only. Rule 1 makes explicit the occurrence of microtemporal fusion: if the content of the new melting pot is complementary with a column (col_{it}) of an existing melting pot (m_i) and if the time-stamp of this column is close enough to t' (i.e., within $\Delta microt$), then microtemporal fusion is performed. Microtemporal fusion may involve undoing a previous fusion. This exception case will be discussed later.

<u>Rule 1</u> Microtemporal fusion (overlap of time intervals)

Given:
- $col_{it} = (p_1, p_2,... , p_j,..., p_n)$:
  one column at time t of an existing melting pot m_i.
- $col_{i't'} = (p'_1, p'_2, ..., p'_j, ..., p'_n)$
  a one column melting pot $m_{i'}$ produced at time t'
- $i \neq i'$

  col_{it} and $col_{i't'}$ are combined if:
- they are complementary:
  Complementary (col_{it}, $col_{i't'}$) is satisfied if:
  $\forall k \in [1..n] : \exists\ info_{ik} \wedge \neg (\exists\ info_{i'k})$
- their time-stamps are temporally close:
  Close (col_{it},$col_{i't'}$) is satisfied if:
  $t' \in [t-\Delta microt, t+\Delta microt]$ (Intersection of time intervals)

Figure 6 illustrates the principles of microtemporal fusion with the example discussed earlier in Figure 2: The user utters the sentence "flights from Boston" at time t_i while selecting "Denver" with the mouse at time t_{i+1}. The melting pot m_i is produced as a result of the selection with its "t_{i+1}" column filled in. Later on, a new melting pot $m_{i'}$ arrives at the Dialogue controller resulting from the speech act. Column t_i of $m_{i'}$ is filled in with the information abstracted from the speech act. m_i and $m_{i'}$ are complementary (their content correspond to distinct

structural parts). In addition, the time stamps of the two columns concerned in m_i and $m_{i'}$ are within $\Delta microt$ (we suppose that $\Delta microt$ is equal to 1 temporal unit). Thus microtemporal fusion can be performed.

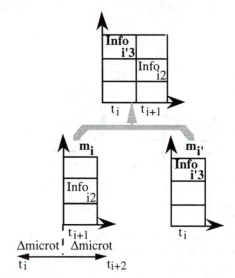

Figure 6: An example of microtemporal fusion.

One particular phenomenon in parallelism is *redundancy* [15]. As shown by the example of Figure 7, a MATIS user may utter the sentence "Flights from Boston" ($Info_{i'1}$ = [Boston]) while selecting "Boston" with the mouse ($Info_{i1}$ = [Boston]). One of the two user's actions must be ignored. i.e., the newly arrived melting pot must be discarded. Redundancy checking is performed before microtemporal fusion is attempted. Rule 2 makes this verification process explicit.

<u>Rule 2</u> Redundancy

Given:
- $col_{it} = (p_1, p_2,... , p_j,..., p_n)$:
  one column at time t of an existing melting m_i
- $col_{i't'} = (p'_1, p'_2, ..., p'_j, ..., p'_n)$
  one column at time t' of a new melting pot $m_{i'}$
- $i \neq i'$

  col_{it} and $col_{i't'}$ are redundant if:
- they contain the same information in the same slots:
  Redundant (col_{it}, $col_{i't'}$) is satisfied if:
  $\forall k \in [1..n] : \exists\ info_{ik} \wedge \exists\ info_{i'k} \wedge info_{ik} = info_{i'k}$
  $\wedge \forall k' \in [1..n] : \neg (\exists\ info_{ik'}) \wedge \neg (\exists\ info_{i'k'})$
- their time-stamps are temporally close:
  Close (col_{it}, $col_{i't'}$) is satisfied if:
  $t' \in [t-\Delta microt, t+\Delta microt]$

Macrotemporal fusion is driven by rules similar to those used for microtemporal fusion where $\Delta microt$ is replaced by temporal windows. Whereas time has a primary role in micro- and macro- temporal fusions, it is not involved in contextual fusion.

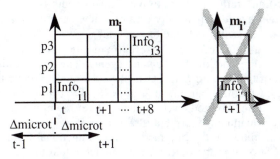

Figure 7: Redundancy: a new melting pot $m_{i'}$ contains information $Info_{i'1}$ equal to $Info_{i1}$ of melting pot m_i produced nearly at the same time as $m_{i'}$.

As described in the above section, *contextual fusion* is the last step in the fusion process. The driving element for contextual fusion is the notion of context. In MATIS, contexts are in a one-to-one correspondence with requests. There is one context per request under specification and the current request denotes the current context. (The user may elaborate multiple requests in an interleaved way.) When a melting pot is complete (all of its structural parts have a value), and its life span expectancy Exp_i expires, it is removed from the set of candidates for fusion. Rule 3 expresses these conditions formally. Exp_i is used for making sure that incorrect fusions have not been performed: when a melting pot is complete, the engine keeps it for a while in the pool of candidates in case the next new melting pots trigger "undo" fusions.

Rule 3 Conditions for removing a melting pot from the list of candidates for fusion:

Melting pot $m_i = (p_1, p_2, ..., p_j, ..., p_n)$ is removed if:
- m_i is complete: $\forall p_j \in m_i, \exists info_{ij}$
- and its span life is over: current date = Exp_i

Undoing erroneous fusions. Because our algorithm favors parallelism, it adopts an "eager" strategy: it does not wait for further information and therefore continuously attempts to combine input data. This approach has the advantage of providing the user with immediate feedback before the functional core is accessed. The drawback is the possible occurrence of incorrect fusions. Incorrect fusion may occur due to the different time scales required to process data specified through distinct languages and devices. As a result, the sequence of melting pots is not necessarily identical to that of the user's actions sequence. For example, in MATIS melting pots that correspond to direct manipulation expressions are built faster than those from voiced utterances. This situation will be illustrated with MATIS in the next section.

A melting pot removed from the fusion pool by the fusion engine is returned to the calling PAC agent for further processing. In the next paragraph we describe how melting pots relate to PAC agents.

The Fusion Engine and the PAC Agents

The PAC agents of the Dialogue Controller are in charge of task sequencing as well as processing the content of the melting pots. This activity is part of the abstraction and concretization processes as described in [3][14]. Abstracting involves multiple processing activities including the use of the fusion engine. When calling the engine, a PAC agent provides a melting pot as an input and receives a list of melting pots as output parameter. Depending on the current candidates in the fusion pool, the content of the input melting pot may or may not be modified by the fusion engine.

Data fusion is one aspect of abstraction. The enrichment of information is also performed by exchanging melting pots across the hierarchy of agents. There is one such hierarchy per task. The set of melting pots are partitioned according to the set of tasks. As a result, an agent hierarchy handles the melting pots that are related to the task it models. In addition the root agent of each hierarchy maintains a mapping function between the melting pots and the PAC agents interested by these melting pots. The benefit of this partitioning is that the fusion engine will not try to combine melting pots that belong to different task partitions. For example in MATIS, if the user utters "Flights from Pittsburgh to this city" while resizing a window, the two melting pots that model the users physical actions does not belong to the same set. As a result, the fusion mechanism does not attempt to combine them.

FROM MODEL TO REALITY: INTERACTING WITH MATIS

In this section we use MATIS to illustrate how PAC agents within the Dialogue Controller operate in conjunction with the fusion mechanism. Figures 8 and 9 show the message passing through the hierarchy of agents and the fusions performed in the context of the following example: the user has already specified the destination slot (i.e., Denver) as well as the departure slot (i.e., Boston) of the current request α. The result of this specification is modelled in Figure 8 as the melting pot m_1 as well as the existence of the *Request α* agent in charge of maintaining a local interaction with the user about this request. The user then utters the sentence "Flights from Pittsburgh" while selecting "TWA" using the mouse.

Because mouse clicks are processed faster than speech input, the mouse selection is first received by the Dialogue Controller through the Presentation facet of the *Tools* agent (<1> in Figure 8). The mouse click is modelled as the melting pot m_2 which contains [TWA]. The Presentation of the *Tools* agent performs a partial immediate feedback by highlighting the selection. Its Control facet calls the fusion mechanism (<2>): the new coming melting pot m_2 is combined with m_1 by contextual fusion. m_1, which now contains [BOS, DEN, TWA], is returned to the *Tools* agent (<3>). In turn, the *Tools* agent, which cannot perform any more processing on m_1, sends m_1 to its parent agent (<4>). As shown in Figure 8, the *Cement* agent which maintains the mapping between melting pots and the agents interested in these melting pots, transfers m1 to the *Request* α agent (<5>). *Request* α agent is then able to update its abstraction facet and its presentation facet (<6>): the request form on the screen (<7>) is updated accordingly with (Boston, Denver and TWA).

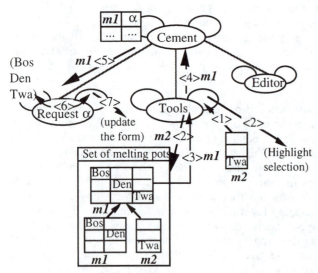

Figure 8: Interacting with MATIS: contextual fusion.

Meanwhile, melting pot m3 which corresponds to the sentence "Flights from Pittsburgh", is received by the *Editor* agent (<1> in Figure 9). The *Editor* agent provides the user with a partial feedback by displaying the recognized sentence while calling the fusion mechanism (<2>). The current set of candidates for fusion is now {m1, m2, m3} (according to rule 3, m1 and m2, which have not reached their life span expectancy, have not been eliminated from the pool). Because the time intervals of m2 [TWA] and m3 [PIT] overlap, they are combined by microtemporal fusion and m2 becomes [PIT, TWA] (rule 1 applies). The previous contextual fusion [BOS, DEN, TWA] is undone: m1 [BOS, DEN] and m2 [PIT, TWA] are returned to the *Editor* agent

(<3>) and reflected back to the *Cement* agent (<4>). The *Cement* agent dynamically creates a new agent *Request* β (<5>)because the new melting pot, m2, [TWA, PIT] has no agent associated with itself (mapping table in the abstraction part of the *Cement* agent). The Presentation facet of the *Request* β agent displays a form containing the state of the new current request (<7>). From now on, the user has elaborated two requests. When completed, the content maintained in the abstract facet of a *Request* agent is transmitted to the Interface with the Functional Core for translation into the SQL format and submitted to the data base maintained in the Functional Core.

SUMMARY AND DISCUSSION

We have presented a software architecture model, PAC-Amodeus, augmented with a fusion mechanism to support the software design of multimodal systems. The platform defined by PAC-Amodeus along with the fusion mechanism fulfills specific requirements of multimodal systems such as data fusion and parallel processing. The fusion mechanism is responsible for combining data specified by the user through different modalities (i.e., a combination of devices and interaction languages). In particular, we have shown the benefits of the symbiosis between the hierarchy of agents of the architectural model and the fusion mechanism. Based on criteria such as time and structural complementarity, the mechanism is generic and reusable. Each melting pot processed may have any number of structural parts (e.g., lines) *that can be filled independently*. Consequently, the PAC-Amodeus model along with the fusion mechanism define a reusable platform for implementing multimodal systems. This property is a distinct advantage over most current tools which are limited in scope.

In a future work, we plan to enrich our fusion mechanism with a confidence factor attached to every slot of a melting pot. The notion of confidence factor provides a simple mechanism for modelling uncertainty and can be usefully exploited for solving ambiguities in deictic expressions. Figure 10 shows the relevance of confidence factors using the example of Figure 2.

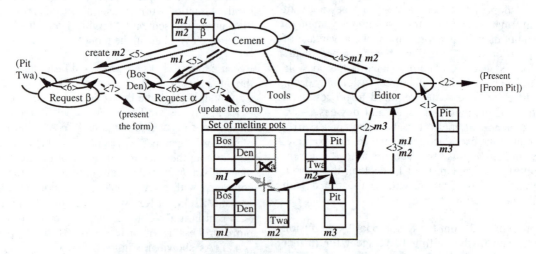

Figure 9: Interacting within MATIS: undoing fusion due to microtemporal fusion.

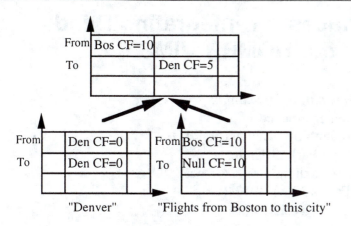

Figure 10: Confidence factor (CF ∈ [1,10]): Example of a deictic expression (see figure 2).

Moreover Δt and Δmicro have been tuned experimentally. One can improve the setting of those parameters by letting the system compute the appropriate values depending on performance of the platform as well as on the behavior of the user. In addition, we will also examine systems that support multiple output modalities. This may lead to the development of a "fission" mechanism as introduced in MSM [3] and suggested in [17].

ACKNOWLEDGEMENTS
This work has been partly supported by project ESPRIT BR 7040 Amodeus II. Many thanks to G. Serghiou for reviewing the paper.

REFERENCES
1. Coutaz, J. PAC, an Object Oriented Model for Dialog Design, in Proc. Interact'87 (Stuttgart, 1-4 Sept. 1987), North Holland, pp. 431-436.

2. Coutaz, J., Balbo, S. Applications: A Dimension Space for User Interface Management Systems, in Proc. CHI'91 Human Factors in Computing Systems (New Orleans, April 27-May 2, 1991), ACM Press, pp. 27-32.

3. Coutaz, J., Nigay, L., Salber, D. The MSM framework: A Design Space for Multi-Sensori-Motor Systems, in Proc. EWHCI'93 (Moscow, Aug. 3-7, 1993), Springer Verlag (Lecture notes in Computer Science, Vol. 753, 1993, pp. 231-241).

4. Coutaz, J. Architectural design for user interfaces, Encyclopedia of Software Engineering, Volume 1 A-N, Wiley-Interscience, 1994, pp. 38-49.

5. Garfinkel, S., Mahoney, M. NeXTSTEP Programming, Springer-Verlag, 1993, 631 pages.

6. Hemjslev, L., Structural Analysis of language, Studia Phonetica, Vol. 1, pp. 69-78.

7. Henry, T.R., Hudson, S.E., Newell, G.L. Integrating Gesture and Snapping into a User Interface Toolkit, in Proc. Symposium on User Interface Software and Technology (Oct. 1990), ACM Press, pp. 112-121.

8. Hill, R.D. The Abstraction-Link-View Paradigm: Using Constraints to Connect User Interfaces to Applications, in Proc. CHI'92 Human Factors in Computing Systems (Monterey, May 3-7, 1992), ACM Press, pp. 335-342.

9. Little, T.D.C, Ghafoor, A., Chen, C.Y.R., Chang, C.S., Berra, P.B. Multimedia Synchronization, IEEE Data Engineering Bulletin, Vol. 14, 3 (Sept. 1991), pp. 26-35.

10. Lunati, J-M and Rudnicky, A. Spoken Language interfaces: The OM system, in Proc. CHI'91 Human Factors in Computing Systems (New Orleans, April 27-May 2, 1991), ACM Press, pp. 453-454.

11. Myers, B. Challenges of HCI Design and Implementation, Interactions, Vol. 1 No. 1 (Jan. 1994), pp. 73-83.

12. Nigay, L. A Case Study of Software Architecture for Multimodal Interactive System: a voice-enabled graphic notebook, TR LGI-IMAG (Oct. 1991), 31 pages.

13. Nigay, L., Coutaz, J., Salber, D. MATIS: A multimodal airline travel information system, SM/WP10, ESPRIT BRA 7040 Amodeus, (Feb. 1993).

14. Nigay, L., Coutaz, J. A design space for multimodal interfaces: concurrent processing and data fusion, in Proc. INTERCHI'93 Human Factors in Computing Systems (Amsterdam, April 24-29, 1993), ACM Press, pp. 172-178.

15. Nigay, L. Conception et modélisation logicielles des systèmes interactifs : application aux interfaces multimodales, PhD dissertation, University of Grenoble, France (Jan. 1994), 315 pages.

16. Norman, D.A. Cognitive Engineering, In User Centered System Design, New Perspectives on Computer Interaction, Hillsdale: Lawrence Erlbaum Associates, 1986, pp. 31-61.

17. Sutcliffe, A., Faraday, P. Designing Presentation in Multimedia Interfaces, in Proc. CHI'94 Human Factors in Computing Systems (Boston, April 24-28, 1994), ACM Press, pp. 92-98.

18. The UIMS Tool Developers Workshop, A Metamodel for the Runtime Architecture of an Interactive System, SIGCHI Bulletin, 24, 1 (Jan. 1992), pp. 32-37.

Developing Dual User Interfaces for Integrating Blind and Sighted Users : the HOMER UIMS

Anthony Savidis and *Constantine Stephanidis*
Institute of Computer Science,
Foundation for Research and Technology-Hellas (FORTH),
Science and Technology Park of Crete,
P.O. Box 1385, GR 711 10, Heraklion, Crete, Greece,
Tel: +30-81-391741, Fax: +30-81-391740
Email: as@csi.forth.gr, cs@csi.forth.gr

ABSTRACT

Existing systems which enable the accessibility of Graphical User Interfaces to blind people follow an "adaptation strategy"; each system adopts its own fixed policy for reproducing visual dialogues to a non-visual form, without knowledge about the application domain or particular dialogue characteristics. It is argued that non-visual User Interfaces should be more than automatically generated adaptations of visual dialogues. Tools are required to facilitate non-visual interface construction, which should allow iterative design and implementation (not supported by adaptation methods). There is a need for "integrated" User Interfaces which are concurrently accessible by both sighted and blind users in order to prevent segregation of blind people in their working environment. The concept of *Dual User Interfaces* is introduced as the most appropriate basis to address this issue. A User Interface Management System has been developed, called HOMER, which facilitates the development of Dual User Interfaces. HOMER supports the integration of visual and non-visual lexical technologies. In this context, a simple toolkit has been also implemented for building non-visual User Interfaces and has been incorporated in the HOMER system.

KEYWORDS: UIMS; Aids for the impaired; Programming environments.

INTRODUCTION

Currently, accessibility to Graphical User Interfaces by blind users is enabled through systems which reproduce the lexical structure of User Interfaces (i.e. interaction objects and their relationships) to a non-visual form. These are characterized as *adaptation oriented* methodologies and the relevant systems, which employ filtering techniques to extract lexical information, are commonly referred to as *screen readers*. One of the main weaknesses of many such screen reader programs is that they explicitly introduce visually oriented concepts to the non-visual interaction. Such concepts come

from the spatial metaphor, which was developed as a result of intensive research efforts based on the human visual information processing capability. Recent approaches aimed at excluding specific types of objects from reproduction and employed 3D audio output techniques for non-visual layout construction [5]. However, such lexical filtering and subsequent lexical transformation of the User Interface suffers from many significant theoretical drawbacks: (i) the original dialogue design has addressed the specific needs and abilities of sighted users, and consequently reproduction to a non-visual form requires syntactic re-structuring according to the needs and abilities of blind users; the translation of visual physical entities to corresponding non-visual structures is like trying to translate text from one language to another by simple substitution of words; also, the role of designers for non-visual dialogues is practically neglected, (ii) no knowledge on the application domain can be extracted and consequently the *idiosyncrasy* of the application is not taken into consideration; for instance, additional error prevention methods would be required for a nuclear reactor control system in comparison to a conventional computer game, (iii) considering that the trend of User Interface software technology is directed towards visual methods such as virtual reality and 3D representation (i.e. visual reality, for instance the *perspective wall* and *cone trees* [1]) it is likely that adaptation oriented techniques will become unrealistic or meaningless, and (iv) there are no facilities for developing non-visual User Interfaces which introduces the serious problem of *lack of tools*; with adaptation oriented methods, design, prototyping, testing and experimentation cycles are completely "vanished" since only some already existing visual applications can be provided to blind users in a fixed non-visual form.

It is therefore evident that there is a need for better quality User Interfaces accessible by blind users and for the corresponding development support. In this context, the *Dual User Interface* concept has been defined as an appropriate basis on which the previous problems can be efficiently addressed. A Dual User Interface is characterized by the following properties: (i) it is concurrently accessible by blind and sighted users, (ii) the visual and non-visual metaphors of interaction meet the specific needs of sighted and blind users respectively (iii) the visual and non-visual syntactic and lexical structure meet the specific needs of sighted and blind users respectively (iv) at any point in time, the same internal (semantic) functionality should be made accessible to both user groups through the visual and non-visual "faces" of the Dual User Interface (v) at any point in time, the same semantic information should be made accessible through the visual and non-visual "faces" of the Dual User Interface. The first requirement is important for enabling cooperation between a blind and a sighted user and thus avoiding further segregation of blind people in their working environment. The second requirement explicitly poses the need of a metaphor specifically designed for blind users in the non-visual environment. The third requirement expresses the need for separate non-visual interface design. Finally, the fourth and fifth requirements indicate that the underlying functionality and information structures should be made equivalently accessible, though via different interactive channels.

A tool has been developed, called *HOMER*, which facilitates the development of Dual User Interfaces. It falls in the domain of language-based User Interface Management Systems (UIMS) and, apart from Dual User Interfaces, it supports the development of separate visual and non-visual User Interfaces as well. The HOMER system can integrate visual and non-visual lexical technologies and it supports concurrently communication with a visual and non-visual lexical technology at run-time. The dialogue specification is facilitated through the *HOMER language*. This language enables abstraction of the physical constructs, which are introduced through lexical technology integration in the HOMER system, to metaphor independent interaction objects. Consequently, sub-dialogues which can be described with such abstract objects are specified with shared descriptions for both the visual and non-visual environments. The Athena and the OSF/Motif widget sets have been integrated as visual technologies, while a simple toolkit for non-visual interface development, called COMONKIT, has been developed and integrated in HOMER. The Dual User Interfaces generated by HOMER can function in two modes: (i) local collaboration, where the visual and non-visual versions run on the same host (support to interface designers for handling keyboard sharing for this case is provided), and (ii) remote collaboration, in which case the visual and non-visual versions run on remote hosts.

RELATED WORK
Solutions which appeared in the past for providing accessibility to Graphical User Interfaces by blind people were based on filtering techniques aiming to reproduce in a non-visual form an internally stored image of the visual display (*off screen model*). Examples of well known commercially available systems are: OUTSPOKEN™ by Berkeley Systems Inc. and SYSTEM 3™ by the TRACE Research & Development Centre at the University of Winsconsin for the Macintosh; SLIMWARE WINDOWS BRIDGE™ by Syntha-Voice Computers Inc., for MS-

WINDOWS™. Research has been also devoted for providing access to the X WINDOWING SYSTEM; two different approaches are discussed in [5]: (i) The one followed by the MERCATOR project at Georgia Institute of Technology, which transforms the User Interface of X-clients to an appropriate auditory representation. (ii) The other followed by the GUIB project, supported by the TIDE Programme of the Commission of the European Union (DG XIII), which addresses both the MS-WINDOWS™ environment and the X WINDOWING SYSTEM. The GUIB approach is based on a proper transformation of the desk-top metaphor to a non-visual version combining Braille, speech and non-speech audio. Additionally, other issues are being investigated by GUIB, including: different input methods which can be used instead of the mouse, the problem of how blind users can locate efficiently the cursor on the screen, issues related to combining spatially localized sounds (both speech and non-speech) with tactile information and design of appropriate non-visual interaction metaphors. Also, the GUIB project has supported two different activity lines: the *adaptation oriented line*, under which the above efforts fall, and the *future development tools line*, in the context of which the work reported in this paper was partially funded.

DUAL USER INTERFACE SPECIFICATION
Interface specification in HOMER is facilitated through the HOMER language which has been designed for supporting formal description of Dual User Interfaces, as well as of separate visual and non-visual User Interfaces. Regarding the visual and non-visual "faces" of a Dual User Interface, the language allows: (i) *metaphor differentiation*, by enabling employment of different metaphors for the visual and non-visual environment, (ii) *structural differentiation*, by facilitating different object hierarchies to be constructed for the visual and non-visual environment, (iii) *syntactic differentiation*, by providing methods for defining separate behaviour for objects and different dialogue sequencing regarding the visual and non-visual environment, and (iv) *shared description of sub-dialogues*, by enabling shared specification of particular syntactic commonalities which can be identified by abstraction methods. Also, the principle of equivalence on represented semantic structures can be preserved through notational facilities for constraining visual and non-visual presentation structures to internal semantic objects.

Virtual interaction objects revisited
The so called *virtual toolkits* aim at enabling platform independent User Interface implementation by providing a common programming interface to various lexical technologies. They deliver collections of interaction objects, usually called *virtual interaction objects*, for interface construction. Since such virtual objects have attributes explicitly related to the desk-top metaphor (e.g. spatial attributes, colours, etc) which forms a specific physical platform model, they are considered mostly *generalisations* of the physical entities rather than *abstractions*. In HOMER, virtual objects can be constructed as pure abstractions, without having lexical attributes; only the abstract behaviour

(e.g. selector, toggle, valuator, etc) and the physically independent state attributes (e.g. state of a toggle object, content of valuator, etc) need to be notationally visualized. Apart from abstraction, generalisation is also supported. For instance, virtual interaction objects can be specified which form generalisations of various platforms implementing the desk-top metaphor. In this case, such desk-top metaphor specific objects are appropriate only for the visual environment; however, they are platform independent and consequently the visual dialogue specification will not need modification for other platforms. In HOMER, the following categories of virtual objects are provided: (i) *Dual* virtual interaction objects, which are pure abstractions of visual and non-visual interaction objects. This type of objects can be used for describing the dialogue for both blind and sighted users in a unified fashion, (ii) *Visual* virtual interaction objects, which are generalisations of visual physical interaction objects. This type of objects can be engaged only on describing the visual dialogue, and (iii) *Non-visual* virtual interaction objects, which are generalisations of non-visual physical interaction objects. This type of objects can be engaged only on describing the non-visual dialogue.

Specification of virtual object genesis and lexical instantiation schemes

There are two levels of interaction objects in HOMER: (i) the virtual level, which concerns the classes of virtual interaction objects, and (ii) the physical level, which consists of the classes of physical interaction objects. Only virtual object classes can be utilised for dialogue specification, while the physical level plays a two-fold role: firstly, it incorporates the description of the interaction objects supported by the lexical technologies which is necessary for technology integration; and secondly, it is used for specification of mapping schemes with which virtual object classes are associated with various physical object classes. The basic relationship between virtual interaction objects in the dialogue is the *parenthood* relationship. Dual objects have two parents, one for the visual and one for the non-visual environment, which need not be the same. Moreover, a Dual object may be parentless for one environment while not for the other. An arbitrary number of attributes and methods can be specified for either virtual or physical object classes; additionally, physical object classes support inherited attributes and methods which are automatically inherited to children objects. During dialogue specification, full notational access is facilitated for virtual interaction objects to the various physical attributes and methods of the associated physical visual or non-visual object class.

A virtual interaction object can be constructed through the genesis mechanism. Construction of objects at the virtual level does not imply construction at the physical level, since no facilities are provided for extending the set of interaction objects supplied by a target platform. The HOMER system provides methods for integrating and utilising the physical constructs which are available. Engagement in dialogue specification of a virtual interaction object is enabled only after a physical instantiation scheme has been defined for the

target environment(-s); for instance, Dual objects require both a visual and a non-visual instantiation scheme to be specified. In Example 1, the genesis of a *Selector* object is demonstrated at the virtual level, supporting one method called *Selected* and two virtual attributes, *totalOptions* and *userOption*. Following the genesis of the virtual class *Selector*, a visual physical object class is introduced, named *VIMenu*; physical attributes like foreground colour, background colour and font are defined.

```
dual genesis    Selector [
        int totalOptions;
        int userOption;
        method Selected;
        constructor (int N) [
                totalOptions=N;
                userOption=-1;
        ]
        destructor []
]
visual object VIMenu [
        int N, userOption;
        string fg, bg, font;
        string* optionsList;
        method VISelected;
        ...
]
```

Example 1: Genesis of virtual objects and specification of physical objects.

```
visual instantiation Selector [
        simpleMenu : VIMenu [
->      {me}.userOption:={me}visual.userOption;
        method Selected : VISelected;
        constructor [
                {me}visual.N={me}.TotalOptions;
        ]
->      ]
        default=simpleMenu
]
1: {x}.attr {x}visual.attr {x}nonvisual.attr
2: method {x} Selected [ stmts ]
3: method {x}visual VIselected [ stmts ]
4: method {x}nonvisual NVselected [ stmts ]
```

Example 2: Specification of instantiation, access to virtual / physical attributes and virtual / physical method specification for virtual interaction objects.

In Example 2, the specification of the visual instantiation for the virtual *Selector* class is demonstrated. Multiple physical instantiation schemes can be defined for each environment. The example scheme is named *simpleMenu* and the physical class in which the object will map is the *VIMenu*. The block indicated between arrows is the description of the instantiation scheme named *simpleMenu*. Attribute associations can be specified either through constraints or monitors. In this example, a constraint enforces the virtual variable *userOption* to always have the value of the physical attribute *userOption*; the qualifier *visual* is used to distinguish physical visual attributes. Following, is the correspondence of virtual to physical methods. The virtual method *Selected* is associated to the physical method *VISelected*; when physical *VISelected* is called, the virtual *Selected* will be called as well. The last four lines show the syntax for accessing virtual or physical attributes and methods. The notation *{X}* provides access to the virtual structure of object *X* while the notations *{X}visual* and *{X}nonvisual* provide access to the visual and non-visual physical structures respectively. Since for each virtual object class it is allowed to specify multiple physical instantiation schemes, it is possible to have objects of the same physical class which are instantiated differently at the physical level. This is the first level of polymorphism (i.e. from the virtual to the physical level) enabled in HOMER. Moreover, the physical level can be modelled in such a way that it forms a generalisation over platforms instead of exactly "visualizing" the conventions used in the particular lexical technology being integrated.

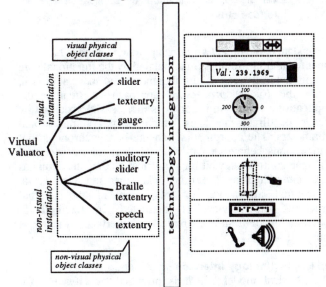

Figure 1: The various intermediate levels for physical realisation of virtual interaction objects in HOMER.

This would practically serve as a second level of polymorphism. Hence, virtual objects are potentially characterized by a meta-polymorphic lexical nature. In Figure 1, the intermediate levels for physically instantiating a Dual virtual object are indicated.

Dialogue agents

The controlling model in the HOMER language is realized via agents; the basic PAC [2] model is adopted, however, it has been enhanced with the introduction of a dual presentation component. Agents in HOMER manage collections of virtual interaction objects, event handlers and owned agents; all managed entities live as long as the owner agent lives. Agents fall in two categories regarding the way in which they are instantiated: (i) agents instantiated *by precondition*, which can not be directly re-used, since they access either global variables or variables belonging to hierarchically higher agents; however, simple preprocessor facilities are usually applied for creating class patterns and for customizing to specific declarations when required, thus enabling a type of generality to agent classes, (ii) agents

instantiated *by parameter passing*, which are explicitly instantiated by statements and they can have an arbitrary number of parameters (either conventional variables or virtual interaction objects); parameterized agents are re-usable and can be applied to easily construct general interface components like dialogue boxes and even domain specific dialogue substructures.

```
1: agent Dbox create if (precondition)
2: agent Dbox (call parameters)
   destroy if (precondition) [
              interaction    objects,    event
              handlers,   methods,   constraints,
              monitors,    functions,    data
              structure definitions, variables,
              agents
       constructor [ stmts ]
       destructor [ stmts ]

   ]
```

Example 3: Specification of dialogue agents enabled by precondition or by parameter passing, and constructs internally to agents.

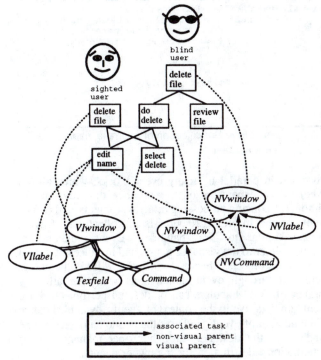

Figure 2: A simple scenario for task structure differentiation.

The skeleton of an agent specification in HOMER is indicated in Example 3. The first line shows the header of an agent instantiated by precondition, while the second a parameterized agent. Agents may optionally have a destruction precondition. Apart from objects, event handlers and children agents, the specification of arbitrary method implementations, constraints, monitors, functions, data structures and variables may take place within the definition body of agents. Also, communication events related to the application interface (e.g. creation of shared objects, receipt of messages in a channel, etc) can be also used as instantiation / destruction preconditions for agents. One way of achieving syntactic differentiation is via dialogue agents and virtual objects. Assume that task-structure differentiation is required for a particular simple task of deleting a file as shown in Figure 2. This scenario is implemented in HOMER via a single agent which owns all the objects indicated in Figure 2. Objects with a prefix *VI*, represent visual objects and those with prefix *NV* represent non-visual objects; the rest are dual objects. The designer defines the hierarchical structure in the context of the owner agent for both environments - notice that dual objects have two parents - and subsequently assigns the desired behaviour for each environments via corresponding method implementation.

Event handlers

The model of event handlers is based on Event Response Systems (ERS) [4] with some modifications and extensions: (i) flags have been substituted by arbitrary preconditions, thus adding more expressive flexibility, (ii) an event may be

```
visual eventhandler DrawArea [
    (pointDragged==true) PointerMoved [
            // arbitrary code here which
            // accesses PointerMoved.x,y
    ]
    (true) PointerMoved [
            !! ButtonPressed (PointerMoved.x,
            PointerMoved.y,
            1);
    ]
]
```

Example 4: Specification of a simple visual event handler.

processed in parallel by many event blocks, which allows orthogonally expansion of the number of event blocks without introducing processing priority problems, and (iii) an event handler in HOMER consists of collections of event blocks (similar to ERL handlers). Event handlers are attached either to the visual or non-visual instances of virtual interaction objects, hence, syntactic differentiation is facilitated through event handlers as well. An arbitrary number of event handlers can be defined per object. Since event handlers live in the context of agents, dynamic change of the number of event handlers attached to an object can be handled by the designer through proper management of instantiation / destruction for the owner agents.

The specification of a simple visual event handler is illustrated in Example 4. Two event blocks are shown for the same event class, thus demonstrating the parallelism for event processing. Also, the first event block has a simple precondition while the second event block indicates the way in which unconditional handling of events can be specified (i.e. setting `(true)` as the precondition). The statement of the second event block (`!! ButtonPressed`) concerns the generation of artificial events (i.e. local event broadcasting).

Application interface

The application interface model combines the shared space model, where arbitrary data structures can be exported, with the model of typed communication channels, where an arbitrary number of direct strongly typed links can be created for information exchange. Even though both combined models are expressively equivalent, the applicational convenience varies depending on the conversation protocol which has to be mapped in the model.

```
enum Commands=[Del,Copy,Ren,Open,Protect];
channel CommandsChannel of Commands;
structEmployee [
        string Name, Surname, Address;
        real Salary;
];
shared Employee;
```

Example 5: Specification of application interface structure.

For instance, the *object oriented* approach which is based on the information structures manipulated by the functional core is more naturally realised through the shared space approach, while the (older) *procedural* approach of treating the application as a semantic server is more easily specified by means of direct messages which carry commands and parameters. The coordination scheme is asynchronous, but synchronous behaviour can be modelled; for instance, the dialogue control can wait for a notification that an action has been completed by the functional core and vice versa. Conceptual abstraction or adaptation of the internal structures and services is facilitated. In Example 5, specification of a simple application interface structure is shown.

Lexical technology interface specification

The physical model which is provided by HOMER for technology integration is centred around the object model. However, with respect to other physical level specification

```
nonvisual        inputevent gestureSelect [
                     objectid obj;
]
visual           inputevent LocatorMoved [
                     int x,y;
]
nonvisual        outputevent Say [
                     string    speechText;
]
visual           outputevent DrawLine [
                     objectid obj;
                     int theGC, x1,y1,x2,y2;
]
```

Example 6: Specification of non-visual and visual input / output events for non-visual and visual technology integration.

models like the one in SERPENT [3] it constitutes a significant extension. All physical constructs can take either the modifier *visual* or *non-visual*. The following classes of physical constructs are supported in HOMER: (i) *interaction objects*, which can have an arbitrary number of attributes and methods; also, *inherited* attributes and methods can be specified, (ii) *input events*, which can model either user generated events or artificial events which are processed locally in the dialogue control and do not involve the technology server, and (iii) *output events*, which can model display functions (always requested in the context of specific interaction objects) and various other miscellaneous functions like resource request, setting resource parameters, requesting resource parameters, releasing resources, etc. In Example 6, specification of some classes of visual / non-visual input / output events is shown.

Dialogue specification

During the design phase of Dual User Interfaces, some common aspects of the visual and non-visual dialogues may be identified. The HOMER language enables expression of such captured syntactic similarities in a compact shared form. One good example of sub-dialogues which help demonstrate this principle are dialogue boxes and message boxes which have been specified by means of re-usable parameterized dialogue agents (see Example 7). The dialogue designer may indirectly handle notification for selection of an item in such sub-dialogues in the following manner (see also Example 8): boolean variables are passed by reference to the re-usable agent, which are set appropriately in this agent according to the selection of commands by the user; on the calling agent, the designer specifies monitors attached to these boolean variables which implement the behaviour for the selected command (e.g. Ok or Cancel).

```
agent   DualDbox (
string msg,
bool* okSelected,
bool* cancelSelected,
string* userText ) [

object VisualPopupContainer viwin();
object NonvisualPopupContainer nvwin();

#define MY_PARENTS      \
: visual parent=viwin   \
: nonvisual parent=nvwin

object DLMessage theMsg(msg) MY_PARENTS;
object DLCommand theOk() MY_PARENTS;
object DLCommand theCancel() MY_PARENTS;
object DLTextEntry theText("") MY_PARENTS;

method {theOk} Activated [
        *okSelected=true;
        destroy myagent;
]
method {theCancel} Activated [
        *theCancel=true;
        destroy myagent;
]
*userText := {theText}.Content;

        constructor   [...]
        destructor    [...]

] // agent DualDbox end
```

Example 7: Specification of a Dual dialogue box.

In Example 7, the prefix *DL* indicates Dual virtual objects. The method *Activated* is a virtual method of the *DLCommand* class, and is called whenever the corresponding visual or non-visual instance is activated (i.e. enabling shared behaviour specification for both environments). As it has been previously discussed, such correspondence between virtual methods and physical

methods is defined as part of the visual and non-visual instantiation schemes for virtual interaction objects.

```
bool delOk=false,delCancel=false;
string file;

create  DualDbox (
        "Specify the file to be deleted",
        &delOk,
        &delCancel,
        &file);

delOk : [       // called when Ok selected
        // behavior here for Ok
]
delCancel : [  // called when Cancel selected
        // behavior here for Cancel
]
```

Example 8: Specification of dialogue for handling Dual dialogue boxes.

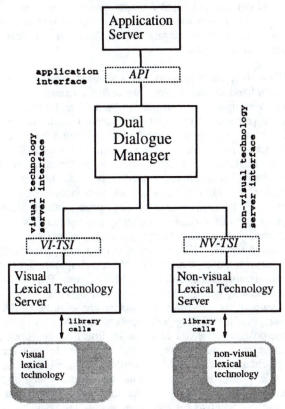

Figure 3: The Dual run-time model.

ARCHITECTURE OF THE HOMER SYSTEM

The HOMER run-time architecture is based on the *Dual run-time model* (see Figure 3) which forms an extension of the Arch model [6] for handling a Dual lexical layer. The Arch model introduces the concept of a generalised "Presentation Component" which stands a level above platforms; however, the concept of "Presentation Objects" as the only entities communicated between the dialogue control and the "Presentation Component" is not considered completely adequate. For instance, direct handling of device input events or even specific output functions which cannot be easily modelled in an object oriented fashion (e.g. the user selects a region of the screen and *copies* it as a bitmap to another location, etc) are not covered by the Arch model. The model which is followed by HOMER, and is reflected in the technology integration notational methods, incorporates explicitly input and output events.

The Application Server, the Dialogue Manager, the application interface (API) and the technology servers are implemented as separate processes. The lexical technology server interfaces (VI-TSI and NV-TSI) are not implemented as separate processes (as it is the case with the API which handles the shared space and message channels). Instead, the communication interface between the Dialogue Manager and the technology servers is realized on the basis of a general and efficient protocol (the *technology server interface protocol*) which explicitly introduces entities such as interaction objects and input / output events. The protocol implementation is provided to technology server developers by means of a class library which hides lower level communication aspects.

The HOMER system supports an integrated and powerful method for handling constraints, preconditions and monitors. Arbitrarily complex expressions can be used and automatic update propagation is efficiently managed. Constraints are uni-directional, however, since *cycle elimination* has been implemented, re-calculations are completely excluded (a cycle is eliminated exactly when it is created) and consequently bi-directional constraints can be safely modelled. Cycle elimination has been also incorporated to monitors. Constrains, monitors and preconditions may engage all types of variables (including aggregate types and pointer variables as well). The programming kernel which is supported in the HOMER language combines a plethora of C and Pascal features, while a flexible and simple model for dynamic memory allocation is provided as well. Regarding inter-component communication, even though the various components of the run-time architecture are separate UNIX processes, the HOMER system supports transparently to the dialogue designer, transfer of dynamic (linear or non-linear) data-structures either via the shared space or through the message channels. The HOMER compiler will generate code to decode / encode dynamic data structures, which are automatically monitored by the run-time system, to / from linear buffers when communication events are exported / imported. This method adds more expressive power to the notational methods for describing communication with either the application component or the technology servers.

NON-VISUAL LEXICAL TECHNOLOGY

The non-visual environment which has been developed and integrated to the HOMER system (as a non-visual lexical technology) is based on a realization of the *rooms* metaphor (see Figure 4) through speech / Braille output and keyboard input. Implementation of this metaphor has been realised by means of a simple non-visual interface development library called COMONKIT. The only container object is the Room class. A Room object may have an arbitrary number of

"children" objects, including objects of the Room class. There are six groups in which a child object may belong: *floor, ceiling, front wall, back wall, left wall* and *right wall*. In case that an object belonging to the vertical walls is of Room class, it is announced to the blind user as a "door"

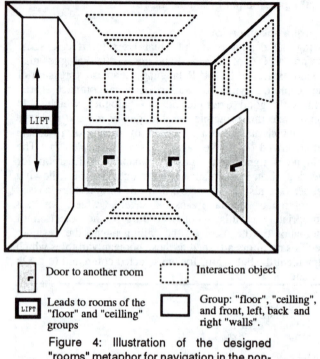

▪ Door to another room	⬚ Interaction object
LIFT Leads to rooms of the "floor" and "ceiling" groups	▭ Group: "floor", "ceiling", and front, left, back and right "walls".

Figure 4: Illustration of the designed "rooms" metaphor for navigation in the non-visual environment.

which leads to another room. Room objects belonging to the "floor" or "ceiling" groups are made accessible through the "lift", which can go down-stairs or up-stairs respectively via blind user control. Other interaction objects which have been implemented include menus, toggles (represented as on / off switches), buttons, text reviewers (represented as books), etc. Particular emphasis was given on providing efficient navigation facilities on the basis of the designed rooms concept.

Using HOMER, many Dual and dedicated non-visual prototypes of interactive applications have been built, like a payroll management system, a personal organizer and a simple "story teller" which includes textual descriptions of pictures and realizes a specifically designed scenario for non-visual picture exploration. A Dual version of the UNIX chess game is also underway. The second version of the non-visual environment, which is currently under development, combines 3D acoustics, 3D pointing and also speech input (recognition of keywords for navigation commands) in addition to the conventional keyboard (see Figure 5). An acoustic sphere illusion is provided to the blind user for navigation purposes which enables a direct perception of the structural aspects of the User Interface in contrast to the highly modal and synchronous approach of the present implementation.

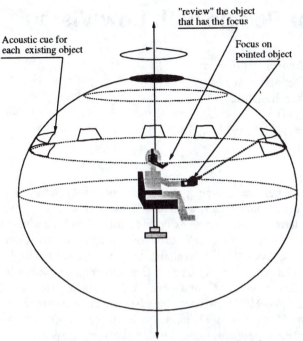

Acoustic cue for
each existing object

"review" the object
that has the focus

Focus on
pointed object

Figure 5: A 3D acoustic non-visual
environment with 3D pointing and keyword
speech recognition.

SUMMARY AND CONCLUSIONS

Existing methods for providing accessibility for blind users
to Graphical User Interfaces suffer from many theoretical
and practical problems. It is argued that the main cause of
the present situation was the lack of tools for supporting
development of User Interfaces meeting the specific blind
user needs. In this context, the concept of Dual User
Interfaces was defined as the most appropriate basis for
integrating blind and sighted user groups. A UIMS has been
developed, called HOMER, which supports the development
of Dual User Interfaces. The HOMER system is platform
independent and it provides methods for integrating visual
and non-visual lexical technologies.

The lack of a non-visual toolkit posed the necessity of
explicitly developing one for subsequent integration in
HOMER. The COMONKIT library was developed, on the
basis of the designed "rooms" metaphor, with particular
emphasis on providing efficient navigation facilities. A
primary requirement for the development of the HOMER
system was the efficient support for interface development
as such, in order to enable high quality Dual, visual and
non-visual User Interfaces to be assembled.

Porting of the HOMER system to MS-WINDOWS™ and
integration of the InterViews library has been planned. Also,
another group will be studying a more effective metaphor
for non-visual environment than "rooms", and will build a
COMONKIT server for handling multiple clients for the
non-visual environment; this will include the implementation
of new efficient non-visual navigator handling all different
clients (the analogy of a visual window manager). The

HOMER system provides an effective framework for Dual
User Interface construction, it fills the gap due to the lack of
tools for non-visual User Interfaces and provides a ground
for further developments and experimentations.

ACKNOWLEDGEMENTS

This work has been partially funded by the TIDE
Programme of the Commission of European Union (DG
XIII), under the projects GUIB (TP 103) and GUIB-II (TP
215). The partners of the GUIB consortium are: IROE-CNR,
Italy; Institute of Computer Science-FORTH, Greece; Vrije
Universiteit Brussel, Belgium; Department of Computer
Science-FUB, Germany; Institute of Telecommunications-
TUB, Germany; IFI, University of Stuttgart, Germany; VTT,
Finland; RNIB, England; F. H. Papenmeier GmbH&Co, KG,
Germany;

REFERENCES

1. Card, S., Robertson, G., and Mackinlay, J.   The
information visualizer, an information workspace. In
*Proceedings of the CHI'91 Conference on Human Factors
in Computing Systems* (New Orleans, Louisiana, April 27-
May 2, 1991). ACM, New York, 1991, 181-188

2. Coutaz, J. Architecture Models for Interactive Software
: failures and trends. In *Engineering for Human-Computer
Interaction*. G. Cockton, Ed. North-Holland, 1990, 137-151

3. CMU/SEI-91-UG-8. *Guide to Adding Toolkits*. Serpent
User's Guide, May 1991

4. Hill, R,D. Supporting Concurrency, Communication, and
Synchronization in Human-Computer Interaction - The
Sassafras UIMS. *ACM Trans. Gr. 5, 3* (July 1986), 179-210

5. Mynatt, E. D., and Weber, G. *Nonvisual Presentation of
Graphical User Interfaces: Contrasting Two Approaches*. In
*Proceedings of the CHI'94 Conference on Human Factors
in Computing Systems* (Boston, Massachusetts, April 24-28,
1994). ACM, New York, 1994, 166-172

6. The UIMS tool developers workshop. A Metamodel for
the run-time architecture of interactive systems. *SIGCHI Bull
24, 1* (January 1992), 32-37.

Improving GUI Accessibility for People with Low Vision

Richard L. Kline and Ephraim P. Glinert
Computer Science Department
Rensselaer Polytechnic Institute
Troy, NY 12180
E–mail: {kliner|glinert}@cs.rpi.edu

ABSTRACT

We present UnWindows V1, a set of tools designed to assist low vision users of X Windows in effectively accomplishing two mundane yet critical interaction tasks: selectively magnifying areas of the screen so that the contents can be seen comfortably, and keeping track of the location of the mouse pointer. We describe our software from both the end user's and implementor's points of view, with particular emphasis on issues related to screen magnification techniques. We conclude with details regarding software availability and plans for future extensions.

KEYWORDS: workstation interfaces, assistive technology, low vision, screen magnification, X Window System

INTRODUCTION

The move towards graphical user interfaces is widely regarded as an advance in human–computer interaction. However, the abandonment of the old–fashioned text based TTY interface presents new challenges to those computer users who are visually impaired [17]. How can we correct this problem?

Our ultimate goal is clear: *To allow the user full and equal access, as if he/she enjoyed normal vision, to any tool or application that he/she may choose to run, whether the output is textual or graphical in nature.*

Much work has been done toward this end for users who are blind [5, 18]. Our own research is intended to assist people who have low vision but who are not blind. To this end, we have attempted to improve the usability of existing graphical interfaces, rather than devise new or alternative interfaces [11, 13]. We have concentrated our efforts on two critical aspects of interaction: selectively magnifying areas of interest so that the contents can be seen comfortably; and finding and controlling the location of the mouse pointer.

In the personal computer market, products that address these issues are now widely available. These span a broad gamut, in terms of both complexity and cost. There are student projects such as Vener's Magnex text editor for the Commodore AMIGA [19] (available from the second author by special request). The CloseView program developed by Berkeley Systems Design is included as a standard feature of the Apple Macintosh operating system; a more elaborate version of this software is marketed by the developers for several hundred dollars under the name inLarge. Specially designed software and hardware produced by companies such as TeleSensory Systems provide the most power, but typically sell for thousands of dollars.

In the comparatively small workstation market, where MIT's X Window System commonly provides the graphics interface engine, there has until now been a paucity of viable solutions for the low vision user community. Old ASCII based aids (such as the large font virtual terminal to UNIX developed by the second author over a decade ago [9]) can sometimes still be run within a window, but are otherwise useless. Because many professionals use workstations rather than personal computers in their work, we chose to implement our collection of programs, known as UnWindows V1, for Sun SPARCstations running X.

A DYNAMIC MAGNIFIER

Magnification is one method commonly employed to help low vision users deal with the small type fonts, illustrations and icons present in much of today's printed media and computer displays. Some features of X reduce, but do not eliminate, the need for a separate magnifier tool. Many text–only X applications allow the user to override the default font settings, but the range of alternative sizes is limited and dependent upon the fonts available on a given display. A more significant problem is that many applications make use of graphical elements as well as text, and it is rare to find an application that will allow the user to specify the displayed size of these nontextual elements. Furthermore, an application's default window might be so large that it would not fit on a display if magnified.

In designing the UnWindows dynamag screen magnification program, we considered two typical uses for physical magnifying glasses. To read the fine print of a legal contract

Figure 1: The control panel for the `dynamag` application, shown much smaller than actual size.

or automobile advertisement, one places the document on a table or other flat surface and *moves* the glass about as needed to inspect different areas of the document. To assemble or repair a small device, on the other hand, an electronics technician positions a glass in a *fixed* location where it will provide the best view of the work area, and then works on the device directly while looking at the image produced by the glass.

In both cases the user can examine and work in areas not under the magnifier without any special effort, and can also reposition the glass, temporarily move it out of the way, or even peer around it to gaze directly at the object of interest, as the need arises. The ability to correlate the magnified view with the reality of the object(s) being viewed is what allows the use of a magnifying glass to become effortless for most of us after a very short learning period. What are the ramifications of our observations to the design of a virtual counterpart to the physical glass?

In the case of a hand–held magnifying glass, the physical separation of the glass from the surface being viewed makes it trivial to keep track of one's place on the page. Where a computer display is concerned, however, the problem of retaining *a sense of global context* manifests itself, because the magnifying glass becomes part of the screen. This problem is most acute in systems which use the entire screen area to draw an enlarged image of a portion of the display. There is no good way to look "around" the magnified view to the unmagnified image "beneath" it, although some intriguing initial efforts have recently been directed at this problem [10].

In theory, this difficulty might be alleviated by imparting to the magnifier the ability to automatically reposition itself in reaction to screen events (e.g., user typing or process output). But on a busy X display, the contents of several windows can change in rapid succession. A naive magnifier that attempted to follow all screen activity would jump around (thrash) hopelessly, imparting nothing but confusion to the user.

Providing global screen context while devoting significant screen real estate to a magnification window are conflicting goals that can only be resolved through compromise. Our solution is to relegate the UnWindows `dynamag` magnifier to a window on the screen, and to support two distinct modes of operation, designed to emulate the real–world examples given above:

- Mobile mode. The magnified area follows the mouse pointer around the screen, dynamically showing whatever lies beneath it. The user may choose:

  a. To have the magnified image be sticky to the mouse pointer (travel around with it).

  b. To have the magnified image remain in a fixed location, yet continuously reflect the current position of the mouse pointer.

- Anchored mode. The magnifier is "anchored" to (associated with) a fixed screen area, then positioned in some (other) arbitrary place and set to continually show an updated view of whatever is in the previously specified area, no matter where the mouse pointer may happen to be.

Note that two of these options break the customary link between the portion of the screen of interest and the position of the magnified image. By affording control over the portion of the screen which is *obscured* as well as that which is *magnified*, we allow the user to minimize the loss of global context on an individual basis and in response to changing circumstances. Some of these ideas have also been recently and independently implemented as "portals" in Bederson and Hollan's Pad++ system [1].

DYNAMAG FEATURES

The `dynamag` program's interface allows the user to easily modify:

- The size of the magnified area.

- The degree of magnification provided.

- The image refresh rate.

The magnifier window is resized using whatever techniques are provided by the window manager being run by the user. The other preferences are changed with the help of a pop up window; cf. Figure 1. New settings are immediately reflected by `dynamag` as they are entered. As with the other UnWindows programs, preferences are stored in a file in each user's home directory.

Let us now examine each of `dynamag`'s two modes of operation in more detail.

Mobile Mode

The screen area which is magnified is not centered around the pointer's location, as one might initially expect. Instead, the area immediately above and to the right of the pointer is magnified. This is done in an effort to reduce the loss of *local* context that results from the magnifier obscuring that part of the display immediately surrounding the pointer. For example, when using our method for the common task of reading a paragraph of text, moving the mouse pointer to the beginning of a new line makes that line (and possibly lines above it as well) visible within the magnifier, while the line immediately below can also be seen (unmagnified).

Because dynamag's screen window, if sticky to the mouse pointer, obscures, once drawn, the very area on the screen which it is magnifying, this window must be removed whenever dynamag requires a new screen image and then redisplayed. This refresh process can prevent dynamag from performing as smoothly as one might wish. The user is therefore given the option of having the application window which displays the enlarged image remain stationary, although the area of the screen to be magnified still is automatically chosen based on the current location of the mouse pointer. When functioning in this way, dynamag's mobile mode performs in a manner that is noticeably smoother (in the absence of frequent window creation and deletion), although we lose the direct analogy with a physical magnifying glass.

Anchored Mode

When the user exits dynamag's mobile mode, the program automatically notes the last screen area that was selected for enlargement. In this way, the magnifier becomes anchored to that part of the display. The dynamag window itself can now be moved to any (other) location on the screen, and the magnified area remains the same. Interesting results are obtained when the dynamag window itself is moved into the area currently being magnified.[1]

Once the magnification window is positioned where desired, the user can interact with the screen as usual, performing work within the window(s) of interest while watching the magnified image being presented in a different location. Interaction with areas of the screen not being magnified does not affect dynamag's operation in any way.

A typical use for this mode of operation is illustrated in Figure 2. The magnification area has been selected (by means of the mobile mode described above) to include the bottom several lines of an xterm window. Once the magnifier has been properly anchored, the dynamag window is moved by the user to a convenient location elsewhere on the screen. With the update interval set to a small value such as 0.25 seconds, the user can type and read from the dynamag window while interacting with the application window.

DYNAMAG IMPLEMENTATION

The dynamag magnifier works by directly polling the X screen, using the Xlib–level XGetImage() routine to find out what is currently being displayed. Obtaining display information at this low level allows dynamag to magnify any image on the screen. We borrowed this method from xmag, a sample application distributed with the X Window System. Each individual pixel in the captured screen image is redrawn within the magnifier's window as a square whose sides are

from 2 to 9 pixels in length, depending on the magnification level selected by the user. At progressively higher magnifications this approach leads to text and graphics that look somewhat "blocky" and unaesthetic, but we believe this is of less importance to the target user community than speed of performance, which would have to be sacrificed if some form of smoothing algorithm were added to the drawing process.

We experimented with moving the entire dynamag application around on the screen when the program was in mobile mode, but found the performance to be insufficient, in that window movement sometimes lagged noticeably behind that of the pointer. Instead, we hide the application window from the screen when entering this mode, using the low level Xlib window management routines for this purpose. The resulting mix of function calls to the two libraries proved to be fragile, and some experimentation was required to arrive at an implementation that correctly processes all of the incoming X events.

The automatic refresh of an anchored magnification window has been implemented through the use of the timeout facility provided by the Xt library calls XtAppAddTimeOut() and XtRemoveTimeOut(). As originally coded, we often noticed a performance lag in the response of dynamag's command buttons when the selected magnification area was large. We determined that interaction between the timeout function calls and processing of other X events was the cause. To solve the problem, we modified our event handler to prevent the backlog of events that we were seeing.

Even with these issues resolved, the sheer amount of pixel data continuously transferred between dynamag and the display server remains the largest hindrance to program performance when the magnification area becomes large. Adding code to take advantage of the shared memory extension to X would improve the program's speed, but only if the display belongs to the workstation running dynamag. At the time of this writing, the Disability Action Committee for X (DACX) is working to complete a screen magnification extension designed specifically to facilitate the writing of screen magnification programs such as dynamag. We hope to rewrite portions of dynamag to take advantage of this new extension when it becomes available.

RELATED MAGNIFICATION SYSTEMS

There is an important distinction between screen magnifiers such as dynamag and special purpose document viewing programs which have built–in "magnifiers" of their own. For example, xdvi, which displays files in the format produced by TEX and LATEX, has a magnification feature which operates very smoothly. Because document viewers such as xdvi have a complete, static representation of the entire image to be displayed when execution starts, some or all of the image can be precomputed. In contrast to this, dynamag must frequently obtain an updated snapshot of the screen display, which can change nearly continuously.

---

[1] The effect is quite similar to that obtained by pointing a video camera at a monitor which is displaying the output of that same camera.

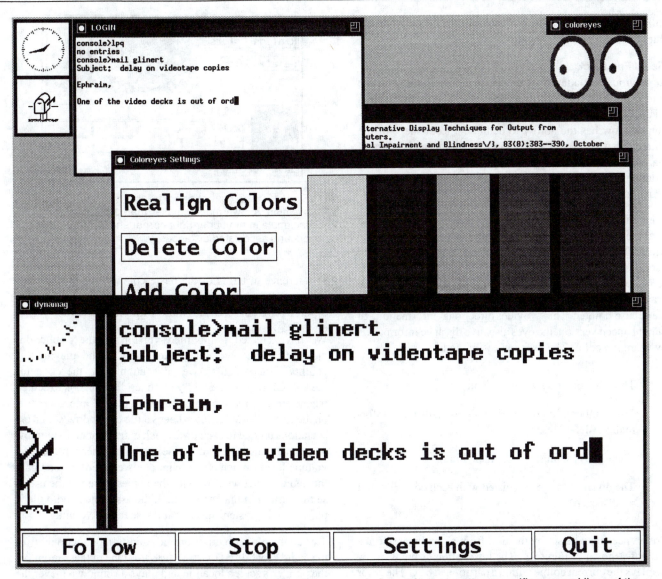

Figure 2: A screen shot of UnWindows in use. The dynamag window, at bottom center, magnifies several lines of the xterm window at top left as well as parts of nearby icons.

Several teams investigating data visualization have recently explored the idea of displaying all of a large graph or document, so as to allow the user to retain global context, while showing full (enlarged) detail for one or several portions of the data being viewed [15, 16]. This is accomplished by shrinking and distorting those parts of the information that are not currently being magnified. It must be noted, however, that this work is again aimed at the display of a single, static document, while a workstation display is very rarely static. We must support an interaction paradigm in which the user may have to refer frequently to areas of the screen away from the current task focus (i.e., the area under magnification) when information on the display changes.

Chin–Purcell's puff program [4] is the only system other than UnWindows of which we are aware that is designed to provide general screen magnification under X. The program operates in a mode very similar to our mobile mode with a stationary display window. It can also be set up to reposition the magnification area automatically in reaction to screen output. The mechanism requires that individual applications register themselves with puff when starting up. This allows puff to ignore screen changes deemed unimportant by the user, such as the redrawing of a clock every minute, but it places the burden on the user to make sure that he/she registers every "important" application and pop–up window that might appear on the screen.

FINDING THE MOUSE POINTER

Knowing where the mouse pointer is situated on the screen is essential to interacting with today's graphical computer interfaces. Yet even users with normal vision often have difficulty in seeing the pointer on a display populated with many windows and icons! UnWindows V1 provides a set of tools that utilize both the visual and aural sensory modalities to convey clues as to the pointer's location.

Visual Feedback

An obvious first attempt at making the pointer easier to see is to make it larger. However, under the X Window System this is difficult to achieve, because each individual application window has the ability to define the local shape of the pointer (that is, how it should appear within the borders of that window). Some of these applications allow the user to redefine the pointer shape (referred to in X as the *cursor*), but many do not.

Our solution is an external visual indicator, a dynamic icon in a fixed location, to assist in highlighting the pointer's position. The UnWindows coloreyes utility is a modified version of the xeyes program that comes bundled with the X Window System software from MIT. The xeyes program draws a stylized pair of eyes which continually gaze toward the mouse pointer. This provides <u>directional</u> information. In our enhanced version, the eyes also give <u>distance</u> information by changing color. The user can easily modify:

- The position and size of the icon.

- The number of colors used to represent distance information (default = 3).

- Their hues (default = red, green, blue).

- The distance range associated with each color (default = 33% each).

Preferences may be set by the user from a graphical settings window, shown in Figure 3. The top portion of the window displays the current colors within a partitioned box. The color representing the area closest to the eyes is situated in the left-most patch of the box. The relative width of each color patch indicates the percentage of relative screen distance that is represented by that color.

The bottom area of the settings window is an RGB color mixer, which allows the user to change existing colors and to set colors used for new partitions. The three sliders represent the intensities of red, green, and blue; the color resulting from the mixture of these values is displayed in the box to their left. Allowing the user to change colors enables him or her to satisfy personal aesthetic preferences. More importantly, however, this feature is essential for users who suffer from various forms of color blindness, or for whom vision is highly dependent upon the contrast between foreground and background.

Aural Feedback

While coloreyes provides pointer location information in a geographic sense (i.e., direction and distance from a fixed point), sound cues are used to provide an alternative frame

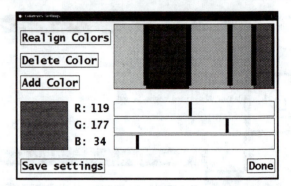

Figure 3: The control panel for the coloreyes application, shown much smaller than actual size. The rectangle in the upper right corner shows current colors and associated distances.

of reference in terms of the basic components of the interface: windows, icons, and the display borders. A collection of three programs provides these audio functions.

We have created a modified version of the public domain twm window manager to add a sound playing capability. In particular, we have augmented the code for the HandleEnterNotify() function, which is executed whenever the mouse pointer has been moved into a window on the screen. UnWindows keeps a file of window/sound associations unique to each user. Each entry contains a window name, the name of an audio file, and a number representing volume level. When a new window is entered, we compare the name of that window with the names stored in the user's settings file. If a match[2] is found, the associated audio file is played on the system speaker at the specified volume.

Two companion programs provide additional functionality to the audio capabilities of UnWindows. One of these monitors the pointer's screen location and plays a sound whenever the pointer moves within a threshold (default = 5 pixels) of the edge of the screen. Each of the four screen edges can be assigned a different sound and/or volume.

The second utility allows the user to create and update his/her personal window/sound association list. Sounds may be previewed, and the desired volume setting for each modified at will. Since recording levels vary from one audio file to another, it is necessary to provide individual playback volume control for each sound. A sample window/sound association list is shown in Figure 4, which shows a portion of the interface for the list maintenance utility program.

An initial question in the design of these tools was the source of sounds to be played. Previous work by Gaver [6] and by Blattner *et al.* [3] in the use of sound allowed their systems

---

[2]When performing name comparisons, we do not check for exact matches. Rather, a window name is considered to match a name in the association list as long as its first characters exactly match an entry in the list. This allows the desired matching to occur even for applications that change their window titles.

Add Sound	Delete Sound	Save Sounds	Quit
0	xterm	train.au	25
1	window	dialtone.au	25
2	xload	gong.au	30
3	oclock	rooster.au	30
4	coloreyes	doorbell.au	30
5	Top_Border	bong.au	60
6	Bottom_Border	bong.au	60
7	Left_Border	bong.au	30
8	Right_Border	bong.au	30
9	dynamag	drip.au	25
10	emacs:	cowbell.au	75
11	xv	ooohh-aahhh.au	35

Figure 4: The window/sound association list maintenance program.

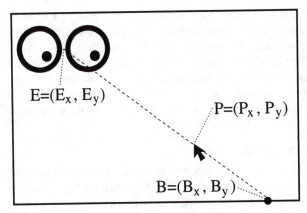

Figure 5: The formula used to determine the color of the eyes in the coloreyes program.

to assign unique sounds to the complete set of objects and actions they wished to identify aurally. In UnWindows V1, however, the set of sounds we needed had to correspond to the application windows which might appear on an X display – a large and ever–growing set. Thus, we feel that our method of letting the user select sounds to represent windows is the most appropriate mechanism for our case.

We do not provide a specific utility within UnWindows V1 to allow the user to record new sounds. However, a number of programs are available for this purpose, including AudioTool, one of the applications which is included with Sun SPARCstations. Although we did not pursue it, the addition of speech synthesis hardware would add further flexibility to the system. Currently, when a window is selected whose name is not in the user's window/sound association list, no sound is played. A speech synthesizer could pronounce the names of those windows not associated with any sound. Unfortunately, speech synthesis is neither common nor inexpensive on workstations running X.

Implementation Issues
Our changes to the original xeyes program consist entirely of additions to the code. The method used to compute the pupils' positions, for instance, remains unchanged. On the other hand, the color in which the pupils and the eyes' outlines is rendered can no longer be set at the beginning of program execution, but must instead be continuously recalculated along with pupil positioning.

The method of computing the color used to render the eyes of coloreyes is illustrated in Figure 5. An imaginary line is drawn from the center of the coloreyes icon E, through the current pointer location P, until it intersects one of the borders of the screen at point B. The resulting ratio of distances is computed as $R = \frac{|P-E|}{|B-E|}$, which will have a value between zero and one. The user's current color settings are then examined and the appropriate hue chosen.

To impart audio output to UnWindows V1, we made some minor modifications to the play program included with the

SunOS operating system in the /usr/demo/SOUND directory, incorporating it into our modified twm. All audio files to be played by UnWindows must be encoded in the Sun standard 8 bit μ–law format which the play program recognizes. Audio files in other formats can be converted into 8 bit μ–law by programs such as the public domain SOX, written by Norskog et al [14].

Unlike the other UnWindows V1 programs, those which play audio files are compatible only with Sun SPARCstations at this time. For the future, we can hope for an audio standard among workstation manufacturers, comparable to X as a graphics standard. For now, however, users wishing to port UnWindows to a new architecture must modify the sound generation function appropriately.

USER FEEDBACK
UnWindows V1 has been released to teams within several organizations, including IBM, Sun Microsystems, DACX, the University of Washington's Adaptive Technology Laboratory, and RPI. In addition to conducting informal small scale user testing with visually impaired individuals (the second author among them), we distributed a user survey to all who requested copies of UnWindows directly from us. Overall, the comments received have been encouraging and positive.

The coloreyes program consistently received positive comments from both visually impaired and normal–sighted users. One visually impaired user commented that while at first he thought of coloreyes as "nothing more than a nifty frill," he found after continued use that it was a natural, "almost subconscious" aid in locating the screen pointer. This same user reported that he found it too distracting to have all of the windows generating sounds; he configured his system to make a "non-intrusive click" only when the pointer approached the screen borders.

Reaction to dynamag has been mixed. Users vary in their preference of magnification mode. Some have noted that, as discussed above, the program's performance can be less than desirable under certain circumstances, and have suggested the addition of mouse and/or keyboard accelerators (similar to those in `puff`) which would allow changes to `dynamag`'s behavior without having to manipulate the program's interface windows directly. Such a feature would have to be designed with great care, however, so as to eliminate (or at least minimize) conflicts with applications that use the same accelerators for different functions.

SYSTEM AVAILABILITY

`UnWindows V1` is freely available via anonymous `ftp`; for more information, please contact the first author. The tools are written in C and utilize only the `Xt` and `Xaw` toolkits provided with the standard X Window System release. The utilities are independent of one another, so that the user can choose to run any or all of them, as required. A detailed exposition of functionality from the user's viewpoint may be found elsewhere [12].

Our programs were developed for Sun SPARCstations running the SunOS operating system. However, users wishing to compile `UnWindows` for other systems should need to modify only those portions of the code that generate audio output. All of the `UnWindows` utilities have been tested under `mwm` and `olwm`, two popular alternative window managers (except our modified `twm`, which is itself a window manager). In addition, users of `UnWindows V1` have compiled `dynamag` and `coloreyes` to run on DEC (both MIPS and Alpha based) and IBM workstations, without modification.

PLANNED ENHANCEMENTS

`UnWindows` continues to evolve. V2, currently under development by the authors and G. Bowden Wise, will help users who are blind, who are hearing impaired, and more [8]. The ultimate objective is to develop transparent interface software which will afford access to (certain categories of) applications without modification. Is this achievable? With the right technology, we believe so. Indeed, our work on `UnWindows` V2 is not aimed solely at people with disabilities. Rather, we are investigating a new approach to multimodal systems in general that we hope will prove broadly applicable.

We seek to develop a new multimedia interaction technology, in which information is not merely regurgitated "as is" but rather is first processed at a high level of abstraction and then distributed among the sensory modalities as required. The hypothesis is that it is impractical for designers of any but the simplest multimedia interfaces to rigidly allocate the output of their systems to specific human sensory channels, because what constitutes acceptable output may depend upon factors which cannot be known when the code is written (e.g., the extra–machine environment, the need to avoid sensory overload due to other applications running concurrently, and of course the need to accommodate a disability).

These observations have led Glinert and Blattner to propose a new class of object in the interface called the *metawidget*

[2, 7]. These abstractions of the widgets with which we are familiar consist of clusters of alternative representations for some information, along with built–in method(s) for selecting among them. The selection methods, as well as the representations themselves, may be time dependent. The metawidget run time environment maintains data on currently active widgets, computes the total cognitive load (according to system specified criteria) to detect overloading, and then posts and/or modifies representations as required. The technology supports a layered approach to multimodal interface construction: a visual or aural toolkit is used to represent information within a modality, while metawidgets constitute the higher level building blocks across modalities.

Properly designing a metawidget's palette of representations, and the mechanism for selecting among them, will clearly be very tricky. Although many open questions remain, we nevertheless hope to have a prototype of `UnWindows V2` that embodies metawidget technology available for distribution and preliminary user testing later this year. The implementation is being carried out in C++ on a platform that consists of an IBM PC with enhanced sound output capabilities running Microsoft Windows. For additional information, please contact the second author.

ACKNOWLEDGEMENTS

This research was supported, in part, by the National Science Foundation under contracts CDA–9015249, CDA–9214887, CDA–9214892 and IRI–9213823.

An early version of `UnWindows` was designed and implemented by Gary Ormsby, who is now with IBM in Austin, Texas.

REFERENCES

1. B. B. Bederson and J. D. Hollan. Pad++: A Zooming Graphical Interface for Exploring Alternate Interface Physics. In *Proc. 7th Annual Symposium on User Interface Software and Technology (UIST'94)*, Marina del Rey, November 2-4, 1994, pages 39–48. ACM Press.

2. M. M. Blattner, E. P. Glinert, J. A. Jorge, and G. R. Ormsby. Metawidgets: Towards a Theory of Multimodal Interface Design. In *Proc. COMPSAC'92*, Chicago, September 22-25, 1992, pages 115–120. IEEE Computer Society Press.

3. M. M. Blattner, D. A. Sumikawa, and R. M. Greenberg. Earcons and Icons: Their Structure and Common Design Principles. *Human–Computer Interaction*, 4(1):11–44, 1989.

4. K. Chin–Purcell. `Puff` computer software. Available (as of this writing) via anonymous `ftp` from `ftp.arc.umn.edu`.

5. W. K. Edwards and E. D. Mynatt. An Architecture for Transforming Graphical Interfaces. In *Proc. 7th Annual*

Symposium on User Interface Software and Technology (UIST'94), Marina del Rey, November 2-4, 1994, pages 39–48. ACM Press.

6. W. W. Gaver. The SonicFinder: An Interface That Uses Auditory Icons. *Human–Computer Interaction*, 4(1):67–94, 1989.

7. E. P. Glinert and M. M. Blattner. Programming the Multimodal Interface. In *Proc. 1st ACM Int. Conf. on Multimedia (MULTIMEDIA'93)*, Anaheim, August 2–6, 1993, pages 189–197. ACM Press.

8. E. P. Glinert, R. L. Kline, G. R. Ormsby, and G. B. Wise. UnWindows: Bringing Multimedia Computing to Users with Disabilities. In *Proc. IISF/ACMJ International Symposium on Computers as Our Better Partners*, Tokyo, March 7–9, 1994, pages 34–42. World Scientific.

9. E. P. Glinert and R. E. Ladner. A Large–Font Virtual Terminal Interface: A Software Prosthesis for the Visually Impaired. *Communications of the ACM*, 27(6):567–572, June 1984.

10. H. Lieberman. Powers of Ten Thousand: Navigating in Large Information Spaces. In *Proc. 7th Annual Symposium on User Interface Software and Technology (UIST'94)*, Marina del Rey, November 2-4, 1994, pages 15–16. ACM Press.

11. E. P. Glinert and B. W. York. Computers and People with Disabilities. *Communications of the ACM*, 35(5):32–35, May 1992.

12. R. L. Kline and E. P. Glinert. X Windows Tools for Low Vision Users. *SIGCAPH Newsletter*, Number 49, pages 1–5, March 1994.

13. M. Krell. LVRS: The Low Vision Research System. In *Proc. First ACM Conf. on Assistive Technologies (ASSETS'94)*, Marina del Rey, October 31-November 1, 1994, pages 136–140. ACM Press.

14. L. Norskog *et al.* SOX (Sound Exchange) computer software. Available (as of this writing) via anonymous ftp from ftp.cwi.nl and other sites.

15. G. G. Robertson and J. D. Mackinlay. The Document Lens. In *Proc. 6th Annual Symposium on User Interface Software and Technology (UIST'93)*, Atlanta, November 3-5, 1993, pages 101–108.

16. M. Sarkar, S. S. Snibbe, O. J. Tversky, and S. P. Reiss. Stretching the Rubber Sheet: A Metaphor for Viewing Large Layouts on Small Screens. In *Proc. 6th Annual Symposium on User Interface Software and Technology (UIST'93)*, Atlanta, November 3-5, 1993, pages 81–91.

17. G. C. Vanderheiden. Nonvisual Alternative Display Techniques for Output from Graphics–Based Computers. *J. Visual Impairment and Blindness*, 83(8):383–390, October 1989.

18. G. C. Vanderheiden, W. Boyd, J. H. Mendenhall, and K. Ford. Development of a Multisensory, Nonvisual Interface to Computers for Blind Users. In *Proc. 35th Annual Meeting of the Human Factors Society*, pages 315–318, 1991.

19. A. R. Vener and E. P. Glinert. MAGNEX: A Text Editor for the Visually Impaired. In *Proc. 16th Annual ACM Computer Science Conference*, Atlanta, February 23-25, 1988, pages 402–407.

Collaborative Tools and the Practicalities of Professional Work at the International Monetary Fund

Richard Harper[1] *& Abigail Sellen*[1,2]

[1]Rank Xerox Research Centre (EuroPARC)
61 Regent St.
Cambridge, CB2 1AB, U.K.
<surname>@europarc.xerox.com

[2]MRC Applied Psychology Unit,
15 Chaucer Rd.
Cambridge, CB2 2EF, U.K.

ABSTRACT

We show how an ethnographic examination of the International Monetary Fund in Washington, D.C. has implications for the design of tools to support collaborative work. First, it reports how information that requires a high degree of professional judgement in its production is unsuited for most current groupware tools. This is contrasted with the shareability of information which can 'stand-alone'. Second, it reports how effective re-use of documents will necessarily involve paper, or 'paper-like' equivalents. Both issues emphasise the need to take into account social processes in the sharing of certain kinds of information.

KEYWORDS

CSCW, work practice, ethnography, paper documents, groupware, professional work, International Monetary Fund

INTRODUCTION

Rank Xerox's Research Centre in Cambridge has a long standing history of examining work settings for the purpose of reasoning about the design of tools and technology in support of collaborative work. These tools and artefacts range from sophisticated, computerised 'groupware' to the more mundane use of paper documents.

One set of these studies has looked at the implications of electronic replacement of paper documents [12]. Others take a broader organisational perspective focusing on such things as the organisational context of office equipment design [1], the willingness of professional staff in research laboratories to allow technology to alter their hierarchical working relations [7], and the social factors underpinning the introduction of a network and associated technologies in the British civil service [2].

This paper reports on an ethnographic study of the organisational practices of professional workers at the International Monetary Fund (IMF). One goal of the study was to understand the role of tools and technology in their current work practices in order to understand what changes might be brought about by introducing new kinds of tools and technologies. We believe that these findings have more general implications for the design of technologies in the support of collaborative work. As will become clear, the collaborative tools we will discuss range from tools and artefacts (including paper documents) which support interaction in face-to-face meetings, to those which are intended to provide access to shared information amongst individuals who are not co-present.

The Nature of Professional Work

Professional workers can be contrasted with secretaries, clerical workers, administrators and technicians, amongst others. They are people who are paid to organise their own work, make judgements and valuations, and who maintain an element of creative control over their own projects. They are of interest to us for at least two reasons.

First, there is the importance of professional workers to organisational activity. It is professionals who are key to decision-making, who have central roles in information production and use, and whose activities are influential in organisational effectiveness.

Second, despite their importance, very little is known about professional workers from a systems design perspective. What is known varies considerably in depth and quality. There is for example, Kidd's remarkable 'The Marks are on the Knowledge Worker' [10] (where knowledge worker is an operationalised term for professional). There, the claim is made that it is the process of making notes and jottings that is key to professional work, rather than re-use of those notes afterwards. This has obvious implications for a whole range of computer devices designed to support or facilitate note-taking. Kidd's study may be contrasted with one which observes that workers 'stack and pile documents on their desks' [13]. In between, one finds the view expressed by the likes of Zuboff [18] and Drucker [4] that manual work is being entirely replaced by professional work. Their claim derives from the belief that information technology enables everyone to access information that they can then process, evaluate and act upon in skilled, that is to say, professional ways. This somewhat gnostic hope is as delightful to contemplate as it is, to date, empirically unverified.

Taken as a whole, it is clear that the work practices of professionals need much more investigation. Theoretical developments from other disciplines, most especially sociology, need to be empirically corroborated, and a great range of empirical material needs to be brought to bear on system design issues.

Needless to say there are many places in which one can examine professional work. We chose the IMF for several reasons: First, nearly a third of its staff have a professional role. Second, these staff are involved in producing highly complex and analytical reports. This work is quintessentially professional. Third, the IMF was willing to allow researchers to examine all aspects of professional work, including the most confidential. It was also willing to allow these examinations over extensive periods of time, enabling the researchers to better understand the nature of professional activity. Thus, in short, the IMF provided an opportunity for a thorough ethnographic examination of professional work. We shall say more about this in the description of the methodological approach.

The Organisational Setting

The IMF, based in Washington D.C., is a financial 'club' whose members consist of most of the countries of the world. Member countries contribute to a pool of resources which can then be used to provide low interest, multi-currency loans should a member find itself facing balance of payments problems.

The IMF has some 3,000 staff, of which 900 are professional economists. These economists analyse economic policies and developments — especially in the macroeconomic arena. They have particular interest in the circumstances surrounding the emergence of financial imbalances (including those that lead to a balance of payments crisis), the policies to overcome such imbalances, and the corrective policy criteria for making loans. This involves going on 'missions' to the country in question. The resulting assessments and criteria of member countries are contained in documents called 'staff reports' which are used by the organisation's executive board for its decision-making.

Methodological Approach

The ethnographic research, carried out by Richard Harper, is ongoing. The first stage consisted of six months field work. The purpose of the field work was to understand what ethnomethodologists, following Garfinkel [5], call 'practical reasoning'. Understanding of this, in turn, enables specification of what may be called the 'logics' of organisational action. In this first stage, the concern was the practical reasoning of the IMF's economists: those skills, methods, techniques, and rules of thumb that

enable IMF economists to produce documents which the organisation itself views as adequate. Because these economists are extensively supported in their activities, the study also included examination of the work and practical reasoning of associated administrative, clerical, and support staff.

The field work centred around the 'life cycle' of staff reports, from the first draft (what is called the 'briefing paper' prepared before a mission commences), through the mission process itself, to the post mission review, and then to translation, printing, and circulation processes. This was accomplished by:

(1) Following a hypothetical staff report around the organisation and interviewing parties that would be involved in its life cycle. Interviews were conducted with desk officers and chiefs who author staff reports, with secretarial and research assistant staff who help in the composition of staff reports, with participants in the review process, including junior economists, and with Front Office chiefs and senior managers (including the deputy managing director). Staff involved in the post authoring and review stages were also interviewed, including the clerical staff who issue and release staff reports once they have been 'cleared', with those who copy and print staff reports, with translators, and finally with archivists.

In all, 138 personnel, including 90 economists were interviewed. These interviews were informal, but consisted of a systematic process of questioning and clarification, whereby the field worker gradually developed a picture of what 'practical reasoning' consisted of in any particular job.

(2) Observing an IMF 'mission' and its allied document production practices. The field worker observed meetings between the mission team before the mission commenced, between the member authorities and the team during the mission, and observed post mission meetings. All related documentation was made available to the field worker. Interviews were also undertaken whenever possible in which observed parties were asked to explain what they were about.

(3) Subsequent to the field work, a set of descriptions were generated and circulated around the IMF. Written comments were gathered and discussed with a number of key 'informants', ranging from senior staff, through to junior economists. A final report was made available to the Fund in 1995.

The results of this research are much more extensive than can be adequately covered here. However, we will focus on two interrelated sets of findings which concern the use of tools in professional collaborative work. We believe these findings have implications for the understanding of activities in other settings and for the generic design and evaluation of collaborative tools, which we will describe.

SHAREABILITY, JUDGEMENT, AND THE DESIGN OF GROUPWARE

Since the invention of organisations, certain types of information have been shared, jointly processed, and used in professional work. With the increasing sophistication and 'user-friendliness' of computer technology, and most especially with the introduction of desktop devices and tools, the capacity for information-sharing was expected to be revolutionised. Not only would organisations be able to provide technological support for what was essentially large scale information-sharing, but now they would be able to support and encourage the sharing of information between individuals and their own, smaller scale, more local information production activities. Thus was born 'groupware' and its associated 'middleware' tools and technologies.

Much of the software originally designed to support groups was designed to provide asynchronous access to the information of others. In other words, it was designed to give access to the work of individuals who, for whatever reason, were not working together in the same physical space at the same time. Examples of this include shared databases and spreadsheets, asynchronous co-authoring tools, and electronic meeting schedulers. Henceforth, it is these kinds of tools we have in mind when we use the term 'groupware' although the term can be and has been used more generally.

Evidence for the success of this kind of groupware has sometimes been contradictory. For example, *Lotus Notes* has been taken up by a great many organisations with the expectation that professional workers would use this software to collaborate more effectively. Some research has found that the use of *Lotus Notes* has not enhanced collaborative work [15], whilst other research has [11]. However, the weight of the evidence suggests that groupware tools frequently fail [e.g., 6]. Numerous explanations have been offered for this. Most have to do with what may be described as social factors: the claim is made that professionals want to hold onto their 'own stuff' so as to preserve an advantage over their colleagues [15]; or that they are unwilling to alter their time-honoured work practices [2].

Our observation of work practice at the IMF, however, shows that differences in the utility of various types of groupware has to do with the type of information intended to be shared within these technologies. Though social factors may be imposed upon it, the fundamental issue is whether information can be shared or whether it cannot. This itself turns on how much individual judgement is used in the creation and management of information.

This can be illustrated by looking at two of the major tasks that professional economists at the IMF carry out: the production of staff reports, and the production of data for a statistical database.

The Production of Staff Reports

Professional workers at the IMF use a variety of tools in their information work. Nearly all use *Lotus 1-2-3* as their spreadsheet tool, an in-house package called *AREMOS* for time series data, and *Word Perfect* for text generation. A number of in-house applications enable the construction of compound documents. Together, these tools are used to generate information that ultimately works its way into what are called *staff reports*. These staff reports are collaboratively authored, and are used by the management of the organisation to make most consequential decisions.

The introduction of a new computer network at the IMF was expected to encourage professional workers to share more of their information throughout the organisation. It was believed that email, the sharing of data files on servers, and the capacity for remote access to information stores would provide new opportunities for information use that previously existing information access and delivery procedures — namely paper mail systems — made difficult. Such broader sharing of information was expected to help create new forms of group collaboration, ones that transcended currently existing group structures. But, as has been found with professional workers in other organisations [2], such changes do not appear to be happening. This can be explained by looking at the practical reasoning, and hence organisational logic, of what professional staff at the IMF do.

Professional economists at the IMF are each responsible for maintaining data on a member country. On an annual basis, or when a request for financial assistance has been made, these individuals are joined by four or five other economists who will work together on the production of a staff report on that particular country. This is known as the mission process. Work on the mission process is divided amongst economists so that one will collect data for, say, the balance of payments, and another for industrial production.

The collection and analysis of data for *Lotus* spreadsheets or *AREMOS* time series is not simply a clerical matter, however. Economists working on a mission are required to make professional judgements about where to fill in missing figures, inconsistencies or to clarify areas where the data seem muddled. This is a natural feature of current economic information. It is never complete, never certain, and always subject to revisions and amendment (sometimes years after the period in question).

Because this work involves judgement, there are checks and balances to ensure that the judgements are correct. These checks and balances consist of the social process that is mission work. For this process involves economists 'working up' their individual data sets iteratively, by corroborating them against the data 'worked up' by their colleagues. Gradually, a commonly specified set of interpretations is agreed upon by the group. This is used to compile the staff reports.

A corollary of this is that data stored on any individual workstation or PC consist of a mixture of judgement and agreed fact whose meaning reflects that stage in the social process of agreement and iteration. Hence at some point in the mission process, the data are rough and incomplete; the social process of figuring out the data being only begun. At another, later stage, the data are more complete, more effectively understood and developed, the social process of which it is part being nearer completion.

We can begin to see the consequences of this by putting things another way. When a team starts work on a mission, when data has just begun to be gathered, and the first meetings have occurred, data are too rough to be shared amongst the team, although each member will have some understanding of other team members' data. Toward the end of the mission cycle, data can be more readily shared, since by that time data will have been more thoroughly assessed and cross validated.

Therefore, at any specific moment in time, the adequacy of data is only visible to participants in a mission team, since it is only they who are aware of what stage of completion the data have reached, and who understands the exact boundaries between judgement and fact. All others, outside the mission team and outside the mission cycle, will find those data opaque and unsuited for use.

A further corollary of this is that outside of the mission process, when economists are working on their own, the data they store will be very difficult for anyone to use but themselves. And it needs to be remembered that these non mission periods are extensive. (As noted, missions typically only occur once a year). For the rest of the time, individual economists are effectively on their own, even though they work closely with their chiefs (and occasionally other colleagues) on the preparation of a variety of small scale data analyses and commentaries.

Taken as whole, these work practices have important implications for groupware. First, when individual economists are working on their own data stores, their data are unsuited for sharing and general use. Those data have not been through the social processes of validation and assessment. Second, when the data have been through such processes, only those within the coterie of a mission team will be able to know at when the data are usable. In addition, only they will know the ins and outs of the data, the difference between judgement and hard fact.

It is for these reasons that professionals at the IMF have not used the network to share information in new ways. This failure to cross information boundaries is not due to physical problems of information distribution, but because there would be no organisational logic in doing so. In summary, then, there are two main findings:

Information involving a high degree of judgement in its production is best interpreted by the producer of that judgement.

At the IMF, individual economists do not use the network to access data created by colleagues working on different countries or in different mission teams. They view those data as being 'individualised' and having a temporally located provenance, therefore being unusable for anyone else's purposes.

Collaborative processes are required to check the judgements used in the production of professionally assessed information.

The IMF's staff reports go through extremely complex and elaborate review, revision, discussion and checking procedures before they are used by the management. These procedures involve numerous personnel and several distinct departments. This is because staff reports contain a great deal of information that derives from individual professional assessment. These individual assessments need to be checked by others.

The Production of Statistics

At the IMF a great deal of effort is put into collecting and publishing statistical data. These data are used for historical analysis but not for management decision-making. Currently, the data are entered into a database known as the 'Economic Information System' running on an IBM 3090 mainframe. This system is about to be replaced with an as yet unspecified server-based environment running on a Novell network.

Irrespective of the system chosen, it is recognised that users will be able and will want to share and access data. They can do so because it is *in the nature of the information itself that it is shareable.* This is because the information that composes the statistical database is strictly only that which derives from standard methods. If any judgement is required to determine vagueness or inconsistency, then those numbers are not added to the database. One consequence of this is that there are numerous omissions in the database. A second is that data can sometimes take years for figures to be agreed and added to the data base. This is particularly troublesome for those involved in policy work and the mission process, since they need up-to-date data. A third consequence, and one we are concerned with here, is that whatever is in this database can be used by anybody. Unlike data which involve judgement, these data are objective and therefore shareable.

Thus, the production of statistics at the IMF stands in contrast to the production of staff reports. For, we find that:

Information that does not require judgement in its production can be more easily shared than information which does.

A straightforward illustration of this is the fact that statistical data on the Economic Information System at the IMF are accessed by individuals other than those who created that information.

Also in contrast to the work practices of the producers of staff reports, we find that:

Social interaction is not as crucial to the sharing of objective information as it is to the sharing of interpreted information.

Professional staff at the IMF will use statistical data without reference to the producers of those data or to other colleagues. Furthermore, documents that derive from the use of those data (and those data alone) are not subject to the same review and assessment procedures that exist in relation to staff reports. This is because there is no need to discuss, check, review, revise and iterate information. The information in question can stand alone and separate from social processes.

Implications for Design

That there is potentially only a small role for information-sharing groupware tools in the production of staff reports, and that there could be greater use for them in the production of the statistical database can be attributed to the extent of judgement used in the production of the two different classes of data. The more judgement used in its production, the less likely that conventional groupware for the asynchronous sharing of information will be useful. The less judgement, the more likely. This relationship can be represented by the schematic graph presented in Figure 1.

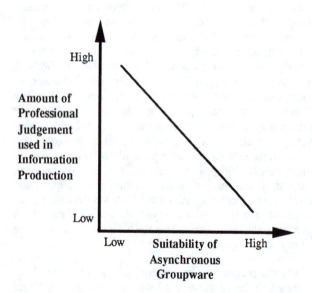

Figure 1. A schematic diagram showing the relationship between the nature of shared information and the suitability of groupware tools.

This is not to say that information which requires high levels of judgement cannot be support by technological tools. However, the implications of these findings are that the tools for supporting this kind of process must also support the social processes of collaboration. This rules out many kinds of conventional groupware technologies which are aimed at information exchange of the work products of others, not the work processes.

However, some technologies are aimed at encouraging and facilitating social processes. Tools to supplement meetings are one class which might be suitable. Professional staff at the IMF spend a great deal of time in meetings and in discussion, exchanging drafts and reviewers' comments.

Media space technologies are another, (i.e., audio-video links in conjunction with access to shared documents). A media space would be suited for those involved in the production of staff reports, but not necessarily useful for those involved in the production of statistics. One wants to support social processes when they are part and parcel of getting the work done, but not necessarily when they are not. (In fact, for a case where a media space was misapplied in this way, see [8])

Finally, it might be objected that the situation one finds at the IMF is unique. In particular, it might be claimed that the information professional workers store and produce does not always involve judgement. Though it is certainly true that the amount of judgement incorporated in the process of creating information will vary amongst different organisations, it must necessarily be the case that all professional work involves some judgement. Otherwise the data collection and production tasks would be a clerical one, and there would be no justification for having highly trained, expert professionals. It is in the nature of all professional work that it involves judgement of one kind or another. Further, the social processes of review can be seen across a variety of organisations: legal briefs are reviewed, medical assessments checked, and auditors' evaluations corroborated.

THE NEED FOR PAPER DOCUMENTS

One of the great hopes of office information systems designers was that their designs would lead to the paperless office. To date the attainment of that goal has not been achieved.

One reason for this is technical: these systems have been characterised by problems of incompatibility, incomplete WYSIWYG editors, and inflexibility, amongst other things. Observation of work practices at the IMF confirms the existence of such problems with their office information systems. However, we can expect that the development of integrated compound document technology (text plus graphics), middleware, and 'open' systems will eventually go some way towards solving these problems.

Another set of difficulties relates to what is sometimes called the cultural preference for paper. It is argued that people like paper, it is familiar to them and they can see no reason why they should stop using it and turn to electronic document forms. This is the cultural inertia argument.

However, our observations suggest that the primary reason for the persistence of paper is that electronic office information systems do not effectively replace the functionalities or affordances of paper. We will argue that users prefer to use paper at certain points in the document life cycle not as an issue of cultural preference, but because these functionalities interact with organisational logic. We will describe an example from the IMF in which the need to re-use documents for ad hoc and unpredicted purposes requires that professional staff re-specify the 'recipient design' of those documents. This process is one which is well supported by the use of paper.

Document Re-Use at the IMF

Professional staff of the IMF produce a whole range of documents with substantive information content. All of these are composed with specific reference to the ultimate user(s) of those documents. Thus, staff reports include all the information that is necessary for IMF executives to make their decisions; briefing papers for a managing director prior to a meeting include only those issues that are pertinent to that meeting; documents for public consumption are designed for a general audience, where the issues of concern are neither politically sensitive or still subject to analytical investigation. Thus, all important documents are 'recipient designed'.

This is important in a setting where there is a great proliferation of documents, each designed for particular sets of purposes and/or audiences. At the IMF, over 70,000,000 pages of documents are copied a year. The executive board receives over 4500 documents, subdivided into a dozen main categories, including approximately 350 staff reports.

However, an important characteristic of the IMF's work practice is that it can also react to unexpected circumstances, and it can do so in part because it can re-use the documents designed for its standard document processes for these unexpected needs.

This effective re-use is achieved by authors redesigning their documents for the new purposes or for the new recipient(s) of their documents. This redesign is especially important in relation to professionally authored documents, since these are likely to contain information that derives from professional judgement. Only the authors will know which information is entirely separate from their assessments and which is not, and thus will be able to assess which of the information is suitable for the unexpected use.

So for example, when members of the IMF's research department wanted to investigate the ratio of military expenditure to private sector growth in underdeveloped economies, they were able to do so, in part, by re-using staff reports. But for this, research department staff needed to get the authors of these reports to explain which of the figures contained in the staff reports derived from calculation and which from professional judgement. These explanations effectively 'redesigned' the reports for the new, unplanned use.

Furthermore, to ensure that these explanations were provided, members of the research department did not access the staff reports electronically but arranged meetings with the respective authors. During these meetings, they were able to outline what they already knew and what they had to find out, and, on this basis, the authors were able to offer the appropriate guidance on using the materials in their reports.

These document-related work practices lead us to three important implications:

For documents to be re-used for unexpected purposes, the authors often need to be 'in the loop', or directly involved in the process of document modification.

At the IMF when staff reports are used for purposes other than executive board activities, such as for gathering research material, the authors of those reports prefer to meet the new recipients. This provides them with an opportunity to learn about what the new recipients need, and to offer them guidance on how to use the information contained in the reports. Thus, this process of document re-use is often essentially social.

In being social another implication is made clear:

To support document re-use, paper documents are preferable to electronic ones. Paper documents support the social mechanisms of document redesign.

Paper documents can be the focus of a face-to-face meeting, can be placed on a desk in view of all parties and each page discussed in turn, and paper documents can be ritually exchanged once an agreement as to its interpretation has been made. In other words, as Luff et al. have noted [12], paper has an 'ecological flexibility' which allows it to be used as a focus for discussion, and for the co-ordination of social interaction. Luff et al. also point out that paper can be more easily interweaved into ongoing collaborative activity, as opposed to screen-based documents which cause interaction to be more localised and fragmented.

So, for example, at the IMF, when the research department wanted to re-use staff reports, meetings were arranged with the area departments that produced those reports. During these meetings, the respective staff sat around a desk and discussed the paper versions of the reports placed in front of them. They made notes on these documents, flicked through them, and drew attention to certain parts. When the authors believed that the research staff had fully understood what they could use the staff reports for, the paper versions were handed over to them.

Accordingly, the third implication is:

There will always be a critical role for paper (or technologies with paper-like qualities) in organisational work practice.

It is important to realise that the above findings will not hold true for all documents. A great many documents can be used again and again for unexpected purposes without the need for authors to be involved. Moreover, certain electronic document tagging applications such as those which use Structured Generalised Mark-up Languages (SGML) can enable users to create their own individually tailored documents. This can be especially useful in relation to such things as technical manuals.

However, for those documents which embody high degrees of professional judgement, the utility of SGML-type applications is reduced. Further, it is difficult to see how such tagging applications could be used to mark out the parts of a document which incorporate judgement from those parts which incorporate just the hard facts. Inevitably these different kinds of information are intertwined so as to convey a cohesive interpretation. This brings us back to the need for keeping the author in the document re-use loop. Only in so doing can correct re-interpretations be devised and agreed.

Implications for Design

A large proportion of the document life cycle in professional work can be carried out electronically. It is largely assumed that it is electronic tools which enable effective document re-use in allowing documents to be easily modified and reproduced. Indeed, electronic tools do support much of standard document-related work practice. However, for an organisation to be flexible, it must be able respond to unexpected circumstances, and it must be able to modify existing documents. In the face of unpredictability, it is best that authors be in the loop of re-use — a process which we have argued is most effectively supported by the use of paper.

This has important implications for designers. Above all, it means that designers should not try to obliterate the use of paper altogether but should attempt to understand and preserve some of its important functionalities.

One way to do this is to integrate or 'interface' the paper and electronic worlds. In this way the need for paper and the benefits of electronic document forms can co-exist. Thus far, technological advancements have mostly benefited moving from the electronic to the paper world. By contrast, moving from the paper to electronic forms is still a cumbersome process in its technological infancy, (albeit with some exciting possibilities [14,16]). However, new developments such as 'glyphs' [9] are representative of advancements made in this direction. Glyphs provide support for the effective moving from paper to electronic and back to paper forms, enabling electronic systems to access (or, more accurately, recognise) important formatting, style, and other information. This can go some way to avoiding the pitfalls of some OCR technology which often destroys

just that information that may be crucial to interpreting and hence using some document.

Another possibility is to consider the use of technologies which possess paper-like qualities. So far, most claims to be developing 'paper user interfaces' refer to the ability to use pen input on flat panel, portable displays, e.g., [3,17]. These technologies have the advantage over paper in being re-imageable and perhaps most interestingly, in providing some interactive capacities, enabling producers of text to manipulate and edit on the 'page' itself.

However, the very fact that the displays are dynamic and reimageable may fundamentally alter the ability for these tools to support the kinds of social processes we have been describing. For example, this may detract from the ability for a group of discussants to easily 'walk through' a document together and gain at-a-glance information by spreading it out on a table. Thus the new affordances offered by alternative technologies need to be set against those offered by paper. The ability for new paper-like technologies to support processes such as sharing, talking over, and exchanging documents between one professional worker and another has yet to be demonstrated.

CONCLUSION

We want to make clear that the preceding discussions and design recommendations are not based on analysis of what is often called 'organisational culture' or, more loosely, 'social factors'. The evidence we have provided turns on the claim that the activities we have described have a fundamental *organisational logic* to them. Professionals prefer paper documents for certain aspects of their work not because they are used to it, but because they afford certain advantages for the achievement of their practical ends. Cultural preferences and other social factors will be superimposed upon this logic, making it obscure to the casual observer. It is only through in-depth, extensive and thorough ethnographic examination that these logics can be uncovered.

Needless to say these 'logics of practice' are considerably more complex and broad than we have been able to convey in this short paper. But those aspects we have presented, relating to the nature of information and the extent to which tools support the social processes involved in the sharing of that information, are we think of fundamental importance in any organisational setting where professional work is undertaken. Therefore, we believe that the recommendations we have offered will have considerable general applicability. If designers take them on board, then we can be confident that we have gone some way towards ensuring that the tools and technologies professional workers will have at hand in the future will be appropriate for their practical requirements.

ACKNOWLEDGEMENTS

We greatly appreciate the help and co-operation of the IMF in allowing the field work to be undertaken and findings to be published. There were a great number of individuals who deserve special thanks. These include

Chris Yandle, Saleh Nsouli, John Hicklin, Alan Wright, Gertrude Long, and Terry Hill. We also thank William Newman and Paul Luff for guidance and comments on this paper.

REFERENCES

1. Anderson, B., Button, G., & Sharrock, W. (1993). Supporting the design process within an organisational context. *Proceedings of ECSCW '93*, (13-17 Sept., Milan), Netherlands: Kluwer, 47-59.

2. Bowers, J. (1994). The work to make a network work: Studying CSCW in action. *Proceedings of CSCW '94*, (22-26 Oct., Chapel Hill, North Carolina), New York: ACM, 287-298.

3. Brocklehurst, E.R., (1991). The NPL electronic paper project. *Int J. Man-Machine Studies, 34*, 69-95.

4. Drucker, P. F. (1973). *Management: Tasks, responsibilities, and practices.* New York: Harper & Row.

5. Garfinkel, H. (1967). *Studies in ethnomethodology.* Englewood Cliffs, NJ: Prentice-Hall, Inc.

6. Grudin, J. (1989). Why groupware applications fail: Problems in the design and evaluation. *Office: Technology and People, 4(3)*, 245-264.

7. Harper, R. H. R. (1992). Looking at ourselves: An examination of the social organisation of two research laboratories. *Proceedings of CSCW '92*, (31 Oct. - 4 Nov., Toronto), New York: ACM, 330-337.

8. Harper, R. H. R., & Carter, K. (1994). Keeping people apart: A research note. *CSCW, 2*, 199-207.

9. Johnson, W., Jellinek, H., Klotz, L., Rao, R., & Card, S. (1993). Bridging the paper and electronic worlds: The paper user interface. In the *Proceedings of INTERCHI '93*, (24-29 April, Amsterdam), 507-512.

10. Kidd, A. (1994). The marks are on the knowledge worker. *Proceedings of CHI '94*, (24-28 April, Boston), 186-191.

11. Korpela, E. (1994). Path to Notes: A networked company choosing its information systems solution. In R. Baskerville, S, Smithson, O. Ngwenyama, & J. DeGross (Eds.), *Transforming organizations with information technology*, North-Holland: Elsevier, 219-242

12. Luff, P., Heath, C. C. and Greatbatch, D. (1992). Tasks-in-interaction: Paper and screen based documentation in collaborative activity. *Proceedings of CSCW '92*, (31 Oct. - 4 Nov., Toronto), New York: ACM, 163-170.

13. Malone, T. (1983). How do people organize their desks? Implications for the design of office information systems. *ACM Trans. Office Info. Systems, 1(1)*, 99-112.

14. Newman, W., & Wellner, P. (1992). A desk supporting computer-based interaction with paper documents. *Proceedings of CHI '92*, (3-7 May, Monterey, CA.), 587-592.

15. Orlikowski, W. (1992). Learning from Notes: Organisational issues in groupware implementation', *Proceedings of CSCW '92*, (31 Oct. - 4 Nov., Toronto), New York: ACM, 362-369.

16. Wellner, P. (1993). *Interacting with paper on the DigitalDesk.* Unpublished Ph.D. dissertation, University of Cambridge, Computer Science Dept.

17. Wolf, C. G., Rhyne, J. R., & Ellozy, H. A. (1989). The paper-like interface. In G. Salvendy & M. J. Smith (Eds.), *Designing and using human computer interface and knowledge-based systems*, Amsterdam: Elsevier, 494-501.

18. Zuboff, S. (1988). *In the age of the smart machine.* New York: Basic Books.

Telephone Operators as Knowledge Workers: Consultants Who Meet Customer Needs

Michael J. Muller*§, Rebecca Carr**, Catherine Ashworth*, Barbara Diekmann*, Cathleen Wharton*, Cherie Eickstaedt**, and Joan Clonts**

* U S WEST Technologies, 4001 Discovery Drive, Boulder CO 80303 USA, +1-303-541-8182 (fax)
** U S WEST Communications, 1801 California, Denver CO 80202 USA, +1-303-896-2563 (fax)

Abstract: We present two large studies and one case study that make a strong case for considering telephone operators as *knowledge workers*. We describe a quantitative analysis of the diversity of operators' knowledge work, and of how their knowledge work coordinates with the subtle resources contained within customers' requests. Operators engage in collaborative query refinement with customers, exhibiting a rich set of skilled performances. Earlier reports characterized the operators' role as an intermediary between customer and database. In contrast, we focus on operator's consultative work in which they use computer systems as one type of support for their primarily cognitive activities. Our results suggest that knowledge work may be a subtle feature of *many* jobs, not only those that are labeled as such. Our methodology may be useful for the analysis of other domains involving skilled workers.

Keywords: Telephone operators, knowledge work, expertise, skilled performance, participatory design, participatory analysis

INTRODUCTION

This paper develops and documents U S WEST's emerging view of telephone operators as knowledge workers. We describe qualitative and quantitative task analyses that show the diversity and depth of operators' skilled performances.

We begin with a brief overview of the work of Directory Assistance (DA) operators, followed by a review of HCI research on telephone operators. We then present data from

§ Muller: michael@advtech.uswest.com, +1-303-541-6564. Carr: rcarr@future.mnet.uswest.com, +1-303-896-8278. Diekmann: diekmann@advtech.uswest.com, +1-303-541-6769. Eickstaedt: ceickst@future.mnet.uswest.com, +1-303-896-5963. Clonts: jclonts@future.mnet.uswest.com, +1-303-778-4023. Ashworth: ashworth@clipr.colorado.edu, +1-303-541-6613. Wharton: cwharton@advtech.uswest.com, +1-303-541-6292.

a detailed task analysis conducted with DA operators at a U S WEST Communications DA office. We describe a variety of types of expertise-based work that operators perform on more than 50% of the calls that they handle. We supplement this major analysis with a second, confirmatory task analysis conducted with DA operators at a second U S WEST Communications site. For comparison purposes, we briefly report a case study of the work of one toll and assistance (TA) operator.

Previous reports characterized the DA operators' role as intermediary between customers and a system or database [10]. Our results support a more human-to-human view of the job, with the important knowledge and expertise vested in the operators, rather than in their computer support tools. Our interpretation is in some ways similar to Kidd's analysis of knowledge work [9]. Our view is aligned with Floyd's process-oriented paradigm (the human work process is the starting point), rather than with the more system-focused, product-oriented paradigm [4].

In our view, DA operators serve as expert consultants. They help customers articulate their needs, engaging in collaborative query refinement at one or more stages of a DA call. Operators serve as experts in a variety of domains of relevance to their customers' lives, helping them to navigate through government agencies, complex business hierarchies, partially remembered geographies, and dynamic changes in their customers' worlds. In a small case study, we find similar themes in the work of TA operators. These interpretations — that is, that operators perform significant knowledge work — have powerful implications for work practices, training, support technologies, and partnerships between management and the labor force.

The Work of Directory Assistance Operators

DA operators are the women and men who look up telephone numbers in response to customers' requests. Different countries have different practices in this area. In the US and Canada, all DA calls go to operators who have access to databases for residential, business, and government telephone numbers (by contrast, in Poland, different operators are responsible for each type of listings). Because of the volume of work and the number of operators who perform it, telephone companies are often concerned to minimize the time required to handle each DA call. Savings of even a tenth of a second per call are multiplied into significant corporate economies.

The hallmark of DA work is fast accuracy. In U S WEST, the average call to a DA operator takes less than half a minute. During this brief time, the operator listens to the customer's request, and often engages in collaborative refinement of that query with the customer. Based on her or his analysis of the customer's needs — and perhaps supplemented by new information that may have been communicated from the business office at the beginning of the work shift — the operator executes one or more searches in a complex set of specialized databases, most of which are internally partitioned. The operator then reports only the relevant subset of the search results to the customer — this is a second opportunity for collaborative refinement. To save work time, the operator often invokes a computerized audio report support tool that delivers the telephone number to the customer. In some localities, this tool may optionally dial the call for the customer.

Previous Research on Telephone Operators

In prior work, Lawrence et al. analyzed the work of DA operators as a special form of mediation between customer and database. In their terms, operators are a particular case of the general class of work involving two humans and a computer: "In these interactions, one person, typically a customer, wants to accomplish some goal with a system but does so by interacting with a human intermediary. The computer operator is, in effect, a 'surrogate user'." [10, p. 399]. They provided examples of several types of translations and inferences that DA operators perform while working in this mode, and continued earlier quantitative research into operators' abilities to time-share their human-to-human and human-to-computer interactions. In this paper, we provide a more rigorous taxonomy of DA operators' cognitive and social work, and quantitative estimates of the occurrence of knowledge work.

Campbell and Velius [2] and Stuart and Gabrys [15] used a modeling approach to predict the impact of experimental technologies on DA calls. Unlike previous successes in the modeling and prediction of Toll and Assistance (TA) operators' work [6,8], both of these papers found large deviations from their models' predictions. These may signal trouble for proposals to automate DA calls [11,12]. Some of the subtle complexities of operators' work, explored in this paper, may help to explain these outcomes.

Conceptions of operators' work are important to U S WEST Communications' Operator and Information Services (OIS) organization and to the operators' union, the Communication Workers of America (CWA), which are together developing knowledge worker descriptions of other job titles [3,7]. It has been important to OIS and CWA to understand operators' work in detail, so that the company and the union can support their work and the quality of service that they provide to U S WEST's customers.

TASK ANALYSIS WITH DIRECTORY ASSISTANCE OPERATORS

Our major task analysis took place in two stages. The first stage was a qualitative study, conducted by subject matter experts (SMEs) from the OIS training organization, in collaboration with a human factors worker. The results of this stage guided the second, quantitative stage. The quantitative task analysis was conducted by one SME and one human factors worker, with ten operator participants who were selected and recruited by CWA. Analyses of the videotaped data were conducted by a team of HCI specialists, in close collaboration with the SME. Interim and final results were validated with representatives of CWA.

Collaborative Qualitative Analysis
Method
The qualitative task analysis used the CARD and PICTIVE techniques from participatory design [13,17]. We emphasize that we were *not*, in this stage, engaged in a participatory activity, because no operators (users) were involved in this preliminary analysis. However, we used participatory techniques in our qualitative analysis.

We applied CARD and PICTIVE in a "bifocal" analytical approach [14]. CARD (which uses card images to represent work tasks and events [16,17]) was used to analyze higher-level task flow issues, and PICTIVE (which uses paper-and-pencil representations of interaction media [13]) was used to analyze lower-level interactions with specific artifacts, including workstation screens. These techniques have been described extensively, and will not be detailed here. However, we note that one aspect of the CARD technique was particularly valuable: This was the ability to include explicit representations of the operator's goals, strategies, and other mental operations as part of the workflow.

Results
Through the qualitative task analysis, we produced an initial description of the work of DA operators, including representative task flows, which were shown in [16]. It quickly became obvious that there were many opportunities for operators to perform a variety of types of knowledge work during these task flows. Figure 1 shows one example of a variety of types of knowledge work (for convenience, two different calls have been combined in the one figure). Specific descriptions of types of knowledge work will be provided in the next section.

We verified these results with representatives of CWA. We then turned to a quantitative task analysis to determine, in part, how often operators engaged in knowledge-work activities, and what resources supported those activities.

Quantitative Analyses
Method
The qualitative task analysis used the method of direct, videotaped observation of one hour of live traffic handled by each of ten operator participants. Calls were recorded during three weekdays during autumn 1993, between the hours of 8:00AM and 8:00PM. Video recordings of the DA calls were supplemented by each operator viewing her or his videotape and providing explanatory comments; these participatory analysis sessions were also videotaped.

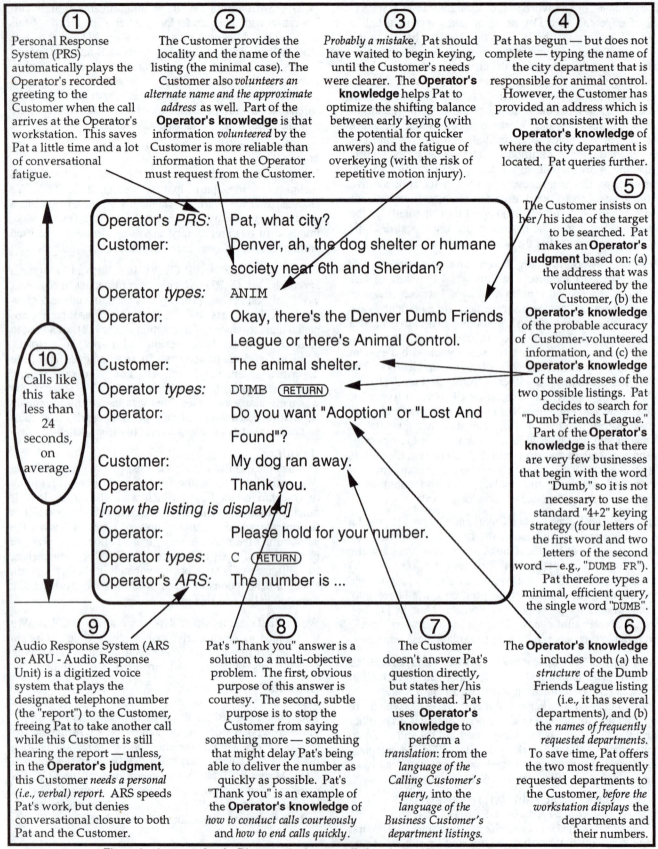

1. Personal Response System (PRS) automatically plays the Operator's recorded greeting to the Customer when the call arrives at the Operator's workstation. This saves Pat a little time and a lot of conversational fatigue.

2. The Customer provides the locality and the name of the listing (the minimal case). The Customer also *volunteers an alternate name and the approximate address* as well. Part of the **Operator's knowledge** is that information *volunteered* by the Customer is more reliable than information that the Operator must request from the Customer.

3. *Probably a mistake*. Pat should have waited to begin keying, until the Customer's needs were clearer. The **Operator's knowledge** helps Pat to optimize the shifting balance between early keying (with the potential for quicker anwers) and the fatigue of overkeying (with the risk of repetitive motion injury).

4. Pat has begun — but does not complete — typing the name of the city department that is responsible for animal control. However, the Customer has provided an address which is not consistent with the **Operator's knowledge** of where the city department is located. Pat queries further.

5. The Customer insists on her/his idea of the target to be searched. Pat makes an **Operator's judgment** based on: (a) the address that was volunteered by the Customer, (b) the **Operator's knowledge** of the probable accuracy of Customer-volunteered information, and (c) the **Operator's knowledge** of the addresses of the two possible listings. Pat decides to search for "Dumb Friends League." Part of the **Operator's knowledge** is that there are very few businesses that begin with the word "Dumb," so it is not necessary to use the standard "4+2" keying strategy (four letters of the first word and two letters of the second word — e.g., "DUMB FR"). Pat therefore types a minimal, efficient query, the single word "DUMB".

Operator's *PRS*:	Pat, what city?
Customer:	Denver, ah, the dog shelter or humane society near 6th and Sheridan?
Operator *types*:	ANIM
Operator:	Okay, there's the Denver Dumb Friends League or there's Animal Control.
Customer:	The animal shelter.
Operator *types*:	DUMB (RETURN)
Operator:	Do you want "Adoption" or "Lost And Found"?
Customer:	My dog ran away.
Operator:	Thank you.
[now the listing is displayed]	
Operator:	Please hold for your number.
Operator *types*:	C (RETURN)
Operator's *ARS*:	The number is ...

10. Calls like this take less than 24 seconds, on average.

9. Audio Response System (ARS or ARU - Audio Response Unit) is a digitized voice system that plays the designated telephone number (the "report") to the Customer, freeing Pat to take another call while this Customer is still hearing the report — unless, in the **Operator's judgment**, this Customer *needs a personal (i.e., verbal) report*. ARS speeds Pat's work, but denies conversational closure to both Pat and the Customer.

8. Pat's "Thank you" answer is a solution to a multi-objective problem. The first, obvious purpose of this answer is courtesy. The second, subtle purpose is to stop the Customer from saying something more — something that might delay Pat's being able to deliver the number as quickly as possible. Pat's "Thank you" is an example of the **Operator's knowledge** of *how to conduct calls courteously* and *how to end calls quickly*.

7. The Customer doesn't answer Pat's question directly, but states her/his need instead. Pat uses **Operator's knowledge** to perform a *translation*: from the *language of the Calling Customer's query*, into the *language of the Business Customer's department listings*.

6. The **Operator's knowledge** includes both (a) the *structure* of the Dumb Friends League listing (i.e., it has several departments), and (b) the *names of frequently requested departments*. To save time, Pat offers the two most frequently requested departments to the Customer, *before the workstation displays* the departments and their numbers.

Figure 1. An example of a Directory Assistance call, focusing on Operators' knowledge work.

Some of the credentials of the operator participants are summarized in Table 1. Five operators were women, and five were men.

Table 1. Backgrounds of Operator Participants

	Min.	Mean	Max.
Time in title (years)	0.3	7.2	22
Education (years)	12	13.4	16
Average work time (seconds)	20.05	23.06	28.8

Fifty calls from each operator's videotape were subsequently analyzed in detail. Of these calls, 73 percent asked about businesses; 18 percent were for residences; 8 percent were for government agencies; and 1 percent were other miscellaneous requests (area code, time of day, etc.). In this brief paper, we cannot present all of our results. The following two subsections focus on operators' knowledge work, and on conditions enabling that work.

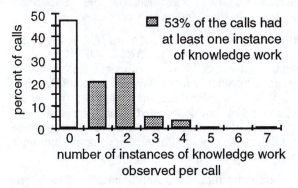

Figure 2. Occurrence of knowledge work in DA calls.

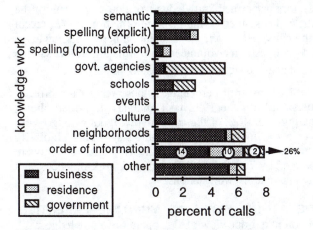

Figure 3. Types of knowledge work in DA calls.

Results: Operators' Knowledge Work

Figure 2 provides an overview of the results relating to operators' knowledge work. Fifty-three percent of the calls involved at least one type of knowledge work. Thirty-one percent involved more than one type.

Figure 3 details the types of knowledge work that we analyzed, based on the qualitative task analysis and on

operators' narratives during the participatory analysis. These included:

- **Semantic.** The operator translated the customer's query into entirely different words that had the intended meaning within the database (e.g., "Mormon" becomes "LDS" [Latter Day Saints]; see also Figure 1). Operators may also volunteer their personal knowledge of changes in the names or locations of businesses, or new namings of government departments. In extreme cases, operators help customers to develop work-around strategies for businesses or agencies that are difficult to reach.

- **Spelling.** The operator keyed a special spelling of the customer's request through knowledge of the database (e.g., "Saint", "Santa", and "San" are all abbreviated "ST" and are compacted onto the following word in the query).

- **Pronunciation.** The operator keyed a special spelling of the customer's request through personal knowledge (e.g., something that sounds like "Berkeley" may be spelled "Buerkle" in Minnesota).

- **Government agencies.** The operator substituted a government department name for the customer's topical query (e.g., "agriculture" for "weed" — or vice versa, depending upon the state and the government agency).

- **Schools.** Some databases list public and private schools in different partitions, or even different files. Operators' knowledge of the database helps them to search the correct file, and to try different naming conventions for particular schools. Operators are also adept at translating approximate addresses into named or numbered school districts — a different database search key.

- **Events.** In some cases (but not in our data), operators are knowledgeable about telephone numbers relevant to current events (sports events, conventions, etc.).

- **Culture.** In some cases, operators use their knowledge to assist customers in finding culturally-relevant numbers (Spanish-speaking, Native American, etc.)

- **Neighborhoods.** Operators can usually translate from localisms — e.g., "uptown," "downtown," "across from that big mall" — into named streets and even specific building numbers on the streets.

- **Order of information.** Part of the operators' "savvy" about using the system involves knowledge of when to try alternative orderings of information in the various search fields. For example, a commercial entity that is named after a person may be searched first-name-first if it is a conventional business, but last-name-first (like a residence listing) if it is a professional practice. Schools named after persons are a particularly difficult case.

Results: Customer-Volunteered Information

What resources do DA operators use to perform this knowledge work? One often-overlooked resource is the information that customers volunteer — that is, information that is over and above the standard *locality* and *name of the listing* information that operators are trained to request, and that technologists design toward [11,12]. When we analyzed customer-volunteered information, we found that it occurred during more than half the calls (Figure 4). Details of the types of customer-volunteered information are provided in Figure 5.

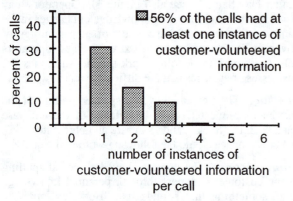

Figure 4. Occurrence of customer-volunteered information (in addition to locality and listing) in DA calls.

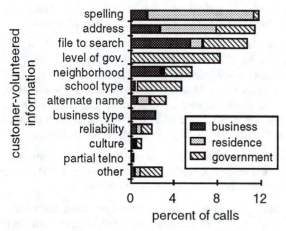

Figure 5. Types of customer-volunteered information (in addition to locality and listing) in DA calls.

Lawrence, Atwood, and Dews also have described some aspects of customer-provided information. They argued that the lack of specificity in the customers' information could be a problem: "They might say that the name is 'something like,' or that 'it might be in Cambridge,' or 'it's near Queens Boulevard'" [10, p. 402]. Our experience has been that operators are adept at making sense and meaning from these approximations, although this process may require an extra conversational turn (see Figure 1, and the knowledge work categories of "Schools," "Events," and "Neighborhoods," above). A customer's approximate query for a government agency provides a similar opportunity for collaborative query refinement, as the operator probes the customer's needs. We have observed operators asking questions such as "Vialsystics? Do you mean birth records?" [vital statistics] and "Well, was it a *moving*

violation?" [to look up the appropriate court] as they helped customers to articulate their needs.

Did operators use the customer-volunteered information for their knowledge work? As Figure 6 shows, knowledge work was significantly associated with customer-volunteered information (χ^2=5.87, p<.02).

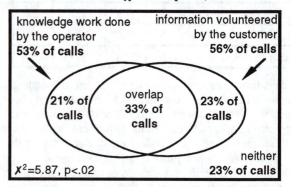

Figure 6. Correlation of knowledge work and customer-volunteered information in DA calls.

CONFIRMATORY TASK ANALYSIS AT A SECOND DIRECTORY ASSISTANCE OFFICE

During the summer of 1994, one of us had the opportunity to conduct a quantitative task analysis at a second DA office in U S WEST. Unlike the first analysis, the second exercise could not use videotaped recordings. Calls were observed and characterized in real-time. Therefore, we suspect that subtle types of knowledge work and customer-volunteered information may have gone undetected.

Eight operators participated. Approximately 50 calls were scored for each operator, using the taxonomy of knowledge work presented above, for a total of 410 calls. DA operators performed knowledge work on 41 percent of the calls, and customers volunteered information on 37 percent of the calls. However, the association between knowledge work and customer-volunteered information in this sample did not achieve significance (χ^2=3.40, p>.05).

In terms of overall percentages, these results are broadly consistent with the 1993 task analysis described above, showing over 40% of the calls involving knowledge work (over 50% in the 1993 task analysis). This similarity occurred despite the fact that the two DA offices were separated by 700 miles, different local work practices, and different vendors' workstations.

OPERATORS' WORK AS MUNDANE EXPERTISE

We compared our view of DA operators' knowledge work with published conceptual analyses of expert performance. DA operators' work fulfilled all seven of the criteria listed by Glaser and Chi [5]:

- *Experts excel mainly in their own domains.* DA operators have detailed domain-specific knowledge that includes the computer systems that they use, the business and geographical domains of their customers, and skills in conducting brief but effective conversations.

- *Experts perceive large meaningful patterns in their domain.* DA operators routinely extract meaning from large numbers of potential answers to customers' queries. Some of these are quite complex — e.g., departmental listings of government agencies, whose naming conventions are often more responsive to the agency's internal logic than to citizens' needs for access.

- *Experts are fast and accurate.* As noted above, DA operators work *quickly*, and with high accuracy.

- *Experts have superior short-term and long-term memory.* DA operators routinely maintain many details of the customer's request in short-term memory. The best keying strategy is often to search on the main name information, and to use address or department information as a cognitive (but not keyed) aid in sorting through the listings returned from the database.

DA operators maintain extensive long-term knowledge of the structure and details of information in their databases. Using this knowledge, they can trim seconds off their calls while providing accurate information. Examples include:

- translating between the customer's language and the language of government departments.

- translating the customer's reference to a neighborhood or a local landmark into the street address information of the database.

- remembering the departments in a particular business, so that they can query the customer for the most likely departments *before* the listings appear on the screen (e.g., Figure 1).

- *Experts represent domain problems at a more principled level.* DA operators possess a detailed structural knowledge of the database, in terms both of its abstract structure (different files, different partitions within files) and in terms of the formal organization of government or business department hierarchies. They are also skilled at diagnosing database entries that have become inaccurate or even subtly misleading, as well as telephone trunks that, while still usable, are in need of maintenance. Their trouble reports and proposed solutions often serve as first-line indications of problems that other offices in the company then attend to.

- *Experts spend time analyzing a problem qualitatively.* While DA operators handle most calls very quickly, they are able to identify customer queries that fall outside of their usual domain of speeded expertise. On this subset of calls, they spend extra time to analyze the problem in detail. In extreme cases, to economize on individual work time, operators transfer the caller directly to a Customer Service Consultant (sometimes called a Service Assistant), whis is assigned to spend the needed time on difficult calls.

- *Experts have strong self-monitoring skills.* DA operators are expected to maintain a certain performance level, usually expressed as average call duration. Failure to achieve this criterion impacts the entire office, forcing other operators to work harder, and causing longer waiting time for customers. Operators maintain a sense of whether they are meeting this criterion. If they are, they can spend more time on a difficult call. If they are not, then they attempt to speed their work, so as regain their targeted performance.

We note that our conceptions of skill and expertise are based on the mundane, every-day skills of an experienced worker (e.g., [9]), and not on the extraordinary skills of an exceptional performer. For a discussion of mundane skills, see [1].

CASE STUDY WITH ONE TOLL AND ASSISTANCE OPERATOR

We were curious to know if our results regarding knowledge work were applicable to another operator job — that of Toll and Assistance (TA) operators. The work of TA operators has been extensively studied and modeled, with spectacular successes for cognitive modeling in real-world settings [6,8]. However, the requirements of these modeling activities appeared to contrast with our knowledge work findings in DA operators. For the modeling techniques to be effective, the modeled behavior must be routine and repeatable. Variations in the behavior should be quantitative, rather than qualitative. For the modeled calls, John wrote: "There is typically no problem solving involved; the [operator] simply recognizes the call situation and executes routine procedures associated with that situation" [8, p. 107]. Similarly, according to Gray et al., "TAOs [toll and assistance operators] recognize each call situation and execute well-practiced methods, rather than engage in problem solving." [6, p. 241].

Could it be that, while DA operators engage in knowledge work on over half of their calls, TA operators do none? One of us worked with one TA operator to explore this question. This section briefly introduces the work of TA operators, and then describes our tentative findings.

The Work of Toll and Assistance (TA) Operators
Operators who work in TA in the US have different responsibilities from those who work in DA. TA operators are primarily concerned with *call completion* — that is, with helping a customer who already knows the number s/he wants to call, but needs operator assistance to place the call. Assistance may take the form of special dialing, but is most often concerned with alternate billing arrangements (collect calls, billing to credit cards or calling cards, or billing to the caller's home telephone number if the caller is traveling). TA operators may also receive nearly all other requests for assistance. These include a great miscellany of problems, ranging from a customer who needs to interrupt someone else's on-going telephone call with an emergency message, to someone who is confused with advanced calling features, to preliminary inquiries about billing, repair, and other telephone company operations that are subsequently handled by other staff at other offices.

Customers' requests for assistance are sometimes the telephone company's first notice of trouble: TA operators therefore serve as first-line trouble diagnosticians, analyzing and reporting system or service problems to other offices in the telephone company. They are aided in their diagnostic work by electronic mail announcements of problems that may affect service. For example, during our observations, the operator received notice of a major forest fire in Montana which had the potential to affect telephone service. The operator used this information to help one customer plan alternate strategies for contacting someone who was on the opposite side of the affected area.

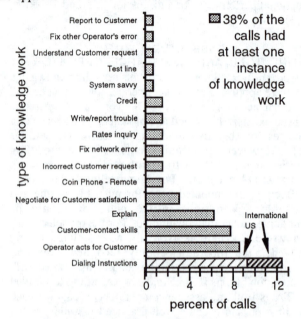

Figure 7. Knowledge work by one TA operator.

Task Analysis of One TA Operator's Work

One of us observed 128 TA calls that were handled by a U S WEST TA operator during one weekday morning in summer 1994. Calls were scored in real-time, without any recording medium. As noted above, this may have led to an undercount of subtle types of knowledge work.

A total of 38 percent of the calls involved significant knowledge work by the TA operator, in areas such as customer contact and negotiation skills, collaborative refinement of the customer's needs and request, dialing instructions suited to the customer's ability to understand (especially in the case of international dialing), remote operation of telephone equipment (coin telephones, line tests), and analytic skills (diagnosing errors by other operators, errors by network elements, database or network errors requiring repair). Quantitative details are in Figure 7.

These results must be interpreted tentatively. They will have to be repeated with a larger sample of operators and of calls. Nonetheless, this case study suggests that the earlier research [6,8] may have been based on only a subset of TA operator work. In all probability, the subset was restricted to call-completion services, in which the TA operator handles alternate billing arrangements, such as credit card,

collect, third-party, and so on. In fact, 46 percent of the calls in our case study involved call completion, and only 17 percent of the call-completion subset (6 percent of the 128 calls in the study) required operator knowledge work.

DISCUSSION

Customers often call operators *not* for mediated access to a database, but rather for expert assistance in finding information that the customers need in order to live and work in their worlds. As we have shown, operators respond to these needs through a mixture of their knowledge of the customers' worlds, of the changing circumstances of those worlds, *and* of the structure and content of their database systems. In this way, we have framed our analysis in terms of the *human processes* in the work, rather than in terms of the software *product* that plays a supporting role for one of those humans (see [4]).

While it is certainly true that operators mediate access between customers and databases [10], this formulation appears to omit important aspects of their work, such as extensive knowledge that is *not* contained within the database, collaborative refinement of customer's queries, and expert diagnostic and problem-solving skills. Similarly, while it is also true that many tasks performed by operators are routine and repeatable [6,8], this formulation, too, appears to omit critical aspects of the application of expert knowledge and analysis, as well as human-to-human skills. Operators are thus a hybrid case between Kidd's *clerical workers* (who rely primarily on external data and resources) and *knowledge workers* (who rely primarily on their own internal representations of information) [9].

We do not know, at present, how these more sophisticated skills contribute to customer's perceptions of the quality of service that they receive, nor to the operators' experience of work that is both pleasant and satisfying. We suspect that these components of knowledge, skill, and expertise are crucial for both of these important considerations.

CONCLUSION

These results have helped U S WEST and CWA to develop new understandings and support strategies for the work of telephone operators. An improved understanding of the sophistication of operators' work and practices has helped us to make informed decisions regarding technology supports for operators' work, and is leading us to consider innovative training approaches. We are also considering changes in work practices and in support artifacts for work, designed to strengthen operators' skills-based and consultative roles. Finally, the area of knowledge work opens new research approaches to improve the supports for operators' work.

We hope that these results will inform CHI work in three ways. First, we encourage others to question assumptions about the claimed simplicity of users' work.

Second, we believe that our work contrasts with the essentialism in Kidd's identification of knowledge work with specific workers [9]: While Kidd would classify operators as clerical workers, we have shown that their clerical job contains significant knowledge work. For our work, Kidd's classifications are more helpful as descriptions of *work components*, rather than descriptions of *workers*.

Third, we offer our two-phase process as a model of effective CHI analysis practices. We postponed formal, quantitative work, beginning our analyses with qualitative, collaborative approaches that were based in the Scandinavian traditions of participatory design and analysis (*collaborative* and *qualitative* are the key concepts here — *not* our specific techniques). We verified our new understandings with the people who were described by those new understandings. The results of these analyses were conceptually compelling, and guided our next steps. Only then did we pursue quantitative analyses, which were compelling for managers and engineers as well as operators.

This two-phase approach allowed us to make discoveries that had previously remained invisible to the more formal approaches used by other CHI researchers [2,6,8,10,12]. Our hybrid model — principled qualitative work, followed by verification with stakeholders, followed by principled quantitative work, followed again by stakeholder verification — provided an effective bridge between the human-process-oriented world of the operators, and the formal-product-oriented world of the engineers. As Floyd has noted [4], negotiating the balance between these worlds is crucial for ethical CHI practice to balance the human as well as organizational needs of all stakeholders.

ACKNOWLEDGMENTS

We thank the following people and organizations for their contributions to our work: Joe Barreda, Susan Barry-Hagen, Kevin Boyle, Communications Workers of America Locals 7201 and 7702, Cindy Darcy, Sandy DeRodeff, Joan Greenbaum, Meg MacRae, Monica Marics, Mike Mase, Judy Olson, Chris Plott, Terry Roberts, Joanie Schifsky, Lynn Streeter, Jeff White, and Mary Wiblishauser.

REFERENCES

[1] Attewell, P. (1990). What is skill? *Work and Occupations* **17**(4), 422-448.

[2] Campbell, J.B., & Velius, G. (1989). On applying speech technologies to directory assistance. *Speech Tech'89*. 130-132.

[3] Communications Workers of America, International Brotherhood of Electrical Workers, and U S WEST (1994). *Job design team future vision*.

[4] Floyd, C. (1987). Outline of a paradigm change in software engineering. In Bjerknes, G., Ehn, P., and Kyng, M. (Eds.) (1987). *Computers and democracy: A Scandinavian challenge*. Brookfield, VT: Gower.

[5] Glaser, R., and Chi, M.T.H. (1988). Overview. In M.T.H. Chi, R. Glaser, and M.J. Farr (Eds.), *The nature of expertise*. Hillsdale NJ: Erlbaum.

[6] Gray, W.D., John, B.E., and Atwood, M.E. (1993). Project Ernestine: Validating a GOMS analysis for predicting and explaining real-world task performance. *Human-Computer Interaction* **8**, 237-304.

[7] Hill, A., Malloy, L, Armbruster, L., Fred, C., Boyle, K., Muller, M.J., Tessicini, L., and Plott, C. (1994). White paper: Integrating design engineering and the workforce. Presented to US National Institute of Science and Technology workshop, Seattle WA, October 1994.

[8] John, B.E. (1990). Extensions of GOMS analyses to expert performance requiring perception of dynamic visual and auditory information. *Proceedings of CHI'90*. Seattle WA: ACM, 107-115.

[9] Kidd, A. (1994). The marks are on the knowledge worker. *Proceedings of CHI'94*. Boston: ACM, 186-191.

[10] Lawrence, D., Atwood, M.E., and Dews, S. (1994). Surrogate users: Mediating between social and technical interaction. *Proceedings of CHI'94*. Boston MA: ACM, 399-404.

[11] Lennig, M. (1990). Putting speech recognition to work in the telephone network. *IEEE Computer*. August 1990. 35-41.

[12] McEwen, S., and Bergman, H. (1993). Automating directory assistance service: A human factors case study. In *Human Factors in Telecommunications: 14th International Symposium*. Darmstadt, Germany: R.v. Decker's Verlag, G. Schenck, 301-309.

[13] Muller, M.J. (1992). Retrospective on a year of participatory design using the PICTIVE technique. In *Striking a Balance: Proceedings of CHI'92*. Monterey CA: ACM, 455-462.

[14] Muller, M.J., Tudor, L.G., Wildman, D.M., White, E.A., Root, R.W., Dayton, T., Carr, R., Diekmann, B., and Dykstra-Erickson, E.A. (1995). Bifocal tools for scenarios and representations in participatory activities with users. Chapter to appear in J. Carroll (Ed.), *Scenario-based design for human-computer interaction*. New York: Wiley, in press.

[15] Stuart, R., and Gabrys, G. (1993). A speech compression proposal for directory assistance operators: GOMS predictions. *INTERCHI'93 Adjunct Proceedings*. Amsterdam: ACM, 159-160.

[16] Tudor, L.G., Muller, M.J., and Dayton, T. (1993). A C.A.R.D. game for participatory task analysis and redesign: Macroscopic complement to PICTIVE. *INTERCHI'93 Adjunct Proceedings*, 51-52.

[17] Tudor, L.G., Muller, M.J., Dayton, T., and Root, R.W. (1993). A participatory design technique for high-level task analysis, critique, and redesign: The CARD method. In *Proceedings of the Human Factors and Ergonomics Society 1993 Meeting*, Seattle WA, October 1993, 295-299.

Ethics, Lies and Videotape...

Wendy E. Mackay
Rank Xerox Research Centre, Cambridge (EuroPARC)
61 Regent Street
Cambridge, UK CB2 1AB
mackay.chi@xerox.com

ABSTRACT
Videotape has become one of the CHI community's most useful technologies: it allows us to analyze users' interactions with computers, prototype new interfaces, and present the results of our research and technical innovations to others. But video is a double-edged sword. It is often misused, however unintentionally. How can we use it well, without compromising our integrity?

This paper presents actual examples of questionable videotaping practices. Next, it explains why we cannot simply borrow ethical guidelines from other professions. It concludes with a proposal for developing usable ethical guidelines for the capture, analysis and presentation of video.

KEYWORDS: HCI professional issues, video editing, ethics, social computing.

INTRODUCTION
The lights dim in the plenary talk at CHI'95. You settle back in your seat to hear from one of the early innovators in HCI - in fact, your former thesis advisor from a decade ago. As expected, he is an entertaining speaker. He quickly has the audience laughing as he shows videos of early interfaces and very perplexed users. Suddenly, you're not laughing. You see a familiar face projected on the 40 foot screen: it's you, ten years ago. You watch in horror as the 2500 members of the audience, now your peers and colleagues, laugh at your 'inept' use of the technology.

Could such a thing happen? It already has. What was the appropriate thing to do? Should the speaker have tried to discover if she were in the audience? Would 'informed consent' given ten years ago have been adequate? What were her rights? What was the audience's responsibility?

These are not easy questions and I won't presume to provide definitive answers. However, I think such examples can raise awareness of the issues facing the CHI community, as we increase our use of video for a wide range of activities. Sometimes, simply being sensitive to the problem is

sufficient; other times, there is no clear course of action. In either case, I contend that we are obligated as a profession to try to deal with these issues as effectively as possible.

As a community, we must educate ourselves about potential misuse and encourage responsible behavior. We must also understand *who* we are trying to protect and the trade-offs in protecting one group versus another. We need comprehensive guidelines to help members of the HCI community make ethical decisions.

The next section challenges the perception that video can be treated as an objective record of events and then presents examples of questionable videotaping practices. I also discuss why the advent of digital video increases the potential for misuse. The subsequent section frames the discussion within a more general ethical framework. I briefly review the perspectives of other professional groups, particularly with respect to their use of video. The last section presents preliminary suggestions for handling video and proposes a strategy for developing more detailed guidelines for the HCI community.

VIDEO: OBJECTIVE OR SUBJECTIVE?
Video is a powerful medium: it can make a point or convince people in ways that other media cannot. Video captures aspects of human behavior, such as gaze and body language, that are not available in any other form. Somehow, video seems "real". Yet, perhaps it is too powerful. Just as statements taken out of context can be very damaging, so can video clips misconstrue events or violate the privacy of the subjects involved.

Researchers often treat videotaped records of human behavior as objective scientific data: they can be viewed repeatedly, individual events can be counted and findings can be verified independently by other researchers. Unfortunately, the appearance of objectivity is just that: an appearance. Someone must choose a location and field of view for the camera, which must include some and exclude other information. The choice of when to press the "record" button also includes and excludes information. More subtly, the context shared by the participants of the videotape may be difficult or impossible to capture and present to subsequent viewers.

The shared context can occur at various levels. For example, Clark & Schaefer (1989) examined conversations between people. If one person is explaining something, she looks to the other person for signs, such as a nod or "uh huh" that he has understood sufficiently well for her to continue. She may not speak clearly but will continue if she is convinced that he is following her. Is she misspeaks, she may see him look puzzled and then smile, indicating that he has understood and she should continue. A camera shot of her face as she speaks will capture the exact words she spoke but not the shared understanding that evolved. The video records only the *fact* that she misspoke. Later, it could be used to "prove" that it was what she "really" meant.

Another problem arises when video captures conversations between people with shared prior experience, who speak in short-hand. In a live setting, an observer might be puzzled by what is meant or ask for clarification. With a video record, the same observer could view it repeatedly, develop a theory about the meaning and become convinced she understands, even if the participants meant something else.

People are used to being able to speak informally in daily conversation. Since both speakers and listeners know their memories can be unreliable, misunderstandings are usually cleared up through further discussion. When casual conversations are recorded, the ways of resolving misunderstandings changes. Suddenly, the speaker can no longer say "I didn't say that"; the videotaped record becomes an independent arbiter of what was said. But what was said is not the same as what was meant. Since people can change their minds over the course of a conversation, statements that seem to establish what the speaker 'really' meant distort the ongoing process of conversation.

Most people (except for politicians[1]) feel uncomfortable being recorded and change their usual behavior; they are not used to speaking "for the record". If electronic mail is notorious for generating misunderstandings due to informal writing, recorded casual speech is worse. Even speaking carefully can be dangerous, since viewers may interpret it as evidence of 'something to hide'. Broadcast media are thus subject to greater restrictions than print media. For example, "Recognizing the particular power of radio and television to influence public opinion, federal legislation was passed limiting the involvement of broadcasters in political camps." (Hall, 1978)

Recording video is only part of the problem. The audience and context in which the video is presented may also affect what is understood. For example, imagine recording a researcher's discussion of a new software interface that 'increases productivity'. This video, shown to employees

who interpret "productivity" as a euphemism for layoffs, suddenly has a very different impact. The infamous "sound bite", in which a short clip is selected to represent a longer event, may distort the original message or make rare events appear representative. "TV news often avoids coverage of the story that doesn't have anything visual and too often makes editorial decisions based on the availability of pictures rather than true news value." (Hall, 1978, p.17)

These examples demonstrate the importance of context and how easily video can be misinterpreted, intentionally or not. Unfortunately, even people who recognize that a videotape is not an objective record find it easy to slip into thinking that it is somehow real. Video is powerful; care is required both in its production and its interpretation. The use of video raises ethical questions: we can look to the literature in ethical theory for help addressing them.

ETHICAL THEORY

The ethical literature is vast, with philosophical discussions dating back to Plato and Aristotle. According to Forester and Morrison (1990), most current professional ethical codes are influenced by three more modern perspectives: ethical relativism (Spinoza), consequentialism or utilitarianism (J.S. Mill) and deontologism (Kant). The latter two are most relevant for computer professionals: "Consequentialism says simply that an action is right or wrong depending upon its consequences, such as its effects on society. [...] By contrast, deontologism says that an action is right or wrong in itself. Deontologists stress the intrinsic character of an act and disregard motives or consequences." (Forester and Morrison, 1990, pp. 16-17)

Older, more established professions, such as medicine and law, provide codes of ethical practice for their members. Their goals are to establish their status as a profession, to regulate their membership and convince the public that they deserve to be self-regulating (Frankel, 1989). Some, such as Ladd (1980), dismiss the notion of organized professional ethics as having few benefits and real potential for harm, while others, such as Bagley (1977), argue that "a written code is a necessity". Luegenbiehl (1992) argues that "Codes of ethics need be neither authoritarian nor designed for the enhancement of a profession. Instead, they should help the professional seeking to engage in ethical practice".

Computer science is a relatively new field but already has a large literature on ethics and computing. (See recent books by Forester and Morrison (1990), Johnson (1994) and Dunlop and Kling (1991).) Martin and Martin (1994) compare four codes of ethics: ACM (1992), IEEE (1992), Data Processing Managers Association (DPMA, 1989) and the Institute for Certification of Computer Professionals (ICCP, 1989). The four codes are similar to each other and to other professional codes because they take a generic approach to ethics. Privacy and confidentiality of data were seen as the only elements that "reflect the unique ethical problems raised by computer technology" (Martin and Martin, 1994). Since video involves both privacy and confidentiality issues, ethical guidelines for HCI must go beyond general ethical codes.

---

[1] I suspect that the reason that political speeches sound so odd is that politicians have learned to speak entirely in "sound bites". Aware that most people will judge a speech from the short clips selected by the media for the evening news, politicians learn to speak in short phrases that will sound good, even when taken out of context. Unfortunately, most people have no experience talking this way and often find themselves looking ridiculous when interviewed.

The ACM Code of Ethics and Professional Conduct, revised in 1992, is generally considered to be the most complete. Anderson et al. (1992) state that the new ACM code "recognized the difficulty that ACM and other societies have in implementing an ethics review system and came to realize that self-regulation depends mostly on the consensus and commitment of its members to ethical behavior". Like Luegenbiehl (1992), they argue that the most important function of a code of ethics is its role as an aid to individual decision-making. They illustrate ethical issues with nine cases that call for individuals to make ethical decisions. Each case has an individual scenario illustrating a typical decision point that relates to sections of the code.

Bok (1982) reported that over 12,000 distinct ethics courses, including law, medicine, business, engineering, liberal arts, research sciences, religion and philosophy, were taught in American academic institutions. Discussing case studies in the class room has been shown to be an effective teaching approach (Dunfee, 1986) and the SIGCAS newsletter regularly presents such ethical case studies for discussion (e.g., Gotterbarn, 1993). Rather than argue about the merits of different ethical philosophies, I have chosen to follow this strategy, presenting scenarios based on real events and proposing guidelines related to the capture, production and presentation of video.

QUESTIONABLE USES OF VIDEO
The following examples of questionable uses of video are based on actual incidents. However, some of the details have been changed to disguise the participants or setting.

Candid Camera?
Linda is preparing her CHI'95 presentation and wants to give an entertaining talk. She looks through her videotapes of user sessions and finds several funny clips of users doing unexpected things. At the talk, she makes a joke and shows the clip; the audience laughs.

Is Linda guilty of perpetuating a "candid camera" approach, in which research videos become transformed into a form of entertainment at the expense of users? Is this an appropriate activity for professionals who purport to support users? On the other hand, does this mean that we can't have entertaining CHI talks or videos?

Lack of permission?
Jane is a trained anthropologist who has just conducted a study of work practices within a corporation. She and her colleagues have videotaped a number of meetings in which sensitive issues, such as determining who should be laid off, have been discussed. The participants are very sensitive about being videotaped and have requested that the videotape not be shown to anyone else in the company. Later that year, Jane presents her work to at a workshop at a CHI conference and includes several clips of video taken from her research.

Is this a violation of her agreement with the participants in her study? Is there a way in which she can disguise the video to prevent any possible feedback from the research audience to the company?

Is the reviewer responsible?
Ralph is reviewing presentations for a workshop he is running. Several of the participants propose to show video of users involved in their work. He decides that it is the responsibility of the authors to obtain the appropriate permissions and does not ask whether the authors have permission to present the tapes in this forum.

What is the reviewer's role? Should he remind the authors of their obligations? Should he go further and request evidence of having obtained appropriate permissions? Under what circumstances should he reject a submission?

Wrong audience?
Fred is developing a technique for combining real data with video simulations to provide training for pilots. He takes data from the flight recorders of planes that have crashed and recreates the situation, including external weather conditions and instrument readings. He plays one of his recreated videos to human factors colleagues, who suddenly find themselves listening to the voice of a real pilot saying: "Oh my God!" followed by a scream and a crash. The audience is stunned. Suddenly the very personal experience of another human being's death was being presented to them, without warning, as a part of a training exercise.

Was it appropriate to show a sensitive video designed for one audience to another? Was this a violation of the dead pilot's privacy? Could he have presented his work to this audience without using the real tape?

Undue influence?
Harry conducts usability studies of new software products for his corporation. He videotapes each usability session and carefully analyzes what causes the user's problems and where they make errors. He then discusses the issues with the software developers. Harry is particularly annoyed by one feature and wants to convince the software developer that it should be changed in a particular way. He shows a video clip of one of the users struggling with the feature as proof that his way is better. He does not show other clips in which users do not experience problems with the feature.

Is Harry taking advantage of people's willingness to think that video is an objective record in order to win an argument? Could Harry provide a more balanced view by presenting an overview of the relevant anecdotes? What would such an overview consist of?

Inappropriate special effects?
John is preparing a video of his new software system for the CHI'95 conference. He carefully records what happens on the screen and then edits out a number of "boring" sections in which the system responds especially slowly. He adds a cut to a separate system, which will eventually be integrated with his, to show what would happen if they were connected.

Under what circumstances is it reasonable to make a system appear faster or more complete than it is? Would a disclaimer, describing the level of editing, be sufficient?

Inappropriate reuse?

Mary is the product manager in charge of a new product being exhibited for the first time at CHI'95. She is proud of their usability lab and shows videotapes of some of the user studies to illustrate how well the interface works. When asked if she had obtained permission from the subjects of the video, she is surprised and says it had not occurred to her to do so. She believes she is safe, legally, since the people in the tape were company employees.

Even if she is not legally liable, does Mary have a responsibility to ask permission from the subjects? When is it appropriate to ask permission? Prior to recording, after the subject has seen the video, or just before each event in which the video will be shown. Is it possible for the subject to really understand what the implications of giving permission are?

Recording without permission?

The XYZ research laboratory allows people in the lab to communicate with each other via live video connections. Privacy issues have been carefully considered and there are a variety of ways for people to select how others may connect to their cameras. A separate program takes snapshots every few minutes from the media space and displays them in a window. One day, one of the participants in the media space walks into a room where a group of her colleagues is laughing at something. She discovers it's a picture of her, with someone giving her a kiss on the cheek (actually, her husband). Since it is impossible to see who the person is, the group laughingly teases her about who it might be.

What is the difference between a temporary record, in which a recently-shot image is displayed, and a more permanent record? Is it acceptable to select segments from an on-going stream of activity and highlight them?

Computing on video

All the previous examples have actually occurred, based on today's technology. We face a potentially much bigger problem with the advent of digital video. At SIGGRAPH '93, a panel of special effects experts showed a "behind the scenes" look at Jurassic Park, in which a stunt woman's image is changed to become that of the main actress. We fully expect special effects in science fiction movies and are amazed by the skill at which dinosaurs can be made to look real. What is less obvious is that special effects are used in most Hollywood movies to create images of reality. These techniques can be used to distort what we see.

Employers already monitor workers through computers. Pillar (1993) surveyed over 300 CEOs and MIS directors and found that 22% searched electronic mail, voice mail, computer files and other networking communications of their employees. Lyon (1994) discusses the role of electronic surveillance in society. Video is increasingly part of that electronic surveillance. For example, Great Britain has a new system that automatically reads the number plates (license plates) of a speeding car and displays the number, together with the excess speed, on a roadside display. The aim at present is to shame the offender, but the next step may be to link the system to a police database. In the past, people had to watch video from electronic surveillance cameras. Now, computers can watch for us.

The above list is not exhaustive, but illustrates problems of varying levels of severity. In most of these examples, the individuals are well-intentioned. In fact, some members of the HCI community will find nothing wrong with some of these scenarios. But this makes the issue problematic: we need to raise the level of awareness and try to establish guidelines that we can agree upon.

GUIDELINES FROM OTHER PROFESSIONS

Since Human-Computer Interaction is a new field, we should learn from other, more established professions. Some research disciplines, particularly the medical and social sciences, have well-established guidelines for using human subjects and include the use of videotaped records in this context. Other disciplines, such as computer science, have no history of using video (or human subjects), leaving HCI members from those fields without any guidance. Unfortunately, even those disciplines that *do* have guidelines for video do not provide sufficient guidance for the diversity of uses of video found in the HCI community. This section briefly summarizes the ethical or legal perspectives of various professions.

Medicine

Physicians have a long history of dealing with ethical issues. The Hippocratic oath urges physicians to "do no harm", i.e. to protect the patient. Key issues include who should choose a patient's treatment plan and how can patients without medical training evaluate risks or give informed consent about procedures. Doctors must present the options and supply all "material" information to the patient, but not necessarily provide full disclosure. Macklin (1987, p.45) describes the evolution of biomedical codes from the *professional community standard*, which asks "what reasonable medical practitioners in similar situations would tell their patients" to the current *reasonable patient standard*: "what the reasonable patient would want to know before giving consent to a recommended therapy." Studies show that poor communication and lack of information make patients more likely to refuse a particular treatment. This standard has helped doctors develop better relationships with their patients, with the accompanying danger that better relationships make it easier to obtain consent. Shannon (1976) and Beauchamp & Childress (1983) provide different views on biomedical ethics. Collste (1992) explores the question of whether computers, particularly expert systems, cause new moral problems.

Social Sciences

Experimental Psychologists who perform experiments with people are expected to follow guidelines established by the American Psychological Association (1991) or the relevant organization in other countries. Individual universities and organizations often publish guidelines, e.g., Queen's University (1989) or UCLA (1987). Most universities also have a committee that reviews research proposals and approves the procedures, e.g., the Massachusetts Institute of Technology Human Subjects Review Committee.

Subjects in Psychology experiments must sign a consent form that describes how any data collected about the subject will be used. After the experiment is completed, the experimenter is expected to "debrief" the subject and explain what occurred. Most guidelines are designed to protect the subject from harm. The APA guidelines were influenced by a famous set of experiments by Milgram (1965). Subjects were told to administer electric shocks to people (actually confederates of the experimenter) if they missed questions on a learning test. Milgram found that subjects followed these orders, even to the extent of believing they had killed the person receiving the shocks. Understandably, the subjects were traumatized by this experience.

Anthropologists and Sociologists work with people in field rather than laboratory settings. Videotape is increasingly used to record people's activities in the context of their daily lives. Both professions have also established ethical guidelines for the protection of their subjects. Critical issues include the problems of how to handle data collected in the field and how to handle naive informants who may not be able to give true informed consent.

Journalism

Hulteng (1985) describes the chief function of journalism as "the communication to the public of a reasonably accurate and complete picture of the world around us [...] The central ruling ethic of journalism [is] to report the news of the world dependably and honestly." (pp. 170-171) Broadcast journalists are thus ethically beholden to their audiences: they "protect" their viewers by presenting an "objective" account of an event. It is ethical to show a person negatively, as long as it is a "truthful" view. However, Hall (1978) explains that the FCC requires journalists to "contact the person attacked, provide a transcript of the charge and allow equal time for a response." Ordinary people (i.e., not celebrities) may not have their images broadcast without permission, unless the event is 'news' that occurred within the past 24 hours.

Hall discusses journalist's rights and responsibilities, from the Fairness Doctrine, which covers libel, slander and invasion of privacy to the Shield and Sunshine laws, which enable journalists to protect their sources. Kronewetter (1988), as well as Hulteng and Hall, discuss journalism ethics and Malcolm (1990) and Alley (1977) provide exposés of ethical violations.

Documentary film-makers do not believe in a single, objective point of view. Their goal is to present a fair perspective, from a particular point of view, through selective shooting and editing. Participants in their films should feel they have been presented fairly, if not always positively.

Marketing Firms

Marketing firms videotape "focus groups" to get customer reactions to new and existing products. Their loyalty is to producer of the products they examine. They must protect their clients, not only from potential lawsuits but also from information leaks to competitors.

Law and Accounting Firms

"Lawyer-client privilege" and "accountant-client privilege" (Causey, 1988) enable clients to speak in confidence to these professionals, another case of protecting the client, both legally and through ethical codes.

Publishers

Publishers must obtain copyright permission from the person who created the videotape before they can distribute it. They are legally responsible for protecting the producer (or copyright holder) of the videotape. Samuelson (1994) discusses legal precedents for the fair use of copyrighted material, including video, e.g., the ability of consumers to videotape broadcast television programs for home use.

Software Developers and Other Corporations

Corporations use video for a variety of purposes, from usability studies to product marketing. Getting permission protects the corporation from lawsuits. Hollywood's Universal Studios obtains global permission from their visitors: a sign informs them that, by entering the park, they have given tacit permission to be videotaped and their images may be used for commercial purposes. People who object are directed to a guest relations office.

Who are you trying to protect?

Trying to understand the goals of each of these professional guidelines reveals a fundamental problem: each is concerned with protecting someone, but they are all different types of people. Some try to protect the person being videotaped. Others try to create an objective view for the benefit of an audience. Some must protect the confidentiality of their clients, while others want to protect the producer of the videotape. The HCI community includes people concerned with each of these situations; our ethical guidelines must somehow address them all.

PRELIMINARY GUIDELINES

Who should the HCI community listen to when developing ethical guidelines for video? We have a diverse (and growing) set of uses of video, both as data about users and technology and as a presentation form for users, customers, management, fellow developers and the HCI research community. What perspective or perspectives should we consider? It is not enough to simply say we should "protect everyone"; we might end up avoiding video all together. We must consider the implications of a variety of uses of video and develop guidelines accordingly.

A good set of guidelines must cover everything from the initial videotaping to its final presentation and address, at least, the following questions: How do we obtain "informed consent"? How should recording of video be constrained? Are restrictions on the analyses performed necessary? Under what conditions should video be presented and to which audiences? Who are we trying to protect? How can people protect themselves and what social structures are needed to ensure that they can? What are the legal and cultural implications of videotaping in different countries? How do we avoid confusing ethics and good taste?

The suggestions presented below are offered as a starting point for discussion, rather than a definitive set of guidelines. They are based on discussions with members of the HCI community and influenced by guidelines from other professions. I encourage people to try them and provide feedback about what does and does not work.

For the purposes of clarity, the term *producer* is used to refer to any person who creates a videotape, including academic researchers, usability specialists and software developers. The term *user* refers to any person in the videotape, including participants in laboratory studies or people being videotaped in the course of their daily activities.

A. Prior to Recording
1. Establish what constitutes informed consent
Prior to recording, obtain *informed consent*[1]: make sure the user understands the implications of being videotaped. The producer must define what constitutes informed consent. This may be difficult, as in the introductory example.

2. Inform people of the presence of live cameras
If a camera is left on, e.g., in a media space or to record an event, let people know when they are on camera and give them the opportunity to avoid being in the camera's view. A sign should state whether or not the video is being recorded. For example, EuroPARC's media space uses a camera in the commons area. A mannequin holds the camera and a sign to let visitors know they are on camera.

3. Ask for permission before videotaping
Tell users that a videotape record will be made and give them the opportunity to speak off the record or stop the recording altogether. Consider if the user feels social pressure to agree and make it clear that saying no is legitimate. Avoiding social consequences may be difficult, e.g., when a meeting is taped and only one person objects.

4. Explain the purpose of the video
Tell users the expected purpose and other potential uses of the video. For example, videotapes from usability studies are sometimes re-used for advertising. Tell users whether separate video clips or the entire session could be used.

---

[1]The principal of 'informed consent' is to ensure that people do not give their permission for something without understanding the consequences. Getting a signature on a piece of paper is not sufficient. The person requesting consent is responsible for explaining the procedures and ensuring that these procedures, as well as the subsequent use of any resulting information, are fully understood.

5. Explain who will have access to the video
Tell users if anyone other than the producer will view the video. Users may not mind a researcher seeing a tape, but may feel uncomfortable if it is shown to colleagues, managers or general audiences, e.g. at a CHI conference.

6. Explain possible settings for showing the videotape
Tell users where the videotape could be shown. For example, at CHI conferences, videotapes may be shown to large audiences during talks, in small videotape viewing rooms, or on the hotel cable TV. In some corporate settings, some video clips may be used for advertising.

7. Explain possible consequences of showing the video
Producers may find it difficult to adequately convey how a user might feel if the video were shown in a certain setting. For example, a video clip shown on a television monitor to colleagues might be acceptable, but highly objectionable when projected on a 40 foot screen to a large audience.

8. Describe potential ways video might be disguised
If the video will be used in unpredictable settings, describe how the user's image will be disguised, e.g., through blurring the user's face. Mantei's (1990) "Strauss Mouse" video is a clever example of avoiding potentially embarrassing use of research videos; she used actors' hands to demonstrate the ways executives misunderstood a 'simple' computer mouse.

B. After Recording
1. Treat videotapes of users as confidential
Do not allow others to view videotapes casually and restrict access to them. This protects producers as well, e.g., if a manager decides to reuse video in ways that violate the original agreement between the user and producer.

2. Allow users to view videotapes
Ideally, give the user the opportunity to view the completed video. If this is not possible, the producer should consider ways in which people can be disguised. For example, some video editing systems can blur or distort a face.

3. If use of the videotape changes, obtain permission again
Asking permission is not a simple matter. Permission can be given before recording or after the user has been taped, or after the user has seen the tape, or just prior to an event in which it will be shown. The user can give blanket approval or approve individual events.

Give users sufficient information to make an informed choice and let them change their minds. For example, in the CHI'89 Kiosk (Soloman, 1990), users who contributed their images for the conference were again asked for their permission when the database was printed on a CD-ROM.

C. Editing Video
1. Avoid misrepresenting data
Producers are responsible for editing videos so as not to imply that particular events are representative if they are not. If video is presented as data, distinguish between anecdotal and representative clips of "typical" events.

2. Distinguish between envisionments, working prototypes and finished products

Clearly label presentations of technology as envisionments, working prototypes or finished products. Envisionments propose or illustrate ideas that have not been fully implemented. Working prototypes have been implemented and should not resort to tricks to make them look more complete. Products are completed commercial systems and must avoid misrepresenting their performance or features. For example, Wellner's (1992) videotape includes clearly labelled envisionments of future ideas contrasted with examples of working software.

3. Label any changes made to enhance technology

Show the actual time it takes for a particular operation or else clearly label cuts designed to improve the pacing of a video presentation. Do not simply cut out the slow sections to make your system appear faster.

D. Presenting Video

1. Protect users' privacy

Hide individuals when possible. For example, shoot over the user's shoulder to see the screen, rather than the user's face. Obviously, this only works if specific characteristics of the user, such as facial expressions, are not an essential part of the record. Consider disguising the user's voice.

2. Do not highlight clips that make users look foolish

Do not show "funny" clips to make users look foolish. This does not mean avoiding all amusing video clips; just be sure that the joke is not at the user's expense.

3. Educate the audience

When giving a presentation, educate the audience: rather than laughing at the user, explain how misconceptions about the technology can lead to breakdowns.

4. Do not rely on the power of video to make a weak point

Be careful when showing video clips to support arguments in favor of particular technology changes. Some video clips may magnify small problems or present a distorted picture.

5. Summarize data fairly

Clearly state the purpose of summaries of video data. Video data can be compressed in a variety of ways. Video clips can provide a shortened version of what occurred in the session or can be used to "tell a story". If clips are presented in random order, they can be combined to show "typical" interactions, highlight unusual or important events, or present collections of interesting observations.

D. Distributing Video

1. Do not use videos for purposes for which they were not intended

Do not allow video of users to be used for purposes that they are not aware of, e.g. for an advertisement.

NEXT STEPS

ACM/SIGCHI has already begun to address a few of the issues relating to video. Every year, attendees ask to videotape CHI conference presentations, often for good reasons, such as non-native speakers who want a video

backup. The SIGCHI executive committee is currently drafting a set of videotaping guidelines to try to balance the needs of audience members with the rights of presenters. The *vision.chi@xerox.com* mailing list has been the forum for the discussion of various drafts and the final version will be published in the SIGCHI Bulletin.

Another policy statement on video appears in the CHI Calls for Participation, e.g. from CHI'95: "Submission of video or pictures of identifiable people should be done with the understanding that responsibility for the collection of appropriate permissions rests with the submitter, not CHI'95." This gives submitters the unfortunate impression that this is solely a legal issue and that once permission has been obtained, the submitter and the conference have no further responsibility in the matter.

The CHI community, given its mix of disciplines and variety of activities, has a unique perspective to offer on the issue of ethics and video. We should take advantage of CHI-sponsored conferences to raise awareness and generate discussions, e.g. Mackay (1989, 1990). We can establish an electronic discussion forum and consider collaborations with other organizations, such as SIGCAS (Computers and Society), CPSR (Computer Professionals for Social Responsibility) and the Electronic Frontier Foundation.

In the late 1980's, SIGCHI sponsored a task force that produced the influential ACM SIGCHI Curricula for Human-Computer Interaction (Hewett et al., 1992). Perhaps the time has come for a similar task force to develop an HCI code of ethics that builds upon the general ACM code and addresses issues unique to HCI, such as video.

CONCLUSIONS

This paper illustrates how easy it is, however inadvertently, to misuse video. Because videotape has become so prevalent in our profession, it is time for us as a community to become aware of the potential dangers and develop guidelines for ethical handling of video. These guidelines must go beyond legal requirements and provide protection for a variety of people involved in the HCI community.

HCI is not the only professional field that uses video. We can learn from other professional ethical codes. However, we cannot blindly adopt other ethical codes. Each profession is concerned with protecting someone: the person in the video, the audience viewing the video, the client paying for the video or the producer of the video. Since the HCI community must address the needs of all of these people, we are uniquely positioned to create a broad-based set of guidelines that help us make informed, ethical decisions about our uses of video. If we are successful, guidelines may influence the wider set of organizations who are struggling with how to handle this powerful new medium.

ACKNOWLEDGEMENTS

The participants in the "Video as a Research and Design Tool" Workshop and attendees of the CHI'90 Discussion Forum on Video Ethics provided many insights and examples. Austin Henderson pointed out the importance of educating audiences when presenting "funny" clips of users. Annette Adler, Sara Bly, and Marilyn Mantei contributed interesting discussions and comments on earlier drafts. I am especially grateful to Michel Beaudouin-Lafon for reading earlier versions of this paper and suggesting the title.

REFERENCES

ACM Code of Ethics and Professional Conduct, (1993) *Communications of the ACM*, vol. 36:2, pp. 100-105.

Alley, R.S. (1977) *Television: Ethics for Hire?* NY: Abington Press.

American Psychological society (1991) The importance of the citizen scientist in national science policy, *APS Observer*, vol. 4, pp. 10-23.

Anderson, R.E., Johnson, D. G., Gotterbarn, D. and Perrolle, J. (1993) Using the new ACM Code of Ethics in decision making. *Communications of the ACM*, vol. 36:2, pp. 98-105.

Bagley, F.J. (1977) Ethics, unethical engineers, and ASME. *Mechanical Engineering*. vol. 99, pp. 42-45.

Beauchamp, T.L. and Childress, J.F. (1983) *Principles of Biomedical Ethics*, NY: Oxford University Press.

Bok, D. (1982) *Beyond the Ivory Tower: Social Responsibility of the Modern University*. Cambridge: Harvard University Press.

Clark, H.H. and Schaefer, E.F. (1989) Contributing to Discourse. *Cognitive Science*, vol. 13, pp. 259-294.

Causey, P.Y. (1988) Duties and Liabilities of Public Accountants. NY: Dow-Jones Irwin.

Collste, G. (1992) Expert Systems in Medicine and Moral Responsibility. *Journal of Systems Software*. Vol. 17, pp. 15-22.

DPMA (1989) DPMA Position Statement Handbook. 505 Busse Highway, Park Ridge, IL 60068.

Dunfee (1986) *Integrating ethics into the MBA core curriculum*. The Wharton School of business. Reseach funded by EXXON Educational Foundation.

Dunlop, C. and Kling, R. (Eds.) (1991). *Computerization and Controversy*. NY: Academic Press.

Forester, T. and Morrison, P. (1990) *Computer Ethics: Cautionary Tales and Ethical Dilemmas in Computing*. Cambridge: MIT Press.

Frankel, M.S. (1989) Professional Codes: Why, how and with what impact? *Journal of Business Ethics*, Vol. 8:2-3, pp.109-116.

Gotterbarn, D. (December 1993) Case Studies: The Captain Knows Best. *Computers and Society*. Vol. 23:3.

Hall, M.W. (1978) *Broadcast Journalism: an Introduction to News Writing*. NY: Comunication Arts Books, Hastings House.

Hewett, T. (1992) *ACM SIGCHI Curricula for Human Computer Interaction*, ACM, New York: NY.

Hulteng, J.L. (1985) *The Messenger's Motives: Ethical Problems of the News Media*. Englewood- Cliffs, NJ: Prentice-Hall .

ICCP (1989) ICCP Code of Ethics. *Your Guide to Certification as a Computer Professional*. ICCP, 2200 E. Devon Ave., Suite 268, Des Plaines, IL 60018.

IEEE (1979) *IEEE Code of Ethics*. IEEE, 345, East 47th St., NY, NY 10017.

Johnson, D.G. (1994) *Computer Ethics*, Second Edition. Englewood Cliffs,NJ: Prentice Hall.

Kronewetter, (1988) *Journalism Ethics*. NY: F. Watts.

Ladd, J. (1980) The quest for a code of professional ethics: intellectual and moral confusion. in R. Chalk, M. Frankel, & S. Chafer, Eds., *AAAS Professional Ethics Project*, American Association for the Advancement of Science, Washington, D.C., pp. 154-159.

Luegenbiehl, H.C. (1992) Computer professionals: Moral autonomy and a code of ethics. *Journal of Systems Software*. Vol. 17, pp. 61-68.

Lyon, D. (1994) *The Electronic Eye: The Rise of Surveillance Society*. University of Minnesota Press.

Mackay, W. (1989) Video as a Research and Design Tool. *Special issue of the SIGCHI Bulletin*, Vol. 21:2.

Mackay W. (1990) Video Ethics discussion forum, in *Proceedings of CHI'90 Human Factors in Computing Systems* (Seattle, WA), ACM Press, pp. 381-386.

Macklin, R. (1987) *Mortal Choices: Bioethics in Today's World*. NY: Pantheon Press.

Mantei, M. (1990) The Strauss Mouse. *SIGGRAPH Video Review*.

Martin, D.C. and Martin, D.H. (June 1994) Ethics Code Review.*Computers and Society*. ACM Press pp.21-25.

Milgram, S. (1965) Some conditions of obedience and disobedience, *Human Relations*, vol. 18, pp. 57-76.

Pillar, C. (July 1993) Bosses with X-Ray eyes: Your Employer may be using computers to keep tabs on you. *Macworld*. pp. 118-123.

Queen's University Department of Psychology (1989) *Ethics Clearance for Research on Human Subjects*, Queen's University, Kingston, Ontario, Canada.

UCLA Faculty Handbook, Supplement A: Policies and Procedures, (1987) University of California at Los Angeles Publications Office, Los Angeles, CA.

Salomon, G. (1990) Designing Casual-Use Hypertext: The CHI'89 InfoBooth. *Proceedings of CHI'90 Human Factors in Computing Systems* (Seattle, WA) ACM Press. pp. 451-458.

Samuelson, P. (January 1994) Copyright's Fair Use Doctrine and Digital Data. *Communications of the ACM*, vol. 37:1, pp. 21-27.

Shannon, T.A. (1976) *Bioethics*. NJ: Paulist Press.

Wellner, P. (1992) The Digital Desk. *SIGGRAPH Video Review*.

Multidisciplinary Modelling In HCI Design
...In Theory and In Practice

*Victoria Bellotti**　*Simon Buckingham Shum*†　*Allan MacLean**　*Nick Hammond*†

* Rank Xerox Research Centre	† HCI Group, Dept. Psychology	*Now at:* Apple Computer Inc.
61 Regent Street	University of York	One Infinite Loop
Cambridge	York	MS: 301-4A, Cupertino
CB2 1AB, UK	YO1 5DD, UK	CA 95014, USA
maclean.cambridge@rxrc.xerox.com	*{sjbs1, nvh1}@unix.york.ac.uk*	*victoria@atg.apple.com*

ABSTRACT

In one of the largest multidisciplinary projects in basic HCI research to date, multiple analytic HCI techniques were combined and applied within an innovative design context to problems identified by designers of an AV communication system, or media space. The problems were presented to user-, system- and design-analysts distributed across Europe. The results of analyses were integrated and passed back to the designers, and to other domain experts, for assessment. The aim of this paper is to illustrate some theory-based insights gained into key problems in media space design and to convey lessons learned about the process of contributing to design using multiple theoretical perspectives. We also describe some obstacles which must be overcome if such techniques are to be transferred successfully to practice.

KEYWORDS: Theory, cognitive modelling, formal methods, design practice, argumentation, design rationale, media spaces, multidisciplinary.

INTRODUCTION

HCI has become a highly eclectic and multidisciplinary field of applied science and design. A fundamental question we now must ask is, how can the theory base of the science keep up with and inform design at the leading edge of innovation? A number of empirical studies have analysed the uptake of analytic techniques in design practice, and report that typically, they have been narrow in scope, have lagged behind innovation and have made little impact upon practice [2, 22, 29]. The case for developing formal methods and systematic analytic approaches grounded in relevant theory has been weakened by this perceived failure to influence design practice.

The time and effort involved in systematic analytic approaches to HCI has always been a disincentive for applying them. Traditionally, the view has been that the costs involved are prohibitive other than for large scale or high risk implementations. However, we can argue that since users dislike having to learn the same thing again and again for each new release of a system, there is always

commercial advantage to be gained from getting the interface and interactive style for core functionality right first time. This is especially true of innovations for which there are likely to be many future releases and upgrades.

There can be little doubt about the benefits of insight into the underlying principles involved in complex user and system characteristics and behaviour which can help us understand how to make an interface better, if not right first time [e.g. 16, 21, 27, 28]. Such an understanding can inform design before prototyping and user testing is possible, and can also highlight the reasons for problems which users might have. However, whilst these benefits appear attractive, the problems with the existing techniques must be overcome if they are to be enjoyed.

If HCI analytic techniques are to support innovative commercial design they must be able to operate in a multidisciplinary fashion to cope with the broad range of issues arising for today's increasingly diverse systems, including multimedia and collaborative applications. Systematic accounts of usability must now transcend the traditional disciplinary boundaries that separate, for example, system, user and design analysis. They must also be able to support non-routine design of more exploratory systems where the problem space is not yet understood.

In this paper, we first describe how HCI analytic techniques provided valuable insights when used in '*multidisciplinary modelling consultancy*' for the exploratory design of a user interface in a digital media space development project. Secondly, we discuss a number of lessons for facilitating the transfer of such techniques to innovative design practice. Since the interaction between modellers, domain and practitioners was of central importance, we describe in some detail how the design context and modelling analyses were communicated.

AMODEUS: AN HCI MODELLING CONSORTIUM

This work is derived from the ESPRIT funded AMODEUS Project, which is now in its sixth and final year. This project has undertaken to develop a number of analytical HCI design approaches, with the goal of bringing theory from the cognitive and computing sciences to bear on the analysis of collaborative and multimodal interaction. Briefly, these techniques are as follows:

System modelling: formally specifying interactive system structure and behaviour [19], also modelling multimodal interfaces as agent architectures with the

Presentation, Abstraction and Control [PAC] framework [17].

Cognitive user-modelling: modelling the use of cognitive resources using the Interacting Cognitive Subsystems architecture for Cognitive Task Analysis (CTA), and the Soar architecture for problem solving using Programmable User Models (PUMs) [7, 26].

Interaction modelling: formal specification of a design in terms of interactions between user and system [1].

Design and integrational frameworks: describing design spaces explicitly; integration of the contributions of the above techniques in relation to design problems [4, 6].

The final component of the project is to *apply and evaluate these approaches* with practitioners [13, 15].

These are ambitious and largely unprecedented aims for a single project. The route to applied, multidisciplinary theory has taken the AMODEUS collaborators through several different kinds of work. Initially, effort was devoted to *establishing mutual understanding* between user-, system- and design-analysts; next, the focus was on *identifying common concepts and interdependency* between very different domains of theory [4]; finally, we have progressed to studies *evaluating the usability* of individual techniques [9, 12], and exercises in which we provide *HCI 'modelling consultancy'* to development projects in industrial [10] and innovative design contexts.

The work reported here falls into the last class of study. We describe a large scale exercise, conducted by 15 researchers for up to a month each (in some cases more), with the aim of demonstrating the following:

- Communication and interdependency between user and system modellers using an integrational design framework.

- Successful communication of modelling analyses to designers using an agreed, shared representation of the design problem and rationale.

- The value of modelling in HCI design analysis.

ECOM: A DIGITAL MEDIA SPACE

One of the design projects with whom AMODEUS partners collaborated was a strongly applied and technologically innovative industrial and academic project called EuroCODE (European CSCW Open Development Environment; Esprit Project 6155). It is developing a range of computer mediated communication and CSCW applications to support collaboration among engineers on the Danish Great Belt bridge and tunnel building scheme. AMODEUS worked with the project team whose job it was to develop the most sophisticated, multimedia applications.

This project team was located at EuroPARC in Cambridge, UK, along with two members of AMODEUS (co-authors of this paper). This meant that there were good opportunities for communication between members of the two projects. The AMODEUS researchers who were co-located with the designers were developing the QOC notation for design space analysis [4, 24]. This succinctly expresses as a semiformal network the key design questions being asked, alternative options being considered, and argumentation about them in the form of trade-offs among criteria. Such a representation shows how different design alternatives have

addressed a common set of problems. As described shortly, design space analysis played a key role in supporting communication between modellers and designers.

We focused on the design of *ECOM*, EuroCODE's media space, a digital audio-visual (AV) communications and messaging system for use over local, high bandwidth and long distance public telecommunications networks. This media space had a well documented design history, based as it was upon at least two existing analogue media spaces, the EuroPARC *RAVE* system [20]; and the Toronto University *CAVECAT* system [25].

PROVIDING HCI MODELLING CONSULTANCY

The modelling exercise involved the following overlapping phases: QOC analysts monitored the ongoing progress of ECOM's design, by interviewing the design team, attending meetings and examining documents and the evolving user interface prototype. Since we wanted to ensure that the AMODEUS modellers were working with real problems specified by designers, rather than issues which they might select to show off the strengths of their analytic techniques, we asked each ECOM designer at EuroPARC what they felt were key issues for the design of the ECOM media space. They raised three important issues for media spaces which became the basis for focusing analytic work in AMODEUS. These revolved around how the system should support:

- Making and breaking AV connections.

- Displaying general levels of availability for communication, and controlling person-specific levels of access for different people to connect to oneself.

- Controlling audio and video parameters (frame rate, quality and bandwidth).

A QOC design space was developed to represent the designers' rationale around these issues. This was based on notes from meetings and discussions with the ECOM designers and another media space designer at EuroPARC, plus research literature and experience of media space use. The design space was verified by the designers as a valid summary of their thinking about the prototype. It consisted of about one hundred QOC Questions with related Options and Criteria. Figure 1 gives an example of one of them.

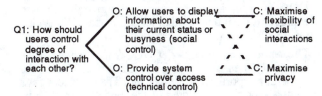

Figure 1: Example of QOC used to represent the designers' own rationale, prior to modelling. Criteria (C) are the bases for deciding among alternative Options (O). Questions (Q) identify issues which structure alternative Options. Solid Option-Criterion links are positive assessments, and dashed links are negative.

An *ECOM design document* was produced to communicate the design to the AMODEUS modellers. The contents of this document included:

- EuroCODE Project background, and a description of the domain in which ECOM would be used (bridge construction).

- A description of target users (construction supervisors and consultants) and a possible ECOM use-scenario.

- Screendumps of the ECOM prototype and other pre-existing, related media space interfaces for comparison.

- Key design issues raised by the designers as the focus for modelling, together with detailed descriptions of how relevant tasks would be accomplished using the ECOM interface or alternative system interfaces.

- The QOC design space showing the reasoning about the three issues with respect to the existing prototype and related systems.

- References to background documents on media space design and relevant literature.

The AMODEUS user-, system- and design-analysts were introduced to ECOM and related media spaces at several project meetings, supplemented by email and telephone conversations. There were also direct email exchanges between modellers and the designers themselves. The modellers were asked to assess the ECOM design from their own perspective and to submit a report of their contributions to assist in refining the prototype. They were instructed to focus on the issues, alternatives and assessments represented in the design space and to provide input in QOC, or QOC compatible terms, in order to avoid misinterpretation of their design contributions by any third party. This enabled the design space analysts to insert these different contributions (with page references back to the analytic reports) into the design space to show how they related to each other and to the ECOM designers' thinking. This *modelling-enriched design space* then served as an overview and index into the accompanying set of AMODEUS analytic reports.

We evaluated the comprehensibility and utility of the modelling and design space integration in several ways. We were especially interested to know if the ECOM design team felt that the modelling input would influence further design, and whether the design space provided a useful summary of that input. Specifically, we assessed the initial design space, the modelling-enriched version and the modelling reports, in the following manner:

- The main ECOM interface designer (D1) used the *initial QOC* (summarising the team's design rationale) to explain his design to another team member (D2) [video recorded session].

- The *modelling-enriched QOC design space* was separately presented to D1 and D2, together with the modelling reports which it indexed. The designers gave feedback on the usefulness of the modelling and QOC [a video recorded session plus a one day workshop].

- Designer D2 used the modelling-enriched design space to help bring a new team member up to speed with important ECOM design issues [video recorded session].

- Modellers presented some of their analyses to a number of media space designers at the CHI'94 conference. Presentations were followed by discussion on the potential of such approaches [3 hour meeting].

The next section provides examples of the lessons learned about specific aspects of the media space designs which were examined. In the following section we step back and describe some of the wider issues related to our experience

of the process of applying multiple theory-based perspectives to a design problem.

VALUE OF THEORY FOR MEDIA SPACE DESIGN

It is important for us to emphasise that there is real value in the output of theoretical analyses. HCI theory is only likely to be accepted by designers, and integrated and applied in practice if we can demonstrate that significant contributions can be made.

The AMODEUS analyses provided several design insights for ECOM which seem to be generalisable to other computer mediated communication systems. In this section we outline three examples of how analysts tackled the design of mechanisms for *displaying general availability for communication* and *setting specific accessibility levels*.

The ECOM interface

ECOM inherited its mechanism for displaying general availability from the Toronto CAVECAT system, and a person-specific access control mechanism from EuroPARC's RAVE system. In CAVECAT general availability levels correspond to iconic door states which the user can select. If they set their door to open, then both *glance* (short, one-way, video-only) and *connect* (two-way AV) connections are possible; if it is half open, then only connect is possible; if it is shut then the recipient of a call must explicitly accept a connection request at the time it is made; and if it is barred, then no connections are permitted. All users are treated equally in terms of whether they can or cannot connect. The RAVE system uses a mechanism by which users set a *permissions list* of people allowed to establish each type of AV connection (*glance, vphone*: a two-way, AV connection, initiated with a 'ringing' protocol allowing the called person to accept or reject the call; and *office-share*: a two-way, AV connection used for very long term links between offices).

CAVECAT's general and RAVE's person-specific access control mechanisms each have their advantages and disadvantages. In ECOM, it was decided that the best of both worlds might be achieved by combining these two kinds of access control, allowing users to select from four *general* availability levels and to choose *person-specific exceptions* for any level.

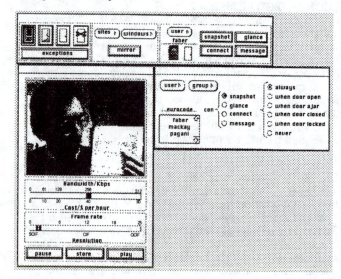

Figure 2*: ECOM media space prototype user interface.*

Figure 2 shows the interface to ECOM's person-specific access control mechanism. By hitting the exceptions button in the main dialogue box (top left), the user brings up the exceptions dialogue box (right hand side of the screen). They then select a user or group (members are displayed in the scrolling window), followed by con, a connection type (one of glance, snapshot, connect and message), and finally a state with which this connection exception is to be associated.

States range from always, through a series of door states through to never. Thus, when the user sets their general availability using one of the door icons in the top left of the figure, any specified exceptions will overrule the default permissions for that door state, for specified people.

The availability of other users can be seen at the time of selecting a connection to them. When making a connection the user selects the target's name from the user menu in the top dialogue box and the target's availability is shown by a door to the right of a small picture of them.

Illustrative modelling insights

There is clearly not space in a paper of this size to do full justice to all the analyses carried out on ECOM. Instead, our aim is to illustrate the kinds of insights we gained from the modelling using three examples:

1. Avoid confusion between the two separate concepts of *general availability* and *person-specific accessibility* (system architecture modelling and cognitive modelling).

2. Users must have *feedback* about their current accessibility settings (cognitive modelling).

3. If access permissions can be applied to groups as well as users, the design must resolve problems of *conflicting permissions* (formal system modelling).

We now describe briefly how the relevant modelling techniques guided the analysts to these conclusions.

The architecture modellers and both of the cognitive modelling groups highlighted the *potential for confusing ECOM's two concepts of general availability and person-specific accessibility*. In CAVECAT, general availability is intended to correspond to the current state of the user. In RAVE, person-specific accessibility corresponds to the user's current permissions settings, i.e., for each other user, which AV connections they request will be accepted and established by the media space.

ECOM attempted to combine these two notions, but in doing so confounded them somewhat, resulting in what was an extremely complex and possibly unworkable control mechanism. The architecture modellers recommended that these concepts should be dealt with separately and the cognitive modellers argued that they arose from two different user concerns which should be supported separately in the interface.

The architecture modellers identified domain, task and system concepts to be dealt with by software handling the making and breaking of connections. They modelled different ECOM connection types and levels of user availability. Their view was that the system architecture would need to maintain a database of *permissions lists* of permitted connections for all users and a representation of

each user's current availability. Each of these would best be handled by different system components.

A person's availability for communication depends on their current activities which may change from time to time. Whilst a user may be allowed to set a level of general availability, as in CAVECAT, this will not always correspond to their desire to be accessible (or inaccessible) to any specific user or group. ECOM gives the user the flexibility to define any set of connections to be permissible at any level. However, this means that the system architecture's representation of availability in ECOM may end up being arbitrarily related to accessibility settings and consequently somewhat meaningless.

The result is highly problematic. A user may forget what access permissions their door state availability levels correspond to. They might also find that they can or cannot connect to person X regardless of X's apparent door state. The system could show a door state which corresponds to the specific accessibility rather than the general availability level of X (this was, we later learned, the designer's intention); however, this defeats the idea of having separate representations for both availability levels and accessibility.

The cognitive modellers showed that ECOM made it hard to separate the task goal of making oneself more or less generally accessible (i.e., available) from that of choosing to be more or less accessible to a particular individual or group. The modellers proposed controlling general and specific accessibility levels separately. Furthermore, the user's current accessibility would be hard to keep track of due to the complexity of the possible exceptions. They recommended that the setting of person-specific exceptions should only be possible at one level of general availability.

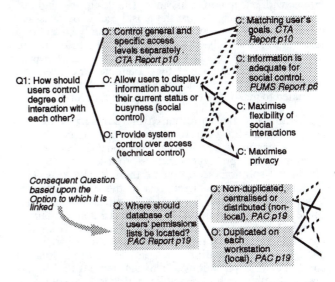

Figure 3: *Example of how the QOC design space was used to summarise modelling analyses of the initial design space. In this example, the analysis in Figure 1 has been added to by the PAC architecture modellers and the CTA and PUMs cognitive modellers.*

Figure 3 shows an example of the way in which modelling input to ECOM's design, and, in this instance, to solving the problem of confusing general availability and person-specific accessibility, were added to the design space as shown in Figure 1. These analytic contributions were always indexed back into their source in a modelling report so that details of the modelling could be examined.

The second contribution we outline is that *users must have feedback about their current accessibility*. This is all the more important if there are to be any exceptions to accessibility settings. Both of the cognitive modelling analyses identified the feedback issue. These suggested that since a user's availability levels were not likely to correspond to permissions settings, they would be unable to remember and therefore unable reliably to predict at any given moment whether any other user could make a particular kind of connection to them. This was compounded by the fact that ECOM did not provide a simple way to find out what any other particular user's exceptions were. One had to select their name and then inspect each connection type to see what the door state settings were.

The third insight we describe came from formal systems modelling (FSM). The complexity of managing media space connections is a good example of an HCI domain warranting the application of formal methods, given their ability to support verification and failure modes analysis. The FSM group formally specified how entities in the system would be defined in terms of their membership of certain sets and the properties they inherited from the sets they belonged to. Thus, users could be members of groups, but users and groups could have different access permissions. The FSM analysis revealed the *implementational requirement to resolve conflicting permissions settings*.

If access permissions are to be applied to groups as well as users, problems arise if users are members of one or more groups with excepted permissions. The designer could opt for the system to apply just the most specific permissions relating to a given user. However, if the only exceptions relevant to a user related to two groups they belonged to, and those two groups had different permissions, the user would reach an impasse if they attempted a connection which they were simultaneously permitted and not permitted to make, due to their group membership.

To summarise, the contribution of the analysts in this design consultancy exercise was to use their approaches to identify usability and implementational flaws in the existing ECOM prototype and to propose a number of alternative solutions to overcome these weaknesses. The modellers' various contributions to the original design space were combined into a single modelling-enriched space, as shown in Figure 3. This space also contained figures (showing example design solutions) from the analytic AMODEUS reports and annotations providing background information such as application domain constraints or explanations of the contributions to help the reader, together with indexing links back to relevant pages in the modelling reports.

Assessment of the value of modelling input

As detailed earlier, the ECOM designers were presented with the modelling-enriched design space and the analytic reports, and were asked to use them in a variety of ways and provide feedback. Overall, their response was that the analytic work had been useful and would certainly influence further design work on the ECOM system. The main designer of the prototype expected that the design space would provide a helpful overview of the issues for subsequent designers, particularly as they might not have ready access to him as he moved to other projects. In this study we did not gather data to be able to make detailed claims about the usefulness of QOC for designers. However, elsewhere we examine the empirical evidence for using design argumentation [11].

Both designers commented that the cognitively motivated design principles and illustrative redesigns of dialogues were helpful, and would be seriously considered for inclusion in future versions of the interface. Since the modelling exercise, the EuroCODE project decided (for reasons unrelated to AMODEUS) to pursue a simpler AV facility than the ECOM prototype we analysed. This means that to date we have not had the opportunity to track the extent to which the modelling recommendations are taken up. However, the issues raised by the designers (controlling multiparty access) will be a core feature of future media spaces, and the insights yielded by modelling on ECOM are not restricted to that particular system.

To judge the value of modelling, designers need to understand it. For example, the CTA modellers experimented with a dual-format report, in which a column of raw modelling notation on one side of the page was accompanied by another column of 'translation' for the designers to follow the analysis and recommendations. This provoked discussion on how much detail modellers should provide to back up their recommendations. ECOM's main designer suggested that it would be useful to have an intermediate-level CTA representation, accessible to a trained designer, to bridge between the raw modelling and analytic conclusions. This is in fact precisely what the CTA approach aims eventually to provide in the form of a tool which semi-automates the modelling process [26]. A trained human factors designer will be able to inspect a high-level representation of the cognitive model. Elsewhere, we report a study investigating the skills which the current prototype tool requires of its users [12].

Since modelling requires extra effort, any discussion of benefits must be balanced by the costs. It is extremely difficult to derive a useful measure of the cost/benefit trade-off in this exercise, since the modelling approaches are in many cases still research tools, being developed through exercises such as the ECOM collaboration. Balanced against *modelling* cost is the cost of generating the ECOM prototype which again is hard to assess, based as it was upon the design and use of other media spaces, and on design meetings involving four other members of the EuroCODE project.

We should also emphasise that we adopted a 'modelling consultancy' paradigm as an effective *research* strategy for investigating HCI modelling techniques, and not as the envisaged *modus operandi* for such techniques in the long term. Ultimately, we envisage modellers based at the same site as the designers, familiar with the design context, and able to provide direct input to the design team. This is the way that both analysts and designers (in this and other design teams with whom we have worked) say that they would like to work, and under such circumstances costs of modelling would drop significantly. In the real world, of course, such constraints are typical of large development projects, where different contributing stakeholders operate under time pressure and often with poor opportunities for communication [3, 18].

A realistic cost/benefit analysis for HCI modelling must therefore be based on modelling tools in use in realistic design contexts by trained practitioners. We have therefore begun to investigate the requirements for training practitioners in a number of these techniques (described further below).

In summary, we cannot yet verify that costs and benefits of HCI modelling were well balanced in this exercise. However, the issues analysed are central to media space design. We maintain that detailed understanding of the usability and implementational implications of design decisions at this early stage will have a pay-off for future systems evolving from the current generation of prototypes. A more detailed report of this modelling exercise is presented in [5].

LESSONS FOR TRANSFERRING THEORETICAL TECHNIQUES TO DESIGN PRACTICE

In addition to the kinds of insights which modelling can offer to designers, we also learned a lot about facilitating the process of applying multiple contributions from the research world to the world of innovative design practice. A key part of our strategy for bringing theory-based techniques into design practice is that significant resources need to be invested not only in the analyses, but also in the delivery mechanisms used to communicate them. This can be thought of as investing effort in the modelling's 'user interface' (in this exercise, the QOC design space and modelling reports) to make the 'underlying functionality' (the modelling) accessible and intelligible to the end-user.

As one might expect with such an ambitious exercise, the envisaged process did not run entirely smoothly. We are now in a better position to understand how this 'interface' can be improved for future exercises. The lessons we draw may assist others seeking to deliver analytic HCI techniques to practitioners. In a number of cases, the points may seem rather unremarkable; however, we suggest that they are comparable to the user-centred orientation which is taken for granted within the usability community—the principles are in some sense 'obvious', but instantiating them is rather more complex, and training others in them even more so. To date, very little attention has been paid to understanding how to make HCI modelling accessible to practitioners.

Making modelling less of a 'black art'

If analytic models are to be delivered to designers, they eventually need to be 'separated' from the expert modellers who developed them. In order to understand the skills needed to render a particular formalism useful, one needs to understand the modelling *process*. Once sufficiently understood, the challenge is to encapsulate this process in appropriate forms to train designers. To date, several studies have been conducted in order to understand the skills which analytic formalisms require from designers [9, 12], and we are developing training to communicate the hands-on practicalities of using techniques [8, 23].

Somewhat ironically, we ourselves failed to anticipate the difficulty which our project partners would have with using the QOC notation to summarise their modelling, even after five years of exposure to it. In retrospect, we should have provided some practical training in using the technique, as we have done for designers over several years. This serves to emphasise that, just as with interactive systems, 'end-user-support' is essential for analytic HCI techniques.

Practically applying an approach draws on very different skills to simply comprehending it in principle.

Provide positive as well as negative design critiques

A notable feature of the modelling analyses which was clearly highlighted by QOC's design argument structures, was the almost exclusively negative character of the assessments made by the analysts of ECOM. This result suggests two things. Firstly, that analysts may, like designers we have observed [4], focus on negative characteristics of designs at the expense of understanding what aspects of the design are positive and should be preserved. Secondly, purely critical feedback from modellers is unlikely to endear them to communities of designers, and they must ensure that they *recognise and reinforce good design decisions as well as point out flaws.*

If we reflect on the 'sociology of method uptake,' it is undeniably the case that new methods and tools are accepted and rejected as much on the basis of the way in which they are presented, and by whom, as on their technical merit. A sense of ownership of the techniques needs to be engendered on the part of the designers. *Practitioners should be respected as fellow professionals, invariably working in more highly constrained environments than researchers.*

Sharing representations: context and participation

There were two main classes of information exchange in the ECOM exercise; describing the design problem to the modellers, and describing the modelling analyses to the designers. We consider that in both cases the effectiveness of the mediating representations could be improved.

Firstly, given the that AMODEUS is pan-European, much communication took place via documentation, email, QOC, or analytic notations. The ECOM design document and related background documents constituted around 250 pages of text and design space. However, the modellers still reported that they were missing background context information concerning design goals, domain and system constraints and characteristics, users, and tasks to be supported. Had the analysts been more closely linked with the designers and the design context, common experience and informal communication would probably have reduced these documentation problems. In the existing literature, *most analytic modelling techniques underplay the importance of access to background information and of participating in the design context prior to formal analytic work.*

With respect to the design space, the designers, although having never encountered the QOC notation before, had little trouble understanding the space when they were asked to verify its contents. When later presented with the modelling-enriched version, they were able to see how the modelling augmented their original rationale. In contrast, the AMODEUS members had not participated in the context and discussions which shaped the initial design space. Whilst they had seen many QOC papers and presentations, they found it very hard to make sense of the argumentation, and relied heavily on the ECOM design document to make sense of it.

A conclusion to draw from this is that for a terse notation like QOC to be accessible outside the original design team, *supplementary detail is needed initially to give context to the analytic representation.* Further research is required to determine the factors which make argumentation and other shared representations intelligible to their users, particularly in situations where projects restrict the opportunities for participation and appropriate contextual experience.

Concretizing theoretical principles

The approaches being developed within AMODEUS place a strong emphasis on the value in design of formal abstractions, whether of task, user cognition, the system, user-system interaction, or the design space and rationale. In contrast, the designers with whom we worked here consistently used concrete representations in the form of interface sketches and mockups, as opposed to the abstract concepts and notations required in modelling.

We have found, both with ECOM and in other modelling consultancy studies, that example sketches of redesigns are an extremely successful way to communicate model-based design insights. Having identified certain principles as critical, the modellers devised simple design examples to satisfy those properties. This provides *something concrete to ground understanding and to reflect upon*, but is not meant to dictate design—having understood the principle at stake, a designer can tailor an example for a specific solution.

Design culture and design formalisms

A commitment to design analysis and representation is obviously compatible with large development projects which are hard to manage without additional representations, and with high risk projects where more formal models support rigorous verification. At present, however, such a design culture is somewhat in contrast to typical media space development teams. By and large, such designers are more comfortable with informal rather than formal analysis, and with constructing working prototypes rather than specifications.

Receptivity to HCI modelling will therefore be moderated by design culture—where a given 'culture' will attract people with particular skills and interests, and discourage others. One obstacle seems to be that the formalisms used by theorists presuppose certain process elements early in the lifecycle, such as a formal requirements specification or task analysis. These do not seem to have been features of the design of ECOM or other media spaces we have considered. A consequence of this 'culture clash' is that designers rooted in the exploratory, design-by-building culture have not been receptive to the idea of using more abstract system specification techniques, even with appropriate training. This was the general reaction at the meeting we held with media space designers at CHI'94, as well as with the ECOM designers. In contrast, work with a software engineering company who already use formal methods has met with much more positive feedback [10].

How is this gulf between cultures to be bridged? A key problem when different cultures meet is lack of mutual understanding. One of the driving forces behind our work within AMODEUS has been to *find ways to communicate to practitioners what modellers seek to do, how, and for whom*. We are acutely aware of the need to provide sufficient background modelling information for designers before attempting detailed discussions. One encapsulation of this background material is a set of executive summaries and short worked examples for each technique, written for designers with whom we are collaborating, and for other interested parties [14].

Of course, education needs to be two-way, and we are working to *understand how HCI modelling techniques can be tailored to the demands of current design practice*. Ultimately, we want to see the fruits of research being exploited by practitioners, not just an elite group comprising those who have developed analytic techniques.

Consequently, we are challenging the conventional notion within the software engineering community that formal methods are only useful if used within a structured development context from the beginning of a project, through refinement, to implementation. This seriously restricts the class of project in which formal methods can be used, cutting out exploratory design projects. The ECOM modelling exercise exemplifies a strategy of *selective* formal specification of particularly complex HCI problems in order to maximise the cost/benefit trade-off.

Selective modelling is also the envisaged strategy of use for cognitive modelling approaches seeking to develop software tools to support rapid re-modelling [7, 26]. Similarly, when training designers to capture and reflect on their rationale using QOC, we emphasise the different ways in which it may fit into their design context, and the importance of using it strategically to maximise the gain from the effort.

In summary, the power and influence of design culture on attitudes to modelling techniques should not be underestimated. In our role as modelling consultants, we explored new ways to communicate modelling to practitioners (by integrating modelling with QOC, and concretising abstract modelling principles). In the longer term, with a view to seeing the techniques in use by designers themselves, we are seeking to *adapt the application and delivery of theory-based modelling techniques to the contexts in which they must be used*.

CONCLUSION

There has been considerable debate in recent years over whether the HCI science base has anything to contribute to real design. In the modelling-design exercise reported here, we have demonstrated how formal modelling techniques grounded in cognitive science and software engineering theory can inform innovative and exploratory user interface design such as that of a digital media space. Moreover, we have explored some of the key issues for communicating such analyses in accessible ways, and for delivering such techniques in forms which are compatible with design practice. We maintain that theoretically grounded HCI design techniques such as these are powerful ways to manage the implementational and cognitive complexity of emergent interface technologies, particularly as they find their way into mainstream and large scale software development. However, *analytic techniques will only achieve uptake if the end-user requirements of design practitioners are properly understood, and the value of such techniques can be demonstrated*. In the exercise reported here, we have made progress in better understanding how to make that match with a design culture which has traditionally shied away from more formally based approaches.

ACKNOWLEDGMENTS

This work is funded by ESPRIT BRA 7040. We are grateful to our AMODEUS colleagues whose varied perspectives have contributed to the work presented here. Our thanks also to the EuroCODE design team for their cooperation, to the BT, EuroPARC, PARC, SunSoft and U. Toronto media space designers for their feedback, and to the CHI reviewers for their comments.

REFERENCES

Note: AMODEUS documents are available by WWW & FTP:
http://www.mrc-apu.cam.ac.uk/amodeus/amodeus.html
ftp://ftp.mrc-apu.cam.ac.uk/pub/amodeus

1. Barnard, P.J. and Harrison, M.D. Towards a Framework for Modelling Human-Computer Interaction. In *Proceedings of EWHCI'92: East-West Conference in HCI*, 1992, pp. 189-196.

2. Bellotti, V. Implications of Current Design Practice for the Use of HCI Techniques. In *Proceedings of HCI'88: People and Computers V*, 1988, CUP: Cambridge, pp. 13-34.

3. Bellotti, V. A Framework for Assessing Applicability of HCI Techniques. In *Proceedings of IFIP INTERACT'90: Human-Computer Interaction*, 1990, Elsevier: Amsterdam, pp. 213-218.

4. Bellotti, V. Integrating Theoreticians' and Practitioners' Perspectives with Design Rationale. In *Proceedings of InterCHI'93: Human Factors in Computing Systems*, 1993, ACM Press: NY, pp. 101-106.

5. Bellotti, V., Buckingham Shum, S. and MacLean, A. Assaying Multidisciplinary Modelling Using QOC as a Mediating Design Expression: The 'EuroCODE' Design Exemplar. Amodeus-2 Project, *Working Paper ID/WP30*, 1994.

6. Bernsen, N.O. Structuring Design Spaces. In *Adjunct Proceedings of InterCHI'93 (Short Papers)*, 1993, ACM Press: New York, pp. 211-212.

7. Blandford, A. and Young, R.M. Developing Runnable User Models: Separating the Problem Solving Techniques from the Domain Knowledge. In *HCI'93: People and Computers VIII*, 1993, CUP: Cambridge

8. Blandford, A.E. and Young, R.M. A tutorial introduction to the PUM Instruction Language. Amodeus-2 Project, *Working Paper UM/WP29*, 1995.

9. Buckingham Shum, S. Analyzing the Usability of a Design Rationale Notation. In *Design Rationale: Concepts, Techniques, and Use*, Moran, T.P. and Carroll, J.M., (Ed.), LEA: Hillsdale, NJ, (in press).

10. Buckingham Shum, S. (Ed.) Preliminary Modelling of the CERD Flight Sequencing Tool. Amodeus-2 Project, *Working Paper TA/WP23*, 1994.

11. Buckingham Shum, S. and Hammond, N. Argumentation-Based Design Rationale: What Use at What Cost? *International Journal of Human-Computer Studies*, 40, 4, 1994, pp. 603-652.

12. Buckingham Shum, S. and Hammond, N. Delivering HCI Modelling to Designers: A Framework and Case Study of Cognitive Modelling. *Interacting with Computers*, 6, 3, 1994, pp. 311-341.

13. Buckingham Shum, S. and Hammond, N. Transferring HCI Modelling and Design Techniques to Practitioners: A Framework and Empirical Work. In *Proceedings of HCI'94: People and Computers IX*, 1994, CUP: Cambridge, pp. 21-36.

14. Buckingham Shum, S., Jørgensen, A., Hammond, N. and Aboulafia, A. (Eds.) Amodeus HCI Modelling & Design Approaches: Executive Summaries & Worked Examples. Amodeus-2 Project, *Working Paper TA/WP16*, 1994.

15. Buckingham Shum, S., Jørgensen, A.H., Aboulafia, A. and Hammond, N.V. Communicating HCI Modelling to Practitioners. In *Proceedings of ACM CHI'94: Human Factors in Computing Systems*, 1994, ACM: New York, pp. 271-272 (Vol.2).

16. Carroll, J.M. and Rosson, M.B. Deliberated Evolution: Stalking The View Matcher in Design Space. *Human-Computer Interaction*, 6, 3&4, 1991, pp. 281-318.

17. Coutaz, J. PAC: An Object-Oriented Model for Dialog Design. In *Proceedings of IFIP INTERACT'87*, 1987, Elsevier: Amsterdam, pp. 431-436.

18. Curtis, B., Krasner, H. and Iscoe, N. A Field Study of the Software Design Process for Large Systems. *Communications of the ACM*, 31, 11, 1988, pp. 1268-1286.

19. Duke, D. and Harrison, M. Abstract Interaction Objects. *Computer Graphics Forum*, 12, 1993, pp. 25-36.

20. Gaver, W., T., M., MacLean, A., Lövstrand, L., Dourish, P., Carter, K. and Buxton, W. Realising a Video Environment: EuroPARC's RAVE System. In *Proceedings of ACM CHI'92: Human Factors in Computing Systems*, ACM Press: New York: Monterey, CA, 1992, pp. 27-35.

21. Gray, W.D., John, B.E. and Atwood, M.E. The Precis of Project Ernestine or an overview of a validation of GOMS. In *Proceedings of ACM CHI'92: Human Factors in Computing Systems*, 1992, ACM Press: New York, pp. 307-312.

22. Gugerty, L. The Use of Analytical Models in Human-Computer Interaction. *International Journal of Man-Machine Studies*, 38, 1993, pp. 625-660.

23. MacLean, A., Buckingham Shum, S. and Bellotti, V. Design Rationale Tutorial. *Tutorials at HCI'92-94: British Computer Society HCI Conferences, UK*, 1992-94

24. MacLean, A., Young, R.M., Bellotti, V. and Moran, T. Questions, Options, and Criteria: Elements of Design Space Analysis. *Human-Computer Interaction*, 6, 3 & 4, 1991, pp. 201-250.

25. Mantei, M.M., Baecker, R.M., Sellen, A.J., Buxton, W.A.S., Milligan, T. and Wellman, B. Experiences in the Use of a Media Space. In *Proceedings of ACM CHI'91: Human Factors in Computing Systems*, 1991, pp. 203-208.

26. May, J., Barnard, P.J. and Blandford, A. Using Structural Descriptions of Interfaces to Automate the Modelling of User Cognition. *User Modelling and Adaptive User Interfaces*, 3, 1993, pp. 27-64.

27. Monk, A., Carroll, J., Harrison, M., Long, J. and Young, R. New Approaches to Theory in HCI: How Should We Judge Their Acceptability? In *Proceedings of IFIP INTERACT'90: Human-Computer Interaction*, 1990, Elsevier: Amsterdam, pp. 1055-1058.

28. Payne, S.J. Understanding Calendar Use. *Human-Computer Interaction*, 8, 2, 1993, pp. 83-100.

29. Rosson, M., Maass, S. and Kellogg, W. The Designer as User: Building Requirements for Design Tools from Design Practice. *Communications of the ACM*, 31, 11, 1988, pp. 1288-1298.

Design Space Analysis as "Training Wheels" in a Framework for Learning User Interface Design

J.W. van Aalst*, T.T. Carey, D.L. McKerlie

Department of Computing and Information Science
University of Guelph
Guelph, Ontario, Canada N1G 2W1
email: tcarey@snowhite.cis.uoguelph.ca
email: mckerlie@snowhite.cis.uoguelph.ca

**Delft University of Technology
Department of Technical Mathematics and Informatics
Delft, the Netherlands
email: jw@snowhite.cis.uoguelph.ca*

ABSTRACT

Learning about design is a central component in education for human-computer interaction. We have found Design Space Analysis to be a useful technique for students learning user interface design skills. In the FLUID tool described here, we have combined explicit instruction on design, worked case studies, and problem exercises for learners, yielding an interactive multimedia system to be incorporated into an HCI design course. FLUID is intended as a "training wheels" for learning user interface design. In this paper, we address the question of how this form of teaching might mediate and extend the learning process and we present our observations on Design Space Analysis as a training wheels aid for learning user interface design.

KEYWORDS: HCI education, Design Space Analysis, design rationale, design skills, interactive multimedia

1. INTRODUCTION:
A TRAINING WHEELS SYSTEM
FOR USER INTERFACE DESIGN

"I think you learn design by apprenticeship. There isn't a recipe we can give but there are some things that everybody picks up, like figuring out what the real problem is, considering the user, ..." [14]

"Teaching as mediating learning involves constructing environments which afford learning about descriptions of the world...to allow more to be learned than is available from experience alone." [12]

Learning about design is a central component in education for human-computer interaction [5]. To teach basic design skills we have to provide learners with opportunities to "learn by doing" on concrete projects, and opportunities to "learn by reflecting" [17] on both their own projects and case studies of other user interface designs.

We have found Design Space Analysis (DSA) [13] to be a useful technique for students learning user interface design skills. Design Space Analysis provides a vocabulary of Issues, Options and assessment against Criteria, and a notation to record the current state of a design in shorthand form. In previous research, we have used adaptations of DSA to promote reflection on design case studies [4, 3]. In the FLUID tool described here, we have incorporated explicit instruction on design and problem exercises for learners, yielding an interactive multimedia system to be incorporated into an HCI course. In the course, the students will work in pairs with FLUID as active preparation for studio sessions involving group design projects.

In the spirit of "teaching as mediated learning" as defined above, we use DSA as a description of the considerations emerging in design. For case studies, this allows us to summarize the important design advances in ways that encourage generalization. For the design exercises, it allows us to support design activities by presenting video clips of advice and war stories. We use the "training wheels"[1] analogy to clarify the role of DSA in the learning context. Here is the explanation given to users of FLUID:

"We will be asking you to construct your design according to a design method called Design Space Analysis (DSA). We use DSA here as a "training wheels" system for learning design. Remember the training wheels on your first bicycle? They helped you to get going and to avoid initial problems. Before long you found you didn't need the wheels any longer because you had internalized the necessary skills. Our requirement that you use DSA is like that: good designers may not have to be so explicit about these steps any more because the processes have become automatic (well, not quite automatic — often we still need to reflect on the design process, so you will find DSA handy later on as well)."

We thus would like to treat Design Space Analysis as a technique for representing design considerations and processes, not as a theory of design practice. However, any notation carries with it implications for the task processes it supports. One interesting aspect of the development of FLUID has been the resulting observations about Design Space Analysis and its implicit model of design.

The case studies in FLUID note that the material has been adapted to serve the training wheels role: the original designers did not necessarily represent their designs in this way. However, design documentation normally rationalizes design history to highlight key decisions [15]. The FLUID tutorial also notes that Design Space Analysis can be useful in other specific roles during design [2].

In the next section we describe a walkthrough of a session with FLUID as a typical learner might experience it. This raises interesting questions about how we learn to design and how teaching might mediate and extend the learning process. In section 3 we discuss the Design Rationale for FLUID, including links to other current research on interactive tutoring systems and reflection about our design based on usability testing. In section 4 we present our observations on Design Space Analysis as a training wheels aid for learning user interface design, including notes on what we learned about DSA through the process of specifying its use in the FLUID Tutorial. The concluding section outlines our current work to deploy and generalize the FLUID system.

---
[1] Our use of the term was inspired by [2], although our training wheels are both prescriptive and restrictive.

2. A WALKTHROUGH OF FLUID USAGE

In this section, we present typical experiences of a learner working through a design exercise in FLUID, an acronym for "Framework for Learning User Interface Design"[2].

FLUID is a learn-by-doing system in which the central learning task is a design exercise. Currently, the learner is asked to be the user interface designer of an interface which makes it easy to place, move and rotate furniture objects in a virtual room, using a standard mouse for direct manipulation. This exercise is adapted from a project report by Stephanie Houde of Apple Computer [9].

Initially, the learning task is presented by a team Manager character appearing in a video window. The learner also meets the client (a furniture shop manager), as well as two other characters who will offer advice during the design process — a Tutor on DSA and a senior Designer. Figure 1 shows the Manager and Tutor pointing out various aspects of FLUID that the learner may wish to explore.

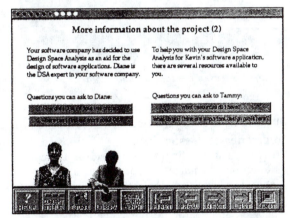

Figure 1. The Tutor and the Manager, introducing the student to the project.

Typically, learners will work through the DSA Tutorial first, although they are free to move back and forth amongst the other components including a Library of worked cases, a Table of Concepts and the design Workbench in which the exercise is carried out (all discussed further below).

The Tutorial introduces the concepts of Design Rationale, DSA, and the DSA notation. In addition, the Tutorial contains some real life "war

---
[2] For those interested in technical specifications, the current FLUID system is implemented in HyperCard 2.2, with video clips (about 150) incorporated as QuickTime™ movies through VideoFusion™.

stories", identifies some disadvantages of DSA, summarizes the different roles that DSA takes on during design, and offers guidelines as to when it is (and is not) advisable to use DSA. The prototype case Library currently contains a single exemplar system, adapted to use the vocabulary and process of DSA. This case is about the design of a device for easy rotation of an object about an arbitrary axis in 3-D space [6]. For the exercise and the worked case, we deliberately chose two related topics to make it easier for the learner to apply the lessons from the case study. Figure 2 shows data from usability testing in the case study (the case study also contains a video clip of a 1981 3-D rotation device [7]).

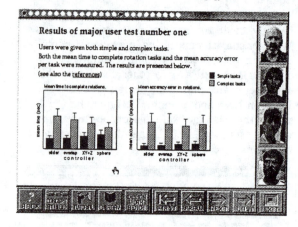

Figure 2. A case from the exemplar Library.

The FLUID design Workbench provides the standard tools of DSA, including a DSA diagram to describe the design Issues[3] , Options, and Criteria, a usage Scenario to describe typical interactions in narrative form, and a design history log which we can use later to show the learner's actions. In addition to the Tutor, Manager, and senior Designer, the learner can also access a group of Users; they report on the usability of the emerging design. The comments of the users are taken from usage data in the published report [9]. A Demonstration is available, to take the learner step-by-step through the basics of filling in a DSA diagram in the Workbench. Figure 3 shows the various navigation and guidance features.

In the introduction of the design exercise, the Manager presents an initial idea for the interface (corresponding to the starting point in [9]). The initial usage Scenario repeats the Manager's description, and the DSA diagram likewise is seeded with this initial information. If learners

request a user test at this point, the Users[4] will tell them that more details will need to be filled in about how the interface works. To add detail or change the current design, learners edit the usage Scenario and the DSA diagram. By selecting a set of design choices, the learner can form a new design Alternative and present it to the Users.

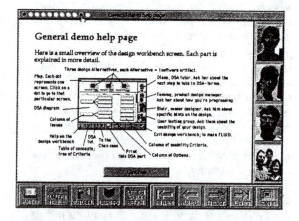

Figure 3. FLUID help screen. The bottom left shows the FLUID components: Table of Concepts, Tutorial, Case library, and Design Workbench.

For example, Figure 4 shows the state of a design after the learner has recognized one of the limitations of the initial idea. In the initial idea (as presented in [9]), a hand-shaped cursor is used to directly manipulate points on the furniture objects, each point corresponding to a different action such as lifting, sliding and rotating. In practice, users had divergent expectations about which part of an object would correspond to a particular movement. The design team introduced "narrative handles", hands which appeared on an object and indicated, by their orientation, the movement invoked by selecting them.

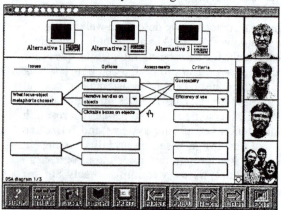

Figure 4. the FLUID Design Workbench.

---

[3] We use the term Issues where DSA uses Questions.

[4] Technically, it should be a User Test group or a Prototyper returning this information, but for simplicity we chose to have the learner interact directly with the Users.

To add this Option as a possible choice for a new design, the learner must edit the usage Scenario to describe a new style of interaction and then edit the DSA diagram to record this choice (these edits can be in either order). In Figure 4 the learner is adding the Option 'Clickable boxes on objects' as a new solution to the Issue called 'What focus-object metaphor to use?'.

Throughout this process, the Tutor, Manager and senior Designer provide help on request. For example, if the learner requests assistance after hearing from the Users that *"the hands don't seem to be in places they expected"*, the Tutor will suggest (in a video clip): *"Think of some other ways to do things (a new Option)"*.

In order to provide this context-sensitive assistance, FLUID must monitor the learner's actions and map them against a state chart which determines what advice to offer. This requires the following features during the process of changing a usage Scenario[5] or DSA diagram:

- learners enter their name for a new Option or Issue in the diagram;
- FLUID provides a list of likely entries in the current state and asks the learners to identify which one is closest to their idea;
- learners can select a part of the usage Scenario to edit, and type in their desired change; and
- FLUID provides a list of likely changes in the current state and requests a selection as above.

We discuss in section 3 the rationale and effects of these features on the learning process.

At the end of the design exercise, the learner can review the project with the various characters. If the design is incomplete, the characters suggest a return to the design Workbench, otherwise they congratulate the learners upon their exercise and suggest follow up reading before the scheduled group design studio exercise.

FLUID was tested with prospective students of the University of Guelph undergraduate HCI course. Completing their DSA diagrams using the design Workbench took on average 59 minutes with a range from 39 to 90 minutes. This does not include the time taken to work through the Tutorial or Library case.

---

5 The Scenario screen is not shown in this paper.

3. DESIGN RATIONALE FOR FLUID
We outline in this section a number of design issues raised by FLUID, and evaluate the choices we made based on preliminary usability testing.

Objectives
Novice user interface designers have to develop expertise in the following basic design skills:

- judging the adequacy of their task knowledge and task analysis;
- recognizing implicit design decisions;
- anticipating user actions/concepts;
- relating observed usage problems to design alternatives;
- envisioning different design choices;
- understanding and balancing design tradeoffs; and
- managing the design process.

The first and last of these objectives are not addressed in the current version of FLUID. Our use of DSA highlights the other objectives for learners, because DSA requires explicit steps corresponding to each of those skills and the creation of an explicit artifact of DSA as a result. This mapping allows us to offer focused advice based on the DSA constructed at any point.

FLUID Conceptual Model
We based FLUID on the Goal Based Scenario concept of Schank and his colleagues [16, 1]. This approach combines principles of active learning, case-based teaching and expert tutoring. In the Goal Based Scenario architecture, learners experience the learning task through a *cover story* (e.g., you are a designer and your manager is giving you an assignment), a *mission* (e.g., design a way to manipulate the furniture in 3-D with a standard mouse) and a *focus* for actually doing the task (e.g., our Design Workbench).

We observed that FLUID users understood the context of their mission in terms of the cover story and got into their roles. However, we need to reinforce the cover story when they encounter the Workbench, since for typical students, the Tutorial and case Library are used after the mission description and before the Workbench.

FLUID Interaction Style
We incorporated in FLUID interaction elements to provide flexibility in learning styles. A Table of Concepts, still under construction, supports learners who want to develop an explicit conceptual model of the knowledge domain (learners can currently access a concept map of usability criteria, containing definitions and

examples from published usability studies. Figure 5 shows one quadrant of the concept map).

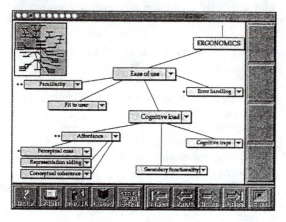

Figure 5. Quadrant of the Table of Concepts.

The Tutorial supports learners who prefer to observe and reflect before launching into a task on their own, and the case Library supports learning by example. In the HCI course, this structure also allows us to relate FLUID to other content: students later do a learning styles assessment and discuss its implications for supporting users with various learning styles.

The Tutorial and case Library appeared to provide users with the background information necessary for learning about DSA. All users explored both interaction elements, suggesting that perhaps these provide complimentary information rather than redundant information. No users referred back to these interaction elements during construction of their DSA in the Workbench.

Choice of Design Exercise

For the exercise which learners would work on, we wanted to adapt a published case study from an industrial environment, so that:

- empirical usage data would be available;
- the design exercise would be perceived as realistic;
- learners would become more aware of the professional literature; and
- learners could read the published paper to contrast the sanitized training wheels version with what actually happened.

We also wanted the design exercise to be novel for the learners, both for motivation and to avoid routine copying from an existing system. Finally, the exercise had to be non-trivial but not require extensive domain knowledge.

The published case study by Stephanie Houde [9] met all of these criteria. A scanned image of the paper is available in FLUID after learners complete the design exercise. The choice of another study to include as a worked case example was largely determined by this choice of design exercise. The study by Chen et al. [6] addressed a similar design problem but with different constraints (i.e., designing a new controller device in [6] versus designing a new way to use an existing device in [9], for similar tasks).

We found that users appreciated the concrete examples provided by both cases. However, there was no evidence that selecting cases from the same domain was advantageous, suggesting that our concern over similarity between cases may have been exaggerated.

Design Workbench

We are not satisfied with the separation between the design Workbench, where learners build a DSA for the exercise, and the other products of design such as the sketches they might make as they work out their designs. The Workbench manipulates DSA, not the entire design. We knew we would not be able to build a design toolkit and also integrate the instructional component, so we accepted the compromise of just developing the DSA Workbench and leaving learners to construct other design products in some other medium (the other activities learners should be doing in parallel with FLUID are described by the senior Designer in the Tutorial). We discuss in section 5 below some of the issues involved in creating a more capable Workbench.

None of the users expressed a need for more creative tools to express their design ideas, although one user took great effort to describe his design idea to the observers of his session with hand gestures. Other users made suggestions for improvement.

Providing Assistance

"For maximum learning to take place, the student should receive just enough of a push or guidance to get past the sticking point without being explicitly told the answer" [8].

It is possible to talk of a future system in which learner input is analyzed for natural content and an appropriate interpretation is made of the meaning of a design choice. There is no such intelligence in the current FLUID system. When users describe a new design Option, they are prompted to identify their choice from a list of preset labels. This mechanism will fail if the list of available choices is not complete enough — i.e., FLUID does not anticipate the users' choices — or if the list presents choices the user has not yet thought of and thereby pre-empts some of the desired learning.

Through pilot testing we established the likely choices, and then structured FLUID to contain nested lists which expose further choices from 'Other' submenus. The final 'Other' submenu notes that the user has produced a choice not included in FLUID and requests feedback by email to the FLUID developers.

The successful aspect of limiting design Options was that users could eventually build an understanding of the set of Options and formulate a meaningful Issue for the DSA diagram. Issues were not constrained in any way, allowing users to create their own labels for each Issue. The well-constructed set of Options provided an opportunity to guide learning about Issues — from previous work, Issues are known to be a difficult concept to teach.

However, test users commented that they felt constrained by these anticipated design Options. With our initial test users, unanticipated Options generated during usage testing (e.g., a 3-D crane instead of the 2-D hands) could have been allowed without jeopardizing the rest of the design space.

There was great variability in use of characters' advice between users during construction of the DSA diagrams. Some users consistently asked for advice from the characters after each addition to the DSA diagram, while others only sought approval after their DSA diagrams were complete.

Roles for Advisor Characters
Ideally, each of the characters providing input in the design exercise should have a distinctive role, so that learners know who to ask for specific needs, and to provide context-sensitive help [8]. We tried to structure the on-line characters so that the DSA Tutor and project Manager provided generic help about the design process, and the Senior Designer provided specific help about the design domain (and occasional reflections on the role of DSA outside the training wheels setting). This proved to also be a useful distinction in terms of future generalizations of the system: for a new design exercise, the specific hints from the Designer would need to be changed but the Tutor's and Manager's video segments are re-usable as is.

We observed that users recognized the differences in the type of advice offered by each of the characters. In general, users preferred the advice from the DSA Tutor for two reasons: the content of the advice seemed most applicable to their task, and the videos for this character tended to be the most engaging. Users consistently commented on the entertainment value of the video clips. These also appeared to be an important motivator for learners to continue to discover new information. However, we noted that in some cases the entertainment value may have been "too engaging" (see also [10]). In some instances the media novelty caused users to miss the important point in the content.

4. DSA AS TRAINING WHEELS FOR USER INTERFACE DESIGN

Issues as Design Dimensions
We found it necessary to emphasize the concept of design Issues in DSA as characterizing dimensions along which designers could look for alternatives. Making this explicit in the Tutorial helped clarify for learners the nature of an Issue in DSA. For example, at a key point in the design exercise, the learners must recognize that their conception of an Issue is too limited: seeing the Issue as "how are users shown selectable parts of an object?" precludes the idea of a manipulable box outside the object itself, which requires the more general conception of the Issue as "how are users shown how to grab the object?".

Role of Scenarios
We had previously found that usage Scenarios were a valuable addition to DSA [4]. Initially, the assumption was made for FLUID that we could easily map changes to the usage Scenarios into changes in the DSA diagram; in the initial FLUID versions this was the only way to change the DSA, to encourage learners to think in Scenario terms. Our assumption proved incorrect — sometimes a new design Option required several changes in the usage Scenario, sometimes a single change in the usage Scenario required alteration of several design Options. We now recognize that the usage Scenarios present a complementary view to the other DSA components. In the current FLUID, users can edit either the Scenario or the DSA diagram. FLUID detects most but not all of the required changes in the other component.

Alternatives Versus Options
We wanted to include usability testing as one of the steps in which learners would engage. Initially, we tried having learners select a set of

Options to present to the users, representing their choices on the various Issues. We found this approach was too abstract, and failed to convey that the design becomes more than just the sum of its parts (see also the point on decomposition below). The explicit notion of an Alternative system design was introduced to fill this gap.

Users were observed to struggle with the notion of an Alternative as a set of selected Options (one from each Issue). This may reflect a weakness in the Case Study and the Tutorial. This will be explored further with students in the HCI design course.

Implications of DSA Notation on Design Processes

As outlined earlier, we found that having to specify the advice to be given to learners in each process step forced us to analyze DSA as a theory of design. We had to develop general heuristics to guide the Tutor's video advice, independent of the particular exercise. For example, in one instance, the tutor says:

"If all your Options are unsatisfactory on a particular usability Criterion, try to identify a new Issue which groups these Options together and suggests a new design dimension to consider."

We also noted the assumption in DSA that an assessment on a specific Criterion can always be related back to the choice of a design Option. This assumption is suitable for the training wheels setting but oversimplifies the design space. For example, in [9] the Option of a hand-shaped cursor satisfies the Criteria of being natural and guessable initially. But once the Option of hands appearing on an object or box is chosen, the assessment of the cursor Options changes. Thus the assessments must be considered on the Alternative system design rather than just the individual Options. The closing section of FLUID uses this example to illustrate how the training wheels experience needs to be extended to cope with more complex situations.

5. CONCLUSIONS AND FUTURE WORK

We found that most learners who were introduced to DSA through FLUID were able afterwards to describe in their own words what DSA is. For example, one user said:

"DSA is a good tool for developing an outline of systematic Options and solutions for most scenarios".

In a pretest questionnaire, we asked test users to describe a previous interface design experience and the decision they made. After the session with FLUID, most were able to suggest ways DSA could have helped them during their design process.

The FLUID system raises questions about what it is that experienced designers know and how others might come to acquire that knowledge most efficiently. We designed FLUID to address particular difficulties which we knew our students encountered as novice designers, in the context of a typical computer science program. The knowledge we gained about DSA in the process of building FLUID was an unanticipated but welcome outcome.

We will be using FLUID in HCI classes in Guelph and Delft in the coming year, probably with pairs of students working together. Further design enhancements are anticipated as we learn more about teaching DSA in this way and more about the usability of FLUID. We have also agreed to distribute versions of the software to other educational institutions and some industrial groups, who agree to furnishing in return both their feedback and additional worked cases and exercises. In this way we hope to accumulate a larger body of exemplar cases. At some point this collection should reach a volume which requires support for learner access to the cases [11], but we do not envision needing this in the short term.

We are also generalizing the FLUID architecture to permit application to additional domains, both within and beyond user interface design. Within user interface design, the next skill to be taught is the design of an evaluation. Outside user interface design, we will be constructing instructional modules for the tasks of problem-based writing and of the role of a field biologist in the Arctic. The key element here is to allow users to engage with the design domain in a way complex enough to seem natural but simple enough to permit monitoring by the environment so that feedback and support can be context-sensitive.

ACKNOWLEDGMENTS

We thank Michael Chen and Stephanie Houde for permission to adapt their work and their comments on the results. Helpful comments were also made by Allan MacLean, Jenny Preece, Richard Jacques, Jonathan Swallow and Charles van der Mast. We would also like to thank the character actors in FLUID who helped bring it to life. This work was supported by the Natural Sciences and Engineering Research Council of Canada, and the University of Guelph's Instructional Development Grants program. We thank the copyright holders who allowed us to incorporate their material — ACM for [9], and the National Research Council of Canada for the videotape related to [7].

REFERENCES

1. Bell, B.L. and R. Bareiss, (1993), Sickle Cell Counselor: Using a Goal-Based Scenario to Motivate Exploration of Knowledge in a Museum Context, in *Proceedings of AI-ED 93*, Association for the Advancement of Computing Education, 153-160.

2. Carey, T.T., A. MacLean, and D. L. McKerlie, Event-based Retrospective Design Rationales, in preparation.

3. Carey, T.T., M. Ellis and M. Rusli, (1993), Re-using User Interface Designs: Experiments with a Prototype Tool and High-level Representations, in R. Winder (ed.), *Proceedings HCI'94 — People and Computers VIII*, Cambridge University Press.

4. Carey, T.T., D. McKerlie, W. Bubie and J. Wilson, (1991) Communicating Human Factors Expertise Through Usability Design Rationales, in D. Diaper and N. Hammond (eds.), *Proceedings HCI'91 — People and Computers VI*, Cambridge University Press, 117-130.

5. Catrambone, R., and Carroll, J.M., (1987), Learning a Word Processing System with Guided Exploration and Training Wheels, in J.M. Carroll and P.P. Tanner (eds.), *Proceedings of CHI+GI'87 Human Factors in Computing Systems and Graphics Interface*, New York: ACM,. 169-174.

6. Chen, M., Mountford, S.J., Sellen, A., (1988), A Study in Interactive 3-D Rotation Using 2-D Control Devices, in *Computer Graphics*, 22(4), ACM, 121-129.

7. Evans, K.B., Tanner, P.P., Wein, M., (1981), Tablet Based Valuators that Provide One, Two, or Three Degrees of Freedom, *Prceedings of SIGGRAPH'81 Computer Graphics* 15(3), 91-97.

8. Feifer, R. and R. Holyoak, (1994), Helping Students Get Unstuck, in proceedings *ED-Media 94*, Association for the Advancement of Computing in Education, p. 645.

9. Houde, S., (1992), Iterative Design of an Interface for Easy 3-D Direct Manipulation, *Proceedings of CHI'92*, ACM, 135-142.

10. Jacques, R., (1994) *Engagement in the Design of Educational Hypermedia*, Ph.D. thesis, Southbank University, U.K., in preparation, 1994.

11. Kolodner, J., (1993), *Case-Based Reasoning*, Morgan Kaufman.

12. Laurillard, D. (1993), *Rethinking University Teaching*, London: Routledge p. 26.

13. MacLean, A., Young, R., Bellotti, V., and Moran, T., (1991), Design Space Analysis: Bridging from Theory to Practice via Design Rationale, in *Proceedings of Esprit'91*, Office for Official Publications of the European Communities, Luxembourg, 720-730.

14. Moran, T., (1994), quoted in Interview with Tom Moran, in J. Preece et al., *Human-Computer Interaction*, Addison-Wesley, p. 349.

15. Parnas, D.L., Clements, P.C., (1986), A Rational Design Process: How and Why to Fake it, in *IEEE Transactions on Software Engineering*, 12(2).

16. Schank, R.C., (1994), Active Learning through Multimedia, *IEEE Multimedia*, 1(1), 69-78.

17. Schoen, D.A., (1987) *Educating the Reflective Practitioner*, Jossey-Bass: San Francisco.

Practical Education
for Improving Software Usability

John Karat
IBM T. J. Watson Research Center
30 Saw Mill River Road
Hawthorne, NY 10532 USA
+1-914-784-7612
jkarat@watson.ibm.com

Tom Dayton
Bellcore
444 Hoes Lane, RRC 4A-1112
Piscataway, NJ 08854 USA
+1-908-699-6843
tdayton.chi@xerox.com

ABSTRACT

A usable software system is one that supports the effective and efficient completion of tasks in a given work context. In most cases of the design and development of commercial software, usability is not dealt with at the same level as other aspects of software engineering (e.g., clear usability objectives are not set, resources for appropriate activities are not given priority by project management). One common consequence is the assignment of responsibility for usability to people who do not have appropriate training, or who are trained in behavioral sciences rather than in more product-oriented fields such as design or engineering. Relying on our experiences in industrial settings, we make personal suggestions of activities for the realistic and practical alternative of training development team members as usability advocates. Our suggestions help meet the needs specified in the recent Strong et al. [21] report on human-computer interaction education, research, and practice.

KEYWORDS: HCI education, technology transfer, participatory design, user-centered design, usability engineering, design problem-solving

INTRODUCTION

What do people need to know to design a usable software tool? Must they be accomplished in some particular field of study? If the design and development of the software tool requires contributions from several individuals (and surely the design of complex software systems remains in this category), should there be a mixture of talents and perspectives, or should teams comprise multiple people all of some single ideal-designer type? These are complex questions that the human-computer interaction (HCI) community is just beginning to confront. In this paper we give our personal (and perhaps controversial) views on these issues.

Whereas 20 years ago software developers may have assumed that developing usable software meant building

systems that met specifications, we have learned that specifications often lack information that influences the usefulness of the system in a real context [23]. Software development teams throughout industry have acquired some sense that issues of technology's use ought to be considered, and even some sense about what should be done [9], but for various reasons they have had difficulty integrating usability activities into software engineering practice (e.g., [10]). Faced with the not-too-well specified problems of software usability, companies either have assumed that usability can be addressed by project personnel using common sense within normal software engineering processes, or have attempted to hire usability specialists.

The educational background preferred by most companies seeking usability specialists is often behavioral science fields (e.g., psychology) that only indirectly train students in skills relevant to the design and development of usable software systems. Being a psychologist who could program seemed to be the perfect mix of knowledge and skill. Recently there has been considerable work to identify a curriculum for software design and development in commercial settings that includes usability as an aspect of project success [11, 21, 22]. To some extent these curricula call for adding traditional behavioral science courses to computer science programs. But our experience is that training software professionals to be sensitive to users and to carry out activities that result in the development of usable software systems involves little that can or should be connected to most current behavioral science training programs. Unfortunately, neither do any other formal educational programs (e.g., computer science) meet the practical needs of the software industry for usability engineers. In this paper we give our personal opinions of what activities are important in usability engineering, who should be doing those activities, what knowledge and skill are required, and how these abilities can be acquired.

WHO SHOULD BE DOING WHAT TO PRODUCE USABLE SOFTWARE?

What Activities?

Many companies are coming to recognize usability as an important component of software quality. But often their quest for that quality is not for the "right" way to do usability engineering, it is for a way that has positive influence on system usability without any offsetting costs of time, money, people, or other resources. They are looking for guidance about how to improve usability, but are unwilling to have the recommendations negatively affect cost and schedule of software product development.

An early path to deciding on usability-enhancing activities was to try applying the methods of behavioral sciences to software development. A number of researchers in a variety of fields have been willing to apply behavioral science to HCI, and many companies hired people with such formal academic training as usability specialists. We two authors are trained as cognitive psychologists and entered the HCI field through this route. The notion was that these HCI specialists had special knowledge and skills, and that this background would transform fairly directly into the skill needed to design user interfaces having good qualities (including usability). In this environment, management made mistakes and HCI specialists made mistakes. Specialists were often called into projects only after the interface was coded, to quickly critique and correct it off the top of their heads without access to users. Some specialists insisted that contributions could be made only through careful (and costly) experimental work or theory building. Engineering approaches to user interface design emerged from the work of some members of the community (e.g., 4), but did not have a broad impact on the way most commercial systems were developed. In our opinion, despite the proliferation of research literature and design guidelines intended to help, industry has not come to view employment of isolated usability specialists as a broadly successful route to usable systems.

New approaches evolved through the 1980s. Aware that the wisdom of psychology and HCI theory were insufficient on their own, usability specialists shifted their focus to rapid iterations of software prototyping on computers and formal evaluation of those prototypes. But workers in industry discovered that few iterations could be fit into the increasingly short development cycles. Formal evaluation of computerized software prototypes is in*efficient* in that is not quick enough for many iterations to occur; problems are found mostly after they have been coded, and even prototype coding is expensive. This approach is also in*effective*—information reported from formal tests of running prototypes (or the real software) tends to be individual problems rather than wholesale design slants, and the human subjects of such tests tend to be more passive sources of problem identification than they are active participants in designing. Thinking is therefore oriented toward fixing individual problems rather than toward the user helping create whole designs that well fit the user and task as a whole.

Attempts to quantify usability have been part of the focus on formal evaluation, and many people took this as a sign of the maturing of HCI. However, workers in industry have realized that numerical summaries of interface problems help little in improving usability, and that customers do not become more inclined to buy an unusable software product just because you can tell them *exactly* how unusable it is. Even *qualitative* formal evaluation of an interface often does not produce good suggestions for fixing or avoiding the problems. In short, evaluation is not how best to influence product usability—*design* is where the action is!

We think the most important activities for developing usable systems are those early in the design process, before substantial resources have been committed to any particular design. These include activities related to the difficult decision of *what* functions to support, not just the decision of *how* to present already-decided-upon functions. Methods that aid understanding the work context and how the new technology might influence it are becoming more important as the HCI community recognizes that usability is more than just screen layout. Understanding users means more than cataloging education profiles or age distributions. It requires involving them—perhaps in new and creative ways—as collaborators in the design process.

The list of activities we think are important is long and not clearly defined; to a large extent we maintain that exactly what activities are carried out depends on the full context of the particular design situation. It includes many activities that contribute to a deeper understanding of the work to be accomplished with the system under design. Many attributes of good usability engineering practice fall under the rubric of *participatory design* [15]. One approach we have found successful is working in early design sessions with a small but representative sample of the stakeholders (including users) in the software product's usability. Such activities are interactive, involving learning on both the designers' and users' parts. These activities are generally aimed at understanding what would constitute a usable system for the domain in question.

Who?

Initially in HCI's history, usability activities were not clearly identified or assigned to particular individuals, since usability was assumed to be attainable by common sense. As usability engineering developed and stressed the importance of specific deliverable products associated with a usability process, some companies recognized that someone in particular should be made officially responsible for usability. But because usability was (and often still is) seen as being attainable largely by common sense ("everyone is a computer user, so anyone can figure out what users need"), the person given that responsibility often was appointed haphazardly from the development team—that is, without requiring any particular knowledge or skill. Appointing someone officially responsible for usability was seen to be necessary merely to provide the motivation and work time for someone to do the required common sense thinking.

Some companies now realize that something beyond common sense is required in usability engineers, but they don't know just what. Out of uncertainty, many companies use the dominant HCI research community as a model, and therefore assume that formal knowledge is required or even sufficient for practicing industrial usability specialists. The further assumption commonly is made that such special knowledge is had almost exclusively by people who have emerged from typical behavioral science education programs.

We agree that usability specialists are needed, but we also insist that placing the responsibility for usability in the hands of a few specialists can never be sufficient [13], regardless of their training. Nearly everyone in software development and marketing organizations should practice usability engineering to some extent, but the degree and type of activity will vary dramatically with the person's particular role and the system under development. For example, marketing people may have plenty of opportunity to discover usability issues from customers and potential customers, and might benefit from knowing what kind of questions they should be asking. Programmers may have no contact with users, but can learn to try to take a users' viewpoint. We maintain that these pervasive kinds of usability engineering activity do not require formal knowledge of behavioral science, though they do require other kinds of uncommon knowledge and skill.

In short, our model is that projects need to have several people involved in varied usability engineering activities. We do not mean that they should all have advanced degrees in behavioral science. We have found that some more appropriately practically-focused training (more than a day though far less than four years) can enable some contributions to come from most participants in a project. These contributions can be facilitated by the work of usability specialists who have better (broader and deeper, formal and informal) backgrounds that are directly relevant to practical usability engineering.

WHAT KNOWLEDGE AND SKILLS ARE NEEDED BY THOSE PEOPLE?

Perhaps in 20 years all software developers (by *developers* we include a variety of people such as system engineers and requirements writers, with backgrounds in computer science and related fields) will have backgrounds that include usability engineering. But for now academic departments are still trying to define what skills contribute to usability engineering, and do not yet value usability engineering enough to make it part of the base knowledge of computer science [21]. We feel that training must help developers see beyond the merely technical issues, to the issues of the technology's impact on the work contexts of users.

There are many attributes needed by a usability engineer, and we feel that most of them are not formal knowledge, but skills or traits. Hewett et al.'s SIGCHI report on curricula [11] gave good consideration to the multidisciplinary nature of the knowledge needed for HCI, but emphasized (in our opinion) formal knowledge. We disagree with that emphasis, though we recognize that it was due partly to the report's charge with focusing exclusively on formal curricula for educational institutions; those institutions are set up to handle formal knowledge more than skills and other informal attributes.

Van Cott and Huey [22] reported skills needed for human factors in general rather than for HCI; we attribute that report's emphasis on formal knowledge partly to a scope that encompassed physical ergonomics. Still, the report also listed several skills outside of normal science program content (e.g., presenting reports). Even more emphasis on practical skills that we consider most relevant to HCI in the real industrial world was present in the Dayton et al. [6] report on attributes needed by user-centered designers. Those attributes included ability to estimate resources, enjoyment of working on teams, and negotiating skill. However, we disagree with [6]'s emphasis on formal knowledge such as theory of cognitive processes as a necessary component of usability engineering skills. We don't think that knowledge of behavioral theory is highly *necessary* for user-centered designers, indeed not even for most people who do most of the broader discipline of HCI. People who do need such knowledge typically work on interfaces for tasks that strain the limits of sensation and perception (e.g., aircraft cockpits).

Granted, knowledge of behavioral science theory is useful—probably more useful than other science backgrounds—for the challenge of critically thinking about HCI design. But such knowledge is not designed to support development of human tools. It helps mostly as a context for thinking about usability, for explaining problems after they have been discovered, for communicating with people who share that theoretical orientation, and maybe for occasionally stimulating ideas of solutions [18]. For example, experienced usability engineers who have no background in formal cognitive theory might nonetheless be able to predict that users will have trouble remembering a set of 20 items on a menu. Naming that difficulty "a short-term memory capacity limitation" does not produce a better prediction than the more provincial "people's heads can only hold so much". Unfortunately, the collection of things we know about human cognition does not form a very useful user-interface design guide.

Some directly relevant knowledge *is* contained in behavioral science programs, but usually only in support of the theory-working that is the focus of such programs. Directly useful formal knowledge includes that of quasi-experimentation; naturalistic field observation; questionnaire construction, administration, and analysis; and descriptive statistics. Knowledge of experimental design and inferential statistics arms usability engineers to critically evaluate usability studies that flaunt the trappings of those tools, such as reporting "$p < .05$". (But usability engineers rarely use those methods themselves.) Awareness that random effects in sample data may masquerade as population differences is quite important. Especially useful are consciousness of the necessity of representative sampling (of users, tasks, work situations, machines) in any kind of activity, and of the many ways that users' behaviors easily can be biased by the observer, the materials, the situation, or the users' own attitudes. Associated with that awareness are skills for carefully dealing with subjects (users) so to avoid biasing them, yet getting their whole-hearted cooperation, all the while treating them ethically. Behavioral science training is also useful for deeply appreciating that people often think differently from each other, and yet that there are fundamental similarities across all people.

We do not intend our emphasis on skill in areas other than scientific theory to suggest that development of behavioral theory of HCI is not worthwhile. On the contrary, we feel that behavioral theory development should be pursued and encouraged, both for its own sake and for its possibly greater applicability in the far future. Card [3, pp. 121–124] is one theoretician who has noted iterative experimental tests of running software interfaces are impractical as a total solution. But his proposed solution was to focus even more on building up the science base of HCI so that the science can be applied. We see that goal as very distant, in fact infinitely distant as a *total* solution. Bits and pieces of science have, do, and will contribute to usability engineering, but always only as part of the solution [5]. Theory has a better chance of being applicable eventually if theoreticians try to cope with the practical realities we describe in this paper.

We are writing here as two individuals who have an interest in furthering a particular kind of activity—the development of computer systems that support human work. We suggest that techniques and training to support this activity need to be very different from the techniques and training for a research career in behavioral science, which necessarily focuses on other theories. More important for the usability specialist, in our view, is a broad range of mostly social, organizational, and artifact-design skills. *Most important is practical experience in the activities of usability engineering.* By this we mean practice in understanding human activity and in designing technology to support it. If theory could help lead to successful design, we would be happy to use it.

HOW TO GET THE KNOWLEDGE AND SKILLS
Our observation is that practical experience in usability engineering, especially in design, is not emphasized to the extent that it could be in training future system designers. Our experience in interviewing job candidates and with new hires is that they rarely have *realistic* experiences of the practical type. Our own academic training once led us, too, to believe that we would conquer the development of usable software through inventive experimental design and data analysis, but we discovered that this belief can lead to difficulties in industrial settings. For instance, someone with that belief might spend so much time on those formal activities that they have too little time for the much more productive cooperative-design activities.

The Hewett et al. [11] report took a step forward by describing the field as "a discipline concerned with the design, evaluation, and implementation of interactive computing systems for human use and with the study of major phenomena surrounding them" (p 5). For the most part this seems appropriate for applied settings, though industry has little direct interest in "the study of major phenomena surrounding them". Skills developed through

the practice of applying knowledge to applied problems, the development of which have long been a part of many engineering programs, are neglected in behavioral science based training.

The Hewett et al. report went on to suggest that educators develop "HCI-oriented" rather than "HCI-focused" undergraduate programs (p. 56). We agree with this suggestion that HCI (or usability engineering) is not a field separate from the base engineering activity, but is a special perspective on that activity. We suggest that both the researcher and practitioner elements of the HCI community need to focus on skills needed to produce usable systems—the design, evaluation, and implementation skills mentioned above, along with the ability to properly consider the use of the system in context. It is not that industry does not want its practitioners to keep current in technical research areas, just that industry elects to invest only small portions of its total expenditures for such research.

Strong et al. [21] extended the Hewett et al. [11] curriculum report in several ways. The extension we think is most relevant to training usability engineers for industry is the emphasis on giving students realistic experience in designing interfaces. Realism is achieved, in our view, by the Strong et al. report's suggestions that students understand the importance of, and gain skill and realistic experience in:

- The entire task and work environment in which the interface will be used
- Constraints and tradeoffs such as limits on resources, and social and organizational pressures
- Work flow, task, and organizational analysis and design
- Involving users in design
- Teamwork
- Reflecting on their practice of usability engineering

Most importantly, these areas should be covered in realistic contexts, by having students participate in several design projects from small to large scale, consult across several projects at once, and experience real software product development, for example through university-industry partnerships.

On the basis of our experiences as usability specialists in several industrial software development environments, we strongly believe that *careful (and usually collaborative) design and informal evaluation, rather than casual design and formal evaluation, are where the leverage is* for producing good interfaces in the real world of industrial software development. Highly skilled people are needed for doing those activities. But current behavioral science education, and indeed any science education, has little to recommend it, because usability engineering is much more craft and engineering than it is applied science [5, 12, 24]. Behavioral science and other educational programs provide substantial knowledge and skill in some areas that are relevant to usability engineering, but fail to provide in other, vital, areas.

There are substantial benefits for applied usability engineering from advanced education in science, but most are not the officially recognized ones such as knowledge of theory and theory-evaluating methods. For instance, getting a PhD in cognitive psychology requires trying to think like other people, which is a valuable skill for usability engineering. Other abilities typically had by someone who successfully completes a PhD in psychology include project management, perseverance, and social skills. A PhD program may not give all those abilities, but at least it tends to filter out people who lack them. So from a pool of job candidates having either traditional computer science or traditional scientific psychology degrees, we two authors strongly favor the psychologists, and among the psychologists we strongly favor those having PhDs.

However, life experiences other than behavioral science training provide similarly important (though often different) skills that are just as or more relevant to usability engineering. Industrial design and engineering educations include many activities and thinking that are relevant to human-computer interface design, even if students worked on artifacts other than human-computer interfaces. Industrial design is a particularly good example, because it forces its practitioners to simultaneously meet demands of utility, esthetics, and the artifact's underlying technology. Human-computer interaction is an applied discipline, by definition. But most educational programs in HCI assume that the mere identity of the topic is enough to make any kind of educational activity, including focus on theory, useful preparation for the real world. Our personal experiences have convinced us that this is clearly untrue. No matter how applied a human factors program is, companies need to train their new hires in the ways of the real industrial world.

Additional training is needed regardless of the educational program (human factors, behavioral science, industrial engineering, industrial design, computer science, whatever), though the areas needing coverage differ. We think several types of training can be used in companies to complement the incomplete abilities that are provided by any of the backgrounds we have mentioned. The importance of any type of practical training is directly related to the amount of one-on-one contact with an experienced usability engineer in the context of a real project. The only advantage of training other than apprenticeship is lower expense. Here are some valuable activities, ordered from those most appropriate early in the new hire's career to those most appropriate after a month of work:

- Formal lecture classes are useful, but often just as preparation for hands-on learning. By "formal" we mean the format rather than the content. The content should always be pragmatic, and peculiar to that company's ways of doing business.
- An internal publication that lists the company's resources for usability engineers, such as which of the company's people are skilled in user contact techniques such as focus groups or contextual inquiry.
- Workshops that provide hands-on practice of methods that are useful for design and evaluation.
- Apprenticeship with experienced usability engineers, working on real projects.

- Mentoring by experienced usability engineers as needed, perhaps daily. Some of this should be in one-on-one lectures, but the lectures should be short and focused on topics relevant to the student's activities for that day or planned for the next.

WHAT WE HAVE TRIED—SOME CASE STUDIES

We two authors have been asked to usability train development team members in a number of settings. These efforts are aimed at creating a larger community of usability activists (people capable of carrying out a range of activities that contribute to software usability), not at providing the levels of knowledge and experience needed to master usability engineering. The following summarizes some of our thoughts on and approaches to this training. In all cases we focus the training on active rather than passive learning, so training tends to last more than a day rather than being short and simple presentation of material.

Usability Advocates for a Medium Size Organization

One of us (Karat) was asked to develop a training program for an IBM software development laboratory staffed by about 200 people. The lab had one full time "usability expert" (a psychologist with about 10 years experience), who was unable to adequately address the usability needs of the approximately 10 projects underway at the lab at any give time. The judgment that current coverage was inadequate came from the expert herself, and was partly a result of the laboratory's efforts to promote quality software development processes. The expert thought it unreasonable to recommend hiring additional experts since the lab was down-sizing, so she proposed establishing "usability advocates"—people who would accept responsibility for product usability—within each project. As the sole expert, she would be a consultant providing formal testing resources and advice to all projects, but clearly the advocates would have to carry out new activities in addition to their current within-project work.

The training program might be summarized as
1. Get management support for developing a usability process, including commitment of reasonable resources.
2. Identify, establish, and train a community of usability advocates.
3. Make the roles of the advocates well known to the entire development community.

To meet a variety of constraints, a program was designed by one of us (Karat) and a collaborator (John Bennett); it included a six-day on-site training program. A two-day advance visit was made to meet with the designated advocates and to gather information about their current projects. Considerable time in this visit was spent meeting with the lab management, to ensure that the instructors and management had similar views of the training. An initial course outline was discussed with management, with the resident specialist, and with the future advocates.

We (Karat and Bennett) attempted to tailor the class exercises so they involved discussion of ongoing real projects. Our focus was on techniques for understanding the work to be supported by the systems under development (techniques such as contextual inquiry and customer interviews), and understanding the impacts of design constraints (e.g., company and international standards) on the system design. We did not attempt to teach the advocates about formal testing, since they knew something about it from the expert. Beyond such "technical content" in the class, we stressed the mutual development of skill by the usability advocates as a team (the advocates were not part of a single department, but reported to different project oriented departments). We focused on shifting the advocates' perspectives, so they viewed usability advocacy as a set of approaches that are not as well defined as engineering professionals normally would like. We did spend about 20% of the class time discussing "partnership" [2] and the roles of different perspectives in design. This proved difficult at first for the advocates, for they felt the class should be focused more on specific procedures they could clearly carry out, whereas we were telling them that the role of usability advocate was going to be one in which listening and teamwork skills were critical. However, in subsequent feedback about the course, the advocates have reported that they now do see the importance of maintaining a community to discuss their usability advocate practice.

We concluded the course by presenting an overview of the usability process to the remainder of the lab (including management), introducing the advocates, and inviting everyone to work with them in improving product usability. We focused on shifting the general lab perspective, so they viewed usability activities as including more than lab evaluations or expert opinions. From the general lab presentations, there were numerous inquiries made and contacts established among parts of the organization that had not previously communicated effectively. Management reported satisfaction that the program had raised the general awareness of usability within the lab, and the advocates expressed satisfaction that people would help them do the job, rather than view them as solely responsible for it.

Consultive Workshops for a Large Organization

The other of us (Dayton) is one of many usability engineers in Bellcore who have been creating and supporting semiformal usability resources. These resources not only help educate new usability specialists, but also educate other people such as managers and developers. There are about 50 usability engineers, user interface designers, and behavioral scientists in Bellcore, and about 7,000 employees of all kinds in the entire company (for an overview see [19]).

Bellcore has had some success with formal (as opposed to informal and semiformal) institutional support for user-centered design (UCD). Although such support has been very helpful, no formal policy can contain all the detailed information that is necessary to understand exactly how to *do* user-centered design. It is just one of the many kinds of activities that are needed for UCD [5, 6, 12]. In keeping with that desirability of diversity in usability practices,

Bellcore provides additional support, ranging from formal to informal. The major resources are

- Corporate policy supporting UCD
- Multiplatform, object-oriented (OO), graphical user interface (GUI) design guide [14, 7]
- Methods for doing participatory, OO, GUI design cheap, early, fast. These include the complementary [7, 16] methods of CARD [16] for designing tasks, PICTIVE [16] for designing interfaces, and newer methods [7] for linking those.
- Suite of lab rooms for usability engineering [8]
- Classes [1]
- Hands-on tutorials [17]
- Consultations
- Consultive workshops [7], described below

One particularly successful resource for helping people learn to practice OO participatory design of multiplatform GUIs has been consultive workshops [7; a partial precursor and related activity is 17]. These are formal insofar as their major goal is to help the participants learn to design. They are informal insofar as only 10 percent of the time is spent on lecture; 90 percent is spent on the participants doing design. The example the participants analyze and design in the workshop is a subset of their own real software product.

The participants in a workshop are the members of a real, intact development team. By *development team* we mean all the stakeholders in the usability of the software product—usability specialists, system engineers, developers, designers, requirements writers, managers, documenters, marketers, subject matter experts, and especially *users* of the software product being designed (or potential users if the product is new). A balanced subset of up to seven members of the team participates in the workshop as a group seated around a small table, on which they manipulate their low-tech materials. (Six is the limit of our real design sessions using this methodology. We use seven in these training sessions so that realistic social problems arise to act as examples to which the meeting facilitation methods can be applied.) But up to four groups can participate in the workshop at once, each at its own table and using its own project as its example. Every group always includes a real end user.

Each group manipulates low-tech materials (paper, index cards, felt-tip pens, sticky notes) that are used to represent the task flow (via a version of the CARD method [16]), the interface design components (via a version of the PICTIVE method [16]), and the task information objects that bridge the task to the interface [7]. Immediately after any of these or other methods is used by the students or the workshop facilitators, the facilitators prompt the students to reflect on the method and why it was used in that instance. The full workshop, from business process analysis through task analysis, task object identification, interface object identification and description, and design of the interface proper, lasts three full days. That may sound like a lot of time, but it is very little time to accomplish the design of a real interface from start to end; the entire workshop runs under severe time pressure—just like real industrial life.

Thus the workshop participants' learning environment is a good approximation of the software development environment in which they must apply their skills after the workshop. They do not return to their job only to be frustrated in explaining the methods to people who were not at the workshop, because most of the members of their real team were in the workshop with them. They do not wonder how to apply the lessons from the workshop to their real project (or wonder *whether* the lessons can be applied), because the lessons were made with their real project instead of an artificial example. Perhaps most importantly, students and managers can justify the time spent in the workshop partly as getting consulting help on their project, rather than as time in class away from their project; this has drastically increased the sign-up rate for our workshops.

CONCLUSION

These two cases from our personal experiences in two companies illustrate only some of the practical ways in which people can be trained in usability engineering, whether they are usability specialists or other members of the development team. There are many similar ways in which people can acquire applicable knowledge and skill relevant to usability engineering in real, industrial software development environments.

In presenting the case for a broader community of usability engineers, and stating that current behavioral science education does not provide sufficient (or perhaps necessary) knowledge and skills for practitioners, we risk understating the indirect benefits of our own backgrounds. However, our personal experience is that the HCI field, and thus education in it, needs to consider more seriously its similarities to other design fields such as architecture and industrial design. Education in those fields acknowledges a large experiential component, utilizes different methods, and focuses on different problems, than does education in more technically encapsulated fields (e.g., [20]).

Part of our message is that usability activities must be practiced by more members of the design and development team than just the usability specialists. Another part is that the knowledge and skills needed for that broader community to contribute in some way to product usability can be obtained in days or weeks rather than years, if the training focuses on directly relevant areas. Many of the same types of educational experiences, but much larger amounts, are needed for usability specialists to acquire appropriately deep and broad expertise to fill their roles. Bringing that larger community up to speed is a challenge, for which we have suggested a few solutions.

We accept that in many ways our presentation in this short paper lacks the specific content that curriculum implementers would need to prepare for tomorrow morning's class. We ourselves are still experimenting with training programs

for usability practitioners of all kinds within our own companies. Our message to our students is that the advice for designing a usable system, like the advice for designing a quality building, cannot be easily codified into technical guidelines. The practitioner in this field must be broadly informed, and must expect each new project to be a learning experience. For these reasons we place our emphasis on learning through interaction more than on learning any specific theory or technique.

ACKNOWLEDGMENTS
We thank John Bennett, Wendy Kellogg, Al McFarland, Ellen White, and several anonymous reviewers who commented on an earlier draft of this paper.

REFERENCES

1. Bellcore. *New Jersey Learning Center Course Catalog.* Bellcore, Piscataway, NJ, 1995.

2. Bennett, J., and Karat, J. Facilitating effective HCI design meetings. In *Proc. of CHI '94* (1994), 198–204.

3. Card, S. K. Human limits and the VDT computer interface. In *Visual Display Terminals.* J. Bennett, D. Case, J. Sandelin, and M. Smith, Eds., Prentice-Hall, Englewood Cliffs, NJ, 1984, pp. 117–.

4. Card, S., Moran, T., and Newell, A. *The Psychology of Human-Computer Interaction.* Lawrence Erlbaum, Hillsdale, NJ, 1983.

5. Dayton, T. Cultivated eclecticism as the normative approach to design. In *Taking Software Design Seriously: Practical Techniques for Human-Computer Interaction Design.* J. Karat, Ed., Academic Press, Boston, 1991, pp. 21–44.

6. Dayton, T., Barr, B., Burke, P. A., Cohill, A. M., Day, M. C., Dray, S., Ehrlich, K., Fitzsimmons, L. A., Henneman, R. L., Hornstein, S. B., Karat, J., Kliger, J., Löwgren, J., Rensch, J., Sellers, M., and Smith, M. R. Skills needed by user-centered design practitioners in real software development environments: Report on the CHI '92 workshop. *SIGCHI Bulletin 25*, 3 (1993), 16–31.

7. Dayton, T., McFarland, A., and White, E. Software development—keeping users at the center. *Bellcore Exchange 10*, 5 (1994), 12–17.

8. Dayton, T., Tudor, L. G., and Root, R. W. Bellcore's user-centred-design support center. *Behaviour & Information Technology 13*, 1 & 2 (1994), 57–66.

9. Gould, J. D., and Lewis, C. H. Designing for usability—key principles and what designers think. *Commun. ACM 28*, (1985), 300–311.

10. Grudin, J. Systematic sources of suboptimal interface design in large product development organizations. *HCI 6*, (1991), 147–196.

11. Hewett, T. T., Baecker, R., Card, S., Carey, T., Gasen, J., Mantei, M., Perlman, G., Strong, G., and Verplank, W. *ACM SIGCHI Curricula for Human-Computer Interaction*. ACM, NY, 1992.

12. Karat, J., and Dayton, T. Taking design seriously: Exploring techniques useful in HCI design. *SIGCHI Bulletin 22*, 2 (1990), 26–33.

13. Lund, A. M. How many human factors people are enough? *Ergonomics in Design*, (Jan. 1994), 36–39.

14. McFarland, A. *Design Guide for Multiplatform Graphical User Interfaces* (LP-R13). Bellcore, Piscataway, NJ, 1994.

15. Muller, M., and Kuhn, S. Participatory design [Special issue of *Commun. ACM*]. *Commun. ACM 36*, 4 (1993).

16. Muller, M. J., Tudor, L. G., Wildman, D. M., White, E. A., Dayton, T., Carr, B., Diekmann, B., and Erickson, E. D. Bifocal tools for scenarios and representations in participatory activities with users. In *Scenario-Based Design for Human-Computer Interaction*. J. M. Carroll, Ed., New York, Wiley, in press.

17. Muller, M. J., Wildman, D. M., and White, E. Participatory design through games and other group exercises [Tutorial at CHI '94, Boston]. *CHI '94 Conference Companion*, (1994), 411–412.

18. Rogers, Y., Bannon, L., and Button, G. Rethinking theoretical frameworks for HCI: Report on an INTERCHI '93 workshop. *SIGCHI Bulletin 26*, 1 (1994), 28–30.

19. Salasoo, A., White, E. A., Dayton, T., Burkhart, B., and Root, R. W. (1994). Bellcore's user-centered design approach. In *Usability in Practice: How Companies Develop User-Friendly Products*. M. E. Wiklund, Ed., Boston, Academic, 1994, pp. 489–515.

20. Schoen, D. *Educating the Reflective Practitioner*. Jossey-Bass, San Francisco, 1986.

21. Strong, G. W., Gasen, J. B., Hewett, T., Hix, D., Morris, J., Muller, M. J., and Novick, D. G. *New Directions in Human-Computer Interaction Education, Research, and Practice*. Drexel Univ., Philadelphia, 1994.

22. Van Cott, H. P., and Huey, B. M., Eds., *Human Factors Specialists' Education and Utilization: Results of a Survey*. National Academy Press, Washington, D.C., 1992.

23. Whiteside, J., Bennett, J., and Holtzblatt, K. Usability engineering: Our experience and evolution. In *Handbook of Human-Computer Interaction*, M. Helander, Ed., Elsevier, Amsterdam, 1988.

24. Wroblewski, D. A. The construction of human-computer interfaces considered as a craft. In *Taking Software Design Seriously: Practical Techniques for Human-Computer Interaction Design*, J. Karat, Ed., Academic, Boston, 1991, pp. 1–19.

Evolution of a Reactive Environment

Jeremy R. Cooperstock

Dept. of Electrical & Computer Engineering
University of Toronto
Toronto, Ontario M5S 1A4
+1-416-978-6619
jer@dgp.toronto.edu

Koichiro Tanikoshi[*], *Garry Beirne, Tracy Narine,*
William Buxton[†]

Computer Systems Research Institute
University of Toronto
Toronto, Ontario M5S 1A4
+1-416-978-0778
{tanikosi, garry, tracyn, butxton}@dgp.toronto.edu

ABSTRACT

A basic tenet of "Ubiquitous computing" (Weiser, 1993 [13]) is that technology should be distributed in the environment (ubiquitous), yet invisible, or transparent. In practice, resolving the seeming paradox arising from the joint demands of ubiquity and transparency is less than simple. This paper documents a case study of attempting to do just that. We describe our experience in developing a working conference room which is equipped to support a broad class of meetings and media. After laying the groundwork and establishing the context in the Introduction, we describe the evolution of the room. Throughout, we attempt to document the rationale and motivation. While derived from a limited domain, we believe that the issues that arise are of general importance, and have strong implications on future research.

KEYWORDS: case studies, CSCW, intelligent systems, reactive environments, home automation, design rationale, office applications

INTRODUCTION

The convergence of computational, communications and audio/video technologies is having an ever-increasing impact on human-machine interaction. The problems introduced by new applications, users and contexts are prompting new ways of thinking about systems and design. "Ubiquitous computing" (Weiser, 1993 [13]) and "Augmented Reality" (Wellner, Mackay & Gold, 1993 [14]) are two examples of this new thinking. The basic tenet of UbiComp and Augmented Reality is that systems should be embedded in the environment. The technology should be distributed (ubiquitous), yet invisible, or transparent. While the theory is appealing, in practice, resolving the seeming paradox arising from the joint demands of ubiquity and transparency is less than simple. This paper documents a case study of attempting to do just that. We describe our experience in developing a working conference room which is equipped to support a broad class of meetings and media for both same place and different

place participation. The work is still "in progress," yet it is sufficiently well advanced that we believe that the timely documentation of our experience will be of benefit to other researchers.

After laying the groundwork and establishing the context, we describe the evolution of the room in more-or-less chronological order. We trace the development of the room from manual control, to manually-driven computer control, to context-sensitive reactive automation -- all the while striving towards the goal of simultaneous ubiquity and invisibility.

What we present is not a simple "show and tell" story. Throughout, we attempt to document the rationale and motivation. Since these derive from the observations of the specifics of use, our story is somewhat bound in the details of the application and technology of the case study. While derived from a limited domain, we believe that the issues that arise are of general importance, and have strong implications on future research.

Background Context

Over the past five years, we have been involved in studying distributed collaborative work, most recently as part of the Ontario Telepresence Project (Riesenbach, 1994 [8]), and earlier as part of the Cavecat Project (Mantei, Baecker, Sellen, Buxton, Milligan & Wellman, 1991 [7]). This work grew out of research at Rank Xerox EuroPARC (Buxton & Moran, 1990 [4]) and the Media Spaces project at Xerox PARC (Bly, Harrison & Irwin, 1993 [3]).

In contrast to work such as Colab (Stefik, Foster, Bobrow, Kahn, Lanning & Suchman, 1987 [10]), our research has focussed mainly on supporting social transactions centered on the offices of the individuals, rather than meeting rooms. Increasingly, however, we have been working towards providing an integrated foundation to support both meeting room and office based collaborative work.

In the process, we were strongly influenced by the emerging ideas of ubiquitous computing and augmented reality. In particular, we were interested in exploring their

[*]Author is visiting from Hitachi Research Laboratory, Hitachi Ltd., Japan.

[†]Author is Principal Scientist, User Interface Research, Alias Research Inc., Toronto, Ontario.

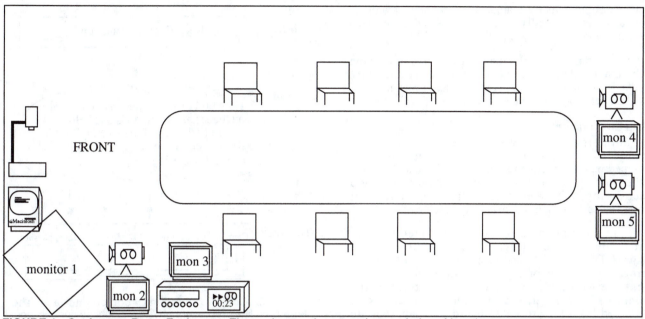

FIGURE 1. Conference Room Equipment. Electronic attendees are given a choice of locations appropriate to the variety of social roles in meeting scenarios.

affordances in preserving the fidelity of conventional social distance/place/function relationships. An early example of this was the use of video "surrogates" employed in our four-way round-the-table video conferencing system, Hydra (Sellen, Buxton & Arnott, 1992 [9]). The driving function here is the notion that for each location in architectural space for which there is a distinct social function, the affordances should be provided to enable that function to be undertaken from that location by any party, be they attending physically or electronically.

We will see examples of this in practice later in the paper. For our purposes now, note the physical distribution of technology that this implies; hence, our interest in the Ubicomp approach. The inevitable problem that arises, however, is as follows: once the equipment implied by this approach is deployed, how can its use and potential ever be accessed within the user community's "threshold of frustration?" To a large extent, the rest of this paper documents our efforts to answer this question. In our attempts, we learned a lot about what worked and what did not, as well as methods of testing ideas and designs.

The Anatomy of the Room

The room which is the object of study is illustrated in Figure 1. It is equipped to support activities such as:

- *Videoconferencing* from the front of the room (permitting remote presentations) or back of the room (permitting remote participants to attend meetings as part of the audience)

- *Video playback* from both local and remote sites

- *Meeting capture* via videotape

- *Electronic collaborative whiteboard* that can be driven locally or remotely, such as described by Elrod et al. (Elrod, Bruce, Gold, Goldberg, Halasz, Janssen, Lee, McCall, Pedersen, Pier, Tang & Welch, 1992 [5]).

- *Support for computer demonstrations*, run either locally or remotely

- *Overhead projection* using video document camera capable of being seen locally and remotely

As the amount of equipment and potential functionality increases, so does the cognitive burden on users. With conventional approaches, presenters must handle multiple remote controls and explicitly establish connections between various devices. Consider, for example, the possible problems of switching from an overhead slide to a videotape sample. On what screen does the video appear? Is it the same one as for the overhead? How is the connection established? Users often complain that control of the equipment is confusing and overly complex. Presenters must either interrupt their talk to manipulate the environment, or simply avoid using the technology because it requires too much effort.

Even if all of these issues are resolved for the local audience, what does the remote person see, and how can one be sure that this is what the presenter intended? These, and a myriad of related problems confront the baffled user. We have not even addressed the basic issue of how the user turned on all of the equipment in the room, initially. Where are all the switches and controls?

While our usage studies indicated that we were trying to incorporate the correct functionality and deploying the components in more-or-less the right locations, our work had not really begun. Regardless of the tremendous potential existing in the room, if the complexity of its use was above the threshold of the typical user, the functionality simply did not exist in any practical sense.

HISTORY

The configuration of hardware and software in the meeting room went through a number of iterations, and continues to evolve. An important concern in our design efforts has been to ensure that users of the room can continue to use whatever tools and techniques with which they are comfortable, for example, using a document camera as an overhead or slide projector, and continuing to have a traditional white board. The underlying design principle is to reduce complexity by enabling users to interact with the room using existing skills acquired through a lifetime in the everyday world.

In this section we describe the design motivation behind each iteration, discuss the solution taken, and evaluate the results. It should be noted that our evaluation was informal, based on personal experiences and anecdotal evidence.

Initial Environment

Our initial room design was intended to allow remote attendees to participate in meetings. The room was equipped with a camera, monitor, microphone and speaker at the front of the room. This equipment functioned as a video surrogate in an existing media space. In short, it corresponded to most basic videoconferencing rooms.

Using this implementation, it was realized through "breakdowns" in meetings that modifications were required. For example, due to the placement of the video surrogate at the front of the room, remote attendees often spent the whole meeting watching the back of the presenter's head. At the same time, local attendees were distracted from the presenter due to the inappropriate location of the remote participant(s) at the front of the room, in the speaker's space. (Note that this situation is the norm in an embarrassingly large number of videoconferencing rooms.) It was clear that different locations of video surrogates were needed for the different social roles of meeting attendees.

First Iteration

The motivation for the first iteration was to allow remote participants to either present, attend or participate in video meetings. This design involved the addition of three video surrogates at the back of the room. These surrogates were placed at the same height as the conference room table so that remote users would be perceived as sitting around the table. Again, an existing media space was used to support this functionality. This design worked well when remote participants were in the appropriate place. However, users could not select their own positions within the room and it was difficult to move from one location to another, such as when the attendee wanted to change roles and become the presenter

At this stage, the user interface consisted of the set of physical connections between devices themselves. This meant that in order for a presenter to realize a goal, such as "record my presentation," it was first necessary to determine which devices to activate, and then make the appropriate connections between them. Figure 2 depicts the user interaction with various devices. The cognitive effort required by the user in order to achieve the high-level goal through direct device manipulation is considerable.

Second Iteration

The next step was the incorporation of an $n \times n$ software-based matrix to implement the patchbay. This is shown in Figure 3.

Each row corresponds to a source device (eg. camera, VCR output) and each column to a destination (eg. visitor view of room, video monitor). By clicking the mouse on entry (i, j), an audio or video switch would make a connection between $source_i$ and $destination_j$. This resulted in considerable time savings, because the user could now establish connections through a graphical user interface,

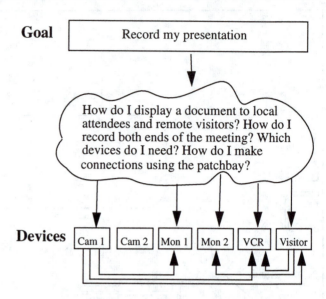

FIGURE 2. Complexity of First Iteration Interface. The inter-device lines represent physical patchbay connections, which the user was required to make.

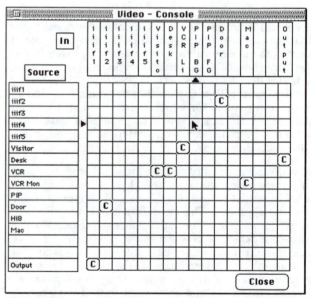

FIGURE 3. Matrix-based interface for controlling equipment (virtual graphical patchbay).

rather than physical wire. However, as depicted in Figure 4, since the user was still responsible for all device connections, the cognitive effort remained high.

Third Iteration

To make the system more efficient and reduce the cognitive burden regarding matrix representations, a provision was added which allowed room administrators to create presets as user selections (see Figure 5). As illustrated in Figure 6, a strong incentive for the development of presets was that they allowed the user to break down a goal into a number of fairly straightforward sub-goals, without concern for the representations of individual devices (Vicente & Rasmussen, 1990 [11]). We found that while this simplified control of the switch, subtle distinctions existed between various presets and users could not decide which ones to choose.

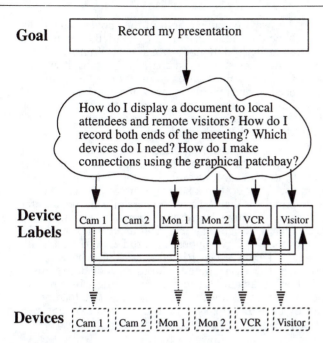

Goal Record my presentation

How do I display a document to local attendees and remote visitors? How do I record both ends of the meeting? Which devices do I need? How do I make connections using the graphical patchbay?

Device Labels | Cam 1 | Cam 2 | Mon 1 | Mon 2 | VCR | Visitor |

Devices | Cam 1 | Cam 2 | Mon 1 | Mon 2 | VCR | Visitor |

FIGURE 4. Complexity of Second Iteration Interface. The solid lines represent user interaction and the dashed lines represent tasks performed by the user interface. Note that the user is still responsible for inter-device connections, now made through the graphical user interface.

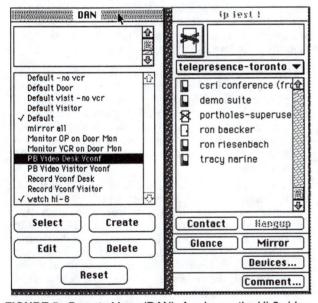

FIGURE 5. Presets Menu (DAN). As shown, the Hi-8 video is currently being viewed and the user is considering the selection of the desktop video deck instead.

At this stage, our work was addressing the problems of control at essentially the same level as many commercial room control systems, such as ADCOM's iVue (ADCOM Electronics Inc., 1994 [1]) and AMX's AXCESS systems (AMX Corporation, 1993 [2]).

Buttons and Lights

Two key problems of the previous iteration were the complexity of the user interface and the lack of diagnostics. As a first step towards reducing the complexity, we are presently constructing a set of button and light modules

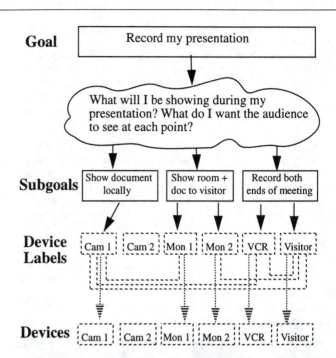

Goal Record my presentation

What will I be showing during my presentation? What do I want the audience to see at each point?

Subgoals | Show document locally | Show room + doc to visitor | Record both ends of meeting |

Device Labels | Cam 1 | Cam 2 | Mon 1 | Mon 2 | VCR | Visitor |

Devices | Cam 1 | Cam 2 | Mon 1 | Mon 2 | VCR | Visitor |

FIGURE 6. Complexity of Third Iteration Interface, using presets. Now, the user can ignore details of device representation and location. However, presets can be confusing, especially when there is more than one way to accomplish a subgoal.

which will be installed on or near each device in the room. Users will be able to make connections simply by pressing the button corresponding to the appropriate source and destination, as shown in Table 1. When the first button is pressed, its module light will flash, indicating that the computer is now waiting for the other end of the connection to be indicated. When the second button is pressed, the second module light will flash momentarily, until the computer has made the connection between each device. At this point, both module lights will turn on. The order in which connections are made is unimportant. Either source or destination can be specified first. Additionally, the source or destination of a connection can be changed simply by pressing the appropriate buttons. A special virtual module is required to represent video surrogates..

Task	Button 1	Button 2
connect source S_i to destination D_i	S_i	D_i
change source S_i to S_j	S_i	S_j
change destination D_i to D_j	D_i	D_j

TABLE 1. Button Action Menu

To illustrate by example, suppose we wish to view a remote participant on monitor 5, and provide this surrogate with the output of our document camera. Pressing the button associated with the surrogate and the button associated with monitor 5 would establish the first connection. The second connection would be formed by pressing the surrogate button and the document camera button. Since the computer knows that monitor 5 is an output-only device and the document camera is input-only, there is no ambiguity as to which connections are intended.

This implementation partly addresses the diagnostics problem of previous iterations through the use of different light states. While the system is working to effect a change, the flashing light indicates to the user that the action is being performed. If a light continues to flash long after a connection has been attempted, a problem exists at a lower level of the system. Obviously, it would benefit us to add diagnostics at these levels as well.

A possible disadvantage of these modules is that they require the user to walk around the room in order to make connections. As an enhancement to this approach, we envision using a laser pointer to point to sources and "drag" them to their destination devices (see Figure 7). As a simple example, one could point to a VCR to select it as input, then drag it to one of the monitors for output. Most of the standard connections necessary during presentations could be accomplished in this manner. In order to provide this capability, we will be installing two calibrated laser detectors to cover the front and back of the conference room. This pointer-based connection process, shown in Figure 7, could provide efficient device selection without the need for the presenter to change location.

FIGURE 7. Conference Room in use. The speaker is using a laser pointer to select a camera view for the remote visitor.

While the buttons and lights modules offer a substantial gain in simplicity, they cannot adequately replace the high-level control of presets provided in the previous iteration. Users may be reluctant to press five buttons (or point to three devices) in order to play a video tape to local and remote conference participants, when a single preset selection would suffice.

REACTIVE ENVIRONMENT

Our next iteration was motivated by three main points:

- We wanted to reduce the overall complexity of operating the room. In particular, we wanted to reduce how much explicit knowledge was required by the user to function effectively.

- We wanted to reduce the intrusion on meetings of managing the operational aspects of the room. If someone wanted to show a videotape, for example, we wanted the user to be concerned only with loading the tape and starting it, not routing the VCR output to appropriate displays.

- One way to achieve both of the above is to simply have the room "driven" by a skilled operator in a computerized room-control system. While this is the norm in most high-end conference rooms, it was not an acceptable solution in our case. Our room was used be several groups, many of whom had no vested interest in the underlying technology. It had to be "walk up and use." The transparent access of Ubicomp had to be achieved.

Our approach to achieving these goals was a variation of the third point, above. The "skilled operator," driving the room in the background would be the technology in the room itself, rather than a human operator. The underlying assumption was that if a human operator was able to infer the user's intentions based on their actions, so should an appropriately designed system. This approach is much like the Responsive Environment of Elrod et al. (Elrod, Hall, Costanza, Dixon & Des Rivieres, 1993 [6]); however, in our case, the environment would collect background information as context to support explicit foreground action. The intent was to reduce the cognitive load of the user by allowing the system to make context-sensitive reactions in response to the user's conscious actions.

To provide a mechanism for such behaviour, the integration of sensors with various devices was required. The output of these sensors allows the computer to determine when certain actions should be taken by the environment, or, in other words, how the environment should react. We call this resulting system a *reactive environment*. Our reactive environment consists of a set of tools, each of which reacts to user-initiated events. For each tool incorporated into our environment, we must keep in mind the following issues of invisibility, seamlessness, and diagnostics:

- How do we make the tool invisible during normal operation? In order for the system to function effectively, users must not perceive themselves to be involved in a two-party communication. Additionally, the rules of interaction must be made explicit to the user, but these rules should seem natural and not require any understanding of the technology.

- How do we provide a seamless "user override" function for those occasions where the intended behaviour of the tool differs from the automatic? There should not be a need to "argue" with the system if it is not behaving according to the desires of the user. If there is a dispute, the tool should seem to disappear. However, some allowance might be made while the system is "learning" the behaviour of a new user. To minimize the possibility of disputes, reactions to user-initiated events should be conservative.

- How do we provide meaningful diagnostics without a graphical user interface? Currently, if something goes wrong with the system, there is no way to find out what has happened. If extra layers of technology are to be added, it is imperative that diagnostics can provide the location of a problem during failures.

The remainder of this section explains the development of our reactive environment in more detail.

Eliminating Remote Controls

A major problem of the current conference room environment is the effort required to turn on all of the equipment. Setting up the room for a video conference typically involves three switches and three to five button presses on multiple remote control units. Since this process tends to be cumbersome, the simple alternative of leaving most of the equipment turned on all of the time is presently taken. Making use of a motion sensor and a computerized infrared transmitter, capable of generating the same infrared signals as any remote control, we can substantially reduce the interaction necessary between user and equipment.

When a user first enters an otherwise unoccupied room, the motion sensor triggers a switch which turns on the lights and activates a transmitter to send the remote control commands necessary to turn on the appropriate devices. If someone is wearing an active badge (Want, Hopper, Falcao & Gibbons, 1992 [12]), for example, the PARC Tab pictured in Figure 8, then the room can identify the user and if appropriate, automatically configure itself for that person.

FIGURE 8. The Xerox PARC Tab.

Through various sensors, the room can detect most actions that will precede the use of a remote control, and issue the appropriate commands itself, using the infrared transmitter.

Since the user does not need to interact with the computer, nor manipulate remote controls to turn on or configure equipment appropriately, the tool which performs these tasks is completely invisible. In our prototype environment, manual use of remote control units is unnecessary, except on rare occasions where the user wishes to override normal system behaviour.

VCR

Manual operation of a VCR is a relatively straightforward task. However, when the additional burden of specifying a video display or camera source is placed on the user, the equipment suddenly becomes complicated. In a conference environment, context can be helpful in determining the intended behaviour of a VCR. For example, if the play button is pressed, video output should appear not only on a monitor in the local conference room, but also on a monitor in any remote site where there are participants in the discussion. Similarly, if the record button is pressed, the VCR should record video locally as well as from the remote site.

Knowledge of whether or not remote participants are involved in a conference is obtained by checking the status of the outside line. VCR functions (eg. play, record, stop) are monitored by polling the VCR interface for user-initiated commands. When a function is selected, our environment can react by establishing the required connections between video sources and VCR inputs, or video destinations and VCR outputs, as appropriate. From the users's perspective, the interface is invisible, since no explicit action beyond pressing the VCR's play or record button was required.

Document Camera

Our conference room has replaced the standard overhead projector typically found in such environments with a document camera, whose output is usually displayed on a large television monitor at the front of the room. Since this monitor is often used for purposes other than viewing documents, presentations involving the document camera can be awkward, especially when a conference presenter wishes the audience to shift its attention from the document to other displays. Even with the buttons and lights interface discussed previously, the need for explicit manual control is too distracting.

Fortunately, selection of the document camera view can be automated easily. Using basic image analysis, we can determine whether or not a document is presently under the camera, and whether or not there is motion under the lens. When either a document or motion is detected, the environment reacts by sending the output from the document camera to the display monitor as well as to any remote participants. If no document is detected over a certain timeout period, then the camera is deselected. Again, the tool is invisible. The simple act of placing a document under the camera is sufficient to activate the device

To provide a mechanism for seamless manual override, we also wanted a method to force the "re-selection" of the document camera. Our solution was very simple. Whenever document motion is detected after a period of inactivity, the document camera is again selected, regardless of its (assumed) current state.

Digital White Board

The large monitor in our conference room is shared by several applications including the document camera and the digital white board, the latter being a Macintosh computer running any interactive application (see Figure 9). Because of the hardware configuration, users of the white board can automatically write or draw with a light pen instead of the mouse. The only special action required is the selection of the Macintosh computer as the input source to the monitor.

Once again, this selection can be automated trivially with the help of a contact sensor on the light pen. Whenever the pen is held, the environment reacts by selecting the Macintosh display automatically and sending this view to remote conference participants as appropriate.

Head-Tracking for Camera Control

By virtue of their location, remote conference participants are currently limited to the view provided by a stationary video camera. In essence, their vision is controlled by a second party, typically the conference presenter, who determines which camera will provide output to the remote site. We considered providing camera selection capability

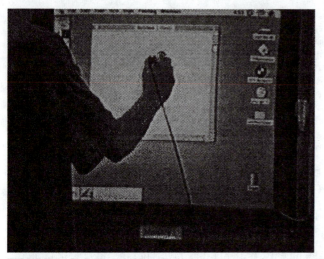

FIGURE 9. The Digital White Board in use.

to the remote end directly, but this solution requires additional computer equipment and communications.

We have adopted a more elegant solution, which requires no additional equipment beyond a video camera and monitor on the remote end, yet which allows the remote participant far more control over the received view. We treat the remote monitor as a window through which the local room can be viewed. Applying a head-tracking algorithm to the video signal, we can determine the position of the remote participant's face in relation to his or her monitor. This position is then used to drive a motorized video camera locally. When the remote participant peers to the left or right, the local camera pans accordingly. Similarly, when the remote participant moves closer to or further from the monitor, the local camera zooms in or out.

Evolution Summary

Table 2 summarizes the evolution of our reactive environment by examining the task space of our conference room environment. The sensors and sensing techniques previously described have now been integrated into our prototype reactive conference room.

CONCLUSIONS

In this paper, we have reported our experience to date. This project is ongoing, however, and there remains a great deal to do. Up to now, we have been exploring the problem space and building prototypes to test designs. However, the system is not yet robust, nor do we have the knowledge to make it so. A number of problems remain.

A standard issue, shared by those working on intelligent agents, is how to deal with exceptions. How do different users specify different behaviours, or how can the system adapt to the evolving desires or expectations of the user?

In another direction, if the room is to make informed decisions based on context, can there be an application-independent architecture for accommodating the shared cross-application knowledge base and heuristics according to which such decisions are made?

Ubiquitous computing holds great promise. For the first time, we have a model that does not confront us with the classic strength vs. generality tradeoff. We no longer have to choose between strong-specific systems and weak-general ones. With Ubicomp, we have the promise of both strength and generality by virtue of the combined power of a family of strong-specific systems working in concert. But the risk is that while any member of the family is easy to use due to its specific nature, complexity and cognitive load may remain the same or increase, by virtue of coordination overhead. In this case, load is simply transferred, not off-loaded.

Our case study attempts to solve this problem. By appropriate design, the complexity of coordination can be relegated to the background, away from conscious action. The intent of this exercise is to begin paving the foundation for an existence proof that useful background processing can be carried out by context-sensitive reactive systems. That being the case, our hope is that this work will stimulate research that will make this capability available sooner rather than later.

Task	Current Interface	Button Selections	Reactive Environment
turn on lights	light switch	"room on" button	motion sensor
turn on equipment	remote/manual controls	"room on" button	motion sensor/PARC Tab
show document	GUI + remote control	doc camera -> dest	document present?
show video	GUI selections	VCR -> dest	VCR play button pressed
revert to document	GUI + remote control	VCR -> doc camera	motion on document camera
show white board	GUI selections	white board -> dest	pen is picked up
let visitor present	GUI selections	monitor A -> monitor B	laser pointer
let visitor look left	GUI selections	camera A -> camera B	remote head tracking

TABLE 2. Task Space for the three different room interfaces. The current graphical user interface is involved in almost every configuration change. However, as we progress to a more reactive environment, explicit interaction with the interface becomes less necessary.

Acknowledgments

Many people have contributed to this project. We would like to thank Bill Gaver for the original head-tracking code, written at the Technical University of Delft and Rank Xerox EuroPARC, Dominic Richens for his work on the early versions of the graphical user interface, and Jie Dai for an early implementation of the infrared transceiver driver. Thanks also to Kim Vicente, George Fitzmaurice, and Kevin McGuire for the useful discussions and feedback.

We also thank Rich Gold, Roy Want, and Norman Adams of Xerox PARC for help with the PARC Tab and Mike Ruicci of CSRI for his outstanding technical support. Special thanks are due to Sidney Fels for the design of the buttons and lights modules, as well as many hours of insightful discussion. Finally, we are greatly indebted to the members of the various research groups who make up the user community of the room. Their patience and feedback has been essential to our work.

This research has been undertaken as part of the Ontario Telepresence Project. Support has come from the Government of Ontario, the Information Technology Research Centre of Ontario, the Telecommunications Research Institute of Ontario, the Natural Sciences and Engineering Research Council of Canada, Xerox PARC, Bell Canada, Alias Research, Sun Microsystems, Hewlett Packard, Hitachi Corp., the Arnott Design Group and Adcom Electronics. This support is gratefully acknowledged.

REFERENCES

1. ADCOM Electronics Inc, iVue: Extending your boardroom around the world, Product Information.

2. AMX Corporation (1993). Advanced Remote Control Systems, Product Information.

3. Bly, S., Harrison, S. & Irwin, S. (1993). Media Spaces: bringing people together in a video, audio and computing environment. Communications of the ACM, 36(1), 28-47.

4. Buxton, W. & Moran, T. (1990) EuroPARC's Integrated Interactive Intermedia Facility (iiif): Early Experience, In S. Gibbs & A.A. Verrijn-Stuart (Eds.). Multi-user interfaces and applications, Proceedings of the IFIP WG 8.4 Conference on Multi-user Interfaces and Applications,

Heraklion, Crete. Amsterdam: Elsevier Science Publishers B.V. (North-Holland), 11-34.

5. Elrod, S., Bruce, R., Gold, R., Goldberg, D., Halasz, F., Janssen, W., Lee, D., McCall, K., Pedersen, E., Pier, K., Tang, J. & Welch, B. (1992). Liveboard: A large interactive display supporting group meetings, presentations and remote collaboration, Proceedings of CHI'92, 599-607.

6. Elrod, S., Hall, G., Costanza, R., Dixon, M. & Des Rivieres, J. (1993) Responsive office environments. Communications of the ACM, 36(7), 84-85.

7. Mantei, M., Baecker, R., Sellen, A., Buxton, W., Milligan, T. & Wellman, B. (1991). Experiences in the use of a media space. Proceedings of CHI '91, ACM Conference on Human Factors in Software, 203-208. Reprinted in D. Marca & G. Bock (Eds.)(1992). Groupware: software for computer-supported collaborative work. Los Alamitos, CA.: IEEE Computer Society Press, 372 - 377.

8. Riesenbach, R. (1994). The Ontario Telepresence Project, CHI'94 Conference Companion, 173-174.

9. Sellen, A., Buxton, W. & Arnott, J. (1992). Using spatial cues to improve videoconferencing. Proceedings of CHI '92, 651-652. Also videotape in CHI '92 Video Proceedings.

10. Stefik, M., Foster, G., Bobrow, D., Kahn, K., Lanning, S. & Suchman, L. (1987). Beyond the chalkboard: Computer support for collaboration and problem solving in meetings. Communications of the ACM, 30(1), 32-47.

11. Vicente, K. & Rasmussen J. (1990). The Ecology of Human-Machine Systems II: Mediating "Direct Perception" in Complex Work Domains, Ecological Psychology, 2(3), 207-249.

12. Want, R., Hopper, A., Falcao, V., & Gibbons, J. (1992). The Active Badge Location System. ACM Transactions on Information Systems, 10(1):91-102.

13. Weiser, M. (1993). Some Computer Science Issues in Ubiquitous Computing, Communications of the ACM, 36(7), 75-83.

14. Wellner, P., Mackay, W. & Gold, R. (Eds.)(1993). Computer-Augmented Environments: Back to the real world. Special Issue of the Communications of the ACM, 36(7).

The High-Tech Toolbelt:
A Study of Designers in the Workplace

Tamara Sumner
Department of Computer Science and Institute of Cognitive Science
University of Colorado
Boulder, CO USA 80309-0430
Email: sumner@cs.colorado.edu

ABSTRACT

Many design professionals assemble collections of off-the-shelf software applications into toolbelts to perform their job. These designers use several different tools to create a variety of design representations. This case study shows how designers *evolve* initially generic toolbelts through a process of *domain-enriching* to make their own *domain-specific design environments*. Comparing this practice with theoretical findings concerning design processes highlights the benefits and limitations of this toolbelt approach. A key benefit is its flexible support for creating and evolving multiple design representations. A key limitation is how it hinders iterative design by making it difficult for designers to maintain consistency across the different design representations. This limitation could be remedied if tools could be extended or "tuned" to support the observed domain-enriching process. Such tuning would enable designers to extend tools during use to: (1) support important domain distinctions and (2) define dependencies between different design representations based on these domain distinctions.

KEYWORDS: Design, Design environments, Domain-orientation, End user modifiability, Iterative design, Interoperability, Tailorability, Task-specificity

INTRODUCTION

In today's workforce, there are many skilled professionals engaged in design activities. These designers often work in emerging high-technology domains such as computer network design, user interface design, or multimedia title design. Domains such as these are characterized by rapid and continual change as underlying technologies evolve. Projections on the workforce of the next century indicate that skilled designers working in rapidly evolving domains will become increasingly prevalent in the years ahead [13]. There are very pragmatic reasons why it is important that the tools used by these designers effectively support their design activities and enable them to create better designs more efficiently. Towards this end, numerous research efforts are focused on creating various types of design support environments [3, 4, 7]. However, a major research challenge is to construct design support tools that are capable of evolving to match the rate of change within dynamic design domains.

The research presented here is concerned with creating evolutionary software tools to support the design activities of skilled professionals in rapidly changing domains. The approach taken has been to observe professional designers in the workplace using the tools of their choice to perform their daily design activities. The assumption is that by analyzing the situations where current practices excel and breakdown, requirements for the next generation of design support tools can be determined. This approach differs from previous work [7] in that the empirical methodology revolves around long-term observations of both designers and their tools solving real design problems.

The case study illustrates how designers rely on collections of off-the-shelf software applications, referred to as "toolbelts," to create a variety of design representations. Over time, designers evolve these initially generic toolbelts through a process of domain-enriching to make their own domain-specific design environments. Domain enriching involves recognizing and incorporating important domain distinctions into design representations.

This paper begins by describing the toolbelt model and comparing it with both theoretical models of design and empirical studies of designers in order to identify its strengths and weaknesses. Next, a long-term case study involving professional designers of phone-based user interfaces is used to illustrate and analyze the toolbelt model. Finally, the implications of these findings for both application designers and creators of design support tools are discussed.

THE TOOLBELT MODEL

A major part of a designer's job is to create and evolve external representations of the design being constructed [15]. These representations occur in many forms such as textual, tabular, or diagrammatic. An important activity for expert designers is figuring out what representations to use or even creating new ones if necessary [1]. For several reasons, more than one external design representation is typically required. First, different design representations at different levels of abstraction support designers to engage in opportunistic design [6]. Second, design activities in the workplace usually involve several different stakeholders from a variety of backgrounds [14]. Often, designers must construct several external representations to facilitate communication and collaboration with each stakeholder group [2].

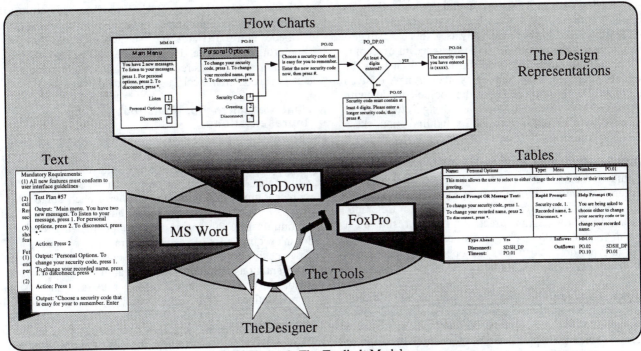

Figure 1: The Toolbelt Model

The designer assembles a collection of software tools and uses them to create different design representations. In the case shown here, the designer used a word processor (MS Word[1]) to create text documents, a flow charting tool (TopDown[1]) to create flow charts, and a database (FoxPro[1]) to create tables.

While some designers rely on "traditional" media such as paper and pencil, increasing numbers rely on software tools. Often, these design professionals assemble collections of off-the-shelf software tools as needed to create the necessary design representations. I refer to these software collections as "high-tech toolbelts" because each designer assembles her personal collection just as a carpenter assembles a collection of hammers, screwdrivers, tape measures, etc. into a personal toolbelt. Figure 1 illustrates the toolbelts used by the designers in this study.

Typical off-the-shelf software tools include word processors, spreadsheets, databases, flow charting tools, and CAD systems. Each tool is good at making a different type of representation. For instance, spreadsheets provide a lot of support for making tabular representations while flow charting packages make it relatively easy to construct node-link types of representations. Thus, *one strength of the toolbelt model is that designers can flexibly pick and choose from a selection of relatively inexpensive, off-the-shelf tools and find ones that provide decent support for making the necessary representations.*

However, designers often end up in a situation where different tools are needed for each different design representation. This situation is problematic because the different representations are usually interrelated; i.e., they are different views of the same information. Now, when the designer changes part of the design in one representation, she must remember to manually carry out the related modifications in all other representations. This makes it very difficult to modify the design because small changes in one representation can trigger a series of changes in other representations. Suddenly, that seemingly small task has blossomed into a large and tedious job. This "blossoming" effect hinders iterative design because designers are reluctant to make changes due to the effort required. Following an iterative design process is not only desirable; it is often necessary, because existing requirements change and new ones are uncovered as design proceeds [1, 6]. Thus, *a weakness of the toolbelt model is that it hinders iterative design.*

Empirical studies indicate that many design errors result from designers' cognitive limitations when managing such dependencies across representations [8]. Besides the danger of introducing errors when making the necessary changes across multiple representations using different tools, the designer must remember to make the changes in the first place. Thus, another weakness of the model is that *the resulting toolbelt provides significantly suboptimal support for the design activity in that it does not help designers to deal with either the cognitive or manual burden of managing dependencies across design representations.*

---

[1] Product Credit and trademark notifications for the products referred to are given here: Excel, MS Word, FoxPro are registered trademarks of the Microsoft Corporation. MacDraw is a registered trademark of the Claris Corporation. MacroMind Director is a registered trademark of the Macromedia Corporation. TopDown is a registered trademark of the Kaetron Corporation. MacFlow is a registered trademark of the Mainstay Corporation.

The following case study serves several purposes. First, its illustrates the toolbelt model and its strengths and weaknesses. Second, a detailed analysis of this case suggest how future software tools might be designed to overcome the weaknesses identified. And third, a look at how the designers' practices, tools, and representations have evolved over time allows us to compare the toolbelt model with alternative task- or domain-oriented design environment models in order to assess the benefits and limitations these different approaches.

CASE STUDY: VOICE DIALOG DESIGN

The study presented here is part of a long-term collaboration between user interface designers at US WEST Advanced Technologies and researchers at the University of Colorado. This analysis of design practices is the result of a combination of workplace observations conducted over a three year period, field notes, and open-ended interviews with members of the design group.

The domain studied here is the design of voice dialog applications; i.e., software applications with phone-based user interfaces. These phone-based interfaces consist of a series of voice-prompted menus requesting the user to perform certain actions; e.g., "to listen to your messages, press 1." The caller issues commands by pressing touch-tone buttons on the telephone keypad and the system responds with appropriate voice phrases and prompts. Typical applications are voice information systems and voice messaging systems. Voice dialog applications are an important technology for many businesses because they reduce the need for human phone operators and provide callers with direct access to information concerning business services. Designing in this domain means specifying the interface for an application at a detailed level.

The Larger Design Process

The designers described in this study are hired by marketing groups to design the functionality and interfaces of voice dialog applications. Once a design has been approved by the marketing group, the designers are responsible for overseeing the implementation of the application by external vendors and testing the final implementation for compliance with the approved design specification. Often, the actual work of testing the interface is contracted out to external consultants. Typically, the market group approaches the design group with an idea for a new product or service and the results from some preliminary market analyses. The designer uses this preliminary information to create one or more initial designs. Programmers construct simulations of the design which are then used by the designers to get feedback from the market group and to perform user testing on the design. The results from using these simulations feed back into the next iteration of the design.

Depending on the size and complexity of the application being developed, this overall design process can take from six months to two years. In summary, the design process is long, complex, highly iterative, and involves a multidisciplinary group of stakeholders (marketers, designers, testers, simulation builders, vendors). Collaboration and communication between the stakeholders is made more difficult by the geographical distance separating them. The designer, marketers, and vendors typically reside in different states and much communication occurs via conference calls using standard telephones.

What Designers Do: Construct Multiple Representations

There are two main facets to the designer's job: constructing the design and communicating the evolving design to all design stakeholders. The two are interrelated in that to communicate the design, the designer must construct design representations that communicate with each stakeholder group. As one designer noted, "the critical problem is communicating the design to other people in a way that doesn't require a lot of specialized training [on their part]." Towards this end, the designers have created different design representations tailored to the special needs of each of the major stakeholder groups. Figure 1 illustrates the various representations that are constructed.

The *flow chart* is the core interface design representation (Figure 2). It consists of different kinds of nodes filled with text. The nodes represents common entities found in the domain, such as audio prompts, voice menus, decisions, and audio messages. These flow charts specify the control flow of the interface, possible user actions, and the text of all audio prompts and messages in the interface. Flow-charting tools such as TopDown or MacFlow[1] are used to create this representation. Flow charts are primarily used to communicate with the marketing group and the simulation builder during design. However, these charts are also used by the vendor when implementing the design and by customer support representatives who must diagnose problems with the application after its release.

Figure 2 shows an excerpt from a flow chart depicting a voice messaging system interface. The chart illustrates that to change a security code, the caller must first navigate through two voice menus (Main Menu and Personal Options) and then listen to an audio prompt requesting the caller to enter 4 digits followed by the # sign. As this example illustrates, the different domain-specific entities have unique appearances and content. These appearances and content are built up by the designer using graphic primitives provided by the flow charting tool, such as text, boxes, and shading. Voice menus consist of a title (the shaded rectangle in Main Menu, Figure 2), an audio prompt (the text under the title), and possible actions associated with touch-tone buttons (the text and small boxes below the text). Each entity in the flow chart is assigned a unique identifier according to a naming scheme the design group has created (e.g. "MM.01" for the Main Menu in Figure 2). This naming scheme facilitates long distance communication involving the marketing group and the vendor. When talking over the phone, the stakeholders can ground their design discussions by referring to "prompt PO.02."

The *table representation* (Figure 2) is constructed primarily for the vendor who will implement the final application on a specialized hardware platform. These tables are constructed using database systems such as FoxPro. Figure 2 shows the table entry for the Personal Options menu. For every entity in the flow chart, there is a corresponding table. Correspondence between the two representations is established via the entity's unique identifier. The table representation is a more detailed elaboration of the entities in the flow chart. It contains both redundant information and additional information to that found in the flow chart. For the example shown in Figure 2, redundant information includes the unique identifier, menu title, and the menu prompt. The descriptions of inflows and outflows is another way of representing the links shown in the flow chart. To create the table, all this redundant information must be manually rekeyed in by the designer or explicitly copied and pasted using direct manipulation. The additional information includes more implementation-oriented details such as pseudo-code descriptions of conditionals and events not shown in the flow chart such as messages spoken in response to help buttons and illegal input conditions.

Test plans are semi-structured text documents created for each task the interface supports. These textual representations are constructed using word processors such as MS Word. The plans detail the actions that must be performed and the output that will be heard along a particular path through the interface. As such, these plans reorganize information already found in the flow chart and table representations. The flow chart shown in Figure 2 would yield several test plans including one for changing the security code and one for changing the recorded greeting. These representations are used by testers who are hired to exhaustively test all features of the final product for compliance with the product's specification.

Other representations containing additional design information not found in the flow chart, table, or test plan representations are also constructed. For the most part, these additional representations are primarily textual in nature and are constructed using word processors such as MS Word.

Iterating the Design: How the Tools (do not) Fit Together

As described above, there are complex relationships between the various design representations, with data in one representation being dependent upon data in another representation. It is important that consistency be maintained between the various inter-dependent representations as the design undergoes modification in response to the demands of marketing, changing technical requirements, and the results of usability testing.

As an example, imagine that usability testing reveals that most callers want to hear their existing security code before they change it. The designer decides to add a new confirmation message between the Personal Options menu (PO.01) and the change security code prompt (PO.02). This requires the designer to adjust all subsequent identifiers to make room for the new message. Additionally, each table corresponding to a changed entity in the flow chart must also be updated with the new identifier and all the inflow /outflow lists must be reworked. Besides being tedious, this change process is fraught with potential errors in that the designer must remember which particular entities in the flow chart changed (a potentially large number) in order that the right tables be updated with the correct information. If the test plans have already been constructed, the designer must also find and update every test plan traversing the path containing the new message.

This example illustrates how seemingly simple design changes can blossom into major jobs involving extensive editing of several design representations created with

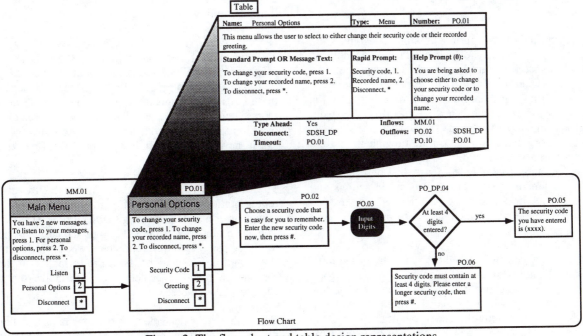

Figure 2: The flow chart and table design representations.

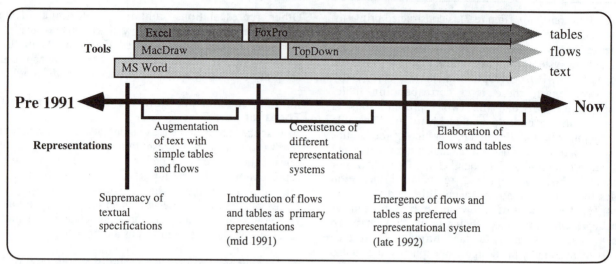

Figure 3: The co-evolution of tools and representations over a three year period.

different software applications. The hardest part is that the burden of the change is placed on the designer - she must remember to make the change and then correctly implement the change without introducing errors or other inconsistencies across the representations. Unfortunately, with existing levels of application interoperability, manual rekeying and copy and paste using direct manipulation are the only methods for maintaining these consistencies! Thus, the cost of iteration (both cognitive and labor) is very high due to the complex dependencies across design representations and the lack of tool support for managing these dependencies.

The Co-evolution of Representations, Tools, and Practices Over Time

The previous description of the major representations is only a snapshot in an ongoing evolutionary design process. As shown in Figure 3, the design representations themselves have undergone a series of major and minor changes throughout the three years that this study encompasses.

At the beginning of the period of study, designs were represented using primarily textual specification documents. These textual specifications distinguished between phrases, prompts, and menus. Important information such as dynamic conditions and help messages were buried in the middle of paragraphs. These textual specifications were augmented with simple tables and flow charts at the request of the vendor organization. The simple flow charts did not contain the full text of all audio prompts and phrases in the interface. The tables augmented the textual description of prompts and menus by depicting all possible actions and responses. MacDraw[1] and sometimes Excel[1] were used to construct the simple tables since early versions of MS Word did not support tables.

In 1991, one designer was asked to create one of the most complex applications to date. He decided it was time for a whole new approach. The textual specifications were getting so large and complex that few people bothered to read them. Those that did had trouble understanding them. This designer established a personal convention of using flow

charts and a new, complex table as the primary design representations. His flow charts contained four different domain distinctions: voice menus, prompts, messages, and decisions. Tables depicted dynamic conditions using pseudocode, showed alternative messages (such as help) and provided new details on error handling. Excel did not support the complex layout needed for the new tables, so the designer switched to a database program called FoxPro. He used MacDraw for the first flow charts simply because that tool was readily available. When he later tried to switch to a flow charting tool, he found that those early applications could not handle large designs, so he continued to use MacDraw.

The two representational systems – textual versus flows and tables – coexisted for about one year. Other designers were reluctant to switch for two reasons. First, several had long-term relationships with their vendors and marketers and were reluctant to switch representations on them. Second, constructing the flow charts in MacDraw was tedious and time-consuming.

By mid-1992, several things happened that led to the emergence of flows and tables as the primary design representations. First, new versions of flow charting tools (e.g., TopDown) had greatly improved their capacity. Using these tools made it much easier to construct the flow chart representation. Second, new designers were hired to embark on a long series of enhancements to an existing product. These new designers readily adopted the flow and table representations.

The designers began to elaborate on the flow and table representations. New domain distinctions emerged such as inputs, error handlers, and digit collectors. A new release of TopDown supported patterned arrows and designers began to use these patterns to differentiate between different types of flow control. The vendor liked the formality of the table representation and suggested ways to formalize the pseudocode parts even more.

Today, due to the dependencies between representations, it is difficult to maintain the table representation when iterating

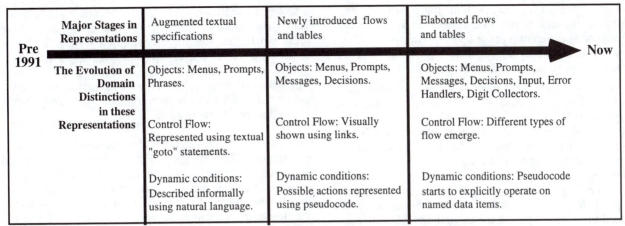

Pre 1991	**Major Stages in Representations**	Augmented textual specifications	Newly introduced flows and tables	Elaborated flows and tables	Now
	The Evolution of Domain Distinctions in these Representations	Objects: Menus, Prompts, Phrases.	Objects: Menus, Prompts, Messages, Decisions.	Objects: Menus, Prompts, Messages, Decisions, Input, Error Handlers, Digit Collectors.	
		Control Flow: Represented using textual "goto" statements.	Control Flow: Visually shown using links.	Control Flow: Different types of flow emerge.	
		Dynamic conditions: Described informally using natural language.	Dynamic conditions: Possible actions represented using pseudocode.	Dynamic conditions: Pseudocode starts to explicitly operate on named data items.	

Figure 4: Domain-enriching is a process whereby domain distinctions emerge and representations evolve to incorporate these emergent distinctions over time.

the design. One designer experienced this in a major way when the market group asked him to redo large parts of a design for which the tables had already been constructed. Most of the tables had to be significantly reconstructed. For this reason, designers are increasingly deferring constructing the table representation until later in the design process.

Not shown in Figure 3 are further signs of major change looming on the horizon. One designer has started working with a new vendor who (so far) has not required the table representation. In another new design, the marketers were dissatisfied with the increasing complexity of the flow charts; they think simplification might be in order.

Looking back over the course of this study, it is clear that several factors influenced the co-evolution of the designers' tools, representations, and practices:

(1) *The various stakeholder groups.* When communication with a stakeholder group broke down, designers changed the representations to overcome the breakdown. Additionally, stakeholders often suggested new ways of representing things.

(2) *The designers.* Over time, better approaches and techniques emerged through practice. As one designer noted, "part of my job is continually thinking of better ways to do things."

(3) *The tools.* Affordances and hindrances of the tools affected both the content of the design representations and the practices of the designers. New features in tools enabled new domain distinctions to emerge and sometimes simply enabled the tool to be used. The flexibility of the tools allowed the representations to undergo quite a bit of modification before a new tool was required. Hindrances, or lack of support for important operations, caused both delayed tool adoption and changes in design practice.

Domain Enriching

This study illustrates how domain distinctions pertinent to voice dialog design emerged over time as a product of the interaction between designers, their tools, and other design stakeholders. In language, "distinctions" are articulated objects and qualities that arise through recurrent patterns of

breakdown in concernful activity [18]. Likewise, domain distinctions in design emerge through recurrent breakdowns between designers, their tools, and other design stakeholders. As the domain distinctions emerged, they were incorporated into design representations. Over time, the representations became more formalized, containing more domain distinctions depicted with greater levels of detail (see Figure 4). The representations also became increasingly interdependent as they shared important distinctions. While the tools themselves were generic in that they had no built-in knowledge of the domain distinctions, the conventions and practices that arose around these tools took on a decidedly domain-specific flavor. The result of this evolutionary process is that the designers created a domain-oriented design environment by enriching their own practices and conventions to reflect important, emerging domain distinctions.

RECOMMENDATIONS FOR IMPROVING THE TOOLBELT

In summary, the primary strength of the toolbelt model is its support for the flexible creation of multiple design representations. The flexibility of the model manifests itself in both large and small scales. In the smaller sense, the flexibility of individual tools supported designers to gradually evolve domain-specific representations. In the larger sense, the ability to change one tool for another without affecting the relationship between other tools and representations enabled designers to radically change design representations when the need arose. However, this pronounced decoupling between different tools and their related representations is also a major weakness of the toolbelt model. It significantly raises the cost of iterative design by forcing designers to manually maintain complex dependencies across representations.

The cost of iteration would be lowered if tools could be extended to automatically manage the dependencies between design representations. The observed process of domain enriching suggests necessary extension mechanisms to allow designers to "tune" their toolbelts to better support their design practices:

• *Within tools, designers need to be able to describe their own layer of named, domain distinctions.* Tools need to be extended to be aware that there are named distinctions, such as prompts, voice menus, and links representing different types of control flow, which have a specific look and content. The contents may also consist of other named domain distinctions.

- *Across tools, designers need to be able to establish automatic mechanisms to manage complex dependencies based on shared named domain distinctions.* For instance, designers should be able to establish a dependency between the flow chart and table representations stating that all prompts with the same identifier should have identical text content. As a second example, designers should be able to establish a dependency that states that whenever a design entity is added to the flow chart, a corresponding table with the same identifier and type should be automatically created and information derivable from the flow chart should be automatically filled in.

The emergence of compound document architectures such as OLE [16] and OpenDoc [12] offer hope that creating tools that provide designers with such mechanisms will soon be possible. These architectures provide a standard way for applications to expose their internal objects for extensibility and programmatic control.

However, these architectures by themselves do not guarantee that the resulting extension mechanisms will be either useful or usable for end users such as the designers in this case study. The prevalent attitude in the software development community is that the key benefit of these architectures is how they enable assembling or bundling an initial collection of smaller components to form a cohesive toolbelt [17]. As this study shows, the benefit is not in the initial bundling; the benefit is in supporting end users to evolve the tools over time. Cohesiveness emerges as end users enrich the tools with knowledge about important domain distinctions and the relationships between domain distinctions. Thus, end users need to be provided with mechanisms supporting the types of extensibility or design-in-use [9] needs uncovered in this study.

RELATED WORK

The research presented in this paper recommends providing designers with flexible generic tools containing mechanisms to support domain enriching. Several other research efforts have also investigated domain-oriented or task-specific design tools.

Fischer [4] has long advocated the construction of knowledge-based domain-oriented design environments. He has noted that one of the key challenges when creating such environments is providing the necessary flexibility to support tasks not envisioned by the creators of the design environment [3]. Several approaches towards providing the necessary flexibility have been pursued, such as providing end user modification substrates [5] and programmable design environments [3]. Both of these techniques have emphasized enabling users to enrich their environment by describing new entities or operations. However, these approaches do not address the need for design representations

to undergo significant evolutionary change, e.g., changing from flow chart to table representations. Thus, these domain-oriented design environments may lack the flexibility needed to keep up with changing practices in rapidly evolving domains. The research presented here suggests that domain-oriented design environments with the required flexibility can be constructed by building on off-the-shelf software components.

Until recently, Nardi had advocated the advantages of task-specific tools [10]. Similar to Fischer, her recent empirical work found that specially constructed task-specific tools are too inflexible to support the activities of professionals [11]. She also discovered that the definition of "task" was highly individual and depended on the fluctuating goals of the people involved. She found that professional slidemakers preferred collections of interoperable components to task-specific tools, and she suggested that such tools be tailored to capture regularities in their daily work [11]. The results of the voice dialog design case study suggest that the ability to describe domain distinctions and establish dependencies across tools based on these domain distinctions are two specific types of tailoring that professional designers require.

While the analysis here is based on a single case study, the basic finding that professionals need the ability to tune generic tools into their personal toolbelts is a general one. Preliminary interviews with designers of a multimedia title (i.e., an interactive book) have yielded similar evidence for the need to tune toolbelts via domain-enriching. The toolbelt used by this design team consisted of MacroMind Director[1], MS Word, and a half dozen other applications for creating visual effects. During the course of several months, the design team evolved a set of domain distinctions and incorporated these distinctions into their representations. They had problems iterating their design due to the dependencies between design representations.

SUMMARY

The voice dialog design case study illustrated how design professionals assemble collections of off-the-shelf software applications into toolbelts necessary for creating the variety of design representations their job demands. The designers evolved initially generic toolbelts through a process of domain-enriching to make their own domain-specific design environments. Iterative design was hindered because the tools provided designers with no support for maintaining consistency across the different design representations. These tools would provide better support for design practices if they could be "tuned" to support the observed domain-enriching process. Such tuning would enable designers to extend tools during use to: (1) support important domain distinctions and (2) define dependencies between different design representations based on these domain distinctions.

ACKNOWLEDGMENTS

I wish to thank Susan Davies, Josh Staller, Mike King, Jason Webb, Lynda Baines, and Bruce Keahy of US WEST Advanced Technologies for their help and support during this project. I also wish to thank the HCC group at the

University of Colorado, John Rieman, Gerry Stahl, Jonathan Ostwald, Marcus Stolze, and Chris DiGiano for helpful discussions on these issues. I also wish to thank the anonymous reviewers for their excellent questions, comments, and suggestions on both this research and this paper. This research was supported by US WEST Advanced Technologies and ARPA under grant No. N66001-94-C-6038.

REFERENCES

1. Curtis, B., H. Krasner and N. Iscoe, "A Field Study of the Software Design Process for Large Systems," *Communications of the ACM,* Vol. 31, pp. 1268-1287, 1988.

2. Ehn, P., *Work-Oriented Design of Computer Artifacts,* arbetslivscentrum, Stockholm, 1989.

3. Eisenberg, M. and G. Fischer, "Programmable Design Environments: Integrating End-User Programming with Domain-Oriented Assistance," *Human Factors in Computing Systems (CHI '94),* Boston, MA (April 24-28), 1994, pp. 431-437.

4. Fischer, G., "Domain-Oriented Design Environments," in *Automated Software Engineering,* Ed., Kluwer Academic Publishers, Boston, MA., 1994, pp. 177-203.

5. Fischer, G. and A. Girgensohn, "End-User Modifiability in Design Environments," *Human Factors in Computing Systems, CHI'90 Conference Proceedings (Seattle, WA),* pp. 183-191, 1990.

6. Guindon, R., "Designing the Design Process: Exploiting Opportunistic Thoughts," *Human Computer Interaction,* Vol. 5, pp. 305-344, 1990.

7. Guindon, R., "Requirements and Design of DesignVision, An Object-Oriented Graphical Interface to an Intelligent Software Design Assistant," *Human Factors in Computing Systems (CHI '92),* Monterey, CA (May 3-7), 1992, pp. 499-506.

8. Guindon, R., H. Krasner and B. Curtis, "Breakdowns and Processes During the Early Phases of Software Design by Professionals," in *Empirical Studies of Programmers: Second Workshop,* G. Olson, S. Sheppard and E. Soloway, Ed., Ablex Publishing Corporation, Norwood, New Jersey, 1987, pp. 65-82.

9. Henderson, A. and M. Kyng, "There's No Place Like Home: Continuing Design in Use," in *Design at Work: Cooperative Design of Computer Systems*, M. Kyng and J. Greenbaum, Ed., Lawrence Erlbaum Associates, Hillsdale, NJ, 1991, pp. 219-240.

10. Nardi, B. A., *A Small Matter of Programming,* The MIT Press, Cambridge, MA, 1993.

11. Nardi, B. A. and J. A. Johnson, "User Preferences for Task-specific vs. Generic Application Software," *Human Factors in Computing Systems (CHI '94),* Boston, MA (April 24-28), 1994, pp. 392-398.

12. Piersol, K., "Under the Hood: A Close-Up of OpenDoc," *BYTE,* Vol. 19, pp. 183-188, 1994.

13. Quinn, J. B., *Intelligent Enterprise,* The Free Press, New York, N.Y., 1992.

14. Rittel, H. and M. M. Webber, "Planning Problems are Wicked Problems," in *Developments in Design Methodology*, N. Cross, Ed., John Wiley & Sons, New York, 1984, pp. 135-144.

15. Schoen, D. A., *The Reflective Practitioner: How Professionals Think in Action,* Basic Books, New York, 1983.

16. Udell, J., "Beyond DOS: Visual Basic Custom Controls Meet OLE," *BYTE,* Vol. 19, pp 197-200, 1994.

17. Udell, J., "Componentware," *BYTE,* Vol. 19, pp. 46-56, 1994.

18. Winograd, T. and F. Flores, *Understanding Computers and Cognition: A New Foundation for Design,* Addison-Wesley, Menlo Park, CA, 1986.

Time Affordances

The Time Factor in Diagnostic Usability Heuristics

Alex Paul Conn

Digital Equipment Corporation
110 Spit Brook Road
Nashua, New Hampshire 03062-2698
(603) 881-0459
alex.conn@zko.mts.dec.com

ABSTRACT

A significant body of usability work has addressed the issue of response time in interactive systems. The sharp increase in desktop and networked systems changes the user's focus to a more active diagnostic viewpoint. Today's more experienced networked user is now engaged in complicated activities for which the issue is whether the system is carrying out the appropriate task and how well it is proceeding with tasks that may vary in response time from instantaneous to tens of minutes. We introduce the concept of a time affordance and a set of principles for determining whether the diagnostic information available to the user is rich enough to prevent unproductive and even destructive actions due to an unclear understanding of progress.

KEYWORDS

Usability engineering, heuristics, time delay, affordances, taxonomy, principles, design rationale, practical guidelines.

1. INTRODUCTION

One of the fundamental problems with systems (computer-based and otherwise), is that users do not always understand what is happening. Even relatively experienced users may be stymied by integrated systems in which the progress of a given task is difficult to assess. When delays are not understood, they can cause dysfunctional behavior even in organizations that are not necessarily computer based, as illustrated by Senge's beer game [11].

For the typical desktop computer system, such as a stand-alone PC, the consequences of not understanding the reason for a given delay may be minimal. However, with the rapid growth in network use and highly integrated systems, the delays can be important indicators of the status of an enterprise information system. With critical applications, such as the control and coordination needed in the trans-

portation and power industries, a wrong conclusion about a delay based on insufficient information can have serious consequences. For example, an improper termination of a computer application could cause a medical operation to be suspended. A computer delay could cause a commuter rail slowdown because the lack of updates could mean that conditions were unsafe. In manufacturing, inappropriate interruptions could result in loss of materials that cool or harden and cannot be manipulated after a delay.

A large body of research exists on various aspects of computer response time [4, 5, 10], providing a good foundation for understanding how rapidly a user's desktop system should respond to requests, or for determining whether a local server can realistically support a potential set of client applications on users' desktops. However, with the world wide web browsers, gophers, combined with increasingly modem-based access, response time is no longer the only focus. The user *knows* a search may take a while. The question facing practical designers is "what constitutes a full set of delay information that users must have in order to understand and respond appropriately to these delays?" In other words, how do designers know, in practice, when their design contains the necessary components of time delay information, so that they can then concentrate on format, style, etc.?

This paper pulls together various existing components of good practice and synthesizes a taxonomy of eight properties in task execution that, if observable, indicate usability along the time dimension. These eight properties together characterize a *time affordance*, an extension of the concept discussed by Don Norman and others [9][3]. We believe the term affordance appropriately captures the *gestalt* impact of the time-delay components, from which users understand what to expect and how to proceed.

Our focus is on the *information*, rather than the presentation format. The precise format of the information should be determined using guidelines such as those described in [7], [8], [12], and [5] and tools and laboratories for developing and evaluating alternatives.

We first introduce four scenarios involving problems with time delays, three of which are roughly based on actual events. We then introduce a taxonomy of time affordance concepts. Next we analyze the scenarios based on the taxonomy, and finally we present a set of principles on which to base a time-affordance interface.

2. SCENARIOS

2.1 Scenario 1: PC with Network Access to an Office Management Information System (MIS).

An office worker uses a well-known GUI-style program to access a database stored on another system in the office. The user attempts to update a field. The action is apparently acceptable because there are no error messages. The only message is the status bar message stating what was going to happen if the user pressed OK. The hour glass appears and stays there for five minutes. Although the mouse will move the hourglass, the PC is otherwise unresponsive (e.g., Alt-Tab will not move between applications).

The user presses `Ctrl-Alt-Delete`. After several seconds, the computer returns with a message saying "Although you can use Ctrl+Alt+Delete to quit an application that has stopped responding to the system, there is no application in this state..." The user should press any key to return to the application or hit Ctrl-Alt-Delete again to reboot system. The user presses a key and Windows returns but still appears to be frozen, and the screen is not properly repainted. In frustration, the user hits `Ctrl-Alt-Delete` twice to reboot.

The status of the update is unclear. Later attempts to do the same update result in the same problems, until the user finds out from a colleague that these kinds of updates always take about 10 minutes. The problem has resulted in a loss of productivity for the better part of a morning.

2.2 Scenario 2: Client Installations

An experienced PC user is trying to install an important update to a package that will allow an existing tool to incorporate a library designed with a new object-oriented approach. Installation is familiar; the graphical progress indicator (which we will call a "thermometer") appears and proceeds slowly but evenly until it reaches 18%. At this point, a message flashes saying "installation complete" but is almost immediately obscured by a message saying "unable to process XYZ because ABC driver not found."

The user repeats the installation several times (in part to try to read the flashing messages but mostly because it said it only had completed 18% of the job). The user then calls a number for help and only after an hour of comparison (file-by-file) do they determine that the installation was indeed complete. In addition, the error message was apparently not important, since the software worked.

2.3 Scenario 3: Networked Manufacturing Floor

A manufacturing floor has recently been integrated by tying together a number of legacy programs over a local area network. The integration required special code to wrap the existing applications and to serve as workflow drivers. Acceptance tests were easily passed, and the system has been working well for over a year. The number and complexity of jobs has been steadily increasing, but monitors have shown that network traffic is still well below saturation.

One day, without warning, a certain piece of computer driven fabrication equipment starts losing communication with data servers. The problem takes days to diagnose, since the network activity (as seen both from the server and the fabrication machine) appears to be normal.

It turns out that to reduce network traffic, a program on the server was designed to compute complex functions output from the legacy application and to send shorthand specifications to the fabrication machine as the results of the computation became available. Recent specifications required so much computation that there was a significant delay between transmission of certain parts of the specification data. The fabrication machine program interpreted the delay as a problem with the server, and went into reset mode, no longer capable of processing the data even when it was ready. On site programmers temporarily fixed the problem by increasing the programmed fabrication machine reset delay value.

However, the delay was originally instituted because the legacy programs occasionally crashed. By timing out, the fabrication machine could become available for fabrication instructions from another server, thus reducing shop floor down time. The dilemma was how to institute a "smart" time-out that distinguished between long computations and malfunctioning legacy programs.

2.4 Scenario 4: A World-Wide Web Browser.

A user employs a world-wide web browser to determine the availability of certain products and request further information from manufacturers. The user has spent considerable time "surfing the web" and can find the necessary information relatively rapidly. The browser is well-designed, presenting many of the reasons for time delay and status information as it happens. There is even an easy way to stop a request if the network traffic is too heavy to allow an acceptable response time.

The hypertext format provides a separate window indicating the network address of a potential request (actually the universal resource locator, or URL, is shown, which contains the network address). However, once a request is made, the URL disappears until the request is completed. Even experienced users often stop perfectly good requests, unable to remember what hypertext items they selected, or afraid that they had accidentally moved the pointer when pushing a mouse button.

Once the user arrives at the menu for ordering information, there is an additional problem. It is easy to read about the desired products and gather together a set of requests, fill out fields to enter a name, address, etc. However, when the user pushes the submit button, there is no way to tell whether anything happened. The user pushes the button multiple times, actually causing multiple requests to be mailed to the manufacturers. Later that day, duplicate automated responses arrive at the user's mailbox.

3. A TAXONOMY OF TIME AFFORDANCE CONCEPTS.

3.1 Time Affordances

A **time affordance** is a presentation of the properties of a delay in a task or anticipated event that may be used by an actor (e.g. user) to determine the need for an interrupting or facilitating action. These properties are often visual but may also have other (e.g., audible) components. As discussed by Graver [3], affordances exist whether the perceiver cares about them or not. "Making affordances perceptible is one approach to designing easily-used systems."

A good example of a time affordance is the "thermometer" on many PC operations, telling visually and with a percent indication how the task is proceeding and how much more needs to be done. Other approaches have been described in [6], [2]. With a good time affordance, a user knows when things are okay and when there are problems, and can generally predict when a task will be completed. A complete time affordance may combine information from more than one area or level of the system in order to provide a visual or other indication of the following eight task properties:

1. **Acceptance**: What the task is and whether it has been accepted with the input parameters or settings.

2. **Scope**: The overall size of the task and the corresponding time the task is expected to take barring difficulties. (Once acceptance and scope are indicated, the task may pause for the user to decide whether to initiate.)

3. **Initiation**: How to initiate the task and, once initiated, clear indication that the task has successfully started.

4. **Progress**: After initiation, clear indication of the overall task being carried out, what additional steps (or substeps) have been completed within the scope of the overall task, and the rate at which the overall task is approaching completion.

5. **Heartbeat**: Quick visual indication that the task is still "alive" (other indications may be changing too slowly for a short visual check).

6. **Exception**: A task that is alive has encountered errors or circumstances that may require outside (i.e., user) intervention.

7. **Remainder**: Indication of how much of the task remains and/or how much time is left before completion.

8. **Completion**: Clear indication of termination of the task and the status at termination. How to terminate the task before completion (whether or not errors have occurred).

Thus, a familiar office elevator may accomplish this goal by (1) illuminating the button that the user pushes to indicate acceptance of the request to go up or down, (2) providing a display showing where the elevator(s) currently are and showing movement of elevator(s), and (3) turning off the local up or down button illumination and/or ringing a

A: **Acceptance:** Computer has accepted input and begun processing.

B: **Initiation** and **Heartbeat:** Blocks are sized so that they always change within the tolerance window. For huge file, this field shows heartbeat.

C: A finer granularity estimate of progress, largely redundant for a fixed rate transfer, but useful as a check on ZMODEM settings.

D: **Scope** and **Remainder:** Computer's estimate of time remaining is believed only if it has been reasonably accurate in the past.

E: **Exception:** User will seek error indications if progress is not being made. If the error rate is sizable, user may take steps to terminate transfer.

F: **Progress** and **Completion:** Visual indication via "thermometer" with percentage indication works only if it is roughly equivalent to time delay. If it is based on non-linear information (e.g., count of files differing in size), the visual indication becomes a source of confusion.

Figure 1: Analysis of a familiar dialog box for ZMODEM

bell when a given elevator is close and about to stop at the requester's floor.

On a PC, modem file transfer implementations frequently provide such information. Figure 1 shows the information provided by a ZMODEM implementation. In addition to the dialog box shown, when the task is complete, the application frequently provides a separate dialog box and/or beep to signal that the transfer is complete.

3.2 Delays

Delays are simply the time period between a request and the completion of the request. E.g., I want an elevator to stop at my floor (going in the direction I want to go); the delay is the time before the elevator actually arrives.

Static delays are periods during which nothing appears to happen even though a task may be proceeding normally. User or machine action during a static delay (and after how long) depends on the time tolerance window, described below. The change of the mouse pointer from an arrow to an hourglass or clock (with no other time affordance) is a classic example of a static delay. If the pointer is animated (e.g., the clock hands move, or the world in the Mosaic interface turns), the result may be a dynamic delay if the animation is a true monitor (heartbeat) for the task in question and is recognized as such by the user.

A **dynamic delay** indicates that although there is a wait, the task has the appearance of proceeding normally and there is no reason for concern. In the early days of computing, console lights indicated something was happening (dynamic delay) and if the lights froze and nothing happened, we knew something was wrong. The flashing "working" message was in general not a dynamic delay because (as described in Section 5.2) most users quickly learned that "working" did not actually monitor progress and did not in reality provide a true "heartbeat" indicator. Note that a dynamic delay may have static components, which can become significant if they result in a delayed heartbeat indication.

3.3 The Time Tolerance window

The **time tolerance window** (or tolerance window for short) is the length of time an actor (user or system) allows before deciding that a task is not making progress or that something must have gone wrong. The time tolerance window is both individual and context sensitive, affected by:

- *Sensitivity or urgency.* A particularly sensitive mission with a chance that things could go seriously wrong may shrink the tolerance window considerably.

- *Estimations.* With properly set expectations, the tolerance window can be increased. (For example, ZMO-DEM warns that the transfer will take half an hour.)

- *Individual differences.* Different actors (humans or machines) have different initial tolerance windows, indicating a need for tailoring (a machine actor may have a different time-out setting stored; users have different levels of patience).

- *Familiarity.* Once a user knows the level of dynamic or static delay in a given interface, he or she will have expectations of what kinds of feedback indicate problems. A machine actor may be programmed using similar experiences.

A time tolerance window will be static if the delay is entirely static (there is no additional information, so the level of tolerance remains unchanged). However, if the delay is dynamic or contains dynamic components, the tolerance window can shrink or grow based on the information available during the dynamic delay. There are actually two kinds of tolerance windows, each of which can be affected by the nature of the dynamic delay:

1. **Progress tolerance window.** Each time the dynamic delay provides new progress information, the actor resets the starting point from which the tolerance window is measured. If the dynamic delay contains static components, the heartbeat must be received within the progress tolerance window in order for the reset to occur (see Figure 2).

2. **Scope tolerance window:** Even if the actor is receiving dynamic information and resetting the progress tolerance window, the overall (dynamic) delay may eventually grow to a point that it exceeds the actor's expectations of the scope of the task (e.g., due to heavy traffic, errors and cretransmissions, etc.). Thus even with a dynamic delay, if the rate of progress is

not high enough and/or the remainder shows that the task will exceed the expected overall scope, the actor will terminate the task.

3.4 The Task Hierarchy

We believe that any task can be viewed as a hierarchy of task steps, each with a corresponding delay. At the highest level is the complete task, which (during the delay after initiation and before completion) is presumably some given percent completed. Below the top level, a task can usually be decomposed into task steps or sub tasks, and these can be further decomposed. Each lower level of decomposition will reveal a set of smaller, and therefore more rapidly completed steps. In theory, by recursing sufficiently deep into the hierarchy, it should be possible to observe task steps that complete well within any progress tolerance window.

A tracking program, which we will call a **hierarchical reporting engine**, could be designed to navigate the task hierarchy. The reporting engine would descend into the task-step hierarchy searching for a granularity of task steps in which change occurs within the progress tolerance window. That is, the reporting engine would move up and down the hierarchy in order to present information for a dynamic delay. If the current level did not provide results within the tolerance window, the engine would descend to a lower hierarchical level for more rapidly changing status information. The engine would need to descend only to a deep enough level where progress could be observed within the tolerance window. The output of a hierarchical reporting engine would thus provide the necessary information so that the updates would occur within the progress tolerance window (see Figure 2).

Computers are now fast enough that whole tasks and/or modules may be completed within an actor's progress tolerance window. Thus in many cases, it is possible to provide dynamic delay information simply by tracking the

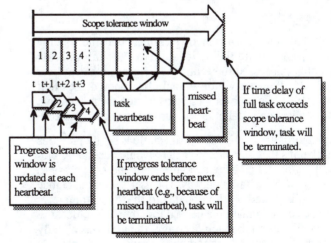

Figure 2: Relationship of progress and scope tolerance window to dynamic delays having heartbeats. The progress tolerance window is elastic and could have been extended with other indications of progress.

completion of modules (e.g., a compressed file is unpacked and copied). If even that information is delayed, the monitoring "engine" often can predict length of task and percent completion at the lower level (e.g., size of file to unpack). Thus a fully-featured engine may not be needed if a dynamic delay can be accomplished by carefully choosing to monitor at a hierarchical level in which feedback occurs within the progress tolerance window.

4. ANALYSIS OF SCENARIOS

In this section, we use the above concepts to characterize and explain the problems encountered in the scenarios.

4.1 Scenario 1

In the update to MIS database scenario, two problems occurred: (1) the user relied on time expectations derived from other experiences, therefore the time tolerance window did not coincide with the actual time needed for the task to complete, and (2) the task required a static window because there was no feedback other than the hour glass to indicate that the computer accepted the input and was proceeding normally.

Two approaches would have reduced the likelihood that a user would give up in this scenario. First, the application might have extended the static window by clearly acknowledging the request [accept], indicating that processing has started [initiate] and setting the expectation for a longer (e.g., ten-minute) period [scope].

Second, the application could have created a dynamic window by providing some kind of feedback indicating progress. Such feedback would answer questions such as: Is something happening [heartbeat], is the right task being accomplished [progress], are there errors requiring intervention [exception]? If it is impossible to predict the percent of the task left to complete [remainder], an alternative approach, such as listing the processing steps and indicating where the system has progressed within the list, might be preferable (as shown in Figure 3).

4.2 Scenario 2

The installation scenario involved both a dynamic window and a problem with the believability of the information presented (discussed in section 5.2). Because the application had a dynamic window, including named files that were copied, the user took no steps to stop the processing. (The application installation was progressing as expected.)

The major issue involved the expectations set by previous installations that the "thermometer" would accurately predict how much had been accomplished and how close to completion the installation was. The user's scope tolerance window, adjusted by the updates to the "thermometer," established a minimal time needed for completion. Because the installation took much *less* time, the user was already concerned that something had gone wrong and erroneously indicated completion. Worse, the error message (essentially unrelated to the bulk of the installation), was sufficient to add enough questions to the process to reject the claim that the installation had completed.

The system used a display object that many users interpret as a linear time indicator to instead indicate a percentage of tasks completed (potentially very non-linear). Had the "thermometer" been replaced by a list of programs that, for example, were grayed out when they were unpacked, the user might have believed the completion indications, especially if he or she could later check that the names at the end of the list were indeed in the directory. See Figure 3.

The role of error messages in reducing confidence is especially important. An error message says something went wrong (perhaps not fully understandable even to experienced users). Just as task-length predictions can increase the tolerance window (because the user knows to wait longer before suspecting problems), error messages have the potential of reducing confidence and thus shortening the tolerance window. The likelihood that the error message would prejudice the overall task depends on how clear

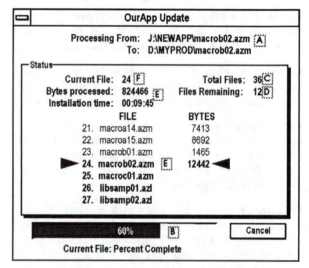

A: **Acceptance:** Computer has accepted input and begun processing.
B: **Initiation** and **Heartbeat:** Because the size of a file cannot be determined at initiation, only the current (and previous) file sizes are known. The "thermometer" serves as a heartbeat indicating the progress of the current file.
C: **Scope:** The total number of files is the only real indicator of scope.
D: **Remainder:** The remaining files tells how much work remains. A thermometer would be misleading because of the implied correspondence between the spatial progression and time.
E: **Progress:** The increasing total bytes processed along with the movement along the list of files indicates progress. The key point is that the list of files processed is clearly indicated so that even if the last 12 are small and take only a second, the user will know that the processing was normal.
F: **Progress** and **Completion:** As the files processed steps toward the number indicated by C, the user understands that progress has occurred, and completion is near. A separate dialog box might be used to indicate true completion.

Figure 3: Possible dialog box for processing files where the size of each file is not known prior to the start of processing.

the indications were during the major task that progress was normal (i.e., how well the eight task affordance properties were presented during the task).

4.3 Scenario 3

In the manufacturing floor fabrication equipment scenario, the fabrication equipment time window was fixed based on programmer expectations when loads were significantly lighter. The design was for a static window, even though a dynamic window capability was possible. For example, the time window was reset each time a buffer or block of primitives was received, not at the beginning of processing of legacy application data.

The driver program that interfaced with the legacy application and built primitives from the data "knew" when it was building a primitive. It could either have (1) determined whether a primitive would take a while and sent an estimate to the fabrication machine [scope], or (2) set appropriate timers to trigger the sending of "status OK" messages to the fabrication machine [heartbeats] if the compu-

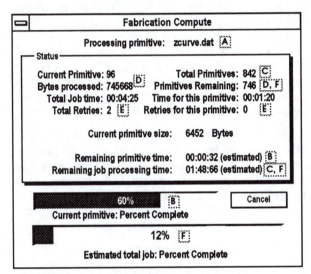

A: **Acceptance and Initiation:** Computer has accepted input and begun processing.

B: **Heartbeat:** Reporting engine uses progress in processing primitive to provide updates within the tolerance window.

C: **Scope:** The total number of primitives indicates scope initially. The program could provide an estimated job time either from a parameter or by using experience from previous runs saved in a data file. As processing proceeds, the estimated remaining processing time is updated based on current job experience.

D: **Progress:** indicated by the increasing bytes, the count of primitives processed (and the thermometers).

E: **Exception:** indicated by total and current retries.

F: **Remainder** and **Completion:** indicated by the estimated remaining time counters, primitives remaining, and the thermometers.

Figure 4: Possible dialog box for hierarchical reporting (displays progress for graphics primitives). This box shows both the overall task remainder (highest abstraction level) and a lower level of abstraction in which something happens within the tolerance window (in this case, the current primitive processing status changes).

tation for a particular primitive was extensive. The same timer could have been used to check on the legacy application (to see if it might have crashed). Figure 4 gives an example of a dialog box showing primitive-by-primitive progress as well as overall task status. Similar information could have been sent in the "status OK" messages to the fabrication machine.

4.4 Scenario 4

The world-wide web scenario illustrates how an otherwise very well designed application can "miss" important components of a time affordance. Once the hypertext item was selected, initiating a transfer, the *progress* component never clearly indicated the goal of the status information. Ideally, the hypertext being executed should visibly change in appearance (e.g., change color) to indicate that the current delay is being caused by selection of that hypertext. In addition, the corresponding universal resource locator (displayed before the hypertext was selected) should remain visible during the time delay.

The problem with the submit button was primarily the lack of an indication of completion, a component of a time affordance, but also important state information even when there are no time delays.

5. BASIC PRINCIPLES AND CONSIDERATIONS

Principle: Every user is engaged in some degree of diagnosis during the operation of an overall task. He or she is prepared to stop the task if something appears to be wrong. The nature of the delay, the context, and the individual's thresholds affect whether their response will be appropriate (i.e., in accordance with the actual state of events).

- Static delays provide no information and are the source of potentially destructive actions by an actor (user or second computer interacting with the system in question).

- Dynamic delays are always preferable to static delays. The quality of the dynamic delay is determined by the information available during the delay (i.e., the time affordances).

- Dynamic delays are generally achievable by engines that can walk the task-step hierarchy seeking status information that is changing at a rate fast enough to hit the progress tolerance window.

- Dynamic delays are also achievable by expanding the time windows using predictive algorithms that set expectations for a longer wait (thus enlarging the progress tolerance window).

- System actors are often programmed with some form of progress tolerance window (e.g., a time-out). By enriching protocols to include the status information in a time affordance, we can reduce or eliminate unnecessary time-outs.

Principle: It is important for the system to provide a good time affordance, ensuring that any delays are dynamic delays wherein progress is observable. A good time affordance will provide all eight properties of a task (acceptance, scope, initiation, progress, heartbeat, remainder, exception, and completion).

5.1 Notions of What is Happening

This paper has focused on the important ingredients in a time affordance. As stated earlier, the guidelines and tradeoffs in the actual format of the information should be explored by work such as [6] and [2], principles [5], and heuristic design principles [8]. Whatever the format, we believe the following to be true:

Principle: An indication that "something is happening" is an important ingredient of a time affordance. Interfaces need to be designed with the necessary "hooks" to provide such feedback within the tolerance window.

The vagueness of the phrase "something is happening" is deliberate. The perception of change that is interpreted as a heartbeat need not be clearly understandable information. From flashing patterns of console lights, we were able to easily recognize the cross-reference stage of an assembler, signaling the nearing completion of a job. Rather than information overload, the lights provided us with interesting patterns from which we could begin to perceive stages of progress. In the same way, Mosaic's byte transfer messages are useful as a heartbeat even though most users do not seriously compute anything from them. We suspect that a starfield view as described in [1] could be designed to provide heartbeat and progress indications similar to console lights.

Because tolerance windows vary between users, time affordance displays may need some kind of "throttle" or tailoring capability, so that the frequency of indication showing that "something is happening" is not distracting.

5.2 The Confidence Factor: Telling the Truth

System designers have for years recognized the response time issue in computer systems as described in [5] and [4]. In an attempt to deal with tasks that do not complete right away, designers have used a number of approaches for dealing with the user. A given user will, based on past predictions and how they were actually realized, adjust their tolerance window. Let us examine the impact of some approaches designers have used to deal with time delays.

Distract: According to John Kemeny, in the original Dartmouth College time-sharing system, the line that printed the name of the program and the starting time was designed not only as an acceptance and initiation indication but also as a delay. The average student's program would compile and run within the time a Model ASR33 Teletype took to print that line. The system therefore seemed to have almost instant response. Thus, at least initially, the system did not require any kind of progress or heartbeat indication for most users. Billboard-style advertising of new features during product installations arguably provides a similar and useful form of distraction.

Mislead: Later time-sharing systems using video display terminals began to incorporate a flashing "working" message. Unfortunately, users soon recognized the message as placating rather than truly indicating heartbeat. The user not only learns to ignore the "working" indication, but worse, may learn to suspect *any* progress or heartbeat indication from the computer, even if valid.

We found similar behavior in a popular streaming tape backup program. Heartbeat was supposed to be shown by a progression of dots across the screen. Yet, on more than one occasion, the dots marched across the screen even when the tape backup was not proceeding at all. (Audible heartbeat indications, described below, that were normally present, such as disk rattles and whining tape, were not present). The dots were thus nothing more than an indication that the computer had not yet crashed!

Surprise: Some implementations provide information that surprises the user. The most common are: (a) progress or remainder indicators that are highly non-linear and therefore do not correspond with true delays. In Scenario 2, the first 18% of the task took much longer than the remaining 82%, causing the user to expend significant time ensuring that the task in fact completed. On the PC, many progress indicators will zoom to 100% only to pause for tens of seconds before any addition heartbeat. Some network copies of small files will show 50% done when the networked computers did not in fact successfully negotiate the protocols and the copy was never completed.

Principle: A system should provide a true estimate of time delay, so that numerical and implied (i.e., spatial) percentages correspond to linear progression over time rather than some other task-related information. If the information is not believed, any time affordance can do very little to extend the tolerance window because the enhanced explanation and forecasting of delays will be suspect.

5.3 Alternatives to visual feedback

A lot of attention has been paid to visual indications of response time and progress indicators. In desktop and laptop computers, hard and floppy disks are often nearby and the noise from seeks and transfers is usually audible. Similarly, streaming tape drives usually whine during transfer and modems provide dialing tones and line noise before connecting. Users quickly learn to interpret such noise a an indication that something useful is happening. While it may not be a good idea to depend on a disk rattle for feedback (especially in a noisy environment), modem designers have apparently standardized on the notion of providing connection noise to convey that initiation and protocol exchange is being accomplished.

We believe that it may be desirable to provide the option for audible indications of heartbeat and progress. In environments where such noise is not objectionable, users might benefit from audible feedback on task progress freeing them to train their visual senses on other aspects of the job.

5.4 Multitasking

As true multi-tasking becomes more commonplace in desktop systems, users will be frequently executing more than one task at a time (although they may focus on one at a time). Users of such systems want to start a task, ensure that everything is going well, and then concentrate on another task. While concentrating on the other task, the user may not require any progress updates for a background task.

In a multitasking environment, the user must be able to easily associate time affordance properties with the correct task. There may also be a need for a foreground mode (with a very active display) and a background mode (once the user is satisfied that the task is successfully underway). In a Windows environment, the modes could change based on whether the display was visible or occluded.

6. SUMMARY AND CONCLUSIONS

A user requests an action of a machine based on a model of the current machine state and the tasks that the machine is expected to be able to carry out. That model is based on knowledge the user either gains explicitly from currently available displays or infers from present and past behavior. The user may lack confidence in that model either because of doubts about the current state or about how to request the desired task.

After initiating a task, the user may thus be looking for indications that the computer's "understanding" of the state and the requested task are the same as his or her own. The time delay during which the task proceeds provides an opportunity for the user to seriously question assumptions about the model and about what is currently happening.

Determining how best to establish a shared man-machine model is the subject of a separate investigation, which will be described in a later paper. However, even with a perfect initial model, enough uncertainty surrounds networked computing that users clearly need help in understanding what is happening during a time delay.

The contribution of this paper is that it pulls together and builds on various principles currently available in the literature related to computer response time and techniques for displaying and controlling task progress. We have developed and presented a taxonomy of time-related properties and a set of guidelines for evaluating information displayed during a time delay. We call this display a *time affordance* because it provides the user with enough information to determine whether to wait or to take action. A time affordance not only describes what information a user might need but also how frequently that information needs to be updated. Without this information, users are likely to take inappropriate steps, either acting too soon to stop a successfully progressing task or waiting too long to terminate a potentially destructive one.

The idea of a time affordance is based on Norman's con-

cepts in [9] and is consistent with the affordances in [3] and percent done indicators in [6]. We believe the ideas are an extension of the concepts of response time, documented in [10] and [4], more completely characterizing the components of a thorough display of time-based information. Enhanced feedback concerning delays can avoid the kinds of inappropriate responses discussed by Senge in [11].

7. ACKNOWLEDGMENTS

Thanks to Geoff Bock for helping establish the context of this work and to fellow members of the Technology and Systems Engineering Architecture and Characterization group at Digital for feedback on many of these ideas.

8. REFERENCES

1. Ahlberg, Christopher and Schneiderman, Ben, "Visual Information Seeking: Tight Coupling of Dynamic Filters with Starfield Displays," CHI '94 Proceedings, pp. 313-317 and 479-480.

2. DeSoi, John F, Lively, William M., and Sheppard, Sallie V., "Graphical Specification of User Interfaces with Behavior Abstraction," CHI '89 Proceedings, pp. 139-144.

3. Gaver, William W., "Technology Affordances," CHI '91 Proceedings, pp. 79-84.

4. Geist, Robert, Allen, Robert, and Nowaczyk, Ronald, "Towards a Model of User Perception of Computer System Response Time" CHI '87 Proceedings, pp. 249-253.

5. Mayhew, Deborah J., *Principles and Guidelines in Software User Interface Design*, Prentice Hall, Englewood Cliffs, New Jersey, 1992. Chapters 9 and 15.

6. Myers, Brad A., "The Importance of Percent-Done Progress Indicators for Computer-Human Interfaces," CHI '85 Proceedings, pp. 11-17.

7. Nielsen, Jakob and Phillips, Victoria L., "Estimating the Relative Usability of Two Interfaces: Heuristic, Format, and Empirical Methods Compared, CHI '93 Proceedings, pp. 214-221.

8. Nielsen, Jakob, Enhancing the Explanatory Power of Usability Heuristics, CHI '94 Proceedings, pp. 152-158.

9. Norman, Donald A., *The Psychology of Everyday Things,* Basic Books, New York, 1988.

10. Schneiderman, Ben, "Response Time and Display Rate in Human Performance with Computers," ACM Computing Surveys, Volume 16, Number 3, September 1984, pp. 265-285. See section 5 in particular.

11. Senge, Peter M., *The Fifth Discipline, The Art and Practice of the Learning Organization,* Doubleday, New York., 1990. See chapters 3 and 5 in particular.

12. Thovtrup, Henrik and Nielsen, Jakob, "Assessing the Usability of a User Interface Standard," CHI '91 Proceedings, pp. 335-341.

RECOMMENDING AND EVALUATING CHOICES IN A VIRTUAL COMMUNITY OF USE

Will Hill, Larry Stead, Mark Rosenstein and George Furnas

Bellcore, 445 South Street, Morristown, NJ 07962-1910

willhill@bellcore.com, lstead@bellcore.com, mbr@bellcore.com, gwf@bellcore.com

http://community.bellcore.com

ABSTRACT

When making a choice in the absence of decisive first-hand knowledge, choosing as other like-minded, similarly-situated people have successfully chosen in the past is a good strategy --- in effect, using other people as filters and guides: filters to strain out potentially bad choices and guides to point out potentially good choices. Current human-computer interfaces largely ignore the power of the social strategy. For most choices within an interface, new users are left to fend for themselves and if necessary, to pursue help outside of the interface. We present a general history-of-use method that automates a social method for informing choice and report on how it fares in the context of a fielded test case: the selection of videos from a large set. The positive results show that communal history-of-use data can serve as a powerful resource for use in interfaces.

Keywords: Human-computer interaction, interaction history, computer-supported cooperative work, organizational computing, browsing, set-top interfaces, resource discovery, video on demand.

INTRODUCTION

With vast stores of multimedia events and objects to choose from, future users of the national information infrastructure will be overwhelmed with choices and human-computer interface designers will be called upon to address the problem. The aim of this research is to evaluate the power of a particular form of virtual community to help users find things they will like with minimal search effort.

Taking *video selection* as an initial test domain, the technique compares a viewer's personal ratings of videos with those of hundreds of others to find people with similar pref-

erences and then recommends unseen videos that these similar people have viewed and liked. The technique outperforms by far a standard source of movie recommendations: nationally recognized movie critics.

Virtual community, not virtual reality nor intelligent agents

The term *community* means "a group of people who share characteristics and interact". The term *virtual* means "in essence or effect only". Thus, by *virtual community* we mean "a group of people who share characteristics and interact in essence or effect only". In other words, people in a Virtual Community influence each other *as though* they interacted but they do not interact. Thus we ask: "Is it possible to arrange for people to share some of the personalized informational benefits of community involvement without the associated communications costs?" Such costs might include for example, the time costs of developing a personal relationship, costs to privacy, costs of synchronous face-to-face communications.

We wish to contrast our idea of virtual community with two popular themes in human interface work: virtual reality and intelligent agents. First we draw the contrast with virtual reality.

Popular future visions of networked computing and infrastructure marry perceptual immersion in virtual reality to high-bandwidth telecommunications. They seek a photo-realistic and real-time "cyber-face to cyber-face" social environment [10]. This immersive vision expects total involvement from participants. The result is what might be called a virtual reality community with its central issues of visual, auditory and temporal fidelity. By virtual community we do not mean virtual reality community. The pitfalls of seeking higher and higher fidelity to face-to-face communication have been well discussed in Brothers et al. [2]. Virtual community is about attempting to realize some of the benefits of community without the associated communications costs.

A second popular vision of networked computing and infrastructure paints scenarios which include a large role for "intelligent agents". The idea is that of semi-autonomous programs somehow endowed with intelligence great enough to impress us with their ability to interpret our needs and their work on our behalf. Our notion of virtual community includes no central role for intelligent agents other than the human participants in the virtual community.

Relation of current work to previous research

Malone et al. [7] propose three types of information filtering activities: cognitive, economic and social. Cognitive activities filter information based on content. Economic filtering activities filter information based on estimated search cost and benefits of use. Social activities filter information based on individual judgments of quality communicated through personal relationships. This paper concentrates upon the computer-assisted mediation of Malone's third type: social filtering activities. However, a basic thesis of this work is that personal relationships are not necessary to social filtering. In fact, social filtering and personal relationships can be teased apart and put back together in interesting new ways. For instance, the communication of quality judgments can occur through less personal, and even impersonal relationships as well as personal relationships. Obviously, people want a satisfying mix of both personal and impersonal relationships.

We have been particularly interested in how social filtering activities can be simultaneously streamlined and enriched through the careful design of communication media. The social relationships in which filtering of information occurs can be streamlined by making them less personal and enriched by making them more personal. For example, adding or removing the communications costs of synchronous face-to-face encounter, anonymity, and choosing a more personal medium such voice or a less personal medium such as text are all means of influencing the personal aspects of communication. Social filtering can be simultaneously streamlined and enriched by making some aspects of a relationship less personal while making other aspects of the relationship more personal.

In the realm of computer-assisted mediation of social filtering, a few HCI experiments sparsely dot the space of possible designs. Goldberg's Tapestry system [3] is a site-oriented email system encouraging the entry of free text annotations with which on-site users can later filter messages. Annotations are rich in high quality information and their successful uses are valuable. However, despite hopes to the contrary, the twin tasks of writing annotations to enter filtering data and specifying queries to use filtering data require significant user effort. Domains where the invested efforts pay off readily are few, but they do exist. In the case of annotations where the method of entering filtering information for the benefit of others has significant user costs,

Grudin's question [4] "Who does the work and who gets the benefit?" becomes noticeably relevant.

Reacting against the trend of interface designers loading additional tasks on users in order to help them find things, the history-enriched digital objects approach (HEDO) [5][6][11] attempts to explore a region of the interface design space that minimizes additional user tasks. Through a combination of automatic interaction history and graphics, depictions of communal history within interface objects hint at their use while user effort is minimized. HEDO techniques record the statistics of menu-selections, the count of spreadsheet cell recalculations and time spent reading documents (e.g., email, reports, source-code,) in a line-by-line manner summing over sections and whole documents. Displays are simple shadings on menus, spreadsheets and document scroll bars. Because the HEDO data are less informative than annotations, they tend to be less useful, but they cost less to gather and use. There is evidently a trade-off here.

One way to think about the trade-off is considering the two approaches to social filtering mentioned so far as two ends of spectrum. On one end of the spectrum we have social filtering interfaces that expect more work from the user and give more value. On the other end of the spectrum we have interfaces that expect no additional work from the user but provide less value. Our thought is that perhaps somewhere in the middle of this spectrum between the two end alternatives, there might lie special niches that offer relatively more filtering value for relatively less filtering work. Such locations on the spectrum, if they existed, we could call design "sweet spots". Figure 1 depicts the spectrum and places a "sweet spot" in the middle.

Figure 1   *Spectrum of Social Filtering Interface Tasks*

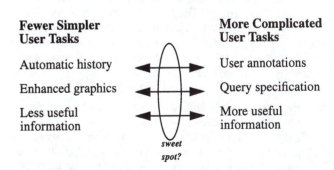

We have in mind the ideal of a community of users routinely entering personal ratings of their interest concerning digital objects in the simplest form possible: a single keypress or gesture. These evaluations are pooled and analyzed automatically in service of the community of use. Members of this community, at their pleasure, receive recommendations of new or unfamiliar digital objects that they are likely to find interesting.

Recommendations might, for instance, take the form of recommendation-enhanced browse-products that tatoo symbols of predicted interest upon object navigation and control points. Later on, Figure 4 shows such a Mosaic Browsing interface with recommendation enhanced hypermedia links and menus.

Of course the question is: does this kind of virtual community work? The answer as we will show is "yes" for videos and probably yes for many other forms of consumer level information items: books (categorized by author), video games, gaming scenarios, music, magazines and restaurants.

Concerning the use of ratings, Allen [1] reported unencouraging results on one of the first investigations (known to us) into personal ratings for HCI-type user-modeling. Recently, Resnick et al. [9] have designed a social filtering architecture based upon personal ratings and demonstrated its application to work-group filtering of Netnews. In a study of eight users reading 8000 Netnews messages, Morita and Shinoda [8] observed strong positive correlations between time spent reading messages and personal interest ratings of those messages. Their work suggests it might be possible for time-on-task measures to stand in for ratings, further reducing user tasks.

Interface Design Goals

In the process of achieving our overall goal of making personal evaluations do significant interface work for a virtual community, our approach illustrates a number of supportive community-oriented design goals:

- *Recommendations and evaluations should simultaneously ease and encourage rather than replace social processes.* Virtual community should make it easy to participate without heavy personal involvement while leaving in hooks for people to pursue more personal relationships if they wish.

- *Recommendations should be information rich, e.g. make apparent their origins.*

- *Recommendations and evaluations should be for sets of people not just individuals.* Of course a set of users who will receive a recommendation or evaluation may contain only a single individual, but the methods should not be limited to that case. This multi-person case of recommending is often important, for example, when two or more people want to choose a video to watch together.

- *Recommendations and evaluations should be from people not a black box machine or so-called "agent".* A particularly useful set of reference users to use in this capacity is the set of people similar to the target users. Equally interesting sets are colleagues or friends of the person or persons for whom the recommendations are computed since existing social context enriches the exchange.

- *Recommendations and evaluations should tell how much confidence to place in them,* in other words they should include indications of how accurate they are.

- *Data entry suffices for data query.* In our case, using a virtual community database to get recommendations requires merely submitting ratings to the community database. In other words, the way to match to others' preferences is to submit your own preferences which in turn may be matched by others at a later time. So use of the database grows the database. Since data entry also queries the database, the work expected of users is cut in half.

Our design also embodies two research tactics.

- *The recommendation and evaluation methods should be generic in the sense of not relying on specific properties or features of items.* This rules out whole classes of approaches, particularly those based on feature analysis where the power comes from regularities among properties specific to the domain of items. This insures a certain generality to the techniques not available in feature-based approaches.

- *The recommendations and evaluations should get their strength from sheer amount of data.* Thus, if the database grows, the quality of recommendations grows. The communal history-of-user approach makes this possible. From the point of view of any given user, a larger database contains at least as many and probably more people with similar preferences than a smaller (i.e., subset) database. Thus overall performance, which is based how similar others' tastes are to the user, improves as a virtual community database grows.

The Research Questions

In order to understand the power of recommending and evaluating choices in a virtual community, we posed three basic questions:

- How well does virtual community perform?

- How reliable are the ratings data?

- How should one measure quality of recommending/evaluating algorithms for incorporation into user interfaces?

- According to the measures, what are the best algorithms for recommending items? For evaluating specific items? What are the trade-offs?

The second and third of these questions deserve further comment. The second question is straight-forward and standard statistical methods apply for answering it. On the third question, no standard measures have emerged as a consensus. At present, we consider two measures: (1) In a split-data test, how well do item ratings predicted by the recommending/evaluating system correlate with actual ratings submitted by users? (2) How do users evaluate the results they see from the algorithms? We report on these measures in the Results section.

METHOD: AN INTERNET CONCEPT TRIAL

Our method was to seed a virtual community in the Internet and to do all the work necessary to exchange high quality recommendations among participants. People participated (and still participate) through an email interface at videos@-bellcore.com. From October 1993 through May 1994 we collected data on how the virtual community functions, how people like it, and how well it performs for participants.

How Virtual Community Technology Works

The virtual community support provided by at videos@-bellcore.com consists of a generic object-oriented database to store and access preference efficiently and give out recommendations and evaluations. It is generic in the sense that one can construct various domains of items: videos, restaurants, books, document pages, and places to visit. In particular, at the time of our analysis, videos@bellcore.com included a data set of 55,000+ ratings of 1750 movies by 291 users. It includes recommending algorithms whose predictions improve as the data grow, and the number of movies, users and ratings and continues to grow daily.

Organization of the Database

The database is organized as set of interrelated instances of object classes. The objects are:

- *person,* which in the videos@bellcore.com case is an Internet subscriber who voluntarily participates by email.

- *item*, which in the videos@bellcore.com case is one of 1750 movies

- *rating*, which is a triple of item, person, and score

- *score,* which is either an integer representing a person's rated preference for an item or a symbol representing the person's relation to this item: *not-interested-in-item, must-see-item, no-previous-use-of-item, item-pending-as-recommendation* or *item-evaluated-for-user*

- *correlation,* which in the case of videos@bellcore.com is a triple of person1, person2, and correlation value, a floating point number between -1.0 and 1.0

The database contains 17 modules. A single high level database interface consisting of the following functions suffices to control it in most circumstances: *load-database, save-database, add-user, erase-user, add-item, erase-item, add-ratings, recommend-items, evaluate-items.*

The Email Interface

Internet participants send a message containing "subject: ratings" to videos@bellcore.com. The system replies with an alphabetical list of 500 videos for the user to evaluate on a scale of 1-10 for the titles they have seen. Rating 1 is low and 10 is high. Users may also rate an unseen movie as "must-see" or "not-interested" as appropriate. Surprisingly, early usability tests showed that it was reasonable to expect self-selected Internet users to rate movies on an alphabetical list of 500 movies. However we do not expect this to be a feature of a deployed system. In order to reduce item/item bias, for every participant 250 of the 500 movies listed are selected randomly. To increase rating hits and to gather a standard set of data for purposes of fair comparison, for every participant the remaining 250 titles are a fixed set of popular movies.

When users return their movie ratings to videos@bellcore.com, an EMACS client process parses the incoming message, and passes ratings data inside a request for a recommendations-text to the server database process. The server process performs *add-user, add-ratings* and *recommend-items.* In the

initial phase of adding ratings for a new user, ratings are added not only in the 1-10, "must-see" and "not-interested" categories, but also in the "unseen" category for titles that the user could have rated but did not. These unseen movies are the first pool from which to compute recommendations.

When a user is new, the database first looks for correlations between the new user's ratings and ratings from a random subsample of known users. We use the random subsample to limit the number of correlations computed to be $O(n)$ rather than $O(n^2)$ in the number of participants. One-tenth of the new user's ratings are held out from the analysis for later quality testing purposes. The most similar users found are used as variables in a multiple-regression equation to predict the new user's ratings. The generated equation is then evaluated by predicting the held out one-tenth of the new user's ratings and then correlating these predictions with the actual ratings.

Once the predication equation exists, it is quite fast to evaluate every unseen movie, sort them by highest prediction and skim off the top to recommend. When recommended, movies are marked in the database as "pending-as-suggestion". A recommendation text is generated and passed back to the EMACS front-end client process where it is mailed back to the user or users.

The Internet email interface is currently a subject-line command interface and there are many commands for specialized actions. Further details are available by sending mail to videos@bellcore.com.

Here is sample reply from the system. Names have been changed to protect anonymity:

Suggested Videos for: John A. Jamus.

Your must-see list with predicted ratings:

> *7.0 "Alien (1979)"*
> *6.5 "Blade Runner"*
> *6.2 "Close Encounters Of The Third Kind (1977)"*

Your video categories with average ratings:

> *6.7 "Action/Adventure"*
> *6.5 "Science Fiction/Fantasy"*
> *6.3 "Children/Family"*
> *6.0 "Mystery/Suspense"*
> *5.9 "Comedy"*
> *5.8 "Drama"*

The viewing patterns of 243 viewers were consulted. Patterns of 7 viewers were found to be most similar.

Correlation with target viewer:

> *0.59 viewer-130 (unlisted@merl.com)*
> *0.55 bullert,jane r (bullert@cc.bellcore.com)*
> *0.51 jan_arst (jan_arst@khdld.decnet.philips.nl)*
> *0.46 Ken Cross (moose@denali.EE.CORNELL.EDU)*
> *0.42 rskt (rskt@cc.bellcore.com)*
> *0.41 kkgg (kkgg@Athena.MIT.EDU)*
> *0.41 bnn (bnn@cc.bellcore.com)*

By category, their joint ratings recommend:

Action/Adventure:

"Excalibur" 8.0, 4 viewers
"Apocalypse Now" 7.2, 4 viewers
"Platoon" 8.3, 3 viewers

Science Fiction/Fantasy:

"Total Recall" 7.2, 5 viewers

Children/Family:

"Wizard Of Oz, The" 8.5, 4 viewers
"Mary Poppins" 7.7, 3 viewers

Mystery/Suspense:

"Silence Of The Lambs, The" 9.3, 3 viewers

Comedy:

"National Lampoon's Animal House" 7.5, 4 viewers
"Driving Miss Daisy" 7.5, 4 viewers
"Hannah and Her Sisters" 8.0, 3 viewers

Drama:

"It's A Wonderful Life" 8.0, 5 viewers
"Dead Poets Society" 7.0, 5 viewers
"Rain Man" 7.5, 4 viewers

Correlation of predicted ratings with your actual ratings is: 0.64 This number measures ability to evaluate movies accurately for you. 0.15 means low ability. 0.85 means very good ability. 0.50 means fair ability.

Instructions are also given for exercising various options in the community. For example, one can order up joint recommendations for more than one person and from a particular set of community members. This second example shows both capabilities at once. Jane and Jim want a joint recommendation of what movie to watch together. They also want recommendations only from Mary and Dick rather than the community at large. Again names have been changed.

Suggested Videos for: Jane Robins, Jim Robins, together.

Your video categories with average ratings:

6.9 "Drama"
6.7 "Mystery/Suspense"
5.8 "Comedy"
5.7 "Children/Family"
5.5 "Action/Adventure"
5.0 "Horror"
4.5 "Musical"

The viewing patterns of 2 viewers were consulted. Correlations with target viewers:

0.21 Mary Divens (no email address)
0.18 Dick Divens (didi@westend.com)

By category, their joint ratings recommend:

Drama:

"Crossroads (1986)" 9.0, 2 viewers
"To Kill A Mockingbird" 10.0, 1 viewer
"Milagro Beanfield War, The" 8.0, 1 viewer

Mystery/Suspense:

"Body Double" 8.0, 2 viewers

"Coma" 7.5, 2 viewers

Comedy:

"After Hours" 9.0, 2 viewers
"Adventures of Shelock Holmes Smarter Brother, The" 9.0, 1 viewer

Children/Family:

"Beauty And The Beast (1991)" 8.0, 1 viewer

Action/Adventure:

"Lawrence of Arabia" 9.0, 1 viewer

Horror:

"Rear Window" 8.0, 1 viewer

Musical:

"West Side Story" 9.0, 1 viewer

THE COMMUNAL HISTORY-OF-USE ALGORITHMS

We have algorithms for two purposes, recommending items and evaluating items. Having tried a few versions of each, we report on the best we have discovered so far. We do not have evidence that these are the best algorithms possible, only that they are good. The algorithms we use for recommending have the following abstract functional form:

recommend-items (n-items, from-items, target-users, reference-users, methods, method-combination, database)

The function name is **Recommend-items** and n-items is how many items to recommend. From-items is the set of items from which to recommend a subset. Target-users is the set of users for which the recommendation is computed. In the case of a recommendation for one user, it is a singleton set. Reference-users is the set of users whose preferences will serve as the basis of the recommendation. For now, we normally compute the reference-users set by finding users whose preferences on the set of from-items correlate most positively with the preferences of the target-users. However, as the second example reply illustrates, community participants may specify a particular set of reference-users (friends, colleagues, whomever) to consider when computing recommendations. The earlier second sample reply showed this. This interesting alternative addresses the idea of supporting rather than replacing social processes. Methods is an optional argument that specifies one or more ways in which information about items, target-users, and reference-users will be analyzed. The Method-combination argument tells how to combine the results of the methods and the database argument specifies the particular data to use. The optional arguments default to standard values.

The function to return an evaluation of a proposed choice looks like this:

evaluate-items (items, target-users, reference-users, methods, method-combination, database)

Here items refers to the set of items to be evaluated for the target-users on the basis of looking at reference-users. The methods, method-combination and database arguments can be used to specify alternative methods for evaluation, how to combine the results and data to use.

RESULTS

The Data

Currently the database consists of 291 participants in the community, 55,000 ratings on a 1-to-10 scale, another 2100 "must-see" or "not-interested" ratings, 64,000 "unseen" and 1200 "pending-as-suggestion" ratings. Of the 1750 movies in the database, 1306 have at least one rating and 739 have at least 3 ratings. 208 movies have more than 100 ratings, and 2 movies have more than 200 ratings. Users rate an average of 183 movies each with a standard deviation of 99. More than 220 of 291 total participants rated more than 100 movies. The database is small, but large enough to conservatively but accurately estimate a number of performance parameters.

For the 739 movies that have three or more ratings. Figure 2 shows the distribution of movies by their mean rating. Notice the slight bias toward positive ratings.

Figure 2   *Distribution of Video Mean Ratings*

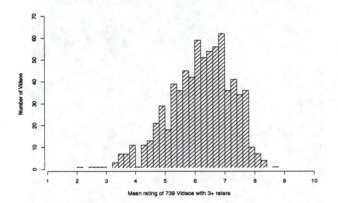

Reliability

Six weeks after they initially tried videos@bellcore.com for the first time by submitting ratings and receiving recommendations, 100 early users were asked to re-rate exactly the same list of movie titles as they had rated the first time. 22 volunteers replied with a second set of ratings. Three outliers were removed from the reliability analysis since they correlated perfectly and were evidently copies of the original ratings rather than second independent sets of ratings. For the remaining 19 users, on movies rated on both occasions, the Pearson *r* correlation between first-time and second-time ratings six weeks apart was 0.83 . This number gives a rough estimate how reliable a source of information the ratings are.

Cross-validated Correlation Study

We held out 10% of every participant's movie ratings to profide a cross-validation test of accuracy. The cross-validated correlation of predicted ratings and actual ratings estimates how well our recommendation method is working. Figure 3 shows that our current best similar viewers algorithm correlates at 0.62 with user ratings. This is a strong positive correlation which means the recommendations are good. How good? We may expect three out of every four recommendations will be rated very highly by a potential viewer. We compared the quality of our virtual community recommendation method to a standard method of getting recommendations, that is, following the advice of movie critics. The ratings of movies by two nationally-known movie critics were entered. Their ratings correlate much more weakly at only the 0.22 level with viewer ratings. Thus the virtual community method is dramatically more accurate, as Figure 3 also shows.

Figure 3   *Two Scatterplots of Actual Ratings by Predicted Ratings. Plot on left shows movie critics as predictor (r=0.22). Plot on right shows virtual community as predictor (r=0.62) (all values are jittered for the purpose of visual presentation, 3269 predictions each for 291 users)*

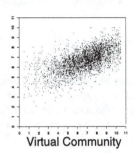

Movie Critics   Virtual Community

User Feedback

Email responses from videos@bellcore.com include a request for open-ended feedback. Out of 51 voluntary responses, 32 were positive, 14 negative and 5 neutral. Here are some sample quotes:

- *Pretty good! Of 8 films you suggested I might enjoy, I agree with 4, disagree with 3, and am uncertain about 1.*

- *The recommendations are so-so. I agree with a few, but the slant of most of them is a suspense, high tension drama. This is not the only thing I watch.*

- *Nice work.*

- *I have seen eight of the ten flicks you recommended and thought they were pretty good; the other two I would definitely like to see.*

- *Your predicted films that I would like were so-so.*

- *Nothing was really off the mark, but I think I prefer a good comedy movie over anything else...*

- *The suggestions hit close to the mark.*

- *None of them were favorites and a few (like werewolf) were real bombs.*

- *It seemed pretty accurate.*

- *I think the recommendations make sense.*

- *I received 5 recommendations. Four looked decent but one was a movie I had already seen and rated low.*

- *Most were things I had seen and had liked, or that seem like things I would enjoy*

- *I wasn't impressed. The list is not what I am interested in seeing.*

- *I think the recommendations were reasonably valid based on the input I gave.*

- *Not bad considering how quickly I filled out my ratings questionnaire.*

- *I'm not convinced that I want to see them.*

- *Some of them were right on (like "Die Hard," and "This is Spinal Tap," I liked both of those) while others were way off (I really don't like the Pink Panther movies and I didn't like Born on the Fourth of July at all.*

- *Intriguing! Of the 10 movies recommended, 8 have been on my "want to see" list for a long time.*

Open ended feedback from users also indicated interest in establishing direct social contacts within their virtual community. Users can participant in either an anonymous or signed fashion. Interestingly, only four users exercised the anonymity option. Wishing to extend the social possibilities of the virtual community, two users asked if they could set "single and available" flags in the community indicating they wanted to use the community as a means of dating. One user found a long lost friend from junior high school. Another wrote that he took the high correlation between his movie tastes and those of someone he was dating as evidence for a long future relationship.

The Upper Limit

One of the standard uses of reliability measures is to put a bound on prediction performance. The basic idea is since a person's rating is noisy (i.e., has a random component in addtion to their more underlying true feeling about the movie) it will never be possible to predict their rating perfectly. Standard statistical theory says that the best one can do is the square root of the observed test-retest reliability correlation. (This is essentially because predicting what the user said once from what they said to the same question last time has noise in at both ends, squaring its effect. The correlation with the truth, if some technique could magically extract it, would have the noise in only once, and hence is bounded only by the square root of the observed reliability). The point to note here is that the observed reliability of 0.83 means that in theory one might be able to get a technique that predicts preference with a correlation of 0.91. The performance of techniques presented here, though much better than that of existing techniques, is still much below this ideal limit. Substantial improvements may be possible.

FUTURE DIRECTIONS

We see a potential for deployment to customers of national information access who will be faced with thousands of possible choices for information and entertainment, in addition to videos.

Virtual Community recommending in the Mosaic Interface to the World Wide Web.

We have instantiated a version of our server where items are World Wide Web URLs (universal resource locators) in place of videos. Figure 4 displays a modified Mosaic browser interface that accepts ratings of WWW pages on a slider widget (near bottom) and reports them to an appropriate virtual community server. When a user clicks on the **Recommend URL** button (near bottom), the browser contacts the virtual community server to get recommended URLs and then fetches the recommended page. It also displays next to every hypertext link, one-half to four stars which represent the virtual community's predicted value of chasing down the hypertext link.

Community Headroom

One direction in which we plan to push the research is toward more individual and social aspects. In particular we are interested in distributed peer-to-peer versions rather than the centralized client/server version that we have now. A wireless deployment of a peer-to-peer version could include wearable PCS devices, pairs of which will, when in close physical proximity, exchange ratings data for local virtual community computation.

Figure 4   *Recommendation-enhanced Browsing in a specialized Mosaic Interface. Note that next to each hyper-link are one to four stars that predict the user's interest in following that link. At the bottom of the window are a recommend button and rating slider to rate the current page.*

Recommend URL   *Rate Links*

CONCLUSIONS

Choice under uncertainty is an opportunity to benefit from other more knowledgeable people. How to support such social filtering with computation has been the topic of this paper. We have demonstrated a *virtual community* method that allows human-computer interfaces to harness the power of a social strategy involving minimal additional work with good utility. We have reported on how it fares in the context of a fielded test case: the selection of videos from a large set. In the case of videos, virtual community recommendations are measurably successful and can be used to recommend or evaluate videos for participants. Virtual communities may also sprout up around other domains such as music, books and catalog products. Targeting both groups and individuals for recommendations and evaluations, it performs well on stringent tests and will continue to improves as the virtual community database grows. When presenting choices in the interface and when a virtual community of users exists to inform those choices, there is no reason to leave users without recommended courses of action. The positive result we have reported suggests that others may want to investigate the power that communal history-of-use data can bring to interfaces.

ACKNOWLEDGMENTS

This work revives a line of research initially pursued jointly with Tim McCandless and David Wroblewski in the late eighties. Their continued interest, suggestions, advice and encouragement concerning the current work have been invaluable. On many occasions, Jim Hollan, Tom Landauer and Bob Allen asked questions and made suggestions that influenced the direction of this work over the past few months. We would also like to thank Ben Bederson, Diane Duffy, Dan Lin, Alan Mcintosh and Kent Wittenburg for many helpful conversation s and suggestions. Finally, without the participation of the Internet community evolving around videos@-bellcore.com, the lessons of this paper would have been impossible to learn. Special thanks to all who made suggestions and found bugs.

REFERENCES

1. Allen, R.B. (1990) User models: Theory, method and practice, *International Journal of Man-Machine Studies*, 32, 511-543.

2. Brothers, L., Hollan, J., Nielsen, J., Stornetta, S., Abney, S., Furnas, G., and Littman, M., Supporting informal communication via ephemeral interest groups. *Proc. ACM CSCW'92 Conf. Computer-Supported Cooperative Work* (Toronto, Canada, 1P4 November 1992), 84-90.

3. Goldberg, D., Nichols, D., Oki, B.M. and Terry, D. (1992) Using Collaborative Filtering to Weave an Information Tapestry. *Communications of the ACM,* 35, 12, pp. 51-60.

4. Grudin, J., Social Evaluation of the User Interface: Who Does the Work and Who Gets the BENEFIT?, Proceedings of IFIP INTERACT'87: Human-Computer Interaction, 1987, 805-811.

5. Hill, W. C., Hollan, J. D., Wroblewski, D., and McCandless, T. (1992) Edit Wear and Read Wear. In: *Proceedings of ACM Conference on Human Factors in Computing Systems, CHI'92.* ACM Press, New York City, New York, pp.3-9.

6. Hill, W.C., Hollan, J.D. (1994) History-Enriched Digital Objects: Prototypes and Policy Issues, *The Information Society,* 10, pp. 139-145.

7. Malone, T.W., Grant, K.R., Turbak, F.A., Brobst, S.A. and Cohen, M.D. (1987) Intelligent Information Sharing Systems. *Communications of the ACM*, 30, 5, pp. 390-402.

8. Morita, M., Shinoda, Y. (1994) Information Filtering Based on User Behavior Analysis and Best Match Text Retrieval, *Proceedings of the 17th Annual International SIGIR Conference on Research and Development*, pp. 272-281.

9. Resnick, P., Iacovou, N., Suchak, M., Bergstrom, P., Riedl, J. (1994) GroupLens: An Open Architecture for Collaborative Filtering of Netnews. *Center for Coordination Science, MIT Sloan School of Management Report WP #3666-94.*

10. Rheingold, H., (1993) *The virtual community: homesteading on the electronic frontier,* Reading Mass: Addison-Wesley.

11. Wroblewski, D., McCandless, T., Hill, W. (1994) Advertisements, Proxies and Wear: Three Methods for Feedback in Interactive Systems, in *Dialogue and Instruction*, Beun, R., Baker, M., and Reiner, M. editors. Springer-Verlag (forthcoming).

Pointing The Way: Active Collaborative Filtering

David Maltz
Dept. of Computer Science
Carnegie-Mellon University
Pittsburgh, PA 15213-3890
dmaltz@cs.cmu.edu

Kate Ehrlich
Lotus Development Corporation
One Rogers St
Cambridge MA 02142
Kate_Ehrlich@crd.lotus.com

ABSTRACT

Collaborative filtering is based on the premise that people looking for information should be able to make use of what others have already found and evaluated. Current collaborative filtering systems provide tools for readers to filter documents based on aggregated ratings over a changing group of readers. Motivated by the results of a study of information sharing, we describe a different type of collaborative filtering system in which people who find interesting documents actively send "pointers" to those documents to their colleagues. A "pointer" contains a hypertext link to the source document as well as contextual information to help the recipient determine the interest and relevance of the document prior to accessing it. Preliminary data suggest that people are using the system in anticipated and unanticipated ways, as well as creating information "digests".

Keywords: Collaborative filtering, information retrieval, hypertext, World Wide Web, Lotus Notes.

INTRODUCTION

Connectivity, networking, and the National Information Infrastructure have become the buzzwords of the day. Underlying the excitement of these times is the promise that each of us will soon have unlimited access to a computer that will cheaply bring us information from sources around the world. We will be in direct contact with all the world's repositories of information -- no matter how small or large -- and in direct contact with the experts and people who create those repositories. Just like the upswell of creation and learning that followed the development of the printing press and the widespread access to information it created, we anticipate a new surge of knowledge to enrich our lives.

Unfortunately, our situation parallels that of the printing press in more ways than one. As the first libraries were built and books became available to larger groups of people, a new problem arose. Once buildings could be filled with more piles of books than any human could possibly read in a lifetime, how could people find books on the topics they wanted? The solution to that problem evolved into an entire field called Library Science. A solution to the problem of finding useful information on a global network promises to be no easier.

Just to put some numbers on the size of problem, it is informative to look at several of the information systems available to users. The World Wide Web allows nearly anyone with a machine on the Internet to create hypertext multimedia documents and link them to other documents at other sites [1]. The Web, started in 1990, had grown to more than 4600 servers in October 1994, and over 10,000 servers in December 1994. Together, these servers make over 1.05 million different documents available for retrieval [11]. Usenet Net News is also growing exponentially. Estimates show that in May 1993 there were over 2.6 million users of Net News at some 87 thousand sites throughout the world. These users generated over 26 thousand new articles a day, amounting to 57 Mbytes of data [13]. Just over a year later, some estimates put the number of users at almost 10 million. The problem gets worse when one considers all the information that organizations are putting on-line in file folders, databases and Lotus Notes amongst other repositories. Not only is the amount of information available to many people far in excess of what can be retrieved through informal browsing methods, but only a very small fraction of the available information will be accurate, up to date and relevant.

Building on the notion of collaborative filtering which originated with the Tapestry project at Xerox PARC, this paper describes a system that supports the informal practice of sharing pointers to interesting documents and sources of information with colleagues. We begin by reviewing existing information filtering systems.

INFORMATION FILTERING

If the problem is that people are swamped by too much information [12, 16], the solution seems to lie in developing better tools to filter the information so that only interesting, relevant information gets through to the user. Many present filtering systems are based on building a user profile. These systems attempt to extract patterns from the observed behavior of the user to predict which items would be selected or rejected. Examples of this form of filtering include LyricTime [9] which compiles a profile of musical preferences, an agent that filters NetNews [15] and INFOSCOPE [4] an interesting hybrid news reader that frames the filtering problem as one of restructuring the information space on a user by user basis to place all the articles relevant to a user in a few known and accessible locations.

However, these systems all suffer from a "cold-start" problem. New users start off with nothing in their profile and must train a profile from scratch. Even with a starter profile, there is still a training period before the profile accurately reflects the user's preferences. During the training period the system can't effectively filter for the user. A better system would allow new users some type of access to the experiences of current users to create a better initial profile. The second drawback is that the user's searches can become circumscribed by the profile. The profile only selects articles similar to the ones the user has already read, and new areas that might be of interest can be missed completely. If the user tries to explore new content areas a new profile must be created that is customized to that subject matter and the user again faces the "cold start" problem.

COLLABORATIVE FILTERING

Most present filtering systems fail to capitalize on a key resource made available by on-line information systems, namely the knowledge and wisdom accumulated as different people find and access documents and form opinions of them. By giving users access to others' prior experience with an information source, we can create a collaborative information filter.

Collaborative filtering systems work by including people in the filtering system, and we can expect people to be better at evaluating documents than a computed function. Current automatic filtering systems attempt to find articles of interest to their user, often using some scoring function to evaluate features of the documents and returning the documents with the highest scores. People can effortlessly evaluate features of a document that are important to other people, but would be difficult to detect automatically. Examples of such features are the writing style and "readability" of a document, or the clarity and forcefulness of an argument the document contains. Imagine the difficulty an automatic filtering system would have figuring out which of two cake recipes is "easier to follow."

Another motivation for collaborative filtering comes from comparing the rich environment of real objects to the much poorer one in which computer users operate. When a user reads a computer file he usually has no way of telling whether he is the first person to ever read it, or if he is looking at the most commonly used reference on the system. Collaborative filtering works in part by associating with computer documents the history of their use. For instance, Hill et al make the observation [7] that the objects we use in everyday life accumulate wear and tear as a normal part of their use: pages in books become wrinkled, bindings creased, and margins smudged with fingerprints. The objects with more wear are the more commonly used ones, and further the wear acts as an index to relevant information inside the object. An example is the way reference books open to commonly used pages when dropped on a desk. Giving searchers access to this usage history lets them take advantage of the type of subtle hints that we commonly use when making read/don't read decisions in the real world.

Tapestry

The concept of collaborative filtering originated with the Information Tapestry project at Xerox PARC [5]. Among its other features, Tapestry was the first system to support collaborative filtering in that it allows its users to annotate the documents they read. Other Tapestry users can then retrieve documents to read based not only on the content of the documents themselves, but also on what other users have said about them. Tapestry provides free text annotations as well as explicit "likeit" and "hateit" annotations so users can pass judgments on the value of the documents which they read.

In its current incarnation, Tapestry suffers from two distinct problems. The first problem is the size of its user base. Because Tapestry is based on a commercial database system it can not be given away freely. Further, Tapestry was not designed for use by large numbers of people at distributed sites. Both these factors combine to limit the pool of potential Tapestry users to researchers at Xerox PARC. Based on anecdotal evidence, this pool does not seem large enough to support a critical mass of users. The vast majority of documents go unannotated, so there is little collaborative information to use when filtering. The second problem with Tapestry is the means by which users enter filters into Tapestry. One common interface to Tapestry requires users to specify requests for information in the form of queries in an SQL-like language. Writing such a query requires the user to have a firm sense of what types of articles he wants to read, which is a hindrance to exploration of new areas and makes it hard to browse the available information for serendipitous hits.

Collaborative Filtering Of Usenet Net News

A further collaborative filtering system for Usenet Net News was created to scale up to a critical mass of users by branching out to more sites, and providing those users with a simpler method for accessing articles [10]. This system modified two common Net News readers to provide collaborative filtering functions: voting and display of articles based on their "popularity". Aside from these extra functions, the news readers appear and handle as they did before. Users of the modified news readers were encouraged to cast votes for or against the articles they read. These votes are used by the system to create a net-wide collective opinion on the usefulness of each article. The news reader clients can then use this information to help future readers of the newsgroup find articles of interest. Users are able to associate their names with their votes or not as they chose.

A positive result of the work on this system was finding that people really will vote. Although no reward or incentive of any kind was offered to the users of one of the modified news readers, 24 of the approximately 40 users of the newsreader voted for articles during the course of one three month study period. However, this number was insufficient to reach critical mass. That is, because of the large number of different documents, this system was dependent on lots of people reading and voting on the same documents. Users reported that when they didn't see any votes in the groups they read, they thought the system must be broken or not working so they didn't bother to vote further because their votes would be lost and wasted. If they knew something useful was happening to their votes, they would have kept voting.

Grouplens

A similar system for NetNews, GroupLens, combines collaboration with user-profiles [14]. Communities of users rank the articles they read on a numerical scale. The system then finds correlations between the ratings different users have given the articles. As viewed by a user called Jane, the goal of the system is to identify a peer group of users whose interests are similar to Jane's, and then to use their opinions of new articles to predict whether Jane will like the articles. Like other filters based on a user-profile, GroupLens suffers from the cold-start problem.

One problem with many of these collaborative filtering systems is that they require a critical mass of users to vote or leave their mark for any aggregate score to be meaningful. Until systems such as the NetNews collaborative filter are well in use, there are uncertain rewards for any user who participates. The lack of a clear reward system can be a major barrier in the acceptance of a groupware application [6]. Some systems also suffer from usability problems due to excessive overhead in either registering a vote, accessing the result of other people's votes, or creating a profile.

"ACTIVE" COLLABORATIVE FILTERING

These three examples of collaborative filtering are well suited to situations where there are a lot of documents in a single database such as NetNews. In these cases, there is a benefit to aggregating the votes of the many readers who access the database. We call this "in-place" or "passive" filtering because there is no direct connection between a person casting a vote and the readers who come later and filter documents based on these aggregated votes.

Another approach to collaborative filtering, and one that we have adopted here, builds on the common practice where people tell their friends or colleagues of interesting documents. We call this "active" collaborative filtering because there is an intent on the part of the person who finds and evaluates a document to share that knowledge with particular people. For example, as part of the World Wide Web system users collect "hotlists" which are effectively lists of hypertext links to the interesting World Wide Web pages that they have found. Several simple systems have been developed to help users format and distribute these hotlists to others, thereby spreading information about good documents on the Web. Another example in the World Wide Web context is Simon [8]. Users of this system create "subject spaces" which are effectively lists of hypertext links to the interesting World Wide Web pages that have been found, and comments on those documents. Individual people can use subject spaces to keep track of their own explorations, but they can also send their subject spaces to a group Simon server. However there is no provision in this system for sending hypertext links to particular individuals.

We see "active" collaborative filtering being useful for distributed information sources such as the World Wide Web, on-line information services, and Lotus Notes databases where users may need help simply finding the source. In "passive" collaborative filtering the system works better the higher the convergence of votes on the same set of documents. In contrast, the benefit of "active" collaborative filtering increases with the divergence of documents that are found.

DESCRIPTION OF THE SYSTEM

The design of the system was informed by the results of a recent study of information retrieval behavior in a customer support group [2]. Contrary to the expectation that support people use on-line or printed documentation to help answer customer's questions, Ehrlich and Cash report that the support people rely on each other to diagnose and solve customer problems, as well as find and interpret formal and informal documentation. Moreover, there was one person in the group who was especially skilled at finding relevant information and applying it to the problem at hand. He was remarkable for his breadth of knowledge of the domain as well as of useful sources of information, general problem-solving, and, communication skills. By virtue of these skills he assumed the role (though not the title) of an "information mediator". That is, he was available as a resource to other

Companies Learning to Sell on the Internet

October 5, 1994

----

 in database <u>IRG Industry Newswire '94 First!</u> on <u>IRG 1</u>   Comments by: Reed Sturtevant , 10/10/94

"More people have and use Internet access than have library cards" claims a consultant teaching businesses how to market themselves using the Internet.

Figure 1: *A sample pointer as seen by a recipient*

---

analysts in the group to help them think through especially troublesome customer problems.

Based on this study we wanted to build a system that supported some level of collaboration and information sharing amongst colleagues. More importantly, we wanted the system to provide tools that would let the "information mediators" in a workgroup easily distribute references and commentary of documents they find. Moreover, we believe that these mediators are important not only for selecting relevant documents but also for selecting reliable sources especially where there are a large number of distributed sources. Because systems such as the World Wide Web or Lotus Notes have few restrictions on authoring and no formal review process, there will be a lot of variability in the accuracy, relevance and completeness of information. Ensuring the quality of information passed on to customers is well known to librarians for whom source validation is an important part of their job [3].

The system described in this paper was developed around three key ideas that we felt were missing from existing informal methods of sharing references.

1. Package contextual information with hypertext links. Existing methods for sharing references to on-line documents are often limited to just the hypertext link, perhaps with a few comments. Yet additional contextual information about the name or location of the source, the date of the document as well as knowledge of the sender's selection biases can be used to judge the relevance of a document prior to reading it.

2. Ease of use. Informal systems are often awkward to use. For instance the ability to add annotations to the hypertext link may not be readily available. Not only should these features be easily accessible but the system in general should return value to the senders and recipients early on to encourage usage. It is through continued and broad adoption that an active filtering system demonstrates its usefulness.

3. Flexible. The main drawback of many of the informal systems is that they work well for a particular situation but are not flexible enough for general use. We intend to explore, for instance, different methods of distributing the package of hypertext link, comments and context to probe the potential extendibility of our system.

Pointers

The basic concept in our system is the "pointer". Modeled after what people do informally, a pointer contains a) a hypertext link to the document of interest; b) contextual information - title and date of document, name of source database, name of sender, and c) optional comments by the sender. A typical pointer is shown in Figure 1.

We implemented our active collaborative filter inside the Lotus Notes environment. Notes is a commercial product that provides a client-server platform for developing and deploying groupware applications. At the user level, Notes consists of documents which are stored in databases on server machines. These documents can be thought of as raw records of field-data pairs. Databases are typically organized around a topic, and the documents they contain can either be created by users inside of Notes, or gatewayed into Notes from external information sources such as Net News, World Wide Web, or clipping services. In addition to documents, each database has a set of forms and views. Database forms can be thought of as document templates containing fields that are filled in during composition time. For instance, the form used to generate the pointer in Figure 1 has a field for each item. Each database has one or more views which are ways of grouping and displaying the list of documents in the database. For instance, documents could be listed in order of document author, document date or category name. The latter being one or more keywords defined by the user to describe a document.

Typical applications in Notes are developed by creating forms, views and databases to support a new type of information, and then adding macros which specify, automate, or regulate the actions that can performed on

205

the information. Notes allows these macros to be embedded directly into the document as buttons whose macros operate on the document's data. Notes' data management infrastructure also provides the equivalent of a hypertext link called a doclink. Doclinks present themselves on the screen as small page icons; clicking on one of these icons opens the referenced document as a new window on the workspace.

Pointers are implemented in Notes as a special form. These forms are accessed from a button we added to the Lotus Notes SmartIcon (TM) bar. Pressing the button while a document is selected will open one of these forms and fill in the document title, creation date and database name of the currently selected document using heuristic rules. The user can add comments of any length. Since all documents in Lotus Notes are semi-structured by the forms used to create them, the information extraction heuristics are, in general, very effective.

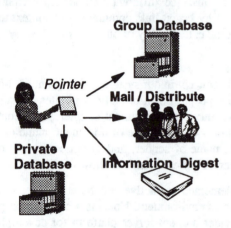

Figure 2: *Different ways of distributing a pointer.*

In addition to information about the pointed to document, the pointer itself includes buttons which help the sender to distribute the pointer to selected people and in a variety of ways --- this distribution process is described below. In order to give pointers some functionality which could not be easily implemented from inside of Notes, our collaborative filtering system also involves a small server process which runs outside of Notes. The buttons on a pointer form communicate with this process which then implements the pointer distribution methods, making it easy to experiment with different distribution methods by altering the server.

Distribution Of Pointers

Once a sender has created the pointer, he/she can save it in a private or public database, email it to one or more people or distribution lists, or save it in a special form we call an "Information Digest". These distribution methods are depicted in Figure 2.

Private or group database. At the simplest level, users can "bookmark" favorite documents by saving a pointer in a private database. We think of these private databases as a scrapbook for saving pointers to interesting documents especially those we might not have time to read immediately. Users can add keywords to their pointers to help organize them. The user can view their list of pointers in various ways including by date of source document, name of database or keyword category. Although not yet implemented, the system allows pointers to be saved in a public database where anyone with the appropriate access rights can add or read pointers.

Email and distribution lists. Pointers can be sent to others using email. Figure 1 shows how a pointer might look when it is opened. Whether from a database or email (which is only another database in Notes), the user opens the referenced document by clicking on the doclink icon. It should be noted that because a pointer is a regular Notes document, anyone running Notes can receive a pointer and follow the hypertext link to the referenced document; they don't need to be running the collaborative filter. Moreover pointers can be forwarded just like any other mail message.

Pointers can also be sent to a distribution list of people who have previously registered themselves as interested in receiving pointers from the evaluating user, similar to a subscription service. This means of distribution gives our system a method of filtering akin to Tapestry's notion of filtering for messages based on who has annotated them. In Tapestry, users can create filters of the form "show me the messages that Joe Blow has annotated." In our system, users can register themselves as being interested in receiving pointers that Joe Blow decides to distribute. Joe himself does not have to worry about who has registered their interest as the system will automatically lookup the registered users and mail out copies of the pointer when Joe presses the Distribute button. We considered calling the Distribute function automatically when any pointer is created, but decided that failed to give the user enough control over who sees their pointers, an especially important concern since we want the system to be useful to people for keeping pointers for their own use.

Information Digests. We have also begun exploring an advanced use in which multiple pointers are saved directly into a pre-designed document containing a combination of original text and pointers. Newsletters, World Wide Web home pages, company profiles, summary reports or even a table of contents, are all examples of this kind of document. A portion of a typical digest is depicted in Figure 3. It consists of a banner title, a subject and several sections. Users can easily add a pointer to any section of a defined digest. The system automatically extracts the information from the pointer and formats it into the digest.

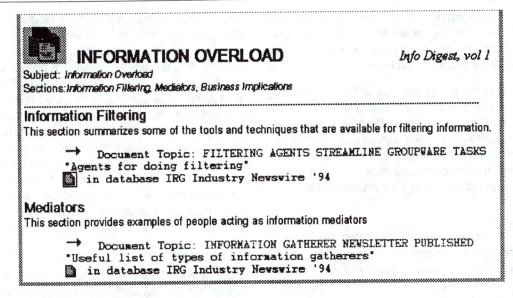

Figure 3: *A sample Information Digest containing pointers formatted by our system and narration text added by the author. The pointers are indicated by arrows while the added text is in Helvetica font.*

As we found in [2], there are people who excel at pulling together bits of information from many places. On Usenet Net News, these people appear as the ones who pull together FAQ's (Frequently Asked Questions) for newsgroups. One form of an information digest might be an on-line newspaper or magazine written by people skilled in selecting, editing, annotating and layout.

Passive filtering. For completeness, our prototype also contains methods to support passive collaborative filtering in addition to active filtering. If a user comes across an interesting document but doesn't know of anyone in particular the document should be brought to the attention of, she can invoke the passive filtering system on the document. The system will then annotate the document in-place so that any future readers of the database can use the annotation to help filter for useful documents. The passive filtering system works by marking the annotated document as having been read by the annotator, and creates a response note to the document which contains any comments the annotator has on the document. Our initial goal was to allow users to mark up the annotated document directly, but this turned out not to be feasible in the current version of Notes.

An active collaborative filtering system similar to ours could be created in the World Wide Web environment quite simply by better integrating Web browsers with email readers. Hypertext links would come almost for free, while information digests could be represented as new Web pages. Private and group databases of pointers might be hard to implement, as no Web clients we know of currently support organizing multiple views of data without external database backends.

Automatically extracting contextual information from Web documents may also pose difficulties. Some contextual information would be easy to obtain, such as the document title. Providing contextual information about the area of Web containing the document would be much harder, since by nature the Web has less structure than Notes databases. Based only on personal experience, it seems like providing both a link to the interesting page and a link to the parent page on which the interesting link was found is a partial solution toward providing recipients with some context on the region of the Web the page was found.

USES OF POINTERS: USER DATA

At the time of writing, the collaborative filter had been distributed to over 50 people at Lotus via email that contained a brief user's guide and the files that made up the system. In some cases the email was sent to an entire workgroup, other times individuals requested it after seeing a colleague use it or getting a demo of it.

Of the people who received the filter, over 50% installed it. It should be noted that people could receive and use pointers sent by others without having to install the filter themselves. The people who didn't install it often did not browse lots of databases themselves and were comfortable relying on others to inform them of interesting or important documents. There is thus an interesting asymmetry between "senders" and "receivers". In a rough survey we found that people in a workgroup of 10 people received 5-10 pointers per week. Approximately 80% of those pointers were sent out by just one person. That person thus serves a similar role to the "information mediator" we identified in [2]. He routinely browsed a few key databases such as the newswire database and sent pointers, including comments, to documents he thought were relevant. People were pleased to receive the pointers and did indeed follow the hypertext link and read the referenced document if the title or comments on it made the document look interesting. As one person said, "I don't tend to browse those databases. I rely on Joe".

Another person in the group used the collaborative filter as a document management tool. She sent pointers to select people in the group alerting them to a particular document in a group-owned database. She saved pointers for herself which she grouped by date (one of the standard views built into the database) and used the Information Digest as a way of creating links between databases.

Design Goals
We designed the prototype around three key ideas: ease of use, contextual information, flexibility. Our preliminary user data suggests that these ideas were critical to the overall design.

Ease of use. It was so easy to create and send pointers - a single click brought up a partially filled pointer form - that people were encouraged to distribute pointers to their friends. We believe that a lot of this information would not have been shared otherwise and this was confirmed by comments we got from senders. One person said that he would have browsed the database and collected up references anyway but probably just waited till a meeting to pass the references on when there may or may not have been an opportunity to tell people about the documents.

Unlike other collaborative filtering systems, our system does not suffer from cold-start problems, the need to create a user profile or even reaching critical mass. In fact people who are content to just receive documents from others can begin participating without even installing the system. And those users who want to use the collaborative filter to save pointers in their private database can get started as soon as the system is installed.

An interesting difference between our system and those which rely on user profiles is that in our system neither senders nor recipients were constrained to particular topics. This meant that people occasionally received pointers to unusual databases or documents somewhat akin to people who forward mail about strange events, poems or facts that they have received or found while browsing the World Wide Web.

Context. Although sharing hypertext links is a standard feature in Notes, several people commented that they were not comfortable following a link unless they had some information about the document at the other end. Thus, people liked getting the additional contextual information in the pointer. They particularly used the document title, and the sender's comments, in judging whether to read the document.

Flexibility. The system was designed to be flexible with respect to methods of distribution (sending by email, saving in a private database or creating an Information Digest). And indeed, we found that people tended to rely primarily on just one of these methods. That is, people who routinely sent pointers to others often did not save pointers for themselves and vice versa. But more significantly, as we look at how the system is used, we see that the simplicity of it and the lack of formal structures had the unanticipated effect of letting group practices evolve. That is, rather than have a particular person (or role) be designated as the one who selects and sends pointers to others, our system was flexible enough to let such a person emerge.

DISCUSSION
Our collaborative filtering system has been used primarily within small workgroups where people know each other's biases and current interests. In fact, the system manifestly trades on making public the identity of the person sending out the pointer. This acknowledgment contrasts with other collaborative filtering systems that work hard to allow the person(s) who contribute to the filtering process to remain anonymous. We believe that knowing something about the person who selected and commented on a document is critical to evaluating the usefulness of that document.

Having the users who find the information also responsible for sending it to colleagues contradicts a common theory on improving information systems. Namely that information finders should be freed from the task of addressing mail and coming up with recipients [16]. Yet, at least one of our studies showed that often users really do have recipients in mind for the information they discover, and we believe our active collaborative filtering system serves users well by allowing them to easily act on this knowledge. Further, since the user who discovers a piece of information is mostly likely the one who knows how it should be fit in with other information, it makes sense that the information finder should have ways of easily writing down that meta-level knowledge.

Based on the informal feedback from users, our system did seem to achieve the goal of providing a simple and hence effective way of sharing knowledge of interesting documents amongst members of a workgroup. Although it has proven useful for that purpose it has some inherent limitations. One limitation is that control over the document selection resides with the sender not with the recipient and puts the recipient in a passive role. This means, for instance, that the recipient cannot use the system to find filtered/reviewed information on a particular subject unless a sender happens to have sent out a pointer to such a document.

On the other hand, the system has the advantage of simplicity and immediacy - no cold-start or critical mass problems to overcome. It also address a need among new users to learn more about the available information space as a whole. By giving everyone in a workgroup the opportunity to save and share pointers, the burden of finding new and interesting relevant documents becomes a shared exercise. There is intrinsic reward in participating for those people who enjoy browsing around looking for information they can share with others. By providing support for an existing means of information sharing, we leverage the best of both worlds: high quality information filtering for recipients plus easy, immediate sharing tools for the senders.

ACKNOWLEDGMENTS

The first sections of this paper borrow from work done at the Xerox Palo Alto Research Center as part of Maltz's SM thesis for the Massachusetts Institute of Technology. We extend our thanks to David Goldberg of PARC and Karen Sollins of MIT who supervised the thesis work and provided valuable comments on this paper, and to Irene Greif and John Patterson for their insightful reading of a previous draft of the paper.

REFERENCES

1. Danzig, P., Obraczka, K., and Li, S-H. Internet Resource Discovery Services, IEEE Computer, September 1993, pp. 8-22.

2. Ehrlich, K. and Cash, D. Turning Information into Knowledge: Information Finding as a Collaborative Activity. In Proceedings Digital Libraries '94 (College Station, 1994), 119-125.

3. Ehrlich, K. and Cash, D. I am an Information Waitress: Bringing Order to the New Digital Libraries. 1994. Lotus Internal Report.

4. Fischer, G. and Stevens, C. Information Access in Complex, Poorly Structured Information Spaces. In Proc. CHI'91 Human Factors in Computing Systems (New Orleans,1991), Addison-Wesley, pp. 63-70.

5. Goldberg, D., Oki, B., Nichols, D., Terry, D.B. Using Collaborative Filtering to Weave an Information Tapestry. Communications of the ACM, December 1992, Vol 35, No12, pp. 61-70.

6. Grudin, J. Why CSCW Applications Fail: problems in the Design and Evaluation of Organizational Interfaces. In Proc. CSCW '88 (Portland, 1988), ACM, pp. 85-93.

7. Hill, W., Hollan, J., Wrobleski, D. and McCandless, T. Edit Wear and Read Wear. In Proc. CHI'92 Human Factors in Computing Systems (Monterey, 1992), Addison-Wesley, pp. 3-9.

8. Johnson, M. Simon Homepage: "Welcome to SIMON," University of London, available via World Wide Web at http://www.elec.qmw.ac.uk/simon/.

9. Loeb, S. Architecting Personalized Delivery of Multimedia Information. Communications of the ACM, December 1992, Vol 35, No 12, pp. 39-48.

10. Maltz, D.A. Distributing Information for Collaborative Filtering on Usenet Net News. MIT Department of EECS MS Thesis (May, 1994). Also available as Tech Report MIT/LCS/TR-603.

11. Mauldin, M., personal communication. see also "The Lycos Home Page:Hunting WWW Information" http://lycos.cs.cmu.edu/.

12. Palme, J. You have 134 unread mail! Do you want to read them now? Proceedings IFIP WG 6.5 Working Conference on Computer-Based Document Services, Nottingham England May 1984, pp 175-184.

13. Reid, B. Usenet Readership Summary Report For May 93, Usenet news.lists, 2 June 1993.

14. Resnick, P., Neophytos, I., Mitesh, S. Bergstrom, P. and Riedl, J. GroupLens: An Open Architecture for Collaborative Filtering of Netnews. In Proc. of CSCW '94: Conference on Computer Supported Cooperative Work (Chapel Hill, 1994), Addison-Wesley.

15. Sheth, B., and Maes, P. Evolving Agents for Personalized Information Filtering. Proceedings of the Ninth IEEE Conference on Artificial Intelligence for Applications, 1993.

16. Terry, D.B. 7 Steps to a Better Mail System. Proceedings IFIP International Symposium on Message Handling System and Application Layer Communication Protocols, October 1990.

Social Information Filtering:
Algorithms for Automating "Word of Mouth"

Upendra Shardanand
MIT Media-Lab
20 Ames Street Rm. 305
Cambridge, MA 02139
shard@media.mit.edu
(617) 253-7441

Pattie Maes
MIT Media-Lab
20 Ames Street Rm. 305
Cambridge, MA 02139
pattie@media.mit.edu
(617) 253-7442

ABSTRACT

This paper describes a technique for making personalized recommendations from any type of database to a user based on similarities between the interest profile of that user and those of other users. In particular, we discuss the implementation of a networked system called Ringo, which makes personalized recommendations for music albums and artists. Ringo's database of users and artists grows dynamically as more people use the system and enter more information. Four different algorithms for making recommendations by using social information filtering were tested and compared. We present quantitative and qualitative results obtained from the use of Ringo by more than 2000 people.

KEYWORDS: social information filtering, personalized recommendation systems, user modeling, information retrieval, intelligent systems, CSCW.

INTRODUCTION

Recent years have seen the explosive growth of the sheer volume of information. The number of books, movies, news, advertisements, and in particular on-line information, is staggering. The volume of things is considerably more than any person can possibly filter through in order to find the ones that he or she will like.

People handle this information overload through their own effort, the effort of others and some blind luck. First of all, most items and information are removed from the stream simply because they are either inaccessible or invisible to the user. Second, a large amount of filtering is done for us. Newspaper editors select what articles their readers want to read. Bookstores decide what books to carry. However with the dawn of the electronic informa-

tion age, this barrier will become less and less a factor. Finally, we rely on friends and other people whose judgement we trust to make recommendations to us.

We need technology to help us wade through all the information to find the items we really want and need, and to rid us of the things we do not want to be bothered with. The common and obvious approach used to tackle the problem of information filtering is *content-based filtering*[1]. *Keyword-based filtering* and *latent semantic indexing* [2] are some example content-based filtering techniques. Content-based filtering techniques recommend items for the user's consumption based on correlations between the *content* of the items and the user's preferences. For example, the system may try to correlate the presence of keywords in an article with the user's taste. However, content-based filtering has limitations:

- Either the items must be of some machine parsable form (e.g. text), or attributes must have been assigned to the items by hand. With current technology, media such as sound, photographs, art, video or physical items cannot be analyzed automatically for relevant attribute information. Often it is not practical or possible to assign attributes by hand due to limitations of resources.

- Content-based filtering techniques have no inherent method for generating serendipitous finds. The system recommends more of what the user already has seen before (and indicated liking). In practice, additional hacks are often added to introduce some element of serendipity.

- Content-based filtering methods cannot filter items based on some assesment of quality, style or point-of-view. For example, they cannot distinguish a well written an a badly written article if the two articles use the same terms.

A complementary filtering technique is needed to address these issues. This paper presents *social informa-*

tion filtering, a general approach to personalized information filtering. Social Information filtering essentially automates the process of "word-of-mouth" recommendations: items are recommended to a user based upon values assigned by other people with similar taste. The system determines which users have similar taste via standard formulas for computing statistical correlations.

Social Information filtering overcomes some of the limitations of content-based filtering. Items being filtered need not be amenable to parsing by a computer. Furthermore, the system may recommend items to the user which are very different (content-wise) from what the user has indicated liking before. Finally, recommendations are based on the quality of items, rather than more objective properties of the items themselves.

This paper details the implementation of a social information filtering system called Ringo, which makes personalized music recommendations to people on the Internet. Results based on the use of this system by thousands of actual users are presented. Various social information filtering algorithms are described, analyzed and compared. These results demonstrate the strength of social information filtering and its potential for immediate application.

RINGO: A PERSONALIZED MUSIC RECOMMENDATION SYSTEM

Social Information filtering exploits similarities between the tastes of different users to recommend (or advise against) items. It relies on the fact that people's tastes are not randomly distributed: there are general trends and patterns within the taste of a person and as well as between groups of people. Social Information filtering automates a process of "word-of-mouth" recommendations. A significant difference is that instead of having to ask a couple friends about a few items, a social information filtering system can consider thousands of other people, and consider thousands of different items, all happening autonomously and automatically. The basic idea is:

1. The system maintains a *user profile*, a record of the user's interests (positive as well as negative) in specific items.

2. It compares this profile to the profiles of other users, and weighs each profile for its degree of similarity with the user's profile. The metric used to determine similarity can vary.

3. Finally, it considers a set of the most similar profiles, and uses information contained in them to recommend (or advise against) items to the user.

7 :	BOOM! One of my FAVORITE few! Can't live without it.
6 :	Solid. They are up there.
5 :	Good Stuff.
4 :	Doesn't turn me on, doesn't bother me.
3 :	Eh. Not really my thing.
2 :	Barely tolerable.
1 :	Pass the earplugs.

Figure 1: Ringo's scale for rating music.

Ringo[7] is a social information filtering system which makes personalized music recommendations. People describe their listening pleasures to the system by rating some music. These ratings constitute the person's *profile*. This profile changes over time as the user rates more artists. Ringo uses these profiles to generate advice to individual users. Ringo compares user profiles to determine which users have similar taste (they like the same albums and dislike the same albums). Once similar users have been identified, the system can predict how much the user may like an album/artist that has not yet been rated by computing a weighted average of all the ratings given to that album by the other users that have similar taste.

Ringo is an on-line service accessed through electronic mail or the World Wide Web. Users may sign up with Ringo by sending e-mail to *ringo@media.mit.edu* with the word "join" in the body. People interact with Ringo by sending commands and data to a central server via e-mail. Once an hour, the server processes all incoming messages and sends replies as necessary. Alternatively, users can try out Ringo via the World Wide Webb (http://ringo.media.mit.edu).

When a user first sends mail to Ringo, he or she is sent a list of 125 artists. The user rates artists for how much they like to listen to them. If the user is not familiar with an artist or does not have a strong opinion, the user is asked not to rate that item. Users are specifically advised to rate artists for how much they like to *listen* to them, not for any other criteria such as musical skill, originality, or other possible categories of judgment.

The scale for ratings varies from 1 "pass the earplugs"' to 7 "one of my favorite few, can't live without them" (Figure 1). A seven point scale was selected since studies have shown that the reliability of data collected in surveys does not increase substantially if the number of choices is increased beyond seven[6]. Ratings are not normalized because as we expected, users rate albums in very different ways. For example, some users only give ratings to music they like (e.g. they only use 6's and 7's), while other users will give bad as well as good rat-

6   "10,000 Maniacs"
3   "AC/DC"
3   "Abdul, Paula"
2   "Ace of Base"
1   "Adams, Bryan"
    "Aerosmith"
    "Alpha Blondy"
6   "Anderson, Laurie"
5   "Arrested Development"
    "Autechre"
3   "B-52s"
    "Babes in Toyland"
    "Be Bop Deluxe"
5   "Beach Boys, The"
    "Beastie Boys"
4   "Beat Happening"
7   "Beatles, The"
1   "Bee Gees"

Figure 2: Part of one person's survey.

ings (1's as well as 7's). An absolute scale was employed and descriptions for each rating point were provided to make it clear what each number means.

The list of artists sent to a user is selected in two parts. Part of the list is generated from a list of the most often rated artists. This ensures that a new user has the opportunity to rate artists which others have also rated, so that there is some commonality in people's profiles. The other part of the list is generated through a random selection from the (open) database of artists. Thus, artists are never left out of the loop. A user may also request a list of some artist's albums, and rate that artist's albums on an individual basis. The procedure for picking an initial list of artists for the user to rate leaves room for future improvement and research, but has been sufficient for our early tests. Figure 2 shows part of one user's ratings of the initial 125 artists selected by Ringo.

Once a person's initial profile has been submitted, Ringo sends a help file to the user, detailing all the commands it understands. An individual can ask Ringo for predictions based upon their personal profile. Specifically, a person can ask Ringo to (1) suggest new artists/albums that the user will enjoy, (2) list artists/albums that the user will hate, and (3) make a prediction about a specific artist/album. Ringo processes such a request using its social filtering algorithm, detailed in the next section. It then sends e-mail back to the person with the result. Figure 4 provides an example of Ringo's suggestions. Every recommendation includes a measure of confidence which depends on factors such as the number of similar users used to make this prediction, the consis-

tency among those users' values, etc. (cfr. [7] for details.) Ringo's reply does not include any information about the identity of the other users whose profiles were used to make the recommendations.

Ringo provides a range of functions apart from making recommendations. For example, when rating an artist or album, a person can also write a short review, which Ringo stores. Two actual reviews entered by users are shown in Figure 5. Notice that the authors of these reviews are free to decide whether to sign these reviews or keep them anonymous. When a user is told to try or to avoid an artist, any reviews for that artist written by similar users are provided as well. Thus, rather than one "thumbs-up, thumbs-down" review being given to the entire audience, each user receives personalized reviews from people that have similar taste.

In addition, Ringo offers other miscellaneous features which increase the appeal of the system. Users may add new artists and albums into the database. This feature was responsible for the growth of the database from 575 artists at inception to over 2500 artists in the first 6 weeks of use. Ringo, upon request, provides dossiers on any artist. The dossier includes a list of that artist's albums and straight averages of scores given that artist and the artist's albums. It also includes any added history about the artist, which can be submitted by any user. Users can also view a "Top 30" and "Bottom 30" list of the most highly and most poorly rated artists, on average. Finally, users can subscribe to a periodic newsletter keeping them up to date on changes and developments in Ringo.

ALGORITHMS AND QUANTITATIVE RESULTS

Ringo became available to the Internet public July 1, 1994. The service was originally advertised on only four specialized USENET newsgroups. After a slow start, the number of people using Ringo grew quickly. Word of the service spread rapidly as people told their friends, or sent messages to mailing lists. Ringo reached the 1000-user mark in less than a month, and had 1900 users after 7 weeks. At the time of this writing (September 1994) Ringo has 2100 users and processes almost 500 messages a day.

Like the membership, the size of the database grew quickly. Originally, Ringo had only 575 artists in its database. As we soon discovered, users were eager to add artists and albums to the system. At the time of this writing, there are over 3000 artists and 9000 albums in Ringo's database.

Thanks to this overwhelming user interest, we have an enormous amount of data on which to test various social information information filtering algorithms. This section discusses four algorithms that were evaluated and

Artist	Rating	Confidence
"Orb, The"	6.9	fair
"Negativland"	6.5	high

Reviews for "Negativland"

They make you laugh at the fact that nothing is funny any more.　　— user@place.edu

"New Order"	6.5	fair

Reviews for "New Order"

Their albums until 'Brotherhood' were excellent. Since then, they have become a tad too tame and predictable.　　— lost@elsewhere.com

"Sonic Youth"	6.5	fair

Reviews for "Sonic Youth"

Confusion is Sex: come closer and I'll tell you.

"Grifters"	6.4	fair
"Dinosaur Jr."	6.4	fair
"Velvet Underground, The"	6.3	low

Reviews for "Velvet Underground, The"

The most amazing band ever.

"Mudhoney"	6.3	fair

Figure 3: Some of Ringo's suggestions.

Tori Amos has my vote for the best artist ever. Her lyrics and music are very inspiring and thought provoking. Her music is perfect for almost any mood. Her beautiful mastery of the piano comes from her playing since she was two years old. But, her wonderful piano arrangements are accompanied by her angelic yet seductive voice. If you don't have either of her two albums, I would very strongly suggest that you go, no better yet, run down and pick them up. They have been a big part of my life and they can do the same for others. — user@place.edu

I'd rather dive into a pool of dull razor blades than listen to Yoko Ono sing. OK, I'm exaggerating. But her voice is *awful* She ought to put a band together with Linda McCartney. Two Beatles wives with little musical talent.

Figure 4: Two sample reviews written by users.

gives more details about the "winning" algorithm. For our tests, the profiles of 1000 people were considered. A profile is a sparse vector of the user's ratings for artists. 1,876 different artists were represented in these profiles.

To test the different algorithms, 20% of the ratings in each person's profile were then randomly removed. These ratings comprised the *target set* of profiles. The remaining 80% formed the *source set*. To evaluate each algorithm, we predicted a value for each rating in the target set, using only the data in the source set. Three such target sets and data sets were randomly created and tested, to check for consistency in our results. For brevity, the results from the first set are presented throughout this paper, as results from all three sets only differed slightly.

In the source set, each person rated on average 106 artists of the 1,876 possible. The median number of ratings was 75, and the most ratings by a single person was 772! The mean score of each profile, i.e. the average score given all artists by a user, was 3.7.

Evaluation Criteria
The following criteria were used to evaluate each prediction scheme:

- The mean absolute error of each predicted rating must be minimized. If $\{r_1, \ldots, r_N\}$ are all the real values in the target set, and $\{p_1, \ldots, p_N\}$ are the predicted values for the same ratings, and $E = \{\varepsilon_1, \ldots, \varepsilon_N\} = \{p_1 - r_1, \ldots, p_N - r_N\}$ are the errors, then the mean absolute error is

$$\overline{|E|} = \frac{\sum_{i=1}^{N} |\varepsilon_i|}{N} \qquad (1)$$

The lower the mean absolute error, the more accurate the scheme. We cannot expect to lower $\overline{|E|}$ below the error in people's ratings of artists. If one provides the same list of artists to a person at different points of time, the resulting data collected will differ to some degree. The degree of this error has not yet been measured. However we would expect the error to at least be ±1 unit on the rating scale (because otherwise there would be 0 or no error).

- The standard deviation of the errors,

$$\sigma = \sqrt{\frac{(\sum (E - \overline{E})^2)}{N}} \qquad (2)$$

should also be minimized. The lower the deviation, the more consistently accurate the scheme is.

- Finally, $\mathcal{T}$, the percentage of target values for which the scheme is able to compute predictions should be maximized. Some algorithms may not be able to make predictions in all cases.

Base Case Algorithm

A point of comparison is needed in order to measure the quality of social information filtering schemes in general. As a base case, for each artist in the target set, the mean score received by an artist in the source set is used as the predicted score for that artist. A social information filtering algorithm is neither personalized nor accurate unless it is a significant improvement over this base case approach.

Figure 5 depicts the distribution of the errors, E. $\overline{|E|}$ is 1.3, and the standard deviation σ is 1.6. The distribution has a nice bell curve shape about 0, which is what was desired. At first glance, it may seem that this mindless scheme does not behave too poorly. However, let us now restrict our examination to the *extreme* target values, where the score is **6** or greater or **2** or less. These values, after all, are the critical points. Users are most interested in suggestions of items they would love or hate, not of items about which they would be ambivalent.

The distribution of errors for extreme values is shown by the dark gray bars in Figure 5. The mean error and standard deviation worsen considerably, with $\overline{|E|} = 1.8$ and $\sigma = 2.0$. Note the lack of the desired bell curve shape. It is in fact the sum of two bell curves. The right hill is mainly the errors for those target values which are **2** or less. The left hill is mainly the errors for those target values which are **6** or greater.

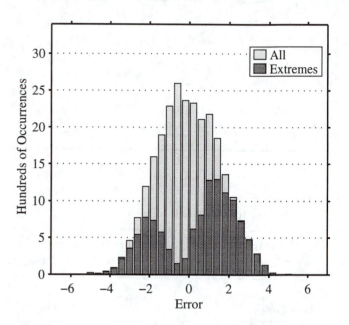

Figure 5: The distribution of errors in predictions of the Base Algorithm.

For the target values **6** or greater, the mean absolute error is much worse, with $\overline{|E|} = 2.1$. Why the great discrepancy in error characteristics between all values and only extreme values? Analysis of the database indicates that the mean score for each artist converges to approximately **4**. Therefore, this scheme performs well in cases where the target value is near **4**. However, for the areas of primary interest to users, the base algorithm is useless.

Social Information Filtering Algorithms

Four different social information filtering algorithms were evaluated. Due to space limitations, the algorithms are described here briefly. Exact mathematical descriptions as well as more detailed analysis of the algorithms can be found in [7].

The Mean Squared Differences Algorithm. The first algorithm measures the degree of dissimilarity between two user profiles, U_x and U_y by the *mean squared difference* between the two profiles:

$$\overline{(U_x - U_y)^2} \qquad (3)$$

Predictions can then be made by considering all users with a dissimilarity to the user which is less than a certain threshold L and computing a weighted average of the ratings provided by these most similar users, where the weights are inverse proportional to the dissimilarity.

The Pearson r Algorithm. An alternative approach is to use the standard *Pearson r* correlation coefficient to measure similarity between user profiles:

$$\frac{\sum (U_x - \overline{U_x})(U_y - \overline{U_y})}{\sqrt{\sum (U_x - \overline{U_x})^2 \times \sum (U_y - \overline{U_y})^2}} \qquad (4)$$

This coefficient ranges from -1, indicating a negative correlation, via O, indicating no correlation, to +1 indicating a positive correlation between two users. Again, predictions can be made by computing a weighted average of other user's ratings, where the Pearson r coefficients are used as the weights. In contrast with the previous algorithm, this algorithm makes use of negative correlations as well as positive correlations to make predictions.

The Constrained Pearson r Algorithm. Close inspection of the Pearson r algorithm and the coefficients it produced prompted us to test a variant which takes the *positivity* and *negativity* of ratings into account. Since the scale of ratings is absolute, we "know" that values below **4** are negative, while values above **4** are positive. We modified the Pearson r scheme so that only when there is an instance where both people have rated an artist positively, above **4**, or both negatively, below **4**, will the correlation coefficient increase. More specifically, the standard Pearson r equation was altered to become:

$$\beta_{xy} = \frac{\sum (U_x - 4)(U_y - 4)}{\sqrt{\sum (U_x - 4)^2 \times \sum (U_y - 4)^2}} \qquad (5)$$

To produce recommendations to a user, the constrained Pearson r algorithm first computes the correlation coefficient between the user and all other users. Then all users whose coefficient is greater than a certain threshold L are identified. Finally a weighted average of the ratings of those similar users is computed, where the weight is proportional to the coefficient. This algorithm does not make use of negative "correlations" as the Pearson r algorithm does. Analysis of the constrained Pearson r coefficients showed that there are few very negative coefficients, so including them makes little difference.

The Artist-Artist Algorithm. The preceding algorithms deal with measuring and employing similarities between *users*. Alternatively, one can employ the use of correlations between *artists or albums* to generate predictions. The idea is simply an inversion of the previous three methodologies. Say Ringo needs to predict how a user, Murray, will like "Harry Connick, Jr". Ringo examines the artists that Murray has already rated. It weighs each one with respect to their degree of correlation with "Harry Connick, Jr". The predicted rating is then simply a weighted average of the artists that Murray has already scored. An implementation of such a scheme using the constrained Pearson r correlation coefficient was evaluated.

Method	All		Extremes		
	$|E|$	σ	$|E|$	σ	T
Base Case	1.3	1.6	1.8	2.0	90
Mean Sq. Diff., $L = 2.0$	1.0	1.3	1.2	1.6	70
Pearson r	1.1	1.4	1.5	1.7	99
Pearson r, $L = 0.35$	1.0	1.3	1.4	1.6	99
Pearson r, $L = 0.5$	1.0	1.3	1.3	1.6	95
Pearson r, $L = 0.65$	1.1	1.4	1.3	1.6	73
Pearson r, $L = 0.75$	1.1	1.5	1.3	1.7	41
Con. Pearson r, $L = 0.5$	1.1	1.3	1.3	1.6	97
Con. Pearson r, $L = 0.6$	1.1	1.4	1.2	1.6	91
Con. Pearson r, $L = 0.7$	1.1	1.3	1.3	1.6	70
Artist-Artist, $L = 0.6$	1.1	1.4	1.3	1.6	89
Artist-Artist, $L = 0.7$	1.1	1.4	1.1	1.5	65

Table 1: Summary of results.

Results

A summary of some of our results (for different values of the threshold L) are presented in table 1. More details can be found in [7]. Overall, in terms of accuracy and the percentage of target values which can be predicted, the constrained Pearson r algorithm performed the best on our dataset if we take into account the error as well as the number of target values that can be predicted. The mean square differences and artist-artist algorithms may perform slightly better in terms of the quality of the predictions made, but they are not able to produce as many predictions.

As expected, there is a tradeoff between the average error of the predictions and the percentage of target values that can be predicted. This tradeoff is controlled by the parameter L, the minimum degree of similarity between users that is required for one user to influence the recommendations made to another.

Figure 6 illustrates the distribution of errors for the best algorithm with the threshold L equal to 0.6. The distribution for extreme values approaches a bell curve, as desired. The statistics for all values and extreme values are $\overline{|E|} = 1.1$, $\sigma = 1.4$ and $\overline{|E|} = 1.2$, $\sigma = 1.6$, respectively. These results are quite excellent, especially as the mean absolute error for extreme values approaches that of all values. At this threshold level, 91% of the target set is predictable.

QUALITATIVE RESULTS

Ultimately, what is more important than the numbers in the previous section is the human response to this new technology. As of this writing over 2000 people have used Ringo. Our source for a qualitative judgment of Ringo is the users themselves. The Ringo system operators have received a *staggering* amount of mail from users— questions, comments, and bug reports. The re-

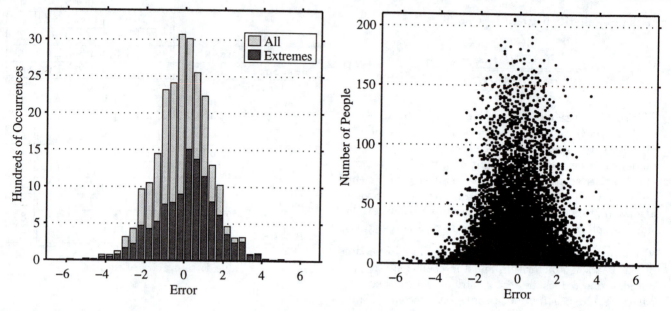

Figure 6: The distribution of errors for the Constrained Pearson r algorithm with $L = 0.6$.

Figure 7: The scatter plot of the error vs. the number of people consulted to make the prediction.

sults described in this section are all based on user feedback and observed use patterns.

One observation is that a social information filtering system becomes more competent as the number of users in the system increases. Figure 7 illustrates how the error in a recommendation relates to the number of people profiles consulted to make the recommendation. As the number of user scores used to generate a prediction increases, the deviation in error decreases significantly. This is the case because the more people use the system, the greater the chances are of finding close matches for any particular user. The system may need to reach a certain *critical mass* of collected data before it becomes useful. Ringo's competence develops over time, as more people use the system. Understandably then, in the first couple weeks of Ringo's life, Ringo was relatively incompetent. During these days we received many messages letting us know how poorly Ringo performed. Slowly, the feedback changed. More and more often we received mail about how "unnervingly accurate" Ringo was, and less about how it was incorrect. Ringo's growing group of regular "customers" indicates that it is now at a point where the majority of people find the service useful.

However, many people are disappointed by Ringo's initial performance. We are often told that a person must do one or two iterations of rating artists before Ringo becomes accurate. A user would rate the initial set, then receive predictions. If the user knows any of the predicted artists are not representative of their personal taste, they rate those artists. This will radically alter

the members of the user's "similar user" neighborhood. After these iterations, Ringo works satisfactorily. This indicates that what is needed is better algorithm for determining the "critical" artists to be rated by the user so as to distinguish the user's tastes and narrow down the group of similar users.

Beyond the recommendations, there are other factors which are responsible for Ringo's great appeal and phenomenal growth. The additional features, such as being a user-grown database, and the provisions for reviews and dossiers add to its functionality. Foremost, however, is the fact that Ringo is not a static system. The database and user base is continually growing. As it does, Ringo's recommendations to the user changes. For this reason, people enjoy Ringo and use it on a regular basis.

RELATED WORK

Several other attemps have been made at building filtering services that rely on patterns among multiple users. The Tapestry system [3] makes it possible to request Netnews documents that have been approved by other users. However, users must themselves know who these similar people are and specifically request documents annotated by those people. That is, using the Tapestry system the user still needs to know which other people may have similar tastes. Thus, the social information filtering is still left to the user.

During the development of Ringo, we learned about the existence of similar projects in a similar state of development. One such example is Grouplens [4], a system applying social information filtering to the personalized selection of Netnews. GroupLens employs Pearson r correlation coefficients to determine similarity between users. On our dataset, the algorithms described in this paper performed better than the algorithm used by Grouplens.

Two other recently developed systems are a video recommendation service implemented at Bellcore, Morristown, NJ and a movie recommendation system developed at ICSI, Berkeley, CA. Unfortunately, as of this writing, there is no information available about the algorithms used in these systems, nor about the results obtained.

The user modeling community has spawned a range of recommendation systems which use information about a user to assign that user to one of a finite set of hand-built, predefined user classes or *stereotypes*. Based on the stereotype the user belongs to, the system then makes recommendations to the user. For example [5] recommends novels to users based on a stereotype classification. This method is far less personalized than the social information filtering method described in this paper. The reason is that in social information filtering, in a sense every user defines a stereotype that another user can to some degree belong to. The number of stereotypes which is used to define the user's taste is much larger.

Finally, some commercial software packages exist that make recommendations to users. An example is Movie Select, a movie recommendation software package by Paramount Interactive Inc. One important difference is that these systems use a data set that does not change over time. Furthermore, these systems also do not record any history of a person's past use. As far as can be deduced from the software manuals and brochures, these systems store correlations between different items and use those correlations to make recommendations. As such the recommendations made are less personalized than in social information filtering systems.

CONCLUSIONS AND FUTURE WORK

Experimental results obtained with the Ringo system have demonstrated that social information filtering methods can be used to make personalized recommendations to users. Ringo has been tested and used in a real-world application and received a positive response. The techniques employed by the system could potentially be used to recommend books, movies, news articles, products, and more.

More work needs to be done in order to make social information filtering applicable when dealing with very large user groups and a less narrow domain. Work is currently under way to speed up the algorithm by the use of clustering techniques, so as to reduce the number

of similarity measures that need to be computed. We are also using clustering techniques among the artist data so as to identify emergent musical genres and make use of these distinct genres in the prediction algorithms.

Finally, we haven't even begun to explore the very interesting and controversial social and economical implications of social information filtering systems like Ringo.

ACKNOWLEDGEMENTS

Carl Feynman proposed the mean squarred differences algorithm and initially developed some of the ideas that led to this work. Karl Sims proposed the artist-artist algorithm. Lee Zamir and Max Metral implemented the WWW version of Ringo and are currently responsible for maintaining the Ringo system. Alan Wexelblat and Max Metral provided useful comments on an earlier version of the paper. This work was sponsored by the National Science Foundation under grant number IRI-92056688 and by the "News in the Future" Consortium of the MIT Media Lab.

REFERENCES

[1] Special Issue on Information Filtering, Communications of the ACM, Vol. 35, No. 12, December 1992.

[2] Scott Deerweester, Susan Dumais, George Furnas, Thomas Landauer, Richard Harshman, "Indexing by Latent Semantic Analysis", Journal of the American Society for Information Science, Vol. 41, No. 1, 1990, pp. 391–407.

[3] David Goldberg, David Nichols, Brian M. Oki and Douglas Terry, "Using Collaborative Filtering to Weave an Information Tapestry", Communications of the ACM, Vol. 35, No. 12, December 1992, pp. 61–70.

[4] Paul Resnick, Neophytos Iacovou, Mitesh Sushak, Peter Bergstrom, John Riedl, "GroupLens: An Open Architecture for Collaborative Filtering of Netnews", to be published in the Proceedings of the CSCW 1994 conference, October 1994.

[5] Elaine Rich, "User Modeling via Stereotypes", Cognitive Science, Vol. 3, pp. 335-366, 1979.

[6] Robert Rosenthal and Ralph Rosnow, "Essentials of Behavioral Research: Methods and Data and Analysis", McGraw Hill, second edition, 1991.

[7] Upendra Shardanand, "Social Information Filtering for Music Recommendation", MIT EECS M. Eng. thesis, also TR-94-04, Learning and Common Sense Group, MIT Media Laboratory, 1994.

A Comparison of User Interfaces for Panning on a Touch-Controlled Display

Jeff A. Johnson

Sun Microsystems

2550 Garcia Ave., MS UMTV12-33

Mountain View, CA 94043-1100

415-336-1625

jeffrey.johnson@eng.sun.com

ABSTRACT

An experiment was conducted to determine which of several candidate user interfaces for panning is most usable and intuitive: panning by pushing the background, panning by pushing the view/window, and panning by touching the side of the display screen. Twelve subjects participated in the experiment, which consisted of three parts: 1) subjects were asked to suggest panning user interfaces that seemed natural to them, 2) subjects each used three different panning user interfaces to perform a structured panning task, with experimenters recording their performance, and 3) subjects were asked which of the three panning methods they preferred. One panning method, panning by pushing the background, emerged as superior in performance and user preference, and slightly better in intuitiveness than panning by touching the side of the screen. Panning by pushing the view/window fared poorly relative to the others on all measures.

KEYWORDS

Touch display, touchscreen, panning, scrolling, navigation.

INTRODUCTION

Often, computer-based displays provide views of a scene or information that is too expansive (wide or tall) to be shown in its entirety. A common solution is to provide a panning or scrolling function, allowing users to control which portion of the subject is visible in the display. If such a function is provided, a user interface -- a way for users to invoke and control panning -- must also be provided.

Panning user interfaces for touch-displays have not been studied much, if at all. Some work has been done on scrolling techniques for more conventional input devices. Informal (and unpublished) investigations at Xerox in the late Seventies focused on whether scrollbars in the Star user interface should treat the document as moving under the window, or the window as moving over the document. Star's designers chose the former [14], but the marketplace for mouse-controlled graphical user interfaces eventually settled on the latter. Bury *et al.* [3] compared scrolling user interfaces for keyboard-controlled character-display terminals and found that subjects performed best when keys indicated the direction of movement of the *viewing area* (e.g., the "up" key moved it upward in the document and hence the document content downward). Beard *et al.* [1] and Glassner [5] described panning and scrolling user interfaces for specific task-domains. Beck and Elkerton [2] compared user interfaces for traversing long lists. MacLean *et al.* used panning user interfaces as an example in their analysis of how to compare alternative user-interface designs rationally [8]. However, all of this work has focused on user interfaces controlled through keyboards and indirect pointing devices such as mice, not touch-displays.

Though empirical studies of touch-display interfaces have been performed [10, 11, 12], these have not included studies of panning and scrolling. Several ways of using touch-displays have been proposed [13] but panning through a flat information space is not among them. Recently, a product that incorporates a user interface for controlling panning has appeared in the marketplace [15], but nothing describing how that design was validated has been published.

The goal of the current study was to determine the best -- meaning most easily used -- user interface for panning and scrolling in a simulated physical world displayed on a touch-controlled display.

Initial experience at FirstPerson (a Sun subsidiary) with a prototype information-space controlled by a touch-display raised in a new form the old issue of whether the scene should be treated as moving under the view-window or the window as moving over the scene. The prototype had been designed such that swiping one's finger across the display caused the scene to shift. The issue was: What should the relationship be between the direction of the swipe and how the scene shifts? Is the user's finger swipe best regarded as pushing the scene background or pushing the "camera" that is viewing the scene? With a Push Background user interface, users swipe in the direction they want the scene to go. This seems completely natural until one considers previous findings on scrollbars and scrolling keys, as well as the fact that to bring information onto the screen that is off-screen left, users would swipe towards the right. With a Push Camera user interface, users swipe in the direction of the off-screen information. This, too, seems natural until one considers that the physical display doesn't actually move, so it is the scene that shifts -- in the opposite direction as the swipe. Of course, both interfaces can't be "natural," since they have opposite effects.

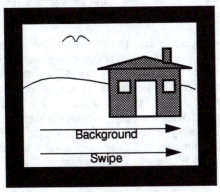

Push Background

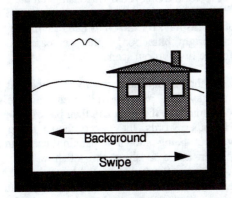

Push Camera

Figure 1. Two contrasting panning methods.

The correct interface depends on how users think about it: do they view themselves as shifting the background or panning the camera? Therefore, the right way to resolve the issue is via user testing. The designers of the initial prototype chose the Push Camera approach, based on an informal paper-and-pencil test they conducted using a few co-workers as subjects. However, experience with the prototype suggested that a significant number of users considered its panning user interface to work "backwards" from what they expected.

In an attempt to resolve this issue, a more formal study was devised, using subjects from outside the company who had no experience with the existing prototype and so were not already familiar with its user interface for panning. The idea of the study was that subjects would first be asked -- before they had tried or seen any panning user interface -- to describe one that would seem natural to them. Subjects would then perform a panning accuracy test with different panning user interfaces. Finally, they would be asked to say which panning user interface they preferred of those they had tried.

As the study was being designed, it was decided that panning user-interfaces other than Push Background and

Push Camera should be considered. One plausible alternative to panning by swiping a finger is panning by touching the sides of the touch-display. In fact, this user-interface for panning was already used in the prototype in certain situations. When a user is dragging an object to a new location that is off-screen, it is impractical to have to drop it, pan the scene, pick the object up, and resume dragging. Therefore, in the prototype, when a user drags an object to a side of the display and holds it there, the "camera" pans towards that side. It is reasonable to consider having panning work that way *all* the time rather than only when the user is dragging something. This interface is best regarded as Touch Edge Camera because the user feels as if hitting an edge of the display pushes the "camera" or "viewing window" in that direction. This interface was added to the set to be compared.

Another possible panning user interface is the opposite of Touch Edge Camera: users press on the side of the display they want the background to move towards. However, because this interface -- referred to herein as Touch Edge Background -- is in the author's experience very counter-intuitive and also seems incompatible with the goal of having panning work well with dragging, it was not included in the comparison.

Regarding the first part of the study, when subjects would be asked to suggest a panning method, it is clear that the appearance of the touch-display can influence what people suggest. This is what Gibson [4] and, later, Norman [9] refer to as an "affordance": when an aspect of an artifact's design suggests how it is to be used. We thought that adding a brightly colored border around the displayed image might suggest "touch here" to users, and might therefore suggest Touch Edge panning (camera or background). We therefore designed the study so that in the initial (elicitation) phase of the study, half of the subjects would see such a border, and half would see the "normal" display (with the displayed scene taking up the entire screen).

METHOD

Subjects

The subjects were six males and six females, ranging in age from 18 to 65 years old. The distribution of subject ages was similar across gender. Subjects were recruited by company employees, and were paid $15 for the one-hour test session. The subjects' occupations include student, homemaker, retiree, clerical, salesperson, and business planner. None are computer engineers, but some do use computers.

Design

The experiment consisted of three tests: Panning Elicitation, Panning Accuracy, and Panning Preference. Since gender differences have occasionally been observed in studies involving hand-eye coordination [7], all three tests were designed to allow gender effects to be distinguished from individual differences.

Panning Elicitation Test

The six subjects of each gender were randomly assigned to one of two display types -- plain or bordered -- in such a way that three subjects of each gender had each display type. The dependent measure was the type of panning user interface the subject suggested (see Procedure and Materials), i.e., a categorical variable. However, as is described in the Analysis and Results section, the response categories were not predetermined, but rather emerged as the study was conducted.

Panning Accuracy Test

Because high intersubject variability was expected in the panning accuracy task, a within-subject design was used for that part of the experiment. Each subject was tested with all three panning user interfaces (Push Background, Push Camera, and Touch Edge Camera), with the order of presentation of the interfaces counterbalanced across subjects both within gender and overall. For each panning user interface, there were twenty-three trials (see Procedure). The following dependent measures were taken on each trial: time, number of moves, and whether the subject started by shifting the scene in the wrong direction (direction errors). Thus, the design of the panning accuracy test was a three (panning user interface) by two (gender) design, with panning user interface varied within subject.

Panning Preference Test

The panning preference test was a very simple design: Ask all 12 subjects (six of each gender) which panning user interface they preferred of the three they had tried, to determine whether any interface is systematically favored or disfavored.

Procedure and Materials

The display was a small (15 cm diagonal) color liquid-crystal screen mounted in a flat case. The case added about 2 cm of border to each edge of the display. A transparent touch-pad was affixed to the front of the display. The display housing was mounted on a pedestal, which held the display about four inches above the surface of a desk (see Figure 2). The display and touch-pad were connected to a Sun™ Sparcstation™ 2 workstation.

display/touch

area

Figure 2. Touch-display apparatus.

Subjects participated in the study individually. A video camera was focused on the display for the entire session, recording what happened on the display as well as the subject's hand when near or on it. Each session was conducted by an experimenter and an assistant (to operate the camera and a timer).

Subjects were told at the start of the session that we (the experimenters) "are testing various design ideas for our product, to make the product easy to use." The experimenter explained that some aspects of the prototype might be hard to use, but if so, that indicated a bad design rather than anything wrong with the subject. Subjects were seated in front of a prototype touch-sensitive display, and their attention was directed to it. The session consisted of three phases: Panning Elicitation, Panning Accuracy, and Panning Preference.

Panning Elicitation

A cartoon scene of a living room was visible on the display. For half of the subjects, the living room display occupied the full area of the display; for the other half, the display had a quarter-inch, bright blue border that decreased the area available for the living room scene.

Subjects were shown that the living room scene was wider than the display (by panning the scene remotely via keyboard commands to the right and then back to the initial position). In other words, they were shown the panning function without being shown a panning user interface. They were then asked to show and tell how they would expect to operate the touch-display to "bring that other part of the scene back into view." Words such as "pan" that might suggest a particular mental model or user interface were intentionally not used in the instructions. Subjects' responses were categorized (see Analysis and Results).

Panning Accuracy

In this phase of the study, the experimenter placed a sticker on the display's bottom bezel that marked three horizontal positions: "A", "B", and "C". Then the experimenter started a program that changed the displayed image from the living room scene to a set of evenly spaced vertical lines, each labeled by a number (0 - 10 from left to right) at its top end (see Figure 3). The experimenter directed the subject's attention to the numbered lines and said that in this part of the test, he would ask the subject to shift the scene so that certain numbered lines were over specified letters. The experimenter explained that the subject would be timed while shifting the scene to the target letter. The experimenter continued: "We'll do that several times, then I'll change the way the scene is shifted and ask you to shift it some more, then we'll change it again and I'll ask you to shift it some more." The subject was reminded that we were testing our designs, not the subject.

The experimenter then demonstrated to the subject how to operate the first of the three panning user interfaces. Subjects were not allowed to practice panning before starting the timed trials. Subjects were instructed to move the line given on each trial to the indicated letter position, and once the line was positioned to his or her satisfaction, to say "OK" so that the assistant would know when to stop the timer. The next trial began after a brief pause, with the scene positioned where it had been left by the previous trial. Twenty three panning trials followed (i.e., all lines moved to all reachable targets) in which the experimenter stated the line number and target letter, the assistant timed the trial, and the experimenter recorded the time on a data sheet. All trials were videotaped to allow rechecking of the times and collection of other data.

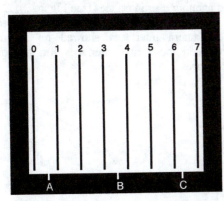

Figure 3. Panning accuracy test display.

After the first set of panning accuracy trials, the experimenter changed the panning method, demonstrated the new panning method, and commenced a second set of trials with a different panning user interface. These were followed by a set of trials with a third panning user interface.

Panning Preference

After completing all three sets of panning accuracy trials, subjects were asked "Which of the three shifting methods do you like best?" Their answers were recorded, both on the data sheet and on videotape.

ANALYSIS AND RESULTS

Panning Elicitation

When subjects were asked to indicate how they would operate the touch-display to bring into view the part of the living room scene that was off-screen-right, they gave a variety of responses. Note that at this point, subjects had not yet seen any actual panning method for the prototype display (though some had used scrolling mechanisms on computers). All they had seen was that the living room scene was wider than could be seen in the display at once, and that the off-screen portion was the right side of the scene. Thus, their responses may be considered to be based upon their prior experience and whatever the prototype display suggested to them.

Subject responses were grouped into five categories, corresponding to the four panning methods under consideration plus an Other category. The panning categories were:

• Push Background: finger motion on screen, shifting the background.

• Push Camera: finger motion on screen, panning the viewer/camera.

• Touch Edge Camera: press edge of screen that camera is to pan towards.

• Touch Edge Background: press edge of screen that background is to shift towards.

• Other: anything else, e.g., scrollbar, physical scrolling buttons on the case.

Six (i.e., half) of the twelve subjects indicated by their actions and words that Push Background was the panning method they first thought of when faced with the prototype touch-display. Three subjects suggested Touch Edge Camera. The remaining three subjects suggested methods that were classified as Other. A statistical test of the skewedness of the observed distribution requires a null hypothesis specifying what distribution would be expected by chance. Because a null hypothesis of equal category probabilities seems naive and no other chance distribution seems any more plausible, a statistical test wasn't feasible. Nonetheless, it is notable (if not "significant" in the formal sense) that half of the subjects suggested Push Background, and that no subject suggested Push Camera or Touch Edge Background (see Figure 4).

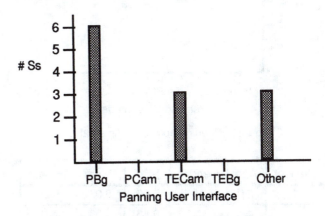

Figure 4. Panning elicitation results.

The foregoing analysis treated all 12 subjects as one group. However, the scene shown to half of the subjects had a bright blue border, which might suggest a Touch Edge panning user interface. Also, orthogonally to this grouping, half the subjects were of each gender. Neither the presence/absence of a blue border nor gender appeared related to subjects' suggested panning user interfaces.

Panning Accuracy

Each subject's time scores were averaged over each trial block, yielding, for that subject, an average time for each of the three panning user interfaces. Similarly, each subject's moves scores were averaged for each of the panning user interfaces. Direction errors were summed for each panning user interface. Tables 1, 2, and 3 show each subject's aggregated data for the time, moves, and error measures, respectively. Note that though treatment means are included in the tables for informal comparison purposes, the table columns are not independent: all three scores in each table-row are from the same subject.

For each dependent measure, a Friedman ranks test for matched scores [6] was used to test for an effect of panning user interface type.[1] To perform this test, each subject's three scores on a given measure were ranked, then the rank scores were submitted to a formula that yields a chi-squared statistic indicating whether any interface had more low or high ranked scores than would be expected by chance.

For all three measures, a significant effect of interface type was observed (time: $x^2(2) = 13.17$, p < .01; moves: $x^2 = 12.17$, p < .01; errors: $x^2 = 15.79$, p < .01; see Figures 5-7). With the Push Background panning user interface, subjects took less time, required fewer moves, and made fewer direction errors than with the other two panning user interfaces. The effect was particularly strong for the direction-errors measure: when using the Push Background, subjects made almost no direction errors, in sharp contrast to the other two panning user interfaces (see Figure 7). A *post-hoc* t-test of difference scores between the Push Camera and Touch Edge Camera interfaces showed no significant difference between those two interfaces for any of the dependent measures.

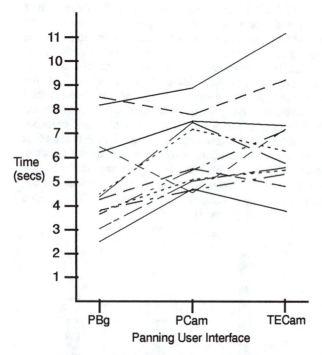

Figure 5. Panning accuracy: time data.

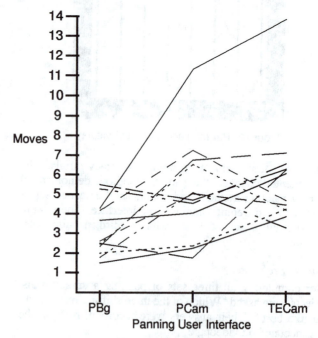

Figure 6. Panning accuracy: moves data.

Subject	Age	Gender	PBg	PCam	TECam
1	21	M	3.06	5.03	5.66
3	43	"	6.48	4.66	7.18
7	29	"	4.58	7.16	6.33
8	26	"	6.29	7.48	7.35
9	65	"	8.14	8.90	11.22
12	45	"	4.44	7.45	5.83
2	23	F	4.27	5.69	4.91
4	65	"	8.57	7.88	9.20
5	18	"	3.88	4.07	4.57
6	39	"	3.71	5.49	7.08
10	50	"	3.81	4.65	5.29
11	19	"	2.53	4.79	3.88
Mean			4.98	6.10	6.54

Table 1. Panning accuracy: time data (avg. over trials).

Subject	Age	Gender	PBg	PCam	TECam
1	21	M	1.48	2.26	3.96
3	43	"	2.57	1.74	6.26
7	29	"	2.39	6.87	7.13
8	26	"	4.26	11.35	13.96
9	65	"	5.57	4.70	6.30
12	45	"	2.43	6.61	4.22
2	23	F	4.13	7.30	4.70
4	65	"	5.22	4.61	6.65
5	18	"	3.74	4.04	6.13
6	39	"	2.57	5.09	4.57
10	50	"	2.09	2.39	4.30
11	19	"	1.87	5.09	3.30
Mean			3.19	5.17	5.95

Table 2. Panning accuracy: moves data (avg. over trials).

---

1. The Friedman test was used in preference to a within-subject ANOVA because of controversy regarding the degrees of freedom for within-subject ANOVAs.

No gender differences appeared on any measure, either for overall performance or effect of panning user interface type.

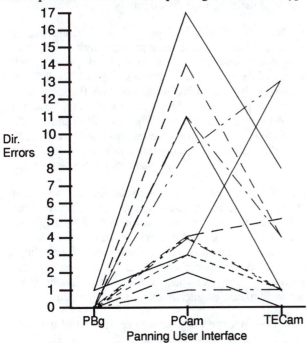

Figure 7. Panning accuracy: error data.

Subject	Age	Gender	PBg	PCam	TECam
1	21	M	0	4	5
3	43	"	0	3	1
7	29	"	0	11	1
8	26	"	1	17	8
9	65	"	0	4	1
12	45	"	0	2	0
2	23	F	0	11	4
4	65	"	0	1	1
5	18	"	0	4	1
6	39	"	1	3	13
10	50	"	0	9	13
11	19	"	0	14	4
Mean			1.67	6.92	4.33

Table 3. Panning accuracy: error data (sum over trials).

One possibility worth checking is that the three panning user interfaces might differ in how performance improves with practice. For example, performance with one user interface might start out worse than with others, but improve faster. Learning effects were examined by comparing, for each subject, performance in the first half of the trials with that in the second half. Since each user-interface-block of trials consisted of an odd number (23) of trials, the "first half" was defined as trials 1-11, and the "second half" was defined as trials 13-23. Trial 12 was ignored.

Overall, there was clear evidence of learning over the trial blocks. Superimposing the trial blocks for the three panning user interfaces, performance in the second half of the trials in a block almost always exceeded that in the first half. Simple sign tests were significant for all three performance measures: time ($p < .01$), number of moves ($p < .05$), and direction errors ($p < .01$).

Since the distance from the starting position to the target position differed from trial to trial, a performance difference between the first and last half-blocks could result from a difference in the distance panned in the two half-blocks. In this case, that explanation can be ruled out, because the distance subjects had to pan in the two half-blocks was almost equal (averaging 102.4 pixels/trial for trials 1-11 *vs.* 103.2 pixels/trial for trials 13-23) and was greater in the second half-block anyway.

To determine whether the learning rate depended on the panning user interface, subject's difference scores (first half-block minus second half-block) were submitted to a Friedman ranks test. For all three performance measures (time, moves, and direction errors), the test showed no significant difference in learning between the three interfaces.[1]

Another possibility worth checking is whether panning time and number of moves depends on the distance panned, and whether the dependency is affected by the panning user interface. It might be, for example, that panning time is directly proportional to distance for one panning user interface but not for another. For each subject, regressions were computed for panning distance *vs.* time and distance *vs.* moves for each of the three panning user interfaces.

Overall, both time and number of moves were positively related to the panning distance. This was determined by a simple sign test on the regression scores: many more of them were positive than would be expected by chance ($p < .01$). On the other hand, Friedman ranks tests applied to the regression scores showed no effect of panning interface type on the strength of the relationship between distance and either time or number of moves.

Panning Preference

Two of the twelve subjects, one male and one female, were accidentally not asked which of the panning user interfaces they preferred of those they had tried. Of the ten remaining subjects, eight preferred Push Background panning, two preferred Touch Edge Camera panning, and none preferred Push Camera panning. A multinomial calculation showed that this distribution is significantly skewed ($p < .01$).

The subjects who preferred Touch Edge Camera panning were subjects 4 and 8. Since neither of these subjects (in fact, no subjects at all; see Tables 1-3) performed better using Touch Edge Camera than with Push Background,

---

1. A learning-based decrease in direction errors was not possible with the Push Background user interface, because direction errors were almost nonexistent.

performance cannot be an explanation for their divergent preference.

Of the ten subjects who were asked to state a preference, five were male and five were female. The distribution of preference scores was exactly the same for males as for females: four out of five preferred Push Background, one out of five preferred Touch Edge Camera. Thus, no gender difference in preference was observed.

DISCUSSION AND CONCLUSIONS

Although the methodology used in the Panning Elicitation test was less formal than that used in the rest of the study, it seems safe to say that Push Background is the most common way people expect to pan a touch display given that they know nothing else about it. Even putting a visible border around the image did not seem to raise the likelihood that subjects would suggest Touch Edge panning instead of Push Background panning.

A possible follow-up to the Panning Elicitation test would be to show subjects a panning scene (without showing them a panning *user interface*, as in the present study) and then ask them: "How do you prefer to think of what just happened: Did the room shift to the left or did the camera view shift to the right?" In other words, subjects would be given a forced choice between two ways of thinking of panning instead of an open-ended question.

Push Background panning was the clear winner of the accuracy test: it resulted in faster times, fewer corrective movements, and fewer initial direction-errors than the other two panning methods tested. This seems to contradict the findings of Bury *et al.* [3] for scrolling keys and the eventual "decision" of the marketplace for scrollbars. Are touch-displays so different from mouse- and keyboard-controlled displays that opposite results make sense? It seems more likely that the important distinction is *not* whether panning/scrolling is controlled *via* keyboard, mouse, or touch. For example, in a touch-controlled system that provided scrollbars or on-screen scrolling "buttons," moving the view might well be more intuitive than moving the background, congruent with the findings of Bury *et al.* In the FirstPerson prototype, there was no separate scrolling control; subjects panned by swiping directly over the displayed scene. Perhaps scrollbars or scrolling keys "afford" camera/view panning (even in touch-displays), while on-scene panning "affords" background panning. Following up on this speculation would be a useful direction for future research.

Even when panning works by swiping directly on the scene (i.e, there is no separate panning/scrolling controller), the intuitiveness and efficiency of the panning user interface presumably depends on other aspects of the application's design, e.g., its user interface for moving displayed objects. Thus, the optimality of Push Background panning may also depend on other elements of the application's gesture-set. This is a another possible topic for future study.

Finally, Push Background was the preferred panning method for most subjects. Though subjects' preference must be influenced by their experience in the Panning Accuracy test,

the fact that two of ten voiced preferences that disagreed with their performance shows that preference for Push Background panning is not based purely on performance.

One advantage of Touch Edge panning (despite poorer performance) is that it requires less hand motion than Push panning does: users can just leave their hands poised above the edges of the screen. It is also possible that Touch Edge Camera wouldn't do so badly against Push Background if the task were grosser, e.g., just moving something offscreen on, or just moving something to the left half of the screen. Touch Edge (Camera or Background) may just be a difficult user interface for panning to an exact position. A less exact panning task is thus a good candidate for a follow-up study.

Touch Edge Background (i.e., touch the side that you want the background to move towards) was not included in this study because informal experience with scrollbars suggested (at least to the author) that it is counter-intuitive. However, the number of direction errors made with Touch Edge Camera and some subjects' comments that it is "backwards" suggest that not everyone would find Touch Edge Background to be a counter-intuitive panning method.

Overall, Push Background seems to be the best of the three panning user interfaces examined in this study. Based on these results, subsequent prototypes developed at FirstPerson used Push Background panning.

This study has shown that it is possible to design an intuitive, highly usable panning user interface for a touch-controlled display without using scrollbars or scrolling keys. Furthermore, the study's results suggests that for such touch-displays, Push Background panning is the right user interface to use unless a strong case can be made for an alternative.

ACKNOWLEDGEMENTS

The study reported in this paper was conducted at FirstPerson, Inc., a subsidiary of Sun Microsystems. The author is grateful for advice and support from his immediate work colleagues: Chuck Clanton, Emilie Young, Joe Palrang, Pat Caruthers, and Cindy Long. Several anonymous reviewers also made suggestions that were helpful in improving this paper.

REFERENCES

1. Beard, D.V., Brown, P., Hemminger, B.M., Misra, R. (1991) "Scrolling Radiologic Images: Requirements, Designs, and Analysis," *Proc. Intl. Symposium on Computer Assisted Radiology (CAR'91)*, West Germany: Springer-Verlag, pages 636-641.

2. Beck, D., Elkerton, J. (1988) "Development and Evaluation of Direct Manipulation Lists," *Proc. ACM Conference on Computer-Human Interaction (CHI'88)*, Washington, DC, pages 72-78.

3. Bury, K., Boyle, J.M., Evey, R.J., and Neal, A.S. (1982) "Windowing versus Scrolling on a Visual Display Terminal," *Human Factors*, 24(4), pages 385-394.

4. Gibson, J.J. (1977) "The Theory of Affordances," in R.E. Shaw and J. Bransford (eds.), *Perceiving, Acting, and Knowing*, Hillsdale, NJ: Erlbaum Associates.

5. Glassner, A.S. (1990) "A Two-Dimensional View Controller," *ACM Transactions on Graphics*, January, 9 (1), pages 138-141.

6. Hayes, W. (1963) *Statistics*, New York: Holt, Rinehart and Winston.

7. Johnson, J. (1991) "Effect of Modes and Mode Feedback on Performance in a Simple Computer Task," *Hewlett-Packard Laboratories Technical Report HPL-91-167*, [also presented as a poster at 1989 ACM Conference on Computer-Human Interaction (CHI'89)]

8. MacLean, A., Young, R.M., Moran, T.P. (1989) "Design Rationale: The Argument Behind the Artifact," *Proc. ACM Conference on Computer-Human Interaction (CHI'89)*, Austin, TX, pages 247-252.

9. Norman, D.A. (1988) *The Psychology of Everyday Things*, New York: Basic Books, Inc.

10. Plaisant, C. and Sears, A. (1993) "Touchscreen interfaces for alphanumeric text display," in B. Shneiderman (Ed.), *Sparks of Innovation in Human-Computer Interaction*, Norwood, NJ: Ablex Publ.

11. Potter, R.L., Weldon, L.J., Shneiderman, B. (1993) "Improving the accuracy of touchscreens: an experimental evaluation of three strategies," in B. Shneiderman (Ed.), *Sparks of Innovation in Human-Computer Interaction*, Norwood, NJ: Ablex Publ.

12. Sears, A., and Shneiderman, B. (1993) "High precision touchscreens: design strategies and comparisons with a mouse," in B. Shneiderman (Ed.), *Sparks of Innovation in Human-Computer Interaction*, Norwood, NJ: Ablex Publ.

13. Shneiderman, B. (1993) "Touchscreens now offer compelling uses," in B. Shneiderman (Ed.), *Sparks of Innovation in Human-Computer Interaction*, Norwood, NJ: Ablex Publ.

14. Smith, D.C., Irby, C., Kimball, R., and Verplank, B. (1982) "Designing the Star User Interface," *Byte*, April 1982, pages 242-280.

15. Sullivan, J. (1993) "Magic Cap," informal demonstration at Sun Microsystems, March.

Pre-Screen Projection:
From Concept to Testing of a New Interaction Technique

Deborah Hix [1,2]

[1] Dept. of Computer Science
Virginia Tech
Blacksburg VA 24061 USA
hix@vt.edu

James N. Templeman [2]

[2] Navy Center for AI
Naval Research Laboratory
Washington DC 20375 USA
templeman@itd.nrl.navy.mil

Robert J.K. Jacob [3]

[3] Dept. of Electrical Engineering
& Computer Science
Tufts University
Medford MA 02155 USA
jacob@cs.tufts.edu

ABSTRACT

Pre-screen projection is a new interaction technique that allows a user to pan and zoom integrally through a scene simply by moving his or her head relative to the screen. The underlying concept is based on real-world visual perception, namely, the fact that a person's view changes as the head moves. Pre-screen projection tracks a user's head in three dimensions and alters the display on the screen relative to head position, giving a natural perspective effect in response to a user's head movements. Specifically, projection of a virtual scene is calculated as if the scene were *in front of* the screen. As a result, the visible scene displayed on the physical screen expands (zooms) dramatically as a user moves nearer. This is analogous to the real world, where the nearer an object is, the more rapidly it visually expands as a person moves toward it. Further, with pre-screen projection a user can navigate (pan and zoom) around a scene *integrally*, as one unified activity, rather than performing panning and zooming as separate tasks. This paper describes the technique, the real-world metaphor on which it is conceptually based, issues involved in iterative development of the technique, and our approach to its empirical evaluation in a realistic application testbed.

KEYWORDS: Interaction techniques, empirical studies, pre-screen projection, egocentric projection, formative evaluation, user tasks, input devices and strategies, interaction styles, input/output devices, Polhemus tracker, visualization, metaphors, user interface component.

INTRODUCTION

Pre-screen projection [16] is a new interaction technique for integrally panning and zooming through a scene. On the surface, developing a new interaction technique such as pre-screen projection sounds as if it should be easy. However, once we devised the initial concept — specifically, panning and zooming a display relative to the user's head position (described later) — we found that designing and implementing it in the context of meaningful tasks in a

CHI' 95, Denver, Colorado, USA
0-89791-694-8/95/0005

realistic setting required a myriad of unanticipated design decisions, which had unobvious consequences. Because the design space for an interaction technique like this is so large, many of these decisions are best made based on very simple, almost continuous *formative evaluation* [9] — early and frequent empirical testing with users to improve a user interface design. We discuss this general design and evaluation process for inventing, implementing, and testing new interaction techniques, using examples of some of the more interesting specific design issues we encountered in developing pre-screen projection.

A key aspect of our work is that we want to go beyond simply creating new technology and assuming it, merely by its novelty, is inherently better for a user than existing technology. Instead, we continually prototype and evaluate with users to assess the effects of our new technology on human task performance. As the HCI community moves beyond WIMP (window, icon, menu, pointer) interfaces, the evolutionary cycle — conceptualization, implementation, and evaluation — we describe becomes essential for producing interaction techniques that are effective and efficient, not merely new and different.

RESEARCH IN INTERACTION TECHNIQUES

An *interaction technique* is a way of using a physical input/output device to perform a generic task in a human-computer dialogue [5]. It abstracts a class of interactive tasks, for example, selecting an object on a screen by point-and-click with a mouse, with the focus on generic tasks. The pop-up menu is an interaction technique. It is not a new device, but rather a way of using a mouse and graphic display to perform selection from a list.

In many applications, a user performs information retrieval and planning tasks that involve large quantities of spatially arrayed data, for example, presented on a map in a military command and control system. Typically, all available or even desired information cannot be legibly displayed on a single screen, so a user must zoom and pan. Zooming and panning are "overhead" tasks that bring the needed information into view; the user then has "real" tasks to perform using the presented information. A goal of our research is to minimize the mental and physical effort required for such overhead tasks, freeing the user to perform the real tasks with minimal distractions. By studying new means of communication to facilitate human-computer interaction, we develop devices and techniques to support these exchanges. Our research paradigm is to invent new interaction techniques, implement them in hardware and software, and then study them empirically to determine whether they improve human performance.

Interaction techniques provide a useful research focus because they are specific enough to be studied, yet generic enough to have practical applicability to a variety of applications. Interaction technique research also bridges a gap between computer science and psychology, especially human factors. But research into interaction techniques *per se* is relatively rare, and research in this area that combines technical innovation with empirical evaluation is unfortunately even rarer.

RELATED WORK

Other researchers have explored the coupling of a dynamic scene to head tracking [4, 17]. In the geometry for this "fish tank" projection, the scene is calculated either behind or within a few centimeters in front of the screen, creating a realistic illusion of 3D scenes. The visual effect of this placement appears very different to a user than pre-screen projection, which exaggerates perspective effects to provide a dramatic way to pan and zoom. Other work on progressive disclosure, a key feature of pre-screen projection, has been done by [6, 8].

Early work on interaction techniques is found in the Interaction Techniques Notebook section of the *ACM Transactions on Graphics*. Examples of other interaction technique work include see-through tools [2], a collection of 3D interaction techniques [15], and a collection of eye movement-based techniques [10]. However, much of this work emphasizes invention of new techniques over evaluation, especially evaluation of human performance. Some work that discusses evaluation includes fish tank virtual reality [17], the alphaslider [1], the Toolglass [12], marking menus [13], and eye tracking [18].

THE CONCEPT UNDERLYING PRE-SCREEN PROJECTION

Real-World Metaphor

The real-world metaphor upon which pre-screen projection is based is the fact that our view of the world changes as we move our head. We are completely familiar with how our view changes as we move about, and we control our view by positioning our head. For example, as we move nearer to objects, they appear larger, and appear smaller as we move away. Pre-screen projection uses a three-dimensional tracking device to monitor head movement and alters the view presented to the user (in a specific way, which we discuss later) based on head position. It is a new technique for integrally panning and zooming through a display, and for progressive disclosure and hiding of information under user control.

Using pre-screen projection, a user wears a light-weight helmet or headband with a three-dimensional Polhemus tracker mounted on the front, as shown in Figure 1. As the user moves from side to side, the display smoothly pans over the world view. As the user moves closer to or further

from the screen, the display smoothly zooms in and out, respectively. (Figure 2 shows a sketch of how the scene on the screen changes as the user moves.) Head tracking is used to control the screen view by making the viewpoint used to calculate a virtual scene correspond to a user's physical viewpoint.

Just as with any zoom or pan technique, pre-screen projection allows a user to selectively reveal different portions of the whole world view on the screen. For example, in a system that updates a geographic display as new information arrives, a user can use pre-screen projection to zoom in when a new situation occurs, to evaluate details, and to zoom out to see an overall view of the situation.

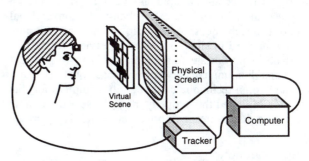

FIGURE 1. Pre-screen projection concept.
Note that the scene actually is displayed on the physical screen, but its dynamic perspective from the user's viewpoint reacts to the user's head movement as if the scene were in front of the screen.

Physical geometry is such that as we move toward a scene, objects in the scene enlarge in our vision; objects closer to us enlarge more rapidly than those in the background. To obtain a dramatic zoom effect in the virtual world presented by the computer, the scene (e.g., a map being panned and zoomed) must be treated as if it were close to the user. We therefore calculate the geometry of pre-screen projection as if the scene were located *in front of* the physical screen (i.e., nearer the user than the physical screen), as Figures 1 and 2 illustrate. This causes the scene to enlarge more rapidly than the screen as a user moves toward it and therefore produces a dramatic zoom. A user, unaware of pre-screen geometry, sees the visible scene displayed on the physical screen. The screen changes continuously (not discretely like some pan or zoom techniques) as the user moves.

In the real world, a person's view naturally pans and zooms simultaneously as the head moves in all three dimensions. However, in many current user interfaces that incorporate panning and zooming, a user performs panning or zooming tasks by moving a mouse or trackball, first to perform one task and then the other, effectively employing two different interaction techniques. This forces a user to perform separately what naturally are two integral tasks [11]. A user typically does not think of zooming or panning as separable tasks, but thinks rather of unified operations like "focus in on that area over there". A strong feature of pre-screen projection is that a user can navigate (pan and zoom) concurrently. We call this operation *integral pan and zoom*, to distinguish it from traditional two step approaches.

Further, because panning and zooming are closely coupled tasks in a user's mind, it is more natural for a user to make a gesture that performs the overall operation, using an integral 3D input device, rather than having to reach for or hold a specific device (e.g., mouse or trackball). Because a user of pre-screen projection does not have to perform a deliberate physical action, such as reaching for a mouse, to acquire the control device (the Polhemus 3D tracker), another feature is that integral panning and zooming is enabled at all times.

Another benefit is the "lightweight" nature of integral pan and zoom obtained using head tracking and rapid, continuous display update. Moving the head toward the screen momentarily to read a portion of the display and then moving away becomes a rapid and natural operation, with less physical and cognitive effort than more conventional zoom or pan techniques. As a result, zooming and panning can now be used in situations where "heavier" techniques would be inappropriate (e.g., zooming in to read a text legend and then immediately zooming back out). Use of a "heavier" zoom technique could be too disruptive for this task. We exploited this property through user-controllable progressive disclosure of information, which is described later.

Visual Perception

Pre-screen projection derives from an understanding of J.J. Gibson's [7] ecological approach to visual perception. Gibson suggests that the visual system is an organ that allows people to acquire useful information about their environment. A person's view of the world changes every time they move their head, and each view discloses new information about both the environment and the user's place in it. People perceive relationships between their movements and the views they see. A knowledge of these relationships is the basis for visual perception. For example, as mentioned earlier, we expect physical objects to appear larger as we move toward them. We use this knowledge to interact with the real world. A dynamic view is the norm in the real world, unlike the static views typically presented on a computer screen. Pre-screen projection carries many dynamic properties of the visual world into the world of computer-generated imagery, because of the real-world metaphor we have adopted. Although it might seem odd to invoke three-dimensional viewing techniques just to shift and scale a flat picture on the screen, by doing so we gain a natural technique for performing head-based integral panning and zooming of the image on the screen.

ITERATIVE DEVELOPMENT OF PRE-SCREEN PROJECTION

Having developed the concept of pre-screen projection, we then began the challenge of developing the interaction technique itself, which includes design, implementation, continuous iterative evaluation, and refinement. As we proceeded with the design, we discovered an enormous design space of possible attributes. Each attribute could have several (and in some cases, many) values, so the combinatorics and interactions of the numerous design decisions quickly became very large. Thus, early design decisions had to be made based on rapid prototyping and quick, informal formative evaluation cycles, rather than on comprehensive experimental evaluation. Below we describe this process for two attributes — scaling and fade-in/out — for which the choice of values was very large and final design decisions were counterintuitive or otherwise unexpected.

Scaling

Scaling — the geometry by which objects in a scene appear larger or smaller as a user zooms (in pre-screen projection, as a function of head movement perpendicular to the screen) — was one of our biggest challenges. Three types of objects in the scene could potentially be designed to scale: the map, icons, and text.

In all design iterations, the map scaled, because the map provides the predominant spatial context and expected visual perspective. Our testbed is a Naval command and control application (briefly described later) in which icons represent military objects such as ships and planes. In our earliest designs, icons did not scale because scaling icons on a map might imply that their size has meaning. In our application, the size of icons conveys no meaning, although it could in other applications. But in formative evaluation we observed that users found it difficult to zoom in on non-scaling icons because their sense of perspective expansion was reduced. So in our final design, icons are scaled.

We chose not to scale text, to maximize the amount of readable text presented to the user at any time. If text is scaled, when the user's head is close to the viewing plane the text can get too large to read, and, conversely, as the user's head moves further away, the text decreases in size and becomes too small to read. We wondered if our design in which the map and icons scale but text does not would be confusing or unnatural to users, but we found in formative evaluation that users were comfortable with this design.

In early designs, we used a scaling factor of real-world linear perspective, $1/z$ (where z is the distance from the user's head to the virtual scene), for the map. When we applied this geometry to make icons scale, they were almost invisible until the user's head was quite close to the viewing plane. That is, a ship in accurate relative scaled size to Europe became a tiny dot when the user had a reasonable portion of Europe in the scene using pre-screen projection. So we redesigned again, by making icons disproportionally large relative to the map.

Yet another issue surfaced with the $1/z$ scaling: a user could not back far enough away (i.e., zoom out) to get the entire world view onto the screen. So we tried various computations for amplifying the scaling as a user moves further away until we found one that worked well for scaling both the map and the icons. We used linear perspective within a 1 foot distance beyond the virtual scene, and beyond that, amplified the scaling by an additional linear factor. In the final design an entire map of Europe and Africa appears when the user is at a distance of about 3 feet from the virtual scene.

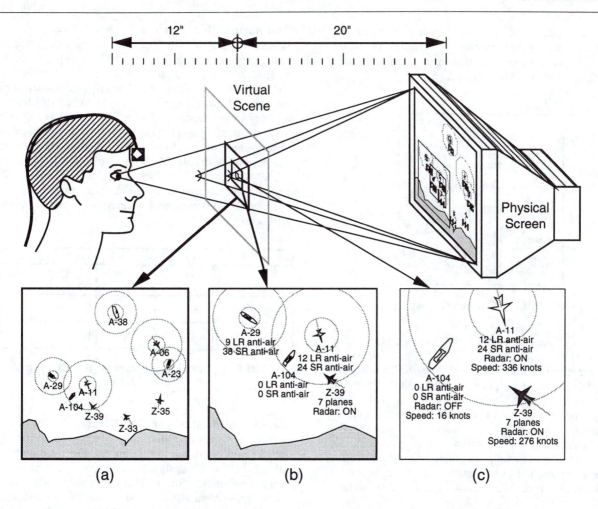

FIGURE 2.   Example of scaling (of map and icons) and fade-in (of text).

Figure 2(a), the first level of fade-in/out, depicts the scene (seen by the user on the physical screen) when the user's head is 12 inches from the virtual scene. At this level, only the military ID is shown for each icon. Figure 2(b), the second level of fade-in/out, shows the screen as the user moves in closer to look at a specific engagement. Additional lines of text appear for each icon. Transition among displays occurs continuously and smoothly, not discretely. In Figure 2(c), the third level of fade-in/out, the user is even closer, to examine further details. (For illustrative purposes here, formats have been altered; in our testbed design, icons take up a smaller portion of the screen, and there are more lines of text.)

Fade-in/out

Fade-in and -out — the gradual appearance and disappearance of objects in a scene — were somewhat easier to design than scaling, but we still made numerous iterations with users before we settled on a design. As with scaling, the same three types of objects — the map, icons, and text — could potentially be designed to fade-in/out.

In all design iterations, the map and icons scaled but did not fade in/out, because of their role in maintaining spatial context. However, text associated with each icon in a scene fades in and out at three different *levels of disclosure*, as a user moves toward and away from the screen. In the first level of fade-in, an icon identifier appears (a military ID), and in each of the next two levels,

additional icon-specific information about weapons, speed, fuel, and so on, appears. The first level of fade-in/out gradually occurs between 13 and 12 inches from the virtual scene; the second level gradually fades in/out between 4 and 3 inches from the virtual scene, and the third level between 2 and 1 inches. A calculated perspective for the virtual scene at 20 inches in front of the physical screen works well for these levels of disclosure. Figure 2 shows a simple example of how scaling and fading in pre-screen projection allow successive disclosure (or hiding) of details as a user moves toward (or away from) the screen.

Figure 3 shows what a display would look like if all information were displayed at once. Pre-screen projection allows a user to control this kind of clutter, by selectively presenting details as the user moves nearer to and further from the screen. Although no one would purposefully design a display to look like this, in a dynamic system this muddle could unpredictably occur.

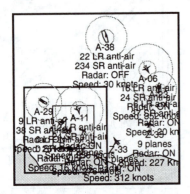

FIGURE 3.   Cluttered display.

We frankly chose distances for the levels of progressive disclosure based on trial and error, but formative evaluation showed that those just described work well. However, we were surprised at the relatively small quantity of text (about 8 to 10 lines per icon) that can be displayed before the text lines for different icons overlap substantially as a user gets close in. We had expected that, because icons move visually further apart as a user gets closer to the viewing plane, there would be space for a rather large amount of information (in our case textual information, but it could also have been graphical) to fade in.

We found that even something as simple as spacing between lines of text must be carefully designed and evaluated. Since text does not scale but the distance between lines of text does, in early designs of fade-in/out, text was too densely packed even after a user zoomed far in. We realized that line spacing must be computed in screen units rather than world map coordinates, even though placement of a block of text is tied to a fixed location on the map. We could have anticipated this during design sessions, but with the large number of design decisions to consider, it is all too easy to overlook or neglect such details. Implementation and user comments quickly make such oversights glaringly obvious.

Also in early stages of design, text at one level faded out as different text from the next level replaced it. We tried this because we wanted to display as much information as possible, which meant re-using screen space. However, users found it hard to determine which level they were in as they looked for specific information. We changed the design to fade in only additional text, beneath (rather than on top of the location of) text displayed at prior levels. We found for our testbed, as mentioned above, that a total of 8 to 10 lines (we use 1 line at the first level and 4 additional lines at both the second and third levels) was the maximum amount that could be displayed and still be readable. This design could vary for different applications.

Counterintuitive Results in Final Design

It is interesting to note that our final design for scaling and fade-in/out of text is exactly the opposite of how scaling and fading naturally work in the real world, as shown below in Table 1. Namely, as a user moves toward and away from text in the real world, that text appears to get larger and smaller (scales) but does not disappear altogether (fade). Based on empirical observations during formative evaluation of pre-screen projection, as discussed above, we found it works best when text fades but does not scale. *This kind of unexpected design decision could only have been corroborated by empirical observations and evaluations with users.*

	In the real world		In our final design	
	Text	Icons/ Map	Text	Icons/ Map
Scaling	Y	Y	N	Y
Fade-in/out	N	N	Y	N

TABLE 1.   Design decisions for scaling and fade-in/out.

Design Attributes for Pre-Screen Projection

While our previous discussion revolved around the attributes of scaling and fade-in/out, there are numerous other factors in the design space for pre-screen projection. Very few interaction techniques are reported in the literature with an attribute list; some notable exceptions are [2, 3, 14]. Some attributes and a few of their possible values are as follows:

- *Scaling* (e.g., type, what is scaled, interaction with fade-in/out)

- *Fade-in/out* (e.g., type, what fades in/out, how many levels, where on screen)

- *Context "world view" presentation* (e.g., miniature map, size, location, information content)

- *Interactions among attributes* (e.g., scaling & fading)

- *Freeze/unfreeze of dynamic screen image* (e.g., how performed by user, relative vs. absolute)

This latter attribute, freeze/unfreeze, is an interesting one for which we are still exploring possibilities. Obviously, with the constantly changing scene afforded by pre-screen projection, a user may, for various reasons, want to momentarily halt movement of the scene. We have incorporated a freeze feature through which, simply by pressing a button, a user can decouple head movement from scene movement. The user can continue to view the static scene as it appears when frozen, while moving the head freely. When the user wishes to unfreeze the scene, re-engaging the coupling between head movement and scene changes, the user again presses a button. However, a freeze breaks the coupled relationship between a user's head and the scene, which raises the question of how to re-establish the relationship when the user performs an unfreeze. We have designed two choices: an absolute unfreeze and a relative unfreeze.

In absolute unfreeze, what a user sees changes to reflect the user's new head position when the unfreeze is performed, resulting in a slow drifting of the scene on the screen to the new perspective. An absolute unfreeze adjusts the virtual scene to that which would have been seen at the new user's head position (at the time of the unfreeze) if dynamic perspective had never been frozen. This maintains the placement and scale of the virtual scene in front of the screen.

In relative unfreeze, what a user sees does not change to reflect the new head position; to accommodate this, the virtual scene is computationally adjusted (shifted and scaled) to maintain its appearance to the user at the new head position. The scene on the screen resumes from where it was when the user performed the freeze, regardless of where the user's head is when the unfreeze is performed. Currently, a user can choose between either of these types of unfreeze, and further evaluation will give us more indications of which is preferable in which situations.

DEVELOPING TASKS, APPLICATION, AND EMPIRICAL STUDIES

As we developed pre-screen projection and incorporated it into a testbed application, we found the major components and relationships in the process of developing an interaction technique to be as shown in Figure 4. Interaction techniques research often stops after the technique has been implemented and tested to make sure that it works. The remaining three components are ignored. We have described the first component, iterative development of an *interaction technique*. However, our work proceeds within the context of this overall development process for interaction techniques, because we want to evaluate effects of new interaction techniques on human performance, not simply see if they operate correctly. It is thus necessary to define *tasks* for a user to perform with the new technique, to set those tasks in an *application* that can be used in an empirical study, and to design the *empirical study* itself. This approach, once again, led us to complexities we had not expected.

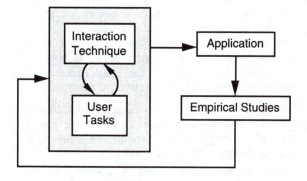

FIGURE 4. Relationship among major components in interaction technique development process.

User Tasks and Application

Panning and zooming are inherently spatial information-seeking activities. A user can, for example, casually browse or actively search for specific information. A user can seek different kinds of information. For example, high-level results of a library search might be displayed as a scatterplot through which a user navigates by panning and zooming to find details about a particular document in the results set. Panning and zooming are pervasive and fundamental activities in many applications, including scientific or information visualization, spatial or geographic presentation, and virtual reality. The way in which a user performs panning and zooming tasks shapes the relationship between that user and a computer environment. To evaluate pre-screen projection, we needed to create user tasks that would capture the essential nature of panning and zooming.

We co-evolved user tasks along with the interaction technique, to produce the best tasks not for the sake of the application, but for evaluating pre-screen projection in a meaningful situation. In fact, we had numerous design iterations of both the technique and the tasks used to evaluate it. We chose Naval command and control (C²) systems as a rich domain relevant to Naval applications. C² systems support the planning, coordination, and execution of military missions. We incorporated pre-screen projection into a C²-like testbed running on a Silicon Graphics Iris 4D/210VGX workstation, and created task scenarios for evaluation of the technique. The purpose of this testbed is to serve as a context for evaluating interaction techniques, not to develop the best C² system.

An interaction technique and goals for its evaluation should drive task development. The task, in turn, influences details of the instantiation of the interaction technique (e.g., in the task of allocating weapons, designing what textual information to display and how to display it for a ship icon). Tasks used for evaluation of interaction techniques are often extremely simplistic (for example, in Toolglass evaluation [12], the experimental task was a simple colorized version of connect-the-dots). This is obviously much easier, but it may be a potentially risky reduction of more complex tasks. Nevertheless, any evaluation is better than none.

We found that important criteria for designing tasks to evaluate pre-screen projection included:

- must have both strong spatial and navigational components (to exploit integral panning and zooming),

- must involve more information than can be presented on a static screen (to evaluate fading and scaling),

- must be simple, quick, and easy to learn (because participants are generally non-military types), and

- must be as deterministic as possible with a minimum of variability caused by individual differences in user strategy.

To meet these criteria, we developed a defensive engagement situation, in which friendly, enemy, unknown, and neutral military units (ships and planes) are displayed on a map of the Mediterranean. We created several scenarios in which multiple enemy, unknown, and neutral units appeared during the scenario, moving along a course to intersect friendly units. Figure 2 gives a flavor of how the screen looks and changes. Circles around each friendly unit (denoted by an A) indicate the range of its missiles. We used realistic (unclassified) military data when feasible in the scenarios.

The user's goals (tasks) were to acquire and maintain an awareness of the evolving situation, and to allocate missiles from friendly units to encroaching enemy planes. Specifically, a user monitors the scene looking for threats. When one is observed, the user determines the number of planes in that threat (by zooming in on textual details that progressively fade-in as described earlier) and the number of missiles the friendly unit has available to fire. The user then uses a slider to enter the number of missiles to shoot and, based on simple "rules of engagement" (see below), then fires those missiles. This series of tasks is time-critical (to detect and eliminate enemy planes as soon as possible), and therefore lends itself well to quantitative metrics such as time to perform tasks and accuracy of task performance. Entering the number of missiles and issuing a fire command are unrelated to panning and zooming, but indicate task closure for measuring human performance.

We formulated our "rules of engagement" by which a user knows how to allocate missiles, to preclude, as much as possible, interference of individual user strategies for performing tasks, as well as the need to understand complicated military rules. For example, one rule that was overly simplified from a real military situation is that each allocated missile will bring down one aircraft with a probability of 100%.

Empirical Studies

Empirical evaluation is a key component in developing a new interaction technique. Best guesses about design are substantiated or refuted by many tight, short cycles of formative evaluation. We have already performed numerous cycles of formative evaluation — some as short as five minutes (these were the really bad designs!), others lasting nearly an hour. Evolution of designs for scaling and fading of the map, icon, and text, which were described previously, as well as many other design details, evolved from numerous rounds of formative evaluation.

Many of our formative evaluation results were previously discussed when we explained the design of scaling and fade-in/out. There were other counterintuitive and surprising results. We expected a slight illusion of three-dimensionality from the perspectively correct visual transformations, but it turned out not to be there. Despite this, users claim that pre-screen projection provides a natural way to navigate a scene. Users are comfortable with the technique and generally can use it effectively within about five minutes. Some users reported that after using pre-screen projection awhile, they found the static screen of a standard system to be "dead".

One interesting experience occurred with a user wearing bifocal glasses. As the user moved her head in close

enough to see details, she also had to tip her head backward to use the lower portion of her bifocals. Obviously, this caused the scene on the screen to shift completely away from where the user was trying to look! We are still considering possible solutions to this difficulty.

An ergonomic issue to be explored is that of fatigue due to carefully controlled head movements over long periods of use of pre-screen projection. Clearly, forcing users to hold their head or upper body at a particular position in order to maintain the desired scene could cause serious problems, even being harmful to users over long periods of time. The freeze/unfreeze feature mentioned previously is one mechanism that we will further investigate as a possibility for overcoming this potential ergonomic problem.

Although this paper discussed our formative evaluation, summative evaluation is also important. *Summative evaluation* [9] is empirical evaluation in which several designs are compared, with primarily quantitative results. In our research, the purpose of summative evaluation is to perform comparative studies of user performance with different interaction techniques for the same tasks. A next step in our research is to perform summative evaluation, comparing user performance and satisfaction using pre-screen projection to other pan and zoom techniques, using the defensive engagement scenarios. These scenarios can be used to evaluate any pan and zoom technique, particularly those involving progressive disclosure of information.

CONCLUSIONS

We learned several important lessons during design, implementation, and evaluation of pre-screen projection. Those of most general interest include:

• When incorporating interaction techniques into an application, we found deliberate violation of user interface design guidelines was sometimes necessary to construct an effective evaluation situation.

For example, we wanted to determine how users use pre-screen projection for navigation when monitoring a situation, so we did not include any audible notification of a new enemy, even if it was off the screen when it appeared. We wanted to see how a user maintains vigilance and develops situational awareness with pre-screen projection. Further, we allowed a user to fire an incorrect number of missiles (more or less than needed), because we wanted to make sure a user can access detailed information (about numbers of missiles, for example).

• We struggled constantly to keep the application testbed simple, finding it all too easy to develop an unnecessarily elaborate testbed design (i.e., C2 system) and to lose sight of evaluating the interaction technique.

The current version of the testbed is so simplistic that most tasks performed by a user could be automated. If all tasks were automated, there obviously would be no need for an interaction technique! We had to keep it simple because real C2 systems take months of training, which is clearly inappropriate for our needs.

- As with any experiment, user performance variability among participants needs to be minimized and/or controlled as much as possible, and experimental tasks need to be deterministic.

This is particularly important for measuring human performance with a novel interaction technique. This constraint limits both the tasks a user can perform during evaluation and also may limit design of the interaction technique itself. Attribute values of the interaction technique must be controllable by the experimenter.

- We often got wildly unexpected results when our continually evolving designs for pre-screen projection were implemented.

We found that even our best guesses, based on expertise, experience, and good design guidelines, did not always work as expected. As discussed, fade-in/out of text was a surprise. The combinatorics of alternative design details (e.g., color, font, spacing, icon shape) were huge. Nonetheless, the devil really is in the details. For example, in an early design, all text was the same color (white), and we found that when icons were close enough for their associated text to overlap, it was not possible to tell which text went with which icon. By tinting the associated text when a specific friendly (blue) and a specific enemy (red) icon were selected, users could then tell which text went with which icon. This small change made a previously impossible task do-able.

- We tried very hard to avoid having issues, constraints, and difficulties with the implementation of pre-screen projection affect design decisions for the technique.

There were a few times when this was unavoidable. For example, using high-resolution data for the map caused (not unexpectedly) the screen to update too slowly and to jitter, so we simplified the map outline. Design of pre-screen projection has led us through many of the same issues involved in designing virtual reality applications.

SUMMARY

Our development of *pre-screen projection* has shown that an interaction technique based on a real-world metaphor can provide a natural, useful means of human-computer interaction. But developing new interaction techniques is difficult, because their design space can be large, implementation complicated, and evaluation time-consuming and resource-intensive. The evolutionary cycle — conceptualization, implementation, and evaluation — we have presented is nevertheless essential for producing interaction techniques that are effective and efficient, not merely new and different.

ACKNOWLEDGMENTS

Kapil Dandekar, Ankush Gosain, Steven Kerschner, and Jon Yi were super-implementers for the testbed. Linda Sibert and Scott Powell contributed many excellent design ideas. Dr. Rex Hartson provided invaluable comments on an early draft of this paper. This work is sponsored by the Decision Support Technology block (RL2C) within the ONR Exploratory Development Program, and is managed by Dr. Elizabeth Wald.

REFERENCES

1. Ahlberg, C., and B. Shneiderman. (1994). The Alphaslider: A Compact and Rapid Selector. *Proc. CHI'94 Conf.*, 365-371.

2. Bier, E.A., M.C. Stone, K. Fishkin, W. Buxton, and T. Baudel. (1994). A Taxonomy of See-Through Tools. *Proc. CHI'94 Conf.*, 358-364.

3. Bleser, T.W., and J.L. Sibert. (1990). Toto: A Tool for Selecting Interaction Techniques. *Proc. UIST'90 Symp.*, 135-142.

4. Deering, M. (1992). High Resolution Virtual Reality. *Computer Graphics*, 26(2), 195-202.

5. Foley, J.D., A. van Dam, S.K. Feiner, and J.F. Hughes. (1990). *Computer Graphics: Principles and Practice*. Addison-Wesley, Reading, MA.

6. Furnas, G.W. (1986). Generalized Fisheye Views. *Proc. CHI'86 Conf.*, 16-23.

7. Gibson, J.J. (1986). *The Ecological Approach to Visual Perception*. Lawrence Erlbaum Associates, Hillsdale, NJ.

8. Herot, C.F., R. Carling, M. Friedell, and D. Kramlich. (1980). A Prototype Spatial Data Management System. *Computer Graphics*, 14(3), 63-70.

9. Hix, D., and H.R. Hartson. (1993). *Developing User Interfaces: Ensuring Usability through Product & Process*. John Wiley & Sons, Inc., New York, NY.

10. Jacob, R.J.K. (1991). The Use of Eye Movements in Human-Computer Interaction Techniques: What You Look At Is What You Get, *ACM Trans. Info. Sys.*, 9(3), 152-169.

11. Jacob, R.J.K., and L.E. Sibert. (1992). The Perceptual Structure of Multidimensional Input Device Selection. *Proc. CHI'92 Conf.*, 211-218.

12. Kabbash, P., W. Buxton, and A. Sellen. (1994). Two-Handed Input in a Compound Task. *Proc. CHI'94 Conf.*, 417-423.

13. Kurtenbach, G., A. Sellen, and W. Buxton. (1993). An Empirical Evaluation of Some Articulatory and Cognitive Aspects of Marking Menus. *Human-Computer Interaction*, 8, 1-23.

14. Mackinlay, J.D., S.K. Card, and G.G. Robertson. (1990). A Semantic Analysis of the Design Space of Input Devices. *Human-Computer Interaction*, 5, 145-190.

15. Snibbe, S., K.P. Herndon, D.C. Robbins, D.B. Conner, and A. van Dam. (1992). Using Deformations to Explore 3D Widget Design. *Computer Graphics*, 26(2), 351-352.

16. Templeman, J.N. (1993). Pre-Screen Projection of Virtual Scenes. *Proc. Fall'93 VR Conf.*, 61- 70.

17. Ware, C., K. Arthur, and K.S. Booth. (1993). Fish Tank Virtual Reality. *Proc. InterCHI'93 Conf.*, 37-41.

18. Ware, C., and H.T. Mikaelian. (1987). An Evaluation of an Eye Tracker as a Device for Computer Input. *Proc. CHI+GI'87 Conf.*, 183-188.

SPACE-SCALE DIAGRAMS:
UNDERSTANDING MULTISCALE INTERFACES

George W. Furnas

Bellcore, 445 South Street

Morristown, NJ 07960-6438

(201) 829-4289

gwf@bellcore.com

Benjamin B. Bederson*

Bellcore, 445 South Street

Morristown, NJ 07960-6438

(201) 829-4871

bederson@bellcore.com

ABSTRACT

Big information worlds cause big problems for interfaces. There is too much to see. They are hard to navigate. An armada of techniques has been proposed to present the many scales of information needed. Space-scale diagrams provide an analytic framework for much of this work. By representing both a spatial world and its different magnifications explicitly, the diagrams allow the direct visualization and analysis of important scale related issues for interfaces.

KEYWORDS: Zoom views, multiscale interfaces, fisheye views, information visualization, GIS; visualization, user interface components; formal methods, design rationale.

INTRODUCTION

For more than a decade there have been efforts to devise satisfactory techniques for viewing very large information worlds. (See, for example, [6] and [9] for recent reviews and analyses). The list of techniques for viewing 2D layouts alone is quite long: the Spatial Data Management System [3], Bifocal Display[1], Fisheye Views [4][12], Perspective Wall [8], the Document Lens [11], Pad [10], and Pad++ [2], the MacroScope[7], and many others.

Central to most of these 2D techniques is a notion of what might be called multiscale viewing. An interface is devised that allows information objects and the structure embedding them to be displayed at many different magnifications, or scales. Users can manipulate which objects, or which part of the whole structure, will be shown at what scale. The scale may be constant and manipulated over time as with a zoom metaphor, or varying over a single view as in the distortion techniques (e.g., fisheye or bifocal metaphor). In either case, the basic assumption is that by moving through space and changing scale the users can get an integrated notion of a very large structure and its contents, navigating through it in ways effective for their tasks.

This paper introduces *space-scale* diagrams as a technique for understanding such multiscale interfaces. These diagrams make scale an explicit dimension of the representation, so that its place in the interface and interactions can be visualized, and better analyzed. We are finding the diagrams useful for understanding such interfaces geometrically, for guiding the design of code, and as interfaces to authoring systems for multiscale information.

This paper will first present the necessary material for understanding the basic diagram and its properties. Subsequent sections will then use that material to show several examples of their uses.

THE SPACE-SCALE DIAGRAM

The basic diagram concepts

The basic idea of a space-scale diagram is quite simple. Consider, for example, a square 2D picture (Figure 1a). The space-scale diagram for this picture would be obtained by creating many copies of the original 2D picture, one at each possible magnification, and stacking them up to form an inverted pyramid (Figure 1b). While the horizontal axes rep-

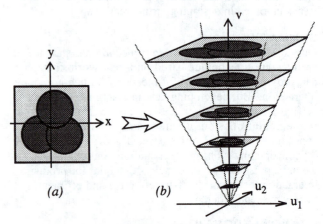

Figure 1. *The basic construction of a Space-Scale diagram from a 2D picture.*

*Current address: bederson@cs.unm.edu, Computer Science Department, University of New Mexico, Albuquerque, NM 87131

resent the original spatial dimensions, the vertical axis represents scale, i.e., the magnification of the picture at that level. In theory, this representation is continuous and infinite: all magnifications appear from 0 to infinity, and the "picture" may be a whole 2D plane if needed.

Before we can discuss the various uses of these diagrams, three basic properties must be described. Note first that a user's viewing window can be represented as a fixed-size horizontal rectangle which, when moved through the 3D space-scale diagram, yields exactly all the possible pan and zoom views of the original 2D surface (Figure 2). This property is useful for studying pan and zoom interactions in continuously zoomable interfaces like Pad and Pad++ [2][10].

Secondly, note that a point in the original picture becomes a ray in this space-scale diagram. The ray starts at the origin and goes through the corresponding point in the continuous set of all possible magnifications of the picture (Figure 3). We call these the *great rays* of the diagram. As a result, regions of the 2D picture become generalized cones in the diagram. For example, circles become circular cones and squares become square cones.

A third property follows from the fact that typically the properties of the original 2D picture (e.g., its geometry) are considered invariant under moving the origin of the 2D coordinate system. In the space-scale diagrams, such a change of origin corresponds to a "shear" (Figure 4), i.e., sliding all the horizontal layers linearly so as to make a different great ray become vertical. Thus, if one only wants to consider properties of the original diagram that are invariant under change of origin, the only meaningful properties of the space-scale diagram are those invariant under such a shear. For example, the absolute angles between great rays change with shear, and so should be given no special meaning.

Now that the basic concepts and properties of space-scale diagrams have been introduced by the detailed Figures 1-4,

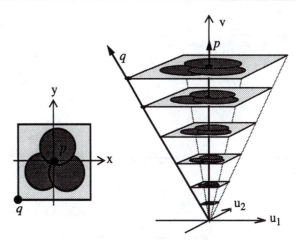

Figure 3. *Points like p and q in the original 2D surface become corresponding "great rays" p and q in the space-scale diagram. (The circles in the picture therefore become cones in the diagram, etc.)*

we can make a simplification. Those figures have been three dimensional, comprising two dimensions of space and one of scale ("2+1D"). Substantial understanding may be gained, however, from the much simpler two-dimensional versions, comprising one dimension of space and one dimension of scale ("1+1D"). It could, for example be a vertical slice from, or an edge on view of, the 2+1D version, or just a space-scale view of a truly 1D world (e.g., a time line). In the 1+1D diagram, since the spatial world is 1D, a viewing window is a line segment that can be moved around the diagram to repre-

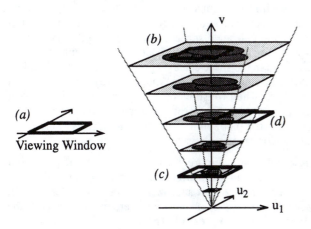

Figure 2. *The viewing window (a) is shifted rigidly around the 3D diagram to obtain all possible pan/zoom views of the original 2D surface, e.g., (b) a zoomed in view of the circle overlap, (c) a zoomed out view including the entire original picture, and (d) a shifted view of a part of the picture.*

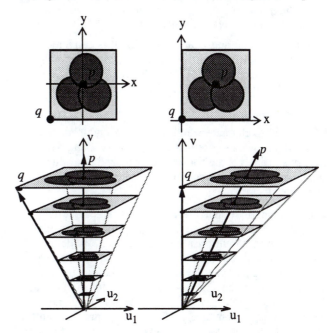

Figure 4. *Shear invariance. Shifting the origin in the 2D picture from p to q corresponds to shearing the layers of the diagram so the q line becomes vertical. Each layer is unchanged, and great rays remain straight. Only those conclusions which remain true under all such shears are valid.*

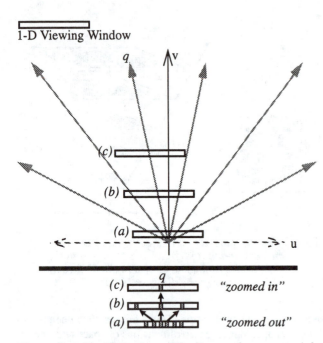

1-D Viewing Window

q   *v*

(c)

(b)

(a)

u

q

(c) — "zoomed in"

(b)

(a) — "zoomed out"

Figure 5. *A "1+1D" space-scale diagram has one spatial dimension, u, and one scale dimension, v. The six great rays here correspond to six points in a 1D spatial world, put together at all magnifications. The viewing window, like the space itself, is one dimensional, and is shown as a narrow slit with the corresponding 1-D window view being visible through the slit. Thus the sequence of views (a), (b), (c) begins with a view of all six points, and then zooms in on the point q. The views, (a), (b), (c) are redrawn at bottom to show the image at those points.*

sent different pan and zoom positions. It is convenient to show the window as a narrow slit, so that looking through it shows the corresponding 1D view. Figure 5 shows one such diagram illustrating a sequence of three zoomed views.

The basic math.

It is helpful to characterize these diagrams mathematically. This will allow us to use analytic geometry along with the diagrams to analyze multiscale interfaces, and also will allow us to map conclusions back into the computer programs that implement them.

The mathematical characterization is simple. Let the pair (x, z) denote the point x in the original picture considered magnified by the multiplicative scale factor z. We define any such (x, z) to correspond to the point (u, v) in the space-scale diagram where $u=xz$ and $v=z$. This second trivial equation is needed to make the space-scale coordinates distinct, and because there are other versions of space-scale diagrams, e.g., where $v=\log(z)$. Conversely, of course, a point (u, v) in the space-scale diagram corresponds to (x, z), i.e., a point x in the original diagram magnified by a factor z, where $x=u/v$, and $z=v$. The notation is a bit informal, in that x and u are single coordinates in the 1+1D version of the diagrams, but a sequence of two coordinates in the 2+1D version.

A few words are in order about the XZ vs. UV characterizations. The (x,z) notation can be considered a world-based

coordinate system. It is important in interface implementation because typically a world being rendered in a multiscale viewer is stored internally in some fixed canonical coordinate system (denoted with x's). The magnification parameter, z, is used in the rendering process. Technically one could define a type of space-scale diagram that plots the set of all (x,z) pairs directly. This "XZ" diagram would stack up many copies of the original diagram, all of the same size, i.e., *without* rescaling them. In this representation, while the picture is always constant size, the viewing window must grow and shrink as it moves up and down in z, indicating its changing scope as it zooms. Thus while the world representation is simple, the viewer behavior is complex. In contrast, the "UV" representation of the space-scale diagrams focused on in this paper can be considered view-based. Conceptually, the world is statically prescaled, and the window is rigidly moved about. The UV representation is thus very useful in discussing how the views should behave. The coordinate transform formulas allow problems stated and solved in terms of view behavior, i.e., in the UV domain, to have their solutions transformed back into XZ for implementation.

EXAMPLE USES OF SPACE-SCALE DIAGRAMS

With these preliminaries, we are prepared to consider various uses of space-scale diagrams. We begin with a few examples involving navigation in zoomable interfaces, then consider how the diagrams can help visualize multiscale objects, and finish by showing how other, non-zoom multiscale views can be characterized.

Pan-zoom trajectories

One of the dominant interface modes for looking at a large 2D world is to provide an undistorted window onto the world and allow the user to pan and zoom. This method is used in [2][3][7][10], as well as essentially all map viewers in GISs (geographic information systems). Space-scale diagrams are a very useful way for researchers studying interfaces to visualize such interactions, since moving a viewing window around via pans and zooms corresponds to taking it on a trajectory through scale-space. If we represent the window by its midpoint, the trajectories become curves and are easily visualized in the space-scale diagram. In this section, we first show how easily space-scale diagrams represent pan/zoom sequences. Then we show how they can be used to solve a very concrete interface problem. Finally we analyze a more sophisticated pan/zoom problem, with a rather surprising information theoretic twist.

Basic trajectories. Figure 6 shows how the basic pan-zoom trajectories can be visualized. In a simple pan (a), the window's center traces out a horizontal line as it slides through space at a fixed scale. A pure zoom around the center of the window follows a great ray (b), as the window's viewing scale changes but its position is constant. In a "zoom-around" the zoom is centered around some fixed point other than the center of the window, e.g., q at the right hand edge of the window. Then the trajectory is a straight line parallel to the great ray of that fixed point. This moves the window so that the fixed point stays put in the view. In the figure, for example, the point, q, always intersects the windows on trajectory (c) at the far right edge, meaning that the point, q, is always at that

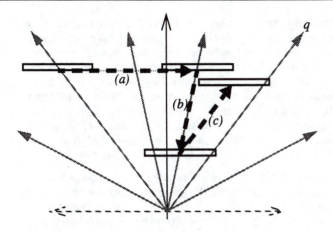

Figure 6. *Basic Pan-Zoom trajectories are shown in the heavy dashed lines:. (a) Is a pure Pan,. (b) is a pure Zoom (out), (c) is a "Zoom-around" the point q.*

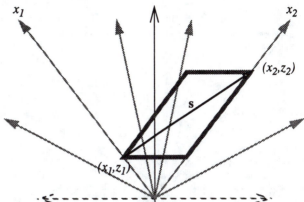

Figure 7. *Solution to the simple joint pan-zoom problem. The trajectory* **s** *monotonically approaches point 2 in both pan and zoom.*

position in the view. If as in this case the fixed point is itself within the window, we call it a zoom-around-within-window or *zaww*. Other sorts of pan-zoom trajectories have their characteristic shapes as well and are hence easily visualized with space-scale diagrams.

The joint pan-zoom problem. There are times when the system must automatically pan and zoom from one place to another, e.g., moving the view to show the result of a search. Making a reasonable joint pan and zoom is not entirely trivial. The problem arises because in typical implementations, pan is linear at any given scale, but zoom is logarithmic, changing magnification by a constant factor in a constant time. These two effects interact. For example, suppose the system needs to move the view from some first point (x_1, z_1) to a second point (x_2, z_2). For example, a GIS might want to shift a view of a map from showing the state of Kansas, to showing a zoomed in view of the city of Chicago, some thousand miles away. A naive implementation might compute the linear pans and log-linear zooms separately and execute them in parallel. The problem is that when zooming in, the world view expands exponentially fast, and the target point x_2 runs away faster than the pan can keep up with it. The net result is that the target is approached non-monotonically: it first moves away as the zoom dominates and only later comes back to the center of the view. Various seat-of-the pants guesses (taking logs of things here and there) do not work either.

What is needed is a way to express the desired monotonicity of the view's movement in both space and scale. This view-based constraint is quite naturally expressed in the UV space-scale diagram as a bounding parallelogram (Figure 7). Three sides of the parallelogram are simple to understand. Since moving up in the diagram corresponds to increasing magnification, any trajectory which exits the top of the parallelogram would have overshot the zoom-in. A trajectory exiting the bottom would have zoomed out when it should have been zooming in. One exiting the right side would have overshot the target in space. The fourth side, on the left, is the most interesting. Any point to the left of that line corresponds to a

view in which the target x_2 is further away from the center of the window than where it started, i.e., violating the non-monotonic approach. Thus any admissible trajectory must stay within this parallelogram, and in general must never move back closer to this left side once it has moved right. The simplest such trajectory in UV space is the diagonal of the parallelogram. Calculating it is simple analytic geometry. The coordinates of points 1 and 2 would typically come from the implementation in terms of XZ. These would first be transformed to UV. The linear interpolation is done trivially there, and the resulting equation transformed back to XZ for use in the implementation. If one composes all these algebraic steps into one formula, the trajectory in XZ for this 1-D case is:

$$z = \frac{z_1 - m\, z_1 x_1}{1 - m\, x} \qquad where \qquad m = \frac{z_2 - z_1}{z_2 x_2 - z_1 x_1}$$

Thus to get a monotonic approach, the scale factor, z, must change hyperbolically with the panning of x. This mathematical relationship is not easily guessed but falls directly out of the analysis of the space-scale diagram. We implemented the 2D analog in Pad++ and found the net effect is visually much more pleasing than our naive attempts, and count this as a success of space-scale diagrams.

Optimal pan-zooms and shortest paths in scale-space.
Since panning and zooming are the dominant navigational motion of these undistorted multiscale interfaces, finding "good" versions of such motions is important. The previous example concerned finding a trajectory where "good" was defined by monotonicity properties. Here we explore another notion of a "good" trajectory, where "good" means "short".

Paradoxically, in scale-space the shortest path between two points is usually not a straight line. This is in fact one of the great advantages of zoomable interfaces for navigation and results from the fact that zoom provides a kind of exponential accelerator for moving around a very large space. A vast distance may be traversed by first zooming out to a scale where the old position and new target destination are close together, then making a small pan from one to the other, and finally zooming back in (see Figure 8). Since zoom is naturally loga-

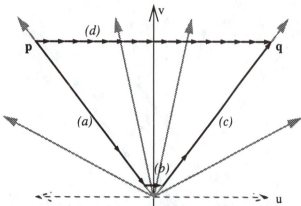

Figure 8. *The shortest path between two points is often not a straight line. Here each arrow represents one unit of cost. Because zoom is logarithmic, it is often "shorter" to zoom out (a), make a small pan (b), and zoom back in (c), than to make a large pan directly (d).*

rithmic, the vast separation can be shrunk much faster than it can be directly traversed, with exponential savings in the limit. Such insights raise the question of what is really the optimal shortest path in scale-space between two points.

When we began pondering this question, we noted a few important but seemingly unrelated pieces of the puzzle. First, one naive intuition about how to pan and zoom to cross large distances says to zoom out until both the old and new location are in the view, then zoom back into the new one. Is this related at all to any notion of a shortest path? Second, window size matters in this intuitive strategy: if the window is bigger, then you do not have to zoom out as far to include both the old and new points. A third piece of the puzzle arises when we note that the "cost" of various pan and zoom operations must be specified formally before we can try to solve the shortest path question. While it seems intuitive that the cost of a pure pan should be linear in the distance panned, and the cost of a pure zoom should be logarithmic with change of scale, there would seem to be a puzzling free parameter relating these two, i.e., telling how much pan is worth how much zoom.

Surprisingly, there turns out to be a very natural information metric on pan/zoom costs which fits these pieces together. It not only yields the linear pan and log zoom costs, but also defines the constant relating them and is sensitive to window size. The metric is motivated by a notion of visual informational complexity: the number of bits it would take to efficiently transmit a movie of a window following the trajectory.

Consider a digital movie made of a pan/zoom sequence over some 2D world. Successive frames differ from one another only slightly, so that a much more efficient encoding is possible. For example, if successive frames are related by a small pan operation, it is necessary only to transmit the bits corresponding to the new pixels appearing at the leading edge of the panning window. The bits at the trailing edge are thrown away. The 1D version is shown in Figure 9a. If the bit density is β (i.e., bits per centimeter of window real estate), then the number of bits to transmit a pan of size d is $d\beta$.

Similarly, consider when successive frames are related by a small pure zoom-in operation (Figure 9b), say where a window is going to magnify a portion covering only $(w-d)/w$ of what it used to cover (where w is the window size). Then too, $d\beta$ bits are involved. These are the bits thrown away at the edges of the window as the zoom-in narrows its scope. Since this new smaller area is to be shown magnified, i.e., with higher resolution, it is exactly this number of bits, $d\beta$, of high resolution information that must be transmitted to augment the lower resolution information that was already available.

The actual calculation of information cost for zooms requires a little more effort, since the amount of information required to make a total zoom by a factor r depends on the number and size of the intermediate steps. For example, two discrete step zooms by a factor of 2 magnification require more bits than a single step zoom by a factor of 4. (Intuitively, this is because showing the intermediate step requires temporarily having some new high resolution information at the edges of the window that is then thrown away in the final scope of the zoomed-in window.) Thus the natural case to consider is the continuous limit, where the step-size goes to zero. The resulting formula says that transmitting a zoom-in (or out) operation for a total magnification change of a factor r requires $\beta w \log(r)$ bits.

Thus the information metric, based on a notion of bits required to encode a movie efficiently, yields exactly what was promised: linear cost of pans ($d\beta$), log costs of zooms ($\beta w \log(r)$), and a constant (w) relating them that is exactly the window size. Similar analyses give the costs for other elementary motions. For example, a zoom around any other point within the window (a *zaww*) always turns out to have the same cost as a pure (centered) zoom. Other arbitrary zoom-arounds are somewhat more complicated.

(a) PAN:

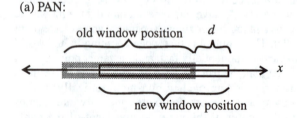

(b) ZOOM:

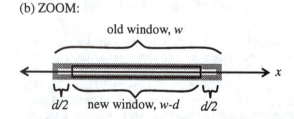

Figure 9. *Information metric on pan and zoom operations on a 1D world. (a) Shifting a window by d requires $d\beta$ new bits. (b) Zooming in by a factor of (w-d)/w, throws away $d\beta$ bits, which must be replaced with just that amount of diffuse, higher resolution information when the window is magnified and brought back to full resolution.*

From these components it is possible to compute the costs of arbitrary trajectories, and therefore in principle to find minimal ones. Unfortunately, the truly optimal ones will have a complicated curved shape, and finding it is a complicated calculus-of-variations problem. We have limited our work so far to finding the shortest paths within certain parameterized families of trajectories, all of which are piecewise pure pans, pure zooms or pure *zaww*'s. We sketch typical members of the families on a space-scale diagram, pick parameterizations of them and apply simple calculus to get the minimal cases. There is not room here to go through these in detail, but we give an overview of the results.

Before doing so, however, it should be mentioned that, despite all this formal work, the real interface issue of what constitutes a "good" pan/zoom trajectory is an empirical/cognitive one. The goal here is to develop a candidate theory for suggesting trajectories, and possibly for modelling and understanding future empirical work. The suitability of the information-based approach followed here hinges on an implicit cognitive theory that humans watching a pan/zoom sequence have somehow to take in, i.e., encode or understand, the sequence of views that is going by. They need to do this to interpret the meaning of specific things they are seeing, understand where they are moving to, how to get back, etc. It is assumed that, other things being equal, "short" movies are somehow easier, taking fewer cognitive resources (processing, memory, etc.) than longer ones. It is also assumed that human viewers do not encode successive frames of the movie but that a small pan or small zoom can be encoded as such, with only the deltas, i.e., the new information, encoded. Thus to some approximation, movies with shorter encoded lengths will be better. (We are also at this point ignoring the content of the world, assuming that no special content-based encoding is practical or at least that information density at all places and scales is sufficiently uniform that its encoding would not change the relative costs.)

To get some empirical idea of whether this information-theoretic approach to "goodness" of pan-zoom trajectories matches human judgment, we implemented some simple testing facilities. The testing interface allows us to animate between two specified points (and zooms) with various trajectories, trajectories that were analyzed and programmed using space-scale diagrams. We did informal testing among a few people in our lab to see if there was an obvious preference between trajectories and compared these to the theory.

For large separations, pure pan is very bad. There is strong agreement between theory and subjects' experience. Theory says the information description of a pure pan movie should be exponentially longer than one using a substantial amount of zoom. Empirically, users universally disliked these big pans. They found it difficult to maintain context as the animation flew across a large scene. Further, when the distance to be travelled was quite large and the animation was fast, it was hard to see what was happening; if the animation was too slow, it took too long to get there.

At the other extreme, for small separations viewers preferred a short pure pan to strategies that zoomed out and in. It turns out that this is also predicted by the theory for the family

piecewise pan/zoom/*zaww* trajectories we considered here. Depending on exactly which types of motions are allowed, the theory predicts that to traverse separations of less than 1 to 3 window widths, the corresponding movie is informationally shorter if it is just a pan.

Does the naively proposed navigation strategy ("zoom out until the starting and ending points are close, then pan in") ever arise in this analysis? At this high level of description, the answer is definitely "yes." The fine points, however, are more subtle. If only *zaww*'s are allowed, the shortest path indeed involves zooming out until both are visible, then zooming in (Figure 10). For users this was quite a well-liked

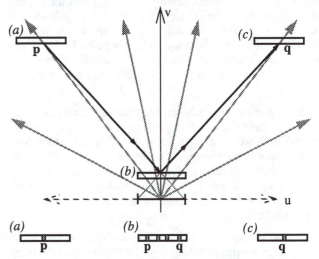

Figure 10. *The shortest zaww path between* **p** *(a) and* **q** *zooms out till both are within the window (b), then zooms in (c). The corresponding views are shown below the diagram.*

trajectory. If pans are allowed, however, the information metric disagrees slightly with the naive intuition. It says instead to stop the zoom just before both are in view, then make a pan of 1-3 screen separations (just as described for short pans), then finally zoom in. The information difference between this optimal strategy and the naive one is small, and our users similarly found small differences in the acceptability. It will be interesting to examine these variants more systematically.

Our overall conclusion is that the information metric, based on analyses of space-scale diagrams, is quite a reasonable way to determine "good" pan/zoom trajectories.

Showing semantic zooming

Another whole class of uses for space-scale diagrams is for the representation of *semantic zooming*[10]. In contrast to geometric zooming, where objects change only their size and not their shape when magnified, semantic zooming allows objects to change their appearance as the amount of real estate available to them changes. For example, an object could just appear as a point when small. As it grows, it could then in turn appear as a solid rectangle, then a labeled rectangle, then a page of text, etc.

Figure 11 shows how geometric zooming and semantic zooming appear in a space-scale diagram. The object on the left, shown as an infinitely extending triangle, corresponds to a 1D gray line segment, which just appears larger as one zooms in (upward: 1,2,3). On the right is an object that changes its appearance as one zooms in. If one zooms out too far (a), it is not visible. At some transition point in scale, it suddenly appears as a three segment dashed line (b), then as a solid line (c), and then when it would be bigger than the window (d), it disappears again.

The importance of such a diagram is that it allow one to see several critical aspects of semantic objects that are not otherwise easily seen. The transition points, i.e., when the object changes representation as a function of scale, is readily apparent. Also the nature of the changing representations, what it looks like before and after the change, can be made clear. The diagram also allows one to compare the transition points and representations of the different objects inhabiting a multi-scale world.

We are exploring direct manipulation in space-scale diagrams as an interface for multi-scale authoring of semantically zoomable objects. For example, by grabbing and manipulating transition boundaries, one can change when an object will zoom semantically. Similarly, suites of objects can have their transitions coordinated by operations analogous to the *snap*, *align*, and *distribute* operators familiar to drawing programs, but applied in the space-scale representation.

As another example of semantic zooming, we have also used space-scale diagrams to implement a "fractal grid." Since grids are useful for aiding authoring and navigation, we wanted to design one that worked at all scales -- a kind of virtual graph paper over the world, where an ever finer mesh of squares appears as you zoom in. We devised the implementa-

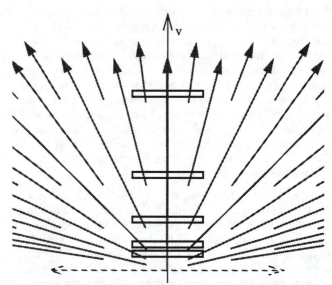

Figure 12. *Fractal grid in 1D. As the window moves up by a factor of 2 magnification, new gridpoints appear to subdivide the world appropriately at that scale. The view of the grid is the same in all five windows.*

tion by first designing the 1D version using the space-scale diagram of Figure 12. This is the analog of a ruler where ever finer subdivisions appear, but by design here they appear only when you zoom in (move upward in the figure). There are nicely spaced gridpoints in the window at all five zooms of the figure. Without this fractal property, at some magnification the grid points would disappear from most views.

Warps and fisheye views

Space-scale diagrams can also be used to produce many kinds of image warpings. We have characterized the space-scale diagram as a stack of image snapshots at different zooms. So far in this paper, we have always taken each image as a horizontal slice through scale space. Now, instead imagine taking a cut of arbitrary shape through scale space and projecting down to the u axis. Figure 13 shows a step-up-step-down cut that produces a mapping with two levels of magnification and a sharp transition between them. Here, *(a)* shows the trajectory through scale space, *(b)* shows the result that would obtain if the cut was purely flat at the initial level, and (c) shows the warped result following.

Different curves can produce many different kinds of mappings. For example, Figure 14 shows how we can create a fisheye view.* By taking a curved trajectory through scale-space, we get a smooth distortion that is magnified in the center and compressed in the periphery. Other cuts can create bifocal [1] and perspective wall [8].

For cuts as in Figure 13, which are piece-wise horizontal, the magnification of the mapping comes directly from the height of the slice. When the cuts are curved and slanted, the geome-

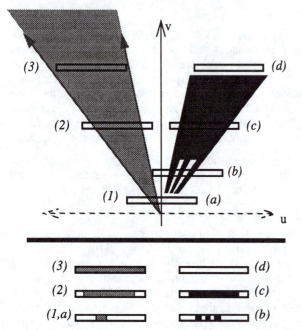

Figure 11. *Semantic Zooming. Bottom slices show views at different points.*

---

* In fact exactly this strategy for creating 2D fisheye views was proposed years ago in [5], p 9,10.

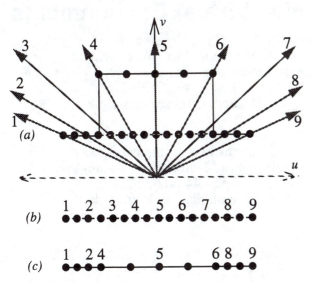

Figure 13. *Warp with two levels of magnification and an abrupt transition between them. (a) shows the trajectory through scale-space, (b) shows the unwarped view, and (c) shows the warped view (notice rays 3 and 7 don't appear).*

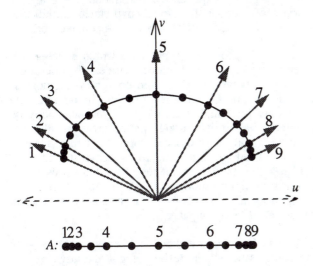

Figure 14. *Fisheye view.*

try is more complicated, but the magnification can always be determined by looking at the projection as in Figure 14.*

CONCLUSION

This paper introduces space-scale diagrams as a new technique for understanding multiscale interfaces. Their defining characteristic and principal virtue is that they represent scale explicitly. We showed how they can aid the analysis of pans and zooms because they take a temporal structure and turn it

---

* Simple projection is only one way for such cuts to create views. For example if one takes the XZ transformed version of these diagrams with cuts, one can use them directly as the magnification functions of [6].

into a static one: a sequence of views becomes a curve in scale-space. This has already helped in the design of good pan/zoom trajectories for Pad++. We showed how the diagrams can help visualization of semantic zooming by showing an object in all its scale-dependent versions simultaneously. We expect to use this as an interface for designing semantically zoomable objects. We also suggested that diagrams may be useful for examining other non-flat multiscale representation, such as fisheye views.

Space-scale diagrams, therefore, are important for visualizing various problems of scale, for aiding formal analyses of those problems, and finally, for implementing various solutions to them.

Acknowledgments

This work was supported in part by ARPA grant N66001-94-C-6039. The authors would like to thank Maria Slowiaczek for her very helpful comments on drafts of this paper.

REFERENCES

1. Apperley, M.D., Tzavaras, I. and Spence, R, A bifocal display technique for data presentation, *Proceedings of Eurographics '82*, pp. 27-43.

2. Bederson, B. B. and Hollan, J.D., Pad++: A zooming graphical interface for exploring alternate interface physics. In *Proceedings of ACM UIST'94*, (1994, Marina Del Ray, CA), ACM Press, pp 17-26.

3. Donelson, W., Spatial management of information. In *Proceedings of ACM SigGraph'78* (Atlanta, GA), ACM press, pp. 203-209.

4. Furnas, G.W., Generalized fisheye views. In *Proceedings of CHI'86 Human Factors in Computing Systems* (Boston, MA, April 1986), ACM press, pp. 16-23.

5. Furnas, G. W., The FISHEYE view: A new look at structured files. *Bell Laboratories Technical Memorandum, #82-11221-22*, Oct 18, 1982. 22pps.

6. Leung, Y.K. and Apperley, M.D., A unified theory of distortion-oriented presentation techniques. In press, TOCHI.

7. Lieberman, H., Powers of ten thousand: navigating in large information spaces. Short paper in *Proceedings of ACM UIST'94*, (1994, Marina Del Ray, CA), ACM Press, pp. 15-16.

8. Mackinlay, J.D., Robertson, G.G. and Card, S.K., The perspective wall: detail and context smoothly integrated. In *Proceedings of CHI'91 Human Factors in Computing Systems*, ACM press, pp. 173-179.

9. Noik, E.G., A space of presentation emphasis techniques for visualizing graphs. In *Proceedings of GI '94: Graphics Interface 1994*, (Banff, Alberta, Canada, May 16-20, 1994), pp. 225-234.

10. Perlin, K. and Fox, D., Pad: An Alternative Approach to the Computer Interface. In *Proceedings of ACM SigGraph `93* (Anaheim, CA) pp. 57-64.

11. Robertson, George. G. and Mackinlay, Jock, The Document Lens. In *Proceedings of ACM UIST'93* (Atlanta, GA), ACM press, pp. 101-108.

12. Sarkar, M. and Brown, M.H., Graphical fisheye views of graphs. In *Proceedings of ACM CHI'92* (Monterey, CA, May, 1992), ACM Press, pp. 83-91.

13. Sarkar, M., Snibbe, S.S., Tversky, O.J., and Reiss, S.P., Stretching the rubber sheet: a metaphor for visualizing large structures on small screens. In *Proceedings of ACM UIST '93* (November 1993), ACM press, pp. 81-91.

User Embodiment in Collaborative Virtual Environments

Steve Benford
Department of Computer Science
The University of Nottingham, Nottingham, UK
Tel: +44-602-514203
E-mail: sdb@cs.nott.ac.uk

Chris Greenhalgh
Department of Computer Science
The University of Nottingham, Nottingham, UK
Tel: +44-602-514225
E-mail: cmg@cs.nott.ac.uk

John Bowers
Department of Psychology
The University of Manchester, Manchester, UK
Tel: +44-61-275-2599
E-mail: bowers@hera.pych.man.ac.uk

Dave Snowdon
Department of Computer Science
The University of Nottingham, Nottingham, UK
Tel: +44-602-514225
E-mail: dns@cs.nott.ac.uk

Lennart E. Fahlén
The Swedish Institute of Computer Science
Stockholm, Sweden
Tel: +46-8-752-1539
E-mail: lef@sics.se

ABSTRACT
This paper explores the issue of user embodiment within collaborative virtual environments. By user embodiment we mean the provision of users with appropriate body images so as to represent them to others and also to themselves. By collaborative virtual environments we mean multi-user virtual reality systems which explicitly support co-operative work (although we argue that the results of our exploration may also be applied to other kinds of collaborative system). The main part of the paper identifies a list of embodiment design issues including: presence, location, identity, activity, availability, history of activity, viewpoint, actionpoint, gesture, facial expression, voluntary versus involuntary expression, degree of presence, reflecting capabilities, physical properties, active bodies, time and change, manipulating your view of others, representation across multiple media, autonomous and distributed body parts, truthfulness and efficiency. Following this, we show how these issues are reflected in our own DIVE and MASSIVE prototype systems and also show how they can be used to analyse several other existing collaborative systems.

KEYWORDS: virtual reality, CSCW, embodiment

INTRODUCTION
User embodiment concerns the provision of users with *appropriate* body images so as to represent them to others (and also to themselves) in collaborative situations.

This paper presents an early theoretical exploration of this issue based on our experience of constructing and analysing a variety of collaborative virtual environments: multi-user virtual reality systems which support co-operative work.

The motivation for embodying users within collaborative systems becomes clear when one considers the role of our bodies in everyday (i.e. non-computer supported) communication. Our bodies provide immediate and continuous information about our presence, activity, attention, availability, mood, status, location, identity, capabilities and many other factors. Our bodies may be explicitly used to communicate as demonstrated by a number of gestural sign languages or may provide an important accompaniment to other forms of communication, helping co-ordinate and manage interaction (e.g. so called "body language").

In our experience, user embodiment becomes an obviously important issue when designing collaborative virtual environments, probably due to their highly graphic nature and the way in which designers are given a free hand in creating objects. However, we believe that many of the issues we raise are equally relevant to co-operative systems in general, where embodiment often seems to be a neglected issue (it appears that many collaborative systems still view users as people on the outside looking in). To go a stage further, we argue that without sufficient embodiment, users only become known to one another through their (disembodied) actions; one might draw an analogy between such users and poltergeists, only visible through paranormal activity. The basic premise of our paper is therefore that *the inhabitants of collaborative virtual environments (and other kinds of collaborative system) ought to be directly visible to themselves and to others through a process of direct and sufficiently rich embodiment*. The key question then becomes how should users be embodied? In other words, are the body images provided appropriate to supporting collaboration? Furthermore, as opposed to merely discussing the appearance of virtual body, we also need to focus on its functions, behaviours and its relation to the user's physical body (i.e. how is the body manipulated and controlled?). Thus, an embodiment can be likened to a 'marionette' with active autonomous behaviours together with a series of strings which the user is continuously 'pulling' as smoothly as possible.

Our paper therefore aims to identify a set of design issues which should be considered by the designers of virtual bodies, along with a set of techniques to support them. These are listed in section two and constitute a diverse, and occasionally conflicting, set of requirements. Designing an appropriate body image will most likely be a case of maintaining a sensible balance between them. Furthermore this balance may be both application and user dependent and will no doubt be constrained by the available computing resources. In the long term it may be possible to refine our initial list of issues into a 'body builder's work-out'. However, we do not yet have sufficient experience to do this. Instead, in section three we describe how the issues are currently reflected in two of our own collaborative virtual environments, DIVE and MASSIVE, giving examples of the bodies we have constructed so far. Section four then uses our list as a framework for analysing how a variety of other collaborative virtual environments and more general CSCW systems tackle user embodiment.

DESIGN ISSUES AND TECHNIQUES

In this section we identify a list of design issues for user embodiments as well as possible techniques for dealing with them. As indicated above, we approach these issues from the perspective of collaborative virtual environments, although we encourage the reader to consider their application to other kinds of collaborative system. We begin with the fundamental issues of presence, location and identity.

Presence

The primary goal of a body image is to convey a sense of someone's presence in a virtual environment. This should be done in an automatic and continuous way so that other users can tell 'at a glance' who is present. In a visually oriented system (such as most VR systems) this will involve associating each user with one or more graphics objects which are considered to represent them.

Location

In shared spaces, it may be important for an embodiment to show the location of a user. This may involve conveying both position and orientation within a given spatial frame of reference (i.e. co-ordinate system). We argue that conveying orientation may be particularly important in collaborative systems due to the significance of orientation to everyday interaction. For example, simple actions such as turning one's back on someone else are loaded with social significance. Consequently, it will often be necessary

to provide body images with recognisable front and back regions.

Identity

Recognising who someone is from their embodiment is clearly a key issue. In fact, body images might convey identity at several distinct levels of recognition. First, it could be easy to recognise at a glance that the body is representing a human being as opposed to some other kind of object. Second, it might be possible to distinguish between different individuals in an interaction, even if you don't know who they are. Third, once you have learned someone's identity, you might be able to recognise them again (this implies some kind of temporal stability). Fourth, you might be able to find out who someone is from their body image. Underpinning these distinctions is the time span over which a body will be used (e.g. one conversation, a few hours or permanently) and the potential number of inhabitants of the environment (from among how many people does an individual have to be recognised?).

Allowing users to personalise body images is also likely to be important if collaborative virtual environments are to gain widespread acceptance. Such personalisation allows people to create recognisable body images and may also help them to identify with their own body image in turn. An example of personalisation might be the ability to don virtual garments or jewellery. Clearly, this ability might have a broader social significance by conveying status or associating individuals with some wider social group (i.e. cultural and work dress codes or fashions).

Activity, viewpoints and actionpoints

Body images might convey a sense of on-going activity. For example, position and orientation in a data space can indicate which data a given user is currently accessing. Such information can be important in co-ordinating activity and in encouraging peripheral awareness of the activities of others. We identify two further aspects of conveying activity: representing user's viewpoints and representing their actionpoints.

A *viewpoint* represents where in space a person is attending and is closely related to the notion of gaze direction (at least in the visual medium). Understanding the viewpoints of others may be critical to supporting interaction (e.g. in controlling turn-taking in conversation or in providing additional context for interpreting talk, especially when spatial-deictic expressions such as 'over there' or 'here' are uttered). Furthermore, humans have the ability to register the rapidly changing viewpoints of others at a fine level of detail (i.e. tracking the movement of other's eyes even at moderate distances). Previous experimental work in the domain of collaborative three dimensional design has already shown the importance of conveying users' viewpoints [8]. In contrast, an actionpoint represents where in space a person is manipulating. Actionpoints typically correspond to the location of virtual limbs (e.g. a telepointer representing a mouse or the image of a hand representing a data glove).

We propose that a user may possess multiple actionpoints and viewpoints. Notice that we deliberately separate where people are attending from where they are manipulating. Although these are often closely related, there appears to be no reason for insisting that they are strictly synchronised; in the real world it is quite possible to manipulate a control while attending somewhere else - indeed, this is highly desirable when driving a car! Representing actionpoints involves providing an appropriate image of a limb driven by whatever device a user is employing. Representing viewpoint involves tracking where a user is attending and moving appropriate parts of their embodiment. Later on we shall see systems that show general body position, head position or even eye position depending on the power of the tracking facilities in use.

Availability and degree of presence
Related to the idea of conveying activity is the idea of showing availability for interaction. The aim here is to convey some sense of how busy and/or interruptable a person is. This might be achieved implicitly by displaying sufficient information about a person's current activity or explicitly through the use of some indicator on their body. This leads us to the further issue of degree of presence. Virtual reality can introduce a strong separation between mind and body. In other words, the presence of a virtual body strongly suggests the presence of the user when this may not, in fact, be the case (e.g., the mind behind the body may have popped out of the office for a few seconds). This is particularly likely to happen with 'desktop' (i.e. screen-based VR) where there is only a minimal connection between the physical user and their virtual body. This mind/body separation could cause a number of problems such as the social embarrassment and wasted effort involved in one person talking to an empty body for any significant amount of time. As a result, it may be important to explicitly show the degree of actual presence in a virtual body. For example, the system might track a user's idle time and employ mechanisms such as increasing translucence or closing eyes to suggest decreasing presence. It might also be possible to put one's body into a suspended state, indicating partial presence to others and perhaps recording on-going conversation to be replayed when subsequently woken up. A suspended body would therefore act as a marker through which one could try to contact its owner in the external world .

As a concrete example of this issue, we cite some of our early experiences with the DIVE system (see below). One of the interesting aspects of DIVE is that a user process that exits unexpectedly often leaves behind a 'corpse' (an empty graphics embodiment). A long DIVE session may produce several such corpses (particularly when developing and testing new applications), which can cause confusion. As a result, two informal conventions have been established among DIVE users. First, on meeting a stationary embodiment, one grabs it and gives it a shake (DIVE allows you to pick other people up). An angry reaction tells you that the embodiment is occupied. Second, bodies that turn out to be corpses are 'buried' (i.e. moved) below the

ground plane. It would be useful to have some more graceful mechanisms for dealing with this problem!

Gesture and facial expression
Gesture is an important part of conversation and ranges from almost sub-conscious accompaniment to speech to complete and well formed sign languages for the deaf. Support for gesture implies that we need to consider what kinds of 'limbs' are present. Facial expression also plays a key role in human interaction as the most powerful external representation of emotion, either conscious or sub-conscious. Facial expression seems strongly related to gesture. However, the granularity of detail involved is much finer and the technical problems inherent in its capture and representation correspondingly more difficult. A crude, but possibly effective approach, might be to texture map video onto an appropriate facial surface of a body image (e.g. the "Talking Heads" at the Media Lab [2]). Another approach involves capturing expression information from the human face using an array of sensors on the skin, modelling it and reproducing it on the body image (e.g. the work of ATR where they explicitly track the movement of a user's face and combine it with models of facial muscles and skin [6] and also the work of Thalmann [10] and Quéau [7]).

This discussion of gesture and facial expression relates to a further issue, that of voluntary versus involuntary expression. Real bodies provide us with the ability to consciously express ourselves as a supplement or alternative to other forms of communication. Virtual bodies can support this by providing an appropriate set of limbs and 'strings' with which to manipulate them. The more flexible the limbs; the richer the gestural language. However, we suspect that users may find ways of gesturing with even very simple limbs. On the other hand, involuntary expression (i.e. that over which users have little control) is also important (looks of shock, anger, fear etc.). However, support for this is technically much harder as it requires automatic capture of sufficiently rich data about the user. This is the real problem we are up against with the facial expression issue - how to capture involuntary expressions.

History of activity
Embodiments might support historical awareness of past presence and activity. In other words, conveying who has been present in the past and what they have done. Clearly we are extending the meaning of 'body' beyond its normal use here. An example might be carving out trails and pathways through virtual space in much the same way as they are worn into the physical world.

Manipulating one's view of other people
In heterogeneous systems where users might employ equipment with radically different capabilities (see MASSIVE below), it will be important for the observer to be able to control their view of other people's bodies. For example, as the user of a sophisticated graphics computer, I may have the processing power to generate a highly complex and fully-textured embodiment. However, this is of little benefit to an observer who does not have a machine

with hardware texturing support. Indeed, the complexity of my body would be counter-productive as the observer would be forced to expend valuable computing resource on rendering my body when it could better be used to render other objects. As a result, the observer should be able to exert some influence over how other people appear to them, perhaps selecting from among a set of possible bodies the one that most suits their needs and capabilities. In short, we propose that it is important for the both the owner and the observers of an embodiment to control how it appears.

This requirement poses a serious problem for most of today's multi-user VR systems - that of subjective variability. Current systems are highly objective in their world view. In other words, all observers see the same world (albeit from different perspectives). A notable exception in this regard is the VEOS system [3]. The ability for people to adopt subjective world views (e.g., seeing different representations of an embodiment) represents a challenge to current VR architectures.

Representation across multiple media
Up to now we have spoken mainly in terms of visual body images. However, body images will be required in all available communication media including audio and text. For example, audio body images might centre around voice tone and quality, be it that of the real-person or be it artificial. Text body images (as used in multi-user dungeons) might involve text names and descriptions or (in a collaborative authoring application) a text-body's 'limbs' might be represented by familiar word processing tools and icons (cursor, scissors etc.).

Autonomous and distributed body parts
We have discussed virtual bodies as if they are localised within some small region of space. We may also need to consider cases where people are in several places at a time, either through multiple direct presence (e.g. logging on more than once) or through some kind of computer agent acting on their behalf (e.g. issuing a database query while browsing an information visualisation).

Efficiency
There will always be a limit to available computing and communications resources. As a result, embodiments should be as efficient as possible, by conveying the above information in simple ways. More specifically, we suspect that approaches which attempt to reproduce the human physical form in as full detail as possible may in fact be wasteful and that more abstract approaches which reflect the above issues in simple ways may be more appropriate. Furthermore, we need to support 'graceful degradation' so that users with less powerful hardware or simpler interfaces can obtain sufficiently useful information without being overloaded. This suggests prioritising the above issues in any given communication scenario. In fact, the real challenge with embodiment will be to prioritise the issues listed in this section according to specific user and application needs and then to find ways of supporting them within a limited computing resource.

Truthfulness
This final issue relates to nearly all of those raised above. It concerns the degree of truth of a body image. In essence, should a body image represent a person as they are in the physical world or should it be created entirely at the whim or fancy or its' owner? We should understand the consequences of both alternatives, or indeed of anything in between. Examples include: truth about identity (can people pretend to be other people?}; truth about facial expression (imagine a world full of perfect poker players); and truth about capabilities (this body has ears on, can they hear me?) On the one hand, lying can be dangerous. On the other, constraining people to the brutal physical truth may be too limiting or boring. The solution may be to specify a *gradient* of body attributes that are increasingly difficult to modify. Those that are easy require relatively little resource. Those that are not require more. For example, changing virtual garments might be easy whereas changing size or face of voice might be difficult. Truthfulness may also be situation dependent (i.e. different degrees may be required for different worlds, applications, contexts etc.). For example, simulation type VR applications may require a very high level of truthfulness.

In summary, we have proposed a list of design issues that need to be considered by the designers of virtual bodies along with some possible techniques for addressing them. The following section now describes how some of these issues have been dealt with in our own DIVE and MASSIVE prototype collaborative virtual environments.

EMBODIMENT IN DIVE AND MASSIVE
The authors have been involved in the construction of two general collaborative virtual environments, DIVE at the Swedish Institute of Computer Science, and MASSIVE at the University of Nottingham. This section considers how the above design issues are reflected in these systems.

Embodiment in DIVE
Virtual reality research at the Swedish Institute of Computer Science has concentrated on supporting multi-user virtual environments over local- and wide-area computer networks, and the use of VR as a basis for collaborative work. As part of this work, the DIVE (Distributed Interactive Virtual Environment) system has been developed to enable experimentation and evaluation of research results [4]. The DIVE system is a tool kit for building distributed VR applications in a heterogeneous network environment. In particular, DIVE allows a number of users and applications to share a virtual environment, where they can interact and communicate in real-time. Audio and video functionality makes it possible to build distributed video-conferencing environments enriched by various services and tools.

A variety of embodiments have been implemented within the DIVE system. The simplest are the 'blockies' which are composed from a few basic graphics objects. The general shape of blockies is sufficient to convey presence, location and orientation (the most common example being a letter 'T' shape). In terms of identity, simple static cartoon-like facial features suggest that a blockie represents a human and the ability for people to personalise their own body images supports some differentiation between individuals (DIVE provides a general geometry description language with which users may specify their own body shapes if they wish). A more advanced DIVE body for immersive use texture maps a static photograph onto the face of the body, thus providing greater support for identifying users in larger scale communication scenarios. This body also provides a graphic representation of the user's arm which tracks their hand position in the physical world via a 3-D mouse.

The display of a solid white line extending from a DIVE body to the point of manipulation in space represents actionpoint in a simple and powerful way and enables other users to see what actions a user is engaged in (e.g., selecting objects). In various DIVE data visualisation applications, each user may also be associated with a different colour which is used to show which data they are accessing (selected objects change to this colour), thereby providing limited peripheral awareness of their activity. Immersive blockies also support a moving head which tracks the position of the user's head in the real world via their head-mounted display (i.e. a six degrees of freedom sensor attached to the top of the user's head). This is very effective at conveying viewpoint, general activity and degree of presence. Finally, video conferencing participants can be represented in DIVE through a video window.

Figure 1 shows a DIVE conference scenario involving a range of embodiments. From left to right we see: an immersed user with humanoid body, textured face and tracked head and arm; a simple non-immersive blockie sporting a humorous propeller hat; a video conferencing participant; and a second immersive user. The scene also shows some DIVE collaboration support tools: a functioning whiteboard which can also be used to create documents and a conference table for document distribution.

Embodiment in MASSIVE

MASSIVE (Model, Architecture and System for Spatial Interaction in Virtual Environments) is a VR conferencing system which realises the COMIC spatial model of interaction [1]. The main goals of MASSIVE are scale (i.e. supporting as many simultaneous users as possible) and heterogeneity (supporting interaction between users whose equipment has different capabilities, who employ radically different styles of user interface and who communicate over an ad-hoc mixture of media).

MASSIVE supports multiple virtual worlds connected via portals. Each world may be inhabited by many concurrent users who can interact over ad-hoc combinations of graphics, audio and text interfaces. The graphics interface renders objects visible in a 3-D space and allows users to navigate this space with a full six degrees of freedom. The audio interface allows users to hear objects and supports both real-time conversation and playback of pre-programmed sounds. The text interface provides a MUD (Multi-User Dungeon)-like view of the world via a window (or map) which looks down onto a 2-D plane across which

users move. Text users are embodied using a few text characters and may interact by typing messages to one another or by 'emoting' (e.g. smile, grimace, etc.).

The graphics, text and audio interfaces may be arbitrarily combined according to the capabilities of a user's terminal equipment. Furthermore, users may export an embodiment into a medium that they cannot receive themselves (thus, a text user can be made visible in the graphics medium and vice versa). The net effect is that users of radically different equipment may interact, albeit in a limited way, within a common virtual world (e.g. text users may appear as slow-speaking, slow moving flatlanders to graphics users). For example, at one extreme, the user of a sophisticated graphics workstation may simultaneously run the graphics, audio and text clients (the latter providing a map facility and allowing interaction with non-audio users). At the other, the user of a dumb terminal (e.g. a VT-100) may run the text client alone. It is also possible to combine the text and audio clients without the graphics and so on. One effect of this heterogeneity is to allow us to populate MASSIVE with large numbers of users at relatively low cost.

MASSIVE graphics embodiments are based on DIVE blockies (although, as with DIVE, users can specify their own geometry via a simple modelling language). Blockies are also automatically labelled with the name of their owner so as to aid identification. In the text interface, users are embodied by a single character (typically the first letter of their chosen name) which shows position and may help identify users in a limited way. An additional line (single character) points in the direction the user is currently facing. Thus, using only two characters, MASSIVE's text interface conveys presence, location, orientation and identity.

Given MASSIVE's inherent heterogeneity, its embodiments need to convey users' capabilities to one another. For example, considering the graphics interface, an audio capable user has ears; a desk-top graphics user (monoscopic) has a single eye; an immersed stereo user would have two eyes and a text user ('textie') has the letter 'T' embossed on their head. Thus, on meeting another user, it should be possible to quickly work out how they perceive you and through which media you can communicate with them (e.g., should you use audio or send text?). Figure 2 shows an example of the graphics interface showing a conference involving five users (the figure shows the view of one of them). We see two non-immersed, audio capable users facing each other across the conference table (ears and a single eye) and a text-only user facing diagonally towards us. We can also see that another non-audio capable user has their back to us.

4. EMBODIMENT IN OTHER SYSTEMS

Next, we briefly analyse the embodiments provided by four further existing technologies, matching them up to the issues identified previously. The four technologies are: dVS, the commercial VR system from DIVISION; ATR's Collaborative Workspace; the multi-user VR game, Doom; and the general use of video as a communication medium. These specific examples have been chosen because of their diversity and because they highlight some interesting aspects of embodiment. Given more space, a wide range of other applications might also have been considered. Indeed, our intention is that designers of future collaborative applications could perform a similar exercise to the following and so gauge the likely effectiveness and limitations of their proposed body images for co-operative work. In order to save space, we only discuss those issues that are actually supported by the chosen examples.

dVS

dVS, from DIVISION Ltd, has been chosen as a typical example of current commercially available VR systems [5]. dVS supports multi-user virtual reality applications running on both DIVISION's own hardware and on Silicon Graphics machines. Users may operate in either immersive or desktop modes. The default embodiment in dVS is a telepointer, although the authors have seen examples involving a disembodied head and a single limb. dVS addresses the following design issues:

- Presence and location - users are directly represented and the use of head and hand tracking support some notion of general location and orientation although the lack of a body linking the two make this difficult to discern.

- Viewpoint and actionpoints - supported through head and hand tracking.

- Gesture - supported through the tracked hand only (though the representation of the hand as a pointer severely limits this ability).

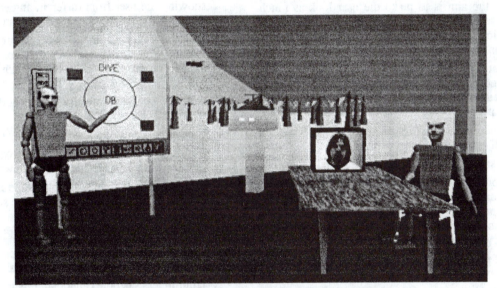

Figure 1: Various embodiments attend a DIVE conference

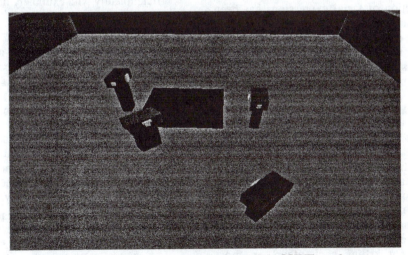

Figure 2: Users show their capabilities at a MASSIVE conference

247

Collaborative Workspace

The ATR lab has been exploring the use of virtual reality to support co-operative work for some years [9]. The main thrust of their research has been on supporting two-party teleconferencing and, in particular, on automatically capturing and reproducing facial expressions. Their collaborative workspace prototype achieves this by attaching a video camera to a head-mounted frame which also supports a position tracker. The use of small reflective disks attached to the user's face allows automatic analysis of their facial movements from the video image. This is then used to animate a texture mapped model of the user's face. Collaborative workspace addresses the following issues:

- Presence - users are directly represented as humanoid looking forms (as realistic as possible).

- Location - as far as we know, the user occupies a relatively fixed overall position (e.g. seated at a table).

- Identity - the aim is to make the user look as much like themselves as possible using a human head model onto which a photographic image of the user is textured and then animated.

- Viewpoint - the user's head position is tracked and represented, as are the positions of their eyes. Thus, this system is one of the very few to convey gaze direction at a very detailed level.

- Actionpoint - the user wears a single data glove and the position of one hand is therefore tracked.

- Gesture - supported through the tracked hand.

- Facial expression - this appears to be the primary focus of this work and a reasonably sophisticated range of facial expressions are possible through the use of tracked mouth, eyebrows and eyes. Both voluntary and involuntary expression are supported.

- Degree of presence - this is not really a problem due to the use of head, eye and hand tracking.

- Efficiency - does not appear to be a key requirement of the project given the super-computers used.

Complimentary, and equally impressive, work on the capture and reproduction of facial expressions has been reported by Thalmann [10]. In this case, the user is not constrained to wearing a head-mounted camera or any facial 'jewellery' or special make-up. The advantage of this is clearly a lack of intrusiveness. However, the disadvantage appears to be the inability to combine facial expressions with head tracking.

Doom

Doom is a multi-user virtual reality game for networked PCs. Doom has been chosen as a representative VR entertainment application intended for mass use and also because it supports many embodiment issues within very limited computing resources. Doom allows up to four users to navigate through a maze of corridors and rooms killing everything that they meet using a variety of weapons. The multi-user version can either be played in death-match mode (i.e. scoring points for killing each other) or, most interestingly, in co-operative mode (i.e. scoring points for killing other things together). Although this may seem far removed from a useful co-operative system, Doom contains several features worth noting. First, the graphics in Doom realise navigable texture mapped environments on a 486 platform. In order to achieve this level of graphics performance, the designers of Doom have placed some constraints on their virtual worlds such as restricting them to use only perpendicular surfaces. Indeed, this is what makes the issue of embodiment in Doom particularly interesting; efficiency is of very great importance. Doom addresses the following design issues:

- Presence - users are directly represented as humanoids.

- Location - each user has a location and a limited number of orientations. Doom portrays users using flat 2-D textures which are always perpendicular to the observer. Swapping between several such textures showing the user from different angles (North, South, East and West) conveys an approximate orientation.

- Identity - other users (player characters in gaming terminology) are distinguished from computer generated monsters (non-player characters). Each user also wears a different colour tunic.

- Activity and availability - the activity of firing weapons is clearly shown.

- Viewpoint - only supported through rough orientation.

- Actionpoint - the impact point of weapons is shown, as is the trace of projectiles for some weapons.

- Facial expression - this is not visible in other people. However, the user does see a separate self image which shows how healthy they are.

- Degree of presence - there is no mistaking a corpse.

- Time and change - not supported except for the user's self image where improvements in health are portrayed.

- Truthfulness - people cannot alter their body images.

- Efficiency - this is where Doom excels; the whole system is an exercise in achieving maximum possible functionality with extremely limited resources.

Video

The use of video in collaborative applications is becoming increasingly widespread and makes an interesting contrast to the above VR based examples. As opposed to considering any specific video conferencing system, we focus on the nature of embodiment within video as a general medium.

- Presence - the presence of the person in front of the camera is clearly represented. However, in situations where there are one way connections (e.g. media space "glances" or surveillance cameras), the presence of the person behind the camera may not be.

- Location - the physical location of a user may be shown to some degree. However, there is no real sense of a common location (i.e. you can't place many people in relation to each other). The same is true of orientation. Other than knowing whether they are facing the camera or not, you cannot tell where someone is looking. First, if they are looking off camera, what are they looking at? Second, in groups of more than two people, who are they looking at if they peer into the camera?

- Identity - is conveyed nearly as well as in the real world (subject to picture resolution problems). Personalisation requires altering your physical self.

- Activity and availability - It may be possible to tell whether someone is busy or not but not what they are doing. Several researchers have investigated techniques for displaying availability to make a video connection (e.g. metaphors such as "doors").

- Viewpoint - not really supported, although you might be fooled otherwise (the orientation issue from above).

- Gesture - supported as in the real world subject to field of view constraints.

- Facial expression - obviously supported (both voluntary and involuntary).

- Truthfulness - generally enforces the brutal truth as there is little chance to break away from the real person's appearance. Some more advanced systems may allow some manipulation of video images.

SUMMARY

The premise of this paper has been that user embodiment is a key issue for collaborative virtual environments (and indeed, for other kinds of collaborative system). Given this assumption, we have identified the following initial list of issues as being relevant to the embodiment of users: presence, location, identity, activity, availability, history of activity, viewpoint, actionpoint, gesture, facial expression, voluntary versus involuntary expression, degree of presence, capabilities, physical properties, manipulating one's view of others, multiple media, distributed bodies, truthfulness and efficiency. We have also shown how these issues are currently reflected in our own DIVE and MASSIVE collaborative virtual environments as well as several others.

We suspect that the importance of any given design issue will be both application and user specific and that the art of virtual body building will involve identifying the important issues in each case and supporting them within the available computing resource. However, at the present time, our list remains only an initial framework for the discussion and exploration of embodiment. In our future work we aim to realise a larger number of these issues within our own DIVE and MASSIVE systems, gaining deeper insights into their relative importance and possible implementation. In the longer term, we would hope to refine our list into a complete 'body builder's work-out', supporting the choice and analysis of the most appropriate designs for the available equipment, application, users, scale and longevity of intended collaborative applications.

ACKNOWLEDGEMENTS
This work has been sponsored by the CEC through the COMIC ESPRIT Basic Research Action and by the UK's EPSRC through the Virtuosi project and its PhD studentship programme.

REFERENCES

1. Benford, S., Bowers, J., Fahlén, L. E., and Greenhalgh, C., Managing Mutual Awareness in Collaborative Virtual Environments, *Proc. Virtual Reality Systems and Technology (VRST) '94*, August, 1994, Singapore.

2. Brand, S., The Medialab - Inventing the future at MIT, Viking Penguin, 1987, ISBN 0-670-81442-3, p. 91-93.

3. Bricken, W., and Coco, G., The VEOS Project, *Presence — Teleoperators and Virtual Environments*, Vol. 3, No. 2, MIT Press 1994.

4. Fahlén, L. E., Brown, C. G., Stahl, O. and Carlsson, C., A Space Based Model for User Interaction in Shared Synthetic Environments, *in Proc.InterCHI'93*, ACM Press, 1993.

5. Grimsdale, C., Supervision - A Parallel Architecture for Virtual Reality, *Virtual Reality Systems*, Earnshaw, R.A., Gigante, M.A and Jones, H. (eds), Academic Press, 1993, ISBN 0-12-227748-1.

6. Ohya, J., Kitamura, Y., Takemura, H., Kishino, F., Terashima, N., Real-time Reproduction of 3D Human Images in Virtual Space Teleconferencing, *Proc.VRAIS'93*, IEEE, Seattle Washington September, 1993, pp. 408-414.

7. Quéau, P., Real Time Facial Analysis and Image Rendering for Televirtuality Applications, *in Notes from Virtual Reality Oslo '94 — Networks and Applications*, eds. Loeffler, Carl E and Søby, Morten and Ødegård, Ola, August 1994.

8. Shu, L., and Flowers, W., Teledesign: groupware user experiments in three-dimensional computer-aided design, *Collaborative Computing*, 1(1), Chapman & Hall, 94.

9. Takemura, H. and Kishino, K., Cooperative Work Environment Using Virtual Workspace, *Proc. CSCW'92*, Toronto, Nov 1992, ACM Press.

10. Thalmann, D., Using Virtual Reality Techniques in the Animation Process, *Virtual Reality Systems*, Earnshaw, R.A., Gigante, M.A and Jones, H. (eds), Academic Press, 1993, ISBN 0-12-227748-1.

Providing Assurances in a Multimedia Interactive Environment

Dorée Duncan Seligmann
Multimedia Communication
Research Department
AT&T Bell Laboratories 4F-605
101 Crawfords Corner Road
Holmdel, New Jersey 07733-3030
(908) 949-4290
doree@research.att.com

Rebecca T. Mercuri
School of Engineering and
Applied Science
University of Pennsylvania
P.O. Box 1166
Philadelphia, Pennsylvania 19105
(215) 736-8355
mercuri@gradient.cis.upenn.edu

John T. Edmark
Multimedia Communication
Research Department
AT&T Bell Laboratories 4F-602
101 Crawfords Corner Road
Holmdel, New Jersey 07733-3030
(908) 949-9223
edmark@research.att.com

ABSTRACT

In ordinary telephone calls, we rely on cues for the assurance that the connection is active and that the other party is listening to what we are saying. For instance, noise on the line (whether it be someone's voice, traffic sounds, or background static from a bad connection) tells us about the state of our connection. Similarly, the occasional "uhuh" or muffled sounds from a side conversation tells us about the focus and activity of the person on the line. Conventional telephony is based on a single connection for communication between two parties — as such, it has relatively simple assurance needs. Multimedia, multiparty systems increase the complexity of the communication in two orthogonal directions, leading to a concomitant increase in assurance needs. As the complexity of these systems and services grows, it becomes increasingly difficult for users to assess the current state of these services and the level of the user interactions within the systems.

We have addressed this problem through the use of assurances that are designed to provide information about the connectivity, presence, focus, and activity in an environment that is part virtual and part real. We describe how independent network media services (a virtual meeting room service, a holophonic sound service, an application sharing service, and a 3D augmented reality visualization system) were designed to work together, providing users with coordinated cohesive assurances for virtual contexts in multimedia, multiparty communication and interaction.

KEYWORDS: auditory I/O, communication, virtual reality, visualization, graphics, teleconferencing, telepresence, user-interfaces.

INTRODUCTION

Advanced teleconferencing is a natural application of multimedia communications systems. No longer are people content to telecommute to a meeting as disembodied voices, they wish to be seen as well as heard. Furthermore, people do not want to telecommute to a meeting empty-handed, they wish to bring electronic media, access services, and control devices for personal referencing or sharing with the group. Although much of the technology is in place to permit such multimodal interaction, traditional systems, applying traditional approaches (e.g., 3kHz monophonic audio, tiled displays, and mouse-driven user-interfaces) fall short in their ability to support and facilitate multiparty communications.

When two parties establish a communication connection, they create a virtual context in which their communication resides, but which coexists with and depends upon the real physical world. We have implemented a model to represent one such virtual context, the virtual meeting room[1] — an interactive environment where conferees may organize, display and manage multiple media streams. The virtual meeting room is an electronic place where users meet and interact with each other and with objects and tools, such as video streams or computer programs. This model establishes a framework on which presentation methods for displaying media outputs to each user are based. It also provides a framework for the design of new media services.

In ordinary phone calls we rely on cues that indicate state (the dial tone, busy signal), connectivity (the other person's voice, static on the line) as well as focus (interjecting "uhuh" or hearing a television in the background). These cues, even if they are not explicitly intended for these purposes, provide information that help assure us about the state and nature of our connection as well as the quality of our interaction. As the complexity of multimedia systems and services grows, so does each participant's uncertainty with regard to the current state of those services, their level and type of connectivity, and the activities of the other participants. The cues we are accustomed to using for two-party connections are not necessarily scalable for multiparty, multimedia systems. For example, it would be unreasonable to expect everyone in a multiparty call to utter "uhuh" to indicate that they are listening.

Different media can be exploited to provide assurances for the other media services. Hence, graphics can, in part, convey presence; while audio can, in part, convey activity. It is in this way that we have enhanced our multimedia interactive environment with synthesized media cues. These cues create assurances that provide context for the conference setting, thus better representing the content of the materials presented, and improving interactions among participants. They also indicate the state of the services provided, as well as the level of participation.

MIXING REALITIES

When two people interact in a simple telephone conversation, it is as if each of them is in two different places simultaneously: their respective physical locations in the world and the virtual place which encompasses their conversation. This notion of coexisting worlds is a fundamental property of the environment in which multimedia, multiparty communication occurs. It is this complex environment that our system depicts. For example, our system represents this mix of real and virtual places and the objects in them, as well as those objects' associations and connectivity. It also represents each participant's equipment and connectivity (which determine the extent of each participant's telepresence and interaction within this virtual place). In the following two examples, 3D graphics and 3D sound are used to depict different aspects of this multimedia, multiparty environment.

Plate 1 shows Rebecca's view of the environment as she browses through it. From this vantage point, she sees the visualization of the physical world below, consisting of some offices at AT&T Bell Laboratories. In the office at the back, John is shown using his computer and phone. In the foreground, Dorée is in her office, likewise using her computer and phone. Here, John and Dorée, and each of their computers and phones, are represented in a ghosted fashion, indicating that they are each telepresent elsewhere. A virtual meeting room hovers above. This room contains John and Dorée and the equipment they have brought with them to share audio and data: their computers and phones. All of the objects in the room are opaque, indicating their telepresence in this virtual place. Connectivity is shown by the cables attaching the devices in the virtual room to their ghosted counterparts in the real world below; data flow is indicated by the animated contents moving through these cables. Rebecca's location and movement in the virtual environment is conveyed through visual and audio cues to the other people in the environment. Rebecca sees the virtual meeting room, who is in it, and what the participants are doing (because this room and that information is explicitly

available to her) and, at the same time, John and Dorée can hear Rebecca as she passes by and they can elect to call her to join their meeting.

Plate 2 shows John's view of a virtual meeting room in which he is telepresent, seated at the round table with a document facing him. To his left is Dorée, to his right is Rebecca. The audio and graphical visualization is constructed from his individual vantage point in this virtual place. The monophonic signals from Dorée's and Rebecca's voices are convolved to correspond to their locations in the virtual meeting room. All three participants are editing a document together. This document is actually a shared X-Window application that is texture-mapped into the 3D environment. Dorée currently has input control, that is, she alone can type into the application, as indicated by the red line connecting her keyboard to the document. Rebecca is currently pointing to a figure in the document, depicted by the cue stick emanating from her direction. Audio cues are generated to indicate activities in the room. Sampled keyboard presses are spatially located near Dorée, while tapping sounds are located near the shared document.

All the information presented originates from the collection of network media services in use: the meeting room service, the shared application program service, the holophonic sound service, and the 3D visualization service. These services together produce the customized integrated multimedia presentation and interfaces to the system for each user.

The above examples illustrate assurances designed to reinforce information about the state, connectivity and focus of objects in the virtual environment. In the remainder of this paper, we emphasize 1) the importance of using a unifying model for virtual context as the basis for multimedia communication and interaction and 2) the types of assurance techniques used within this model. We will also describe how our underlying infrastructure and shared protocols for multimedia services have enabled us to implement a variety of assurances presented by cooperating media services. Users of our system need not have the same equipment or software in order to interact with each other. We will show examples from several different configurations.

THE VIRTUAL MEETING ROOM

The virtual meeting room model can enhance data stream delivery by establishing or imposing relationships among the individuals and objects in the conference. We have addressed some of these issues in our work with the Rapport multimedia communication system[1,6] and in the development of its user-interfaces.[7] Plate 3 shows Bob's view of the virtual meeting room with phone, video, and shared application services. The graphical interface dynamically generates coordinated presentations and control mechanisms for shared resources, connectivity, and presence. Controls and state information are presented in sets and every user is presented with his own customized view of the meeting rooms. The graphical interfaces indicate the current level of participation of each user and what other media capabilities are available. The representation of each participant includes his name, a picture or live video, and various indicators of presence and connectivity. The binoculars below Sid's live video feed notify Bob that Sid is viewing Bob's talking head. The application program window frames and corresponding iconic representations indicate who is currently providing input, who *can* provide input, and where the program is executing. This information is conveyed using the state information transmitted by the individual media services, not by a central controlling module.

A virtual meeting can be as simple as an audio conference call. Here, the ability to identify individual data streams is essential to effective interaction. When a participant starts talking, we need to know who she is and how to recognize her. In a two-party communication, once the initial identification of the talkers are made, no further identity tags are needed for the duration of the call. When a third person is added, the dynamics of the conversation necessarily change. The auditory cues that we use to keep from interrupting one another in a two-person interaction are not sufficient as more parties join the conference. Less-aggressive individuals are often left out, and the other participants may even wonder (or query) if they are still "on the line." If a group is added (say via a speaker-phone), there may be a sense that unidentified listeners are in the room, silently monitoring the discussion, which may be disconcerting to some parties. Various participants may want to put themselves "on hold" in order to carry on private discussions, and then rejoin the group at a later point. It is useful to all parties to know who is still around and just observing quietly, and who is temporarily absent from the session. The virtual meeting room can be used to reveal and clarify this information, augmenting the virtual interaction by providing cues that are associated with face-to-face meetings.

THE MR ARCHITECTURE

The types of assurances we wish to provide depend on a fairly complete representation of the state of the system. This is difficult to accomplish in multimedia systems if the various media services that comprise the system do not share basic structures and naming conventions, or simply do not represent the connectivity or activity over the connections.

Our multimedia communication system is built on an architecture we call MR (Meeting Room). Briefly, MR is a platform-free, transport-free, device-free, hardware-free infrastructure on which (cooperating) network-based media

services are built. The rich representation maintained by each module in the system facilitates the implementation of assurances.

Each media service (i.e. video, audio, etc.) is comprised of a network Server and a local Manager for each individual user. The Server may have associated servers and devices, and the Manager may have local clients (such as an interface), devices and servers. The Server and Managers each maintain representations of the following base classes: virtual meeting rooms (the persistent contexts), conferees (persons associated with the room), materials (objects in the room), and connections.

Figure 1 shows a partial view of MR. In the network, the MR Server maintains the state of all the virtual meeting rooms and also the states of the associated participants and media services. The MR Server has access to a name server that describes registered objects. On the local site shown, a Conversation Manager (CM) communicates with the MR Server. It has a user-interface (UI) that allows the user to issue meeting room commands (creating, entering, leaving room; associating and disassociating media services; inviting or calling people to join, etc.).

Virtual meeting rooms are *persistent*, serving as electronic rendezvous places for people to meet, acting as depositories for media objects, and providing structure for reestablishing services. Yet the system is not connection-based, as meeting rooms can exist even when devoid of users and/or objects. The MR Server maintains a representation for each room, in the network, until it is explicitly torn down. Each local site corresponds to one user. A user can move physical locations, change hardware configurations, and still access the people and objects in any given virtual meeting room. Connections are simply dropped when not needed and (re)established to bring media services into a room at any time. A user can access a meeting room from any point as long as there is a local CM that can communicate with the MR Server. Each

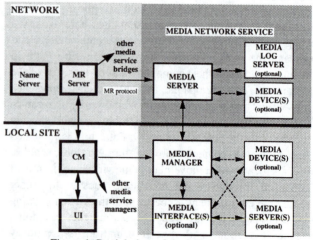

Figure 1. Partial view of the MR Architecture

network service is handled by local Managers that communicate to network Servers. For example, a user can first enter a virtual meeting room from a location via a phone. His phone Manager communicates with the phone Server in order to add him to the conference call. (Note that this location need not be registered in advance; local sites are dynamically established and associated with particular users.) Similarly, a user can create a virtual meeting room from her office and begin to execute a program within it. She then leaves the virtual room (although it continues to exist) and her physical office, and travels to a remote location. Now, with a different piece of hardware, she may reestablish contact with the MR Server and join the virtual room where the still-executing program is located.

We can view the sharing of a medium within the context of a virtual meeting room in two ways: first, as shared objects, and second, as modes of interaction. In an audio service, an example of a shared object is a musical CD that everyone in the virtual meeting room can hear, while one mode of interaction is shared voice. In a video service, an example of a shared object is a video stream that everyone in the virtual meeting room can see, while one mode of interaction is talking heads. Similarly, we define two types of sharing of a windowing system. A shared object could be a program with which everyone in the virtual room can interact, while one mode of interaction is a pointer placed within the context of a program's display.

ASSURANCES

Assurance cues are useful for establishing a sense of connectivity and focus in the virtual meeting. *Connectivity assurances* provide information about the status of the connections between and amongst the participants during the course of communication. They reflect the integrity of the system in operation, and do not simply refer to the physical connections but include the logical ones as well. *Focal assurances* provide information about the nature of each participant's involvement with the system. They reflect the activities and configurations of each media stream (representing users, objects, devices, etc.) in the system during interaction.[11]

Graphical simulation can be used to provide visual feedback about the virtual meeting and the connectivity among the services, while simultaneously recreating the real world settings in which the users and equipment reside. Texture mapped human models can transcend simple iconified user figures through enhancement with back channel response cues, such as head nods, smiles, arm gestures, and so on, in order to reinforce an active sense of connectivity and presence. For example, if a participant's attention is distracted from the meeting (perhaps by a call to another meeting), the displayed head may momentarily turn away or the eyelids may close.[4]

Holophonic audio (monophonic sound streams to which transforms are applied in order to generate a 3D pair) can be used to simulate a consistent virtual acoustical display where sounds are provided with directional context relative to a listener's vantage point.[5,15,16] Spatial tracking may also be used to correlate each participant's motions in the real world with their position in the virtual environment.[14] Experiments by Alexander Graham Bell[3] in the late 1800's, and in the 1940's at Bell Laboratories by Koenig[10] have long indicated that the spatial localization provided by binaural listening is important for discriminatory processing of audio. Simple stereo pairing allows listeners to subjectively localize sounds in the 3D audio space; reduces the perception of reverberation, background and impulse noises; lends greater ease in differentiation of multiple speech streams; and enhances comprehension of speech in negative signal-to-noise environments. Shimizu[12] observed that, within teleconferencing settings, stereophonics enabled listeners to more easily identify speakers with whom they were unfamiliar.

Although our virtual meeting rooms shall exist in the cold electronic void of cyberspace, they need not be bland and sterile. They should be warm places with character and individuality, where natural conversations are facilitated. When we go to a restaurant, the sights and smells that greet us as we proceed through the door, prepare us for the feast ahead. As we approach a lecture hall, the murmuring of the crowd (or their snoring) indicates to us the level of anticipation of the gathered group. So too, when we enter a virtual meeting room, we should be presented with cues that aid us in understanding behaviors and expectations for the meeting we have joined. These cues can also provide contextual assurances for system state and connectivity levels.

A virtual presence can be applied to provide a further transcendental context to the meeting room environment. Companies establish a corporate image that is consistently conveyed through the appearance of their products, physical plant, logo, and publicity. Individuals create unique atmospheres for themselves in their workspaces and homes. Our virtual settings can similarly use ambient sounds, texture-mapped displays, and carefully designed interfaces in order to establish or enhance a desired mood (i.e. energy can be motivated with bright colors and up-tempo music, calm encouraged by muted background visuals and soft environmental noises). In this way, a sensory experience can be created that contributes positively to the dynamics of the meeting and reinforces the memories taken away by the participants.

Our media services operate within the virtual meeting room model to help support various paradigms, as follows:

Expectation: a sense of anticipation or prior knowledge of activities and surroundings, and the behaviors related to proper and efficient manipulation of the environment. (I know where I am and what to do here.) Sounds within the room may be heard from outside, or as you approach; differently sized and shaped rooms replicate the view and acoustics that would occur in a real room; mood can be augmented with visual and audio ambiance.

Identification: a clear view of who and what is sharing the virtual place is established. (I know who is talking, who is typing, what room we are in.) Each room can have its own image and sound color; people in the room are identifiable by texture-mapped images, location, timbre, volume and diction; audio and visual metaphors for user feedback and control can be applied to spatially placed objects in the environment.[2,9]

Association: relationships between data streams and the individuals who generated or share them are established and may be modified dynamically. (I understand which voice goes with what picture, which program just beeped.) The connection between sounds and users or objects can be reinforced with aural cues, such as the clicking of keys on the keyboard or the whir of a printing device.[13]

Differentiation: similar items must be distinguishable from each other. (I know there is a difference between a participant's voice and the voice in the movie we are watching.) Here too, spatial location, timbre, volume, and virtually applied acoustics and images encourage distinction among objects.

Memory: events have a temporal context. (I remember who was the last person speaking, what was the last thing we did.) Sound and visual imagery will increase retention of the meeting experience and the relationships among the participants. In addition, the persistent character of the

virtual meeting room allows for a sense of continuity between sessions, permitting familiar visual and aural elements in the room to trigger memories from previous sessions.[8]

Reference: items exist in the environment within some common context understood by all of the participants. (I am looking at the movie in the back of the room, you are sitting to the left of me and I am on your right.) A global sense of locality ensures that all objects retain their relative positions as the space is observed from different points.

Process: awareness and understanding of one's relationship to the environment, and vice versa, is enhanced. (I know who is listening to me, and can comprehend their reactions to what I do.) Directionality increases the awareness of one's existence within the space, and improves the sense of immediacy of the communication.

Attention: a variety of indicators (verbal, gestural, graphical, etc.) for use in emphasis and articulation should be available across the various media. (I am pointing at you, this text needs to be cut out of that document, that last remark was directed toward me). Volume (whisper, shout) and gesturing can give verbal emphasis; sound metaphors can be used as audio pointers, and color and intensity metaphors can be used as visual pointers.

THE HOLOPHONIC SOUND SERVICE

In this section we describe how we use the holophonic sound service to reinforce the presentation of the virtual meeting room and environment. The entire environment exists in a 3D coordinate system and each object is initially assigned a spatial location in this virtual place. For example, participants in a virtual meeting room are positioned at a round table whose dimensions expand as new members join. Shared objects are located on virtual shared surfaces. We will describe a virtual meeting room with two active media

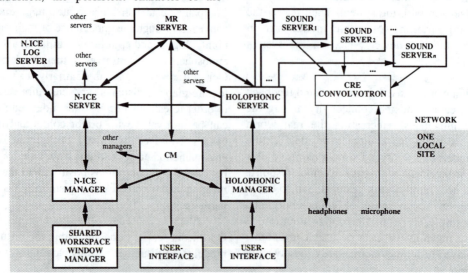

Figure 2. Partial representation: virtual meeting room service, 3D sound and application program sharing

services: N-ICE (Networked Interactive Collaborative Environment, an application sharing service) and the holophonic (3D audio) service. Figure 2 shows how these media services are interconnected with the MR Server. In the network, the MR Server sends MR protocol events to both the N-ICE and holophonic servers. The N-ICE and holophonic servers transmit messages to each other indicating the state of each service. On each local site, each of the related Managers (CM, N-ICE, Holophonic) creates a customized view of the state of the system for the user. Sounds corresponding to the participants and their tools are convolved to reflect their assigned spatial locations from each participant's own perspective.

We have implemented the following categories of audio cues to provide assurances:

3D Realtime Voice. The holophonic Server convolves the monophonic voice signals for each participant. This is not just restricted to the interiors of the virtual meeting rooms. Persons browsing the hallways can also hear the spatially located audio from their vantage points as they wander through the virtual environment.

Generic Events. The MR protocol involves the creation, destruction and use of the virtual meeting rooms. A user is advised of changes in state of the virtual meeting room by messages that are spatially located near the user's ear (such as a whispered briefing). For example, when a user enters a virtual meeting room, she is advised of the people present, and each participant is advised of the new arrival. Sounds for hallway or room events are also assigned to global locations. Broadcast messages, such as audio cues to indicate that a meeting is about to adjourn or commence, are provided.

Interaction with Objects. Audio assurances are used to indicate activity in the room, such as input to shared application programs. N-ICE allows for different input protocols, including chaotic mode, in which anyone in the virtual meeting room can provide input to a particular application program. When simply viewing an application program's windowed displays, a participant may be uncertain (in the absence of additional cues) as to who produced the events that are changing the display. The holophonic service maps selected input events to sampled audio cues and spatially locates them near the representation of the person from whom they originated. For example, we use sampled keyboard clicks for key press events. Using application-specific knowledge, the holophonic service selects different audio cues for similar events. For example, a mouse motion event in a drawing program is presented aurally as a pen scratching noise; the same event in a CAD/CAM program is represented by sampled drafting tool sounds. Inherently collaborative media services, such as a shared whiteboard application, can provide more extensive

information about events to aid in mapping to audio cues.

Objects. N-ICE reports the state of each application program as it changes. When a registered application is executed, the holophonic sound service maps successive events to audio cues. For example, when an image viewer opens a window, sampled sounds of a slide projector are played; for document editors, sampled sounds created with paper are used.

Interaction-based events. N-ICE also supports a set of windowed interaction devices, such as pointers and annotators. The holophonic Server uses a tapping sound to accompany pointing.

Participants need not adopt the same set of media services in a virtual meeting room. A participant may choose not to use (or may not be able to use) the N-ICE service. In this case, the holophonic service can provide different aural assurances vis-à-vis N-ICE-specific events (such as the opening of a new application program, someone has just typed a "t", etc.). Furthermore, the MR architecture allows for *any* media service to provide assurances cues for the same events. Thus, the same event can be presented by combinations of cues, such as visual captions, bridged audio, or synthesized speech, in addition to those described in this paper.

USER FEEDBACK

Although we have not yet had the chance to conduct any formal usability tests, we have received feedback from different trial sites and casual testers. Almost everyone has expressed a desire to control both the volume and mapping of each audio cue. It was noted that at higher volume levels the sampled keyclicks became annoying over extended periods of time. Also, many of the cues were repeated if they were mapped to window mapping events; these need to be filtered perhaps via user experience levels. On the positive side, everyone reacted very favorably to the 3D spatialization of the conferees and to the advisor messaging. It was agreed that even without an accompanying 3D visualization, speaker identification was vastly improved. Users in an ongoing trial using N-ICE particularly like two features of the the shared window manager: it clearly differentiates shared application windows from private ones appearing on the same display, and it prevents people not in the virtual meeting room from either seeing or modifying the shared applications. Users also like the enhanced control of their personal video stream offered by the meeting room model, where only the people in the room can access a user's stream, and then only if explicitly made available by that user. Tests will be conducted to determine problems that arise when users enter the virtual meeting room with different equipment, ranging from simple desktop phones to the full 3D visualization and holophonic audio system. As the system matures, formal usability tests should be performed and analyzed.

CONCLUSIONS

We have described how the virtual meeting room provides a useful framework to model new media services and create a cohesive presentation of connectivity and activity. In particular, we have described how separate media services can be enhanced to coalesce disjoint data streams and information into rich representations of people, objects, and places in an environment that is part real and part virtual. These services combine synergistically to produce a compelling experience of telepresence (of oneself, others, and objects) and place.

Multimedia systems are for *people* to use. While we search for better transport algorithms to guarantee data arrival rates and synchronization, new compression techniques and data formats, we must also seek new methods for organizing and humanizing the presentation of this information. Context enhancements within the meeting room model are a step toward providing a seamless transition from the real world to that of the virtual and back again. Architectures, such as MR, make it possible to support persistent, flexible and extensible virtual contexts that facilitate the communication and interaction process.

IMPLEMENTATION

All code was written in C++ and executes on workstations running UNIX® and using the base classes and protocols of the MR system. The holophonic sound service employs Crystal River Engineering, Inc. Convolvotron hardware. N-ICE executes on native X-Window Servers using unaltered X-Window applications. The 3D visualization system is written with SGI's OpenInventor C++ libraries and OpenGL. Plates 1 and 2 show displays that were generated on a SGI Crimson Reality Engine1.

ACKNOWLEDGEMENTS

Sid Ahuja and J. R. Ensor have been collaborators since 1986, when we first developed Rapport. MR was implemented by Murali Aravamudan and Babu Ramakrishnan. Cati Laporte designed parts of the 3D visualization system. May Pack designed and implemented portions of N-ICE; Dave Weimer wrote the N-ICE window manager. We also thank Sandy Thuel and Pierre Welner for their comments on this paper.

REFERENCES

1. Ahuja, S. R., Ensor, J. R., and Horn, D. N. "The Rapport Multimedia Conferencing System." In *Proceedings of the Conference on Office Information Systems*, Palo Alto, California, March 1988.

2. Beaudouin-Lafon, Michel, and Gaver, William W. "Eno: Synthesizing Structured Sound Spaces." In *Proceedings of UIST'94*, November 2-4, 1994, Marina del Rey, California, ACM Press, 1994.

3. Bell, A. G. "Experiments Relating to Binaural Audition." In *American Journal of Otology*, 1880.

4. Cassell, J., Pelachaud, C., Badler, N., Steedman, M., et al "Animated Conversation: Rule-based Generation of Facial Expression, Gesture & Spoken Intonation for Multiple Conversational Agents." In *Proc. ACM SIGGRAPH '94*, Orlando, FL, July 24–29, 1994.

5. Cohen, Michael "Throwing, Pitching and Catching Sound: Audio Windowing Models and Modes." In *International Journal of Man-Machine Studies*, Volume 39, 1993.

6. Ensor, J. R., Ahuja, S. R, Connaghan, R.B., Pack, M., and Seligmann, D. D. "The Rapport Multimedia Communication System." In *Proceedings of ACM SIGCHI '92 Human Factors in Computing Systems*, Monterey, California, May 3–7, 1992.

7. Ensor, J. R., Ahuja, S. R, and Seligmann, D. D. "User Interfaces for Multimedia Multiparty Communications." In *Proceedings of IEEE International Conference on Communications ICC '93*, Geneva, Switzerland, May 23–26, 1993.

8. Gabbe, J., Ginsberg, A., Robinson, B. "Towards Intelligent Recognition of Multimedia Episodes in Real-Time Applications." In *Proceedings in ACM Multimedia '94*, San Francisco, California, Oct. 15-20, 1994.

9. Gaver, W. W., Smith, R. B., and O'Shea, T. "Effective Sounds in Complex Systems: The ARKola Simulation." In *Proceedings of ACM SIGCHI '91 Human Factors in Computing Systems*, New Orleans, Louisiana, April 27–May 2, 1991.

10. Koenig, W. "Subjective Effects in Binaural Hearing." In *Journal of the Acoustical Society of America*, Volume 22, Number 1, January, 1950.

11. Seligmann, D. D., and Edmark, J. "User Interface Mechanisms for Assurances During Multimedia Multiparty Communication." In *1st International Workshop on Networked Reality in Telecommunication*, Tokyo, Japan, May, 1994.

12. Shimizu, Y. "Research on the Use of Stereophonics in Teleconferences." In *Business Japan*, March, 1991.

13. Takala, Tapio, and Hahn, James "Sound Rendering." In *Proceedings of ACM SIGGRAPH '92*, Chicago, Illinois, July 26-31, 1992.

14. Teodosio, L. A. and Mills, M. "Panoramic Overviews for Navigating Real-World Scenes." In *Proceedings of ACM Multimedia '93*, Anaheim, California, August 1–6, 1993.

15. Wenzel, Elizabeth M., Wightman, Frederic L., and Kistler, Doris J. "Localization with Non-Individualized Virtual Acoustic Display Cues." In *Proceedings of ACM SIGCHI '91 Human Factors in Computing Systems*, New Orleans, Louisiana, April 27–May 2, 1991.

16. Wenzel, Elizabeth M. "Three-Dimensional Virtual Acoustic Displays." NASA Ames Research Center, July, 1991.

A Virtual Window On Media Space

William W. Gaver[*†]   *Gerda Smets*[†]   *Kees Overbeeke*[†]

[*]Royal College of Art
Kensington Gore, London SW7 2EU, U.K.
gaver@rca-crd.demon.co.uk

[†]Technische Universiteit Delft
Jafalaan 9, 2628 BX Delft, The Netherlands
c.j.overbeeke (or) g.j.f.smets@io.tudelft.nl

ABSTRACT

The Virtual Window system uses head movements in a local office to control camera movement in a remote office. The result is like a window allowing exploration of remote scenes rather than a flat screen showing moving pictures. Our analysis of the system, experience implementing a prototype, and observations of people using it, combine to suggest that it may help overcome the limitations of typical media space configurations. In particular, it seems useful in offering an expanded field of view, reducing visual discontinuities, allowing mutual negotiation of orientation, providing depth information, and supporting camera awareness. The prototype we built is too large, noisy, slow and inaccurate for extended use, but it is valuable in opening a space of possibilities for the design of systems that allow richer access to remote colleagues.

KEYWORDS: CSCW, groupwork, media spaces, video

INTRODUCTION

Media spaces are computer-controlled networks of audio and video equipment designed to support collaboration [2, 4, 6, 14, 17, 25]. They are distinguished from more common videophone, video-conferencing, and video broadcasting systems in that they are continuously available environments rather than periodically accessed services. Because maintaining high-bandwidth connections is costly, current video services are typically used for planned and focused meetings. Media spaces, in contrast, assume a future in which broadband networks are commonplace, the data rates needed for high fidelity video and audio are a trivial fraction of the total available, and thus the systems can be left "on" all the time. In practice, this is usually simulated in-house, using dedicated networks of analog audio and video cables leading from central computer-controlled switches to equipment in offices and common areas.

The constancy of connections that characterizes media spaces has several implications for how they are used. Because they are not associated with special events (e.g. meetings), they become part of the everyday work environment. It is common always to have a video connection somewhere, often to a common area, if only because such

views are more pleasant than blank monitors. This implies, finally, that the proportion of time spent using the media space for meetings is relatively low, and instead they often are used to support a more informal, peripheral awareness of people and events. Again, this distinguishes them from more commonly encountered video systems.

Trouble in Media Space

While there is anecdotal evidence that media spaces can support professional activities [2, 6, 13, 14, 17], and particularly long-term collaborative relationships [1, 3], quantitative data supporting the value of these technologies have been more difficult to find. Typically studies show that adding video to an audio channel makes no significant effect on conversation dynamics or on the performance of tasks that do not rely heavily on social cues [7, 19, 20]. Even in a more naturalistic media space setting, Fish et al. [4] found that people usually used the Bellcore system as a prelude to physically co-present meetings, and concluded that it did not clearly add to the functionality provided by telephone or email systems.

These results seem relevant primarily for the role of video in supporting relatively focused interactions. But one of the motivating intuitions behind early media space research was that they might help create and sustain a more informal sense of shared awareness [e.g., 2, 17, 25]. The fact that much recent research meant to assess media spaces seems focused on more formal uses may be because of the difficulty of finding quantitative data that addresses their informal possibilities. In any case, it has made it difficult to assess these original intuitions.

Access to the Task Domain

Observational and analytic studies, on the other hand, have suggested limitations on the ability for media spaces to support informal shared awareness. For instance, Heath and Luff [9] described how co-present collaborators shape their activities and utterances for their partners, and contrasted this with the difficulties they observed in organizing these sorts of visually mediated activities in and through media spaces [10]. They concluded that while a great deal of everyday collaboration is mediated by access to colleagues in the context of their tasks, this sort of access is often not provided by current media spaces.

A similar point was made by Nardi et al. [15] in their study of video used during neurosurgery within operating rooms and remote offices. Video is important in such settings because it allows visual access to events that are otherwise inaccessible (e.g. cameras are pointed into the head during brain surgery), and thus provides the awareness necessary for coordination. The emphasis of this sort of application is on visual access to tasks, not faces;

thus Nardi et al. [15] recommend "turning away from talking heads" and instead focusing on "video-as-data."

This work is valuable in emphasizing the ability for video to support awareness of task-related artifacts. It is less convincing as a case against face-to-face video, however, since giving access to tasks is not incompatible with giving access to people. From our perspective, the point is not that cameras should be focused away from people towards workbenches (or skulls, as the case may be), but that the narrow focus of video itself must be broadened.

Extending Affordances

Gaver [5] analysed the affordances of media spaces to understand how the technologies shape perception and interaction. This analysis emphasized several limitations on the visual information media spaces convey:

- Video provides a restricted field of view on remote sites.

- Video has limited resolution.

- Video conveys a limited amount of information about the three-dimensional structure of remote scenes.

- There are discontinuities (or "seams," [13]) on the edges of scenes and between views from different cameras.

- There are also discontinuities between local and remote scenes and their geometries.

- The medium is anisotropic: the discontinuities between local and remote geometries are not reciprocal (and thus not predictable).

- Movement with respect to remote spaces is usually difficult or impossible.

Each of these attributes has implications for collaboration in media space. But the inability to move with respect to remote spaces may be most consequential of all. As Gibson [8] emphasized, movement is fundamental for perception. We move towards and away from things, look around them, and move them so we can inspect them closely. Movement also has implications for the other constraints produced by video. If we can look around, we increase our effective field of view. Moving can compensate for low resolution [21]. It provides information about three-dimensional layout in the form of movement parallax [8, 16, 22]. Finally, movement might allow people to compensate for the discontinuities and anisotropies of current media spaces.

Allowing Movement in Remote Spaces

One approach to approximating movement within remote sites was explored using the MTV (for Multiple Target Video) system, which employed several switched video cameras in each of two offices [7]. Observations of six pairs of partners collaborating on two tasks indicated that the increased access was indeed beneficial. Participants used all the views, and were often creative, finding unexpected ways to gain access to their colleagues and their working environments. In fact, they accessed face-to-face views for much less time than views that included places and objects relevant for the tasks. This supports suggestions that access to task domains may be more useful than access to colleague's faces [10, 15]. However, participants did seem to rely on quick views of their colleagues as a way to assess attention and availability; looking times may

be misleading as a basis for judging the importance of these views.

Though multiple cameras provided valuable visual access for collaboration, a number of problems with this strategy became clear. Despite the proliferation of cameras (and associated clutter), there were still significant gaps in the visual coverage provided. In addition, participants seemed to have problems establishing a frame of reference with one another, and in directing orientation to different parts of their environments. One result was that the video images themselves became the shared objects, rather than the physical spaces they portrayed, and participants would point at these images rather than the offices themselves. In general, the greater access provided by multiple cameras seemed outweighed by the addition of new levels of discontinuity and anisotropy.

Despite these problems, increasing visual access to remote environments seems a clearly desirable goal. In this paper, we describe another approach, involving the creation of a Virtual Window that allows true visual movement over time, rather than a series of views from static cameras. By providing an intuitive way to move remote cameras, we believe we can overcome many of the limitations of video for supporting peripheral awareness without introducing the problems that come with multiple cameras.

THE DELFT VIRTUAL WINDOW

The basic idea of the Virtual Window is that moving in front of a local video monitor causes a remote camera to move analogously, thus providing new information on the display (see Figure 1). To see something out of view to the right, for instance, the viewer need only "look around the corner" by moving to the left; to see something on a desk, he or she need only "look over the edge," and so forth. The result is that the monitor appears as a window rather than a flat screen, through which remote scenes may be explored visually in a natural and intuitive way.

Movement Parallax and Depth Television

The Delft Virtual Window was invented originally as a means for creating *depth television*, allowing information for three-dimensional depth to be conveyed on a two-dimensional screen [16, 22]. The system creates the self-

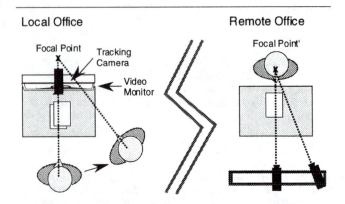

Figure 1. The Virtual Window: Local head locations are detected by a tracking camera and used to control a moving camera in the remote office. The effect is that the image on the local monitor changes as if it were a window.

generated optic flow patterns that underlie *movement parallax*. As the head is moved around a focal point (shown in Figure 1), objects appear to move differently from one another depending on their distances (this is easy to see by moving one's head around an object while focusing on it: objects in the background seem to move parallel with the head, while those in the foreground move against it). Movement parallax is well suited for depth television because it does not require different images to be presented to both eyes. Indeed, similar methods have been used for computer graphics [12], but the Delft Virtual Window is the first system that provides movement parallax around a focal point for realtime video [23].

The Virtual Window has been tested experimentally by comparing people's accuracy at judging depth in remote scenes when they were viewed from static cameras, from moving cameras that they did not control, and from the Virtual Window system [16, 22]. A clear advantage was found for the Virtual Window system over static views, and a significant decrease in variability of depth judgements when compared with those made from passively viewed moving scenes. The experimental evidence thus supports the intuitive impression that the Delft Virtual Window can do a good job of conveying depth information.

Affordances for Increased Access

It is difficult to implement Virtual Window systems with the speed and accuracy necessary to give very good impressions of depth. But the technique gives rise to a number of other, serendipitous affordances that make even less-ambitious versions potentially beneficial for media spaces.

Field of View Because the camera moves around a focal point, it provides access to a much larger area of the remote scene than stationary cameras do (see Figure 2). The distance of the focal point from the camera determines the effective field of view. If it is set at infinity, for example, the camera moves only laterally and relatively little is added to the field of view. At the opposite extreme, if the focal point is set at the front of the camera itself, there is

no lateral movement and the camera movement is equivalent to that provided by a pan-tilt unit. The field of view is greatly expanded, but parallax information for depth is lost.

Resolution As Gaver [5] pointed out, for static cameras there is an inherent tradeoff between field of view and resolution. This conflict does not exist for moving cameras: Not only is the effective field of view increased by allowing movement, but Smets et al. [21] have shown that information for fine details can be obtained over time from a moving camera: effective resolution is increased as well.

Continuity of the Remote Scene Although the greater field of view offered by the Virtual Window must be accessed over time, new views are linked continuously. Instead of jumping from one view to another, one moves smoothly among views, making it easy to understand how they relate to one another. This contrasts with the MTV system, in which jumps among views introduced gaps and discontinuities that seemed to impede orientation [7, 11].

Continuity with the Local Scene If visual movement within the remote scene appears continuous with movement-induced shifts of perspective on the local one, the sense of continuity in and through media space should be increased. The glass screen of the video monitor will continue to act as a barrier between local and remote spaces, of course, but no longer spaces with different physics (i.e., in which head movements produce different visual consequences).

Control and Coordination: Finally, local control over remote cameras has several implications for perception and interaction in media space. Not only does it imply a larger field of view, but one available for active exploration rather than one depending on passive presentation. This may help support coordination with remote colleagues. As a simple example, it is common to hold something up to show a remote colleague, only to misjudge and hold it partially off-camera. Correcting the error usually requires explicit negotiation ("a little to the left...no, *my* left!"). The Virtual Window system allows the remote viewer to compensate for his or her partner's mistake simply by moving, without requiring any explicit discussion about the mechanics of the situation.

The combination of these affordances – the ability to expand the field of view, to raise the effective resolution, to increase the continuity within and between spaces, to support control and coordination, and to provide depth information – make the Virtual Window concept appealing for media space research. In the following sections, we describe our approach to implementing such a system and our experiences with the prototype we built.

EXPERIENCES WITH A VIRTUAL WINDOW

We collaborated to design, build, and assess an instantiation of the Virtual Window system. Most of the design, implementation, and initial programming were done at the Delft University of Technology. Two of the three devices were then installed, the software ported and developed, and the results tested at Rank Xerox Cambridge EuroPARC.

There are three separate aspects involved in instantiating a Virtual Window system:

- *Head-tracking* The location of the viewer's head with respect to the monitor must be determined.

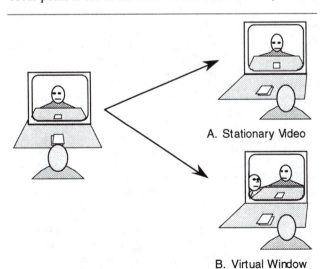

A. Stationary Video

B. Virtual Window

Figure 2. If cameras are stationary (A), local movements do not change the field of view, but do introduce discontinuities between local and remote spaces. In the Virtual Window system (B), local moves can provide a greater field of view continuously with local visual changes.

- *Camera-moving* The camera must be moved in the remote site.
- *Mapping* The head location must be mapped to a desired camera location.

A number of approaches may be taken to these issues [16, 22]. The prototype we built depended on a combination of idealistic goals (e.g., hands-free operation) tempered – sometimes betrayed – by pragmatic realities (e.g., cost of implementation). In the end, the process of designing, building, and trying it ourselves taught us, at least as much as watching it in use, both about the fundamental issues at stake and about the realities of implementation. Here we describe our tactics in some detail, and discuss some of the implications for our experiences with the system.

Head-tracking

We decided at the outset of the project that head-tracking should be accomplished without requiring users to wear any special devices or clothing. This seemed crucial if the Virtual Windows were to be used as casually as the rest of EuroPARC's media space. However, this precluded the use of commercially available devices such as Polhemus sensors or infrared trackers. Instead, our version of the Virtual Window uses image processing on a video signal to determine head location.

For our implementation, a "tracking camera" is mounted on the local video monitor (Figure 1) and the incoming video stream is processed to extract the viewer's head location. The basic image processing strategy is shown in Figure 3. First, a single frame is digitized from the head-tracking camera when nobody is in view; this is used as the *reference image*. While the system is running, the reference image is subtracted from each incoming video frame, leaving a *difference image* that is processed to find an area of large differences assumed to be the viewer's head.

Finding such an area is at the heart of the image processing algorithm. First the differences along the rows of the image are summed, giving a difference profile for the height of the image. A threshold is set between the overall average of the differences and the greatest difference, and the top of the head is taken to be the first row of the image from the

top that crosses the threshold (the head is assumed to be upright in the image). Then a horizontal difference profile is taken from a row on or just below the supposed top of the head, and a new threshold is set. The first cells to exceed this threshold from the right and left are assumed to be the sides of the head, and the center of the head to be halfway between the two.

A number of small variations can be used to improve this basic algorithm. For instance, it is useful to set a threshold for the minimum distance required before moving the camera. This helps to avoid spurious camera jitter caused by small fluctuations between successive frames.

This algorithm is simplistic in a number of ways. For instance, it does not recognise a head per se, but only areas where the incoming image is very different from the reference image. This means that the algorithm will track any source of change, such as a moving hand. It also means that the algorithm is very sensitive to changes in the ambient light, since these tend to introduce spurious differences between the incoming and reference images. Finally, it implies that more than one source of difference – such as two people in the tracking camera's field of view – may cause it to return inaccurate values (it tends to track whoever is higher in the tracking image, and returns an average horizontal value if they are at the same level). This is a manifestation of the more fundamental problem of scaling the Virtual Window to provide the correct visual information to more than one viewer.

Nonetheless, the algorithm works surprisingly well for all its simplicity. When conditions are good, the algorithm produces generally accurate values allowing a viewer's head to be tracked even against a cluttered background. Clearly there are more sophisticated approaches that might be used for this task, but there are severe constraints on the amount of processing that can be done while maintaining reasonable system latency. Even using this simple algorithm, we only achieved rates of about 3 - 7 frames per second on a Sparcstation 2; more accurate algorithms might not be worth still slower rates.

Camera-Moving

To move a camera around a focal point, recreating the optics of looking through a window, it is necessary both to rotate it and to move it laterally. This means that commercially available pan-tilt units are inadequate, unless the focal point is set to the front of the camera and no lateral movement is required.

We constructed our camera-moving apparatus from two A3 size flat-bed plotters that originally used software-controlled stepper motors to move pens over paper. We modified them extensively, cutting away most of the flat bed to reveal the basic frame, moving the control boards, and mounting them together so they would stand vertically (see Figure 4). The two pen transports are used to move the front and back of a Panasonic thumb-sized camera separately; each is powered by two stepper motors controlled over an RS232 link by the host computer. Though we had originally planned to use the built-in hardware and software to control the motors, this produced only instantaneous acceleration and deceleration, which led to unacceptably shaky camera movement. We hired an electronics contractor to develop new control hardware and

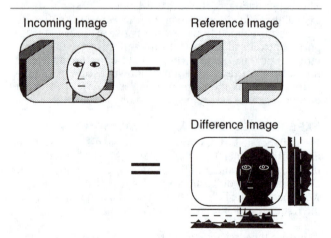

Incoming Image　　　　Reference Image

Difference Image

Figure 3. Head-tracking is accomplished by looking for values over threshold in a difference image produced by subtracting a reference image from each incoming frame.

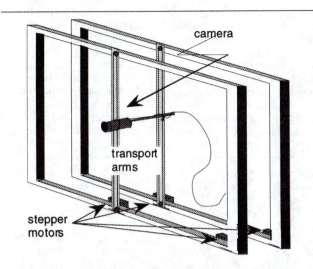

Figure 4. The camera transport mechanism uses two transport arms to move the front and back of a thumb camera separately.

software, which greatly enhanced the system by allowing smooth acceleration and deceleration of each motor separately.

The camera-moving devices are successful in being able to move a camera relatively quickly and smoothly over an area of about .35 X .2 meters. However, when two of them were moved from the large workshop in Delft where they had been designed and initially tested to the smaller, quieter office environment in Cambridge, it quickly became apparent that they are far too large and noisy to be acceptable for office use. Each of the devices takes up a volume of about .7 X .5 X .2 meters, and has a footprint of roughly .8 X .5 meters, larger than most of the video monitors being used. In addition, the motors cause audible vibrations in the frame. When we changed the system to allow each of the four motors to accelerate and decelerate independently, as described earlier, the noise problem was greatly exacerbated because each motor introduced its own independently changing frequency component. The resulting noise, though sounding impressively like a science fiction sound effect, is clearly too intrusive to be used in an office environment. In sum, the camera-moving devices have been adequate for our initial research, but a different design would be necessary for longer-term use.

Mapping Head Location to Camera Movement

A final issue for implementing a Virtual Window is the mapping between head and camera location. We discuss two aspects of this here: the determination of the focus point and errors caused by the expression of location as a point in the tracking camera's picture plane.

Determining the Focal Point One difficulty in implementing the virtual window system is in determining the focal point about which the remote camera is to move. Ideally, the viewer's actual focus could be determined by measuring gaze direction, convergence and accommodation. In practice, this seems difficult at best, not clearly necessary depending on the aims of the system, and almost certainly unfeasible if the system is to be used casually.

For our prototype, then, the focal point was set by the user using a simple graphical interface. We assumed that the focal point is always on a line extending from the center of the camera moving device. By taking the origin of our movement coordinates at that point, we can express the focal point simply as the ratio of front and back camera movements (see Figure 5). If the ratio is 1, the focal length is infinite, the front of the camera moves as much as the back, and the effect is one of lateral movement with no rotation. If the ratio is 0, the focal point is the front of the camera, and the camera only rotates without moving laterally, just like a pan-tilt unit. Intermediate ratios give intermediate focal lengths.

Angular Locations and Visual Information For our prototype, we simply mapped the pair of coordinates returned by the head-tracking software to a new location for the back of the camera so that the maximum values of each would map to one another. This seems satisfactory in practice, but in reality it leads to systematic differences from the optical changes that movement in front of a window would make. In Figure 6, for instance, the two heads are both on the edge of the tracking camera's field of view, and so would return the same head locations and receive the same view from the remote camera. But if the monitor were really a window, the views would be different, as indicated by the lines of sight shown in the figure. This disparity arises because the edges of the tracking camera's image plane do not map to the edges of the monitor. The practical consequences of this disparity are unclear – again, our simple mapping seems satisfactory – but the issue bears consideration.

As we suggested earlier, implementing a Virtual Window that can move a remote camera with the speed and accuracy necessary for veridical depth perception is difficult; some of the issues we have just discussed should make clear why this is so. We relaxed a number of the requirements for our prototype, since we were less interested in producing convincing depth information than we were in exploring the other affordances offered by the Virtual Window. Nonetheless, in many cases the changing scene provided by our implementation does evoke a good impression of depth (albeit at the wrong scale: often the remote office seems like a relatively small box). More importantly, the prototype has allowed us to explore some of the possibilities of

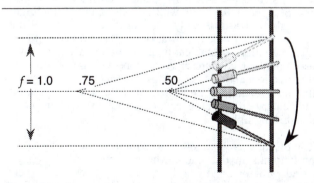

Figure 5. The focal point, *f*, can be expressed as the ratio of front to back movement. When *f* is 1, the focal point is at infinity and the camera only moves laterally. When *f* is 0, the focal point is at the front of the camera and the effect is like a pan-tilt device. Here the camera is shown as it moves around an *f* of .5 from top to bottom.

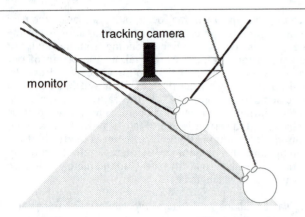

Figure 6. Equal locations in the tracking camera's picture plane should sometimes map to different camera positions.

using the Virtual Window to provide greater access to remote sites.

Observing the Virtual Window in Use

To observe the system in use, we had six pairs of participants use it in pursuing two simple collaborative tasks. Subjects sat in separate offices, each controlling camera movement in his or her partner's office using the Virtual Window. The first task was called the *Room-Draw Task*, and required each participant simply to draw a floor-plan of his or her colleagues' office. The second task was the *Overhead Projector Design Task*, which asked the partners to redesign an overhead projector so that the lens-carrying arm would not block the audience's view. These tasks were modelled after similar ones used previously to assess collaboration in media spaces [7, 11]. They are designed to be simple, easily understood and motivated, and to focus on participants' access to their remote colleagues' environment.

Our observations tended to confirm the advantages, and emphasize the deficiencies, that we had noticed in developing the system. In the following, we briefly describe the problems that participants had with the system, then the advantages it provided.

When It Was Bad...

The first two pairs of participants used the system on a beautiful spring day, with white clouds racing over a bright blue sky. Unfortunately, this provided a compelling demonstration of the head-tracking algorithm's susceptibility to variations in ambient light. The reference images we used could not be representative of the wide ranges of room illumination, and so the cameras often moved erratically as the head-tracking algorithm located the areas of greatest momentary difference, even though these were often due to the shifting light.

The results were extremely puzzling and frustrating to the participants in the study, who had not used the Video Window before, and who for the most part were relatively naive about media spaces in general. The movements of the view were only partially related to their own movements, and it seemed that because they were new to the system they had little comprehension of what or whether anything was going wrong. In effect, they became passive rather than active observers of the remote scene, a situation that has been shown experimentally to produce worse performance than if no motion were provided [21].

In any case, there was little that participants could do to correct tracking problems except to take a new reference image, which required ducking under the table so that they would not be in view of the tracking camera. On occasions when the view would show an area of the remote office that was useful, participants would often freeze in an attempt to keep the camera from moving. Ironically, in these circumstances a stationary camera would have given the participants better access to the remote site than a moving one – a point to which we return.

But When It Was Good...

Fortunately, the remaining participants were tested on cloudy days more typical of England, which meant that the systems were relatively accurate and stable. In these conditions, several advantages of the Virtual Window became clear. For example, there were several instances in which a participant would move slightly to achieve a better view on something his or her partner was displaying; thus, as we had expected, the system appeared to allow subjects mutually to negotiate orientation. In addition, there were occasions in which the system seemed to help participants maintain awareness of their partner's field of view, by increasing their awareness of the camera and its orientation (though this may in part have been due to the salience of the camera-moving device).

Most importantly, though, the Virtual Window did succeed in allowing participants to explore their partner's office visually, and the mapping between local movements and remote views appeared natural to the users. It seems difficult to convey the force of this result because of its simplicity. For instance, when one participant wanted to look down and to the side, he simply stood up and moved to the side. This sort of observation seems easy to overlook in the midst of the many difficulties people had with the current system. But the fact that this is possible at all, and that it seemed so natural, is a major success of the Virtual Window system.

CONCLUSIONS

Providing the ability to move with respect to remote spaces seems a clearly desirable goal. But our experiences with the Virtual Window, as well as with the earlier MTV system [7] suggest that the vague notion of "remote movement" should be decomposed. From this perspective, experiencing a monitor as a window requires:

- user access to new views of the remote site
- linked continuously in space and time
- produced by local head movement
- with enough speed and accuracy for movement parallax.

This decomposition is useful in comparing strategies for providing greater access to remote scenes. For instance, the original MTV system [7] provided new views of remote sites, but they were not linked continuously in space or time. A later version, which replaced switching with multiple monitors [11], allowed continuous access over time, but there were still discontinuities (gaps) in spatial coverage. Pan-tilt-zoom units provide both sorts of continuity, but are typically controlled by joysticks and similar devices. Finally, the Virtual Window we built enables head-tracked camera movement, but not true movement parallax.

Though the prototype we built is too slow and inaccurate to provide good movement parallax, and too large and noisy for everyday use, many of the problems we encountered seem less like inherent failings of the concept and more like challenges for iterative design. We may have been too ambitious in our design, rejecting reliable off-the-shelf equipment and using less-reliable custom solutions in an attempt to avoid compromising our ideals about how the system should work. Nonetheless, the prototype does illustrate some of the potential advantages of the Virtual Window approach. In addition, it opens a space of possibilities for the design of systems that allow much richer access to remote sites.

For instance, the inaccuracy of the head-tracking algorithm was clearly due to its reliance on an accurate reference image. There are several possibilities for increasing the robustness of this algorithm. If the overall differences between the incoming and reference pictures are consistently large, for example, it might be assumed that the reference image is out of date and the user could be notified. More fundamentally, greater accuracy can be obtained if the reference frame is updated adaptively [e.g., 24]. Another possibility is to replace the reference image with the results of low-pass filtering the current stream of images; this would have the effect of blurring out any movement (e.g., of the head) and helping to compensate for shifts in light. Finally, other head-tracking techniques might fruitfully be explored, including those which require users to wear special devices.

Similarly, we might expect that further iterations of the camera-moving system would greatly help with its size and noise. One possibility is to shift priorities from providing movement parallax towards providing a greater field of view. This would imply that lateral movement is unnecessary and allow the use of a commercially available pan-tilt-zoom unit. An additional advantage of using an off-the-shelf unit would be the opportunity to incorporate zoom as well, so that leaning towards the monitor might cause the camera to enlarge the image around the focal point. In fact, we are currently exploring such a system with Koichi'ro Tanikosi, Hiroshi Ishii, and Bill Buxton at the University of Toronto.

A more radical design option is to avoid moving a camera at all, and instead to produce a shifting view on remote scenes by moving a window over, and then undistorting, the view from a fish-eye lens. Apple Computer has developed a similar strategy for creating Quicktime "virtual reality" [18], but not for use with realtime video. The processing demands of such a strategy are quite high, but it has a number of advantages. Not only would it eliminate the difficult problems of mechanically moving the mass of a camera very quickly with no discernible vibrations, but it would also do away with the problem of scaling the system to deal with multiple, distributed remote viewers. It is not clear that the strategy could be extended to produce lateral as well as rotary camera movement, but it seems well worth further investigation.

Finally, it is also desirable to design for the enduring differences between Virtual Windows and real ones. For example, a clear finding of our user study was the need to distinguish and allow separate control over movement in local and remote spaces. Once participants had achieved good views of remote spaces, they often seemed reluctant to move for fear of losing them. This problem is partially an effect of the current system's limitations. When working in front of a real window, moving away to achieve some local goal is easily reversed simply by moving back again. Using the current implementation of the Virtual Window, in contrast, moving back is no guarantee of recovering the original view. Though future versions should alleviate this problem, it may actually be desirable to maintain the dissociation. A foot pedal could be added to the system, for instance, allowing people to stop the Virtual Window so that local movement would not disturb a good view of the remote site.

In sum, the prototype Virtual Window is useful in opening up a wide space for the design of new video systems. Perhaps none will succeed in fully creating the experience of looking through a window into an office thousands of miles away, but many are likely to be useful in overcoming the limitations of existing systems. In the end, perhaps the most important contribution the Virtual Window makes is as a concrete reminder that media spaces need not be constrained to single, unmoving cameras left sitting on top of video monitors.

ACKNOWLEDGEMENTS

We thank Rank Xerox Cambridge EuroPARC—particularly Bob Anderson and Allan Maclean—and the Faculty of Industrial Design Engineering at Delft TU for supporting this collaboration. Pieter Jan Stappers was an invaluable guide to Virtual Window design, particularly the head-tracking algorithm. We thank Ronald Teunissen for work on the camera-moving apparatus and Jeroen Ommering for the "Cameraman" motor-control software. Finally, we are extremely grateful to Abi Sellen for helping with the study reported here, and to her and Christian Heath, Paul Luff, Anne Schlottmann, Paul Dourish, Sara Bly and Wendy Mackay for valuable discussions about this work.

REFERENCES

1 Adler, A., and Henderson, H. (1994). A room of our own: Experiences from a direct office share. *Proceedings of CHI'94.* ACM: New York, 138 - 144.

2 Bly, S., Harrison, S., and Irwin, S. (1993). Media spaces: Bringing people together in a video, audio, and computing environment. *Communications of the ACM,* 36 (1), 28 - 47.

3 Dourish, P., Adler, A., Bellotti, V. and Henderson, A. (1994). Your place or mine? Learning from long-term use of video communication. Working Paper, Rank Xerox Research Centre, Cambridge Laboratory.

4 Fish, R., Kraut, R., Root, R., and Rice, R. Evaluating video as a technology for informal communication. *Proceedings of CHI'92.* ACM, New York, 37 - 48.

5 Gaver, W. The affordances of media spaces for collaboration. *Proceedings of CSCW'92* .

6 Gaver, W., Moran, T., MacLean, A., Lövstrand, L., Dourish, P., Carter, K., and Buxton, W. Realizing a

video environment: EuroPARC's RAVE system. *Proceedings of CHI'92*. ACM, New York, 27 - 35.

7   Gaver, W., Sellen, A., Heath, C. and Luff, P. (1993). One is not enough: Multiple views on a media space. *Proceedings of INTERCHI'93*. ACM: New York, 335 - 341.

8   Gibson, J. J. (1979). The ecological approach to visual perception. Houghton Mifflin, New York.

9   Heath, C., and Luff, P. (1992a). Collaboration and control: Crisis management and multimedia technology in London underground line control rooms. *CSCW Journal*, 1 (1-2), 69 - 94.

10 Heath, C., and Luff, P. (1992b). Media space and communicative asymmetries: Preliminary observations of video mediated interaction. *Human-Computer Interaction*, 7, 315 - 346.

11 Heath, C., Luff, P., and Sellen, A. (1994). Rethinking media space: The need for flexible access in video-mediated communication. Rank Xerox Research Centre technical report.

12 Hodges, L., and McAllister, D. (1987). True three-dimensional CRT-based displays. *Information Display*.

13 Ishii, H., Kobayashi, M., and Arita, K. (1994). Iterative design of seamless collaboration media. *Communications of the ACM*, 37 (8), 83 - 97.

14 Mantei, M., Baecker, R., Sellen, A., Buxton, W., Milligan, T., and Wellman, B. Experiences in the use of a media space. *Proceedings of CHI'91*. ACM, New York, 203 - 208.

15 Nardi, B., Schwarz, H., Kuchinsky, A. Leichner, R., Whittaker, S. and Sclabassi, R. (1993). Turning away from talking heads: The use of video-as-data in neuro-

surgery. *Proceedings of INTERCHI'93*. ACM: New York, 327 - 334.

16 Overbeeke, C., Smets, G., and Stratmann, M. (1987). Depth on a flat screen II. *Perceptual & Motor Skills, 65*.

17 Root, R. (1988). Design of a multimedia vehicle for social browsing. In Proceedings of the CSCW'88. ACM, New York. 25-38.

18 Rose, H. (1994). QuickTime VR: Much more than "virtual reality for the rest of us." *Converge*, August.

19 Sellen, A.. Speech patterns in video-mediated conversations. *Proceedings of CHI'92*. ACM, New York, 49 - 59.

20 Short, J., Williams, E., and Christie, B. *The social psychology of telecommunications*. London: Wiley & Sons, 1976.

21 Smets, G., Overbeeke, C., and Blankendaal (1995, submitted). Movement induced visual perception and resolution for product design. Submitted to *Automatica*.

22 Smets, G., Overbeeke, C., and Stratmann, M. (1987). Depth on a flat screen. *Perceptual & Motor Skills 64*, 1023 - 1034.

23 Smets, G., Stratmann, M., & Overbeeke, C. (1988). Method of causing an observer to get a three-dimensional impression from a two-dimensional representation. *US Patent* 4, 7575, 380.

24 Stappers, P. (1995, submitted). Tracking head movements in front of a monitor. Submitted to *Behaviour Research Methods, Instruments, and Computers*.

25 Stults, R. (1986). Media space. Xerox PARC technical report.

Virtual Reality on a WIM:
Interactive Worlds in Miniature

Richard Stoakley, Matthew J. Conway, Randy Pausch

The University of Virginia
Department of Computer Science
Charlottesville, VA 22903
{stoakley | conway | pausch}@uvacs.cs.virginia.edu
(804) 982-2200

KEYWORDS

virtual reality, three-dimensional interaction, two-handed interaction, information visualization

ABSTRACT

This paper explores a user interface technique which augments an immersive head tracked display with a hand-held miniature copy of the virtual environment. We call this interface technique the *Worlds in Miniature* (WIM) metaphor. In addition to the first-person perspective offered by a virtual reality system, a *World in Miniature* offers a second dynamic viewport onto the virtual environment. Objects may be directly manipulated either through the immersive viewport or through the three-dimensional viewport offered by the WIM.

In addition to describing object manipulation, this paper explores ways in which *Worlds in Miniature* can act as a single unifying metaphor for such application independent interaction techniques as object selection, navigation, path planning, and visualization. The WIM metaphor offers multiple points of view and multiple scales at which the user can operate, without requiring explicit modes or commands.

Informal user observation indicates that users adapt to the *Worlds in Miniature* metaphor quickly and that physical props are helpful in manipulating the WIM and other objects in the environment.

INTRODUCTION

Many benefits have been claimed formally and informally for using immersive three-dimensional displays. While virtual reality technology has the potential to give the user a better understanding of the space he or she inhabits, and can improve performance in some tasks [17], it can easily present a virtual world to the user that is just as confusing, limiting and ambiguous as the real world. We have grown accustomed to these real world constraints: things we cannot reach, things hidden from view, things beyond our sight

and behind us, and things which appear close to each other because they line up along our current line of sight. Our virtual environments should address these constraints and with respect to these issues be "better" than the real world.

In particular, we notice that many implementations of virtual environments only give the user one point of view (an all-encompassing, immersive view from within the head mounted display) and a single scale (one-to-one) at which to operate. A single point of view prohibits the user from gaining a larger context of the environment, and the one-to-one scale in which the user operates puts most of the world out of the user's immediate reach.

Figure 1: *The World In Miniature (WIM) viewed against the background of a life-size virtual environment.*

To address these two concerns, we provide the user with a hand-held miniature representation of the life-size world (figure 1). This representation serves as an additional viewport onto the virtual environment and we refer to it as a *World in Miniature* (WIM). The user can now interact with the environment by direct manipulation through both the WIM and the life-size world. Moving an object on the model moves the corresponding life-size representation of that object and vice versa.

Previous work has explored the use of miniature three-dimensional models for visualization and navigation. Our contribution is in extending this work and integrating it with object selection and manipulation under the unifying *Worlds in Miniature* metaphor. As an adjunct to the WIM, we have explored the advantages and disadvantages of grounding the user's perception of the model with a physical prop; in this case, a clipboard.

The rest of this paper discusses related previous work, a description of our WIM implementation, the basic interaction techniques we have used to demonstrate the effectiveness of the WIM concept, and the importance of asymmetric two-handed interaction for this metaphor. We conclude with results from informal user observation of the WIM interface and a discussion of future work.

PREVIOUS WORK

Many researchers have dealt with the open questions of three-dimensional object manipulation and navigation in virtual environments. The *World in Miniature* metaphor draws on these previous experiences, and attempts to synthesize an intuitive, coherent model to help address these questions. Previous work falls into three categories: object manipulation, navigation, and object selection.

Previous Work in Object Manipulation

Ware's Bat [22] interface demonstrates the use of a 6 degree-of-freedom (DOF) input device (a position and orientation tracker) to grab and place objects in a virtual environment. In this work, Ware used the bat to pick up and manipulate the virtual objects themselves, not miniature, proxy objects. Ware found that users easily understood the 1:1 mapping between translations and rotations on the input device and the object being manipulated. This study was a unimanual task and did not place the user's hands in the same physical space as the graphics.

In Sachs's 3-Draw [19], we see two hands used asymmetrically in a three-dimensional drawing and designing task. In addition to this, Sachs used props for each of the user's hands and found that relative motion between hands was better than a fixed single object and one free mover. 3-Draw was not implemented in an immersive, head-tracked environment and the system did not provide multiple, simultaneous views. The input props controlled the point of view by rotating the object's base plane.

Hinckley's [13] work with props exploited the asymmetric use of hands, which follows from work by Guiard [12]. This work showed how a prop in the non-dominant hand can be used to specify a coordinate system with gross orientation, while the user's preferred hand can be used for fine grain positioning relative to that coordinate system. This work is also three-dimensional but non-immersive and directly manipulates an object at one-to-one scale in a "fishtank VR" paradigm.

3DM [3] was an immersive three-dimensional drawing package, but provided only one point of view at a time and required the user to change scale or fly explicitly to manip-

ulate objects which were currently out of arm's reach. Butterworth states that users sometimes found the scaling disorienting.

Schmandt's [20] early explorations of Augmented Reality (AR) used a half-silvered mirror over a stationary drafting tablet in order to specify both a base plane and a slicing plane in computer generated VLSI models. He found this surface invaluable in constraining the user's input to a plane. The system was not immersive and presented a single scale at any given time.

Aspects of Bier's *Toolglass and Magic Lenses* [2] work, closely resemble a two-dimensional analog to the WIM. For example, when using a magnification lens, objects can easily be manipulated at a convenient scale, while maintaining much of the surrounding context.

Previous Work in Navigation

We define "navigation" to cover two related tasks: movement through a 3D space and determining orientation relative to the surrounding environment.

Darken's [7] discussion of navigating virtual environments enumerates many important techniques and compares their relative strengths and weaknesses. Several of the navigation techniques presented were WIM-like maps, but were primarily two-dimensional in nature. Through the WIM interface, some of these techniques have been extended into the third dimension.

Ware [23] explored the possibilities of holding the three-dimensional *scene* in hand for the purpose of quickly navigating the space. He found this *scene in hand* metaphor particularly good for quickly viewing the bounding-cube edges of a scene. The scene in hand task was a unimanual operation which employed ratcheting to perform large rotations.

The work most closely resembling the WIM interface was Fisher's *map cube* in virtual reality [9]. The NASA VIEW system used a three-dimensional miniature map of the immersive world to help navigate. In addition, the VIEW system used multiple two-dimensional viewports to jump from one place in the virtual environment to another. A user's manipulation of the "map cube" was unimanual. A similar map-cube concept was referred to as the God's-eye-view in the Super Cockpit project [11].

Previous Work in Object Selection

Many researchers have explored methods for selecting objects in a virtual world. Common approaches include raycasting [10] [22] and selection cones [14]. Both of these techniques suffer from object occlusion and therefore need to be tied closely with some mechanism that can quickly establish different points of view.

Put-That-There [4] used selection via a combination of pointing and naming (or description). Pointing in this two-dimensional application is analogous to raycasting in virtual environments.

SYSTEM DESCRIPTION

To explore the benefits and limitations of the WIM metaphor, we built a simple three-dimensional modeling package that could be used as a design tool for a traditional architecture design project called a *Kit of Parts*.

We provide the user with an immersive point of view through a headmounted display (HMD) tracked by a Polhemus 6 degree of freedom tracker. To control the position and orientation of the WIM, we outfitted the user's non-dominant hand with a clipboard attached to another Polhemus position sensor. The surface of the clipboard represents the floor of the WIM. By looking at the clipboard, the user observes an aerial view of the entire scene.

In his or her other hand, the user holds a tennis ball modeled after the UNC cueball, in which we have installed two buttons and another Polhemus sensor. This buttonball was used as the selection and manipulation tool for all of our user observations and WIM development. By pressing the first button on the buttonball, users can select objects; the second button was left open for application-specific actions (figure 2).

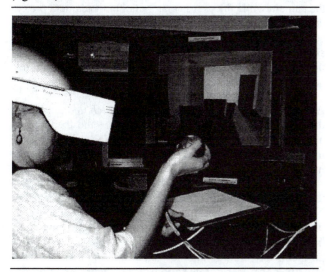

Figure 2: *A user manipulates the* WIM *using the physical clipboard and buttonball props.*

The WIM graphics attached to the clipboard are nothing more than a miniature copy of all the surrounding graphics in the immersive environment. Each of the objects in the WIM copy are tied to their counterparts in the immersive environment through pointers and vice versa at the point of WIM creation. In this way, when an object responds to a method call, the object has enough information to ensure that the same method gets called on its "shadow" object. Thus the user can manipulate the objects in the WIM and the objects in the world will follow (video figure 1 - The WIM Interface). The environment itself (in miniature) *becomes its own widget* for manipulating objects in the environment [26].

Software and Equipment

The Kit of Parts modeler was implemented using the Alice Rapid Prototyping System [6] running the simulation on a Sun Microsystems Sparc 10™ and rendering on a Silicon Graphics Onyx Reality Engine²™. Typical rendering rates were about 25 frames per second (FPS), while simulation rates were typically 6 FPS. A Virtual Research Flight Helmet™ was used for the display and was tracked with a Polhemus Isotrak™ magnetic tracker. The buttonball and clipboard each carried a Polhemus tracker sensor for position and orientation information.

INTERACTION TECHNIQUES USING THE WIM

In this section, we discuss basic application independent WIM-based interaction techniques.

Quickly Changing the POV

Being able to see objects from many different angles allows us to quickly remove or reduce occlusion and improves the sense of the three-dimensional space it occupies [21]. Because the WIM is a hand-held model, the user can quickly establish different points of view by rotating the WIM in both hands. Note that this form of "WIM fly-by" can often give the user all of the information that he or she needs without destroying the point of view established in the larger, immersive environment. We believe that this interaction technique can establish a new viewpoint more quickly and with less cognitive burden than a technique that requires an explicit "flight" command and management of the flight path.

Object Selection: Overcoming Range and Occlusion

If the virtual, immersive environment is sufficiently large, there will be objects that are out of physical arm's reach. If the user must touch an object to select it, the user would have to employ a separate flying mechanism, which means moving the camera; a sometimes disorienting or otherwise inappropriate approach. Armed with a *World in Miniature*, the user now has the choice of selecting objects either by pointing to the object itself (as before) or by pointing to its proxy on the WIM. By turning the model in his or her hands, the user can even view and pick objects that are obscured by his or her current line of sight from the immersive camera viewpoint. The WIM provides a second (often "bird's eye") point of view from which to examine the scene.

Object Manipulation

Once objects are selected, the WIM allows us to manipulate those objects at either the scale offered by the WIM or the one-to-one scale offered by the immersive environment. If the scale of the WIM is smaller than that of the immersive world, manipulating objects on the WIM necessarily gives the user far-reaching coarse grained control of objects.

The WIM can also display objects at a greater than one-to-one scale, implementing a three-dimensional magnifying glass of sorts. This gives the user very fine grain control of objects through the WIM at the expense of range. Though we have not implemented zooming in our current system, we clearly see the need for allowing the user to get more detail on the WIM or to zoom out to view more context. We are currently pursuing this avenue of research.

We speculate that because the WIM is clearly a model attached to the user's hand, it is seen as something separate from the rest of the immersive environment. The WIM therefore naturally offers two different scales to the user without requiring explicit modes or commands.

An Example: Hanging a Picture

Putting these ideas together, we can consider an example task: hanging a picture on a wall. This task is typical of a broad class of two-person tasks in which the proximity required to manipulate an object interferes with the desire to see those effects in a larger context. With a WIM, a single user can stand at a comfortable distance to view the picture in context, while at the same time reaching into the WIM to manipulate it.

Of course, the user could choose to use the WIM the other way around: fly close to the wall to stand next to the picture, then use the WIM to view the entire room in miniature to determine if the picture is straight. Examining relative strengths and weaknesses of each of these approaches is an area for further study.

There are other common ways to approach the manipulation of objects at a distance. For example, a ray can be used to select and manipulate an object. Unlike the WIM, raycasting suffers from an occlusion problem and a lever arm problem, as mentioned before.

Mixing Scales and Operations

Viewing, selection, and manipulation are independent operations. Because the WIM gives the user another scale at which to operate, the user can choose the most appropriate scale for any given subtask, and even switch scales in the middle to suit the requirements of the task. For example: the user can reach into the WIM to *select* a distant object (taking advantage of the greater than 1:1 scale of the WIM), and then reach out to the immersive world to *move* the WIM-selected object *at a distance* in 1:1 scale [22] [14] all the while *viewing* the scene in the WIM.

Rotation

Our current implementation allows users to rotate objects, through *ratcheting* (repeated grabbing, rotating and releasing) [24] and is therefore more awkward than a rotation done with just the fingers [14]. Interestingly, some users found it just as effective to grab the object and to counter-rotate the entire WIM.

In our current implementation, rotation is gridded to 30 degree increments, primarily to assist in aligning rectilinear objects [14]. We found that if the rotation grid is too course (greater than about 45 degrees), some people assume that they cannot rotate at all and if set to 15 degrees or less, users report that aligning rectilinear objects is very difficult.

Navigation: Flight with a WIM

To make the view travel through the immersive environment, the most common user interface technique in virtual environments is probably "flying." If the WIM includes some representation of the user as an object in the scene, the user can simply reach into the WIM and "pick himself up" to change his location in the environment. This raises the question of when to update the immersive world as objects in the WIM are manipulated. We enumerate three possibilities.

Updating after Manipulation: immediate, post-facto and batch

When changes are made on the WIM, we usually move the real object and the proxy object simultaneously, something we refer to as *immediate update*. Under some conditions, immediate update is either not desirable (due to visual clutter or occlusion) or impossible (the complexity of the object prevents changes to it from being updated in real time). In these situations, we use *post-facto update*, where the immersive environment updates only after the user is finished with the WIM interaction and has released the proxy.

A good special case of post-facto update is the case of moving the user's viewpoint. We find that immediate update of the camera while the user is manipulating the camera proxy is highly disorienting, instead we wait until the user has stopped moving the camera, and then use a smooth slow-in/slow-out animation [15] to move the camera to its new position. This animated movement helps maintain visual continuity [14].

Another useful form of update delay is *batch update*. Here, the user makes *several* changes to the WIM and then issues an explicit command (e.g. pressing the second button on the buttonball) to cause the immersive environment to commit to the current layout of the WIM. This is useful for two reasons. First, before the user commits his or her changes, the user has two independent views of the environment (the "as-is" picture in the immersive world and the "proposed" picture in the WIM). Secondly, it might be the case that moving one object at a time might leave the simulation in an inconsistent state, and so "batching" the changes like this gives the user a transaction-like *commit* operation on the changes to objects in the scene (with the possibility of supporting *rollback* or *revert* operations if the changes seem undesirable halfway through the operation).

VISUALIZATION

The *Worlds in Miniature* metaphor supports several kinds of displays and interaction techniques that fall loosely under the heading of visualization. These techniques exploit the WIM's ability to provide a different view of the immersive data with improved context. It would seem that the WIM is good for visualization for all the same reasons that a map is good for visualization:

Spatially locating and orienting the user: the WIM can provide an indicator showing where the user is and which way he or she is facing relative to the rest of the environment.

Path planning: with a WIM we can easily plan a future camera path in three-dimensions to prepare for an object fly-by. The user can even preview the camera motion before committing him or herself to the change in the larger, immersive viewpoint.

History: if the user leaves a trail behind as he or she travels from place to place, the WIM can be used like a regular 2D map to see the trail in its entirety. Dropping a trail of crumbs is not as useful if you cannot see the trail in context [7].

Measuring distances: the WIM can be configured to display distances between distant (or very closely spaced) points that are difficult to reach at the immersive one-to-one scale. The WIM also provides a convenient tool for measuring areas and volumes.

Viewing alternate representations: the immersive environment may be dense with spatially co-located data (i.e. topological contours, ore deposits, fault lines). The user can display data like this on the WIM, showing an alternate view of the immersive space in context. The improved context can also facilitate the observation of more implicit relationships in a space (e.g. symmetry, primary axis) and can display data not shown in the immersive scene (e.g. wiring, circulation patterns, plumbing paths). Here, the WIM might act as a filter; similar to a three-dimensional version of Bier's "magic lenses" [2] or one of Fitzmaurice's "active maps" [10].

Three-Dimensional Design: the WIM serves many of the same functions that architectural models have traditionally served (e.g. massing model, parti diagram).

MULTIPLE WIMS

Until now, we have considered only a single instantiation of a WIM in a virtual environment, but clearly there might be a reason to have more than one such miniature active at a time. Multiple WIMs could be used to display:

- widely separated regions of the same environment
- several completely different environments
- worlds at different scales
- the same world with different filters applied to the representation
- the same world displayed at different points in time

This last option allows the user to do a side by side comparison of several design ideas. A logical extension of this notion is that these snapshots can act as *jump points* to different spaces or times, much the same way hypertext systems sometimes have thumbnail pictures of previously visited documents [1]. Selecting a WIM would cause the immersive environment to change to that particular world [9] (video figure 2 – Multiple WIMs).

Multiple WIMs enable users to multiplex their attention much the same way Window Managers allow this in 2D. These multiple views into the virtual world, allow the user to visually compare different scales and/or different locations [8].

MANIPULATING THE WIM

Through the exploration of the previous interfaces, several issues arose concerning the interface between the human and the WIM tool.

The Importance of Props

One of our early implementations of the WIM work did not use physical props; the user grasped at the WIM graphics as he or she would any other graphical object in the scene. As long the user continued the grasping gesture, the WIM followed the position and orientation of the user's hand and when released, it would remain hovering in space wherever it was dropped. While this was sufficient for many tasks, we found that rotating the WIM without the benefit of haptic feedback was extremely difficult. Invariably, users would contort themselves into uncomfortable positions rather than let go of the WIM to grab it again by another, more comfortable corner.

After Sachs [19], we decided to use physical props to assist the user's manipulation of the WIM itself. To represent the WIM, we chose an ordinary clipboard to which we attached a Polhemus 6 DOF tracker for the user's non-dominant hand. For the user's preferred hand, we used a tennis ball with a Polhemus tracker and two buttons.

Props: The Clipboard

This prop allows the user to rotate the WIM using a two-handed technique that passes the clipboard quickly from one hand to the other and back when the rotation of the WIM is greater than can be done comfortably with one hand. Interestingly, some users hold the clipboard from underneath, rotating the clipboard deftly with one hand. Both of these techniques are hard to imagine doing in the absence of the haptic feedback provided by a physical prop.

Props: The Buttonball

Before we settled on the buttonball as our primary pointing device, we experimented with a pen interface to the WIM. This technique is most appropriate for manipulation of objects when they are constrained to a plane [20] (the base plane being the default). When manipulation of objects in three-dimensions is called for, a pen on the surface of the clipboard does not appear to be expressive enough to capture object rotation well.

Two-Handed Interaction

Our implementation of the WIM metaphor takes advantage of several previously published results in the field of motor behavior that have not been fully exploited in a head tracked virtual environment [2] [12]. The most important of these results state that a human's dominant (preferred) hand makes its motions relative to the coordinate system specified by the non-dominant hand, and the preferred hand's motion is generally at a finer grain [12]. In our case, the non-dominant hand establishes a coordinate system with the clipboard and the dominant hand performs fine grained picking and manipulation operations.

While the dominant hand may be occupied with a pointing device of some kind, it is still sufficiently free to help the other hand spin the WIM quickly when necessary.

Shape of Props

Like all real world artifacts, the shape of the props and the users' experience suggest things about the usage of the props [16]. For example, the shape of the clipboard says something to users about its preferred orientation. The cursor's physical prop is spherical, indicating that it has no preferred orientation, and in fact it does not matter how the cursor is wielded since rotation is relative to the plane specified with the non-dominant hand, which holds the clipboard.

The clipboard also provides a surface that the user can bear down on when necessary. This is similar to the way an artist might rest his or her hand on a paint palette or a guitarist might rest a finger on the guitar body.

PROBLEMS WITH PHYSICAL PROPS

The use of the clipboard as a prop presents some problems of its own.

Fatigue

Holding a physical clipboard, even a relatively light one, can cause users to fatigue rather quickly. To overcome this problem, we created a simple clutching mechanism that allows the user to alternately attach and detach the WIM from the physical prop with the press of a button. When detached, the WIM "floats" in the air, permitting the user to set the prop down (video figure 3 – Prop Clutching). This clutching mechanism extended well to multiple WIMs: when the user toggles the clutch, the closest WIM snaps to the user's clipboard. Toggling the clutch again disengages the current WIM and allows the user to pick up another WIM.

Another technique for relieving arm stress is to have the user sit at a physical table on which the clipboard could be set. Users can also rest their arms on the table while manipulating the model. The presence of the table clearly presents a mobility problem for the user because it prevents the user from moving or walking in the virtual environment and so may not be ideal for all applications.

Limitations of Solid Objects

In our experience, one of the first things a user of the WIM is likely to try is to hold the WIM close to his or her face in order to get a closer, more dynamic look at the world. Users quickly discover that this is an easy, efficient way to establish many different points of view from inside the miniature. Unfortunately, many times the physical prop itself gets in the way, preventing the user from putting the tracker in the appropriate position to get a useful viewpoint. Fortunately, the ability to disengage the WIM, leaving it in space without the clipboard helps alleviate this problem.

INFORMAL OBSERVATION

We observed ten people using the WIM. Some had previous experience with virtual reality, and some had three-dimensional design experience (e.g. architecture students). Users were given a simple architectural modeler and asked to design an office space. We proceeded with a rapid "observe, evaluate, revise" methodology to learn about the *Worlds in Miniature* interface.

The user was given a miniature copy of the room he or she was standing in, and a series of common furniture pieces on a shelf at the side of the WIM (figure 3). The user was then asked to design an office space by positioning the furniture in the room.

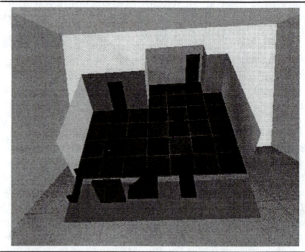

Figure 3: *The starting configuration of the Kit of Parts* WIM. *The parts themselves are at the bottom, outside the regular working area.*

In many ways, this design task replicates the traditional architectural design project known as a *Kit of Parts*. The furniture pieces (half-height walls, shelves, tables and table corner pieces) represent the kit of manipulable objects. Moving the traditional kit of parts project into virtual reality was particularly appealing for several reasons:

- It constrains the space in which the user can work.
- Since users are designing a life-size space, clearly seeing that space at 1:1 scale is helpful.
- The WIM succinctly replaces the plans, elevations and models typically associated with this type of design project.

The WIM that we used was a 1/4" scale version of the immersive world, with a snap spacing of 1/8" (0.5 scale feet). In addition to the translation snap, rotation was constrained to be about the Z axis in increments of 30 degrees.

User Reactions

We observed users in order to see how viable a solution the WIM interface was to several types of tasks. While it was not our intention for the study to produce concrete numbers, we were after what Brooks refers to as interesting "Observations" [5]. We hoped to gain some sense of:

1) How quickly do users take to the WIM metaphor?
2) How do users like the weight and maneuverability of the physical clipboard?
3) Do users like clutching? Do they take to it easily?
4) How do users feel about moving themselves via the camera proxy?

None of the users expressed problems establishing the mapping between objects on the WIM and objects in the life-size virtual world. Several users were even able to grab and manipulate objects on the WIM without explicitly looking at the WIM itself. This was the strongest evidence that people had developed an internal model of the three-dimensional space established by the WIM in their hands.

Arm fatigue was not a major problem. Subjects developed a pattern of raising the WIM, making the manipulations and then lowering the WIM out of view to look at the life-size world. Also, subjects were sitting, which allowed them to rest their arms on their legs and adjust their gaze to look down at the WIM. The weight of the head-mount was the only fatigue complaint we received, although most users were not immersed for more than fifteen minutes.

The participants generally understood and liked the clutching mechanism. They seemed to have no problem associating the second button on the buttonball with the clutching of the clipboard.

Users found that rotating the camera with immediate update of the user view was very disorienting. Not surprisingly, we also found that the animation of the user's view from a starting point to it's ending point, needed to be as smooth as possible. Problems arose when rendering frame rates were insufficient due to environment complexity and when users tried to move great distances in a short period of time, causing large translation deltas for any given rendering frame.

Additional Observations

Intentional misregistration: Interestingly, none of our users noticed that the position of the WIM is about a ten centimeters above the physical location of the clipboard. This was done to allow users to place the center of the buttonball on the floor of the WIM. Without this offset, the closest a user could get to the floor is the radius of the buttonball. When asked whether the miniature world was in registration with the clipboard, all the users said yes, even though they often moved the cursor through the base plane of the virtual model during placement tasks. This of course should be impossible because the clipboard would prevent the physical buttonball from penetrating its plane. We believe this to be a good example of the perceptual phenomena known as "visual capture", (roughly: the visual perception system dominating the proprioceptive system) [18].

Selection by trolling: Pressing the buttonball's selection button and *trolling* through objects in order to select them seemed a much more effective selection technique than the traditional "position-then-select" technique. Because the world was sparsely populated with objects, users had few problems accidentally passing through (hence selecting) the wrong object. Few users discovered the trolling technique on their own.

IMPLEMENTATION ISSUES

We found it extremely useful to take advantage of the well known *backface culling* option on the SGI hardware. By modeling rooms such that the polygons making up interior walls face into the room, the viewer could look at the WIM from any angle without the exterior walls occluding the

interior while leaving the walls on the opposite side of the WIM visible (video figure 4 - Backface Culling). We later decided that this *backface culling* greatly improved the usefulness of the WIM for many tasks, and had the hardware not performed this, we would have implemented it ourselves. Interestingly, none of our users in our informal studies noticed this effect until it was pointed out to them.

Another issue related to our software implementation was that the lighting effects on the model do not represent the lighting effects seen on corresponding objects in the virtual world. This may not be a problem for many applications, but will probably be an issue for classically trained architects who traditionally use models to test lighting.

FUTURE WORK

Because of the encouraging experience we have had with the current WIM implementation, we intend to continue exploring some of the issues behind WIM-based interaction.

Value of the Prop: While anecdotal evidence suggests that the WIM interface with the clipboard prop is faster than manipulating the WIM with a virtual hand, an informal study is planned to determine to what extent the presence of the prop enhances the WIM interface.

Scrolling: At some point, scaling issues cause a straightforward WIM interface to break down. For example: if the WIM's scale is very small, selection and manipulation become to difficult. One solution is to maintain a reasonable scale, and allow the user to scroll the world into the working volume of the WIM.

Clipping: Clipping a large WIM in X and Y seems intuitive, (clip to the comfortable limits of the user's reach) but the clipping in Z remains an issue yet to be explored.

Zooming: When the immersive world is large, we need to adjusting the scale of the WIM to keep the working volume within the user's physical reach. We intend to explore the best ways of extending this functionality to the WIM.

Multiple Participants: The WIM may serve as a useful tool for multiple user interaction. Any number of users could manipulate their own personal WIM and all participants could watch the effects in the life-size environment and on their own WIM's.

Three-Dimensional Design: We would like to give the WIM to more architecture students to determine whether it is a more effective design medium for some tasks than traditional pencil, paper and foam core modeling board.

Two-Dimensional Widgets in VR: Even though the environment is three-dimensional, it seems clear that there is still a place for some of the 2D widgets to which we have become accustomed (e.g. buttons). We have speculated that the clipboard, and possibly a table would be viable surfaces on which to project standard 2D GUI widgets, whenever those controls would make sense. We see this as an interesting opportunity to explore ways of "recovering" decades of

2D GUI results in the relatively new arena of virtual environment interfaces and head tracked displays [25].

CONCLUSION

The WIM interface gives the user of an immersive three-dimensional environment a chance to operate at several different scales through several different views at once without engaging explicit modes. From our informal user observations, we believe some of the most important features of this technique to be:

- intuitive use of the WIM metaphor
- multiple, simultaneous points of view
- multiple scales for selecting and manipulating objects
- easy manipulation and selection of objects at a distance
- improved visual context
- removal of occlusion and line of sight problems
- general applicability across a wide range of application domains

The most encouraging result from our informal observations was that users fluidly used multiple capabilities of the WIM. For example, users often manipulated objects in the *World in Miniature* while simultaneously changing their point of view. This level of integration implied to us that users were able to focus on the task without being distracted by the interface technology.

ACKNOWLEDGEMENTS

We would like to thank Rich Gossweiler, Ken Hinkley, Tommy Burnette, Jim Durbin, Ari Rapkin, and Shuichi Koga for their help in the construction of the WIM interface, for the creation of the supporting video tape, and for their many invaluable ideas. We would also like to thank the SIGCHI reviewers for their many thoughtful suggestions and improvements.

REFERENCES

[1] Apple Computer, Inc., *Hyperscript Language Guide: The Hypertalk Language.* Addison-Wesley 1988.

[2] Eric A. Bier, Maureen C. Stone, Ken Pier, William Buxton, Tony D. DeRose, *Toolglass and Magic Lenses: The See-Through Interface.* SIGGRAPH '93, pp. 73-80.

[3] Jeff Butterworth, Andrew Davidson, Stephen Hench, Marc Olano, *3DM: A Three Dimensional Modeler Using a Head-Mounted Display.* Proceedings 1992 Symposium on Interactive 3D Graphics, pp. 135-138.

[4] Richard A. Bolt. *Put-That-There.* SIGGRAPH '80, pp. 262-270.

[5] Frederick P. Brooks, *Grasping Reality Through Illusion: Interactive Graphics Serving Science.* SIGCHI '88, pp. 1-11.

[6] Robert DeLine, Master's Thesis. *Alice: A Rapid Prototyping System for Three-Dimensional Interactive Graphical Environments.* University of Virginia, May, 1993.

[7] Rudy Darken, John Sibert, *A Toolset for Navigation in Virtual Environments.* UIST '93, pp. 157-165.

[8] Steven Feiner, Clifford Beshers, *Visualizing n-Dimensional Virtual Worlds with n-vision.* Proceedings 1990 Symposium on Interactive 3D Graphics, pp. 37-38.

[9] Scott Fisher, *The AMES Virtual Environment Workstation (VIEW).* SIGGRAPH '89 Course #29 Notes.

[10] George Fitzmaurice, *Situated Information Spaces and Spatially Aware.* Communications of the ACM, 36 (7), 1993, pp. 39-49.

[11] Dr. Thomas A. Furness, III, *The Super Cockpit and Human Factors Challenges.* Human Interface Technology (HIT) Laboratory of the Washington Technology Center, September 1986

[12] Yves Guiard, *Asymmetric Division of Labor in Human Skilled Bimanual Action: The Kinematic Chain as a Model.* The Journal of Motor Behavior, pp. 486-517, 1987.

[13] Ken Hinckley, Randy Pausch, John C. Goble, Neal F. Kassell, *Passive Real-World Interface Props for Neurosurgical Visualization.* SIGCHI '94, pp. 452-458.

[14] Jiandong Liang, Mark Green, *JDCAD: A Highly Interactive 3D Modeling System.* 3rd International Conference on CAD, pp. 217-222, August 1993.

[15] John Lassiter, *Principles of Traditional Animation Applied to 3D Computer Animation.* SIGGRAPH '87, pp. 35-44.

[16] Donald Norman, *The Design of Everyday Things.* Doubleday. 1988.

[17] Randy Pausch, M. Anne Shackelford, Dennis Proffitt, *A User Study Comparing Head-Mounted and Stationary Displays.* Proceedings '93 IEEE Symposium on Research Frontiers in Virtual Reality, pp. 41-45.

[18] Irvin Rock, *The Logic Of Perception.* Cambridge, p.70. MIT Press, 1983.

[19] Emanuel Sachs, Andrew Robert, David Stoops, *3 Draw: A Tool for Designing 3D Shapes.* '91 IEEE Computer Graphics and Applications, pp. 18-25.

[20] Chris Schmandt, *Spatial Input/Display Correspondence in a Stereoscopic Computer Graphic Work Station.* SIGGRAPH '83, pp. 253-262.

[21] Ivan Sutherland, *A Head-Mounted Three Dimensional Display.* Fall Joint Computer Conference, pp. 757-764, 1968.

[22] Colin Ware, Danny R. Jessome, *Using the Bat: A Six-Dimensional Mouse for Object Placement.* '88 IEEE Computer Graphics and Applications, pp. 65-70.

[23] Colin Ware, Steven Osborne, *Exploration and Virtual Camera Control in Virtual Three Dimensional Environments.* Computer Graphics 24 (2), pp. 175-183, 1990.

[24] Colin Ware, *Using Hand Position for Virtual Object Placement.* The Visual Computer 6 (5), pp. 245-253. Springer-Verlag 1990.

[25] David Weimer, S. Ganapathy, A synthetic visual environment with hand gesturing and voice input. CHI '89, pp. 235-240.

[26] Robert C. Zeleznik, Kenneth P. Herndon, Daniel C. Robbins, Nate Huang, Tom Meyer, Noah Parker, John F. Hughes, *An Interactive 3D Toolkit for Constructing 3D Widgets.* SIGGRAPH '93, pp. 81-84.

The "Prince" Technique: Fitts' Law and Selection Using Area Cursors

Paul Kabbash
Input Research Group
Computer Systems Research Institute
University of Toronto
Toronto, Ontario,
Canada M5S 1A4
Tel: +1 (416) 978-6619
kabbash@dgp.toronto.edu

William Buxton
Alias Research, Inc.&
University of Toronto
c/o Alias Research, Inc.
110 Richmond St. East
Toronto, Ontario,
Canada M5C1P1
Tel: +1 (416) 362-9181
buxton@alias.com

ABSTRACT

In most GUIs, selection is effected by placing the point of the mouse-driven cursor over the area of the object to be selected. Fitts' law is commonly used to model such target acquisition, with the term A representing the amplitude, or distance, of the target from the cursor, and W the width of the target area. As the W term gets smaller, the index of difficulty of the task increases. The extreme case of this is when the target is a point. In this paper, we show that selection in such cases can be facilitated if the cursor is an area, rather than a point. Furthermore, we show that when the target is a point and the width of the cursor is W, that Fitts' law still holds. An experiment is presented and the implications of the technique are discussed for both 2D and 3D interfaces.

KEYWORDS: Input techniques, graphical user interfaces, Fitts' law, haptic input.

INTRODUCTION

Although the traditional method of selection in direct-manipulation systems is generally effective, there are certain conditions where it breaks down. One of these is when the target is very small. The extreme case of this is when the target is a point. The reason for the problem can best be explained by Fitts' law [3, 8], expressed as:

$$ID = \log_2(A/W + 1). \qquad (1)$$

From this formulation we see that the index of difficulty (*ID*) of a target acquisition task is a function of the amplitude (*A*), or distance, of the target from the cursor, and the width of the target (*W*). The index of difficulty rises as the width of the target gets smaller.

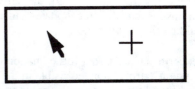

Figure 1: Two typical "point" cursors. Selection with conventional GUIs is literally "pointing," for example, with the point of the arrow or the intersection point of the cross-hair.

With most conventional GUIs, the selection tool is a point, such as represented by the point of an arrow shaped cursor, or the intersection of the lines in a cross-hair cursor (Figure 1). In the extreme case, therefore, we are selecting a point with a point.

The purpose of the research described in this paper is to explore an alternative approach whereby (in the 2D case) the cursor is represented by an *area*, rather than by a point. Just as the area of a fly-swatter makes it easier to swat a small fly, likewise the area of such a cursor should make it easier to select small targets and points.

More formally, it is our claim that selecting a small target with an area cursor can be modeled by a slight twist of Fitts' law, namely, that the *W* term now applies to the width of the cursor, rather than the width of the target. Figure 2 illustrates the approach using an area cursor, as well as the traditional approach.

In what follows, we report on an experiment that demonstrates the applicability of Fitts' law to selecting point targets with an area cursor. We follow this with a discussion of the design implications of our findings to other 2D and 3D tasks.

Finally, due to the similarity of their benefits, we name the use of area cursors after the first manufacturer of oversized tennis rackets: the *Prince* technique.

THE EXPERIMENT

Subjects

Twelve students from the University of Toronto participated as paid volunteers. All had experience using the mouse and were strongly right handed based on the Edinburgh Handedness Inventory [10].

Apparatus

Equipment was an Apple *Macintosh IIfx* with 13-inch RGB monitor. Subjects performed the tasks using their right hand and a standard mouse. The control/display ratio of the mouse was adjusted to the second fastest setting on the *Macintosh* Control Panel. Since even small lags (75 ms) in display response have been found to degrade human performance on Fitts' law tasks [5, 9], the software was optimized to ensure that drawing updates did not delay movement of the Prince cursor.

Before the experiment, we tested for animation delays by making the system "arrow" pointer visible in the center of the paddle and found that it was not possible to shake the paddle from the arrow, even at movement speeds likely to be much faster than those encountered during the experiment.

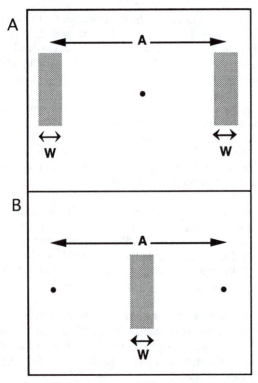

Figure 2: Two versions of Fitts' reciprocal aiming task. The conventional approach is shown in 2(a). Here a standard "point" cursor is moved amplitude *A* between two targets of width *W*. In 2(b) we see the variation, where an area cursor of width *W* is moved between two point targets separated by distance *A*.

Procedure

Subjects performed a reciprocal point-select task using both the Fitts and Prince techniques (Figure 2). They were given written instructions and several warm-up trials prior to data collection. In addition, they performed one practice session on each technique.

For each technique, two targets appeared on either side of the monitor. Subjects moved the cursor back and forth between the targets and selected each target by pressing and releasing the mouse button. They were instructed to balance speed and accuracy for an error rate around 4%, and an error beep sounded if selection occurred outside the target. Results of movement time and error rate were given to subjects at the end of each session.

The cursor and the target were represented in the two techniques using different objects. In the Fitts condition the cursor was a small black dot with radius 2 pixels, and the targets were rectangles having width *W* and height 200 pixels. The Prince condition reversed these objects exactly (see Figure 2b), so that subjects controlled a rectangular cursor (width *W*, height 200 pixels) and used it to capture two target dots (radius 2 pixels). In both conditions, the rectangles were unframed and shaded light blue. The Prince cursor was transparent, so that the targets could be clearly seen beneath it. When the target and cursor overlapped, their appearance in the two techniques was nearly indistinguishable, the primary difference being which object moved or was stationary.

Design

A fully-crossed, within-subjects factorial design with repeated measures was used. Factors were movement amplitude (64, 128, 256, and 512 pixels), target or cursor width (8, 16, 32, and 64 pixels), and technique (Fitts and Prince). The amplitude and width conditions yielded seven levels of task difficulty, ranging from 1 bit to 6.02 bits. The *A-W* conditions were presented in random order with a block of ten trials performed at each condition. A session consisted of a sequence of sixteen blocks covering all *A-W* conditions. After training, ten sessions were performed in all, alternating between the Fitts (five sessions) and the Prince (five sessions) techniques. The order of techniques was counterbalanced, with half of the subjects beginning with the Prince technique and the other half with the Fitts technique. Subjects took about one hour to complete the experiment.

Dependent variables were movement time (*MT*), error rate (*ER*), constant error (*CE*), and variable error (*VE*). The latter two measures were used to describe the quality of placement of the response selections [11]. *CE* was measured in the horizontal axis, as the signed distance between target and cursor centers at the moment a selection occurred, and was used to detect systematic trends towards undershooting or overshooting the target center. *VE* captures the endpoint variability of responses and corresponds to effective target width ($W_e = 4.133 \times SD_x$; see [8]).

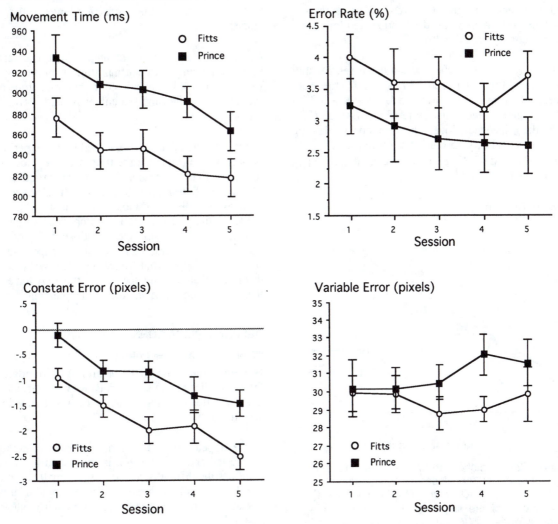

Figure 3: Means (with standard error bars) for movement time, error rate, constant error and variable error, decomposed by technique and session.

Results

There were significant main effects of technique on all four dependent variables. While subjects were slower using the Prince technique (900 ms vs. 841 ms; $F_{1,10} = 273.5$, $p < .001$), they produced fewer errors (2.8% vs. 3.6%; $F_{1,10} = 15.7$, $p < .005$) and tended to aim nearer to the target centers ($CE = -.922$ vs. -1.787 pixels; $F_{1,10} = 27.0$, $p < .001$). Subjects therefore appear to have been more careful when making selections with the Prince technique. This did not however provide an advantage in terms of motor response variability. With both techniques, VE was very close to the average nominal width of 30 pixels. However, VE was 30.9 pixels in the Prince technique and 29.5 pixels in the Fitts technique, so subjects were somewhat less variable in endpoint placement using the Fitts technique ($F_{1,10} = 9.40$, $p < .02$).

This suggests that the bias towards accuracy rather than speed in the Prince technique was due to its unfamiliarity. We investigated this possibility further, by examining performance in relation to learning phase over the five sessions (Figure 3). The analysis revealed a main effect of

session on MT ($F_{4,40} = 11.8$, $p < .001$) representing a small improvement for both techniques, in total less than 8% from sessions 1 to 5. A significant effect of session on CE ($F_{4,40} = 10.4$, $p < .001$) suggests that subjects also increasingly undershot the targets as they grew more confident with the task. They were able to do so without incurring greater errors or endpoint variability, as evidenced by the lack of session effects on ER ($F_{4,40} = 1.07$, $p > .05$) and VE ($F_{4,40} = .543$). However, the two-way interaction of technique x session was not significant for any of the four dependent variables ($F_{4,40} < 1.06$, $p > .05$), implying that the speed-accuracy tradeoff in the Prince technique did not change relative to the Fitts technique even as subjects progressed through the trials. Thus, it is unclear from the present data to what extent further practice with the Prince technique would have altered its accuracy bias.

Fit of the Model

Since there were no significant interactions on MT between session effects and those of the other independent variables, regression lines were fitted to the MT data averaged over sessions 1 to 5 (Figure 4). The regression analyses were performed on the data normalized for errors, using the method described in [9].

As predicted by Fitts' law, there was a linear relationship and high correlation between MT and ID for both techniques. The fitted equations were $MT = 198 + 204\ ID$ ($r = .978$) for the Fitts technique and $MT = 267 + 203\ ID$ ($r = .965$) for the Prince technique, with the model explaining 95.6% of the variability in the Fitts technique ($F_{1,15} = 302.5$, $p < .001$) and 93.2% in the Prince technique ($F_{1,15} = 191.4$, $p < .001$).

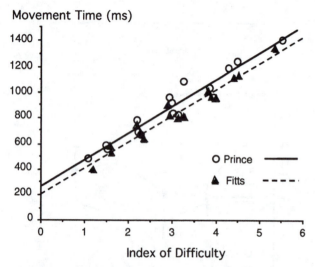

Movement Time (ms)

Figure 4: Scatter-plots of the MT-ID relationship in the Fitts and Prince techniques. The equation fitted was $MT = a + b\ ID$, where $ID = \log_2(A/W_e + 1)$. The reciprocal of the slope of each line gives the index of performance ($IP = 1/b$) for the technique.

As is evident in Figure 4, the slopes of the two regression lines did not differ. Each yielded an index of performance (IP) of about 4.9 bits/s, which is comparable to earlier experiments [7]. Because the techniques had different intercepts, however, the MT-ID relationship in the Fitts technique was displaced slightly downward with respect to the Prince technique.

Individual subject regressions also were computed. These exhibited similar trends with no differences between techniques for slope ($F_{1,10} = .416$), but a significantly greater positive intercept in the Prince technique ($F_{1,10} = 22.1$, $p < .001$). The correlation coefficients obtained for individual subjects were, with one exception, greater than .90 and revealed no differences between techniques ($F_{1,10} = 1.04$, $p > .05$). Hence the results for the aggregated regressions given in Figure 4 appear to have been consistently exhibited within each subject.

Discussion

While the results clearly demonstrate that Fitts' law applies to the Prince technique, a question remains as to why the performance differences between techniques were reflected

in the intercept of the regression lines, rather than the slope.

One interpretation of the intercept is that it represents time spent on the targets rather than time spent moving between them [13, p. 146]. Considered in this way, "time on target" includes only the time the cursor is held motionless over the target.

For the reciprocal aiming task used in this experiment, time on target would include the time necessary for the subject to verify that the cursor is over the target, the time to execute the button press itself, as well as preparation time to program the next movement, as in [1]. There is evidence that the verification component, in particular, is sensitive to the accuracy demands and objectives of the task [1, 6, 12]. Thus, if subjects in our experiment were in fact being somewhat more careful with the Prince technique than the Fitts technique, this may have increased their verification time and hence the intercept.

DESIGN IMPLICATIONS

"The Prince and the Pointer"

Having established that the area and standard cursors follow similar prediction models, we now consider some of the properties of using an area cursor as a positioning and selecting tool.

There appear to be two main benefits of applying the Prince technique. The first is illustrated in Figure 5 and contrasts the difficulty of acquiring a small target using the standard cross-hair cursor and a rectangular area cursor. For such tasks, the area cursor approach is clearly much easier.

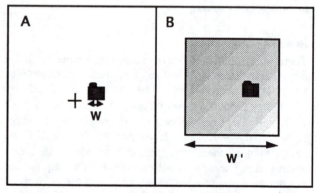

Figure 5: In 5(a) the target is selected using a standard "cross-hair" cursor. The difficulty of the task is limited by the size of the target (W). In 5(b), an area cursor with width W' surrounds the target to select it. The difficulty of this task is a function of W'.

We can use the results from the experiment to quantify the difference between the two approaches. Assume the target has width W, the area cursor width W', and the distance moved in both cases is A. Then, when A/W is large (i.e., the task is hard), the difference in index of difficulty (ID, equation 1) for the two tasks will approach $\log_2(W'/W)$.

For example, if W is 6 pixels and W' is 96 pixels, the Prince technique represents a savings of about 4 bits as rated by Fitts' law. Given the performance level arrived at in the experiment (4.9 bits/s), this translates to a movement time savings of roughly .75 s per mouse selection. In the case of A = 384 pixels, for instance, this is a 93% reduction. (Of course, the movement time savings will be even larger with a device that does not perform as well as the mouse; e.g., using IP = 1.5 bits/s, reported in [7] for trackball performance during a dragging task, the predicted savings are well over 2 s per selection.)

A second capability of the area cursor is that it may function as a "net." Used in this way, an area cursor can group and select a collection of points or small objects with a single pointing movement, much as the "lasso" tool is used in drawing applications like MacDraw. This capability, however, also serves to illustrate a drawback with using the area cursor as the only selection tool in a GUI. This is that the Prince technique is inappropriate for fine positioning tasks, because selections may become ambiguous when displays are cluttered.

Our belief is that an effective way of exploiting the Prince technique is to combine it with the traditional point-cursor approach. Where fine positioning is not required, it may be possible to replace it by coarse positioning and the Prince technique. Furthermore, by dynamically switching between Prince and point-cursor positioning techniques, the difficulty of positioning tasks can be matched more closely to task context.

We illustrate this using three examples.

Example 1: Dragging a File into a Folder

Figure 6 gives an example of how use of the area and point cursors can be combined in traditional 2-dimensional GUIs. It illustrates how task difficulty can be matched to the accuracy demands of the task, through only a small modification of current practice. In this case, the cursor switching is effected by the system rather than the user.

The task is moving a file to a folder, as in the *Macintosh* Finder (Figure 6). Typically, an outline of the file is displayed beneath the mouse pointer as it is being dragged towards the folder. The drag outline can be interpreted as an area cursor. We can then define the task of acquiring the folder in two ways, with the system responding either to the location of the pointer (Figure 6a) or the icon of the file being dragged (Figure 6b).

The second representation of the task (Prince technique) will be easier whenever the file icon is larger than the folder icon. (The second task may also be more intuitive. e.g., Why do people miss the Trash icon so often? Perhaps it's because we're attending to the file we're moving, rather than the location of the pointer.) If ambiguity results with this technique, as in Figure 6c, users can either reposition the file cursor to remove the ambiguity or they can revert to positioning with the point cursor. That is,

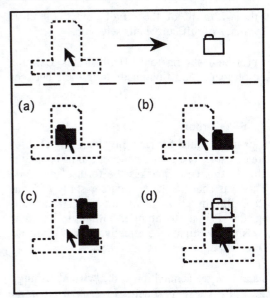

Figure 6: Moving a file into a folder. The user acquires a file and begins to drag its outline toward the folder. The folder will be highlighted to indicate selection. This occurs in (a) when the pointer moves inside the folder, and in (b) when the folder and the file overlap. In (c), the selection is ambiguous because the file overlaps with two folders. This can be resolved as in (d), by repositioning the pointer in one of the folders.

fine positioning need only be invoked as a last resort (when displays are cluttered), not as the default.

Example 2: Toolglass & Magic Lenses

The Prince and point-cursor techniques may be used in combination. This is seen, for example, in the Toolglass and Magic Lens techniques introduced by Bier, Stone, Pier, Buxton, & DeRose [2]. These techniques illustrate an elegant solution to the "display clutter/ambiguity" problem raised in the previous example.

With them, the area and point cursors are represented as separate objects and distributed between the two hands. The area cursor—that is, the Toolglass and Magic Lens widgets themselves—is controlled by the nondominant hand, while the dominant hand manipulates the point cursor.

When a task does not need fine positioning, the widgets can be used alone. An example is using a magnifying magic lens. Because of the widget's size, it is easy to position over the desired regions of the screen with the nondominant hand, and there are few if any negative consequences of "spill over" to other objects. However, when fine positioning is required, the technique can take advantage of the interaction between the two hands. An example is *clicking-through* the magnification widget with the point cursor in order to select one of the small objects being magnified. In this example, the synergies work to our advantage since the magnification widget itself increases the width W of the target object, therefore reducing the index of difficulty in its selection.

The head prop and slicing tool [4] illustrates a similar distribution of coarse and fine positioning tools between hands.

Example 3: Silk Cursor

All of our examples thus far have involved 2D selection using an area cursor. The Prince concept can be applied in 3D as well. In this case, it takes the form of a *volume* cursor. This was seen in the silk cursor study of Zhai, Buxton, & Milgram [14]. This was our first study employing the Prince technique, although its main objective was to investigate the effectiveness of occlusion cues in 3D selection, rather than Fitts' law.

A 3D dynamic target acquisition task, "virtual fishing," was designed for the experiment. In each trial of the experiment, an "angel fish" with random size and color appears swimming randomly within a 3D virtual environment. The subjects were asked to move a 3D volume cursor to envelop the fish and "grasp it" when the fish was perceived to be completely inside of the cursor.

In the experiment, only one fish was presented at a time since the technique breaks down in crowded waters.

A partial solution worth studying is to give the net a size operation (e.g., perhaps performed by the other hand). If you're swimming in open water, the net can be large. But when you swim into a school of fish, then you'd scale the net to make it smaller. This makes it easier to swim between the fish. At some point, however, two fish are going to be so close that they can't easily be distinguished even with a small net. The size operation by itself is also not ideal, since important occlusion information about the targets is lost when the net is too small.

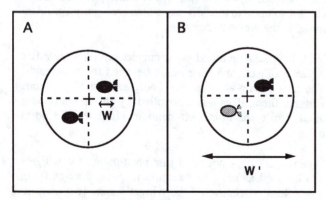

Figure 7: The silk cursor with a homing grid. The depth occlusion cues in 7(a) tell the user that two fish are caught in the net. Hence, the homing grid must be used to select one of the fish. In 7(b), one fish is shown within the net while the other is in front.

The software can help here by drawing a "homing grid" inside the net (with a cross-hair cursor showing the center of mass of the net). The homing grid can always be

shown, even when the net is large. Presumably, there will be a cross-over point in terms of what the user pays attention to. When tasks are coarse, selections can be made with reference to the surface of the net. When the display is cluttered, attention shifts more to the homing grid. But even here the silk net can help with positioning by providing occlusion information about the environment (see Figure 7).

CONCLUSION

An alternative approach to pointing, called the *Prince* technique, was investigated and found to be comparable to traditional pointing methods. Because the Prince technique uses a cursor of large area or volume, it is suitable for tasks that are normally difficult with the standard pointer, such as acquiring small targets or points. We feel that the Prince technique may be especially valuable when used in conjunction with traditional pointing techniques, where it can be used to tailor task difficulty more closely to the accuracy demands of the task. The examples presented three distinct methods suggesting how this might be accomplished.

The current study is an initial probe into a rich design space. Many questions and issues remain. We investigated selection tasks involving one width parameter, either the target or the cursor. What happens when there are two width parameters, defined by moving and stationary objects? The whole issue of "grasping" isolated objects from among a close cluster requires much more investigation. Likewise, the 3D case of the volume cursor deserves study. It would also be worthwhile to compare and/or combine the technique with gravitational "snapping" techniques. Finally, for the full potential of the technique to be realized, it is likely that new affordances (such as supporting "grasping") need to be built into input devices, such as mice. This also requires further study.

ACKNOWLEDGMENTS

This research has been undertaken under the auspices of the Input Research Group (IRG) of the University of Toronto. The authors gratefully acknowledge the contribution of the members of the group to this work. We would especially like to acknowledge the contribution of Shumin Zhai. We would also like to thank Abigail Sellen of Rank Xerox EuroPARC, Scott MacKenzie of the University of Guelph, and Mark Tapia of the University of Toronto for their contributions.

The IRG is generously supported by the Information Technology Research Centre of Ontario (ITRC), the Natural Sciences and Engineering Research Council of Canada (NSERC), Xerox PARC, Alias Research Inc. and Hitachi Corp. This support is gratefully acknowledged.

REFERENCES

1. Adams, J. J. (1992). The effects of objectives and constraints on motor control strategy in reciprocal aiming movements. *Journal of Motor Behavior, 24,* 173–185.

2. Bier, E. A., Stone, M. C., Pier, K., Buxton, W., & DeRose, T. D. (1993). Toolglass and magic lenses: The see-through interface. *Proceedings of SIGGRAPH '93* (pp. 73–80). New York: ACM.

3. Fitts, P. M. (1954). The information capacity of the human motor system in controlling the amplitude of movement. *Journal of Experimental Psychology, 47,* 381–391.

4. Hinckley, K., Paush, R, Goble, J. C., & Kassell, N. F. (1994). Passive real-world interface props for neurosurgical visualization. *Proceedings of the CHI '94 Conference on Human Factors in Computing Systems* (pp. 452–458). New York: ACM.

5. Hoffman, E. R. (1992). Fitts' Law with transmission delay. *Ergonomics, 35,* 37–48.

6. Jagacinski, R. J., Repperger, D. W., Moran, M. S., Ward, S. L., & Glass, B. (1980). Fitts' law and the microstructure of rapid discrete movements. *Journal of Experimental Psychology: Human Perception and Performance, 6,* 309–320.

7. MacKenzie, I. S., Sellen, A., & Buxton, W. (1991). A comparison of input devices in elemental pointing and dragging tasks. *Proceedings of the CHI '91 Conference on Human Factors in Computing Systems* (pp. 161–166). New York: ACM.

8. MacKenzie, I. S. (1992). Fitts' law as a research and design tool in human-computer interaction. *Human-Computer Interaction, 7,* 91–139.

9. MacKenzie & Ware. (1993). Lag as a determinant of human performance in interactive systems. *Proceedings of the CHI '93 Conference on Human Factors in Computing Systems* (pp. 488–493). New York: ACM.

10. Oldfield, R. C. (1971). The assessment and analysis of handedness: The Edinburgh inventory. *Neuropsychologia, 9,* 97–113.

11. Schutz, R. W., & Roy, E. A. (1973). Absolute error: The devil in disguise. *Journal of Motor Behavior, 5,* 141–153.

12. Walker, N., Meyer, D. E., & Smelcer, J. B. (1993). Spatial and temporal characteristics of rapid cursor positioning movements with electromechanical mice in human-computer interaction. *Human Factors, 35,* 431–458.

13. Welford, A. T. (1968). *Fundamentals of skill.* London: Methuen.

14. Zhai, S., Buxton, W., & Milgram, P. (1994). The "silk cursor": Investigating transparency for 3D target acquisition. *Proceedings of the CHI '94 Conference on Human Factors in Computing Systems* (pp. 459–464). New York: ACM.

Applying Electric Field Sensing to Human-Computer Interfaces

Thomas G. Zimmerman, Joshua R. Smith, Joseph A. Paradiso, David Allport[1], Neil Gershenfeld

MIT Media Laboratory - Physics and Media Group
20 Ames Street E15-487
Cambridge, Mass 02176-4307
(617) 253-0620
tz,jrs,joep,dea,neilg @media.mit.edu

ABSTRACT

A non-contact sensor based on the interaction of a person with electric fields for human-computer interface is investigated. Two sensing modes are explored: an external electric field shunted to ground through a human body, and an external electric field transmitted through a human body to stationary receivers. The sensors are low power (milliwatts), high resolution (millimeter) low cost (a few dollars per channel), have low latency (millisecond), high update rate (1 kHz), high immunity to noise (>72 dB), are not affected by clothing, surface texture or reflectivity, and can operate on length scales from microns to meters. Systems incorporating the sensors include a finger mouse, a room that knows the location of its occupant, and people-sensing furniture. Haptic feedback using passive materials is described. Also discussed are empirical and analytical approaches to transform sensor measurements into position information.

KEYWORDS: user interface, input device, gesture interface, non-contact sensing, electric field.

INTRODUCTION

Our research on electric field (EF) based human-computer interfaces (HCI) grew out of a project to instrument Yo-Yo Ma's cello [8]. We needed to measure bow position in two axes with minimum impact on the instrument and its playability. In this paper we discuss two types of EF sensing mechanisms: *shunting*, where an external EF is effectively grounded by a person in the field; and *transmitting*, where low frequency energy is coupled into a person, making the entire body an EF emitter. The benefits of each sensing mechanism are

1. Visiting scientist from HP Labs, Bristol, England.

presented along with comparisons to other sensing means. We report on several EF systems and applications, designed by arranging the size and location of EF transmitters and receivers, and suggest some future applications.

Since electric fields pass through non-conductors, passive materials that apply force and viscous friction may be incorporated into EF sensing devices, providing haptic feedback. We have constructed a pressure pad and a viscous 3-D workspace based on this principle.

PREVIOUS WORK

The first well-known use of EF sensing for human-machine interface was Leon Theremin's musical instrument. Two omnidirectional antennas were used to control the pitch and amplitude of an oscillator. Body capacitance detunes a resonant tank circuit [7]. The effect of body capacitance on electric circuits was well known to radio's pioneers, who saw the effect as an annoyance rather than an asset.

As the need for electronic security and surveillance increases, there is growing use of remote (non-contact) occupancy and motion detectors. Sensing mechanisms include capacitance, acoustic, optoelectronic, microwave, ultrasonic, video, laser, and triboelectric (detecting static electric charge) [5]. Many of these mechanisms have been adapted to measure the location of body parts in three dimensions, motivated by military cockpit and virtual reality (VR) applications [15].

Acoustic methods are line-of-sight and are affected by echoes, multi-paths, air currents, temperature, and humidity. Optical systems are also line-of-sight, require controlled lighting, are saturated by bright lights, and can be confused by shadows. Infrared systems require significant power to cover large areas. Systems based on reflection are affected by surface texture, reflectivity, and incidence angle of the detected object. Video has a slow update rate (e.g., 60 Hz) and produces copious amounts of data that must be acquired, stored, and processed.

Microwaves pose potential health and regulation problems. Simple pyroelectric systems have very slow response times (>100 msec) and can only respond to changing signals. Lasers must be scanned, can cause eye damage, and are line-of-sight. Triboelectric sensing requires the detected object to be electrically charged.

Mathews [14] developed an electronic drum that detects the 3-D location of a hand-held transmitting baton relative to a planar array of antennas by using near-field signal-strength measurements. Lee, Buxton, and Smith [13] use capacitance measurement to detect multiple contacts on a touch-sensitive tablet. Both systems require the user to touch something.

Capacitive sensors can measure proximity without contact. To assist robots to navigate and avoid injuring humans, NASA has developed a capacitive reflector sensor [22] that can detects objects up to 30 cm away. The sensor uses a driven shield to push EF lines away from grounding surfaces and towards the object. Wall stud finders use differential capacitance measurement to locate wood behind plaster boards by sensing dielectric changes [6]. Linear capacitive reactance sensors are used in industry to measure the proximity of grounded objects with an accuracy of 5 microns [4]. Electrical impedance tomography places electrode arrays on the body to form images of tissue and organs based on internal electric conductivity [21].

Weakly electric fish (e.g., Gymnotiformes, sharks, and catfish) are very sophisticated users of electric fields [1]. These fish use amplitude modulation and spectral changes to determine object size, shape, conductivity, distance, and velocity. They use electric fields for social communication, identifying sex, age, and dominance hierarchy. They perform jamming avoidance when they detect the beating of their field with an approaching fish: the fish with the lower transmit frequency decreases its frequency, and the fish with the higher frequency raises its frequency. Some saltwater weakly electric fish have adapted their sensing ability to detect EF gradients as low as 5nV/cm.

Given this long history of capacitive measurement, one might wonder why EF sensing is not common in human-computer interfaces. But it is only recently that inexpensive electronic components have become available to measure the small signals produced by EF sensors. Also non-uniform electric fields have made it difficult to transform these signals into linear position coordinates. Our research addresses these issues to help make EF sensing more accessible to interface designers.

It will be shown that EF sensors provide ample resolution and that converting the EF signal strength into position is the more challenging task.

MODES OF OPERATION
The Human Shunt

An electrical potential (voltage) is created between an oscillator electrode and a virtual ground electrode (Figure 1). A virtual ground is an electrical connection kept at zero potential by an operational amplifier, allowing current I_R to ground to be measured. The potential difference induces charge on the electrodes, creating an electric field between the electrodes. If the area of the electrodes is small relative to the spacing between them, the electrodes can be modeled as point charges producing dipole fields. The dipole field strength varies inversely with distance cubed. In practice the measurable field strength extends approximately two dipole lengths (distance between the transmitter and receiver electrodes). As the electrodes are moved farther apart, a larger electrode area is required to compensate for the decrease in signal strength.

When a hand, or other body part, is placed in an electric field the amount of displacement current I_R reaching the receiver decreases. This may seem counter-intuitive since the conductive and dielectric properties of the hand should increase the displacement current. However, if an object is much larger than the dipole length, the portion of the object out of the field serves as a charge reservoir, which is what we mean by "ground". The hand intercepts electric field lines, shunting them to ground, decreasing the amount of displacement current I_R reaching the receiver.

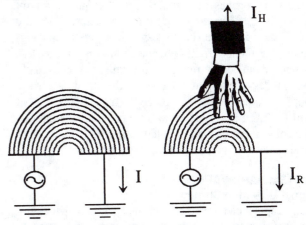

Figure 1. An electric dipole field created between an oscillating transmit electrode and virtual ground receiver electrode is intercepted by a hand. Displacement current to ground I_R decreases as the hand moves further into the dipole field.

The Human Transmitter

Low frequency energy is capacitively coupled into a person's body, making the entire person an EF emitter (Figure 2). The person can stand on, sit on, touch, or otherwise be near the oscillator electrode. One or more receiver electrodes are placed about the person. The displacement current into a receiver I_R increases as the person moves closer to that receiver. At close proximity, the person and the receiver electrode are modeled as ideal flat plates, where displacement current varies with the reciprocal of distance. At large distances, the person and the receiver electrode are modeled as points, where displacement current varies with the reciprocal of distance squared.

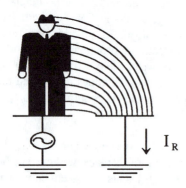

Figure 2. Energy from an oscillator is coupled into a person standing on the transmit electrode making the person an electric field emitter. As the person moves any body part closer to the grounded receive electrode, the displacement current into the receiver I_R increases.

Mode Crossover

When a hand (or any large object relative to the dipole length) approaches the dipole field of Figure 1 (shunt mode), the displacement current I_R decreases. When the hand gets very close (much less than a dipole spacing) the displacement current I_R begins to increase; the system changes from shunt mode to transmit mode. Actually both modes occur simultaneously, the hand is always coupling some field to the receiver (transmit mode) but until the hand is very close to the electrodes, the amount of displacement current shunted away from the receiver exceeds the amount coupled into the receiver.

SYSTEM HARDWARE
Signal Detection Strategy

Many capacitance detection schemes [5, 6, 13] measure the charging time of a resistor-capacitor (RC) network. The capacitance and displacement currents for EF sensing are on the order of picofarads (10^{-12} farad) and nanoamps (10^{-9} amps), requiring more sophisticated detection strategies. A synchronous detection circuit (Figure 3) is used to detect the transmitted frequency and reject all others [10], acting as a very narrow band-pass filter. Other detection methods include frequency-

modulation chirps (as used in radar), frequency hopping, and code modulation (e.g., spread spectrum).

The displacement current can be measured with approximately 12 bits accuracy (72 dB) using the components shown in Figure 3. There is a trade-off between update rate (sample rate/number of samples averaged) and accuracy (signal-to-noise ratio). The signal-to-noise ratio increases as the square root of the number of samples averaged. For example, averaging 64 samples increases the signal-to-noise a factor of eight (+18 dB), with a corresponding 1/64 update rate.

Information can be coded in the modulated transmitter signal. A multitude of small EF sensing devices can be scattered about a room, like eels in a murky pond, transmitting measurements to neighboring devices with the same EF used to measure proximity. The jamming avoidance mechanism of weakly electric fish [1] suggests that such devices can adjust their transmission frequencies autonomously when new devices are introduced into the sensing space.

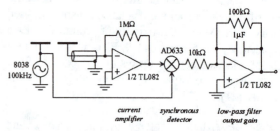

Figure 3. Synchronous detection circuitry.

Small displacement currents require good shielding, however the capacitance of shielded coaxial cable is orders of magnitude greater that the capacitance between electrodes. Cable capacitance low-pass filters the received signal, typically limiting the operating frequency to 30 kHz, and introduces a phase shift that is compensated for in the synchronous detector (not shown). Placing the current amplifier at the receiver electrode allows higher frequencies, limited by the amplifier's slew rate. For example, attaching the receive electrode directly to the TL082 current amplifier allows an operating frequency of 220 kHz.

Transmitter Power

The frequency range we use for EF HCI is 10 kHz to 200 kHz. Below this range, displacement currents and update rates are too small. Above this range FCC power regulations become more stringent [3]. The distance between electrodes is a fraction of a wavelength, so no appreciable energy is radiated. The only power consumed by the transmitter is the energy required to charge the capacitance of the transmitter electrode to the oscillating voltage.

In practice the transmitter power is less than a milliwatt. This allows the design of very low power systems with no radio interference. By adding an inductor, the transmitter can be driven into resonance, decreasing energy dissipation and increasing the transmitter potential, for example 60 volts from a 5 volt supply. A larger transmitter potential increases the strength of the received signal and therefore the signal-to-noise ratio, producing greater spatial resolution. Transmit signal strength can be increased until the current amplifier is saturated.

"Fish" Evaluation Board

To assist researchers in exploring EF HCI, our group has produced a small microprocessor-based EF sensing unit, supporting one transmitter and four receivers (Figure 4). It is called a "fish" after the amazing EF abilities of weakly electric fish, and because fish can navigate three dimensions while a mouse can navigate only two. The evaluation board supports MIDI, RS-232, and RS-485 serial communication protocols. We are currently designing a "smart fish," a second generation EF evaluation board utilizing a digital signal processor to allow automatic calibration and the exploration of more complex detection strategies, such a spread spectrum. The smart fish also measures the power loading of the transmitter to disambiguate mode crossover. Transmitter loading is monotonic; the current drawn from the transmit electrode always increases as an object approaches the transmit electrode.

ELECTRIC FIELD GEOMETRY

The value returned by a sensor is unfortunately *not* directly proportional to the distance between a hand and

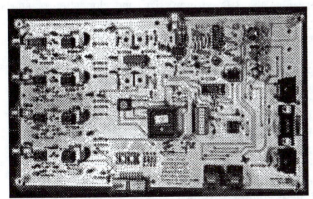

Figure 4. Fish electric field sensing evaluation board.

the sensor. Recovering information such as the *(x,y,z)* position of a hand from three sensor values *(r,s,t)* is a non-trivial problem. Solving the problem requires a

model of the electric field geometry. The absolute signal strength depends on the coupling of the person to a reference (ground for shunt mode and the transmit electrode for transmit mode). This coupling acts as a global system gain. The relative signal strength of the sensors contains the position information. For this reason normalized sensor readings are used to calculate position information.

There are two basic strategies for creating this model. In the analytical approach, knowledge of electrostatics (Laplace equation) is used to derive, for a given sensor geometry, an expression for the signals received as a function of hand position. The expression is then inverted analytically or numerically. In the empirical approach, signals are measured for a variety of known hand positions, and a function (e.g., radial basis function) that converts sensor values to hand positions is fit to the resulting data set. The analytical approach provides insight into the behavior of the sensors and does not require a training phase. However, any given analytical solution is applicable only for a particular sensor geometry, and different sensor geometries require new solutions. The empirical approach is more flexible, because changes in the sensor layout or environment can be accommodated by retraining.

Since our measurements occur within a fraction of a wavelength, we are in the near-field limit where the electric field is the gradient of the potential across the electrodes, so we can treat the situation as an electrostatics problem [12]. The same physics applies for electrode spacing that ranges from microns to meters. Small electrode spacing has been used to measure position with micron resolution [19]; large electrode spacing has been used to measure the location of a person in a room. We are not interested in the absolute values of sensor values; we care only about their functional dependence on the position of the body part we are measuring. Since the human body is covered with conductive, we treat the body as a perfectly conducting object. The hand is treated as a grounded point in space. In practice, the finite area of a hand and its connection to an arm serves to blur or convolve the ideal point response. But this point approximation usually works well as long as the real hand is a constant shape, the same convolution is being applied everywhere, and so the basic functional form of the hand response will be the same as that of the point response.

In-Plane Measurements

Figure 5 shows a contour plot of the predicted received signal, calculated using the classic dipole field expression [12] for a hand moving around a Z plane 0.9 dipole units above the dipole axis. A dipole unit is the distance between the transmit electrode and receive electrode. The predicted contour compares well to data collected by moving a grounded cube (2.5 cm on each side) across the plane.

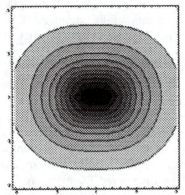

Figure 5. Contours of electric field strength in plane Z = 0.9 (measured in units of the dipole spacing) predicted by analytical model.

Out-Of-Plane Measurements

The relationship between hand proximity Z and displacement current I_R is measured using an electrical equivalent of a hand and arm suspended above the center of a dipole. The term proximity is used to emphasize that EF sensing measures the integrated (convolved) effect of an object in the electric field. When a hand is placed near a dipole, the hand, arm, and body attached to the arm all affect the field, though each contributes less as they are progressively farther away from the dipole.

The surrogate hand and arm combination is an aluminum tube 7.6 cm in diameter and 48.3 cm long and is grounded through a suspending wire for shunt mode and connected to an oscillator for transmit mode. The transmit and receiver electrode, each measuring 2.5 cm x 2.5 cm, are 15.2 cm apart on center. A least squares fit of the data reveals the following functional form for both shunt and transmit modes;

$$I_R = A + \frac{B}{\sqrt{Z}}$$

where A and B are constants determined by electrode geometry, detection circuit gain and bias, oscillator frequency and voltage, and Z is distance above the dipole. For shunt mode B is negative since displacement current I_R decreases as the object moves closer to the dipole.

Proximity resolution is expressed as the change in distance Z that produces a 6 dB change in displacement current I_R over the noise floor (two times the noise floor). The resolution is dependent on the signal-to-noise ratio of the detection system, which is a function of integration time. The longer the data is averaged, the greater the

proximity resolution, albeit with a corresponding slower update rate. The fish evaluation board used in these measurements has an integration time constant of 10 milliseconds. Figure 6 plots proximity resolution as a function of distance Z for shunt mode. At 85 mm distance, proximity resolution is 1 mm.

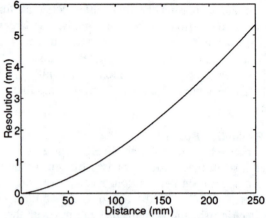

Figure 6. Proximity resolution of a surrogate arm in shunt mode plotted as a function of distance from dipole to arm. Resolution is the change in distance that produces a 6dB change in signal over noise.

Imaging: Converting Signals To Position

Each dipole measures a degree of freedom, either object position or size. A single dipole cannot distinguish a close small object from a large distant object, as both might block the same number of field lines. A second dipole operating on a longer length-scale (greater electrode spacing) can be used to distinguish these two situations, or to measure two spatial coordinates of a single fixed-size object. Three dipoles can measure the 3-D position of an object of fixed size, or determine the 2-D position and size of an object. Four dipoles can determine the size and 3-D position of an object. Five dipoles can determine the 3-D position, size, and elongation of an object. We are working on the continuum limit of adding more dipoles, to perform low-resolution imaging.

Optimal Sensor Placement

Each receiver measurement constrains the position of a small object (relative to dipole spacing) to an ellipsoid centered on the dipole axis (see Figure 5). The dipoles should be oriented orthogonally in order to minimize the sensitivity of the solution (x,y,z) to errors in (r,s,t). The problem of inverting the sensor readings is equivalent to the geometrical problem of finding the intersection points of these ellipsoids. Often additional constraints (prior knowledge) must be imposed to select one solution from the many symmetric cases that are consistent with the data. For example, to make a two-dimensional mouse using only two dipoles, we must impose the constraint that the hand is on one side of the dipoles.

synthesizers and sound sources, creates a complex sonic terrain based on the location of the person, allowing navigation of a sonic environment.

Smart Chair

A chair is fitted with one transmitter in the seat and four receivers: two located in the headrest to measure head rotation, and one at each armrest to measure hand proximity. A person in the chair navigates multiple audio channels by head and hand placement [16]. The sensors are mounted underneath the chair fabric, so they are invisible to the user. Smart chairs may be used to control radio functions in a car, home audiovisual equipment, or simply to turn off a computer monitor when a user leaves a workstation.

In another application, a transmitter is installed in a chair to allow the magicians Penn & Teller to perform music by waving their arms near four receivers. Hand position controls various sound parameters produced by computer-controlled sound synthesizers.

Haptic Feedback in 3-D Space

A foam pad is placed on top of a dipole pair. Pressing on the foam produces a force feedback. Since force is proportional to position (Hooke's law), and finger position is measured by EF sensing, finger force is measured. A passive piece of foam on an EF sensor is a pressure sensor.

A plastic box is fitted with electrodes on three sides to measure hand position in 3-D. The box is filled with bird-seed (millet) to provide a viscous medium for haptic feedback. The seed allows users to rest their hand in space, reducing fatigue, and provides something to grab. Slight compression of the seed increases viscosity. Perhaps a computer-controlled piston, bearing on a movable wall of the box, could provide a simple way to simulate an environment with variable viscosity.

FUTURE APPLICATIONS

Researchers are currently exploring direct manipulation of instrumented real objects to facilitate 3-D orientation and manipulation [9, 17]. Electric field sensors may be incorporated in objects to measure object deformation, position, and orientation.

The Tailor project [20] allows disabled individuals to run computer applications by mapping the unique anatomical movement ability of each individual to control signals. Combining EF sensing with such mapping techniques could provide a person in a wheelchair with individually tailored, unobtrusive, invisible, low-power, and low-cost computer and machine interfaces.

Hermetically sealed EF sensors in a palm top could determine when the case is open, when the unit is being held, and could create a large control space around the small device. Foam EF buttons could provide force and tactile feedback, detect finger approach and finger pressure, and distinguish between slow and fast presses.

Multiple transmitters and receivers, multiplexed in time, frequency, or by coding sequence, could be placed under a carpet to determine the number and location of people in a room. When an electrode under a person is activated, that person becomes the EF source. Smart floors can be used for multi-participant VR simulations without the burden of wires or the complexities of video cameras.

Attempts have been made to instrument whiteboards using video cameras [11] and optoelectronics [2]. Both systems require rear imaging to record stylus movement. A conventional plastic whiteboard can be fitted with an array of EF sensing electrodes to measure the location of a metal-cased marker in the hand of a shunting or transmitting person.

Watches have a very small workspace and very little energy capacity. An EF sensor can be used to create a large workspace over a small watch face. Such watch controllers can be used to search through audio databases.

CONCLUSION

We have discussed some HCI systems and future applications of EF based sensing. The near-field nature of low-frequency electric fields allows the same detection scheme to be scaled from microns to meters. EF sensing provides high resolution proximity information. The difficulty is converting proximity to position. We have worked out an analytical method to correct for the non-uniform nature of dipole fields. Empirical methods may be used to compensate for complex field distortion caused by dielectrics or conductors in the field. Some of EF sensing's greatest qualitative appeals are the sense of magic, simplicity, and "naturalness" it brings to an HCI. The abilities of weakly electric fish to perform object detection, communicating, and jamming avoidance demonstrate what is possible with EF sensing. The authors know of no other sensing mechanism or system that can deliver non-contact sensing with millimeter resolution at kilohertz sample rates and millisecond lag times for a few dollars a channel. As computing power leaps off the desk and into a multitude of small battery-powered devices, the need for low-power unobtrusive interfaces grows. It is our belief that EF sensing can make a significant contribution to the sensing abilities of computing machines.

COMPARISON TO OTHER SENSOR TECHNOLOGIES

Electric field sensors detect a bulk effect, integrating the body's interception of EF. Unlike optical system, the effect does not depend on object surface texture and reflectivity. The data from EF sensors is continuous with a resolution limited by transmission strength and noise rejection. There is an economy of data; only three channels are required to locate a hand in 3-D. In comparison, a video camera produces an abundance of data, on the order of 75 megabits per second, while updating at 60 Hz. An EF system operating at 100 kHz can average 100 samples, provide a 1 kHz update rate, with 1 millisecond lag time.

Electric field systems can be extremely small, light-weight and low power, as required by the ever shrinking real estate and energy capacity of lap, palm, and watch based computers. Since electric fields penetrate non-conductors, sensors can be hidden, providing protection from weather and wear, as well as adding an element of magic to the interface.

SYSTEM CONFIGURATIONS AND APPLICATIONS

The transmit method provides large receive signals, operates over large areas, and can distinguish multiple persons. Capacitively coupling energy into a person requires continuous close contact with the person. We have used transmit electrode ranging from 5 to 150 square cm, depending on proximity to the person. The transmit electrode can be incorporated into the seat of a chair, a section of a floor, the back of a palm computer, or a wristwatch band. Direct conductive contact with the person's skin requires a much smaller electrode area (<5 square cm). Asymmetric placement of receiver electrodes helps decouple signal strength from position calculations.

The shunt method does not require close contact with a person. For each dimension, a minimum of one receiver is required. Prototyping interfaces is basically an "arts and crafts" project, consisting of cutting out electrodes, typically aluminum foil and copper tape, taping them down, and wiring them up to the fish evaluation unit.

Figure 7. Two-dimensional finger-pointing mouse.

2-D Finger-Pointing Mouse

We have implemented a two-dimensional finger-pointing mouse on a laptop computer (Figure 7). The input device is activated by touching a small transmitter electrode with the fourth (little) finger of the left hand. Energy is coupled into the person, and the EF emitted from the pointing finger is sensed at two receiving electrodes. A thin uniform copper strip running across the top of the screen senses Y position, and a tapered strip along the side of the screen senses X position. The taper renders the electrode more sensitive to the EF emitted by the pointing finger and less sensitive to the field emitted by the arm. The shaped electrode physically implements an analog spatially varying signal gain. A third small receiving electrode, placed below the spacebar, allows the thumb of the left hand to generate click signals.

The pointing finger does not need to be in contact with, or even close to the screen, thereby avoiding screen smudges and occlusion of the cursor by the pointing finger. Position sensing is easily disabled by lifting the forth finger off the transmitting electrode, the equivalent of lifting and putting down a mouse, facilitating relative position control.

Smart Table

To demonstrate the concept of "smart furniture," a co-linear dipole pair (i.e., receiver, transmitter, receiver) is placed underneath a wooden table to measure hand gestures. A computer screen displays an electronic newspaper whose pages are flipped forward and backward by sweeps of a hand across the table (X-axis). Placing the hand down on the table (Z-axis) advances to the next section, lifting the hand up displays the previous section. Gestures are detected by applying a threshold to the X and Z velocities. Position in the X-axis is approximated by differencing the two receiver signals; position in the Z-axis is approximated by the sum of the receiver signals.

An array of dipoles can turn a table into a multidimensional digitizing and gesture input device. Such an EF sensing matrix may substitute for or augment a video camera for video desk applications [18]. Perhaps visual ambiguities and occlusions could be arbitrated by EF sensing, indicating hand location to the video analysis system.

Person-Sensing Room

In an installation piece at the MIT Media Lab, a single transmitter electrode covers the entire floor of a room, coupling energy into a person walking on the floor. Four receiver electrodes, located on the walls, measure relative signal strength, indicating the location of the person. A computer program, controlling a multitude of

ACKNOWLEDGMENTS
The authors would like to thank Henry Chong for his programming assistance, the team of Teresa Marrin, Pete Rice, Edward Hammond, John Crouch, Ryan Christensen, and Alexander Sherstinsky for their work on the person-sensing room, and the Hewlett-Packard Corporation and the "News in the Future" consortium for their support.

REFERENCES

1. Bullock, T.H. Electroreception. Ann. Rev. Neuroscience, Vol. 5 (1982), pp. 121-170.

2. Elrod, S., et. al. Liveboard: A Large Interactive Display Supporting Group Meetings, Presentations and Remote Collaboration. CHI'92 Human Factors in Computing Systems (Monterey, May 3-7, 1992), ACM Press, pp. 599-607.

3. Federal Communications Commission, Part 15 Radio Frequency Devices. FCC, Washington, D.C.

4. Foster, R. Linear Capacitive Reactance Sensors for Industrial Applications. SAE Technical Paper 890974, 40th Annual Earthmoving Industry Conference, Peoria, Ill. (April 11-13, 1989).

5. Fraden, J. AIP Handbook of Modern Sensors, American Institute of Physics, New York, 1993, pp. 312-346.

6. Franklin, R. Fuller, F., Electronic Wall Stud Sensor. U.S. Patent No. 4,099,118, July 4, 1978.

7. Garner, L. For That Different Sound, Music a'la Theremin. Popular Electronics (November 1967), pp. 29-33,102-103.

8. Gershenfeld, N. Method and Apparatus for Electromagnetic Non-Contact Position Measurement with Respect to One or More Axes. U.S. Patent No. 5,247,261, Sept. 21, 1993.

9. Hinckley, K., et. al. Passive Real-World Interface Props for Visualization. CHI'94 Human Factors in Computing Systems (Boston, April 24-28, 1994), ACM Press, pp. 452-458.

10. Horowitz, P., Hill, W. The Art of Electronics, Cambridge University Press, New York, 1989, p. 889.

11. Ishii, H., Kobayashi, M. ClearBoard: A Seamless Medium for Shared Drawing and Conversation with Eye Contact. CHI'92 Human Factors in Computing Systems (Monterey, May 3-7, 1992), ACM Press, pp. 525-540.

12. Jackson, J. Classical Electrodynamics. John Wiley & Sons, New York, 1975, p. 138 and 392.

13. Lee, S., Buxton, W., Smith, K. A Multi-touch Three Dimensional Touch-Sensitive Tablet. CHI'85 Human Factors in Computing Systems (1985) ACM Press, pp. 21-24.

14. Mathews, M. Three Dimensional Baton and Gesture Sensor. U.S. Patent No. 4,980,519, Dec. 25, 1990.

15. Meyer, K., Applewhite, H., A Survey of Position Trackers. Presence, Vol. 1, No. 2 (Spring 1992), MIT, pp. 173-200.

16. Mullins, A. AudioStreamers: Browsing Concurrent Audio Streams. Master's Thesis, MIT Media Lab, June 1995 (forthcoming).

17. Murakami, T. Nakajima, N., Direct and Intuitive Input Device for 3-D Shape Deformation. CHI'94 Human Factors in Computing Systems (Boston, April 24-28, 1994), ACM Press, pp. 465-470.

18. Newman, W, Wellner, P. A Desk Supporting Computer-Based Interaction with Paper Documents. CHI'92 Human Factors in Computing Systems (Monterey, May 3-7, 1992), ACM Press, pp. 587-598.

19. Paradiso, J. A Simple Technique for Measuring Axial Displacement in Stretched-Wire Alignment Systems. (May 1994), Draper Technical Report GEM-TN-94-607, Cambridge, Massachusetts.

20. Pausch, P., et. al. One-Dimensional Motion Tailoring for the Disabled: A User Study. CHI'92 Human Factors in Computing Systems (Monterey, May 3-7, 1992), ACM Press, pp. 405-411.

21. Webster, J.G. Electrical Impedance Tomography, Adam Hilger, Bristol, 1990, pp. 1.

22 Vranish, J. McConnell, R., Driven Shielding Capacitive Proximity Sensor. U.S. Patent No. 5,166,679, Nov. 24, 1992.

Learning to Write Together Using Groupware

Alex Mitchell, Ilona Posner and Ronald Baecker
Dynamic Graphics Project, University of Toronto
10 Kings College Road
Toronto, ON M5S 1A1, CANADA
Tel: 1-416-978-5473
E-mail: alex@dgp.toronto.edu

ABSTRACT

Most studies of collaborative writing have focused on mature writers who have extensive experience with the process of writing together. Typically, these studies also deal with short, somewhat artificial tasks carried out in a laboratory, and thus do not extend over a period of time as real writing usually does.

This paper describes an ethnographic study of collaborative writing by two groups of 4 grade six students using synchronous collaborative writing software for one hour per week over a 12 week period. Despite initially having little appreciation of what it means to write together, and no experience in synchronous collaborative writing, both groups produced nearly one dozen short collaboratively conceived, written, and edited documents by the end of the study.

A careful analysis of video tape records, written documents, questionnaires, and interviews demonstrated the importance of concepts such as awareness, ownership, and control in the writing process, and highlighted many examples of strengths and weaknesses in the writing software.

KEYWORDS: CSCW, groupware, group work, collaborative writing, learning to write, novice writers, ethnography.

INTRODUCTION

Writing together is difficult. We have been carrying out research designed to advance our understanding of how people write together, how they learn to write together, and what kinds of computer-based tools could aid this process [1, 11, 14, 19, 20]. Other investigators have also studied how groups write together [2, 5, 9]. Most of these studies tend to take the form of surveys or questionnaires, and have largely focused on mature writers. However, some have dealt with collaborative writing in the classroom [3], identifying the difficulties facing novice writers in collaborative situations.

Figure 1: Cover of the magazine, by Ryan Fields, age 12.

The above research has provided insight into the collaborative writing process. A number of theories of collaborative writing have been developed [5, 20, 23] to characterize this process. There have also been a number of tools designed to support the collaborative writing process. Most notable among the many systems are GROVE [6], PREP [16], Quilt [10], SASSE [1], and ShrEdit [17].

Studies have been conducted into the use of several of these collaborative writing tools [12, 18]. These types of studies, while valuable, do not provide much insight into how collaborative writing tools would be used in real, extended scenarios.

Based on this previous research and our own experience with SASSE [1], we felt useful insight could be gained from a detailed study of the use of a collaborative writing tool by novice writers over an extended period of time. Although collaborative writing is a common practice, novice writers are uncertain as to how to proceed, and often have difficulty even understanding what is meant by collaborative writing. Studying the use of a collaborative writing system provides the opportunity to observe how novice writers learn to write together, and reveals much about the special needs of novice writers. We felt that the use of a collaborative writing tool would help with the process of learning to write together, and would give us insight into the strengths and weaknesses of the tools, providing clues as to the requirements for design of future systems.

Our study involves two groups of 4 grade six students working together to produce a magazine on prejudice (see Figure 1). Through the course of the twelve week study, they learned how to use Aspects[1], a commercial synchronous collaborative editor, and developed the skills necessary to successfully write together. The students all developed different levels of expertise, and were able to share that expertise with each other to help the group accomplish its task. The study had several objectives. We wanted to observe extended use of a synchronous collaborative editor, in a situated context, with inexperienced writers, in the hope of both validating the concept of a synchronous shared editor and deriving useful insight into the design of future systems. We also wanted to see if the use of a synchronous collaborative editor would benefit the writing and learning processes.

This paper presents preliminary results of the study in the form of qualitative observations and suggestions for the design of future collaborative writing tools. We will begin by giving an overview of the study, the setup and activities involved, and the data collection and analysis methods. We will then briefly explain the way in which the students developed expertise at using Aspects and writing together. The rest of the paper will focus on the way in which the students used the collaborative writing tool, and the design recommendations that can be derived from these observations. A companion paper [21] will provide more details into the learning process, how the use of collaborative writing tools affected the product, and the impacts the experience had on the participants.

THE PREJUDICE PROJECT

The Prejudice Project took place at the Huron Street Public School in Toronto between January and May 1994. The goal of this project was for grade six students to learn about prejudice while collaboratively writing and producing a magazine on that subject.

We conducted an ethnographic study of two groups of students preparing the written material for this magazine. Eight students were selected, with the assistance of their teachers, from 14 volunteers out of two grade 5/6 classes. The students were experienced with the Computer Supported Intentional Learning Environment (CSILE) shared knowledge building system [22], but were not familiar with synchronous collaborative work on a computer.

Through the course of the study the students learned to write together using Aspects on networked Macintosh computers. Aspects allows multiple users to work

concurrently on shared documents. It uses a replicated architecture, provides various locking mechanisms, and minimal consistency control. It is a fairly conservative but stable system as opposed to research systems which may provide many useful features but may lack the robustness necessary for serious extended use.

Study Setup

Each group met once a week for one hour after school. The students worked in a classroom, sitting at adjacent Macintosh computers (see Figure 2). Each networked computer ran a version of Aspects, with documents shared between all machines. The students' seating arrangements were changed each week to reduce the possibility of subgroup formation and other influences of physical placement.

Figure 2: Physical setup of the study

During the twelve-week period, the students were given training in the use of the Aspects system, introduced to the concepts and skills necessary for collaborative writing, and then given the freedom to use those skills as they saw fit.

The first five weeks were highly structured in order to cover various topics related to prejudice. The format of group work was guided by the instructor (Ilona) in an attempt to introduce a variety of ways of working collaboratively. The activities included writing a poem and a story as a group. Students were exposed to the various mechanisms provided by Aspects, giving them the tools necessary to perform their tasks. The assigned tasks were designed to expose the students to a variety of writing styles and approaches [19, 20, 23]. They started with *scribe/consultant* writing, where a *scribe* enters text in a document, and one or more *consultants* provide ideas but do not actually enter them in the document. They were also given tasks involving *parallel writing,* with writers individually entering text at the same time in the same document, but in different regions, and *joint writing*, with writers working closely on one section of the document.

[1]Aspects was developed by Group Technologies Inc.

In the remaining weeks the students were free to choose what they wanted to work on and how they would work together. Their work included doing research, writing articles, creating artwork, and editing some of the materials created in the first five weeks to be incorporated into the magazine.

Throughout the study, a balance had to be maintained between the amount of training given and our desire to see how the students would use the technology without guidance.

Data Collection and Analysis
During each weekly session all of the group's interactions were videotaped. The recording setup included two cameras covering the students working on the computers, and two cameras capturing screen images. In addition, time-indexed notes were made using the Timelines [7] video annotation software while the sessions were in progress.

Exploratory analysis was conducted on the video records of the sessions by two people using Timelines. This analysis identified common problems, incidents and trends in the data. More in-depth discourse analysis and coding will be done in the future to explore in detail some of the areas discussed in this paper.

A number of other measures were used to provide a rich view of the sessions. The students' teachers were asked to evaluate the students' abilities, and to provide blind evaluations of a selection of work, both individual and group. The evaluations were intended to provide information which could be used to give insight into the effects of the use of groupware technology on the students' learning experience and the quality of the document. The results from this data were used in the companion paper [21].

We also conducted a halfway and a final questionnaire and a final individual interview with each student. A few weeks after the end of the project, we came back and conducted a group discussion about the project, which gave additional insight into the students' experience.

OBSERVATIONS
As a result of the analysis of this data, we have compiled a number of observations that can be drawn from the study and contribute to the understanding of collaborative writing and collaborative writing tools. Through the process of creating their magazine, the two groups of students were confronted with a number of challenges, most notably learning to write together and learning to use the tools provided. We also observed many activities directly related to the collaborative task: awareness, ownership, and access control.

Having never worked on a group project that demanded such close collaboration, the students had to learn both what it means to write together and how to do that successfully. In addition to learning about the task of writing together, the students had to learn to use the tools provided. The students had no problem working with the computers. All had extensive experience with computers both in the classroom and at home. In fact, the students were so comfortable with the computers that they took a cruel pleasure in causing the software to crash.

The first major difference encountered by any user when moving from single-user to multi-user software is the notion of a shared space in which other people are working. This experience was illustrated in many ways. The students developed awareness of themselves and each other. Similarly they developed patterns for determining who would control the shared space. Finally, they negotiated access to that space.

Learning to Write Together
From the start, both groups were not sure how to approach the task of "writing together". When encouraged to work together, they claimed that they didn't know what this meant. When the groups first began using Aspects, they instinctively shifted into a parallel writing style, all working independently with little communication. Although they were all able to enter text into the document at the same time and see each other's entries, they weren't initially comfortable with this. It took some time for them to get used to the fact that the document was shared.

When they found that entering duplicate text without consultation didn't work very well, the group switched to a consultant/scribe mode of writing. Despite the ability to access the document synchronously, one person would do all the typing with the others suggesting ideas. This style of writing leads to interesting control and ownership problems, as will be discussed later. This form of writing most closely resembles the way groups work together synchronously using traditional technology [20], and thus involved the least adaptation on the part of the students.

Later, as they became more comfortable with the task and the system, the students began to make full use of the synchronous editing capabilities of Aspects. The following is a brief example of how the students made effective use of the shared workspace. While composing a story during the fifth week of the study, both groups used a consultant/scribe writing style. One student "drove" the writing, eliciting ideas and entering them in the document, while the others contributed ideas and followed on their own screens. However, one student frequently moved around in the document, rereading the story from the start and suggesting changes. She also pointed out errors using a telepointer. Interestingly, though, she did not actually make changes. This notion of control over the document by the scribe will be discussed further in the section on control below.

Part of the problem with the notion of what it means to write together is the question of what it means for a piece of writing to be a group document. The students weren't sure whether a document written in parallel could ever be considered a unified piece of writing. They felt that such a document should be rewritten from scratch by one person.

This came up when the first group was editing a poem with stanzas written in parallel by each group member.[1]

```
Sue     I think we should write the whole
        thing over because it sounds like
        3 poems stuck together.
Hope    Yeah, I think one person has to
        rewrite it.
Liz     Yeah.
```

This same difficulty was encountered with several other documents that were written in parallel. It was only when a document was written jointly with one person acting as scribe that the group acknowledged that the document was coherent.

This problem faced by our novice writers highlights an ongoing question faced by researchers of collaborations: when should a document be considered to have been truly *written together*.

Learning To Use the Technology

Collaborative writing is a very difficult task. To support it successfully, the tools provided must not add to that complexity. Experienced writers often get distracted from the content of their writing when composing on a computer [8]. The vast number of fonts and styles available tend to encourage a focus on format and layout. This is especially true for novice writers. The students found it hard enough to stay on task without technological distractions. Tools such as chat boxes and cute telepointer shapes were often more distracting than useful. An interface that supports gradual disclosure of features would allow users to be comfortable with the system at all stages of learning (a).[2]

The feedback provided by the system was often obscure and confusing, leading the students to ignore it. When the messages were critical, such as when document consistency was lost, this led to later problems (b). The fact that documents are shared, yet replicated, caused continuous confusion to both the students and the experimenters. A number of documents were lost and had to be reentered because the various contributors all assumed that someone else had saved the document. The location of the document was not at all obvious from the interface (c).

However, as the students became more familiar with the concept of shared access to a common document, they developed working patterns that took advantage of the technology. As will be discussed below, the students came to realize how the technology could be used in different situations.

---

[1]Note that the names of the students have been changed, but gender has been preserved.

[2] Throughout our discussion of the study we will use letters such as (a) to link our observations to the list of design recommendations appearing near the end of this paper.

Collaboration and Awareness

The importance of collaborator awareness mechanisms has been well recognized [1, 4, 14]. When dealing with novice writers, this is especially important.

Self-Awareness

Even after becoming familiar with the notion of a shared workspace and having worked on the system for six weeks, the students still had difficulty determining where they were and what they were doing.

Where Am I? There are several things that the students were asking when they asked "Where am I?" The simplest is location in the shared document. However, it is also important to provide some feedback as to whether the user is in a conference or working alone, and whether the current document is private or shared with others (d).

It was common for people to become confused as to whether their work would be seen by others or not, and whether they would be able to see others' work. This was especially true when someone stopped working closely with the group, and then later returned.

In one case, it turned out that the students who had been working together to edit a document had in fact been working in separate copies of the document (e).

```
Sue     [moving her mouse and looking at
        Hope's screen]   Why doesn't it
        [her pointer] show up?
Ilona   Why don't I have a copy of that?
Dan     [moves mouse, looks at Sue's]
```

This situation wasn't discovered until Sue happened to glance at Hope's screen and notice that her telepointer was not showing up.

What Am I Doing? Similarly, in a multi-user conference with multiple documents it isn't always obvious what you are doing at a given moment. This can range from confusion as to whether you are telepointing or not, a simple interface problem, to more subtle concerns, such as whether you are interfering with someone else's attempts to edit text. The system needs to make the information about your relative location and influence on others readily available (e).

Collaborator Awareness

Collaborator awareness is always important, but even more than usual when dealing with users who are learning to work together. Awareness not only includes awareness of where people are within a document, but who is present for collaboration and who is potentially present.

Where Are You? Lack of any reminder of where others are makes it easy to forget that there is a shared workspace (f):

```
Group   [they start to divide the task up
        by questions - each group member
        tells the others what they wrote
        as if the others can't see it]
```

```
Ilona    Everyone can see the same thing.
         [goes over and scrolls Sally's to
         show them]
```

What Are You Doing? When working individually, it is easy to lose track of what others are doing. This problem is often overcome by resorting to physical pointing and glancing at each other's screens (g):

```
Sally    I'm getting confused. Rob, what
         are we changing here?
Rob      [makes changes, points to screen
         to indicate what he is doing]
Carol    [watches what Rob is doing]
Sally    [doesn't notice Rob's gesture] Rob
         what are we doing here?
Rob      Its gonna look like a poem.
Sally    [sees gesture and looks over] Okay
```

Who Did That? With synchronous shared access to the document, it is possible to enter text or to delete someone else's text without that person's knowledge. This can lead to confusion:

```
Dan      [deleting something]
Liz      No no don't erase it DON'T! who
         erased that?
Hope     Not me I just got in.
Liz      [looks over at Sue] Sue?!?
Dan      [looks around, says nothing]
```

Tracking of where other people are and what they are doing can come in many forms. Aspects provides bars alone the side of the document indicating that someone has control of a region of text. This tells you that someone is there, but not who it is.

Pointing and Gesturing

Users may also want to explicitly provide information to others about their actions. Aspects provides a simple telepointing mechanism, allowing each user to gesture with a remote cursor of a user-selected shape. However, the students often found it easier to use physical pointing and gestures:

```
Sally    Rob, can you show me what you're
         trying to do?
Rob      Take a look here. [points to her
         screen]
```

Telepointers were too limiting because they were both unable to draw collaborators' attention and lacked information about the person who was pointing. All this information is available in a simple hand gesture (g).

The telepointers also tended to be rather distracting; the students often ended up chasing each others' pointers around the screen. However, one group did learn to use them effectively when proofreading and editing:

```
Liz      Meant is spelled M-E-A-N-T [points
         at her screen]
Dan      Where is it?
Liz      Its, I'll mark it, there I've
         marked it.  See?  That's where it
```

```
         is [uses telepointer] where my
         little annoying thingy is.
```

Having discovered this function, Liz explains it to the others:

```
Liz      Say he spelt birth wrong [points
         with finger, Sue looks] you go to
         that [moves telepointer there] and
         go like that. [wiggles it]
Sue      OHHH.
```

The shared workspace encourages this type of consulting and collaborative learning.

Effect of Physical Placement

The students tended to take advantage of the physical placement of the computers to aid in their awareness of group activities. The computers were placed in a row, allowing each student to glance around at the other students and at their screens. From the start, they tended to glance around a lot, anxious to stay aware of what the rest of the group is doing. This ability to look at each other's computers also led to shifts between working independently or together on separate computers and working huddled around one machine.

The physical placement of the machines also allowed people to notice when someone is looking at their work; this is useful for encouraging and facilitating collaboration and consultation:

```
Group    [they start entering comments]
Liz      [glances at Hope's screen for
         confirmation she's doing it right]
Hope     [notices, gives her advice]
```

However, the physical placement can also lead to formation of subgroups and exclusion of peripheral group members. To minimize this, we rearranged seating patterns each session.

Document Ownership

The perception of who has a claim to ownership of a section of text, or over the entire document, was independent of how ownership was represented by the technology. Aspects doesn't provide any explicit indication of who wrote a section of text. However, the students would often assume, especially in the early parts of the study, that the person who typed a section was the only person who could change it.

Similarly, there was a connection between who typed in a section of text and who got credit for the ideas contained in the text. We observed that the scribe usually provided fewer ideas than the rest of the group. Despite this, the scribe occasionally took credit for the content of the document. For example, in the first group, the person who typed a story claimed the next day that it was his story:

```
Dan      I wrote the story.
Sue      No I did.
Liz      I did.
Hope     We all did.
Dan      The one about the...
```

```
Hope    I made up Tiger Lily.
Ilona   I thought everybody wrote it.
Dan     I wrote it most cause I typed
        everything.
```

Although the system did not provide explicit ownership information, the students tended to continue to identify text with the person who typed it. This showed up most clearly when the second group was editing one of the documents. In this case, the sections of the document were all written in parallel, with each section easily identifiable as belonging to a different person. Two of the group members were suggesting changes, but refused to make the final alterations until the entire group gave permission.

However, when the group was working together on a document that had already been edited, the group members had no reservations about arbitrarily deleting someone else's text without telling them (h). This suggests that at this point the group members had come to regard the text as shared, rather than just owned by the person who typed it.

Document Control

The system's assumptions about control over the document, both in terms of the ability of group members to edit sections of text and to access documents, had several effects on the collaborative writing process

The students in the study used Aspects in paragraph locking mode. This allows each user to gain control of a paragraph of text and make changes within that paragraph. As long as the user doesn't move the selection point out of the paragraph, other users are locked out. This granularity of locking led to some interesting behaviour.

When working with physical documents, the students were able to gain control over a paper by grabbing it if necessary. In Aspects there was no way to force a shift in control. One student understood the technology to the extent that he deliberately kept an entire document as one paragraph to keep control of the changes being made to the text, even when encouraged to add paragraph breaks:

```
Ilona   Can I suggest you put some more
        returns in there? [gets up and
        puts returns in so Dan isn't
        locking the whole document]
Liz &   [start typing like mad as Ilona
Sue     puts in spaces]
Dan     I didn't want to or else they'll
        start doing funny things with it.
```

However, the group soon learned to overcome the limitations of the locking mechanism by simply using the other person's computer rather than trying to get control of the document from within the system.

```
Dan     [takes Sue's mouse when she's not
        looking, moves her out of a
        paragraph]
Sue     [looks back, sees Dan]
Dan     One sec... stop, let me work on
        this part.
```

Control over the text also had an effect on the roles taken on by the group members. When deciding who would be the scribe, the group would either take a vote or argue until someone managed to get control of the text. For example, when the first group was composing a story, Dan ended up gaining control of the text. The group went along with this, everyone dictating while he types; ideas were passed around, negotiated, and the final decision was made by the scribe.

Although the system gave the scribe explicit control of the document, the other members of the group were still able to make significant contributions and provide feedback which affected the contents of the document. In the above example, where Dan had control of the document, there were several occasions where other students tried to make changes. Kim would attempt to alter a sentence, and failing that would ask Dan to make the change, using the telepointer to indicate the change. So, in spite of the control mechanisms provided by the technology, the entire group was able to influence the document (i).

Synchronous Access and Collaboration

The fact that everyone could access the workspace influenced the style of collaboration. Initially, everyone wanted to type just because they could. However, as they gained experience with the technology and with group writing, the students became more selective in their choice of writing style. For example, when working on the last day on the magazine's introduction, the second group shifted between scribe and independent parallel writing to solve a consensus problem. Sally was acting as scribe, but the group couldn't decide on the wording of one section of the document. To solve this, they all entered their own ideas, then all read them and selected the best. They then shifted back to scribe mode, and continued (h).

Working synchronously but on separate sections of the document worked well in a task that lends itself to division. On the tenth day, the two members of the first group were preparing the questions for an interview – they discussed the content, then split the task up, but talked back and forth while entering the text. Then, after the interview, they worked in a scribe fashion, one student dictating the answers while the other typed.

However, it was not always clear who had been given the role of scribe. In fact, the assignment of roles tended to change dynamically, since the technology lets anyone take control as long as the previous scribe is willing to relinquish control. In general the groups were able to adapt to this shifting of control. If someone's ideas were not being accepted or they were being ignored, that person would sometimes go off and start entering the ideas independently in a different section of the document.

The students were able to adapt their use of the system to suit their working patterns, and to take in to consideration the social and group interactions taking place as they worked. It was possible because the system did not attempt to impose strict roles and patterns of usage on the students (j).

DESIGN RECOMMENDATIONS

The above observations stress the importance of maintaining an awareness of both the shared space in which you are working, and the fact that there are other people working in that shared space. Dourish and Bellotti [4] indicate that the use of shared feedback, the notion of providing implicit, peripheral information about everyone in a shared space, is a promising approach. Examples of this approach can be seen in ShrEdit [12] and SASSE [14]. ShrEdit allows you to request that the system find and track movements of others; SASSE provides peripheral information in the form of colour-coded, shared scrollbars, audio cues, a document overview (or gestalt), and a tracking mode.

Another important problem introduced by collaborative tools is the need to keep track of changes in a shared document. One way to provide the necessary information about changes in the document is through the display of differences in the document, often called "diffs", either as change bars [15], through the use of annotations [14], or through more active notification [13].

Ownership of and access to the shared workspace are also important considerations. As we have seen, the way in which both of these are handled by the system have an influence on the behaviour of the group. While some systems such as Grove [6] and PREP [16] assign roles to collaborators, most systems leave this up to the group. This, along with flexible access and floor control mechanisms, allows social interactions, not the system, to determine working patterns and group behaviour.

The above discussion suggests a number of design recommendations, which we will summarize below:

(a) provide tools appropriate to the users' level of expertise; avoid distracting tools; use gradual disclosure

(b) make sure the system's feedback is simple and concise

(c) provide a clear and accurate mental model of the system

(d) provide self-awareness in terms of location in the shared workspace, and potential actions in that location

(e) provide awareness of the user's effects on others

(f) provide awareness of the presence of others in the shared workspace to encourage discussion and negotiation

(g) provide collaborator awareness in terms of shared feedback and explicit information such as gestures

(h) provide flexibility in terms of the representation of ownership information to allow for changes over time

(i) allow flexibility in terms of document control to allow for shifting roles at different stages of the writing process

(j) avoid imposing patterns on natural social interactions

CONCLUSION

The observations we have made of the students working together using Aspects are very encouraging. Over the 11 hours they were working with Aspects, they developed distinct, mature strategies for working together. Despite having never worked with synchronous collaborative writing software, both groups managed to produce coherent documents which they felt reflected the work of the entire group. Together they successfully produced a 32 page magazine which will be on display early next year as part of an exhibit at the Ontario Science Centre.

The students testified to having written the magazine "together", something that they did not even know the meaning of at the start of the study. Perhaps this in itself is the best definition of group writing – the perception that the results of your work are the result of the group's work, rather than the work of the individual members of the group. This achievement validates the concept of a synchronous shared text editor, and provides promise for the use of such technology in education and writing in general.

The group jointly learned about an important topic, prejudice. As one student said in the final group discussion:

```
Liz    Everyone is different, we all have
       different beliefs, and we should
       respect that. this is what we
       learned.
```

They had become comfortable with the idea of writing together, and confident with the technology. From not knowing what it means to write together, they had progressed to feeling that they were able to succeed at, and enjoy, group writing. During the final group discussion, one student volunteered the following:

```
Rob    The best thing was learning how to
       work with everybody, we weren't
       too good at that before.
```

By observing this learning process, we have been able to gain an insight into the nature of group writing, and identify some of the effects of the use of collaborative writing tools on this process.

We have seen that the technology has a distinct effect on the way in which novice writers approach the collaborative writing task. However, at the same time we have seen that writers, as they become familiar with both the task and the technology, are able to exploit the features of the system and use it to their advantage in creative ways. The observations we have made of the problems students have learning to write together are very similar to those experienced seen in adult writers. From these observations we have drawn a series of recommendations for future design.

Our study stretched over twelve weeks. With a task domain as complicated and unfamiliar as collaborative writing, it is important that any observations be made over an extended period, in situ, and in a situation where users are allowed to approach their tasks as freely as possible. This allows usage patterns to develop naturally, and provides the time needed to learn about the task and the technology. Although this type of ethnographic study is harder to run and much more time-consuming to analyze than a traditional lab study, the type of real usage that we have observed could never be seen in a usability laboratory.

ACKNOWLEDGMENTS

The authors would like to thank the teachers and students at the Huron Street Public School for their time and enthusiasm, and the researchers at OISE, Marline Scardamalia, Peter Rowley, and Jim Hewett, for their help with the project. Ben Smith-Lea and Russell Owen provided essential technical support. Hiroshi Ishii, Sara Bly and our reviewers provided valuable comments. We would also like to thank those who provide funding for our research, especially the Natural Science and Engineering Research Council of Canada.

REFERENCES

1. Baecker, R.M., Nastos, D., Posner, I.R., and Mawby, K.L. The User-centred Iterative Design of Collaborative Writing Software, *Proceedings of InterCHI'93*, ACM, 1993, 399-405, 541.

2. Beck, E. A Survey of Experiences of Collaborative Writing. In Sharples, M. (Ed.), *Computer Supported Collaborative Writing*, Springer-Verlag, 1993.

3. DiPardo, A. and Freedman, S.W. *Historical Overview: Groups in the Writing Classroom*, Technical Report No. 4, Centre for the Study of Writing, University of California, Berkeley, 1987.

4. Dourish, P. and Bellotti, V. Awareness and Coordination in Shared Workspaces, *Proceedings of CSCW'92*, ACM, 1992, 107-114.

5. Ede, L. and Lunsford, A. *Singular Texts/Plural Authors: Perspectives on Collaborative Writing.* Southern Illinois University Press, 1990.

6. Ellis, C.A., Gibbs, S.J., and Rein, G.L. Groupware: Some Issues and Experiences, *CACM* 34(1), 1991, 38-58. Reprinted in Baecker, R.,M. *Readings in Groupware and Computer-supported Cooperative Work: Facilitating Human-Human Collaboration*, Morgan Kaufmann, 1993.

7. Harrison, B., Owen, R., and Baecker, R.M. Timelines: An Interactive System for the Collection and Visualization of Temporal Data, *Proceedings of Graphical Interface '94*, Morgan Kaufmann, 1994.

8. Hass, C. Does the Medium Make a Difference? Two Studies of Writing with Pen and Paper and with Computers. *Human-Computer Interaction*, 4, 1989, 149-169.

9. Kraut, R.E., Egido, C., and Galegher, J. Patterns of Contact and Communication in Scientific Research Collaborations. In Galegher, J., Kraut, R.E., and Egido, C. (Eds.), *Intellectual Teamwork: Social and Technological Foundations of Cooperative Work*, Erlbaum, 1990, 149-171.

10. Leland, M.D.P., Fish, R.S., and Kraut, R.E. Collaborative Document Production Using Quilt, *Proceedings of CSCW'88*, ACM, 1988, 206-215.

11. Mawby, K. Designing Collaborative Writing Tools, M.Sc. Thesis, Department of Computer Science, University of Toronto, 1991.

12. McLaughlin Hymes, C. and Olson, G.M. Unblocking Brainstorming Through the Use of a Simple Group Editor, *Proceedings of CSCW'92*, ACM, 1992, 99-106.

13. Minor, S., and Magnusson, B. A Model for Semi-(a)Synchronous Collaborative Editing, *Proceedings of ECSCW'93*, 1993, 219-231.

14. Nastos, D. A Structured Environment for Collaborative Writing, M.Sc. Thesis, Department of Computer Science, University of Toronto, 1992.

15. Neuwirth, C.M., Chandhok, R., Kaufer, D.S., Erion, P., Morris, J., and Miller, D. Flexible DIFF-ing in a Collaborative Writing System, *Proceedings of CSCW'92*, ACM, 1992, 147-154.

16. Neuwirth, C.M., Kaufer, D.S., Chandhok, R., and Morris, J.H., Issues in the Design of Computer Support for Co-authoring and Commenting, *Proceedings of CSCW 90*, 183-195. Reprinted in Baecker, R.,M. *Readings in Groupware and Computer-supported Cooperative Work: Facilitating Human-Human Collaboration*, Morgan Kaufmann, 1993.

17. Olson, J.S., Olson, G.M., Mack, L.A., and Wellner, P. Concurrent Editing: The Group's Interface. *Proceedings of Interact '90*, 835-840.

18. Olson, J.S., Olson, G.M., Storrøsten, M., and Carter, M. How a Group-Editor Changes the Character of a Design Meeting as well as its Outcome, *Proceedings of CSCW'92*, ACM, 1992, 91-98.

19. Posner, I.R. A Study of Collaborative Writing, M.Sc. Thesis, Department of Computer Science, University of Toronto, 1991.

20. Posner, I.R. and Baecker, R.M. How People Write Together, *Proceedings of the Twenty-fifth Annual Hawaii International Conference on System Sciences*, 1992, 127-138. Reprinted, with slight modification, in Baecker, R.,M. *Readings in Groupware and Computer-supported Cooperative Work: Facilitating Human-Human Collaboration*, Morgan Kaufmann, 1993.

21. Posner, I.R., Mitchell, A., and Baecker, R.M. Learning to Write Together using Computers - Product and Experiences, submitted for review.

22. Scardamalia, M., Bereiter, C., McLean, R., Swallow, J., and Woodruff, E. Computer Supported Intentional Learning Environments, *Journal of Educational Computing Research* 5(1), 1989, 51-68.

23. Sharples, M., Goodlet, J.S., Beck, E.E., Wood, C.C., Easterbrook, S.M., and Plowman, L. Research Issues in the Study of Computer Supported Collaborative Writing. In Sharples, M. (Ed.), *Computer Supported Collaborative Writing*, Springer-Verlag, 1993, 9-28.

Electronic futures markets versus floor trading: Implications for interface design

Satu S. Parikh
Finance Department
The Wharton School of the University of Pennsylvania
1300 Steinberg Hall, Philadelphia, PA 19104-6366
LITZ2@finance.wharton.upenn.edu
(215) 382-7749

Gerald L. Lohse
Operations and Information Management Department
The Wharton School of the University of Pennsylvania
1326 Steinberg Hall, Philadelphia, PA 19104-6366
Lohse@wharton.upenn.edu
(215) 898-8541

ABSTRACT

The primary concern in designing an interface for an electronic trading system is the impact on market liquidity [9]. Current systems make use of efficient order-execution algorithms but fail to capture elements of the trading floor that contribute to an efficient market [9]. We briefly describe tasks conducted in futures pit trading and current off-hours electronic trading systems. Understanding the tasks helps define key components to an interface for electronic trading. These include visualization of the market and its participants, a trading process which allows active participation and price discovery as well as concurrent interaction among each of the participants.

KEYWORDS: Futures trading, automated exchange, trading pits, interface design, electronic markets

INTRODUCTION

In the futures pits of the Chicago Mercantile Exchange (CME), tens of thousands of people crowd into 70,000 sq. ft. and trade in excess of 550,000 contracts a day by using their voices and hands. Even this volume is pale in comparison to the total dollar volume of all futures contracts worldwide, which are over $500 billion a day [12]. In comparison, the electronic automated counterpart to the Exchange, the GLOBEX Trading System, trades only 6,000 contracts daily. "[GLOBEX] is just nickel-and-dime small investor retail business so far," says a portfolio manager at a large New York investment firm. Another person familiar with exchanges and trading, Mr. Dale Lorenzen, first vice-chairman of the Chicago Board of Trade, readily admits, "I just don't think GLOBEX is ready to take the volume that we can generate with open outcry." Why an electronic system cannot effectively replicate the human component of floor trading, highlights some important interface design issues.

The CME currently has some of the most crowded futures pits in the world. Imagine you are a trader in such a futures pit, and your task is to sell 100 soybean contracts at a given price. You scream in as loud a voice as possible, "100 for [price]" while simultaneously touching a closed fist to your forehead with palm facing outwards. Within seconds a person wishing to buy a hundred soybean contracts

indicates a desire to buy but wishes to lower the price. Some negotiation takes place, and once an acceptable price is found, the trade is executed. The average trader on the CME executes 60 trades an hour.

There are a number of existing attempts to automate some of the most liquid financial and commodity markets in the world, and the results have been less than desirable. GLOBEX, a joint venture between Reuters Holdings PLC and the CME, is one example. Two years after its introduction, the system has been described as restrictive, slow, and inefficient relative to the open outcry method of trading, and there has been a great deal of debate about why this is so, or rather why this *should* be so [14]. In spite of the many reasons why a floor-based system is "better" than its screen-based counterpart, from an interface standpoint the superiority of the trading pit structure is readily apparent.

The trading pits are where all of the dealing in a particular futures instrument is legally required to take place [2, 10]. More specifically the Commodity Futures Trading Commission's Regulation 1.38 states the following [6]:

> All purchases and sales of any commodity for future delivery, and of any commodity option . . . shall be executed openly and competitively by open outcry or posting of bids and offers by other equally open and competitive methods in the trading pit or ring or similar place provided by the contract market.

In these pits, the traders' shouts and hand signals form a complex network of communication that may seem uncontrolled and chaotic, but which is, in fact, remarkably orderly. One may postulate that with a computerized trade execution system, communication between traders can be provided by showing the best bids and offers in the market for a specific contract at any particular moment, thus eliminating any need for hand signals and/or verbal communication. This is true, but what needs to be considered and explained is that by removing the human element in a trade, the efficiency of price matching may increase but only at the expense of market liquidity, which is defined as the ability to execute an order of any size without a significant discrepancy to the market price [3]. Thus, the electronic automation of futures markets has definite interface design issues, namely effectively replicating the communications bandwidth of floor-based futures trading in an electronic securities market. It is important to note, however, that the intent of this research is not to provide decision support for those traders engaged in open outcry. Rather, it is directed towards addressing interface design problems and issues in screen-based futures trading.

To understand why the existing computerized trade execution systems fail to substitute for an open outcry market, one must first understand the basic market principles as well as some details about how different orders are executed. Thus, section 1 begins with an overview of the futures markets. Section 2 delves into aspects of screen-based futures trading systems currently in operation with a primary focus on GLOBEX. While other systems are currently in operation [12], GLOBEX is the highest volume electronic market and is the most widely used for after-hours trading [9].

Section 3 looks primarily at the interface considerations of both the floor and screen-based trading systems. Here the different problems are stated and explained, and possible ways to increase the communications bandwidth in an electronic futures market are explored.

Section 4 describes the components of a prototype we are developing that represents the beginnings of a solution to this very difficult interface design problem.

FUTURES MARKETS

A futures contract is simply an agreement to buy or sell a given amount of a commodity or financial instrument at a particular price on a stipulated future date. Common contracts are crude oil, precious metals, agricultural goods, interest rate instruments, and currencies. Each is traded on an exchange, and some are traded on more than one. Futures contracts are used for hedging, speculation, and market making.

Hedging is used to offset the risk exposure from potential future cash market transactions in the product. For example, large multinational corporations may use currency futures to hedge against a future payment in foreign currency or lenders may use interest rate futures to hedge against a future fall in rates.

Obviously futures are very useful in reducing market risk, but they may also be used as speculative instruments. Arbitrage is exploiting small discrepancies in different markets. By having positions in both the cash market and the futures market, one could earn a riskless profit if market inefficiencies arise.

Market making is not specific to futures, but the way it is implemented is unique to the futures markets. Making a market involves taking both sides of a trade and profiting off of the spread, which is the difference between the buy (bid) and sell (ask) price. In most futures exchanges, market makers, also called floor traders or speculators, are those trading for their own account as opposed to filling an order for a customer. Thus, it is the market maker who supplies much of the capital that ensures a liquid futures market.

Order Execution

The actual trade is executed in a ring shaped pit. The buyer, wishing to invest a given amount of money in a particular contract, places an order with a member firm (e.g., Lind-Waldock Futures Inc., Merrill Lynch, etc.). In placing the order, the buyer may set certain restrictions on the order such as price limit, market, or opening only [16]. The member firm communicates the order via telephone or computerized order-entry to its clerks on the exchange floor. The order is time-stamped for order-clearing purposes, and then a runner delivers it to the firm's floor broker in the trading pit. Using open outcry, the floor broker shouts the order into the pit. As there are many brokers and traders shouting in these pits, shouting alone may not accurately convey intentions to buy or sell. Thus, hand signals are an integral way in which a pit trader or broker communicates with the rest of the pit participants.

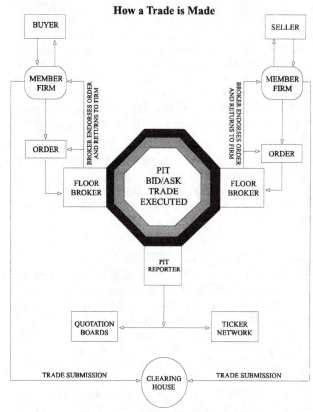

How a Trade is Made

Figure 1 How a trade is made [4]

Once the broker has indicated the buy order, other brokers with sell orders or market makers wishing to create or close out a position may choose to sell the required contracts to the buyer-represented broker. If the broker wishes to buy a large quantity of contracts, she may enter into a number of trades with different brokers in order to fill the order. At any rate, once the order has been filled, the broker writes the seller's identification number and the details of the trade such as the time, price, and quantity on a trade ticket. Once the trade is executed, exchange employees relay the price to the exchange's computerized price reporting system. In relaying the information, the exchange's employees use a variety of methods. The CME uses mobile phones; the New York Cotton Exchange uses hand-held computers; and the Coffee, Sugar and Cocoa Exchange employees use hand signals to report prices to other employees entering the price data into a computer. This information is then transmitted to market observers and participants worldwide, thereby allowing people around the globe to trade with relative ease.

The last step in the execution process is confirming the trade. A runner returns the filled order to the member firm's communications desk, where it is once again time-stamped. Confirmation of the futures purchase is then sent to the customer.

Order Clearing

After an order is filled it must be cleared by the respective exchange's clearing agent. The CME's Clearing House is the guarantor of performance for each futures contract traded. This means that the timely delivery by the seller of the exact quantity and quality of the commodity or financial instrument and the full and timely payment by the buyer, are fully ensured by the Clearing House. In addition to monitoring the physical delivery process, the clearing agent oversees the daily settlement of all accounts. This means that if an unliquidated (open) position exists at the end of the day, it must be marked-to-market. Marking to market involves recalculating the value of the position using that day's settlement price. Any loss in value is treated as a debt to the clearing agent which must be repaid, and any increase in value is credited to the customer's account.

In understanding the different interface implications in automating the futures market, one must be familiar with how and why such a market functions. The next section explains the GLOBEX system and other automated securities markets.

AUTOMATED SECURITIES EXCHANGES

Rapid development in technology has permitted the move from floor-based markets to screen-based systems, and one such system currently in operation is the Global Exchange (GLOBEX) Trading System. More specifically, GLOBEX is an international, automated order entry and matching system that is operated by Reuters Holdings PLC and the CME. The GLOBEX Network extends to ten financial centers, including New York, Chicago, London and Tokyo. At present, more than 325 trade terminals are in operation. The system operates after the close of the regular trading hours and closes just prior to market open [7, 9].

Types of Orders Available

Primarily, two types of orders are given, market and limit [11]. A market order instructs the floor broker to execute the trade at the current market price or better with no restrictions on price. While a limit order, restricts the broker to a given price or better in her bidding or offering.

Many view GLOBEX to be restrictive in placing different types of trades, since GLOBEX only permits limit orders to be entered. A trader at a GLOBEX terminal chooses the contract he wishes to trade in, and then enters an order, which is ranked by a price/time priority algorithm and stored in a limit order "book." The trade is executed when the prices cross.

GLOBEX offers two trade options, hit and take, which are akin to market bids and offers, but the GLOBEX alternatives differ from market orders distinctly. Hit orders, when entered, place a sell order at the best bid price, and take orders place a buy order at the best offer. Both are supposed to duplicate market orders. The difference,

however, lies in reconciling the quantity of contracts specified in the orders. On the floor, market orders are either filled in entirety or in part. If they are only partially filled, the broker can continue to trade the remaining contracts at the current market prices. On GLOBEX, hit/take trades are filled to the quantity of the standing order, and the remaining balance is automatically canceled.

Another way a market order is constructed on GLOBEX, is by entering an order at a better price than the best standing order. Thus, the new order is immediately executed at the best bid or offer. If the newly entered order is for a large quantity, the system moves through the book, providing fills at the best prices, until the order is filled or the limit price is reached.

PROBLEM DEFINITION

The problem in electronic securities exchanges is the fact that by automating the trade process, the human element is forgone. While the efficiency of trade clearing and order matching may increase, the added liquidity, provided by those willing to trade on the human element (i.e., noise traders), is sacrificed. The question at hand is how to improve the human interaction in these automated trade systems.

Current GLOBEX Interface

The current interface for GLOBEX (Figure 2) uses a windowing technique which subdivides the area of the screen into windows: trading window, response window, full ticker window, request for quote (RFQ) and alert window, monitor window, and the limit order book. While this setup may be adequate in displaying one window of information, it is difficult to view information across multiple windows. Further, the interface slows the entry of trades relative to open outcry and it would be difficult to enter, let alone execute, 60 trades per hour using GLOBEX.

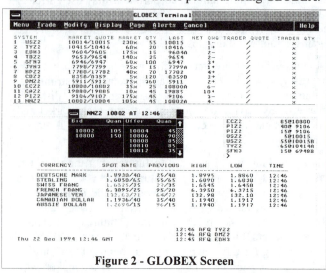

Figure 2 - GLOBEX Screen

The trading window is where a trader finds the best bid and offer available for a GLOBEX product. As it contains the most information, it is the largest of all the windows, comprising nearly a third of the entire screen. It contains the current contracts being traded as well as their best bids and offers. The quantity and last trade price are also displayed along with the high, low, and volume for the current trading session. The trader may modify a portion of this window in order to show the total number of his orders entered as well as his last bid/offer.

In addition to displaying the market for a given contract, the trading window is where trades are entered and modified. Assume the trader's intentions are to place an order to sell a notional amount of a given product listed on GLOBEX. First, he must depress the number key representing the contract or use the mouse to select it. Once this is done, a dialog box appears giving the trader a number of alternatives. More specifically, he may bid/offer the contract thus making a market, separately bid or offer the contracts, hit/take the best bid/offer, change or cancel an existing order, request for quote, or display the order book which only shows the five best bids/offers. Assuming this trader wishes to sell his contracts, he would use the keyboard's "arrow" keys coupled with the "enter" key or the mouse to select the "offer" button. The trader must now grapple with another dialog box requesting the details of the trade, such as the offer price and quantity. After all information has been entered as desired, the trader clicks the "OK" button, thus entering the trade into the limit order book. The trade is subsequently ordered by GLOBEX's price/time priority structure and is advanced in the execution queue as trades are executed.

Trades take five seconds to enter for hit/take orders and ten seconds for regular bid/offer limit orders. Use of the mouse in trading generally speeds up the order entry process. By allowing traders to quickly select choices instead of typing, order entry using a mouse is 30% faster than a keyboard.

The response or trading mailbox window is smaller than the trading window. This window enables the trader to receive messages from the central or host computer, including order acknowledgments, executions, and respective order numbers.

The ticker windows lies to the right of the trading mailbox. It displays the most recent market movements providing an updated message each time a trade occurs and each time there is a change in the price or size of the best bids and offers. The ticker employs a vertical scrolling technique in updating market information. Each time there is relevant information, it is displayed from the top of the windows and each successive new piece of trade information is updated below. Thus, the window does not actually scroll but rather updates and overwrites old information.

GLOBEX also utilizes a monitor window, which contains information from the cash (spot) markets. An option allows the trader to show those futures' and/or options' prices from those traded in the trading windows. The trader may also view a number of indices and market composites which are provided by GLOBEX as well.

The Request For Quote (RFQ) and Alert windows alerts the trader when the price of a selected instrument has reached a preset level. GLOBEX permits alerts in all of its listed products. Another feature, which is appealing for thinly traded contracts, is the RFQ. If a contract in which a trader wishes to transact in is not being traded, the trader may

"ask the market" for a quote by clicking the appropriate button on the particular contract's aforementioned dialog box. This is broadcast to GLOBEX screens around the world. Others wishing to trade or perhaps make a market in the contract may then choose to respond with a bid or offer in the requested market.

GLOBEX is efficient in clearing and matching trades according to its price/time priority algorithm. However, the system is inadequate in maintaining an active and liquid market. This is partly due to the nature of after-hours trading, but much more attributable to the system's inability to recreate the human-to-human interaction present in the trading floor environment. This interaction is forgone (1) by GLOBEX's method of order entry and matching, and (2) by the system's interface.

Order Execution Algorithm

The rules of an automated trade-execution process are referred to as an *order-execution algorithm* [11]. For example, GLOBEX uses an automated continuous double auction, in which transactions take place when limit prices cross. For the continuous double auction process, a standing price/time priority structure is used. That is, bids (offers) are assigned priority in the limit order "book" from the highest (lowest) price to the lowest (highest) price. Orders of equal prices are given priority according to the time of entry. When order prices cross, the trade occurs at the lower of the two quantities specified in the orders.

The following brief example illustrates this process. Assume the following chronologically successive bids and offers for a specific futures contract:

Time	Offer	Quantity	Time	Bid	Quantity
10:35	1.59	50	10:36	1.63	45
10:33	1.63	35	10:33	1.605	95
10:34	1.63	70	10:32	1.57	100
10:30	1.64	100	10:30	1.55	85
10:31	1.65	25			

At 10:30 the best buy order is at the price of 1.55 while the best sell order is for 1.64 per contract. Since order prices do not cross, no trade execution is possible. Note that the spread between the best bid and offer is .09. This simply signifies that the cost of buying while simultaneously selling a notional amount, would cost .09 per contract. At 10:35 the best offer is entered at 1.59 per contract, while the best bid stands at 1.605. Since the best offer is less than or equal to the best bid and, conversely, as the best bid is greater than or equal to the best offer, the trade is executed at the standing order's price of 1.605. Since the quantities are not equal, only 50 of the 95 contracts are executed at the bid price. The balance, at the trader's request, may remain as a standing order or simply be canceled. At 10:35, immediately after the trade has been executed, the order book would look as follows:

Time	Offer	Quantity	Time	Bid	Quantity
10:33	1.63	35	10:33	1.605	45
10:34	1.63	70	10:32	1.57	100
10:30	1.64	100	10:30	1.55	85
10:31	1.65	25			

A new best bid is entered at 10:36 at the price of 1.63. Since this crosses with the best offer, another trade is executed. However, as there are two offers at the price of 1.63, GLOBEX makes use of the time priority to distinguish the orders. Since both orders occurred at different times, the offer entered first is given priority. Thus the new bid is executed at standing price of 1.63, and since the existing offer is for 35 contracts, this is the number of contracts traded. Assuming the trader entering the order did not want the balance canceled, the remaining 10 contracts would be executed against the new best offer of the same price for 70 contracts. Thus, the order book would look as follows after the new order-entry at 10:36 and the two new trades occurring immediately after 10:36.

Time	Offer	Quantity	Time	Bid	Quantity
10:33	1.63	35	10:36	1.63	45
10:34	1.63	70	10:33	1.605	45
10:30	1.64	100	10:32	1.57	100
10:31	1.65	25	10:30	1.55	85

Time	Offer	Quantity	Time	Bid	Quantity
10:34	1.63	70	10:36	1.63	10
10:30	1.64	100	10:33	1.605	45
10:31	1.65	25	10:32	1.57	100
			10:30	1.55	85

Time	Offer	Quantity	Time	Bid	Quantity
10:34	1.63	60	10:33	1.605	45
10:30	1.64	100	10:32	1.57	100
10:31	1.65	25	10:30	1.55	85

As new orders are entered, they are prioritized in the book as described, and if matching is possible, the trade is executed at the standing price. Reconciling the quantity depends on the trader's instructions. If the balance is left as a standing order, it remains in the order book and is executed as illustrated above; otherwise it is simply canceled.

Orders may be modified on GLOBEX. Traders may change the price and/or quantity of an existing order if it has not yet been executed. However, by modifying an existing an order, time priority is lost. The system views order modification as equivalent to canceling the existing order and entering a new one. This may prove costly in a high volume market, where many orders exist for a given price.

Discussion of Order Entry and Matching

An understanding into why GLOBEX hinders an efficient and liquid market thus requires discussion into some of the previously published ideas about the inadequacy of trade automation. The two primary reasons explored are the importance of traders or "locals" and the psychology of a futures pit[14].

Locals and Traders

The trader is integral to a properly functioning market. Trading for their own accounts, they have a great deal of flexibility in deciding how their money is allocated. Some choose to assume no risk by taking both sides of the market and profiting from the spread (i.e., scalper or market

maker). Others incur some risk by playing natural market movements (i.e., spreader, arbitrageur, or position trader). Whether some or a great deal of risk is borne by traders, they are infusing the markets with a large supply of liquid reserves, and it is these liquid funds which result in commitments to a newly listed futures contract or option as well as the long term viability of an established market.

Although GLOBEX allows market making it does not have the communications bandwidth necessary to trade actively in a liquid market. As an illustration, assume two markets, GLOBEX and a floor based exchange. In addition assume a market maker wishes to place a bid and an offer for a highly liquid contract. On the floor this trader using the open outcry method of trading can indicate her desire for both orders in tandem. Using hand signals she can place the order in a few seconds. On GLOBEX the trader would have to grapple with the contract's dialog box which does allow a "BID/ASK" option. After entering a specific bid, offer, price and quantity, the order is entered into the limit book where it is ranked according to the price/time priority algorithm. Clearly, this would take much longer than simply waving and shouting one's desires. Speed, however, is not the only issue of import in distinguishing between the two systems. Versatility is equally substantial. If the market begins to move away from the trader, on the floor she can simply modify her hand gestures and shouting to account for this. On GLOBEX it is not as easy. The trader can clearly change her order, but not without relinquishing time priority in the queue. In a liquid and volatile market, forfeiting an advantageous place in the order queue may result in an unconsummated order.

One also has to be aware of the aforementioned communications bandwidth present on the floor of an exchange. In a specific pit, a trader can view the whole market in front of him. He can select an appropriate price and trade immediately. In addition, it is not imperative that he trade with someone in physical proximity, the trader can choose someone across from him or anywhere in the pits. This is because the system of hand gestures is so efficient, traders can signal intentions and desires at an instant to a large number of people.

Pit Psychology

Less quantifiable than a high volume and liquid market are the psychological issues unique to the trading pit and how these are to be replicated on a screen-based trade system. Many traders, experienced with both the pit structure as well as GLOBEX, assert that on the floor there is information meaningful to the negotiation process which cannot be found on GLOBEX. One example is noise or information trading. This refers to the increased noise in an extremely active market. Traders sometimes choose which pit and its respective contract to risk their money in by the noise associated with it. On a screen based system such as GLOBEX there is no way to hear price movement. There is no replication of the hundreds of shouting voices, waving hands, or stamping feet that one commonly sees on the trading floor. Thus, it can be argued that this lack of auditory and visual stimuli on GLOBEX may preclude some involvement by traders seeking profit.

Another example of the pit psychology is available in Leo Melamed's *Critique of Automation* [14]:

"There is yet another type of liquidity which is exclusively a product of the floor, or rather of the pits.... This liquidity is the product of pit psychology. For example, pit trader A, a larger trader who trades solely for his own account, is about to enter the pit. Pit trader A is known to have 'bought the market' earlier that day or the previous day. The market has fallen during the last thirty minutes. As pit trader A enters the pit, but before he has indicated any bid or offer, he is seen by pit traders B, C, D, and E, who also trade for their own accounts. Pit trader B, a fairly large local trader, is short the market. His split second reaction is that A will no doubt continue buying the market and perhaps cause the rally. Thus, he immediately buys to offset some of his shorts in an attempt to secure some of his profit. Pit trader C, a fairly large local trader, who is long the market and is presently with losses and nervous about his position, reacts entirely differently. He feels that trader A is about to sell and liquidate his entire long position causing the market to fall further. Thus C, to minimize his losses, immediately begins to sell the available market bids. Pit trader D, a small local 'scalper' believes that A will not sell out his position, but rather that he will buy more. Therefore, D tries to buy one or two contracts for his own account, so that he can 'scalp them a few seconds later at the higher price he expects will soon develop. Pit trader E, a 'spreader' between two or more different contract months, has just taken a short position in the contract month where A is about to enter the pit. E had intended to stay in an 'unhooked' position for thirty seconds or so before completing his spread by buying the other contract month. Upon seeing A, however, he decides quickly to 'hook up' his spreads in the other contract month so that he does not unnecessarily expose himself, since he is unsure of what A will do.

Actually it may turn out that A entered the pit, looked around, and left without trading; or he may have done what some of the traders expected. However it really does not matter what A did. All the bids, offers, and transactions in reaction to A's entering the pit accounted for additional market liquidity. In fact, they created new market positions which, in turn, would have to be offset, probably that day, thereby creating further market liquidity.

The reasons for the trades are unimportant to this discussion. It is important that there was a visual and psychological interaction which cannot be duplicated by computer. Without the pit psychology, this form of liquidity is lost."

Interface Design Issues

The primary concern in creating a screen-based system is the impact on market liquidity [9]. Many factors on the trading floor contribute to an efficient market, and these may not be able to be fully replicated on a computer [10], but *full* replication should not be the immediate goal. Rather, a system developer should be concerned with the implementation of an interface which incorporates *some* of the defining characteristics of a floor-based market. These include visualization of the market and its participants, a trading process which allows active participation, and the

human interaction between each of the participants. The last characteristic clearly is not easy to create. How does one duplicate the traders' pitch and tone of voice and level of urgency in order execution? The answer partially depends upon the state of technology.

Currently, virtual reality (VR) systems are available for viewing financial data. For example, VR applications exist for showing multidimensional financial data to aid foreign exchange currency option traders using head-mounted display helmets, data gloves, and stereoscopic images[15]. Another example is a VR room with bobbing icons, each with different visual attributes corresponding to different financial data, through which a broker can accurately view a client's portofolio holdings[8]. This technology in its refined state could greatly facilitate the trade process, but current technological limitations make these early systems impractical for trading use. Other technologies with application in the area include an eye tracking device for object selection, voice recognition, multimedia, video conferencing, and touch screen.

Design distinctions must be made between the symbolization of information, defined as data relevant in decision-making, and tasks (i.e., bidding, offering, canceling, etc.). Information can be further classified as either quantitative in nature (i.e., prices, quantities, spot data, etc.) or qualitative or psychological in nature. The latter characterization is illustrated by the pit psychology example. It is difficult to transpose this information asymmetry, present in psychological information, onto a screen-based system, but a number of ways exist to incorporate the available quantitative figures onto such a system.

The Pit

In creating an automated trading environment, it is necessary to establish a central object which will represent the place where "virtual" traders will reside and bid and offer contracts. A pit structure, such as those on the CME or COMEX, may be chosen as an appropriate central object. As the pit is static in nature, it may be constructed using a graphics program or bitmap image. It is worthwhile to consider that on the different floor exchanges, the actual physical characteristics of the pit differ. This suggests that they serve little purpose other than to provide a central area where efficient and open exchange is possible. However, in creating this image, it is necessary to indicate leveling. An example of the aforementioned psychological information, the grading of the pit exhibits a hierarchy among traders residing there. Figure 3 illustrates the idea.

Generally the above relationships hold, but, since traders are continuously gleaning profits and losses from the market, movement between "rings" is common [1]. In addition, there is not and should not be a definitive line on the pits demarcating the different classes of traders.

Traders

Traders view each other as information segments. Not only do they help relay public and private information to the marketplace, they signal their intentions by their bid/offer price and quantity. Furthermore, there are psychological ramifications to their every action. Thus, by having a list of the best bids and offers, as GLOBEX does, one does not see the entire market and the intentions of its participants. By modifying GLOBEX to show the entire limit order book, this would be solved, but at the significant cost of feasibility since these limit books are generally very large. Another way is to use an icon representation of each trader. Since some markets are very large with respect to the number of participants, there is an issue of available screen "real estate." A solution to this would be to dynamically display those traders who actually have standing bids and offers. Unfortunately, this avoids the psychological component of viewing the entire market, but it is feasible given the current state of screen technology.

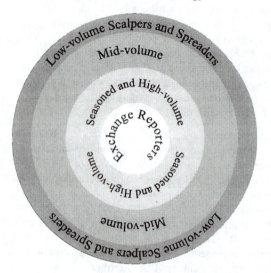

Figure 3 Trading Pit Hierarchy

Using icons gives users symbols of actual human traders. With the advent of video conferencing it may soon be possible to show a bit mapped or even real-time image of each participating trader. While this may be farfetched, the idea should be transparent: to give the trader the human element which exists on the floor. Using traders' initials as the icon is more practicable, and coupled with price and quantity data, the icon would combine quantitative as well as qualitative aspects into one package. It is also easy to view the icons superimposed onto a bit mapped image of a trading pit. To account for the hierarchical issues discussed above, icons can move to different regions on the pit depending on some moving average of that specific trader's total dollar volume of the current day (initial positions can be determined from the previous day's volume figures). Figure 3.1 shows an example of a trader-populated electronic pit. Each icon represents a trader. Those willing to buy are represented by a white icon and those willing to sell are depicted by a black icon. The icon for a small trader or scalper V, would indicate an offer for 5 contracts at 1.53, while an experienced, high-volume trader may be trader AE who is located towards the center of the trading ring bidding 1.37 for 200 contracts.

Trading Process
Optimizing the process by which trades are executed has important implications for interface design. It is necessary to note that this process is primarily governed by the order-execution algorithm rather than the interface. However, the ease of trading on a screen-based system is directly

dependent on the user's perception of the information and the facility of her cognitive processes in interpreting it.

An advantage of using the icons in the screen design is their functionality. Not only do they serve as visual representations of human traders, they also can be used as order execution vehicles. By using the mouse to click on the desired trader's bid/offer, the user can quickly choose her best price and quantity. This is assuming traders are actively involved in the market as opposed to simply entering limit orders. After viewing the marketplace and it's best prices, a trader can quickly execute the trade by clicking the respective trader's icon. Thus, communications bandwidth is maintained as is the speed of order entry and execution.

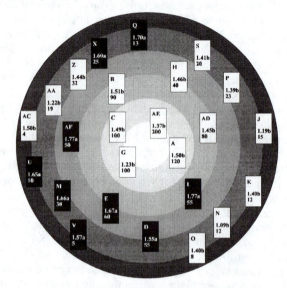

Figure 3.1 Electronic Pit Representation

London International Financial Futures Exchange's (LIFFE) Automated Pit Trading (APT) system [13] uses a 10 second limit on all orders after which they must be refreshed. This added feature forces traders to get more involved in the trading process. On the floor, traders must shout and gesticulate continuously in order to place a trade. Thus, on a theoretical system as proposed above, such a feature could be incorporated to keep the traders' orders executable as long as their "breaths are still warm[13]."

Other Design Issues
Other market information such as spot and forward data, should be included as well. These quantitative facts may be displayed in separate windows. Thus, this element of the configuration would be very similar to GLOBEX. Since they are solely quantitative in nature, there is no need to be overly intricate in designing its interface. One technique would be to make use of scrolling which GLOBEX currently uses or create a substitution routine to simply replace the current data.

This system must also allow orders to be changed easily. Currently, GLOBEX can cancel an order and reenter it with a time priority loss. A new such system would have to allow order modifications as dynamically as possible, without any such loss of time priority.

CONCLUDING REMARKS

Many other issues affect the interface design. The Commodity Futures Trading Commission (CFTC) continuously assesses the impact of an electronic trading system to the jobs, prices of exchange seats, as well as the open and competitive provision in Regulation 1.38 (see Section I). Automation could led to abandonment of floor-based trading. This would reduce the value of a seat on the exchange and eliminate numerous jobs. Thus, the push towards automation is likely to face great opposition from traders, clerks, and others employed by floor exchanges.

However, automation can be advantageous in certain cases (i.e., the deregulation of the London Stock Exchange (LSE) in 1986 led to introduction of electronic trading - Big Bang[5]). Essentially by using screen-based trading as a replacement for floor trading, LSE gained competitive advantage over other stock exchanges. After Big Bang, London saw average daily share turnover increase from £643 million to £1,161 million. In addition the liquidity seemed to increase evidenced by the number of market makers in operation before and after deregulation.

While London's Big Bang clearly seemed to be advantageous, an analogous information technology upheaval at a floor based futures exchange would most likely have adverse affects on liquidity and volume. Part of the difference is a function of different fundamentals underlying trading in both markets. Trading stock is different from trading futures contracts. One difference which is interesting to note is that in trading stock there is more emphasis on fundamental analysis (i.e., balance sheet strength, price/earnings ratio, and return on equity) rather on overall price trends, and this is understandable since a share of stock represents direct ownership of a company. By contrast, trading futures contracts is based largely on expectations. Thus, a trader of futures is much more concerned with the overall trend in expectations than in the fundamentals for that particular contract. On the floor expectations are largely visible and price trends can be easily perceived.

Current interest in automating the $500 billion a day futures exchanges throughout the world insures continued research into methods of replicating floor trading on a screen. However, until additional communications bandwidth can be added to the interface of electronic trading systems, the acceptance of such methods will likely be limited. We are currently developing such a prototype that incorporates the design elements discussed in the paper.

ACKNOWLEDGMENTS

This material is based upon work supported by the Information, Robotics, and Intelligent Systems Division of the National Science Foundation under Grant No. IRI-9209576. We are grateful to John Santos and Robert Push, Brown Brothers Harriman & Co., Brain Sayler, Commodity Clearing Corporation and Elizabeth Block, Chicago Mercantile Exchange, for their assistance with this project.

BIBLIOGRAPHY

1   Baker, Wayne E., The Social Structure of a National Securities Market, *American Journal of Sociology*, 1984. pp. 775-811.

2   Becker, Lopez, Berberi-Doumar, Cohn, and Adkins, *Automated Securities Trading*, 1992.

3   Black, Fischer., Toward a Fully Automated Stock Exchange, *Financial Analysts Journal*, 1971. pp 29-45.

4   Chicago Mercantile Exchange, *A World Marketplace*, 1990.

5   Clemons E.K. and Weber B.W., London's Big Bang: a Case Study of Information Technology, Competitive Impact, and Organizational Change. *Journal of Management Information Systems*, 1990, 6(4), 41-60.

6   Corcoran, Andrea M. and Lawton, John C., Regulatory Oversight and Automated Trading Design: Elements of Consideration, *The Journal of Futures Markets*, 1993. Vol 13, No 2, pp. 213-222.

7   Diamond, Barbara and Kollar, Mark., *24-Hour Trading - The Global Network of Futures and Options Markets*, New York, NY, 1989.

8. Display Technology: City Lights, *The Economist*, September 3, 1994, p.80-81

9   Domowitz, Ian, Automating the Price Discovery Process: Some International Comparisons and Regulatory Implications, *Journal of Financial Services Research*, 1992. pp. 305-326.

10 Domowitz, Ian, Equally Open and Competitive: Regulatory Approval of Automated Trade Execution in the Futures Markets, *Journal of Futures Markets*, 1993. Vol 13, No 1, pp. 93-113.

11 Domowitz, Ian, The Mechanics of Automated Trade Execution Systems, *Journal of Financial Intermediation*, 1990. pp.167-194.

12 Domowitz, Ian, A Taxonomy of Automated Trade Execution Systems, *Journal of International Money and Finance*, 1993. pp. 607-631.

13 The London International Financial Futures and Options Exchange, *APT - A Trading System for the Future*, 1994.

14 Melamed, Leo, The Mechanics of a Commodity Futures Exchange: A Critique of Automation of the Transaction Process. *Hofstra Law Review*, 1977, 6, 149-172.

15 Nielson, G. M., Rosenblum, L. J., Visualization Comes of Age, *IEEE Computer Applications and Graphics*, 11(3), p.15-19

16 Stoll, Hans R., Principles of Trading Market Structure, *Journal of Financial Services Research*, 1992. pp. 75-107.

Dinosaur Input Device *

Brian Knep
Industrial Light and Magic
San Rafael, CA

Craig Hayes
Tippett Studios
Berkeley, CA

Rick Sayre
Pixar
Richmond, CA

Tom Williams
Industrial Light and Magic
San Rafael, CA

ABSTRACT

We present a system for animating an articulate figure using a physical skeleton, or *armature*, connected to a workstation. The skeleton is covered with sensors that monitor the orientations of the joints and send this information to the computer via custom-built hardware. The system is precise, fast, compact, and easy to use. It lets traditional stop-motion animators produce animation on a computer without requiring them to learn complex software. The working environment is very similar to the traditional environment but without the nuisances of lights, a camera, and delicate foam-latex skin. The resulting animation lacks the artifacts of stop-motion animation, the pops and jerkiness, and yet retains the intentional subtleties and hard stops that computer animation often lacks.

KEYWORDS: Entertainment applications; Motion capture; Animation

INTRODUCTION

Motivated by the large amount of high-quality computer graphics animation called for by the film *Jurassic Park* [8] and by the desire to use the talent of experienced traditional animators, we built a system that allows them to animate computer graphics characters easily. We wanted the animators to be able to use their developed skills without first climbing the learning curves of computers and computer graphics, and in particular, without learning to use a complex interface such as that of a large commercial animation package. The result is a *stop-motion-capture device* called the Dinosaur Input Device, or DID. The DID is a highly intuitive three-dimensional (3D) input device, and many of its features are applicable to ordinary motion capture and to 3D input generally.

To accomplish our goal, we use a tool familiar to stop motion: the *armature*, an articulate skeleton often made of aluminum and steel rods, hinges, and swivels. In traditional stop-motion, the armature is posed by the animator and photographed onto a single frame of film. The armature is then moved incrementally, and photographed again onto the next frame of film. After many poses have been photographed, the resulting film will appear to show a moving armature. Instead of using film, our system captures the poses of an armature and uses them to position and animate a computer model.

To produce high-quality animation, the system must be very precise. One of the advantages of stop-motion animation over computer animation is the subtle movements and hard stops that are often smoothed away by computer interpolation. An imprecise or error-prone system would destroy these subtleties. The system must also be fast so the animators can immediately see their results composited over a background picture. This allows them to animate to the background, resulting in better integration of the live and computer-generated footage. Finally, the system cannot impair the movement of the armature or hinder the animators.

The most commonly used motion-capture devices are image based, such as SuperFluo's ELITE Motion Analyzer [3] and Motion Analysis's ExpertVision [7]. They rely on two or more high-quality CCD cameras, several reflective markers, and some specialized hardware. The markers are placed on the objects or actors and the hardware uses the images produced by the cameras to find the 3D position of each marker. The cameras and other hardware make these systems expensive. They are also imprecise—the calculated position varies over the surface of the marker—and slow—once the positions have been found, they need to be tracked and correlated with the computer joints. If the actors' motion causes markers to cross, the systems often get confused and the user must intervene to identify markers. They also require a large, unobstructed space to set up the cameras. These systems are useful for capturing the real-time gross motion of an object or actor, but lose the high-detailed motion and introduce too much noise into the data.

There are also systems such as the Polhemus Fastrak and the Ascension Bird that use electromagnetic transmitters and receivers to determine the location and orientation of an object. The object is connected to the computer via a cable tether. These systems work in real time, but they are noisy and imprecise, especially around metal objects like the armature.

Then there are systems that use potentiometers to record joint rotations. Examples are the Dextrous Handmaster [9], the Compact Master Manipulator [4], and the body suit used in the feature film *Toys* [6] [10]. These systems are fast and cheap, but the potentiometers are noisy and imprecise.

The above approach is the closest to what we need, and we overcome its shortcomings by using optical encoders rather than potentiometers. In the Armature section, below, we

---

*This device can of course be used to animate non-dinosaurs, but the name has stuck.

describe these encoders in detail and compare them against potentiometers.

The system we built, the DID, allows an animator, either a traditional stop-motion artist or a computer-graphics artist, to create computer animation by posing a physical armature. Digital sensors and specialized hardware monitor the global position and orientation of the armature and the orientations of the armature's joints and make them available to a computer. The computer uses these orientations to pose and display a 3D computer model composited over a background frame. Once the animator is satisfied with the pose, the computer records it and the animator moves on to the next pose. The recorded animation can be treated like any other computer animation, such as those produced on the SoftImage, Alias, Wavefront, and TDI modeling and animation systems. The animation curves can be edited, and the animation can be rendered with a high-quality renderer using motion blur. These are things that cannot be done with traditional stop-motion animation.

As an input device, the DID is very intuitive: To put the model in a particular pose, users pull and push physical joints; to look at a different view, users walk around the model or move their heads. This is much simpler than a hierarchy of nodes with rotate, translate, and scale values at each node, accessed by menu choices, sliders, and mouse movements. The DID is also rock solid: the computer model doesn't move unless the armature does.

The DID is composed of three parts: the physical armature, the controller, and the computer software. The next three sections describe each of these parts in detail.

ARMATURE
The armature is similar to those used in traditional stop-motion animation but with important differences: It is bare, with no foam-latex skin or adornments; its joints are monitored by digital sensors; it is larger than the traditional armature; and it uses a different set of joints than the traditional armature (see Color Plate 1).

No Skin
Armatures are usually covered in painted foam-latex skin and adorned with hair, nails, and teeth. The DID armature is bare: Animators manipulate the joints directly. The advantages are that animators don't have to worry about damaging delicate details, they can find a joint easily if it needs to be tightened or loosened (necessary for certain movements), they can make tiny micro-movements that otherwise wouldn't be possible, and they can see the range of movement left in a joint before it reaches its mechanical bounds.

Sensors
Each joint in the armature is monitored by a set of sensors. When a joint moves, each of the corresponding sensors sends out a signal describing the motion. Sensors also detect movements of the rig on which the armature is mounted. This rig controls the position and orientation of the entire armature.

For sensors we use optical encoders. They are encased in plastic boxes measuring about three-quarters of an inch on each side with a cylindrical shaft projecting from one side and wires projecting from the opposite side. When the shaft

turns, the amount of rotation is digitally encoded and sent out along the wires.

The raw encoders can detect rotations as small as one-third of one degree. Using reduction gearing we can attach an encoder to a joint so that a single revolution of the joint results in many revolutions of the encoder shaft. This lets the encoders detect joint rotations much smaller than one-third of one degree. The gears, however, introduce an upper limit on the precision due to *mechanical backlash*, the amount a gear can move before neighboring gears move. We have found that with our armatures one-third of one degree is precise enough for high-quality animation.

Two types of optical encoders are common: absolute and relative. An absolute encoder has a code for its entire resolution etched onto an internal disk. As the disk turns, a detector reads this code and gives an absolute and instantaneous readout of the encoder's current rotation. Relative encoders have an internal disk with simple alternating stripes (255 stripes in our encoders). Two out-of-phase detectors generate equal-period pulse trains when the shaft is rotated; counting pulses tells how far the disk has turned, while the relative phase of the two signals tells the direction. Absolute encoders are by necessity larger than relative encoders of equivalent resolution, and considerably more expensive. We therefore use relative encoders. Unfortunately, this means we have to maintain the absolute rotations of the joints elsewhere (see the Controller section below).

We use optical encoders rather than potentiometers and an analog-to-digital converter for several reasons. First, potentiometers are noisy even if perfectly shielded from electromagnetic interference. Optical encoders are much less susceptible to interference because they produce signals in the digital domain. This makes expensive, bulky, shielded cables unnecessary. Second, potentiometers are highly nonlinear, making accurate and consistent measurement of small incremental movements impossible. Optical encoders are very linear (on the order of fractions of the encoder sensitivity). Third, potentiometers produce a noisy spike when turned more than 360 degrees, making reduction gearing and multi-revolution motions impractical. Optical encoders do not have this limit. Finally, potentiometers are not perfectly *repeatable*—they might give different values at the same physical rotation. Optical encoders are perfectly repeatable.

Size
In order to fit the encoders on the armature without hindering its range of motion, we had to enlarge the underlying skeleton. Our armatures are about three feet from the head to the tip of the tail, which is about 30% larger than they would be in a traditional stop-motion environment. We were concerned that the larger size would impair the animators, but once they got used to it they found that the larger armature was in fact easier to manipulate.

Joints
Each encoder can monitor only a single axis of rotation, so each joint is either a hinge, a swivel, or a universal joint that we developed that allows three encoders to monitor all possible rotations around a central point. Unfortunately, the

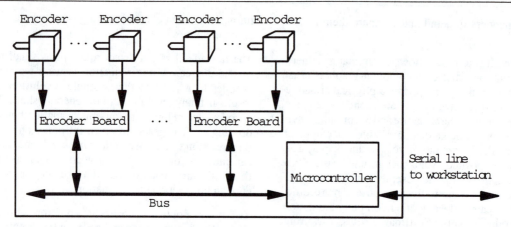

Figure 1: A simplified schematic of the controller.

universal joints are large and can only be used in uncluttered areas, like the neck and ankles. There are no ball joints in the armature.

CONTROLLER

Because we use relative encoders, we need to maintain the absolute rotations somewhere. We built specialized hardware to maintain these values and feed them to a host computer. We call this hardware the *controller*.

The controller is highly modular. An embedded microcontroller communicates with the workstation over a high-speed serial link with a bus designed to hold various interface cards (see Figure 1). We built interface cards that work with the optical encoders, but the controller can handle other cards that support different types of sensors. For example, an analog-input card could be built for strain gauges for measuring very small movements, such as those of armature toes. These interface cards plug into the bus, making it easy to configure a system with support for the desired number and type of sensors. The controller manages the cards, and allows the host to read the monitored armature joints and reset or preload their values for calibration.

Each interface card is capable of supporting sixteen encoders. There is one 24 bit counter for each encoder. By observing the leading and trailing edges of both encoder pulses, the hardware can increase the available resolution by a factor of four. These extra bits of precision are not guaranteed to be strictly linear, but they are guaranteed to be strictly monotonic and repeatable, and the non-linearity is consistent between defined counts. This therefore lets us track 16,384 revolutions of the encoder shaft (2^{24} values / 1024 pulses-per-revolution).

In order to calibrate the counters, we first pose the armature in a *zero-position* with the spine and legs straight and perpendicular to each other. Mechanical stops help us set this pose. A metal cube, for example, helps ensure a right angle between the legs and the spine. Although this is a completely unnatural pose, it is much easier to form and verify than a neutral, or relaxed pose. We then reset the counters in the controller to zero.

SOFTWARE

The software reads the encoder values from the controller and uses them to pose and display a 3D model of the object being animated. By compositing the model over a background picture, and by using a simulated camera whose motion has been matched to the camera that filmed the background pictures, the animator can see the pose in context as it will be seen on film. In other words, the system is *WYSIWYG*—what you see is what you get.

We can also display 3D geometry representing objects in the scene, such as tables and lamps. Animators use these to gauge the spatial relations between the animated object and the other objects. We can make sure, for example, that a foot is solidly on top of a table. To aide this, the camera can be snapped to the x, y, and z axes and moved around the scene using standard camera controls.

We can display either the computer model or a simple ball-and-stick model depicting the joints and rods of the armature. For prototyping speed, information about the armature is stored in a file that is interpreted at run time. This file describes the armature hierarchy, the names of the joints, the axes around which they rotate, and the ratio of joint revolutions to encoder revolutions. An example of a leg as described in this file is in Figure 2.

The whole process—reading the rotations from the controller, matching the computer model to the armature, and displaying the computer model composited over a background plate—takes one second on a Silicon Graphics Indigo R4000 and thus gives the animator quick feedback. Once he or she is satisfied with the pose, the software records it and moves to the next frame—moving the camera and bringing in a new background plate if necessary. The software can also generate a quick flipbook-type rendering of all the recorded animation for preview.

Matching

Due to software, animation and physical constraints, the skeletons of the physical models and the computer models often do not match. Software and animation constraints force us to build our computer models with joints in specific locations while physical constraints do not allow us to match

```
node {
  name rightKneeBend      # joint name
  axis    1 0 0           # joint axis
  sensor 5                # sensor index
  scale  25               # gear ratio
  node {
    name    rightCalve    # rod name
    length 4              # rod length
    node {
      name rightAnkleSwivel
      axis 0 0 1
      sensor 6
      scale -10
      node {
        name rightAnkleBend
        axis    1 0 0
        sensor 7
        scale  25
      }
    }
  }
}
```

Figure 2: An excerpt from an armature description file.

these joint locations in the armature. Each joint in the computer model is a universal joint: it can rotate about the x, y, and z axes. As explained in the Armature section, not all of the joints on the armature are universal, and so we cannot make a one-to-one match between them and the joints on the computer model—where the computer model has one joint the armature might have two joints, one for x rotation and one for y rotation, separated by a short distance of perhaps one inch. The models have different numbers of joints often in different locations and bending in different directions. Posing the computer model based on the physical joint data is therefore a non-trivial task.

This problem is over-constrained. The poses of the two different skeletons cannot be matched exactly—the best we can do is get a good pose based on a chosen metric. Our concern is that key joints, or *anchors*, match up as closely as possible; the rest of the joints need to approximate the pose but need not follow it exactly. The anchors are the leg joints, the hips, the head, and the tip of the tail (see Figure 3)

To match the models, we first group the joints into chains with the anchors as the endpoints of the chains. In our example there's a chain between the head and the hips and between the hips and the tip of the tail. There are also chains between the leg anchors, but since they have no intermediate joints, these chains are simply straight lines.

Each chain of joints in the physical model represents a curve in 3-space. Our goal is to match this curve with the corresponding chain of joints in the computer model. This is analogous to fitting a spline to a curve [5] [2]. In our case, however, the distances between the control points of the spline (the joints of the chain) must remain constant.

The method we use to match the chains is simple. We translate the anchor at one end of the computer chain to the position of the corresponding anchor of the physical chain. We then rotate the translated anchor until the second joint in the chain touches the physical chain. We then rotate the second joint until the third joint touches the physical chain. We continue until we reach the end of the computer chain. This process is illustrated in figure 4.

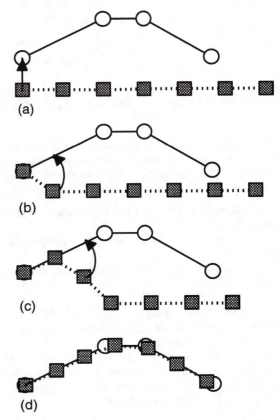

Figure 4: Matching a computer chain (grey squares) to a physical chain (white circles).

This method is easy for users to understand and has very predictable behavior. We get a perfect match for one of the anchors and a good fit for the chain, but because we push the error out toward the far anchors, these anchors may not match well. In practice, our chains are similar enough that this error is not a problem. We also don't put the armature into very extreme poses that could make it difficult to find a good match. We start at the hips and match toward the head (pushing the error toward the head), then match each leg (no error since the legs match one-to-one), and finally match the tail (pushing the error toward the tip of the tail). Since all the joints in the legs are anchors and match up one-to-one, we skip the curve-fitting stage in the legs.

For prototyping speed, the spline-matching information is stored in a file that is interpreted at run time. This file describes which joints are anchors, and which chains to match.

RESULTS

The DID is precise enough for high-quality animation, fast enough for interactive feedback, compact enough not to hin-

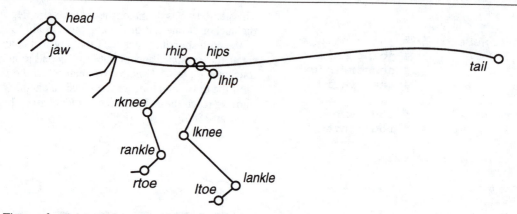

Figure 3: The *anchor* joints. When matching the computer model to the armature, we try to match these joints exactly. The remainder of the joints are matched using a simple curve-fitting technique.

der the movement of the armature, and easy to use. Animators with stop-motion experience are able to start animating immediately, and usually prefer this setup over the traditional setup. Here are some advantages they note:

- They don't have to worry about lights, cameras, or other stage impediments.
- They don't have to worry about hiding the support rods that hold up the armature.
- They are working with a naked armature, and thus can manipulate the joints directly.
- They can edit the resulting animation.
- They can render the resulting animation using computer graphics techniques for more natural motion-blur, textures, and integration with other elements in the scene.
- They can generate a quick, flipbook-type rendering of the animation composited over a background sequence, so they can see how the motion looks in the shot right away, instead of having to wait for film to return from a lab.

Jurassic Park

The DID was used with much success on *Jurassic Park* [8]. Of the 52 shots with computer animation, 15 were animated with the DID. Some of these shots had two creatures, making a total of 20 DID-animated creatures.

Two sequences were composed mostly of DID shots: the main-road sequence, where a tyrannosaur breaks out of her paddy and attacks the park tourists and destroys their jeeps, and the kitchen sequence, where two velociraptors hunt the two children in a large kitchen. To animate these shots we built four functional systems: two for the tyrannosaur and two for the velociraptors. The largest has 74 sensors, each with four wires—two for control, one for power, and one for ground—making a total of 296 wires. In comparison, the body suit used to create the war-room effects in the feature film *Toys* [6] uses only 24 sensors, all potentiometers, each with two wires [10]. The Dextrous Handmaster [9] and the Compact Master Manipulator [4] also both use a small number of potentiometers.

As the animators became more familiar with the DID, they began experimenting with *keyframing*—posing only every fifth or tenth frame and letting the computer interpolate be-

tween the poses. They found the resulting motion too smooth, however, so they used keyframing only to experiment with different movements before animating the final shot, which they did frame by frame.

For more on the human-interest side, see the *Cinefex* article on *Jurassic Park* [1].

CONCLUSIONS

The DID is easy to use because the physical device corresponds directly to the computer model it is controlling. Movements of the armature correspond to the same movements in the computer model, and the armature gives tactile feedback and spatial cues that match the computer environment. This is an example of a physical device that is easier to control than a virtual one.

The DID is distinct from a puppeteering device because of the direct correspondence it has with the computer model. The ideas we present, however, could be used to build devices that don't correspond directly to the models they control. For example, we could build a generic bipedal armature and use it to animate several bipedal computer models, each with different proportions. The matching algorithm would have to map the movements of the generic armature onto the computer models, perhaps by scaling or warping the movements. Although the correspondence wouldn't be as direct as the DID, the user would still enjoy the advantages of manipulating a physical 3D input device.

ACKNOWLEDGMENTS

We would like to thank Phil Tippett, Bart Trickel, Tom St. Amand, Stuart Ziff, Adam Valdez, and Randal M. Dutra from Tippett Studios; Dennis Muren, Janet Healy, and the software staff and technical directors at Industrial Light and Magic.

REFERENCES

1. Jody Duncan. The beauty in the beasts. *Cinefex*, 55:42, August 1993.

2. Gerald Farin. *Curves and Surfaces for Computer Aided Geometric Design*, chapter 23, pages 293–300. Academic Press, 1988.

3. Giancarlo Ferrigno and Antonio Pedotti. ELITE: A digital dedicated hardware system for movement analysis via real-time TV signal processing. *IEEE Transactions on Biomedical Enginerring*, 32:943, November 1985.

4. Hiroo Iwata. Aritificial reality with force-feedback: Development of desktop virtual space with compact master manipulator. In *Proceedings of the ACM SIGGRAPH, Computer Graphics*, volume 24(4), page 165, August 1990.

5. Peter Lancaster and Kestutis Salkauskas. *Curve and Surface Fitting: An Introduction*. Academic Press, 1986.

6. Barry Levinson. *Toys*. 20th Century Fox/Baltimore Pictures, November 1992. Motion picture.

7. MotionAnalysis Corporation, Santa Rosa, CA. *System Specifications*, 1993.

8. Steven Spielberg. *Jurassic Park*. Universal Studios, July 1993. Motion picture.

9. Dave Sturman. *Whole-hand input*. PhD thesis, Media Arts and Sciences, Massachusetts Institute of Technology, 1992.

10. Mark Cotta Vaz. Toy wars. *Cinefex*, 54:54, May 1993.

Dynamic Stereo Displays

Colin Ware
Faculty of Computer Science
University of New Brunswick
P.O. Box 4400, Fredericton
New Brunswick, Canada E3B 5A3
Email: cware@UNB.ca

ABSTRACT

Based on a review of the facts about human stereo vision, a case is made that the stereo processing mechanism is highly flexible. Stereopsis seems to provide only local additional depth information, rather than defining the overall 3D geometry of a perceived scene. New phenomenological and experimental evidence is presented to support this view. The first demonstration shows that kinetic depth information dominates stereopsis in a depth cue conflict. Experiment 1 shows that dynamic changes in effective eye separation are not noticed if they occur over a period of a few seconds. Experiment 2 shows that subjects who are given control over their effective eye separation, can comfortably work with larger than normal eye separations when viewing a low relief scene. Finally, an algorithm is presented for the generation of dynamic stereo images designed to reduce the normal eye strain that occurs due to the mis-coupling of focus and vergence cues.

KEYWORDS: Stereo displays, Virtual reality, 3D displays.

INTRODUCTION

The stereoscopic depth cue consists of relative differences or *disparities* between parts of the images available to the two eyes. In normal circumstances this information is effective only for objects less than 25 meters away, and it is optimal for objects that are much closer.

In computer graphics, some objects have no inherent spatial size, others may be representations of mountains or microscopic entities. One way of obtaining an appropriate stereo view is to scale the scene and bring it to an appropriate viewing distance. Another is to change the effective eye separations dynamically. An interesting research question is whether it is possible to create a system in which the stereo disparity is changed

dynamically so as to create near optimal disparities for perceiving depth information no matter whether the graphical object is at a great distance or close. The question addressed here is the extent to which changing disparities in real-time is perceptually disturbing. Two new experiments and two demonstrations are reported that show that human perception of depth through stereopsis is highly flexible, large distortions of the correct perspective geometry are possible and these distortions may be changed dynamically without undue perceptual ill effects.

An algorithm is presented which is designed to take advantage of this perceptual flexibility to allow the real-time adjustment of stereo disparity values as the user moves through the image space. The goal is to create a system in which the disparity values are always comfortable.

Stereo Vision

First some terminology and basic facts relating to stereo vision. Figure 1 illustrates the simplest possible stereo display. The eyes are fixated on the vertical line **a.** A second line **b** is closer to **a** in the right eye's image than in the left eye's image. The brain resolves this discrepancy by perceiving the lines as being at different depths as shown.

Retinal disparity is the difference between the angular separation of a and b in the two eyes (disparity = $\alpha - \beta$). *Vergence* is the degree to which the two eyes converge to fixate a target (this is also called phoria).

If the disparity between the two images becomes too great then *diplopia* occurs. This is the appearance of the doubling of part of the image. Another way of putting this is that the images are no longer *fused*. Whether two images can be fused or not and the area within which fusion occurs is called *Panum's Fusion Area*. However the size of Panum's fusion area is highly dependent on a number of visual display parameters such as the exposure duration to the images and the size of the targets. It is also true that depth judgments can be made despite diplopia, in other words, outside of the fusion area, although these are less accurate. For an excellent introductory review of stereo vision from a human factors perspective see [7].

Virtual Eye Separation

In stereo photogrammetry and in certain kinds of range finders it is common to create stereo images which have an effective eye separation much larger than any actual eye separation [5]. The reason for this is obvious; human eyes are only placed approximately 6.3 cm apart, which means that stereo information is only a useful depth cue up to 30 meters or so. However, if we can effectively change the eye separation then far more distant objects can be resolved by stereopsis. In viewing a mountain 10 km distant a virtual eye separation of 1 km might be appropriate. If viewing an object at 1 cm (as in a stereo microscope) a virtual eye separation of 1 mm will be more suitable.

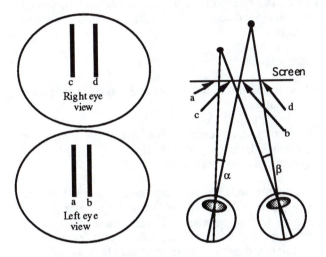

Figure 1. An illustration of some of the basic geometry relating to stereoscopic viewing.

Depth Cue Conflict

Occlusion is one of the major depth cues. It is a perceptual rule that says that closer objects always occlude (i.e. cover up) further objects. The problem is that when disparity information causes an object appear in front of a screen display the edge of the screen may appear to occlude that object and since occlusion is the stronger depth cue, the conflict is resolved perceptually in favor of occlusion, destroying the illusion of depth.

A more subtle depth cue conflict can arise if we try to change the stereo separation dynamically while moving through a scene. One of the most important depth cues comes from the dynamic flow of information across the retina, and evidence suggests that this is more important to 3D space perception than stereopsis [1,2]. When we are driving along a highway, we have a very strong sense of space yet almost all objects are likely to be outside the range at which stereo disparity is an effective depth cue. Figure 2 illustrates the kind of visual flow field that result from forward motion. This depth cue is also called motion parallax or, in some cases, the kinetic depth effect.

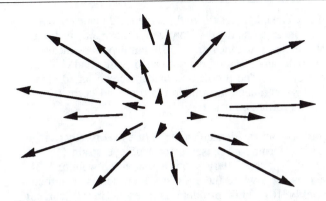

Figure 2 The kind of visual flow field that results from forward motion through a 3D environment.

Dynamically changing disparities, should cause changes in the relative depths in a scene. However, if other depth cues are stronger this effect may not be apparent. There is some evidence that the other dynamic depth cues, such as motion parallax, so dominate space perception that altering the effective disparities will be invisible. The perceptual question is, can we fly around a scene dynamically changing the effective eye separation without the users perceiving a rubbery distortion of the scene? Distortion should occur if the brain is a perfect geometry processor. Also, if rubbery distortion does appear is the effect disturbing or is it an acceptable price to pay for optimal stereopsis?

A piece of indirect evidence for the relative weakness of the stereo depth cue comes from a paper by Wallack [10]. In his study Wallack increased the effective eye separation in a telestereoscope which more than doubled the effective eye separation of the subjects. His subjects viewed a rotating wire object. The point that is relevant here is that before the actual experiment Wallack had to discard half his subjects because they failed to perceive any size distortion of the object as it rotated, whereas the disparity-vergence information should have made the object appear to stretch greatly in depth as it rotated. Clearly for those subjects that were discarded the kinetic depth effect (perhaps combined with object rigidity assumptions) completely dominated the percept. What is more after a short period of exposure the shape distortion of the object appeared much reduced even for those subjects who passed the initial test.

The Vergence focus problem

When we fixate objects at different depths, two things happen: the degree of convergence of the eyes changes (called vergence) and the focal length of the lens in the eye changes to create a sharp image on the retina. The vergence and the focus mechanism are known to be coupled in the human visual system. In fact if one eye is covered the vergence of that covered eye changes as the uncovered eye focuses on objects at different distances.

In a screen all objects lie in the same focal plane no matter what the apparent depth. However, the eye may be fooled into thinking that they are at different depths by means of stereo display that provides accurate disparity and vergence information. The problem is that in screen based stereo displays vergence information is provided correctly but focus information is not.

There is some evidence that the failure to correctly change focus information causes a form of eye strain [5]. A recent Japanese study showed that after watching 3D images for a while the eyes lose their ability to refocus quickly [6]. This problem is present in all current generations of stereoscopic head mounted displays and with monitor based stereo displays.

In another study it has been shown that the coupling of accommodation and vergence can be changed [4] and that this change can persist for some time. There appears to be considerable flexibility in the visual system regarding the coupling of focus and vergence. Anytime that a person dons a pair of reading glasses her visual system is forced to make an adjustment to a fixed change in focal length of her eye. This forces a change in the focus vergence relationship. With bifocals this re-adjustment must be continuously effected.

In view of the above observations how may we reduce the problems associated with the decoupling of focus and vergence in stereo display? One solution that seems obvious is to try to make images lie in the vicinity of the monitor screen, to reduce the parallax. This will minimize the focus vergence discrepancy. Valyrus [8] found experimentally that the accommodation vergence discrepancy should not be more than 1.6 degrees. He proposed a guideline based on parallax, which he defined as the spatial discrepancy on the screen between homologous image points from the two eyes. His guideline states that

$$P \leq 0.03 \cdot D$$

where P is the parallax and D is the viewing distance, otherwise diplopia will occur. Veron et al. [9] used this formula to derive the guideline that screen based stereo displays should be placed 2.3 metres from the viewer to give an image that it should always be possible to fuse. They assumed that the virtual object would always be placed behind the screen .

Based on a different analysis of the problem Williams and Parrish [12] concluded that a practical viewing volume falls between -25% and +60% of the viewer to screen distance. They proposed a method whereby objects at different depths can optimally use the available disparity range and show how objects at two or more different distances can be brought into the useful viewing volume. Their scheme parcels out the available disparity so that certain depth ranges are enhanced stereoscopically, while others are reduced in terms of the stereo depth. For example, in a scene with two objects, the distance between the front and back of each objects is allocated a large disparity range, while the empty space between them is

made devoid of disparity. What is interesting here is that this approach assumes that disparity is more important for seeing the local 3D shape of the individual objects rather than the 3D relationship between the two objects. Disparity becomes only local depth cue and not a global depth cue. Whether or not this is appropriate it assumes that the brain can tolerate inconsistencies between disparity information and other depth cues.

Summary Of Major Points So Far

• Having a different virtual eye separation for scenes of different sizes is a good idea.

• Having something that is always just behind the screen will also tend to be a good idea (to avoid occlusion conflicts).

• Large disparities result in diplopia

• Flow information will dominate dynamic disparity information.

• The focus and vergence mechanisms of the human eye are coupled, but only for high frequency changes. The system can re calibrate itself.

A reasonable working hypothesis is that most of our understanding of 3D space comes from depth cues such as occlusion, motion parallax and linear perspective. Stereo disparity provides additional, rather local information about relative depths. Therefore it may be reasonable to devise algorithms to dynamically adjust disparity information so that it is optimal for a particular situation, because the fact that depth cue conflicts will result is unlikely to be noticed.

The following series of demonstrations and experimental studies were all devised to test the validity of this hypothesis and explore the usefulness of dynamic stereo adjustments.

Changing Effective Eye Separations

It is possible to change the effective eye separation by a number of means.

1) By changing the eye separation parameter in a computer graphics stereo display the scene can be flattened or depth enhanced. Given the correct viewing position it is possible to construct a stereo view of a 3D scene such that the images presented to the two eyes are correct for an object in the vicinity of the monitor [2]. We assume that an eye separation parameter is set to 1.0 for "correct viewing" and 0.0 when both eyes get the same image, as in single viewpoint graphics. Clearly, it is also possible to set this parameter to intervening values, or even outside of the range 0-1. If it is negative then depth relationships are inverted (which in not likely to be useful), if greater

than 1.0 then stereo depth is enhanced. Thus this parameter has the effect of flattening or depth enhancing the image in terms of disparity cues.

2) By scaling an object and bringing it closer or moving it futher away, the separation of the eyes in relative to the object's size can be changed. This if a 5 km mountain viewed at 10 km distance is shrunk to 0.5 meter and moved to a viewing distance of 1 meter the eyes will be effectively 10,000 times further apart relative to the mountains original size.

3) By use of mirrors or prisms it is possible to actually change the optical separation of the pupils of the eyes. We will not be concerned with this method here.

Equipment Used

All of the studies, except for the last used an Indigo^2 Extreme with the Cyberscope^{TM}. The Cyberscope consists of a hood that can be placed over a small monitor allowing for the stereo viewing of properly constructed images. The Cyberscope uses front surface mirrors to displace and rotate the images presented to the two eye as shown in Figure 3

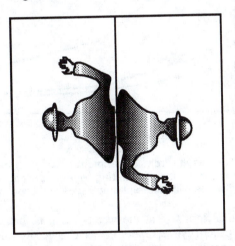

Figure 3. The cyberscope optically rotates the images from the two halves of the screen, 90 deg clockwise and counter clockwise respectively, and superimposes them. This is done using front surface mirrors to provide perfect optical clarity.

DEMONSTRATION 1: Does Motion parallax dominate stereo disparity when they are in conflict?

This study was directed at the relationship between optical flow information spatial cues, and stereo disparity cues. A special display was created in which the flow of visual information was consistent either with a continuously approaching surface, or with an inflating surface, This scene was constructed to be self similar at all scales of

resolution, and it was designed to be constantly expanding about a center at the plane of the screen (Illustrated in Figure 4 and in Color Plate 1, Ware,). Constantly inflating objects are not common but an expanding flow field is present whenever we move forward through the environment. Thus, based on experience observers might be expected to perceive movement of the scene towards them.

However, the scene was viewed in stereo and the stereo depth cues should have been enough to tell the observers that the scene was in fact expanding away from them.

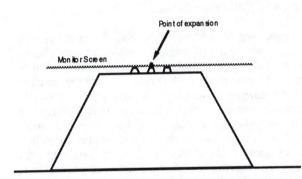

Figure 4. A schematic cross section of the recursively defined scene. The truncated pyramid in the center of each group of three, has a group of three truncated pyramids on top of it. The scene is self similar through scale transformations about the dot. The scene was viewed in stereo from above while continuously expanding.

The issue was, would subjects perceive the scene as inflating - which was the only geometrically consistent way of perceiving the pattern - or would they perceive something constantly coming towards them, as would be more consistent with everyday experience.

Observers were asked to look at this display and comment on what they saw. The general consensus on observing this display is that it shows a scene "coming up towards me". This impression lessens somewhat with time, and sometimes the rate of advance appears to slow and speedup, depending on the stage in the animation cycle. However, none of the observers reported expansion, and none of them reported seeing the scene moving away from them, as was in fact happening. This suggests a very powerful dominance of the optical flow information over stereo information.

EXPERIMENT 1: How fast can we change disparity cues without the effect being noticeable ?

In the introduction to this paper a case was made that the perceptual motor system is capable of re calibrating the disparity depth cue mechanism in the presence of other depth cues, such as motion parallax. Another way of describing it is that the disparity mechanism is insensitive to low frequency change. This study is directed to asking the question of how fast this re calibration can take place by measuring the frequence of disparity changes that are just detectable.

Method

In order to investigate this problem a scene was constructed in which a moving carpet dotted with truncated pyramids moved perpetually towards the observer. An image configured for the Cyberscope is illustrated in Color Plate 2 (Ware).

The scene was viewed in stereo and the effective eye separation was changed sinusoidally with an accelerating frequency. To describe this transformation it is useful to examine the extremes. With zero eye separation we have the same images presented to the two eyes but kinetic information consistent with a 3D scene (like looking at a moving television picture). A separation of 6.3 cm is normal and results in a correct stereo display for which disparity cues and the motion parallax are consistent with a 3D scene (making the normal perceptual assumptions about rigidity). Sinusoidal changes in eye separation should result in a sensation of oscillating depth if the brain were to rely primarily on disparity information but this would be in conflict with the rigidity assumption given the linear perspective and motion flow information.

On each trial the change in separation was started slowly and gradually sped up until it became noticeable. This speedup was such that after 50 seconds the eye separation was being changed at 1 Hz. At 100 seconds the frequency would be 2 Hz. There was also a random offset in time to the start of oscillation so that subjects could not anticipate this in their responses. The actual eye separation did not oscillate through the full range but varied with different amplitudes under different conditions. The amplitudes of oscillation were 10%, 20%, 30% The eye separations also varied 6.3 cm, 4.2 cm and 2.1 cm. Thus there were nine viewing conditions given by the product of these sets of settings.

One of the problems with this study was the difficulty in describing to subjects what they were supposed to look for. Subjects do not report sinusoidal depth changes. Instead a kind of paradoxical sideways movement is perceived; it is paradoxical because it is going in both directions at once. Subjects had to be trained to be able to see this phenomenon. Once this was achieved they were instructed to push a mouse button as soon as the oscillation became noticeable. This had the effect of recording the result and initiating the next trial.

Nine subjects who were all undergraduate or graduate students were used as observers.

Results

The results showing the mean frequency at which subjects detected the paradoxical motion are plotted in Figure 5. They show that the oscillation frequency that is detectable varies inversely with the amplitude of oscillation, and inversely with the effective eye separation. Both of these are to be expected since both eye separation and amplitude increase the amount of disparity change over time. The worst case was for a the maximum amplitude and the maximum eye separation, in which case the time to average frequency was 0.3 Hz. This is a remarkably rapid rate of adaptation to changing disparity ratios.

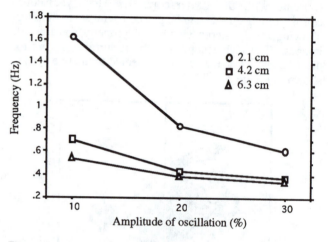

Figure 5. Results from Experiment 1 showing how the frequency at which oscillating eye separations are detected varies with different amplitudes and base values.

In practical terms what the results mean is that with a moving pattern eye separation can be changed dynamically as long as it is done gradually, taking several seconds to smoothly change. In this case viewers are unlikely to notice that anything unusual has happened.

EXPERIMENT 2: How do observers adjust their eye separation ?

The second experiment was initially designed to address the issue whether or not there is an obviously correct eye separation setting that is consistent with the geometry of a scene. However, in our pilot study we found that subjects had very little idea of what the "correct setting" was, therefore we changed the task and asked the users to create "the maximum comfortable setting" in terms of eye separation. Subjects were given control over the effective eye separation and instructed to increase the eye separation until diplopia occurred and then move it back to a comfortable value. The moving carpet display was used again for this study.

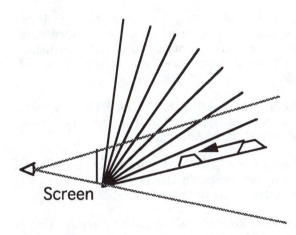

Figure 6. The moving carpet of truncated pyramids was presented in stereo and at different angles to the vertical plane of the monitor screen.

Subjects were given controls that allowed them to adjust the eye separation by depressing one of two keys, one of which increased the eye separation, the other of which decreased the eye separation. Subjects did this with the computer graphics model of the moving carpet set at 8 different angles with respect to the monitor (as shown in Figure 6) and they repeated the procedure twice to provide two settings at each angle.

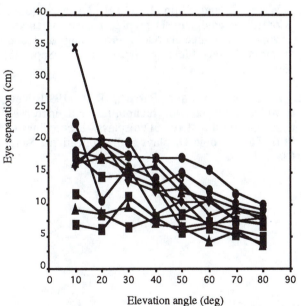

Figure 7. The results 10 subjects participating in Experiment 2

The results are shown in Figure 7 They indicate that subjects could comfortably tolerate much greater disparities with scenes having little depth in them, and small disparities with scene that contained a lot of depth.

This suggests that automatically changing the effective eye separation information about depth in a scene is probably a good idea even if it means breaking the rules of consistent geometry to do so. It is also clear that there are large individual difference with respect to the amount of disparity that can be tolerated suggesting that users of stereo displays should be able to customize a disparity parameter for their own comfort.

AN ALGORITHM FOR DYNAMIC STEREO ADJUSTMENT

The following algorithm was created to allow for the viewing of any scene with automatic adjustment so the stereo values would be in a reasonable range and could change dynamically. This algorithm has three steps.

Step 1: measure the closest portion of the displayed image. (This can be done by sampling the Z buffer).

Step 2: scale the scene about a mid point between the observer's two eyes in such a way that the closest point lies just behind the screen.

Step 3: render this modified scene in stereo using the normal methods for constructing off axis perspective views [3].

The transformation is illustrated in Figure 8.

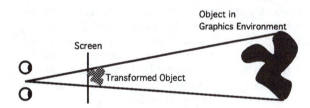

Figure 8. Schematic illustration of the effects of the stereo adjustment algorithm.

This algorithm achieves the following things.
- It reduces the focus-vergence discrepancy for far objects because the vergence will be that for an image at the screen.
- It enhances the disparities for far objects. For near objects that would normally lie in front of the screen disparities are reduced, thus reducing the likelihood of diplopia.
- It does not change the screen image of the scene components by a large amount and therefore if it is implemented dynamically the motion flow information is preserved.

DEMONSTRATION 2: Stereo adjustment algorithm with a large screen

Representatives of the UNB Ocean Mapping Group showed several thousand people the dynamic stereo display using a large format, 60 in diagonal screen, Electrohome projector at the CeBIT trade show in Germany. They all viewed a digital elevation map showing a line of undersea volcanoes in the South Pacific. This is shown as a monocular image in Color Plate 3, Ware. The eye separation was set to be considerably larger than usual, about 24 cm, in order to get appropriate disparities given the approximately 3 meter viewing distance. The interface allowed people to "fly" over the terrain using a six-degree of freedom input device [11]. In general viewers were very impressed by the large format, high resolution stereo display. None reported that the scene appeared to be expanding and contracting as they moved through the artificial landscape.

CONCLUSION

All of the evidence presented here is consistent with the hypothesis that the disparity depth cue is a highly flexible depth enhancement, rather than the primary determinant of 3D space perception. What this means is that in the absence of evidence to the contrary, hyper stereo adjustments are a useful tool in information display. We apparently do not need to be careful about matching the stereo geometry with the actual eye geometry. Rather what is important is to create stereo displays which maximize disparity gradients while maintaining them at a level below that at which diplopia sets in. Given this interpretation it seems to be worth artificially changing scenes so that the stereo information about relative depths is optimized, even though this stereo information may be in conflict with the other depth cues available, such as linear perspective and motion flow. The two advantages to such manipulations will be that disparities can be optimized for depth discrimination in a given scene and vergence-focus conficts can be reduced - which has the effect of reducing long term eye-strain.

References

1. Arthur, K., Booth, K.S. and Ware, C., (1993) Evaluating 3D Task Performance for Fish Tank Virtual Worlds. ACM Transactions on Information Systems. 11(3) 239-265.

2. Cutting, J.E. (1986) Perception with an eye for motion. MIT Press, Cambridge, Mass.

3. Deering, M. (1992) High Resolution Virtual Reality. Proceedings of SIGGRAPH '92. Computer Graphics, 26, 2 (July 1992), pp. 195-202.

4. Judge, S.J. and Miles, F.A. (1985) Changes in the coupling between accomodation and vergence eye movments induce in human subjects by altering the effective interocular separation. Perception 14, 617-629.

5. Lippert, T.M. and Benser, E.T. (1987) Photointerpreter Evaluation of Hyperstereographic Forward Looking Infrared (FLIR) Senso Imagery.

6. Noro, Kageyu, (1993) Industrial Aplication of Virtual Reality and Possible Health Problems. Japanese Journal of Ergonomica, v. 29, 126-129.

7. Patterson, Robert (1992) Human Stereopsis. Human Factors, 34(2) 669-692.

8. Valyrus, N.A. "Stereoscopy" Focal Press, London, 1966.

9. Veron, H. , Southard, D.A. Leger, J.R. and Conway, J.L. 1990 Stereoscopic Displays of Terrain Database Visualization, Proceedings of SPIE Stereoscopic Displays and Applications, Santa Clara, 124-135.

10. Wallack, H, and Karsh, E. (1963) The modification of stereoscopic depth perception based on oculomotor cues. Perception and psychophysics. 11, 110-116.

11. Ware, C. and Slipp, L. (1991) Using Velocity Control to navigate 3D graphical environments: a comparison of three interfaces. Proceedings of Human Factors Society Meeting, San Francisco, Sept. 35, 300-304.

12. Williams, S.P, and Parrish, R.V. 1990 New computational control techniques and increased understanding for stereo 3D displays. Proceedings of SPIE Stereoscopic Displays and Applications, Santa Clara, 73-82.

Transparent Layered User Interfaces: An Evaluation of a Display Design to Enhance Focused and Divided Attention

Beverly L. Harrison [1,3]

[1]Dept. of Industrial
Engineering
University of Toronto
Toronto, Ontario, Canada
M5S 1A4
beverly@dgp.utoronto.ca
benfica@ie.utoronto.ca

Hiroshi Ishii [2]

[2]NTT Human Interface Lab
1-2356 Take, Yokosuka-Shi
Kanagawa, 238-03 Japan
ishii@chi.xerox.com

Kim J. Vicente [1]

[3]Alias Research Ltd.,
110 Richmond St. East
Toronto, Ontario, Canada
M5C 1P1

William A. S. Buxton [3,4]

[4] Dept. of Computer Science
University of Toronto
Toronto, Ontario, Canada
M5S 1A4
buxton@dgp.utoronto.ca

ABSTRACT

This paper describes a new research program investigating graphical user interfaces from an attentional perspective (as opposed to a more traditional visual perception approach). The central research issue is how we can better support both focusing attention on a single interface object (without distraction from other objects) and dividing or time sharing attention between multiple objects (to preserve context or global awareness). This attentional trade-off seems to be a central but as yet comparatively ignored issue in many interface designs. To this end, this paper proposes a framework for classifying and evaluating user interfaces with *semi-transparent* windows, menus, dialogue boxes, screens, or other objects. Semi-transparency fits into a more general proposed display design space of "layered" interface objects. We outline the design space, task space, and attentional issues which motivated our research. Our investigation is comprised of both empirical evaluation and more realistic application usage. This paper reports on the empirical results and summarizes some of the application findings.

KEYWORDS: display design, evaluation, transparency, user interface design, interaction technology

INTRODUCTION

This paper describes results from an experiment used to evaluate transparent user interfaces against a proposed attentional model. The central research issue is how we can better support both focusing attention on a single interface object (without distraction from other objects) and dividing or time sharing attention between multiple objects (to preserve context or awareness).

The *technological problem* addressed by transparent interfaces is that of screen size constraints. Limited screen real estate combined with graphical interface design has

resulted in systems with a proliferation of overlapping windows, menus, dialog boxes, and tool palettes. It is not feasible to "tile" computer workspaces to facilitate keeping track of things. There are too many objects. Overlapping opaque objects obscure portions of information we may need to see and therefore may also be undesirable. Transparent interfaces address these issues, but may also introduce new challenges for designers.

The associated *psychological problem* we are addressing is that of focused and divided attention. When there are multiple sources of information we must make choices about what to attend to and when. At times, we need to focus our attention exclusively on a single item without interference from other items. At other times, we may need to time share or divide our attention between two (or more) items of interest. In this case, we rapidly switch attention back and forth between the items (necessitating minimal "switching costs"). There is a trade-off between these attentional requirements (depicted in Figure 4).

The *need* for focused or divided attention is largely determined by the demands of the user's task. However, our *ability* to successfully focus or divide (share) attention can be enhanced or degraded by the display design choices we make. For example, opaque overlapping window designs are problematic for divided attention (some information cannot be seen) but facilitate focused attention (the hidden background window cannot create visual interference). The interaction between the task characteristics and the design characteristics determine the attentional requirements and performance (Figure 1).

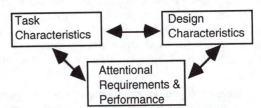

Figure 1. Design, Task, and Attentional Performance

Task characteristics largely determine attentional requirements and minimum acceptable performance levels. These task characteristics are pre-determined based on the nature of the work. Design characteristics (e.g., level of transparency) facilitate or prevent the task goals from being attained, limiting attentional performance. Our approach is: given an understanding of the task, can we manipulate the design characteristics to produce the necessary attentional performance?

Several key design issues need to be investigated if users are expected to focus on or divide attention between two superimposed images. Can users selectively attend to a chosen "layer" without visual interference from the other? Are there certain display characteristics or task properties which facilitate or preclude overlapping displays? How do these design choices affect attentional performance?

DISPLAY DESIGN CHARACTERISTICS

The small amount of display real estate available relative to the amount of data to be displayed presents a real challenge to user interface design. To date, two main strategies have been applied to the problem. In the first, the screen is partitioned, or *tiled*, into a number of non-overlapping windows. This we refer to as the *space multiplexed* strategy. In the second, windows lie on top of one another. Only the top one is visible at any given time, but a mechanism is provided to rapidly change which window is visible (temporal sequencing). This we refer to as the *time multiplexed* strategy. Most frequently, a hybrid of the two is used. What we propose in this paper, however, is a third strategy. Through the use of transparency in the background of windows, the contents of windows underneath others is visible, or at least partially visible. This "new" strategy we refer to as *depth multiplexing.*

On the one hand, the depth multiplexing approach offers the best of both worlds: windows need not be tiled to be visible. Hence, ideally, less information is obscured. On the other hand, the potential for content of one window interfering with another above or below it is introduced. Our prototypes show clearly that in some situations the technique works well, while in others there are real problems. The objective of our research agenda, of which the current paper is a part, is to develop a more formal understanding of the constraints of such an approach.

We propose a design space that captures the above three strategies and applies, in general, to foreground and background interface layers (Figure 2 and Figure 3). This design space allows us to methodically categorize and investigate both existing technologies and more novel technologies.

In one dimension (upon which this paper focuses), we vary the level of transparency/opacity between the two displays. *Fully opaque objects* reflect traditional window, palette, and menu design in current graphical user interfaces. *Fully transparent designs* reflect some of the more advanced interfaces such as those used in Heads Up Displays (HUDs)

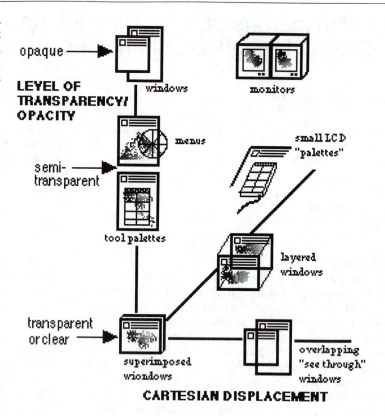

Figure 2. Design Space Dimensions

in aviation [12, 18] or in the Clearboard system [5]. In HUD design, aircraft instrumentation (a graphical computer interface) is superimposed on the external real world scene, using specially engineered windshields. In the Clearboard work, a large drawing surface is overlayed on a video image of the user's collaborative partner. *Semi-transparent designs* include such things as video overlays (like those used in presenting sports scores while the game is playing), "3-D silk cursors" [19] or Toolglass–like tool palettes [2,7].

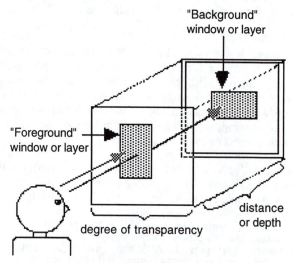

Figure 3. Concept of Layered Displays

Along another dimension we can vary the perceived depth of the planes between two displays, where one image appears closer to the user while the other is in the background. This can be accomplished using half-silvered mirrors, polarizing filters, or special transparent LCD displays (creating binocular disparity or stereopsis). In this case, the user looks through the display presented in the foreground to see the display presented in the background (e.g., [10]). Layers on this axis are distinguished by both transparency and depth. There are limited examples of such systems. Knowlton [9] used graphical overlays projected downwards onto half-silvered mirrors over blank keyboard keys to dynamically re-label buttons and functions keys (e.g., for telephone operators). Schmandt [16] built a system to allow users to manually manipulate and interact with objects in a 3-D computer space using a 3-D wand. Again a half–silvered mirror was used to project the computer space over the user's hand(s) and the input device. Disney has also developed a product called the "ImaginEasel" for animators and artists. ImaginEasel keeps the user's hand and input device in the workspace (using mirrors).

The proposed design space provides us with a means of categorizing both existing technologies and new technologies. However, the utility of any particular design will depend upon how well it supports the task characteristics and goals.

TASK CHARACTERISTICS

A number of situations arise as part of our day-to-day work which require us to focus or divide our attention. A number of such situations are outlined below, reflecting the diverse range of possible applications.

Focused attention examples:
- working on a document when a dialog box or warning message interrupts
- a pull-down menu (or pie menu) may temporarily block part of our current window. (The selected menus items may go on to create further dialog boxes of their own.)
- using a help system which displays a window of information and we would like to remember the context from which we requested help

Divided attention examples:
- using tear-off tool palettes (which behave as tiny overlapping windows)
- collaborating with a partner shown in a video window and we want to glance at both our partner and the work in progress (e.g., a drawing or document)
- viewing a live video conversation with one person while monitoring several Portholes-like connections to others for peripheral awareness of their availability
- using an interactive dialog box to change the drawing or modeling characteristics of an underlying image, model, or animation

These situations all share a common attentional problem: we need to be visually aware of multiple objects which overlap and obscure each other. All of these scenarios have two (or more) "tasks". In some cases we wish to time-share the two tasks (divided attention), while in other cases we selectively attend to one task excluding the other (focused attention). By their very nature, many of the proposed task pairs have an implicit active and passive task, We need a *peripheral awareness* of the passive task while we temporarily divert most of our attention to the active ask. The extent of this awareness determines the extent to which we must divide or focus our attention. We also must consider the visual contents and distinctiveness of the two layers within the task. How similar are they? What is the information density and level of detail of each? This determines how much interference may result when we focus our attention on one object. These characteristics may be unique for each task. A detailed task analysis is required to determine them (and hence the appropriateness of transparent design solutions within a particular domain).

DIVIDED AND FOCUSED ATTENTION

We are concerned with three critical attentional components: the ability to divide attention between two items, the ability to separate the visual characteristics of each source and focus on any single item with minimal interference from other items, and the switching cost (time, mechanism, learning, awareness) of shifting attention from one item to another.

To facilitate focused attention (ignoring information from the background layer while focusing on the foreground) we want to make the attributes of the information on foreground objects as different from the background as possible. We also wish to reduce the visibility of the background objects. This will minimize interference. By contrast, for divided attention (being able to see both foreground and background layers), we need to support simultaneous visibility of both layers. However, the user must still be able to separate which features belong to the foreground and which to the background in order to accurately perceive the objects.

There are many ways of achieving differentiation between layers (with varying success), such as different colors, content attributes – analog (images or graphics) versus verbal (text based), font sizes or styles, etc. Many of these features are pre-determined by the task. The level of transparency effects visibility of the background. Low degrees of transparency (more opaque) distinguishes the appearance of the foreground and background object, allowing the user to easily focus attention on the foreground. For divided attention, a high degree of transparency is desirable to support higher visibility of both layers.

Clearly there is a trade-off between these two goals. We need to support this trade-off since most real world jobs require *both* focused and divided attention. We have characterized the trade-off in Figure 2 which provides a framework for this research. We have used level of transparency as the visibility control variable. From this analysis, we can predict that the optimal degree of transparency is determined by the trade-off of supporting both focused and divided attention. As degree of transparency increases, it gets easier to divide attention between information on the top object and information on the background object but more difficult to focus attention on either object exclusively. The optimal transparency (OT) is a result of a trade-off. The curves and the location of optimal transparency in the figure are hypothetical but may reveal the trend. The non-linear nature of the curves is also proposed but appears to be supported from our preliminary experimental work.

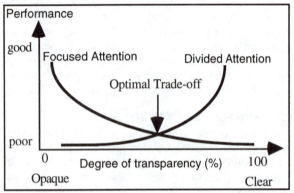

Figure 4. A simple model of transparency selection.

PREVIOUS RESEARCH IN ATTENTION AND DISPLAY DESIGN

Research in selective and divided attention, selective looking, and display design suggest that transparency is a promising method of presenting foreground and background layered information.

Kohler [11] originally investigated selective looking (monitoring dual tasks) by building headgear using half-silvered mirrors which presented the scene of the world in front of him superimposed on the scene of the world behind him. He reported that he could easily switch between these two views; the unattended scene seemed to "disappear" from sight.

Motivated by this work, further studies were carried out [15, 1] using two superimposed video images presented on a single monitor. In the first study [15] the tasks were visually distinctive: a hand slapping game and a ball tossing game. In the later study [1] both tasks were visually similar ball tossing games; the tasks were differentiated by the color of the shirts worn by the players. In both cases, subjects were asked to monitor one task and indicate the irregular occurrence of target events in this task. Meanwhile, bizarre events were sporadically presented in the non-monitored task. Subjects were easily able to monitor the target task to the exclusion of the unattended task. Subjects did not notice the bizarre events, even when the experiment was stopped during or immediately after the bizarre event occurred and the subjects were asked about it. This result still held when the bizarre event was presented in the exact same visual location where the target event

occurred (i.e., within foveal range). This seems to indicate that the intentionally unobserved task goes virtually unnoticed. A number of alternative explanations for this phenomenon were discussed and discounted. This work suggests that two superimposed *video* tasks can be easily monitored with minimal interference. However, the extent of simultaneous task awareness is unclear.

Similar results in selective looking have been found in studies of dual task monitoring in Heads Up Displays typically used in aircraft control and navigation tasks. Specific advantages cited include improved flight performance, superior object tracking, [12, 18]. The primary disadvantage is "attentional tunneling" – fixation on the HUD to the exclusion of events in the real world, particularly unexpected events (or vise-versa) [18]. Again subjects are easily able to differentiate either display layer easily. Practice seems to improve *simultaneous* monitoring performance.

This previous research, though not applied directly to graphical user interface design per se, suggests promising evidence for the use of superimposed transparent displays. Based on these results, one would anticipate reduced switching time and improved awareness by minimizing head and eye movement and re-focusing. Also, one can reasonably anticipate that users will be able to treat the sources separately and voluntarily attend to one or the other (with varying degrees of interference).

As in most interface designs, one can anticipate some inappropriate applications and pitfalls as well. In cases where "missed observations" have a high cost, reducing visibility through transparency might be undesirable. Also if both tasks must be simultaneously monitored and both have high attentional demands, the attentional tunneling problems might arise. Finally, while this would seem feasible for distinctive types of information, we must evaluate how well this technique works for visually similar information types.

RESEARCH METHODOLOGY

We are taking *two* complementary approaches to study transparent designs: formal experiments and realistic field studies. This paper emphasizes our empirical results.

To reveal how focused and divided attention changes, i.e. how the curves in Figure 4 are shaped, we are conducting formal experimental studies with well controlled models and simulations. By varying the degree of semi-transparency in between the two layers, the experimental results provide us with precise performance measures on how well the user can see both foreground and background information and on how high the interference is between the two "layers".

However, we realize that controlled experimental paradigms address a restricted set of design dimensions only. Real applications consist of a much richer task space. We have also developed several prototype systems which are more representative of real world applications. We are evaluating these systems and observing user behavior to gain further insights into the design of transparent user interfaces. This combined research program allows us to further formulate research issues while remaining confident that our research results have external (real-world) validity. The two approaches are conducted in parallel.

EXPERIMENT — TRANSPARENCY EFFECTS ON TASK INTERFERENCE AND LEGIBILITY

Our first set of formal experiments used a very simple but robust task to measure interference between two layers called the Stroop Effect [17]. In traditional Stroop tasks, a series of words are presented in randomly chosen colors (e.g., red, green, blue, yellow). Subjects must name the *ink color* while ignoring the word. Some of the words are neutral (e.g., uncle, shoe, cute, nail); other words are the names of conflicting colors (e.g., yellow, blue, green, red). Consistent, significant performance degradation occurs when conflicting color words are used and subjects attempt to name the color of the ink (e.g., the word "red" appears in green ink; the correct response is green). In later studies (e.g., [8]), a consistent and significant Stroop Effect was found even when the word was printed in black ink, presented adjacent to a color bar. It is virtually impossible to consciously block or prevent the Stroop Effect in selective looking tasks, despite numerous experimental permutations (over 700 articles – for reviews see [6, 13]).

Our experiments test how varying *transparency* effects interference between the displayed word and the color target, using a traditional Stroop test. The Stroop test was used to evaluate interference because it provides an sensitive, extreme measure of the extent of interference. As such, it should suggest worst case limitations. In our experiment, the word is seen by looking "through" the color patch. At high levels of transparency (e.g., 100% - clear) we anticipate that users will experience high levels of interference from the word when they try to name the color (difficulty in focused attention). As the color patch becomes more opaque the interference from the word should decrease (making focused attention easier). This would support the focused attention curve in Figure 4.

We used the word naming component of the Stroop Test to test the divided attention curve proposed. In this case users are asked to ignore the color patch and read the word in the background layer. This experiment reflects more of a legibility test, necessary for divided attention. The color patch in the foreground is always clearly visible and perceived. By reading the background word the user is, in effect, creating a divided attention task. At high levels of transparency (e.g., 100% - clear) it should be very easy to read the background word (divided attention is easy). At more lower levels (opacity increases) it should become progressively more difficult or impossible to read the word (loss of ability to divide attention).

When combined, results from the two experiments suggest interface design parameters where interference is minimized and the word is still fairly legible (awareness is preserved).

Hypotheses (stated as null hypotheses)

H1:　As transparency level increases (i.e., the word is more visible through the color patch) the response time and errors will be unchanged in the color naming task.

We anticipate more interference as transparency increases and therefore reduced performance as shown in Figure 4. Furthermore we anticipate a leveling-off point where performance does not continue to degrade.

H2:　As transparency increases the response time and errors will be unchanged for the word naming task.

We anticipate that as transparency increases the word gets easier to see and is therefore faster and more accurate to read.

Experimental Design: Color Naming Experiment and Word Naming Experiment

We used 4 colors: red, blue, green, and yellow. Words (helvetica, 78 point, uppercase) appeared "through" the colored rectangular patch. We used neutral words UNCLE, NAIL, CUTE, and FOOD in addition to the four color names. Transparency levels were varied as: 0% (baseline condition - only one of the word or color shows), 5%, 10%, 20%, 50%, 100% (clear - both the word and color show). Task order (color naming versus word naming) was counter–balanced and spaced one day apart. No cross task interference is anticipated [14]. The word naming experiment baseline condition was a word only – presented with no color patch. The color naming experiment baseline condition was a color patch only – presented with no word. There were no other differences between the two experiments. (The word naming experiment should not have any Stroop effects but performance should be affected by the visibility of the word.)

A fully randomized, within subject, repeated measures design was used. There were 4 conditions: non-conflict or neutral (the word was a neutral word), incongruent color (a conflicting color word was present), congruent color (the color word matched the color of the patch), and baseline (color or word only). Transparency levels of 0%, 5%, 10%, 20%, 50%, 100% were used for all word-color combinations for a total of 180 unique combinations. For each of 16 subjects, three sequences of the entire set of 180 images were shown. Trials were presented in random order at 5 second intervals. Each experiment lasted about 45 minutes. Verbal responses were logged within 1 msec of accuracy. Errors in response were recorded. Error trials were removed from subsequent analysis of response times.

Experimental System Configuration

The experiments were run using the PsyScope software and hardware [3] with a headset microphone on a Macintosh IIfx. Audio levels were adjusted before each subject was run. Subjects sat at a fixed distance of 100cm from the screen. All sessions were video taped.

Procedure

Subjects were given 20 practice trials. These trials were randomly selected from the set of 180 possible combinations. Following this, subjects were shown three sequences 180 combinations (15 minutes per set), with rest breaks in between each set.

Subjects were debriefed at the end of the experiment. Open ended comments were recorded and the experiment was video and audio taped for analysis purposes. Response times and errors were logged by the computer.

Subjects

A total of 16 students from the University of Toronto were run as subjects They were pre-screened for color-blindness. Subjects were paid for their participation and could voluntarily withdraw without penalty at any time.

RESULTS – COLOR NAMING TASK

A univariate repeated measures ANOVA was carried out on the data. As hypothesized, significant main effects were found for transparency $F(5, 719)=11.12$, p< .0001 and word type $F(3, 719)= 36.19$, p < .0001. This suggests that the Stroop Effect was present and that transparency may indeed dilute the interference. Not surprisingly, color also showed a significant main effect $F(3, 719)=15.51$, p < .0001, suggesting that saturation or luminance might dilute the interference (i.e., affects word legibility - see below). There were no significant interaction effects across factors.

Post-hoc analyses were carried out to compare means for the transparency and word type (Student-Newman-Keuls test with alpha levels = .05). Response times for transparency levels occurred in four statistically significant groupings: 100%+50%+20%, 10%, 5%, and 0% (baseline condition). As expected, word types were grouped according to the predicted Stroop Effect: incongruent (color name conflicted with color word), neutral+congruent, and blank (color only - baseline condition). Our primary interest is in the effect of transparency under maximum interference conditions (incongruent word). The mean response times of primary interest are shown in Figure 5.

At 5% transparency (word was only slightly visible) the means across all word types are not statistically different from 0% (no interference/Stroop effect). At levels above 10%, three groupings of means occur (as the Stroop effect would predict): blank, congruent+neutral words, and incongruent words. Interference peaked at 50% – increasing transparency did not degrade performance.

Subject errors in response occurred only occasionally (average of 4 per 540 trials) and almost exclusively on the color-incongruent trials. Errors were approximately evenly distributed across all levels above 5% (5% showed few errors). Error trials were not used in the above analysis.

RESULTS – WORD NAMING TASK

A univariate repeated measures ANOVA was carried out on the data. As hypothesized, a significant main effect was found for transparency $F(5, 8614)=25.94$, p< .0001. Word type and color also showed significant main effects: word type $F(3, 8614)=16.06$, p < .0001 and color $F(3, 8614)=26.55$, p < .0001. Additionally there was a significant interaction between transparency and color $F(15, 8614)=4.36$, p < .0001. This suggests that word legibility is affected by not only level of transparency (i.e., visibility) but also the properties of the color used (i.e., saturation and luminance). (Figure 5 shows overall mean response times.)

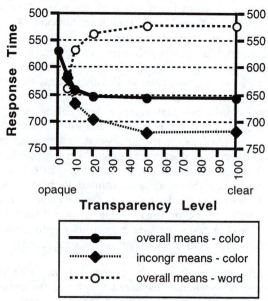

Figure 5. Mean response times

Post-hoc analyses were carried out to compare means for the transparency and transparency * color interactions (Student-Newman-Keuls test with alpha levels = .05). Transparency levels occurred in three significant groupings: 5%, 10%, 100%+50%+20%+0%. The baseline word only condition (0%) was not statistically different from the 100% condition (word with color background). Analysis of word type showed an unexpected Stroop Effect (despite counter-balancing order with the color naming experiment).

For levels of transparency of 5% subjects reported great difficulty in seeing the word, about 15% of the trials were errors. (Subjects reported "none" when they could not make out the word.) At 5% and 10% levels, certain colors produced better performance (lower response times, fewer errors) than others. Yellow was "easiest" followed by green (by post-hoc analysis of means). Blue and red were "hardest" and not statistically different. For transparency levels above 10%, subjects made virtually no errors and performance was consistent across colors. At 20% levels and higher, all words were easily read and there were no significant differences in response times.

DISCUSSION

For the color naming experiment, we have found that degree of transparency dilutes the interference/Stroop effect in a seemingly logarithmic fashion, with performance leveling off at 50% transparency. At levels of 5% (and likely less) minimal or no interference seems to take place (using means comparison tests). This supports our proposed focus attention curve, with performance cut off points at 5% (lower) and 50% (upper). The error rates also seem to support this: errors only occurred below 5%. .

Word naming seems highly error prone at levels of 5%. At levels of 10% subjects could accurately name most of the words, though they seemed to perform slightly better, depending upon what the background color was. It seems that there was an interaction between saturation/luminance and legibility. This suggests that certain colors might be more profitably used in transparent windows or interfaces - though this remains to be tested. Word naming performance improved more dramatically than hypothesized, with performance leveling off at 20%. Our hypothesized divided attention curve seems to underestimate the effect of increased transparency. Also we did not observe the hypothesized continual performance improvement but rather saw performance roughly peak and remain constant from 20% transparency to 100%.

The Stroop test was used to evaluate interference between transparent layers because it provides an sensitive, extreme measure of the extent of interference. As such, it should suggest worst case limitations. Our results suggest that for divided attention tasks, substantial performance gains occur within the first 20-25% transparency, but may not occur from 20% to 100%. Levels of 5% or less do not seem usable. For focused attention tasks, there is a rapid performance degradation between 5% and 50% transparency. At 50% performance is at it worst and does not deteriorate substantially with further increases in transparency.

Clearly, different tasks will have different levels of error tolerance and acceptable performance limits. Also the legibility of layers will be determined by visual distinctiveness in addition to overall transparency levels.

CURRENT WORK IN REAL APPLICATIONS

The above experiment tested one of the most stringent interference tasks possible and gave us insights into both the proposed attention model and some of the upper and lower threshold values for transparency. In addition to the empirical work, we wish to evaluate our theories of attention, performance, and interface design in more realistic prototype and application domains. This work is briefly summarized here (see [4] for more detail).

We installed transparency into some interactive dialog boxes within a 3-D modeling/animation system. In this system the user needs to see a potentially large model (full screen, background) while changing various attributes of the model or of the drawing tools (using windows in the foreground), resulting in a divided attention problem. Typically, a user might have 3 or 4 such interactive dialog windows open at all times.

We had several users of varying levels of expertise evaluate the transparent windows. We also asked users to select a "personal favorite" transparency level using a slider bar. Substantial in-depth investigation is still being conducted. However, several insightful comments have already been noted.

The degree of visual distinction between the two tasks strongly influences the extent of possible interference and perceived difficulty. Users found transparent windows (text, buttons) were easier to use over solid models/images than those superimposed over wire frame drawings. Higher levels of "opacity" seemed to partially compensate in the more difficult task situation (by minimizing interference as in the Stroop experiment). This suggests that level of detail or information density might also be a determining factor when choosing transparency levels.

As familiarity with the interactive window layout improved, users preferred corresponding increases in transparency. They preferred to see "less" of the interactive dialog boxes and more of the underlying image. The dialog box items were needed only as outlines to target selections - the actual legibility of the text was substantially less important. This suggests that border of windows and buttons and data entry areas might be handled in a different way than the actual names and labels. Performance improvements are similar to Heads Up Display research findings. However, this suggests new and intriguing possibilities for dynamically evolving interfaces based on increased expertise.

We additionally developed anti-interference (AI) outlines for text and borders of objects, based on feedback from prototyping (Figure 6) [4]. These AI graphics use an opposing contrast level outline to encircle the object or letter (e.g., white objects have black border outlines). This has dramatically improved visibility and distinctiveness of items in transparent foreground menus and windows. Work and evaluation in this area is on-going.

Figure 6 (a). Plain font style, 20% transparency
(b). "Anti-interference" (AI) font style, 20% transparency

CONCLUSIONS

We have illustrated a method of empirically testing our proposed design space dimensions and the proposed attentional framework using well-established theoretical measures. We are now evaluating focused and divided attention and performance with more complex visual information in a variety of real world tasks using transparency in menus, dialog boxes, and windows. We are additionally experimenting with dynamically evolving interfaces for example, whether transparency level should automatically change for the entire window (or a portion of the window) when the cursor is moved over it.

We believe that interface designers can take advantage of both the intrinsic properties of the task and of an understanding of human visual attention to design new display techniques and systems. The design space proposed in this paper supports the idea of active/passive tasks by providing users with an awareness of one task while they focus on the other. In this way, inherent characteristics of the task are supported in the interface while providing enhanced functionality. We believe that results thus far show promising advantages for creating new user interfaces and interaction techniques. We are exploiting possibilities of new technology in a way that is sensitive to both psychological and task constraints.

ACKNOWLEDGMENTS
Support for our laboratory is gratefully acknowledged from the Natural Sciences and Engineering Research Council (NSERC), Alias Research Inc., Apple Computer. the Information Technology Research Centre (ITRC), Xerox PARC, and the Ontario Telepresence Project. We would also like to thank Dr. Allison Sekuler and the Psychology Dept., Dr. Colin MacLeod, Dr. Chris Wickens, Shumin Zhai, and members of the Graphics Lab and Cognitive Engineering Lab.

REFERENCES

1. Becklen, R. and Cervone, D. (1983) Selective looking and the noticing of unexpected events. *Memory and Cognition*, 11 (6), 601-608.

2. Bier, E. A., Stone, M. C., Pier, K., Buxton, W., and DeRose, T. D. (1993). Toolglass and magic lenses: The see-through interface. *Proceedings of SIGGRAPH'93*. Anaheim, CA.

3. Cohen J.D., MacWhinney B., Flatt M. & Provost J. (1993). PsyScope: A new graphic interactive environment for designing psychology experiments. *Behavioral Research Methods, Instruments & Computers*, 25(2), 257-271.

4. Harrison, B. L., Zhai, S., Vicente, K. J., Buxton, B. (1994). Semi-transparent User Interface Object: Supporting Focused and Divided Attention. *Cognitive Engineering Lab Technical Report CEL-94-08*, Dept. of Industrial Engineering, Univ. of Toronto, November, 1994.

5. Ishii, H. and Kobayashi, M. (1991). Clearboard: A seamless medium for shared drawing and conversation with eye contact. *Proceedings of CHI'91*, Monterey, CA, 525-532.

6. Jensen, A. R. & Rohwer, W. D. (1966). The Stroop color-word test: A review, *Acta Psychologica*, 25, 36-93.

7. Kabbash, P., Buxton, W. A. S., and Sellen, A. (1994). Two-handed input in a compound task. *Proceedings of CHI'94*, Boston, MA., 417-423.

8. Kahneman, D. and Chajczyk, D. (1983) Tests of the Automaticity of Reading: Dilution of Stroop Effects by Color-Irrelevant Stimuli. *Journal of Experimental Psychology: Human Perception and Performance*, 9 (4), August, 1993, 497-509.

9. Knowlton, K. C. (1977). Computer displays optically superimposed on input devices. *Bell System Technical Journal*, Vol. 56, No. 3, March, 1977. 367-383.

10. Kobayashi, M. and Ishii, H. (1994). DisplLayers: Multi-Layer Display Technique to Enhance Selective Looking of Overlaid Images. Poster from *CHI'94 Conference*, April, 1994, Boston MA.

11. Kohler, P. A. (1972) *Aspects of Motion Perception*. New York: Pergamon.

12. Larish, I and Wickens, C. D. (1991). Divided Attention with Superimposed and Separated Imagery: Implications for Head-Up Displays. *University of Illinois Institute of Aviation Technical Report* (ARL-91-4/NASA HUD-91-1).

13. MacLeod, C. M. (1991). Half a Century of Research on the Stroop Effect: An Integrative Review. *Psychological Bulletin*, Vol. 109, No. 2, 163-203.

14. MacLeod, C. M. and Dunbar, K. (1988). Training and Stroop-like interference: Evidence for a continuum of automaticity. *Journal of Experimental Psychology: Learning, Memory, and Cognition*, 14, pp.126-135.

15. Neisser, U. and Becklen, R. (1975) Selective Looking: Attending to Visually Specified Events. *Cognitive Psychology*, 7, 480-494.

16. Schmandt, C. (1983). Spatial input/display correspondance in a stereoscopic computer graphic workstation. *Computer Graphics*, Vol. 17, No. 3., 253-259.

17. Stroop, J. R. (1935). Factors affecting speed in serial verbal reactions. *Journal of Experimental Psychology*, 18, 643-662.

18. Wickens, C. D., Martin-Emerson, R., and Larish, I. (1993). Attentional tunneling and the Head-up Display. *Proceedings of the 7th Annual Symposium on Aviation Psychology*, Ohio State University, Ohio, 865-870.

19. Zhai, S., Buxton, W., and Milgram, P. (1994). The "silk cursor": Investigating transparency for 3D target acquisition. *Proceedings of CHI'94*, Boston, MA., 459-464.

User-Centered Video: Transmitting Video Images Based on the User's Interest

Kimiya Yamaashi, Yukihiro Kawamata, Masayuki Tani, Hidekazu Matsumoto

Hitachi Research Laboratory, Hitachi, Ltd.
7-1-1 Omika, Hitachi, Ibaraki, 319-12 Japan
Phone: +81-294-52-5111
E-mail: yamaashi@hrl.hitachi.co.jp

ABSTRACT

Many applications, such as video conference systems and remotely controlled systems, need to transmit multiple video images through narrow band networks. However, high quality transmission of the video images is not possible within the network bandwidth.

This paper describes a technique, User-Centered Video (UCV), which transmits multiple video images through a network by changing quality of the video images based on a user's interest. The UCV assigns a network data rate to each video image in proportion to the user's interest. The UCV transmits video images of interest with high quality, while degrading the remaining video images. The video images are degraded in the space and time domains (e.g., spatial resolution, frame rate) to fit them into the assigned data rates. The UCV evaluates the degree of the user's interest based on the window layouts. The user thereby obtains both the video images of interest, in detail, and the global context of video images, even through a narrow band network.

KEYWORDS: Networks or communication, Digital video, Compression, User's interest, Computing resources

INTRODUCTION

Most image compression techniques, such as H.261 [1], JPEG (Joint Photographic Experts Group) [2] and MPEG (Moving Pictures Experts Group) [3], are not user-centered. These techniques compress a video image in the same way without considering what a user is watching in the video image. Furthermore they cannot allow transmission of multiple video images through a narrow band network such as a public network like ISDN (Integrated Services Digital Network). For example, the MPEG technique allows only one video image to be transmitted through the ISDN-1500 (1.5 Mbits/second) network [3]. Since the technique degrades all the video images uniformly to fit them into the network, a user cannot obtain video images of individual desired quality.

Many applications, such as video conference systems, remotely controlled systems and security systems, rely on transmission of multiple video images through public networks. For example, users of video conference systems want to look at multiple video images showing all participants in the conference at one time. Operators of remotely controlled systems always look at multiple video images of remote spots because problems may occur anywhere. In these applications, each site is a long way from the others. In video conference systems, sites may be several hundred miles from each other. In remotely controlled systems, remote spots may be several miles from the control room. These systems need to use a public network to transmit multiple video images because it is very expensive to build a private network between such remote locations.

Users are not interested in all the video images at a time. They are watching a part of the video images carefully and seeing the remaining video images peripherally to obtain a global context of the situation. For example, when an operator in a remotely controlled system is inspecting a machine, the operator is carefully watching the video image that shows the machine. The operator sees peripherally the remaining video images only to grasp whether problems are occurring there or not.

This paper describes a technique for transmitting multiple video images, based on a user's interest, through a network. The technique, User-Centered Video (UCV), compresses video images of interest with high quality and degrades the remaining video images according to the user's interest. That is to say, the UCV assigns a larger part of a network data rate to a video image of interest than the remaining video images. The UCV transmits multiple video images through the following steps:

1) The UCV evaluates a user's interest dynamically based on window layouts.

2) The UCV degrades video images according to the evaluated user's interest by changing the frame rate and the number of pixels of video images to fit them into the network bandwidth.

There are several approaches to control display [4-6] or image compression [7,8] according to a user's interest. For

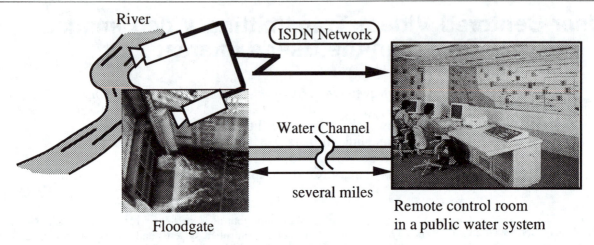

Figure 1 A public water system needs to transmit multiple video images through a public network

example the Generalized Fisheye Views approach enables a user to recognize a long program list by displaying the part of interest in detail, while giving the whole list in a global context [4]. These techniques control only screen space among the computing resources. The UCV assigns network data rates to multiple video images according to a user's interest. The image query system developed by Hill et al. [7] transmits a rough image of the whole image first and then transmits a detailed image of the region specified by the user. But, they discuss only transmission of a static image. The UCV transmits multiple continuous media (video images) through a network. Plompen et al. [8] proposed a video conference system that tracks movement of participants' heads. They assumed that participants in one site are interested in the movement of participants' heads in the other site. As they determined a user's interest statically, the technique cannot adapt to dynamic changes of the user's interest. The UCV evaluates the degree of the user's interest dynamically so that it can adapt to the dynamic changes.

In this paper, first we describe the concept of the UCV. Then we describe the UCV in detail; the architecture, evaluation of the degree of a user's interest, and determination of video parameters (frame rate and number of pixels). Finally we demonstrate the UCV with a prototype system.

CONCEPT OF USER-CENTERED VIDEO
A typical application of the UCV is to a remotely controlled system, such as a public water system shown in figure 1. The system locates several video cameras around an intake tower to watch conditions (e.g., at the floodgate for the intake, on the river, at gates to the water channel, etc.). The system usually has several intake towers that are several miles from the control room, so it is cost-desirable to transmit multiple images around each intake tower through a public network.

The MPEG technique allows only one video image at 30 frames/second through the ISDN-1500 network. So when the system transmits 8 video images, MPEG transmits all the video images at only 3-4 frames/second. It is hard for an operator to control and watch the floodgate because smooth motion of the floodgate is lacking with such a degraded video image.

The system does not need to transmit the video images with the same quality. During opening of a floodgate for intake the operator is focusing on the video image of the floodgate to examine the exact position and motion of the floodgate. The operator needs to obtain the video image of the floodgate with high quality. At the same time the operator looks at global context of the remaining video images rather than the details. For example, the operator is seeing peripherally the video images of gates that are not being handled. Because rough images of the gates allow the operator to grasp whether the gates are still or not, the operator does not need details such as the exact positions of the gates.

The user's interest changes dynamically according to the situation. For example, when some problems begin to occur in a video image, the operator changes the user's interest to the video image in which the trouble is occurring.

The UCV transmits video images of interest with high quality by degrading the remaining video images, because it is useless transmitting video images of little interest with the same quality as video images of interest. The UCV evaluates the degrees of the user's interest of video images dynamically. The UCV assigns a network data rate to each video image according to the evaluated degree of the user's interest. The UCV changes quality of each video image to fit them into the assigned data rate.

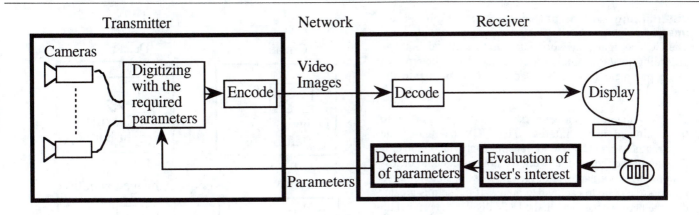

Figure 2   A schematic diagram of User-Centered Video technique

USER-CENTERED VIDEO

Architecture of UCV

A schematic diagram of the UCV is shown in figure 2. Only the receiver gets the user's conditions. In the UCV the transmitter not only sends video images to the receiver, but also receives information about the user's interest from the receiver. The UCV adds two steps to the conventional video transmitting steps: (1) dynamic evaluation of the degree of the user's interest; and (2) determination of video parameters (frame rate and number of pixels of each video image) of each video image.

The UCV can also use a conventional video compression technique (JPEG, MPEG and H.261) in the encode and decode steps in figure 2. The UCV compresses video images more efficiently than conventional video compression techniques.

Evaluating user's interest

The UCV evaluates a user's interest based on window layouts. For example, when a user compares two video windows, the user locates these video windows close to each other to look at both at one glance. We can estimate the user's interest based on the distance between windows.

The degree of a user's interest decreases with the distance from a focused window as shown in figure 3. When the user is focusing on a region, user's visions are clear at the center of the view and fuzzy in the surrounding regions. This is because the retina of a human eye is hierarchically decomposed into a foveal region that perceives details and a surrounding low resolution region [5].

In this paper the UCV formalizes DOI (Degree Of Interest) of each video window as the following function:

$$DOI(x) = \exp(-d)$$

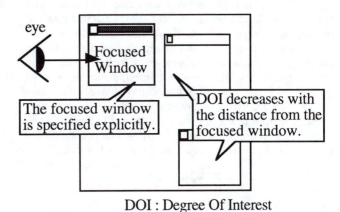

DOI : Degree Of Interest

Figure 3   Window layouts represent user's interest

where $\exp()$ is an exponential function and d is the distance between a window x and the focused window. We define DOI of the focused window as one (maximum). DOI of a video window is equal to 0 when the distance from the focused window is infinite.

In the UCV a user specifies the focused window explicitly like an active window. There are many ways to detect the focused window. For example an eye tracker can detect the spot on which the user is focusing. We adopt a simple way such as the user's explicit specification because most users dislike wearing special devices such as glasses for an eye tracker.

Determining video parameters of each video image

The UCV assigns a network data rate to each video image according to the evaluated user's interest and determines video parameters of each video image to fit it into the assigned network data rate.

The UCV determines a network data rate in proportion to the evaluated user's interest. The UCV calculates the required network data rate to transmit each video image. If the sum of the required network data rate is equal to or less than the network bandwidth, the UCV assigns the required network data rate to each video image. If the sum is greater than the network bandwidth, the UCV divides the bandwidth into the network data rates for video images in proportion to the required network data rate multiplied by the DOI of each video image.

The UCV determines the frame rate and the number of pixels of each video image based on the assigned network data rate. The data amount of a video image is expressed by the mathematical product of the frame rate and the number of pixels. The assigned network data rate determines only the value of the mathematical product. The frame rate and the number of pixels are traded-off. There are many ways to determine the frame rate and the number of pixels. In this paper the UCV calculates a temporary frame rate based on the assigned network data rate assuming that the size of each video image is equal to the displayed size of the video image. The UCV then decreases the number of pixels by decreasing the number of rows and columns sequentially and it increases the frame rate until the frame rate becomes more than the minimum frame rate that is specified in advance.

UCV IN A VIDEO IMAGE

Specification of a region of interest

The UCV transmits a video image of interest with higher quality than without the UCV. When a user wants to look at the video image of interest with higher quality, the UCV allows the user to obtain the regions of interest in the video image with higher quality by degrading the remaining region.

There are regions of interest and regions of no interest in a video image. For example, during opening of the floodgate, an operator is carefully watching the region of the floodgate, rather than the other places, such as the background wall. The operator wants to look at the region of the floodgate with higher quality even if the remaining region is degraded.

In the UCV a user can create, delete and move a rectangular region and change the resolution of the specified region as shown in figure 4. The UCV transmits the specified region of interest with higher quality by degrading the remaining region (the background region).

Video parameters of each region

The UCV changes the resolution of each region according to the user's specified resolution of each region. The UCV

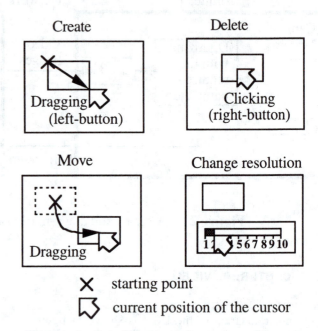

Figure 4 Specification of a region of interest

assigns the required network data rate to transmit each region with the specified resolution, to the region from the assigned network data rate. The UCV then degrades the background region to fit it into the remaining data rate of the assigned network data rate.

We might try to change the frame rate to change the data amount of each region, but if the frame rates of regions are changed, seams might appear at the edge of each region because the digitizing time of each region is different. The UCV does not change the frame rate of regions.

EXAMPLES WITH A PROTOTYPE SYSTEM

Example for ISDN-1500

We developed a prototype system that simulates views of video images transmitted by the UCV. The system digitizes and displays multiple video images while changing the frame rate and the number of pixels of each video image so that the total data amount of the video images is equal to the assumed bandwidth. This prototype system can simulate views of video images through any network by changing the bandwidth.

We assumed that the prototype system's bandwidth was 23 Mbits/second, which can display a video image (320 columns and 240 rows, 24 bits/pixels) at 13 frames/second. The views of the system correspond to views of video images transmitted with JPEG through ISDN-1500 networks, because JPEG can compress a video image at the compression rate of 1.5-2.0 bits/pixel with indistinguishable quality from the original [2].

Yamaashi_PLATE 1 shows a view of 4 video images (displayed size is 320 columns and 240 rows) with the UCV. VIDEO 1 is the focused window designated FO. The title bar of each video window shows the name and the frame rate and the zoom rate of columns and rows. The frame rate was adjusted to more than 5 frames/second in determining the frame rate and the number of pixels. The zoom rate is the rate of the displayed size and the digitized size of a video image. For example, the zoom rate of a column of 3 means that a digitized pixel is magnified to 3 pixels in the column direction when it is displayed.

Returning to our example, we see that an operator can inspect details of a floodgate with smooth motion. Without the UCV the operator can see all video images at only the average frame rate (3 frames/second). This frame rate is too low to observe effectively the motion of the floodgate. The UCV can show the focused video image of the floodgate with about three times smoother motion (8 frames/second). The assigned ratio of the network bandwidth to the focused video images is 59 %.

The UCV degrades the remaining video images according to the distance from the focused video image. For example VIDEO 3 is displayed with the zoom rate of 2x2 at 8 frames/second. The assigned ratio of the network bandwidth for VIDEO 3 is 16 %. VIDEO 2, which is the farthest from the focused window, is displayed with the zoom rate of 3x3 at 6 frames/second. The assigned ratio of the network bandwidth for VIDEO 2 is only 5 %.

The operator can obtain the global motion of video images even when they are degraded. For example, VIDEO 2 shows carp in the water quality monitoring tank. The abnormal motion of the carp means that the water is polluted with something. The operator needs to examine the carp motion to grasp whether the water quality is normal or not. VIDEO 2 does not show the carp in detail (e.g., the pattern of carp's bodies), but the global context which allows the operator to judge whether the carp are swimming normally or not.

This example shows that the UCV allows an operator to obtain a smoother video image of interest with a global context of video images.

Example for ISDN-64

We simulated a video transmission with the ISDN-64 network. We assumed the bandwidth of the prototype system was 1.9 Mbits/second, which can transmit a video image (320 columns and 240 rows, 24 bits/pixels) at 1.1 frames/second. The ISDN-64 network has 2 data lines (64 Kbits/second) and 1 control line (16 Kbits/second). We can use 2 data lines to transmit video images. The network bandwidth is 128 Kbits/second. The views of this prototype system correspond to views of video images with JPEG through the ISDN-64 network.

Yamaashi_PLATE 2 shows a view of video images with the UCV. Yamaashi_PLATE 3 shows a magnified image of the focused video image. An operator can understand the global motion of each video image, but the network

bandwidth is too low to examine the focused video image in detail. The zooming rate of the focused video image is 3x3. The operator can not read the water level with the numbers on the water gauge to inspect the water level.

In Yamaashi_PLATE 3 the operator specifies a region of the water gauge as a region of interest, then the UCV shows the video image of Yamaashi_PLATE 4. The specified region of interest is shown with the red rectangle. The UCV shows the region of interest (80 columns and 45 rows) at full resolution (zooming rate 1x1), while the resolution of the background is 1.8 times rougher than Yamaashi_PLATE 3. In Yamaashi_PLATE 4 the operator can read the numbers on the water gauge.

This example shows that the UCV allows an operator to look at numbers on a gauge by specifying the region as a region of interest, while the numbers cannot be seen without the UCV.

CONCLUSION

We have proposed a User-Centered Video (UCV) that transmits multiple video images with a narrow band network. The UCV assigns network data rates to video images and regions of a video image according to the user's interest. The UCV evaluates the degree of the user's interest from window layouts (distance from the focused window) and direct specification of regions of interest in a video image.

We demonstrated the UCV using examples that simulated views of multiple video images with ISDN-1500 and ISDN-64 networks. The example for the ISDN-1500 network demonstrated that the UCV allows a user to get the focused video with much smoother motion than without the UCV and the global context of video images is obtained even while the remaining video images are degraded. The example for the ISDN-64 network illustrated that the UCV allows a user to look at the details of regions of interest by degrading the remaining regions, although the user cannot look at the details without the UCV.

We assigned network data rates to video images according to the user's interest. In the future, we would like to assign computing resources also according to user's interest. For example, graphic power of three dimensional (3D) graphic hardware is a limited computing resource. When a user displays multiple 3D graphics, the required graphic power overwhelms the power of the graphic hardware. It is desirable to assign the graphic power according to the user's interest.

REFERENCES

1. Liou, M. Overview of the px64 kbits/s video coding standard, Communications of the ACM, 1991, ACM Press, Vol. 34, No. 4, pp. 59-63.

2. Wallace, G.K. The JPEG still picture compression standard, Communications of the ACM, 1991, ACM Press, Vol. 34, No. 4, pp. 30-44.

3. Gall, D.L. MPEG: A video compression standard for multimedia applications, Communications of the ACM, 1991, ACM Press, Vol. 34, No. 4, pp. 46-58.

4. Furnus, G.W. Generalized fisheye views, in Proc. CHI'86 Human Factors in Computing Systems (Boston, April 13-17, 1986), ACM Press, pp. 16-23.

5. Mackinlay, J.D. Robertson, G.G. Card, S.K. The perspective wall: detail and context smoothly integrated, in Proc. CHI'91 Human Factors in Computing Systems (New Orleans, April 27-May 2, 1991), ACM Press, pp. 173-179.

6. Stone, M.C. Fishkin, K. Bier, E.A. The movable filter as a user interface tool, in Proc. CHI'94 Human Factors in Computing Systems (Boston, April 24-28, 1994, ACM Press, pp. 306-312.

7. Hill Jr., F.S. Walker Jr., S. Gao, F. Interactive image query system using progressive transmission, Computer Graphics (July 1993), ACM Press, Vol. 17, No. 3, pp. 323-330.

8. Plompen, R.H.J.M. Groenveld, J.G.P. Booman, F. Boekee, D.E. An image knowledge based video codec for low bitrates, SPIE, Advances in Image Processing, 1987, Vol. 804, pp. 379-384.

Visualizing Complex Hypermedia Networks through Multiple Hierarchical Views

Sougata Mukherjea, James D. Foley, Scott Hudson
Graphics, Visualization & Usability Center
College of Computing
Georgia Institute of Technology
E-mail: sougata@cc.gatech.edu, foley@cc.gatech.edu, hudson@cc.gatech.edu

ABSTRACT

Our work concerns visualizing the information space of hypermedia systems using multiple hierarchical views. Although overview diagrams are useful for helping the user to navigate in a hypermedia system, for any real-world system they become too complicated and large to be really useful. This is because these diagrams represent complex network structures which are very difficult to visualize and comprehend. On the other hand, effective visualizations of hierarchies have been developed. Our strategy is to provide the user with different hierarchies, each giving a different perspective to the underlying information space, to help the user better comprehend the information. We propose an algorithm based on content and structural analysis to form hierarchies from hypermedia networks. The algorithm is automatic but can be guided by the user. The multiple hierarchies can be visualized in various ways. We give examples of the implementation of the algorithm on two hypermedia systems.

KEYWORDS: Hypermedia, Overview Diagrams, Information Visualization, Hierarchization.

INTRODUCTION

Overview diagrams are one of the best tools for orientation and navigation in hypermedia documents [17]. By presenting a map of the underlying information space, they allow the users to see where they are, what other information is available and how to access the other information. However, for any real-world hypermedia system with many nodes and links, the overview diagrams represent large complex network structures. They are generally shown as 2D or 3D graphs and comprehending such large complex graphs is extremely difficult. The layout of graphs is also a very difficult problem [1]. Other attempts to visualize networks such as Semnet [3], have not been very successful.

In [13], Parunak notes that: "The insight for hypermedia is that a hyperbase structured as a set of distinguishable hierarchies will offer navigational and other cognitive benefits that an equally complex system of undifferentiated links does

not, even if the union of all the hierarchies is not itself hierarchical." Neuwirth et al. [12] also observed that the ability to view knowledge from different perspectives is important. Thus, if different hierarchies, each of which gives a different perspective to the underlying information can be formed, the user would be able to comprehend the information better. It should be also noted that unlike networks some very effective ways of visualizing hierarchies have been proposed. Examples are *Treemaps* [7] and *Cone Trees* [15].

This paper proposes an algorithm for forming hierarchies from hypermedia graphs. It uses both structural and content analysis to identify the hierarchies. The structural analysis looks at the structure of the graph while the content analysis looks at the contents of the nodes. (Note that the content analysis assumes a database-oriented hypermedia system where the nodes are described with attributes). Although our algorithm is automatic, forming the "best" possible hierarchy representing the graph, the user can guide the process so that hierarchies giving different perspectives to the underlying information can be formed. These hierarchies can be visualized in different ways.

Section 2 presents our hierarchization process. Section 3 shows the implementation of the algorithm in the **Navigational View Builder**, a system we are building for visualizing the information space of Hypermedia systems [10], [11]. This section discusses the application of the algorithm on a demo automobile database and a section of the *World-Wide Web*. Section 4 discusses how the hierarchies can be transformed to other forms of data organizations. Section 5 talks about the related work while section 6 is the conclusion.

THE HIERARCHIZATION PROCESS

New Data Structure

For our hierarchization process we use a data structure which we call the **pre-tree**. A pre-tree is an intermediate between a graph and a tree. It has a node called the *root* which does not have any parent node. However, unlike a real tree, all its descendants need not be trees themselves - they may be any arbitrary graph. These descendants thus form a list of graphs and are called *branches*. However, there is one restriction - nodes from different branches cannot have links between them. An example pre-tree is shown in Figure 1. Note that pre-tree is another data structure like *multi-trees* [4] - it is not as complex as a graph but not as simple as a tree. Also note that although the term "pre-tree" has not been used before, this data structure has a long history in top-down

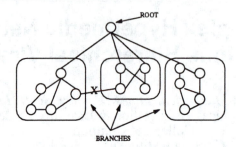

Figure 1: An example pre-tree. It has a root node which does not have any parent. The descendants of the root node are graphs. However, none of these graphs have any links between them. Our hierarchization algorithm tries to identify the best pre-tree to represent the given graph. The final tree is formed by calling the algorithm recursively for the branches.

clustering techniques [5]. Top-down clustering would often be halted when new divisions were not auspicious, leaving a final structure which is essentially a pre-tree.

Hierarchization Algorithm

The algorithm tries to identify a suitable pre-tree from a given graph. Thus a root node is identified and the other nodes are partitioned into branches. This root node forms the root of the final hierarchy. The algorithm is recursively called for each of the branches and the trees formed by these recursive calls become children of the root of the final hierarchy. The recursion stops if a branch has very few nodes or the required depth of the final tree has been reached. It may also happen that for certain branches, no suitable pre-trees can be formed. In these cases, the nodes of the branches become children of the parent of the branch. (This case generally occurs for branches with very few nodes).

For identifying potential pre-trees both content and structural analysis are used.

- *Content analysis:*

For content analysis, for each attribute, the nodes of the graph are partitioned into branches based on the attribute values by *Content-based Clustering*. The clustering algorithm is explained in [11]. If too many or too few branches are formed, the attribute is not suitable for forming a pre-tree. Otherwise a new pre-tree is formed with these branches. The root of the pre-tree is a *cluster* representing all the nodes of the graph.

- *Structural analysis:*

A pre-tree is formed for nodes in the graph which can reach all other nodes. These nodes are designated as the roots of the pre-trees. The branches are the branches of the spanning tree formed by doing a breadth-first search from the designated root node. [1]

Both content and structural analysis can identify several potential pre-trees. A *metric* is used to rank these pre-trees. The metric consists of the following submetrics:

- Information lost in the formation of the pre-tree: When the nodes are partitioned for forming the branches, all links joining nodes in different branches are removed. Thus valuable information is lost and a submetric calculates the ratio of the number of links remaining in the branches to the total number of links in the original graph to rank the pre-trees in

order of the least amount of information lost.

- "Treeness" of the branches: Since our overall objective is to form trees, it is advantageous if the branches of the pre-tree are already close to trees. If all the branches only consisted of trees, there would be a total of $n - c$ links where n is the total number of nodes in the branches and c is the number of connected components. Thus a submetric which calculates the ratio $(n - c)/l$ where l is the total number of links is an indication of the "treeness" of the branches.

- "Goodness" of the root: For a structural pre-tree the goodness of the root is determined by the sum of the distances of the shortest path from the root to all other nodes. A "good" root will reach all other nodes by following only a few links so that the resulting tree is not very deep. A deep tree is not desirable since it will force the user to follow a long and tedious path to reach some information. For content analysis the goodness of the root is determined by the relevance of the attribute. (For example, for an automobile database, the manufacturer of the cars is a more relevant attribute than the number of doors of the cars).

Each submetric returns a number between 0 and 1. The overall metric is calculated by a weighted sum of the submetrics where the weight is determined by the relative importance of the submetrics.

The Role of the User

By default, the entire process would be automatically forming the "best" hierarchical form for the original graph. However, the user can guide the process both during the translation of the graph to a tree and during the visualization of the tree.

- Translation phase:

– The users can control the various variables that are used in the translation process. For example, they can control the variable which specifies the maximum possible depth of the tree (the recursion stops when this depth is reached).

– The user can control the relative importance of the various submetrics in the overall metric that is used to rank a given pre-tree. For example the user can specify that the "goodness' of a root is not a useful criteria for judging pre-trees. The user can also assign different weights to different link types to influence the submetric calculating the amount of information lost.

– The algorithm generally selects the best possible pre-tree at each level. However, the user can choose the pre-tree instead. The user is shown the possible pre-trees that can be selected ranked by the metric and the user can choose one of them. The user can specify to what level of the hierarchy

---

[1] A detailed analysis is omitted for the purpose of brevity. The algorithm is explained with examples in the next section.

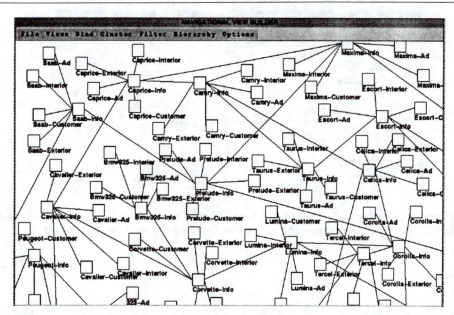

Figure 2: An overview diagram of an automobile database. The diagram is very difficult to comprehend.

the pre-trees would be chosen. By choosing different pre-trees during different runnings of the algorithm, different hierarchies, giving different perspectives to the data can be formed.
• Visualization phase:
– Besides a 2D tree, the hierarchy can also be visualized as Cone Trees, Treemaps or as a Table of Contents of a book (which is formed by listing the nodes in the order of a depth-first search).
– Different visual attributes can be bound to information attributes in the views. This is an extension of the work reported in [10].

IMPLEMENTATION

The algorithm has been implemented in the Navigational View Builder, a system for forming overview diagrams of hypermedia systems. Figure 2 represents an overview diagram of an automobile database. There are a lot of inter-connected nodes showing, for example, textual information about the cars, images of the cars, TV advertisements and audio of previous buyers' comments. There are also links to other cars with similar price and other models by the same manufacturer. From this complex network a hierarchy can be formed automatically. The top-level root of this tree and its children are shown in the left hand screen of Figure 3. In this case, the attribute *Price* was used to form the initial partitioning and the root represents a cluster for all the nodes.

The user can form different hierarchies by selecting other pre-trees. For example, if the user wanted to select the pre-tree at the initial level, the dialog box shown in Figure 4 pops up. If the user wants to partition based on the attribute *Country*, the tree shown in the right hand screen in Figure 3 is formed. In this figure some of the children represents clusters for countries. For example the node labeled Japan represents all the Japanese cars and its children are shown in the left hand screen of Figure 5. Here the partitioning is done by the attribute *Manufacturer*. For some other countries the nodes in the cluster formed a tree. In these cases the roots of the

tree were identified by structural analysis and they became the children of the overall root. Thus for Sweden, Saab-Info is the root of the tree for all nodes related to Swedish cars. Its children are shown in the right hand screen of Figure 5.

Color plate 1 shows a 3D Tree view of this hierarchy. In this view, the colors of the nodes represent various countries and the colors of the links represent link types. Various zooming and filtering operations that are mentioned in [15] are possible for this 3D tree. Moreover, smooth animation is used so that the view changes are not abrupt and allow the user to see the changes easily. (Note that the implementation is done using C++, Motif and Open Inventor [18].)

Forming Hierarchies in the World-Wide Web
Let us now look at an example from perhaps the most popular hypermedia system, the *World-Wide Web*. For input to the Navigational View Builder, information was automatically extracted from the WWW about the various files, their authors, their links to other files and other information by parsing the HTML documents using the method described in [14]. Figure 6 shows an unstructured overview diagram of the WWW pages about the research activities at the Graphics Visualization & Usability (GVU) Center at Georgia Tech.[2] Obviously, this information is very complicated.

The left hand screen of Figure 7 shows the top level of the hierarchy automatically created for the data by the system. The file *research.html* which lists the various research activities of the GVU Center is the root. It has branches to the major research area as well as to *gvutop.html*, a file containing general information about GVU. The right hand side of Figure 7 shows a view of a section of this hierarchy where the nodes are listed as a table of content of a book.

A major drawback of the World-Wide Web is that very few useful semantic attributes are defined for the pages. To create some other meaningful hierarchies, attributes like the topic

---

[2]URL: http://www.gatech.edu/gvu/gvutop.html

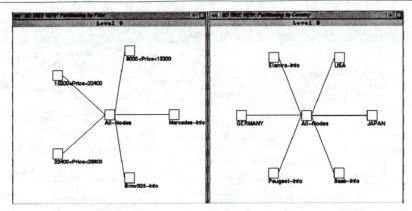

Figure 3: The left hand screen shows the default tree formed for the automobile database. The top-level partitioning is by the attribute *Price*. The right hand screen shows the tree formed if the top-level partitioning is done by the attribute *Country*.

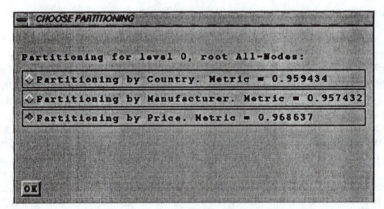

Figure 4: At each level various pre-trees can be used. A metric ranks these pre-trees. By default the pre-tree with the best metric is selected. However, the user can select others using the above menu.

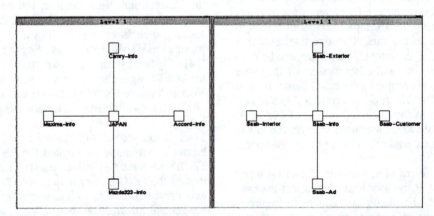

Figure 5: Examples of Content and Structural analysis for forming pre-trees. The left hand screen represents the nodes for Japan. The root is a cluster representing all Japanese cars. The nodes are partitioned by the attribute *Manufacturer*. The right hand screen is for Swedish cars. These nodes form a tree with the node Saab-Info as the root.

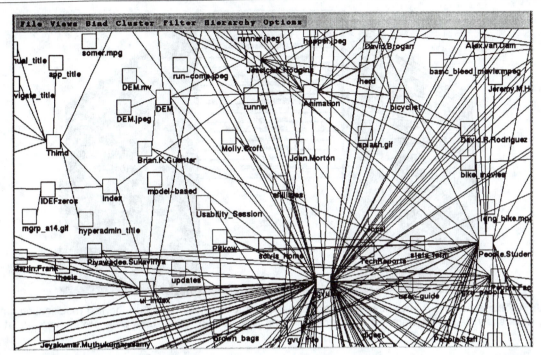

Figure 6: An overview diagram of the World-Wide Web pages about the research activities at GVU. It indicates clearly why traditional overview diagrams are useless for real-world hypermedia systems.

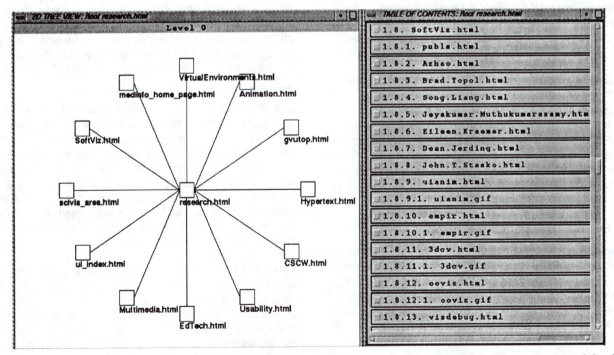

Figure 7: The left hand screen shows the top level of the default hierarchy formed for the GVU WWW pages. *research.html* is the root and the major research areas are shown. The right hand screen shows a book view of a portion of this hierarchy showing research in Software Visualization.

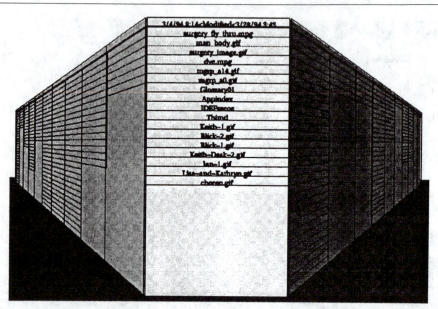

Figure 8: A Perspective Wall view showing a linear arrangement of the files based on the last modification time. The different walls show files which were last modified in different time frames. Only some walls are in the focus at a given time.

of the page (whether it is a research page or a personal page, etc.) were inserted manually. (Efforts are underway to incorporate metadata into WWW and hopefully in the near future we can extract all useful information from the WWW automatically.) The left hand screen of Color plate 2 represents a treemap view of a hierarchy formed when the initial partitioning is done by the topic of the page. The colors are used to represent the kind of users who created the pages. Green is used to represent Phd students and the color plate indicates that the Phd students are the primary authors of the pages.

Multiple hierarchies, each giving a different perspective to the underlying information space can be formed. If a user selects a node in one view, its positions in the other views are also highlighted. Thus, these views help the user in comprehending the data. It should be also noted that the user can go directly to the corresponding WWW page for the selected node. Thus in the Treemap view, the node *visdebug.html* is highlighted. The corresponding WWW page is shown on the right hand screen of Color plate 2.

GENERATING OTHER VIEWS

Once a hierarchy is formed from the original graph structure, the hierarchy can be transformed to other data organizations as well. Visualizations can be formed for these data organizations also. For example, if the original partitioning for forming the hierarchy was done by a quantitative attribute, a linear structure sorted by that attribute can be formed from the subtrees of the root node.

Figure 8 represents a perspective wall [9] view of a linear arrangement of the GVU WWW pages sorted by the last modification times of the files. From the hierarchy whose initial partitioning was by the attribute *last-modified-time*, the files were divided into partitions based on the time when they were last modified. These partitions were arranged on walls. Only some walls are in the focus at a given time. The

user can easily control the walls which are in focus through a scrollbar. Similarly, for the automobile database a Perspective Wall view can be formed where the cars are sorted by the attribute *Price*.

Other views can also be generated. For example, a tabular view showing information like average price, mileage, etc. for various car models and also such useful statistics for different manufacturers of the cars can be formed by a depth-first traversal of the hierarchical structure whose partitionings are done by the attributes *Manufacturer* and *Car-Model*.

RELATED WORK

Our structural analysis is similar to that described in [2] for identifying hierarchies from hypermedia structures. Although using just structural analysis to identify hierarchies works for hypertext systems with simpler underlying structures, identifying meaningful hierarchies by structural analysis alone is difficult for real-world systems. Content analysis is also essential as is evident from the paper. [6] describes a method to linearize complex hyper-networked nodes to facilitate browsing using a book metaphor. However, this work also uses structural analysis only.

This paper is also related to systems that deal with graphical presentation of information automatically or semi-automatically. Examples include APT [8] and SAGE [16]. However, our information domain is different from these systems - these systems deal with highly structured information. The views that we want to develop are also different. The previous systems generally produced bar diagrams, scatter plots and such graph views.

CONCLUSION

One of the best ways to comprehend a large complicated information structure is to form multiple simpler structures each highlighting different aspects of the original structure. Our work tries to use this philosophy to make a complex hypermedia system understandable to the user. We believe that by forming various effective views of the underlying space, we would allow the user to better understand the complex information. We give examples of the hierarchization process from two complicated hypermedia systems to illustrate our point. These examples show that our algorithm was able to extract meaningful hierarchies which gave better insights into the complex information spaces.

Future work is planned along the following directions:

- *Visualizing Larger Databases:* Although a detailed complexity analysis is beyond the scope of this paper, it can be shown that the major bottleneck of the algorithm is the structural analysis to identify roots. [2] uses an $O(n^3)$ algorithm to identify roots. On the other hand we use $O(n^2 + nl)$ algorithm to identify roots (by calling the breadth-first search for each node). Although in the worst case $l = O(n^2)$, on average $l = O(n)$ and our algorithm will perform better. For the WWW database with about 400 nodes and 800 links our algorithm took about 7 seconds on a SGI reality engine. Although this is acceptable, we will face problems for larger databases. We are investigating ways to enhance the performance by improving the efficiency of the code and using probabilistic algorithms to identify roots. Moreover, even cone trees and treemaps are not able to visualize larger databases effectively. New visualization techniques are needed.

- *Usability Studies:* A limitation of our system is that no evaluation of how useful our views really are have been done so far. We plan to do serious usability studies in the near future. These studies may give us new insights that will help to improve our system.

ACKNOWLEDGEMENT

This work is supported by grants from Digital Equipment Corporation, Bell South Enterprises, Inc. and Emory University System of Health Care, Atlanta, Georgia as part of the Hypermedia Interface for Multimedia Databases project. We would also like to thank the reviewers of this paper for their useful comments.

REFERENCES

1. G. Battista, P. Eades, R. Tamassia, and I. Tollis. Algorithms for Drawing Graphs: an Annotated Bibliography. Technical report, Brown University, June 1993.

2. R. Botafogo, E. Rivlin, and B. Shneiderman. Structural Analysis of Hypertexts: Identifying Hierarchies and Useful Metrics. *ACM Transactions on Office Information Systems*, 10(2):142–180, 1992.

3. K. Fairchild, S. Poltrok, and G. Furnas. Semnet: Three-dimensional Graphic Representations of Large Knowledge Bases. In R. Guindon, editor, *Cognitive Science and its Applications for Human-Computer Interaction*. Lawrence Erlbaum, 1988.

4. G. Furnas and J. Zacks. Multitrees: Enriching and Reusing Hierarchical Structures. In *Proceedings of the ACM SIGCHI '94 Conference on Human Factors in Computing Systems*, pages 330–336, Boston, Ma, April 1994.

5. J. Hartigan. *Clustering Algorithms*. John Wiley and Sons, 1975.

6. S. Ichimura and Y. Matsushita. Another Dimension to Hypermedia Access. In *Proceedings of Hypertext '93 Conference*, pages 63–72, Seattle, Wa, November 1993.

7. B. Johnson and B. Shneiderman. Treemaps: A Space-filling Approach to the Visualization of Hierarchical Information. In *Proceedings of IEEE Visualization '91 Conference*, pages 284–291, San Diego, Ca, October 1991.

8. J. MacKinlay. Automating the Design of Graphical Presentation of Relational Information. *ACM Transactions on Graphics*, 5(2):110–141, April 1986.

9. J. D. Mackinlay, S. Card, and G. Robertson. Perspective Wall: Detail and Context Smoothly Integrated. In *Proceedings of the ACM SIGCHI '91 Conference on Human Factors in Computing Systems*, pages 173–179, New Orleans, La, April 1991.

10. S. Mukherjea and J. Foley. Navigational View Builder: A Tool for Building Navigational Views of Information Spaces. In *ACM SIGCHI '94 Conference Companion*, pages 289–290, Boston, Ma, April 1994.

11. S. Mukherjea, J. Foley, and S. Hudson. Interactive Clustering for Navigating in Hypermedia Systems. In *Proceedings of the ACM European Conference of Hypermedia Technology*, pages 136–144, Edinburgh, Scotland, September 1994.

12. C. Neuwirth, D. Kauffer, R. Chimera, and G. Terilyn. The Notes Program: A Hypertext Application for Writing from Source Texts. In *Proceedings of Hypertext '87 Conference*, pages 121–135, Chapel Hill, NC, November 1987.

13. H. Parunak. Hypermedia Topologies and User Navigation. In *Proceedings of Hypertext '89 Conference*, pages 43–50, Pittsburgh, Pa, November 1989.

14. J. Pitkow and K. Bharat. WEBVIZ: A Tool for World-Wide Web Access Log Visualization. In *Proceedings of the First International World-Wide Web Conference*, Geneva, Switzerland, May 1994.

15. G. G. Robertson, J. D. Mackinlay, and S. Card. Cone Trees: Animated 3D Visualizations of Hierarchical Information. In *Proceedings of the ACM SIGCHI '91 Conference on Human Factors in Computing Systems*, pages 189–194, New Orleans, La, April 1991.

16. S. Roth, J. Kolojejchick, J. Mattis, and J. Goldstein. Interactive Graphic Design Using Automatic Presentation Knowledge. In *Proceedings of the ACM SIGCHI '94 Conference on Human Factors in Computing Systems*, pages 112–117, Boston, Ma, April 1994.

17. K. Utting and N. Yankelovich. Context and Orientation in Hypermedia Networks. *ACM Transactions on Office Information Systems*, 7(1):58–84, 1989.

18. J. Wernecke. *The Inventor Mentor: Programming Object-Oriented 3D Graphics with Open Inventor*. Addison-Wesley Publishing Company, 1994.

SageBook: Searching Data-Graphics by Content

Mei C. Chuah, Steven F. Roth, John Kolojejchick, Joe Mattis, Octavio Juarez

School of Computer Science
Carnegie Mellon University
Pittsburgh, PA, 15213, USA
Tel: +1-412-268-2145
E-mail: mei+@cs.cmu.edu; steven.roth@cs.cmu.edu

ABSTRACT

Currently, there are many hypertext-like tools and database retrieval systems that use keyword search as a means of navigation. While useful for certain tasks, keyword search is insufficient for browsing databases of data-graphics. SageBook is a system that searches among existing data-graphics, so that they can be reused with new data. In order to fulfill the needs of retrieval and reuse, it provides: 1) a *direct manipulation, graphical query interface*; 2) a *content description language* that can express important relationships for retrieving data-graphics; 3) *automatic description* of stored data-graphics based on their content; 4) search techniques sensitive to the *structure and similarity* among data-graphics; 5) *manual and automatic adaptation* tools for altering data-graphics so that they can be reused with new data.

KEYWORDS: Data-visualization, Data-graphic design, Automatic presentation, Intelligent interfaces, Content-based search, Image-retrieval, Information-retrieval

INTRODUCTION

Our approach to supporting the creation of data-graphics is to view their design as two complementary processes: design as a constructive process of selecting and arranging graphical elements, and design as a process of browsing and customizing previous cases. SageBook supports the latter process by enabling users to find, browse, and apply previously created data-graphics to the construction of new ones that reflect current data and design preferences.

Current data-graphic design tools, particularly those provided with spreadsheets, do not support these processes well because they do not enable people to combine diverse information in a single graphic. They are unable to integrate different kinds of graphical objects, properties, or chart types to show the relationships among many data attributes. Instead, isolated graphical styles must be selected individually from a lengthy menu (e.g. charts with bars, charts with lines, charts with plot points, etc.).

There are *constructive* tools that enable users to assemble or sketch combinations of graphic elements flexibly [5,8]. These tools support a vast number of different data-graphics based on the combination and organization of many graphical elements (e.g., those in Figures 2, 4, 7, 9). Nevertheless, constructing a data-graphic, especially one that contains a lot of information, still requires a user to have substantial design expertise. Even expert designers may need ideas when working with new data sets, and a good source of ideas exists in other users' successful visualizations of similar data.

One of our approaches to providing expertise has been to give users access to a library of data-graphics, created by users of a constructive system called SageBrush or created automatically by a related knowledge-based system called SAGE [8]. Since searching a portfolio of hundreds of data-graphics can be laborious, we created SageBook, a content-based search and browsing tool that enables users to retrieve data-graphics based on their appearance and/or the properties of the data they present.

In [8], we gave an overview of the three components of our system (SAGE, SageBrush, and SageBook), but primarily focused on SageBrush. In this paper, we focus on SageBook's browsing interfaces and mechanisms for content-based search and reuse. SageBook's goal is to provide content-based retrieval facilities in the context of supporting user-directed, data-graphic design. To fulfill this goal, we identified five crucial needs:

1. *A direct manipulation graphical query interface* - a flexible and intuitive query interface with which users sketch graphics similar in appearance to those they want to browse. Alternatively, users may select subsets of their data to retrieve graphics that display similar data. SageBrush serves as SageBook's query interface (Figure 2 Bottom).

2. *A content description language* - an expressive vocabulary for describing the graphical and data relationships contained in data-graphics, so that they can be searched by content. In addition, it is necessary to translate user queries into this vocabulary, so that users can communicate them unambiguously (i.e. without vocabulary mismatches). The problem of vocabulary mismatch is well summarized by Lesk [1].

3. *Automatic description* - automatic indexing of stored data-graphics, so that the data-graphic library can be easily populated, maintained, and organized for efficient

search. By indexing we mean the categorization of data-graphics using the content description language.

4. *Structural and similarity-based search* - a mechanism for matching queries and stored data-graphics, based on the spatial organization and structural relationships among graphical elements, and the characteristics of and dependencies among data attributes. This mechanism supports retrieval based on partial matches (i.e., based on *similarity* between query and graphic). Structural search is more powerful than keyword search because the latter is not expressive of the relationships among multiple data and graphical components.

5. *Manual and automatic adaptation* - facilities to help users alter the data-graphics retrieved by SageBook's search strategies, so that they can be applied to a user's current task.

We are aware of no other approaches that address these needs for data-graphics. Some have been addressed in systems for retrieving photographs or images, but none have provided a solution that takes into account all of them. Garber [2] developed a retrieval system for advertising photographs based on a study of art directors. Queries are posed by typing in keyword descriptions of objects or travel locations (e.g. man, dog; Florida, New England). Users can select a level of similarity for defining the degree of relaxation allowed in retrievals. Photographs are ordered according to how close they match the keywords in the query (based on the implementors prestored judgments of similarity). The use of keywords in the query system makes the process susceptible to vocabulary mismatches (i.e., the descriptions specified by the user may not match those used to describe the stored photographs). In addition, the photograph library has to be manually indexed; thus populating it is laborious and error-prone.

Nishiyama et al. [6] described an image-retrieval system that searches based on the relative position of objects in a photograph, and on some object attributes. Queries are graphical sketches, so users need not learn a keyword system (thus reducing the mismatch problem). However, the content description language and query interface are limited to six object types. As was shown in their evaluation, this is insufficient to describe the space of pictures that might be in the library. As in Garber's system, pictures in the library are manually indexed.

TRADEMARK [3] is an image-retrieval system that does matching based on physical features (e.g. colors, lines) of images. Using image analysis techniques, TRADEMARK can automatically index or sort its library. However, this type of search (characterized as "machine-oriented" in [6]) does not produce a content description beyond the surface features of the image. Therefore, it is unable to search for concepts like "person" or "beach". Furthermore, the interface requires users to create a detailed query also at the surface feature level.

ART MUSEUM [3] is an image-retrieval system for art pieces. Its search criteria are graphical features and keywords of artistic impressions. The search for graphical features is based on the physical appearance of the pictures (e.g. color, texture) and has the same limitations as TRADEMARK. The artistic impressions associated with each picture have to be manually entered. Furthermore, the search done on artistic impressions is a keyword matching process, making it especially sensitive to vocabulary mismatches.

None of these systems provide adaptation tools because they were created for the task of image-retrieval only. In data-graphic design, reuse is a primary user task, thus adaptation facilities are of the utmost importance. Reuse involves extracting the design that was inherent in an existing data-graphic and reapplying it to the design of a new data-graphic.

We have designed a system that directly supports the five needs of a retrieval and reuse facility for data-graphic design. Our system provides users with a direct manipulation interface (shown in Figures 2 and 7) to pose complete or partial data and graphic queries. A query is translated into a content description language, which has also been used to express automatically-generated descriptions of the data-graphics in SageBook's library. SageBook compares the query with these descriptions and retrieves a set of data-graphics that fulfills its similarity tests (Figure 7). Users can then manually or automatically adapt these data-graphics as desired. We first give an overview of the interactions and information flow among system components, and then we discuss how we deal with the needs of retrieval and reuse.

SYSTEM OVERVIEW

SageBook is integrated with two other modules: SageBrush and SAGE. SageBrush is a tool for sketching data-graphics from primitive graphical elements; as such, it can be used both as a design space and query interface. SAGE is an automatic presentation system. Details on SAGE and SageBrush can be found in [8].

A retrieval transaction emphasizing the relations among SageBook and the other modules is shown in Figure 1.

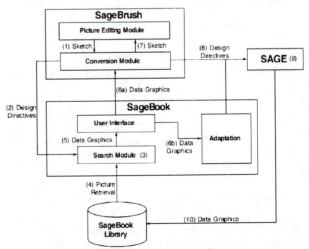

Figure 1: The flow of typical transactions among SageBook, SageBrush, and SAGE.

1. A user creates a data or graphic query using SageBrush.
2. The query is converted by SageBrush into *design directives*, which are then passed to SageBook. Design directives are partial specifications of a data-graphic, expressed in terms of the system's content description language.
3. SageBook's search module uses the design directives to locate matches between the query and stored data-graphics.
4. The matching items are retrieved from SageBook's data-graphic library.
5. The data-graphics found are then sent to the browser in SageBook.
6. From the browser, the user may pick one or more data-graphics to be (a) manually modified in SageBrush or (b) automatically modified in SageBook.
7. To manually modify a data-graphic, SageBrush first converts it into a sketch and displays it. This sketch can then be adapted by the user. Figure 2 (Top) shows an example data-graphic that has been retrieved by SageBook. Figure 2 (Bottom) shows a sketch of the data-graphic when it is brought into SageBrush for manual adaptation.
8. After editing the sketch, a user may generate a new graphic (i.e. direct SageBrush to convert the sketch back into design directives and send it to SAGE).
9. SAGE automatically generates the new data-graphic.
10. The user may then save the data-graphic in SageBook's library so that it may be reused later.

The example above only shows one possible sequence of actions. A user is not restricted to executing exactly these actions, and can combine the different functionalities of the three modules flexibly.

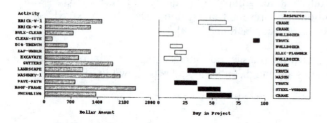

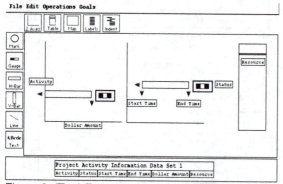

Figure 2: (Top) Example data-graphic retrieved by SageBook. (Bottom) SageBrush interface, showing a sketch of the data-graphic created by SageBook.

The process of retrieval and reuse described above can be divided into four phases, each emphasizing the different needs of information retrieval and data-graphic design.

- How do I tell the system what I want (easily and without any ambiguity)?
- How can the content of queries and data-graphics be expressed?
- How does the system find what I want?
- How can a data-graphic be adapted for new data after retrieval?

The following sections describe each of these phases in detail and explain how we dealt with the retrieval and reuse needs that were previously raised.

QUERY INTERFACE: HOW DO I TELL THE SYSTEM WHAT I WANT?

Queries are constructed in SageBrush by assembling graphical sketches or by selecting *data-domains* (i.e. database attributes) to be visualized. Interface details are provided in [8]. Whether querying based on graphical or data content, users do not need to know a complex vocabulary for describing that content. They do not have to learn the terms the system uses internally to refer to axes, map spaces, interval bars, gauges, indented text, etc. Instead with SageBrush, they can select and arrange *spaces* (e.g. charts, tables), the objects contained within those spaces (e.g. *marks*, *bars*), and the objects' properties (e.g. *color*, *size*, *shape*, *position*). Likewise, users do not have to learn the terms for describing the characteristics of data, like *scale of measurement* (nominal, ordinal, quantitative) or relationships among data-domains (*functional dependency*, *interval*, *2D coordinate*). Instead, they simply load the data-sets that they wish to peruse into SageBrush's data area, and select the data-domains that they wish to visualize. This is in contrast to previous systems [2,3,10] that require users to specify the characteristics of the query object via keywords. Systems that do not provide direct-manipulation query interfaces force users to learn an underlying object description language.

SageBrush contains methods to convert a data or graphical query into a language (design directives) that is understood by SageBook and SAGE. When users select a data set of current interest, the system extracts the characteristics of each selected data-domain (attribute) and reformulates the query in terms of underlying data properties.

SageBook does require data objects to be characterized when the data is first created. Currently, this characterization must be provided by database creators. We expect to be able to build modules that extract this characterization either by examining the information typically stored in databases (e.g. relation schemes), by examining the data itself, or by interacting with users. However, once data is characterized and stored, users need not be aware of the characteristics or the language that is used to describe them.

In addition to serving as a query interface, SageBrush can also be used to construct data-graphics and to manually adapt retrieved data-graphics. Because of SageBrush's multiple functionality, any data-graphic that can be constructed can also be queried.

REPRESENTATION: HOW CAN THE CONTENT OF QUERIES AND DATA-GRAPHICS BE DESCRIBED?

A common data and graphic representation is used by all the modules of our system. It provides a vocabulary that is capable of expressing the syntax and semantics of data-graphic designs, and of characterizing the data contained within them. It is able to express the spatial relationships between graphical objects, the relationships between data-domains, and the various graphic and data attributes. Through this language, the content of data-graphics can be fully described.

A query specified by the user with data and graphical symbols is first translated into this internal representation before it is passed to SageBook for processing. This common language allows the user and the different modules of the system to communicate without any vocabulary mismatches. In addition, all data-graphics generated by SAGE are described using this language. SageBook, in turn, uses the description associated with each data-graphic as an index for its search strategies. As a result, all data-graphics in the SageBook library are automatically indexed by SAGE when they are first generated. This is a significant advantage compared to other visual search systems [2,3,6], which require the descriptions of images in the graphic library to be manually entered as keywords.

The data characterization has been described in [9] and is not repeated here. It includes the scales of measurement (nominal, quantitative, ordinal), structural relationships among data (such as between the endpoints of ranges and between the two domains of a geographic 2D coordinate), and the dependencies among domains (e.g. whether a person has one or more birthdates, residences, or children). However, we will briefly describe the main structures of the graphical representation that relate to SageBook in order to facilitate an understanding of the search procedures.

Graphic Representation

Each data-graphic is described as a *design specification*, which consists of several *spaces*. Each space represents a grouping of graphical elements that are positioned according to a single *layout discipline*. There are many types of layout disciplines; some examples are shown in Figure 3.

Figure 3: Different types of layout disciplines.

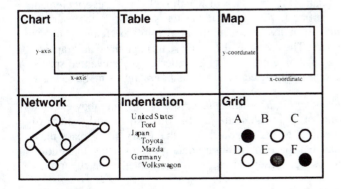

Within each space there may be several objects called *graphemes*. Examples of graphemes are marks, bars, text, lines, and gauges. Each grapheme uses different *properties* to define its appearance. Some of these properties may be used to encode data-domains or distinguish different relations shown in the same space. For example, Figure 4 shows a data-graphic of steel-factory data. This graphic was designed using SAGE and it uses the *size* of the marks in the first space to encode *billet-thickness* and the *color* of the bars in the second space to distinguish between *materials-cost* and *labor-cost*. Attributes not encoding domains or relations have default values (e.g. the *color* of the marks).

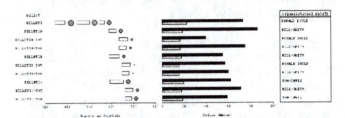

Figure 4: Steel-factory data.

Figure 5 expresses the data-graphic in Figure 4 in terms of its constituents. The data-graphic contains three horizontally aligned spaces. Two of the spaces use the *chart* layout discipline and one the *table* layout discipline. Within the first space are two sets of graphemes: *marks* and *interval bars*. The *position* of the interval bars is used to express the furnace schedule for the different billets, and the *size* of the marks is used to express *billet-thickness*. The second space contains two sets of *bar* graphemes that use the *color* property to distinguish the two *cost* data attributes that the bars encode. Their *lengths* encode the data values. The last space has a set of *text* graphemes whose *lettering* encodes data.

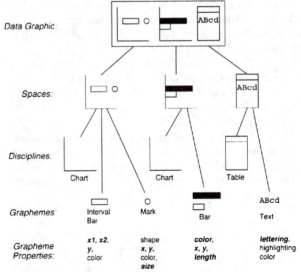

Figure 5: Constituents of data-graphics.

Thus, the content description language describes: classes of objects, the spatial relationships among spaces and graphemes, the graphical properties of objects, and the way that those properties are assigned to data.

SEARCH: HOW DOES SAGEBOOK FIND WHAT I WANT?

The process of matching a user query to the SageBook library is carried out by two components of the search module: the data-matcher and graphic-matcher. The graphic-matching component has three alternative match strategies and the data-matching component has four. The different match strategies provide different degrees of relaxation on the search criteria based on the degree of overlap between the library data-graphic and the user query. Each retrieves a different number of data-graphics depending on its degree of relaxation. Partial overlap matching or similarity matching was shown to be important and useful in Garber's photograph retrieval system [2].

A typical reason for relaxation is to find compromises in lieu of finding exactly what one wants. Additionally, similarity-based relaxation finds items that are equally desirable but that would otherwise not match because of insignificant feature differences. Most importantly, supporting data-graphic design suggests an additional function of relaxation: giving users ideas for how to integrate additional graphical elements and properties with partial designs they have created. The latter answers questions such as: How can additional graphemes be added to the space I've created and integrated with the graphemes I've already included? How have previous data-graphics used additional properties of these graphemes? How can other spaces or graphemes be substituted for the ones I've selected to express the same data? Enabling users to answer questions like these motivated the choice of match criteria that evolved in SageBook. Finally, our choice of criteria reflected the fact that it was easy for users (or the system) to remove extra spaces, graphemes, and properties when adapting the design for new data.

The search strategies in SageBook are based on the structural properties of the graphical and data elements in a data-graphic. Structural search is more robust and powerful than keyword search because:

- It recognizes positional relationships among graphical elements and the functional relationships in data. For example, structural search would be able to distinguish between Figure 6a and Figure 6b because it can tell that in Figure 6a the mark is in the same space as the bar, while in Figure 6b the mark is in the same space as the line. Similarly, it can also tell that the color property is used to encode data for the bar in Figure 6a, whereas in Figure 6b the color property is used to encode data for the mark.

(a)   (b)

Figure 6: Data-graphics that are distinguishable by structural search but not by keyword search.

- It can separate graphical and data elements and relations into different classes and weight the classes differently in its search criteria. For example, the layout discipline (chart, map, table, etc.) could be weighted more heavily than the type of grapheme (mark, bar, line, etc.).

A. Graphic-Matching Strategies

Figure 7 shows a graphical query (i.e. sketch) and the data-graphics retrieved with that query, using a moderately-relaxed matching strategy. SageBook provides the following three alternative graphic-matching strategies.

Close Graphic-Matching: This strategy searches for library data-graphics that have the same number of spaces as the query. In Figure 7, this strategy would have retrieved the first four stacks of similar data-graphics (i.e. the stacks outlined in black). These data-graphics only contain one space because the query has only one space.

For a *space* in the query to match a *candidate space* in a library data-graphic, both must employ the same layout discipline, and every grapheme in the query space must match a grapheme in the candidate space (i.e. the candidate space may contain unmatched graphemes, even though the query space may not). For a *grapheme* in the query space to match a *candidate grapheme*, both must have the same *grapheme-class* (e.g. bar, line, mark), and every *property* specified in the query grapheme (e.g. color, shape, size) must be used by the candidate grapheme. Using this search strategy, the query in Figure 7 will retrieve only those data-graphics consisting of a single *chart* that contains at least one *mark* grapheme. Note that only the positional properties of the mark were specified in the query; thus retrieved data-graphics may use additional grapheme properties that the query did not specify.

Subset Graphic-Matching: This strategy is more inclusive than close graphic-matching. In subset matching, a library data-graphic may contain more spaces than the query, as long as every query space matches a space in the data-graphic. This strategy retrieves all of the stacks of data-graphics in Figure 7. The stacks are sorted according to their degree of similarity to the query, based on the match criteria. For example, in Figure 7, all one-space matches are shown first, followed by all two-space matches, etc.

Subset matching supports a process resembling a library search. First, the user enters a query and retrieves a super-set of data-graphics, each of which will contain every element specified in the query. If the set is too large, the user can narrow it by adding more constraints or features to the query. The user may then browse through the data-graphics, and pick one based on other criteria. Any unwanted spaces can be easily deleted from the data-graphic using SageBrush.

Overlap Graphic-Matching: Subset matching may exclude data-graphics that are useful but fall slightly short of meeting the match criteria (i.e. that every query space must match a space in the library data-graphic). Thus, in addition to a strict subset search, we implemented a match strategy that sets upper and lower bounds around the number of query spaces that need to be matched. These bounds are set to be percentages of the total number of spaces in the query.

B. Data-Matching Strategies

Data-Relation Matching: This strategy searches for library data-graphics that contain every relation that was specified in the query. This matching strategy is useful when sets of daily or weekly data must be redisplayed in a consistent style. This also suggests an additional use for data-graphic retrieval - searching for information (rather than just graphic displays) stored as graphic media.

Close Data-Matching: This strategy enables users to find graphics showing data that has similar characteristics to their current data. Given a list of domains (i.e. the query) and their characteristics, the close data-matching algorithm tries to find a mapping from the query domains to the domains in a library data-graphic. For a query domain to match a candidate domain in a data-graphic, they must have the same *data-type* (nominal, ordinal, quantitative) and *frame of reference* (quantitative/valuation, coordinate), and must participate in the same kinds of *functional-dependencies* and *complex types*. Figure 8 shows an example of this data-matching process. *Activity* matches *houseID* (both have *nominal* data-types), and *materials-cost* matches *number-of-rooms* (both have *quantitative* data-types). *Start-date* and *end-date* match with *date-on-market* and *date-sold*, since they both have the same frame-of-reference (*coordinate*) and belong to the same complex-type (*interval type*). This matching process ensures that the domains in the library data-graphic and the query are equal

in number, and match one-to-one, as Figure 8 illustrates.

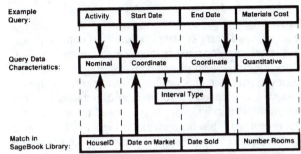

Figure 8: Close data-matching process.

Unlike the relation matching strategy, which requires the query and library data-graphic to contain the very same relations, the close data-matching strategy only requires that the domains have *similar* data characteristics and interrelationships. Thus, this strategy is not a keyword search, but rather is a search based on a *similarity* of structure between data-sets.

Subset Data-Matching: The idea behind this strategy is analogous to that of subset graphic-matching. Subset data-matching is like close data-matching, except that instead of requiring a *bijective* (i.e. *one-to-one* and *onto*) mapping between domains in the query and the library data-graphic, subset matching allows the library data-graphic to contain more domains than the query, as long as every query domain matches a domain in the data-graphic.

Overlap Data-Matching: As with graphic-matching, a variant of the subset data-matching strategy was created that sets upper and lower bounds around the number of query

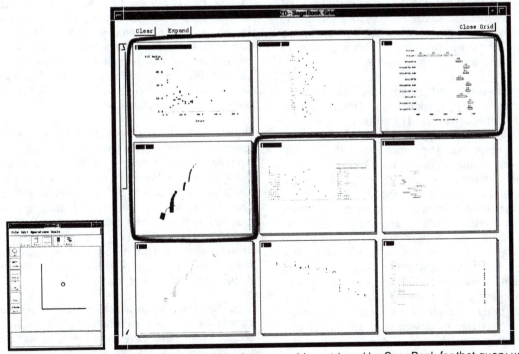

Figure 7: A graphical query, and the grid of data-graphics retrieved by SageBook for that query using a subset criterion.

domains that need to be matched, instead of using a strict subset rule.

C. Browsing

If the SageBook library contains hundreds of data-graphics, some queries may retrieve a large set of items. In such cases, the cognitive load placed on users to browse through the retrieved data-graphics would be significant. To support browsing, we developed a scrollable, grid-like interface that enables multiple data-graphics to be viewed at once (Figure 7). Our recent work has been on exploring ways to enhance browsing efficiency by grouping similar data-graphics into a *stack* in one cell of the grid. The number of data-graphics in a stack is indicated by the length of a black bar at the top of each cell. The *expand* operation can be used to distribute members of any stack into a new grid. An interesting challenge has been to develop effective grouping strategies (i.e. similarity criteria) for organizing a large number of data-graphics into a small number of meaningful stacks. The formal representation of data-graphics provides a framework for grouping strategies, as it did for graphic and data queries.

Since SageBook's purpose is primarily to help users' get design ideas, we defined four criteria that increased design differences between stacks by grouping similar data-graphics together. The method names reflect the aspect of the data-graphics within a stack that can be different. *Data-only* groups into a stack all those data-graphics that have the same number and types of spaces, ordering of aligned spaces, types and number of graphemes within each space and properties of graphemes. Effectively, these are cases in which the same design was saved for different data. The *spaces-order* method groups together the same data-graphics as the *data-only* method, but in addition, it includes data-graphics that have identical designs except for the ordering of aligned spaces. For example, data-graphics like the one in Figure 4 would be stored in the same stack regardless of the left-to-right ordering of the three spaces.

The two techniques mentioned group together data-graphics that show the same design approaches. Other methods differentiate design alternatives. The *grapheme-property* method groups together data-graphics that meet the *data-only* criterion, except that graphemes may use different properties. For example, data-graphics like the one in Figure 4 would be stored in the same stack regardless of the properties of the circles that were used (e.g. color, shape, size). The *grapheme-number* method groups data-graphics that have the same types of graphemes, and uses the same properties for each type, in each space. However, the number of each grapheme type in a space may differ. For example, this groups bar charts with one, two, or more bars per axis element in the same stack or maps with points containing a single label or multiple labels in the same stack. Finally, other methods are possible that group graphics based on *styles* of design (e.g. aligned charts, clustered graphemes, networks, tables, etc.).

We are exploring the different possibilities of providing these methods as individual options or combined sequentially to form a hierarchical classification of graphics within each stack. Our current implementation groups data-graphics using a four-tier hierarchy, consisting of the *data-only* (bottom), *space-order*, *grapheme-property*, and *grapheme-number* (top) categorization methods. Expanding a stack is equivalent to removing a constraint for that particular stack so that members of the stack can be viewed in greater detail. A stack can be expanded into a series of stacks which can be further expanded until the bottom of the hierarchy is reached.

REUSE AND ADAPTATION: HOW CAN A DATA-GRAPHIC BE ADAPTED FOR NEW DATA AFTER RETRIEVAL?

The existence of similarity search strategies opens up the possibility that some of the data-graphics retrieved by SageBook may not fully conform to what the user desires. In such cases our system provides manual adaptation capabilities through SageBrush and automatic adaptation capabilities through SageBook.

The automatic-adaptation module does the mapping between data-domains in the query to data-domains in the retrieved data-graphic based on their characteristics. When there are data-domains in the retrieved data-graphic that cannot be mapped to domains in the query, the adaptation module will discard graphical objects from the data-graphic as necessary. When it is forced to do this, the adaptation module tries to preserve spaces first, graphemes second and grapheme properties last.

Figure 9 (Top) shows a data-query and an example data-graphic that is retrieved by that query. This data-graphic shows a supply-network with supply routes/paths (indicated by the lines) and demand units (indicated by the marks). The data-graphic was retrieved because it contains "paths" which are defined by the geographic coordinates of their end-points. This exactly matches with the data-domains *start-location-n/s start-location-e/w, end-location-n/s* and *end-location-e/w* in the query data.

Figure 9 (Bottom) shows the new data-graphic that is generated from the query data after automatic adaptation has been performed on the data-graphic in Figure 9 (Top). Note that the *marks* in Figure 9 (Top) were discarded in Figure 9 (Bottom) because the old domains which it expressed (geographic location of demand units and the quantity required by those units) could not be mapped to any of the new domains in the query (i.e. *temperature* and *troop-movement-size*). This is because *temperature* and *troop-movement-size* are properties of the "paths", whereas the demand units are totally separate objects.

When there are additional data-domains in the query data that cannot be mapped to the retrieved data-graphic, the adaptation module leaves it to SAGE to add them into the new data-graphic. In the example adaptation shown in Figure 9, SAGE additionally encoded *temperature* by using *color* and *troop-movement-size* by *line thickness*. In general, we have developed and equipped SAGE with knowledge-based design techniques that can complete partial design specifications [8]. Partial specifications may be constructed either by SageBook's automatic adaptation module or by the user. We have explained how Figure 9 (Bottom) can be constructed automatically through SageBook; [7] shows how it can be constructed by the user through SageBrush.

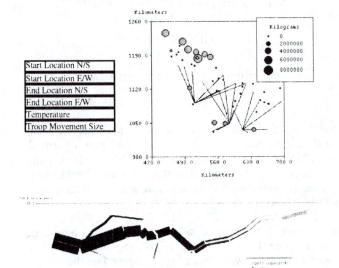

Figure 9: (Top) A data query and an example data-graphic retrieved by that query. (Bottom) A new data-graphic generated from the query data and the data-graphic design after automatic adaptation.

SUMMARY AND FUTURE WORK

We have designed and implemented a content-based search system, SageBook, which provides users with design expertise by giving them access to a database of prior data-graphics. Unlike prior image-retrieval systems, the goal of SageBook is to provide content-based retrieval facilities in the context of supporting data-graphic design. In order to fulfill this goal:

- We have designed and developed a *graphical direct-manipulation interface* (SageBrush) from which users can specify requests to the system with ease. SageBrush can also be used to manually adapt previous data-graphics and to construct new ones.
- We have formalized a *content description language* in order to characterize the graphical elements and data-domains that are contained within a data-graphic as well as the relationships among them. User requests are translated into this vocabulary before they are passed on to SageBook or SAGE so that they can be conveyed to the system without ambiguity.
- We have implemented an automatic presentation system, SAGE, which creates data-graphics specified by users and automatically stores a description of the data and graphical characteristics with each image. This provides SageBook with a growing library of data-graphics that has been *automatically described* when they are first produced.

- We have provided strategies to search for data-graphics based on their content (i.e. *structure*) and their *similarity* to the user query.
- We have constructed *manual and automatic adaptation* tools to aid users in the process of reusing retrieved data-graphics for their new data.

Currently, we are running a series of user tests to see what effect SageBook has on measures like ease of creating designs and the quality and diversity of graphics that are created. Another area of future work is validating the utility of the match criteria, especially our assumptions of the important criteria for judging similarity. Finally, we are exploring ways to base search on the *information-seeking goals* that the graphics are being designed to support rather than just the data that is being visualized [8,9].

REFERENCES

1. Borgman, C.L., Belkin, N.J., Croft, W.B., Lesk, M.E., and Landauer, T.K. Retrieval Systems for the Information Seeker: Can the Role of the Intermediary be Automated? *CHI'88 Human Factors in Computing Systems*, ACM, April 1988, p.51-53.
2. Garber, S.R. and Grunes, M. B., The Art of Search: A study of Art Directors. *Proceedings CHI'92 Human Factors in Computer Systems*, ACM, May 1992,p.157-163.
3. Kato, T., Kurita, T., and Shimogaki, H., Multimedia Interaction with Image Database Systems. *SIGCHI Bulletin* 22, 1 (July, 1990), p. 52-54.
4. Mackinlay, J.D. Automating the Design of Graphical Presentations of Relational Information. *ACM Transactions on Graphics*, 5, 2 (Apr 1986),p.110-141.
5. Myers, B., Goldstein J., Goldberg, M.A. Creating Charts by Demonstration. *Proceedings CHI'94 Human Factors in Computing Systems*, ACM, April 1994, p.106-111.
6. Nishiyama H., Kin, S., Yokoyama, T. and Matsushita Y. An Image Retrieval System Considering Subjective Perception. *Proceedings CHI'94 Human Factors in Computing Systems*, ACM, April 1994, p.30-36.
7. Roth, S.F., Kolojejchick J., Mattis J., Chuah M., SageTools: An Intelligent Environment for Sketching, Browsing, and Customizing Data-Graphics. *Proceedings CHI'95 Human Factors in Computing Systems*, ACM, May 1995.
8. Roth, S.F., Kolojejchick J., Mattis J., Goldstein J., Interactive Graphic Design Using Automatic Presentation Knowledge. *Proceedings CHI'94 Human Factors in Computing Systems*, ACM, April 1994,p.112-117.
9. Roth, S.F. and Mattis J. Data Characterization for Intelligent Graphics Presentation. *Proceedings SIGCHI'90 Human Factors in Computing Systems*, Seattle, WA, ACM, April, 1990, p. 193-200.
10. Tou, F. N., Williams, M.D.Fikes R., Henderson, A., & Malone, T. RABBIT:An Intelligent Database Assistant. *AAAI-82 Proceedings of the National Conference on Artificial Intelligence*, 1, August. 1980, p.314-318.

Finding and Using Implicit Structure in Human-Organized Spatial Layouts of Information

Frank M. Shipman III, Catherine C. Marshall

Department of Computer Science
Texas A&M University
College Station, TX 77843-3112
(409) 862 - 3216
{shipman, marshall}@cs.tamu.edu

Thomas P. Moran

Xerox Palo Alto Research Center
3333 Coyote Hill Road
Palo Alto, CA 94304
(415) 812 - 4351
moran@parc.xerox.com

ABSTRACT

Many interfaces allow users to manipulate graphical objects, icons representing underlying data or the data themselves, against a spatial backdrop or canvas. Users take advantage of the flexibility offered by spatial manipulation to create evolving lightweight structures. We have been investigating these implicit organizations so we can support user activities like information management or exploratory analysis. To accomplish this goal, we have analyzed the spatial structures people create in diverse settings and tasks, developed algorithms to detect the common structures we identified in our survey, and experimented with new facilities based on recognized structure. Similar recognition-based functionality can be used within many common applications, providing more support for users' activities with less attendant overhead.

KEYWORDS: emergent structure; spatial diagrams; spatial structure recognition; informal systems; hypermedia.

INTRODUCTION

Many kinds of software, ranging from operating systems to graphics editors to collaboration substrates to hypermedia applications, provide users with two or two-and-a-half dimensional spaces in which they can organize information in the form of graphical objects. These manipulable objects include icons or other visual symbols representing larger pieces of information, as well as complete chunks of information. People tend to follow certain conventions when they lay out these objects to indicate relationships among them. We are investigating methods of recognizing such implicit spatial structure and using it to support users' activities.

We first examined the kinds of spatial layouts that people produce in both computational and noncomputational media. Based on these results, and the results of analogous

studies of the way people organize materials in their physical environments [4][6][7], we developed a "spatial parser" to recognize common structures. The parser is built to be portable and extensible so that it can be added to new systems or augmented to recognize new kinds of structures with little effort.

We integrated the parser with VIKI [11], a spatial hypertext system, to explore the use of recognized structure. VIKI currently includes two types of support based on the parser. First, VIKI provides users with ready access to implicit structures. Second, at user initiative, VIKI suggests formally represented structures based on the implicit structure.

The next section summarizes our study of user-constructed spatial arrangements and the conventions we found. We then describe the architecture developed for recognizing these common spatial structures. Finally, we discuss the application of structure recognition to support users performing information management and analysis tasks in VIKI.

LOOKING AT PRACTICE

Experiences with the use of Aquanet [9] led us to recognize the importance users place on implicit spatial structures. This realization prompted us to perform a survey of spatial structures people created in diverse hypermedia systems and in non-computational settings (such as notecards and Post-its stuck on a wall) [10].

As part of the survey, we analyzed nine layouts, each the result of a long-term information management or analysis task. Three of the layouts were created in NoteCards [2], one in the Virtual Notebook System [14], and three in Aquanet [8]. The two non-computational layouts in our survey involved wall-sized arrangements of 3" x 5" cards, Post-its, and other pieces of paper. Table 1 shows the task, source, and number of objects for each of these layouts.

To perform the analysis, we introduce three analytic abstractions. The first is the notion of graphical *information objects* already discussed. Also, we use the idea of *object type*. Here, type is used to reflect an object's distinguishing visual characteristics and functional role within a given spatial layout. We also use the idea of *spatial structure*,

task	source	# of objects	# of types
comparing editors	paper	72	3
product design analysis	paper	284	8
anthropological analysis	NoteCards	36	3
anthropological analysis	NoteCards	20	3
anthropological analysis	NoteCards	50	3
bug tracking (VNS)	VNS	63	6
bug tracking (Aquanet)	Aquanet	130	6
analysis of tech. field	Aquanet	1506	14
analysis of software	Aquanet	193	6
writing paper	Aquanet	74	3

Table 1: Spatial layout task, source, and data characteristics

which is a perceptually distinct group of objects. Table 1 includes the number of object types identified in our analysis.

Following this analytic framework, we encoded the data in a canonical form which recorded spatial and visual aspects of the graphical information objects, including the relative planar location and the extent of each object. Each object was also assigned a type based on its system type (if it had one) and distinguishing visual characteristics (for example, font or color). Discussions with the people who created the layouts helped us to understand the intended structures in each example and their meaning. We discovered that a small set of common structures, like stacks and lists, were common across the layouts, even though they were created in a variety of systems and in service of different tasks.

Analysis of the content and function of the objects used in these structures revealed that people used spatial layouts to represent different types of relationships among constituents. First, spatial structure was used as a means of categorization or to build up sets. Second, spatial proximity was used to indicate specific relationships among objects or among types of objects. Finally, spatial arrangement was in some instances dictated by the way in which objects were used together in a task.

The following two examples, one computational, one non-computational, illustrate the kinds of conventional structures we found, and preview the issues that arise in developing heuristics for recognizing these structures and using them to support users' activities.

Example from Aquanet

Figure 1 is an example of the kind of layouts we gathered for our survey; the layout is the result of a group effort to write a paper in the Aquanet hypertext system. Each long thin gray rectangle represents a major topic heading;

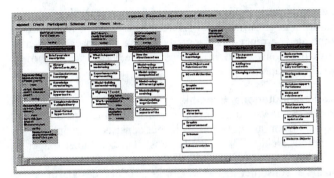

Figure 1: Example of spatial layout from Aquanet

outlined rectangles represent subtopics (and the underlying paragraphs explicating them); and thicker gray rectangles are annotations, notes from one author to the others. In Aquanet, these visual properties are solely dependent on the object's type in the system. The other computational sources, however, did not have this notion of object type in the system but the authors had improvised, creating visually apparent object types using different fonts and colors for objects.

It is readily evident to human perception that there is a significant amount of structure in the spatial layout in Figure 1. Yet very little of this structure is expressed in such a way that it is accessible to the system in which the layout was created (even though the system provided users with a mechanism for expressing just such structure).

Figure 2 is a close-up of one of the structures in Figure 1. The arrangement includes a list with six similar elements (of type subtopic); it is a good example of the kind of structure people use to express categories or sets. This list is part of a higher-level structure that includes the list and its heading, which is of a different type. We refer to this kind of structure (one consisting of a regular pattern of different object types) as a *composite*. The list of annotations on the right side of the list refers to a portion of the list. Unlike the rest of the structure, without examining contents, this sort of reference is more idiosyncratic and remains ambiguous to human perception. In our analysis of the layouts for common structures, we resolved such ambiguities by looking at textual content and talking to the original authors.

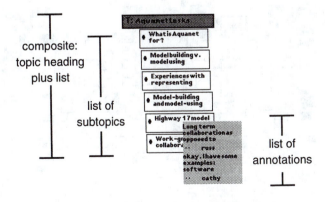

Figure 2: Close-up of structure with description

Referring back to Figure 1, we can also see that this list is part of a larger structure of six such lists, some of which are annotated, arranged horizontally across the space. It is a structure constructed in service of the larger activity of writing a paper.

Non-computational Example

Figure 3 shows a diagram of one of the non-computational layouts from our survey. This layout consists of around two hundred eighty 3"x5" cards, Post-Its, and other paper covering a good-sized wall. The arrangement was used to analyze consumer behavior for a product design in an industrial design firm.

In this non-computational arrangement, color (and kind) of paper and marker are used to visually reflect a notion of object type similar to that supported by Aquanet and improvised in the other systems. Objects shown in the diagram in Figure 3 have this information encoded in their shade of gray. While the majority of cards contain handwritten text, there are also some pictures and diagrams; these cards were assigned a different type because they were so visually distinct.

As in the prior example from Aquanet, there is an apparent structure to this diagram. The cards on the top and to the left act as labels resulting in an incomplete matrix of lists. A distinct horizontal row of cards divides the matrix at its center. Annotations take the form of cards and Post-Its attached to or next to cards they discuss.

In summary, our data analysis uncovered sufficient regularity to support the idea that automatic recognition of implicit spatial structure was possible, and that it could identify several useful kinds relations.

COMMON SPATIAL STRUCTURES

Based on our survey of spatial arrangements, we developed an initial set of primitive spatial structures: lists (aligned objects of the same type), stacks (overlapping objects of the same type), composites (repeated arrangements of alignment between objects of different types), and heaps (overlapping objects of different types.) These four primitives may be composed into higher-level structures,

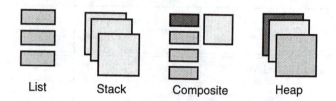

List   Stack   Composite   Heap

Figure 4: Examples of primitive spatial structures

like lists of composites. Figure 4 shows abstract examples of these four types of structure.

Lists. Lists are objects of the same type that are vertically or horizontally aligned; they were by far the most common spatial structure we identified in our survey, found in almost every example. Lists were used primarily for categorization: membership in a particular list indicated elements shared common features. Some of the lists in the diagrams included internal ordering principles, such as importance or temporal sequence. Also, some lists included groupings of more closely related items. Because these additional types of structure relied on analysis of object content, they are not represented in our list of primitive spatial structures.

Stacks. Stacks are objects of a single type that overlap significantly; they tended to be used for categorization when there was not enough space for lists. By arranging objects into stacks, the user has decided to favor compactness over ease of access since even partially obscured objects are more difficult to manipulate or view.

Composites. Composites are regular spatial arrangements of different types of objects; they were the most interesting kind of structure in our survey because they were used to denote higher level abstractions, relationships among types of objects. For example, in our survey, we found a repeated pattern of "implementor" objects above "system" objects; this pattern suggested an implicit "developed" relation. Similarly, labels often appeared above lists. In both cases, the spatial arrangement represents an implicit semantic relationship among the objects; it is a one-to-many relationship in the label/list case.

Heaps. Heaps are overlapping collections of different types of objects. They are usually created during the course of a

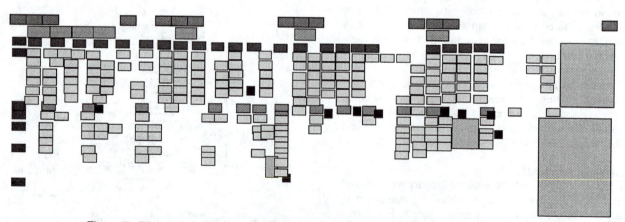

Figure 3: Diagram of physical layout of paper created during product design meeting.

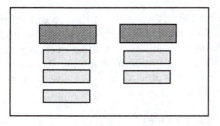

Figure 5: Sample graphic layout

sorting task as a strategy for managing a crowded space. People group dissimilar objects with the intention of spreading them out again once space becomes available.

RECOGNIZING SPATIAL STRUCTURE: AN EXAMPLE

We have implemented a spatial parser to recognize the common spatial structures identified in our observations. This section uses an example layout (shown in Figure 5) and its resulting parse to motivate our discussion of automatic recognition. The recognition architecture and its use will be described in later sections of this paper.

The layout shown in Figure 5, a simplified version of the arrangement in Figure 1, contains seven individual objects of two distinct types. The initial structures, described using the spatial primitives, are two vertical lists of smaller light gray objects. These lists, in turn, are part of two instances of a composite type that consists of a larger darker gray object over a list of light gray objects. Because these are instances of the same type of composite and they are aligned, a horizontal list with two elements is the next result of the parse.

Figure 6 shows the final parse tree of the spatial layout from Figure 5. Non-terminal nodes of the graph reflect the primitive and intermediate structures that may be identified through analysis of spatial layout, including Horizontal list, Vertical list, and Composite. Terminal nodes show graphic depictions of the original types from the layout.

The recognition algorithms were not tuned to any particular style of layout, and use only the location (x and y position), extent (width and height), and system-supplied type of the information objects. This limited set of characteristics was used so that the parser could be easily adapted to a variety of systems. For systems that do not include a notion of type, the visual characteristics of objects may be used to

automatically assign them a type. This mapping of visual features to implicit object type is system dependent since different visual features are modifiable in different systems.

Using our survey examples as test cases, we found that this kind of parsing was fairly successful. At a low level, such as the labelled lists in Figure 1, the parser is quite accurate at determining what elements are part of the same structure; thus the small irregularities in alignment and spacing seem to be handled correctly. Ambiguous structures were not always identified as authors intended. For example, the fourth labelled list from the left in Figure 1 is identified as two separate lists (even though the author saw them as one) because of the large gap between the fragments.

Higher level structures were sometimes missed. For example, the six labelled lists in Figure 1 are parsed as two separate structures. In this example, the parser recognizes the first three lists as one structure, and the second three lists as another, because each of the lists on the left side of the diagram has an annotation directly above it. This is a second example of how layouts may also be ambiguous to humans without access to the semantics of the situation (if a reader did not realize the annotation functioned as such, he or she would make the same error as the parser). In most of our test cases, in spite of omissions and inaccuracies, the structure identified was often consistent with the authors' intent.

RECOGNITION ARCHITECTURE

We wanted to develop a flexible and easily extensible architecture that would allow us to tailor recognition performance to match different conventions, tasks, user preferences, and system-dictated layout constraints (such as gridding). As a result, the recognition architecture includes a reconfigurable pipeline of spatial structure *specialists* that share a blackboard of information and a recognition *strategist* that configures the specialists.

The strategist begins the recognition process by determining the order in which the pipeline of recognition specialists will be applied. This ordering is based on a statistical assessment of layout features. The specialists then begin a bottom-up parse of the layout; each specialist is responsible for identifying a particular type of structure. If the specialists define new types (as a result of encountering composites or heaps), they add these to the blackboard and recompute usage statistics to reflect the new structures. Figure 7 diagrams this process. The strategist, the blackboard, and recognition specialists are described in more detail below.

Strategist. The strategist performs some initial analysis of the overall space to determine the order in which the specialists will be applied. The strategist uses an initial calculation of overall object alignments to determine whether the arrangement seems to favor a vertical or horizontal orientation, or neither. According to these results, the specialists are ordered in the pipeline. (In the future work section, we discuss the potential for more complex or cooperating strategists.)

Blackboard. The blackboard contains global information shared by the specialists. This information includes user-

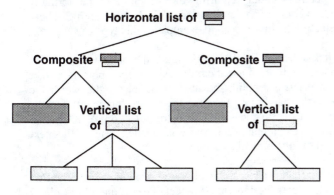

Figure 6: Spatial parse tree for sample layout

Initial Object Position, Extent, and Type Information

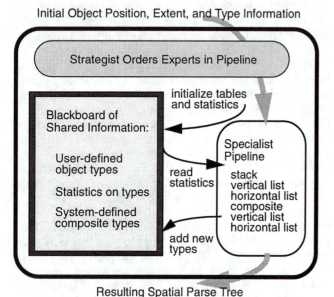

Resulting Spatial Parse Tree

Figure 7: Architecture for spatial recognition

defined object types, system-defined composite types, and statistics on each type's use. Initial statistics are the average width and height for objects of a given type.

Specialists. Recognition of implicit structure is performed by a pipeline of computerized specialists. Each specialist is designed to recognize a single type of structure. Besides the information in the blackboard, the specialists are provided with the spatial display and its current parse. Because each specialist produces a new parse as output, the specialists can be pipelined together in any order and a single specialist can operate multiple times; this enables the recognition of complex structures like the horizontal list of composites used as an example in the last section.

To keep the algorithms system-independent, the specialists use only the visual and planar characteristics described earlier: object position, extent, and type. Specialists that identify heterogeneous structures (heaps of dissimilar objects and composites) generate new unique types for later specialists called by the parser; structures formed from homogeneous objects assume the type of their constituents. By repeated application of the recognition algorithms on parse results, higher levels of structure are recognized (like lists of composites, or composites which include lists).

SUPPORTING THE USER WITH RECOGNIZED STRUCTURE

Why is there a need to develop heuristic algorithms to find structure that is already perceived by the users?

First, they can help authors interact with ad hoc organization; the found structures can be used as the basis for supporting simple but repetitive information management tasks. Second, if a more formal knowledge base is a desired outcome of the task, recognizing structures is an important method for helping people notice and express the regular structure of their domain and maintain its consistency [15].

We are investigating these uses of the recognized structure in the context of VIKI [11]. With the addition of the spatial parser, VIKI assists authors in both interaction with ad-hoc structures and formalization of emerging structures. Before describing this support in detail we will give a brief overview of VIKI and its goals.

Brief Description of VIKI

VIKI is a spatial hypertext system that supports the emergent qualities of structure and the abstractions that guide its creation; the tool is intended for exploratory interpretation, making sense of a body of electronic information collected in service of a specific task such as analysis, design, or evaluation. Users manipulate graphical objects within a hierarchy of information workspaces (collections), using the visual properties of the objects along with a more formal types mechanism to express a highly nuanced interpretation of the collected materials.

VIKI gives users the ability to work with three kinds of elements: objects, collections, and composites. Each graphical *object* is a visual symbol that refers to an underlying piece of semi-structured information. Each *collection* is a subspace that can reflect semantic categorization or a task-oriented subset of the information space. *Composites* are two or more objects which are used together to make up a meaningful higher-level unit of structure. These three kinds of elements allow users to build up the same kinds of structures we observed in practice.

VIKI collections act as clipping regions so users can see information at multiple levels in the hierarchy of workspaces. Similar to Boxer [1], clicking on the border of a collection causes that collection to fill the VIKI window. Figure 8 shows before and after images of collection traversal in VIKI.

VIKI uses the results of spatial parsing two ways: it supports interaction based on implicit structure (where implicit structure remains undeclared) and it helps people use the object-collection-composite data model by supporting the transition from implicit to declared structure. Examples of each of these uses are described below.

Access to Implicit Structures

The spatial parser allow users to interact with implicit structure without requiring the structure to be formally defined or even anticipated. VIKI's hierarchic click-selection facility is a good example of this type of support.

Click-selection in VIKI works much the same way as it does in a text editor. In a text editor, a single click puts the cursor at a particular point; the next click selects the word; the next, the paragraph; the next the entire document; and the next returns to the single point of selection. VIKI uses a similar technique: each successive click selects the next level of hierarchical structure.

Figure 9 shows hierarchic click selection in action. The first click selects the individual object. The second click selects the list of three objects. The third adds the label for that list

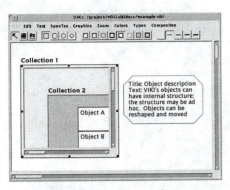

a.) before traversal

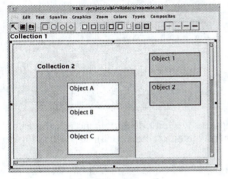

b.) after traversal

Figure 8: Screen A shows a VIKI collection acting as a clipping region prior to traversal. Screen B shows the view after navigating into the collection.

and the fourth selects a similar labelled list above the first list.

Users realize two immediate advantages from hierarchic click-selection. First, users may select objects that are part of partially hidden structures without having to scroll or traverse from the current view. This is especially important in VIKI since collections act as clipping regions, displaying only portions of their contents. Thus, selecting a structure in one collection and moving it to another collection, a fairly frequent action, can be accelerated through use of click selection.

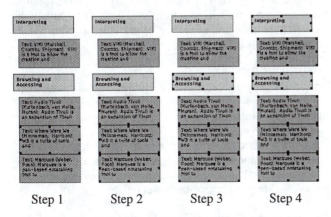

Step 1   Step 2   Step 3   Step 4

Figure 9: Four steps in VIKI's recognition-based hierarchic click-selection

Second, users may select partial structures in areas where objects are too densely packed for sweep-selection. In VIKI, such situations arise when users sort through large numbers of references to external documents--a frequent activity in the type of analysis tasks VIKI is designed to support.

In each of these cases, users tend to select objects this way for non-destructive operations, such as move, because the entire extent of the selection may not be visible.

Support for Creating Formal Structures
While undeclared structure can be the basis for certain kinds of interaction, formal or declared structure can facilitate other kinds of support.

Experiences with a variety of information management tools point to the difficulty users have in creating and using formal structure [16]. This difficulty led us to support a process of "incremental formalization", in which information is initially entered in an informal representation and later can be formalized when needed.

Incremental formalization aims, first of all, to eliminate the cognitive costs of formalization that inhibit user input. Secondly, it aims to reduce the burden of formalization by distributing it, and making it demand driven.

To further lower the cost of formalizing information, we are investigating techniques for using the recognized structure from the spatial parser to support incremental formalization. VIKI uses the results of the spatial parser to provide formalization suggestions to the user. This work builds on our experience with supporting incremental formalization based on the recognition of textual cues of inter-object relations in the Hyper-Object Substrate [17].

Collection Suggestion
By supporting incremental formalization VIKI helps users bridge the gap between their activities and the system's data model. For example, at a user's initiative, VIKI will suggest collections -- apparent subdivisions of materials for starting new subspaces. Collection suggestion is an accelerator: it greatly reduces author effort in creating new collections and moving existing, visually structured materials into them.

To suggest collections in VIKI, we look for the highest level of contiguous structure. These higher level structures correspond to the task-oriented workspaces we observed in our survey. In determining which top-level structures to suggest as collections, structures which greatly overlap in space (and thus would obscure one another if made into collections independently) are combined. We limit the number of extraneous small collections that VIKI suggests by requiring a minimum number of constituent objects.

Collection suggestion uses a standard spelling checker as a model of interaction. In our interface, shown in Figure 10, the user can iterate through the list of suggestions and accept those that are appropriate. VIKI displays these suggestions to the user by selecting all objects and collections that will become part of the new collection and outlining its extent

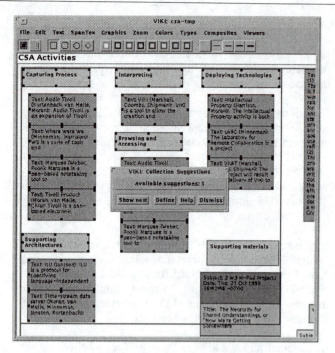

Figure 10: VIKI suggests a new collection based at the user's request

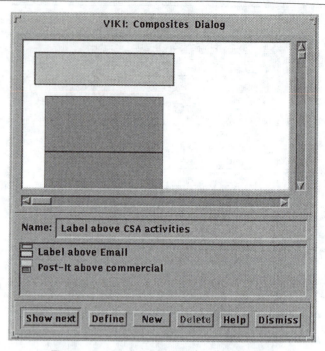

Figure 11: Composite definition dialog with suggestion based on the structure shown in Figure 10

with a dark band. Users can modify suggestions by interactively changing which objects are selected and thus will be included in the new collection.

Composite Suggestion

Composite suggestion is an example of how VIKI can use recognition results to support not only the formalization of instances of relations, but also new classes of relations. The recurrent spatial patterns of object types that are identified by the spatial parser are used as suggestions for VIKI composite types.

Users can ask for VIKI to suggest potential composite types from within the composite definition dialog. Users may either accept the suggestions as is, modify them, or start from scratch to develop new composite types. Figure 11 shows the composite definition dialog with a suggested composite based on a recurrent pattern in the user's work. The suggestion appears as a set of abstract objects in an editable workspace with a composite name based on the names of the constituent objects and their arrangement.

RELATED WORK

Work on the recognition of spatial regularities has been a long-standing part of vision processing. But our work emphasizes the recognition of patterns in layouts of discrete, declared objects generated by purposeful human activities.

Document analysis and recognition shares our goal of identifying structures implicit in the layout of information. It differs, however, in its emphasis on the recognition of presentational structures common to known genres rather than identifying the more dynamic, idiosyncratic structures that evolve in the process of manipulating information

objects. Thus, while some basic techniques may be shared, they are apt to diverge due to this crucial difference.

Spatial recognition and parsing is also found in work on visual languages, but has different goals and assumptions which influence the types of recognition algorithms produced. Unlike Lakin's visual language parsing in vmacs [5], our algorithms do not assume that we can unambiguously recover their underlying syntactic structure. Unlike Pictoral Janus [3], a visual programming environment that bases connectivity on assessments of inside, connected, or touching, our purpose is not to "debug" formal visual/spatial structures, but rather to tease out some implicit partially-framed structure.

Our goal of providing better support for users' already apparent desire to work with implicit structure has influenced the spatial parser's design from the start. Saund's perceptually-supported sketch editor [13] and Moran's support using implicit structure in Tivoli [12], although focussing on recognition in pen-based sketches, are based on similar goals.

FUTURE WORK

This work may proceed in several different directions to improve our understanding of the recognition and use of implicit spatial structure.

From our experience, it is apparent that new specialists are required for some kinds of recognition to proceed. Outlines, matrices and tables were found in our survey but are not parsed well by the current implementation.

Because of the idiosyncratic and ambiguous nature of the spatial structure, in cases where there is a high overhead for incorrect parses, interaction will be important in guiding this kind of recognition. Up to now, we have emphasized uses of recognized structure that are tolerant to inaccurate parsing; uses where the results are lightweight and used to provide alternatives for actions already possible through other means. Future work on interaction methods for correcting incorrect parses would be required before error-intolerant uses of recognized structure can be evaluated in real-use situations.

More generally, experience with enabling and supporting incremental formalization in VIKI shows promising results. This work leads to more issues to be explored in both methods of producing and interfaces to suggesting possible formalizations. We need more experience before we can answer questions like when suggestions should be provided to the user and what good interfaces are for providing such suggestions.

CONCLUSIONS

Our survey showed that people use the visual and spatial characteristics of graphical layouts to express relationships between objects, icons, or other representations of underlying information. We discovered a small set of underlying primitives that capture the conventions illustrated by the layouts in our survey and implemented a set of heuristic algorithms to recognize these common structures. Because these structures are readily perceived by people, but remain implicit (and therefore opaque) to the systems in which they were created, we postulated that they would be a good basis for providing new system services.

Our extensible, tailorable architecture for performing spatial structure recognition uses a pipeline of specialists, configured by a strategist, that communicate through a blackboard. The specialists, which may be reapplied as many times as necessary, perform a bottom-up parse of a spatial layout to identify multiple levels of implicit structure. This architecture not only allows new heuristics to be added to the spatial parser, but also allows the parser to be modified to work with different systems or tools.

We have demonstrated the utility of this approach by integrating the parser with VIKI, a spatial hypertext system. In VIKI, we use spatial structure recognition as the basis for user interaction with implicit forms and as a means of suggesting more formal representations to the user at his or her initiative. Already the parser-based support provides advantages for VIKI users; as use continues, we believe that we will be able to enhance both the algorithms, the way they are configured, and the kinds of system features we can provide based on the results of their analysis.

ACKNOWLEDGEMENTS
We would like to thank Frank Halasz and Jim Coombs, who aided in the development of the ideas and systems described in this paper. This work was done while at Xerox PARC.

REFERENCES
[1]   diSessa, A., Abelson, H. "Boxer: A Reconstructible Computational Medium." *Communications of the ACM*, 29, 9, (1986), pp. 859-868.

[2]   Halasz, F.G., Moran, T.P., Trigg, R.H. "NoteCards in a Nutshell." *Proc. of the ACM CHI + GI Conference*, (Apr. 5-9, 1987), pp. 45-52.

[3]   Kahn, K. "Concurrent constraint programs to parse and animate pictures of concurrent constraint programs." *Proc. of the International Conference on Fifth Generation Computer Systems*, June 1992.

[4]   Kidd, A. "The Marks are on the Knowledge Worker." *Proc. of CHI '94*, (Apr. 24-28, 1994), pp. 186-191.

[5]   Lakin, F. "Visual Grammars for Visual Languages." *Proc. of AAAI '87*, (July, 1987), pp. 683-688.

[6]   Malone, T.W. "How do People Organize their Desks? Implications for the Design of Office Information Systems." *ACM Transactions on Office Information Systems*, 1, 1 (Jan., 1983), pp. 99-112

[7]   Mander, R., Salomon, G., Wong, Y.Y. "A 'Pile' Metaphor for Supporting Organization of Information." *Proc. of CHI '92*, (May 3-7, 1992), pp. 627-634.

[8]   Marshall, C.C., Halasz, F.G., Rogers, R.A., Janssen, W.C., Jr. "Aquanet: a hypertext tool to hold your knowledge in place." *Proceedings of Hypertext '91*, (Dec. 16-18, 1991), pp. 261-275.

[9]   Marshall, C.C., Rogers, R.A. "Two Years before the Mist: Experiences with Aquanet." *Proc. of ECHT '92*, (Dec. 1-4, 1992), pp. 53-62.

[10]   Marshall, C.C., Shipman, F. M. "Searching for the Missing Link: Discovering Implicit Structure in Spatial Hypertext." *Proc. of Hypertext '93*, (Nov. 14-18, 1993), pp. 217-230.

[11]   Marshall, C.C., Shipman, F. M., Coombs, J.H. "VIKI: Spatial Hypertext Supporting Emergent Structure." *Proc. of ECHT '94*, (Sept. 18-23, 1994), pp. 13-23.

[12]   Moran, T.P., Chiu., P., vanMelle, B., Kurtenbach, G. "Implicit structures for pen-based systems within a freeform interaction paradigm." *Proc. of CHI '95*, (May 7-11, 1995).

[13]   Saund, E., Moran, T.P. "A Perceptually-Supported Sketch Editor." *Proc. of UIST '94*, 1994.

[14]   Shipman, F.M., Chaney, R.J., Gorry, G.A. "Distributed Hypertext for Collaborative Research: The Virtual Notebook System." *Proc. of Hypertext '89*, (Nov. 5-8, 1989), pp. 129-135.

[15]   Shipman, F.M. *Supporting Knowledge-Base Evolution with Incremental Formalization*. Ph.D. Dissertation, Technical Report CU-CS-658-93, Dept. of Computer Science, Univ. of Colorado, 1993.

[16]   Shipman, F.M., Marshall, C.C. "Formality Considered Harmful: Experiences, Emerging Themes, and Directions." Xerox PARC technical report ISTL-CSA-94-08-02, 1994.

[17]   Shipman, F.M., McCall, R. "Supporting Knowledge-Base Evolution with Incremental Formalization." *Proc. of CIII '94*, (Apr. 24-28, 1994), pp. 285-291.

A Comparison of Face-To-Face and Distributed Presentations

Ellen A. Isaacs, Trevor Morris, Thomas K. Rodriguez, and John C. Tang

SunSoft

2550 Garcia Ave., MTV 21-225

Mountain View, CA 94043-1100, USA

(415) 336-1167, (415) 336-3097, (415) 336-4918, (415) 336-1636

ellen.isaacs@sun.com, trevor.morris@sun.com, tom.rodriguez@sun.com, john.tang@sun.com

ABSTRACT

As organizations become distributed across multiple sites, they are looking to technology to help support enterprise-wide communication and training to distant locations. We developed an application called Forum that broadcasts live video, audio, and slides from a speaker to distributed audiences at their computer desktops. We studied how distributed presentations over Forum differed from talks given in face-to-face settings. We found that Forum attracted larger audiences, but the quality of interaction was perceived to be lower. Forum appeared to provide more flexible and effective use of slides and other visual materials. On the whole, audiences preferred to watch talks over Forum but speakers preferred to give talks in a local setting. The study raises issues about how to design this technology and how to help people discover effective ways of using it.

KEYWORDS: Distributed presentations, distance learning, computer-supported cooperative work (CSCW), video conferencing, multimedia, organizational communication.

INTRODUCTION

Over the past few years, many organizations have been trying to cope with an increasingly distributed workplace. Corporations, universities, and governments are forming partnerships with organizations in other parts of the world, and many organizations are distributing their own operations across many sites. At the same time, business consultants are emphasizing the importance of communication to help workers make informed, well-coordinated decisions that help further the goals of the organization [Peters, 1987; Smith, 1991]. In addition, many organizations and individuals are increasing their demand for training to keep themselves competitive in a global marketplace [Peters, 1987].

This combination of distributed workplaces and increased need for communication and training is leading many organizations to look to technology for support. In particular, the growth of networked computers can help

them "shrink the workplace." So far, the CSCW community has explored two approaches to the problem. It has studied ways to use electronic mail [Sproull and Kiesler, 1988] and databases [Liberman and Rich, 1993] to help large groups communicate efficiently across time and distance. And it has experimented with the use of video and audio to provide a sense of teleproximity to small distributed groups who need to tightly coordinate their efforts [Olson and Bly, 1991; Tang and Rua, 1994]. There has been less exploration, however, of the use of video and audio to help support the communication needs of large distributed organizations.

There have been a few attempts to use multimedia to support large distributed group communication. For example, members of the Internet community have conducted live video presentations over the Multicast Backbone (MBone) [Macedonia, 1994]. Usually, video is multicast from a speaker and the audience participates through an audio channel. A shared whiteboard program may also be used. The growing use of MBone conferences demonstrates the need to connect large groups of people spanning across continents.

Our group has developed an application called Forum that broadcasts video-based presentations to audiences sitting at their workstations distributed across a network [Isaacs, et.al, 1994]. Like MBone conferences, Forum multicasts live audio, video, and slides from a presenter, and it provides a mechanism for audiences to ask questions. Unlike the MBone, Forum enables the speaker to poll the audience anonymously, and it allows audience members to send written comments to the speaker and to each other. Forum is also distinguished by its emphasis on making the user interface easy and compelling.

As Forum and similar applications are used to improve communication and training, it is important to understand how technology-supported arenas differ from lecture halls and classrooms. Such an understanding will help us improve the design of these applications and it will help us set realistic expectations for successful use of the technology. To compare these two environments, we conducted a study comparing a set of presentations that were given both over Forum and face-to-face. In this paper, we describe the Forum application, how we conducted the study, and what we found. We also discuss ways to improve the design of distributed presentation technology and how people might adapt presentations to most effectively use the technology.

FORUM

Figure 1 shows the interface used by Forum audiences. Participants watch the speaker's image in the left portion of the main (upper left) window and view the speaker's slides in the window below. Pre-recorded videos also can be shown in the video region, replacing the speaker's image. By default, the slide window tracks the speaker's slides, but audiences can also move through the slides independently by clicking on the thumbnails to the right of the main slide. They can also get back in sync at any time. The speaker and each audience member can type or draw on the slides (shown in Figure 1), but only the speaker's marks are seen by everyone. Audiences can save the slides (with or without annotations) for later reference.

Audience members interact with the speaker through a live poll, a spoken question, or a written comment. Speakers use the poll meter to ask the audience multiple-choice questions (shown as a yes/no poll in Figure 1). As each person clicks on a choice, everyone sees the poll meter update. To ask a question, audience members press the Ask to Speak button, which puts their name in the In Line to Speak list. When called on, they press the Speak button and speak into their microphones. Everyone else hears the question and sees a small picture of the person that appears while she speaks. In Figure 1, for example, Ellen Isaacs is asking a question and three others are waiting to speak. Audience members can also send the speaker written comments. The Comments button brings up a window in which they can type a note

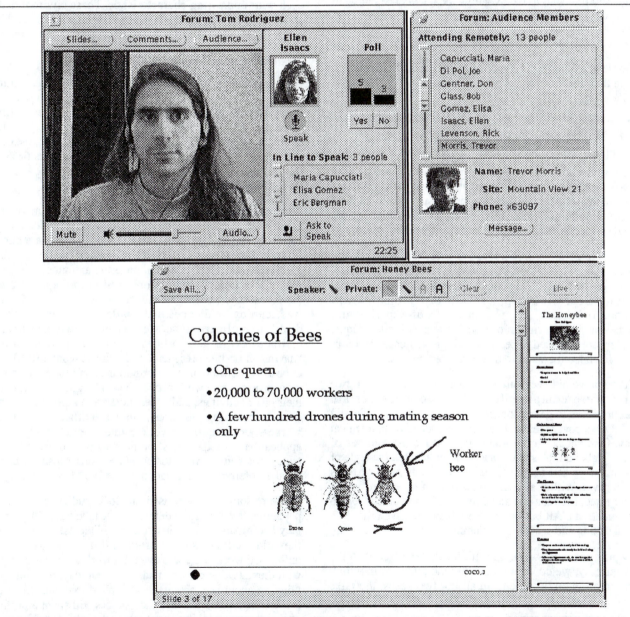

FIGURE 1. Forum's audience interface. Audience members watch the video and interact with the speaker in the upper left window. They view the slides in the lower window, using the thumbnails to view slides independently of the speaker. The Audience Window shows a list of attendees and allows audience members to send each other text messages.

and then send it in. Only the speaker sees these comments, unless she chooses to present one to the audience as a slide. Everyone can see an audience list (upper right window), and audience members can send each other text messages. (See Isaacs et al., 1994, for a complete description of Forum.)

To give a presentation, speakers sit in front of a workstation and speak to a camera. Only one person, a member of the Forum development team, is usually in the room with the speaker to help manage any unexpected problems.

METHOD

Seven pairs of presentations were observed over the course of about eight months, once given over Forum and once in either a lecture hall or conference room. All seven were originally prepared to be given to a local audience but were also given over Forum at our request.

The data consist of the following:

- Surveys asking both audience members and speakers to rate their reactions to the talks. (Audience survey response rate averaged 42% per talk.)
- Videotapes of all talks, in most cases from both the speakers' and the audiences' perspectives.[1]
- Computer logs of the speakers' and audiences' Forum use.

Because the adoption of Forum will be determined by how people perceive it, we wanted to understand their reactions. For this reason, we rely heavily on the surveys. Since user perceptions do not necessarily reflect actual differences, we use the videotapes and logs to help explain users' reactions.

We attempted to observe a range of presentation topics and styles. Presentation topics included: a company executive's vision of the future, a new human resources policy, a career development workshop, a review of a project team's accomplishments, two presentations of experimental studies, and a description of a design evaluation technique. Three were designed to be given to large audiences and four were intended for small to medium-sized groups.

Although we did everything we could to make sure the two talks were comparable, we had to make certain compromises to gain cooperation with users. The talks did not always cover identical material explained in an identical way. One speaker combined material from a number of related talks on the subject and another adapted the style of presentation to Forum's capabilities. Four pairs of talks were given within a week or so of each other, but in the other three, between one and three months separated the two presentations. All but one of the talks were given first in a local setting and second over Forum.

Because we were working with real talks offered to the community, we could not control who attended in which presentation environment. In all but two cases, the talks

---

1 We were unable to videotape one of the seven pairs of talks, and we could not collect surveys from a different pair of talks, so the survey and videotape analyses are each based on six pairs of talks.

were announced without indicating they would be presented in another format. Since users usually did not choose between formats, presentation environment played a role only to the extent that it influenced a person's decision to attend. Nonetheless, since audiences were self-selected, any results must be attributed to a combination of the technology and the type of people it attracted, not the technology alone.

FINDINGS

Based on the videos, surveys, and logs, we found that Forum talks differed from local talks in size, audience attentiveness, perceived quality of presentation, use of visual materials, perceived quality of interaction, and degree of audience awareness. Although some of these differences were subtle, they gave rise to two distinct user experiences, each with interesting characteristics. These environments appealed to speakers and audiences to different degrees. The following sections describe these differences in more detail.

Attendance and attention

Analyses of variance showed that Forum attracted larger audiences than did local talks, but the audiences were more likely to split their attention with other activities during Forum talks. Forum audiences averaged 141 people, compared with 60 for local talks ($F(1,5) = 22.42$, $p<.01$). Audiences reported paying attention to Forum talks 65% of the time, compared with 84% of the time in local talks ($F(1,409) = 7.07$, $p<.001$).

These results are not surprising because Forum is one application in an environment where other activities on the computer and in the office compete for the user's attention. First, Forum makes it easy for audiences to attend talks. Audiences received an e-mail message announcing a talk, they clicked on an "invitation," and Forum appeared on their screens just before the talk began. Second, once Forum was running, audiences had available to them all their desktop applications as well as other resources (and distractions) in their offices. As one user commented, "having an on-line presentation like this is wonderful! I can choose to focus on areas of importance and tune out to get more worth-while work accomplished." When asked about their other activities, audiences most often reported reading e-mail, answering the phone, talking to officemates or co-workers, or doing other work. Some reported using other applications to help understand the material, for example by taking notes in a text editor or by browsing Mosaic when a speaker referred to information on the World Wide Web.

We were somewhat surprised that local audiences reported paying far less than full attention to the talk, especially since they had relatively few distractions. When asked what else they did, audiences most often said they thought about unfinished work and some even did work they had brought with them. (Other common answers were daydreaming and thinking about how the material applied to them.) So even those who decided to leave their offices still spent a portion of their energy thinking about the work they left behind.

Forum talks also tended to run longer than local talks, 1:10:31 vs. 1:01:33, but this difference was not significant ($F(1,5) = 4.86$, $p<.079$). Speakers did report losing track of time during Forum talks, perhaps because they could not see when audience members became restless or left as the scheduled end time approached. Also, some speakers said the Forum clock was not prominent enough.

Presentation Quality

On the whole, both speakers and audiences thought speakers presented the material better in local settings than they did over Forum. On a five-point scale (with 5 high), audiences rated the quality of the speaker's presentation at 4.19 in local talks vs. 3.95 over Forum. This difference was significant, ($F(1, 411) = 6.40$, $p<.05$). This effect may in fact be larger because most speakers gave their local talk first, and so had more practice for their Forum talk.

Speakers perceived a wider difference in their performance than the audience, rating themselves 4.42 in local settings and 3.42 over Forum ($F(1,5) = 15.00$, $p<.05$). Although we cannot compare speakers' and audiences' ratings statistically, it is interesting to note that relative to audiences' ratings, speakers appeared to overestimate their performance in local settings and underestimate their performance over Forum.

Audiences had a slight tendency to find the material more interesting in a local setting (4.21) than over Forum (4.17) ($F(1,415) =5.37$, $p<.05$), but this difference may be due to the fact that people who made the effort to attend in person were more likely to be interested in the material. It is in fact encouraging that the difference in interest levels between Forum and local audiences was so small.

Speakers also felt that certain features of Forum enhanced their presentations compared with local talks, but other features detracted from their talks. Speakers said Forum enhanced their presentation more than the equipment available in local talks did, 3.00 vs. 2.00 ($F(1,5) = 7.50$, $p<.05$). They said it did so by providing the ability to annotate their slides, anonymously poll the audience, easily show videos, and reach a wider audience. However, they felt Forum detracted more than a local talk environment, 3.17

vs. 1.83 ($F(1,5) = 10.00$, $p<.05$) because it did not provide an image of the audience and it forced them to stay seated. "I like to move," one explained, "and my voice gets more interesting when I speak to a large room with lots of people." Some were distracted by the video preview of themselves and others said they had trouble referring to their notes, which they had to keep on their laps because Forum did not provide on-line support for speakers' notes.

Visual Materials

Forum provided more flexibility in the use of visual materials, which both speakers and audiences appreciated. During the Forum presentations, all speakers used slides, three presented pre-recorded videos, and one presented a live demo of a hand-held device. During the local talk, the live demo was replaced with a videotape of the demo because there was no equipment to project the small device onto a big screen. Audience members especially liked having their own copy of the slides, which they could peruse at their own pace, annotate, and save for later reference. Although audiences praised the availability of visual materials, many complained about video quality. They were distracted by the slow video frame rate (4 fps) and lack of audio-video synchronization.

Audiences said they thought the visual materials helped them understand the information better over Forum (4.17) than in a local setting (3.58) ($F(1,376) = 15.44$, $p<.001$). This finding may be due to the fact that the slides window took up a large portion of the screen (see Figure 2) and that each person received a close-up and unobstructed view. The logs show that an average of 68% of each audience took advantage of its ability to view slides independently of the speaker, indicating that participants were using the slides to understand the material at their own pace. In addition, an average of 35% of each audience saved their own on-line copy of the slides. Audiences reported taking fewer notes during Forum presentations ($F(1,406) = 28.02$, $p<.001$), again probably because the slides were made available. Only two speakers handed out their slides to local audiences, and one of those did not provide a complete set.

 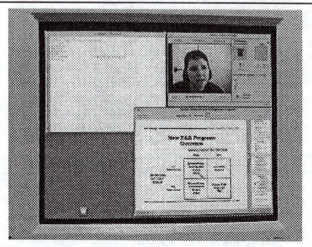

FIGURE 2. The same speaker giving a talk in a local setting (left) and over Forum (right). In both cases, she points to the upper left box on the slide, in the local talk by gesturing on the projector and over Forum by using the pointer.

The videotapes showed that speakers pointed to the slides during Forum talks an average of 26.8 times per presentation compared with 20.8 for local talks, but this difference was not significant $(F(1,5) = 0.51, ns)$. (Figure 2 shows the same speaker pointing in each setting.) They also drew on the slides more often during Forum talks, 15.8 vs. 0.3 times per talk, but this difference was not significant because of large variances $(F(1,5) = 3.83, p<.11)$.

Interactivity

So far the discussion has focused on the material flowing from the speaker to the audience. However, the biggest difference between Forum talks and local talks was the quality of interaction with and among the audience. Forum's polling and questions mechanisms supported a similar *quantity* of interaction as local talks, but the *quality* of that interaction was seen as reduced. Forum did not provide sufficient support for the subtle cues that speakers use to monitor and adjust to audiences.

The videotapes showed that significantly more spoken questions were asked in local talks (19.5) than Forum talks (2.8) $(F(1,5) =12.05, p<.05)$. But when written questions are included (0.0 in local settings vs. 9.3 over Forum), there was no significant difference in the number of questions asked in local talks (19.5) vs. over Forum (12.2), $(F(1,5) = 1.60, ns)$. In addition, more poll questions were asked over Forum (2.2) than in face-to-face talks (0.8) $(F(1,5) = 7.27, p<.05)$.

However, interactions in local talks appeared to be richer in a number of ways. First, people who asked questions were more likely to ask follow-up questions or clarifications, which created more of a sense of a discussion rather than a lecture. Audiences asked follow-up questions 8.7 times per local talk compared with only 1.0 times per Forum talk. This difference was significant $(F(1,5)=5.66, p<.05)$.

Second, speakers were more likely in local settings to incorporate other people into their presentation, most commonly drawing out audience members who had also worked on the material being discussed. During the six talks, there were 48 instances when other knowledgeable people volunteered information to follow up on the speaker's point or the speaker asked an audience member to help answer a question. One talk was in fact given by two people and it had a very different feel in the two settings. Over Forum, each speaker gave his part of the presentation separately, but in the conference room, each called out comments throughout the other person's section. Although it was possible over Forum for attendees to volunteer information or the speaker to draw out information, it happened only three times. Two of those times occurred through the written comments, which have a built-in delay. It was also more risky for a speaker to ask someone in the audience list to help them answer a question because they had no guarantee that the person was paying attention or that they had a microphone that would enable them to respond.

When asked to rate how well speakers handled questions, audiences rated Forum speakers at 3.93, compared with 4.29 for local talks $(F(1,400) = 10.21, p<.01)$. Speakers also rated

themselves lower for Forum talks (2.67) than for local talks (4.17) $(F(1,5) = 45.0, p<.001)$. In this case, speakers appeared to underestimate their performance in both settings relative to audiences' perceptions, but they thought they did a much poorer job of answering questions over Forum than audiences did. One speaker mentioned a common problem in her response, "you don't get the immediate feedback of seeing the employee you are answering a question for. You can't see them shake their head, that yes they understand or no they need more info. I sometimes felt like I repeated myself too many times trying to answer the question because I could not see if I had made my point." Speakers also rated the quality of the questions poorer over Forum than in local talks (3.00 vs. 4.25), but this difference was not significant $(F(1,5) = 4.75, ns)$.

Audience Awareness

Just as people felt it was difficult for the speaker and audience to interact over Forum, they also felt a weaker sense of the audience's reaction during a presentation. Much awareness of an audience comes from knowing who is there and from the questions they ask, but there is another more subtle aspect that comes from seeing the audience, hearing its spontaneous reactions, and chatting with people before and after a talk.

The videotapes showed that the local talks included an average of 16.3 instances of spontaneous laughter or chuckling among the audience and an average of 1.2 instances of applause. Since Forum had no such mechanisms, there were no instances of either laughter or applause. Speakers said the lack of such feedback gave Forum a more sterile feeling. The videotapes showed that audience members sometimes did in fact applaud and laugh in their offices, but they had no means of conveying those emotions to the speaker or the rest of the audience. Some audience members were frustrated by the lack of such a channel and some went so far as to send in written comments explicitly praising the talk and thanking the speaker for doing a good job.

Both audiences and speakers reported having a lower sense of the audience's reaction to the talk. Audiences rated their awareness of each other at 3.08 at local talks but only 2.30 at Forum talks $(F(1,397) = 57.91, p<.001)$. Speakers felt an even more dramatic gap, rating audience awareness at 4.50 at local talks and 1.83 over Forum $(F(1,5) = 64.00, p<.001)$.

We also asked audiences to indicate how much they interacted with others before, during, or after talks in a way that related to the talk. Forum audiences rated their degree of interaction at 1.69 compared with 2.14 in local talks $(F(1,398) = 38.59, p<.001)$. We found it interesting that although audiences interacted with each other more at local talks, they did not report interacting much in either setting. Speakers also said they interacted with audience members before and after the talk more often in a local setting than over Forum, 3.67 vs. 1.33 $(F(1,5), p<.01)$.

The videotapes showed that audience interaction at the local talks consisted of participants chatting with each other before and after the talks as well as whispering to their neighbors during the talk. Some talks included more novel types of interaction. In one small-group presentation, each person briefly introduced himself or herself. In another talk, the person who had sponsored the presentation gave a pitch for her group's services before the talk began and then handed out material at the end. At the end of two other talks, the speaker passed around a prototype of a hand-held device and people discussed it as they passed it around.

We asked Forum audiences how they had interacted and they reported watching with others, talking with officemates or passersby, chatting in the halls with others who had also watched, having lunchtime discussions about the talk, and sending each other text messages. The logs indicated that an average of 13% of each audience sent an average of 2.9 messages per person. Although Forum audience members heard each others' questions and sometimes interacted with other attendees both near and far, those isolated interactions did not replace the sense of community that is conveyed among a large group in a single room.

Preferences

This combination of advantages and disadvantages led speakers to prefer giving talks in local settings, where they could see and respond to the audience. But audiences preferred attending over Forum, where they could receive the information while getting other work done.

When speakers were asked which format they preferred, three chose a local setting, two said it depended on the material, and one chose Forum. The one who chose Forum did not actually prefer using it; instead she felt it was most important to reach a broad audience. (She described a new human resources policy, which had to be communicated to everyone in the company.) Speakers also said they enjoyed giving their talks in a local setting (4.25) more than they did over Forum (2.92) (F(1,5) = 10.00, p<.05). As one speaker reported, "face-to-face is still the most natural communication form and I feel that I can express myself better as a speaker when I can use my entire body and can see the audience." Nonetheless, most speakers made a point of saying that they enjoyed using Forum and appreciated being able to reach a wider audience. One explained, "I prefer face-to-face for giving a talk, but I would happily use Forum if I wanted to reach an audience who I didn't think would attend a real talk."

Audiences, on the other hand, strongly preferred Forum. When we asked audience members how they would attend next time, we found that Forum audiences preferred Forum by a wider margin than local audiences preferred local talks (χ^2 = 7.13, df = 1, p<.01). Ninety percent of Forum audiences preferred Forum, whereas 72% of local audiences said they would attend again in person.

When we asked audiences to evaluate the extent to which the talk was worth the time and energy they spent to attend, they rated Forum and local talks almost equally (4.01 vs. 4.04) (F(1,410) = 2.45, ns). Since local talks take more time and energy to attend, this result shows that people who attended in person felt that the extra effort was worthwhile.

Those who preferred Forum offered many explanations. They liked the convenience of attending, the ability to do other work, the ability to tune in and out depending on their interest level, and the lack of travel time. Others also mentioned always getting "the best seat in the house," being able to talk with others, being able to leave talks politely, and being accessible to co-workers during the talk. Those who preferred local talks liked being able to interact and influence the discussion more easily, concentrate better, and see and hear the speaker with higher fidelity.

Effects of Presentation Topic

Since we did not collect examples of previously determined categories of presentations, we have so far factored out the effects of topic and examined the differences due solely to presentation environment. However, when we focused on differences between talks, some notable differences appeared. Perhaps the most effective Forum talks were the presentation about the new human resources policy (HR talk) and the executive's vision of the future (vision talk). The least effective Forum talk was one discussing an experimental study (experimental study talk).

The HR policy talk provided information about a relatively important but mundane topic. It was the only one at which the Forum audience thought the material was more interesting (3.42 vs. 3.00) and presented better (3.73 vs. 3.54) than the local audience did (although these differences were not significant). It was also the only one in which the Forum audience thought they had made significantly better use of their time (3.46 vs. 2.62, F(1,38) = 7.77, p <.01). When asked which way they would attend in the future, even the local audience chose Forum by a 2:1 margin. Audience members commented that Forum was ideal for this kind of talk because they could get the information without devoting an uninterrupted hour to it.

The vision talk made good use of Forum because it reached a wide audience (250 people), whereas attendance at the local talk was restricted due to room size. People also appeared to feel more comfortable writing rather than speaking their questions to a high-level executive; he handled 24 (written) questions over Forum but only five (spoken) in person. Unlike at other pairs of talks, the Forum and local audiences' reported no significant difference in their sense of the audience (2.73 vs. 2.67), and the amount they interacted with each other (1.65 vs. 1.68). In most other talks, local audiences rated themselves significantly higher on these measures than Forum audiences.

At the other extreme, the experimental study talk did not seem to translate well to Forum. This talk was designed for a lunchtime discussion series among a small group of colleagues. Compared with the Forum audience, the local audience of this talk found the material significantly more interesting (4.44 vs. 3.33) (F(1,26) = 9.16, p<.01), thought the presenter did a much better job of presenting it (4.11 vs. 2.97) (F(1,6) = 17.49, p<.001), and was more satisfied that their time had been well spent (4.56 vs. 2.88) (F(1,25) = 19.30, p<.001). Audience members had a much greater awareness of each other (4.00 vs. 1.70) (F(1,25) = 54.94, p<.001) and interacted with each other to a far greater degree than they did at the Forum talk (3.22 vs. 1.61) (F (1,26) = 16.34, p<.001), probably because many of its members knew each other. Far more questions were asked (44 vs. 10) and there were more instances of laughter in this talk than any other (35).

This talk was also given by two presenters who frequently interacted and added to each other's explanations. It was very difficult for the speakers to translate this type of intimate, highly interactive, and informal presentation to Forum, and as a result, they strongly preferred the local setting over Forum (rating them 5 vs. 1). As one audience member said, "[This talk] series is as much of a social event as it is a technical talk. I would not want to miss the social aspects of a room full of interesting people."

DISCUSSION

These findings indicate that remote presentations reach a wide audience, enhance the effectiveness of visual materials, and provide for the basic exchange of questions and answers, but they do not support subtle, ongoing interaction cues among speakers and audiences. The lack of these cues detracts from a speaker's ability to present the material, answer questions effectively, and adapt to the audience's reactions. It also generally reduces satisfaction with the experience of giving a presentation.

Although audiences also perceive lower quality interaction, they still prefer attending remotely because of the convenience, the ability to get other work done during the presentation, and the ability to easily view and save the slides. They generally believe they learn the material effectively and much more efficiently. As one person explained, "I actually liked the ability to look over the slides at my own pace, to continue doing work if I so choose. Great time savings and I got the information I needed." As a result, people attend a wider range of presentations remotely than they do when forced to attend in person. Several Forum audience members mentioned that they would not have attended the presentation if they had been required to travel to a conference room.

It is worth considering how Forum and other related tools might be designed to better support presentations. On the other hand, it is also reasonable to discuss how speakers might learn to adapt their presentations to better overcome the limitations of reaching remote audiences.

Improving the Design

The most obvious difference between Forum and local talks is that the audience is not visible in Forum talks. This difference is probably a major reason for the reduced level of interaction and audience awareness. To improve these factors, video of audience members could be provided, perhaps as a matrix of images that update slowly and, in the case of large talks, rotate through the audience. At the very least, video of the audience member asking a question could be shown. Speakers could then use visual cues to help determine whether they had answered questions adequately.

When developing Forum, we decided not to provide video of the audience because of limited network bandwidth and a lack of commonly available video equipment. We were concerned that current networks would be overloaded if video were transmitted from even a portion of the audience. We also did not want to require audience members to have cameras because it would radically reduce the size of available audiences. We expect that network bandwidth will increase over time, however, and more people will have cameras built into their computers. At that point, it will be more practical to provide video of audience members.

Another approach is to allow a small group to attend locally as the talk is transmitted to a remote audience. We are in the process of building a studio to do just this. The local audience should provide a sense of at least part of the audience and it should create a more intimate atmosphere even when many people are attending. Speakers will be able to stand and move around, which some said would help them better express enthusiasm. And it will facilitate tight interaction among multiple presenters. One potential pitfall is that the remote audience may feel less involved in the talk. Care must be taken to carefully integrate the two audiences, especially during interactive portions of talks.

Less radical changes may also be possible. We found that much of Forum's sterile atmosphere was due to the lack of ongoing audio information from the audience. The lack of laughter and applause were notably missing, as were other spontaneous reactions, such as groaning at a bad joke and quick "quips" from the audience. Forum did not allow each audience member to have an "open mic" because doing so would have caused an unacceptable level of audio feedback. If improvements were made in echo cancellation and silence detection software, it would be possible to allow a more open environment in which spontaneous noises could be captured. Barring that, we could provide more overt mechanisms to convey applause and possibly amusement, although it would be important to make the interface natural and the sounds realistic. We might also enable audience members who asked a question to easily indicate when they were satisfied with the answer.

Another useful feature would be support for on-line speaker's notes. As mentioned, speakers often looked at notes on their laps because they were not available on line. Speakers could focus more of their attention on the material if they could prepare annotations that only they could view.

Adapting to the Tool

Even if we could provide an ideal remote presentation tool, speakers would still have to learn to tailor their presentations to use it effectively. All seven speakers in this study gave their first-ever Forum talk, and only four had previously attended as a Forum audience member. Not surprisingly, all of them felt they would do a better job next time. Most of their proposed modifications were intended to improve interaction with the audience. In addition, we felt that over time speakers would learn to take advantage of the fact that audiences were watching from their desktops, with access to other desktop tools.

Many speakers said that if they gave another presentation, they would plan to use the poll more often or ask provocative questions. Two speakers, including the one who did adapt his presentation to Forum, posed a question and asked audience members to respond with written comments. After responses arrived, they read a series of answers to the audience. This technique seemed to work quite well because it gave them a sense of their audience, got the audience used to responding, and kept everyone interested.

One person said he would ask audience members to identify themselves, which would give him a better idea of who was attending and make the audience more comfortable using their microphones. A few said they would look at the camera more, hoping to create more sense of intimacy for the audience. And one person said he would put less on his slides to create more drama as he built to a conclusion. He disliked the audience peeking ahead at his punch lines.

Although none of the speakers mentioned using other computer desktop resources, we think this is a promising possibility, especially because it makes a virtue of the fact that audiences often split their attention during talks. During a presentation, audiences could be encouraged to explore more detailed on-line information, for example by using Mosaic. Doing so may encourage participants to ask more informed questions that raise the level of discussion. During a technical discussion or a training session, participants could be asked to retrieve examples from their own work to focus the discussion on issues most relevant to them.

Another approach is to keep users' attention focused by making Forum talks more compelling. Just as broadcast television uses attractive imagery and clever composition to capture viewers' attention, Forum speakers might draw in audiences by creating presentations with high production values. Of course, doing so requires more effort, but such investment may be appropriate for certain presentations.

CONCLUSIONS

This study gives insight into the kind of environment we can provide through media-supported distributed presentation tools. Distributed presentations can effectively convey information to large groups of people who might not otherwise participate. Visual materials appear to be more important in distributed presentations, and attention should be paid to making those materials useful and engaging. Attempts should be made to exploit audiences' ability to use other desktop tools and to pace themselves. A distributed environment can provide adequate support for asking and answering questions, but it is not ideal for encouraging active audience participation or providing fine-grained feedback from the audience.

With improved technology and designs, audience interaction will become easier and more natural, but even today, companies and universities can successfully use video-based presentation tools to supplement face-to-face instructional environments and text-based on-line tools. Companies can keep more people informed and

synchronized in their efforts, and schools can reach out to those who cannot attend in person. It is important to recognize, however, that remote presentation environments have their advantages and disadvantages. Presenters should use the environment that best matches their purpose and adapt their presentation accordingly.

ACKNOWLEDGEMENTS

We would like to thank the rest of the Collaborative Computing group at SunSoft for their help in developing Forum and setting up the presentations for this study. We thank Tom Pratt for helping set up a connection between two Sun campuses, making Forum accessible to more people. We thank our Information Resources department for their cooperation. We are very grateful to the audiences who attended the talks, filled out the surveys and stuck with us when we "experienced technical difficulties." And we especially thank the speakers who were willing to broadcast low-resolution images of themselves to hundreds of people, and then gave us their insightful reactions and suggestions.

REFERENCES

1. Isaacs, E.A., T. Morris, and T.K. Rodriguez, A Forum For Supporting Interactive Presentations to Distributed Audiences, *Proceedings of the Conference on Computer-Supported Cooperative Work (CSCW '94)*, October, 1994, Chapel Hill, NC, pp. 405-416.

2. Liberman, K. and J.L. Rich, Lotus Notes Databases: The Foundation of a Virtual Library, *Database*, 1993, 16(3), pp. 33-46.

3. Macedonia, M.R. and D.P. Brutzman, MBone Provides Audio and Video Across the Internet, *IEEE Computer*, April, 1994, pp. 30-36.

4. Olson, M.H. and S. A. Bly, The Portland Experience: A Report on a Distributed Research Group, *International Journal of Man[sic]-Machine Studies*, 1991, 34 (2), pp. 211-228.

5. Peters, T., *Thriving on Chaos: Handbook for a Management Revolution*, 1987, New York: A. A. Knopf, Inc.

6. Smith, A.E., *Innovative Employee Communication: New Approaches to Improving Trust, Teamwork, and Performance*, 1991, Prentice Hall: Englewood Cliffs, NJ.

7. Sproull, L. and S. Kiesler, Reducing Social Context Cues: Electronic Mail in Organizational Communication, in *Computer-Supported Cooperative Work: A Book of Readings*, I. Greif (Ed.), 1988, Morgan Kaufmann: San Mateo, CA, pp. 683-712.

8. Tang, J.C. and M. Rua, Montage: Providing Teleproximity for Distributed Groups, *Proceedings of the Conference on Computer Human Interaction (CHI '94)*, April, 1994, Boston, MA, pp. 37-43.

What Mix of Video and Audio is Useful for Small Groups Doing Remote Real-time Design Work?

Judith S. Olson, Gary M. Olson,
Collaboratory for Research on Electronic Work (CREW)
The University of Michigan
701 Tappan Street
Ann Arbor, MI 48109-1234
(313) 747-4948   olsons@crew.umich.edu

David K. Meader
Department of Management Information Systems
Karl Eller Graduate School of Management
430 McClelland Hall
University of Arizona
Tucson, AZ 85721
(602) 621-3600   dmeader@bpa.arizona.edu

ABSTRACT

This study reports the second in a series of related studies of the ways in which small groups work together, and the effects of various kinds of technology support. In this study groups of three people worked for an hour and a half designing an Automated Post Office. Our previous work showed that people doing this task produced higher quality designs when they were able to use a shared-editor to support their emerging design. This study compares the same kinds of groups now working at a distance, connected to each other both by this shared editor and either with high-quality stereo audio or the same audio plus high-quality video. The video was arranged so that people made eye contact and spatial relations were preserved, allowing people to have a sense of who was doing what in a way similar to that in face-to-face work. Results showed that with video, work was as good in quality as that face-to face; with audio only, the quality of the work suffered a small but significant amount. When working at a distance, however, groups spent more time clarifying to each other and talking longer about how to manage their work. Furthermore, groups rated the audio-only condition as having a lower discussion quality, and reported more difficulty communicating   Perceptions suffer without video, and work is accomplished in slightly different manner, but the quality of work suffers very little.

KEYWORDS

Group support system, remote work, concurrent editing, small group behavior, desktop video.

INTRODUCTION

Meetings are a central component of collaborative work in organizations. They can range from formal meetings that are scheduled in advance with a pre-defined agenda to informal, ad hoc meetings where the members of a work group get together to work interactively on some problem on the spur of the moment. Traditionally, all forms of what we think of as meetings took place face-to-face, in

meeting rooms, commons areas, and lunchrooms. But as we all know, groups no longer need to meet in the same location; new technologies are allowing us to relax the constraint of co-location. Modern telecommunications make available an interesting array of options such as teleconferences, video conferences, and synchronous interactions over computer networks. These alternatives to face-to-face interactions have distinct properties, and have not, in any real sense, replaced what it is possible to do in face-to-face interactions. However, they offer organizations and communities of practice additional flexibility in how teams are structured and deployed. To do this effectively, we need to learn more about both the opportunities and constraints offered by these new modes of synchronous interaction.

This paper grows out of a line of research whose aim is to understand these issues. We have focused on synchronous interactions among small teams working on design problems. We chose small teams because they are such a widespread and enduring form for working on projects in organizations, and are a hallmark of such new organizational forms as adhocracies [1,28]. We chose the task of design because it is a representative ill-structured problem solving task [22,25] that interleaves many subprocesses such as planning, creativity, decision-making and cognitive conflict (dimensions in McGrath's [12] task taxonomy). Design is also usually a collaborative task [e.g., 11].

In this line of research, we began with field studies of groups in real organizations doing software system design [17]. Our goal was to understand better what small group behavior was like for design tasks, and what opportunities existed for supporting this activity with technology. We learned much about both, leading to our developing a simple shared editor called ShrEdit [13] that we felt had properties that would be useful for groups doing these kinds of tasks. It provided the members of a group with an electronic workspace in which they could all enter and edit their ideas. We took ShrEdit into the laboratory to assess this. We created a design task that elicited design behavior similar to what we had seen in the field [19], and compared real groups of three people (i.e., people who knew each other and had worked together before they came into the laboratory) using ShrEdit with groups working with the more traditional meeting room media of whiteboard, and paper and pencil. The groups using ShrEdit produced higher quality designs, though they were somewhat less satisfied with their work. To our surprise, they produced higher quality designs by exploring fewer ideas rather than more.

The next step in this research, reported in this paper, was to study comparable groups working with ShrEdit but no longer physically co-located. In designing ShrEdit we assumed that groups would have other communication channels available to them. In a face-to-face setting, of course, the groups can talk and gesture in their usual interactive ways, and indeed, the groups in our studies engaged in extensive discussion while using ShrEdit as a workspace to capture and revise their emerging ideas. So in looking at the use of ShrEdit by distributed groups we provided them with other communication channels for talking and interacting.

We decided to provide communication for our groups that was as ideal as we could make it given their distributed set-up. We wanted a baseline for later studies that looked at other kinds of communication, such as digital desktop video. In the present study we focused on how groups of three performed when they have a shared workspace tool and *ideal* remote communication.

Many investigators have pointed out that shared workspace tools are important sources of coordination in collaborative problem solving tasks [4,7,27,30]. We know from our prior work [19] that ShrEdit is an effective shared workspace tool for the kind of design task we have used.

High quality audio is also very important to remote synchronous work [6,20,27]. So we had half our groups work with high quality audio in addition to the shared workspace. Our audio was full duplex, directional for both input and output, and of far better quality than found in teleconferencing or most commercial video conferencing systems.

More controversial is whether video adds significant value for groups doing distributed problem solving. While the research record is quite mixed [5,27], many theories [e.g., 3,23,24,29] and most people's intuitions are that video should add substantial value to such work. Thus, the other half of our groups had our good quality audio plus high quality analog video connections to each of their colleagues. The video was arranged in an optimal fashion to create the feeling of sitting around a table with one's colleagues, with the shared workspace in the center. We took more care than usual to create what we felt would be the best possible video conferencing set-up.

We were interested in how these video/audio groups using ShrEdit would compare to face-to-face groups using ShrEdit from our earlier study [19] on a range of measures: quality of the work product, satisfaction, and characteristics of the group process. We also compared these audio/video groups to the audio-only groups to assess the added effect of the video. What distinguishes our study from previous investigations is the use of an established workspace tool of known value for sharing the work, and the care we took to ensure that the audio and video were of the highest quality we could get with present communication technology.

Another function of this study was to establish a baseline from which we could conduct later studies of a variety of less-than-optimal communication technologies. Multimedia desktop conferencing systems that run over the Internet or ISDN lines are generally quite constrained in the quality of the audio and video they can provide. With our baseline data we can assess these situations in future investigations.

METHOD
Subjects
The subjects were 36 existing groups of three professionals. All of the groups consisted of three MBAs[1] each enrolled in the Michigan Business School. In all groups, the members had worked together before in class or work projects and all knew at least one Macintosh or Windows application.

The Setting and its Communication Support
The group members were seated in three separate rooms made to look like offices, all part of the Collaboration Technology Suite (CTS) [16]. As shown in Figure 1, in these rooms, a workstation with a large screen was centered on a desk, with two 13" video monitors on each side of the screen. A camera and microphone were mounted on each video monitor, with the camera placed at the center of the top of the monitor so that when the participants faced each other, they appeared to each other to be making eye contact.[2] Furthermore, when the other two remote participants were facing each other, their images projected to the receiving participant made them look as if they were looking at each other.

The microphones and speakers were similarly situated to either side of the central screen, corresponding to the person shown on the video screen. They were open full-duplex channels that additionally projected a sense of spatial location. Indeed, in the audio-only condition, group participants moved their heads to face the speaker boxes of the people they were addressing. The audio condition used the identical microphones and speakers of the video condition; the only difference was that the video monitors were turned off.

Technology Support
The technology used here was one we designed and built, called "ShrEdit," which stands for *Shared Editor* [13] For reasons outlined in the previous paper [19], ShrEdit is a simple text editor which allows all participants to view and change the same simple text document, with all participants being able to type simultaneously within one character position of each other. Although the individuals' views of

---

[1]Ninety percent of the MBAs at Michigan have significant work experience before coming back to school. These are professionals with practical group experience.

[2]Eye contact was not perfect. Participants reported that the other person appeared to be looking at their throat when they looked into their eyes.

Figure 1. A diagram of the audio and video configuration in our remotely connected offices.

the document are normally independent (each can scroll to a different part of the document or arrange the windows on the screen in a unique way), the views can also be locked together if the discussion calls for it. ShrEdit does not support layout and font features that make the document appear in final format; this was originally an expedient for development, but in the end proved an asset [4] in that such formatting operations are not the kinds of things a group should spend its group time doing. ShrEdit is a simple editor, learnable by our subjects within a 15 minute period, during which we focus on the aspects of the interface that are unique to the fact that a group is using this simultaneously, such as the fact that people bump into each other and can find where each other is working in a long document.

Task

We chose to study the task of group design, that of early requirements definition, because it is both important and interesting. Our observations of design in the field show a seemingly erratic sequence of these subtasks, with associations, implications and evaluations taking the group from one topic to the next and back. But by categorizing details of the activity in the episodes into core design activities and coordination activities, we found interesting, systematic patterns of behavior [18]. In particular, groups engaged in periods of design activity, punctuated by coordination episodes. The behavior within each kind of activity showed strong regularities, and there was little intermixing. Once designing, the group designed; when they needed coordination, they did that for a time, and then jumped back into the regular design pattern.

In this study, all groups were instructed to draft the initial requirements for an Automatic Post Office (APO), a collection of postal services offered through a stand-alone device similar to an ATM for which a prototype could be built by their fictitious company of 30 people in a year. They were instructed to determine the core services they would offer, some of the required equipment, the rough cost/benefit analysis, and a list of things they would like to investigate before the next time they would (hypothetically) meet. They were given 1-1/2 hours to complete the assignment, producing meeting notes that could be read by a (fictitious) additional group member who could not attend that day's meeting.

Procedure

The subjects came to the CTS for a single three-hour period, broken into two 1-1/2 hour portions with a 15 minute break in between. In the first half, the subjects filled out background questionnaires and permission forms, and learned to use ShrEdit in a 20 minute training session. The instructions demonstrated the system's capabilities and how to control them, but did not prescribe how to use these features to support work. After the training session, the group members were taken to their individual remote offices. After a few minutes of acclimation to set camera angles and audio levels, they were asked to solve two small problems, of 20 minutes each. These tasks served both to allow the subjects to warm up to the task situation and to learn the software and adjust to the software's capability for simultaneous editing. Following a 15 minute break, the groups did the APO task in a single 1-1/2 hour sitting. All groups filled out a questionnaire after the session.

In total, we ran 39 groups, 37 of which survived the full 3 hours without a fire alarm going off (an unrelated event to the conduct of the experiment). One group was eliminated because they spent nearly half their time digressing. Thirty six groups were included in the final analysis. Eighteen of the groups used the full video and audio technologies to communicate; 18 had only the audio. All groups used ShrEdit.

Study Design

Groups were recruited en masse through various MBA classes, and encouraged to sign up at convenient times, either morning, afternoon, or evening sessions for three hours. The groups were assigned to conditions at random as they came in with the proviso that at the end of each week the conditions were balanced for time of day and each of the three experimenters served in an equal number of video and audio conditions.

Measures

Our goal was to assess three things: the quality of the product, the participants' satisfaction with the process, and the process of design and coordination.

Outcome.

We assessed how the technology affected the *outcomes* of the meetings, the quality of the design as reflected in the final document of each group. The final document was intended to be notes readable by a fictitious group member who was not present at that meeting but who would be present at the next.

We used the same quality measure as was used in our earlier study [19]. This measure was developed after extensive discussion with both designers and researchers of design. Three major aspects of the groups' output were scored: how completely the output covered all the aspects of the assigned task, the ease of understanding of the ideas reported in the document, and the judged quality of the post office design, including the feasibility of producing a prototype of the suggested post office within the stated time and manpower constraints, the coherence of the ideas, and the judged success of the ideas if marketed. Each aspect was then detailed further and a rating form constructed. Six researchers then coded the output from the same six meetings. The average pairwise correlation between raters was .85. Since this was well above acceptable range for reliability of measures, one researcher then coded the quality of the remaining meetings' outputs using the same instrument. Out of a possible score of 80 points, the quality of the meetings ranged from 40-74.

Satisfaction

To assess satisfaction, we constructed a post-session questionnaire that asked the participants to

a) rate their satisfaction with the *process* that they used (adapted from [9,10]) as well as with the design *result* [10],

b) assess the evenness of the participants' contributions [9] and c) identify a leader if one emerged.

c) rate how easy it was to understand the other participants and be understood.

The first two sets of questions were identical to those asked of the groups run in the companion face-to-face study; the remaining 23, which focused on various details of the communication media, were new to this study.

Process

All sessions were video taped. with the three locations displayed in a split screen format. From audio tapes which were collected at the same time, we transcribed the verbal reports. In addition, we captured timed keystrokes from ShrEdit; we integrated this typing activity into the verbal transcript. These transcripts were then coded for what kinds of activity were taking place at each moment, using the categories devised and tested in our study of field design meetings and the earlier related lab study [17, 19] We identified times when the participants stated the issue on the table, when they generated alternatives, when they critiqued the ideas. These categories have origins in the Design Rationale literature [13]. We also catalogued the time it took the participants to organize themselves (an activity we call "meeting management"), to clarify their ideas, talk through difficulties with the technology, or engage in side digressions. The division of activity into task and process management was inspired by work in the literature [21,31].

Several new categories were required to account for the work surrounding producing the output: times when they would plan the organization and wording or dictate the words, called Plan and Write. In the supported groups, we required yet another two categories: times when they were confused about something having to do with the technology, and other comments about the placing their work into the windows on the screen. We called these Technology Confusion and Technology Management, respectively.

Inter-rater reliability of the core 22 categories were measured in two ways. A strict measure shows the correspondence of categorization, second by second; our inter-rater reliability

is .68%, with a Cohen's K = .64. If we look at the summary measures used, the correlation between the two raters' summary statistics on time in category was .97.

RESULTS

The results are organized by the major classes of measures as described above: outcome, satisfaction, and process.

Outcome differences

Figure 1 arrays the quality scores for the two new conditions (Remote Audio and Remote Video) and displays them in the context of the comparable conditions in the companion study [19] in which the same kind of groups doing the identical task worked face-to-face, half of them using ShrEdit, half using whiteboard and paper and pencil. This table shows that the average judged quality of the groups supported by audio plus video was *not* significantly higher than that for groups supported by audio only $(t(36) = 1.41, p<.16)$.

FTF Unsup.		Remote Audio		Remote Video		FTF Supported
54.7	=	56.7	=	61.5	=	64.4

|--------------p<.01----------------|
|----------------------p<.01----------------------|
|------------(p<.09)------------|

Figure 2. Mean Ratings of the Quality of the Output for groups in four conditions, with indications of which pairs of conditions were significantly different from each other.

Although these quality differences were not significant, they showed a pattern of differences with the face-to-face conditions that are interesting. All four quality ratings were not significantly different from their adjacent values (FTF Unsupported = Remote Audio; Remote Video = FTF with ShrEdit). However, FTF with ShrEdit was significantly higher than the Remote Audio $(t(35) = 2.67, p<.01)$ and FTF Unsupported $(t(36) = 2.71, p<.01)$. Remote Audio was not significantly higher than FTF Unsupported $(p<.09)$.

In sum, if there are quality differences when one takes away video connections, they are small and do not always overcome the large between-group variances in performance. But more interesting, the quality of the output of work with remote high-quality video is not significantly different from that of face-to-face work. Remote work without video is not as good as face-to-face.

Satisfaction

Analysis of the questionnaires revealed that without video, the participants reported that the quality of the discussion was significantly *poorer* $(t(106) = 2.32, p<.02)$. Furthermore, remote work with video was judged to be as high quality as face to face with ShrEdit support. The normal way of working, however, face to face with whiteboard and paper and pencil was the highest of all. This is not surprising because all of the modes of working with ShrEdit were new and may have temporarily unsettled people.

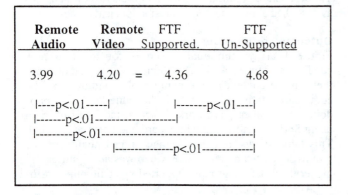

Remote Audio	Remote Video	FTF Supported.	FTF Un-Supported
3.99	4.20 =	4.36	4.68

Figure 3. Mean Ratings of the Perceived Quality of the Discussion for groups in four conditions, with indications of which pairs of conditions were significantly different from each other.

In other questions, the remote group with audio only reported being less able to tell how their other group members were reacting to things said (t(106) = 2.28, p<.025). They also reported that the communication system got in the way of their being able to persuade others about their ideas (t(106) = 3.52, p<.001) or to resolve disagreements (t(106) = 2.15, p<.03).

Process differences

We now look at how the groups conducted their work, in particular how they spent their time and how they moved among activities in the course of design.

We asked the groups to spend an hour and a half working on the APO problem. Overall, the mean time spent in the meetings was 87 minutes, which was not significantly different over the four conditions.

Furthermore, the groups talked a great deal during this 87 minutes; they were not just silently typing. On average the groups spent 64 minutes talking. The three groups that had ShrEdit talked significantly less than the group using whiteboard paper and pencil, by 13 minutes (F(3,70) = 6.06, p<.001). There was no difference among the groups supported by ShrEdit, whether they be FTF or remote, or supported by video or only audio.

As described above, we coded the transcripts of the spoken parts of the meetings, noting the kinds of activities they engaged in, and then summarized how much total time was spent in each activity as well as the flow between activities. Figure 4 shows a view of the flow of activities, the face-to-face supported groups on the left, the audio only ones in the middle, and the video groups on the right. In these diagrams, each category of activity is represented by a circle, the area of which represents the total time the group spent in that activity, aggregated over the whole meeting.

White portions of the circle represent the direct introduction of the idea; the black wedges represent the time spent clarifying that topic. The arrows denote the transitions between them, the width of which reflects the likelihood of going from one category to the next.[3]

The groups in both conditions spent their time in almost identical ways. Furthermore, they are almost identical to the way in which face-to-face groups worked with the same shared editor tool these groups used. The differences that were significantly different included:

Video groups spent less time than audio groups **stating and clarifying the Issues** (t(34) = 2.54, p<.02; t(34) = 2.25 p<.03).

Remote groups (both Video and Audio) spent significantly more time **managing their meeting** than the FTF group using ShrEdit (Video: t(35) = 2.92, p<.006; Audio: t(35) = 3.18, p<.003).

Remote groups (certainly Audio and marginally Video) spent significantly more time clarifying what they meant to each other (all categories combined) than the FTF group using ShrEdit (Video: t(35) = 1.81, p<.08; Audio: t(35) = 2.31, p<.003).

CONCLUSIONS

With high quality communication (both audio and video) and a shared workspace tool, distributed groups can produce work that is indistinguishable in quality from face-to-face groups using the same workspace tool. Taking away the video from distributed groups leads to poorer quality designs when compared to face-to-face groups. The audio-only groups were marginally different from the video/audio groups. Thus, high quality group intellectual work is possible under distributed conditions, and video appears to add some value.

The perceptions of the users, however, is that video adds value. The groups working at a distance without video do not like it as much as those that have the video. They were less able to tell how their other group members were reacting to things said. They also reported that the communication system got in the way of their being able to persuade others about their ideas or to resolve disagreements. Tang and Isaacs [27] found that groups in a field setting who were offered video in addition to shared workspace and audio used the system more heavily than those who had audio and workspace tools, suggesting that the satisfaction differences we saw in our study are probably a harbinger of usage pattern differences if these capabilities were discretionary.

However, judged by how people used their time, distributed work does require greater process overhead. The remote groups spent more time managing their work and clarifying what they meant than the face-to-face groups. Working

---

[3]To make this diagram less "busy," we include here only those transitions that occurred at least 1% of the time.

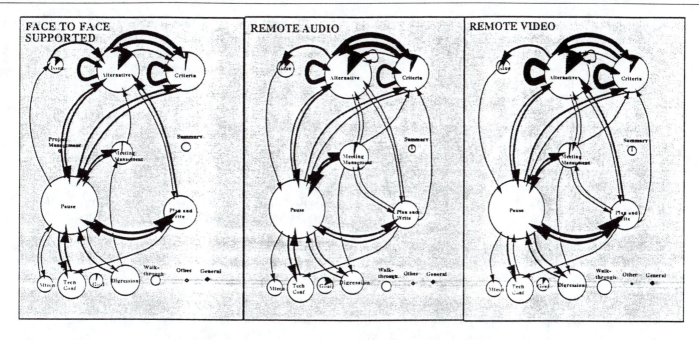

Figure 4 The user of time for various activities along with the pattern of transitions between the activities for the groups that had Video (right) Audio (middle) and those Face to Face.

under distributed circumstances is not equivalent to working face-to-face. Perhaps there is more sense of what others are doing and what they mean when we are face-to-face than can be presented via even very good video channels.

These results are important. We confirm the results of others [2, 14, 26] in that remote work can be done without loss of quality. This study has added to this body of findings, however, in that it uses intact groups doing a more realistic task, and uses measures of process as well as quality and satisfaction. In addition, we found that remote work takes extra effort to manage the group and clarify things. Adding video to remote work has some value in terms of the work accomplished by the groups, and has a clearer effect on the satisfaction of the group members. People like to see each other. Video makes them feel more able to *communicate* with each other, to persuade and resolve issues. For work that extends over long periods of time, these preferences are very likely to be important, as shown in the Tang and Isaacs [27] field study.

ACKNOWLEDGMENTS

This work has been supported by the National Science Foundation (Grant No. IRI-8902930), and by grants from Ameritech, the Ameritech Foundation, and AT&T. Many people participated in the collection and analysis of the data reported here, including Mark Carter, Stacey Donahue, Sue Schuon, Barb Gamm, Patsy Gore, Arona Pearson, Sidney Levy, Shawn Salata, Michael Walker, Rodney Walker, David Sisson, and Isabelle Byrnes.

REFERENCES

1. Bennis, W.G. (1968) The temporary society. In *The Temporary Society*, W.G. Bennis & P.E. Slater (Eds.), Harper & Row, New York.

2. Chapanis, A., and Ochsman, R. N. (1972) Studies in interactive communication: I. The effects of four communication modes on the behavior of teams during cooperative problem-solving. *Human Factors*, 14: 487-509.

3. Daft, R.L., & Lengel, R.H. (1986) Organizational information requirements, media richness and structural design. *Management Science, 32*, 554-571.

4. Dourish, P, and Bellotti, V. (1992) Awareness and coordination in shared workspaces. *Proceedings of the Conference on Computer Supported Cooperative Work.*

5. Egido, C. (1990) Teleconferencing as a technology to support cooperative work: Its possibilities and limitations. In J. Galegher, R.E. Kraut, & C. Egido (Eds.), *Intellectual teamwork: Social and technological foundations of cooperative work.* Lawrence Erlbaum Associates, Hillsdale, NJ. 351-371.

6. Fish, R.S., Kraut, R.E., & Chalfonte, B.L. (1990) The VideoWindow system in informal communication. *In Proceedings of CSCW '90.* ACM, New York.

7. Fish, R.S., Kraut, R.E., Root, R.W., & Rice, R. (1993) Video as a technology for informal communication. *Communications of the ACM, 36*, 48-61.

8. Gallupe, R., DeSanctis, G., and Dickson, G. (1988) Computer-based support for group problem solving: An experimental investigation. *MISQ,* 12, 277-296.

9. Gouran, D. S., Brown, C., and Henry, D. R. (1978) Behavioral correlates of perceptions of quality in decision-making discussions. *Communication Monographs.* 45(1), March. 51-63.

10. Green, S. G., and Taber, T. D. (1980) The effects of three social decision schemes on decision group process. *Organizational Behavior and Human Performance.* 25, 97-106.

11. Herbsleb, J.D., Klein, H., Olson, G.M., Brunner, H., Olson, J.S., & Harding, J. (in press) Object-oriented analysis and design in software project teams. *Human-Computer Interaction.*

12. McGrath, J. E. (1984) *Groups: Interaction and performance.* Englewood Cliffs, NJ: Prentice-Hall.

13 McGuffin, L., and Olson, G. M. (1992) ShrEdit: A shared electronic workspace. CSMIL Technical Report #45, The University of Michigan.

14. Minneman, S., and Bly, S. (1991) Managing a trois: A study of a multi-user drawing tool in distributed design work. *Proceedings of the Conference on Computer Human Interaction,* ACM Press, 217-224.

15. Moran, T. and Carroll, J., (Eds.) (in preparation) *Design Rationale.*

16. Olson, G. M., Olson, J. S., Killey, L., Mack, L. A., Cornell, P., and Luchetti, R. (in press) Flexible facilities for electronic meetings. In S. Kinney, R. Bostrom, and R. Watson, (Eds) *Computer Augmented Teamwork: A Guided Tour.* Van Nostrand Reinhold.

17. Olson, G. M., Olson, J. S., Carter, M., and Storrøsten, M. (1992) Small group design meetings: An analysis of collaboration. *Human Computer Interaction,* 7, 347-374.

18. Olson, G. M., Olson, J. S., Storrøsten, M, Carter, M., Herbsleb, J., and Rueter, H. (in press) The structure of activity during design meetings. In T. Moran and J. Carroll (Eds.) *Design Rationale.*

19. Olson, J.S., Olson, G.M., Storrøsten, M., & Carter, M. (1993) Groupwork close up: A comparison of the group design process with and without a simple group editor. *ACM Transactions on Information Systems, 11,* 321-348.

20. Pagani, D.S., & Mackay, W.E. (1993) Bringing media spaces into the real world. In G. De Michelis, C. Simone, & K. Schmidt (Eds.), *Proceedings of ECSCW '93.* Milan, Kluwer Academic, 341-356.

21. Putnam, L. L. (1983) Small group work climates: A lag-sequential analysis of group interaction. *Small Group Behavior,* 14(4), 465-494.

22. Reitman, W. R. (1965) *Cognition and Thought* New York: Wiley.

23. Rutter, D.R. (1984) *Looking and Seeing: The Role of Visual Communication.* Wiley, New York.

24. Short, J.A., Williams, E., & Christie, B. (1976) *The Social Psychology of Telecommunications.* Wiley, New York.

25. Simon, H.A. (1981) *The Sciences of the Artificial* (2nd. Ed.) MIT Press, Cambridge, MA.

26. Smith, R. G., O'Shea, T., O'Malley, Cl, Scanlon, E., and Taylor, J. (1989) Preliminary experiments with a distributed multi-media problem solving environment. *Proceedings of the First European Conference on Computer Supported Cooperative Work,* 19-34

27. Tang, J.C., & Isaacs, E. (1993) Why do users like video? Studies of multi-media supported collaboration. *Computer Supported Cooperative Work (CSCW), 1,* 163-196.

28. Toffler, A. (1980) *The Third Wave.* Morrow, New York.

29. Weick, K.E., & Meader, D.K. (1993) Sensemaking and group support systems. In L. Jessup & J. Valacich (Eds.), *Group support systems: New Perspectives.* Macmillan, New York.

30. Whittaker, S., Geelhoed, E., & Robinson, E. (1993) Shared workspaces: How do they work and when are they useful? *International Journal of Man-Machine Studies.*

31. Zigurs, I., Poole, M. S., and DeSanctis, G. L. (1988) A study of influence in computer-mediated group decision making. *MISQ,* 12, 645-665.

Designing SpeechActs:
Issues in Speech User Interfaces

Nicole Yankelovich, Gina-Anne Levow†, Matt Marx‡

Sun Microsystems Laboratories
Two Elizabeth Drive
Chelmsford, MA, USA 01824
508-442-0441
nicole.yankelovich@east.sun.com

ABSTRACT

SpeechActs is an experimental conversational speech system. Experience with redesigning the system based on user feedback indicates the importance of adhering to conversational conventions when designing speech interfaces, particularly in the face of speech recognition errors. Study results also suggest that speech-only interfaces should be designed from scratch rather than directly translated from their graphical counterparts. This paper examines a set of challenging issues facing speech interface designers and describes approaches to address some of these challenges.

KEYWORDS: Speech interface design, speech recognition, auditory I/O, discourse, conversational interaction.

INTRODUCTION

Mobile access to on-line information is crucial for traveling professionals who often feel out of touch when separated from their computer. Missed messages can cause serious inconvenience or even spell disaster when decisions are delayed or plans change.

A portable computer can empower the nomad to some degree, yet connecting to the network (by modem, for example) can often range from impractical to impossible. The ubiquitous telephone, on the other hand, is necessarily networked. Telephone access to on-line data using touch-tone interfaces is already common. These interfaces, however, are often characterized by a labyrinth of invisible and tedious hierarchies which result when menu options out-number telephone keys or when choices overload users' short-term memory.

Conversational speech offers an attractive alternative to keypad input for telephone-based interaction. It is familiar,

† Normally resides at the MIT AI Laboratory.

‡ Normally resides at the MIT Media Laboratory.

requires minimal physical effort for the user, and leaves hands and eyes free. And since physical space presents no constraint for a speech system, the number of commands is virtually unlimited.

Implementing a usable conversational interface, however, involves overcoming substantial obstacles. Error-prone speech recognizers require the system to emphasize feedback and verification, yet the time it takes to identify and repair errors can be tiring. Further, conversational interfaces are young, and transferring design principles from other media such as graphics can lead to unusable systems. Despite these problems, we, along with others [6, 9, 10, 11], believe the technology good enough and the promise exciting enough to make experimentation worthwhile.

In the SpeechActs project, we seek to identify principles and challenges of conversational interface design and to pinpoint limitations of current technology. In so doing, we hope to define useful avenues for research and suggest strategies for addressing the difficult problems facing speech user interface designers.

In this paper, we first describe the functionality of the SpeechActs system. We then explain our methodology, including usability testing and iterative redesign. We conclude with speech user interface design challenges and strategies for meeting those challenges in a speech-only environment.

THE SpeechActs SYSTEM

SpeechActs is a research prototype that integrates third-party speech recognition and synthesis with telephony, natural language processing capabilities, and other tools for creating speech applications. For an overview of the SpeechActs architecture, see [13]. To date, the system includes speech-only interfaces to a number of applications including electronic mail, calendar, weather, and stock quotes.

With the **mail** application, which uses Sun's Mail Tool backend, users can hear their messages, skip forward or backward from one header to the next, fax a message, reply to a message, or initiate a new message to a person on their *short list* of names known to the speech recognizer. To fax messages, users can say the name of a predefined location (i.e., work, home) or can specify a fax number by using telephone keypad input. When sending a message, the user has the option of including the current message and/or including a recorded audio attachment. Following is an example of a typical interaction with the mail application:

SpeechActs: You have 14 new messages and 47 old messages. The first new message is from Eric Baatz regarding "report completed."

User: Let me hear it.

SpeechActs: "The first draft is ready for your comments. Eric."

User: Reply and include the current message.

SpeechActs: Preparing message to Eric Baatz. Begin recording after the tone. When finished, pause for several seconds.

User: Eric, I'll get those to you this afternoon.

SpeechActs: Say cancel, send, or review.

User: Send.

SpeechActs: Message sent. What now?

User: Next message.

SpeechActs: New message two from Stuart Adams, subject "Can we meet at 3:00 today?"

User: Switch to calendar...

The SpeechActs **calendar** interface, based on Sun's Calendar Manager application, allows users to browse their own calendar as well as the calendars of other users on their short list. When the user requests information, the application reads them all the events on a selected day. Typical calendar queries include:

What do I have tomorrow?

What about Bob?

What did he have last Wednesday?

And next Thursday?

What was Paul doing three days after Labor Day?

The **weather** application provides an interface to the University of Michigan's on-line Weather Underground forecasts. Users can call up and ask for weather for states and for major cities around the country. For example, the user can say:

What's the weather in Seattle?

How about Texas?

I'd like the extended forecast for Boston.

Like the weather application, the **stock quotes** application provides a speech interface to a dynamic data feed. The user is able to ask for the prices of selected stocks, ask about their highs, lows, and volume, or ask for the prices of stocks in their portfolio (a stored list of stocks). Sample queries include:

What's the price of Sun?

What was the volume?

Tell me about IBM.

How's my portfolio doing?

As with multiple graphical applications running in the same environment, SpeechActs supports a standard set of functions that are always available in any application. For example, the user may always switch to a different application, ask for help, or end a session by saying "good bye."

USER STUDY / ITERATIVE DESIGN

Before the SpeechActs software was written, we conducted a survey and a field study [12] which served as the basis for the preliminary speech user interface (SUI) design. Once we had a working prototype, we conducted a usability study in which we adhered to Jakob Nielsen's formative evaluation philosophy of changing and retesting the interface as soon as usability problems are uncovered [8]. As a result, the formative evaluation study involved small groups of users and a substantial amount of iterative redesign.

Formative Evaluation Study Design

Fourteen users participated in the study. The first two participants were pilot subjects. After the first pilot, we redesigned the study, solved major usability problems, and fixed software bugs. After the pilots, nine users, all from our target population of traveling professionals, were divided into three groups of three. Each group had two males and one female. An additional three participants were, unconventionally, members of the software development team. They served as a control group. As expert SpeechActs users, the developers provided a means of factoring *out* the interface in order to evaluate the performance of the speech recognizer.

After testing each group of target users, we altered the interface and used the next group to validate our changes. Some major design changes were postponed until the end of the study. These will be tested in the next phase of the project when we plan to conduct a longer-term field study to measure the usefulness of SpeechActs as users adapt to it over time.

Tasks

During the study, each participant was led into a room fashioned like a hotel room and seated at a table with a telephone. They were asked to complete a set of 22 tasks, taking approximately 20 minutes, and then participate in a follow-up interview.

The tasks were designed to help evaluate each of the four SpeechActs applications, as well as their interoperation, in a real-life situation. To complete the tasks, participants had to read and reply to electronic mail, check calendar entries for themselves and others, look up a stock quote, and retrieve a weather forecast.

Instead of giving explicit directions, we embedded the tasks in the mail messages. Thus the single, simple directive "answer all new messages that require a response" led to the participants executing most of the tasks desired. For example, one of the messages read as follows: "I understand you have access to weather information around the country. If it's not too much trouble, could you tell me how warm it is going to be in Pittsburgh tomorrow?" The participant had to switch from the mail application to the weather application, retrieve the forecast, return to the mail application, and prepare a reply.

Although the instructions for completing the task were brief, participants were provided with a "quick reference card" with sample commands. For example, under the heading "Mail" were phrases such as "read me the first message," "let me hear it," "next message," "skip that one," "scan the headers," and "go to message seven." In addition, keypad commands were listed for stopping speech synthesizer output and turning the recognizer on and off.

Summary of Results

After testing the first group of users, we were able to identify the main problems in the interface. Each of our users bemoaned the slow pace of the interaction, most of them thought the computer gave too much feedback, and almost everyone insisted that they be able to interrupt the speech output with their voice. Most egregious was our inappropriate translation of the Sun Mail Tool message organization into speech. A technique that worked well in the graphical interface turned out to be confusing and disorienting in the speech interface. Details about this problem with message organization along with other design-related study results are woven into the discussion on design challenges in the following section.

In the study, our main aim was not to collect quantitative data; however, the data we gathered did suggest several trends. As hoped, we noticed a marked, consistent decrease in both the number of utterances and the amount of time required to complete the tasks from one design cycle to the next, suggesting that the redesigns had some effect. On average, the first group of users took 74 utterances and 18.5 minutes to complete the tasks compared to the third group which took only 62 utterances and 15 minutes (Table 1).

Participants	Utterances	Time (minutes)
Group 1	74	18.67
Group 2	63	16.33
Group 3	62	15.00
Developers	43	12.33

Table 1. Average number of utterances and time to complete tasks.

At the start of the SpeechActs project, we were aware that the state of the art in speech recognition technology was not adequate for the conversational applications we were building. One of our research questions was to determine if cer-

tain types of interface design strategies might increase users' success with the recognizer. Unfortunately, none of our redesigns seemed to have an impact on recognition rates—the number of utterances that resulted in the system performing the correct action. They remained consistent among the groups, with the developers showing about a 10% better rate than the first-time users. More significant than the design was the individual; for instance, female participants, on average, had only 52% of their utterances interpreted correctly compared to 68.5% for males. Even with these low recognition rates, the participants were able to complete most of the 22 tasks. Males averaged 20 completed tasks compared to 17 for females (Table 2).

Participants	Recog. Rates	Tasks Completed
Female	52%	17
Male	68.5%	20
Developers	75.3%	22

Table 2. Average recognition rates and number of tasks completed..

Paradoxically, we found that recognition rates were a poor indicator of satisfaction. Some of the participants with the highest error rates gave the most glowing reviews during the follow-up interview. It is our conclusion that error rates correlate only loosely with satisfaction. Users bring many and varying expectations to a conversation, and their satisfaction will depend on how well the system fulfills those expectations.

Moreover, expectations other than recognition performance colored users' opinions. Some participants were expert at using Sun's voice mail system with its touch-tone sequences that can be rapidly issued. These users were quick to point out the slow pace of SpeechActs; almost without exception they pointed out that a short sequence of key presses could execute a command that took several seconds or longer with SpeechActs.

Overall, participants liked the concept behind SpeechActs and eagerly awaited improvements. Barriers still remain, however, before a system like SpeechActs can be made widely available. The next section provides a more in-depth discussion of the challenges inherent in speech interfaces as well as solutions to some of these suggested by our users' experience with SpeechActs.

DESIGN CHALLENGES

In analyzing the data from our user studies, we have identified four substantial user interface design challenges for speech-only applications. Below is a description of each challenge along with our approach to addressing the challenge.

Challenge: Simulating Conversation

Herb Clark says that "speaking and listening are two parts of a collective activity" [1]. A major design challenge in creating speech applications, therefore, is to simulate the role of speaker/listener convincingly enough to produce successful communication with the human collaborator. In designing our dialogs, we attempt to establish and maintain what Clark calls a *common ground* or shared context.

To make the interaction feel conversational, we avoid explicitly prompting the user for input whenever possible. This means that there are numerous junctures in the conversational flow where the user must take the initiative. For example, after a mail header is read, users hear a prompt tone. Almost all users comfortably take the lead and say something appropriate such as "read the message," or "skip it." In these cases, we adequately establish a common ground and therefore are rewarded with a conversation that flows naturally without the use of explicit prompts.

When we engaged users in a subdialog, however, study participants had trouble knowing what to say, or even if it was their turn to speak, when the subdialog concluded. The completion of a subdialog corresponds to a *discourse segment pop* in the discourse structure terminology described by Grosz & Sidner [3]. When the subdialog is closed, the context returns to that preceding the subdialog. For example, the user might read a string of messages and then come across one that requires a response. In the reply subdialog, the user has to decide whether or not to include the current message, has to record the new message, and, perhaps, has to review the recording. When finished, the user is back to a point where he or she can continue reading messages. In the Mail Tool graphical user interface (GUI), the reply sequence takes place in a pop-up window which disappears when the user sends the message, and their previous context is revealed. We found that we needed an analogous signal.

Our first attempt to provide a discourse pop cue—a prompt tone at the end of the subdialog—failed. We considered the use of an intonational cue, which is one technique used by human speakers. Since our synthesizer could not produce a clear enough intonational cue, we included an explicit *cue phrase*—"What now?"—to signal the discourse pop. Surprisingly, this small prompt did, in fact, act to signal the subdialog's completion and return the user to the main interactional context.

Prosody. Prosody, or intonation, is an important element in conversations, yet many of the synthesizers available today do a poor job reproducing human-sounding intonational contours. This means that many types of utterances used by humans cannot be employed in the speech interface design. For example, as an alternative to the phrase "What did you say?", we tried to use "hmm?" and "huh?", but could not reproduce the sounds convincingly.

Despite the lack of good prosodics, most of our study participants said the speech output was understandable. On the other hand, many complained that the voice sounded "tinny," "electronic," or "choppy."

Pacing. Another important aspect of conversation involves pacing. Due to a variety of reasons, the pacing in SpeechActs applications does not match normal conversational pacing. The pauses in the conversation resulting from recognition delays, while not excessively long by graphical interaction standards, are just long enough to be perceived as unnatural. One user commented: "I had to get adjusted to it in the beginning...I had to slow down my reactions."

In addition, the synthesizer is difficult to interrupt due to cross-talk in the telephone lines which prevents the speech recognizer from listening while the synthesizer is speaking. In the implementation used by study participants, users had to use keypad input to stop the synthesizer from speaking. Unfortunately, as Stifelman also found [11], users had a strong preference for using their voice to interrupt the synthesizer. A user said: "I kept finding myself talking before the computer was finished. The pacing was off."

We have identified several strategies to improve pacing. First, we are experimenting with a *barge-in* technique that will allow users to interrupt the speech synthesizer using their voice. Second, we would like users to be able to speed up and slow down the synthesized speech. This way they could listen to familiar prompts and unimportant messages quickly, but slow the speaking down for important information. We are also considering adding keypad short-cuts for functions common to all applications (e.g., next, previous, skip, delete, help, etc.). This will allow advanced users to move more quickly through the information, skipping prompts when appropriate. Another potential aid for advanced users, which Stifelman recommends [11], is replacing some of the spoken prompts with auditory icons or sounds that evoke the meaning of the prompt.

Challenge: Transforming GUIs into SUIs

Since one of the goals of the SpeechActs project is to enable speech access to existing desktop applications, our initial SUI designs were influenced by the existing graphical interfaces. Our user studies, however, made it apparent that GUI conventions would not transfer successfully to a speech-only environment. The evolution of our SUI design shows a clear trend towards interpersonal conversational style and away from graphical techniques.

Vocabulary. An important aspect of conversation is vocabulary. We discovered early on that the vocabulary used in the GUI does not transfer well to the SUI. As much as they may use a piece of software, users are not in the habit of using the vocabulary from the graphical interface in their work-related conversations. Here is one of many examples from our pre-design field study where we analyzed human-human conversations relating to calendars: On the telephone, a manager who is a heavy user of Sun's calendar GUI, asked his assistant to look up information on a colleague's calendar:

Manager: Next Monday—Can you get into John's calendar?

To access another user's calendar in the GUI, the assistant had to select an item (johnb@lab2) from the Browse menu. In his request, the manager never mentioned the word "browse," and certainly did not specify the colleague's user ID and machine name. Also note his use of a relative date specification. The graphical calendar has no concept of "next Monday" or other relative dates such as "a week from tomorrow" or "the day after Labor Day." These are not necessary with a graphical view, yet they are almost essential when a physical calendar is not present.

It turned out that the assistant could not, in fact, access John's calendar. She received the error message: "Unable to access johnb@lab2." Her spoken reply was:

Assistant: Gosh, I don't think I can get into his calendar.

In designing each of the SpeechActs applications, we tried to support vocabulary and sentence structures in keeping with users' conversational conventions rather than with the words and phrases used in the corresponding graphical interface. The field study as well as the formative study both indicate that it is unlikely users will have success interacting with a system that uses graphical items as *speech buttons* or spoken commands.

Information Organization. In addition to vocabulary, the organization and presentation of information often does not transfer well from the graphical to the conversational domain. The difficulties we encountered with the numbering of electronic mail messages illustrates the translation problem. In Sun's Mail Tool GUI, messages are numbered sequentially, and new messages are marked with the letter "N." Thus, if you have 10 messages and three are new, the first new message is number 8. The advantage of this scheme is that messages retain the same number even when their status changes from new to old. The "N" is simply removed after a message is read.

We initially used the same numbering scheme in the SUI, but with poor results. Even though the start-up message told the user how many new and old messages they had, users were uniformly confused about the first new message having a number greater than one. When asked about their concept of message numbering, users generally responded that they expected the messages to be organized like Sun's internal voice mail where new messages start with number 1. No one alluded to the Mail Tool organization of messages.

We improved the situation by numbering new messages 1 to n and old messages 1 to n. Of course, this introduced a new problem. Once a message was read, did it immediately become old and receive a different number? Since we wanted users to be able to reference messages by number (e.g., "Skip back to message four."), remembering the messages seemed unwise. Instead, we added the concept of

"read messages," so if users revisited a message, they were reminded that they had already read it, but the message numbers stayed constant until the end of the session. Following the changes, users consistently stated that they knew where they were in the system, and specifically mentioned the helpfulness of the reminder messages.

Information Flow. Just as one way of organizing information can be clear on the screen and confusing when spoken, so it is with information flow. A frequently used flow-of-control convention in GUI design is the pop-up dialog box. These are often used to elicit confirmation from the user. A typical example is a Yes/No or OK/Cancel dialog box that acts as a barrier to further action until the user makes a selection. The pop-up is visually salient, and thus captures the user's attention. The closing of the dialog box also serves as important feedback to the user.

We attempted to create speech dialog boxes. For example, we wanted a confirmation from the user before sending a new message (e.g., "Your message is being sent to Matt Marx. Okay?"). The only acceptable answers to this question were "yes," "okay," "no" and some synonyms. Users were highly non-compliant! Some seemed confused by the question; others simply ignored it. Some of the confusion was understandable. Occasionally, users had said something other than "send." If this happened, users often repeated or rephrased their command (e.g., "review") instead of answering the question with a "no." Even without recognition problems, only a few users answered the yes/no question directly. Instead, many simply proceeded with their planned task (e.g., "Read the next message."). Sometimes they added "yes" or "no" to the beginning of their phrase to acknowledge the prompt. This phenomenon was also observed by researchers at NTT [5].

When considered in the context of spoken dialog, this behavior is actually quite natural. As the classic example "Do you have the time?" illustrates, yes/no questions rarely *require* yes/no answers. The listener frequently has to infer yes or no, or pick it out from the context of a larger utterance.

Not being able to count on reliable answers to yes/no questions can be problematic from a design standpoint since designing for errors is a necessity in the speech arena. We handled this problem in a number of different ways. First, we removed as many of these spoken dialog boxes as possible. Where we felt confirmation was necessary, we allowed users to preface commands with yes or no. If they did not, we treated a valid command as an implicit request to "do the right thing." For example, in the case of the exit dialog, "Did you say to hang up?", we treated any valid input as an implicit "no." In the few rare cases where we wanted to be absolutely sure we were able to understand the user's input, we used what Candace Kamm calls *directive prompts* [4] instead of using a more conversational style. For instance, after the user has recorded a new mail message, we prompt them to "Say cancel, send, or review."

Challenge: Recognition errors

Ironically, the bane of speech-driven interfaces is the very tool which makes them possible: the speech recognizer. One can never be completely sure that the recognizer has understood correctly. Interacting with a recognizer over the telephone is not unlike conversing with a beginning student of your native language: since it is easy for your conversational counterpart to misunderstand, you must continually check and verify, often repeating or rephrasing until you are understood.

Not only are the recognition errors frustrating, but so are the recognizer's inconsistent responses. It is common for the user to say something once and have it recognized, then say it again and have it misrecognized. This lack of predictability is insidious. It not only makes the recognizer seem less cooperative than a non-native speaker, but, more importantly, the unpredictability makes it difficult for the user to construct and maintain a useful *conceptual model* of the applications' behaviors. When the user says something and the computer performs the correct action, the user makes many assumptions about cause and effect. When the user says the same thing again and some random action occurs due to a misrecognition, all the valuable assumptions are now called into question. Not only are users frustrated by the recognition errors, but they are frustrated by their inability to figure out how the applications work.

A variety of phenomena result in recognition errors. If the user speaks before the system is ready to listen, only part of the speech is captured and thus almost surely misunderstood. An accent, a cold, or an exaggerated tone can result in speech which does not match the voice model of the recognizer. Background noise, especially words spoken by passersby, can be mistaken for the user's voice. Finally, out-of-vocabulary utterances—i.e., the user says something not covered by the grammar or the dictionary—necessarily result in errors.

Recognition errors can be divided into three categories: rejection, substitution, and insertion [10]. A *rejection* error is said to occur when the recognizer has no hypothesis about what the user said. A *substitution* error involves the recognizer mistaking the user's utterance for a different legal utterance, as when "send a message" is interpreted as "seventh message." With an *insertion* error, the recognizer interprets noise as a legal utterance—perhaps others in the room were talking, or the user inadvertently tapped the telephone.

Rejection Errors. In handling rejection errors, we want to avoid the "brick wall" effect—that every rejection is met with the same "I didn't understand" response. Based on user complaints as well as our observation of how quickly frustration levels increased when faced with repetitive errors, we eliminated the repetition. In its place, we give

progressive assistance: we give a short error message the first couple of times, and if errors persist, we offer more assistance. For example, here is one progression of error messages that a user might encounter:

What did you say?

Sorry?

Sorry. Please rephrase.

I didn't understand. Speak clearly, but don't overemphasize.

Still no luck. Wait for the prompt tone before speaking.

As background noise and early starts are common causes of misrecognition, simply repeating the command often solves the problem. Persistent errors are often a sign of out-of-vocabulary utterances, so we escalate to asking the user to try rephrasing the request. Another common problem is that users respond to repeated rejection errors by exaggerating; thus they must be reminded to speak normally and clearly.

Progressive assistance does more than bring the error to the user's attention; the user is guided towards speaking a legal utterance by successively more informative error messages which consider the probable context of the misunderstanding. Repetitiveness and frustration are reduced. One study participant praised our progressive assistance strategy: "When you've made your request three times, it's actually nice that you don't have the exact same response. It gave me the perception that it's trying to understand what I'm saying."

Substitution Errors. Where rejection errors are frustrating, substitution errors can be damaging. If the user asks the weather application for "Kuai" but the recognizer hears "Good-bye" and then hangs up, the interaction could be completely terminated. Hence, in some situations, one wants to explicitly verify that the user's utterance was correctly understood.

Verifying every utterance, however, is much too tedious. Where commands consist of short queries, as in asking about calendar entries, verification can take longer than presentation. For example, if a user asks "What do I have today?", responding with "Did you say 'what do I have today'?", adds too much to the interaction. We verify the utterance implicitly by echoing back part of the command in the answer: "Today, at 10:00, you have a meeting with..."

As Kamm suggests [4], we want verification commensurate with the cost of the action which would be effected by the recognized utterance. Reading the wrong stock quote or calendar entry will make the user wait a few seconds, but sending a confidential message to the wrong person by mistake could have serious consequences.

The following split describes our verification scheme: commands which involve the presentation of data to the user are verified implicitly, and commands which will destroy data or set in motion future events are verified explicitly. If a user asks about the weather in Duluth, the system will indicate that it is the report for Duluth before reading the contents. The user is then free to regain control of the interaction by interrupting the synthesizer (unfortunately using a touch-tone command in our current implementation). If, on the other hand, the user wants to fax a 500 page mail message, the system will check to make sure that's what was really meant.

Although not its primary purpose, the SpeechActs natural language component, called Swiftus, helps to compensate for minor substitution errors [7]. It does so by allowing the application developer to convert phrases meaning the same thing into a canonical form. For example, the following calendar queries will all be interpreted the same way:

What does Nicole have May sixth?

What do Nicole have on May six?

What is on Nicole's schedule May sixth?

This means that some substitution errors (e.g., "Switch to weather," misrecognized as "Please weather") will still result in the correct action.

Insertion Errors. Spurious recognition typically occurs due to background noise. The illusory utterance will either be rejected or mistaken for an actual command; in either case, the previous methods can be applied. The real challenge is to prevent insertion errors. Users can press a keypad command to turn off the speech recognizer in order to talk to someone, sneeze, or simply gather their thoughts. Another keypad command restarts the recognizer and prompts the user with "What now?" to indicate that it is listening again.

Challenge: The Nature of Speech

Current speech technologies certainly pose substantial design challenges, but the very nature of speech itself is also problematic. For users to succeed with a SUI, they must rely on a different set of mental abilities than is necessary for successful GUI interactions. For example, short-term memory, the ability to maintain a mental model of the system's state, and the capacity for visualizing the organization of information are all more important cognitive skills for SUI interactions than for GUI interactions.

Lack of Visual Feedback. The inherent lack of visual feedback in a speech-only interface can lead users to feel less in control. In a graphical interface, a new user can explore the interface at leisure, taking time to think, ponder, and explore. With a speech interface, the user must either answer questions, initiate a dialog or be faced with silence. Long pauses in conversations are often perceived as embarrassing or uncomfortable, so users feel a need to respond quickly. This lack of think time, coupled with nothing to look at, can cause users to add false starts or "ums" and "ahs" to the beginning of their sentences, increasing the likelihood of recognition errors.

Lack of visuals also means much less information can be transmitted to the user at one time. Given a large set of new mail messages or a month's worth of calendar appointments, there is no quick way to glance at the information. One user said: "Not being able to view it—I was surprised at the level of frustration it caused."

To partially compensate for the lack of visual cues, we plan to use both scanning and filtering techniques. For example, during the iterative redesign we added the ability to scan mail headers. We also plan to add functionality so that users can have their mail filtered by topic or by user, and their calendar entries summarized by week and by month. This way, important messages and appointments will be called out to the user first, eliminating some of the need to glance at the information.

Speed and Persistence. Although speech is easy for humans to produce, it is much harder for us to consume [10]. The slowness of the speech output, whether it be synthesized or recorded, is one contributing factor. Almost everyone can absorb written information more quickly than verbal information. Lack of persistence is another factor. This makes speech both easy to miss and easy to forget.

To compensate for these various problems, we attempted to follow some of the maxims H.P. Grice states as part of his *cooperative principle* of conversation [2]. Grice counsels that contributions should be informative, but no more so than is required. They should also be relevant, brief, and orderly.

Because speech is an inherently slow output medium, much of our dialog redesign effort focused on being brief. We eliminated entire prompts whenever possible and interleaved feedback with the next conversational move so as not to waste time.

We also eliminated extraneous words whenever possible. By using a technique which we call *tapered presentation*, we were able to shorten output considerably in cases where we had a list of similar items This technique basically involves not repeating words that can be implied. In the stock quotes application, for example, when a user asks for his or her portfolio status, the response is something like:

Currently,	Sun is trading	at 32, up 1/2 since yesterday.
	SGI is	at 23, down 1/4.
	IBM is	at 69, up 1/8.

With the first stock, we establish the pattern of how the data is going to be presented. With successive stocks, we streamline the presentation by eliminating repetitive words.

Also in the pursuit of brevity and in an attempt not to stress user's short-term memory, we avoid the use of lists or menus. Instead, we use conversational conventions to give users an idea of what to say next. In the calendar application, for example, we always start with "Today, you have..." By initiating the conversation and providing some common ground, it seems natural for users to respond by saying, "What do I have tomorrow?" or "What does Paul have today?"

Ambiguous Silence. Another speech-related problem, also observed by Stifelman [11], is the difficulty users have in interpreting silence. Sometimes silence means that the speech recognizer is working on what they said, but other times, it means that the recognizer simply did not hear the user's input.

This last problem is perhaps the easiest to overcome. Clearly, the user needs immediate feedback even if the recognizer is a bit slow. We plan to add an audio cue that will serve the same purpose as a graphical watch cursor. This will let users know if the computer is working on their request, leaving silence to mean that the system is waiting for input.

CONCLUSIONS

Based on our experience designing SpeechActs, we have concluded that adhering to the principles of conversation does, in fact, make for a more usable speech-only interface. Just as in human-human dialog, grounding the conversation, avoiding repetition, and handling interruptions are all factors that lead to successful communication.

Due to the nature of speech itself, the computer's portion of the dialog must be both as brief and as informative as possible. This can be achieved by streamlining the design, using tapered presentation techniques, providing short-cuts that make use of another medium (such as touch-tones), and making verification commensurate with the cost of the action.

As with all other interface design efforts, immediate and informative feedback is essential. In the speech domain, users must know when the system has heard them speak, and then know that their speech was recognized correctly.

Finally, we have strong evidence to suggest that translating a graphical interface into speech is not likely to produce an effective interface. The design of the SUI must be a separate effort that involves studying human-human conversations in the application domain. If users are expected to alternate between modalities, care must be taken to ensure that the SUI design is consistent with the corresponding graphical interface. This involves consistency of concepts and not a direct translation of graphical elements, language, and interaction techniques.

While interface challenges abound, we hope that working with speech technology at this stage in its development will provide speech vendors with the impetus to make the improvements necessary for the creation of truly fluent speech interfaces.

ACKNOWLEDGEMENTS
The SpeechActs project is a collaborative effort. Eric Baatz and Stuart Adams have implemented major portions of the

framework while Paul Martin and Andy Kehler are responsible for the natural language components. Special thanks to Bob Sproull for his contributions to the architectural design of the system.

REFERENCES

1. Clark, Herbert H. *Arenas of Language Use.* University of Chicago Press, Chicago, IL, 1992.

2. Grice, H. P. "Logic and Conversation," *Syntax and Semantics: Speech Acts,* Cole & Morgan, editors, Volume 3, Academic Press, 1975.

3. Grosz, Barbara, and Candy Sidner. "Attention, Intentions, and the Structure of Discourse," *Computational Linguistics,* Volume 12, No. 3, 1986.

4. Kamm, Candace. "User Interfaces for Voice Applications," *Voice Communication Between Humans and Machines,* National Academy Press, Washington, DC, 1994.

5. Kitai, Mikia, A. Imamura, and Y. Suzuki. "Voice Activated Interaction System Based on HMM-based Speaker-Independent Word Spotting," *Proceedings of the Voice I/O Systems Applications Conference,* Atlanta, GA, September 1991.

6. Ly, Eric, and Chris Schmandt. "Chatter: A Conversational Learning Speech Interface," *AAAI Spring Symposium on Intelligent Multi-Media Multi-Modal Systems,* Stanford, CA, March 1994.

7. Martin, Paul and Andrew Kehler. "SpeechActs: A Testbed for Continuous Speech Applications," *AAAI-94 Workshop on the Integration of Natural Language and Speech Processing,* 12th National Conference on AI, Seattle, WA, July 31-August 1, 1994.

8. Nielsen, Jakob. "The Usability Engineering Life Cycle," IEEE *Computer,* March 1992.

9. Roe, David, and Jay Wilpon, editors. *Voice Communication Between Humans and Machines,* National Academy Press, Washington, DC, 1994.

10. Schmandt, Chris. *Voice Communication with Computers: Conversational Systems,* Van Nostrand Reinhold, New York, 1994.

11. Stifelman, Lisa, Barry Arons, Chris Schmandt, and Eric Hulteen, "VoiceNotes: A Speech Interface for a Hand-Held Voice Notetaker, ACM *INTERCHI '93 Conference Proceedings,* Amsterdam, The Netherlands, April 24-29, 1993.

12. Yankelovich, Nicole. "Talking vs. Taking: Speech Access to Remote Computers," ACM *CHI '94 Conference Companion,* Boston, MA, April 24-28, 1994.

13. Yankelovich, Nicole and Eric Baatz. "SpeechActs: A Framework for Building Speech Applications," *AVIOS '94 Conference Proceedings,* San Jose, CA, September 20-23, 1994.

Integrating Task and Software Development for Object-Oriented Applications

Mary Beth Rosson and John M.Carroll
Virginia Polytechnic Institute and State University
Blacksburg, VA 24061, USA
Tel: 1-703-231-6470
E-mail: <lastName>@cs.vt.edu

ABSTRACT

We describe an approach to developing object-oriented applications that seeks to integrate the design of user tasks with the design of software implementing these tasks. Using the Scenario Browser — an experimental environment for developing Smalltalk applications — a designer employs a single set of task scenarios to envision and reason about user needs and concerns and to experiment with and refine object-oriented software abstractions. We argue that the shared context provided by the scenarios promotes rapid feedback between usage and software concerns, so that mutual constraints and opportunities can be recognized and addressed early and continuingly in the development process.

KEYWORDS: Prototyping, Design tools, Scenarios, Object-oriented programming, Software engineering, design rationale

INTRODUCTION

In contemporary design practice for applications and user interfaces, task scenarios are a medium for reasoning about user activities; they are used for requirements capture and analysis, user-designer communication, system envisionment and development, documentation and training, usability evaluation. Scenarios have raised the hopes of many in human-computer interaction for an integrated design framework.

Nevertheless, scenario-based design methods are not a mature technology; currently they are not even a coherent methodology (see, e.g., the discussion in the April 1992 issue of *SIGCHI Bulletin*). For example, there is little consensus about the appropriate level or levels at which to couch scenario descriptions: is it the level of self-conscious human activity, the level of input and output transactions, or is it both? What implications arise from couching scenario descriptions at different levels? What are the procedures and representations for

doing this? It is more than likely that at the current time no developer or practitioner of scenario-based methods would render the same answers to these questions.

Scenario-based methods also raise a serious — indeed classic — system development issue: Will scenario design representations *increase* the "gap" between specification and implementation? While much work in software engineering in past decade has been directed at narrowing this gap [9, 16], it seems prima facie that a focus on detailed narratives of user interaction may widen it anew.

Researchers studying object-oriented design (OOD) have developed scenario-based approaches that complement those used in the design of human-computer interaction: scenarios are often used in developing a problem domain model, which is then further elaborated and refined to build an object-oriented design [6, 10, 15, 17]. By using scenarios as a source of important problem objects and their interactions, these approaches narrow the gap between specification and implementation. However, the OOD methods are not focussed on reasoning about user tasks — they use the scenarios to gain insight into possible *software* models.

We are trying to integrate these complementary uses of scenarios in design. In this paper we describe our work with the Scenario Browser, an experimental design environment for Smalltalk applications, in which a single set of task scenarios is used both to envision and reason about user needs and concerns, and to model and refine object-oriented abstractions that implement these tasks. For concreteness, our discussion focuses on an example application (a bibliography utility) developed using the tool. We begin with an overview of how the applications are developed in the Scenario Browser environment, then consider in more depth the issues of scenario evolution, completeness, consistency, task prototyping and object-based task models that arise in this approach to application development.

SCENARIO-BASED DEVELOPMENT

We assume a view of scenarios as couched at the level people construe their work to themselves, the level at which tasks become meaningful to the people who carry them out. For a person building our example

(a)

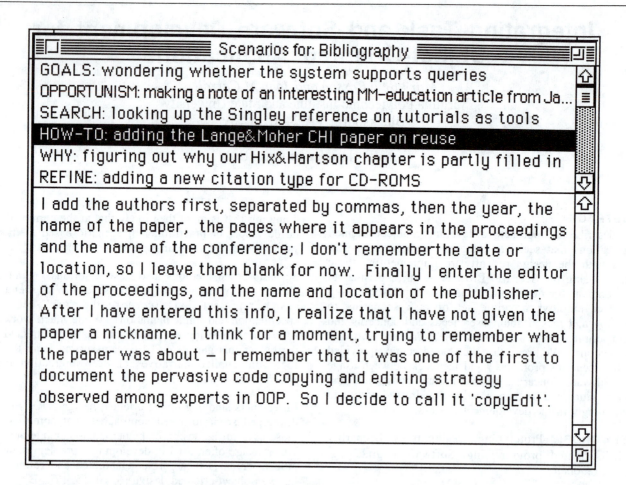

(b)

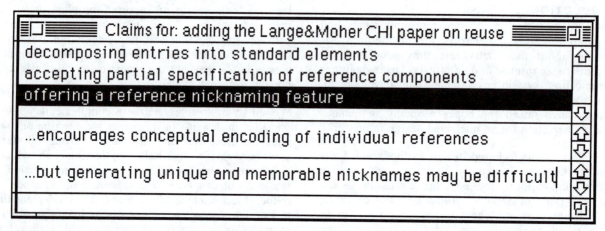

Figure 1. Task specification and rationale in the Scenario Browser environment. The designer develops a textual narrative for candidate scenarios in the Scenario View (a), and records hypothesized usability consequences implied by the tasks in the Claims View (b).

bibliography utility, example scenarios might be "making a note of an interesting multi-media education article from Jack", "looking up the Singley reference", "adding the Lange&Moher CHI paper on reuse", and so forth. Our design method involves generating scenarios — narrative descriptions of specific task episodes — to document requirements, problems, and goals, and from this to systematically envision reason about, and debate design solutions [4].

In the Scenario Browser environment, the designer uses a Scenario View to create and elaborate textual versions of the task scenarios (see Figure 1a). In doing this, the designer might start with a few general situations, using them as a heuristic for generating a broad range of user concerns [4]. So, for example, a common situation involves the carrying out of straightforward procedures. The addition of a reference would be normally be such a task (in Figure 1, this is indicated by the prefix HOW-TO); in contrast, browsing someone else's reference for interesting article reflects opportunistic behavior.

Figure 1b displays a Claims View of the bibliography tool under development. Designers may use this view to record usability rationale associated with individual scenarios, in the form of *claims* [4]. A claim is a causal hypothesis about some aspect of the scenario; it relates features of the system in use to one or more possible consequences for the user (positive consequences appear in the middle pane, negative in the bottom). Early in development, these claims are likely to be the result of analytic work by the designers; later on, they would incorporate observations obtained through usability testing. At any point in development, however, the claims analyzed for a scenario represent the designers' current best guesses about the usability issues pertaining to that particular use situation.

In the example scenario — adding the Lange and Moher CHI'89 paper — we considered that users might want to create a personal nickname for the reference. As the scenario suggests, this feature could have various consequences for a user: The user might benefit from being able to make use of personal experience and article content to build a more meaningful citation name. On the other hand, people are often unreliable at designing names, and the user might have difficulties, might even create a quirky nickname that is rapidly forgotten, necessitating wasteful search. By recording these hypothesized consequences, we identified this scenario as one context for considering usability issues associated with the nicknaming feature.

Our design approach concentrates on developing a coherent set of basic tasks first and only then developing an effective user interface to these tasks. One consequence of this is that most of the early scenarios are of the "How-to" variety — this situation includes the basic tasks a user can perform with a system. The other situations emerge once the system has been elaborated enough to evoke or facilitate particular goals or problem-solving strategies in the user. For example, "Why" scenarios normally describe situations in which the user is trying to understand an unexpected or problematic response from the system; envisioning such situations requires that a fairly detailed progression of user actions and system responses has been specified.

As soon as a task has been described — at any level of detail — the designer can begin building software to implement this specification. In Smalltalk, this consists of creating and refining an object model of the scenario: Smalltalk objects are created, connected to one another, and given scenario-relevant behaviors. Depending on the degree of elaboration provided by a scenario, this object model may be more or less complete. However, even if just one object can be modeled, the designer can begin experimenting with this part of the scenario implementation, sending test messages to the object, and refining its characteristics and behavior.

Incremental software development in the Scenario Browser environment is supported by the Implementation View and the Bittitalk Browser (see Figure 2). The designer can either instantiate classes already existing in the system (e.g., creating an instance of OrderedCollection to serve as the "refs" object) or create and instantiate new classes to represent new abstractions (e.g., a Name class that structures the text components of author names). The objects created to implement a scenario are graphed in the lower pane of the Implementation View, so that the designer can see a visual representation of the various cooperating Smalltalk objects and their relation to one another (Smalltalk objects "point" to one another via instance variables, and these are graphed as arcs). The Bittitalk Browser provides access to the class definitions behind the instances in the object model — this tool is identical to the standard Smalltalk/V® class browser, but instead of providing access to the entire (and very large) Smalltalk/V® class hierarchy, it lists only the classes currently being used to implement a scenario [14].

From the perspective of object-oriented design, an important characteristic of our scenario-based approach is that the software development is *instance-centered*: the designer develops new abstractions (classes) and behaviors (methods) by creating and experimenting with instances playing various roles in a scenario, rather than focussing on the more abstract task of writing class definitions and methods in a class hierarchy browser [7]. The testing and refinement of objects in the model is driven by the needs of specific scenarios. So, for example, in working on the conference paper scenario, we first modeled the authors' names as simple strings (i.e., 'Beth M. Lange'); as we elaborated the scenario to include visual feedback to the user, these string implementations were replaced by instances of the new class Name, which contained more intelligence about the structure of names.

The concrete situations provided by individual task scenarios promote the analysis of object responsibilities, a fundamental aspect of OOD [6, 15, 17]. But from the perspective of usability, they also create a focus on *tasks* during the implementation of an application. Thus any software design issues or alternatives that arise can

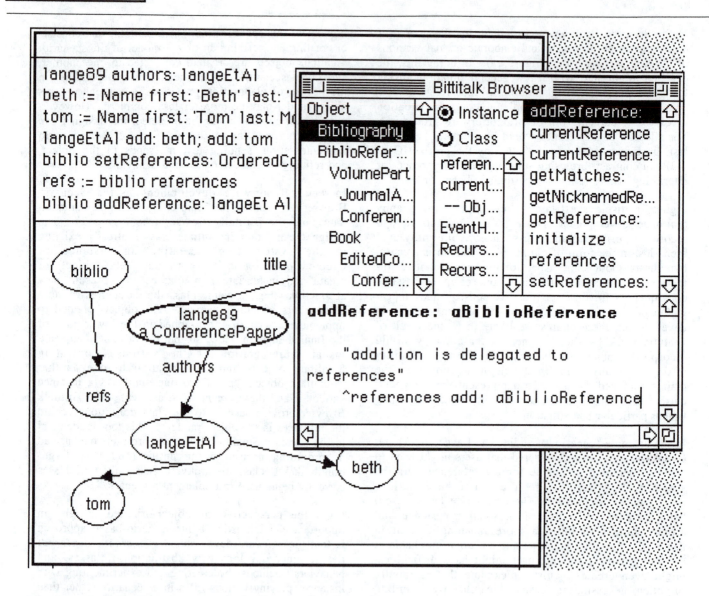

Figure 2. Software development and rationale in the Scenario Browser environment. The designer instantiates and experiments with Smalltalk objects in the Implementation View (on left), and creates new classes and behaviors in the Bittitalk Browser (on right). Software quality claims can also be recorded in a Claims View analogous to that shown for usability claims in Figure 1a.

the implementation of an application. Thus any software issues or alternatives that arise can be considered in light of specific tasks that they might impact, *at the time* that the issues first surface. Designers can use a Claims View analogous to that in Figure 1a to encode an explicit analysis of such issues. Thus a designer might record software claims — which are analogous to usability claims, but with features and consequences bearing on issues of software quality — to document his or her best guesses as to how specific software design features seem likely to impact the "goodness" of the object model developed for a particular scenario.

The scenario framework also supports communication between the designer and "user" of the bibilography (the

first author used the Scenario Browser to build the bibliography utility, with requirements obtained in conversation with the second author, an expert on bibliographic forms). Designer-user communication has long been recognized as a problem in development, in that designers tend to think and talk about their design objects, while users think and talk about their tasks [8]. A number of researchers have argued that building object-oriented design models should ease this communication, in that the main elements of the software model will be objects from the problem domain [1, 12]. We found that the task scenarios did provide an concrete medium for questions and clarifications concerning the user's needs and preferences: bits of functionality could be described or demonstrated and alternatives considered (with reference to the current usability or software rationale), with the relevant scenario serving as a shared context for discussion.

To sum up, object-oriented applications are developed in the Scenario Browser environment through an integrated process of scenario specification and implementation. Tasks are specified and analyzed for usability issues in parallel with the development and analysis of object models that implement the tasks. We now turn to some of the general issues raised by this approach to application development.

SCENARIO EVOLUTION

Developing an application through the creation and elaboration of task scenarios is an evolutionary process in which analysis and experimentation with individual scenarios reveals increasingly more about task and software requirements. In the simplest cases, scenarios develop through accretion — successively more of the problem requirements are specified and implemented. For the task specifications, this involves adding textual description. But for the scenario implementations, it means adding or altering computational objects.

As long as the designer has anticipated the nature of the relevant software abstractions, additional objects can be instantiated and integrated seamlessly into the developing object model for a scenario. However, designers cannot anticipate all of the elements needed in an object model, and some refinements will require changing the abstraction (class) behind one or more objects already in the model; in Smalltalk/V®, such a modification is difficult.

For example, our initial object model of the conference paper scenario did not include the date and location of the conference. Later, while working on a formatting scenario, we recognized that some reference formats require this information, and so needed to elaborate the conference paper object accordingly. In the standard Smalltalk environment, we would have been forced to destroy the conference paper object already in the model, add the new structural information (to the class definition for that object), and then re-build the object. The Scenario Browser environment eased this object model evolution by allowing us to add to (or remove) an existing object's attributes, so that we were able to add elements to a scenario without destroying and rebuilding the existing context.

The Scenario Browser also supports cases in which design objects are re-engineerd (i.e., not merely added to an object model). Any object in the implementation of a scenario can "mutate" to a new kind of object (i.e., class), without disrupting its current role in the model. For example, the conference paper was originally modeled as a relatively simple object, whose state consisted largely of strings and collections of strings (e.g., title, authors, proceedings title and editors). At a later point, we reconceptualized conference papers as

objects that contain information specific to a paper (e.g., title, authors), but whose conference information (e.g., proceedings title, editor) is held by a second reference object, a conference proceedings. Again, in most object-oriented environments, the original object model would have been thrown away and re-implemented to reflect the new abstraction. However, we were able to use the tools in the Scenario Browser environment to "split" the original conference paper object into two objects — instances of the revised conference paper class and the new conference proceedings class.

COMPLETENESS

A development process that relies heavily on scenarios must be very sensitive to issues of completeness: How many scenarios are sufficient as a specification for a system? Should we expect that the scenario set covers all of the intended functionality? In our earlier work, we have have found that a dozen or so key scenarios have provided a good specification for systems under design. [4]. Clearly, however, we would not recommend that designers *implement* exactly and only these dozen or so tasks; the software solution to a problem must be a general one.

In our bibliography utility scenarios, the object model we built for any given scenario was much more general than its specification. This appears to be a natural consequence of moving to a more symbolic and abstract representation (i.e., the domain of algorithmic entities). Whereas a task narrative focusses on the detailed goals and experiences of the user *in a particular situation*, the object model implicitly covers many variations of a scenario, for example, strings of varying length, typing errors, missing data and so on. In this sense, the "scenario" that we implemented can be seen as a kind of use case in the sense of Jacobson et al. [10]. Some of the considerations bearing on this generalization process are captured in a scenario's software claims; others are captured more informally as textual notes (see, e.g., the "notepad" in the upper right of the Implementation View in Figure 2).

CONSISTENCY

Another critical issue for scenario-based design methods is maintaining consistency across scenarios. Scenario elaboration can be viewed as a bottom-up approach to design problem-solving, with an emphasis on the rich context provided by specific task situations. However, modeling the details of each particular task setting might introduce inconsistencies in the overall design. For example, we began our work on the conference paper object in the context of reference-entry tasks, and the view we developed reflected our analysis of how users think of references when adding them to a database (e.g., with components ordered in the fashion they are normally seen, with the option of a personalized encoding). In contrast, a designer working initially on another scenario — browsing OOD design papers perhaps — might have come up with a rather different concept of a reference, one that was heavily weighted with attributes representing relations among references.

An explicit integration of scenario descriptions and their implementations can promote consistency in the overall design. For example, an important aspect of OOD is the development of abstraction hierarchies — abstract classes are used to define shared characteristics of related objects; subclasses then "inherit" the shared behavior, leading to a more efficient design. For designers modeling an object in a specific scenario, then, one design issue concerns the attributes or behavior this object might share with objects in other scenarios. Our first bibliography scenario modeled just one reference item, a conference paper. When we moved to other scenarios involving different kinds of reference items, however, we began to create a "BiblioReference" hierarchy (see the Bittitalk Browser in Figure 2); as additional scenarios were designed, this hierarchy was expanded and refactored.

The Scenario Browser environment also supports the reuse of pieces of object models developed in different scenarios. At any point, the designer can "grab" any object modeled in an existing scenario implementation, and incorporate it into a new scenario. Thus, several of our object models used the Lange and Moher conference paper, but in different roles (e.g., creating it, citing it in a document, updating it). Rather than building a new conference paper instance for each case, we simply carried our original object from scenario to scenario, elaborating it as needed for the new settings. Through such reuse we ensured that we always had a single, consistent implementation for this problem entity.

Of course, changing the implementation of an object in a second scenario can cause a scenario implemented earlier to "break": the other objects in this earlier scenario may no longer be able to communicate correctly with the changed object. When we "split" the Lange & Moher conference paper object into two objects (the paper and its proceedings), we went back to other scenarios whose implementations used this object and updated them as necessary. The Scenario Browser anticipates such a re-engineering process by providing direct access to all other scenario implementations using ann given object.

A more analytic approach to consistency management is to withdraw periodically from work on individual scenarios, to take a more cumulative view of the developing application. In the Scenario Browser environment, views of the objects, classes, and rationale developed across all scenarios are available. The designer can analyze a cumulative object graph for related but inconsistent objects, or for missed opportunities for abstraction. Our bibliography application was developed by one designer, but a system developed by a group of designers, or over a more extended time period, would be more prone to inconsistencies and redundancies (e.g., two different instantiations of the "same" conference proceedings object, one named chi89, the other chi1989). By analyzing the objects in a cumulative graph and the roles that they are playing, a designer may discover objects whose functionality is identical or overlapping, and carry out appropriate modifications (e.g., breaking up one object into two or more others, replacing one object with another, combining one or more objects into a single new object).

TASK PROTOTYPING

User tasks and the software designed to support them are interdependent. Tasks set requirements for new systems; as systems are developed, their software and hardware characteristics create constraints or opportunities for the tasks [2]. Many development projects manage this mutual interdependency through tight feedback loops between task specification and implementation, codifying specifications in prototypes early and repeatedly to "try out" on users [11]. The approach we have described can be seen as a refinement of rapid prototyping: it encourages early and continued testing of a design specification, even before a conventional "user interface" has been implemented. However it goes beyond the purely empirical feedback of prototyping with its parallel creation and evolution of design rationale for the tasks and supporting software.

An important characteristic of our approach is that we prototype *tasks*. We are seeking to support a process in which designers experiment early and often with software abstractions, but always in the context of specific task requirements. So, for example, when generating an object model for the conference paper scenario (see Figure 1), we worked directly from a usage description and analysis reflecting hypotheses about the *user's* concept of a reference (e.g., references contain components such as authors and title, and reference-entry consists of specifying these parts). This motivated and rationalized a software design in which these components were easily identifiable and manipulable. This is a common sort of dependency between user tasks and the software supporting them, where some aspect of the user's task structure must be respected or reified in the software model.

From the perspective of object-oriented software engineering, our approach can be seen as an extension of an emerging family of scenario-based analysis and design methods [6, 10, 15, 17]. These methods argue that analysis of task scenarios is a useful technique for developing a problem domain model, a fundamental task in object-oriented design. Like these researchers, we have argued that scenarios are the right starting point for constructing a model of cooperating problem objects [13]. However, these scenario-based approaches to OOD have focussed only on the goal of developing good software models. Our work extends this goal to include parallel development of useful and usable *tasks* [3, 4].

Most development projects will uncover tradeoffs between usability and implementation concerns, and the discovery and resolution of these tradeoffs is an important aspect of task prototyping. The formatting scenarios provide an example from the bibliography project. One of our first descriptions of a formatting scenario described an author citing references in a familiar style, with the added ability to use nicknames. When implementation began, however, a number of issues arose dealing with how the document was to be parsed (e.g., recognizing a variety of citation forms, distinguishing between citations and other parenthesized material, recognizing and responding appropriately to prior mentions of a reference author). These issues were then considered, but in the context of the affected scenarios, so that consequences for usability and for implementation could be considered jointly. The usability and software claims developed at this point documented these concerns, marking the issue as one to be carefully tracked.

OBJECT RESPONSIBILITIES IN A TASK

Most developers are familiar with the constraints that user requirements place on software and vice versa. A less familiar — but perhaps even more important — form of interaction between tasks and software occurs when reasoning about a software model suggests new *opportunities* for users' tasks. Users conceptualize their tasks in certain ways and their understanding of task requirements typically reflects this conceptualization. However, computerizing a task can significantly change the way it is carried out, enabling or suggesting activities that had been difficult or even impossible to achieve [2]. If designers hope to generate innovative solutions to users' needs, they must recognize, exploit and refine such opportunities early and continuingly in application development.

Application of the object-oriented pardigm is one vehicle for discovering possible extensions to existing tasks. In OOD, problems are decomposed into computational objects and their responsibilities; when a problem entity is conceptualized as an "active" agent, it is often given responsibilities normally carried out by humans or other agents [12]. Thus our object-oriented analysis of the bibliography suggested that the various reference types should be "responsible" for their own formatting rules: a conference paper should contain knowledge about how to format itself in IEEE style, in ACM style and so on. This is in contrast to the non-computerized version of the task, where formatting rules are stored in the author's head, or in a reference manual of some sort. This object-oriented view of the formatting task tied formatting knowledge intimately to the object being formatted. It suggested a task model in which *doing* the formatting was integrated with specifying the *rules* for the formatting; a user should have access to (and thus be able to experiment with) an object's format at any time, rather than being forced to leave a

particular formatting job and move to a distinct formatting "rules" task.

An example of how we documented such task extensions can be seen in our reasoning about conference proceedings. Our initial model of a conference paper did not include a separate component for the proceedings containing the paper; authors do not normally cite an entire proceedings or journal issue, so items such as these did not appear in our task analysis. But as we developed the object models for the various conference paper scenarios, we recognized that an individual conference paper should not be responsible for information about its containing volume. This pointed to the need for separate volume objects, which in turn suggested a task enhancement: creation or modification of a volume reference through work on any of the papers it contains. We immediately explored usability and software rationale for such an enhancement in the context of the affected scenarios. Our conclusion was that while the user's conceptual model would now be more complex (journal articles, book chapters and conference papers would be "compound" references), the additional power provided by the new model warranted the change. We recorded the usability and software claims that that we had developed in reasoning about these tradeoffs; an explicit link between the two sets of claims was created to document this input from the software model to the task design.

These examples argue for particular contributions of the object-oriented paradigm to scenario-based development. Certainly there is nothing in principle that prevents the interleaved development of task scenarios and a *non-*object-oriented software design. However, we believe that the integration of task and software development will produce the greatest payoff in the design of object-oriented systems: the use-oriented descriptions offer a rich and concrete context for the analysis of object responsiblities, while at the same time this responsibility analysis can suggest new ways of envisioning elements of users' tasks [12].

DISCUSSION

Our work on the Scenario Browser embodies a new perspective on the gap between task specification and implementation. Rather than rely only on the empirical feedback available through a rapid prototyping process, we propose that tasks and implementation be developed jointly, within an analytic framework that supports recognition and management of their interdependencies. Specifically, we have created a design tool that might integrate use-oriented reasoning with reasoning about software abstractions, using scenarios as the unit or context of analysis and design. In this we are integrating two converging areas of research, that on scenario-based design of human-computer interaction, and that on the use of scenarios in object-oriented analysis and design.

Our research has been guided by the principle that an application's usefulness and usability is inherently constrained by the problem domain model embodied in the application software [5, 12]. It may be possible to "paper over" (i.e., via a user interface optimized through continued iteration) slight mismatches wuth the software's task model, but if a system is to evolve and grow as a user's requirements and skills evolve, it must be founded on well-engineered software abstractions that reflect a task's structure. The most fundamental stage in object-oriented analysis is the enumeration of the problem domain objects, their individual responsibilities, and their collaborations [17]. Our design approach proposes that this enumeration can also play a key role in understanding and extending user tasks.

In sum, our work investigates the proposition that scenarios of the *same sort* we have been collecting and constructing to guide envisionment of the "external" system, the system as directly experienced by the user, can provide a use-oriented context for object-oriented analysis and design. Our focus is on design situations in which the task model is evolving, situations in which the designer is trying (or at least willing) to get beyond merely making a good model of an extant situation. We want to help designers develop a first-pass analysis of the "right" set of task scenarios, and the "right" set of objects and behaviors. We want to support a fluid exchange of reasoning between the task level and the software level, such that user tasks can inspire software models and software models can extend user tasks.

ACKNOWLEDGEMENT

A partially overlapping discussion of this research will appear in the proceedings of a workshop on scenario-based design.

REFERENCES

1. Bruegge, B., J. Blythe, J. Jackson, and J. Shufelt. Object-oriented system modeling with OMT, in *OOPLSA '92: Object-Oriented Programming, Systems and Applications*. 1992. New York, NY: ACM, pp. 359-376.

2. Carroll, J. M. and R. L. Campbell. Artifacts as psychological theory: The case of human-computer interaction. *Behavior and Information Technology 8*, (1989), 247-256.

3. Carroll, J. M. and M. B. Rosson. Human-computer interaction scenarios as a design representation, in *HICSS-23: Hawaii International Conference on System Sciences*. 1990. Los Alamitos, CA: IEEE Computer Society Press, pp. 555-561.

4. Carroll, J. M. and M. B. Rosson. Getting around the task-artifact cycle: How to make claims and design by scenario. *ACM Transactions on Information Systems 10*, (1992), 181-212.

5. Fischer, G. Domain-oriented design environments, in *KBSE'92: The 7th Annual Knowledge-Based Software Engineering Conference*. 1992. Los Alamitos, CA: IEEE Computer Society Press, pp. 204-213.

6. Gibson, E. Objects — Born and bred. *Byte* (1990), 245-254.

7. Gold, E. and M. B. Rosson. Portia: An instance-centered environment for Smalltalk, *OOPSLA '91: Proceedings* 1991. New York, NY: ACM, pp. 62-74.

8. Greenbaum, J. and M. Kyng. *Design at Work: Cooperative Design of Computer Systems*. 1991, Hillsdale, NJ: Lawrence Erlbaum Associates.

9. Harrison, M. and H. Thimbleby, ed. *Formal Methods in Human-Computer Interaction*. 1989, Cambridge University Press: New York, NY.

10. Jacobson, I., M. Chriserson, P. Jonsson, and G. Overgaard. *Object-oriented Software Engineering: A Use-case Driven Approach*. 1992, Reading, MA: Addison-Wesley.

11. Myers, B. A. and M. B. Rosson. Survey on user interface programming, in *CHI'92: Human Factors in Computing Systems*. 1992. New York, NY: ACM, pp. 195-202.

12. Rosson, M. B. and S. R. Alpert. Cognitive consequences of object-oriented design. *Human-Computer Interaction 5*, (1990), 345-379.

13. Rosson, M. B. and J. M. Carroll, Extending the task-artifact framework: Scenario-based design of Smalltalk applications, in *Advances in Human-Computer Interaction*, ed. H.R. Hartson and D. Hix. 1993, Ablex: Norwood, NJ. pp. 31-57.

14. Rosson, M. B., J. M. Carroll, and R. K. E. Bellamy. Smalltalk scaffolding: A case study in Minimalist instruction, in *Human Factors in Computing Systems, CHI '90 Conference*. 1990. New York, NY: ACM, pp. 423-429.

15. Rubin, K. S. and A. Goldberg. Object behavior analysis. *Communications of the ACM 35*, 9 (1992), 48-62.

16. Wasserman, A. I. Rapid prototyping of interactive information systems. *ACM Software Engineering Notes 7*, (1982), 171-180.

17. Wirfs-Brock, R., B. Wilkerson, and L. Wiener. *Designing Object-oriented Software*. 1990, Englewood Cliffs, NJ: Prentice Hall.

Using Computational Critics to Facilitate Long-term Collaboration in User Interface Design

Uwe Malinowski

Siemens Corporate R&D
ZFE ST SN 5
D 81730 München, Germany
+49 (89) 636-2969
malinow@zfe.siemens.de

Kumiyo Nakakoji

Software Research Associates Inc.
Software Engineering Lab.
1-1-1 Hirakawa-cho, Chiyoda-ku
Tokyo 102, Japan

Department of Comp. Science
University of Colorado
Boulder, CO 80309-430, USA
+1 (303) 492-3912
kumiyo@cs.colorado.edu

ABSTRACT

User interface design and end-user adaptation during the use of the system should be viewed as an ongoing collaborative design process among interface designers and end-users. Existing approaches have focused on the two activities separately and paid little attention to integration of the two by supporting their asynchronous collaboration over a long period of time throughout the evolution of the interface design. Our knowledge-based domain-oriented user interface design environments serve both as design media and as communication media among interface designers and end-users. An embedded computational critiquing mechanism not only identifies possible problematic situations in a design for user interface designers and end-users but also facilitates asynchronous communication among stakeholders. The presentation of critiquing messages often triggers designers and end-users to articulate design rationale by describing how they responded to the critiques. The recorded design rationale mediates collaboration among end-users and user interface designers during the end-user adaptation and redesign of the interface by providing background context for a design decision.

KEYWORDS: Usability Engineering, Collaborative Design, Design Rationale, User Interface Design Environments, Critiquing Systems, End-user Adaptation, Process Control.

INTRODUCTION

Issues of concern to user interface development are twofold: (1) knowledge necessary for interface design is distributed among interface designers and end-users; and (2) design tasks are ill-defined and open-ended by nature.

First, design knowledge necessary for interface design is distributed [11]. Designing a user interface requires two types of knowledge: knowledge about user interface design and knowledge about the application domain. Initially, there is a symmetry of ignorance between interface designers and end-users in which (1) the understanding required to solve the design problem is distributed, and (2) there is no com-

mon language that allows the stakeholders to communicate their understanding to each other [11]. Thus, not only the thin spread of application domain knowledge among user interface designers [2] but also little understanding about computer technology among end-users [3] causes communication breakdowns between end-users and interface designers [5].

Approaches have been made to improve collaboration among interface designers and end-users. A participatory design approach encourages end-users to have a stake in the user interface design process from the very beginning [3]. Another approach is to encourage interface (system) designers to participate in end-users' work settings [12]. Although we acknowledge the importance of such face-to-face collaboration among interface designers and end-users, a challenge we see is that interface designers are not available all the time during the use of the system. Once the system is released, designers are rarely assigned to maintenance and it is difficult for end-users to contact designers directly in order to communicate their requests. End-users and interface designers need to collaborate asynchronously over a long period of time throughout the life-cycle of a system.

Second, user interface design is an ill-defined and open-ended problem-solving task. Understanding a design problem and solving it are intertwined tasks; one cannot completely specify the problem before starting to solve it [23, 24]. Starting with a vague and incomplete problem description, people sketch out a partial solution. By seeing the partial solution, they identify portions of the problem that have not yet been understood, gain an understanding of the problem, and then refine the solution [25]. By iterating this process, understanding of the problem gradually emerges. Design problems are open-ended in that the knowledge necessary for a design can never be completely articulated a priori [23]. Traditional knowledge acquisition approaches that aim at capturing expert knowledge will not work because some part of expert knowledge always remains tacit - people know more than they can tell [21].

Thus, end-users cannot foresee the many different ways in which they will use the system. Furthermore, the system might change the way tasks are performed, and changes to the environment might make a different design necessary. Specific needs always emerge during the system usage, not due to sloppy requirements analysis but due to the very nature of design tasks. In response to this, ways to enable the end-user to adapt the user interface have been proposed [15].

To cope with these challenges, we present our approach to view user interface design as an evolving collaborative design process by interface designers and end-users of the system over a long period of time [9]. The process starts with the development of a user interface by user interface designers while collaborating with end-users. During the use of the system, end-users should be enabled to customize the system to their specific needs, which cannot be determined a priori but emerges only during use. After a while, user interface designers should be able to redesign the interface because end-users may have introduced inconsistencies and inefficient modifications during the adaptation.

Knowledge-based, domain-oriented design environments that facilitate asynchronous collaboration among different types of stakeholders in various design domains have been studied [7]. A computational critiquing component of such an interface design environment [8] supports the interface design and end-user adaptation as an evolutionary design process. The embedded critiquing component:

- supports both user interface designers and end-users by identifying potentially problematic situations in a design and presenting design knowledge relevant to the situation; and
- facilitates long-term collaboration among interface designers and end-users by triggering them to record design rationale, which helps them to understand the underlying context of a certain design decision.

In this paper, we first describe a computational critiquing mechanism embedded in a domain-oriented user interface design environment as a framework of our approach. We then present a scenario to illustrate how the critics support user interface development as a collaborative evolutionary design process using the domain of traffic management systems as an *object-to-think-with*. We describe the prototype system and discuss necessary improvements.

FRAMEWORK:
EMBEDDED CRITICS IN A DESIGN ENVIRONMENT

Domain-oriented knowledge-based design environments that cope with the issues discussed in the previous section have been studied in various domains, including kitchen floor plan designs and network design [7]. We have applied the design environment approach to user interface design. In the application domain of integrated traffic management systems, for example, the system provides knowledge not only about interface design (e.g., a guideline in designing a pop-up menu) but also about the domain (e.g., necessary functions for a train station).

Figure 1 illustrates a cycle of design, user adaptation, and subsequent redesign as a process model for user interface design supported with a design environment. The system is initially designed by interface designers in collaboration with end-users. Although not shown in Figure 1, this collaboration is very important for both design process and result. The application is equipped with mechanisms that allow end-users to adapt the interface during the use of the system. By this means, the interface "gradually evolves." Finally, the system is occasionally "redesigned" by interface design-

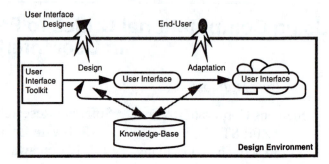

Figure 1: User interface design as an evolutionary process

ers in collaboration with the end-users to reorganize changes made by the end-users that might have introduced inconsistencies.

Critiquing is the exchange of a reasoned opinion about a product or action, which triggers further reflection on or changes to the artifact being designed [8]. Critiquing has been found to be an effective and natural method people use in collaborative problem solving [10, 17]. The critiquing mechanism embedded in a design environment helps people engage in design activities by:

- monitoring the current status of user interface design and informing them of possible breakdowns in what they have done;
- providing them with knowledge (either about user interface design or about the application-domain) relevant to the identified breakdowns; and
- encouraging them to articulate design rationale by motivating them to describe how they responded to the critiquing messages.

The first two aspects of the mechanism support user interface designers and end-users to exploit the system's stored knowledge by delivering the task-relevant knowledge even when they do not know about the existence of the knowledge, or when they do not realize the need to know the design knowledge. A part of such knowledge can be accumulated through the third aspect of the mechanism: design rationale.

Previous efforts [8, 14] have reported that embedded computational critiquing mechanisms in design environments have proven to be effective in supporting user interface designers by providing application-domain knowledge while developing a user interface.

We also use this critiquing mechanism to support end-users to adapt their interface to their needs by providing user interface design knowledge. Computer-aided adaptation has been proposed to enable users to adapt the user interface [13]. Experiments with prototypes [16] showed that enabling is not enough - end-users need support during the adaptation process.

We have identified that the critiquing approach is an effective knowledge-elicitation technique by motivating people to articulate design knowledge [19]. In the study of critiquing components in various design environments, we have observed that users often articulated why they violated the principles when being critiqued. This information was accumulated as design rationale. For example, in the study of the KID kitchen design environment [18], when a subject used a double-bowl sink when designing a kitchen for a single-person household, and KID presented a critiquing message saying that the subject should use a single-bowl sink if the design was for a small family, the subject articulated: "Well, ... Do you think a college student will do dishes every day? I don't think so... I wanna stack up dishes in one sink and I need another one."

In the study of a critiquing mechanism of another system called VDDE (Voice-Dialogue Design System) [1], when the system informed subjects that a design violated a certain design guideline, they stated that another conflicting rule was more important, therefore, they had to violate the indicated guideline.

The above evidence demonstrates that critiquing mechanisms are effective methods to elicit knowledge from users and encourage them to articulate design rationale.

Design rationale plays a crucial role in supporting long-term asynchronous communication among stakeholders in design tasks [6]. Design rationale provides descriptions and explanations for why a certain design decision is being made [6, 26]. Because the context of certain design decisions is not always explicit, explicitly articulated design rationale that describes why a design decision is made with regard to which design intention helps people to have a better understanding of the design. In user interface design, for example, design rationale recorded by interface designers will help end-users adapting the user interface to understand why the user interface has been designed that specific way. A rationale for adaptation recorded by end-users will help user interface designers to redesign the interface. This rationale can later be translated into knowledge usable by the critiquing component. Thus, triggered by critiquing mechanisms, the system serves as communication media that supports collaboration between user interface designers and end-users. Figure 2 illustrates this process.

THE SCENARIO

In this section, we present a scenario to illustrate our approach in the domain of user interface design for a traffic management system to allow traffic managers to monitor and influence the traffic flow on streets as well as that of railway trains. An interface design environment for the domain is provided for user interface designers. It consists of a user interface toolkit, a critiquing component and an argumentation component. The scenario focuses on how the order of functions in pop-up menus has been refined by designers and by end-users throughout the design life cycle, supported by the system's critiquing.

(1) Initial Requirements

The user interface is required to show the graphical representation of the objects, including streets, intersections, railway tracks, and stations. Figure 13 illustrates an example interface consisting of several streets and railroad lines, two intersections with the main street ("IS-1" and "IS-2") and two railway stations ("ST-A" and "ST-B"). The required functionality for each of the objects is depicted in Figure 4.

(2) Initial Design by Interface Designers

Using the given requirements, interface designers start designing pop-up menus for the functionality of the objects under consideration. Since no order criterion is specified, they choose to organize the pop-up menus alphabetically. However, *Display Video* is placed in the last position in the menu for the intersection IS-1 in order to keep the consistency among objects of the same type, i.e., with IS-2, which does not need the functionality because video cameras are not installed in IS-2 (Figure 5).

(3) System's Critique of the Initial Design

The critiquing component analyzes the design. The rule "Consistency between pop-up menus" fires, saying that the pop-up menus for stations and intersections are not consistent. Having chosen alphabetical order and consistency between pop-up menus for objects of the same type as the ordering criteria, the designers do not understand why the system complains about this. So the designer decides to inspect the related argumentation, which describes why this rule fired in the current situation. The argumentation component explains that in this domain, the functions *Schedule Screen* and *Traffic Light* are semantically analogous to end-users (see Figure 9).

This leads the designer to exchange the positions of *Schedule Screen* and *Display Video* on the menus for ST-A and ST-B so that *Schedule Screen* and *Traffic Lights* appear in the same position in the menus (Figure 6). The designers record this design rationale in the argumentation component

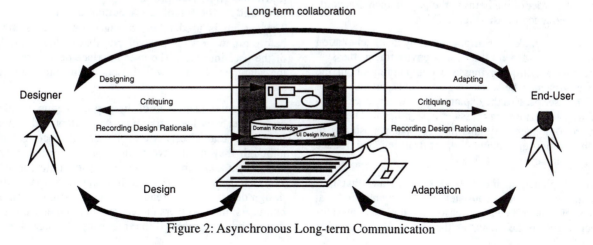

Figure 2: Asynchronous Long-term Communication

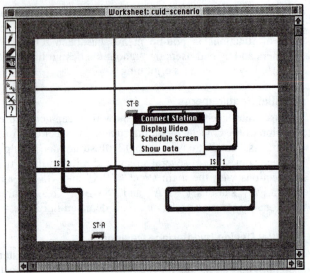

Figure 3: User Interface for Traffic Management

- **Show Data**: Display all available data for this station, such as typical number of passengers at this day/time for each direction, current number of passengers waiting for each direction, current schedule of trains for all directions, and observed or expected problems.
- **Display Video**: Provide video pictures of platforms and railroad lines from video cameras installed in the station.
- **Connect Station**: Setup a video and audio link to a station manager to discuss problems.
- **Schedule Screen**: Reschedule trains, add additional trains, and change tracks for trains when problems occur.

(a) For Each Station

- **Traffic Lights**: Change the traffic light program for the intersection or switch traffic lights "manually".
- **Connect Police**: Set up audio connection to the police when an accident occurs or the traffic lights fail.
- **Display Video**: Display pictures from a video if video cameras are installed in the intersection (Cameras are installed in IS-1 but not in IS-2).

(b) For Each Intersection

Figure 4: Required Functionality

ST-A and ST-B	IS-1	IS-2
Connect Station	Connect Police	Connect Police
Display Video	Traffic Lights	Traffic Lights
Schedule Screen	Display Video	
Show Data		

Figure 5: Menus after the Initial Design

ST-A and ST-B	IS-1	IS-2
Connect Station	Connect Police	Connect Police
Schedule Screen	Traffic Lights	Traffic Lights
Display Video	Display Video	
Show Data		

Figure 6: Menus after Redesign Based on Critiquing

ST-A and ST-B	IS-1	IS-2
Show Data	Display Video	Traffic Lights
Schedule Screen	Traffic Lights	Connect Police
Connect Station	Connect Police	
Display Video		

Figure 7: Menus after Adaptation by the End-users

ST-A and ST-B	IS-1	IS-2
Show Data	Traffic Lights	Traffic Lights
Schedule Screen	Connect Police	Connect Police
Connect Station	Display Video	
Display Video		

Figure 8: Menus after Redesign Based on Adaptations

to describe why they switched the positions of *Schedule Screen* and *Display Video*. The design is now released to end-users for real use.

(4) End-users' Adaptation and the System's Critiquing

After using the system for some time, one of the users thinks that the order of the functions in the pop-up menus is inadequate. In a discussion with colleagues, the decision is made to change the order of the items in the pop-up menus according to the frequency of use (Figure 7).

When end-users make this arrangement by using adaptation mechanisms provided by the design environment, the system presents a critiquing message complaining about the lack of consistency between pop-up menus, as mentioned above. The user presses the explain button to get an explanation of why the critiquing message is significant. The argumentation component explains that consistent positions of functions in pop-up menus support learning and reduce search time and error rate (Figure 10).

With this explanation, the end-users understand that it is principally good to have consistent pop-up menus and that this rule must have been the reason for the user interface designer to set up the menus as they were. However, after

discussing this conflict between the users' needs and the rule from the interface guidelines, the users decide to change the order as planned (Figure 7) conscious of the violation of the rule. The users document this decision in the argumentation tool, explaining that it seems to be more important for them that the menus fit their needs defined by their tasks than the user interface guidelines (Figure 11).

(5) Redesign by Designers - Informed by the Recorded Argumentation

The system is used in the changed setup for some time before a new version is developed that includes a redesign of the user interface. The user interface designers examine the adaptations made by end-users during the usage of the system.

Inspecting the adaptation of the pop-up menus, the designers wonder why the users might have adapted the interface in a way that clearly violates user interface guidelines, although they must have been informed by the critiquing component about the rule concerning semantic consistency. The designers examine the argumentation available for this design decision. After reading the explanation given by the end-users, the designers understand that the end-users think that usability increases if the frequency of usage is reflected

---

What is Schedule Screen?

Schedule Screen allows the user to reschedule trains, add additional trains, and reschedule trains on different tracks if problems occur.
The functionality is similar to **Traffic Lights** operations at an intersection.

domain-distinction:
    semantic-correspondence
        ("schedule screen" "traffic lights")

Figure 9: Argumentation for Designers

How to arrange menu items?
- semantic consistency
  [pro] semantically consistent order in menus supports learning and reduces search time and error rate.
  domain-distinction: **semantic-correspondence**
- alphabetical order
  [pro] searching is very fast
  domain-distinction: **alphabetical-order**

Figure 10: Argumentation for End-Users

How to arrange menu items?
- frequency
  The order of operations according to the usage frequency speeds up work!
  Frequency for Train Stations:
  1. *Show Data*   2. *Schedule Screen*
  3. *Connect Station*   4. *Display Video*
  Frequency for Intersections:
  1. *Traffic Lights*   2. *Connect Police*

Figure 11: Design Rationale Recorded by End-User

How to arrange menu items?
- usage-frequency order
  [pro] Fast access to those operations most often used.
  domain-distinction: **usage-frequency**

What is the sequence of operations according to usage frequency?

For **Train Stations**: 1. *Show Data*   2. *Schedule Screen*
                  3. *Connect Station* 4. *Display Video*
domain-distinction:
    usage-frequency ("Show Data" "Schedule Screen"
                  "Connect Station" "Display Video")
For **Intersections**: 1. *Traffic Lights*   2. *Connect Police*
domain-distinction:
    usage-frequency ("Traffic Lights" "Connect Police")

Figure 12: Knowledge-Base Update by a Knowledge Engineer based on the Design Rationale of the End-User

---

in the order of the menu items. This was not part of the initial requirements, nor had end-users detected this before using the system on a regular basis.

Therefore, the designers decide to design the pop-up menus in a partly consistent way, so that the menus for one type of objects is consistent, but not necessarily consistent with the menus of different objects by changing the order of functions in the menu for IS-1 and IS-2. This design also allows consideration of the frequency of usage of the functions (Figure 7). The solution is discussed with and accepted by the end-users who made and explained the adaptation of the pop-up menus.

Besides informing the user interface designer about the application domain, the domain knowledge implemented by the end-users in the argumentation tool is used in a second way. A knowledge engineer analyzes all the information and implements it in the knowledge base of the critiquing component. In this example, the frequency of usages of the functions is added to the knowledge base (Figure 11).

DISCUSSION OF THE SCENARIO

The scenario illustrates how the embedded computational critiques support the evolution of the user interface design by:
- helping interface designers to become aware of domain knowledge such as semantic correspondence between "Schedule Screen" and "Traffic Lights" operations;
- helping end-users to learn about user interface guidelines, for example, that pop-up menu items should be ordered with semantic consistency; and
- informing interface designers of the rationale behind the change end-users have made, i.e., it was more important for the end-users to arrange pop-up menus according to the frequency of usage rather than the semantic consistencies.

Our emphasis in using critiquing mechanisms in a design environment is on providing related argumentation and explanation of why the critiquing rule is significant, rather than on simply notifying users or designers of possible problematic situations. Different stakeholders need different descriptions for the same critiquing rule. That is, critics for user interface design are related to both user interface guidelines and domain semantics. In the scenario a critique fired using these two types of knowledge:

(1) user interface guideline: the order of operations in pop-up menus has to be semantically consistent; and

(2) domain knowledge: "schedule screen" for a train station and "traffic lights" for an intersection are semantically corresponding operations.

For example, in the scenario, the same critiquing rule fired to both the end-user and the designer: "pop-up menus need to be ordered semantically consistent." User interface designers needed to know which operations are semantically corresponding to each other in this specific domain. End-users needed to know what "semantic consistency" is and when and why it is important in user interface design.

Thus, the critiquing mechanisms need to combine both domain knowledge and user interface design knowledge. Our implementation separates the two types of knowledge by using the notion of "domain-distinction," which is further described below. By integrating these two types of knowledge-bases rather than using a single knowledge-base for "user interface design for traffic management systems," the same user interface guideline knowledge can be combined with another application domain knowledge to produce a domain-oriented design environment.

THE PROTOTYPE

The current prototype of a user interface design environment for integrated traffic control systems is implemented on a Macintosh computer using MCL. The prototype has been built on top of Agentsheets, a substrate for building

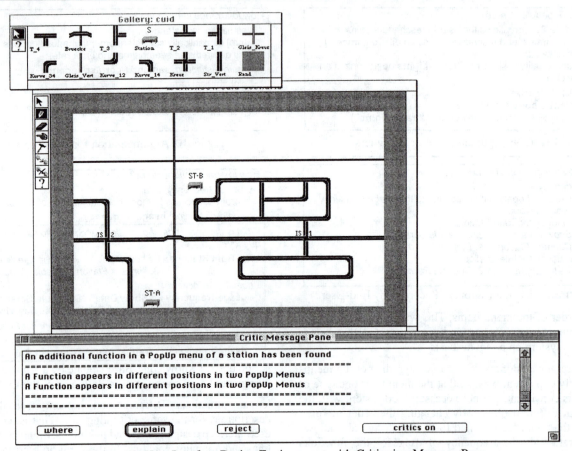

Figure 13: User Interface Design Environment with Critiquing Message Pane

visual design environments [22]. Agentsheets supports the creation of galleries of design units that can be used to build visual representations in a work area.

In this prototype system, tools for adding, erasing, and moving objects can be selected from a toolbar. The user interface objects are presented in the gallery and can be copied into the artifact. Currently, the application-specific objects include streets, intersections, railroad lines, bridges, and stations. Additionally, pull-down and pop-up menus are available (Figure 13).

The critiquing component consists of a critiquing manager and a set of rules. The manager decides which rules are applicable in the current situation, based on the action of the designer and on the type of the objects involved in the current design activity. The rules themselves are of the form

> if *condition* then *reaction*

with condition describing a situation that is an error or a problem that requires a warning. Reaction defines the message that will be given to the designer.

The critiquing component monitors the actions of the designer. Every time an interface object is placed, moved, or deleted, or the definition of a menu is changed, the critiquing manager identifies all rules that are applicable in this situation. The conditions of all rules are inspected. The messages defined in the rules where the condition is true are displayed in the message window. This means that all information is delivered to the designer immediately as a response to the design activity. Additionally, the designer

can select a "critique complete design" function. This is necessary because some conditions can be checked in a meaningful way only when the design is finished; otherwise, they would deliver the same information after each design step. An example involves the completeness rules (see below).

The messages are presented in a message window. In this window, the "explain" button gives access to the explanations of rules. Additionally, rules can be disabled for the current situation ("reject" button), allowing the designer to define an exception situation. A rule disabled for a specific situation will not be checked in the future for this part of the design, but remains active for the rest of the artifact. Pressing the "where" button highlights all objects related to any currently displayed message. Selecting a single message highlights only the objects related to this message.

The set of currently implemented rules covers different groups of rules. Some rules represent user interface design principles, like a maximum number of items in a menu or the maximum depth of cascaded menus. The second group of rules checks the completeness of the design according to a given specification. This currently covers the required objects and required functions for the system and for the individual objects. Another group of rules represents the consistency requirements, for example, that functions with the same name always appear in the same location in all pop-up menus. Finally, a group of rules that represent design knowledge that is strictly application domain specific was implemented. These rules describe the topological relationship of user interface objects that are allowed in this domain.

As mentioned above, our prototype system integrates two knowledge-bases: user interface guideline knowledge and application domain knowledge. Figure 14 illustrates an applicability condition. It checks whether the operations in the same position of two different menus (menu-A and menu-B) are semantically consistent. Note that the function "menu-consistency" is independent from the application domain. Providing another "domain-map" will check semantic consistency in a different domain.

In order to link critiquing rules with argumentation, we use the notion of domain-distinctions (see Figures 9, 10, and 12). The argumentation-base is structured as a network of nodes consisting of issues, answers, and arguments based on IBIS [4]. An issue-answer pair represents a design decision in terms of function, structure, or behavior at various levels of abstraction. In our system, issue-answer pairs and arguments are associated with domain-distinctions. Whereas the domain-distinctions come from the vocabulary of user interface design, the arguments are part of the application domain vocabulary. Each domain-distinction is related to a LISP function that checks the current interface design. For example, "Semantic-correspondence (menu-A menu-B)" is related to two predicates: first to determine which functions (operations) are semantically corresponding (i.e., domain knowledge), and second, to determine whether semantically corresponding operations such as "schedule screen" and "traffic lights" occur in different positions in different pop-up menus.

DISCUSSION AND FUTURE DIRECTIONS

The use of critiquing techniques for user interface design is not new [14], and supporting end-user adaptation by a critiquing component is implicitly suggested by the CAA approach [13]. However, none of the existing approaches combine the two and use the technique to facilitate long-term collaboration among end-users and user interface designers. Our approach is to integrate the two activities with a design environment, where the knowledge-base gradually evolves through the two activities.

```
(defun menu-consistency
#II checks whether operations in menu-A and operations in menu-B are
ordered semantically consistent. Semantic consistency among operations
for different types of object is provided by "traffic-control-map" II#
        :menus (menu-A menu-B)
        :domain-map #'traffic-control-map
        (for each (item-a item-b) in (menu-A menu-B) do
        (let
            ((consistent
                (semantic-check
                item-a
                item-b
                traffic-control-map))
            (if (not consistent)
(                fire-critic-rule #'menu-consistency item-a item-b)))

(setq traffic-control-map
#II a list of semantically corresponding (type operation) paris. II#
        #'( (Trainstation 'Schedule-Screen) (Intersection 'Traffic-Lights)
        .....)
```

Figure 14: An Applicability Condition Rule for
Menu Consistency

As of now, the critiquing component has been implemented, and ways to implement a design rationale component and to combine it with a critiquing component have been studied in different domains, including voice dialog design and kitchen design [20]. A hypermedia system has proven to be a useful basis for the implementation.

Critiquing and design rationale in user interface design differ from other domains that have been investigated, as different types of critiquing and explanation are necessary for users with different background knowledge, i.e., user interface designers and end-users. The need for different support is obvious, but further investigations are necessary to identify how to consider these differences.

To make the critiquing component most effective for the designer, only information relevant to the current situation should be delivered. The designer needs the ability to customize the set of applicable rules. A first step is made by allowing the designer to disable rules for specific situations. Additionally, the rules need to be organized in different rule sets that can then be enabled or disabled by the designer. The need for rule sets occurs, for example, when a designer creates user interfaces for different customers and each of them has corporate design guidelines. Additionally, different domains require different rule sets.

Furthermore, it will be interesting to explore how different mechanisms to support the designer or the design process can be combined in a way that is most effective for the designer. Mechanisms that share the goal of supporting the designer in creating a usable interface include constraints, implemented in the user interface design toolkit; critics, as described in this paper; and agents, which also support the designer in an active way.

Finally, integrating a specification component that allows stakeholders to state their design requirements with a critiquing mechanism allows further tunes of the critics. For example, if the user interface design is a walk-up-and-use type of system, the operations of a pop-up menu should be ordered alphabetically, or, if the user interface needs to be internationalized, then the operations should not be ordered alphabetically because different languages may cause conflicts among the orders. Those conflicting guidelines will support designers and end-users to further refine their understanding of the design intention [20].

In this paper, we have presented our approach of using critiquing mechanisms to support user interface design. We have discussed problems inherent in user interface design and claimed that such design should be facilitated as an ongoing, evolving, and collaborative design process between user interface designers and end-users. Critiquing mechanisms can support both end-users and interface designers by providing design knowledge relevant to their task at hand, and invoke them to record design rationale. Such stored design rationale plays a crucial role in supporting long-term asynchronous collaboration among interface designers during redesign and end-users during adaptation.

Acknowledgments

We would like to thank Gerhard Fischer and the HCC group at the University of Colorado, who contributed to the conceptual framework and the systems discussed in this paper. This research was conducted at the University of Colorado and supported by Software Research Associates, Inc. (Tokyo, Japan) and Siemens Corporate Research and Development (Munich, Germany).

References

1. N.Bonnardel, T.Sumner: From System Development to System Assessment: Exploratory Study of the Activity of Professional Designers. in Proceedings of the 7th European Conference on Cognitive Ergonomics, 1994, forthcoming.

2. B.Curtis, H.Krasner, N. Iscoe: A Field Study of the Software Design Process for Large Systems. Communications of the ACM, Vol. 31, No. 11, November, 1988, pp. 1268-1287.

3. P. Ehn and M. Kyng: The Collective Resource Approach to Systems Design. In P.Ehn, G.Bjerknes, M. Kyng (eds.): Computers and Democracy – A Scandinavian Challenge. Avebury, Aldershot, UK, 1987.

4. J. Conklin and M. Begeman: gIBIS: A Hypertext Tool for Exploratory Policy Discussion, Transactions of Office Information Systems, Vol. 6, No. 4, 1988, pp. 303-331.

5. G. Fischer: Communications Requirements for Cooperative Problem Solving Systems. In The International Journal of Information Systems (Special Issue on Knowledge Engineering), Vol. 15, No. 1, 1990, pp. 21-36.

6. G. Fischer, J. Grudin, A.C. Lemke, R. McCall, J. Ostwald, B.N. Reeves, F. Shipman: Supporting Indirect, Collaborative Design with Integrated Knowledge-Based Design Environments. Human Computer Interaction, Special Issue on Computer Supported Cooperative Work, Vol. 7, No. 3, 1992, pp. 281-314.

7. G. Fischer, K. Nakakoji: Beyond the Macho Approach of Artificial Intelligence: Empower Human Designers - Do Not Replace Them. Knowledge-Based Systems Journal, Vol. 5, No. 1, 1992, pp. 15-30.

8. G. Fischer, K. Nakakoji, J. Ostwald, G. Stahl, T. Sumner: Embedding Critics in Design Environments. The Knowledge Engineering Review Journal, Vol. 4, No. 8, 1993, pp. 285-307.

9. G. Fischer, R. McCall, J. Ostwald, B.N. Reeves, F. Shipman: Seeding, Evolutionary Growth and Reseeding: Supporting Incremental Development of Design Environments. Human Factors in Computing Systems, CHI'94 Conference, (Boston, MA), ACM, pp. 292-298.

10. G. Fischer, B.N. Reeves: Beyond Intelligent Interfaces: Exploring, Analyzing and Creating Success Models of Cooperative Problem Solving. Applied Intelligence, Special Issue Intelligent Interfaces, Vol. 1, No. 1992, pp. 311-332.

11. J. Greenbaum, M. Kyng: Design at Work: Cooperative Design of Computer Systems. Lawrence Erlbaum Associates, Hillsdale, NJ, 1991.

12. K. Haltzblatt, H. Beyer, Making Customer-Centered Design Work for Teams, Communication of the ACM, Vol.36, No.10, October, pp. 93-103.

13. T. Kühme: A User-Centered Approach to Adaptive User Interfaces. In Knowledge -Based Systems Vol. 6, No. 4, December 1993, Butterworth-Heinemann, Oxford, 1993.

14. A.C. Lemke, G. Fischer: A Cooperative Problem Solving System for User Interface Design, Proceedings of AAAI-90, Eighth National Conference on Artificial Intelligence, AAAI Press/The MIT Press, Cambridge, MA, August 1990, pp. 479-484.

15. U.Malinowski, T.Kühme, H.Dieterich, M.Schneider-Hufschmidt: A Taxonomy of Adaptive User Interfaces. In A. Monk, D. Diaper, M. D. Harrison (Eds.): People and Computers VII, Proceedings HCI '92. Cambridge University Press, Cambridge 1992.

16. U.Malinowski, T.Kühme, H.Dieterich, M.Schneider-Hufschmidt: Computer-Aided Adaptation of User Interfaces with Menus and Dialog Boxes. In G.Salvendy, M.J.Smith (eds.): Human-Computer Interaction: Software and Hardware Interfaces. Proceedings of HCI Int. 1993, Orlando, Florida. Elsevier, Amsterdam, 1993.

17. N. Miyake: Constructive Interaction and the Iterative Process of Understanding. Cognitive Science, Vol. 10, No. 1986, pp. 151-177.

18. K. Nakakoji: Increasing Shared Understanding of a Design Task between Designers and Design Environments: The Role of a Specification Component. Dissertation Thesis, University of Colorado at Boulder, 1993.

19. K.Nakakoji, G.Fischer: Intertwining Knowledge Delivery and Elicitation: A Process Model for Human-Computer Collaboration in Design. In Special Issue of Knowledge-Based Systems on Human-Computer Collaboration, forthcoming in 1995.

20. K.Nakakoji, T. Sumner, B.Harstadt: Perspective-Based Critiquing: Helping Designers Cope with Conflicts Among Design Intentions. In J.Gero (ed.): Artificial Intelligence in Design '94. Butterworth-Heinemann 1994.

21. M. Polanyi: The Tacit Dimension. Doubleday, Garden City, NY, 1966.

22. A. Repenning: AGENTSHEETS: A Tool for Building Domain-oriented Dynamic, Visual Environments. University of Colorado at Boulder, Department of Computer Science, Technical Report CU-CS-693-93, 1993.

23. H. Rittel, M.M. Webber: Planning Problems Are Wicked Problems. In N. Cross (ed.), Developments in Design Methodology. John Wiley & Sons, New York, 1984.

24. H.A. Simon: The Structure of Ill-Structured Problems. Artificial Intelligence, Vol. No. 4, 1973, pp. 181-200.

25. A.S.Snodgrass, R. Coyne: Is Designing Hermeneutical? Technical Report, Department of Architectural and Design Science, University of Australia, 1990.

26. K.Yakemovic, E.Conklin: Report on a Development Project Use of an Issue-Based Information System. In Proceedings of the Conference on CSCW, Los Angeles, October 1990, pp. 105-118.

A Theoretically Motivated Tool for Automatically Generating Command Aliases

Sarah Nichols & Frank E. Ritter

ESRC Centre for Research in Development, Instruction and Training
Department of Psychology
University of Nottingham
University Park
Nottingham
NG7 2RD
Tel: ++44 (0115) 951-5151
E-mail: ritter@psyc.nott.ac.uk

ABSTRACT

A useful approach towards improving interface design is to incorporate known HCI theory in design tools. As a step toward this, we have created a tool incorporating several known psychological results (e.g., alias generation rules and the keystroke model). The tool, simple additions to a spreadsheet developed for psychology, helps create theoretically motivated aliases for command line interfaces, and could be further extended to other interface types. It was used to semi-automatically generate a set of aliases for the interface to a cognitive modelling system. These aliases reduce typing time by approximately 50%. Command frequency data, necessary for computing time savings and useful for arbitrating alias clashes, can be difficult to obtain. We found that expert users can quickly provide useful and reasonably consistent estimates, and that the time savings predictions were robust across their predictions and when compared with a uniform command frequency distribution.

KEYWORDS

HCI design tools; Keystroke-Level Model; design problem solving.

INTRODUCTION

Card, Moran and Newell [1] noted that the action is in the design. While some work has gone into tools for design that incorporate their and others' theories (see, for example, Casner and Larkin [3]), most of interface design is still done by hand, and without direct recourse to known HCI theory.

The Rationale for Theoretically Motivated Design

This situation is unfortunate, for there are now numerous

HCI results that could be applied, for example, about good and bad HCI interfaces [1, 11], about problem solving and interface use [7, 8], and about perceptual processes [3]. These results offer the opportunity to improve interface design at an early stage. However, it may be beyond the grasp of a designer with a deadline to apply all these results, particularly if the designer has to find the results and then apply them by hand.

The Need For a Tool to Apply Theory to Design

The appropriate approach, it appears to us, is to incorporate theories into a tool or set of tools to either design interfaces or to quickly inspect designs. For an HCI tool to be beneficial, the utility has to be easy to use, cheap, automatic, honest, and complete for the task at hand. The tool must also be flexible to incorporate additional results. This affords fast implementation of psychological theory in an accessible and testable environment.

One often used and easily approached way to improve interfaces is by introducing *aliases* — abbreviations of command names. The use of aliases reduces the work required to execute a command and thus also lowers the possibility of error in expert users (for errors that are simply related to workload). It is also an area where there are known constraints that can inform design, both from empirical research and from HCI theories [1, 7].

Command line interfaces are a worthy area to serve as an example application of such a tool. They are ubiquitous, and as we shall show, the payoff can be quite large. The commands are often needlessly long, and typing mistakes can require a lengthy command to be re-entered. We also chose this type of interface because it is easy to work with and, in this case, has practical use in our daily work.

The use of aliases is not a new concept, but as far as we are aware, there have been no tools put forward to create aliases automatically. It would be useful to have tools that help in HCI related tasks in general — some screen layout tools are available [3] but they are not used much. If this lack of application is caused by the difficulty in applying the theory to a task at hand easily and quickly, a step toward remedying the lack of application of theory to design is to create a tool to make this process easier.

We have created a tool incorporating several known psychological constraints for creating theoretically motivated aliases for command line interfaces. The tool described here has been applied to the command line interface of Soar, a cognitive modelling system [8]. A set of principles for abbreviation method have been incorporated into an alias generating function, and the predicted execution times of the created aliases have been compared with the existing commands using the Keystroke-Level Model.

THE THEORIES IN THE TOOL

We incorporated two sets of psychological theories into our alias design tool. The first predicts how long it will take to enter the command set when they are weighted by frequency. The other is used to derive a set of aliases. They are brought together in a programmable spreadsheet.

The Keystroke-Level Model

We used the Keystroke-Level Model [1] as a measure of design efficiency as it predicts the time taken for a set of commands or aliases to be executed based on sub-task speeds and frequency of tasks. It is a useful and practical simplification of GOMS (Goals, Operators, Methods and Selection) analysis at the level of individual keystrokes [1], when the user's interaction sequence can be specified in detail, as it can here. The Keystroke-Level Model has been shown to predict user performance with between 5 and 20% error [2, 5].

There are several basic assumptions and restrictions associated with the Keystroke-Level Model that this domain and our approach satisfy:

*   The acquisition time and execution time for a task are independent.

*   The users are experts.

*   No account is taken of errors or error recovery.

The total time taken for one task is considered to consist of the time taken to acquire the task (which is not addressed here) and the execution time. The main components of command execution that are considered here are the physical motions required to input the command (in the case of a command line language this is keystrokes) and the number of mental operators that are necessary to remember and correctly execute the commands. Our current model is adaptable for typing speed, but does not include the effect of different typing speeds for different letters. The inclusion of mental operators is governed by heuristics specified by Card et al. [1], specifically rule 2 in Figure 8.2.

An important question to be considered is the increase in burden on memory and mental operators that the introduction of aliases might impose. The Keystroke-Level Model does make several suggestions about how the aliases should be generated consonant with the generation guidelines below. If a rule set is used to generate the aliases, once the initial (very small) rule set has been learned only a single mental operator is required

to enter a command alias and that behaviour can become automated more quickly — rather than having to learn by rote each alias, a rule can be applied [1].

The Keystroke-Level model does not specify where the rules or keystrokes come from, or how they are mapped onto the task. This leaves the interface builder with a classic design problem of creating a set of aliases that are easily learned and quickly entered.

Existing Guidelines for Alias Set Generation

While the Keystroke-Level Model does not tell us how to create commands themselves, there are some results that make strong suggestions for how to create aliases of existing commands. There are two main techniques commonly used for abbreviating command names — *truncation* and *contraction* (although other rules have also been used [7]). Truncation involves deleting the last few letters of the original command name, and contraction involves removing letters from the middle and end of the word. In order to avoid confusion, the abbreviation length can be governed by a *minimum-to-distinguish* system, where a sufficient number of letters are used to form the abbreviation to ensure that the abbreviation is unique. However, this means that users must know how many letters are required for each individual command. Research into techniques of abbreviation has shown that abbreviations formed by truncation are more easy to encode (i.e., produce the abbreviation on presentation of the word) than abbreviations formed by contraction [4]. It has been suggested that this is because it is easier to have consistency between commands abbreviated by truncation (e.g., with the rule *always use the first three letters*) as opposed to contraction. A GOMS analysis [7] has also shown that two letter truncation and minimum-to-distinguish are the most efficient forms of abbreviation.

Watts [11] emphasises the need for consistency of abbreviation format within a system, and notes that it is vital to avoid confusion between aliases. If all commands are to be shortened to single letters, then although time would undoubtedly be saved at a keystroke level the memory burden would be greatly increased as the user would have to remember which of the commands were abbreviated to a single letter and which were exceptions. Ehrenreich and Porcu [4] state that if abbreviations are generated by a simple rule then the memory load is greatly reduced. If rules are learned then once the rules are known by the user then their access becomes automated and there is no increase in mental operator time.

Payne and Green [9] have proposed a theory of learnability based on writing a task action grammar (TAG). In their case, different types of subtasks would have different syntax, which would provide a basis for using different abbreviation mechanisms or rules for different types of subtasks. This should decrease the number of clashes and make the resulting aliases more memorable. Aliases presumably would then indicate their category, which would lengthen them. However, if exceptions do occur, then they have to be dealt with using rules similar to the rules above, such as vowel deletion. Since different rule generating mechanisms can be used across subtasks, the user must remember which category the command is in to retrieve the exception rule. An experiment that they performed compared aliases generated for two interfaces, one with a fairly uniform TAG, and with a less uniform TAG. The aliases for the more uniform interface yielded fewer errors, a lower rate of use of a help facility, and a more efficient solution of the problem than the less uniform interface. But their paper by no means states the method supported by Ehrenreich and Porcu [4] produces bad aliases, merely that the command language itself should be regular as well.

In the development and application of this tool we were not prepared to include the redesign of command sets, which the application of TAG encourages, and which Payne and Green note may have to be done before alias creation. We are not sure how easily the Soar commands would lend themselves to being divided up into different types of tasks, as would be necessary if a TAG approach was taken. TAG grammars and semantics (such as destructive commands) could be included as a future extension, by providing a column to indicate a semantic category of each command, and a function that uses that category when automatically creating the alias set.

It can be difficult to apply these principles consistently and time-consuming to implement them by hand. The Keystroke-Level Model notes that command frequency data needs to be gathered, new aliases need to be devised, and in order to assess how useful the aliases are, the time savings need to be computed. Therefore the next step would appear to be to provide a flexible and extendible facility to create and assess aliases automatically.

Dismal — a Motivated Tool for HCI

Dismal [10] is a spreadsheet that was explicitly developed for manipulating psychology data. Dismal is written in GNU-Emacs Lisp, making extensions and modifications such as computations based on the Keystroke-Level Model easy to incorporate. This approach could be used within any programming language, but we prefer a spreadsheet to display the aliases. This visual presentation and the use of functions to compute and display the expected times makes the process easy to follow and provides updates automatically to the designer. We believe that recent versions of commercial spreadsheets now often include a full programming language providing the necessary functionality, but we would be less able to distribute the complete tool, because Emacs is free and can now be run on both UNIX systems and Macs.

The HCI theories were implemented by using two Lisp functions which were added to the Dismal spreadsheet. The first — *key-val* — took a command and calculated the number of mental operators and keystrokes used in the execution of the command. It then used the estimated typing speed to calculate a time prediction for the execution of the command according to the Keystroke-Level Model. The second function — *make-alias* — used the specific rules noted below to automatically generate

commands aliases. However, human input was still needed to decide the best way to arbitrate clashes between aliases given the large command set examined here.

EXAMPLE APPLICATION

The command set initially optimised with this tool was taken from Soar [8], a unified theory of cognition realised as a cognitive modelling language. It has previously been command line driven, and over 50 commands are available. In the past year, a Soar Development Environment (SDE) [6] has been developed, which is menu and keystroke driven. The command line interface remains an important part of SDE however, as some members of the Soar community still prefer a command line interface. Also, some Soar users cannot use SDE due to the limitations of the hardware that they use to run Soar.

The ideas behind the alias creation could in fact be applied to any command set, but Soar was a suitable candidate for modification for several reasons.

- The 'alias' command recently added to Soar allows aliases to be added quickly, easily and cheaply.

- The Soar language is currently being used within our local environment and therefore it is worthwhile enhancing the system for both local users and the Soar community world-wide.

- There is potential to get command use frequency information and usability feedback from the local users.

- It is fair to assume that most Soar users can be considered "expert users" and are highly familiar with the commands.

- Some of the commands that exist within Soar are quite long — this suggests that users of the Soar Command Line interface would benefit greatly from the introduction of aliases.

Devising the Aliases

The aliases were devised with the generating function noted above incorporating the alias generation guidelines in the literature (primarily truncation, which means that the abbreviation is the first two letters of the original command). The function aimed to minimise keystrokes while avoiding ambiguity. The characteristics of the command language itself were also considered — particularly with reference to the fact that many of the commands consisted of several words. Therefore, we added the following additional guidelines to create a consistent alias set:

- Include in the alias the first letter of each word in the case of multi-word commands.

- If the length of the command is 5 letters or less, a one letter alias may be provided if clashes do not occur as a result of the new abbreviation. In general this means that already short, abbreviated commands (e.g., *pwd*), do not get shortened to one letter, but short one word commands (e.g., *watch*) do get abbreviated to a single letter.

The value for the mental operator M was taken from Card, Moran, and Newell [1] as 1.35 s. The speed of typing was assumed to be 40 wpm (average non-secretary typist), or 280 ms/keystroke.

Other Aliases Included in Our Set

The alias set generated automatically was augmented for distribution with aliases generated with several additional principles in mind. Computation of time saved, however, was done solely with the main set.

Aliases for experts _and_ novices. While most of our expected users will be experts, every expert used to be a novice, so aliases should be available for use by both. A rule based system for developing aliases is appropriate for both of these categories of users. However, for the purposes of analysis, the times are only applicable to the experts' behaviour, as the Keystroke-Level Model can only be used to predict the speed of an expert completing a task. The alias set is still useful for novices, but this cannot be quantitatively shown using the Keystroke-Level Model.

For experts, once the rules have been learned, no additional knowledge is required in order to be able to use the aliases and so time is instantly saved in the form of keystroke reduction without an increase in mental load from having different abbreviations to remember for different commands

For novices, the meanings of the original commands can still be retained — the aliases are both syntactically and semantically compatible. This can be contrasted with control keystroke command type of aliases, which although reducing time, remove any semantic component of the original command. This is particularly important when considering Soar, which is both a theory and a language, but is also relevant when considering other command line languages.

Aliases — reducing errors. We can reduce how the long commands within Soar can lead to errors in two ways. (a) The number of keystrokes required means that errors are more likely to occur than with shorter commands. Therefore by shortening the command names we expect both to save time and reduce the likelihood of simple typing errors. (b) We included several common misspellings of commands (e.g., *init-saor* for *init-soar*).

Estimations of Task Frequency

The Keystroke-Level Model uses task frequency to balance the time it takes to do each task in a set of unit-tasks. An initial analysis of approximately two hours of actual subject data was performed to compute these frequencies. In this session, commands were used while a specific problem was solved. Perusal of additional transcripts suggested that command usage was highly dependent on the task, there were large individual differences, and in order to generate meaningful frequency

data enormous amounts of keystroke logs would be required (we estimate on the order of hundreds of hours). Steps have been taken to log this data in the latest version of the Soar Development Environment [6].

In an attempt to generate useful frequencies more quickly, four expert Soar users, all with more than three years experience working with Soar, were asked to provide frequency estimates for their own use of the original command set. These were easy to provide, although they do not correlate very well with each other (mean pair-wise correlation of 0.49). In future work we expect to validate this approach as an approximation to complete logs.

RESULTS

The automatic function generated 80% of the final aliases, where the rules could be strictly adhered to, and the exceptions were manually adjusted according to frequency information and the conventions shown in Table 1. The complete alias set is shown in Table 2.

Table 1 - How Exceptions Were Dealt With.

The letters here are included as footnotes in Table 2, indicating how the various aliases that could not have automatic aliases created were adjusted.

(a) These commands were already of length <= 3 letters and if they had been shortened any more then clashes would have occurred

(b) The alias *q* rather than *e* is used to prevent accidental exit from Soar as so many other commands have the initial letter e. If *e* is typed then the *echo* command will be executed — this is not a dangerous situation.

(c) The commands *load* and *log* are both short, but are abbreviated by contraction rather than truncation to avoid confusion between the two — if the abbreviation *lo* is typed then a warning is echoed to the screen advising the user to type *ld* or *lg*. The same principle applies to *warnings* and *watch* with a warning being displayed if *wa* is typed.

(d) The command *memory-stats* is shortened to *mems* rather than *ms* to avoid confusion with the already existing Soar command *ms* (which means "list the rules that match"). However, if the user follows the rules then the result is not dangerous.

Figure 1 shows the estimated time savings based on the time to perform the original command set and the alias set balanced for each individual's frequency distribution. The time taken to execute the original command set is represented by 100%. This total time, based on the normalised command frequency distributions, varied between 245 s (flat frequency) to 458 s (User 3). When the keystroke values of the commands and aliases were calculated and weighted using a uniform command distribution (i.e., assuming all commands were used equally often), it was found that the aliases provided a 55% saving in time over the original command set.

Table 2 - Predicted Execution Times for Original Commands and Aliases.

See Table 1 for full explanation of exceptions noted as footnotes. Data imported from Dismal spreadsheet, values rounded to two decimal places.

	Original Command	Predicted time (s)	Alias	Predicted time(s)	Time Saving
1	add-wme	4.87	aw	2.17	56%
2	agent-go	5.13	ag	2.17	58%
3	cd (for chdir)	2.17	cd[a]	2.17	0%
4	chunk-free-problem-spaces	12.42	cfps	2.70	78%
5	d	1.90	d	1.90	0%
6	default-print-depth	9.46	dpd	2.43	74%
7	echo	2.70	e	1.90	30%
8	excise	3.24	ex	2.17	33%
9	excise-all	5.67	ea	2.17	62%
10	excise-chunks	6.49	ec	2.17	67%
11	excise-task	5.94	et	2.17	64%
12	exit	2.70	q[b]	1.90	30%
13	firing-counts	6.49	fc	2.17	67%
14	go	2.17	g	1.90	13%
15	help	2.70	h	1.90	30%
16	init-soar	5.40	is	2.17	60%
17	learn	2.98	le	2.17	27%
18	list-chunks	5.94	lc	2.17	64%
19	list-help-topics	8.64	lht	2.43	72%
20	list-justifications	8.10	lj	2.17	73%
21	list-productions	7.60	lp	2.17	70%
22	load	2.70	ld [c]	2.17	20%
23	log	2.43	lg [c]	2.17	11%
24	matches	3.51	ma	2.17	38%
25	max-elaborations	7.30	me	2.17	70%
26	memory-stats	6.21	mems [d]	2.70	57%
27	ms	2.17	ms [a]	2.17	0%
28	object-trace-format	9.46	otf	2.43	74%
29	p (for print)	1.90	p	1.90	0%
30	pgs	2.43	pgs [a]	2.43	0%
31	preferences	4.59	pr	2.17	53%
32	print-all-help	8.10	pah	2.43	70%
33	print-stats	5.94	ps	2.17	64%
34	ptrace	3.24	pt	2.17	33%
35	pwd	2.43	pwd [a]	2.43	0%
36	quit	2.70	q	1.90	30%
37	r (for run)	1.90	r	1.90	0%
38	remove-wme	5.67	rw	2.17	62%
39	rete-stats	5.67	rs	2.17	62%
40	schedule	3.79	sc	2.17	43%
41	select-agent	6.21	sa	2.17	65%
42	soar-news	5.40	sn	2.17	60%
43	sp	2.17	sp [a]	2.17	0%
44	stack-trace-format	9.19	stf	2.43	74%
45	stats	2.98	s	1.90	36%
46	time	2.70	t	1.90	30%
47	unptrace	3.79	un	2.17	43%
48	user-select	5.94	us	2.17	64%
49	version	3.51	ve	2.17	38%
50	warnings	3.79	wr	2.17	43%
51	watch	2.98	wt	2.17	27%
52	wm	2.17	wm [a]	2.17	0%
TOTAL	(Using flat weighting)	245.17		132.58	55%

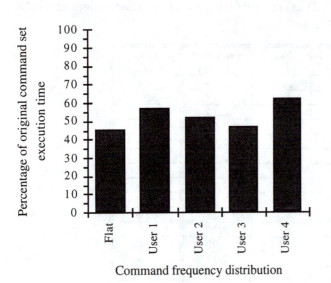

Figure 1 - Comparison of Predicted Benefits of Generated Aliases for Different Command Frequency Distributions.

Although each user had different preferences regarding which commands they used (or thought they used) to perform tasks, when savings were calculated with these frequency estimates, the time savings for individual users for the complete command set ranged between 38 and 53%. When the unchanged command names were omitted from the analysis, the average time savings across all the distributions increased to 62%.

DISCUSSION AND CONCLUSIONS

Overall, this tool appears to be a robust and inexpensive way of applying simple HCI theories to design to reduce command execution time. Aliases can be constructed and tested very easily, and, with the use of an HCI tool, the principles behind the forms of command aliases can be applied routinely and uniformly. Savings estimates can also be documented directly and used to guide design by hand when it is necessary or desirable. The use of a system using guidelines to generate aliases means that the alias forms are easy to learn and can generally be predicted without the particular alias being specified. The local Soar user group has found the alias set to be generally useful, and the aliases have also been distributed to the Soar community at large. Aliases improve the interface at quite modest cost, there appears to be no reason not to take this efficiency gain.

Successes and Implications

Several successes can be drawn from this initial attempt at creating and applying a tool that incorporates an HCI theory. Some of these concern the tool and the theory it incorporates, and others have implications for such theory-based design tools in general.

Comparison with automatic command completion. An analysis using Dismal, Emacs and the existing alias set suggests that automatic command completion would be inferior to this alias set. Command completion would not reduce the keystrokes in this set of commands as much as the aliases do, and would not decrease the mental operators at all.

Implications from the Keystroke Model — The effect of mental operator time. One of the factors that increased the Keystroke-Level Model's time predictions for the original command names was the number of mental operators in the original commands, one for each word. Using single word or acronym aliases are faster not only because they are shorter, but because they require fewer mental operators to perform. In the future, it may be more appropriate to include different types of mental operators for different parts of the task as occurs in the GOMS models [1].

Implications for the Keystroke Model — Task frequency generation and use. Command frequency distributions are used to arbitrate alias clashes and compute expected savings. Ideally, one might assume that these logs should come from actual users. For large command sets, this appears to be unnecessary and considerably expensive. All four of our expert users had different patterns of command use but they all showed a considerable and nearly equal decrease in execution time when aliases where used. The effect of the new command set appears to be robust across the various frequency predictions, and indeed, even if a flat distribution is used. In order to make this approach more tractable, we suggest using frequency estimates instead.

The Effect on Errors. No users type perfectly, and decreasing command length, if it is consistent, may also decrease errors for expert users (where expert users are those who are familiar with Soar, the command abbreviation rules and the aliases themselves). While we do not have empirical evidence of this yet for our alias set, we can derive a simple prediction. Our alias set reduced the number of keystrokes by 68%. (This is greater than the time savings because the alias generation algorithm removes more keystrokes than mental operators.) The simplest prediction is that the simple typing errors are related to the number of keystrokes, no matter what the users' typing accuracy or speed. Although words are inherently easier to type, these aliases should be fairly easy to type. The semantic component of the command is retained and they are generated in a regular way. This suggests that the number of simple typing errors should significantly decrease as well.

Remaining Problems

While we were generally pleased with the tool and the alias set it generated, three problems remain that need to be acknowledged.

Could the Tool Have Generated More Aliases Automatically? If the rules for devising the aliases had been increased in size then more of the aliases could have been generated automatically. However, the trade-off between increase in the size of the rule set and decrease in the number of exceptions must be considered. If an additional rule had been introduced (e.g., "if truncation to the first two letters of the command leads to a clash, use the first and third letters as the alias instead"), then some of the exceptions would have been automatically generated, but this would have been at the expense of an increase in rules to be learned. It is beyond the scope of this paper to consider the different cognitive burdens that these would impose.

One cannot enumerate all the causes of exceptions — we suspect that this is an open set. We found several here (e.g., usage frequency, clashes, common existing conventions for usage) and do not feel that we have even started to enumerate all the possible causes. This is a problem that is likely to remain, and so while the use of automatic generation should be maximised, the ability for easy manual intervention, such as is provided here with the use of an accessible and flexible spreadsheet, must be retained.

Could the Alias-Generating Tool Have Been More Powerful? As noted above, there are some differences in opinion about which alias generation rules are best, and we have ignored the possibilities of command frequency forcing inconsistencies as well. The point of this paper and tool was not to provide the definitive alias generating function but to show how such a system would work, and provide the means for creating such a function. In other words, the focus should be on our approach, not the specific details of our situation. Since aliases are now cheap and easy to customise, it is not difficult to create additional alias sets for individual users to either adapt our set or create a new set based on their preferred generating rule based on their own cognitive style.

Some points about the Soar aliases do need to be clarified. The Soar command set is a fairly small command set, which has two major implications. First, there were not many clashes formed by the automatic alias generation rules, and second, expert users, for whom aliases are created for, are likely to know all the commands in the set. When a larger command set is being considered, then the need for use of frequency data as a method of arbitration between alias clashes is increased.

Overall effect of improving the command line interface. While we can present a strong case for decreasing typing time, and these aliases appear to increase user satisfaction and reduce errors (although, we have only obtained anecdotal evidence only so far), this result may have a smaller effect on overall task time than might be implied for two reasons. (a) Since the Keystroke-Level model was being used as the analytic tool, the higher level processing was not considered. In the user logs we examined, only a small proportion of time was spent typing (again, this varies largely across users and time, and requires extensive user logs to measure). (b) It is also important to note that several of the commands take arguments, such as files to load or productions to examine. These are not considered here — from a practical point of view, computing their average length is a difficult problem where individual and site differences will reign. Including some estimate remains a necessary step, and the burden that we anticipate it to be suggests that providing them automatically would be a particularly fruitful action to include in the interface.

Future Directions
It is possible that this technique could be directly (although perhaps not simply) extended to analyse menus and other forms of input, as both the Keystroke-Level Model and GOMS analysis could handle this level of complexity. Indeed, the integrated and examinable nature of the menus in the spreadsheet we used (or the other Soar interface) suggests that this would be an appropriate and reachable next step.

Simple Lisp functions were used to implement the command set generation and their analysis. While not a problem in this tool, incorporating additional knowledge and extending the range of tasks that can be examined will become a serious problem quite soon given this representation. A higher level language for representing the constraints on good design and a more complete time prediction system (as could be provided in the future by an extension of Soar itself), will also be necessary steps.

ACKNOWLEDGEMENTS
Support for this work has been provided by the DRA, contract number 2024/004, and by the ESRC Centre for Research in Development, Instruction and Training. Erik Altmann graciously provided access to transcriptions of Soar users. Four anonymous expert Soar users kindly provided us with command frequency estimates. Paul Tingle, Matt Southall and Rob Jones performed a preliminary analysis suggested by Kate Cook that inspired this work. The anonymous CHI reviewers contributed comments that lead to greater clarity.

HOW TO GET DISMAL INCLUDING THESE EXTENSIONS
The latest versions of Dismal and the Soar alias command set are available via anonymous FTP from host unicorn.ccc.nott.ac.uk (128.243.40.7) in the directory "/pub/lpzfr".

REFERENCES

1. Card, S., Moran, T. and Newell, A. The psychology of human-computer interaction. 1983, Hillsdale, NJ: LEA.

2. Card, S.K., Moran, T.P. and Newell, A. The keystroke-level model for user performance time with interactive systems. Communications of the ACM, 1980. 23(7): p. 396-410.

3. Casner, S. and Larkin, J.H. Cognitive efficiency considerations for good graphic design. In Proceedings of

the Annual Conference of the Cognitive Science Society. 1989. Hillsdale, NJ: LEA.

4.   Ehrenreich, S.L. and Porcu, T. Abbreviations for Automated Syatems: Teaching Operators The Rules, In Directions in Human/Computer Interaction, A. Badre and B. Shneiderman, Editors. 1982, Ablex: Norwood, NJ.

5.   Haunold, P. and Kuhn, W. A. Keystroke Level Analysis of a Graphics Application: Manual Map Digitizing. In CHI '94, Human Factors in Computing Systems. 1994. Boston, MA: ACM.

6.   Hucka, M, The Soar Development Environment, 1994. Artificial Intelligence Laboratory, University of Michigan: Ann Arbor.

7.   John, B.E. and Newell, A. Predicting the time to recall computer command abbreviations. In CHI'87 Conference on Human Factors and Computing Systems. 1987. New York: ACM Press.

8.   Newell, A. Unified Theories of Cognition. 1990, Cambridge, MA: Harvard University Press.

9.   Payne, S.J. and Green, T.R.G. Task-action grammars: A model of the mental representation of task languages. Human-Computer Interaction, 1986. 2: p. 93-133.

10. Ritter, F.E, Lochun, S, Bibby, P.A. and Marshall, S. Dismal: A free spreadsheet for sequential data analysis and HCI experimentation. In Computers in Psychology '94. 1994. York (UK): CTI Centre for Psychology, U. of York.

11. Watts, R.A. Introducing Interactive Computing. 1984, Manchester: NCC Publications.

A Focus+Context Technique
Based on Hyperbolic Geometry
for
Visualizing Large Hierarchies.

John Lamping, Ramana Rao, and Peter Pirolli

Xerox Palo Alto Research Center
3333 Coyote Hill Road
Palo Alto, CA 94304
<lamping,rao, pirolli>@parc.xerox.com

ABSTRACT

We present a new focus+context (fisheye) technique for vi-sualizing and manipulating large hierarchies. Our technique assigns more display space to a portion of the hierarchy while still embedding it in the context of the entire hierarchy. The essence of this scheme is to lay out the hierarchy in a uniform way on a hyperbolic plane and map this plane onto a circular display region. This supports a smooth blending between fo-cus and context, as well as continuous redirection of the focus. We have developed effective procedures for manipulating the focus using pointer clicks as well as interactive dragging, and for smoothly animating transitions across such manipulation. A laboratory experiment comparing the hyperbolic browser with a conventional hierarchy browser was conducted.

KEYWORDS: Hierarchy Display, Information Visualization, Fisheye Display, Focus+Context Technique.

INTRODUCTION

In the last few years, Information Visualization research has explored the application of interactive graphics and animation technology to visualizing and making sense of larger informa-tion sets than would otherwise be practical [11]. One recur-ring theme has been the power of focus+context techniques, in which detailed views of particular parts of an information set are blended in some way with a view the of the overall structure of the set. In this paper, we present a new technique, called the hyperbolic browser, for visualizing and manipulat-ing large hierarchies.

The hyperbolic browser, illustrated in Figure 1, was origi-nally inspired by the Escher woodcut shown in Figure 2. Two salient properties of the figures are: first that components di-minish in size as they move outwards, and second that there

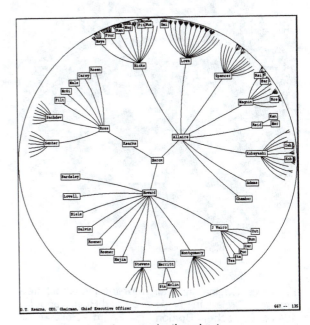

Figure 1: An organization chart.

is an exponential (devilish) growth in the number of compo-nents. These properties—"fisheye" distortion and the ability to uniformly embed an exponentially growing structure—are the aspects of this construction (the Poincaré mapping of the hyperbolic plane) that originally attracted our attention.

The hyperbolic browser initially displays a tree with its root at the center, but the display can be smoothly transformed to bring other nodes into focus, as illustrated in Figure 3. In all cases, the amount of space available to a node falls off as a continuous function of its distance in the tree from the point in the center. Thus the context always includes several generations of parents, siblings, and children, making it easier for the user to explore the hierarchy without getting lost.

The hyperbolic browser supports effective interaction with much larger hierarchies than conventional hierarchy viewers and complements the strengths of other novel tree browsers.

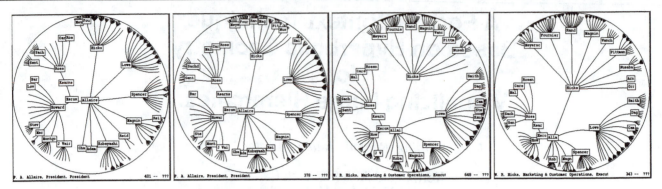

Figure 3: Changing the focus.

Figure 2: Original inspiration for the hyperbolic tree browser. Circle Limit IV (Heaven and Hell), 1960, (c) 1994 M.C. Escher / Cordon Art – Baarn – Holland. All rights reserved. Printed with permission.

In a 600 pixel by 600 pixel window, a standard 2-d hierarchy browser can typically display 100 nodes (w/ 3 character text strings). The hyperbolic browser can display 1000 nodes of which about the 50 nearest the focus can show from 3 to dozens of characters of text. Thus the hyperbolic browser can display up to 10 times as many nodes while providing more effective navigation around the hierarchy. The scale advantage is obtained by the dynamic distortion of the tree display according to the varying interest levels of its parts.

Our approach exploits hyperbolic geometry [2, 9]. The essence of the approach is to lay out the hierarchy on the hyperbolic plane and map this plane onto a circular display region. The hyperbolic plane is a non-Euclidean geometry in which parallel lines diverge away from each other. This leads to the convenient property that the circumference of a circle on the hyperbolic plane grows exponentially with its radius, which means that exponentially more space is available with increasing distance. Thus hierarchies—which tend to expand exponentially with depth—can be laid out in hyperbolic space in a uniform way, so that the distance (as measured in the hyperbolic geometry) between parents, children, and siblings is approximately the same everywhere in the hierarchy.

While the hyperbolic plane is a mathematical object, it can be mapped in a natural way onto the unit disk, which provides a means for displaying it on an ordinary (Euclidean) display. This mapping displays portions of the plane near the origin using more space than other portions of the plane. Very remote parts of the hyperbolic plane get miniscule amounts of space near the edge of the disk. Translating the hierarchy on the hyperbolic plane provides a mechanism for controlling which portion of the structure receives the most space without compromising the illusion of viewing the entire hyperbolic plane. We have developed effective procedures for manipulating the focus using pointer dragging and for smoothly animating transitions across such manipulation.

We have implemented versions of the hyperbolic browser that run on Unix/X and on Macintoshes. We conducted an experiment with 4 subjects to compare the hyperbolic tree browser with a conventional browser on a node location task. Though no statistically significant performance difference was identified, a strong preference for the hyperbolic tree browser was established and a number of design insights were gained.

PROBLEM AND RELATED WORK

Many hierarchies, such as organization charts or directory structures, are too large to display in their entirety on a computer screen. The conventional display approach maps all the hierarchy into a region that is larger than the display and then uses scrolling to move around the region. This approach has the problem that the user can't see the relationship of the visible portion of the tree to the entire structure (without auxiliary views). It would be useful to be able to see the entire hierarchy while focusing on any particular part so that the relationship of parts to the whole can be seen and so that focus can be moved to other parts in a smooth and continuous way.

A number of focus+context display techniques have been introduced in the last fifteen years to address the needs of many types of information structures [7, 15]. Many of these focus+context techniques, including the document lens [13], the perspective wall [8], and the work of Sarkar et al [14, 16], could be applied to browsing trees laid out using conventional 2-d layout techniques. The problem is that there is no satisfactory conventional 2-d layout of a large tree, because of its exponential growth. If leaf nodes are to be given adequate spacing, then nodes near the root must be placed very far apart, obscuring the high level tree structure, and leaving no nice way to display the context of the entire tree.

The Cone Tree[12] modifies the above approach by embedding the tree in a three dimensional space. This embedding of the tree has joints that can be rotated to bring different parts of the tree into focus. This requires currently expensive 3D animation support. Furthermore, trees with more than approximately 1000 nodes are difficult to manipulate. The hyperbolic browser is two dimensional and has relatively modest computational needs, making it potentially useful on a broad variety of platforms.

Another novel tree browsing technique is treemaps [5] which allocates the entire space of a display area to the nodes of the tree by dividing the space of a node among itself and its descendants according to properties of the node. The space allocated to each node is then filled according to the same or other properties of the node. This technique utilizes space efficiently and can be used to look for values and patterns amongst a large collection of values which agglomerate hierarchically, however it tends to obscure the hierarchical structure of the values and provides no way of focusing on one part of a hierarchy without losing the context.

Some conventional hierarchy browsers prune or filter the tree to allow selective display of portions of the tree that the user has indicated. This still has the problem that the context of the interesting portion of the tree is not displayed. Furnas [3] introduced a technique whereby nodes in the tree are assigned an interest level based on distance from a focus node (or its ancestors). Degree of interest can then be used to selectively display the nodes of interest and their local context. Though this technique is quite powerful, it still does not provide a solution to the problem of displaying the entire tree. In contrast, the hyperbolic browser is based on an underlying geometry that allows for smooth blending of focus and context and continuous repositioning of the focus.

Bertin[1] illustrates that a radial layout of the tree could be uniform by shrinking the size of the nodes with their distance from the root. The use of hyperbolic geometry provides an elegant way of doing this while addressing the problems of navigation. The fractal approach of Koike and Yoshihara [6] offers a similar technique for laying out trees. In particular, they have explored an implementation that combines fractal layout with Cone Tree-like technique. The hyperbolic browser has the benefit that focusing on a node shows more of the node's context in all directions (i.e. ancestors, siblings, and descendants). The fractal view has a more rigid layout (as with other multiscale interfaces) in which much of this context is lost as the viewpoint is moved to lower levels of the tree.

There have been a number of projects to visualize hyperbolic geometry, including an animated video of moving through hyperbolic space [4]. The emphasis of the hyperbolic browser is a particular exploitation of hyperbolic space for information visualization. We don't expect the user to know or care about hyperbolic geometry.

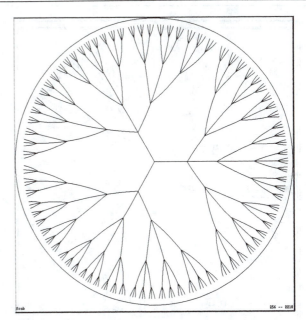

Figure 4: A uniform tree of depth 5 and branching factor 3 (364 nodes).

HYPERBOLIC BROWSER BASICS

The hyperbolic browser replaces the conventional approach of laying a tree out on a Euclidean plane by doing *layout* on the hyperbolic plane and then *mapping* to the unit disk (which is straightforwardly mapped to the display). *Change of focus* is handled by performing a rigid transformation of the hyperbolic plane, moving the laid out tree in the process. Thus layout is only performed once. Space for displaying *node information* is also computed during layout and automatically transformed with each change of focus.

The implementation of points and transformations on the hyperbolic plane is briefly discussed in the appendix. The rest of this section presumes an implementation of the hyperbolic plane and discusses higher level issues.

Layout

Laying a tree out in the hyperbolic plane is an easy problem, because the circumference and area of a circle grow exponentially with its radius. There is lots of room. We use a recursive algorithm that lays out each node based on local information. A node is allocated a wedge of the hyperbolic plane, angling out from itself, to put its descendants in. It places all its children along an arc in that wedge, at an equal distance from itself, and far enough out so that the children are some minimum distance apart from each other. Each of the children then gets a sub-wedge for its descendants. Because of the way parallel lines diverge in hyperbolic geometry, each child will typically get a wedge that spans about as big an angle as does its parent's wedge, yet none of the children's wedges will overlap.

The layout routine navigates through the hyperbolic plane in terms of operations, like moving some distance or turning through some angle, which are provided by the hyperbolic plane implementation.

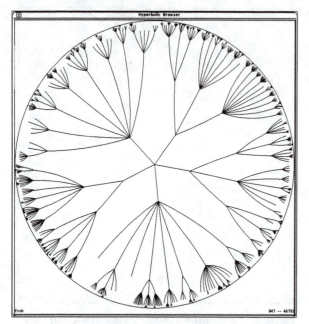

Figure 5: The initial layout of a tree with 1004 nodes using a Poisson distribution for number of children. The origin of the tree is in the center.

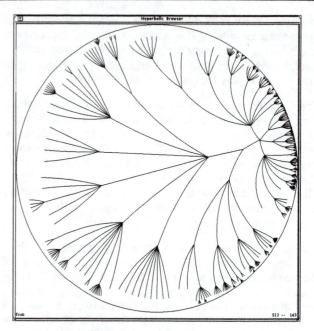

Figure 6: A new focus.

Figure 4 shows what the layout of a uniform tree looks like. Notice how the children of each node span about the same angle, except near the root, where a larger wedge was available initially. To get a more compact layout for non-uniform trees, we modify this simple algorithm slightly, so that siblings that themselves have lots of children get a larger wedge than siblings that don't (the wedge size grows logarithmically). This effect can be seen in Figure 5, where, for example, the five children of the root get different amounts of space. This tends to decrease the variation of the distances between grandchildren and their grandparent.

Another option in layout (in contrast to all examples so far illustrated) is to use less than the entire 360 degree circle for spreading out the children of the root node. With this option, children of the root could all be put in one direction, for example to the right or below, as in conventional layouts. An example of this option, discussed below, appears in Figure 8.

Mapping and Representation

Once the tree has been laid out on the hyperbolic plane, it must be mapped in some way to a 2-d plane for display. (We can barely imagine the hyperbolic plane, not to mention see it.) There are two canonical ways of mapping the hyperbolic plane to the unit disk. In both of them, one vicinity in the hyperbolic plane is in focus at the center of the disk while the rest of the hyperbolic plane fades off in a perspective-like fashion toward the edge of the disk, as we desire. We use the conformal mapping, or Poincaré model, which preserves angles but distorts lines in the hyperbolic space into arcs on the unit disk, as can be seen in the figures. The other possibility, the projective mapping, or Klein model, takes lines in the hyperbolic plane to lines in the unit disk, but distorts angles. You can't have it both ways.

We tried the Klein model. But points that are mapped to near the edge by the Poincaré model get mapped almost right on the edge by the Klein model. As a result, nodes more than a link or two from the node in focus get almost no screen real-estate, making it very hard to perceive the context.

Change of Focus

The user can change focus either by clicking on any visible point to bring it into focus at the center, or by dragging any visible point interactively to any other position. In either case, the rest of the display transforms appropriately. Regions that approach the center become magnified, while regions that were in the center shrink as they are moved toward the edge. Figure 6 shows the same tree as Figure 5 with a different focus. The root has been shifted to the right, putting more focus on the nodes that were toward the left.

Changes of focus are implemented as rigid transformations of the hyperbolic plane that will have the desired effect when the plane is mapped to the display; there is never a need to repeat the layout process. A change of focus to a new node, for example, is implemented by a translation in the hyperbolic plane that moves the selected node to the location that is mapped to the center of the disk.

To avoid loss of floating point precision across multiple transformations, we compose successive transformations into a single cumulative transformation, which we then apply to the positions determined in the original layout. Further, since we only need the mapped positions of the nodes that will be displayed, the transformation only needs to be computed for nodes whose display size will be at least a screen pixel. This yields a constant bound on redisplay computation, no matter how many nodes are in the tree. And, the implementation of translation can be fairly efficient; we require about 20 floating point operations to translate a point, comparable to the cost of rendering a node on the screen.

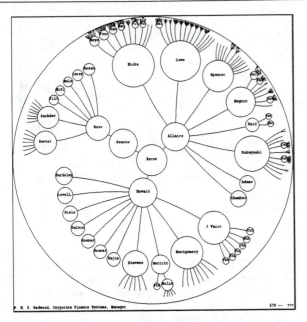

Figure 7: The display regions of nodes.

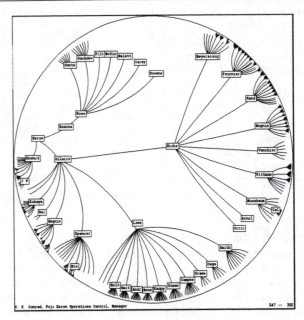

Figure 8: Putting children toward the right.

Node Information

Another property of the Poincaré projection is that circles on the hyperbolic plane are mapped into circles on the Euclidean disk, though they will shrink in size the further they are from the origin. We exploit this property by calculating a circle in the hyperbolic plane around each node that is guaranteed not to intersect the circle of any other node. When those circles are mapped onto the unit disk they provide a circular display region for each node of the tree in which to display a represenation of the node. This can be combined with a facility that uses different representations for nodes depending on the amount of real space they receive. Figure 7 shows the same tree as Figure 1, with the display region of each node indicated.

PRESERVING ORIENTATION

Orientation presents an interesting issue for the hyperbolic browser, because things tend to get rotated. For example, most nodes rotate on the display during a pure translation. There is a line that doesn't rotate, but the farther nodes are on the display from that line, the more they rotate. This can be seen in the series of frames in Figure 3. The node labeled "Lowe", for example, whose children fan out to the upper right in the top frame ends up with its children fanning out to the right in the bottom frame. These rotations are reasonably intuitive for translations to or from the origin. But if drags near the edge of the disk are interpreted as translations between the the source and the destination of the drag, the display will do a counter-intuitive pirouette about the point being dragged.

There is a fundamental property of hyperbolic geometry that is behind this and that also causes another problem. In the usual Euclidean plane, if some graphical object is dragged around, but not rotated, then is always keeps its original orientation— not rotated. But this is *not true* in the hyperbolic plane. A series of translations forming a closed loop, each preserving the orientation along the line of translation, will, in general, cause a rotation. (In fact the amount of rotation is proportional to the area of the closed loop and is in the opposite direction to the direction the loop was traversed.) This leads to the counter-intuitive behavior that a user who browses around the hierarchy can experience a different orientation each time they revisit some node, even though all they did was translations.

We address both of these problems by interpreting the user's manipulations as a combination of both the most direct translation between the points the user specifies and an additional rotation around the point moved, so that the manipulations and their cumulative effects are more intuitive. From the user's perspective, drags and clicks move the point that the user is manipulating where they expect. The additional rotation also appears natural, as it is designed to preserve some other property that the user expects. The user need not even be particularly aware that rotation is being added.

We have found two promising principles for adding rotations. In one approach, rotations are added so that the original root node always keeps its original orientation on the display. In particular, the edges leaving it always leave in their original directions. Preserving the orientation of the root node also means that the node currently in focus also has the orientation it had in the original image. The transformations in the examples presented so far all worked this way. It seems to give an intuitive behavior both for individual drags and for the cumulative effect of drags.

The other approach we have taken is to explicitly not preserve orientation. Instead, when a node is clicked on to bring it to focus, the display is rotated to have its children fan out in a canonical direction, such as to the right. This is illustrated in Figure 8 and also in the animation sequence in Figure 9. This approach is aided when the children of the root node are all laid out on one side of that node, as also illustrated in the two figures, so that the children of the root node can also fan out in the canonical direction when it is in focus.

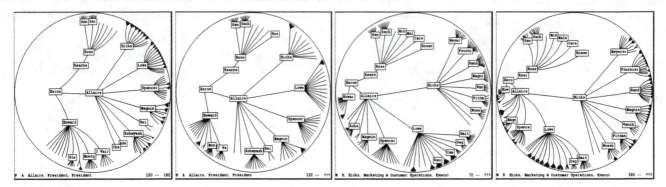

Figure 9: Animation with compromised rendering.

ANIMATED TRANSITIONS

As demonstrated by recent work on information visualizations, animated transitions between different views of a structure can maintain object constancy and help the user assimilate the changes across views. The smooth continuous nature of the hyperbolic plane allows for performing smooth transitions of focus by rendering appropriate intermediate views.

Animation sequences are generated using the so-called "nth-root" of a transition transformation, i.e. the rigid transformation that applied n times will have the same effect as the original. Successive applications of the "nth-root" generate the intermediate frames. The sequences in Figure 3 and Figure 9 were generated this way.

Responsive display performance is crucial for animation and interactive dragging. This can be a problem for large hierarchies on standard hardware. We achieve quick redisplay by compromising on display quality during motion. These compromises provide options for use in a system that automatically adjusts rendering quality during animation, e.g. the Information Visualizer governor [10] or Pacers [17]. Fortunately, there are compromises that don't significantly affect the sense of motion. Figure 9 shows an animation sequence with the compromises active in the intermediate frames. Unless specifically looked for, the compromises typically go unnoticed during motion.

One compromise is to draw less of the fringe. Even the full quality display routine stops drawing the fringe once it gets below one pixel resolution. For animation, the pruning can be strengthened, so that descendants of nodes within some small border inside the edge of the disk are not drawn. This tremendously increases display performance, since the vast majority of nodes are very close to the edge. But it doesn't significantly degrade perceptual quality for a moving display, since those nodes occupy only a small fraction of the display, and not a part that the user is typically focusing on.

The other compromise is to draw lines, rather than arcs, which are expensive in the display environments we have been using. While arcs give a more pleasing and intuitive static display, they aren't as important during animation. This appears to be the case both because the difference between arcs and lines isn't as apparent in the presence of motion, and because the user's attention during motion tends to be focused near the center of the display, where the arcs are already almost

straight.

One other possible compromise is to drop text during animation. We found this to be a significant distraction, however. And text display has not been a performance bottleneck.

EVALUATION AND FUTURE WORK

A laboratory experiment was conducted to contrast the hyperbolic browser against a conventional 2-d scrolling browser with a horizontal tree layout. Our subjects preferred the hyperbolic browser in a post-experimental survey, but there was no significant difference between the browsers in performance times for the given task, which involved finding specific node locations. The study has fueled our iterative design process as well as highlighted areas for further work and evaluation.

The two browsers in the study support mostly the same user operations. "Pointing" provided feedback on the node under the cursor in a feedback area at the bottom of window. "Clicking" moved a point to the center. "Grabbing" any visible point allowed interactive dragging of the tree within the window. The 2-d scrolling browser provides conventional scrollbars as well.

The experiment was based on the task of locating and "double-clicking" on particular nodes in four World-Wide-Web hierarchies identified by their URLs (the application intent being that a Web browser would jump to that node). Though this particular task and application were adequate for a rough baseline evaluation, there are problematic aspects. Typically, this task would better be supported by query-by-name or even an alphabetical listing of the nodes. Furthermore the WWW hierarchy (based on breadth-first flattening of the network) contained many similarly-named nodes and the hierarchy wasn't strongly related to a semantic nesting.

After pilot trials, we added to both browsers a feature to rotate the names of children of a pointed-to node through the feedback area and then jump to the current child. We also added a toggleable "long text" mode in which all nodes beyond an allocated space threshold disregard their boundaries and display up to 25 characters. Despite the overlapping of the text, this leads to more text being visible and discernible on the screen at once (see Figure 10).

The experiment used a within-subject design with four subjects, and tested for the effects of practice. We found no

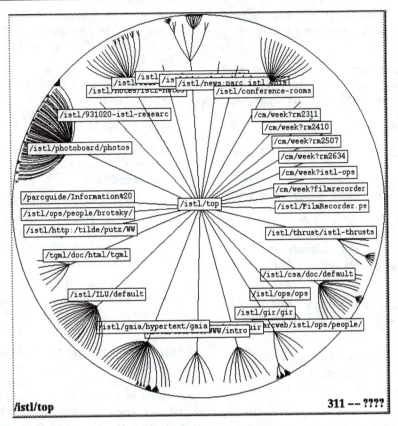

Figure 10: Hyperbolic browser in long text mode in World Wide Web hierarchy utilized in laboratory experiment.

significant difference in time or number of user actions in performing the tasks across the browsers. There was a significant practice effect in which practice produced a decrease in the number of user actions required to perform the task for both browsers, but there was no practice effect on task completion time for either browser. These practice effects did not differ significantly between the browsers.

Our post-experimental survey showed that all four subjects preferred the hyperbolic browser for "getting a sense of the overall tree structure" and "finding specific nodes by their titles," as well as "overall." In addition, specific survey questions and our observations identified relative strengths and weaknesses of the hyperbolic browser. Three of the subjects liked the ability to see more of the nodes at once and two mentioned the ability to see various structural properties and a better use of the space.

The amount of text that the hyperbolic browser displays was a problem. The experimental task was particularly sensitive to this problem because of the length and overlap of URLs, and the ill-structured nature of the WWW hierarchy. Though the long text mode was introduced before the study, none of the subjects used this feature during the study, preferring to point at the parent and then rapidly rotate the children through the much larger feedback area.

Problem areas mentioned by one subject were that the hyperbolic browser provide a weaker sense of directionality of links and also of location of a node in the overall space (because

shapes changed). Layout in a canonical direction as shown in Figure 8 addresses the first of these problems, but may worsen the second. In particular, for applications in which access to ancestors or to the root node is particularly important, this layout makes it easy to find and navigate toward these nodes.

A number of refinements may increase the value of the browser for navigating and learning hierarchies. For example, landmarks can be created in the space by utilizing color and other graphical elements (e.g. we painted http, gopher, and ftp links using different colors). Other possibilities are providing a visual indication of where there are nodes that are invisible because of the resolution limit, using a ladder of multiscale graphical representations in node display regions, and supporting user control of trade-off between node display region size and number of visible nodes (i.e. packing). The effective use of these variations are likely to be application or task dependent and so best explored in such a design context.

CONCLUSION

Hyperbolic geometry provides an elegant solution to the problem of providing a focus+context display for large hierarchies. The hyperbolic plane has the room to lay out large hierarchies, and the Poincaré map provides a natural, continuously graded, focus+context mapping from the hyperbolic plane to a display. The hyperbolic browser can handle arbitrarily large hierarchies, with a context that includes as many nodes as are included by 3d approaches and with modest computational requirements. Our evaluation study suggested this technique could be valuable, and has identified issues for further work.

We believe that the hyperbolic browser offers a promising new addition to the suite of available focus+context techniques.

ACKNOWLEDGEMENTS

We would like to thank the reviewers, our four subjects, and the colleagues who made suggestions for the prototype. Xerox Corporation is seeking patent protection for technology described in this paper.

REFERENCES

1. J Bertin. *Semiology of Graphics*. University of Wisconsin Press, 1983.

2. H. S. M. Coxeter. *Non-Euclidean Geometry*. University of Toronto Press, 1965.

3. George W. Furnas. Generalized fisheye views. In *Proceedings of the ACM SIGCHI Conference on Human Factors in Computing Systems*, pages 16–23. ACM, April 1986.

4. C. Gunn. Visualizing hyperbolic space. In *Computer Graphics and Mathematics*, pages 299–311. Springer-Verlag, October 1991.

5. B. Johnson and B. Shnedierman. Tree-maps: A space-filling approach to the visualization of hierarchical information. In *Visualization 1991*, pages 284–291. IEEE, 1991.

6. Hideki Koike and Hirotaka Yoshihara. Fractal approaches for visualizing huge hierarchies. In *Proceedings of the 1993 IEEE Symposium on Visual Languages*. IEEE, 1993.

7. Y.K. Leung and M.D.Apperley. A review and taxonomy of distortion-oriented presentation techniques. *ACM Transactions on Computer-Human Interaction*, 1(2):126–160, June 1994.

8. J. D. Mackinlay, G. G. Robertson, and S. K. Card. The perspective wall: Detail and context smoothly integrated. In *Proceedings of the ACM SIGCHI Conference on Human Factors in Computing Systems*, pages 173–179. ACM, April 1991.

9. E. E. Moise. *Elementary Geometry from an Advanced Standpoint*. Addison-Wesley, 1974.

10. G. G. Robertson, S. K. Card, and J. D. Mackinlay. The cognitive coprocessor architecture for interactive user interfaces. In *Proceedings of the ACM SIGGRAPH Symposium on User Interface Software and Technology*, pages 10–18. ACM Press, Nov 1989.

11. G. G. Robertson, S. K. Card, and J. D. Mackinlay. Information visualization using 3d interactive animation. *Communications of the ACM*, 36(4), 1993.

12. G. G. Robertson, J. D. Mackinlay, and S. K. Card. Cone trees: Animated 3d visualizations of hierarchical information. In *Proceedings of the ACM SIGCHI Conference on Human Factors in Computing Systems*, pages 189–194. ACM, April 1991.

13. George G. Robertson and J. D. Mackinlay. The document lens. In *Proceedings of the ACM Symposium on User Interface Software and Technology*. ACM Press, Nov 1993.

14. Manojit Sarkar and Marc H. Brown. Graphical fisheye views of graphs. In *Proceedings of the ACM SIGCHI Conference on Human Factors in Computing Systems*, pages 83–91. ACM, April 1992.

15. Manojit Sarkar and Marc H. Brown. Graphical fisheye views. *Communications of the ACM*, 37(12):73–84, December 1994.

16. Manojit Sarkar, Scott Snibbe, and Steven Reiss. Stretching the rubber sheet: A metaphor for visualizing large structure on small screen. In *Proceedings of the ACM Symposium on User Interface Software and Technology*. ACM Press, Nov 1993.

17. Steven H. Tang and Mark A. Linton. Pacers: Time-elastic objects. In *Proceedings of the ACM Symposium on User Interface Software and Technology*. ACM Press, Nov 1993.

APPENDIX: IMPLEMENTING HYPERBOLIC GEOMETRY

We use the Poincaré model for our underlying implementation of the hyperbolic plane, because that makes translation between the underlying representation and screen coordinates trivial. We represent a point in hyperbolic space by the corresponding point in the unit disk, which is represented by a floating point complex number of magnitude less than 1. Rigid transformations of the hyperbolic plane become circle preserving transformations of the unit disk.

Any such transformation can be expressed as a complex function of z of the form

$$z_t = \frac{\theta z + P}{1 + \overline{P}z}$$

Where P and θ are complex numbers, $|P| < 1$ and $|\theta| = 1$, and $\overline{P}$ is the complex conjugate of P. This transformation indicates a rotation by θ around the origin followed by moving the origin to P (and $-P$ to the origin).

The composition of two transformations can be computed by:

$$P = \frac{\theta_2 P_1 + P_2}{\theta_2 P_1 \overline{P_2} + 1} \qquad \theta = \frac{\theta_1 \theta_2 + \theta_1 \overline{P_1} P_2}{\theta_2 P_1 \overline{P_2} + 1}$$

Due to round-off error, the magnitude of the new θ may not be exactly 1. Accumulated errors in the magnitude of θ can lead to large errors when transforming points near the edge, so we always normalize the new θ to a magnitude of 1.

As an aside, on graphics hardware that has fast support for 3×3 matrix multiplication, it might be faster to use the Klein model, as done in [4], because rigid transformations can then be expressed in terms of linear operations on homogeneous coordinates. Display then requires computing the Poincaré mapping of points represented in the Klein model, which is just a matter of recomputing the distance from the origin according to $r_p = r_k / (1 + \sqrt{1 - r_k^2})$.

GeoSpace: An Interactive Visualization System for Exploring Complex Information Spaces

Ishantha Lokuge and Suguru Ishizaki

Visible Language Workshop, Media Laboratory
Massachusetts Institute of Technology
20 Ames St, Cambridge, MA 02139
ishi@media.mit.edu, suguru@media.mit.edu

ABSTRACT

This paper presents a reactive interface display which allows information seekers to explore complex information spaces. We have adopted information seeking dialogue as a fundamental model of interaction and implemented a prototype system in the mapping domain—GeoSpace—which progressively provides information upon a user's input queries. Domain knowledge is represented in a form of information presentation plan modules, and an activation spreading network technique is used to determine the relevance of information. The reactive nature of the activation spreading network, combined with visual design techniques, such as typography, color and transparency enables the system to support the information seeker in exploring the complex information space. The system also incorporates a simple learning mechanism which enables the system to adapt the display to a particular user's preferences. GeoSpace allows users to rapidly identify information in a dense display and it can guide a users' attention in a fluid manner while preserving overall context.

KEYWORDS: Interactive techniques, intelligent interfaces, cartography, multi-layer, graphics presentation, activation spreading network

1. INTRODUCTION

The exploration of complex data spaces in an age where both technology and information are growing at exponential rates is a challenging task. Recent developments in interactive media with high quality graphics have provided interface designers a means of creating more comprehensive environments for visualizing complex information. However, most interactive information systems provide collections of discrete visual presentations, in that they do not relate one presentation to another. Consequently, they fail to support a user's continuous exploration of visual information and gradual construction of understanding.

In order to create a more responsive visual information display, we have focused on developing an interactive visualization system that embodies the following characteristics:

- **A reactive display:** The system provides users with a responsive visual environment. When a user specifies a query, the display reacts in real time to reflect the input request.
- **Conversational interaction:** The use of an information seeking dialogue as a model of interaction enables a continuous exploration of the information space.
- **Context preservation:** The system maintains a notion of current state and presents the requested information within this context. This prevents one from getting lost in the complex information space.
- **Visual clarity:** Dynamic use of various visual design techniques proposed in [5][13] are integrated to enhance the clarity of the display by reducing users' cognitive load.
- **An adaptive knowledge base:** The knowledge base can be customized for specific user preferences.

Two main areas of research have influenced the work presented in this paper. The first area of research involves visual techniques and direct manipulation as a means of exploring complex information. One such approach is the use of overlapping multiple layers of information in which individual layers are accessible [3]. Belge et al. proposed a layering system in which a user can select and pull out layers by directly accessing visual elements on them. While their approach provided users with more control over the composition of layers, it does not provide semantic access to the information; nor does it visually organize the display. Colby et al. proposed a multi-layer map composting system [5] where users specify the importance of relevant information using sliders. Based on the importance values, their system automatically adjusts transparency, focus, and intensity of visual elements using simple rules encoded by a graphic designer. Although their system allows users to focus on the semantics (importance) of information rather than directly manipulating the graphics, the interaction becomes cumbersome when the number of information layers increases (i.e., the importance of each layer must be adjusted by the user). Most multi-layer approaches provide users with an interface that controls the display based on layers in order to simplify the interaction. However, they only allow limited access to specific graphic components within a given layer in order to avoid complex interaction.

While the above approaches emphasize direct manipulation and visual techniques, other interface displays have been proposed that incorporate domain and presentation knowledge (e.g.,[7][11][12]). For example, Maybury introduces an interactive visual presentation method that considers visual presentation as communicative acts (e.g., graphical, auditory, or gestural) based on the linguistic study of speech acts [11]. A multimedia explanation is achieved by using rhetorical acts, which is a sequence of linguistic or graphical acts that achieve a certain communicative goal such as identifying an information entity. Rhetorical acts are represented in a form of a plan, which is similar to our representation. Although these intelligent presentation systems enables sophisticated presentation based on a user's single request, they do not provide a mechanism to maintain a model of the user's information seeking goals from one query to another.

In this paper, we propose an information visualization system based on a model of an information seeking dialogue, where an information seeker incrementally asks questions and an information provider gradually answers the questions. Geographic information is chosen as an example domain which involves highly complex information and it is used to illustrate the proposed technique. Our ultimate goal is to apply this approach in various other domains of complex information such as news spaces and other abstract information domains.

In section 2, we describe an interaction model on which our system is based. Section 3 presents a prototype system, GeoSpace, in the geographic information systems domain. In Section 4, we outline our technical framework for implementing GeoSpace. Finally, in section 5 we discuss potential directions in which our research can be extended.

2. USER INTERACTION MODEL

We have adopted information seeking dialogue as a fundamental model of interaction, since the information space is hard to comprehend by a single query [4]. Most users find it difficult to formulate their information seeking goals in one request. Hence, an information display that gradually augments this process would greatly enhance the user's comprehension.

Our interaction model based on our informal observation is as follows: The first query by the information seeker (IS) makes the information provider (IP) guess what is important to show. After the IP provides information based on the first query, the IS may ask the second query based on what is provided. The IP then determines what is important to show next considering both the first and the second queries. The information seeking dialogue may continue until the user is satisfied.

A user's information seeking process can be top-down, bottom-up, or a combination of both. For example, suppose the IS is trying to look for a new apartment. On one hand, the IS may start a dialogue by stating that s/he is looking for an apartment. This can be considered top-down since the IS provided the ultimate goal of the dialogue. In this case, the IP is not certain about what kind of detailed information the IS is aware of. On the other hand, the IS may ask for a particular location. This can be considered bottom-up, since it targets a specific item of data. In this case, the IP is not certain about what the IS's ultimate goal is.

We consider the IP to be an expert in both domain information and visual presentation and the IP's knowledge is canonicalized in a form of reactive patterns. Instead of deliberately reasoning about what to present every time the user asks a question, the IP simply reacts to it by using canonical presentation techniques.

Based on this model, we have developed: (1) a knowledge representation scheme for representing domain knowledge together with visual design knowledge, (2) a computational mechanism whereby the system reacts to a series of user requests while maintaining overall context, and (3) a learning mechanism that enables the system to be molded according to user profiles, or to particular projects.

There have been approaches that use queries coupled with graphical displays both for narrowing down information to be presented [2][8] and for supporting users' exploration of the data space [6]. The information seeking model used in GeoSpace emphasizes the latter in its purpose.

3. A TYPICAL SCENARIO

This section presents GeoSpace—a prototype system which embodies the information visualization technique proposed in this paper. Figure 1 shows a snapshot of the initial state of the display. The visual complexity of this map display makes it hard for users to discern specific information while interacting without getting lost. GeoSpace allows the user to progressively ask questions in order to acquire appropriate information.

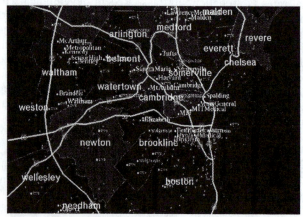

Figure 1. Map of Boston area showing the dense nature of the display

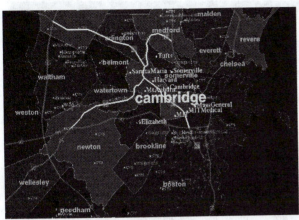

Figure 2. "Show me Cambridge"

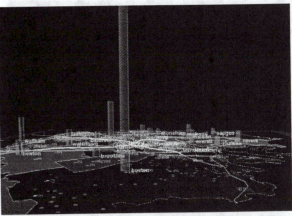

Figure 4. 3D view of Figure 3

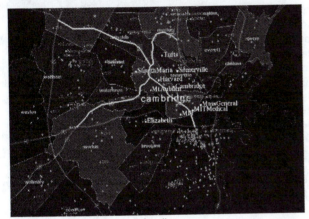

Figure 3. "Show me crime data"

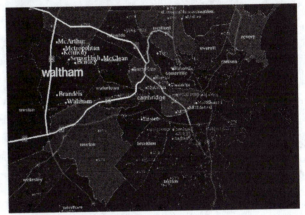

Figure 5. "Show me Waltham"

GeoSpace provides users with two types of interaction. First, the user can use text or speech to enter a query to which the system responds using the mechanism described above. Second, the user can use the mouse to zoom, pan, and move around the two and three dimensional map display.

The following is a scenario in which a person new to the Cambridge area tries to explore the information around the area (e.g., looking for a place to live). Having heard of the perilous life styles of people in Cambridge, suppose that the person is interested in relative crime level and accessibility to hospitals in the neighborhood.

First, the user asks the system *"Show me Cambridge."* Then, the label Cambridge increases in opacity to bring this information to the user's attention (Figure 2, Color plate 1). The typographic size also changes accordingly, resulting in a sharper focus of Cambridge. Notice also that related information such as hospitals, highways and colleges around Cambridge are also identified visually to a slightly lesser degree. This demonstrates the reactive nature of discerning information rapidly in a visually dense environment.

The user then asks, *"Show me crime data."* This shows a spatial distribution of crime data for the greater Boston area (Figure 3). By rotating the plane of the map, the user

obtains a three dimensional view of crime data in the form of a bar graph as shown in Figure 4 (Color plate 3). GeoSpace currently treats crime data as a unit of information. However, it is also possible to construct the domain knowledge such that the crime data is broken down into finer units. This type of decision is important when designing an application.

After looking at crime data, the user found that there are fewer crimes around Waltham and asks, *"Show me Waltham."* The label *Waltham* and other information relevant to Waltham, such as highways, hospitals and schools, increase in opacity. The typographic size also changes accordingly (Figure 5, Color plate 2). While providing information on Waltham, GeoSpace still distinguishes Cambridge from other regions in the map about which information was not sought. This is accomplished by a gradation in transparency between Cambridge and other areas of the map. This demonstrates the feature of maintaining the previous context using translucency.

Now, the user is interested in hospitals and asks, *"Show me Hospitals."* All the hospitals are indicated by employing the visual techniques discussed above (Figure 6). Imagine the user's expectation of seeing pharmacies was not realized by the system. The user can then explicitly change the system to a learning mode and say *"Show me pharmacies*

411

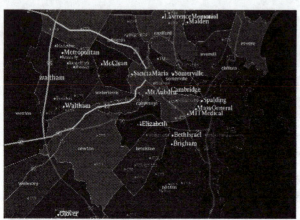

Figure 6. *"Show me hospitals"*

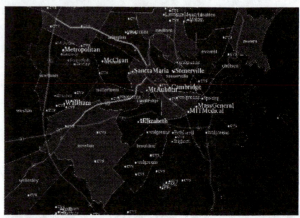

Figure 7. *"Show me pharmacies"*

too." Then, the system behaves as it would when asked to display pharmacies (Figure 7). Furthermore, it will have learned the relationship between hospitals and pharmacies. Thus, if the user asks, *"Show me hospitals"* in future interactions, the system would display hospitals and pharmacies simultaneously. This demonstrates the ability of the system to adapt to a user's preferences.

GeoSpace is implemented in C++ and GL graphics language on a Silicon Graphics Onyx workstation with Reality Engine graphics.

4. APPROACH

Our approach incorporates domain knowledge and presentation knowledge together in the form of presentation plans, and uses an activation spreading network technique to determine relevant presentation plans to be executed. The reactive nature of the activation spreading network, combined with visual design techniques, such as typography, transparency and color, enables the system to support the information seeker to explore the complex information space. The following section describes the components that constitute the proposed approach embodied in GeoSpace.

4.1. Domain Knowledge

Information seeking goals and presentation plans are the basic components of GeoSpace. A plan consists of a list of sub-plans, a list of conflicting plans, and a list of effects. The effect-list contains a set of goals that are achieved by executing the presentation plan. The sub-plan list contains a set of goals that must be achieved in order to accomplish goals in an effect-list. The conflict list contains a set of goals that are either semantically irrelevant or visually conflicting with the plan. Since our interest is to determine what the user is interested in, it is important to recognize when a subject of interest changes. Knowledge about semantic conflicts helps the system to identify a shift of interest. When a large amount of data exist in a database, it is often the case that the same visual features (such as color, typeface, orientation, or motion) are used by more than one visual element. Knowledge about visual conflicts helps the system to identify visually confusing situations.

Figure 8 shows typical presentation plans. Plan (a) says, in order for a user to know about transportation, a user must know about bus routes, subways, and place names. The plan also indicate that hospitals and bookstores are not relevant when a user wants to know about transportation. In the current representation semantic and visual conflicts are not distinguished. Plan (b) is much simpler; it has neither sub-plans nor conflicts. The activation level specifies the threshold energy required to realize the plan.

4.2. Activation Spreading Network

Plan:	{Show_Transportation}
Sub-Plans:	{Know_Place_names, Know_Bus_routes, Know_Subways}
Conflicts:	{Know_Hospitals, Know_Bookstores}
Effects:	{Know_Transportation}
Realization:	ø
Activation:	0.8 (a)

Plan:	{Show_Bus_map}
Sub-Plans:	ø
Conflicts:	ø
Effects:	{Know_Transportation}
Realization:	#<bus_map-object>
Activation:	0.3 (b)

Figure 8. *Typical presentation plans*

The system uses an activation spreading network [1][9] to determine priorities of plans based on the user's request. A plan module's activation level is changed by the user's immediate goals, and when their activation level exceed the threshold, positive and negative activation energy is sent to other plan modules connected by hierarchical links and conflicting links respectively. The current system iteratively injects a small amount of constant energy to fluidly change the overall activation state. In every iteration, activation levels of all the plan modules are normalized to the most active plan. This also results in the gradual decay of plans whose links are not explicitly specified.

For example, when the user specifies a query such as "Show me transportation", *know_transportation* becomes the current information seeking goal. The system then injects activation energy to the plans that contain *know_transportation* in the effect-list. When a plan module's activation level reaches a certain threshold, it spreads energy to the plans in the sub-plan list. A plan also spreads activation energy upwards to the higher level plans that contains *know_transportation* in their sub-plan list. This upwards activation results in activating indirectly related information. Figure 9 shows a simple example of an activation spreading process.

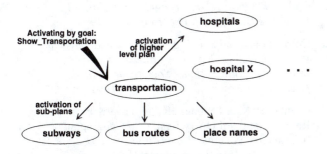

Figure 9. Schematic diagram of typical activation spreading

An activation spreading network not only presents the immediately relevant information, but it can also preserve the user's previous states of exploration. When a user requests new information, the system seamlessly transforms the previous state into the new state. The network can also prepare for the user's future request by activating plan modules that are potentially relevant in the following interactions. This could greatly assist users to formulate subsequent queries towards satisfying a particular goal.

4.3. Visual Design
The map display involves many layers of information each of which corresponds to a different set of data. The system is intended to incorporate various visual techniques, such as translucency and focus which helps clarify visual information without losing overall context [5][13]. We have incorporated these new techniques along with traditional graphical design techniques in the design of the map display. In the current implementation, translucency is particularly important in visually organizing the dense information space without losing a larger context. The most important information is displayed with a higher level of opacity, and related information is displayed with medium translucency. Irrelevant information is displayed almost transparent. Since, the display can show secondary information using translucency, the user has a chance of realizing a new question to ask next. Also, previously displayed information can be shown with medium to high translucency so that the user can maintain a continuous dialogue.

Plans may or may not have a graphical presentation. For example, a plan for highways does not have a graphical representation, but each highway has a graphical representation. Those plans that have a graphical representation change their graphical style according to their activation levels. Currently, the energy levels are mapped to transparency values and/or typographic sizes on the cartographic display. The mapping from the activation levels to graphical styles is achieved by simple procedures that are implemented according to design principles. In other words, visual design knowledge is embedded in those procedures and presentation plans. Thus, the quality of visual presentation, such as legibility, readability, and clarity are significantly enhanced.

In the map display, much of the spatial layout of the various graphic elements was inherently determined by geographic location. But when large amounts of information are involved, the layout often becomes a serious design problem. We did not use cartographic layout algorithms in GeoSpace; rather, we took an alternative approach that incorporates a mechanism that prioritizes information, and a set of dynamic visual techniques in order to avoid the complex layout problem.

4.4. Learning Mechanism
GeoSpace incorporates a simple learning mechanism in order for users to customize the domain knowledge. This is important since the initial domain knowledge is constructed by a particular designer and in some cases the system might behave in ways that do not reflect users preferences. The learning mechanism allows the user to personalize the response of the information display. Consider the user who wishes to see hospitals but the system does not know to show pharmacies when hospitals are requested. In other words, *know_pharmacies* is not included in a sub-goal list of the plan to show hospitals. In such a case, a user may want to customize the system so that the system can associate pharmacies and hospitals for future interaction. In the current system, the user must explicitly tell the system what to learn. The system accomplishes the above task by adding the goal *know_pharmacies* to the sub-plan list of the plan to show hospitals. However, an ideal system should be able to detect a user's interaction patterns and automatically learn that pharmacies are associated with hospitals. The current implementation does not have a mechanism to detect users' interaction patterns. Statistical methods to identify a user's interaction behavior is currently being explored to enable implicit learning capabilities.

5. FUTURE DIRECTIONS
Currently, our approach has been examined in the domain of geographic information. We intend to explore other domains that involve more abstract information, such as online news and financial data. In order to further examine the power of this technique, we are also increasing the size of the current database by adding more data.

Currently, the domain knowledge base is built manually by a designer. Further research will include the development of (1) a graphical interface for building domain knowledge and (2) a mechanism that automatically constructs the initial domain knowledge base for certain types of information provided that presentation plans can be implied from the database.

The activation network is fairly sensitive to the amount of activation energy spread. This will control the pace of the transitions from one display to another; hence, it is important to determine the optimal activation values. We are experimenting with varying energy levels to find the optimal network.

Finally, we are experimenting with the use of weighted links [9] for the activation spreading network to employ an implicit learning mechanism. While the current explicit learning mechanism provides rapid adaptation, implicit learning will provide a natural means for incorporating users preference over longer periods of time.

6. CONCLUSIONS

We have presented a reactive interface display for interactively exploring complex information spaces. We have shown that the knowledge representation scheme using presentation plans and information seeking goals, combined with the activation spreading network, provides the information display with a reactive capability. The mechanism can implicitly chain presentation plans by hierarchically spreading activation energy, and can respond to an immediate shift of interest by spreading negative energy to conflicting plans. The system can also direct a user's attention in a fluid manner without losing overall context, by gradually changing the states of activation. Dynamic use of various visual techniques, such as translucency, type size and color, are directly associated with activation levels of plans and visually clarify the display. We have also presented a learning mechanism as an integral part of the system, which allows users to customize the information display. These features make a user's exploration of complex information spaces a more dynamic experience.

ACKNOWLEDGEMENTS

We would like to thank Prof. Muriel Cooper, Ron MacNeil, and Dave Small for their continued support and advice at all times during the course of this project. Special thanks to Prof. Whitman Richards, Louis Weitzman, Hasitha Wickramasinghe, and Yin Yin Wong for commenting on drafts of this paper. We would also like to thank the other members of the MIT Media Laboratory's Visible Language Workshop for providing valuable suggestions as we developed our ideas.

This work was in part sponsored by ARPA, JNIDS, NYNEX and Alenia. However, the views and conclusions expressed here are those of the authors and do not necessarily represent that of the sponsors.

REFERENCES

1. Anderson, John. A Spreading Activation Theory of Memory. *Journal of Verbal Learning and Verbal Behavior* 22, pp.261-295, 1983.

2. Ahlberg, Christopher and Shneiderman, Ben. Visual Information Seeking: Tight Coupling of Dynamic Query Filters with Starfield Displays. *Proceedings of SIGCHI Human Factors in Computing Systems,* 1994.

3. Belge, Matt, Lokuge, Ishantha and Rivers, Dave. Back to the Future: A Graphical Layering System Inspired by Transparent Paper. *INTERCHI'94 Conference Companion,* 1993.

4. Carberry, Sandra. *Plan Recognition in Natural Language Dialogue,* A Bradford Book, 1990.

5. Colby, Grace and Scholl, Laura. Transparency and Blur as Selective Cues for Complex Visual Information. *International Society for Optical Engineering Proceedings,* Vol 1460, 1991.

6. Egenhofer, Max J. Manipulating the Graphical Representation of Query Results in Geographic Information Systems, *Proceedings of the IEEE Workshop on Visual Languages,* 1990.

7. Feiner, Steven and McKeown, Kathleen. Automating the Generation of Coordinated Multimedia Explanations. In: *Intelligent Multimedia Interfaces,* ed. Mark T. Maybury, AAAI Press/The MIT Press, 1993.

8. Goldstein, Jade and Roth, Steven. Using Aggregation and Dynamic Queries for Exploring Large Data Sets. *Proceedings of SIGCHI Human Factors in Computing Systems,* 1994.

9. Maes, Pattie. Situated Agents Can Have Goals, *Designing Autonomous Agents: Theory and Practice from Biology to Engineering and Back,* ed. P. Maes, MIT Press/Bradford Books, 1990.

10. Maes, Pattie. Learning Behavior Networks from Experience, *Towards a Practice of Autonomous Systems: Proceedings of the First European Conference on Artificial Life,* ed. F.J. Varela & P. Bourgine, MIT Press/Bradford Books, 1992.

11. Maybury, Mark. Planning Multimedia Explanations Using Communicative Acts. In: *Intelligent Multimedia Interfaces,* ed. Mark T. Maybury, AAAI Press/The MIT Press, 1993.

12. Roth, Steven and Hefley, William. Intelligent Multimedia Presentation Systems: Research and Principles. In: *Intelligent Multimedia Interfaces,* edited by Mark T. Maybury, AAAI Press/The MIT Press, 1993.

13. Small, David, Ishizaki, Suguru., and Cooper, Muriel. Typographic Space. *SIGCHI Conference Companion,* 1994.

Enhanced Dynamic Queries via Movable Filters

Ken Fishkin, Maureen C. Stone
Xerox PARC, 3333 Coyote Hill Rd., Palo Alto CA 94304
E-mail: {fishkin, stone} @parc.xerox.com

ABSTRACT

Traditional database query systems allow users to construct complicated database queries from specialized database language primitives. While powerful and expressive, such systems are not easy to use, especially for browsing or exploring the data. Information visualization systems address this problem by providing graphical presentations of the data and direct manipulation tools for exploring the data. Recent work has reported the value of dynamic queries coupled with two-dimensional data representations for progressive refinement of user queries. However, the queries generated by these systems are limited to conjunctions of global ranges of parameter values. In this paper, we extend dynamic queries by encoding each operand of the query as a Magic Lens filter. Compound queries can be constructed by overlapping the lenses. Each lens includes a slider and a set of buttons to control the value of the filter function and to define the composition operation generated by overlapping the lenses. We demonstrate a system that supports multiple, simultaneous, general, real-valued queries on databases with incomplete data, while maintaining the simple visual interface of dynamic query systems.

Key Words: viewing filter, lens, database query, dynamic queries, magic lens, visualization

INTRODUCTION

Traditional database query systems require the user to construct a database query from language primitives [19]. Such systems are powerful and expressive, but not easy to use, especially when the user is unfamiliar with the database schema. Information visualization systems [15] provide graphical display of database values and direct manipulation tools for exploring relationships in the data. These systems have many advantages over language-based systems, including a visual representation that provides an intuitive feel for the scope of the data, immediate feedback, and incremental, reversible interactions. There is a tension in such systems, however, between providing expressive power and ease of use.

In this paper, we present a new direct manipulation technique for exploring a database displayed as a two dimensional set of

points. With it, users can incrementally construct full boolean queries by layering queries encoded as Magic Lens™ filters [5,6,17]. Our work builds directly on two techniques for information visualization that were presented at CHI '94. The first, the *starfield display* [2], supports interactive filtering and zooming on scatterplot displays. The second, the *movable filter* [17], supports multiple simultaneous visual transformations and queries on underlying data. In this paper we combine the two techniques, enhancing the starfield display by augmenting it with the flexibility and functionality of the movable filter. The advantages include: a direct manipulation mechanism for creating general boolean queries, multiple simultaneous views, and a uniform mechanism for providing alternate views of the data. We also demonstrate how our query mechanism was extended to support general, real-valued queries on databases containing missing fields.

After describing related work, we describe using magic lens filters to generate boolean queries over graphically displayed data. We then discuss how they provide a uniform mechanism for generating multiple views of the data. We then extend our model to support real-valued queries over incomplete data, and present our conclusions and plans for future work. Concrete examples taken from an application for exploring US Census data are used throughout the paper.

RELATED WORK

Scatterplots or *thematic maps* are well-established techniques for displaying data as points in a 2-D field [4,18]. Such displays can encode large amounts of data and can provide an intuitive way to visualize groups of related data items. Transformations of the display produce patterns that are easy to interpret, even for high-dimensional datasets [14].

Dynamic queries [1] apply direct manipulation techniques to the problem of constructing database queries. A selector (e.g. a slider) or set of selectors is used to control the range of values required of a particular attribute or set of attributes. When combined with a graphical representation of the database such as a scatterplot, users can rapidly explore different subsets of the data by manipulating the selectors [2]. However, the number of attributes that can be controlled is limited by the number of selectors that can be easily applied to the data. The effect of combining slider filters is strictly conjunctive; disjunctive queries may only be performed by performing each sub-query sequentially. The effects of the selectors are global; there is no way to limit the scope to only a portion of the data except by zooming in on it. Finally, the number of selectors, and hence the number of possible queries, is fixed in advance.

The Aggregate Manipulator [11] as well as XSoft's Visual Recall™ [20] allow somewhat more powerful (but non-interactive) queries, supporting disjunctive and limited compound composition via textual, hierarchical, menu-driven interfaces. The conceptual prototype of Egenhofer [10] is a hybrid technique. Query operands are typed in using a database language, but those operands are visually composed, with modes that support disjunctive, conjunctive, and subtractive composition.

Magic lens filters [6,17] are a user interface tool that combine an arbitrarily-shaped region with an operator that changes the view of objects viewed through that region. The operator can be quite general, accessing the underlying data structures of the application and reformatting the data to generate a modified view. The filters are spatially bounded, may be parameterized, and are interactively positioned over on-screen applications. Filters may overlap, in which case they compose their effects in the overlap region.

Magic lens filters provide a number of advantages desirable for data visualization. Since they are spatially bounded, they can perform their operation within a restricted focus while maintaining global context. Since they can overlap, compositional semantics can be defined and controlled.

BOOLEAN QUERIES BY COMPOSITION

Given some mechanism for displaying data in scatterplot format, we can use a set of lenses to create dynamic queries on the data. Each lens acts as a filter that screens on some attribute of the data. A slider on the filter controls a threshold for numeric data. Buttons and other controls on the lens can control other functions. When the lenses overlap, their operations are combined. This provides a clean model for building up complex queries. This physical, rather than merely conceptual, composition of multiple sliders appeals to existing user intuitions that spatial overlaps imply a composition, as shown in Venn diagrams and color circles.

To create boolean queries by composing magic lens filters, we need to provide a way to specify how the filters are combined. To provide full boolean functionality, we need to provide a mechanism for the AND, OR, NOT and grouping or parenthesizing operations.

We define for each lens a filtering function and a composition mode that describes how the result of the filtering function is combined with the output of lenses underneath. More formally, a lens L=(F, M), where F is a filter and M is a boolean operator. The filter, F, describes the output calculation for the filter on some datum. The mode, M, describes how that output is combined with the output from lower filters. For example, given L1=(F1, OR) and L2=(F2, AND), the result of positioning L1 over L2 is (F1 OR F2). Conversely, the effect of positioning L2 over L1 is (F2 AND F1). We implement the composition mode as a button on the lens, making it easy to change the mode as needed.

The NOT operation can be encoded as a lens whose filter inverts the sense of the data coming in. Using the formalism of the previous paragraph, an inverting lens N = (NULL, NOT). That is, N applies a NOT to the output of lower filters, and has no intrinsic filter function. For example, consider the query (F1 OR NOT F2), where F1 and F2 filter for various attributes. To implement this query, the user would lay down filter F2, then the NOT filter N, then filter F1 with its composition mode set to OR.

To incorporate grouping, we need a mechanism for encapsulating the expression defined by a stack of lenses. To do this, we provide an operation that replaces a stack of lenses with a single *compound* lens that is semantically equivalent. The user creates such a compound lens by selecting a point on a stack of lenses through a click-through button, a partially transparent button on an overlaying lens [5]. All lenses beneath this point are combined to create a compound lens that generates the same query as the stack. The resulting lens also has a composition mode that defines how it is combined with other lenses. This new compound lens can be manipulated by the user just as any other lens, providing a simple way to encapsulate a complex query into a single conceptual unit.

To create the query (F1 AND F2) OR (F3 AND F4), for example, we create compound lenses for the values in parentheses: C1=(F1 AND F2) and C2=(F3 AND F4). By giving lens C1 a composition mode of OR, we can create the desired expression by positioning C1 over C2. Compound lenses may contain other compound lenses, allowing queries to grow to arbitrary complexity.

Since we can incorporate AND, OR, NOT, and grouping, we can represent any boolean query. Complex queries can be incrementally built up by direct manipulation of the lenses and their modes. Useful queries can be saved as compound lenses. The resulting lenses can then be used to build up more complex queries in a completely uniform manner. While we have implemented only three common boolean composition modes, this model supports any of the boolean operators.

EXAMPLES

To demonstrate these ideas we created an application to browse a database of US census data [7]. In this database, each row represents a city and the columns describe the city along various census metrics: population, crime rate, property tax rate, and so forth. We chose this database because it is publicly available, lends itself to fairly interesting queries, and the data elements have an intuitive mapping to the 2D plane, namely the physical location of the city on a map.

In this implementation, each lens is implemented as an X window [16]. A lens manager server extends the X window system to support magic lens functionality. Therefore, the lenses can be manipulated using the regular window manager interface. Lenses can display their output using the X graphics library, or, upon request, they can display it in PostScript® [3] form. Using this facility, the application can generate PostScript pictures describing its screen appearance. We used such generated PostScript for all the black and white figures in this paper. Choosing this form, instead of a screen snapshot, allows us to customize the presentation to fit the space and color limitations of the proceedings. The figures in the color plates are screen snapshots that show the screen appearance of the running application.

Color plate 1 shows a typical lens filter in action. Each city is displayed as a blue-rimmed white box at a point proportional to its latitude and longitude. The data attribute associated with the lens filter covering the center of the country, "1991 crime index," is displayed in the window header. Below that on the left is a slider used to control the threshold value for the query, and a label showing the current slider value (12158.5 in this example). The buttons to the right of the slider control whether the user is screening for data less than (<) or greater than (>) the slider value. Cities shown in red, rather than white, pass the filter. For example, in this case, cities in the center of the country with a crime rate greater than 12158.5 are shown in red. Buttons along the right edge of the application are used to spawn new lenses (13 different types are presently implemented), and for application housekeeping.

Figure 1 shows the effects of applying different composition modes to the composition of two filters. The composition mode for a lens is defined by a group of four radio buttons. The buttons labeled **AND** and **OR** set the composition mode accordingly. The **SELF** button causes the lens to display only the effect of its own filter, ignoring other lenses, and the **NOP** button sets the filtering function to NULL, disabling the effect of the lens. These two modes are useful when interpreting complex queries. In figure 1(a) we look for cities which have high annual salaries and low taxes. We can make the query less demanding by using the **OR** button to change the composition mode (figure 1(b)), to see cities which have either high salaries or low taxes. The **SELF** and **NOP** buttons can be used to examine these components separately, to determine which cities have high salaries and which have low taxes.

An alternative interface for setting the composition mode is to use a click-through tool. In this interface, a single button on the lens indicates the current mode. A click-through tool contains buttons that can be used to change the mode by clicking through them to the lens below. The advantage of this interface is that it supports a large set of composition modes without visual clutter The disadvantage is that it requires more steps to change the mode. Our application supports both interfaces; users can choose the interface they prefer. An example of a lens with a single mode button is shown in color plate 2.

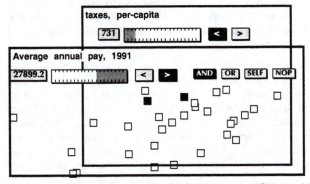

Figure 1(a)   High salaries AND low taxes. Cities with partial data are shown in figure 1(b) but not in 1(a).

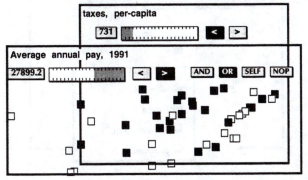

Figure 1(b) High salaries OR low taxes. Both conjunctive (AND) and disjunctive (OR) queries are incorporated in our system.

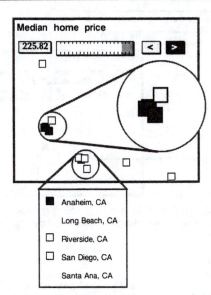

Figure 2. Semantic filters can be augmented with visual filters. Here, a magnifying lens and a call-out lens show clumped cities while maintaining context elsewhere.

MULTIPLE VIEWS

Simultaneous multiple queries can be performed by positioning different lenses over different parts of the display. Each lens, and hence each query, can be independently customized by manipulating the controls on the lens. We can also use lenses to generate alternate views of the data, such as magnifying lenses, callouts, sorted views, and other such transformations. The user interface of these visual transformations is identical to that of the semantic transformations (the filters).

For example, a scatterplot representation of the data will often have clumps, where some set of data points map to nearby points on the display or even overlap. Figure 2 shows two examples of this clumping, and uses multiple visual filters to aid in viewing the data. A small magnification lens is positioned over northern California, letting us see that there are four cities clumped into the bay area. Over southern California, a callout lens is placed. This lens displays the cities as a list, making it easy to separate them. The rest of the map is displayed in the usual manner. This allows easy identification of cities in the dense region while maintaining global context. The boxes next to the city names on the callout are active, so filters placed over them act exactly as if they were placed over the original map. Cities listed without boxes are missing the data selected by the filter, "Median home price."

In the star-field paradigm, a user manipulates a set of selectors, and observes their filtering effect on the data. By associating each selector with a single movable filter, we gain the ability to pose multiple simultaneous queries, each with its own set of parameters. For example, suppose we wish to determine which cities in each region of the country have relatively low housing prices. We have data available for the average housing price per city. However, the range of values for this attribute is wide, and varies geographically. For example, houses on the west coast are typically more expensive than houses in the midwest. Therefore, we need to filter on a higher threshold on the west coast than in the midwest. Figure 3 shows two filters with two different threshold values positioned over California and over Texas.

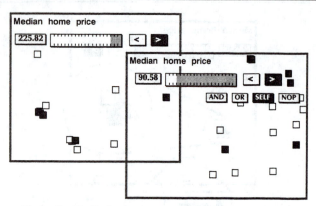

Figure 3. To find relatively high housing prices in California and Texas, two different filters are positioned simultaneously.

EXTENSIONS

In the previous section, we discussed how movable filters can be used to perform general boolean queries on scatterplot data. In this section, we discuss further extending the power of our filters such that they support real-valued queries and undefined data values.

Real-valued Queries

To support more powerful queries, we extended our system such that filters assign a real-valued score on the range [0...1] to each datum. Data with a score of 0 fails the filter entirely, data with a score of 1 succeeds entirely, and data with intermediate scores partially satisfy the filter. This provides more information about the value returned by the query. In general, the scoring function could be arbitrary—a filter might prefer data with extreme values, or assign a gaussian falloff to data with the highest score going to data of a certain value, or other similar metrics. Our implementation currently supports only linear and step scoring functions, but this is not due to any limitation imposed by the technique. We present the score visually by showing each datum as a partially filled-in square; the higher the score, the more of the square is filled in.

For example, in figure 4(a) a boolean filter is screening for crime rate, and both Dallas and Fort Worth are seen to have high crime rates. When we switch to real-valued filters (figure 4(b)) we see that Dallas has a lower crime rate than Fort Worth. The values displayed were computed as follows. As in the boolean case, the filter slider value defines a threshold. In figure 4, for example, the threshold is 12838. When screening for data greater than the threshold (>), data below the threshold is assigned a score of 0, as is the case for Arlington in Figure 4. The city in the database with the greatest value for the data (in the case of crime data, it happens to be Atlanta, GA, with a crime index of 18953) is assigned a score of 1.0. Cities with values between the threshold and the maximum are assigned a simple linear score equal to their value divided by the maximum value. When screening for values less than the threshold (<), values below the threshold are assigned non-zero scores in a similar manner.

Computations can be performed on the output of real-valued filters. For example, in figure 5 we have placed a sorting lens over a real-valued crime rate filter. Only the cities under the sorting lens are sorted. We can see that in Florida, Jacksonville has the lowest crime rate and Miami has the highest.

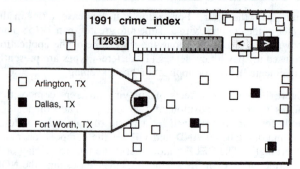

Figure 4(a) boolean query on crime rate for three cities in Texas.

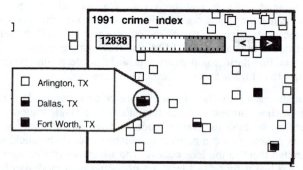

Figure 4(b) Real-valued query on crime rate for the same cities. Extending our filters from boolean-valued to real-valued allows distinctions to be maintained.

Real-valued filters require real-valued composition modes. The real-valued MIN and MAX composition modes correspond to the boolean AND and OR as in fuzzy logic [21]. That is, MIN and MAX work the same on 0.0 and 1.0 as AND and OR do on 0 (False) and 1 (True), but can also incorporate values in-between. A real-valued NOT filter returns 1.0 minus its input value.

Just as boolean filters can be composed with an arbitrary boolean operation, not just AND and OR, so too can real-valued filters be composed with an arbitrary numerical operation, not just MIN and MAX. These operations can be statistical transformations used to change the data distribution (sqrt, arcsin, etc.), fuzzy logic operators used to tune the query ("very," "somewhat," "more or less," etc.), or mathematical operators used for general manipulation (difference, log, etc.). Color

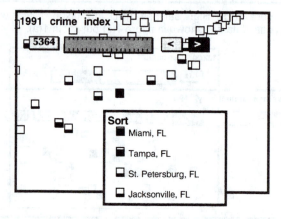

Figure 5. A sorting lens sorts cities by crime rate in Florida.

plate 2 shows some of these composition modes in use to visualize the correlation between poverty and crime rate. The DIFF composition mode is an example of a mathematical composition operator: it is defined as the absolute value of the difference between its two operands. The filter above that, the "Very" filter, is an example of an operator from fuzzy logic: Very(x) is defined as x squared [8]. By overlaying a real-valued NOT filter above that, we ask where crime rate and poverty rate are NOT VERY DIFFerent. The higher the score, the redder the city's box, the greater the correlation. We can see that poverty and crime rate are highly correlated in most cities.

Missing Data

Databases in general, and the US census database in particular, do not always have all data fully defined. In the case of the US census database, for example, some cities may have population figures included but not crime data, or vice versa. A robust information presentation system must address this problem to give an accurate view of the data. There are two issues: how such data is visually presented and how queries based on that data are performed.

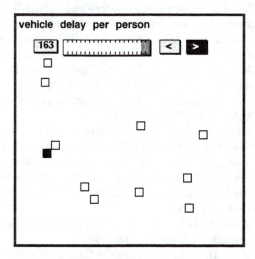

Figure 6(a). A filter finds only one city (San Francisco) with a high score.

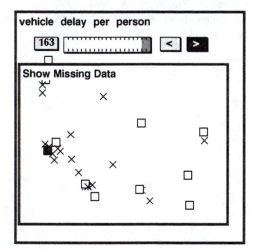

Figure 6(b). A missing data lens shows that attribute values are missing for many cities. Cities with missing data are marked with an 'X'.

We visualize missing data by use of a special lens. Normally, any city whose value as a result of a query is missing, or undefined, is simply not shown. The "Show Missing Data" lens makes these cities visible. For example, figure 6(a) shows a query, centered on the West Coast, filtering for cities with severe traffic delays. Only San Francisco is shown to have such delays. By adding a "Show Missing Data" lens, however, as shown in figure 6(b), we see that most of the other California communities don't have data for this category—perhaps San Francisco is not as unattractive as it first appeared.

We incorporated missing data into the semantics of composition by considering missing data to be analogous to the IEEE concept of a non-signaling Not-A-Number (NaN), a value that is defined to "afford retrospective diagnostic information inherited from invalid or unavailable data and results" [12]. In accordance with common practice [13], we then define:

$$OR(a, NaN) = a, \qquad MAX(a, NaN) = a$$
$$F(a, NaN) = NaN, \quad \text{for all other composition modes } F.$$

Note that a city that is invisible because of missing data may re-appear when another filter is laid upon it, if that filter's composition mode is MAX or OR.

CONCLUSIONS

Applying magic lens filters to the problem of exploring and visualizing data creates a user model that is expressive yet easy to understand. Placing a lens over the data evokes the physical model of using a lens for filtering or enhancing the view of the data beneath it. Including a slider on the lens allows the user to dynamically vary the query by adjusting the filter threshold.

Overlapping the filters creates a natural metaphor for creating compound queries. By providing explicit composition modes and a mechanism for grouping expressions, the full power of a boolean query language can be expressed graphically. We have also demonstrated how this language can be expanded to include real-valued functions that gracefully handle missing values. Powerful queries can be built up incrementally, then preserved for future exploration. In this way, the lenses provide a mechanism for capturing the result of a data exploration session.

Magic lens filters provide a natural mechanism for visual transformations as well as semantic transformations of the data. We have shown examples of callout, magnification, missing data, and sorting lenses. Other possible functions include alternate representations of score such as color or size, and overlaying geographic information such as city or state boundaries.

In our application, magic lens filters have been combined with click-through tools to implement a wide range of user-interface operations. These can be used to manipulate the data, the filtering operations, or the tools themselves. For example, click-through tools are used to create compound lenses, generate figures, provide additional parameters for the filtering operations, and to change the layout of the buttons and sliders.

In summary, magic lens filters provide a uniform, powerful and extensible mechanism for data visualization and exploration applications.

FUTURE WORK

Within the particular application of analyzing city census data, we are working to include a wider variety of scoring functions, to explore other visual techniques (e.g. blurring and/or translucency as discussed in [9]) for expressing queries and their results, and to make the user interface smoother, faster and more sophisticated. This will enable us to perform user studies to help quantify the value of this technique.

We are also interested in extending this work to apply to other types of data and other forms of information visualization.

ACKNOWLEDGMENTS

We thank Jock Mackinlay for his enthusiasm, encouragement, and shared realization that magic lens filters could support disjunctive queries. We thank Eric Bier for his user-interface and illustration design expertise, and Ken Pier for his help in the preparation of the paper. Finally, we thank Xerox PARC for its continuing support.

Trademarks and Patents: Magic Lens and Visual Recall are trademarks of the Xerox Corporation. PostScript is a trademark of Adobe Systems, Inc. Patents related to the concepts discussed in this paper have been applied for by the Xerox Corporation.

REFERENCES

1. Christopher Ahlberg, Christopher Williamson, and Ben Shneiderman. Dynamic Queries for Information Exploration: an Implementation and Evaluation. *Proceedings of CHI '92,* 1992. pp. 619-626.

2. Christopher Ahlberg and Ben Shneiderman. Visual Information Seeking: Tight Coupling of Dynamic Query Filters with Starfield Displays. *Proceedings of CHI '94,* (Boston, MA, April 24-28) ACM, New York, (1994), pp. 313-317.

3. Adobe Systems Incorporated. *PostScript® Language Reference Manual, second edition.* Addison-Wesley, 1990.

4. Jacques Bertin. Semiology of Graphics. University of Wisconsin Press. 1983.

5. Eric A. Bier, Maureen C. Stone, Thomas Baudel, William Buxton, and Ken Fishkin. A Taxonomy of See-Through Tools. *Proceedings of CHI '94,* (Boston, MA, April 24-28) ACM, New York, (1994), pp. 358-364.

6. Eric A. Bier, Maureen C. Stone, Ken Pier, William Buxton, and Tony D. DeRose. Toolglass and Magic Lenses: The See-Through Interface. Proceedings of Siggraph '93 (Anaheim, CA, August), *Computer Graphics* Annual Conference Series, ACM, 1993, pp. 73-80.

7. Bureau of the Census. Statistical Abstract of the United States 1993. Washington DC, 1993.

8. C.L. Chang. Interpretation and Execution of Fuzzy Programs. In Fuzzy Sets and Their Applications to Cognitive and Decision Processes. Academic Press. New York. 1975. pp. 191-218.

9. Grace Colby and Laura Scholl. Transparency and Blur as Selective Curs for Complex Visual Information. Proceedings of SPIE '91 (San Jose, Feb). SPIE, 1991, pp. 114-125.

10. Max J. Egenhofer. Manipulating the Graphical Representation of Query Results in Geographic Information Systems. *Proceedings of the 1990 IEEE Workshop on Visual Languages.* IEEE Computer Society Press, Los Alamitos CA, 1990, pp. 119-124.

11. Jade Goldstein and Steven F. Roth. Using Aggregation and Dynamic Queries for Exploring Large Data Sets. *Proceedings of CHI '94,* (Boston, MA, April 24-28) ACM, New York, (1994), pp. 23-29.

12. IEEE. The IEEE Standard for Binary Floating-Point Arithmetic. IEEE, New York, 1985.

13. W. Kahan. How should Max and Min be defined? University of California, Berkeley. May 25, 1989.

14. Daniel A. Keim and Hans-Peter Kriegel. VisDB: Database Exploration Using Multidimensional Visualization. IEEE CG&A, **14**(5), September 1994. pp. 40-49.

15. George G. Robertson, Stuart K. Card, and Jock D. Mackinlay. Information Visualization Using 3D Interactive Animation. Communications of the ACM, **36**(4), April 1993. pp. 57-71.

16. Robert W. Scheifler, James Gettys, and Ron Newman. X Window System. Digital Press, Bedford MA, 1988.

17. Maureen C. Stone, Ken Fishkin, and Eric A. Bier. The Movable Filter as a User Interface Tool. *Proceedings of CHI '94,* (Boston, MA, April 24-28) ACM, New York, (1994), pp. 306-312.

18. Edward R. Tufte. The Visual Display of Quantitative Information. Graphics Press. 1983.

19. Jeffrey D. Ullman. Principles of Database Systems. Computer Science Press, 1980.

20. XSoft Corporation. Visual Recall for Windows. Xerox Corporation. Palo Alto, CA. 1994.

21. Lotfi Zadeh. Fuzzy Sets. Information Control, vol. 8, pp. 338-353. 1965.

Turning Research into Practice:
Characteristics of Display-Based Interaction

*Marita Franzke**
Institute of Cognitive Science
University of Colorado, Boulder, CO 80304

ABSTRACT

This research investigates how several characteristics of display-based systems support or hinder the exploration and retention of the functions needed to perform tasks in a new application. In particular it is shown how the combination of the type of interface action, the number of interaction objects presented on the screen, and the quality of the label associated with these objects interact in supporting discovery and retention of the functionality embedded in those systems. An experiment is reported which provides empirical evidence for Polson & Lewis's CE+ theory of exploratory learning of computer systems [11]. It also extends this theory and therefore leads to a refinement of the cognitive walkthrough procedure that was derived from it. The study uses an experimental method that combines observations from realistically complex task scenarios with a detailed analysis of the observed performance.

KEYWORDS: exploration, retention, display-based systems, direct manipulation, cognitive theory, cognitive walkthrough, experimental method.

INTRODUCTION

Researchers agree today that learning by guided exploration, is not only a preferred mode of knowledge acquisition, but also one of the more successful ones [3,13]. The study reported here builds on this assumption and investigates the process of task-oriented exploration at a small grain size. It asks how design decisions embedded in commercially available display-based systems assist in exploratory search, how exploratory performance is related to forgetting, and how systems could be designed to support exploration and retention better.

Characteristics of discretionary use of software

At present, a growing number of users of new software are computer-literate. These users will have used a variety of computer applications before encountering a new one, and are therefore more likely to attempt learning new functionality by exploration rather than by reading manuals and tutorials [13,14]. These discretionary users of systems may also use systems sporadically, since many applications support very specialized functions that are not needed every day. Graphing applications are one example of these specialized systems. If applications are only used occasionally, retention of once discovered functionality comes to be a usability issue. Hence, two activities, discovery and retention of functionality, are two important characteristics of discretionary use of software.

This means that design of application software should support exploratory activities of computer literates, even for applications that are more complex than the often cited example for walk-up-and-use systems, ATM-machines. Secondly, interfaces should provide ample cues to casual users, so that once-discovered functionality can be easily remembered or rediscovered after a longer interval of non-use.

Graphical user interfaces (GUI's), or display-based systems, in our terminology, were once hailed to solve these two problems of explorability and retainability simply by way of displaying relevant information to the user [e.g. 15]. Early empirical work, however, showed that exploration of display-based systems may be difficult, if not impossible, at least for novice users [2]. A cognitive model of explanatory behavior with computer systems will help to explain why these computer users failed to discover important functionality.

Exploration

Polson & Lewis's [11, 12] CE+ addresses the problem of explorability of interfaces in the framework of a theory of problem solving. It is assumed that task-centered exploration of a new interface is guided by general task goals, such as "make a new graph from the data in spreadsheet". Having formed such a goal, users will search the interface for an object that promises progress towards that goal. This search is guided by label-goal matches. If an interface object (such as a menu item, a button, or an icon) displays a label that matches the goal (for example, 'graph', or 'chart'), users will select it, given that they know how to take an action on this particular object. After executing an action, users evaluate the system feedback. If they conclude that they made progress towards their goal, the search-action cycle continues, until they have completed their goal in this manner.

There is empirical evidence for both, the importance of well-matched labels [e.g., 4], as well as the importance of differentiating and recognizing interface objects, and knowing which actions are available on them [2].

The CE+ theory identifies four critical points at which the exploratory search can fail: (1) Users can form an inadequate goal, (2) they might not find the correct interface object (because of poor label match), (3) users may not know how to execute an action on the relevant interface object, and (4) they may receive inappropriate and misleading feedback. The experiment reported here focuses on points two and three. It asks whether the discovery of the appropriate interface object is also dependent on the number of distracting interface objects, and on the type of

*The author can now be reached at:
U S WEST Technologies,
4001 Discovery Drive, Boulder, CO 80303
e-mail: mfranzk@advtech.uswest.com

action that will be necessary. In particular, its aim is to identify whether the goodness of the label, the type of action, and the number of interface objects interact. The result of the experiment should lead to a refinement of the model of the search process.

Retention
Several empirical studies have shown that recall of important parts of display-based interfaces is poor even for frequent users of these system, who can demonstrate virtually flawless performance in using these same features in the context of the application [9, 10]. Theoretical accounts of display-based interaction are therefore based on the assumption that recognition rather than recall of commands and interface objects drives the interaction [e.g., 7, 8, 10]. The experiment described here compares performance at a short and a longer retention interval. This will answer the question of whether display-based computer skill is indeed robust against forgetting. It also concerns itself with the question of whether the particular design features listed above (label match, number of interface objects, and type of action) will have an effect on the retainability of particular interactions.

Turning research into practice I: the cognitive walkthrough
The cognitive theory of Polson and Lewis [11] discussed above has previously been extended into a method for the evaluation of walk-up-and-use interfaces, the cognitive walkthrough [12, 17]. This method helps interface designers and engineers to evaluate a system or system specification by decomposing various user goals into chains of interface actions (such as menu selections, button presses, etc.). For each goal and for each action in accomplishment of a goal, the evaluator considers a series of criteria, derived from the four-step cognitive model: (1) whether the user will have trouble forming an appropriate task goal, (2) whether an action is clearly available and whether the label associated with the action matches the users' representation of the task goal, (3) whether the user will know how to execute an action, and (4) whether the system feedback is clearly interpretable.

The method has been introduced into industrial use and has received various criticisms in the literature [e.g. 16,17]. Most of these reviews center on procedural aspects of the method which have been taken into account in newer versions of the cognitive walkthrough [17]. Some other criticisms concern the methods' ability to detect a broad range of usability problems. One of the points brought forward [16] is that the method is too narrowly focused on identifying linguistic difficulties, namely mismatches between the users' (linguistic) representation of a task goal, and the (linguistically expressed) label of the goal. The method gives little guidance in determining whether a graphical layout of the screen design is more or less conducive to finding an important object. Furthermore, it is difficult to estimate whether a particular interface action (a button press, for example) will be known to a group of users. This is especially problematic when the new application is for a large and relatively diverse user-group.

The experimental results, reported in this paper, by extending CE+, can also be used to add more specific evaluation criteria to the walkthrough procedure. In particular, the results will inform us about how the number and grouping of interface objects affects search, and will provide us with information about the relative difficulty of a range of different interface actions. The results of the

retention trials will also show whether difficulties identified with the walkthrough procedure will only be problematic during exploration or also for application after longer time delays. This information will be integrated into the walkthrough in the discussion section.

Overview of the study
In the experiment, familiar users of Macintosh systems learned a new application, one of four graphing systems (see below). They were asked to create a graph and do several modifications to the default graph that the system brought up. The subjects participated in two of these trials, each time with different data and superficially different modifications. The first of these trials was the *exploration trial*, here the subjects had to discover the necessary functions on their own. The second trial (*experienced performance*) was administered either after a short (a few minutes) or a long (one week) retention interval. Comparison between these two delay conditions allowed for an observation of the effect of forgetting on performance.

METHOD
Subjects
Thirty-three males and forty-three females participated in this experiment. The subjects had an average of 2.8 years of Macintosh experience, and were familiar with an average of 3 different Macintosh applications. The majority of subjects (72%) had additional experience with PC's. On the average, subjects had 1.6 years experience with PC's, and knew 1.8 PC applications. None of the subjects had used any graphing applications before. The age range was 15 to 44 years, with an average of 25 years. Subjects were paid $15 for participation. The data of four subjects were excluded from the analysis, because of failures to complete the task in the exploration phase, for a total of 72 subjects.

Design and Materials
Subjects were randomly assigned to one of eight experimental groups (four interfaces by two delay conditions). The interfaces were CGI[1], CGIII[2], EXC tool[3], and EXC menu[4]. Subjects were assigned to one interface and used it throughout the whole experiment. Half of the subjects performed the second (experienced) trial after a short delay (approximately ten minutes), and the other half after a long (one week) delay. Subjects worked on a Apple Macintosh II cx with a 13'' color monitor, set to black and white. The screen interactions were videotaped over the subjects' shoulders and an audio-track was recorded.

Task and Procedure
Tasks: Subjects were provided with a HyperCard stack of instructions for both tasks. They were told which subgoals to complete in which order. Specifically, they had to (1) create a line graph from data in a file provided to them. Then they received a sample graph and instructions on what formatting changes to perform to match the default graph to the sample graph. Subjects were instructed to (2) move the legend to a different location, (3) change the font size of the legend text, (4) change the line and symbol style of the plotted graph, (5) change the font and style of graph title, (6/7) change the font and style of both axes, and (8/9) edit the title and x-axis title content. The instructions provided subjects with detailed information on *what* to do, but no hints on *how* to accomplish it. The tasks in the

[1]CA Cricket Graph, version 1.3.2, 1989.
[2]CA Cricket Graph III, version 1.01, 1992.
[3]MS EXCEL, version 3.0, 1990, toolbar usable.
[4]MS EXCEL, version 3.0, 1990, toolbar disabled.

exploratory and experienced session were isomorphic, but subjects were provided with different sets of data and different sample graphs in both cases. The instructions, and sample and default graphs were identical in all interface conditions. The instructions were provided in a HyperCard stack, which overlapped with the application window. If the subjects wanted to read the instructions, they had to click on the stack to bring it to the front (see Figure 1). This procedure allowed us to account for the time subjects spent in reading the instructions.

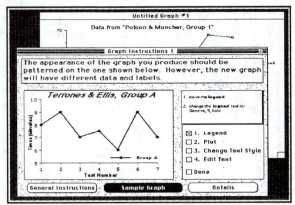

Figure 1. Example of instruction card.

Procedure: On arrival, subjects filled out a brief questionnaire about their computing background. After this they completed a simple editing task in which they were warmed up to the window switching procedure involving the instructions. They then started the first graphing task. The experimenter stayed in the room with the subject during these tasks and provided brief, action-oriented hints if subjects had not made any progress toward the next correct action for more than two minutes. After completion, half of the subjects received another editing task (as distracter), and the second graphing task (the experienced trial). The other half of the subjects did these two tasks (the second editing and graphing tasks) one week later.

For each correct action step in fulfillment of the tasks we will report on the action time (time to find the correct action - time to view instructions) and number of hints needed.

The results are reported in two sections: First, global results that concern overall performance measures are summarized. Second, results from the detailed analyses are reported. In that section a description of the coding of the design parameters and their effect on the subjects' performance is given. A more complete set of analyses can be found in [6].

RESULTS
Global Results
Effects of Training and Delay: Mixed two-factorial MANOVA's (trial, repeated; condition, between-subjects) with a covariate controlling for Macintosh experience were performed on action times and the number of hints. See Figure 2 for an illustration of the group means underlying these analyses.

There was a main effect of trial for action times (F $(1,59)$ = 95.83, p.<.01) and for hints (F$(1,54)$ = 63.76, p.<.01). Subjects' performance showed a sharp improvement between the first and the second trial across interfaces and delay conditions (see Figure 2). Overall, subjects were able to cut their action times in half, from an average of about

fifteen to an average of about seven minutes for the whole task. This results shows that interface literate users were able to discover functionality in a new system in a reasonable amount of time, without extensive use of external help, and were able to use the discovered methods efficiently in a second trial.

There were no significant main effects for delay for action times (F$(1, 59)$ = .24, p. > .05), nor for hints (F$(1, 54)$ = 2.21, p. > .05). However, the interaction between trial and condition was significant for action times (F$(1, 59)$ = 4.64, p. < .05). This result shows that the overall performance time decrease was influenced by the delay. The performance time decrease was smaller when the second trail happened after a one week delay (see Figure 2). There was no significant interaction between trial and delay condition for the number of hints (F$(1, 54)$ = .15, p. > .10).

These results indicate that while learning effects are strong, as can be expected on the background of theories of skill acquisition [e.g., 1], forgetting plays a surprisingly little role for performance with display-based systems. If users have not used a new system for a longer time period, they need more time to perform the same tasks. However, the observed forgetting effects are relatively small when compared to the large savings between the first and second trial and their significance could be debated on practical grounds. In a previous study that investigated performance on a transfer task, we had found that forgetting may play a role only in complex interactions, but not in simple ones [5]. The analyses reported below investigate the exact locus of the delay effects, and try to relate the difficulties in exploration and the observed forgetting effects to several design parameters.

Overall performance analyses also showed significant differences between the four systems, but these differences disappeared, when the number of action-steps for the task were controlled for [6].

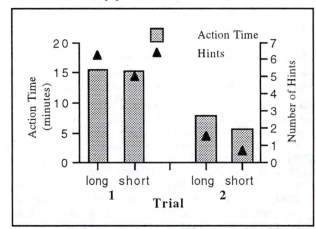

Figure 2. Effect of delay on second task. Long = one week delay, short = ten minutes delay between first and second task.

Detailed Analyses and Results
If the current task artifact analysis is to inform further designs of display-based applications, we need to refine our level of analysis. For any type of design recommendations we need to know which type of interactions were difficult to discover during the exploration phase (first task), and which interactions were responsible for the forgetting effects observed in the global results.

Analysis by Subgoals: To answer these questions, the analysis was first taken to the level of subgoals. For this level of analysis we will only report results for the action times to reduce complexity in the presentation.

Simple ANOVA's on action times associated with trial 1 (exploration) and on differences between the delay conditions on trial 2 show significant effects due to subgoals, $F_{(9,630)} = 49.00$, p. $<.01$ for exploration times and $F_{(9,621)} = 3.41$, p. $<.01$ for the differences between delay conditions. For an illustration of this effect see Figure 3. Separate analyses of variance were performed on the overall action times associated with each subgoal, testing whether there were differences between the two delay conditions. The arrows in Figure 3 point to the subgoals where delay differences were statistically as well as practically significant, $F_{(1,70)} = 13.27$, p. $<.01$ for 'create graph', $F_{(1,70)} = 11.15$, p. $<.01$ for 'change legend', and $F_{(1,70)} = 10.95$, p. $<.01$ for 'edit title 1'.

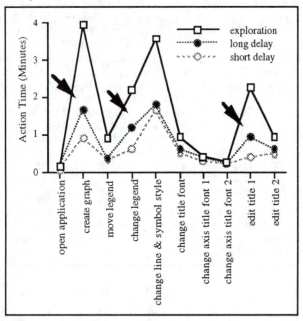

Figure 3. Action times during task 1 (exploration) and task 2. Arrows indicate significant delay effects.

An inspection of the graph in Figure 3 illustrates that these subgoals were also associated with particularly long exploration times. All three subgoals introduced situations in which subjects had to discover and learn a completely new method. They had never created a graph before, they had to acquire a general method for modifying objects in the graph (change legend) and they had to discover another method for editing text associated with graph objects. Along these lines we suggest that long exploration times and some forgetting effects due to long retention delays may appear in situations where completely new methods need to be discovered. In all other subgoals where transfer of old (move) or newly learned methods (e.g. change title font) were possible, exploration times were lower, and no forgetting due to the delay appeared. One exception to this rule appears to be the subgoal 'changing line and symbol style', where performance on the second task was very poor for both conditions.

Analysis by Action Steps: To determine exploration and retention difficulties further, the analysis was finally taken down to the level of individual action steps. For this, we

included individual action steps that comprised the three subgoals of interest in the regression analyses described below. Figure 4(a/b) provides an example for the level of detail of this analysis. Figures 4a displays the first action step for the subgoal 'create graph' for Cricket Graph III: select menu-bar item 'graph'. Figure 4b shows the second action-step: select menu-bar item 'new graph'.

For each action step (e.g. 'click on menu-bar item 'graph'', Figure 4a) three variables were encoded, (1) the type of action, (2) the semantic distance between goal and label, and (3) the number of objects competing for attention. For the *type of action* we recorded what type of interaction needed to be performed by the subject (e.g.: menu bar selection, button click, move operation, tool selection, etc.).

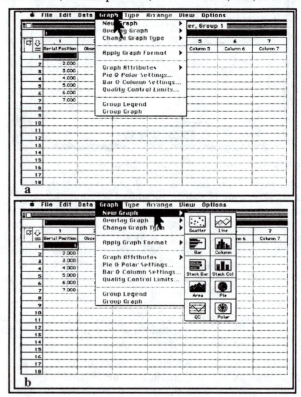

Figure 4. First two action-steps to create a graph in CGIII.

For the *semantic distance* it was assumed that subjects represented their active goal in terms of the task description. For example, if the assumed goal was to 'create a graph', then a menu item with the label 'graph' was defined as semantic difference of 0, a menu item with the label 'chart' as a value of 1 (for synonym), the label 'drawing tools' as a value of 2 (semantically related, but inference required), the label 'file' received a value of 3 (no direct semantic link, connection has to be learned). Finally, for the *objects competing for attention*, each object in the relevant object group was counted, as well as the number of competing object groups (e.g. for Figure 4a: 0 competing menu items (greyed out) + 1 for the menu bar + 1 column labels in spreadsheet + 1 for spreadsheet entries = 3).

Exploration and Experienced Performance. The design parameters derived above were used in three sets of simple and multiple regressions, to explore their effects on the action times. Details on these analyses and their results are reported in [6]. Here, we will focus on the illustration and discussion of the significant effects.

Figures 5-9 demonstrate that the three coded design parameters indeed had the predicted effects on subjects' action time performance. Figure 5 shows, that there was an effect of label quality, so that action times were longer with a larger semantic difference between the labeling of an object and the subjects' representation of the goal.

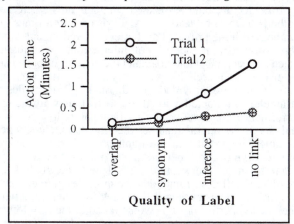

Figure 5. Time per action step by label quality and trial.

Similarly, the more objects were displayed at any given time, the longer the action times to access the correct object (Figure 6). Additionally, in both cases these main effects interacted with trial, so that action times during the exploration trial were more affected than action times during the second trial (see Figures 5 and 6). The number of objects to search and the difficulties involved in evaluating potential operators both affect search time during exploration. During trial 2, however, where subjects had access to knowledge of successful interactions, the number of objects to search and the goodness of the semantic match did not matter as much.

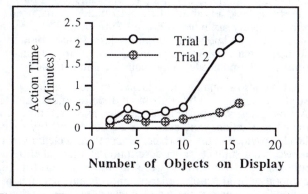

Figure 6. Time per action by number of objects and trial.

Furthermore, there was a three-way interaction between these two design parameters (number of objects and semantic difference) and trial, so that search time during trial one due to semantic difference was elevated additionally if there were many objects to search (Figure 7). If the semantic differences between labels and goals were small, there is no effect of the number of objects on display, and in fact, the action times during trial one did not seem to be very different from the action times during trial two.

There was also a significant effect of the type of interactions on action times that varied with trial. During trial one (exploration) some interactions are especially difficult to discover; Figure 8 displays the types of interactions in order of difficulty. Some of the more difficult interactions were

dragging and dropping of items, clicking on tool icons, or double- and single-clicking on objects.

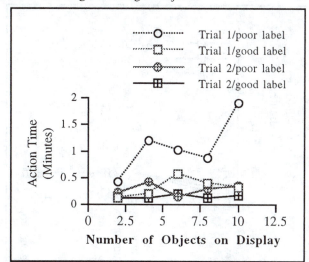

Figure 7. Time per action by number of objects, quality of label and trial.

On the low end of the action time spectrum were selections of buttons, selections of menu items and of lists in dialog boxes. None of the more difficult interactions were labeled in any meaningful way. The easier interactions however, were associated with labels. It seems that in searching a new application for methods, subjects oriented themselves towards reading of labels, rather than blindly trying direct manipulation operations on objects or icons. The order of difficulty of these interaction types also explains the strong correlation between semantic distance and interaction type that was reported in the beginning of this section.

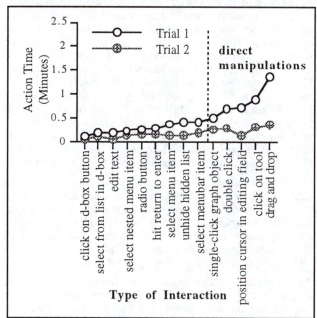

Figure 8. Time per action by type of interaction and trial.

Finally, we observed a triple interaction of type of interaction, the number of objects on display, and trial. If there were many objects to search, action times during trial one were inflated especially for interactions that were difficult to discover. Figure 9 displays this effect for several interaction types. Unfortunately the combinatorial set of all action types with all numbers of display objects was not complete in our sample.

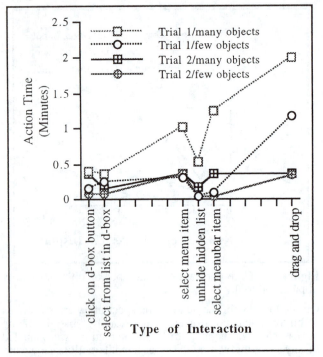

Figure 9. Type of interaction by number of objects and trial.

Exploration Summary: We found that the three coded design parameters, semantic difference, number of objects on display, and type of interaction, had effects on action times. Action times during the exploration trial were affected more strongly than times during later performance (trial two). Once subjects had acquired a new action they seemed to be able to use this knowledge immediately in further interactions. Additionally, there were interactions between some of the design parameters. The pattern of these interactions suggests that subjects first search interface objects with labels, and only consider unlabeled choices (e.g. iconic tools) later. Second, well-labeled objects will be considered and found faster, and the number of objects in the display does not seem to affect search times for such situations. If the relevant object is labeled poorly, more items will be considered in depth, so that the number of objects to search will elevate the action times even further.

Delay Performance. In the analyses by subgoals we noted that performance on subgoals where new methods had to be discovered was more prone to performance decrease due to the delay manipulation. The following analyses investigate these effects at a lower level of detail. Only the actions from subgoals (3) create graph, (4) change legend, and (9) edit title 1, from trial two were included in these analyses.

As with exploration performance, we found strong effects of all three parameters on the differences between the two delay conditions. The same factors that lead to poor performance during search also lead to poorer performance after a week delay between the exploration and second session. Subjects were more prone to forgetting when they had to use direct manipulation techniques, when the action was poorly labeled, and when there were many objects to search. In no case was this performance close to the long search times during exploration, however. In other words, the more correct interpretation of the display depends on retrieval of

specifically learned information from memory (learned meanings of labels, learned interaction techniques), the stronger the influence of forgetting on performance.

Overall Summary: We were able to show that exploration and delay performance varied by subgoal, and the results suggest that longer exploration times and poorer delay performance are both associated with subgoals where new methods have to be learned. Analysis of action times and numbers of hints provided at the level of individual action steps showed that the type of interaction, the semantic match between label and goal, and the number of objects on the display all are reliable predictors of action times and the numbers of hints provided. Furthermore, all three design parameters interacted with trial, so that they influenced performance most clearly during the exploration trial. An analysis of the interactions between the parameters, where possible, suggested that the number of objects to search impairs exploration performance only when a simple semantic match between the label and the goal is not possible. Finally, we found that the three design parameters also influence performance at a week-long delay. After a week delay, performance is worse (action times are longer, and more hints have to be provided) when interaction objects are not or poorly labeled, and when many objects have to be searched.

DISCUSSION

In this section we will first discuss the theoretical implications of the results. Then the extended model of exploratory search will be used to refine the cognitive walkthrough evaluation criteria.

Lessons about exploratory search

First, we have found that discretionary users of display-based systems are able to learn a new application by task-oriented exploration in a reasonable amount of time, if support in the form of occasional hints is available. On the average, the subjects in our study were able to get through the first task in about fifteen minutes, and needed approximately six hints to do so. Furthermore, they were able to use the knowledge acquired during this exploratory sequence in a second task, where they cut their performance time in half, and needed only about one hint per trial.

We have also found ample evidence for the label-following strategy suggested by Polson and colleagues [4, 11]. Action times, which can be interpreted as search times for the correct action, were clearly related to the quality of the labels provided in the interface. If there was a clear overlap between the subjects' goal description and the label in the interface, search times were minimal, even for the exploration trial (in the order of ten to twenty seconds per action). However, if the match between the label and goal was less evident, search times were greatly increased (up to 90 seconds).

Furthermore, we saw that this result is modified by the number of simultaneously displayed objects that the users have to search. If the label matches the goal representation well, the number of objects competing for attention do no alter the search time significantly. Even with as many as ten objects on screen, search times are no longer than thirty seconds. If the label match is poor, however, the number of objects on screen considerably changes the search times, so that a combination of poor matches and many objects can produce search times up to two minutes.

Finally, there was evidence to suggest that the particular type of interaction required also influences the ease with which an action will be discovered. The subjects in this study, who all had previous MS Word experience, considered labeled actions, such as menu-interactions, button-presses and the like, first, with search times under 30 seconds. They took longer to discover direct manipulation interactions on unlabeled objects, such as double-clicks on graph objects, drag-and-drop operations, etc., and the more alternative objects were presented on the screen, the longer were the search times (up to two minutes).

This suggests the following scan-search process for exploratory search of interfaces (an extension on the step 2 in Polson & Lewis' CE+): After forming a task goal, users will quickly scan the interface for interface objects containing a semantically 'promising' label. If such a label is provided, subjects will find it quickly; it seems to 'pop-out' at the user, no matter how many distracting objects are on the screen. If no good match with any provided label is possible, a slower, more careful evaluation of the interpretation of labels has to follow. This process is sensitive to the number of options to consider. Finally, when all labeled choices have been considered and found inappropriate, the users in this study turned their attention to unlabeled action choices and try direct manipulation operations on display-objects. Again, this is a slow process that is sensitive to the number of simultaneously displayed options.

Lessons about retention
Our results show that retention of newly discovered functions is indeed quite good in interaction with display-based applications, even after a week long delay between the first and the second use of the system. Subjects were able to cut their performance times in half, and even though there were differences in performance dependent on the length of the delay, these were relatively small in comparison to the great overall improvement from the first to the second trial. The detailed analyses showed that forgetting occurs mainly in the context of completely new subgoals, and can be related to the failure to retrieve newly learned information from memory [8].

From research to praxis II: Refinement of the cognitive walkthrough
In the introduction we pointed out how the four-step processing model of Polson and Lewis was translated into a four-step evaluation method of human-computer-interaction. We also pointed to some of the criticisms of the cognitive walkthrough method, namely that it does not provide guidelines about what makes an action 'clearly available' to a user, and which types of actions will be known and considered by a broad class of users.

The results offer answers to these two concerns. First, the goodness of the label match really *does* determine, whether an action will seem readily available to a user. Good label matches assure that users will discover the correct action choice. This points to the necessity of good participatory design: The better the designers understand their users' task goals and language, the better they will be able to provide labels that matches the users' expectations. If a label match can not be insured (because of a large variability in the user population), the number of options to consider should be kept to a minimum. The more options the user has to ponder, the longer it will take him or her to find the correct choice, and the more likely it is that he or she will make the wrong choice. If a design team is therefore unsure about the labeling of an action, it should try to consult the user

population for suggestions, or try to minimize the alternative choices for this action-step of the interaction.

Second, direct manipulation actions will be considered last by a user who is familiar with standard off-the shelf systems such as MS Word or MS Excel, which are mainly menu-driven. If a design team finds that an action sequence necessitates the use of direct manipulation, they should ask again carefully, how many alternative actions are available at the time. As a general guideline, missing labels (direct manipulation), or poor label-goal matches, *in combination with* a number of alternative actions to search, will lead to long search times and user frustration. In our case, the experimenter had to intervene in situations like this, because the subjects were unable to make any progress on themselves. Design teams should also ask careful questions about the characteristics of the respective user-group that they are designing for. We have additional evidence [6] to suggest users might be more willing to consider direct manipulation operations when they are working in a system context that uses this type of interaction frequently.

Figure 10 summarizes these evaluation recommendations for each one of the steps in a action sequence Writing in plain print states the walkthrough questions as summarized in [17]. Bold text refers to the refining questions derived from this study.

---

1) Will the user try to achieve the right effect (form the right goal)?

2) Will the user notice that the correct action is available?
   - **Is the action labeled or unlabeled?**
   - **If the action is labeled, is there a poor match between the label and the users' representation?**
     - **If in doubt, talk to your users!**
   - **Are there more than 10 screen actions to be considered at this time?**
   - **Is there a combination of labeling problems and a large number of objects?**

   The more questions are answered with NO, the more 'obvious' will the correct choice appear to the user.

3) Will the user associate the correct action with the effect trying to be achieved?

4) If the correct action is performed, will the user see that progress is being made?

Figure 10. Original walkthrough questions and refinements.

Finally, proofing an interface with the cognitive walkthrough in this way should not only help identify and fix problems that occur during exploration (the walkthrough was originally designed for this), but it should also help to smooth intermittent use of systems. Actions that were easy to discover in the first place will either be easily remembered after long time delays, or they will be trivial to rediscover.

CONCLUSIONS
The current study presents new insights into the nature of display-based interaction, namely that learning by task-oriented exploration is a possibility for interface-literate users, both in terms of the initial learning phase, as well as the use of this knowledge in later trials. We showed that forgetting plays a relatively small role in the use of display-based systems, and pointed to some of its possible causes. Furthermore, we presented evidence for a scan-search

strategy in service of exploration of new applications, that is dependent on the good labeling of actions, as well as on the number of action options that need to be considered. Direct manipulations on unlabeled objects proved to be especially difficult to discover. These results were discussed in the context of the CE+ theory of exploration with computer systems, and the findings were integrated into the suite of evaluation criteria embedded in the cognitive walkthrough method.

ACKNOWLEDGMENTS
Thanks to Peter Polson, Clayton Lewis, John Rieman, Evelyn Ferstl, Sharon Irving, and the members of the CHI'94 doctoral consortium for supporting this work in various ways. The anonymous reviewers of the CHI'95 conference provided appreciated editorial help as well as many thoughtful comments. This work was supported by NSF grant IRI 9116640.

REFERENCES
[1] Anderson, J.R. (1982). Acquisition of cognitive skill. *Psychological Review*, 89, 369-406.

[2] Carroll, J.M., & Mazur, S.A. (1986). LisaLearning. *IEEE Computer*, 91, 35-49.

[3] Charney, D., Reder, L., & Kusbit, G. (1990). Goal setting and procedure selection in acquiring computer skills: a comparison of problem solving and learner exploration. *Cognition and Instruction, 7,* 323-342.

[4] Engelbeck, G. (1986). Exceptions to generalizations: implications for formal models of human-computer interaction. Unpublished master's thesis, University of Colorado, Department of Psychology, Boulder.

[5] Franzke, M. & Rieman, J. (1993). Natural Training Wheels: Learning and Transfer between two Versions of a Computer Application. In T. Grechenig & Tscheligi (Eds.). *Lecture Notes in Computer Science: Human Computer Interaction.* Vienna Conference VCHCI. Berlin, FRG: Springer-Verlag.

[6] Franzke, M. (1994). Exploration and experienced performance with display-based systems. Unpublished dissertation, University of Colorado, Department of Psychology, Boulder.

[7] Howes, A. (1994). A model for the acquisition of menu knowledge by exploration. *In Proceedings of CHI '94*, Boston, MA: ACM, 445-451.

[8] Kitajima, M. & Polson, P.G. (1994). A model-based analysis of errors in HCI. *In Conference Companion of CHI '94*, Boston, MA: ACM, 301-302.

[9] Mayes, J.T., Draper, S.W., McGregor, A.M., & Oatley, K. (1988). Information flow in a user interface: the effect of experience and context on the recall of MacWrite screens. In Jones, D.M, Einder, R. (Eds.): *People and Computers IV*, Cambridge UK: Cambridge University Press, 191-220.

[10] Payne, S. (1991). Display-based action at the user interface. *International Journal of Man-Machine Studies, 35,* 275-289.

[11] Polson, P., & Lewis, C. (1990). Theory-based design for easily learned interfaces. *Human-Computer-Interaction, 5, 191-220.*

[12] Polson, P., Lewis, C., Rieman, J., & Wharton, C. (1992). Cognitive walkthroughs: A method for theory-based evaluation of interfaces. *International Journal for Man-Machine Studies, 36,* 741-733.

[13] Rieman, J.F. (submitted). A field study of exploratory learning strategies. *Submitted to ACM Transactions on Human Computer Interaction.*

[14] Santhanam, R., & Wiedenbeck, S. (1993). Neither novice nor expert: the discretionary user of software. *International Journal of Man-Machine Studies, 38,* 201-229.

[15] Shneiderman, B. (1983). Direct manipulation: a step beyond programming languages. *IEEE Computer*, 57-69.

[16] Wharton, C., Bradford, J., Jeffries, R., and Franzke, M. (1992). Applying cognitive walkthroughs to more complex interfaces: Experiences, issues, and recommendations. *Proceedings CHI'92 Conference*, Monterey, CA, 381-388.

[17] Wharton, C., Rieman, J.R., Lewis, C., and Polson, P. (1994). The cognitive walkthrough method: A practitioner's guide. In J. Nielsen and R. Mack (Eds.), *Usability Inspection Methods,* John Wiley, NY.

Learning and Using the Cognitive Walkthrough Method: A Case Study Approach

Bonnie E. John
Departments of Computer Science and Psychology
and the HCI Institute
Carnegie Mellon University
Pittsburgh, PA 15213
(412) 268-7182
bej@cs.cmu.edu

Hilary Packer
Master of Software Engineering Program
Carnegie Mellon University
Pittsburgh, PA 15213
(412) 268-3759
hpacker@cs.cmu.edu

ABSTRACT

We present a detailed case study, drawn from many information sources, of a computer scientist learning and using Cognitive Walkthroughs to assess a multi-media authoring tool. This study results in several clear messages to both system designers and to developers of evaluation techniques: this technique is currenlty learnable and usable, but there are several areas where further method-development would greatly contribute to a designer's use of the technique. In addition, the emergent picture of the process this evaluator went through to produce his analysis sets realistic expectations for other novice evaluators who contemplate learning and using Cognitive Walkthroughs.

KEYWORDS: usability engineering, inspection methods, Cognitive Walkthrough.

INTRODUCTION

In the last few years, there has been growing interest in studying different usability evaluation techniques to understand their effectiveness, applicability, learnability, and usability. Developers looking want to know which techniques will be suited for their system, budget and time frame. Universities want to know what to teach in the growing number of HCI courses. Researchers developing techniques need to have feedback to allow them to improve and expand the techniques in useful directions.

Some work has been done that compares different techniques (e.g., [3], [4], [7], [8]). This work has taken the form of experiments, formally comparing performance outcomes of different techniques. The dependent variables are typically quantitative: the number (and type) of usability problems identified, how closely a given technique predicts a user's behavior, the time it takes to perform the evaluation, the labor costs involved.

Perhaps the biggest difficulty with these studies is that they provide no data about what people actually *do* when they are using these techniques. Without process data, it is difficult to understand how the technique itself leads the analyst to identify usability problems, as opposed to the analyst simply being clever. Without process data, it a developer does not know what to expect when setting out to use a new technique. Finally, without process data, it is difficult to provide meaningful feedback to the method-developers so they can improve their techniques. This situation is akin to a usability study of a system that only provides numerical data about performance time or total number of errors; what we need is the equivalent of a think-aloud protocol for assessing usability techniques.

In his book *Case study research: Design and methods*, Robert Yin [19] states that a case-study approach has an advantage over surveys, experiments, and other research strategies "...when a 'how' or 'why' question is being asked about a contemporary set of events over which the investigator has little or no control." ([19], p. 9) This seems to describe the situation facing the field of HCI when evaluating methods. We are asking *how* a given technique can be used to predict usability problems, *why* it works in some situations and not in others, and we have virtually no control over how an analyst learns or uses a technique.

The essence of the case-study approach is to collect many different types of data and use them "in a triangulating fashion" ([19] p. 13) to converge on an explanation of what happened. When multiple sources of information converge, it boosts our confidence that we have understood the series of decisions occurring in a case; how these decisions were made, how they were implemented, and what result was achieved. This deep understanding should allow us to know whether these processes and results are likely to reoccur with other developers or in the next design project.

This paper presents the first in a series of five case studies of novice analysts learning and applying different usability assessment techniques to a single interface.[1] These studies analyze several sources of evidence: problem description forms filled out as the analysts worked, discussions with

---

[1] In all, data on five cases were collected with analysts using Cognitive Walkthrough, Claims Analysis, User Action Notation, GOMS, and Heuristic Evaluation. However, due to space limitations, this paper will report the details of only the first case.

the analysts, final reports by the analysts, and, to obtain detailed process data, diaries kept by analysts while they learned and applied these techniques. This case-study approach allows us to look at what the analysts did, the confusion and insights they had about the techniques, as well as more traditional performance measures like the number and type of usability problems identified.

THE ANALYSIS SITUATION

It is not uncommon for a development team to ask one of its members (or an outside person) with an interest in HCI to make recommendations about a user interface design (in fact, it was just such a request that led the first author to switch fields from mechanical engineering to HCI 15 years ago). When this happens the analyst must figure out what assessment technique he or she wants to use, find books or papers describing how to use it, learn the technique from these materials, and apply it to the system design. In this study, we set up a similar situation in the following way.

Figuring out what technique to use

Five volunteer analysts were given a 30 minute lecture on HCI assessment techniques. This lecture was the methods part of the *Introduction and Overview of Human-Computer Interaction* tutorial given at the CHI conference for the last three years [2]. The analysts heard this lecture from the same lecturer (the first author) and received the same tutorial notes and bibliography as tutorial participants, which is the way scores of professionals get their first introduction to usability assessment methods each year. One week later, each analyst chose the method they wished to use.

The system and specifications

The analysts were given two documents with which to do their analyses: the user interface specification of the ACSE multimedia authoring system and a target multimedia document (described below). The analysts read these documents at home for one week and then were given two 1-hour sessions with the head developer of the system to clarify any points of misunderstanding. The results of these sessions were written up and sent to each analyst via e-mail. The analysts had access to the developer via e-mail if any other difficulties arose understanding the documents, and all such e-mail conversations were sent to all analysts.

The ACSE Multi-media Authoring System The Advanced Computing for Science Education (ACSE) project has built a multi-media science learning environment to teach the skills of scientific reasoning [12]. This software system provides an author with a structured document framework and a set of tools for constructing science lessons containing text, still graphics, movies, and simulations. The system provides the science student with tools for navigating through the lesson, viewing the movies, and manipulating and running the simulations.

Based on several years experience with this system, the ACSE project recently designed a second version of it's authoring tool, called the VolumeViewBuilder (henceforth, *Builder*). The redesign was a group effort recorded in a user

interface specification written by a software developer and a technical writer, both of whom had recently sat in on an HCI design class [5]. The implementation of the Builder was begun at the same time as our case studies and continues at the time of this writing.

The ACSE project gave us their user interface specification for analysis. The specification included an introduction to the system and the goals of the document (4 pages), the interface for science students (the VolumeViewExplorer, 9 pages) and the interface for the Builder (22 pages). The specification included 37 figures of the screens (1 in the introduction, 11 in the Explorer, 24 in the Builder) which ranged from small pictures of specialized cursor icons to full-page figures of the entire screen (e.g., Figure 1).

The Example Multi-Media Document: The target multi-media document was a biology lesson about Drosophila development adapted from an advanced undergraduate biology course using the ACSE system for laboratory sessions. This volume is 55 pages long, and in addition to text contains 23 high resolution images and figures, 3 movies, 7 simulations, 10 fragments of simulation code,

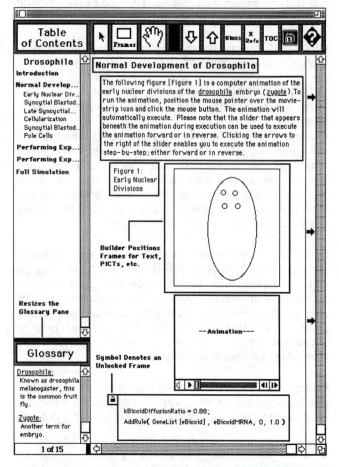

Figure 1. Example of an illustration in the *Volume View Interface Design* [5, p. 16]. In that document, this is a full-page illustration of the screen of the Builder application

and 10 review questions. The original volume was produced with the first version of the Builder and did not included a table of contents, a glossary, or hyper-links, because those features were not included in that version of the Builder. The first author modified this lesson to include these features and produced a hard copy target document. This document did not run on a computer, but the links were explicitly indicated in supplementary lists, i.e., all the table of contents entries and their page numbers were in a list, all the glossary terms were in a list, all the hyperlink "hot phrases" and where they would point were in a list. This document and the lists were given to the analysts as a typical document created with the new Builder.

THE ANALYSIS PROCESS

The analysts worked primarily on their own for an elapsed time of ten weeks. They used two forms for recording their work on an ongoing basis: a structured diary and a problem description form (described below). The group met once or twice a week to discuss the analysis process. Each analyst produced both a verbal and written final report of their analyses. A questionnaire assessed the analysts educational and professional background. All of these data were analyzed to produce the case reported in this paper.

The Diary Form

The diary form was adapted from [14]. The analyst used it to record his or her activity each half hour while working on the analysis. The analyst wrote a short description of the activity and then placed it in one of six different categories: literature search, reading for "what-it-is", reading for "how-to-do-it", reading/analysis (when reading and analysis are so intertwined as to be inseparable), analysis (when the analyst knows the technique well enough to analyze without reference to the literature), and unrelated (e.g., copying papers). The form also had columns for recording a difficulty with the technique, an insight into the technique, a problem with the design, a solution to a design problem, and "other." At any time, the analyst could make a note on the form and write an extended explanation of what he or she was thinking in a separate, free-form diary.

The Problem Description Form

The problem description report (PDR) was adapted from [7]. The PDR provided an area for describing the problem, an estimate of the severity and frequency of the problem, and an assessment of whether these judgments came from the technique itself, as a side effect of the technique, or from some form of personal judgment. Each PDR had a reference number that also appeared in a diary's column for recording a design problem.

The Analysts' Discussions

The discussions with the analysts were structured around several open-ended questions: What did you do in your analysis in the last couple of days? Did you have any difficulty with the technique? Did you have any insights into the technique? Did you discover anything else notable about the technique? Thus, these discussions centered around the process, not the content of the analyses. That is, they discussed things like problems getting or understanding papers, problems making the techniques applicable to the Builder, types of information their techniques needed or provided, but not specific usability problems with the Builder (e.g., that the method of creating hyperlinks was awkward or that the menu items were in the wrong place). The first author took notes during the discussions, which contributed to the case in this paper. These discussions were also audio taped, but the tapes have not yet been analyzed and do not contribute to this case.

The Final Report

Each analyst presented a videotaped verbal and a written report that included a brief summary of the technique, an annotated bibliography of the papers used to learn and apply the technique, any modifications made to the technique to allow it to be used for the Builder specification, any areas of exceptional doubts or confidence about using the technique, suggestions for improving the technique, and the three most important problems that the designers should fix in the Builder. Again, the written artifacts contribute to the case in this paper, but the tapes have not yet been analyzed.

THE COGNITIVE WALKTHROUGH ANALYSIS
The analyst, A1

A1 was a researcher in the School of Computer Science. He had taken over a dozen courses in CS, considered himself fluent in two programming languages, and had worked professionally as a programmer before taking part in this evaluation. He had taken one cognitive psychology course, but none in HCI. A1 received graduate-course credit for participating in this analysis.

The choice of a technique

A1 spent 12 hours finding and reading papers about several HCI techniques and reading the Builder specification before deciding which analysis technique to choose. He considered PUMs [6] but rejected it because it required getting the PUM simulation code. He considered heuristic evaluation as defined by Jeffries and colleagues [7] but decided he did not have sufficient UI design experience. He considered using standards or guidelines but rejected it because it meant "a lot of reading though thick books." He read Chapters 5 (Usability Heuristics) and 8 (Interface Standards) in Nielsen's *Usability Engineering* [11]. Despite some doubts about whether he had the prerequisite training, he chose Cognitive Walkthrough (CW) based on the initial HCI methods overview lecture (above) and the summary of the CHI'92 workshop on usability inspection methods [10].

CW is an usability analysis technique similar in rationale and execution to requirements or code walkthroughs. CW focuses on the ease of learning a new interface. This method has been evolving since its introduction in 1990 [9] and it's current form [18] was eventually used by A1. Inputs to a CW are a description of the interface, a task scenario, assumptions about the knowledge a user will bring to the task, and the specific actions a user must perform to accomplish the task with the interface. The group, or individual, performing the CW then examines

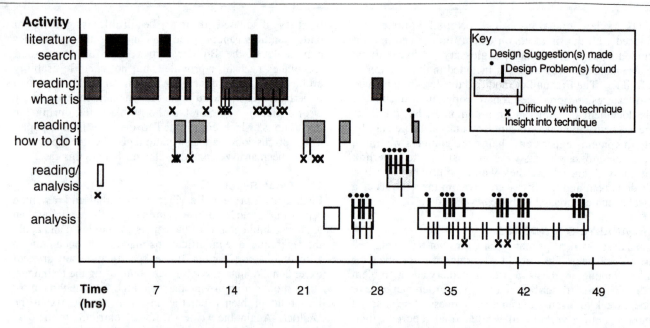

Figure 2. Timeline of the activities of analyst A1 as he learned and applied the Cognitive Walkthrough inspection method to the Builder multimedia authoring tool, as recorded in his diary forms.

each step in the correct action sequence asking the following four questions. "(1) Will the user try to achieve the right effect? (2) Will the user notice that the correct action is available? (3) Will the user associate the correct action with the effect that the user is trying to achieve? (4) If the correct action is performed, will the user user see that progress is being made toward solution of the task?" ([18] p. 106) If a credible success story cannot be told for each question at each step, then a usability problem has been identified and CW suggests ways of fixing the problem.

The time course of the analysis
A timeline of the activities of A1 appears in Figure 2. After choosing CW, A1 read several other papers [1, 9, 13, 15, 16] before finding [18] (which at the time was only a University of Colorado technical report, but is now publically available). In his final report, A1 stated that [18] is the only reference really needed to learn and use the technique and strongly recommends it to other analysts because of its clear, step-by-step description and concise examples. When he found this paper he read it with special emphasis on learning how to do the technique (total 3 hrs).

A1 then spent a total of 4.5 hours setting up task scenarios with correct action sequences in preparation for doing the walkthrough. To do this, A1 examined the functionality of the Builder and the frequency of elements (e.g., PICT frames, movies, glossary terms) occurring in the Drosophila target document. In doing so, A1 discovered that the Drosophila document did not include several of the features included in the Builder interface document, e.g., it did not include graphical representations of simulation code like radio buttons for a fixed set of mutually exclusive parameters or slide-bars for changing numerical parameters in a simulation. Noting this deficit in the target document in the diary for later discussion, A1 chose two task

scenarios, the first being a short one (creating pps. 1-3 of the target document) to test his understanding of the walkthrough technique, and the second being much longer (creating pps. 24-35; 104 user actions).

A1 then determined the correct sequence of 36 user actions for the short task (1.5 hours) and performed the walkthrough (2 hours). He found six usability problems and six associated design suggestions during this self-imposed practice walkthrough. In addition to usability problems, this short analysis pointed out many gaps in the Builder interface specification. After this, A1 read [17], which he found interesting, but not useful for applying CW.

A1 then began to apply the CW method in earnest. Since he did not have ready access to designers of the system, he performed the walkthrough himself, rather than in a group. He first prepared the correct action sequence for this longer task (104 actions). He began the walkthrough with frequent reference to the [18], but after 2.5 hours of tightly intertwined reading/analysis and a second reading of the Builder specification, he did not need to refer to the papers anymore. He then spent 16 hrs doing the walkthrough, that is, for each user action he asked the four questions recommended by CW and either wrote a success story or filled out a problem description report for each question. During this phase he reported 36 usability problems and 35 design suggestions. His final report cites seven more general psychology and HCI papers, which his diary does not record as being read during this time period (presumably read during his cognitive psychology course).

An example of the content of A1's walkthrough
Figure 3 shows a portion of the action sequence A1 prepared for himself for the second task scenario. He listed

---

```
Action  Sequence
Task B: Create 24-35 in Drosophila  document

USER  ACTION                 SYSTEM   FEEDBACK

Reopen Drosophila volume and go to p. 23
1) Click on Drosophilia icon    brings up simulation

2) Choose "Show Volume"        brings up Drosophila volume
   from the "Windows" menu      with page 1 on top
```

Figure 3. Sequence of correct actions for a portion of the long task scenario used by A1.

both the action a user would have to take to accomplish the task and the system's response for each step.

Figure 4 shows A1's walkthrough of the actions specified in Figure 3. For each action, he enumerated the CW questions and either checked off a question if he considered it a success and recorded the reason for his judgement, or justified why he thought the question indicated a failure of the interface to support learnability. If there was a failure at a action, A1 prepared a PDR (Figure 5) and put the number of the PDR in his CW. In addition, A1 often recorded a suggestion for redesigning the system to prevent the failure.

The usability problems identified

A1 submitted 42 PDRs, all of which were written while doing tightly-coupled reading/analysis or analysis. These PDRs included 48 unique usability problems and 4 duplicate problems (there were 9 PDRs with more than one problem on them) as judged by consensus between the authors. These problems covered all aspects of the Builder application: bookmarks, frames, cross-references, glossary, table of contents, cursor, help system, pages, undo, volume, and the entire application as a whole; but not problems with the Drosophila target document. Of these PDRs, A1 said he found 61% of them directly from using the technique, 12% as a side effect of using the technique, 15% simply from reading the Builder design specification, and 10% from other sources (2% were left blank). Of the "other" reports, the source of these problems were the group discussions.

The PDRs asked for a judgment of how frequently the problem would occur to a user of the Builder. A1 judged that none of the problems he reported would occur only once to a user, 22% to occur rarely, 37% to occur occasionally, 29% to occur often, 10% to occur constantly, (2% were left blank). None of these judgments came directly from using the technique, 39% came from personal judgment, 56% came from "other" sources (5% were left blank). The other source of frequency information was usually the frequency analysis of the target document that A1 had performed when selecting the task scenarios, whereas a few were attributed simply to "common sense."

The PDRs also asked for a judgment of the severity of the usability problem on a scale of 1 to 5, where 1=trivial,

```
Cognitive  Walkthrough
Task B: Create 24-35 in Drosophila  document

1 .1 ͜  part of task, Mac experience
  .2 ͜  icon visible
  .3 ͜  Mac experience
  .4 ͜  simulation comes up
  success
2 .1   problem here: Mac experience tells you that you
       immediately get into the application by clicking on
       an icon representing an instance of this application
  .2 ͜  accessible [through] menu
  .3   again problem here
  .4 ͜  loads Drosophila volume ‡ visible in window
  failure ‡ 26
```

Figure 4. A1's answers to the four CW questions for the actions shown in Figure 3. The arrow pointing to "26" indicates that this failure is reported in PDR #26 shown in Figure 5.

```
Problem Description Form

Reference number in diary: 26
Brief description of the problem: Starting a volume in builder
  mode: Necessity to choose "Windows - Show Volume" to
  get into builder mode with the current volume, contradicts
  Mac experience that you get into the correct application
  just by clicking on a file created by this application
How did you find this problem? Using my technique
How frequently will users encounter this problem? Often
How did you estimate frequency? "Other" - [this will occur]
  each time the builder wants to edit an existing volume
How serious is this problem? Serious
How did you estimate severity? Personal Judgement
Other comments? Design suggestion: Redesign procedure
  to open an existing volume. 1) click on icon representing
  the volume, 2) a dialog box comes up asking whether this
  volume shall be opened in builder or explorer mode, 3)
  application loads volume and displays first page of it
  without any further action.
```

Figure 5. PDR#26 reporting the learnability problem indentified in the CW of step 2 (Figure 4).

3=moderately serious and 5=must be changed for software to be usable! On this scale, A1 judged none of the problems he found to be trivial problems, 12% to be 2, 29% to be moderately serious (3), 49% to be 4, and 5% must be changed for the software to be usable at all (5% were left blank). Again, none of these judgments were attributed to the CW technique, whereas 95% were personal judgment (5% were left blank).

In his final report, A1 said that the three most important problems to fix were the icons in the tool palette, the items in the menus, and the relationship of the glossary and table-of-contents panes to the main-media pane of the Builder window. The first two of these problems had many PDRs associated with them (6 and 5, respectively, about 13% of the total number of PDRs for each one). This is not surprising, as two of CW's four questions focus on the availability and meaningfulness of cues in the system, both of which point to issues with icons and menus. The last problem appears in only one PDR. However, A1 justifies the importance of this problem with his frequency analysis of the Drosophila document: glossary terms were the most frequently occurring feature of that document.

Difficulties with CW and insights into its use
In all, A1 recorded 87 notes in his diary which he labeled either a difficulty with the CW technique or an insight into its use. Figure 2 shows that the difficulties and insights occur not only when initially reading about the technique or when learning how to use it, but throughout the analysis. These notes fall into several major content categories. Also, some concerns disappear as A1 learns more or becomes more familiar with CW, whereas others persist for the duration of the analysis.

Background and training required of the analyst (3 notes). When reading about the CW technique, A1 was initially concerned about the amount of training necessary for the analyst because of statements made in [9] and [10]. However, this concern disappeared from his diary notes after the 16th hour. By the final report, A1 was "very comfortable with learning and applying the technique" and asserted that "little or no experience in either user interface evaluation or cognitive psychology is required of the user [of the CW technique]."

Applicability of CW to the Builder application (5 notes). A1 wondered whether CW would scale up to the complex Builder interface. However, reading [16] seemed to allay these concerns. In the final report, A1 says he "...wondered whether the assumptions of the walkthrough...would apply to the evaluation of the Volume View. But the assumptions about the user population (novice users with Mac experience) turned out to be consistent with these assumptions and the more recent versions of the CW are not restricted to walk-up-and-use interfaces."

Underlying theory (13 notes). A1 expressed many initial concerns about the underlying theory of CW. He was concerned about the validity of CE+ [13] and what part the CE+ theory played in the actual walkthrough. He was concerned about his lack of knowledge about the underlying theory, and how that would impact his ability to conduct an effective walkthrough. However, these concerns disappear while reading [18] because A1 reasoned that CW's assumption that the user's actions would largely be guided by the interface applied to the Builder. In his final report, A1 feels confident in his ability to conduct a walkthrough, and states that "Little or no experience in...cognitive psychology is required... The main strength of the CW is that the technique is very simple and easy to learn."

Task scenarios (13 notes). A1 had two types of concerns about the task scenarios CW requires. The first concern is with his ability to develop "good" task scenarios. He felt that in order to come up with realistic task scenarios, he would have to talk to the users of the Builder, but this application was too new to have many users or a history of use, so he had to settle for performing a frequency analysis on the Drosophila example document. He continued to have insights about the choice of task scenarios throughout the analysis process, right up until the very end when he decided that task scenarios for a design tool like the Builder should include modification tasks and recovery from error as well as the create-from-scratch tasks he actually analyzed (this last insight came from the group discussions, brought to the table by the analyst using GOMS).

A1's second concern was about not having the designers accessible to dictate the action sequences for the tasks as the CW papers suggest. This remained a concern throughout the analysis because he believed he needed to know more about the Builder than the specification contained, as evidenced by the fact that his diary contains 23 notes about gaps in the specification. However, A1 turned this problem into a virtue by the final report; he stated, "determining the sequence of correct actions for longer task scenarios can take quite a lot of time; and this can be quite costly if being done by a designer. Therefore, I suggest that the sequence of actions is determined by the evaluator himself. Another reason for this modification is that it puts the evaluator naturally in a situation of learning by exploration. After all, this is the main focus of the Cognitive Walkthrough." A1 recognizes, however, that if the evaluator determines the sequence of actions, he or she must be given an opportunity to clarify questions with the designer.

The process of doing a CW (9 notes). A1 had several concerns about the process of performing a CW that arose from reading the early research papers about the technique [1, 9, 13, 15, 16]. He was concerned about filling out structured forms, how to handle design suggestions, and how to record usability problems that were not directly connected to the task scenarios. However, all of these concerns disappeared when A1 read [20].

During the practice walkthrough, A1 found the process of stepping through re-occurring sub-procedures to be very tedious. To relieve this burden, A1 introduced macros for tasks that occur frequently in the same context in the task scenarios. That is, he would perform the walkthrough for the first occurrence of a sub-procedure at the lowest level of granularity, but for subsequent occurrences, he used a macro to symbolize the detailed steps.

One process concern recorded by A1 was justified when the PDRs were examined closely. Early on, A1 was concerned about how to keep track of many usability problems particularly when the walkthrough is done over an extended period of time. Of the 52 problems reported by A1, we (the authors) agreed that 4 of these problems were duplicates. This amounts to 8% of the problems, but this problem might escalate rapidly with larger systems.

The basic capabilities of the CW method (7 notes) A1 voiced an early concern, raised in [18] that CW could not address global design issues. Evidently nothing in his experience with CW helped to quell that concern, for in his final report, A1 reiterates, "through the focus on the narrow path determined by the sequence of correct actions, global design issues are not explicitly addressed." Furthermore, A1 says the focus on correct sequences of actions prevents CW from addressing the ease of recovery from error. His suggestion for explicitly including error-recovery task scenarios is an effort to overcome this problem.

Finally, A1 was concerned that the technique does not provide guidance for rating the frequency and severity of usability problems. Indeed, even if this concerned had not been articulated evidence from the PDRs is overwhelming. On the 42 PDRs submitted, *none* of the estimates of either frequency or severity were accredited to the CW technique.

The quality of A1's walkthroughs

All of A1's products were examined in detail by John Rieman, a developer of the CW method. Rieman's opinion of the work was that "A1's final paper showed an excellent understanding of the...technique...[His] walkthroughs were fairly good...but they had some flaws." (personal communication, 16 Dec 1994) Rieman thought that A1 often failed to recognize steps where the user may not have the right goal (question 1). In addition, Rieman thought A1 was too strick in his failure criterion for question 3. For instance, A1 listed a failure when the user's goal was to create a glossary entry, but the button was labelled "Gloss." Rieman would have called that a successful instance of label-following rather than a failure. The first flaw would miss usability problems in the interface and the second flaw would produce false alarms. In future empirical usability studies of the interface, we will be able to assess Rieman's predictions of the effectiveness of A1's walkthroughs.

DISCUSSION
What this case study says about the Cognitive Walkthrough evaluation technique

The over-riding sense of the experiences of A1 is that the Cognitive Walkthrough evaluation technique is learnable and usable for a computer designer with little psychological or HCI training. There are also some specific lessons in the details of this case study for different interest groups.

For computer designers, the strongest message is that reading the early research papers about CW can be confusing and raise many concerns about the theory and practice of CW. (It is unclear from this case whether this is true for everyone, or only for analysts without formal psychology training.) However, the practitioner's guide [18] cleared up most of those issues for A1, suggesting that designers should read only that chapter. This case study alone cannot make such a recommendation with confidence because it is impossible to determine whether A1 benefitted from [18] because it is sufficient to learn the technique or only because he had read the other papers. To investigate this further, the first author assigned only [18] to her undergraduate HCI class and lectured only from the content of that paper. The walkthroughs conducted by the six undergraduate teams were examined in detail by Rieman and

considered "all fairly good and some were excellent." (personal communication, 16 Dec 1994) Where there were problems they often involved question 1 (right goal?) and question 3 (good label?) as was the case with A1 . This experience strengthens our recommendation to avoid the early papers and read only the practitioners' guide if you want to use CW. However, pay particular attention to the explanations and examples of questions 1 and 3.

Another message to designers is that CW itself will not give you much guidance about how to pick task scenarios. (This same message results from the GOMS and Claims Analysis cases as well.) Scenario generation still seesm to be a black art. However, this case suggests that including modification and error-recovery tasks in the scenarios may be important and easily overlooked. Similarly, CW will not give any guidance as to the frequency or severity of usability problems (a deficit shared by all five evaluation methods studied). These gaps in HCI techniques call to the developers of evaluation techniques to fill a crying need.

A more specific question to the developers of CW concerns the modification suggested by A1 that macros be used to reduce tedium. Before recommending this procedure to designers, the impact of macros should be assessed. True, they may decrease analysis time. However, they may also focus attention away from tedious tasks that users themselves would find objectionable, just as the analysts do. The developers of CW, or others who want to extend its use, may view the assessment of macros as another opportunity for contribution.

A comment on the case-study approach

This case study approach used many different types of information to converge on a picture of the process A1 went through to produce his analysis. This story itself can help designers learn and use CW by providing realistic expectations about the process, the difficulties and insights, and the length of time it takes to perform an analysis. Such expectations are missing from a textbook description of a technique. However, without them, new analysts can become unnecessarily discouraged, or never even consider embarking on such an analysis.

FUTURE WORK AND AN INVITATION

An important element missing from this analysis is a measure of how effective A1 was in predicting the usability problems that occur in the ACSE Builder application. The traditional benchmark against which to compare usability assessment is empirical usability testing. Unfortunately, the Builder has not yet been implemented to specification, so empirical evaluation must remain as future work. Another interesting benchmark would be to compare the design suggestions made through the CW technique to the actual implementation of the Builder when it is finished. It is possible that some portion of the usability problems associated with the design document would be fixed in the normal software development process as the programmers think hard about the details of implementation.

In addition to the lessons learned from a single case study such as this, the potential for many more lessons is inherent in multiple case studies. The data from the other four evaluation techniques (Claims Analysis, User Action Notation, GOMS, and Heuristic Evaluation) will contribute to our knowledge of learning and using different methods for HCI evaluation, especially as we begin to compare the experiences using the converging evidence of the case-study approach. However, even more can be learned with replication and expansion of the conditions under which these cases were performed. Therefore, the first author invites any practitioner or researcher who wishes to participate to contact her about collecting process data for case-studies. The PDRs and diary forms are available for distribution, as are the Introduction to HCI Methods lecture materials, the Builder interface design document, and the Drosophila example document.

ACKNOWLEDGMENTS
This work was supported by the Advanced Research Projects Agency, DoD, and monitored by the Office of Naval Research under contract N00014-93-1-0934. The views and conclusions contained in this document are those of the authors and should not be interpreted as representing the official policies, either expressed or implied, of ARPA, ONR, or the U. S. Government. We thank Robin Jeffries for providing her original PDR and helpful suggestions, "A1" for his thoughtful and thorough CWs, and John Rieman for his evaluations of the CWs discussed herein.

REFERENCES
1. Bell, B., Rieman, J., and Lewis, C. (1990) Usability Testing of a Graphical Programming System: Things We Missed in a Programming Walkthrough. In *CHI'90 Proceedings*, Seattle, WA, ACM, NY.

2. Butler, K, Jacob, R. J. K., & John, B. E. (1993) *Introduction and Overview of Human-Computer Interaction*. Tutorial materials, presented at INTERCHI, 1993 (Amsterdam, April 24 - April 29), ACM, New York.

3. Cuomo, D. L. & Bowen, C. D. (1994) Understanding usability issues addressed by three user-system interface evaluation techniques. *Interacting with Computers, 6, 1*, 86-108.

4. Desurvire, H. W. (1994) Faster! Cheaper!! Are usability inspection methods as efficient as empirical testing? In Jakob Nielsen and Robert L. Mack (eds.) *Usability Inspection Methods*. New York: John Wiley.

5. Gallagher, S. & Meter, G. (1993) *Volume View Interface Design*. Unpublished Report of the ACSE Project, School of Computer Science Carnegie Mellon University, May 4, 1993.

6. Howes, A., and Young, R.M. (1991) Predicting the Learnability of Task-Action Mappings. In *CHI'91 Proceedings*, New Orleans, LA, ACM, NY.

7. Jeffries, R., Miller, J.R., Wharton, C., and Uyeda, K.M., (1991) User interface evaluation in the real world: A comparison of four techniques, *CHI'91 Proceedings*, New Orleans, LA, ACM, NY.

8. Karat, C.-M. (1994) A comparison of user interface evaluation methods. In J. Nielsen and R. L. Mack (eds.) *Usability Inspection Methods*. New York: John Wiley.

9. Lewis, C., Polson, P., Wharton, C., Rieman, J. (1990) Testing a walkthrough methodology for theory-based design of walk-up-and-use interfaces. In *CHI'90 Proceedings*, Seattle, WA, ACM, NY.

10. Mack, R., Nielsen, J. (1992) Usability Inspection Methods: Summary Report of a Workshop held at CHI'92.. *IBM Technical Report* #IBMC-18273. IBM T.J. Watson Research Center, Yorktown Heights, NY.

11. Nielsen, J., (1993) *Usability Engineering*, San Diego: Academic Press Inc.

12. Pane, J.F., & Miller, P.L. (1993). The ACSE multimedia science learning environment. Proceedings of the 1993 International Conference on Computers in Education, Taipei, Taiwan.

13. Polson, P.& Lewis, C. (1990) Theory-based design for easily learned interfaces. *Human Computer Interaction*, 5, 191-220.

14. Rieman, J. (1993) The diary study: A workplace-oriented research tool to guide laboratory efforts In *Proceedings of INTERCHI', 1993* Amsterdam, ACM, NY.

15. Rowley, D. E., & Rhoades, D. G. (1992) The Cognitive Jogthrough: A fast-paced user interface evaluation procedure. In *CHI'92 Proceedings*, Monterey, CA, ACM, NY.

16. Wharton, C., Bradford, J., Jeffries, R., Franzke, M. (1992) Applying Cognitive Walkthrough to more complex user interfaces: Experiences, issues and recommendations. In *CHI'92 Proceedings*, Monterey, CA, ACM, NY.

17. Wharton, C., Lewis, C. (1994) The role of psychological theory in usability inspection methods. In J. Nielsen and R. L. Mack (eds.) *Usability Inspection Methods*. New York: John Wiley.

18. Wharton, C., Rieman, J., Lewis, C., & Polson, P. (1994) The Cognitive Walkthrough Method: A practitioner's guide. In J. Nielsen and R. L. Mack (eds.) *Usability Inspection Methods*. New York: John Wiley..

19. Yin, R. K. (1994) Case study research: Design and methods (2nd ed., Applied Social Research Methods Series Vol. 5). Thousand Oaks, CA: Sage Publications.

What Help Do Users Need?: Taxonomies for On-line Information Needs & Access Methods

A.W. Roesler, S.G. McLellan

Austin Systems Center
Schlumberger Well Services
8311 North FM 620 Road
P.O. Box 200015
Austin, Texas 78720-0015
roesler@austin.asc.slb.com

ABSTRACT

The feasibility of using a general on-line help taxonomy scheme as the starting point for our interactive graphical applications' on-line help specifications was investigated. We assumed that using such a taxonomy would make it easier for users of the help system, regardless of the application used. The literature, software conferences, trade shows, and the like point to enormous differences of opinion about what help even IS, much less how it should be designed, accessed, displayed, stored, or maintained. While much research described sound design principles and access methods, very little was available on WHAT to organize or access. Our effort on defining a taxonomy for on-line help was based upon three tests:

- Test1, a Wizard-of-Oz usability study of an application that identified what types of on-line help our interactive software users actually ask for;
- Test2, a test that validated a general taxonomy for on-line help content for help providers, based on the results of Test1, and a general taxonomy of access methods derived from these content types; and
- Test3, a repeat of Test1, substituting a prototype help system for Wizard-of-Oz help that successfully validated the usability of both on-line help content and access taxonomies for help users.

This paper summarizes the results of all three tests, highlighting the proposed taxonomies and key findings about them from Test2. Together, the results from all tests indicate that a general taxonomy of information needs and the taxonomy of access methods to particular information types make it easy both for help providers to understand what information they need to supply and for help users to find the help they need quickly.

KEYWORDS

On-line help, taxonomy, user interface, usability, empirical evaluation, methodology.

THE PROBLEM

Early in 1993, as part of the requirements for a large commercial system of over 55 complex interactive graphics applications comprising some 16 million lines of code, an effort was begun to determine the feasibility of using a general on-line help taxonomy scheme as the starting point for all applications' on-line help specifications. The current help facility, utilizing a book mental model and related table of contents and indices as the primary navigational aids, had proven inadequate--for example, for the application chosen for the present study, only 32% of help requests from users could be answered or found in the current documentation. "One striking thing about a survey of the current literature on help systems," Mayhew [1] indicates, "is what is missing, rather than what is available. In particular, no paper or study surveyed here directly addresses the basic question of what the content or subject matter of a user assistance program should be. Few researchers seem to have posed and answered the question, 'What kind of questions do novice and expert users actually ask when using a system?' All the fancy formatting, navigational ease and ease of access in the world will not be of much use if the information contained in the help system is not the information that users seek." Likewise, Elkerton and Palmiter [2] find that "knowledge about help content is underdeveloped, whereas research on help presentation and access is more widely available." In his 1993 book on help systems, Duffy [3] identifies the primary challenge in designing information for on-line help as "match[ing] the information provided to users with the different kinds of knowledge that they require." Clearly, what is required is a practical examination of the content of help that users need and if it is possible to categorize this content for reuse from application to application.

THE STRATEGY TOWARDS A SOLUTION

Our short-term efforts at defining a taxonomy for on-line help were based upon three basic tests:

- Test1, a Wizard-of-Oz usability study of an application that identified what types of on-line help our interactive software users actually ask for,
- Test2, a test that validated a general taxonomy for on-line help content for help providers, based on the results of Test1, and a general taxonomy of access methods derived from these content types, and

• Test3, a repeat of Test1, substituting a prototype help system for Wizard-of-Oz help that successfully validated the usability of both on-line help content and access taxonomies for help users.

The longer-term goals of our work were to establish empirically-based guidelines that would:

• make it simple for help developers to determine and place required help information so that help users can easily find the information where they expect it.
• explicitly cover all questions indicated by Test1.
• implicitly cover all similar specific questions that could arise, given a different set of tasks.
• minimize user navigation for information.

Before presenting either taxonomy of information or access methods, let's examine what the first and third tests revealed about the kind of information users required and how the prototype help system satisfied these requirements.

Comparing Wizard-Of-Oz Help (Test1) & Prototype On-line Help (Test3)

A Wizard-of-Oz technique [4], whereby a person, the Wizard, received the test subjects requests for help and supplied that help as required, was chosen for Test1 to capture how users formulated their requests for help as they encountered obstacles or questions as they worked with an application. A specific interactive application, which allows users to edit graphical displays of complex geological data, was selected. A set of 15 specific tasks was formulated from a real-world use of the program in the field. Five clients with varying experiences using the application were asked to bring up the application and to think aloud as they performed these prescribed tasks. They were also told that they could rely on help at any time from an expert, the wizard, sitting alongside [5]. Results were videotaped to capture the entire session with user commentary, errors, questions, and the like.

Some months later, a usability test was run to validate if users could successfully use the prototype on-line help system that derived its specific taxonomy of help content and specific taxonomy of access methods from the questions asked and the errors made in Test1. Three users performed this Test3, the same as Test1 with one exception: the wizard used in Test1 was replaced by the prototype help system in Test3. At the beginning of each session, test subjects were shown how to use help ("Help on Help" entry from Help Menu in Figure 1). Again, user performance was videotaped. Though we are still analyzing the data, our observations to date show the following:

• All test subjects easily learned how to use the help system and referenced it as needed. No subject found it necessary to access help on help during the duration of the test.
• Subjects in Test3 committed none of the three critical errors subjects in Test1 did. One reason was that all the critical need-to-know information was provided by the help system as a whole, whereas the wizard only provided information about the single problem a subject in Test1 was asking about. Typically, the sub-

ject in Test1 then committed the next critical error and asked the wizard for help.
• There were 44% fewer overall errors than in the first test (44 for 3 users versus 131 for 5 users). In part, this resulted because repeated errors were less frequent in Test3. Videotapes reveal that test subjects repeatedly utilized prompt messages, reminding them about what components were (What Is) and what actions to perform and how to perform them (What Next) correctly.
• No test subject failed to find the information he or she needed from the help system.
• Even though our help documentation was terse, test subjects made comments that they would like for it to be even more terse. In particular, they are not interested in rationalizations for actions they have to perform being part of what they FIRST see in the help system. In short, they wanted some control of whether they saw any rationalizations or not.

Defining a Taxonomy of Help Information (Test2)

The taxonomy of help content was determined by starting with existing taxonomies in the literature [6,7,8,9,10,11], then tailoring them to a generalization scheme that incorporated our findings from Test1. In particular,

• Users posed their requests for help as questions. *Example*:
"Which button here controls the fill of the box?"
• Based on the information requested and errors committed, required information included some unexpected types: for example, what an application CAN'T do and how to use Motif conventions (despite users indicating an expert level of Motif knowledge).
Examples:
"What about a double ended arrow?"
"How do I change the colors of a curve?"
A user lost text he typed into a text field dialog because the default button was changed to the Cancel button and the user pressed the Return key.
"Is there any difference between the Close and Exit buttons?"
• Repeated and severe errors pointed to a need to provide users up front with cues to critical, need-to-know-first information to avoid early setbacks.
Examples:
Many users selected the wrong file to display after they saved data.
All users did not initially set parameters to display all of the data they needed.
• Since some information (for example, encountered bugs and workarounds) cannot always be anticipated ahead of time, alternative information about assistance (for example, Customer Support) must be available to users.
Example:
"Looks like the program is broken. What happened?"
• Since all users, new or experienced to an application, experienced repeated impasses and needed reminders about HI components or next required actions, part

of the taxonomy must include methods and components directly in the application interface that provide application-directed information when appropriate.
Examples:
"What are these arrow buttons"
"How do you close a polygon again?"

- Any taxonomy of information needs includes information that is specific to the application and information that may span multiple applications.
Examples:
"What are the right color and pattern to indicate oil?" "How do I get help on Motif?"

The taxonomy we derived from an examination of the questions is given in Table 1.

In Test2, a set of six test subjects, consisting of application developers and technical writers, were given

- all of the questions from Test1 (a total of 209), along with the errors (a total of 131), in the context of the tasks that the original Test1 subjects were asked to perform, and
- the taxonomy of proposed information types.

They were asked to place a checkmark under the help category (1-10) to indicate the location in help where they would go FIRST to find the answer to the question asked. Two additional categories were provided: "Don't Understand Question" if they did not understand the question at all and "Don't Know Category" if they understood the question but did not know what help category to go for help first. The test subjects were not shown the actual video recordings of interactions but were asked to infer the context for each question as best they could. Below is a summary of our results for Test2:

- Test subjects took an average of 45 minutes to classify the 209 questions.
- Test subjects were able to classify 80% of queries for help in the same way, even without the benefit of knowing the explicit context in which the earlier users actually asked the questions.
- When test subjects were given additional context for the remaining 20% of queries, results climbed to 97%.
Example question: Is it beginning tail (to draw arrow)?
Revised question: I'm in arrow drawing mode. Is the head of the arrow at the first click or the tail?

At the end of the test, subjects were asked this question: "Based upon questions users asked and what errors users made, what information would you classify as critical information (to be included in the "What Must I Know First" help category)?" Responses confirmed how easy it is for help providers, with the aid of usability results, to point to the same list of critical information.

Defining a Taxonomy of Help Access methods (Test2)

The taxonomy of information needs was only half of the solution for help providers. What was still needed was to map this information uniformly to the access methods to

them. Specifically, we speculated how we might minimize the number of times users had to leave an application and minimize the amount of navigation for help.

Implicit to our taxonomy of information were two general help categories: user-directed and application-directed. User-directed help means that a user determines the kind of help desired. Application-directed help means that an application determines the kind of help provided. Examining the percentage of answers placed in the "What Is" and "What Next" categories, we noticed that 37% of queries for help could be answered within the application with the help of a message box and two different access methods:

- Button 1 down over an application's human interface primitive or composite component (e.g., button, menu item, menu bar) to display in the Message Box user-directed "What Is" information about the component
- Button 1 up (activation) over a primitive to display in the Message Box application-directed "What Next". information about the next possible actions required or allowed.

This was confirmed in Test3, where 44% of the help that test subjects used was contained in the one line "What Is" and "What Next" messages.

A third access method, Button 1 down plus the "Help" key, would allow users to bring up the help application with additional context-sensitive information on the component without going to the Help menu in the menu bar and searching for help. This help would include more information about what the component is and how it relates to other components, what next actions are expected, including critical need-to-know-first information, and a link to the master lists of application components (What Is), possible states and actions (What Next), and tasks that users can and can't do (How Do I).

A fourth method, the help button in any HI window, would provide access to critical need-to-know-first information (What Must I Know First) about using this window.
A fifth method, help in the Help Menu, would provide access to master lists of help specific to the application (What Must I Know First, What Is, What Next, and How Do I), as well as access to help common to a set of applications (Meaning of Terms, Help on Help, etc.).

Our access taxonomy was, then, derived from our taxonomy of information, not the other way around. This taxonomy of access methods is given in Table 2. We also implemented the same help menu and access methods in the help window for the application. This would allow users to obtain additional help in the same way as they did within the application itself.

Test3 demonstrated the usefulness of an access strategy that is simple and that utilizes messaging within an application. Specifically,

- No test subject got lost navigating in the help system throughout the test.
- 44% of the help information that test subjects used was found in the one line "What Is" and "What Next" messages within the application. This was

comparable to the 37% of questions whose answers were placed in either the "What Is" or "What Next" categories in Test2.

- In almost all cases the test subjects used the minimum amount of navigation to the information they needed outside the application (only 35% of accesses required more than 2 actions, excluding help found in "What Is" and "What Next" messages).

WHAT WE LEARNED

Overall, the results from our tests indicate that a general taxonomy of information needs and the taxonomy of access methods to particular information types make it easy both for help providers to understand what information they need to supply and for help users to find the help they need quickly (See Figure 1). Our tests suggest that

- Usability studies are needed to find and verify information in several help categories (e.g., critical information in the "What Must I Know First" category and tasks that can't be performed in the "How Do I" category).

- Subjects could not distinguish whether the information they needed was related to Motif or the specific application, despite their indication that they knew Motif behavior (e.g., how to delete the contents of a text field). Help was constructed so that subjects did not have to know whether their question was a Motif-related question or an application-specific question.

- The use of two types of prompts (e.g., for "What Is" and "What Next" information) directly in the application substantially reduces the occurrences of leaving an application for help.

- Constructing application-specific help around two levels (e.g., master lists of "What Next", "What Is", and "How Do I" versus specific "What Next" prompts, "What Is" prompts, and "How Do I" task descriptions) minimizes the amount of navigation of help outside an application.

ACKNOWLEDGEMENTS

The authors would like to thank Allison Elliott from Georgia Tech University and Ruven Brooks from Schlumberger Austin Research for their technical assistance in setting up the usability tests, compiling, and analyzing the results.

REFERENCES

1. Mayhew, D. J. (1992). *Principles and Guidelines in Software User Interface Design*, Englewood Cliffs: PTR Prentice Hall.

2. Elkerton, J., & Palmiter, S. L. (1989). Designing Help Systems Using the GOMS Model: An Information Retrieval Evaluation, *Proceedings of the Human Factors Society 33rd Annual Meeting*, 1989, pp. 281-85.

3. Duffy T. & others (1993). *On Line Help: Design and Evaluation*. Norwood: Ablex Publishing Corp.

4. Hill, W. C. (1993). A Wizard of Oz Study of Advice Giving and Following, *Human-Computer Interaction*, 1993, Vol. 8, pp. 57-81.

5. Hill, W. C., & Miller, J. R., (1988). Justified Advice: A Semi-Naturalistic Study of Advisory Strategies, *CHI '88 Proceedings*, May 1988, pp. 185-90, ACM.

6. Sondheimer, N. K., Relles, N. (1982). Human Factors and User Assistance in Interactive Computing Systems: An Introduction, *IEEE Transactions of Systems, Man and Cybernetics*, Vol. SMC-12 (March-April 1982), pp. 102-107.

7. Hartley, J. R., & Smith, M. (1988). Question answering and explanation giving in on-line help systems. *Artificial intelligence and human learning: Intelligent computer-aided instruction* (pp.338-360). London: Chapman and Hall.

8. Miyake, N., & Norman, D. A. (1979). To ask a question one must know enough to know what is not known. *Journal of Verbal Learning and Verbal Behavior*, 18, 357-364.

9. Sellen, A. & Nicol, A. (1990). Building user-centered on-line help. In B. Laurel (Ed.), *The art of human-computer interface design* (pp.143-154). Reading, MA: Addisoon-Wesley.

10. Robertson, S., & Swartz, M. (1988). Why do we ask ourselves questions? *Questioning Exchange*, 2, (1), pp. 47-51.

11. Harris, R. A., & Hosier, W. J. (1991). A Taxonomy of Online Information. *Technical Communication*, 1991, 38(2), pp. 197-209.

Table 1. Taxonomy for On-line Information Needs

Help Menu No.	Help Item & Type	Help Item Description
1	What Must I Know First?	Short overview of application, concepts, & methods
		Critical need-to-know-first info, bugs, workarounds
2	How Do I?	Tasks that can & can't be done
3	What Is?	Description of HI control(s) & component(s)
4	What Can I Do Next?	Application state(s) and next possible action(s)
5	Help on Help?	How to get help
6	Meaning of Terms?	Alphabetized glossary of terms
7	Mouse & Keyboard Conventions?	HI conventions, selection, activation, navigation, etc
8	Related Product Info?	Related info/docs on system, current application (user, reference, tutorial), discipline (e.g., geology)
9	Customer Assistance?	How to get outside expert assistance
10	Version Info?	Version number, other pertinent info for software administrator or customer support

Table 2. Taxonomy for On-line Access Methods

Help Method No.	Help Access Method	What Help Is Displayed & Where
1	Button 1 Down	Displays 1-line description of HI control or component (What Is) in application's Message Box.
2	Button 1 Up	Displays 1-line description of current state & next available actions (What Can I Do Next) in application's Message Box.
3	Button 1 Down + Help Key	Brings up separate Help screen with more detailed information on what component/control is & how it relates to other components (What Is), what next actions are expected (What Can I Do Next), including critical need-to-know-first information (What Must I Know First), as well as a links to 3 master lists: application components (What Is), possible states & actions (What Can I Do Next), & tasks that can & can't be performed (How Do I).
4	Button on Help Button in HI	Brings up separate Help screen with more detailed information on critical need-to-know-first information (What Must I Know First).
5	Select from Help Menu	Brings up particular master list of application-specific info: critical info (What Must I Know First), tasks that can & can't be performed (How Do I), HI components (What Is), possible states & next actions (What Can I Do Next). Or brings up info common to set of applications: Help on Help, Meaning of Terms, Mouse & Keyboard Conventions, Related Product Info, or Customer Assistance.

Bricks: Laying the Foundations for Graspable User Interfaces

George W. Fitzmaurice

Dynamic Graphics Project
CSRI, University of Toronto
Toronto, Ontario,
CANADA M5S 1A4
Tel: +1 (416) 978-6619
E-mail: gf@dgp.toronto.edu

Hiroshi Ishii

NTT Human Interface Lab.
1-2356 Take,
Yokosuka-Shi, Kanagawa,
238-03 JAPAN
Tel. 81-468-59-3522
E-mail: ishii.chi@xerox.com

William Buxton

University of Toronto &
Alias Research, c/o CSRI
University of Toronto
Toronto, Ontario,
CANADA M5S 1A4
Tel: +1 (416) 978-1961
E-mail:buxton@dgp.toronto.edu

ABSTRACT

We introduce the concept of Graspable User Interfaces that allow direct control of electronic or virtual objects through physical handles for control. These physical artifacts, which we call "bricks," are essentially new input devices that can be tightly coupled or "attached" to virtual objects for manipulation or for expressing action (e.g., to set parameters or for initiating processes). Our bricks operate on top of a large horizontal display surface known as the "ActiveDesk." We present four stages in the development of Graspable UIs: (1) a series of exploratory studies on hand gestures and grasping; (2) interaction simulations using mock-ups and rapid prototyping tools; (3) a working prototype and sample application called GraspDraw; and (4) the initial integrating of the Graspable UI concepts into a commercial application. Finally, we conclude by presenting a design space for Bricks which lay the foundation for further exploring and developing Graspable User Interfaces.

KEYWORDS: input devices, graphical user interfaces, graspable user interfaces, haptic input, two-handed interaction, prototyping, computer augmented environments, ubiquitous computing

INTRODUCTION

We propose a new paradigm, Graspable User Interfaces, which argues for having some of the virtual user interface elements take on physical forms. Traditional graphical user interfaces (GUIs) define a set of graphical interface elements (e.g., windows, icons, menus) that reside in a purely electronic or virtual form. Generic haptic input devices (e.g., mouse and keyboard) are primarily used to manipulate these virtual interface elements.

The Graspable UIs allow direct control of electronic or virtual objects through physical artifacts which act as handles for control (see Figure 1). These physical artifacts are essentially new input devices which can be tightly

coupled or "attached" to virtual objects for manipulation or for expressing action (e.g., to set parameters or to initiate a process). In essence, Graspable UIs are a blend of virtual and physical artifacts, each offering affordances in their respective instantiation. In many cases, we wish to offer a seamless blend between the physical and virtual worlds.

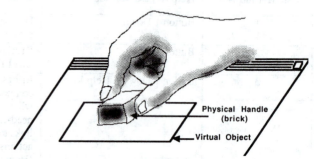

Figure 1. A graspable object.

The basic premise is that the affordances of the physical handles are inherently richer than what virtual handles afford through conventional direct manipulation techniques. These physical affordances, which we will discuss in more detail later, include facilitating two handed interactions, spatial caching, and parallel position and orientation control.

The Graspable UI design offers a concurrence between space-multiplexed input and output. Input devices can be classified as being *space-multiplexed* or *time-multiplexed*. With space-multiplexed input, each function to be controlled has a dedicated transducer, each occupying its own space. For example, an automobile has a brake, clutch, throttle, steering wheel, and gear shift which are distinct, dedicated transducers controlling a single specific task. In contrast, time-multiplexing input uses one device to control different functions at different points in time. For instance, the mouse uses time multiplexing as it controls functions as diverse as menu selection, navigation using the scroll widgets, pointing, and activating "buttons." Traditional GUIs have an inherent dissonance in that the display output is often space-multiplexed (icons or control widgets occupy their own space and must be made visible to use) while the input is time-multiplexed (i.e., most of our actions are channeled through a single device, a mouse, over time).

Only one task, therefore, can be performed at a time, as they all use the same transducer. The resulting interaction techniques are often sequential in nature and mutually exclusive. Graspable UIs attempt to overcome this.

In general, the Graspable UI design philosophy has several advantages:

- It encourages two handed interactions [3, 7];
- shifts to more specialized, context sensitive input devices;
- allows for more parallel input specification by the user, thereby improving the expressiveness or the communication capacity with the computer;
- leverages off of our well developed, everyday skills of prehensile behaviors [8] for physical object manipulations;
- externalizes traditionally internal computer representations;
- facilitates interactions by making interface elements more "direct" and more "manipulable" by using physical artifacts;
- takes advantage of our keen spatial reasoning [2] skills;
- offers a space multiplex design with a one to one mapping between control and controller; and finally,
- affords multi-person, collaborative use.

BASIC CONCEPTS

Graspable UIs allow direct control of electronic objects through physical artifacts which we call *bricks*. The bricks, approximately the size of LEGO™ bricks, sit and operate on a large, horizontal computer display surface (the Active Desk, described later). A *graspable* object is an object composed of both a physical handle (i.e., one or more bricks attached) and a virtual object (see Figure 1).

The bricks act as specialized input devices and are tracked by the host computer. From the computer's perspective, the brick devices are tightly coupled to the host computer — capable of constantly receiving brick related information (e.g., position, orientation and selection information) which can be relayed to application programs and the operating system. From the user's perspective, the bricks act as physical handles to electronic objects and offer a rich blend of physical and electronic affordances.

One Handle

In the simplest case, we can think of the bricks as handles similar to that of graphical handles in computer drawing programs such as MacDraw™ (see Figure 2a). A physical handle (i.e., a brick) can be attached to an object. Placing a brick on the display surface causes the virtual object beneath it to become attached (see Figure 2b). Raising the brick above the surface releases the virtual object. To move or rotate a virtual object, the user moves or rotates the attached brick (see Figure 3). Note that the virtual object's center of rotation is at the center of the brick.

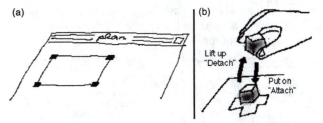

Figure 2. (a) Traditional MacDraw-like application which uses electronic handles to indicate a selection. (b) Selecting using a brick.

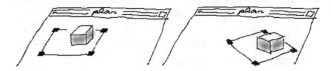

Figure 3. Move and rotate virtual object by manipulating physical brick which acts as a handle.

A simple example application may be a floor planner (see Figure 5a). Each piece of furniture has a physical brick attached and the user can arrange the pieces, most likely in a rapid trial-and-error fashion. This design lends itself to two handed interaction and the forming of highly transient groupings by touching and moving multiple bricks at the same time.

Two Handles

More sophisticated interaction techniques can be developed if we allow more than one handle (or brick) to be attached to a virtual object. For example, to stretch an electronic square, two physical bricks can be placed on an object. One brick acts like an anchor while the second brick is moved (see Figure 4).

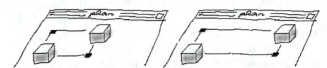

Figure 4. Two bricks can stretch the square. One brick acts like an anchor while the second brick is moved.

Placing more than one brick on an electronic object gives the user multiple control points to manipulate an object. For example, a spline-curve can have bricks placed on its control points (see Figure 5b). A more compelling example is using the position and orientation information of the bricks to deform the shape of an object. In Figure 6, the user starts off with a rectangle shaped object. By placing a brick at both ends and rotating them at the same time, the user specifies a bending transformation similar to what would happen in the real world if the object were made out of a malleable material such as clay. It is difficult to imagine how this action or transformation could be expressed easily using a mouse.

One key idea that the examples illustrates is that the bricks can offer a significantly rich vocabulary of expression for input devices. Compared to most pointing devices (e.g., the mouse) which only offers an x-y location, the bricks offer multiple x-y locations and orientation information at the same instances of time.

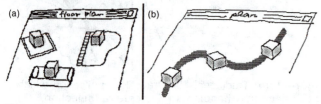

Figure 5. (a) Proposed simple floor planner application. (b) Many physical bricks are used for specifying multiple control points for creating a spline curve.

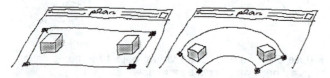

Figure 6. Moving and rotating both bricks at the same time causes the electronic object to be transformed.

RELATED RESEARCH

Some research and commercial systems have been developed with a similar graspable theme. In some sense, many of these emerging systems exhibit the property of ubiquitous computing [14] in which computation is embedded in many physical artifacts and spread throughout our everyday environment. The following systems illustrate the push towards ubiquitous computing, physical manipulation interfaces and merging physical and virtual artifacts.

The LegoWall prototype (developed by A/S Modulex, Billund Denmark in conjunction with the Jutland Institute of Technology in 1988) consists of specially designed LEGO blocks that fasten to a wall mounted peg-board panel composed of a grid of connectors. The connectors supply power and a means of communication from the blocks to a central processing unit. This central processing unit runs an expert system to help track where the blocks are and what actions are valid.

The behavior construction kits [9] consist of computerized LEGO pieces with electronic sensors (such as light, temperature, pressure) which can be programmed by a computer (using LEGO/Logo) and assembled by users. These LEGO machines can be spread throughout the environment to capture or interact with behaviors of people, animals or other physical objects. The "programmable brick," a small battery-powered computer containing a microprocessor, non-volatile ROM and I/O ports is also being developed to spread computation.

The AlgoBlock system [13] is a set of physical blocks that can be connected to each other to form a program. Each block corresponds to a single Logo-like command in the programming language. Once again, the emphasis is on manipulating physical blocks each with a designated atomic function which can be linked together to compose a more complex program. The system facilitates collaboration by providing simultaneous access and mutual monitoring of each block.

Based on a similar philosophy of the 3-Draw computer-aided design tool [11], Hinckley et al. has developed passive real-world interface props [5]. Here users are given physical props as a mechanism to manipulate 3D models. They are striving for interfaces in which the computer passively observes a natural user dialog in the real world (manipulating physical objects), rather than forcing a user to engage in a contrived dialog in the computer-generated world.

Finally, the DigitalDesk [15] merges our everyday physical desktop with paper documents and electronic documents. A computer display is projected down onto a real physical desk and video cameras pointed at the desk use image analysis techniques to sense what the user is doing. The DigitalDesk is a great example of how well we can merge physical and electronic artifacts, taking advantage of the strengths of both mediums.

STAGE 1: EARLY BRICK EXPLORATIONS

A series of quick studies was conducted to motivate and investigate some of the concepts behind Graspable UIs. Having decided on bricks, we wanted to gain insights into the motor-action vocabulary for manipulating them.

LEGO separation task

The first exploratory study asked subjects to perform a simple sorting task as quickly as possible. The basic idea was to get a sense of the performance characteristics and a range of behavior people exhibit while performing a task that warrants rapid hand movements and agile finger control for object manipulation. Subjects were presented with a large pile of colored LEGO bricks on a table and were asked to separate them into piles by color as quickly as possible.

We observed rapid hand movements and a high degree of parallelism in terms of the use of two hands throughout the task. A very rich gestural vocabulary was exhibited. For instance, a subject's hands and arms would cross during the task. Subjects would sometimes slide instead of pick-up and drop the bricks. Multiple bricks were moved at the same time. Occasionally a hand was used as a "bulldozer" to form groups or to move a set of bricks at the same time. The task allowed subjects to perform *imprecise* actions and interactions. That is, they could use mostly ballistic actions throughout the task and the system allowed for imprecise and incomplete specifications (e.g., "put this brick in that pile," which does not require a precise (x, y) position specification). Finally, we noticed that users would enlarge their workspace to be roughly the range of their arms' reach.

Domino sorting task

The second exploratory study asked subjects to place dominos on a sheet of paper in descending sorted order. Initially, the dominos were randomly placed on a tabletop and subjects could use the entire work surface. A second condition was run which had the dominos start in a bag. In addition, their tabletop workspace was restricted to the size of a piece of paper.

Once again this sorting task also revealed interesting interaction properties. Tactile feedback was often used to grab dominos while visually attending to other tasks. The non-dominant hand was often used to reposition and align the dominos into their final resting place while, in parallel, the dominant hand was used to retrieve new dominos. The most interesting observation was that subjects seemed to inherently know the geometric properties of the bricks and made use of this everyday knowledge in their interactions without prompting. For example, if 5 bricks are side-by-side in a row, subjects knew that applying simultaneous pressure to the left-most and right-most end bricks will cause the entire row of bricks to be moved. Finally, in the restricted workspace domino condition we observed one subject taking advantage of the "stackability" of the dominos and occasionally piled similar dominos on top of others to conserve space. Also, sometimes a subject would use their non-dominant hand as a "clipboard" or temporary buffer while they plan or manipulate other dominos.

Physical manipulation of a stretchable square

To get a better sense of the issues for manipulating physical versus virtual objects, we designed a "stretchable square" constructed out of foam core. This square looks like a tray with a one inch rim around each side. Users could expand or collapse the width of the square (see Figure 7). We displayed an end position, orientation and scale factor for the physical square and asked subjects to manipulate the square to match the final target as quickly as possible. A variety of cases were tested involving one, two or all three transformation operations (translate, scale, and rotate).

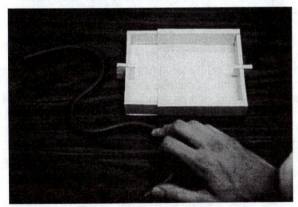

Figure 7. Flexible curve and stretchable square.

We found that each subject had a different style of grasping the stretchable square for position and orientation tasks. This served to remind us that physical objects often have a wide variety of ways to grasp and to manipulate them even given natural grasp points. In addition, subjects did not hesitate and were not confounded by trying to plan a grasp strategy. One subject used his dominant hand to perform the primary manipulation and the non-dominant hand as a breaking mechanism and for finer control.

Perhaps the most salient observation is that users performed the three operations (translation, rotation and scaling) in parallel. That is, as the subjects were translating the square towards its final position, they would also rotate and scale the square at the same time. These atomic operations are combined and chunked together [1].

Comparison Using MacDraw Application

The same matching tasks were then done using virtual objects and a stylus on a large, horizontal drafting table with a computer display projected on the writing surface. Using the MacDraw II™ program, subjects were asked to move a virtual object on top of a target virtual object matching position, orientation and scale factors.

We observed that even when we factor out the time needed to switch in and out of rotation mode in MacDraw, task completion time was about an order of magnitude longer than the physical manipulation using the stretchable square. We noticed a "zoom-in" effect to reach the desired end target goal. For example, subjects would first move the object on top of the target. Then they would rotate the object, but often be unable to plan ahead and realize that the center of rotation will cause the object to be displaced. Thus, they often had to perform another translation operation. They would repeat this process until satisfied with a final match.

The MacDraw user interface, and many other interfaces, forces the subject to perform the operations in a strictly sequential manner. While we can become very adept at performing a series of atomic operations in sequence, the interface constrains user interaction behavior. In effect, the interface forces users to remain novices by not allowing them to exhibit more natural and efficient expressions of specifying atomic operations in parallel.

Curve Matching

Continuing to explore our skills at physical manipulations, we asked subjects to use a flexible curve (see Figure 7) to match a target shape. The flexible curve, often used in graphic design, consists of a malleable metal surrounded by soft plastic in the shape of a long (18 inch) rod. The inner metal allows the curve to hold its shape once deformed.

We found that users quickly learned and explored the physical properties of the flexible curve and exhibited very expert performance in under a minute. All ten fingers were often used to impart forces and counterforces onto the curve. The palm of the hand was also used to preserve portions of the shape during the curve matching task. We observed that some subjects would "semantically load" their hands and arms before making contact with the flexible curve in anticipation of their interactions. The semantic loading is a preconceived grasp and manipulation strategy by the user which, in order to execute properly, the arms, hands and fingers must start in a specific, sometime uncomfortable, loaded position. This process often allowed the subject to reach the final target curve shape in one gestural action.

STAGE 2: MOCK-UP AND SIMULATIONS

Next, we mocked-up some sample brick interactions using a prototyping tool (Macromind Director) and acted them out on the Active Desk. By using a few LEGO bricks as props and creating some basic animations using the prototyping tool, we could quickly visualize what the interactions would look like and begin to get a sense of how they will feel. These sample interactions were video taped and edited. We were able to mock-up many of the primary ideas such as: attaching and detaching bricks from virtual objects; translation and rotation operations using one brick; using two bricks each attached to separate virtual objects, and finally two bricks attached to a single virtual object to specify stretching and simple deformations.

All of these exploratory studies and mock-ups aided us to quickly explore some of the core concepts with minimum set-up effort. Finally, the video tapes that we create often serves as inspirational material.

STAGE 3: PROTOTYPE

After the mock-ups, we built the bricks prototype to further investigate the Graspable UI concepts. The prototype consists of the Active Desk, a SGI Indigo2 and two Ascension Bird receivers (see Figure 8).

Active Desk

The Active Desk is a large horizontal desktop surface which has a rear projected computer screen underneath the writing surface (see Figure 8). Modeled after a drafting table, the dimensions of the surface are roughly 4.5' by 3.0' on a slight 30 degree angle. The projected computer screen inset has a dimension roughly 3' by 2'. A Scriptel transparent digitizing tablet lays on top of the surface and a stylus device may be used for input. The LCD projection display only has a 640x480 resolution so the SGI screen is down converted to an NTSC signal and sent to the LCD display.

Bricks

To prototype the graspable objects (bricks), we use the Ascension Flock of Birds™ 6D input devices to simulate the graspable objects. That is, each receiver is a small 1 inch cube that constantly sends positional (x, y, and z) and orientation information to the SGI workstation. We currently have a two receiver system, which simulates two active bricks that operate on top of the Active Desk. More receivers can be added to the system but the wires attached to the receivers hinder interactions. Nevertheless, the two receivers offer us an initial means of exploring the design space in a more formal manner.

GraspDraw — A simple drawing application

A simple drawing application, GraspDraw, was developed to test out some of the interaction techniques. The application lets users create objects such as lines, circles, rectangles and triangles (see Figure 8). Once created, the objects can be moved, rotated and scaled. GraspDraw is written in C using the GL library on an SGI Indigo2.

The two Bird receivers act like bricks and can be used simultaneously to perform operations in parallel. One of the bricks has a push button attached to it to register additional user input. This button is primarily used for creating new objects. Grasps (i.e., attaching the brick to a virtual object) are registered when a brick is near or on the desktop surface. To release a grasp, the user lifts the brick off of the desktop (about 2 cm).

To select the current tool (select, delete, rectangle, triangle, line, circle) and current draw color, we use a physical tray and an ink-well metaphor. Users dunk a brick in a compartment in the tray to select a particular tool. A soft audio beep is heard to act as feedback for switching tools. Once a tool is selected, a prototype shape or tool icon is attached to the brick. The shape or icon is drawn in a semi-transparent layer so that users may see through the tool.

Figure 8. GraspDraw application and ActiveDesk.

The concept of an *anchor* and *actuator* have been defined in interactions that involve two or more bricks. An anchor serves as the origin of an interaction operation. Anchors often specify an orientation value as well as a positional value. Actuators only specify positional values and operate within a frame of reference defined by an anchor. For example, performing a stretching operation on a virtual object involves using two bricks one as an anchor and the other as an actuator. The *first* brick attached to the virtual object acts as an anchor. The object can be moved or rotated. When the *second* brick is attached, it serves as an actuator. Position information is registered relative to the anchor brick. If the first anchor brick is released, the actuator brick is promoted to the role of an anchor.

STAGE 4: COMMERCIAL APPLICATION

In following the goals of user centered design and user testing, there are some real problems when working with new interaction techniques such as the Graspable UI. First, in order to conduct formal experiments, one must generally work in a restricted controlled environment. The demands of experimental control are often at odds with human performance in the more complex context of the real world. Secondly, University researchers typically do not have access to the source code of anything but toy applications.

Therefore, testing and demonstrations of innovative techniques like the Graspable UI are subject to criticisms that "it's fine in the simple test environment, but it won't work in the real world."

The first point can be dealt with by careful experimental design and differentiating between controlled experiments and user testing. Our approach to the second is to partner with a commercial software company that has a real application with real users. In so doing, we were able to access both a real application and a highly trained user community.

Hence, we have implemented a critical mass of the Graspable UI into a modified version of Alias Studio™, a high-end 3D modeling and animation program for SGI machines. Specifically, we are exploring how multiple bricks can be used to aid curve editing tasks. Although we have just begun this stage of research, we currently have two bricks integrated into the Studio program. The bricks can be used to simultaneously edit the position, orientation and scale factor for points along a curve. Future investigations may use bricks to clamp or freeze portions of the curve. This integration process and evaluation will further help us to refine the Graspable UI concepts.

DISCUSSION

We have conducted some preliminary user testing of the bricks concept using the GraspDraw application. All of the approximately 20 users who have tried the interface perform parallel operations (e.g., translate and rotate) at a very early stage of using the application. Within a few minutes of using the application, users become very adept at making drawings and manipulating virtual objects. Some users commented on the fact that the bricks were tethered, which hindered some of their interactions.

One could argue that all Graphical UI interactions, except perhaps touch (e.g., touchscreens) are already graspable interfaces if they use a mouse or stylus. However, this claim misses a few important distinctions. First, Graspable UIs make a distinction between "attachment" and "selection." In traditional Graphical UIs, the selection paradigm dictates that there is typically only one active selection; Selection N implicitly causes Selection N-1 to be unselected. In contrast, when bricks are attached to virtual objects the association persists across multiple interactions. Selections are then made by making physical contact with the bricks. Therefore, with Graspable UIs we can possibly eliminate many of the redundant selection actions and make selections easier by replacing the act of precisely positioning a cursor over a small target with the act of grabbing a brick. Secondly, Graspable UIs advocate using multiple devices (e.g., bricks) instead of channeling all interactions through one device (e.g., mouse). Consequently, not only are selections persistent, there can be one persistent selection per brick. Thirdly, the bricks are inherently spatial. For example, we can temporarily arrange bricks to form spatial caches or use them as spatial landmarks for storage. By having more spatial persistence, we can use more of our spatial reasoning skills and muscle memory. This was exhibited during the LEGO and Domino exploratory studies. Clearly, the bricks are handled differently than a mouse.

One may suggest to eliminate using bricks and instead use only our hands as the physical input devices. While this may be useful for some applications, in general using a physical intermediary (i.e., brick) may be more desirable. First, tactile feedback is essential; it provides a way of safeguarding user intent. The bricks supply tactile confirmation and serve as a visual interaction residue. Secondly, hand gestures lack very natural delimiters for starting and stopping points. This makes it difficult to segment commands and introduces lexical pragmatics. In contrast, the affordances of touching and releasing a brick serve as very natural start and stop points.

There are many open design issues and interaction pragmatics to research. For example, should we vary the attributes of a brick (shape, size, color, weight) to indicate its function? Should all the bricks have symmetrical behavior? How many bricks can a user operate with at the same time? Do the bricks take up too much space and cause screen clutter (perhaps we can stack the bricks and they can be made out of translucent material)? For fine, precise pointing, do bricks have a natural hot spot (perhaps a corner or edge)? Sometimes it is more advantageous to have a big "cursor" to acquire a small target [6].

Inter-Brick behaviors

Much of the power behind the Bricks is the ability to operate and interact with more than one brick at the same time. Our interaction techniques need to be sensitive to this issue and define consistent inter-brick behaviors for one-handed (unimanual) or two-handed (bimanual) interactions. Moreover, we will need to develop a new class of techniques that use combinations of unimanual and bimanual interactions during the life span of a single technique. For instance, a technique may be initiated using one hand, transfer to using both hands and then terminate back to using one hand. The key point is that we need to provide for seamless transitions within a single interaction technique that switches between unimanual and bimanual interactions. As we noted earlier, the Anchor/Actuator behavior serves as one example.

Our goal has been to quickly explore the new design space and identify major landmarks and issues rather than quantify any specific subset of the terrain. The next phase of our evaluation will include a more detailed evaluation at places in the design space that have the most potential.

DESIGN SPACE

We have developed an initial design space for bricks which serves to lay the foundation for exploring Graspable UIs. Table 1 summarizes the design space. The shaded region in the table represents where our current Bricks prototype fits in the design space. Each of the rows in the table represent dimensions of the design space which are described below.

Brick's internal ability — Does the brick have any internal mechanisms (physical or electronics) that generates additional information or intelligence? Inert objects have no internal mechanisms, only external features (color, shape, weight). Smart objects often have embedded sensors and microprocessors.

Input & Output — What properties can be sensed and displayed back to the user (or system)?

Spatially aware — Can the brick sense the surroundings and other bricks? Bricks can be unaware (work in isolation); mutually aware (aware only of each other); or be aware of their surroundings (primitive sensing of its environment and other bricks) [4].

Communication — How do the bricks communicate among themselves and to host computers? The mechanisms range from wireless (such as infra-red), tethered (requiring wires or cables) and grid board (a specialized operating surface with pluggable connecting parts).

Interaction time span — Given a task, are users manipulating the bricks in quick bursty interactions (sometimes gesturing in fractions of a second); using a set of bricks, accessing them within seconds or minutes (an interaction cache); or are the interactions long term running days, months, years between interactions (e.g., an archive)?

Bricks in use at same time — Do users manipulate one brick at a time (one handed interactions), two at a time (two handed interactions), or more than two? Users could manipulate 5 to 10 bricks at a time (e.g., bulldozer) or perhaps even 50 to 100 at a time.

Function assignment — How frequently and by what mechanism do the bricks get assigned their functions? Permanent assignment means that each brick has one function or role for its lifetime. With some effort, programmable assignment allows bricks to have their function reassigned. Transient assignment allows for users to rapidly reassign the brick's function.

Interaction representations — Is the system designed to have a blend of physical and virtual artifacts? When there is a mix, are the dual representations equal (i.e., functions can be performed using either physical or virtual artifacts), complimentary (i.e., one medium can perform a subset of the functionality that the other medium cannot) or combinatoric (together both offer functionality that either one could not provide alone).

Physical & Virtual layers — Are the layers direct (superimposed) or indirect (separated)?

Bond Between Physical & Virtual layers — Tightly coupled systems have the physical and virtual representations perfectly synchronized, the physical objects are tracked continuously in real time. Loosely coupled systems allow for the representations to become out of synchronization for long stretches of time (i.e., minutes, days) and updated in more of a batch mode.

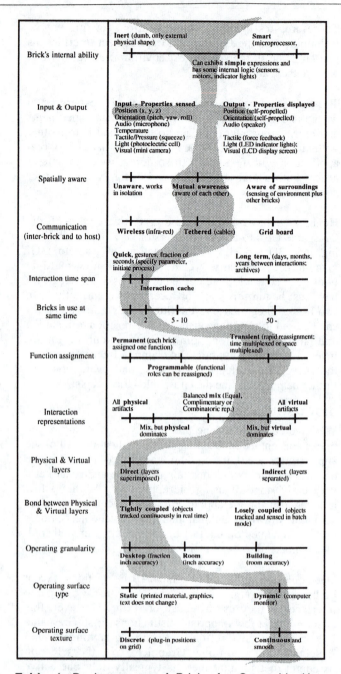

Table 1. Design space of Bricks for Graspable User Interfaces. Gray region shows where current Brick prototype fits into design space.

Operating Granularity — What is the range of space that the bricks operate in and at what sensing resolution? For example, the devices may operate at a *building* level (e.g., capable of determining what room they are currently in), *room* level (e.g., capable of determining, within an inch accuracy, position and orientation information inside a room), and *desktop* level (e.g., micro accuracy within 0.1 in of position and orientation information on a desktop).

Operating surface texture — What granularity or texture do the bricks operate on? A discrete texture requires that the bricks be plugged into special receptors (e.g., a grid board) while a continuous texture allows for smooth movement or dragging (e.g., tabletop).

Operating surface type — Do the bricks operate on a static surface (e.g., a tabletop) or a dynamic surface which can be changing constantly (e.g., Active Desk)?

It should be noted that this is not an exhaustive parsing of the design space. Robinett [10], however, proposes a more formal taxonomy for technologically mediated experiences which may aid our investigation. Yet, the many dimensions of our design space exhibit its richness and provides a more structured mechanism to explore the concepts behind Graspable UIs.

FUTURE WORK

There are many future directions we would like to explore. First, we will conduct more formal evaluation measures on the GraspDraw program. Next we will investigate other regions of the design space including developing techniques in 3-D as well as to operate on 3-D virtual objects. In addition, we hope to develop multiple, untethered bricks. Two promising areas are computer vision techniques [12] and electric-field sensing [16].

CONCLUSIONS

In this paper we have introduced a new technique, the Graspable User Interface, as a means of augmenting the power of conventional Graphical User Interfaces. In so doing, we have attempted to go beyond a simple "show and tell" exercise. Through the methodology described, we have attempted to both explore the overall design space effectively, and tease out the underlying human skills on which we could build our interaction techniques.

The Graspable User Interface is an example of "radical evolution." It is evolutionary in the sense that it builds upon the conventions of the GUI. Hence, both existing technology and human skill will transfer to the new technique. However, it is radical in that the incremental change that it introduces takes us into a radically new design space. Assuming that this new space is an improvement on what preceded it, this combination gives us the best of both worlds: the new and the status quo.

From the experience gained in the work described, we believe these new techniques to be highly potent and worthy of deeper study. What we have attempted is a proof of concept and exposition of our ideas. Hopefully this work will lead to a more detailed exploration of the technique and its potential.

ACKNOWLEDGMENTS

This research was undertaken under the auspices of the Input Research Group at the University of Toronto and we thank the members for their input. We especially would like to thank Mark Chignell, Marilyn Mantei, Michiel van de Panne, Gordon Kurtenbach, Beverly Harrison, William Hunt and Kim Vicente for their input. Thanks also to Ferdie Poblete who helped design and build the stretchable square prop. The Active Desk was designed and built by the Ontario Telepresence Project and the Arnott Design Group. Our research has been supported by the Information Technology Research Centre of Ontario, the Natural Sciences and Engineering Research Council of Canada, Xerox PARC and Alias Research.

More material including dynamic figures can be found on the CHI'95 Electronic Proceedings CD-ROM and at URL: http://www.dgp.utoronto.ca/people/GeorgeFitzmaurice/home.html

REFERENCES

1. Buxton, W. (1986). Chunking and phrasing and the design of human-computer dialogues. *Proceedings of the IFIP World Computer Congress*. pp. 475-480.
2. Eilan, N., McCarthy, R. and Brewer, B. (1993). Spatial Representation. Oxford, UK: Blackwell.
3. Guiard, Y. (1987). Asymmetric Division of Labor in Human Skilled Bimanual Action: The Kinematic Chain Model. In *Journal of Motor Behavior*, 19(4), pp. 486-517.
4. Fitzmaurice, G.W. (1993). Situated Information Spaces and Spatially Aware Palmtop Computers, *Communications of the ACM*. 36(7), pp. 38-49.
5. Hickley, K., Pausch, R., Goble, J. C. and Kassell, N. F. (1994). Passive Real-World Interface Props for Neurosurgical Visualization. *Proc. of CHI'94*, pp. 452-458.
6. Kabbash, P. and Buxton, W. (1995). The 'Prince' Technique: Fitts' Law and Selection Using Area Cursors, To appear in *Proc. of CHI'95*.
7. Kabbash, P., Buxton, W. and Sellen, A. (1994). Two-Handed Input in a Compound Task. *Proc. of CHI94*, pp. 417-423.
8. MacKenzie, C. L. and Iberall, T. (1994). The Grasping Hand. Amsterdam: North-Holland, Elsevier Science.
9. Resnick, M. (1993). Behavior Construction Kits. In *Communications of the ACM*. 36(7), pp. 64-71.
10. Robinett, W. (1992). Synthetic Experience: A Proposed Taxonomy. *Presence*, 1(2), pp. 229-247.
11. Sachs, E., Roberts, A. and Stoops, D. (1990). 3-Draw: A tool for the conceptual design of three-dimensional shapes. CHI'90 Technical Video Program, ACM SIGGRAPH Video Review, Issue 55, No. 2.
12. Schneider, S.A. (1990). Experiments in the dynamic and strategic control of cooperating manipulators. Ph.D. Thesis, Dept. of Elec. Eng., Stanford Univ.
13. Suzuki, H., Kato, H. (1993). AlgoBlock: a Tangible Programming Language, a Tool for Collaborative Learning. Proceedings of 4th European Logo Conference, Aug. 1993, Athens Greece, pp. 297-303.
14. Weiser, M. (1991). The computer for the 21st Century. In *Scientific America*, 265(3), pp. 94-104.
15. Wellner, P. (1993). Interacting with paper on the DigitalDesk. In *Com. of the ACM*. 36(7), pp. 86-96.
16. Zimmerman, T., Smith, J.R., Paradiso, J.A., Allport, D. and Gershenfeld, N. (1995). Applying Electric Field Sensing to Human-Computer Interfaces. To Appear in *Proceedings of CHI'95*.

Situated Facial Displays:
Towards Social Interaction

Akikazu Takeuchi
Sony Computer Science Laboratory, Inc.
3-14-13 Higashi-Gotanda
Shinagawa-ku, Tokyo 141, Japan
TEL: +81-3-5448-4380
takeuchi@csl.sony.co.jp

Taketo Naito
Computer Science Department
Keio University
3-14-1, Hiyoshi, Kohoku-ku
Yokohama, Kanagawa 223, Japan
TEL: +81-45-560-1150
naito@mt.cs.keio.ac.jp

ABSTRACT

Most interactive programs have been assuming interaction with a single user. We propose the notion of "Social Interaction" as a new interaction paradigm between multiple humans and computers. Social interaction requires that first a computer has the multiple participants model, second its behaviors are not only determined by internal logic but also affected by perceived external situations, and finally it actively joins the interaction. An experimental system with these features was developed. It consists of three subsystems, a vision subsystem that processes motion video input to examine an external situation, an action/reaction subsystem that generates an action based on internal logic of a task and a situated reaction triggered by perceived external situation, and a facial animation subsystem that generates a three-dimensional face capable of various facial displays. From the experiment using the system with a number of subjects, we found that subjects generally tended to try to interpret facial displays of the computer. Such involvement prevented them from concentrating on a task. We also found that subjects never recognized situated reactions of the computer that were unrelated to the task although they unconsciously responded to them. These findings seem to imply subliminal involvement of the subjects caused by facial displays and situated reactions.

KEYWORDS: User interface design, multimodal interfaces, facial expression, anthropomorphism, subliminal involvement

MOTIVATION AND INTRODUCTION

Recently, there is growing interest in an agent/agency model [15]. Maes and her colleagues are developing interface agents [6] which are (semi-)automatic/autonomous and help users to perform various tasks. An autonomous system has the ability to control itself and to make its own decisions. Autonomy is essential to surviving in a dynamically changing world. It is the subject of research in many areas including robotics, artificial life and artificial ecosystems.

However, although autonomy is vital to surviving in the real world, it is only concerned with "self." It is selfish by nature. It does not seem to work well in human society. Socialness is a higher-level concept defined above the concept of an individual, and is the style of interaction between the individuals in a group. Socialness can be applied to the interaction between humans and computers, and possibly to that between multiple computers. However, most interactive programs have been assuming interaction with a single user. Programs of this type all share the following features:

- Dialogical: Only two participants, a human (asker) and a computer (answerer), are assumed. Turn-taking is trivial (alternate turns).
- Transformational: A computer is regarded as a function that receives an inquiry and outputs an answer.
- Passive: A computer does not voluntarily speak.

The dialogical and transformational views are well suited to applications such as question-answering on databases and drawing. It is true that there are many domains in which single-user-oriented systems fit very well.

However, daily conversation is not always functional. One example is the co-constructive conversation studied by Chovil [2]. Co-constructive conversation is that of a group of individuals in which, say, people talk about the food they ate in a restaurant a month ago. There are no special roles (like a chairperson) for the participants to play. They all have the same role. All participants try to contribute to recalling the food by relating his or her memories about the food, adding comments, and correcting the others' impressions. Turn-taking is controlled by eye-contact, facial expression, body-gestures, voice tones, and so on. The conversation includes many subconversations, some of them existing in parallel and dividing the group into subgroups. The conversation terminates only when all the participants are satisfied with the conclusion.

Co-constructive conversation closely approximates day-to-day conversation. Conversation is a social action. Suchman said that communication is not a symbolic process that happens to go on in real-world settings, but a real-world activity in which we make use of language to delineate the collective relevance of our shared environment [10]. To create a computer that can participate in social conversations such as the co-constructive conversation described above, is our research goal. To this end, we propose the notion of "Social Interaction" as a new interaction paradigm between humans and computers.

SOCIAL INTERACTION AND SITUATEDNESS

In contrast to the single-user-oriented view, social interaction has the following features:

- Multiple participant model: Interaction involves more than two participants that are humans or computers. A computer identifies every participant together with his/her/its character. There are no fixed roles. Turn-taking is flexible, and highly dependent on the interaction situation.
- Situatedness: Actions of a computer are not only controlled by internal logic, but also affected by situations which are perceived through various sensors, that is, multimodally. A computer observes the situation, recognizes participants there and detects their conversational actions. Some actions can be defined as reactions to situations perceived. Turning to look at a person suddenly taking some action is an example of such actions. These play the role of unconscious indication of attention, which can be used, for example, as a cue for turn-taking or as a sign of interruption.
- Bi-directional: A computer actively joins the interaction: that is, it takes every opportunity to make an action. Computers and humans interact in a symmetric fashion. Symmetry comes from knowing each other, i.e. an ability to identify partners. Bi-directional interaction implies that computers voluntarily engage in actions with others.

Face-to-face communication is a real-world activity, in which we use various communication modalities, including verbal and nonverbal ones. Multimodal signals are consciously or unconsciously directed to other participants in social interaction. People can perceive these signals and utilize them to coordinate social interaction among a group. Participants' actions such as body gestures, eye-contact, facial expressions, and coughing are major resources in a situation. Social interaction is multimodal interaction, hence it is essentially situated interaction.

We are attempting to bring facial displays into computer human interaction as a new modality that makes the interaction tighter and more social [12]. The system we developed has a synthesized face capable of various facial

displays and shows an appropriate display depending on the current conversational situation. In this paper we describe a new system extended towards social interaction. Figure 1 illustrates an overview of the new system. The new system can handle the multiple user model to some extent. It is situated in the sense that it can perceive an external world through a video camera and its behaviors are determined by internal logic as well as perceived situations. Its behaviors are expressed as facial displays, eyeball movements, and/or head motions. Since it uses a situated behavior model, a set of possible facial displays are called *situated facial displays*. The system is not a transformational question-answering system, but an autonomous system observing users' actions and performing appropriate actions in appropriate situations.

RELATED WORK

Within the context of CSCW, especially in videoconferencing research, social interaction among human participants are studied, and the importance of gaze/eye-contact is reported (for example [5]). There is also an attempt to develop such a videoconference system that can support local eye-contact between two participants [9] to promote social interaction. Our work is in parallel with these activities, and tries to develop a synthetic agent capable of social interaction with humans and other agents.

Takeuchi et al. reported that communicative facial displays help interaction between a human and a computer [12]. Walker et al. studied how facial expression affected users' performance/productivity and reported that a stern face is good for productivity, but creates a bad impression [13]. It implies that handling a human face, or humanity in general, is not a simple task since it brings up various unclear human factors at the same time. Brennan et al. reported that interaction with an anthropomorphized agent takes more effort [1]. Nass et al. reported that even computer experts respond to computers as if computers were social actors [7].

There are several application domains in which the notion of social interaction is essentially useful. These domains share the following features:

- Computers' autonomous/automatic actions help to solve or improve the current problematic situation.
- There is a clear role separation among participants so that a computer can understand the social situation easily.

One example is a car navigation system where it is important for a computer to take an action when the driver takes a wrong path. There is also a clear distinction between the driver and the human navigator. "Backseat Driver" is a successful example in this application domain [8]. It talks to the driver whenever the car comes close to or passes by an intersection where the car should turn.

Another example is a tutoring system, where there is a clear distinction between the tutor and the student. A computer can actively help both of them by watching the situation and giving appropriate advice.

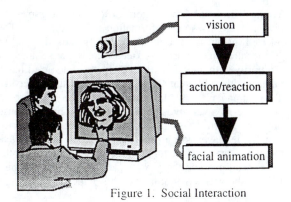

Figure 1. Social Interaction

SYSTEM ARCHITECTURE

The new system was developed for verifying the notion of situated facial displays and social interaction. The system consists of three subsystems: a vision subsystem that processes motion video input, an action/reaction subsystem that determines the system's behavior based on internal logic and vision input, and a facial animation subsystem that generates a three-dimensional face with various facial displays (Figure 1). Currently, all subsystems are running on an SGI 320VGX equipped with VideoLab as a live video grabber.

Vision

The vision subsystem gets image data through a video camera that is placed on top of the display (see Figure 1). There is a steep trade-off between the processing speed and the content of the processing. The more information, the slower the processing. Since slow reactions are essentially useless in a real-world setting, we have to force ourselves to reduce time-consuming processing. In the current implementation, the subsystem can detect and track several users' positions simultaneously.

The vision subsystem keeps as a reference frame one still frame, which is the image of the scene with no human. By detecting differences between incoming images and the reference frame, and segmenting detected regions, moving objects are extracted in real-time. Assuming that the only moving objects are humans in the room, we can determine 2D positions of the humans in the image. Using the camera position and direction, the positions are translated into 3D orientations, which can be applied to eyeball rotation and face rotation when drawing a 3D face. Since this process is fast enough, the subsystem can keep track of several users' positions at 30 frames per second.

Action/Reaction

The outputs of the vision subsystem are representations of external events observed visually, and these are sent to the action/reaction subsystem. Although the information obtained is limited, we can extract the following:

- each person's position,
- the speed of each person's movement when they move

The system's behaviors are expressed in terms of facial displays, eyeball movements and head motions. The system only has a synthesized human head. The face can express various facial displays, including those listed in [12]. In addition to them, the eyeballs and head itself can also express various communicative expressions such as gaze, wink, nod and so on. A "gaze" can establish eye-contact, which is a powerful communication signal. Eye-contact/gaze is especially important when several individuals are involved in social interaction, since eye-contact can indicate who a message is directed to.

In [12], computer's facial displays are determined by a correspondence between conversational situations and communicative facial displays, where "conversational situations" means different logical contexts for processing a user's inquiry. The current system was extended to incorporate the idea of "situatedness." Instructions to the facial animation subsystem are first determined by the current logical context. This forms a basic behavior. The other factor influencing the system's behavior is the "physical situation." The physical situation means various information, including information about users' position, users' gestures, and users' facial displays (not all of them are handled by the current system). For example, when one person moves quickly, the action "look at that person" is invoked. Actions invoked in this way are called *reactions*. In contrast, actions invoked based on logical contexts are simply called *actions*. In general, reactions can always override actions. Reactions are quick motions and never last long.

Facial Animation

The face is modeled three-dimensionally. The current face is composed of approximately 500 polygons. The face may be rendered using a skin-like surface material by applying a texture map taken from a photograph or a video frame. In 3D computer graphics, a facial display is realized by the local deformation of the polygons representing the face. Waters showed that deformation that simulates the action of muscles underlying the face looks more natural [14]. We use the numerical equations defined by Waters to simulate muscle dynamics. Currently, 16 muscles and 10 parameters, controlling mouth opening, jaw rotation, eye movement, eyelid opening, and head orientation are incorporated. The facial modeling and animation system is based on Takeuchi and Franks' work [11].

Gaze control is also implemented in the facial animation subsystem using a video camera fixed on top of a display. Using this, eye-contact between a user and a computer are achieved, although it is not real three dimensional eye-contact since the computer face is projected onto a 2D display screen. Real eye-contact is selective in the sense that only one pair of participants gets the feeling of eye-contact. In contrast, eye-contact with a computer face on the screen is like eye-contact with a TV news announcer. As long as the announcer looks straight ahead, every person watching the TV screen has the feeling of eye-contact.

EXPERIMENT

Method

Using the current prototype system, we performed the experiment under several different conditions to investigate the effect of situatedness and to study how much socialness was achieved.

Task. The task of the system is to watch a game and give nonverbally some information to participants. The selected game is a card matching game similar to "Concentration" where sixteen cards are arranged arbitrarily with their faces turned down. Each card has a pattern with a different shape and color on the underside. Two human players play alternately. In one turn, a player turns over two cards sequentially by a mouse pointer. When the shapes on the underside of two cards match, the player gets a point and the cards are removed. Otherwise the cards are turned over again and placed in the same positions and the next player takes a turn. Color Plate 1 shows a screen image of this game.

Table 1 The behaviors of the current system: the first column lists names of actions and reactions; the second lists their descriptions; the third shows how they were recognized by the subjects as a percentage.

Game actions		
Looking at a player	It indicates the current turn by looking at the corresponding side.	62.5
Looking at a pair of cards	Since the computer keeps records of cards already turned over, it can inform its ally of which cards should be turned over.	62.5
Smiling at a card	The computer smiles at a card when an ally player touches with the mouse one of the cards that are indicated to be open.	75
Smiling on an ally	The computer smiles on an ally player when he/she has turned over a card the computer suggests.	
Disappointed	An expression shown when an ally player has not followed the computer's suggestion.	37.5
Happy	An expression shown when an ally has got a point.	100
Shrug	An expression shown when an opponent has got a point.	100
Smiling	In the beginning of an ally turn, the computer smiles on an ally if an ally is ahead.	0
Win	An expression shown when an ally wins a game.	0
Even	An expression shown when a game results in a draw.	0
Lose	An expression shown when an ally lose a game.	0
Situated reactions		
Tracking a player	The computer's face and eyes track the current player.	75
Tracking a card	The computer's face and eyes track a card on which the mouse pointer is placed.	50
Question	An expression shown when the current player does not play for a certain amount of time.	0
Looking at a player	The computer looks at the player who is not the current player and moves quickly.	0
Looking for a player	The computer tilts its synthesized head and appears to look into the distance when the current player disappears from its sight.	0

At the beginning of the game, the computer chooses one player as its ally and the other as its opponent. As the computer is watching the game, it performs two types of behaviors. The first is a *game action:* this is an action that is associated with the current logical context of the game. The second is a *situated reaction:* this is a reaction triggered by the current physical situation.

Table 1 lists both game actions and situated reactions implemented in the current system.

Table 2. Four conditions

	Situated reaction	No Situated reaction
with face	SF	NF
with arrow	SA	NA

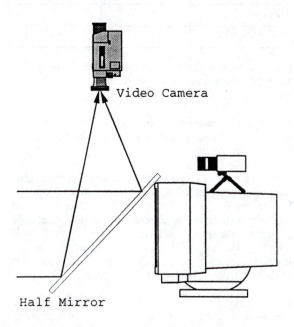

Video Camera

Half Mirror

Figure 2. Video recording configuration

Subjects. The prototype system was tested on 7 volunteer subjects from a group of university-level computer science students. The average age of the subjects was 22.

Other conditions. In addition to the experimental condition described so far, which is denoted by SF, three other conditions, NF, SA and NA were prepared. They are all listed in Table 2. The intention in designing these four conditions is first to examine the effect of situatedness by turning on and off the *situated* reactions. In conditions SF and SA, situated reactions were on, but in NF and NA they were turned off. The second is to examine the effect of a realistic face by comparing it with an arbitrary 3D object, in this case, a 3D arrow. A 3D arrow was considered as equivalent to a face without expressions (Color Plate 2). This arrow could perform actions mainly expressed by the head's direction such as "looking at" and "tracking". One exception is "smiling at a card". In this case, the arrow can flash its color to indicate whether or not an ally touches the correct card. SF and NF denote the conditions with a face, while SA and NA denote the conditions with an arrow.

Seven subjects were grouped into four couples (one subject played twice). Each couple was requested to perform the experiment under four different conditions. The experiment was recorded by a video camera so that we could extract scenes in which a subject looked at the computer face or a subject reacted to a computer's action/reaction. Figure 2 illustrates a video recording setting. With a half mirror, we could obtain the view

Table 3 Usefulness and entertainment evaluation

	useful, reliable	entertainment, fun
[SF] Situated Face	3.00	1.75
[SA] Situated Arrow	1.50	2.63
[NF] Face	3.25	2.00
[NA] Arrow	1.88	3.00

from the center of the screen so that analysis of the subjects' eye movements was made easy.

Result
Although the number of subjects is not enough to conclude something statistically, we obtained a number of interesting observations.

The computer shows various actions and reactions. The third column in Table 1 shows how they were recognized by the subjects. It shows that the subjects noticed relatively well those actions/reactions which were closely related to the game such as "Happy", "Shrug" and "Smiling at a card", while they never noticed some of the situated reactions although they unconsciously responded to those reactions as shown in Color Plate 3.

Color Plate 3 shows two diagrams which illustrate behaviors of human players and the computer face during the experiment under SF condition and SA condition. They were created by analyzing a video recording the game. In the picture, the vertical axis is the time axis. Horizontal gray stripes indicate opponent turns except the first one that is the opening scene in the game, while white zones between gray zones are ally turns. In each diagram, there are three columns; blue, red and green with yellow. The blue column represents the ally's behavior where short branches from a fat trunk indicate periods when the ally looked at the computer's face. Similarly, the red column represents the opponent's behavior with left branches indicating the opponent looked at the computer's face. The green column represents the computer face's behavior, especially in which direction it was looking. It consists of three vertical subcolumns, and they represent "looking at the ally," "looking at cards," and "looking at the opponent", from left to right, respectively. Yellow branches included in the left and right subcolumn indicate "Looking at a player" reactions. The picture shows eye-contact as line-contact between blue and green (yellow) branches, or red and green (yellow) branches. By close examination of the picture, we found that:

- Subjects look at the computer face more often during their turn to read some information from the face.
- They also look at the computer face often around the turn-changing point.
- Situated reactions such as "Looking at a player" appeared many times. In many cases, it triggered a players' reaction and eventually established eye-contact.
- Human players look at the computer face more often than an arrow.

Table 3 shows subjects' evaluation of usefulness and entertainment factors for each condition. Smaller numbers in each cell indicate better scores. As the table indicates, an arrow was recognized as a useful and reliable tool, while a face was accepted for entertainment or fun.

DISCUSSION AND FUTURE DIRECTIONS
From the analysis of the results of the experiment, we have the following observations:

- Subjects tend to try to interpret facial displays and head behaviors. Such involvement prevents them from concentrating on the game.
- A face is considered entertaining, but less useful than an arrow.
- A face attracts human eyes more often than an arrow.
- Some situated reactions such as "Question", "Looking at a player" and "Looking for a player" were never identified by subjects, although they often responded to them.
- A situated arrow is most helpful and reliable in winning the game. A simple arrow is second. This is because an arrow directly shows useful information. For example, color-flashing when a subject's mouse touches the correct card.

It is interesting that subjects never recognized situated reactions that were not directly related to the game although they unconsciously responded to them. This implies a *subliminal involvement*. People respond to systems with a human-like face in clearly different way from another system with an arrow. The difference is their attitude. When people face a human-like system, they try to read subtle signals, interpret them, and respond to them. All this happens consciously and unconsciously

Facial displays are subtle expressions by nature. People naturally try to interpret them. This causes slow response and unconscious reactions in systems with human-like faces. The same evidence was reported by Brennan [1]. Anthropomorphism has been criticized for being inefficient and requiring more effort [3]. However, this is because people appreciate human images and try to interpret them. Such involvement is not a negative effect. We surmise that once people are accustomed to synthesized faces, performance becomes more efficient, and a long partnership further improves performance. Human-like characterization is one good form of autonomous agents, because people are accustomed to interact with other humans.

It is clear that facial displays are useful in entertainment applications. In addition, facial displays can be used in more subtle, complex and therefore sophisticated situations. GUIDE system is a successful example [4]. Although it uses the whole human video image, it exemplifies the usage of human images as indications of cultural background.

The present paper suggests that people's attitude towards computers with human-like faces is completely different from that to ordinary computers with desktop-style interfaces. Understanding this difference is important when designing a computer system capable of social interaction with humans. Situated reaction seems to affect users' behavior and to get participants involved in interaction, but we do not know how it can contribute to the content of interaction. We need more research on application domains.

ACKNOWLEDGMENTS

We thank Steve Franks for his early contribution to the facial animation subsystem. Special thanks to Keith Waters for his original animation system. We thank Mario Tokoro and our colleagues at Sony CSL for their encouragement and discussion.

REFERENCES

1. Brennan, S. E. and Ohaeri, J. O. Effect of Message Style on Users' Attributions toward Agents, In CHI'94 Conference Companion Human Factors in Computing Systems (Boston, April 24-28, 1994), ACM Press, pp. 281-282.

2. Chovil, N. Discourse-oriented facial displays in conversation. Research on Language and Social Interaction, 25 (1991/1992) 163-194.

3. Don, A. and Brennan, S. and Laurel, B. and Shneiderman, B. Anthropomorphism: from Eliza to Terminator 2, In Proc. CHI'92 Human Factors in Computing Systems (Monterey, May 3-7, 1992), ACM Press, pp. 67-70.

4. Don, A. and Oren, T. and Laurel, B. GUIDES 3.0, In Proc. CHI'91 Human Factors in Computing Systems (New Orleans, April 27-May 2, 1991), ACM Press, pp. 447-448.

5. Ishii, H. and Kobayashi, M. and Grudin, J. Integration of Interpersonal Space and Shared Workspace: ClearBoard Design and Experiments.ACM Trans. on Information Systems, 11, 4 (Oct. 1993) 349-375.

6. Maes, P. and Kozierok, R. Learning interface agents, In Proc. AAAI'93 (1993), MIT Press, Cambridge, pp. 459-465.

7. Nass, C. and Steuer, J. and Tauber, E. R. Computers are Social Actors, In Proc. CHI'94 Human Factors in Computing Systems (Boston, April 24-28, 1994), ACM Press, pp. 72-78.

8. Schmandt, C. M. and Davis, J. R. Synthetics speech for real time direction giving, In Digest of Technical Papers, IEEE ICCE, 1989, pp. 288-289.

9. Sellen, A. and Buxton, B. Using Spacial Cues to Improve Videoconferencing, In Proc. CHI'92 Human Factors in Computing Systems (Monterey, May 3-7, 1992), ACM Press, pp. 651-652.

10. Suchman, L. A. *Plans and Situated Actions*, Cambridge University Press, Cambridge, 1987.

11. Takeuchi, A. and Franks, S. *A Rapid Face Construction Lab*. Tech. Report. SCSL-TR-92-010, Sony Computer Science Laboratory, Inc., Tokyo, 1992.

12. Takeuchi, A. and Nagao, K. Communicative Facial Displays as a New Conversation Modality, In Proc. INTERCHI'93 Human Factors in Computing Systems (Amsterdam, April 24-29, 1993), ACM Press, pp. 187-193.

13. Walker, J. and Sproull, L. and Subramani, R. Using a Human Face in an Interface, In Proc. CHI'94 Human Factors in Computing Systems (Boston, April 24-28, 1994), ACM Press, pp. 85-91.

14. Waters, K. A muscle model for animating three-dimensional facial expression, In Computer Graphics 21, 4 (July 1987), 17-24.

15. Special Issue on Intelligent Agents, Communications of the ACM 37, 7 (July 1994).

Glove-TalkII: An Adaptive Gesture-to-Formant Interface

Sidney Fels
Department of Computer Science
University of Toronto
Toronto, ON, Canada, M5S 1A4
ssfels@ai.toronto.edu

Geoffrey Hinton
Department of Computer Science
University of Toronto
Toronto, ON, Canada, M5S 1A4
hinton@ai.toronto.edu

ABSTRACT

Glove-TalkII is a system which translates hand gestures to speech through an adaptive interface. Hand gestures are mapped continuously to 10 control parameters of a parallel formant speech synthesizer. The mapping allows the hand to act as an artificial vocal tract that produces speech in real time. This gives an unlimited vocabulary, multiple languages in addition to direct control of fundamental frequency and volume. Currently, the best version of Glove-TalkII uses several input devices (including a Cyberglove, a ContactGlove, a polhemus sensor, and a foot-pedal), a parallel formant speech synthesizer and 3 neural networks. The gesture-to-speech task is divided into vowel and consonant production by using a gating network to weight the outputs of a vowel and a consonant neural network. The gating network and the consonant network are trained with examples from the user. The vowel network implements a fixed, user-defined relationship between hand-position and vowel sound and does not require any training examples from the user. Volume, fundamental frequency and stop consonants are produced with a fixed mapping from the input devices. One subject has trained for about 100 hours to speak intelligibly with Glove-TalkII. He passed through eight distinct stages while learning to speak. He speaks slowly with speech quality similar to a text-to-speech synthesizer but with far more natural-sounding pitch variations.

KEYWORDS: Gesture-to-speech device, gestural input, speech output, speech acquisition, adaptive interface, talking machine.

INTRODUCTION

Many different possible schemes exist for converting hand gestures to speech. The choice of scheme depends on the granularity of the speech that you want to produce. Figure 1 identifies a spectrum defined by possible divisions of speech based on the duration of the sound

for each granularity. What is interesting is that in general, the coarser the division of speech, the smaller the bandwidth necessary for the user. In contrast, where the granularity of speech is on the order of articulatory muscle movements (i.e. the artificial vocal tract [AVT]) high bandwidth control is necessary for good speech. Devices which implement this model of speech production are like musical instruments which produce speech sounds. The user must control the timing of sounds to produce speech much as a musician plays notes to produce music. The AVT allows unlimited vocabulary, control of pitch and non-verbal sounds. Glove-TalkII is an adaptive interface that implements an AVT.

Translating gestures to speech using an AVT model has a long history beginning in the late 1700's. Systems developed include a bellows-driven hand-varied resonator tube with auxiliary controls (1790's [16]), a rubber-moulded skull with actuators for manipulating tongue and jaw position (1880's [1]) and a keyboard-footpedal interface controlling a set of linearly spaced bandpass frequency generators called the Voder (1940 [4]). The Voder was demonstrated at the World's Fair in 1939 by operators who had trained continuously for one year to learn to speak with the system. This suggests that the task of speaking with a gestural interface is very difficult and the training times could be significantly decreased with a better interface. Glove-TalkII is implemented with neural networks which allows the system to learn the user's interpretation of an articulatory model of speaking.

The obvious use of an AVT is as a speaking aid for speech impaired people. Clearly, the difficulties encountered with this application include the extreme motor demands and time required to learn to use the device compared to other speech prostheses which require less control. Additionally, users must be able to hear to effectively use the device which further limits the potential user group. Of course, care must be taken when considering these criticisms since AVTs potentially provide a much richer speech space than other coarse granularity systems which may be preferable for some people. And, just as children who are learning to speak are willing to

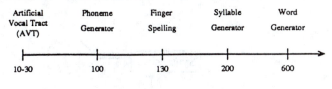

Figure 1: Spectrum of gesture-to-speech mappings based on the granularity of speech.

spend the large amount of time required to control their vocal tracts, it is not unreasonable to expect users to spend on the order of 100 hours to learn to speak with an AVT like Glove-TalkII. Besides the obvious application of Glove-TalkII, the neural network techniques used successfully can be applied to other complex interfaces where adaptation between a user's cognitive space and some objective space is required, for example; musical instrument design and telerobotics.

This paper first describes the Glove-TalkII system and then the experience of a single subject as he learned to speak with Glove-TalkII over 100 hours. Quantitative analysis of Glove-TalkII only provides a rough guide to the performance of the whole system. Observation of the single subject allows for qualitative analysis of Glove-TalkII to determine its effectiveness as a gesture-to-speech device.

OVERVIEW OF GLOVE-TALKII

The Glove-TalkII system converts hand gestures to speech, based on a gesture-to-formant model. The gesture vocabulary is based on a vocal-articulator model of the hand. By dividing the mapping tasks into independent subtasks, a substantial reduction in network size and training time is possible (see [5]).

Figure 2 illustrates the whole Glove-TalkII system. Important features include the three neural networks labeled vowel/consonant decision (V/C), vowel, and consonant. The V/C network is a 12-10-1 feed forward neural network with sigmoid activation functions[1]. The V/C network is trained on data collected from the user to decide whether he wants to produce a vowel or a consonant sound. Likewise, the consonant network is trained to produce consonant sounds based on user-generated examples from an initial gesture vocabulary. The consonant network is a 12-15-9 feed forward network. It uses normalized radial basis function (RBF [2]) activations for the hidden units and sigmoid activations for the output units. In contrast, the vowel network implements a fixed mapping between hand-positions and vowel phonemes defined by the user. The vowel network is a 2-11-8 feed forward network. It also uses normalized RBF hidden units and sigmoid output units[2].

---

[1]See [13] for an excellent introduction to neural networks and how they can be trained.

[2]Quantitative analysis of each of the various neural networks on

Eight contact switches on the user's left hand designate the stop consonants (B, D, G, J, P, T, K, CH), because the dynamics of such sounds proved too fast to be controlled by the user. The foot pedal provides a volume control by adjusting the speech amplitude and this mapping is fixed. The fundamental frequency, which is related to the pitch of the speech, is determined by a fixed mapping from the user's hand height. The output of the system drives 10 control parameters of a parallel formant speech synthesizer every 10 msec. The 10 control parameters are: nasal amplitude (ALF), first, second and third formant frequency and amplitude (F1, A1, F2, A2, F3, A3), high frequency amplitude (AHF), degree of voicing (V) and fundamental frequency (F0).

Once trained, Glove-TalkII can be used as follows: to initiate speech, the user forms the hand shape of the first sound she intends to produce. She depresses the foot pedal and the sound comes out of the synthesizer. Vowels and consonants of various qualities are produced in a continuous fashion through the appropriate co-ordination of hand and foot motions. Words are formed by making the correct motions; for example, to say "hello" the user forms the "h" sound, depresses the foot pedal and quickly moves her hand to produce the "e" sound, then the "l" sound and finally the "o" sound. The user has complete control of the timing and quality of the individual sounds. The articulatory mapping between gestures and speech is decided *a priori*. The mapping is based on a simplistic articulatory phonetic description of speech [10]. The X,Y coordinates (measured by the polhemus) are mapped to something like tongue position and height[3] producing vowels when the user's hand is in an open configuration (see figure 3 for the correspondence and table 1 for a typical vowel configuration). Manner and place of articulation for non-stop consonants are determined by opposition of the thumb with the index and middle fingers. Table 1 shows the initial gesture mapping between static hand gestures and static articulatory positions corresponding to phonemes. The ring finger controls voicing. Only *static* articulatory configurations are used as training points for the neural networks, and the interpolation between them is a result of the learning but is not explicitly trained. For example, the vowel space interpolation allows the user to easily move within vowel space to produce dipthongs. Ideally, the transitions should also be learned, but in the text-to-speech formant data we use for training [11] these transitions are poor, and it is very hard to extract formant trajectories from real speech accurately.

---

typical training data can be found in [6]. As is typical with speech research though, care must be taken when using quantitative analysis of the networks performance to judge the performance of the whole system. For this reason, qualitative analysis of the single user is important.

[3]In reality, the XY coordinates map more closely to changes in the first two formants, F1 and F2 of vowels. From the user's perspective though, the link to tongue movement is useful.

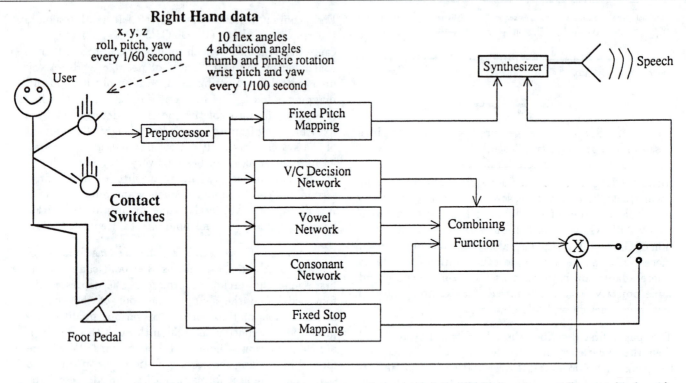

Right Hand data

x, y, z
roll, pitch, yaw
every 1/60 second

10 flex angles
4 abduction angles
thumb and pinkie rotation
wrist pitch and yaw
every 1/100 second

Figure 2: Block diagram of Glove-TalkII: input from the user is measured by the Cyberglove, polhemus, keyboard and foot pedal, then mapped using neural networks and fixed functions to formant parameters which drive the parallel formant synthesizer [14].

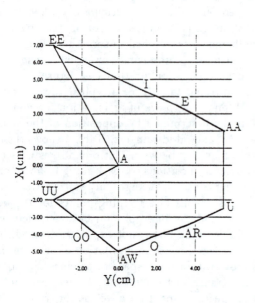

Figure 3: Hand-position to Vowel Sound Mapping. The coordinates are specified relative to the origin at the sound A. The X and Y coordinates form a horizontal plane parallel to the floor when the user is sitting. The 11 cardinal phoneme targets are determined with the text-to-speech synthesizer.

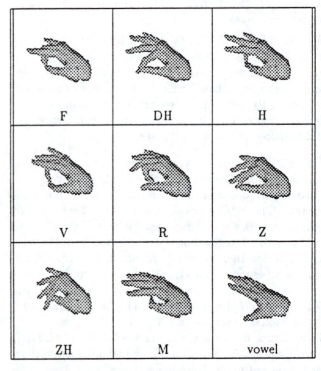

Table 1: Examples of static gesture-to-consonant mapping. Note, each gesture corresponds to a static *non-stop* consonant phoneme generated by the text-to-speech synthesizer and the neural networks provide the continous interpolation.

Hardware and Software Tools

There are five main pieces of hardware used in Glove-TalkII, four for input and one for output. The glove input device is a Cyberglove. This device has 18 flex sensors embedded inside a lightweight glove. The flex sensors respond linearly to bend angle and are placed strategically in the glove. These angles are measured at a frequency of about 100Hz.

The second input device is a polhemus sensor which measures the X,Y,Z, roll, pitch and yaw of the hand relative to a fixed source. The small sensor is mounted on the back of the Cyberglove on the user's forearm; thus, the six parameters are independent of the user's wrist motion. The device measures the parameters at a frequency of 60Hz.

The third input device is a ContactGlove. This device measures contact between points on the fingers to the thumb which are mapped to stop consonants.

The final input device used is a foot pedal. This device has a variable resistance which is an approximately linear function of foot depression. The variable resistance is used in a voltage divider circuit. The variable voltage is sampled by the A/D circuitry included with the computer at its lowest frequency of 8kHz. Additionally, several elastic bands have been attached to the base to provide some force feedback and also to return the foot pedal to the fully undepressed position when the user's foot is lifted.

The output device is a Loughborough Sound Images (LSI) parallel formant speech synthesizer. The device requires 16 speech parameters at 100 Hz to operate. The parameters are quantized to 6 bits (integer range [0,63]). The ten main parameters are these:

- ALF - low frequency amplitude; logarithmic scale
- F1 - first formant frequency; 115 to 1060 Hz in increments of 15 Hz
- A1 - first formant amplitude; logarithmic scale
- F2 - second formant frequency; 730 to 2620 Hz in increments of 30 Hz
- A2 - second formant amplitude; logarithmic scale
- F3 - third formant frequency; 1510 to 3400 Hz in increments of 30 Hz
- A3 - third formant amplitude; logarithmic scale
- AHF - high frequency amplitude; logarithmic scale
- V - degree of voicing
- F0 - fundamental frequency; 25 to 417 Hz using logarithmic scale

The first eight are called formants parameters and can can be thought of as resonances of the vocal tract. The last two represent glottal controls. The parameters are sent to the synthesizer using a parallel port. Control of these parameters is sufficient to produce high quality speech. A text-to-speech synthesizer [11] is available which outputs formant parameters to drive the formant synthesizer and provide formant targets for training the neural networks.

All the software runs on a Silicon Graphics Personal Iris 4D/35. The Xerion Neural Network libraries simulate neural networks and run all the hardware devices [9]. After all the preprocessing and data collection, there is enough computing power remaining in each 10 msec interval to simulate networks with up to 1000 floating point weights which is sufficient for Glove-TalkII to operate without significant interruption. Glove-TalkII requires about 200,000 floating point operations per second.

LEARNING TO SPEAK WITH GLOVE-TALKII

One subject has been trained extensively to speak with Glove-TalkII. The subject is an accomplished pianist who can speak. It was anticipated that his skill in forming finger patterns for playing the piano and his musical training would transfer positively to aid his learning to speak with Glove-TalkII.

The subject went through 8 learning phases during speech acquisition. The phases are:

1. initial set-up
2. initial network training
3. individual phoneme formation within simple words and CV/VC pairings
4. word formation and interword pitch control
5. short segment formation with suprasegmental pitch control and singing
6. passage reading
7. fine tuning; movement control and phrasing
8. spontaneous speech

During his training, Glove-TalkII also adapted to incorporate changes required by the subject. Of course, his progression through the stages is not as linear as suggested by the above list. Some aspects of speaking were more difficult than others, so a substantial amount of mixing of the different levels occurred. Practice at the higher levels facilitated perfecting more difficult sounds that were still being practiced at the lower levels. Also, the stages are iterative, that is, at regular intervals the subject returns to lower levels to further refine his speech. An interesting research issue would be to determine how adaptation by the user interacts with adaptation by the interface.

Initial Set-up

The first phase was initializing the system and familiarizing the subject with the system. The subject's hand parameters were calibrated[4] using the graphical hand and the displayed finger angles. A scale file was created from recorded hand data which is used to scale the input data between 0 and 1 for input to the neural networks. The subject was familiarized with putting on the glove and setting up Glove-TalkII.

Initial Network Training

The second phase was initial network training. The subject familiarized himself with the initial mapping. A complete set of training data was collected from the subject. The typical data collection scheme for a single phoneme is as follows:

1. A target consonant plays for 100 msec through the speech synthesizer.
2. The user forms a hand configuration corresponding to the phoneme.
3. The user depresses the foot pedal to begin recording. The start of the recording is indicated by a green square appearing on the monitor.
4. 10-15 time steps of hand data are collected and stored with the corresponding formant targets and phoneme identifier; the end of data collection is indicated by the green square turning red.
5. The user chooses whether to save data to a file and whether to redo the current target or move to the next one.

The first training set consisted of 2830 examples of static consonants for training the consonant network and 3502 examples (2830 consonants and 672 vowels) to train the V/C network. These data were used to train Glove-TalkII's neural networks to map the subject's interpretation of the initial gesture vocabulary. During data collection, the subject memorized the static hand configuration to static consonant mapping. In addition, he provided hand configurations most suited to his hand that approximated the initial mapping. The simplicity of the data collection procedure and the ease with which the networks train are important for Glove-TalkII to be a useful adaptive interface.

Simple Words

The third phase involved the subject learning to say individual sounds within consonant–vowel pairs (CV), vowel–consonant pairs (VC), and simple word contexts reliably. This was the first time the subject had spoken with the system. The speech produced was unintelligible at this time; however, individual sounds were recognizable. Most importantly, the subject began to recognize different types of phoneme sounds within the sounds he produced. Phoneme recognition helps provide audio feedback for the subject to adapt his motions to produce desired effects. Much of the subject's practice time was spent repeating each of the consonant and vowels

---

[4]Calibration is performed infrequently due to the robustness of the Cyberglove sensors.

sounds, for example "we", "el", "I". After several hours of practice the subject determined which sounds were most difficult to produce. For these difficult sounds, more training data were collected and the networks retrained with the new data added to the original data. This process was repeated several times in an attempt to improve the consonant mapping. As the subject became more proficient at producing static hand configurations (about 5 more hours of practice) it became clear that some of the difficulty producing individual sounds was that often a mixture of vowels and consonants was being produced. This effect was caused by the V/C network having an output slightly larger than zero for consonant sounds causing the output of the vowel network to mix with the consonant sound. Vowel data and consonant data were collected to retrain the V/C network (the consonant network was also retrained with the new data). The subject found this version of the V/C network enabled him to speak individual sounds very reliably. The only phoneme not completely satisfactory was ZH (as in *pleasure*) which rarely occurs in English.

Words and Pitch

The subject proceeded to stage 4: word formation and word pitch control. The subject spent time practicing common English words (from [3]). This exercise provided practice forming transitions the subject would be likely to encounter during conversational speech. While practicing these words some of the difficult transitions became evident; for example, the N→L transition in the word "only". For difficult transitions such as this one, an inverse mapping was used to assist the subject in finding the correct hand gesture timing to make the word intelligible [7]. One of the phonemes the subject found difficult to say was the R sound in words like "are". Using the inverse mapping and a pseudo-spectrum of the R sound it was observed that this sound is actually a vowel sound with dropping pitch. Thus, the subject experimented with the effects of pitch control on individual words to make them intelligible.

Stages 3 and 4 are less distinct than suggested above. The subject practiced individual words and individual sounds simultaneously. This amalgamation became particularly prominent as the subject became more proficient with individual sounds. Data were collected for improving the consonant sounds during the many hours of practice during phases 3 and 4.

Glove-TalkII was retrained about 10 times during these initial phases, sometimes with more data for particular phonemes and other times with replacement data. For future subjects, good performance of the V/C network must be a key focus in the early stages of learning. Several retraining sessions were probably unnecessary since the phoneme errors were caused by mixtures of vowels and consonants caused by poor vowel/consonant distinctions.

Three more significant adjustments were made after the V/C network was performing properly. First, the I position on the vowel mapping was shifted to (5,0) from (4.5,1) which is midway between EE and E (see figure 3). This modification was necessary because the subject had difficulty saying the I phoneme as in "is" reliably. This phoneme occurs frequently in English causing significant intelligibility problems. This problem was probably due to the I and E sound being placed relatively close to each other on the initial mapping, correspondingly, after the vowel network was trained, the area in the X-Y plane which produces the I sound was too small relatively. Second, the subject created another complete training set for every static phoneme sound once the V/C network performed well. The consonant network was trained with this new data set plus the data set used to train the good V/C network. Third, the entire vowel space was compressed by a factor of 0.75 since the subject found that he had to move his hand extensively in X,Y plane to speak. A factor of 0.5 was also tried but was found extreme. Another interesting attempt to provide a better vowel space was to form a radial representation of the static phonemes. Using A as the centre, the remaining 10 vowel phonemes were placed at equidistant positions along a ring 5 cm away from the A. Training data were generated by partitioning the plane into sectors formed by the mid-points between phonemes on the ring, and by specifying phoneme targets for each of the sectors sampled evenly with 60 points out to a radius of 10 cm. The subject found that already after 15 hours of training on the original vowel space the new vowel space was too different to integrate into his speech quickly. In comparison, shifting the I phoneme was easily integrated. From this observation, it appears that users can adapt relatively quickly to the first mapping, after which it becomes difficult to alter the mapping radically without significant performance penalties.

At this point, Glove-TalkII was relatively stable allowing the subject to produce static phonemes in sequence reliably. The subject could intelligibly say simple words that had been practiced. He was proficient at manipulating pitch within a word as well as getting difficult phoneme transitions, especially stop-to-vowel or stop-to-non stop consonants.

Phrases

The next phase of the subject's learning involved saying short segments. He would combine practiced words into meaningful utterances like "Hello, how are you?", the alphabet song ("a-b-c-d-e-f-g... now, I've said my abc, next time won't you sing with me?"), and excerpts from [15] (i.e. "Sam I am." and "I do not like green eggs and ham."). Much practice was required to get the word transitions correct so that they were intelligible. Further, pitch control was practiced over the whole segment to further improve intelligibility. His proficiency

with pitch control was such that his version of the alphabet song was actually *sung*. By the end of this phase, the subject could say any simple utterance (1 to 2 syllables) intelligibly after only a few attempts. However, individual words were spoken slowly at 2.5 to 3 times slower than normal speaking rate[5]. Pronunciation difficulties were mostly in the following areas:

1. producing correct vowel phonemes reliably
2. producing stop consonant clusters
3. timing the R sound precisely

First, it is very important for vowel phonemes to sound correct to achieve proper enunciation of slow speech. With Glove-TalkII, it is difficult to know exactly which vowel will be produced until the foot pedal is depressed since there is poor absolute hand position feedback. Second, timing stop consonants is difficult since the stop phonemes are produced within 100 msec. Small timing errors produce unintelligible stop sounds. Third, if the R sound is sustained for too long, the speech produced sounds muffled and its intelligibility is impaired. Forty milliseconds should be a typical duration of the R sound, but this short timing is difficult to achieve since the static gesture required is hard to produce quickly (see table 1). Notice that when making the R sound, the index finger is very bent. To extend the finger requires a fairly large motion which must be made quickly to achieve the necessary transition. One technique to achieve the necessary transition speed is to form the R sound partially instead of completing the finger trajectory. This technique requires a large degree of finger control since the subject's index finger does not oppose the thumb in this case. Another alternative for *some* R sounds is to use one of the R sounding vowel sounds with a drop in pitch as in "ar" in the British pronunciation of "farther". Examples of R's that can be made in this fashion include UR, AR and ER as in "curious", "are", and "curd" respectively. This type of R sound is much easier to produce quickly. The difficulty for the subject is learning to know automatically which way to make the R sound. The subject uses a combination of a pitch drop and a short R burst as a safe alternative for unknown R contexts.

Reading

The subject reached the sixth stage next; learning to read lengthy passages. Some of the passages include [15], [12], and the "Little Miss Muffet" nursery rhyme. As the subject progressed through the stages, the length of speaking time without excess fatigue increased. In the first few stages, one hour of continuous practice was exhausting. By the reading stage he was comfortable enough that 1-2 hours of continuous practice was possible. Reading exercises helped improve intonation control. For example, the children's story

[5]Normal rate is defined by the text-to-speech synthesizer rate for unrestricted text.

"Green Eggs and Ham" [15] has two voices with different intonation i.e.

Would you like them in a house?
Would you like them with a mouse?
I would not like them in a house.
I would not like them with a mouse.

In addition, reading caused improvement in the three most difficult areas for producing intelligible speech: reliably producing vowel sounds, stop consonant clusters and R technique.

Several distinguishing features of the subject's speech were observed in informal listening tests. First, a strong contextual effect occurred. In particular, when a listener hears the subject speak for the first time, she sometimes does not understand a single word; rather perceives a long slurred speech-like utterance. However, once the listener is told what the utterance was and hears the subject say it again, the words become intelligible and distinguishable. Subsequent novel speech also becomes intelligible. This effect is similar to the adaptation people make when listening to speakers with strong accents or speech impairments. For familiar utterances, the subject's speech is very intelligible; for example, counting and saying the alphabet were never misunderstood even by listeners whose first language is not English. Second, the subject speaks slowly. Third, by using appropriate pitch control the subject produces some relatively natural-sounding speech compared to the text-to-speech synthesizer. As shown through interword pitch variation, proper control of pitch improves intelligibility of the subject's speech. Fourth, even with considerable practice some stops (i.e. P, T, K) are still difficult to discriminate in all contexts. While the R sound still sounds a bit muffled, after considerable practice (approximately 50 hours) the AR, ER, and UR sounds are made reliably in appropriate R-contexts, which alleviated the need for the consonant hand configuration for R to be used in these cases.

Fine Tuning

The next phase the subject reached while learning to speak is the fine tuning stage. The subject designed exercises to help overcome problem areas and make his speech more easily understood and natural-sounding. The exercises fell into two categories: speed, and phrasing. Speed exercises involved saying individual words and phrases as quickly as possible without compromising intelligibility. The phrases used increased in length from two words up to 5 words as he became more proficient at the exercise. During these exercises the subject tried to minimize his hand motion through vowel space. An interesting artifact of speaking faster is that vowel accuracy is not vital for intelligibility. The speed exercises help overcome the poor R quality as well as improving stop onset and offset timing appropriately.

The phrasing exercises involved saying utterances with appropriate pauses controlled by the foot pedal. Pitch control is further refined by synchronizing the phrasing of the speech with the intonation. It is interesting to note that when phrases in a sentence are said quickly in chunks, separated by bringing the foot pedal to the full upright position (i.e. volume turned off), the speech quality improves greatly. Together, these exercises help the subject speak better.

Conversation

The final phase is spontaneous speech. Practicing while conversing with someone helped improve the whole spectrum of skills required for intelligible natural speech. Conversations with unaccustomed listeners are particularly useful since they had not adapted to the peculiarities of Glove-TalkII speech and forced the subject to speak well.

Some of the stages of learning the subject progressed through are similar to the stages encountered while learning to play a musical instrument. The stages can also be categorized according to Fitts' three stages of learning [8]: cognitive, associative and autonomous. Using Fitts' levels, stages 1–4 correspond to the cognitive level, stages 5–7 the associative level and stage 8 the autonomous level. One of the key features discovered while the subject was at levels 3 and 4 was that the V/C network must work well for the user to get adequate feedback about which phonemes he produces.

After 100 hours of practice the subject progressed from simple, barely speech-like noise to intelligible somewhat natural-sounding speech. The subject exhibits two levels of performance, one for rehearsed speech and one for unrehearsed. Rehearsed speech sounds similar to slow text-to-speech synthesized speech with natural intonation contours. For unrehearsed speech the subject still has difficulty pronouncing polysyllabic words intelligibly. However, with a few tries he can say any utterance found in the English language. Additionally, he can sing and make non-vocal sounds. The subject can also speak other languages. Even though Glove-TalkII has been designed for English speech sounds, it is a relatively simple matter to modify Glove-TalkII to produce speech sounds from other languages.

SUMMARY

The initial mapping for Glove-TalkII is loosely based on an articulatory model of speech. An open configuration of the hand corresponds to an unobstructed vocal tract, which in turn generates vowel sounds. Different vowel sounds are produced by movements of the hand in a horizontal X-Y plane that corresponds to movements of the first two formants which are roughly related to tongue position. Consonants other than stops are produced by closing the index, middle, or ring fingers or flexing the thumb, representing constrictions in the vocal tract. Stop consonants are produced by pressing keys on the keyboard. F0 is controlled by hand height and speaking intensity by foot pedal depression.

Glove-TalkII learns the user's interpretation of this initial mapping. The V/C network and the consonant network learn the mapping from examples generated by the user during phases of training. The vowel network is trained on examples computed from the user-defined mapping between hand-position and vowels. The F0 and volume mappings are non-adaptive.

One subject was trained to use Glove-TalkII. After 100 hours of practice he is able to speak intelligibly. The subject passed through 8 distinct stages while he learned to speak. His speech is fairly slow (1.5 to 3 times slower than normal speech) and somewhat robotic. It sounds similar to speech produced with a text-to-speech synthesizer but has a more natural intonation contour which greatly improves the intelligibility and naturalness of the speech. Reading novel passages intelligibly usually requires several attempts, especially with polysyllabic words. Intelligible spontaneous speech is possible but difficult.

ACKNOWLEDGEMENTS

We thank Peter Dayan, Sageev Oore and Mike Revow for their contributions. This research was funded by the Institute for Robotics and Intelligent Systems and NSERC. Geoffrey Hinton is the Noranda fellow of the Canadian Institute for Advanced Research.

REFERENCES

[1] A. G. Bell. Making a talking-machine. In *Beinn Bhreagh Recorder*, pages 61–72, November 1909.

[2] D. Broomhead and D. Lowe. Multivariable functional interpolation and adaptive networks. *Complex Systems*, 2:321–355, 1988.

[3] G. Dewy. *Relativ Frequency of English Speech Sounds.* Harvard University Press, Cambridge, Mass, 1950.

[4] Homer Dudley, R. R. Riesz, and S. S. A. Watkins. A synthetic speaker. *Journal of the Franklin Institute*, 227(6):739–764, June 1939.

[5] S. Fels and G. Hinton. Glove-Talk: A neural network interface between a data-glove and a speech synthesizer. *IEEE Transaction on Neural Networks*, 4:2–8, 1993.

[6] S. Fels and G. Hinton. Glove-TalkII: Mapping hand gestures to speech using neural networks. In D.S. Touretzky, editor, *Advances in Neural Information Processing Systems*, volume 7, Denver, 1995. Morgan Kaufmann, San Mateo.

[7] S. S. Fels. Glove-TalkII: Mapping hand gestures to speech using neural networks, August 1994. Dissertation.

[8] P. M. Fitts. Perceptual–motor skill learning. In A. W. Melton, editor, *Categories of human learning.* Academic Press, New York, NY, 1964.

[9] Connectionist Research Group. *Xerion Neural Network Simulator Libraries and Man Pages; version 3.183.* University of Toronto, Toronto, ON, CANADA, 1990.

[10] P. Ladefoged. *A course in Phonetics (2 ed.).* Harcourt Brace Javanovich, New York, 1982.

[11] E. Lewis. A 'C' implementation of the JSRU text-to-speech system. Technical report, Computer Science Dept., University of Bristol, 1989.

[12] H. A. Rey. *Curious George.* Houghton Mifflin Company, Boston, 1941.

[13] D. E. Rumelhart, J. L. McClelland, and the PDP research group. *Parallel distributed processing: Explorations in the microstructure of cognition. Vols. I and II.* MIT Press, Cambridge, MA, 1986.

[14] J. M. Rye and J. N. Holmes. A versatile software parallel-formant speech synthesizer. Technical Report JSRU-RR-1016, Joint Speech Research Unit, Malvern, UK, 1982.

[15] Dr. Seuss. *Green Eggs and Ham.* Beginner Books, New York, 1960.

[16] Wolfgang Ritter von Kempelen. *Mechanismus der menschlichen Sprache nebst Beschreibungeiner sprechenden Maschine. Mit einer Einleitung von Herbert E. Brekle und Wolfgang Wild.* Stuttgart-Bad Cannstatt F. Frommann, Stuttgart, 1970.

Pictures as Input Data

Douglas C. Kohlert and Dan R. Olsen, Jr.
Computer Science Department
Brigham Young University
Provo, UT 84602
dkohl@tim.cs.byu.edu   olsen@cs.byu.edu

ABSTRACT

This paper suggests that there exists a large class of inherently graphical applications that could use pictures as their primary input data. These applications have no need to store input data in any other format and thus eliminate the need to do conversions between input data and a graphical representation. Since the graphical representation is the only representation of the data, such applications allow users to edit an application's input data by manipulating pictures in a drawing editor. Such an environment would be ideal for users of pen-based machines since data would not have to be entered via a keyboard, instead a gesture based drawing editor could be used. CUPID, which is a tool for **C**reating **U**ser-Interfaces that use **P**ictures as **I**nput **D**ata, is presented.

KEYWORDS VISUAL LANGUAGES, PICTURE PARSING, PICTURE-BASED APPLICATIONS

INTRODUCTION

Much of user interface design focuses on how to effectively represent application data in such a way that it can be easily comprehended as well as manipulated by the user. For many application domains, this requires converting an awkward and unnatural internal data representation to a more comprehensible graphical form. Some applications are inherently graphical in the way users visualize information. For example, the relations in a pedigree tree for a genealogical application can be quite difficult to understand if presented in a textual format, however, the same tree presented in a graphical form can be quickly grasped and understood by the user.

This paper suggests that inherently graphical applications need not try to do conversions between the internal representation and the displayed graphical representation of data such as is done by TRIP3 [9].

Instead, application input data can be stored and displayed as graphical objects with relations between objects based on geometric relations. This allows arbitrary pictures to become an application's input data. A visual language could be defined for each application domain that would determine a "picture syntax" that could be used to "parse" application data sets. This syntax would determine how objects in a picture are related and thus could be used by the application to access and retrieve information from various parts of the picture.

Using pictures to represent input data has five major advantages. First, the data that is input into the application is easily understood by the user. Pictures of data also have certain affordances that make it clear how to change relationships within the picture [3], such as by moving the position of a line. Second, "drawing pictures" to represent application data is a task that most users are familiar with. Third, modifying application data is reduced to simply editing an already existing drawing or picture. Fourth, with the ever increasing number of gesture based machines, it is very advantageous to be able to draw an application's data rather than input it in a textual form. Although a gesture-based drawing editor similar to GEdit [8] could be used for sketch recognition, only a palette-based drawing editor has been used in this work. Fifth, picture-based applications give the user significant freedom over fixed object systems. Unlike fixed object or dialogue based systems, the user has more freedom to place a variety of objects in more locations in a picture as long as the picture syntax is not violated.

The remainder of this paper will present the CUPID visual language tool for **C**reating **U**ser-Interfaces that use **P**ictures as **I**nput **D**ata. However, before presenting CUPID, we will first present some example applications that have been built with CUPID. We will then discuss how these types of applications would be built in a traditional approach. Next, we present how these applications can be quickly constructed using CUPID. Finally, the results of using CUPID will be presented, followed by our summary.

EXAMPLE PICTURE-BASED APPLICATIONS

To help illustrate how useful picture based applications can be, we will present three picture based applications that have been built with the CUPID tool kit. The Image Composition program shown in Figure 1 represents how pictures can be used to represent image data as well as operations to be performed on that data.

This application allows the user to construct an image expression tree composed of image names and operations to be performed on those images. The image resulting from parsing the expression is displayed at the top area of this window.

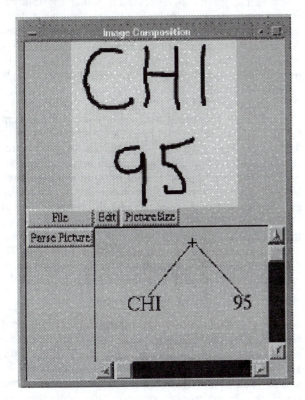

Figure 1

The leaf nodes in an expression tree are the names of files that contain images. The interior nodes represent image operations that operate on their children. All operations require one or more image parameters. An image parameter is specified by drawing a line from the operator node to the root of the subtree that computes the image parameter's value. Having constructed an image expression tree, the user can select any node in the tree and request that the image represented by the expression rooted at that node be computed and displayed. For example, in Figure 1 the user has entered an expression that adds the "CHI" and "95" images to produce the "CHI 95" image.

With this application a user can quickly construct an image expression tree to build a rather complex image by drawing lines between operators and image names, instead of having to type in a series of commands. The user explores possible combinations of image operations by simply rearranging the drawing. The interface becomes a workspace rather than an expression entry tool. Unfortunately current tool kits make such interfaces very difficult to write.

Another type of application that can be built with CUPID is one that can convert pictures of data to another representation such as a data file for another application.

This might be done to create a more intuitive interface for another application. We constructed a genealogy application that parses an arbitrary simple pedigree tree as shown in Figure 2 into a corresponding GEDCOM file that can be used by other genealogy programs.

A pedigree tree is composed of individuals with lines between them to represent parental/child relationships. Thus a family is defined as individuals connected by lines. This makes building a family in our application a very simple task. To specify the father of an individual, the user would simply draw an arrow from the "Father:" text in the individual, to the connecting circle on the actual father individual as show in Figure 2.

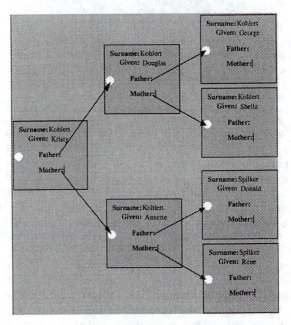

Figure 2

Although creating an individual and adding them to a family in a program such as *Personal Ancestral File* is relatively simple, changing relationships between individuals is a rather difficult task. This requires specifically removing the individual from the current family, and adding them to the desired family. To do this requires the user to know the RIN, (Record Identification Number), of the individual to be removed. If the RIN is not known, the user must search the database for the desired individual so the RIN number can be retrieved. However, in our application the lines between individuals have an affordance to be moved from individual to individual to change the appropriate relationship. To change the father of an individual in our application the user would simply change the endpoint of the line from the old father to the new father.

The UNIX Shell Script Builder, Figure 3, is an example of an application built with our system that can be used as an interface to the UNIX shell. A shell script picture consists of five different type of objects.

- a Prompt object to specify where the script is to begin,

- a File object for specifying filenames,
- an object for specifying a command line,
- a Printer object to represent a printer, and
- a line object for connecting other objects.

The UNIX Shell Script Builder allows the user to draw a picture of a UNIX shell script. Lines between objects represent either a redirection or a pipe depending on the context of the line. When the user's picture is parsed, a corresponding shell script is executed.

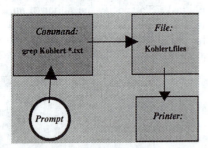

Figure 3

With this approach, rather complex scripts can be built without having to know about all the various ways to pipe things around in UNIX. For example, the output of a command can be simultaneously piped or redirected to any number of other objects by simply drawing multiple lines out of the object to the other target objects. In UNIX one would have to learn about and use the tee command and temporary files to achieve the same result.

Although these examples are rather simple, they are only a small sample of a rather large class of applications that include logic diagrams, PERT-charts, organizational charts, multimedia synchronization diagrams, equation editors, genetic inheritance diagrams, circuit diagrams, musical notation, molecular modeling, object-oriented class structure and inheritance diagrams. All of these applications are characterized by diagrammatic representations of information.

PICTURE-BASED APPLICATION BY HAND
Building a picture-based application such as the UNIX Shell Script Builder file in Figure 3 is a non-trivial task. A specialized direct manipulation editor must be built to allow the user to create and manipulate pictures representing user input. This editor would need to supply template objects of all possible components in an acceptable picture, such as a prompt object or command object. These template objects would be used to create objects in the picture.

As the user creates a new object from a template, the system would create an internal representation of the object. When the user modifies the contents of an object, the corresponding data structure would also be changed. To do this, the editor would have to continually monitor the user's actions in order to keep the internal representation up to date. For example, if the user were to create a command object with its default command line of

"command line", the system would create an internal representation of a command with the value "command line". If the user changes the command line of the command object to say "ls -l *.pic", the system must change the contents of the internal command to match. Keeping the graphical representation and the internal representation synchronized requires some sort of mechanism for mapping between the two representations. Also required is the input handler to perform the edits on objects. Typically tool kits don't allow the user to directly edit a text value, instead a dialogue box is presented for editing.

Creating links between objects also requires the system to internally manage what objects are connected to what. This is a somewhat more difficult task than simply representing a picture object internally because changing connections between objects requires managing at least three internal objects. Two of these objects are the originally connected pair, and the third object is the newly connected object. For example, in Figure 3, the user may decide that he wants to send the output of the "grep Kohlert *.txt" command to the printer rather than the file "Kohlert.files." This would cause the system to change not only the command and file objects, but also the printer object.

Creating objects and maintaining an internal representation of the picture is one problem that must be overcome when building a picture based application. However, there are many small problems that must be solved such as constraint problems that will constrain graphical objects to each other so that when they are dragged they stay connected. For example, in Figure 3, if the user were to drag the entire Command rectangle, both the lines connected to it would also move, thus eliminating the need for the user to reattach the lines. However, there are systems such as Unidraw [11] that can help solve this problem.

Although creating such an editor is not extremely difficult, it does require some serious thought and design. To create such a specialized editor would take several weeks to program, not to mention design and testing time.

BUILDING PICTURE-BASED APPLICATIONS WITH CUPID
The focus of this section is to describe how a picture-based application is built with CUPID. To do this we will use the UNIX Shell Script application as our target application. Our approach to this problem is to start with a fully functional constraint-based drawing editor with additional picture parsing primitives. We will then restrict the editor to the target domain rather than build up an editor from tool kit primitives.

Applications Built With CUPID
All applications built with the CUPID tool kit have two windows. The main window is based on a drawing editor. The other window is a palette of draw objects or

templates that can be drawn in the drawing window. For example, the UNIX Shell Script Builder created with CUPID is shown in figure 4.

The window on the right is the main drawing window with an already existing shell script picture. The palette window on the left shows the various template objects that are used to draw script pictures.

The drawing editor used by CUPID, and the applications created with CUPID, is primitive-based similar to MacDraw. Each primitive, or draw object, is created and manipulated by control points. To create a new object in the drawing window, one of the template objects in the palette must first be selected. To then draw the object, the user clicks in the drawing area once for each control point. For example, if the selected template object was a command line object from figure 4, the user would click in the drawing window once for one corner of the rectangle, and once for the opposite corner.

Once on object has been drawn, it can be edited in many ways. An object's location may be changed by selecting the object and dragging it to the desired location. The size and possibly shape of an object can be changed by selecting one of an object's control points and dragging it to the desired position.

Placing or dragging a control point on top of an already existing object constraints that point to that object. Such control points are constrained to that object. Thus if the underlying object is moved, the control point will also move. If the object is resized, the control point will remain proportionately in the same location. For example, in the sample script picture in figure 4, the line between the "rev" command and the "sort" command are mapped to those objects. If the user grabbed the end of the line over the "sort" command and moved it to be over the "outfile" file name object, then that line control point would then be mapped to the file name object instead of the command line object.

Other attributes that can be edited on appropriate draw objects include line width, line color, fill color, font face, point size, and font style.

The drawing editor also allows for grouped objects. A group object is one or more objects that are treated as a single object. There are two different types of groups supported by this editor. One type simply takes any number of selected objects and wraps an invisible wrapper, rectangle, around them to make them one object. This type of group is not scalable since the objects have no real relation among them.

The other type of group takes a selected object, and groups all other objects that are mapped to it through transitive closure, into its own parametric space. Thus when the base object is resized, all the subobjects will also be resized. All of the interactive behavior described so far is all built into the general drawing editor, and is thus, not built by the tool designer.

Each of the template objects in the palette window in figure 4 with the exception of the arrow-line are primitive objects that have been grouped. The command line, file name, and printer objects are grouped around their base rectangles, while the prompt is grouped around the base circle.

Once an input data diagram has been drawn in a CUPID generated application, the picture may be parsed by pressing a "Parse Picture" button. Parsing a picture means to interpret the picture within the application's domain. For example, the UNIX Shell Script Builder knows how to parse and interpret pictures of UNIX shell scripts created within itself. However, the shell script builder does not know how to parse or interpret pedigree charts, or even pictures of shell scripts created in other applications.

In short, applications built with the CUPID tool kit are based around a drawing editor which allows the user to draw pictures of input data from a small, but complete, set of template objects designed especially for the application. Once the picture is drawn, the user can tell the application to interpret the picture by pressing a "Parse Picture" button.

Now that the reader is familiar with the workings of applications built with CUPID, the next section will discuss how to build such an application.

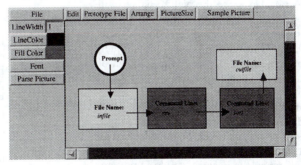

Figure 4

How To Build

In order to build a picture based application with CUPID, there are three things that must be done: template objects must be created, a picture parsing routine must be defined, and user-changeable attributes must be decided. After all of these things have been done, CUPID can generate the code for the desired application.

Templates. Template objects are used by the user to create new draw primitives. Template objects contain most of the pertinent information needed by the desired application. For example, a file object in figure 4, contains the name of a file to be used in a shell script.

Templates are used for three different reasons. First, template objects eliminate the need for users to draw complex objects by hand. Second, templates help the user draw more easily parsed pictures by limiting their choices of drawable objects. If all draw primitives were available, without any application specific template objects, the user would be less likely to produce a parseable picture. Third, template objects can be used to enforce geometric relationships that can be utilized by the picture parsing routine.

Template objects are built in a drawing editor just like the one the end user will use except a richer interface is provided. It provides a full palette of primitive draw objects including circles, rectangles, parallelograms, arcs, splines, text and various types of lines.

A template object is created by combining one or more draw primitives together. For example, to draw the command line object template in figure 4, a base rectangle object is drawn first. Next, a text object is placed on this rectangle and its value edited to be "**Command Line:**." Another text object is then placed on the rectangle and its text is edited to be the default value of "*Command* ." Now the base rectangle can be selected and made into a group, thus making it and the two text objects one group object.

Although the command line template built above would work, there is some danger in using this template. The main danger is that all drawing objects are free to be edited by the user. Although the user should be able to edit the "*Command*" text object to be actual commands, they should not be able to change the "**Command Line:**" text since this is what will be used to identify a command line object. Also, since the picture parser will be defined using geometric relations (this will be described later) and since the "*Command*" text object needs to remain below the "**Command Line:**" text object, the user should not be allowed to change the position of these text objects.

To remedy this problem, the "**Command Line:**" text object can be made locked so that it may not be edited. Also, the positions of these text objects can be locked so that they may not be moved by themselves. They can then only be moved when the base rectangle is moved.

With the template object constrained this way, the end user will not be able to move either of these text objects and thus will only be able to edit the "*Command*" text object or move the base rectangle.

To make a template object available to the user in the final application, it must be put into the template palette. To do this the desired template object is copied into the template palette window of the CUPID tool kit. CUPID will only use the objects in this window as templates for the final application. Thus, nothing else that is drawn in the main drawing editor can be used as a template. Also, none of the primitive draw objects are made available in the final application unless the application designer draws one and copies it into the template palette window.

In short, building template objects involves three main concepts: drawing the object with primitive draw objects, locking and constraining the appropriate pieces of the object, and placing the object in the palette window.

Parsing Routine. The next step is to define a parsing routine to extract pertinent information from the picture. Traditional applications that use graphical representations of internal data structures such as TRIP3, do not actually parse pictures, but rather maintain an internal representation as the picture is manipulated. Applications built with CUPID on the other hand actually retrieve all information directly from the picture itself by means the NIC [6] scripting language NICScript [7].

The NICScript is similar to other scripting languages such as Tcl [10] and SuperTalk [1]. However, the NICScript language has been extended to parse pictures created by the drawing editor described above.

The extensions made to the NICScript language for this work include geometric parsing methods on draw primitives. These parsing methods are based on the object's bounds, control points, and a selection criteria.

With NICScript, the designer may specify how to parse a picture by means of geometric relationships such as *Right*, *Above*, *Touches* and *ConnectedTo*. These are the same types of relations that are used by other visual language parsers [2][4]. Each of these relations has a corresponding method that has been defined on the base draw object class. These methods include:

- Right - *finds all objects lying anywhere to the right of the target object*
- DirRight - *finds all objects that are directly right of, but not above or below of the target object*
- Left - *finds all objects lying anywhere to the left of the target object*
- DirLeft - . . .
- Above, DirAbove - . . .
- Below, DirBelow - . . .
- Contains - *finds all objects within the target object's bounds*
- ContainedIn - *finds all objects that the target object lies within*

- Touches - *finds all object touching the target object*
- Parent - *if the target object is part of a composite object, the base or parent object is returned*
- Children - *if the target object is a composite object, the subobjects or children are returned*

Geometric Searches. These methods can be used to find objects relative to the target object. For example, in Figure 5, if object A were the target object and the *DirRight* (Directly Right) method were called, then the set of objects {a, b, B, E, G} would be returned.

In contrast to the *DirRight* method, the *Right* method is used to access all objects that lie somewhere to the right of the object. Thus, if the *Right* method were called on object A, then the set {a, b, c, B, D, E, F, G} would be returned.

Each of these methods simply uses a bounding box to locate other objects. This bounding box is an extension to the target object's bounds. Any object that intersects the defined bounding box is returned in a sorted list.

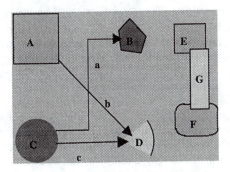

Figure 5

For example, in Figure 5, if we called the *DirLeft* method on object F, then a rectangle extending from the left edge of F with the same height would be extended to the left edge of the drawing. Objects D, b, c, and C would be returned in a list in that order. All of the left methods return objects sorted right-to-left. All of the right methods sort the list left-to-right, while the above methods sort bottom-to-top and the below methods sort top-to-bottom. The other methods, such as *Touches*, are unsorted.

Although there are methods for right, left, above, and below, we sometimes need to find objects that are below and to the left of an object. To accomplish this, each of the parsing methods returns a Set object containing the objects found. The Set class has methods defined on it for doing union, intersection and difference, allowing the user to access an object to the lower-left by calling the Left and Below methods on the desired object and then intersecting the resulting sets. The Set class also has methods defined for sorting the set in a particular direction such as right-to-left, or top-to-bottom. This

allows the user to access objects closest to or farthest away from the current object in the desired direction.

Selection. The parsing methods as described so far, allow the user to find all objects that have a certain geometric relation to the target object, however, the script author may only be interested in certain types of objects, say circles. To allow the user to specify what type of objects should be returned a selection pattern is passed as an argument to the various parsing methods. Only objects that match the pattern are returned.

One type of pattern is simply a class name. With a class name pattern, only objects that are of the specified class are returned. For example, in Figure 5, if object A is the target object and the pattern :PolygonDO (polygon draw object) is passed as an argument to the *Right* method then only the set containing object B will be returned. Patterns can be used to match any attribute of an object such as color, line width etc. A complete description of NIC patterns can be found in the NIC documentation [5].

Connections. Direct relations between objects are usually represented by drawing lines between the objects. There are three methods that can be used to discover these relations: *Connected*, *ConnectedTo*, and *ConnectedFrom*. These methods find objects that are connected to the target object in various ways.

Before these methods are explained, it is necessary to explain that two draw objects can be connected by any other draw object. For example, in Figure 5, the objects A and D are connected by line b, objects C and B are connected by a, objects C and D are connected by c, and objects E and F are connected by object G. In order to connect two draw objects *x* and *y*, another draw object *w* must have its first control point within *x*, and its second control point within *y*. In our example picture, object A is connected to object D by arrow line b. Thus the first control point of b is in A, and the second control point of b is in D. Control points are numbered by the order they are specified. Thus the first control point specified is numbered 1, while the second point specified is numbered 2, etc.

It is important to note that connectedness is a pictorial relation to be discovered and not simply stored internally. Each of the methods used to find connections have, in addition to a selection pattern, a connection pattern as a parameter. The connection pattern is similar to the selection pattern but it is used to determine what types of objects can be considered as connectors. For example, a connection pattern of :LineDO, will only consider lines to be connectors, while a pattern of (| :LineDO :RectangleDO), where | is the OR operator, will accept lines or rectangles as connectors.

The *Connected* method simply returns all objects that are connected to the target object. The *ConnectedTo* method returns all objects that are connected to the target object when the first control point of the connector is within the target object and the second control point is within the other object. In the example picture, A is connected to D, C is connected to B and D, and E is connected to F. The *ConnectedFrom* method is just the opposite of the *ConnectedTo* method in that the second control point of the connector must be within the target object and the first control point must be within the other object. Thus in the example picture, D is connected from A and C, B is connected from C, and F is connected from E.

Writing Scripts. Most of writing the parsing script is simply calling various parse methods on draw objects. In the UNIX shell script builder, the "Prompt" object specifies where the script is to begin. To find the prompt object we can call the Children method on the root group draw object of the drawing editor

(:= (!RootGroup Children :CircleDO) !Prompt)

passing :CircleDO as the selection pattern since the prompt is the only circle object in the picture.

From the "Prompt" it is necessary to find the first command to run. Calling of the "Connected" method on the "Prompt" similar to

(:= (!Prompt Connected :ArrowLineDO :RectangleDO) !CmndObj)

will return all :RectangleDO objects connected to the "Prompt" object by an :ArrowLineDO.

In the example UNIX shell script picture in Figure 4, this will be the rectangle object containing the text objects "File Name" and "infile." To get the "File Name" text object from within this object we can call

(:= (!CmndObj Children (:! (& :TextDO
                    ([Value "File Name"])
)))
    !FileNameLabel)

This will return all text objects that have the text "File Name" and are children of the !FileObj, where !FileObj is the rectangle object found above. To actually get the name of the file we want we could then call

(:= (!FileNameLabel DirBelow :TextDO) !FileName)

where !FileNameLabel is the "File Name" text object found above. This call will find the text object that lies directly below the !FileNameLabel object. In this case this will be the text object whose value is "infile."

It is important to note at this point that this example does not find the "infile" text object in the most efficient manner. We choose to use this example to give the reader a better feel for how the geometric parsing can be used.

Editable Attributes. The final step in building a picture-

based application with CUPID is to determine which attributes the user should be able to edit in the final application. This is important because as was mentioned earlier, the parsing routine may use selection patterns that are based on an object's attributes. For example, the parsing routine of the shell script builder may require that all Printer objects be purple.

To prevent the user from making the mistake of changing the color attribute on such an object, the application designer can tell CUPID to not put the interface components necessary to change a certain attribute into the final application. CUPID provides a checklist of attributes that can be removed from the final application. Thus the designer simply selects the attributes that are to be included in the final application. For example, if the designer specified that the FillColor attribute should not be editable in the UNIX Shell Script Builder, then CUPID would generate the same application shown in Figure 4 except the FillColor interface components below the LineColor interface components would be missing.

After completing these steps CUPID can generate the NICScript code for the desired application.

Generating an Application. Because CUPID and its applications are built upon NIC, the data model, scripting language, pattern language, and user interface tools are all integrated into one data model. Thus the templates, and parsing routines, built with CUPID are nothing more than NIC data objects.

CUPID uses these data objects to modify an existing template of a generic application. This template user interface looks just like the UNIX Shell Script Builder in Figure 4 without the template objects and example picture. Inserting the template objects into a new application is achieved by adding a statement to the template interface's initialization code to load a template file which CUPID creates. The parse routine defined in CUPID can then be attached to the "Parse Picture" button by setting the Action attribute on that button with the desired routine.

Removing the user interface components that edit attributes that should not be editable is done by finding those components in the interface description and removing them from the interface.

Generally the interface generated by CUPID is sufficient. However, if the designer is not satisfied with the design of the default interface, or simply needs to add more to the interface, the interface may be loaded into NIC's interface editor to be edited as desired.

Summary of Building an Application with CUPID. Designing a CUPID application takes three steps:
- Draw template objects, locking parts that the user should not change. Once drawn, the template should be copied into the template palette window.

- Write a parsing routine should be defined using the geometric parsing methods defined in NICScript. The methods used in this routine are almost always the same ones that would be use if you were telling someone over the phone how to find the important pieces of a picture.

- Specify which attributes in the final application should be editable.

USER EXPERIENCE

To evaluate the usefulness of the CUPID tool kit, we asked a couple of individuals that were familiar with NIC and the NICScript programming language but have had no experience with picture parsing, to use CUPID to build the UNIX Shell Script Builder.

The script builder was required to handle the redirection of file input and output as well as the piping of commands. It also needed to support the printing of files and the output of commands.

Each individual worked independently with minimal verbal explanation of the tool. The first subject spent approximately 5 hours building the script builder. Of this 5 hours, approximately 1.5 hours went to reading the documentation., and 3.5 hours to building the tool. His finished picture-parser consisted of 81 lines of script code.

The second subject spent approximately 10 hours building the tool. Of these hours, 2 were spent reading the documentation, 1 hour was spent playing with the CUPID tool kit to see how it worked, and 7 hours was spent actually designing and implementing the script builder with the CUPID tool kit. This subject's finished picture-parser consisted of 180 lines of script code.

Another individual used CUPID's geometric parsing methods to build the Image Composition application in Figure 1. With these parsing methods, the individual was able to create the parsing script in approximately 1 hour.

Both of these experiences show that new users to the CUPID tool kit can rapidly learn to use the tool and actually build a useful application.

SUMMARY

The CUPID tool kit presented in this paper makes it possible to quickly build applications that use picture-based input. To design such an application only requires the designer to know the various different geometric methods that can be used to find objects relative to other objects. By using geometric relations to reference other objects, a whole new class of picture-based applications can be easily built.

Implementing such an application only requires the knowledge of primitive-based drawing editors, and the NIC programming language. The designer of the CUPID tool kit was able to build the genealogy application

discussed above in approximately 10 hours. Most of that time was learning how generate GEDCOM files rather than how to parse the pedigree charts.

Once a picture-based application has been built, users can quickly create complex input structures by merely drawing and editing pictures in a rather simple drawing editor. This requires minimal knowledge and expertise of the end user, while still giving the user the freedom to create arbitrarily complex data structures within the confines specified by the designer. Being able to change data structures by merely changing graphical images is straight forward and very self explanatory. There is no need to go through unnecessary dialogues that are prominent in traditional systems.

REFERENCES

1. Calica, B., A. Himes, and C. Mascis, *SuperCard, The Personal Software Toolkit.* 1989, Silicon Beach Software, Inc.

2. Costagliola, G., M. Tomita, and S.-K. Chang. A Generalized Parser for 2-D Languages. in *IEEE Workshop on Visual Languages.* 1991. Kobe, Japan: 98-104.

3. Gaver, W.W. Technology Affordances. in *Human Factors in Computing Systems (CHI '91).* 1991. New Orleans, Louisiana: ACM Press, 79-84.

4. Helm, R., K. Marriot, and M. Ordersky. Building Visual Language Parsers. in *Human Factors in Computing Systems (CHI '91).* 1991. New Orleans, Louisiana: ACM Press, 105-112.

5. ISSL, NIC Pattern Language. 1993, WWW URL: file://issl.cs.byu.edu/docs/NIC/NICPat.intro.html.

6. ISSL, NIC Programming Documentation. 1993, WWW URL: file://issl.cs.byu.edu/docs/NIC/home.html.

7. ISSL, NICScript. 1993, Mosaic document: WWW URL is file://issl.cs.byu.edu/docs/NIC/NICScript.html.

8. Kurtenbach, G. and B. Buxton, GEDIT: A Test Bed for Editing by Contiguous Gestures. *SIGCHI Bulletin*, 1991. **23**(2): p. 22-26.

9. Miyashita, K., S. Matsuoka, and S. Takahashi. Declarative Programming of Graphical Interfaces by Visual Examples. in *Proc. of ACM User Interface Software and Technology (UIST '92).* 1992. Monterey, California: ACM Press, 107-116.

10. Ousterhout, J.K. Tcl: An Embeddable Command Language. in *Proc. of the 1990 Winter USENIX Conference.* 1990. 133-146.

11. Vlissides, J.M. and M.A. Linton, Unidraw: A Framework for Building Domain-Specific Graphical Editors. *Transactions on Information Systems*, 1990. **8**(3): p. 237-268.

Planning-Based Control of Interface Animation

David Kurlander and Daniel T. Ling

Microsoft Research
One Microsoft Way
Redmond, WA 98052

(206) 936-2285
E-Mail: djk@microsoft.com

ABSTRACT

Animations express a sense of process and continuity that is difficult to convey through other techniques. Although interfaces can often benefit from animation, User Interface Management Systems (UIMSs) rarely provide the tools necessary to easily support complex, state-dependent application output, such as animations. Here we describe Player, an interface component that facilitates sequencing these animations. One difficulty of integrating animations into interactive systems is that animation scripts typically only work in very specific contexts. Care must be taken to establish the required context prior to executing an animation. Player employs a precondition and postcondition-based specification language, and automatically computes which animation scripts should be invoked to establish the necessary state. Player's specification language has been designed to make it easy to express the desired behavior of animation controllers. Since planning can be a time-consuming process inappropriate for interactive systems, Player precompiles the plan-based specification into a state machine that executes far more quickly. Serving as an animation controller, Player hides animation script dependencies from the application. Player has been incorporated into the Persona UIMS, and is currently used in the Peedy application.

KEYWORDS: Animation, planning, User Interface Management Systems, UIMS, user interface components, 3D interfaces.

INTRODUCTION

With the advent of inexpensive graphics rendering hardware and faster computers, interface animation will become commonplace. One building block of many such interfaces will be a User Interface Management System (UIMS) component that controls the sequencing of animation

actions. We have developed such a component, called Player, that simplifies the task of coordinating these animations. Player supports a convenient plan-based specification of animation actions, and compiles this specification into a representation that can be executed efficiently at run time.

To better motivate the need for this capability in a UIMS, consider the Persona UIMS currently being developed by our group, and a prototype agent-based interface called Peedy, built on Persona, and demonstrated at CHI '94 [1]. Peedy's visual representation is that of a 3D animated parrot. Peedy has a rich repertoire of animated bird-like and human-like behaviors, and responds to spoken natural language requests for musical selections. The Peedy application is shown in a video clip accompanying electronic distributions of this paper.

Figure 1a illustrates the slice of the Persona UIMS's architecture that handles output. The application sends control events to the animation controller. This controller interprets the incoming events according to its current internal state, informs the low level graphics system (called Reactor) what animations to perform, and adjusts its own current internal state accordingly.

For example, consider the path of actions when the user asks Peedy "What do you have by Bonnie Raitt?" This is illustrated in Figure 1b. First the application interprets the

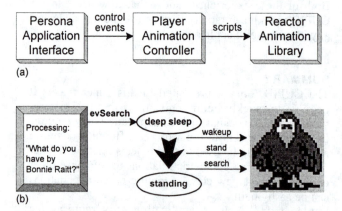

FIGURE 1. Persona and Peedy. (a) architecture of the output component of the Persona UIMS; (b) an example of its use in the Peedy application.

message, and sends an **evSearch** event to the animation controller, to have Peedy search for the disc. The animation controller knows that Peedy is in his "deep sleep" state, so it sequentially invokes the wakeup, standup, and search animations. It also changes Peedy's current state (as represented in the animation controller) to standing, so that if another **evSearch** event is received immediately, Peedy will forego the wakeup and standup animations, and immediately perform a search.

One can view the animation controller as a state machine, that interprets input events in the context of its current state, to produce animation actions and enter a new state. Originally we specified the animation controller procedurally as a state machine, but as new events, actions, and states were added, the controller became unwieldy, and very difficult to modify and debug. It became clear that we needed a different manner of specifying the controller's behavior. One of the difficulties of specifying this behavior is that graphical actions make sense only in limited contexts for either semantic reasons (Peedy cannot sleep and search at the same time) or animation considerations (the search script was authored with the expectation that Peedy would be in a standing position).

In traditional UIMSs, the programmer must specify all these transitions, which can be a tedious and error-prone process. The Player component of the Persona UIMS calculates these transitions automatically, freeing the implementer from part of the chore of constructing animated interfaces. To accomplish this, Player relies on *planning*, a technique traditionally used by the AI community to determine the sequence of operators necessary to get from an initial start state to a goal state. In our system, the operators that affect system state are animation scripts, and the programmer declares preconditions and postconditions that explain how each of the scripts depend on and modify state. One of the major problems with planning algorithms is that they are computationally intensive. Animation controllers, however, have to operate in real time. Our solution is to precompile the conveniently specified planning notation into an efficient to execute state machine.

There are several contributions of this work. We describe a UIMS component that isolates animation dependencies from the application, and identify a valuable use of planning technology within this component. To satisfy the real-time constraints of a user interface, we present an algorithm suitable for converting a plan-based specification into a state machine. Unlike traditional AI planning techniques, this algorithm must find transitions to a goal from any possible state, not just a single start state. We describe a technique for doing this efficiently, exploiting coherence in the search space. Our language for specifying animation controllers is described in this paper. It includes provisions for handling goal-oriented behaviors (such as speaking), as well as autonomous actions (such as snoring). Our system employs a novel state hierarchy to simplify the task of specifying preconditions and postconditions. A final contribution of this work is an implementation within the Persona UIMS,

and its use within the Peedy application, as a proof of concept of this research.

The next section describes other research that shares some of the same goals or uses related techniques. Following that, we present our language for specifying the behavior of the animation controller. The subsequent section explains Player's planning algorithm. Additional implementation issues are described following this, and then the paper presents our conclusions and possible future directions.

RELATED WORK

One way to discuss UIMSs is in terms of the Seeheim model [14], presented in Figure 2. The user interacts directly with the *presentation* component, which passes interaction events on to the *dialogue control*. The dialogue control component then determines which application services should be requested through the *application interface model*. To produce output, the application interface model can drive the presentation component to display graphics and play sound. Alternatively, it can send events to the dialogue control, which in turn can drive the presentation. Note that the Persona output architecture of Figure 1a can be mapped directly to the Seeheim model of Figure 2, with the animation controller serving as the dialogue control, and Reactor handling presentation services.

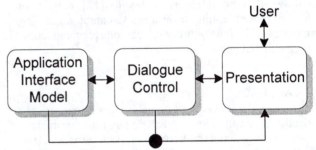

FIGURE 2. The Seeheim UIMS model (adapted from [14]).

UIMSs have employed several different techniques for dialogue control. Green compares the state machine, grammar, and event-based approaches in [9]. However, these techniques have focused on interpreting sequences of input, producing graphics only incidentally in the process. Olsen writes, "The notion of dialog control having an influence over the presentation of application data is signified in the diagram [Figure 2] by the small circle around the path from the application interface to the presentation. This data display facet of the model has not been fully realized in any UIMS" [14]. It is this facet that Player attempts to address.

The UIDE system also relies on preconditions and postconditions for dialogue control [5] [8], using them to determine when to enable application actions, and to support interface transformations. However, UIDE does not use planning to automatically sequence visual presentations. An exception to this is Sukaviriya's Cartoonist system, that extends the

UIDE model to automatically provide context-sensitive help animations [16]. UIDE also uses planning to generate context-sensitive textual help [4]. However, the UIDE work does not use planning to sequence general animation presentations, and does not employ precompilation techniques to maintain real-time constraints.

Many researchers have addressed other aspects of incorporating animation into the interface. For example, PARC's Cognitive Co-processor deals with timing issues [2]. Several interfaces incorporate traditional animation effects [3] [10]. Virtual reality research frequently coordinates simulation and novel interface devices with animation [13].

Researchers in artificial intelligence have studied planning issues extensively, and their work is summarized in several surveys [7] [6]. The planning technique described here is a simple variant of goal regression, and is not intended to contribute to the planning literature. However, we do believe that this paper presents a novel application for planning technology. Our approach differs from traditional planning by precompiling all necessary plans prior to execution. Schoppers also precompiles plans, though his methods differ somewhat from ours and he applies his work to the robotics domain [15].

Others have applied planning to computer animation. For example, Lengyel uses graphics hardware to accelerate path planning, and then animates his results [12]. Koga uses special purpose planning techniques to compute grasping animations [11]. None of this work exploits precompilation, nor is integrated in a UIMS.

THE LANGUAGE

This section describes the language used to express the desired run-time behavior of the animation controller. Presenting this language here serves two purposes. The language helps us communicate the capabilities of the controller. Also the language has been refined over time to make behavior specification easier, so our design choices should be of general interest.

There are five components to the language. Recall that the animation controller accepts high-level animation events and outputs animations scripts. So the language must contain both event and script definitions. The language also contains constructs for defining state variables that represent animation state, autonomous actions called autoscripts, and a state class hierarchy that makes defining preconditions easier. Each of these language constructs will now be described in turn.

State variables

State variables represent those components of the animation configuration that may need to be considered when determining whether a script can be invoked. State variable definitions take on the form:

(state-variable name type initial-value <values>)

All expressions in the language are LISP s-expressions (thus the parentheses), and bracketed values represent optional parameters. The first three arguments indicate the name, type, and initial value of the variable. State variables can be of type boolean, integer, float, or string. The last argument is an optional list of possible values for the variable. This can turn potentially infinitely-valued types, such as strings, into types that can take on a limited set of values (enumeration types). Examples of state-variable definitions are:

(state-variable 'holding-note 'boolean false)

(state-variable 'posture 'string 'stand '(fly stand sit))

The first definition creates a variable called **holding-note**, which is a boolean and has an initial value of false. The second creates a variable called **posture**, which is a string that is initialized to **stand**. It can take on only three values (**fly, stand,** and **sit**), and this should be expressed to the system because in some cases the system can reason about the value of the variable by knowing what it is not.

There is a special class of state variable, called a time variable. Time variables are set to the last time one of a group of events was processed.

Autoscripts

Autoscripts make it easy to define autonomous actions, which are actions that occur typically continuously when the animation system is in a particular set of states. Examples of this would be having an animated character snore when it is asleep, or swing its legs when it is bored. Autoscripts are procedures that are executed whenever a state variable takes on a particular value. For example, to have the **snore** procedure called when a variable called **alert** is set to **sleep**, we write the following:

(autoscript 'alert 'sleep '(snore))

The third argument is a list, because we may want to associate multiple autonomous actions with a given state variable value. Note that though we typically bind autoscripts to a single value of a state variable, we could have an autoscript run whenever an arbitrary logical expression of state variables is true, by binding the autoscript to multiple variable values, and evaluating whether the expression is true within the autoscript itself before proceeding with the action.

Event definitions

For every event that might be received by the animation controller, an event definition specifies at a high-level what needs to be accomplished and the desired timing. Event definitions take on the form:

(event name <directives>*)

The term <directives>* represents a diverse set of statements that can appear in any number and combination. The :state directive tells the controller to perform the sequence of operations necessary to achieve a particular state. The single argument to this directive is a logical expression of state variables, permitting conjunction, disjunction, and negation.

This high-level specification declares the desired results, not how to attain these results. In contrast, the :op directive instructs the system to perform the operation specified as its only argument. The animation controller may not be in a state allowing the desired operation to be executed. In this case, the controller will initially perform other operations necessary to attain this state, and then execute the specified operation.

For example, the evBadSpeech event is received by Player whenever our animated agent cannot recognize an utterance with sufficient confidence. Its effect is to have Peedy raise his wing to his ear, and say "Huh?" This event definition is as follows:

```
(event 'evBadSpeech :state 'wing-at-ear :op 'huh)
```

When an evBadSpeech event comes over the wire, the controller dispatches animations so that the expression wing-at-ear (a single state variable) is true. It then makes sure that the preconditions of the huh operator are satisfied, and then executes it. Note that wing-at-ear could have been defined as a precondition for the huh operator, and then the :state directive could have been omitted above. However, we chose to specify the behavior this way, because we might want huh to be executed in some cases when wing-at-ear is false.

By default, the directives are achieved sequentially in time. Above, wing-at-ear is made to be true, and immediately afterwards huh is executed. The :label and :time directives allow us to override this behavior, and define more flexible sequencing. The :label directive assigns a name to the point in time at which it appears in the directives sequence. The :time directive adjusts the current time in one of these sequences. An example follows:

```
(event 'evThanks
       :op 'bow
       :label 'a
       :time '(+ (label a) 3)
       :op 'camgoodbye
       :time '(+ (label a) 5)
       :op 'sit)
```

As defined above, when the animation controller receives an evThanks event, Peedy will bow. The label a represents the time immediately after the bow due to its position in the sequence. The first :time directive adjusts the scheduling clock to 3 seconds after the bow completes, making this the time that the camgoodbye operator executes, moving the camera to the "goodbye" position. The second :time directive sets the scheduling clock to 5 seconds after the bow, and then Peedy sits. If Peedy must perform an initial sequence of actions to satisfy the sit precondition, these will begin at this time, and the sit operation will occur later. Note that these two timing directives allow operations to be scheduled in parallel or sequentially.

Four additional directives are used, albeit less frequently. The :if statement allows a block of other directives to be executed only if a logical expression is true. This allows us, for example, to branch and select very different animation

goals based on the current state. Occasionally it is easier to specify a set of actions in terms of a state machine, rather than as a plan. The :add and :sub directives change the values of state variables, and in conjunction with the :if directive, allow small state machines to be incorporated in the controller code. The :code directive allows arbitrary C++ code to be embedded in the controller program.

Operator definitions

Scripts are the operators that act on our graphical scene, often changing the scene's state in the process. Operator definitions are of the following form:

```
(op opname <:script scriptname>
           <:precond precondition>
           <:add postcondition>
           <:sub postcondition>
           <:must-ask boolean>)
```

This creates an operator named opname associated with the script called scriptname. The operator can only execute when the specified precondition is true, and the postcondition is typically specified relative to this precondition using :add or :sub. Since operators typically change only a few aspects of the state, relative specification is usually easiest. The :must-ask directive defaults to false, indicating that the planner is free to use the operator during the planning process. When :must-ask is true, the operator will only be used if explicitly requested in the :op directive of an event definition. An example script definition appears below:

```
(op 'search
    :script 'stream
    :precond '((not holding-note) and ...)
    :add 'holding-note)
```

This defines an operator named search, associated with a script called stream. The precondition is a complex logical expression that the state class hierarchy, described in the next section, helps to simplify. The part shown here says that Peedy cannot be holding a note before executing a search. After executing the search, all of the preconditions will still hold, except holding-note will be true.

Though we have so far referred to operators and scripts interchangeably, there are really several different types of operators in Player. Operators can be static scripts, dynamic scripts (procedures that execute scripts), or arbitrary code. In the latter two cases, the :director or :code directives replace the :script directive.

We can also define macro-operators, which are sequences of operators that together modify the system state. As an example, the hard-wake macro-operator appears below:

```
(macro-op 'hard-wake
          :precond '(alert.snore and ...)
          :add 'alert.awake
          :seq '(:op snort :op exhale :op focus))
```

The above expression defines a macro-operator that can only be executed when, among other things, the value of alert is snore. Here, the '.' ("dot") comparator denotes equality. Afterwards, the value of alert will be awake. The effect of invoking this macro-operator is equivalent to executing the snort, exhale, and focus operators in sequence, making Peedy snort, exhale, then focus at the camera in transitioning from a snoring sleep to wakefulness in our application. The :time and :label directives can also appear in a macro definition to control the relative start times of the operators, however, our system requires that care be taken to avoid scheduling interfering operators concurrently.

State class hierarchy

In the last two examples, the preconditions were too complex to fit on a single line, so parts were omitted. Writing preconditions can be a slow, tedious process, especially in the presence of many interdependent state variables. To simplify the task, we allow programmers to create a state class hierarchy to be used in specifying preconditions. For example the complete precondition for the search operator defined earlier is:

```
((not holding-note) and alert.awake and
    posture.stand and (not wing-to-ear) and
    (not wearing-phones))
```

Since this precondition is shared by five different operators, we defined a state class (called **standing-noteless**) that represents the expression, and is used as the precondition for these operators. This makes the initial specification easier, but also subsequent modification, since changes can be made in a single place.

Class definitions take the following form:

```
(state-class classname states)
```

State class hierarchies support multiple inheritance. Here, **states** is a list of state variable expressions or previously defined state classes. A state-class typically inherits from all of these states, and in the case of conflicts, the latter states take precedence. State hierarchies can be arbitrarily deep. The **stand-noteless** class is not actually defined as the complex expression presented earlier, but as:

```
(state-class stand-noteless
        '(stand-op (not holding-note)))
```

In other words, the **stand-noteless** class inherits from another class called **stand-op**. Figure 3 shows most of the state class hierarchy used in the Peedy system, with extra detail given for the descendents of **ack-op**. It shows the expressions that each class adds to those inherited from its parents.

We have found that the semantics of an application and its animations tend to reveal a natural class hierarchy. For example, for our animated character to respond with an action, he must be awake, and for him to acknowledge the user with an action, he must not have his wing to his ear as if he could not hear, and cannot be wearing headphones. These three requirements comprise the class **ack-op** (for

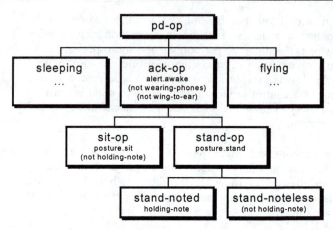

FIGURE 3. The state class hierarchy used in the Peedy system. Class names are in the larger font. Classes inherit state components from their ancestors.

acknowledgment operation), from which most of our operations inherit, at least indirectly.

ALGORITHM

Typical planning algorithms take a start state, goal state, and set of operators, compute for a while, then return a sequence of operators that transforms the start into the goal. Since our animated interface must exhibit real-time performance, planning at run-time is not an option. Instead, Player precompiles the plan-based specification into a state machine that has much better performance. This places an unusual requirement on the planning algorithm—it must find paths from any state in which the system might be to every specified goal state.

A naive approach might apply a conventional planner to each of these states and goals independently. Fortunately, there is coherence in the problem space that a simple variation of a traditional planning algorithm allows us to exploit. Our planning algorithm appears in Figure 4, and like other goal regression planners, works by beginning with goals and applying operator inverses until finding the desired start state (or in our case, start states). The algorithm is a breadth-first planner, and is guaranteed to find the shortest sequence of operators that takes any possible start state to a desired goal.

The basic algorithm consists of two procedures: Main and an auxiliary routine called SubPlan. Main begins by iterating over the goals specified in the event definitions (line 2). These include each argument to the :state directives, and the preconditions of the operators specified by the :op directives. Often the same goal appears in multiple event definitions, so we remove duplicates after putting the goals in a canonical form that makes identifying these duplicates easy.

Next, the algorithm sets up two queues, called ResultQ and WorkingQ. These queues both hold plans, which are pairs whose first element is a state, and whose second element is

```
1    Procedure Main ():
2        for each goal state, G, specified in an event definition do
3            WorkingQ := MakeEmptyQueue()
4            ResultQ := MakeEmptyQueue()
5            Enqueue([G, NULL], ResultQ)
6            Enqueue([G, NULL], WorkingQ)
7            SubPlan(WorkingQ, G, ResultQ)
8            RecordPlan(G, ResultQ)
9        end for
10   end procedure

11   Procedure SubPlan(WorkingQ, SolvedStates, ResultQ):
12       while (NotEmpty(WorkingQ)) do
13           RootPlan := Pop(WorkingQ)
14           RootState := First(RootPlan)
15           RootOps := Second(RootPlan)
16           if (Length(RootOps) >= MAXDEPTH) exit loop
17           for each state transforming operator O do
18               if (not(RootState => not(Postconditions(O))) do
19                   NewState := TransformState(RootState, O)
20                   if (not (NewState => SolvedStates)) do
21                       NewOps := Concatenate(O, RootOps)
22                       Enqueue([NewState, NewOps], ResultQ)
23                       Enqueue([NewState, NewOps], WorkingQ)
24                       SolvedStates := SolvedStates OR NewState
25                   end if
26               end if
27           end for
28       end while
29   end procedure
```

FIGURE 4. Regression-based planner, which finds plans from all possible states, leading to specified goals. The algorithm is explained in the text.

a sequence of operators that transform this state to the current goal. ResultQ will hold all the plans (ordered from shortest to longest) found for a given goal, and WorkingQ holds those plans that perhaps can still be expanded to form other plans. Both of these queues are initialized to contain the empty plan, representing the fact that when the start state is the goal state, no operators need be executed (lines 3-6).

The SubPlan procedure performs regression-based planning, beginning with the empty plan, sequentially adding operators to it in a breadth-first manner. It takes as arguments the two queues and the parameter SolvedStates, an expression representing the disjunction of the start states for which we have already found solutions. WorkingQ represents the current positions in the search. While there are still plans in WorkingQ, we remove the first element, call it RootPlan, call this plan's start state RootState, and its operator sequence RootOps (lines 12-15). It is from these roots that we try to build the next plans. If the length of RootOps is greater than a predefined MAXDEPTH constant, then we truncate the search (line 16). This guarantees that the algorithm will terminate when presented with an infinite search space.

The procedure now considers, in sequence, concatenating each operator onto the front of RootOps (line 17). Actually, it need only consider those operators that can change Root-State in some way. If the operator does not change Root-State, then we have already found a plan (namely RootPlan), that transforms RootState to the goal state in fewer operations. We need not consider operators with the same preconditions as postconditions. Also, operators whose :must-ask flag is true (see the Language section) need not be considered, since they cannot participate in the planning process.

For each of the operators that we need to consider, we ensure that its postconditions might possibly be true in RootState (line 18). If so, we calculate the effect of applying the inverse of the operator to RootState, and call this NewState (line 19). If NewState is not subsumed by Solved-States, then we need to build a plan for it (line 20). First, we construct the operation sequence that converts NewState to the goal state, by concatenating the new operation onto the front of RootOps to form NewOps (line 21). Next, we enqueue the plan represented by NewState and NewOps onto the ends of ResultQ and WorkingQ (lines 22-23). Finally, we express that we have now found a solution for NewState by setting SolvedStates to be the disjunction of the previous SolvedStates and NewState (line 24).

When SubPlan returns, ResultQ contains plans that take any possible state to the goal state (except those whose operation sequences would be longer than MAXDEPTH). Back in the Main procedure, these plans are recorded for this particular goal (line 7), and then Main continues to solve for other goals.

IMPLEMENTATION

The next step, after the planning algorithm finishes, is to build the actual state machine. Our system generates C++ code for the state machine, which is compiled and linked together with the Reactor animation library and various support routines. The heart of the state machine has already been calculated by the planner. Recall that plans are (state conditional, action sequence) pairs, which the planner computed for every goal state. These plans can readily be converted to if-then-else blocks, which are encapsulated into a procedure for their corresponding goal. These procedures also return a value indicating whether or not the goal state can be achieved. We refer to these procedures as *state-achieving procedures*, since they convert the existing state to a desired state.

Next, the system outputs *operator-execution procedures* for every operator referenced in event definitions. These procedures first call a state-achieving procedure, attempting to establish their precondition. If successful, the operator-execution procedures execute the operator and adjust state variables to reflect the postcondition. When multiple operators share the same precondition, their operator-execution procedures will call the same state-achieving procedures.

Finally, we generate *event procedures* for every event definition. These procedures, called whenever a new event is received from the application interface, invoke state-achieving procedures for each :state directive, and operator-execution procedures for each :op directive in the event definition. The :time directive produces code that manipulates a global variable, used as the start time for operator dispatch. The :label directive generates code to store the current value of this variable in an array, alongside other saved time values.

The planner and ancillary code for producing the state machine are implemented in Allegro Common Lisp, and run under Microsoft Windows NT. Our animation controller specification for the Peedy demo contains 7 state variables (including 1 time variable), 5 autoscripts, 43 operators, 9 state classes, and 35 event definitions. The planner had to solve for 13 unique goals, and the plans tended to be short (all were 5 operators or less). The system took about 4 seconds to generate a state machine from this controller specification on a 90 MHz Pentium.

It is important to note that in our Peedy application, not all animation is scheduled via planning. We have found that low-level animation actions, such as flying or blinking, are conveniently implemented as small procedural entities or state machines that are invoked by the higher-level animation planner. These state machines can be activated through autoscripts and the :director directive, and they can maintain their own internal state, or reference and modify the animation controller's state variables at run-time. As mentioned earlier, state machines can also be embedded into the animation controller using the event definition's :if directive. Our experience suggests that planning-based specification should not entirely replace procedurally based specification. The two techniques can best be used together.

CONCLUSIONS

Animation is of increasing interest to people building user interfaces, and tools must be created to facilitate incorporating animation in the interface. Animation actions or scripts often have strict dependencies on the state of the scene, so one of these tools should be concerned with tracking these dependencies and guaranteeing that they are satisfied. We have built such a tool, called Player, that uses a precondition and postcondition based specification to encode these dependencies, and automatically determine the sequence of animation actions that must be executed when a high-level animation event is received.

Effectively, this raises the level of the protocol used by the application interface to put graphics on the screen. On the input side, traditional UIMSs rely on a dialogue control subsystem to mediate between low-level input events from the presentation component and higher-level application services. However, on the output side, the application interface typically controls the screen state directly through low-level calls to the presentation interface. For complex presentations, such as animations, this low-level output dialogue is

inadequate, since it requires that the application concern itself with the detailed requirements of the animation scripts. The animation controller described here raises the level of the output protocol, hiding presentation details from the application, by serving as an output dialogue control component. By acting as a dialogue control for output, Player establishes a link from the dialogue component to the presentation component of the Seeheim model, which though called for by the model, is typically ignored by UIMSs.

Initially we implemented an animation controller specified entirely as a state machine, but as the controller grew, the state machine became difficult to maintain and enhance. The plan-based specification described here has proven easier to work with, in part because the system automatically calculates necessary transitions between animation states. Player has been incorporated within the Persona UIMS, and is employed by the Peedy application. In building Player, we made a number of design decisions to simplify specifying the animation. We have found the need to incorporate autonomous actions, as well as goal-directed behaviors. A state class hierarchy has simplified the process of specifying operator preconditions. In event definitions, we have found it useful to provide directives for both establishing new states, and invoking particular operators. Timing directives provide necessary control over the overlap and spacing of animation actions. It has proven convenient to allow small state machines and procedures to be embedded in the plan-based specification, providing a greater degree of expressivity within the easily-specified plan-based framework.

To provide real-time interaction, particularly as planning time increases, we believe it necessary to precompile the plan-based behavior specification into a state machine. This requires an algorithm that computes plans from all possible start states to all goal states expressed in event definitions. The paper describes such an algorithm, and the process for converting the resulting plans into a state machine. We have been happy with both the ease of specifying animation control, and the performance of this approach, for controllers of the complexity described earlier.

Although we have referred to Player as an animation controller, it truly can be applied to a wide range of presentation media. In the Peedy application, Player controls the sequencing of several media types, including animation, sound, and text. It could also be used to sequence static graphics, and general multimedia systems could benefit from this technology.

FUTURE WORK

This work suggests a number of important topics for subsequent research. Planning algorithms tend not to scale well, and planning research has investigated ways of dealing with this problem. We would like to determine the realistic maximum specification size and state space complexity that can be handled by our approach, and explore acceleration techniques that others have developed for traditional planning that might be also applicable to our plan precompilation method.

We have not adequately dealt with the problem of animation parallelism across events. In the current system, people can specify that two animation actions dispatched by a single event definition should execute in parallel, but they cannot indicate how two parallel event executions should affect each other. For example, if the system receives two events in rapid succession, one that requires Peedy to fly to a given location, and another requiring him to sit, there are several ways to interpret this sequence. The actions can be interpreted sequentially, allowing Peedy to sit only after he reaches his destination. The second action might preempt the first, making Peedy start to fly, and then sit instead. Alternatively the first action might override the second, circumventing the sit operation entirely. The actions could also be interleaved, with the character starting his flight, deciding to sit instead, and then reconsidering and completing his flight. A more complete animation controller would allow these alternatives to be expressed. Currently our application interface never sends conflicting events to Player simultaneously, however this should be allowed.

Several software engineering issues should be considered. We have no simple mechanism for automatically duplicating the name space of event, operator, and state variable definitions, to allow multiple identical agents to be controlled in the same scene. It would be helpful to have graphical tools for debugging and performing regression testing on plan specifications. We would like to build an animation authoring environment in which planning is performed dynamically during the development cycle, compiling plans only after the interface is debugged.

Projects are currently under way to add collision detection, inverse kinematics, and dynamics into our graphical environment. We hope to make our animation controller work seamlessly with these techniques. In addition, we would like to explore more sophisticated planning techniques, including temporal planning.

ACKNOWLEDGMENTS

The Persona and Peedy projects have benefited from the efforts of many group members. Andy Stankosky and David Pugh wrote Reactor. Tim Skelly modeled and animated Peedy. Maarten Van Dantzich wrote our graphical file format translator. David Thiel added sound to Peedy's environment, and Gene Ball dealt with natural language issues and wrote much of Persona's and Peedy's substrates.

REFERENCES

1. Ball, J. E. et al. Reactor: A System for Real-Time, Reactive Animations. In *CHI '94 Conference Companion.* (Boston, MA, April 24-28). ACM, New York. 1994. 39-40.

2. Card, S. K., Robertson, G. G., and Mackinlay, J. D. The Information Visualizer, an Information Workspace. In *CHI '91 Proceedings.* (New Orleans, LA, April 27-May 2). ACM, New York. 1991. 181-188.

3. Chang, B. and Ungar, D. Animation: From Cartoons to the User Interface. In *UIST '93 Proceedings.* (Atlanta, GA, November 3-5). ACM, New York. 1993. 45-55.

4. de Graaff, J. J., Sukaviriya, P., and van der Mast, C. A. P. G. Automatic Generation of Context-Sensitive Textual Help. Tech. Report GIT-GVU-93-11. GVU Center. Georgia Institute of Technology. 1993.

5. Foley, J. et al. A Knowledge-based User Interface Management System. In *CHI '88 Conference Proceedings.* (Washington, DC, May 15-19). ACM, New York. 1988. 67-72.

6. Genesereth, M. R., and Nilsson, N. J. *Logical Foundations of Artificial Intelligence.* Morgan Kaufmann, San Mateo, CA. 1988.

7. Georgeff, M. P. Planning. In *Readings in Planning,* edited by J. Allen et al. Morgan Kaufmann, San Mateo, CA. 1990. 5-25.

8. Gieskens, D. F., and Foley, J. D. Controlling User Interface Objects through Pre- and Postconditions. In *CHI '92 Conference Proceedings.* (Monterey, CA, May 3-7). ACM, New York. 1992. 189-194.

9. Green, M. A Survey of Three Dialogue Models. *ACM Transactions on Graphics 5,* 3. (July 1986). 244-275.

10. Hudson, S. and Stasko, J. T. Animation Support in a User Interface Toolkit: Flexible, Robust, and Reusable Abstractions. In *UIST '93 Proceedings.* (Atlanta, GA, November 3-5). ACM, New York. 1993. 57-67.

11. Koga, Y. et al. Planning Motions with Intentions. In *SIGGRAPH '94 Proceedings.* (Orlando, FL, July 24-29). ACM, New York. 1994. 395-408.

12. Lengyel, J. et al. Real-Time Robot Motion Planning Using Rasterizing Computer Graphics Hardware. In *SIGGRAPH '90 Proceedings.* (Dallas, TX, August 6-10). ACM, New York. 1990. 327-335.

13. Lewis, J. B., Koved, L., and Ling, D. T. Dialogue Structures for Virtual Worlds. In *CHI '91 Proceedings.* (New Orleans, LA, April 27 - May 2). ACM, New York. 1991. 131-136.

14. Olsen, D. R. Jr. *User Interface Management Systems: Models and Algorithms.* Morgan Kaufmann, San Mateo, CA. 1992.

15. Schoppers, M. J. Universal Plans for Reactive Robots in Unpredictable Environments. In *IJCAI '87 Conference Proceedings.* Vol. 2. (Milan, Italy, August 23-28). Morgan Kaufmann. 1039-1046.

16. Sukaviriya, P. and Foley, J. D. Coupling a UI Framework with Automatic Context-Sensitive Animated Help. In *UIST '90 Conference Proceedings.* (Snowbird, UT, October 3-5). ACM, New York. 1990. 152-166.

Bridging the Gulf Between Code and Behavior in Programming

Henry Lieberman
Media Laboratory
Massachusetts Institute of Technology
Cambridge, Mass. USA
lieber@media.mit.edu

Christopher Fry
Harlequin, Ltd.
1 Cambridge Center
Cambridge, Mass. USA
cfry@harlequin.com

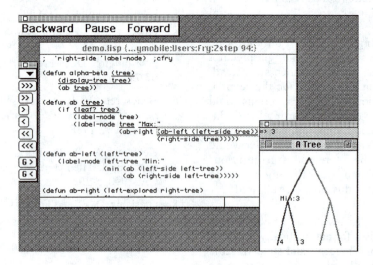

ABSTRACT

Program debugging can be an expensive, complex and frustrating process. Conventional programming environments provide little explicit support for the cognitive tasks of diagnosis and visualization faced by the programmer. *ZStep 94* is a program debugging environment designed to help the programmer understand the correspondence between static program code and dynamic program execution. Some of ZStep 94's innovations include:

- An animated view of program execution, using the very same display used to edit the source code

- A window that displays values which follows the stepper's focus

- An incrementally-generated complete history of program execution and output

- "Video recorder" controls to run the program in forward and reverse directions and control the level of detail displayed

- One-click access from graphical objects to the code that drew them

- One-click access from expressions in the code to their values and graphical output

KEYWORDS: Programming environments, psychology of programming, debugging, educational applications, software visualization

INTRODUCTION

Debugging accounts for about half of the estimated $100/line cost of a programmer's time [13], a major expense in the $92 billion US software market [6]. While the interface community has directed much attention toward improving interfaces for end users of applications, surprisingly little attention has gone toward improving the human interface of program debugging tools. Applying widely recognized human-computer interface principles to the problem can result in dramatic improvements in the effectiveness of debugging interfaces.

Donald Norman [12] refers to the *Gulf of Evaluation* and the *Gulf of Execution* as the gaps that often occur between the user's intent, the actual effect of a command given to the computer, and the result. In programming, this manifests itself as the difficulty of understanding the dynamic behavior of a program from the static source code of the program. The primary cognitive task in debugging is forming a mental model of the correspondence between code and behavior.

SUPPORTING THE COGNITIVE TASKS IN PROGRAMMING

Programming is the art of constructing a static description, the program code, of a dynamic process, the behavior which results from running the program. In that sense, it is analogous to composing music. The program code is like a musical score, whose purpose it is to cause the performer [in the programming case, a computer] to perform a set of actions over a period of time.

What makes programming cognitively difficult is that the programmer must imagine the dynamic process of execution while he or she is constructing the static description, just as a composer must "hear the piece" in his or her head, while composing. This puts a great burden on the programmer's short term memory. What makes programming even more difficult than composing is that a musical composition usually specifies a single performance, whereas a program may be executed in a wide variety of conditions, with different resulting behavior.

Many, if not most, routine program bugs result from a discrepancy between the programmer's imagining of the desired behavior in a given situation, and the actual behavior of the program code in that situation. [This would not be true only in the case of deep conceptual misunderstandings, where the program may actually perform as originally intended, but the programmer realizes that the original intention does not solve the problem.]

For the programmer, the problem of translating intent into program code corresponds to what Norman calls the *Gulf of Execution*. Interactive tools such as on-line context-sensitive help systems and syntax-directed editors can provide intelligent assistance in bridging that gap by relieving reliance on the programmer's memory of programming language details.

Once program code is written, the problem remains of verifying that the code written actually expresses the programmer's intent under all circumstances of interest. This is the *Gulf of Evaluation*. Interactive tools such as debuggers and program visualization systems can be invaluable in bridging that gap. Instead of trying to imagine how the events in a program unfold over time, why not have the machine show them to you?

We have designed a program debugging environment to explicitly support the problem solving methodology of matching the expectations of a programmer concerning the behavior of code to the actual behavior of the code. This environment is called *ZStep 94*, a descendent of the stepper described in [8].

PROBLEM SOLVING PROCESSES IN DEBUGGING: LOCALIZATION AND INSTRUMENTATION

Two principal activities in debugging that can be assisted by tools in the programming environment are *instrumentation* and *localization*. Instrumentation is the process of finding out what the behavior of a given piece of code is, the software analog of attaching oscilloscope probes to a hardware component. Traditional tools that assist instrumentation are trace, breakpoints, and manually inserted print statements: trace instruments all calls to a function, a breakpoint or print statement instruments a specific function call. The problem with using trace and breakpoints in debugging is that they require some plausible hypothesis as to where the bug *might* be, so you know where to place the instrumentation. They are not of much help when you have no idea where a bug might be, or there are too many possibilities to check individually.

Localization is the process of isolating which piece of code is "responsible" for some given undesirable behavior of a buggy program, without any prior knowledge of where it might be. Among traditional tools, a *stepper* is potentially the most effective localization tool, since it interactively imitates the action of the interpreter, and the program can in theory be stepped until the error is found.

However, traditional steppers have a fatal interface flaw: they have poor control over the level of detail shown. Since the programmer's time is too valuable to make looking at every step of an evaluation feasible, only those details potentially relevant to locating the bug should be examined. Typical steppers stop before evaluation of each expression and let the user choose whether or not to see the internal details of the evaluation of the current expression. But how can the user make the decision about whether to see the details of an expression if he or she doesn't know whether this expression contributes to the bug or not? This leaves the user in the same dilemma as the instrumentation tools -- they must have a reasonable hypothesis about where the bug might be before they can effectively use the debugging tools!

20-20 HINDSIGHT: REVERSIBLE CONTROL STRUCTURE

The solution adopted by ZStep 94 is to provide a reversible control structure. It keeps a complete, incrementally generated history of the execution of the program and its output. The user can confidently choose to temporarily ignore the details of a particular expression, secure in the knowledge that if the expression later proves to be relevant, the stepper can be backed up to look at the details. Thus, ZStep 94 provides a true localization tool.

There has been a considerable amount of past work on reversible program execution [1,3,8,9,11,15], but this work has concentrated on the details of minimizing the space requirements of the history and tracking side effects to data structures, both of which are important, but secondary to the user interface aspects which are the emphasis of this paper. It is important not only to "back up" variables to their previous values, but also to "back up" a consistent view of the user interface, including static code, dynamic data, and graphical output so that the user "backs up" their mental image of the program execution.

The reversible control structure aspects of ZStep 94 are discussed in more detail in [9]. We will address the issue of the computational expense of the history-keeping mechanism later.

ZStep 94's main menu uses a bi-directional "video recorder" metaphor. The single-arrow "play" and "reverse" correspond to single-step in a traditional stepper, and the "fast-forward" and "rewind" operation go from an expression to its value and vice versa, without displaying details.

Go to end of program ——————————— >>>

Show value of expression, without stopping —— >>

Single step ————————————————— >

Single step backwards ——————————— <

Back up from value to expression ————— <<

Go to beginning of program —————————— <<<

Single step "graphically" —————————— G >

Single step backwards "graphically" ———— G <

ZStep 94's "control menu"

It is important to note that ZStep 94's expression-by-expression stepping is *not* the same as statement-by-statement or line-by-line stepping found in many steppers for procedural languages. Individual lines or statements may still contain complex computations, and it is a severe limitation on the debugger if the granularity cannot be brought down finer than a statement or line.

The two "graphic step" operations G> and G< are an innovation that lets the user step the *graphic output of the program* back and forth, rather than control the stepper in terms of the expressions of the code. This will be discussed below.

ZStep 94 also has a "cruise control" mode, in which the stepper can run continuously, in either direction, without user intervention. The distance of the cursor from the center of the panel controls the speed. The user can stop it at any point when an interesting event appears, and run the stepper in either direction at that point.

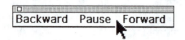

ZStep 94's "cruise control"

KEEPING THE DEBUGGING PROBLEM IN CONTEXT

A crucial problem in designing an interface for program debugging is *maintaining the visual context*. Because programming is an activity in which many items of interest have complex temporal and spatial relationships to other items, it is important to present each item with its context clearly identified.

Items such as the expression currently being evaluated, the value of a variable, or graphics drawn by the code may have almost no meaning outside their proper context. The programmer wants to know *where* in the code that expression was evaluated, *which* instance of the code was it, *when* did the variable have that value, *how* did that graphic appear on the screen?

If the item and its visual context are spatially or temporally separated, a new cognitive task is created for the user -- matching up the item with its context. This new cognitive task creates an obstacle for debugging and puts additional burden on the user's short term memory. Linear steppers or tracers that simply print out the next expression to be evaluated create the task of matching up the expression printed to the place in the code where that expression appears. "Follow the bouncing ball" interfaces that point to an expression and print out a value in another window lead to "ping-ponging" the user's attention between the code display and the value display.

FOLLOW THE BOUNCING WINDOW

Because most programmers input their code as text in a text editor, the primary mental image of the program becomes the text editor's display of the code. Thus, to preserve the WYSIWYG property, ZStep 94 always use the text editor's display to present code during debugging. To maintain visual continuity, it is important that the exact form of the user's input be preserved, including comments and formatting details, even if they are not semantically significant. As the interpreter's focus moves, the source code of the expression that is about to be evaluated, or has just returned a value, is highlighted.

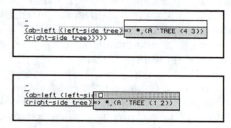

The value window moves through the code

Steppers always have the problem of how to show code expressions and their values simultaneously. In ZStep 94, as the editor's focus moves from expression to expression, we use a floating window to display the value. The display of the value is always exactly aligned with the expression to which it corresponds, so that the visual association between an expression and its value is apparent. The floating value window is colored light green to indicate if the expression is about to be evaluated, light blue to indicate a return value, or yellow if it has caused an error.

Earlier versions had used the idea of *substituting* the value for the code in place. This keeps the user's attention focused on the point of execution, but loses the original expression. Another version maintained two windows, one with the original code, the other with the substituted values, but this was affected by the "ping-pong" problem. All these are valid approaches, but we prefer the floating value window as the best compromise between visibility and maintaining visual continuity. We might also explore making the values window *translucent*, to reduce the effect of obscuring the code underneath.

WHAT DID THAT CODE DO?

Our approach is to integrate instrumentation tools directly into the stepper. We provide two facilities for pointing at a piece of code and inquiring about the behavior of the code. Rather than inserting breakpoints or print statements into the actual code and resubmitting the code to the interpreter or compiler, we let the user simply point at the desired expression, then run the stepper until that expression is reached. This is called *Step to Mouse Position*. This is like a breakpoint, but an advantage is that the stepper is runnable both forward and backward from the point where the program stops, and all information about the computation remains available.

```
ZStep Current Form
Eval Defun for ZStep
_____Modes_____
Show Value Under Mouse
Step to Mouse Pos
_____Navigation_____
Go to First Stepped Form
Go to Current Stepped Form
Go to Last Stepped Form
_____Misc_____
✓Auto-position Value Window?
Print Event Trace?
Current Form History
Make Values Filter
Delete Events After Current
Inspect Current Event
Help
```

ZStep 94's pull-down menu

Even more dynamic is *Show Value Under Mouse*, which is like a continuously updated Step to Mouse Position. The user simply waves the mouse around the code, without clicking, and the expression currently underneath the cursor displays its value window. Unlike Step to Mouse Position, this works only for values that have been previously computed and are quickly retrievable, and does not run the stepper past the current execution point.

WHAT HAS THAT CODE DONE?

We also provide a facility to track the behavior of a given expression over different execution histories. The operation *Current Form History* allows the user to point at an

expression and bring up a menu of the past values of that expression.

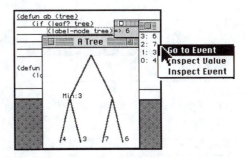

The history of values of an expression

Another history facility is the *Values Filter*, which brings up a menu of all returned values up to that point satisfying a condition. Clicking on one of the values returns the stepper to the corresponding event. We could also provide a filter on the expression executed, which would correspond to a traditional trace.

WHAT CODE DID THAT?

One of the most essential, but also most difficult, tasks in debugging is being able to reason backward from the manifestation of some buggy behavior to the underlying cause. This is especially problematic when the program in question itself has a graphical user interface. The programmer must work backwards from an incorrect user interface display to the code responsible. Traditional tools do not make any special provision for debugging programs with graphic output; worse, the user interface of the debugger often interferes with the user interface of the target program itself, making it impossible to debug!

ZStep 94 maintains a correspondence between events in the execution history and graphical output produced by the current expression. Considerable care is taken to assure that the graphic output always appears consistent with the state of execution. When the stepper is run forward or backward to a certain point in the execution, the graphic display is also moved to that point.

Furthermore, individual graphical objects on the display also are associated with the events that gave rise to them. We allow the user to click on a graphical object, such as a tree node in our example, and the stepper is automatically positioned at the event which drew that node. Just as in our other operations like Step to Mouse Position, the stepper is active at that point, and the program can be run forward and backward from that point.

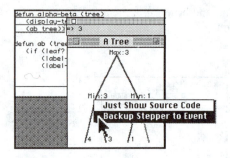

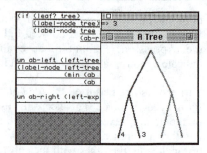

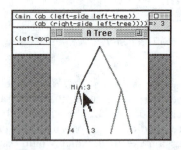

Clicking on a graphical object backs up the stepper to the event which drew it

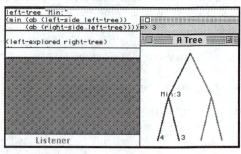

Four successive "graphic steps"

LET'S SEE THAT AGAIN, SLOWLY

In reasoning from the behavior of the program to the code, it is useful to be able to *step the behavior* rather than step the code. The user conceptualizes the behavior of the program as a set of graphic states that unfold over time, as the frames of an animation. The increments of execution should be measured in terms of the animation frames rather than execution of code, since events that happen in the code may or may not give rise to graphic output.

ZStep 94 provides two operations, Graphic Step Forward and Graphic Step Backward, that run the stepper forward or backward, respectively, until the next event happens that results in significant graphic output. Below, each graphic step results in an exploration of the next branch of the tree.

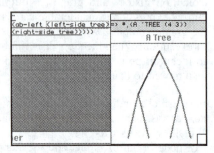

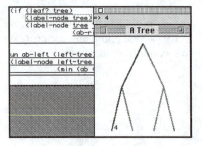

Each graphic step runs the stepper forward or backward until it is pointing at the event which was responsible for the graphic output, and the stepper remains live at all times. While stepping non-graphic code, the effects of previous graphic operations remain visible, just as they do in a normally-running program.

We could also provide graphic step operations analogous to "graphic fast forward" and "graphic rewind". Because the stepper can be run from either the code or the graphics at any point in time, the user can easily move back and forth between the different points of view.

ERROR CONDITIONS

ZStep 94 has a unique approach to dealing with execution errors. Traditionally, errors during program execution are disruptive. They either print an error message and halt execution, or put the user into a breakpoint loop, from which special commands can examine the error, and the stack can be inspected. In either case, errors disrupt execution and often lose information about partial computations which may have finished correctly.

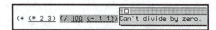

ZStep 94 displays an error message

In ZStep 94, an expression which results in an error simply displays the error message in place of the value of the subexpression most relevant to the error. The value window is colored yellow to indicate the error condition. The stepper remains active, all intermediate values are preserved, and the program can be run backward to examine the history that led up to the error.

The current version of ZStep 94 has no independent display of the stack, though the user can flip through events representing containing computations manually, whether or not an error has occurred. Previous ZStep versions had a stack display that was updated continuously with each event, animated in tandem with the source code animation. Each stack frame was itself a menu item, and clicking it would return you to that frame.

What happens once an error is found? ZStep 94 facilitates the repair phase, since it leaves you in the text editor with the cursor pointing to exactly the code in need of repair. Further, you are then just one click away from restarting the entire computation after the edit. However, we cannot support restarting the computation from any point in ZStep's history after the edit, because editing a running program cannot guarantee consistency of the event data structures. However, we could imagine techniques that would allow conservative restart of the code, or at least warn you of potential difficulties.

THERE'S NOTHING SLOWER THAN A PROGRAM THAT DOESN'T WORK YET!

By now, many readers will be thinking: isn't all this history-keeping and use of special-purpose interpreters ridiculously expensive? Histories eat up enormous amounts of storage, and interpreting code is slow. The answer is: yes, it can be expensive. But does it matter?

First of all, we have to keep in mind that the purpose of the stepper is to debug programs that don't work yet, and so worrying about optimizing execution is silly. Even in large programs, judicious testing can frequently quickly isolate a buggy example involving short executions that can be feasible to debug using complete history-keeping.

Nevertheless, we admit that there may be bugs that appear only after long runs involving large amounts of code, and so our techniques may be inappropriate in these cases. However, we conjecture that the vast majority of bugs are relatively shallow, and the productivity improvements from finding simple bugs quickly will far outweigh slower execution during testing.

Second, the key to making these techniques feasible over a wider range of programs are tools for selectively turning on and off history-keeping mechanisms. A simple way to automate selective processing of history in a common case would be to run the program normally until an error occurs. Then a program could use the stack inspector to determine what functions were involved in the error, and history-keeping would be turned on selectively for those functions. The problem would then be run again from the beginning.

Another common objection to our approach is that it is not guaranteed to work in the presence of programs that have side effects. First, we should observe that certain kinds of shallow side effects do in fact work without any special

provisions. Incrementing a variable will, because of the history, preserve the values both before and after the operation. More complex side effects involving shared data structures will not work, but one is no worse off than with conventional debugging techniques. More elaborate history-keeping mechanisms such as those studied in the simulation literature could alleviate this problem.

One of the principal uses of our stepper could be in educational applications, where execution efficiency is not of much concern, but interactive control over speed is important. One of the best ways to teach programming to a beginner would be to have the student step through example programs. In general, a stepper is an excellent tool for understanding code written by others.

IMPLEMENTATION

ZStep 94 was implemented in Macintosh Common Lisp 2.0. It is a prototype, and not a production implementation in several respects. First, it works with only a subset of Common Lisp. Second, no attempt was made to optimize speed or space constraints. Third, we did not perform any formal user testing, besides getting feedback from both experienced and novice programmers. Our main goal was to experiment with novel interfaces to dynamic program visualization.

Adapting ZStep 94 concepts to C would be possible, but challenging. A complete parser and unparser for C syntax would be required and care would have to be taken to assure the C interpreter or compiler kept enough type and run-time information.

The lesson for design of languages and environments is to consider debuggability as a primary criterion. If the goal of a new environment is to make programmers more productive, nothing could contribute to this goal more than introspective features that provide the foundation for sophisticated debuggers.

RELATED WORK

We're sad to report that there is not as much related work in this area as there should be. Even recently implemented programming environments seem to provide only the same set of tools that have been around for the past 30 years: trace, breakpoints, stack inspection, and perhaps a line-by-line stepper.

A notable exception has been the Open University in England, which has long been a source of innovative programming environments. The Transparent Prolog Machine [5] provides an innovative graphical view of program execution, and an interface carefully designed with Prolog's more complex execution model in mind. An innovative stepper for Lisp which shares some of the principles described here was recently implemented by Watt [14].

Many of the elements of our approach do have a long history. Reversible debuggers have been explored as far back as 1969 [3], and more recently by Moher [11].

However, these debuggers did not provide reversible animations of both the code and its graphical output, nor connections between individual code expressions, values and individual graphical objects.

The field of *visual programming* [7] uses graphical objects to represent the elements of the program, such as variables, functions and loops. The best of these environments also provide some animation of the graphical representation during execution. The pioneering work on animated visualization of program code in a single stepper was done by Ron Baecker [2].

Animation of visual representations of data manipulated by programs often appears under the name *algorithm animation* [4] [or *scientific visualization* if the algorithm represents a physical process]. Animations of data help a programmer visualize the dynamic behavior of a program as it runs. But most visual programming and algorithm animation systems confine themselves to visualizing and animating either the code or the data, but not both.

No one of these approaches -- reversiblity, animation of code, or animation of data -- by themselves will lead to a satisfactory set of debugging tools. ZStep 94's contribution is to integrate reversibility, animation of code, and correspondence between code expressions, values and graphic output, all under unified interactive control. Using the control structure of a stepper to control visualization of data helps solve one of the fundamental problems of software visualization: establishing the correspondence between data that looks faulty and determining the code that corresponds to the error. Adding data visualization facilities to the code visualization that steppers provide solves the problem of determining the effect of a particular piece of code upon the [sometimes complex] program state.

ACKNOWLEDGMENTS
Support for Lieberman's work comes in part from research grants from Alenia Corp., Apple Computer, ARPA/JNIDS, the National Science Foundation, and other sponsors of the MIT Media Lab, and for Fry's work from Harlequin, Inc. The authors would like to thank Marc Brown, John Stasko and Blaine Price, who ran last year's CHI Workshop on Software Visualization, and John Domingue for helpful suggestions.

REFERENCES
[1] Hiralal Agrawal, Richard deMillo, and Eugene Spafford, An Execution-Backtracking Approach to Debugging, IEEE Software, May 1991.

[2] Ron Baecker, Two Systems Which Produce Animated Representations of the Execution of Computer Programs, SigCSE Bulletin, February 1975.

[3] Robert Balzer, EXDAMS, EXtensible Debugging and Monitoring System, Spring Joint Computer Conference, 1969.

[4] Marc Brown, Perspectives on Algorithm Animation, Proceedings of ACM CHI'88 Conference on Human Factors in Computing Systems, p. 33-38.

[5] Marc Eisenstadt and Mike Brayshaw, The Transparent Prolog Machine: An Execution Model and Graphical Debugger for Logic Programming, Journal of Logic Programming 5(4), p.1-66, 1988.

[6] W. Wayt Gibbs, Software's Chronic Crisis, Scientific American, Sept 1994.

[7] E.P. Glinert, ed, Visual Programming Environments: Applications and Issues, IEEE Press, 1991

[8] Henry Lieberman, Steps Toward Better Debugging Tools for Lisp, ACM Symposium on Lisp and Functional Programming, Austin, Texas, August 1984

[9] Henry Lieberman, Reversible Object-Oriented Interpreters, First European Conference on Object-Oriented Programming, Paris, France, Springer-Verlag, 1987.

[10] Henry Lieberman, A Three-Dimensional Representation for Program Execution, in [7]

[11] Thomas Moher, PROVIDE: A Process Visualization and Debugging Environment, IEEE Transactions on Software Engineering, Vol. 14, No. 6, June 1988.

[12] Donald Norman, Cognitive Engineering, in Donald Norman and Stephen Draper, eds. User Centered System Design, Lawrence Erlbaum Assoc. 1986.

[13] Dennie van Tassel, Program Style, Design, Efficiency, Debugging and Testing, Prentice-Hall, 1974

[14] Stewart Watt, Froglet: A Source-Level Stepper for Lisp, Human Cognition Research Laboratory, Open University, Milton Keynes, England, 1994.

[15] M. V. Zelkowitz, Reversible Execution, Communications of the ACM, Sept. 1973.

Implicit Structures for Pen-Based Systems
Within a Freeform Interaction Paradigm

Thomas P. Moran, Patrick Chiu, William van Melle, Gordon Kurtenbach**

Xerox Palo Alto Research Center
3333 Coyote Hill Road
Palo Alto, CA 94304

{moran,chiu,vanmelle}@parc.xerox.com
gkurtenbach@alias.com

ABSTRACT

This paper presents a scheme for extending an informal, pen-based whiteboard system (Tivoli on the Xerox Live-Board) to provide a structured editing capability without violating its free expression and ease of use. The scheme supports list, text, table, and outline structures over handwritten scribbles and typed text. The scheme is based on the system temporarily perceiving the "implicit structure" that humans see in the material, which is called a *WYPIWYG* (What You Perceive Is What You Get) capability. The design techniques, principles, trade-offs, and limitations of the scheme are discussed. A notion of "freeform interaction" is proposed to position the system with respect to current user interface techniques.

KEYWORDS: freeform interaction, implicit structure, pen-based systems, scribbling, whiteboard metaphor, informal systems, recognition-based systems, perceptual support, list structures, gestural interfaces, user interface design.

INTRODUCTION

Our goal is to create computational support for the informal collaborative processes of small groups working together in real time. We are concerned especially with "generative" tasks (creating and assessing new ideas and perspectives, discussing them, playing with them, organizing them, negotiating about them, and so on). Human interaction in such situations is informal, freewheeling, rapid, and subtle.

Computational systems are typically ill-suited to such situations, because they force users to create and deal with more-or-less formalized representations. The overhead of using such representations inhibits the very processes they are meant to support [7,12]. One of the big challenges for current HCI design is to create systems for informal interaction.

Pen-based systems that allow scribbling on wall-size displays or notepads can support whiteboard or shared notebook metaphors for interacting with informally scribbled

material. The free, easy, and familiar expression permitted by such systems make them a promising class of tools to support informal interaction.

Our base tool is a large, shared, pen-based electronic display device called the *LiveBoard* [4]. We have developed a software system, called Tivoli [9], that simulates whiteboard functionality on the LiveBoard.[1] (There is a commercial version of Tivoli called MeetingBoard.[2]) This paper presents and discusses a new scheme that we have designed and implemented in Tivoli to extend its editing power, while yet remaining simple, natural, and consistent with the informal nature of the tool.

This paper begins by proposing the notion of "freeform interaction" to help pin down what we mean by "informal." Then we describe the extended editing scheme, which is based on the system perceiving the "implicit structure" that humans see in the material. This is followed by a discussion of the design principles, trade-offs, limitations, and comparison to other systems.

FREEFORM INTERACTION

The notion of "informal interaction" is somewhat vague, and so we define a more operational notion. A graphical editing system allows a user to manipulate graphical objects (GOs) that have defined positions in a 2-D space. A *free graphical object* (freeGO) is a GO that has no constraints or structural relations with other GOs; it can be freely operated upon independently of any other GOs in the space. Any kind of GO — such as an ink stroke, a text character, an icon, or a composite GO — can be a freeGO. Typical operations are drawing, erasing, wiping, dragging, and gesturing (for both selecting and operating). A representation consisting solely of freeGOs is a *freeform representation*. The unconstrained

---

1. Tivoli is written in C++ and runs under Unix and X Windows on Sun-based LiveBoards and on Sun workstations.

2. MeetingBoard is being developed and marketed by Xerox's LiveWorks, Inc. It runs under Microsoft Windows on PC-based commercial LiveBoards and on PC workstations.

---

*Patrick Chiu is with LiveWorks, Inc., A Xerox Company, 2040 Fortune Drive, San Jose, CA 95131. Gordon Kurtenbach is now with Alias Research Inc., 110 Richmond Street East, Toronto, Canada M5C 1P1.

interaction enabled with such a representation is *freeform interaction.*[3]

Scribbling is a prime example of freeform interaction: In scribbling, strokes can be created (drawn) anywhere without affecting existing strokes. Any strokes can be changed or erased without affecting any other strokes.[4]

In contrast, traditional text editing is not freeform, because there is an underlying string structure among the characters; e.g., deleting a character affects the positions of all characters later in the string. To be freeform, characters would all have to be freeGOs, i.e., have no underlying string structure. Erasing some characters would not cause any other characters to move. Such a seemingly limited model of text would have a crucial advantage: characters and strokes could be freely intermixed.

IMPLICIT STRUCTURES

During freeform interaction, users create material on the display in order to react to it and see new relationships and to try new things with it by manipulating the material. There is — *in users' minds* — quite a bit of structure. The structure may be deliberate or it may emerge from the interaction. It may be partial and ill-formed. Because interaction is freeform, there is no constraint to remain within the conventions of a particular structure. There is often a mixture of structures on the display. Users' commitment to particular struc-

---

3. We are aware that the notion of freeform needs much more discussion. We present it here briefly to introduce a seemingly useful notion and to position the specific scheme described in the paper.

4. Strokes also have the property of being *freehand* (being able to take on arbitrary shapes based on the movement of the creating instrument). This is not to be confused with their being freeform (a property of the ensemble), which is the relevant property here.

---

tures is often tentative and transitory, with the material being "re-registered" as different kinds of structures.

Such structure is *implicit* in the sense that it is perceived by the user but not by the system, because it is not defined or declared to the system. Keeping structure implicit is the essence of freeform interaction. The benefit is that users have the freedom to treat the material any way they want at any time; the cost is that the system cannot take advantage of the implicit structure in supporting the users' operations. Therefore, we would like the system to automatically perceive structure in the material in order to support a *WYPIWYG* (What You Perceive Is What You Get) capability [11]. But it is crucial to have the system perceive structure in the material only when the user needs support, and to keep the interaction freeform otherwise.

Our experience with LiveBoards and whiteboards is that list-like structures are ubiquitous. Thus, we set out to support the manipulation of four kinds of list-related structures:

- **lists** (vertically aligned sequences of items, which are horizontally clustered sets of GOs),
- **text** (horizontally aligned sequences of GOs),
- **tables** (arrays of elements, which are clustered GOs, aligned both horizontally and vertically),
- **outlines** (lists with indented items).[5]

The system need only perceive enough structure to support the "naturally expected" behavior of each of these structures.[6]

---

5. Because of space limitations, we will say little about outlines in this paper. The design principles can be illustrated by our treatment of the first three structures.

6. "Full" recognition, such as handwriting recognition or "parsing" by a "visual grammar" (e.g., [6]), is not required.

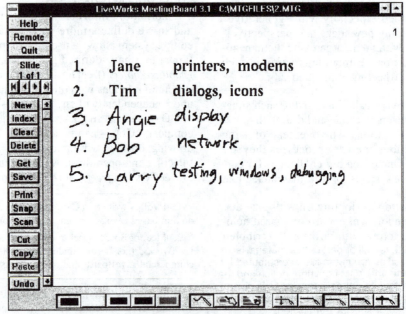

Figure 1. The commercial version of Tivoli: the Xerox LiveWorks MeetingBoard display.

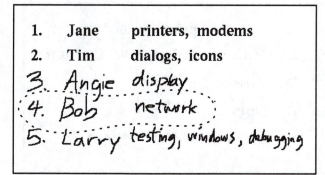

Figure 2a. Freeform move: User makes a loop gesture to select item 4, then drags the selection to before item 2.

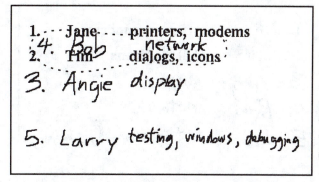

Figure 2b. Result: The selection remains exactly where the user left it, with no further adjustments.

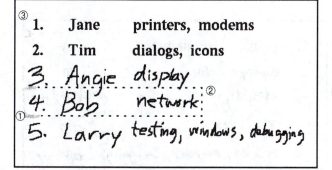

Figure 3a. Structured list move: User makes a left bracket gesture ①, which the system projects rightward into a structured selection ②. The wedge gesture ③ indicates the desired destination of the move.

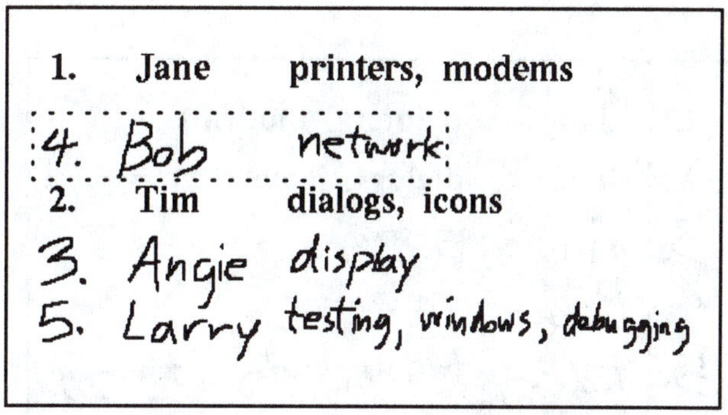

Figure 3b. Result: Items 2 and 3 are moved downward to open up space; item 4 is moved into the space.

The general design technique is to embed *ephemeral perceptual support* within freeform interaction: Whenever the user takes an action that implies a structural interpretation, the system temporarily perceives the structure in the material, carries out the current operation according to the expected behavior of that structure, and then returns to freeform interaction. Before discussing specific design techniques, we illustrate how our scheme works.

HOW THE BASIC SCHEME WORKS

Figure 1 shows the MeetingBoard display on the LiveBoard, which contains a mix of handwritten and typewritten material. The material is freeform, i.e., stroke and character freeGOs, but you can clearly perceive a lot of structure in it. It can be seen as a list of five items. Each item is numbered. After each number is a name. The names can be seen as a column. Horizontal clusters of characters and of strokes can be seen as words.

Let us see how to manipulate this material. Figure 2 shows a freeform move. First, a segment of material is selected by drawing a loop around it (Figure 2a); then it is dragged dynamically to a new location (Figure 2b). Because it is a freeform move, the dragged material stays where it is dropped, and none of the material around is adjusted. This is not satisfactory if the material is regarded as a list.

To deal with the material as a list (Figure 3a), the user first indicates the intent by using a structural selection gesture: a bracket ("[") gesture at the left of the list item. The system projects the legs of the bracket to the right to enclose the whole list item. The resulting rectangular selection enclosure signals that the system regards this as a structural selection. Then the user makes a wedge (">") gesture to tell the system where to insert the selected item. After the wedge gesture is made, the system opens up space for the item, moves the item, and closes the space where the item used to be (Figure 3b). The system animates all of the movement, so that the user, and the other people the user is working with, can easily track and understand the move.

To move a phrase, left and right bracket gestures are made to select the phrase (Figure 4a). Then a caret ("^") gesture is made to indicate a textual insertion. The system animates the moving of the phrase as well as the closing and opening up of spaces (Figure 4b).

To move a column, the material must be treated as a table. A top bracket indicates a column, the legs of the bracket projecting downward to select it (Figure 5a). A caret gesture shows where to move the column. Again, the system animates the move. All of the material to the right of the selected column is regarded as a column and moves left to close up the vacated space (Figure 5b).

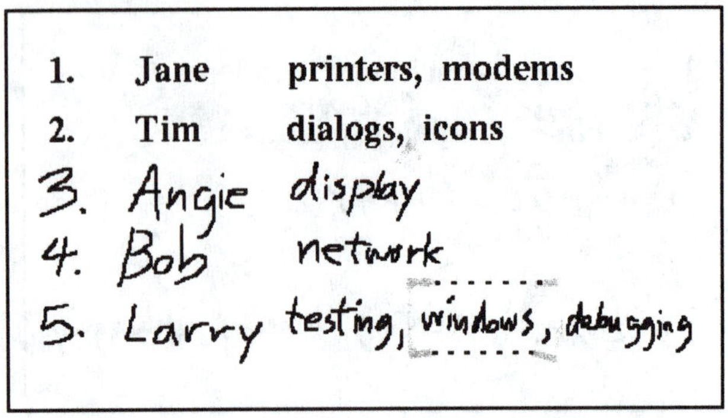

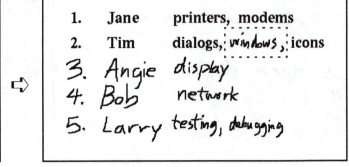

Figure 4a. Text move: User selects a word by making a left bracket gesture followed by a right bracket. The caret gesture indicates the desired destination of the move.

Figure 4b. Result: Space is opened up in item 2, the selected word is moved into the space, and the vacated space is closed up in item 5.

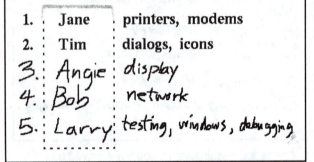

Figure 5a. Column move: User makes the "top bracket" gesture, then a caret gesture to indicate the destination of the move.

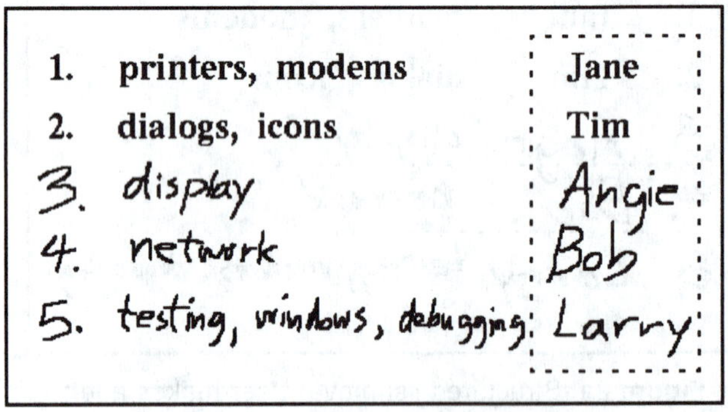

Figure 5b. Result: The selected column is moved to the right, and the space it vacated is closed up by moving the remaining objects leftward.

Any of the structural moves can also be made by dragging. To move a list item, the item is selected (as in Figure 3a) and dragged to a location near to where it is to be inserted (as in Figure 2b, except with a rectangular selection). Then the system animates the opening of space for the item, moving it from its dragged position, and closing the space where it started (the result being like Figure 3b).

Wedge and caret gestures indicate whether to insert a selection as an item or as text. In the case of dragging, the type of insertion is determined by where the selection is dragged to. If it is dragged to the gap between two lines, it is inserted as a list item; if it is dragged to a point within an item, then it is inserted as text.[7]

DISCUSSION OF DESIGN FEATURES

In developing the implicit structure scheme, we have been led to many unusual design decisions by our goal of working within a freeform interaction paradigm. Most of these decisions were arrived at during a process of iterating and exploring many alternatives. In this section we discuss some of the important features of the scheme.

Ephemeral Perception

The defining feature of implicit structure is that the system's perception is ephemeral — it is only in force for the duration of the immediate structural operations. The user interface issue is how to temporarily evoke the perceptual mechanisms in an effortless manner. Evoking structure is done by *gestural triggering* — making a structural selection gesture. The rectangular shape of the selection enclosure[8] makes it visually apparent that the system is ready to treat the next operation as a structural operation. After the user operates on a structural selection (e.g., moves it), the structural selection remains for further operations. The user can revert to freeform by simply beginning to draw strokes, and the selection is "dismissed" as a side effect.

We might argue that these transitional actions are costless for the user, since the user has to do them in any case. But, before making a selection the user must decide whether to make a structural or a freeform selection. The cost is mental (a choice has to be made [2]); and occasionally users make errors (e.g., selecting freeform but expecting a structural move to occur). We feel the benefits outweigh the costs.[9]

---

7. The drag "point" is the location of the stylus or cursor at the end of the drag.

8. In contrast, a freeform selection is enclosed by a freehand loop.

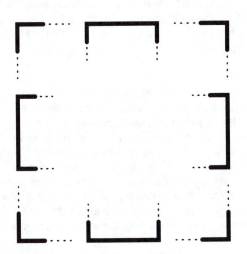

Figure 6. The eight structural selection gestures. Each gesture can be used to select the material enclosed by the projection of the legs, or to modify (extend or shrink) an existing structural selection.

Composite Structural Model

Each of the list-like structures we support has unique features; but they also share many common features, such as horizontal-vertical alignment, sequential elements, compactness (i.e., space preservation between items), and so on. Thus, we treat these four structures as a single *composite structural model*, rather than as four different models. There are two reasons: uniform selection and delayed commitment.

There is a uniform set of structural selection gestures, brackets and L-shaped gestures (Figure 6). These gestures work by projecting from their legs to define a rectangular region. This reduces structural selection to simple geometry, i.e., defining a rectangle. The user does not have to commit to a particular structure when a selection is made; the structure is not determined until an operation on the selection is invoked. For example, when an item in a list is selected, it can be either moved to another position in the list (with space opened vertically to make room for it) or to a place within another item (with space opened horizontally to make room for it).

Character-Based Structure

Characters are freeGOs in this scheme. Characters are created either by being imported from a text source or by being typed in.[10] In either case, they align nicely and look like structured text. But there is no underlying structure. If an eraser is swept through a neatly-aligned array of characters (such as those on this page), the characters touched by the

eraser would be deleted, but none of the other characters would move; and the text-like appearance would be damaged.

Character freeGOs can be made to behave like text by invoking implicit structures. Structurally selecting a horizontal sequence of characters and then moving them will cause space to be opened and closed in a text-like manner. Typing is treated as an implicit structure operation. Making an ink dot on the display creates a type-in point; if the dot is near some characters, it will "snap" into a position so that the typed characters align with existing characters. If there are characters immediately to the right of the type-in point, they are moved to the right to accommodate the newly-typed characters.[11]

Generic Commands

Most of Tivoli's basic editing commands are generic (polymorphic); the same commands can be applied, with somewhat different but appropriate results, to different freeGO types and structures. For example, the same pigtail (delete) gesture applies to both strokes and characters, in either freeform or structural selections (in the freeform case the selected freeGOs are simply deleted, but in the latter case there is a further side-effect of moving surrounding freeGOs to close up the space). Erasing is always interpreted to be a freeform operation, and typing commands (character-creating and spacing commands) are always interpreted to be structural operations.

Borders & Multiple Structures

Consider a display in which a sketch is made to the right of a list. Structural operations are not possible on the list, because the system cannot perceive how much of the material on the display belongs to the list. If a left bracket gesture is made, it will project all the way to the right into the sketch. If the resulting "item" is then moved, the sketch will be altered in a nonsensical way (since it is not a list).

This situation is handled by the concept of *borders*. Very long strokes are considered to be borders that divide the display into regions. Borders delimit structures. Thus, structure operations stay within the confines of borders. In the exam-

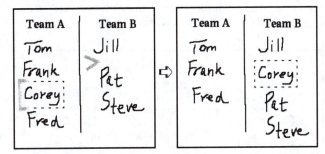

Figure 7. A "border" stroke confines the structural interpretation of selection and movement commands to one side of the border.

---

9. There is another important distinction for the user in Tivoli: The user must press a pen button to indicate that a stroke is a gesture. This sometimes causes errors. The benefit is that the user is totally free to draw anything without worrying whether some kinds of ink strokes are "reserved" as gestures.

10. Although it would be consistent with our scheme for characters to be recognized from handwriting, we do not support this, because it is not appropriate in the group work context of Live-Board use.

11. Similarly, the keystrokes Space, Backspace, Tab, and Return evoke familiar positioning operations. There are, of course, no "formatting character" (e.g., Space, Tab, Return) freeGOs in this scheme.

ple above, if a vertical stroke is drawn between the list and the sketch, then operations on the list are confined to the left side of the border, and the sketch will not be disturbed.

The most common use of borders is to divide the display into columnar regions for multiple lists. Structural operations occur independently in the different regions. For example, an item can be moved from one list to another across borders. Figure 7 shows two lists with an item selected in the left list. When the user moves the item to the right list, it will be inserted there; and the resulting opening and closing of spaces will be confined by the border so they don't interfere with each other.[12]

Cleanup Operations

Structures emerge from freeform interaction. Because freeform interaction is not constrained, material is not always rendered precisely with respect to a given structure; and thus the system may not perceive it in the same way as the user does. This can lead to some unpleasant perplexities.

To deal with this problem, we provide *cleanup operations*, which neaten up the alignment of material. In carrying out a cleanup operation, the system must decide whether elements are aligned or not. It is not so important what the system decides; what is important is that it makes the system's perception clear to the user. The user can then adjust those elements that were misperceived.

For example, the horizontal cleanup operation is useful for tables. Consider the table in Figure 8a, which is taken from our user test data. The user created the table column-wise, and hence the rows are not well-aligned horizontally. Note that it is impossible to select the first row of the table because it dips and the elements are crowded. The horizontal cleanup operation analyzes a table column by column, identifying the items in each column, finds correspondences between the items in the different columns, and decides

---

12. We also permit selections to be extended across borders. For example, in a table with vertical lines delimiting its columns, a "[" gesture selects up to the nearest border. Then a "]" gesture on the other side of the border extends the selection across it. In carrying out the operation on the extended selection (such as moving a row of the table), the border lines are treated as "ambient" in that they remain fixed during the operation.

what is in each row; it then respaces the elements to make the spacing between rows clear. The result in this example is shown in Figure 8b, where it can be seen that the first row is now easily selected.

Animation

Structural operations, especially moves, cause not only the selected material to move but also some of the non-selected material. It can be confusing when many objects are jumping to new locations. This is especially critical in the Live-Board situation, where one person is making the edits and other people are watching (the "demo phenomenon"). The operations on the display need to be *socially perceptible*. Our solution to this is to *animate* all the movements so that anyone watching can visually track the changes. (See [3,10] for a more detailed rationale for the utility of animation.)

Undo

The implicit structure scheme has evolved to be quite simple and natural. But this does not mean that anyone can walk up and use it without any practice. There are a lot of subtleties of interaction that must be learned through experience. Our experience is that it takes users about 10 minutes of practice to learn the basic Tivoli gestures plus about 15 minutes more to feel confident and comfortable with the implicit structure scheme. In LiveBoard situations, where users are in meetings when they first decide to use the structural operations, they cannot take time out to practice. Therefore, there is often a lot of initial fumbling. In these situations, we have found our unlimited *Undo operation* makes a huge difference. Even experienced users are occasionally surprised when the system perceives materials differently than they do (an inherent issue in this kind of technique), and the Undo reduces these potential disasters to manageable glitches.

THE SYSTEM'S PERCEPTUAL MECHANISM

The whole implicit structure scheme rests on the ability of the system to quickly perceive structure. The mechanism we use is based on computing a "projected ink profile." Figure 9 shows graphically the profile computed in both horizontal and vertical dimensions for the material from Figures 2 through 5. The peaks and valleys of the profile curves are analyzed to identify the structural elements and the spaces between them. While more sophisticated clustering techniques could also be used, this computational technique is

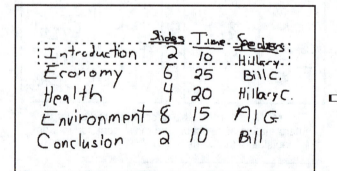

Figure 8a. This hand-drawn table has evolved to the point where the rows are too skewed and crowded to be selectable by a structured gesture.

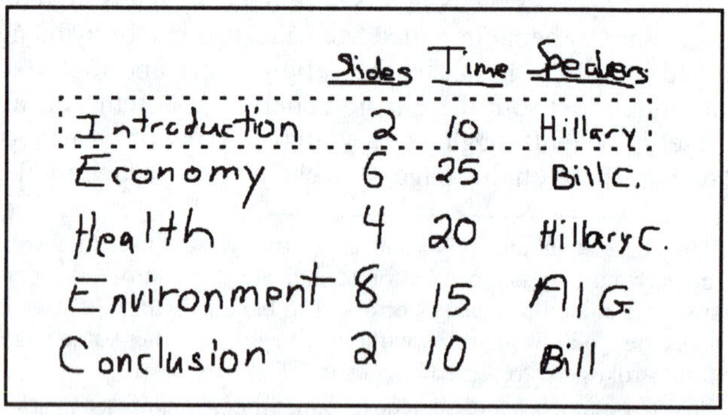

Figure 8b. After applying a "table cleanup" operation, the rows have been straightened out and spread apart, making them easily selectable.

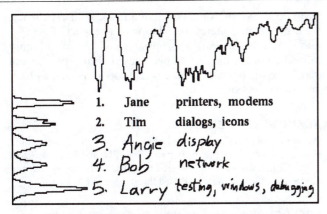

Figure 9. The projected-ink profile graphs at the left and top show the system's measure of the ink density in the x and y dimensions.

well suited to our purposes and has served us well. It is good at picking up the alignment relations that are inherent in list-like structures, while tolerating some degree of overlap between elements (e.g., the strokes in items 3 and 4 in Figure 9). It is efficient enough that structures can be computed on demand, which is critical for supporting ephemeral perception.

DESIGN HISTORY

Designing and iteratively refining the implicit structure scheme took place during a period of a little over two years. A variety of empirical tests were employed: experiences of the developers and close colleagues, independent user tests, and real use situations.

At first we implemented a "lined paper" method for handling lists (which is what other pen-based systems, such as [1,8], do). We set aside that scheme when we were satisfied that we could recognize lists on a blank surface. It took almost a year of iteration and refinement, with lots of testing within the Tivoli group before we had a version without "obvious" flaws. There were many design issues at this point that needed empirical evidence to help us address.

We conducted a small set of tests with independent users. In each session of about an hour, a user was trained until he/she was confident; then the user was given a range of tasks by playing the role of "scribe" in a simulated meeting. It took only four users before enough major problems were raised that we had to address. The problems involved confusion among the various structures (rows, columns, segments, blocks), which at that time were treated as different kinds of selection. During the next few months we redesigned and reimplemented the conceptual model and user interface: The structures were simplified to the composite structural model, and dragging and animation were added.

Another example of a problem raised in the user tests was that our early design for L-gestures was much too confusing to users. The original L-gestures are shown in Figure 10. The L-gestures were powerful operations for opening and closing spaces. Their design was perfectly logical: the order in which an L was drawn was significant; the first leg of an L indicated where the space was to be adjusted, and the sec-

ond leg indicated the direction and extent of the adjustment. But users could not remember this abstract logic. Therefore, we abandoned these operations[13] and made all L-gestures be simple projective selection gestures (Figure 6). Users have no trouble with these.

Once these changes were made, we had an opportunity to use the system in support of a series of real meetings, whose function was to collaboratively reorganize the priorities among large sets of items. To do this they used the ability to manipulate multi-column lists. These meetings went on for several months, and prompted the refinement of several minor aspects of the interface and the workings of the underlying recognition algorithms.

Finally, a subset of the implicit structure scheme (most of the aspects described in this paper) was chosen to be "hardened" and incorporated into LiveWorks' MeetingBoard product.

RELATION TO OTHER SYSTEMS

There are hints of implicit structure in current user interfaces, although the principles are not articulated or developed. For example, most word processors implicitly recognize words, and yet allow users to deal with text at the level of characters. Emacs [14] carries this idea the farthest. Emacs uses the character string as the base representation (analogous to freeGOs), and allows modes to be applied to the base (e.g., text, Lisp, or C++ modes). The structures are embedded in recognition routines and not data structures, and thus different modes can be applied to the same text. Lakin took inspiration from Emacs and extended this idea to the visual realm [6]. But structure in Emacs is not ephemeral; the user must explicitly declare modes. Further, modes are applied wholistically to the entire character string.

Pen-based systems promise informal interaction. Yet they mostly use the pen to input characters and then treat the text in the standard way. That is to say, they are not freeform in the sense defined in this paper. Perhaps the most notorious pen-based system today is the Apple Newton MessagePad [8]. The Newton uses two basic structures, character strings and structured graphics. However, once handwriting or drawing is interpreted, the interpretation is permanent. Thus

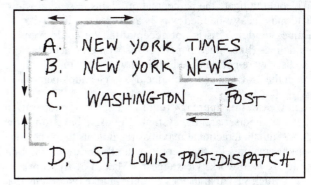

Figure 10. A set of structured command gestures, no longer used in the present system.

---

13. There are other ways to open and close spaces.

handwritten text cannot be treated in a structural manner, and strings of characters generally cannot be freeform. Also, characters and graphics cannot be manipulated together in a structural manner; e.g., a graphic object located within a character string will not be moved when characters prior to it are deleted.

The a*ha*! InkWriter [1] is perhaps the closest system to ours in its basic objectives. Its main goal is to treat handwriting as text. It supports text paragraphs and lists of handwriting or characters. Graphics are treated as separate paragraphs. The system compromises on freeform interaction, because it uses a "lined paper" background. Strokes are interpreted as handwriting only if they occur between the lines, and strokes are interpreted as graphics only if they are more than two lines high (and at least one space from a handwritten paragraph). Once input is interpreted, it remains as either text or graphics. Neither system exhibits the fluidity or flexibility that we feel is necessary for a truly usable informal (i.e., freeform) system.

LIMITS OF THE IMPLICIT STRUCTURE APPROACH

We should put the notion of implicit structure within freeform interaction in perspective. Freeform interaction is appropriate only in situations where constraints would inhibit rather than support a process. There are times when constraints are helpful. We would not be against "freezing" (making explicit) an implicit structure in such a situation. What seems needed is a way to transition from freeform to structured interaction [12]. Treating structure implicitly and ephemerally is useful in early, formative stages of a process.

But it should be understood that there are limits to the implicit structure approach.[14] Given the inherent freedom of expression in freeform interaction, it is difficult for users to stay within the confines defined by particular "visual grammars." Even if users mentally stick to particular structures, there are the manual problems of neatness, in which users vary considerably. These make system perception difficult in general. That is why we have chosen not to implement elaborate recognition grammars, but rather to "perceive" simple visual features (e.g., alignment) that are useful across different structures.

We could mitigate the perceptual problems in several ways. The most heavy-handed way is a structure editor (like an Emacs mode). This is not acceptable. We could provide guidelines (like "lined paper"). This would be acceptable if the user were not confined in rigid ways to them. We have taken the "softest" approach: the use of cleanup operations.

In any case, this approach requires a spirit of cooperation between the user and the system. The users have to follow good-faith "interactional maxims" (analogous to the conversational maxims [5] that people naturally follow) if implicit structure is to work. We suggest a Maxim of Appropriateness, which says that users will only invoke operations that are appropriate given the material at hand (e.g., they will not

try to do a list move on a sketch of a face). With experience, the user becomes more attuned to what to expect of the system (i.e., the sense of "appropriateness" becomes highly refined), and the interaction becomes skillful. The user can then have the benefits of structural support as well as freedom of expression.

ACKNOWLEDGMENTS

We would like to thank many colleagues — Eric Saund, Frank Halasz, Kim McCall, Steve Harrison, Sara Bly, and other members of the Collaborative Systems Area at PARC — for trying out and discussing this work as we iterated through various designs.

REFERENCES

[1] *aha! InkWriter Handbook* (1993). Mountain View, CA: aha! software corporation.

[2] Card, S. K., Moran, T. P., & Newell, A. (1983). *The Psychology of Human-Computer Interaction*. Hillsdale, NJ: Lawrence Erlbaum Associates.

[3] Chang, B. W., and Unger, D. (1993). Animation: from cartoons to the user interface. *Proceedings of UIST'93*, 45-55. New York: ACM.

[4] Elrod, S., Bruce, R., et al. (1992). LiveBoard: A large interactive display supporting group meetings, presentations and remote collaboration. *Proceedings of CHI'92*. New York: ACM.

[5] Grice, H. P. (1975). Logic and conversation. In P. Cole & J. Morgan (Eds.), *Syntax and semantics 3: speech acts*. New York: Academic Press.

[6] Lakin, F. (1987). Visual grammars for visual languages. *Proceedings of AAAI'87*, 683-688.

[7] Moran, T. P. (1993). Deformalizing computer and communication systems. Position Paper for the *Inter-CHI'93 Research Symposium*.

[8] *Newton MessagePad Handbook* (1993). Cupertino, CA: Apple Computer, Inc.

[9] Pedersen, E. R., McCall, K., Moran, T. P., & Halasz, F. G. (1993). Tivoli: An electronic whiteboard for informal workgroup meetings. *Proceedings of Inter-CHI'93*, 391-398. New York: ACM.

[10] Robertson, G. G., Card, S. K., MacKinlay, J. D. (1989). The cognitive co-processor architecture for interactive user interfaces. *Proceedings of UIST'89*. New York: ACM.

[11] Saund, E., & Moran, T. P. (1994). A perceptually-supported sketch editor. *Proceedings of UIST'94*. New York: ACM.

[12] Shipman, F. M., & Marshall, C. C. (1993). Formality considered harmful: experiences, emerging themes, and directions. Technical Report, Department of Computer Science, University of Colorado.

[13] Shipman, F. M., Marshall, C. C., & Moran, T. P. (1995). Finding and using implicit structure in human-organized spatial information layouts. *Proceedings of CHI'95*. New York: ACM.

[14] Stallman, R. (1985). *GNU Emacs Manual*. Cambridge, MA: Free Software Foundation.

---

14. In fact, these limits are probably inherent in all recognition-based systems.

Back To The Future: Pen And Paper Technology Supports Complex Group Coordination

Steve Whittaker
Lotus Development Corporation
One Rogers St
Cambridge MA 02142
whittaker@crd.lotus.com
+1 (617) 693 5003

Heinrich Schwarz
Science, Technology, Society Program
Massachusetts Institute of Technology
Cambridge MA 02139
schwarz@mit.edu
+1 (617) 497 5827

ABSTRACT

Despite a wealth of electronic group tools for co-ordinating the software development process, instead we find many groups choosing apparently outmoded *"material"* tools in critical projects. To understand the limitations of current electronic tools, we studied two groups, contrasting the effectiveness of both kinds of tools. We show that the *size*, *public* location and *physical* qualities of material tools engender certain crucial group processes that current on-line technologies fail to support. A large wallboard located in a public area promoted group interaction around the board, it enabled collaborative problem solving, as well as informing individuals about the local and global progress of the project. Furthermore, the public nature of the wallboard encouraged greater commitment and updating. However, material tools fall short on several other dimensions such as distribution, complex dependency tracking, and versioning. We believe that some of the benefits of material tools should be incorporated into electronic systems and suggest design alternatives that could bring these benefits to electronic systems.

KEYWORDS:

CSCW, ethnography, group work, co-ordination, group memory, interpersonal communications, media, software development.

THE PAPER PARADOX

There are strong intuitive reasons why software should be useful in the co-ordination of large teams in complex domains [8,15]. On-line tools allow rapid distribution of information supporting electronic access to project data and schedules. They enable updates for individual project members, allowing the schedule to be kept current and reducing time spent in meetings reporting progress. On-line scheduling tools also support computation for: (a) compiling and planning complex schedules; (b) generating

multiple views of data; (c) exploring dependencies between project subtasks; (d) hypothetical reasoning for planning.

Despite these compelling advantages, we found that in the company that we studied, a number of groups had *abandoned* on-line tools in favour of what seem to be outmoded technologies. Instead of on-line tools, we observed the use of large, publicly located, *wallboards*, with individual tasks and milestones written on paper slips which were then pinned to the boards (see Fig 1). These groups had access to, and expertise with, on-line project scheduling tools, email and workflow applications, but nevertheless preferred these "outmoded techniques" for managing software development.

Our empirical case study set out to address this paradox. We begin by examining the planning, co-ordination and communication processes involved in this type of complex multiperson project. We then contrast the effects of using *electronic* versus *material* co-ordination tools by comparing the software development process in two different groups. The first group relied on *electronic* scheduling tools and the other on the *wallboard* tool just described.

THE DEVELOPMENT GROUPS AND ORGANISATIONAL CONTEXT

The two projects involved large multidisciplinary teams of about twenty people, including software developers, user interface designers, quality assurance, documentation writers, and their managers. Both projects worked closely with external groups: in one case software components were being provided by an external group; in the other, the group itself produced software components for another product. Both groups also co-ordinated with other corporate entities such as marketing and upper management.

One group was developing a presentation package. For their development process they used a mixture of project bulletin boards, electronic mail, a scheduling tool (MS-Project™) and face-to-face meetings. Electronic bulletin boards and face-to-face meetings were used to compile design specifications for product features. These specifications were broken into a list of development tasks. These tasks, along with developer estimates of how long they would take to implement, were then combined into the scheduling tool and a schedule generated. Ideally, each week, individual developers reported their progress to their project manager via email. The manager entered the data into the on-line schedule. This data was then compiled and distributed via email, or hardcopy, to both project members and external groups. The output each person received consisted of the list of tasks for the entire project, with each task allocated to an individual, with start and completion dates attached, along with progress on each task.

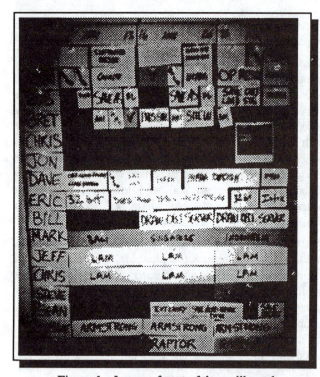

Figure 1: *Image of part of the wallboard*

The second group was writing software components to allow different software programs to interact together. This group had initially used an on-line scheduling tool, but had abandoned this, in favour of pen-and-paper procedures. During initial planning, project members entered both task descriptions on a large wallboard (Fig 1). The board depicted the names of all the people in the project vertically and a time-line horizontally. Against each person's name was a series of labelled pieces of card each representing different tasks: the length of the card indicated the expected duration of the task.

There were different coloured pieces of card to represent different task types, and the labels described the task. Each individual handwrote the labels for their own tasks. There were also various symbols to represent deadlines and review points. Unallocated tasks and an envelope containing notes about unforeseen tasks or potential problems were located in neutral areas of the board. People also pinned up photographs showing prior versions of the board, as well as cartoons, postcards and joke tasks.

The board was large: it was about 5ft high and 25ft long. It was located in a prominent public location in the hallway, so that team members walked past it 8-10 times a day, on their way to the printer, lunch and the bathrooms. It was also visible to external groups passing by the project area. Ideally, each week, group members reported their progress face-to-face, either with their project manager or in small project groups. They stood in front of the board, discussing the progress of old tasks already on the board for the last week, and planning the next week. They did this by moving around and cutting up the different pieces of card, to represent new tasks they had done or intended to undertake. In addition, some replanning meetings among subgroups of the project took place in front of the board.

The groups were similar in size, and both had participatory and open styles of management. One difference lay in their history: most of the presentation package group had worked together for over two years, whereas the components group had only been in existence for 6 months. Both groups predominantly communicated externally with other development groups, management and marketing in face-to-face meetings. The wallboard group sometimes met in front of the board for these meetings. There was no corporate requirement to produce an on-line schedule, although the presentation group distributed their electronic schedule to relevant external groups.

METHOD

We interviewed members of the two groups and their management individually using semi-structured questions. We asked participants to describe: (a) the software development process and their individual roles and responsibilities; (b) the planning and communication methods used by their group, with a focus on temporal co-ordination. We examined different artifacts they produced, eg. schedules, specification documents, bug reports and on-line project bulletin boards, and asked people to explain how these were used. We also analysed some opportunistic interactions among group members, and talked to participants as we watched their daily activities over the 14 week period of the study. Altogether between 10 and 20 hours of observation were carried out. We interviewed 30 people for between one and two hours. Our analysis is mainly based on transcripts of these recorded interviews, with the two initial groups, supplemented by interviews of other groups using similar tools. Most of the interviews were conducted after each group had completed a significant project goal, although observation work was carried out before and after this.

There are few established methodologies for case studies and the analysis of semi-structured questionnaires. We first analysed our interviews for descriptions of the software development process itself, its individual phases and the problems that arose in executing those phases. We collected every user comment about each phase, to see how the phase was affected by the medium. Thus for each phase, for both the *wallboard* and the electronic schedule, we collected the perceived benefits and drawbacks of the two scheduling methods. We present representative quotes from participants about these benefits and drawbacks, as well as participants' explanations about why they occurred. Where there was substantial disagreement or inconsistency between participants' opinions, then we present quotes representing alternative points of view.

THE SCHEDULING PROBLEM

Complex software development is a difficult process to manage and schedule accurately. In the software industry missed deadlines and late product releases are frequent. Nevertheless, schedules remain out of control, often to an extent beyond that in other industries:

"If this were a space mission and you underestimated by a factor of 2, you'd be off. But this is not. This is software design. Being off by a factor of 10 is bad, but not out of the realm of possibilities."

Our informants identified four major issues with scheduling their projects:

Coordination: There are multiple dependencies within and between projects, necessitating careful sequencing of tasks, and frequent communication about progress. This is to avoid people duplicating or redoing tasks, or being idle waiting for prerequisite tasks to be completed. Constructing a *shared schedule* is a technique for group co-ordination. It informs people inside and outside the group about what tasks should be completed by specific dates, allowing people to co-ordinate their actions accordingly. It therefore serves as a *plan*, about what the group and its members have committed to do. For a schedule to be an effective co-ordination tool, however, people must be able to understand it, and they must be committed to it. This requires careful work in generating, maintaining and modifying the schedule.

Initial planning: Planning realistic schedules for novel software is inherently difficult. The true extent of coding tasks may only emerge once coding has begun. Features may also need to be altered part way through development, in the light of marketing or UI feedback. Furthermore, given the competitiveness of the software market, there is political pressure for short (and sometimes unrealistic) deadlines. These factors mean that there is often a *discrepancy* between what was planned in the schedule for a given date and the actual state of the developed code, when development has begun. Repeated experiences with trying to execute unrealistic schedules can lead people to ignore the schedule, further reducing its utility as a co-ordination tool.

Updating: Given the uncertainty of the development process, discrepancies between schedule and reality are inevitable. The best that can be done is therefore to ensure that the schedule is kept *up-to-date.* This way at least people will be clear about what has been done and hence what remains to be completed in the original schedule. Without regular updates it may be unclear exactly where the project actually stands.

Replanning: In the course of development, new features are sometimes added to the initial design, often in response to competing products. Combined with the emerging discrepancies between schedule and reality, this means that *replanning* is a constant feature of the development process. It is therefore crucial as part of this process, to support *learning*: access to previous project data should enable people to be more accurate about the replanning process, by casting their new plans in the light of prior experience.

In what follows, we compare the utility of *material* versus *electronic* schedules for group *co-ordination*, based on their capacity for supporting *initial planning, updating* and *replanning.*

THE IMPACT OF DIFFERENT TOOLS ON THE DEVELOPMENT PROCESS

An immediately striking fact was that board was considered more "real" and "credible" than traditional electronic schedules.

"Somehow, it seems more real when it's on the board, I have to say. Perhaps, when I see it on the screen, I think here's another attempt to automate this stuff. The board somehow seems more real".

In contrast, weekly paper printouts of schedules generated from software tools lacked reality:

"Maybe because you get a new printout like every week and it's a little bit different. I don't know, it's like a copy. It's not the master database, it's not like you're looking at the real thing".

And because the software schedule is not considered real, it fails in its function as a tool for communicating information and co-ordinating actions:

"managers really sort of use the electronic software tools to derive schedules for the groups that they manage, and that, whether you intend it or not, that information doesn't get shared as well as it could ... The information conveyance seems to be somewhat inconsistent. Even though you distribute it to everybody... people may file it. Or people may not share. It just doesn't seem to sort of get out there in any sort of consistent manner... It's dead on arrival. "

Why then, was the board perceived and used differently from electronic tools? We will argue that three physical characteristics of the board afford a different style of co-ordination usage. First the board is *large* and *visual.* This promotes a straightforward representation of the global aspects of the complex scheduling problem. Secondly, the board is *public,* which means it can serve as a site for communication and collaborative planning, as well as engendering commitment to the schedule. Finally, the board is *material:* the fact that people physically manipulated and moved task cards leads to a more reflective style of scheduling activity. We now discuss the impact of these factors in the context of the scheduling problems described earlier: namely the generation, updating and replanning of the schedule, as well as its use for communication and co-ordination.

Initial Planning

The initial planning process is crucial for the accuracy and credibility of the entire schedule. Accurate planning is difficult, however, because of unforeseen dependencies between tasks, the iterative nature of code design, and pressure for short deadlines. One of the major differences between a *material* scheduling tool such as the board and a software schedule, lies in the way any kind of scheduling manipulation is performed. The board requires manual manipulation of paper artifacts, eg. cutting pieces of paper, writing on them, or taking them from their current location and putting them into a new location. These manual processes seem to encourage more thorough reflection about what one is doing, and the impact of one's actions on others. In contrast, the simplicity of changing numbers in electronic scheduling tools reduces this thoughtful handling of the estimations.

"in the mechanical process of cutting a task bar, maybe you're thinking of, 'Jeez, did I cut it right to really represent the duration?' And that may cause you to think about, 'Did I really represent this duration realistically? Did I talk to all my people about it?' And then in the act of posting. . . So all of that takes a certain amount of time, and in your mind, you could be thinking ... about some questions that relate to this. So, it tends to draw you in -- at least me -- a little bit more than doing it electronically, where I can just sort of read from my notes what it is, type it in, recalculate, and consider myself done with it. This [board] forces me ... It gives me the opportunity to think about the schedule a little bit more, because the process of building it requires more time."

The electronic form appears to make the process less concrete, and less reflective. It seems to promote a more cursory style of planning:

"And they'll say, Look this over and change some things. So you're not even looking at the date because it's too hard to try to really think about, so you're just looking at the numbers. Cross it out, you know, make 5, make that a 7, make this 3 a 2, and then you hand it back. That's how the process goes."

Furthermore, the concrete character of the wallboard representation gives the planned tasks more reality and permanence:

"having a pile of cards that have to be placed, it becomes clear what is possible and what not: you can't cheat. The cards don't go away when you wait a couple of minutes."

The public nature of the wallboard is another crucial characteristic that distinguishes it from electronic tools. This has at least two effects. First, transactions performed with it, are potentially done in a social situation. Second, the displayed information is permanently publicly visible. Together, these encourage collaborative planning which promotes awareness of the plans of others: this is vital given the number of dependencies between team members.

"the board ... seems pretty clearly to be a matter of different people's work, and that board couldn't have come out of one person sitting in his office and throwing something together that looks pretty fancy in an hour or two and that has things arranged in a certain way."

"[with the board] if you're leaving stuff out, it's possible to see it. Somebody will say, we forgot."

Public activities at the board also seems to have social consequences. Awareness of the public nature, as well as social interaction, seem to encourage reflection and commitment.

"there was something about the board and the arrangement of the tasks that said the developer ... was on the carpet discussing with the manager what these various things were and making certain contracts. I'm less impressed with that fact when I see you've got another Windows application doing it."

There are major limitations of the board for planning, however. Software can better represent the complex dependencies among tasks, that need to be included in initial planning. The current instantiation of the wallboard does not provide a way of showing the relations between tasks.

Updating

The utility of the schedule as a co-ordination tool is much reduced when it is not accurately updated, and there is a discrepancy between schedule and reality. Accurate schedules depend on frequent and honest updates. Updating includes reporting the state of completion of scheduled tasks, justifying slippage, and informing about what has actually been worked on during the last week, as opposed to what was planned. Updating requires work and motivation, however. Without any benefits from, or belief in the schedule, updating will be neglected, done superficially or used to conceal possible delays. When we first observed the uses of the board, the updating process was done meticulously. Emphasis was placed on recording what was really worked on and what the slippages were, reducing discrepancies between schedule and reality.

"This got changed a lot last week to be more of a reflection of how reality had progressed instead of what it was supposed to have been".

Several conditions can support or impede the updating process. As with planning activities, the material, social, and public qualities of the wallboard mean that people treat the updating process differently from software updating.

"just by changing the number on a spreadsheet with a couple of key strokes, as opposed to walking out the door, going down the hall, out in the public, where I move my date out... The public stuff seems to have heavier weight to it, in terms of the import."

The public and visible nature of the board also motivates updates because people want to be seen to be working:

"people .... got really nervous that they didn't have things after their name. They got jumpy.... So they're eager to get up there"

With software the process is organized in a way that seems to promote less responsibility and interest in updating:

"whereas the developers in my group couldn't care less about the dates. Really they see it as my role to track that and to wave a flag if the dates are going haywire or whatever."

The material nature of the board makes it harder, however, to do detailed updating. Every change means producing a paper strip of the right length, naming it, putting it on the board. There is a minimal size of the paper that limits tasks to one-day-length, so updating brief tasks becomes a laborious process. This may reduce update accuracy and frequency.

As against this, the public nature of the board promotes accuracy in other ways: Often updating happened between the developer and the manager, or in small groups. They would stand in front of the board, look at the last week's tasks and talk about what was actually worked on, changing the board accordingly. These conversations not only revealed where the individual developer really was, but also why delays had occurred and which unplanned additional tasks may have come up. The board therefore reflected the past and present state of the project after this type of collaborative updating. Furthermore, the updating process on the board was seen as the responsibility of the developers, since they had access to the schedule themselves. This resulted in project organisation being devolved from the manager:

"I like the fact that the developers are getting involved. They're starting to take some ownership of .... the way their tasks are... they've taken responsibility for all the tasks they're working on together, and they're beginning to reallocate them via each other as necessary."

Replanning

The frequent need for recalibrating the original schedule, and the fact that new features are often added in the middle of the development cycle, mean that replanning is a constant feature of the development process. Many of the benefits of the board in initial planning are equally true for replanning. One developer describes a situation where they had to change the plan in order to meet an upcoming

deadline. The interactive, negotiating, context-related, manual, and visual aspects afforded by the board become clear:

"we were rating tasks that were on the board, everything out past the deadline wasn't in, and we were able to pull things in and rearrange them... Everyone had a main task list in hand. We put them on the board, pushed everything out, got everything lined up, tried to get estimates from people on when they thought they could make it, realigned task estimates on things, added new tasks, added tasks for things like the stuff we discussed, filling in .... adding integration [time] .... saying, no, I've got too much on my stack, can you do it? Oh, Chris is doing this and that's related to this, wouldn't it be better if he took that and I gave this to Dave who's got something else that's kind of like that or related to that? There is a lot going on in half an hour. A lot of last-minute load levelling assignments, task definition, changing the size of the tasks, all happening in a very face to face [manner]."

Contrast the above process, with a description of replanning activity among two managers of a group that uses electronic management tools. The discussion takes place in the absence of the development teams. If the former situation can be called collaborative, this one can be titled private:

"we move tasks around between developers and I decide which tasks we move and how we move them ..I try to concentrate on moving tasks that are at the end of the schedule and tasks that, you know, different people could do equally well ... if this person has too many things and I can move this to this person, but not to this person, and then I can move something else from this person to this person and that sort of thing happens."

The board was also more flexible. People tacked up new unplanned tasks that occurred to them in the course of their current work, but which were unforeseen in the original plan. These were placed in neutral areas of the board. They could then be invoked in updating, which encouraged the board being used as a continuous replanning tool.

Given the uncertainty of the development process, learning from project history is essential when replanning during development. This process can be supported by a tool that helps recalibrate the original plan in the light of the actual work that was done and actual time needed. The board supports registering what actually happened. It can therefore give a systematic sense of which tasks were originally underestimated, which kind of overheads forgotten, and where in the schedule that delays occurred. The plan can be adjusted and refined once some of these factors are better understood.

"We're scheduled three days for it [a certain task]. We don't know if the three days are accurate, but we know it's a task that's going to be out there at the time. So we put an

estimate out there that's on the board ... that gives us a record so that we're learning from this as we go."

However, despite attempts to do this manually, using the board, a major advantage of software scheduling tools lies in their ability to keep track of the *set* of changes a schedule goes through over time. Different *versions* of the plan can be saved without much effort, reflecting different states of the project, while a material schedule loses older versions after each change.

"I know of no other way to capture the state of the board at a given frame of time other than doing it photographically. So, while it seems hokey, it may be the only practical thing to do."

Coordination and Communication

The central function of the schedule is to co-ordinate group collaboration. People need to be able to determine who is working on what task and when they will complete that task. So how does the board promote this inside and outside the group? Clearly one of the major advantages of the board schedule is its visibility. The board is big and colourful, offering a better project overview than with software printouts:

"I'd say that that's a big weakness of [MS-Project] ... the whole communication .... of a visual higher picture is totally missing"

"it's harder to flip through pieces of paper when you've got a large number of people, than it is to go out here [to the board] and see what everybody's up to and how we're converging on them [our goals]."

It also provides easy public access to the group's goals as well as individual's tasks and commitments, serving a crucial co-ordination function:

"it was good to see all of us up there from the beginning, and an idea of what we were going to work on and when. Because you came in and you knew what you were going to work on."

The board was used as a co-ordination tool between a local developer W, and another developer, X, who is collaborating from a remote site:

"[W] is sitting here -- he's phoning [X], who's also up on the board -- saying, well, this is where we are now, you've got this coming up and I'll get that -- why don't I take this and you take that? And then he [W] will get up and move them around on the board. And -- it's working real well as a -- a mechanism for communicating information about the project."

This same developer worked part time. He frequently used the board when he came back to work after a four day absence, to re-orient himself. He would look at last week's work, and his work for the coming week. He said it also helped him to get initial information about the work and progress of his collaborators before he talked to them in person.

Having access to this kind of information is a prerequisite for developers assisting each other. According to the requirements of the situation, developers may also help each other spontaneously.

"[the developers] are beginning to reallocate [the tasks] via each other as necessary. If somebody finds that they have more work coming up all of a sudden, something happened, if they had to go up and do something else, then they're working well to reallocate it"

Electronic schedules in contrast seem to be less a method for group co-ordination than private progress checking:

"so once a week, they'll send us an Email going, the new updated development schedule is there. I look at that probably every couple of weeks.[...] I just check every couple weeks to make sure I -- you know, I didn't forget something big"

The privacy of the electronic process has consequences for individual's awareness about the progress of others. Developers using electronic schedules reported little knowledge of how the project in general was progressing, nor about the details of others' work. This may arise from the immense amount of detail in electronic schedules, making it difficult to obtain a general overview. One schedule had 1200 tasks allocated between 19 project members, organised on a person-by-person basis, extending to 25 pages when printed out. Developers typically updated on-line by accessing their own tasks, which were often spread over several screens. The complexity of accessing their own tasks and the sheer amount of detail meant that they were unlikely to serendipitously notice the progress of others while updating.

The benefits of the board for internal communication partially extend to external communication processes. The visibility and public nature of the board, seem to promote certain types of external communications in a way that is harder to achieve with software:

"Bob from Product Z is ... scheduled to integrate our code. I came in two mornings ago and found him standing there in front of the board looking at it to see where things were, when he's going to be able to start doing his integration."

"I've seen a lot of schedules here which are on PCs and then get sort of -- never really make it to management; they never really see the project. They get reports -- But they don't see what that schedule [is] -- how it looks."

The central location of the board also offers a place for opportunistic communication. Although this doesn't happen frequently, activities around the board can draw the attention of other people and invite opportunistic conversations. Updating situations encourage people to watch and join in conversations.

The biggest limitation of a material schedule for external co-ordination however lies in distribution. The board, as one manager pointed out, *"is not very portable"*. This is not a problem for collocated team members, but for team members at remote locations, the difficulty of distributing the board becomes a major disadvantage.

(about a group member in another city): "He can't see this here. The best he can do is imagine what it might look like. I could maybe mail him a Polaroid [of the board], or I could scan one and send it in bit-map. [...] So he's getting his information coming very static on a one week granularity. That's not good.

Furthermore the information on the wallboard is difficult to integrate with other electronic information:

"I tried to get some [deadline] dates out of the [board] group, ... and their response to me was, come down and look at the wall..... the down side of it is the communications outside of the group, and because we're all so dependent on each other these days"

Other negative aspects to the board's public nature, are that sensitive or confidential information may be publicly displayed.

THEORETICAL AND DESIGN ISSUES FOR GROUP CO-ORDINATION TOOLS

The crucial function of a schedule is to facilitate *asynchronous group co-ordination* over extended periods of time. In general, attempts to create *electronic* tools to explicitly support *asynchronous co-ordination* have not been successful [1,4]. Our study suggests explanations for this, as well as ways to address these failures. As in previous research [1,4], individuals using electronic schedules perceived little direct benefit for the additional effort of maintaining the group tool. As a consequence, the electronic schedule was not carefully maintained, and its ability to serve as a co-ordination tool was compromised. In contrast, a key benefit of the board was that it converted the schedule from a tool that was used mainly by management, into a valuable personal and group resource. Why was the board used and perceived differently? Firstly, the board's ready *visual, public* availability made it easy to employ for personal and group reminding. It also provided a place and focus for synchronous group planning. In addition, there was greater belief in, and commitment to, the wallboard schedule. Belief and commitment arose from the *visual, public* nature of the board encouraging greater responsibility and the *material* interaction promoting reflective planning. Thus, the effort of updating and maintaining the wallboard schedule was outweighed by its benefits in enabling people to better plan and co-ordinate their work with others, based around a credible schedule.

Other research work has identified the importance of synchronous interpersonal communications in long term collaboration [6,13]. The *public* forum of the board provided an ideal physical location for arranged and opportunistic meetings. The board was not only a *place* to meet, however. An important secondary function of the board was in providing tools and support for such meetings. Our results extend other work on *synchronous collaboration* processes and tools. In synchronous meetings, the wallboard served as an *artifact* in promoting group problem solving, it helped record progress in complex tasks and provided *material, visual* conversational props to aid reasoning and discussion. Research on shared workspaces shows they offer some of the same benefits for distributed collaboration [9,10,12,14]. Work on support for collocated design meetings is also consistent with our findings here: it shows the importance of *large* recording surfaces, and some of the advantages of pen and paper over computer techniques in complex planning tasks [5,7].

Our current data, mainly based on interviews, indicate a higher degree of satisfaction, and greater use of the wallboard schedule[1]. This group completed code development to *deadline*, whereas the electronic group reported slips against the schedule. The two groups were developing different code however, so we should be careful about the generality of the conclusions we draw. Given the limitations of this case study, we need to supplement these observations with more systematic long term evaluations of the impact of the technologies.

We also need to look at different types of groups using the scheduling technology. This will help distinguish the effects of *inherent media characteristics* from the ways in which groups *chose to use the technologies*. In our current study these are hard to separate because we only looked at two groups. Thus for example, the big/visual characteristic of the board seems to directly promote overview information. In contrast, the fact that people chose to hold replanning meetings in front of the board is not the immediate result of media properties, but is part of the group operating principles. The public character of the board may facilitate such meetings, and it may have increased the likelihood that update meetings took place at the board, but people could clearly have decided to update and replan elsewhere. More research investigating groups with different culture and habits using these technologies is needed to disentangle media effects from group operating principles. Furthermore we have to understand the *types of* groups in which this type of technology is most useful. Is the tool as effective for groups where there are fewer dependencies between the activities of the members, and where tight co-ordination is less important? We also suggest a number of different potential benefits of the

---

[1] The perceived success of the board is supported by the fact that in the course of our four month study, of the six groups we contacted who were using MS-Project when the study started, a further two had abandoned software in favour of the board.

wallboard, and future experimental work should evaluate which of these benefits are most significant for the types of groups and activities we studied.

What are the implications of this work for the design of both scheduling tools and groupware in general? We have identified several limitations of electronic media when compared with existing paper based processes. Electronic tools may: (a) fail to promote face-to-face communications which have been shown to be critical for group collaboration [6,13]; (b) decrease awareness of the actions of other group members [1,3], which may reduce helping behaviour and collaboration; (c) suffer from lack of visibility and permanence which may compromise their function for reminding and co-ordination [11,12]; (d) fail to engender belief and commitment [4]. However, despite their benefits, material tools also fall short on several dimensions such as distribution, versioning and complex dependency tracking.

How then can we combine the benefits of electronic and material systems? One possibility is to use a pen-based large display such as [2] in a public location, running a modified scheduling application, that is updatable by all group members. This would satisfy the requirements of *size* and *visibility,* and allow *public* interactions. The pen-based interface would allow handwritten inputs, with tasks being placed and moved using direct manipulation techniques, to simulate *material interaction*. The benefits of software would be that the information could be remotely distributed. Coworkers at remote locations could also have Liveboards with fast audio (and possibly video) connections, so that collaborative planning and updating meetings could be held "in front" of the board at short notice. Furthermore the computational capabilities of the application could be used to depict dependencies, save previous versions of the "board" to facilitate learning, and integrate the data with other corporate information.

In future, we should design electronic group tools with the benefits of the material wallboard. Our electronic tools need to be *public*, to promote commitment and conversation; *material* in affording engagement and reflective use of the tools; and they need to simulate the dimensions of *size* and *visibility* in supporting ready access to complex information.

ACKNOWLEDGMENTS

Thanks to the two groups for participating in the study and to Irene Greif, Candy Sidner and Lyn Walker for discussions of this work.

REFERENCES

[1]Bowers, J. The work to make a network work. In *Proceedings of Conference on Computer Supported Co-operative Work*, 287-298, 1994.

[2]Elrod, S., Gold, R., Goldberg, D., Halasz, F., Janssen, W., Lee, D., McCall, K., Pedersen, E., Pier, K., Tang, J. and Welch, B. Liveboard: a large interactive display supporting group meetings, presentations and remote collaboration. In *Proceedings of CHI'92 Human Factors in Computing Systems*, 599-607, 1992.

[3]Gaver, W., Moran, T., Maclean, A., Lovstrand, L., Dourish, P., Carter, K ., and Buxton, W. Realizing a video Environment: Europarc's RAVE system. In *Proceedings of CHI'92 Human Factors in Computing Systems*, 27-35, 1992.

[4]Grudin, J. Why CSCW applications fail. In *Proceedings of Conference on Computer Supported Co-operative Work*, 85-93, 1988.

[5]Karat, J., and Bennett, J. Supporting effective and efficient design meetings. In J. Carroll (Ed.), *Designing Interaction*. Cambridge, 1990.

[6]Kraut, R., Fish, R., Root, B. and Chalfonte, B. Informal communication in organizations. In R. Baecker (Ed.), *Groupware and Computer Supported Co-operative Work*, 287-314, Morgan Kaufman, 1992.

[7]Kyng, M. Designing for a dollar a day. In *Proceedings of Conference on Computer Supported Co-operative Work*, 178-188, 1988.

[8]Malone, T., and Crowston, K. What is coordination theory and how can it help design cooperative work systems? In R. Baecker (Ed.), *Groupware and Computer Supported Co-operative Work*, 287-314, Morgan Kaufman, 1992.

[9]Nardi, B, Schwarz, H, Kuchinsky, A, Leichner, R, Whittaker, S., and Sclabassi, R. Turning away from talking heads: an analysis of "video-as-data". In *Proceedings of CHI'93 Human Factors in Computing Systems*, 327-334, 1993.

[10]Tang, J. Findings from observational studies of collaborative work. *International Journal of Man Machine Studies, 34*, 143-160, 1991.

[11]Walker, M. Experimentally evaluating communication strategies: The effect of the task. In *AAAI94*, 1994.

[12]Whittaker, S., Brennan, S., and Clark, H. Co-ordinating activity: an analysis of computer supported co-operative work. In *Proceedings of CHI'91 Human Factors in Computing Systems*, 361-367, 1991.

[13]Whittaker, S., Frohlich, D., and Daly-Jones, O. Informal communication: what is it like and how might we support it? In *Proceedings of CHI'94 Human Factors in Computing Systems*, 130-137, 1994.

[14]Whittaker, S., Geelhoed, E., and Robinson, E. Shared workspaces: how do they work and when are they useful? *International Journal of Man-Machine Studies, 39*, 813-842.

[15]Winograd, T and Flores, F. *Understanding computers and cognition*. Addison-Wesley, 1986.

Recognition Accuracy and User Acceptance of Pen Interfaces

Clive Frankish[1], Richard Hull[2], and Pam Morgan[1]

[1] University of Bristol
Department of Psychology, 8 Woodland Road
Bristol BS8 1TN, U.K.
Tel: 0272 288559
email: C.Frankish@uk.ac.bristol

[2] Hewlett Packard Research Laboratories
Filton Road, Stoke Gifford
Bristol BS12 6QZ, U.K.
Tel: 0272 799910
email: rh@uk.com.hp.hpl.hplb

ABSTRACT

The accuracy of handwriting recognition is often seen as a key factor in determining the acceptability of hand-held computers that employ a pen for user interaction. We report the results of a study in which the relationship between user satisfaction and recogniser performance was examined in the context of different types of target application. Subjects with no prior experience of pen computing evaluated the appropriateness of the pen interface for performing three different tasks that required translation of handwritten text. The results indicate that the influence of recogniser performance on user satisfaction depends on the task context. These findings are interpreted in terms of the task-related costs and benefits associated with handwriting recognition. Further analysis of recognition data showed that accuracy did not improve as subjects became more practised. However, substantial gains in accuracy could be achieved by selectively adapting the recogniser to deal with a small, user-specific subset of characters.

KEYWORDS: Pen-based input, handwriting recognition

INTRODUCTION

The development of pen interfaces is a key element in strategies for increasing the market for lightweight, hand-held computers. The small size of these devices precludes the use of conventional keyboards, and many are intended to appeal to users who do not have keyboard or computing experience. The preferred technical solution for products such as Personal Digital Assistants (PDAs) is often a graphical user interface in which a pen can be used for pointing and selection functions, drawing, and text entry. In this context, the pen is at least as effective as a mouse for direct manipulation of screen objects. In a well-designed pen interface, screen prompts, user action, and

system feedback will be integrated in an immediate and intuitive manner. Direct graphical input, such as freehand drawing, is also much easier with a pen than with a mouse. However, the virtues of handwriting recognition as a means of entering text are less certain. The idea of handwritten input is attractive, particularly for inexperienced computer users, but is offset by the need to correct recognition errors. There are also dissimilarities between human perception and machine recognition which can cause frustration when handwriting recognisers behave in ways the user does not understand.

In assessing the potential for pen based computing, the unreliability of handwriting recognition is the most obvious limiting factor. For system developers, improved accuracy is therefore a major goal. However, it is far from clear what we should set as a realistic target for this effort. There are also very few empirical data to indicate what gains we might expect in user acceptance for a given increase in recognition performance. One aim of the study described here was to provide some quantitative information about the relationship between recognition accuracy and user satisfaction, for different types of pen application. The results confirmed that this relationship is highly task-dependent. In the light of these findings, error data from the test applications were further analysed to assess gains in recognition accuracy that might be achieved as users become more practised, or through limited adaptation of the recogniser to individual users.

QUESTIONS ADDRESSED IN THE STUDY

How important is recognition accuracy in determining user acceptance of pen interfaces?

Few could disagree with the assertion that "the value of a handwriting recognition system is dependent on the degree to which the system can accurately interpret handwritten characters" [6]. This might be taken to imply that high targets should be set for acceptable levels of system performance, and this appears to be supported by data from some evaluation studies in which text copying tasks have been used to assess recogniser performance. In one such study, the general opinion of subjects who had achieved a mean accuracy of 93.2% was that the recogniser would not be of practical significance unless it could be made both faster and more accurate [8]. The impact of recognition errors must also be considered both in terms of the additional costs associated with error correction, and the acceptability of transcription errors in the completed text. In the present evaluation study, it was assumed that all recognition errors would be corrected. However, for note-taking or memo applications, this may not be an absolute requirement. Uncorrected text that is 97% correct may be judged acceptable for personal notes or for passing information to a secretary or colleague, although for communications to a superior, a higher criterion of 99% accuracy or better may be required [1].

User evaluation of handwriting recognition accuracy is not necessarily a reliable indication of the acceptability of an application in which the recogniser is used. In most pen applications, translation of handwritten text will be only one component of user interaction. Users are likely to judge the value of an application primarily in terms of its appropriateness for completing a task, rather than their satisfaction with this one aspect of pen function. The impact of recognition errors will therefore depend upon factors such as the amount of text entry required, the acceptability of uncorrected recognition errors in the final text, and the benefits of using handwritten text input as compared with other available methods of data entry. The importance of recognition accuracy can therefore only be assessed in a broader context that takes into account the nature of the task that users are trying to perform [4]. Within this context, evaluation studies should attempt (a) to establish minimum levels of acceptable performance, and (b) to assess the extent to which further gains in accuracy above this minimum level will result in increased user satisfaction.

To what extent is recognition accuracy under users' control?

The performance of handwriting recognisers is partly determined by individual differences in neatness, consistency, and usage of idiosyncratic letter forms. These attributes will also determine the gains in accuracy that can be achieved by using handwriting samples to adapt the recogniser to individual users. Recogniser 'training' will generally produce greatest benefits for users who produce letter forms that are idiosyncratic, consistent, and mutually distinctive.

For most individuals, handwriting characteristics are well established and stable. Indeed, one of the attractions of pen interfaces is the prospect of using familiar pen and paper skills for interacting with computers. However, handwriting skills do not transfer completely from paper to screen. New users must adjust to differences in the pen and writing surface, and more importantly, to constraints imposed by the recogniser. The amount of learning that occurs, the time it takes, and the effects on recognition

accuracy are all important in determining how potential users should be introduced to pen interfaces. Adjustment of motor skills is likely to be fairly rapid. However, deliberate modification of writing styles to improve recognition accuracy is more problematic, even for motivated users. Feedback from the recogniser could in principle allow users to develop an internal model of the recognition process, and to adapt their writing styles accordingly. If this were happening, recognition accuracy should improve with practice. Available data on practice effects suggest that this improvement is perhaps rather marginal. One reported study found no evidence that experience with a recognition device caused subjects to change their writing styles [8]. In this case no differences were detected between 'ink' generated by subjects during an initial period of pen familiarisation, and that produced after extended practice with the recogniser. In a longer study, the frequency of misrecognition errors remained constant over a sequence of 14 test sessions, although subjects did manage to reduce segmentation errors by improving their control of letter spacing [6]. Finally, there is some evidence that users will be more successful in learning to control their writing styles if some form of explicit instruction or support is provided. learning For example, recognition accuracy might only improve with practice if subjects are allowed to inspect their training prototypes during the test session [7].

In these studies, the amount of explicit instruction and pre-trial practice was variable and not always clearly reported. One aim of the present investigation was to monitor the performance of an untrained recogniser from subjects' first attempts at handwritten input, through an extended period of use. The extent of user adaptation to the system was measured both in terms of overall accuracy and the relative accuracy of first and second attempts at recognition. First attempts may fail because of execution errors which cause characters to be poorly formed, or because users revert to earlier writing styles. However, second attempts at entering misrecognised characters are much more carefully controlled. Users who have developed an accurate model of the recognition process should be able to utilise this model to improve the accuracy of their re-entry attempts. If this were happening, we would expect to find that when all other factors are controlled for, second attempts at data entry are generally more successful than first attempts.

How can recogniser performance be improved?

For recognisers designed to identify discrete, handprinted characters, the problem is one of discriminating between patterns within a limited set of possible alternatives. Because discrimination will generally be easier for smaller set sizes, system designers can improve recognition accuracy by designing interfaces in which input fields will only accept restricted subsets, such as digits or lower case letters. This is an effective strategy for some types of application, particulary those that involve form-filling tasks where the format of information is well defined.

With unrestricted input of handprinted characters, the performance of the current generation of recognisers approaches the accuracy achieved by humans, which might be taken as a theoretical limit. One much-quoted study reports an accuracy of 96.8% for human identification of carefully handprinted upper case letters, viewed in isolation [5]. Mean levels of accuracy at around these levels have been reported in laboratory tests, using copying tasks [6], [2]. However, these very high levels of accuracy tend to be obtained for trained recognisers, often with large training sets that have been elicited from users under supervision. For devices intended for inexpert users, there is inevitably a question whether lengthy procedures for recogniser training will be fully or appropriately completed. In the present study, untrained recognition data from individual subjects was examined in order to assess the gains in accuracy that might be achieved by a selective retraining procedure targeted on a small subset of poorly recognised characters.

THE EXPERIMENTAL STUDY

For the purposes of this study we devised three test applications, representing different types of task that might be accomplished by means of a pen-based system. Each was implemented on an IBM-compatible PC, using the Microsoft Windows for Pen software environment, with a Wacom PL-100V pen tablet. For the section of the study from which the present data are taken, the recognisers were set to operate in 'boxed, discrete' mode; i.e. all recognised input was written as discrete characters, with the positions of individual characters defined by 'comb' guidelines. The data reported here were obtained from a total of 24 subjects, each using one of three available recognisers. Since differences in the performance of these recognisers were marginal, and are not germane to the results reported here, no distinction will be made between data obtained from different recognisers.

The three test applications were designed to contrast in the following ways:

- the amount of handwriting recognition required for successful task completion,

- error tolerance; i.e. the extent to which the task demanded that all handwritten text be correctly recognised

- the balance between pen use for input of recognised text and other pen functions, such as pointing, menu selection, and creation of non-recognised 'ink'.

The three applications were:

Fax/memo In this application there were three input fields on the display. Two were used for entering recognised handprinted text; one for the name of the message recipient, the other for a six-digit telephone number.

Subjects were required to correct any recognition errors within these fields. A third and larger 'scribble' field was used for writing the message itself. Pen traces appearing in this area were not passed to the recogniser. Completed messages were despatched by a pen tap on a 'send' screen button.

Database This task was organised around a database containing approximately 1500 simulated student records, indexed by name. Access to a particular record was achieved by entering handprinted text in surname and first name fields. The recognised text was matched against the database and the corresponding index region displayed in a scrollable window. Best-fit matching meant that successful access could be achieved by incomplete or partially misrecognised input. Final selection and display of the desired record was achieved by using the pen in point-and-click mode, after which new data (three two-digit examination scores) were entered into the record, again as recognised handprinted text.

Diary This was a standard type of diary application, with a month overview that could be expanded using point-and-click responses to open the appropriate appointments page. Subjects were given a series of brief scenarios, from which they created diary entries in their own style. These entries were entered as recognised handprinted text, and subjects were instructed to correct any recognition errors before closing the diary page.

The three experimental tasks were entirely accomplished by means of the pen interface, and in all three cases the pen was used for point-and-click selection, as well as for input of recognised text. For the fax/memo application, these pen functions were embedded in a task context that focused on the creation of an unrecognised ink trace. In the database task, the combination of point-and-click and recognition of handwritten input provided an economical and effective means of accessing records. The requirement for completely correct recognition was confined to entry of a limited amount of new data. Finally, in the diary task the main focus was on completely correct entry of short text notes (typically around five or six words). In addition to these three simulated applications, the test session also included two sessions of a copying task ('Pen Practice'). This was based on entry of a set of single words or five-digit strings which included the complete set of letters and digits, and was used to obtain a controlled measure of recognition accuracy.

In the experimental session, subjects with no previous experience of pen computing were first shown the pen and tablet, and given a brief description of the principles of pen interaction and handwriting recognition. They then immediately filled in a questionnaire which dealt with their expectations of this type of pen based system. They then had an average of one and a half hours experience with the system. This began with a brief instruction in the use of simple editing procedures, which were practised without the need for character entry. This was followed by the Pen Practice task, and then sessions with the each of the three test applications, each involving completion of a series of predetermined tasks. The order of these sessions was counterbalanced across subjects. Subjects then completed an evaluation questionnaire which sampled various aspects of attitudes to the pen interface. Finally, recognition accuracy was again assessed using a modified version of the initial Pen Practice task. In the whole course of the experimental session, subjects entered an average of 1023 characters to the recogniser.

DATA

Data logged from the pen tasks included ink traces, all user control actions, and recogniser output. Because subjects occasionally performed incorrect actions (e.g. mis-spellings) scoring of recognition accuracy was based on a transcribed record of the session, in which user inputs were verified by inspection of the corresponding ink trace. The following analysis directly addresses the questions about users' attitudes to recognition accuracy raised in the opening discussion.

How important is recognition accuracy in determining user acceptance of pen interfaces?

Mean recognition accuracy was 87.0% (range 76.2 - 94.6%). The distribution of individual scores is shown in Figure 1. Further analysis indicated that letter identification was rather better than these figures might suggest. Mean recognition rates for lower and upper case letters were 90.9% and 76.1%, respectively. The relatively

high error rate for upper case letters was largely accounted for by case errors, with frequent lower case substitutions for letters such as C, O, S, V, etc., which have identical lower and upper case forms. When operating in 'boxed' mode, recogniser assignment of case is based on the size of the character relative to the comb guide. When case errors were excluded, mean recognition accuracy for upper case letters was 89.4%.

Figure 1: Mean recogniser performance (24 subjects)

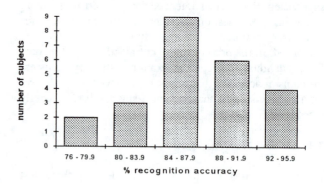

Seven questionnaire items were designed to elicit general attitudes towards the pen interface. Subjects indicated their responses to these items on a ten point scale, which was then scored so that 10 represented the most favourable, and 1 the least favourable response to each item. A mean rating of 5.5 thus represents the midpoint of this scale. The seven questionnaire items are listed in Table 1, together with the mean ratings, and the correlations between subjective rating and mean recognition accuracy.

Table 1:   General appraisal and task appropriateness items from the subjective evaluation questionnaire.

Questionnaire item	mean rating	correlation with recog.
General appraisal		
1. How much do you like the idea of using a pen as a means of working with computers?	6.2	0.21
2. Would you use this type of computer again?	6.0	0.15
3. Did you enjoy using the pen based computer?	6.4	0.33
4. Did you find the pen system you have just used: frustrating vs. satisfying	4.3	0.41
5. difficult to use vs. easy to use	5.8	0.23
6. difficult to understand vs. easy to understand	8.8	0.17
7. time consuming vs. time saving	4.1	0.35
Mean rating, items 1 - 7	5.7	0.43
Task appropriateness		
How appropriate was the pen-based system for completing the tasks?		
8. Appropriate for fax	6.6	0.36
9. Appropriate for records	8.9	0.22
10. Appropriate for diary	4.8	0.45
Mean rating, items 8-10	6.7	0.48

Average ratings across all 24 subjects suggest a mixed attitude to the pen interface. A high rating for ease of understanding confirms the intuitiveness of the pen interface. Several of the remaining attitude responses were neutral or marginally positive. However, consistent ratings of the pen applications as frustrating and time consuming suggest that recognition errors were perceived as a significant problem. If this is generally the case, we would expect that subjects who managed to achieve high overall recognition accuracy would be more satisfied than those who were less successful. But although the correlations between recognition performance and subjective appraisal are all positive, the relationship is not a strong one. The only correlations with overall accuracy that are statistically significant are for ratings of frustration/satisfaction, and for scores averaged over all seven appraisal items (critical value for $r(23) = 0.41$, $p<.05$).

The second set of questionnaire items shown in Table 1 asked subjects to rate the appropriateness of the pen interface for completing each of the three experimental tasks. The records application received a very high rating on this scale, as compared with the mildly favourable assessment of the fax application, and low evaluation of the diary. To what extent are these ratings determined by recogniser performance? This question can be answered by examining the correlations between ratings of task appropriateness and recognition accuracy. These show a positive, but relatively weak relationship between the two measures, and the correlation is statistically significant only for the diary task. In Figure 2, mean appropriateness ratings are plotted for each of the three test applications as a function of recognition accuracy. From these data it can be seen that the records application was assessed favourably even by subjects whose handwriting was poorly recognised. In contrast, the diary task was seen as appropriate only by those who achieved high levels of recognition accuracy.

Figure 2: Task appropriateness and recognition accuracy

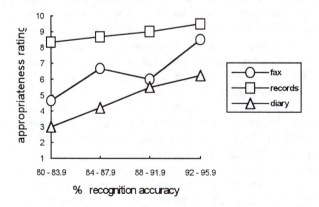

These findings suggest that recogniser performance is a factor in determining the success of pen applications, but that its influence is heavily task-dependent. Some pen applications will be successful despite relatively poor recognition accuracy. The records task used in this study was designed to model a typical information retrieval activity (a target area for PDA applications), where a modest amount of recognised pen input allows the user to achieve significant subgoals within the overall task. In contrast, users may perceive only a marginal advantage in the translation of handwritten diary entries, and this advantage must be weighed against the costs incurred in achieving correct recognition. By reducing these costs to the user, improvements in recogniser performance will be a significant factor in extending the range of potentially successful applications.

How much control do users have over recognition accuracy?

Motivated and compliant users of pen based systems will make considerable efforts to improve recognition accuracy. In the present study, some subjects were clearly experimenting with different writing styles and letter forms. Questionnaire responses reflected a general belief that recognition accuracy did improve with practice. However, the fact that subjects also felt that they had a poor understanding of the causes of recognition failures implies that they did not have an explicit model of the processes involved in machine recognition.

This study was specifically designed to assess the extent to which subjects were able to adapt to the recogniser in the initial stages of practice. The first copying task ('Pen Practice') was completed at the beginning of the experimental session, and was for all subjects their first experience with a recogniser. A matched test was also presented at the end of the session, after subjects had completed a series of trials with each of the test applications. In the event, there were no significant differences in mean recognition accuracy for the first and second Pen Practice sessions (85.3% and 86.4%, respectively; $t(23) = 1.21$).

This finding has two immediate practical implications. First, it suggests that for discrete character recognition, 'out of the box' performance of an untrained recogniser may come close to the standard achievable by a more practised user. The second implication is that unless explicit guidance is given, new users may be unable to modify their writing styles in ways that would allow them to obtain the best possible performance from a particular recognition device.

At another level, the absence of practice effects implies that our subjects were unable to use their experience with the recogniser to develop an internal model of the recognition process. This type of understanding is crucial in enabling users to respond effectively to input failures. In performing tasks with the pen interface, even the most careful writer will sometimes enter characters that are poorly formed. As with keyboarding errors, these occasional failures need not detract from the fluency of the interaction, provided that the user is able to make an efficient repair. This will be easier if the user understands what was wrong with the first attempt, and knows how to enter a more acceptable form. Thus the accuracy of re-entry may be more sensitive than practice effects as an indicator of the extent to which users possess an accurate model of the recognition process.

In the event, we found no evidence that subjects knew how to modify their writing when recognition had failed. The analysis of first and second entry attempts reported here is for lower case letters only; upper case letters are of less general interest, because the preponderance of case errors is a characteristic of boxed input. To avoid selection bias (letters occurring as second attempts tend to be those that are poorly recognised), accuracy was calculated for each subject by first determining error rates for individual letters, and then taking the average of these values. For mean error rates expressed in this way, there was no evidence of a significant difference in the success of first and second attempts at entry (10.8% and 9.7%, respectively; $t(23) = 1.14$).

Taken together, the lack of improvement with practice and the inability to increase recognition accuracy on re-entry attempts both indicate that subjects' behaviour was not directed by an appropriate model of the machine recognition process. This does not mean that they were incapable of functioning in this way, only that they did not spontaneously use their experience to construct the necessary model. We might therefore conclude that some form of explicit instruction is likely to be necessary if users of pen interfaces are to achieve higher recognition accuracy as they become more practised at using the system.

How can recogniser performance be improved?

There are two main reasons why an untrained recogniser can fail to identify handprinted characters correctly. One is simply the difficulty of discriminating between highly confusable pairs, such as "t" and "f", or "2" and "Z". The other is that individuals may use written forms that are quite discriminable, but do not match the prototypes used to build the recogniser. The best way to deal with the first problem will almost certainly be to introduce 'top down' procedures that use contextual cues to resolve ambiguity. Dictionary lookup is one strategy that is likely to prove effective for many applications. The second problem requires either that users modify their written forms to match recogniser prototypes, or that the recogniser is designed to adapt to individual users. The evidence presented here suggests that users are generally unwilling or unable to adapt their writing styles to an untrained recogniser. On the other hand, laboratory studies have shown that training the recogniser does significantly

improve performance. The disadvantage of this approach is that new users may be required to complete a lengthy training procedure before pen applications can be used.

In order to assess the merits of various approaches to recogniser training, we need more information about costs and benefits to the user. This in turn requires more detailed analysis of the pattern of recognition failures for individual users. For example, lengthy pretraining of the entire character set will not repay the effort involved if either (a) most errors are due to confusable pairs within the recognition set, and discrimination of these pairs does not improve with training, or (b) untrained recognition of most characters is acceptable, but each user has a few idiosyncratic letter forms that are poorly recognised.

The following analysis illustrates the problems that must be addressed, and provides some preliminary answers. The data are based on first attempts at recognition of lower case letters. To reduce sampling errors in estimation of error rates, the data set was further restricted by including only those letters that appeared an average of 12 or more times (range 12.3 - 92.0) in protocols from individual subjects. This excluded b, j, k, q, v, w, x, and z, leaving a total of 18 letters.

For each subject, mean error rates for these 18 letters were calculated, and arranged in rank order. We can then identify the letters that were most problematic for that individual by looking at the top four in this list. Data for the group as a whole give some idea of the extent to which these difficulties are idiosyncratic. If recognition failures were completely due to difficulty in discriminating similar letter forms (e.g. "f" and "t"), then all 24 'top four' lists would be identical. If on the other hand, error patterns were completely idiosyncratic, then the 'top four' lists might look like random selections from the entire set of letters.

The experimental data fall into a pattern that is, not unexpectedly, somewhere between these two extremes. Table 2 shows the number of 'top four' lists that contained each letter, together with the average error rate for the group as a whole. There is clearly some consistency across subjects, and this reflects the difficulty of discriminating letter pairs such as f/t, l/I, h/n, and r/v. But of these, only "f" was a particular problem for a clear majority of subjects. Elsewhere, it is evident that individual subjects had difficulty with letters that were generally quite well recognised. The degree of failure in these cases was often quite extreme; average error rates for the first and second letters in individual ranked lists were 38% and 27%. Overall, these data suggest that for discrete recognition of handprinted characters, there may be advantages in devising a selective procedure for recogniser training, targeted on a small subset of items that are problematic for the individual user.

Table 2: Letter-by-letter analysis for individual subjects, and mean error rates

letter	'top four' count	mean % error	letter	'top four' count	mean % error
f	18	26.4	y	5	6.2
l	13	12.3	d	3	5.9
h	10	13.9	s	3	5.5
r	10	12.5	o	2	4.8
t	9	10.8	a	1	3.4
g	6	7.6	p	1	2.3
c	5	8.9	n	0	1.8
i	5	10.0	e	0	1.6
u	5	7.8	m	0	1.2

CONCLUSIONS

The aim of this study was to collect data on user acceptance of pen interfaces experienced in simulated task contexts. Within these contexts we monitored both recognition accuracy and subjective evaluation. The general picture that emerges is that recogniser performance is a significant factor in determining user satisfaction, but that its impact depends on the nature of the task being performed, and the functional advantage of translating 'ink' traces into recognised text. This is essentially a cost/benefit relationship; users will accept the costs associated with recognition errors if (as in our record updating task) there is a substantial payoff in terms of achieving task goals. In this context, variations in recognition error rates that are generally within the range of 5-20% are only of minor importance in determining levels of user satisfaction. In tasks where there is a smaller benefit, users become more sensitive to the costs associated with handwriting recognition. In setting targets for recogniser performance, it is therefore inappropriate to think in terms of a fixed target that would make pen interfaces a viable option. It is rather the case that progressive increases in accuracy will extend the range of applications that users will find acceptable.

These findings are consistent with other studies which have concluded that at present, handwriting recognition is unlikely to be an effective method for unconstrained text input [3]. For routine tasks, people seem to prefer implementations in which the pen interface is highly structured, with screen prompts linked to clearly defined fields for text entry. In the short to medium term, the most effective strategy for extending the range of pen applications might therefore be to develop interfaces in which the task is adapted to this structured format, rather than trying to build recognisers capable of handling free text input.

In the pursuit of improved recognition performance, we have also considered the relative contributions that might be made by users and system designers. One striking finding in this study was that recognition accuracy did not improve with practice. Given the immediate linkage between user behaviour and potentially rewarding system feedback, we might have expected to find some evidence of behaviour modification. However, informal observation suggests that many individuals are either unwilling or unable to modify writing styles that involve stable, long-established motor skills. We might perhaps expect such changes to occur in the long term, but this assumes that users will find the pen system sufficiently attractive to reach this level of practice.

If users cannot adapt to the recogniser, the best (and widely used) strategy is to adapt the recogniser to the user. Analysis of the pattern of recognition errors for individual subjects indicates that in cost/benefit terms the best strategy will be to target the training procedure on a small subset of characters that are poorly recognised. This leaves open the question of how we can best identify this subset for individual users.

Finally, we should perhaps conclude by noting that most of our experimental subjects found the idea of a pen interface attractive. This enthusiasm was only slightly dimmed by their experience of the realities of unreliable recognition. Provided that pen applications are appropriately matched to recogniser performance, the future prospects for pen computing appear promising.

ACKNOWLEDGEMENT

This research was funded by Hewlett Packard Research Laboratories, Bristol, U.K.

REFERENCES

1. LaLomia, M.J. User acceptance of handwritten recognition accuracy. *Proceedings ACM CHI '94 Human Factors in Computing Systems Conference, Boston, Mass.*, (1994), p. 107.

2. Neisser, U., and Weene, P. A note on human recognition of hand-printed characters. *Information and control*, **3**, (1960), pp 191-196.

3. Oviatt, S.L., Cohen, P.R., & Wang, M. Toward interface design for human language technology: Modality and structure as determinants of linguistic complexity. *Speech Communication*, **15**, (1994), pp 283-300.

4. Rhyne, J.R., & Wolf, C.G. Recognition-based user interfaces. In R. Hartson and D. Hix [eds.] *Advances in human-computer interaction, vol. 4.* (1993). New York: Ablex, Chapter 7, pp. 191- 250.

5. Santos, P.J., Baltzer, A.J., Badre, A.N., Henneman, R.L., and Miller, M.S. On handwriting recognition performance: Some experimental results. In: *Proceedings of the Human Factors Society 36th Annual Meeting*. Santa Monica CA: Human Factors Society (1992), pp 283-287.

6. Schoonard, J.W., Gould, J.D., Bieber, M., and Fusca, A. A behavioral study of a hand print recognition system. *IBM Research Report RC 12494*. Yorktown Heights, NY: T.J. Watson Research Center, 1987.

7. Tappert, C.C., and Jeanty, H.H. A study of several accuracy-improvement methods for a handwriting recognition system. *IBM Research Report RC 15373*. Yorktown Heights, NY: T.J. Watson Research Center, 1990.

8. Wolf, C.G. Understanding handwriting recognition from the user's perspective. In: *Proceedings of the Human Factors Society 34th Annual Meeting*. Santa Monica CA: Human Factors Society, (1990), pp 249-253.

Designing the PenPal: Blending Hardware and Software in a User-Interface for Children

Philippe P. Piernot[†], Ramon M. Felciano, Roby Stancel, Jonathan Marsh and Marc Yvon[†]
Stanford University

Philippe P. Piernot
Knowledge Systems Laboratory
Stanford University
701 Welch Road, Bldg. C
Palo Alto, CA 94306
Tel: 1-415-323-2531
E-mail: piernot@ksl.stanford.edu

Ramon M. Felciano
Section on Medical Informatics
Stanford University School of Medicine
MSOB X-215
Stanford, CA 94305-5479
Tel: 1-415-725-3398
E-mail: felciano@camis.stanford.edu

ABSTRACT

As part of the 1994 Apple Interface Design Competition, we designed and prototyped the PenPal, a portable communications device for children aged four to six. The PenPal enables children to learn by creating images and sending them across the Internet to a real audience of friends, classmates, and teachers. A built-in camera and microphone allow children to take pictures and add sounds or voice annotations. The pictures can be modified by plugging different tools into the PenPal, and sent through the Internet using the PenPal Dock. The limited symbolic reasoning and planning abilities, short attention span, and pre-literacy of children in this age range were taken into account in the PenPal design. The central design philosophy and main contribution of the project was to create a single interface based on continuity of action between hardware and software elements. The physical interface flows smoothly into the software interface, with a fuzzy boundary between the two. We discuss the design process and usability tests that went into designing the PenPal, and the insights we gained from the project.

KEYWORDS

Hardware and software integration, user-centered design for children, Internet and multimedia application, educational application, portable computing.

INTRODUCTION

In January 1994, Apple Computer organized the Third Interface Design Competition. The theme was to design a *portable* device that uses the *Internet* to foster *learning*,

† visiting from LIAP-5, Université René Descartes, Paris, France.

using technology that should be commonly available within 2-3 years [1]. Competing teams had to submit a documented interface design, including prototypes of the main aspects of the interface and video footage of user studies. Eleven schools from around the world were invited to participate in the competition. At Stanford University, the competition was integrated into an interdisciplinary course on Human-Computer Interaction and Product Design taught by Terry Winograd, David Kelly, and Brad Hartfield. The authors designed the PenPal (Figure 1), a visual communications device to introduce children between the ages of four and six to creating digital media and exchanging it across the Internet. The PenPal project won an award for "Best Hardware/Software Integration and User Involvement" and an award for "Best Presentation".

Figure 1. The PenPal.

We decided to design for children of kindergarten and preschool age because their special needs and interests are not addressed in today's typical software and hardware solutions. In particular, we wanted to let children interact with other children at distant locations without requiring adult assistance. The communication capabilities of computers could help children contact and collaborate with other children around the world. To address the special needs of this user group, we developed two central principles that are present throughout the design. These principles were to (a) use physical interactions to take advantage of childrens' affinity for tactile manipulations, and (b) blur the boundaries between hardware and software to provide a single fluid interface.

The PenPal meets the thematic requirements of the competition because:

- The PenPal is a *communication tool*. The Internet is already being used by older children to communicate. For example, KidLink is an Internet service that offers to connect children with e-mail correspondents [8]. The PenPal expands the richness of conventional electronic communications by focusing on the use of pictures and sounds, thus lowering the age and skill level required for participation.

- The PenPal is a *learning tool*. At the young end of our user group, learning comes from the process of creating and sending a message—the process provides the educational value [6]. For older children, plug-in PenPal tools can help them learn to read, to tell time and dates, to plan ahead, to speak a new language, or to budget their allowance.

- The PenPal is *portable*. The PenPal can be carried by hand or on a shoulder strap, so that learning can happen alone, in a small group, or with the whole class—wherever the child happens to be.

We designed all aspects of the PenPal user-interface, but due to temporal and financial constraints (10 weeks from initial design to final presentation, with a limited budget), we were only able to fully prototype the main elements of the interface and we didn't do a formal evaluation. In particular, we prototyped and tested the seamless integration of hardware and software, which we believe is an important contribution of this project. We first present our user group and our design principles, then we describe the design process, usability tests, and details of our solution, and we conclude with a discussion of the insights gained.

USER GROUP

Our goal was to combine three activities children enjoy: creating, communicating and learning. Our approach was framed by some unique design constraints imposed by our chosen user group of four to six year old children:

- Children may be *unable to read*. Literacy is not a requirement for using the PenPal, but text remains present in the user interface to help children as they learn to read. More directed learning could come in the form of PenPal tools that teach children to read.

- Children have *limited symbolic reasoning*, so we decided not rely on the ability to process the graphical abstractions present in many current user interfaces. The PenPal doesn't use a desktop metaphor with icons to represent applications and files. Instead, interactions are tied to physical actions: manually plugging a tool into the PenPal launches the software.

- Children have *limited planning capabilities* and may have trouble carrying out tasks that require several steps. The PenPal uses auditory and visual coaching to help users carry out sequences of operations, and provide positive feedback for their actions.

- Children have a *shorter attention span* than adults. The PenPal provides an immediacy of results (e.g., as soon as one takes a picture, it is placed on the screen and ready to manipulate), as well as an entertaining interface to keep the user's attention.

Clearly, these constraints present a worst case scenario, since some of our users could read or had previous experience using computers. Indeed, the challenge presented by these constraints was compounded by the fact that children of this age have widely differing real world knowledge, experiences, and abilities. We decided to design within these confines so that the interface would not be too complicated for our youngest users.

DESIGN PRINCIPLES

To address these issues, we developed two principles, inspired by the nature of our user group, that guided our design:

- The PenPal features a *Physically Manipulable Interface*. Many toys for children of this age contain physical elements for children to operate (e.g., knobs, buttons, levers). We decided the PenPal should have a similarly tactile interface to take advantage of this natural interest in touching and manipulating such objects. For example, users switch into drawing mode by plugging in the PaintSet tool, not clicking on an icon.

  Because functionality is tied to physical actions and objects, the features of the PenPal are always visible and available to the user. The focus on physical interactions allows us to use direct manipulation to the greatest extent possible in order to get what we call *WYTIWYG* ("What You Touch Is What You Get"). We believe physical modes and actions afford natural mappings for our users, and therefore lead to a more intuitive user-interface.

- We wanted the PenPal to have *a single interface that consisted of both the physical interactions and the software ones, with a purposefully fuzzy boundary between the two*. For example, when one slides a MessageCard into a slot on the PenPal, a life-sized picture of the card slides onto the screen next to the slot: the card is both a physical and a software interface element. Every physical interaction with the PenPal produces similar feedback, and creates a *spatial and temporal continuity of action* between the physical interface and the software interface.

  Images on the screen are not iconic or abstract in character, but instead are realistic representations of real-world objects. By tightly coupling on-screen images and interactions with physical actions, we hope to reduce the *semantic and articulatory distance* required for the child to understand the meaning and use of the image [7].

In addition to these basic design principles, we had some overall design goals in mind:

- The PenPal should be *personal* and *fun to use*. Children should perceive the PenPal as being built exclusively for them, rather than a computer for adults that has been retrofitted for younger users. We provide a number of ways to customize the hardware-software interface: a personal address book with pictures of other users, a custom software startup sequence that runs when the PenPal is turned on, and exchangeable hardware screen frames with different colors and visual themes.

- The PenPal should be *expandable*. Although we wanted to keep the interface simple, we wanted to take advantage of the computer's ability to perform many different tasks. The PenPal must be able to grow with the children, expanding functionality as they learn to do more with the tool.

DESIGN PROCESS

The Stanford course focused on teaching user-centered design. The first phase involved brainstorming and exchanging ideas with the goal of forming teams and elaborating product concepts. The course faculty emphasized the unique advantages of interdisciplinary design [9]; our five-person team has members from France, Germany, and the United States, with backgrounds in Computer Science, Product Design, Art and Engineering.

The rest of the course was broken down into user observation and scenario generation, hardware and software prototyping, and usability testing. Oral design critiques were performed every two weeks. The user-centered design cycle iterated through several prototypes and usability tests (12 boys and girls between the ages of two and a half and six years old) [5]. The process was supported by market observations, expert interviews (parents, teachers, education experts), user interviews, and visits to schools and toy stores. Because we were unable to build a fully integrated prototype, each test involved a hardware and a software phase.

Figure 2. Previous hardware prototypes.

We built two series of prototypes in parallel. The first series of prototypes explored the appearance, ergonomics and function of the physical unit. We generated several life-sized prototypes using materials from clay and cardboard, to laser-cut acrylic (Figure 2), and conducted usability tests to determine which form-factors worked best for our users.

The second series of prototypes helped us develop the interactive aspects of the PenPal. In order to understand the important issues in merging physical and screen interfaces, we conducted a *Wizard of Oz* experiment with software prototypes developed in Macromedia Director. To model the physical interactions, we taped a touch-screen onto a computer monitor and taped a large-scale cardboard mock-up of the PenPal on top of the touch-screen (Figure 3) [3]. The cardboard mock-up included the main hardware features of the PenPal: two tool slots, and a coaching button at the top of the screen. The software was controlled through a combination of behind-the-scenes manipulation by a project member and of direct user interaction with the touch-screen.

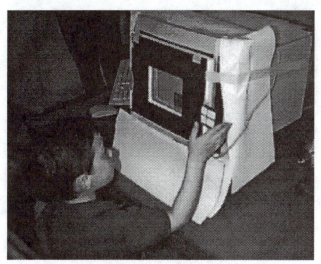

Figure 3. Setting for the software user tests.

PHYSICAL DESIGN

In recognition that a "user interface is more than software" [10], we discuss the design of the physical product as well as its software elements.

Form Factor

We determined the initial shape of the PenPal by asking one of our users to take modeling clay and "wrap" a small pad of paper. The pad was wrapped irregularly and in several colors of clay, indicating that the PenPal didn't necessarily need a conventional rectangular viewing area. We used this fact to create fun and colorful screen frames that occasionally overlap the screen.

Our final hardware design features a landscape-oriented color liquid-crystal display (LCD) touch-screen, about 3 inches by 5 inches in size. The screen is embedded in a slim frame with two barrel-shaped grips. On either side of the screen are slots for plug-in tools or credit-card-sized MessageCards (Figure 4). A camera is embedded at the top of the left grip, and a microphone and speaker in at the top of the right grip. A single button is positioned at the top of the screen, and brings up the PenPal's prompting and coaching feature. The back of the unit holds the battery storage area, clips for a shoulder strap, and various contacts for connection to the Dock.

Figure 4. PenPal, MessageCard and PaintSet tool.

When lying on a flat surface, the PenPal is tapered towards the user so that the screen slopes slightly upward to face the user for more ergonomic usage. We added this taper after users repeatedly placed one of our clay prototypes in a portrait orientation rather than the intended landscape orientation, because the left side of the prototype was higher than the right side. Users placed the prototype sloping toward them rather than toward the right, even though this orientation made controls and tools

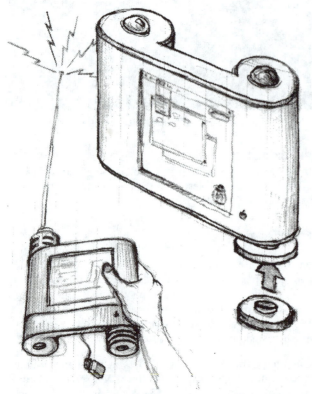

Figure 5. Memory coin.

asymmetrical. This challenged our assumption about the best orientation for the PenPal; we did not have enough time to perform additional tests to determine which orientation our users preferred.

Data Storage

A major design challenge was to find a way for children to store pictures and sounds, and exchange them with other users. We developed several concepts such as "data marbles" that would contain a single image, or "memory coins" that plugged into the grips of the PenPal (Figure 5).

However, feedback from teachers, parents, and other students in the course led us to believe that users would have difficulty thinking about coins or marbles as items that could be sent to someone, so we settled for the concept of a postcard. In the final design, camera images or drawings are stored on MessageCards, credit-card-sized cards that have six LCD fields on one side and an addressing area on the other (Figure 6). The LCD fields are magnetized to indicate how full the card is.

Figure 6. MessageCards.

The PenPal Dock

To provide an inexpensive connection to the Internet, we developed a docking station rather than a wireless connection in the PenPal itself (Figure 7). The PenPal Dock has a build-in modem, a MessageCard slot for sending and

Figure 7. PenPal Dock.

receiving messages, and a flashing light for indicating when a new message has arrived. The dock sits on the wall or on a table, and also serves as a storage rack and recharging station for the PenPal.

Customization

Both teachers and parents emphasized the importance of including personal elements in the PenPal. This is achieved by customizing their address book with friends' addresses, adding important names and dates, or choosing from a variety of PenPal screen frames that have different visual themes such as the interior of a car or a tropical jungle (Figure 8).

Figure 8. Screen frames.

The screen frames occasionally intrude upon the screen area. For example, the Jungle screen frame has a few trees and bushes that overlap from the frame onto the screen area; this again blurs the hardware-software separation. Although we did not prototype this idea, we envision that screen frames could have software elements as well: the jungle scene, for example, could change the animated coaching character to a monkey or parrot.

INTERACTION DESIGN

We now turn to the design of the software elements in the PenPal, concentrating on how they relate and blend with the hardware features.

Data Storage

MessageCards contain six fixed-size images. We chose fixed-size images to avoid the need for scrolling, and a fixed number of images per card to simplify browsing. To verify that users could create on a canvas the size of the PenPal screen, we stuck sheets of paper onto our hardware prototypes and had our users draw on them.

Inserting a MessageCard into a PenPal slot slides an on-screen image of the card into view next to the slot. This virtual MessageCard shows a reduced-size version ("thumbnail") of each image on the card; tapping on one of these six thumbnails zooms it to full size and plays any sounds associated with the image (Figure 9). Because of the speed limitations of our testing setup (touch-screen and Director prototype), users often tapped repeatedly until the image zoomed to full size. We are convinced that with a better implementation, the system will react quickly so that users will understand that a single tap is sufficient.

Addressing

Once we chose MessageCards as the data storage mechanism, we realized we could use the address side of the card to embody both the hardware and software aspects of addressing the card. The back of a MessageCard looks like the address side of a postcard with an address area and a stamp; all but two users recognized the visual similarity between MessageCards and postcards after we added a stamp. If a MessageCard is inserted into the PenPal with the address side facing up, a personal Address Book with faces of other correspondents appears on the screen (Figure 10). Touching one of the faces moves it to the address area of the on-screen card, and a spoken voice reinforces the action ("This card is addressed to Pierre"). To send an addressed MessageCard across the Internet, the user inserts it into the slot on the PenPal Dock the same

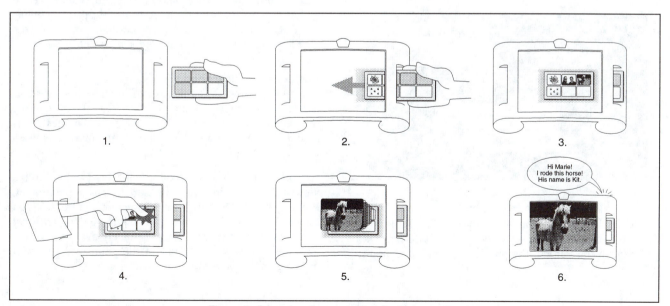

Figure 9. Browsing the contents of a MessageCard.

way one would drop a real postcard into a mailbox.

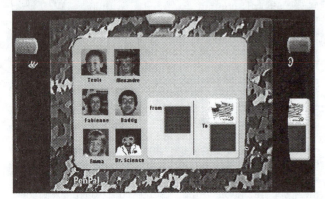

Figure 10. Addressing a MessageCard.

In our initial design, touching a face moved it directly to the addressing area. Users commented that it behaved differently than the data side of the MessageCard, where touching on images zoomed them to full size. In the final design, touching a face in the Address Book zooms it to full size, then zooms it back down into the addressing area. Once we added this animation, users seemed to feel more comfortable with the interface: they recognized pictures of their friends, and understood that they could now send the MessageCard to these friends.

Tool Design

We tested a number of different methods for plugging in tool modules, including extensions to the grips and various cabled plug-ins. Most of these ideas were discarded because they appeared to make the PenPal awkward to hold and use. We decided to use the same slots for the tools as for the MessageCards: all plug-in / pull-out interactions consistently occur in conjunction with one of these two slots that are clearly visible from the top of the unit.

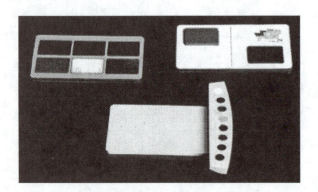

Figure 11. PaintSet and MessageCards.

While children can use drawing programs like KidPix [2], desktop computer operating systems are not designed for children. In the PenPal, the physical aspects of a plug-in

tool govern its usage. For example, plugging in the PaintSet tool (Figure 11) switches the software into paint mode: a voice confirms the new functionality (e.g., "This is the PaintSet! Touch the screen to draw on it!")."Dipping a finger" into one of the colored patches on the PaintSet switches colors while the PenPal pronounces the name of the new color. Users can draw with the PenPal as if finger-painting.

Tool Slots

Early tests showed that users had difficulty finding the slots. Because the slots were located on the side of the prototypes and not easily seen from above, users tried to place the cards directly onto the screen. We made the slots clearly visible from the top, and added distinctive color differences on the handles (Figure 12).

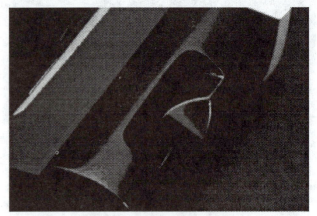

Figure 12. PenPal tool slot.

We ran into some difficulties when we tried prototyping different MessageCard sizes and PenPal slot designs at the same time. As a result, some cards simply would not fit into the slots. Users might have noticed the slots but not tried to insert the cards because the card they happened to be holding was clearly too large to fit into that prototype. Once we settled on a fixed size card and slot, users were able to perform the action repeatedly without trouble.

Camera and Microphone

The built-in camera is activated by sliding back a lever thus revealing the snapshot button on the front, and uncovering the camera lens on the back (Figure 13). The gesture is similar to winding up a photo camera, and our young users enjoyed playing with the lever.

As soon as the lever is turned, the display becomes a live viewfinder, revealing what the camera lens is pointing at. Pressing the shutter button takes a picture and freezes it on the screen. This is similar to taking a conventional photograph, but the result is immediately available for the user to work with (e.g. plug in the paint tool and start drawing on the picture).

Figure 13. Camera lever.

For self-portraits, the camera lever can be rotated 180° so that the lens faces the user; in this case, pressing the snapshot button starts a 10-second, audible countdown instead of taking the picture right away. A similar mechanism controls the microphone on the right side: after the lever is pushed to start recording, it slowly rotates back into place like a kitchen timer to indicate the amount of recording time left.

An earlier design placed the camera on a pivoting arm; lowering the arm would activate the camera. We discarded this idea after users pointed out that it seemed quite fragile. However, their reactions were probably biased by the fragility of our hardware prototype: several users accidentally pulled the arm off the unit while trying to rotate it, after which they handled all interactions with the unit with an exaggerated delicacy. Our final prototype was much sturdier and the camera lever fit more snugly into the unit, making it harder to remove accidentally.

Coaching and Help

Educators we spoke with emphasized the importance of auditory feedback and reinforcement during a child's learning period. We added auditory and animated coaching to the PenPal to reinforce what users were doing and to guide them. Usability tests showed that children occasionally needed prompting to suggest what options were available to them. A "help me" button above the screen activates an animated character that gives users options about what they might like to do next (Figure 14).

Figure 14. Animated character.

Whenever possible, the animations attempt to reinforce the fuzzy border between hardware and software by making the character aware of surrounding hardware such as slots and levers. For example, in our usability test, the animated character suggests that the user plug in a PaintSet tool: the character drops down from under the "help me" button that the user just pressed, walks across the screen and *points* to the physical slot (located next to the right edge of the screen) as she talks.

INSIGHTS

Constrained Design Space

We ascribe our accomplishments primarily to the course requirements, and the constrained design space imposed by both our user group and the prerequisites of the competition. The course curriculum focused on a by-the-book, user-centered design methodology, and forced us to explicitly articulate our design for bi-weekly presentations to a group of peers and faculty for criticism. Selecting a well-focused user group permitted an in-depth understanding of users' needs, abilities, and daily lives. As the computing industry moves towards special purpose devices that perform a narrower range of tasks than the typical desktop computer, designing for such narrowly-defined user groups will be an increasingly common goal.

Designing for Children

Usability testing. Performing usability studies with children required tailoring conventional testing methods [4]. Adult users are often asked to evaluate features by rating them according to a scale. Some of our users found this difficult, so we experimented with simple, directed questions or task statements that attempted to extract useful feedback without influencing users' responses. For example, when testing our early designs for the tool slots, we found that questions that required even simple hypothetical thinking on the part of the user were less useful than simple task assignments. Thus, we told users to "Put this card into the PenPal" instead of asking them "Where would you insert this card in the PenPal?" The most useful tests were brief and straightforward, and involved intriguing tasks to keep users interested.

Children had difficulty with explicit "thinking aloud" techniques, and we sometimes struggled to understand the verbal feedback they gave during post-test interviews, due either to their inability to articulate their thoughts, or our inability to understand them. They were also more likely to verbalize positive reactions than negative ones, and the tendency to seek approval from the tester that is often seen in adult usability tests appeared exaggerated in our young users. Therefore, we relied more on user observation than on interviews to find flaws in our design; often the most valuable information came from interpreting users' facial expressions and spontaneous reactions. We found it vital to have at least two cameras running at all times: one videotaping the user's face, the other taping the interaction with the user interface, typically over the user's shoulder.

Customization. The customization and "personalization" of the PenPal turned out to be a very important factor in making it appealing to our user group. Our users loved the animated start-up sequence, the address book filled with familiar faces of friends, and the colorful screen frames. The success of colorful mouse pads and animated screen-saver software for desktop computers would suggest that this kind of customization appeals to adults as well, yet it is rarely integrated into the basic design of the computer.

Animation at the Interface

Developing a design featuring hardware and software working in fluid unison was one of the major challenges in this project. The use of sound and animation proved to be a valuable tool in developing our solution. When a MessageCard is inserted on the right side of the PenPal, the user moves the card from the right to the left as it slides into the slot. The on-screen image of the card mimics this dynamic motion so that it appears at the same location and moves in the same fashion as the physical card. Through the integrity of this illusion, the PenPal serves as a magic magnifier to allow users see what is on their MessageCards.

While other interface systems use graphical feedback about physical actions (e.g. a floppy disk icon appearing on the Macintosh desktop after a disk is inserted into the drive), the PenPal provides a "continous representation of the object of interest" that requires less cognitive interpretation on the part of the user [7]. The spatial and temporal continuity of this parallelism results in physical modes and actions that appeared to be more natural for our users than software-only manipulations.

ACKNOWLEDGMENTS

We thank Terry Winograd, David Kelley, Brad Hartfield, Harry Saddler, Deanna Marsh, Jean-Michel Moreau and Angel Puerta as well as our young users for their contributions. Special thanks to Joy Mountford and Greg Thomas for organizing the Apple Design Competition, and to Vicky O'Day for providing valuable assistance in performing our usability tests. Certain computing services were provided by Apple Computer, Inc. and the CAMIS resource at Stanford, NIH grant No. LM-05305.

REFERENCES

1. Apple Computer, design competition brief, 1994: `<http://www-pcd.stanford.edu/hci/apple.txt>`

2. Brøderbund Software, *KidPix User Manual*, 1991.

3. Ehn, P. and Kyng, M. Cardboard Computers: Mocking-it-up or Hands-on the Future, in Greenbaum, J. and Kyng, M., eds., *Design at Work: Cooperative Design of Computer Systems*. Lawrence Erlbaum Associates, Hillsdale, New Jersey, 1991, pp. 169-195.

4. Gomoll, K. Some Techniques for Observing Users, in Laurel, B., ed., *The Art of Human-Computer Interface Design*. Addison-Wesley Publishing Company, Inc., Reading, Massachusetts, 1990, pp. 85-90.

5. Gould, J.D. and Lewis, C. Designing for Usability: Key Principles and What Designers Think. *Communications of the ACM*. 28, 3 (March 1985), pp. 300-311.

6. Hakansson, J. Lessons Learned from Kids: One Developer's Point of View, in Laurel, B., ed., *The Art of Human-Computer Interface Design*. Addison-Wesley Publishing Company, Inc., Reading, Massachusetts, 1990, pp. 123-130.

7. Hutchins, E., Hollan, J. and Norman, D. Direct Manipulation Interfaces, in D. Norman and S. Draper, eds., *User Centered System Design*. Lawrence Erlbaum Associates, Hillsdale, New Jersey, 1986, pp. 88-124.

8. KidLink Home Page: `<http://kidlink.ccit.duq.edu:70/0/kidlink-general.html>`

9. Kim, S. Interdisciplinary Cooperation, in Laurel, B., ed., *The Art of Human-Computer Interface Design*. Addison-Wesley Publishing Company, Inc., Reading, Massachusetts, 1990, pp. 31-44.

10. Vertelney, L. and Booker, S. Designing the Whole-Product User Interface, in Laurel, B., ed., *The Art of Human-Computer Interface Design*. Addison-Wesley Publishing Company, Inc., Reading, Massachusetts, 1990, pp. 57-63.

Amazing Animation™: Movie Making for Kids
Design Briefing

Shannon L. Halgren   *Tony Fernandes*

Deanna Thomas

Interface Design Group
Claris Corporation
Santa Clara, CA 95052
(408) 987-3912, email: shannon_halgren@claris.com

ABSTRACT

The development of the interface for Amazing Animation was a challenging, unique, and a rewarding experience for our Interface Design Group at Claris.   Given the constraints of a very tight timeframe and working with a user population we were unfamiliar with, our group was able to make numerous improvements which had a tremendous impact on the product's usability.   This having been our first time designing for and testing children, we learned volumes about this unique user population.   Design assumptions and testing methodologies used in adult products must all be reworked for kids. This paper describes the progression of Amazing Animation interface and points out the lessons learned about testing and designing for kids along the way.

KEYWORDS

Interface Design, Kids Software, Designing for Children, Testing Children

PRODUCT DESCRIPTION

Kids ages 5-14 can use Amazing Animation to create movies, video games and interactive school projects. Kids stamp animated characters and sounds onto pre-drawn backgrounds to make stories about space, dinosaurs, underwater life, and many other topics.   Amazing Animation automatically records characters movement across the screen to be replayed over and over. New characters and backgrounds can be drawn using the drawing tools or clip art can be imported and animated. Text and sounds are used to narrate the movies. Frame-by-frame editing is possible, as are special effect transitions between frames. A frame from a typical Amazing Animation movie is presented in Figure 1.

Figure 1. Frame from a sample Amazing Animation movie.

THE ORIGINAL STATE OF AFFAIRS

The Claris Interface Design Group was initially approached by the Amazing Animation product marketing lead and asked to provide interface design support to the development team and possibly do some usability testing. The program was already in Alpha which meant time was of essence and interface changes involving significant recoding was going to take a lot of convincing. However, the development team was extremely open to design and usability input which allowed positive interface changes to occur.

The Design Group immediately dedicated an interface designer and a usability specialist to the project. These two individuals worked on both the testing and design phases of the project. After the first round of usability testing, a visual designer was added to the team to help with the design phases of the project.

THE FIRST STEP: USABILITY TESTING

All parties involved agreed that usability testing would be the first step towards interface improvement of Amazing Animation.   Being that this was the first kids' product our team had tested, we had to quickly shift from an adult to a kid world. Our Usability Studio at Claris was not appropriate for testing children; the desk was too high, the chairs too big, and the modern office decor did not shout Sesame Street. Therefore, we decided the best option for us on a limited timeframe would be to perform a field test.

After some negotiation, we were allowed to perform our usability test at an interactive learning computer lab. This was a private business where children could come in after school or during the summer and work on basic reading and math skills using interactive software. We set up a computer and video camera in a corner of the busy computer lab. When the kids had a breaking point in their lessons, they were allowed to come over to our station and play with our product. The switch from math to movie making seemed to be a welcome change.

USABILITY RESULTS & LESSONS LEARNED

We're Not in Kansas Anymore: Testing Kids is Not Like Testing Adults

After the first few children had come and left our testing station, we were left with the unrefutable conclusion that observing kids interact with software is in no way similar to observing adults interact with software; we were in an entirely new world of computer usage. Our carefully prepared test scripts and questionnaires were left unused. Our empirical methods and procedures crashed to the floor when our questions and prompts to stay on track were repeatedly ignored. The kids were much more interested in such unpredictable tasks as hitting the mouse button repeatedly to create 10s of new screen characters by the second or dragging a character around the screen in an endless, frantic loop.

Eventually, we were taught by the kids that observation with some guidance was the preferred method of conducting a kids usability test. We suggested things we'd like to see the children do with our product (like make a movie), but had to remain flexible and follow the child's preferred path through our product. In the end, we probably learned a lot more about how kids will use our product from this observational method than from any test script we could have created.

Over the course of the study, our ability to interact with the kids also improved. Kids seemed to have two distinct modes of interaction with us. They either told us *everything* that was on there mind about our product and the world in general, or they told us *nothing* which left figuring out their thought process one big puzzle. One thing we began to learn was that an unanswered question

appeared to mean "I don't know" or "no". If the child knew the answer or the answer was positive, they usually proudly shared the answer.

Follow the Yellow-Brick Road: Provide a Clear Path Through the Interface

Most kids' interaction with our prototype was not the streamlined movie-making session that was hoped for. Figure 2 displays Amazing Animation as it first appeared upon launching. Kids' typical first path through our prototype is listed on the screen.

Figure 2. Typical Sequence Using the Opening Screen of Our Prototype

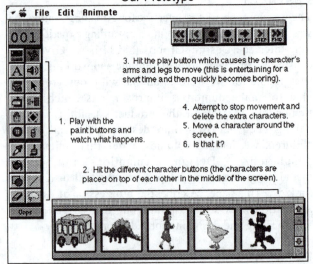

The complexity of the interface and the overwhelming number of controls made it difficult to keep on a movie-making track. Kids know how to push buttons and they do it with great glee and disregard for the consequences. Our kids quickly got themselves trapped in advanced paint and movie-making modes which surpassed their expertise. They also loved clicking on the character buttons at the bottom of the screen. Each click of a character button placed a new character on the center of the screen. Before the child knew what was happening, a mass of arms, legs, and bodies was assembled in center-screen and waiting for action.

The next hurdle for kids to jump was to figure out how to record a character's movement and then watch the animation. The intended path for this task was to hit the record button, move a character around the screen, hit Stop, hit Rewind, hit Play to view the animation, and finally, hit Stop to stop the animation which played in an endless loop. Our kids, however, preferred to hit only the Play button. To them, "Play" meant having fun, not starting a movie. When the Play button was hit before any recording was performed, all the characters' arms and legs began to move (an automatic feature of characters), but their bodies remained glued to center-stage. This was wildly amusing at first, but became quickly boring. After futilely trying to stop the movement and/or delete any extra characters, kids eventually became tired with our prototype and came to the conclusion we had a boring product.

When redesigning the interface for this product, we paid close attention to making the process of moving through the interface much simpler. The first step towards this end was to remove many of the controls from the opening screen. Only the very basic movie controls were left. Advanced paint and movie controls were moved out of sight to pull-out drawers located at the bottom of the screen. We felt drawers offered a metaphor that kids could relate to, contrary to abstract tool palettes containing numerous icon controls found in adult products. Kids have most likely had experience storing objects such as toys or tools in drawers that can be opened and closed.

Selecting the up and down arrow controls to the right of the drawers displayed different sets of drawers containing different tools. We were hoping that by hiding the advanced tools, the novice users would not stumble onto them and get lost in their functionality. Rather, only the advanced users who might want the advanced tools would go looking for more options. This redesign allows the product to be engaging and usable for a wider range of ages and abilities. For reference, our redesigned screen is pictured in Figure 3.

The process of creating a movie was made more intuitive by displaying only tools directly related to movie making on the default opening screen. The most efficient process for creating a movie is to select a background, select desired characters and sounds, record the characters movement and finally playback the movie. The three drawers at the bottom of the screen are placed in the order that they should be used; scenes or backgrounds first, stamps or characters next, and finally, sounds.

Contrary to the old design, recording of movement in the new design is done automatically when the character is dragged around the screen. The old design required the pressing of the record button. We found kids did not recognize the button labeled "REC" as record and even so, didn't realize that recording was a key element in movie-making. To provide feedback when the product is recording, a melody is played and the frame counter located below the Play button increases as new frames are recorded. If no action is taken, after a few moments an

auditory prompt provides the next step towards making movie. While we found that the auditory prompt was a useful way of providing help to kids, it does provide problems when localizing and therefore, isn't the ideal solution. More thought will be given to this for version 2.

After the character has stopped moving across the screen, and thus recording has stopped, the letters on the play button move up and down. This draws attention to the button. When pressed, the movie will automatically rewind to frame one and play once through eliminating the rewind and stop steps necessary in the original design.

Figure 3. Opening Screen of the Redesigned Prototype.

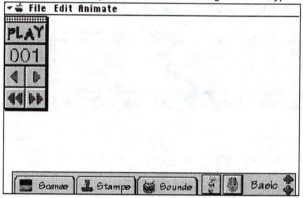

Don't Push an Adult Interface on Kids

Observing kids use our prototype very clearly showed us where assumptions were made that kids use a computer like adults. Adult Graphical User Interface (GUI) controls such as scroll bars, window close boxes, menus, and paint tools were all problematic for most kids. Some of the older kids we tested were computer savvy and had worked with these controls, but all the younger ones were at a loss in situations like figuring out how to close an open window which only displayed the typical title line close box.

In general, our redesigned interface took out adult interface controls. Drawers of tools which automatically closed replaced typical toolbars and window close corners. The drawers were labeled not only with an icon, but also with text to increase recognition. Drawers were also given ample space so scrolling for more options was kept to a minimum. Standard paint tools were either cut because of their complexity (e.g., the eyedropper and lasso tools) or presented in a more kid-friendly manner. For example, the text tool previously placed on the tool palette using the standard "A" icon, was moved to a drawer with an ABC icon and label stating "Letters". The new drawer presents different font, style and size options in labeled sections (see Figure 4).

Figure 4. The text tool location in the original and redesigned prototype.

Original Prototype - Standard text tool brings up a text palette

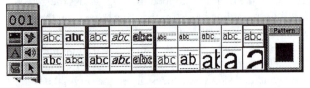

Redesign - Text drawer is accessible only by switching to the tools mode using the arrow buttons. Drawer label is more kid-friendly.

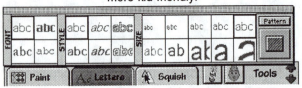

Don't Push Adult Metaphors on Kids

The original prototype we tested used two strong metaphors that may have been appropriate for adults, but were usually lost on kids. The first metaphor, that of making a movie that contains individual, sequenced frames was too abstract for children. Kids understand animation only at a cursory level - animation is things moving. The features that allowed kids to edit individual frames were also too easy to stumble into and get confused about. Again, in the redesign, these advanced features were taken off the basic, default opening screen.

The VCR metaphor used for the movie controls in the original prototype was also missed by kids. For the most part the movie controls were not recognized as familiar and so were not used correctly. The rewind, fast forward, and step frame controls were rarely touched. Play and Stop, however, were selected a great deal, but often at inappropriate moments.

Our redesign tried to reduce the dependence on these adult metaphors. The movie metaphor was still present, but only visibly when using the advanced controls. Large colorful, animated Stop, Start, and Record buttons which were only displayed when it was appropriate to use them, replaced the standard VCR control panel (see Figure 5).

Figure 5. The text tool location in the original and redesigned prototype.

Original Prototype - Movie controls are presented using a standard VCR control metaphor.

Redesign - Movie controls are simplified by enlarging and animating buttons which are only displayed when relevant.

Target Age is Critical: Do Your Home Work

The older kids who tested our prototype were quick to tell us that our characters and backgrounds were too babyish. In some situations it was clear that they were a little embarrassed to be using the product in the social atmosphere where we were testing. Ideas for new characters and backgrounds were quite different for the different age groups and between girls and boys. This finding forced us to step back and critically think through which ages we wanted to market this product towards and select content and features based on that decision. We also increased the number, content, and quality of backgrounds and stamps available.

ROUND 2 OF TESTING & REDESIGN

After some negotiating with the development team and making a few modifications, our redesigned interface was quickly implemented and within a few weeks we were ready for another usability test. We went back to the same location for testing and were fortunate enough to have some of the same kids as well as different ones look at our new design. We were pleased to find that many of the usability issues we found in the original prototype had been solved with our redesign.

Drawer Mode Redesign

Some problems still existed in the interface which were addressed in subsequent design changes. For example, the original buttons for switching between the modes of drawers at the bottom of the screen were too big and attention-grabbing. Little kids were still pushing them and then getting lost in the advanced features. We reduced the button size and toned down their appeal. Additionally, more sets of drawers containing advanced features were added. We moved from having to sets (beginning and advanced) to four; Basic, Movies, Effects & Tools. These four categories which were labeled were used to group the drawers into meaningful groups which allow users to more accurately guess drawer contents. The final buttons used are the ones shown in Figures 3 and 4.

Move, Resize, & Rotate Tools Redesign

Another group of controls that required iterative design were the move, resize, and rotate controls. These tools could be used during recording to record the movement, growth or reduction of character size, or character rotation. On the other hand, they could also be used to get the character in the desired state and position before recording started. In the prototype tested in our first study, these more advanced editing controls were displayed in the tool palette (see Figure 6, Stage 1). The icons for these tools were not recognized by kids nor were the subtleties of using each in the two different record states understood.

In the first redesigned interface, the controls were moved to a "Squish" drawer located in one of the more advanced modes, Tools (see Figure 6, Stage 2). The icons were redesigned so that they all contained a consistent element, the triangle. Because in the new design, recording was made automatic when a character is dragged across the screen, additional controls had to be added that allowed moving, resizing, and rotating the character to be done while recording or while paused before recording. In this design the controls in the left column were static and represent manipulations that will not be recorded. The icons in the right column were animated to suggest that while using them actions will be recorded. The animations of the right column icons were also used to help enhance icon meaning. The top move icon pictured the triangle being dragged in a diagonal path across the icon. The middle resize icon displayed the triangle being expanded from very small to the size of the icon. Finally, the bottom rotate icon displayed the triangle being turned around and around. These animations continually play while the Squish drawer is open.

The Stage 2 redesign was an improvement over the original design, Stage 1, because young kids did not stumble onto these tools and get lost in their functionality. Older kids who found the tools, however, had difficulty figuring out what they did. The final design of this drawer shown in Figure 6, Stage 3, regrouped the icons and added labels for enhanced meaning even further.

Figure 6. Stages of Move, Resize and Rotate Tool Design.

Stage 1 - Abstract adult icons in large tool palette

Stage 2 - Tools in drawer only accessible in Tools mode. Icons contain consistent elements and are animated to enhance icon meaning.

Stage 3 - Labels are added and icons are grouped to enhance meaning even further.

ONE LAST MEETING WITH KIDS

Unfortunately, after our second round of usability testing and redesign the members of the interface design team needed to move on to different projects and thus their involvement with Amazing Animation lessened. Nonetheless, members from the development and marketing teams felt strongly about meeting with our target market one last time. This time a beta version of the product was dropped off at the computer lab of a local middle school. The lab instructor agreed to let some classes use the product during lab time. Students were given incentive to use the product by being allowed to participate in a Claris-sponsored contest to find the best kid-created movie. Contest-winning movies were later used for PR and marketing purposes. One week after drop off a team went back to see what kids had created with Amazing Animations. Each lab class which used the product was given a brief presentation by the product developer and were allowed to ask questions about the product or his job. Afterwards, the kids broke up into small groups to continue making their movies. Groups were observed and interviewed about the product. Unfortunately, no one from the Design Group was able to attend these classes, so no specific data was gathered about the success of our redesigns such as the text drawer or move, resize and rotate buttons. We can only hope the redesigns enhanced usability as we expected.

CONCLUSION

The design process of creating the Amazing Animation interface was unique, challenging, and fun on a variety of levels. Initially, this seemed like an impossible mission. Our team was brought in post-alpha, initially told we could make few, if any changes which would require major recoding, and given roughly a month's time to work our magic. Nonetheless, the developer's willingness to go along with our changes and the incredibly quick turn around times we were all able to make, allowed significant, positive changes to happen.

The product interface was significantly reworked to adapt to the playful, experimental, exploratory method of use that children brought to the product. We focused on improving usability and engagability for a wide age range of users by simplifying many features and processes, making some features more visible, and by hiding advanced features and tools in drawers to prevent users

who are just messing around from stumbling onto them.

This project was an all too rare example of the ideal user-centric development cycle. The product team collectively agreed to put the effort forth to gather user's input and then to really listen to that input. All redesign that was done was based on information gathered from real users; the kids. The developers and design team worked cooperatively to develop a usable product in a timely fashion. This project was truly a design success story!

Finally, the experience of designing for and testing children was a welcome challenge from the world of adult products which now feels a bit mundane. Our team learned a great deal from this project about working with kids and our enthusiasm for working on more kids products has greatly increased.

PRESENTER'S BACKGROUND AND PROJECT INVOLVEMENT

Shannon L. Halgren

Shannon is a Usability Engineer in the Interface Design Group at Claris Corporation. Her job is to manage usability testing at Claris. She earned her Masters and Ph.D. degrees in Human-Computer Interaction at Rice University. Shannon designed and helped conduct the usability tests associated with this project and participated in many of the interface design decisions.

Tony Fernandes

Tony is manager of the Interface Design Group at Claris. The group is responsible for the UI design of Claris' Windows and Mac products. He participated in the field testing of this product as well as in the development of design solutions to the problems encountered.

Deanna Thomas

Deanna is a Visual Designer in the Interface Design group at Claris Corporation. Her job is to create visual information be way of icons, navigation in dialogs and product identities for Claris. She received her Bachelor of Fine Arts in Studio at the San Francisco Art Institute and her Master of Fine Arts at University of North Texas. A six year veteran of the Apple community, her involvements with visual interface also include (of late) PowerTalk & AppleSearch. Deanna provided the visual interpretation of our design brainstorming sessions and gave Amazing Animation much of its look and feel.

Drag Me, Drop Me, Treat Me Like an Object

Annette Wagner, Patrick Curran, and Robert O'Brien

SunSoft, Inc.

2550 Garcia Ave., MTV 21-225

Mountain View, CA 94043-1100, USA

(415) 336-5427

Email: annette.wagner@sun.com

ABSTRACT

This design briefing covers the major human interface design issues encountered in the development of the Common Desktop Environment Drag and Drop Convenience Application Programming Interface. The presentation will walk through the icon development, user testing and the different problems and solutions that arose during development.

Keywords: Computer-human interface, direct manipulation, drag and drop, Common Desktop Environment, icons, drag icons, Motif 1.2.

BACKGROUND

In the spring of 1993, several major players in the UNIX software market joined in an agreement which became known as the Common Open Software Environment (COSE). One of the results of this agreement was to join forces to create a unified UNIX-based desktop, referred to as the Common Desktop Environment (CDE).

CDE is based on Motif 1.2 and includes a file manager, mailer, calendar, text editor and other standard applications. In addition it has a Front Panel for quick access and launching of applications and support for multiple workspaces.

What is Drag and Drop?

To the user, "drag-and-drop" on the desktop is being able to pick up the latest newsletter and put it away in the news folder.

A more technical way of describing drag and drop would be: The user positions the mouse pointer over the object in question and presses a mouse button and then moves the mouse, thus dragging the selected object across the screen. The user can drag the selected object to whatever place they want to drop it. This can mean dragging the object over a folder which might then highlight to indicate to the user the folder is an "ok" place to drop the object. When the user finds an acceptable place to put the object the user then

releases the mouse button and the object does whatever operation is appropriate given the context of the drop.

The Problem

In CDE, one of the human interface goals was to support pervasive Drag aNd Drop (DND) throughout the desktop. To accomplish this, there had to be programming support that allowed applications to drag and drop objects between them without having to know intimate details about the other application. The code to facilitate this kind of application integration was written as an API (Application Programming Interface) which any application can access.

DND on our desktop is based on Motif 1.2 DND which meant everyone on the team needed to be familiar with how DND worked in Motif 1.2. Motif 1.2 supports the ability to drag text pretty freely. For example, users can drag text from static labels on buttons - something that we discovered users found pretty useless and we turned off.

The ability to drag icons in a file manager type application was inherited from a previous version of the file manager used in CDE. To drag an icon required the file manager to have code internal to the file manager that could not be easily re-used by other applications.

Our task was to extend the base Motif API to provide support for applications to easily integrate DND into their human interface. This meant extending the set of drag cursors in Motif 1.2 to support dragging icons, not just text. We were also asked to improve the usability of the text drag icons where possible.

The resulting work was incorporated into the CDE Drag-and-Drop Convenience Libraries, an API that built on the base Motif API to allow developers to integrate DND into their applications.

The Team

The team consisted of two software engineers (O'Brien and Tuthill), one of whom was the team leader for the project, two human interface engineers, and one documentation writer (Winsor). I was the main human interface engineer on the project and was responsible for the design, specifications, and coordination of the human interface work. The other human interface engineer, Patrick Curran, was responsible for running the usability evaluations.

Terminology

The team used the OSF Motif 1.2 Style Guide terminology since we were basing our work on the Motif 1.2 API.

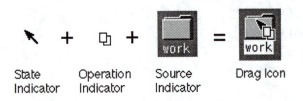

| State Indicator | Operation Indicator | Source Indicator | Drag Icon |

Figure 1   *Starting on the left, the parts of a drag icon are the state indicator, operation indicator, source indicator which together build the drag icon.*

There are three parts to a drag icon (see Figure 1). The state indicator tells the user if the drag is valid or invalid. The operation indicator tells the user if the drag operation is a move, copy, or link. The source indicator tells the user what is being dragged. The source indicator is usually the icon the user has selected.

Files, buffers, and text all had special meanings in DND. Text refers to text selections that the user can select and drag to do accelerated cut, copy and paste. Files are literally files that exist in the file system; the user sees these as glyphs or icons in a file manager. Buffers are bits that do not exist independently in the file system but, when dragged, should still be represented as glyphs or icons to the user. Displaying a calendar appointment attached to a e-mail message is one example of a buffer that a user sees. Files and buffers are effectively the same and will be referred to as files, meaning the icons the user sees on the computer screen.

MAJOR DESIGN ISSUES

Constraints

The Common Desktop Environment project was a highly constrained project from the viewpoint of resources, scheduling, and coordination. The amount and kind of changes we could make to the Motif toolkit were considered carefully before being approved. This should sound familiar to anyone shipping products. The two biggest impacts on the design process were the large amount of time spent coordinating across several companies and the schedule changes.

The entire team had one constraint in common - we were all starting from scratch with regard to Motif knowledge. The first task I had was to familiarize myself with Motif 1.2 DND and determine what changes we could make to improve usability and to integrate the functionality of the new API.

Motif 1.2 DND has a set of bitmaps that it uses to build the text drag icons. There is a bitmap for each state and each operation indicator. We had the option of changing the bitmap designs but not the method of building the drag icons or the

size of the bitmaps. The state and operation bitmaps are in black and white; there is no support for color at this time.

Design Process

Since this was a small team we did much of our work in the team leader's office on the white board. Most interactions took place between the two software engineers and the main human interface engineer. We would sketch out problems in the center of the board, API calls on the edges, and to-do lists on the sides. Everything was color coded in purple and green marker colors (the official team colors).

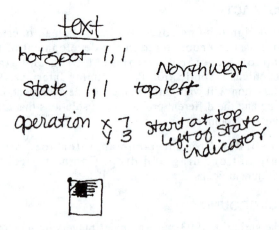

Figure 2   *An example of the notes from our meetings.*

I took copious drawn and written notes (Figure 2). Depending on what was discussed I would re-write the notes as an email and send it around, make copies of the notes for others, or just keep them for my own records.

Since the code was written as a base library, and base libraries have to be done first so other applications can depend on them, the schedule was very tight. For two to three months we met practically everyday for an hour or more.

Initially designs were done on paper and on the white board. A fair amount of time was spent figuring out why Motif did what it did. Understanding the behavior was critical to understanding what could be changed.

I used the CDE Icon Editor to generate the bitmap files. The software engineers developed a test program that could be used to swap in the drag icons as they were designed and would let us try out different parameters for building the drag icons. We used this program to visually evaluate the drag icons on screen and to try out dragging text and icons. This enabled us to do fairly rapid turnaround on drag icon designs.

A word on the metaphor...

Drag icons are a mixed bag of metaphors. The arrow is usually there to represent the mouse on the screen. The operation indicator tells the user if they are making a move or a copy or a link. This is usually done with very abstract icons and few pixels. The source indicator is perhaps the most clear because, if done right, it is the actual icon the user selected. The basic idea is "put arrow on what you want to move and move it".

Moving right along...

One of the first issues we dealt with was the move cursor. Motif 1.2 uses a small document glyph attached to an arrow for the drag icon displayed in a move operation (see Figure 3). Most GUIs use a plain arrow pointer for move. Anecdotal evidence from users of Motif indicated most folks thought the move drag icon to be confusing - they were not sure what the little document attached to the arrow meant especially as it is used in the context of moving text, not moving a document.

Drag Icon with Move feedback.   Drag Icon with Copy feedback.   Drag Icon with Link feedback.

Figure 3   These are the Motif 1.2 text drag icons as per the OSF Motif 1.2 Style Guide.

I proposed to the team that we leave the bitmap for the move operation indicator empty. This would result in a move drag icon that only displayed a plain arrow similar to other GUIs. The response from the team was positive: they had not thought about resolving the issue this way but once they explored this proposal they realized it worked well without breaking any Motif code.

We went ahead and tried this solution and an alternate solution in which I put the arrow head in the valid state indicator bitmap and the tail of the arrow in the move operation bitmap.

The reason we tried this was that we had theorized that the drag icon was built by drawing the state indicator on top of the operation indicator on top of the source indicator. This made logical sense to us. If this were true then separating the two pieces into different bitmaps should have resulted in a better looking drag icon.

In the process of testing with the sample program, we discovered that Motif 1.2 builds the drag icon in a different order than we had originally theorized (Figure 4). The *operation* indicator is drawn on top of the state indicator on top of the source indicator. This ordering is hard-coded and cannot be changed (however illogical it appeared to us).

Given that the drag icon was built in this order, it meant that the operation indicator always overwrote what was under it. So, the proposal in which the arrow and its tail were separate bitmaps was not necessary. We put the arrow with its tail into the valid state indicator bitmap and left the move operation bitmap empty. Though the state indicator is always overwritten with the operation indicator, in the move case it is an empty bitmap so the arrow and tail appear as they should.

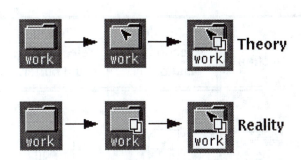

Figure 4   The top row shows the order in which we thought the drag icons were built. The bottom row shows the actual order in which they are built.

The main point is that I had to understand the design of the drag icons from the perspective of a software engineer as well as a human interface engineer. It is less likely that the final solution would have come out as well had I only looked at this problem as a visual design issue. And because I had taken the time to understand the engineering issues that affected the drag icon design, the team was more receptive to my proposals.

What is that you have there?

The chart in Figure 5 shows some of the various drag icons in our desktop. Different drag icons result depending on what the user has selected. There were two basic drag icon types, one for text and one for files. In this section we discuss the design of the file drag icons.

The source indicator bitmap for files is handed off to the DND Convenience API by the application. This is optional: if the application does not supply a source indicator then the DND Convenience API uses the default source indicator seen in Figure 5, the plain document icon.

The file drag icons presented an interesting design difficulty in that the design had to work regardless of what the source indicator bitmap design or size was.

Optimally, the source indicator is the actual icon the user selected. In the case of the CDE File Manager, the file manager icons were used directly as source indicators. Icons could be 16X16 or 32X32 pixels. An icon is created within a "bounding box". The file manager icons in our desktop do not always use the entire bounding box for the actual icon design. Typically the rest of the bounding box is filled up with a transparent color.

The DND Convenience API could have figured out the size of the icon being passed in, but it was too expensive in terms of performance to figure out what actual bits were being used in the bitmap. Knowing the actual bits used would have enabled us to crop the source bitmap used in the drag icon and would have facilitated better placement of the state and operation indicators.

Given the trade-off of exact placement of the state and operation indicators versus immediate feedback when the

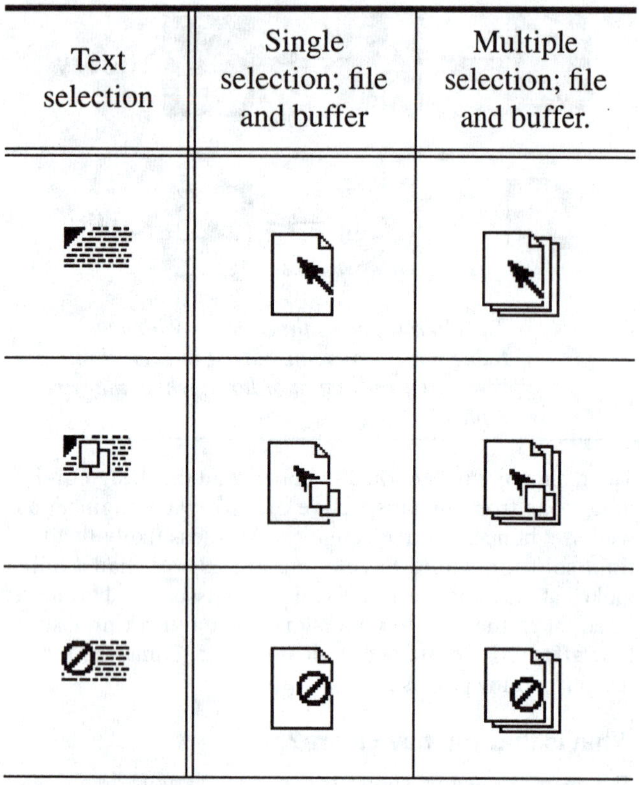

Text selection	Single selection; file and buffer	Multiple selection; file and buffer.

Figure 5　This chart shows examples of the final drag icons. Since we cannot do screen shots of the actual drag icons, this artwork is approximate.

user starts dragging, immediate feedback is more critical. If users can't tell they are dragging something, they will never get to the point of dropping it. So making a change that might seriously degrade system response time was not an option.

All of this meant that the file drag icons had to be designed so that they worked with 16X16 and 32X32 sized bitmaps regardless of where the actual bits were or what size icon was passed in.

We altered the 32X32 icons used in the file manager so that icons would be left bottom justified in their bounding boxes. The 16X16 icons typically use the full bounding box. This gave us a starting point for determining placement. We began user testing with the arrow placed at the top left corner of the source indicator, regardless of size.

Drag 20 pixels to the left, then...

The act of DND cannot be tested on paper. The very nature of dragging and dropping requires that it be tested interactively. If DND is working correctly, users never notice the drag icons or have any problems with them. To actually point out the drag icons to the user and ask their opinion would be akin to pointing out the engine of a car and asking a user how they liked the design. Unless the user is techno-literate, they don't care about the engine, they just want the car to start in the morning.

I did run an early paper study on the operation indicator bitmap designs for copy and link. Ten users were asked what the various designs meant and which one they would use for copy. We got clear feedback on which icon meant copy and which meant link. This feedback determined the icon design we chose.

The drag icons were all tested interactively. We compared the Motif 1.2 text drag icons and the file drag icons from early versions of the CDE File Manager to the new designs of the text and file drag icons in two rounds of user testing.

We had a very limited amount of time to do user testing of the drag icons. At one point we didn't think we would be able to run any testing as no applications were using the library yet. Thankfully, the engineering team produced a 'hack' to the Motif library that allowed us to swap back and forth between the Motif drag icons and the new drag icons when we restarted the system. The hack worked with only two applications: early versions of the text editor and the file manager.

We came up with a series of tasks: selecting text and dropping it in a new location, selecting file icons and moving the icons inside a folder window, dragging them to a printer. We had participants try out both types of layout (As Placed, Use Grid) in the file manager using both sizes of icons (16X16, 32X32).

There were 4 participants in the first round and 6 participants in the second round. In each test participants were asked to do the tasks in one environment and then the other. The test was counter-balanced.

The results of the first round of testing showed that users had trouble positioning with the file icons. We had used file drag icons that were built similar to the text drag icons where the arrow was in the top left corner, see Figure 6, top row. When the user dropped an icon however it did not place exactly where the user dropped it. One participant stated, with regard to placing icons versus text positioning, "visually I prefer the arrow in the middle better, but for positioning I prefer it to be in the outer edge."

When the user drags a piece of selected text to a new place in a document, the user has to position the text very exactly because the text is effectively pasted where it is dropped. There is no highlighting or other visual indicator of the drop site because only the user knows where they are going to drop the text. Conversely, when the user drags a document icon typically the icon is either being re-positioned in a window or getting dropped on a large drop site. The amount of precision needed to be successful is what makes the two tasks different.

To aid users in these two tasks, we ended up designing two different sets of drag icons. In the text drag icon the arrow is placed in the top left corner of the text source indicator. In the file drag icon, we placed the arrow pointer more in the center of the source indicator. Placing the arrows in the two drag icons in these two positions allowed users to more accurately predict where something would get dropped or

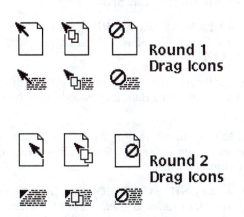

Round 1
Drag Icons

Round 2
Drag Icons

Figure 6   *The top row shows the drag icons used in Round 1 of user testing. The bottom row shows the redesigned drag icons used in Round 2.*

placed when they released the mouse button (Figure 6, bottom row).

In the second round of user testing participants preferred the new drag icons and their error rates were better using the new drag icons. (We did not put a large emphasis on the error rates for either study as users were not given much training and it was impossible to differentiate whether errors were due to a new desktop GUI or specifically due to DND.)

Drop Point versus Insertion Point

We realized early on that we had two very different tasks that had to be accommodated by the drag icons. We thought we had designed the two varieties of drag icons to match the tasks the user would be doing across the desktop. But then, in the course of implementing functionality in the CDE Mailer and Text Editor applications, we realized we had a problem.

This unfortunately fell very late in the development schedule. An unfortunate circumstance of working on an API is that this type of code must be stable and frozen very early on in the project so that other applications can then write code using it. One implication of this is that the DND API was frozen *before* any application had fully integrated it. While the plan is to cover all possible uses of an API, reality does not always match the plan no matter how good your team is.

As the CDE Mailer team was implementing the ability to drag and drop files/buffers, we realized that we needed to support dropping files onto the text editor window with the end result that the file was inserted into the document open in the text editor. The question was *where* should it be inserted: at the drop point or the insertion point?

We had designed the drop behavior for the text drag icon such that text was always inserted at the drop point, not at the insertion point. This matched users expectations for a task that mirrored cut, copy and paste. At first it seemed obvious that we should insert file at the drop point. The problem was that we had designed the text drag icons to

work for inserting at the drop point, but the file drag icons were **not** designed to work for inserting at the drop point.

Redesigning the file drag icons was not an option as the DND library was frozen. I had to have a recommendation within two days. We did not have time or resources for running a user test. So, I did the next best thing I could think of: I went back to the Open Windows Mail Tool application which supported the same functionality and reviewed the human interface design and the user testing reports. I went around and asked current users what they thought of the functionality. I brought the issue up in the joint company human interface conference call and reviewed my ideas and findings.

What I found was that dragging and dropping a file or buffer onto a mailer compose window or onto a text editor window was really an accelerator for File: Include operation. The File: Include menu item brings up a File Selection Box which allows the user to choose a file to include in the current document. File: Include always inserts the chosen file at the insertion point.

This behavior had never been documented as a problem: it appeared to match users expectations. From informal questioning, most Open Windows users were very comfortable with this behavior and with the difference between text drags and file drags. Most responded by saying things like, "it does what it's supposed to do."

Given the information I gathered, I proposed that we implement file drags such that they always insert at the insertion point, not the drop point. So far, this appears to be the right decision.

IF I HAD TIME TO DO IT RIGHT?

I always have difficulty with this question. It has no reality in a product development environment. You NEVER have time to do it right. You only have time to make compromises and the best use of your time is to spend it making sure you are making the best compromise you can.

That said, there are a few things I wish could have been implemented differently.

- The arrow should be appear on the source icon whereever the user places it upon button down.

- Having the option to crop icons to only the actual bits in use would facilitate better arrow placement and hopefully better error rates.

- When the user is dragging text a little "bouncing" caret should follow along with the text telling the user where the text will end up. (Users asked for that.)

- Get rid of the copy and link feedback. Further exploration would be needed before doing that.

- Explore the use of color in the valid and invalid indicators.

- I think drop site feedback is under utilized in applications and is something that can aid users and make the system seem much more predictable. I would like to explore more and better uses of this in the future.

ACKNOWLEDGMENTS

The author would like to acknowledge Robert O'Brien, Will Tuthill, Patrick Curran, and Janice Winsor, the members of the Drag and Drop team. And many thanks to John Fowler, Maria Capucciati, Kim Donner, and Rick Levine for their support on the design issues as they came up.

REFERENCES

As is typical on a product with tight schedules, we did not have time to reference everything we would have liked to. I have listed here the books we did reference and the systems we reviewed for comparative analysis.

1. Hewlett-Packard Company, International Business Machines Corp., Sun Microsystems, Inc., Novell, Inc. *Common Desktop Environment Style Guide.* Unpublished manuscript (publishing date Spring, 1995).

2. Hewlett-Packard Company. *HP Visual User Environment 3.0 User's Guide*, Hewlett-Packard Company, Corvallis, Oregon, 1992

3. Open Software Foundation, Inc. *OSF/Motif Style Guide* (revision 1.2). Prentice Hall, Englewood Cliffs, New Jersey, 1993.

4. Sun Microsystems, Inc. *OPEN LOOK Graphical User Interface Functional Specification*. Addison-Wesley, Menlo Park, California, 1989.

We reviewed DND in OPEN LOOK in OpenWindows 3.1 on the Sun platform, in System 7.1 on the Macintosh, in MSWindows 3.1, Hewlett-Packard VUE, on an Apple Newton, and AIX on the IBM platform.

The Effects of Practical Business Constraints on User Interface Design

Debra Herschmann
Sigma Imaging Systems
622 Third Avenue, 30th Floor
New York, New York 10017
(212) 476-3000

ABSTRACT

In a business environment, resource, budget and schedule constraints profoundly affect a product's user interface design. This paper describes the design of a graphical workflow application as it was affected by compromise between management, design and development during the product life cycle.

The product is tracked from its initial implementation as a highly functional utility with a non-standard user interface, to its brief life as a prototype representing the ultimate workflow tool. Primary focus is on the third, most recent version, and the design problems that arose in delivering a highly usable interface within practical, real world constraints.

KEYWORDS

Iterative design, resource constraints, compromise, prototyping, usability testing.

PROJECT HISTORY

RouteBuilder 1.0, released in 1990, was the first graphical workflow product on the market, allowing businesses to define the routing of work between users for processing. The goal was to enable business managers to create and maintain department workflows without knowledge of programming or scripting languages. Developed by a small group of programmers with limited user interface expertise, RouteBuilder 1.0 focused on utility rather than usability. The ability to build a workflow using drawing tools was revolutionary, setting the standard subsequently copied by other software companies. Still, major portions of the interface were difficult to use. While the RouteBuilder 1.0 development team had a thorough understanding of the users' needs, no formal usability testing was performed before release.

After release of the initial product, developers worked closely with users for three years to understand how they utilized and interacted with the product. Problems were corrected, features added incrementally, and recommendations gathered for future iterations. This field experience, together with trends in the marketplace, formed the basis of later versions.

RouteBuilder 2.0, a prototype, focused on an appealing user interface, and illustrated future product direction. The goal was to identify the scope of work necessary to implement a new front end incorporating added functionality, and to surpass other companies beginning to release competing products. The front end was prototyped using visual programming tools and designed without consideration of practical constraints. The cost of actually implementing expensive user interface features was disregarded, encouraging over-featuring of the product. RouteBuilder 2.0 was created by a user interface designer and a product manager, with limited input from development. This version, demonstrated at trade shows and conferences, was very well received.

RouteBuilder 3.0, currently in the development process for release in 1995, strikes a balance between the two previous versions, focusing on both functionality and usability. The goal was to design a powerful, usable product that could be implemented within budget and schedule, while laying the groundwork for future versions. The product was designed by a larger, more varied team, applying a more structured approach to software development The design cycle included requirements definition, functional specification, prototyping, usability testing and quality assurance testing.

THE DESIGN TEAM

The RouteBuilder 3.0 design team consisted of a core of developers with a range of GUI and workflow experience, two project leaders responsible for resource management and scheduling, a user interface designer, and a marketing person familiar with existing users and industry trends. Periodic input was solicited from documentation, quality assurance and graphic design experts. This multidisciplinary approach encouraged discussion and allowed many perspectives and areas of expertise to be incorporated into the final product. However, it also brought together people with different, sometimes conflicting agendas.

Project management's agenda was to deliver a product that satisfied the functional requirements, within schedule, with the allocated development staff. Marketing wanted to provide functionality and a user interface competitive in the marketplace. The interface designer's objective was to provide an elegant, easy to use interface camouflaging complex underlying technology. Development's goal was to implement a stable, technically powerful, state of the art product, optimizing programming efficiency by reusing programming code across the application. The system designers, much to their credit, rarely allowed implementation issues to cloud the front end design.

Disagreement between team members was resolved in several ways. In most cases, project management's agenda took priority. They weighed necessity against affordability and either included or cut features. In several instances, upper management vetoed decisions made by project management. One figure in upper management, heavily invested in interface design, became the user interface champion, salvaging components that he deemed necessary for the product to be competitive by allocating additional resources. Issues were occasionally decided by personal factors, such as how persistent and energetic each team member was, and who argued the most eloquently or screamed the loudest.

This team approach to design resulted in a well balanced application. The interface model was preserved, though with less front end functionality than originally specified. Practical constraints resulted in features being cut, especially those provided solely to enhance ease of use. Multiple techniques for performing the same task were eliminated. Interface behavior was occasionally revised to allow reuse of program components or to increase performance. For example, when the algorithm for drawing workstep connection lines proved too slow, the visual design was changed to optimize performance. Some functionality, eliminated during the estimation process, was restored after testing proved it necessary for ease of use. In addition, features supported at low cost by a new class library, Microsoft Foundation Classes (MFC), were exploited. This boosted the quality of the user interface and its compliance with Windows GUI standards.

These design modifications affected departments outside of development. Documentation, quality assurance and marketing departments began writing user guides, test scripts and promotional materials. They soon discovered that they were documenting a moving target. As this problem became clear, they were informed of upcoming design revisions more quickly. While the iterative design cycle slowed their documentation process, frequent communication helped avoid duplication of effort.

The Role of the Designer
In this development environment, the role of the designer became to design the most reasonable interface within practical constraints, eliminating or revising features and exploiting easily supported functionality. It also became an exercise in revising, refining and documenting highly detailed designs, and communicating these changes to development, documentation and quality assurance departments. Multiple versions of design documentation were maintained, to be used when cut features were restored, and to provide input for future product versions.

A primary responsibility was to argue for features that, if eliminated, would jeopardize product usability or preclude future enhancements. However, the designer was occasionally too reluctant to abandon designs and could have spent less time on detailed definition of features likely to be cut. The design cycle was a lesson in compromise, paring down the ultimate yet impractical front end to achieve a scaled down, usable and implementable interface.

ROUTEBUILDER USER INTERFACE DESIGN
RouteBuilder 3.0 includes several unique and innovative features that distinguish it from other graphical workflow tools. The evolution of these features, including graphical route creation, dynamic and intelligent rule creation, prepackaged worksteps and legends, is described below. Specific design goals for RouteBuilder 3.0 included tightly integrating visual elements with underlying behavior, camouflaging the complex workflow engine, eliminating redundant tasks, and complying with GUI standards and state of the art interface techniques.

Prototyping and Usability Testing
Prototyping was advocated by management from the beginning of RouteBuilder 3.0. Prototypes met with varying degrees of success based on the tools used, the suitability of the features selected for prototyping, and the technical skill of the individual building the prototype.

Usability testing included paper prototypes, on-screen prototypes built using Visual Basic, Visual C++ and MFC, and customer presentations and reviews. RouteBuilder 3.0 was coded in segments, allowing features to be tested and revised on an ongoing basis, rather than waiting for testing of a complete product.

GRAPHICAL ROUTE CREATION
Window structure and components vary between the three product versions, based on design decisions, implementation constraints and changes in GUI standards. The main windows of the three product versions appear in figures 1-3.

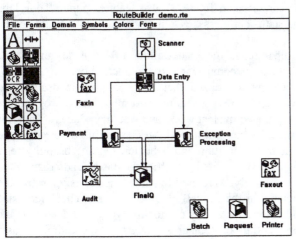

Figure 1. RouteBuilder 1.0

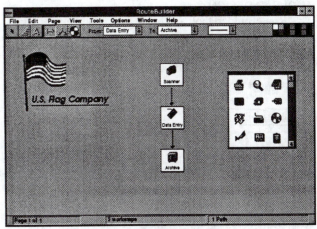

Figure 2. RouteBuilder 2.0

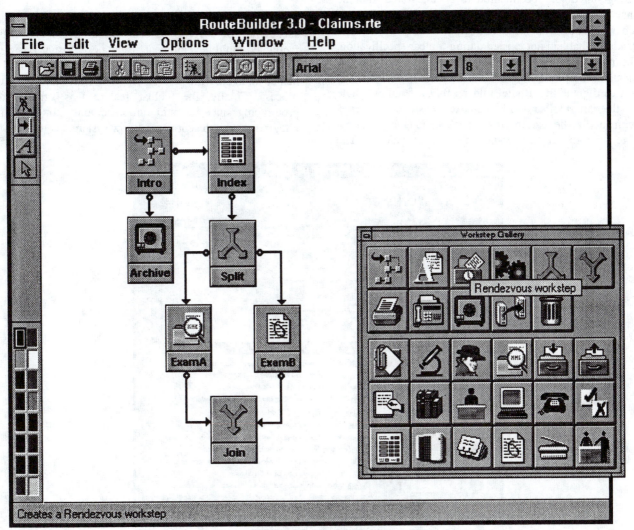

Figure 3. RouteBuilder 3.0

TOOLBAR DESIGN

The RouteBuilder toolbar changed dramatically between product versions. RouteBuilder 1.0's static toolbar was replaced by RouteBuilder 2.0's dynamic toolbar (figure 4), whose elements changed as different tools and route objects were selected.

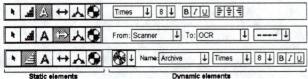

Figure 4. RouteBuilder 2.0 dynamic toolbar

RouteBuilder 3.0 initially called for the highly dynamic toolbar prototyped in version 2.0. The 3.0 design was revised several times due to tradeoffs between costly user interface behavior and practical budget constraints. Deemed too costly to implement, the dynamic toolbar was replaced by one with static buttons. Microsoft Foundation Classes allowed advanced interface behavior scaled down during functional specification to be integrated into the final product. MFC facilitated the implementation of behavior compliant with emerging Windows interface standards that had not existed during toolbar specification. The result was not exactly like the 2.0 toolbar, but was equally usable and had additional capabilities, including user configurability. RouteBuilder 3.0 now features a state of the art toolbar with context sensitive elements. Segments can be rearranged and docked along any window edge, or torn off to become free floating palettes. The toolbar (seen in figure 3) also features popup Tool Tips and status bar messages to provide users with on-line help.

Toolbar icons for RouteBuilder 1.0 were designed by a graphically competent programmer. For version 3.0, management realized that hiring a graphic designer to create an icon set would not cost more than a programmer doing the equivalent work, and would produce far superior results. Visual Basic and MFC were used to prototype toolbar composition and behavior. On-line usability tests (figure 5), created in Visual Basic and distributed electronically, were used to validate icon design, select the contents of the RouteBuilder icon library, check icon size and color on a variety of monitor types, and determine the composition of toolbar segments. In addition to user responses, hardware information, such as monitor resolution, was captured automatically. The toolbars were initially tested in-house, then distributed to customers in the United States and finally tested internationally. Testing resulted in 25% of the original icons being totally redesigned, and another 30% modified slightly.

Prototyping and usability testing of the toolbar were successful for several reasons. Visual Basic and MFC facilitated rapid development at low cost. Visual Basic, in particular, did not require much technical expertise. It also helped automate the test distribution and data gathering process. Finally, the toolbar feature was suitable for prototyping since it could run standalone, not requiring integration with the entire workflow system.

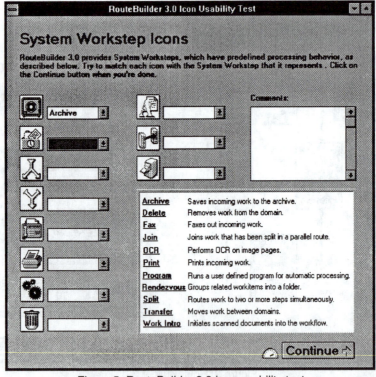

Figure 5. RouteBuilder 3.0 icon usability test

DYNAMIC RULE CREATION

Each RouteBuilder workstep represents a processing activity. User generated rules determine the routing of items between worksteps. Unlike other workflow tools, RouteBuilder provides a graphical interface for rule creation, and does not require users to program.

RouteBuilder 1.0 provided a linear technique for rule creation (figure 6), requiring navigation through multiple modal dialog boxes, with rules created and viewed individually. Complex rules could be created, yet the interface was cumbersome.

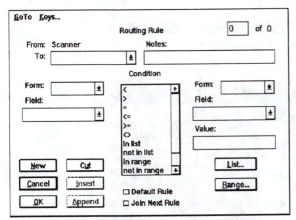

Figure 6. RouteBuilder 1.0 Rules dialog

The RouteBuilder 2.0 Rules dialog (figure 7) retained the basic rule creation structure, but replaced multiple modal dialogs with a single dialog box. Rules were set via controls displayed dynamically based on context. Users could view and manipulate multiple rules simultaneously.

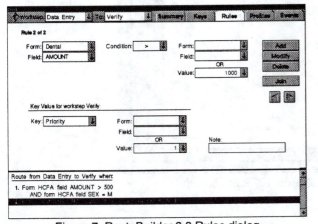

Figure 7. RouteBuilder 2.0 Rules dialog

RouteBuilder 3.0 expanded the scope of the rules, requiring the addition of interface elements to support new functionality. This motivated total revision of the rule creation interface (figures 8-10). Controls were originally specified to appear dynamically within the rule, providing context sensitive selections and allowing direct manipulation (figure 9). While this seemed usable as

documented, and even as tested with paper prototypes, it was visually disconcerting on screen. When the design problems could not be resolved within the time allocated, the moving prompt was revised to be displayed in a fixed location (figure 10). The resulting interface is effective for experienced users, but daunting to novices. Features to assist users, including context sensitive help and support of novice and expert modes, are contingent on scheduling.

Rules prototyping was originally intended to be done with MFC, with prototype code being used for the actual front end. This plan failed because inexperience with the new tools, including Visual C++ and MFC, made turnaround time too slow to be effective for rapid prototyping and revision. More successful was the use of Visual Basic for quick, disposable prototypes demonstrating individual aspects of rule behavior. Simple behavior was demonstrated effectively using third party applications not usually considered as prototyping tools. For example, formatting and line wrap were shown using tables in Microsoft Word. Cut, copy, paste and delete were demonstrated using Excel macros. These prototypes were useful in validating design decisions, communicating ideas between team members, and raising design details for resolution.

The RouteBuilder 3.0 rule design process was challenging, intensely detailed and the cause of much anxiety. The commitment to solving and implementing this design reflects on its high degree of support by upper management, determined to provide a unique interface suitable for novices and experts. Were it not for this support, the design would have been initially resolved in a less risky manner. Of all the specified RouteBuilder functionality, it was most difficult to envision the final look and feel of the rule interface. It was understood from initial design that the end result may be revolutionary for workflow applications, or may prove to be difficult to use. The interface is being refined and is subject to change.

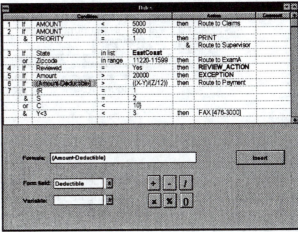

Figure 8. A preliminary RouteBuilder 3.0 Rules dialog

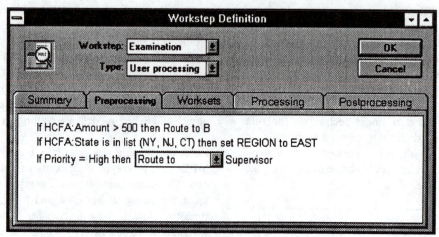

Figure 9. An early Rules dialog with in-place editing

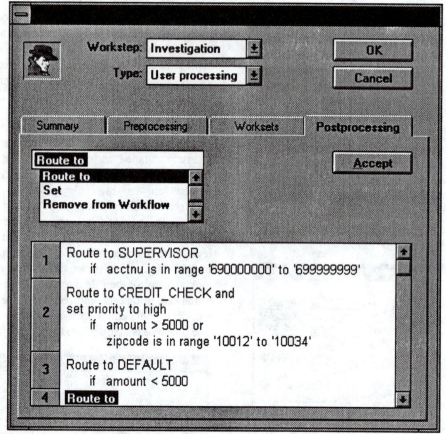

Figure 10. The most recent RouteBuilder 3.0 Rules dialog

PACKAGED WORKSTEPS

Years of field experience showed that similar workstep types were being created by many customers. This resulted in a new prepackaged workstep feature (figure 11), making the creation of predefined workstep types instantaneous. While in previous versions, workstep icons were purely visual, prepackaged worksteps couple icons with predefined functionality.

Packaged worksteps highlight the value of examining product usage and providing functionality that automates common tasks. Conceptualized by team members with extensive customer experience, knowledge of workflow system functionality and an understanding of how to package that functionality, this feature would not have been realized by a less diverse team.

Figure 11. RouteBuilder 3.0 Workstep Gallery with Packaged Worksteps

LEGENDS

The RouteBuilder 3.0 Legend feature (figure 12) allows line styles and colors to represent different types of work flowing through a route. Motivated by the complex, spaghetti-like route maps generated by users of the initial product, Legends allow hiding and displaying of connections by style, filtering the map. The original concept was to tightly integrate visual representation with underlying behavior. By associating rules with Legend styles, users could quickly generate complex rule sets by simply drawing a connection between worksteps.

Due to time constraints, the Legend feature will provide visual mapping of styles to user defined descriptions, and toggling of individual styles. The ability to associate rules with a Legend style has been postponed.

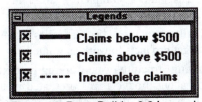

Figure 12. RouteBuilder 3.0 Legends

LESSONS LEARNED

Below are lessons learned as a designer working in a business environment.

Keep the ultimate product vision in sight and define base functionality which cannot be compromised. Learn to distinguish between features that are critical and shape the product's future direction from features that can dropped or added incrementally after release.

Create a detailed and explicit user interface functional specification, and maintain multiple versions as features are revised. On the RouteBuilder 3.0 project, documentation of all features was saved even after features

were revised or eliminated. When features were resurrected, it was simple to provide developers with the necessary specifications. Such documents also provide ideas for future versions of the application.

Develop rapid and disposable prototypes, rather than time consuming code, which management may feel committed to use after expending resources. Prototype individual features until a complete product is available for testing. Simple behavior can be demonstrated quickly using third party applications such as Word and Excel, not usually associated with prototyping.

During the estimation process, beware of questions that start with "How much would you mind if.." and "How bad would it be if..."

Communicate design revisions with departments external to development affected by the changes. Overlooking this factor early in the RouteBuilder design revision process caused quality assurance to rewrite test scripts. Informing documentation, training, Q.A. and marketing of modifications in a timely manner helps avoid duplication of effort.

Track the user interface as it progresses through all phases of development. Test interim versions of the software. Discuss interface behavior with the developers implementing each feature. This helps avoid misinterpretation or oversights by managers and developers. It also allows clarification of details missing from the functional specification and resolution of design problems arising during implementation.

Most importantly, realize that a user interface designer, developing a product within practical constraints, after negotiation and compromise, is part of the creative process.

REPLACING A NETWORKING INTERFACE "FROM HELL"

Roxanne F. Bradley
Hewlett-Packard Company
19420 Homestead Road, MS 43LE
Cupertino, CA 95014
408/447-4240
roxanne_bradley@hp6600.desk.hp.com

Linn D. Johnk
Hewlett-Packard Company
19483 Pruneridge Avenue, MS 48NA
Cupertino, CA 95014
408/447-0804
ljohnk@cup.hp.com

ABSTRACT

A multidisciplinary design team at Hewlett-Packard (HP) has successfully designed a new user interface for a network troubleshooting tool. Users felt that the new interface let them focus on the task of network troubleshooting, thus freeing them from the details of the interface and its underlying implementation. The design team believes that the success achieved is due to the process used and the multidisciplinary aspect of the team.

This design review describes the process followed by the design team, the difficulties encountered, the results obtained from a comparative evaluation of the new and existing product interfaces, and the lessons learned.

KEYWORDS

User-centered design, usability release criteria, usability inspections, comparative usability testing.

INTRODUCTION

Networking products are typically difficult to learn and use. This is because, by definition, networking products span multiple systems, multiple communication methodologies, and, probably, multiple geographic areas. Representing these complex concepts and data structures makes developing interfaces for networking products quite difficult.

This project involved redesigning the existing interface for a network troubleshooting tool -- Network Tracing and Logging (NetTL). NetTL is a facility that is used by many of Hewlett-Packard Company's (HP's) networking products. It provides the user with a consistent way to gather information about networking products to solve problems with the network. A survey of both external and internal users found that using NetTL was so difficult that many users refused to use it or used only a small fraction of its

capabilities. In fact, some users have referred to this interface as the interface "from hell."

By analyzing the users of NetTL and the tasks that NetTL needs to support, the design team found that:

- Users are under pressure when solving network problems.
- Problem resolution must be as quick as possible.
- Users need to use this tool only infrequently.
- Users do not have knowledge of the underlying networking products' structure (i.e., subsystems).

The team found that the current interface (a command line) did not meet the needs of the user because of its steep learning curve and high cognitive (or mental) load. To illustrate, the NetTL user had to understand:

- Two cryptic command line interfaces with multiple, non-intuitive options (there was one interface for capturing data and a completely different one for reporting data).
- The nettl file structure, because files have to be modified to get products to work as wanted.
- The underlying structure of networking products since information about products was presented in terms of subsystems with no references to the associated products.

Figure 1 shows the steps, using the current interface, to perform a fundamental task -- changing the type of information gathered about a networking product (in this case, OSI) from Disaster and Error messages to Disaster, Error, and Warning messages.

Because of the steep learning curve and heavy mental load placed upon the user when using the current interface, the user had no time to focus on the networking problem. The team agreed that a well-designed graphical user interface would decrease the learning curve and lessen the user's mental load and thus significantly improve the accuracy and speed with which the user could get networking information. Figure 2 shows the steps, using the graphical interface, to perform the same fundamental task as outlined in Figure 1.

1. The user must **remember or look up** the command "nettl -ss".

2. The user types "nettl -ss" and receives the output as shown below.

3. The user must **remember or look up** the subsystems associated with OSI. NOTE: At this point the user very often would give up and call HP. Subsystems are: ACSE_US, CM, EM, HPS, SHM, ULA_UTILS.

4. The user must **remember or look up** the command to change the logging levels:

   nettl [-l class [class...] -e subsystem [subsystem...]]

5. The user must type the command.

   nettl - l disaster error -e acse_us cm em hps shm ula_utils

6. To ensure the changes are correctly made, the user must **remember** and type "nettl -ss" and **remember** the subsystems to look for changes on the display.

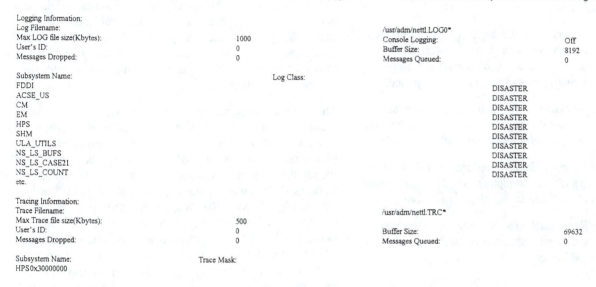

```
Logging Information:
Log Filename:                                        /usr/adm/nettl.LOG0*
Max LOG file size(Kbytes):           1000            Console Logging:               Off
User's ID:                           0               Buffer Size:                   8192
Messages Dropped:                    0               Messages Queued:               0

Subsystem Name:                          Log Class:
FDDI                                                                     DISASTER
ACSE_US                                                                  DISASTER
CM                                                                       DISASTER
EM                                                                       DISASTER
HPS                                                                      DISASTER
SHM                                                                      DISASTER
ULA_UTILS                                                                DISASTER
NS_LS_BUFS                                                               DISASTER
NS_LS_CASE21                                                             DISASTER
NS_LS_COUNT                                                              DISASTER
etc.

Tracing Information:
Trace Filename:                                      /usr/adm/nettl.TRC*
Max Trace file size(Kbytes):         500
User's ID:                           0               Buffer Size:                   69632
Messages Dropped:                    0               Messages Queued:               0

Subsystem Name:              Trace Mask:
HPS 0x30000000
```

Figure 1. Current NetTL Interface Steps for Changing Logging

1. The user must **remember or look up** the command "nettladm".

2. The user types "nettladm." On the resulting display (see below), the user finds the product of interest in the window, highlights all of the subsystems associated with the product of interest, and selects "Modify Logging..." from the "Actions" menu.

3. In the resulting dialog box (see below), the user selects the appropriate logging levels, and selects "OK". NOTE: The user can change the console notification here as well. With the current interface, the user would have to use a different tool *and* modify a file to change the console notification.

4. The logging screen is returned, where the user can quickly verify that this information is correct.

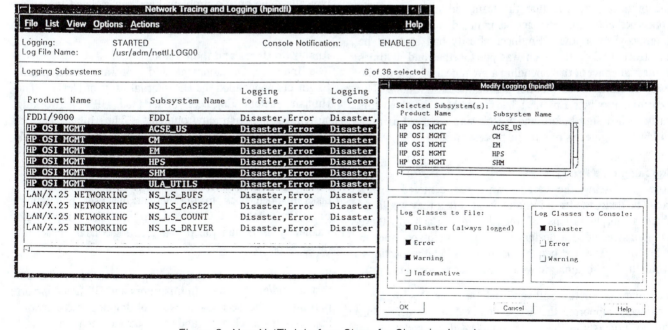

Figure 2. New NetTL Interface Steps for Changing Logging

THE DESIGN PROCESS

Creating a Team

The design team consisted of a Human Factors Engineer, a Learning Products Engineer, an Interface Development Engineer, and a Software Development Engineer. The Human Factors Engineer was responsible for leading the team in applying the user-centered design process. During this process each member of the team was responsible for different points of view:

Human Factors Engineer	Responsible for applying Motif interface design rules and HP's internal Motif design implementation, and ensuring that human factors principles were applied.
Learning Products Engineer	Responsible for ensuring that the design required minimal, if any, hardcopy documentation and minimal online help.
Interface Development Engineer	Responsible for ensuring that the design could be implemented with the tools chosen and for developing prototypes.
Software Development Engineer	Responsible for ensuring that the design could be implemented given the current NetTL technology.

Although the entire team was committed to making the new interface easy to use, only the Human Factors Engineer had any experience with the user-centered design approach. Each member of the team began the design process with a different perspective about what the process and their role in the process would be. The Software Development Engineer believed that the interface was independent from the product code and that the team did not need to be concerned about any changes being made to the code. The Interface Development Engineer already had a prototype constructed when the team was put together and at times expressed concern that spending a great deal of time on the users and their tasks was not getting the design done. The Learning Products Engineer had never been involved this early in a design before and wasn't sure what a Learning Products Engineer's role in the process was.

Agreeing on a Process

After discussing the various ways user-centered design could be done, the team initially decided that they would do three major iterations of the design. Each of these iterations included a design/redesign, a prototype, and an evaluation. The first two evaluations would be inspections, while the final evaluation would be a user performance test.

The manager of the Software and Interface Development Engineers supported the user-centered design process, but did not understand that code would be developed later in the process than anticipated. Several times throughout the process the manager had to be reminded that a user-centered design approach focused on design first and then coding.

Understanding the Product

The Software Development Engineer explained to the team that NetTL is a facility that can be used for tracing and logging by HP-developed networking products. Networking products consist of several subsystems. During operation, these subsystems may encounter some activity that is designated as an exception to the normal rules of operation. Since exceptional or abnormal events may indicate that the product is not operating as it should, the user needs to be notified of these events.

The purpose of logging is to record or log events as they occur. Events fall into four major categories: Disaster (always logged), Warning, Error, and Informative. To obtain notification of desired event types, the user must specify the types of events about which he or she wishes to be notified. The user can also specify how that notification should take place. Events are recorded in a log file; notification of events may also be sent to the console monitor if desired. Once information has been recorded in a log file, the user may retrieve all or part of it by setting up appropriate search criteria. Capturing and retrieving the right information is a critical first step in the troubleshooting process.

Tracing differs from logging in that it is not triggered by an event. Rather, tracing records the activity of a given subsystem. As with logging, the tracing user can specify what information is captured for later perusal, as well as the search criteria to apply to the captured data. Tracing is typically done once a problem has been isolated to one or more subsystems.

Analyzing Users and their Tasks

The Interface Development and Software Development Engineers understood the HP internal users of NetTL. The Human Factors Engineer understood HP's networking customers and their environments. The Human Factors Engineer began leading the team through the user and task analysis by facilitating discussions of: a) who all the users in a networking environment were, b) what the networking tasks for those users were, and c) how the NetTL-specific tasks mapped back to the networking user population. From that analysis the team determined the primary and secondary users and the primary and secondary tasks for those users.

The primary users of the logging portion of NetTL are system and/or network administrators and HP field support personnel. The secondary users of logging are network application developers and HP development and call support personnel. The experience with networking for these primary and secondary users ranges from little or none to considerable. All these users primarily use logging to obtain notification of desired event types and to troubleshoot problems.

For the tracing portion of NetTL, the primary users are network application developers and HP development and call support personnel, while the secondary users are system and/or network administrators and HP field support personnel. The secondary users of tracing are most likely to use that portion of the facility under the guidance of a primary tracing user. Troubleshooting is the primary task that tracing supports. Tracing facilitates troubleshooting by helping the user find out as much as possible about what is happening on the network.

Since it had been determined in a survey that the primary users of logging were completely unsuccessful using NetTL, the design team decided that simplifying logging tasks for these users would be the primary goal of the design team. Thus, the design team determined the key task objectives for the first release of the new interface were:

- Ability of the primary logging user to log events of interest.
- Ability of the primary logging user to retrieve logging data necessary to perform basic troubleshooting tasks.
- Ability of the primary logging user to identify and correct problems resulting from loss of messages by the logging facility.
- Ability of the primary tracing user to guide a secondary tracing user successfully through a trace.

The team learned to use this information about the users and their tasks to make design decisions. As the team became more skilled, the role of usability advocate increasingly became a shared one, rather than the responsibility of the Human Factors Engineer alone.

Developing Release Criteria

Release criteria are used to set goals for products. These goals help the team make decisions about added functionality, interface design trade-offs, and whether the product is ready to be released. Release criteria also describe what the user experience should be like when performing tasks using a product.

Before developing the release criteria, the team evaluated what impact the knowledge of the users and their tasks could have on the design of NetTL. The team agreed that: 1) notifying users of networking problems is critical, 2) enabling users to solve networking problems in a timely manner is critical, and 3) enabling users to solve problems without having to call HP support is critical.

The team understood that, for the new NetTL interface to be successful in meeting user needs, the learning curve, the

mental load while performing tasks, and user time on task would all need to be reduced. At the same time, the user's chance of success and level of satisfaction with the product would have to be increased. Based on this understanding, the team set the release criteria in Table 1.

Measure	Target	Minimum Acceptable
Time on task	≤ 2 minutes	≤ 10 minutes
Success rate	100%	80%
Percentage using hardcopy	0%	20%
Satisfaction rating	≥ 7 on 9-point scale	≥ 5 on 9-point scale

Table 1. Release Criteria

Target values represent ideal or "stretch" goals, while the minimum acceptable is the level of performance that must be achieved in order to release the product. Because these were product release criteria, this meant that, before the product could be released, at least the minimum acceptable criteria had to be met for each key task objective.

Most of these measures are well understood within HP as user needs but had never been used as product release criteria. Because each key task had to meet each of these release criteria, the criteria were considered very aggressive, if not impossible, by other HP development teams in similar business areas.

Iteration 1

The first iteration involved designing the interface, developing a paper prototype, and evaluating that prototype.

The design. For this design, the team relied upon the user and task analysis to make design decisions. Each member of the team contributed to the design in unique ways based upon their expertise. For example, the Learning Products Engineer evaluated each design idea and determined what the documentation requirements would be. If hardcopy documentation would be required to support the design idea, the Learning Products Engineer informed the team and the team modified the idea or devised a new approach.

As the team designed, a paper prototype was drawn with descriptions of the fields and navigation that would take place in the context of performing key logging and tracing tasks.

The evaluation. For the evaluation of the design, the team performed a usability inspection. A usability inspection is an evaluation process in which people review the interface using typical task scenarios. The inspectors raise concerns about how the interface or product works, and these concerns, as well as any ideas or suggestions are logged. The design team then reviews the concerns and determines the root cause for each concern. These root causes are then examined and the interface redesigned based upon what the team believes will fix the root cause. The inspection team for the paper prototype consisted of internal users of the existing product.

Most of the inspectors' concerns with the interface were about terminology and the purpose of an action or screen. These concerns indicated a mismatch between the interface and the user's mental model of NetTL. It was obvious that there were some major problems with the design. In fact, most of the inspectors were unable to complete the task scenarios within the allotted time.

Iteration 2
The second iteration involved redesigning the interface, developing a prototype, and evaluating the design.

The redesign. The root cause of many concerns raised during the inspection was that information was not being presented in a format users could understand. Another root cause was that the design too closely imitated the existing product by arbitrarily separating tasks, such as capturing the data from retrieving it. The team agreed that the new graphical user interface should hide more of the complexities of the existing product.

For example, the configuration of logging to the log file and logging to the console was one example where the original design arbitrarily separated tasks by too closely following the existing product. In the redesign, these configuration activities were "integrated" by modifying the interface so that the two tasks could be performed on the same screen.

The Prototype. After many design meetings in which the team discussed the user's mental model and iteratively redesigned each screen, the redesign was complete. The redesign was documented and a prototype developed. Navigation was possible in the prototype, and, although data could be manipulated on the screen, it was not saved, nor did data manipulations affect other screens. With this design, the user was able to see the objects of interest upon entry into the product. Also, the actions allowed were related to the objects on the screen.

Development of the prototype is another example of where the multidisciplinary team was of value. As the Interface Development Engineer developed the prototype and found technological problems implementing the design, the Interface Development Engineer brought solutions to these problems to the team. The team discussed the proposed solutions and either agreed to them or modified them.

The evaluation. Once the prototype was complete, another usability inspection was performed. The inspectors from the first inspection participated.

With the redesigned interface, the inspectors were able to perform more of the tasks successfully. There were fewer concerns during this inspection than in the first, and the concerns focused more on the functionality than on the terminology. From an analysis of the concerns raised by the inspectors, it became clear the interface still needed to provide the user an obvious way to access and accomplish frequently performed tasks.

Iteration 3
The third iteration involved redesigning the interface and updating the prototype.

The redesign. Addressing the concerns from the second inspection did not necessitate a complete redesign as was the case with the first inspection. These concerns caused the team to rethink some of the information on the screen and how it was presented. One of the root causes for the concerns identified was the underlying operation of the product. For example, console notification was determined by setting the events for all products. If Disaster events were being reported to the console, and the user wanted to be notified of Disaster and Error events for a particular subsystem, Disaster and Error had to be turned on for all subsystems. The team agreed that console notification should be determined on a subsystem basis. This required modifications to NetTL by the Software Development Engineer.

The Software Development Engineer was willing to make the changes to console notification so that it could be turned on for individual subsystems, as well as other modifications necessary to improve usability. This need to make internal changes to improve usability and navigation underscores the fact that a user interface is not something added on to the product as an afterthought or at the end of a development cycle.

The remainder of the changes involved fine tuning the interface. This was done fairly quickly for most screens.

The Prototype. Based on the comments received, the team decided that no more major iterations were necessary, it was time to evaluate the product in terms of the release criteria. For the evaluation of the release criteria, HP customers would use the prototype. The prototype reflected the new design and simulated the manipulation of real data.

MEETING THE RELEASE CRITERIA
The design team decided that a timed usability test was the best way to determine whether the criteria had been met. The team concluded a timed test could determine how users of the graphical and command line interfaces would

perform relative to the release criteria and whether there would be a difference in user performance based on the type of interface used. To avoid the risk of introducing bias into the testing, the design team selected a Human Factors Engineer who was not part of the design team as the test administrator.

Test Design

Test Participants. Twenty system and/or network administrators were recruited from the HP customer base to participate in the test sessions. All test participants indicated they were at least sometimes responsible for monitoring network products and doing preliminary troubleshooting. These individuals were either novice or occasional users of the existing NetTL interface.

Half of the participants performed logging and tracing tasks using the new graphical user interface (GUI); the other half of the participants performed the same tasks using the existing command line interface (CLI).

Test Environment and General Procedures. The tests were conducted in the HP Cupertino Site Usability Lab. Participants were videotaped while performing representative tasks. Since these were timed tests, the think-aloud method was not employed. Users were told at the outset that if difficulties were encountered, they should attempt to work through them on their own as best they could, but not to the point of frustration. A maximum of ten minutes was allowed per task (based on the minimum acceptable criteria for time on task).

Three logging and one tracing task comprised the test session. The tasks were as follows:

1. Changing the logging level for a particular networking product from Disaster to Disaster, Warning, and Error.
2. Monitoring the log file to determine that the appropriate information was being logged for the networking product of interest.
3. Determining if there was any problem with logging information being "dropped" from the log file and, if such a problem existed, resolving it.
4. The tracing task was a scripted, simulated support call.

Following completion of the final task scenario, a posttest attitude survey was administered to measure reactions to the product. After finishing the questionnaire, participants were debriefed, both to gain an understanding of the participant's ratings as well as to obtain retrospective feedback on the experience of using the product and any suggestions for improvements.

Data Analysis. To assess performance relative to the release criteria, both quantitative and subjective measures were obtained. The following three measures of user performance were taken and analyzed for each task tested:

1) time required to complete tasks, 2) successful completion, and 3) whether hardcopy documentation was required to perform a task, where merely reaching for the manual was counted as requiring it.

Satisfaction ratings were collected and analyzed for the interface in general. The ratings were made on a 9-point scale and included the following:

- Separate averages for each of seven attributes [Easy to learn, Easy to use, Flexible, Power (i.e., adequate functionality to accomplish tasks), Stimulating, Satisfying, Wonderful].

- Composite average of the seven attribute scores, representing an overall perception of usability.

Test Results

On all measures, performance with the GUI exceeded the minimum acceptable release criteria while performance with the command line interface seldom did, specifically:

- The GUI met the minimum acceptable criteria for all tasks. It also met the target criteria for satisfaction and hardcopy usage for all tasks.

- The CLI met the minimum acceptable criteria for time and success on only one task: the simulated support call task (a task users did not have to perform entirely on their own). For the Power attribute of the user ratings, the CLI also met the minimum acceptable.

Time Results. Time on task using the GUI (4.1 minutes averaged over all test tasks) met or exceeded the minimum acceptable criteria of ≤ 10 minutes. Participants using the command line interface required more time to complete tasks. In fact, except for one task, all command line users required more than the allotted 10 minutes, so the CLI did not meet the minimum acceptable time criteria.

The following comments from representative users reflect the difference in time:

- CLI user: "After an hour [on all test tasks] I still couldn't get it to work. I felt another hour wouldn't help. It looked like it would take days."

- GUI user: "The time it took to do things was reasonable; it only took a minute or two."

Success Results. Participants using the GUI were considerably more successful than CLI users (GUI -- 78% success rate averaged over all test tasks; CLI -- 16% success rate averaged over all test tasks). The GUI success rate is not statistically different from 80% so the GUI met the minimum acceptable success criteria despite the presence of two critical defects. Those defects have been fixed, and subsequent user feedback suggests the success rates for the GUI would now be much higher. The CLI success rate value reflects the one relatively simple task that some participants were able to complete successfully.

These user comments describe the difference in the perception of task success:

- CLI user: "Nothing was easy. I don't think I succeeded in any task."

- GUI user: "It was satisfying to use. I felt confident about getting on the road to the solution."

Hardcopy Results. On this measure, participants using the GUI met the target criteria of 0% requiring hardcopy. Participants using the command line interface were more likely to require hardcopy documentation in their efforts to perform tasks. Eight of the ten CLI users reached for the manual at least once during the test session, and, on average, CLI users required hardcopy on over half the tasks. While a few of the CLI participants reached for the manual as a matter of course, most first attempted to accomplish their tasks before resorting to the manual.

The following comments reflect the difference:

- CLI user: "I couldn't do anything. I needed to use the documentation and manpages a lot. The time that took was frustrating."

- GUI user: "I could figure out where everything was at. I didn't need to look at the manual. Playing led to learning."

Satisfaction Results. Satisfaction ratings for the GUI exceeded those for the command line interface on all usability attributes and the resulting composite measure, see Figure 3. The GUI ratings exceeded the minimum acceptable criteria (≥ 5 on a 9-point scale) on all attributes and were not statistically different from the target criteria (≥ 7 on a 9-point scale) on five of the seven individual attributes, as well as the composite. Ratings for the command line interface met the minimum acceptable criteria on only the "Power" attribute. Even there, the GUI score exceeded the command line result.

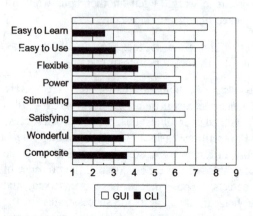

Figure 3. Satisfaction Ratings

These user comments reflect the differences in user perceptions of the two interfaces:

- CLI user: "Working with this was like doing carpentry through a keyhole."

- GUI user: "I believe it [the product] would keep me from having to make a support call."

Implications of the Results. During the design effort, the team found that a sizable volume of networking-related support calls involve helping customers use NetTL, thus significant support cost savings are anticipated. Cost savings are also anticipated from a reduced need for hardcopy documentation. Taken together, these savings were very instrumental in helping the design team justify implementation of their design to management. Harder to

measure than the savings to HP but of potentially greater significance are anticipated customer benefits from increased productivity and confidence.

LESSONS LEARNED

Design Process

Multidisciplinary Design Team. Having a multidisciplinary design team helped to make the design process a success. Everyone contributed to the design from their own perspective. The team believes that these different perspectives helped them to develop an interface that met the release criteria. This is because no one view of the product or interface could dominate the design. The members of the team believe that all design teams should be multidisciplinary.

Experience. Having only one person on the team with user-centered design experience does work if that person has the leadership skills and techniques necessary to, in essence, provide on-the-job training in user-centered design. Some of the many challenges this person faces are:

- Frustration that the team has to move slower than desired. This is because the team has to learn as they go.
- Continually keeping the team focused on the process.
- Answering doubts about the process and the value of the process from the team as well as management.

The Process. Product team members are familiar with identifying who their users are, but not defining their users and determining how this information may impact the design. Team members are also relatively unfamiliar with analyzing tasks from a user's perspective. They tend to adopt a product perspective instead. When the design begins and the user profile and task analysis information is used to make design decisions, then the value of the effort becomes clear to the team.

However, even at this stage there can be doubt about the process. Most doubts disappear once the first inspection is complete. The developers are particularly struck by the thoroughness of the inspectors and the amount of criticism and suggestions generated. For perhaps the first time, they realize that what they thought was intuitive isn't necessarily so for their users. The inspection feedback also convinces the team that several iterations of design and evaluation may be needed before the release criteria can be met. In fact, in early inspections, surface structure problems and mental model mismatches can hide fundamental problems with product functionality and/or structure.

Testing Process

By the time the testing of the release criteria happens, the design team is enthusiastically supportive of the process and understands the value of the evaluation. They are eager to see external customers use their design, so it is important to encourage them to observe as many test sessions as possible. During these sessions they can see ways to improve their contribution to the product. However, the test administrator needs to discourage team members from making changes until all test sessions are complete and the data has been analyzed.

There are several areas in the testing process where the test administrator must exert a controlling influence due to the risk of introducing bias. These include: a) interacting with test participants, b) analyzing test data, c) determining whether release criteria have been met, and d) identifying defects and assessing their severity. Because of this, it is important to have an unbiased tester, that is, someone not involved in the product design.

As mentioned previously, two critical defects were uncovered during the user testing. This demonstrates that inspections cannot entirely take the place of user tests. Nonetheless, the effectiveness of usability testing can be enhanced when a product has undergone one or more iterations of usability inspections. Based on experience with both inspected and uninspected products, the numbers of critical and serious defects found during usability testing is consistently less for products that have undergone inspections. Like more and more development teams at HP, the NetTL team determined they would not release their product until all critical and serious defects had been resolved. By finding and fixing defects as early as possible, the NetTL team has been able to freeze the design much earlier and move through the implementation phase more rapidly than is the case with traditional design processes.

Usability tests that validate release criteria can greatly enhance a design team's confidence in their product. Although this type of testing is expensive and should probably not be used to test every design, it does determine whether release criteria have been met. These quantitative results can also be used in many ways, such as to persuade management to support user-centered design or continue funding of the project.

CONCLUSION

Based upon the experience gained through this effort and many others, we believe that following the user-centered design process will result in products that users can use and that as more teams use this process, the value gained will become known. The success of the new NetTL interface has generated new interest and enthusiasm for user-centered design within the local networking products division. Several product teams are now working through the same process with the support of Human Factors. In addition, we are becoming user-centered design advocates by taking the NetTL success story to other local divisions, as well as to the wider HP audience and to upper management. We feel it is important for all user advocates to support this process and to ensure that all teams gain experience in user-centered design.

In conclusion, we believe the following are the keys to the success demonstrated in the NetTL design story:

- Establishing a multidisciplinary team.
- Making a commitment to user-centered design.
- Maintaining that commitment despite periods of doubt and frustration.
- Sharing the role of user advocate.
- Setting aggressive but attainable release criteria.
- Allowing those criteria to help drive design decisions.
- Getting user feedback early and often and using that feedback to refine the design.
- Moving beyond the product perspective to a user perspective.
- Validating that the product has met its release criteria.
- Quantifying anticipated company and user benefits due to improved product usability.

ADDITIONAL READINGS

1. Gould, J.D. How to Design Usable Systems. In M. Helander (Ed.), Handbook of Human-Computer Interaction. North-Holland, New York, 1988, pp. 757-789.

2. Nielsen, J. and Mack, R.L. (Eds.). Usability Inspection Methods. Wiley & Sons, Inc., New York, 1994.

3. Norman, D.A. and Draper, S.W. (Eds.). User Centered System Design. Lawrence Erlbaum Associates, Hillsdale, N.J., 1986.

User-Centered Development of a Large-Scale Complex Networked Virtual Environment

Thomas W. Mastaglio
Loral Federal Systems
12461 Research Parkway, Suite 400
Orlando, FL 32817
(407) 823-7345
tmast@greatwall.cctt.com

Jeanine Williamson
Dynamics Research Corporation
12461 Research Parkway, Suite 400
Orlando, FL 32817
(407) 823-7345
willt@greatwall.cctt.com

ABSTRACT

An integrated development team comprised of industry engineers, government engineers, and user community representatives is developing a large-scale complex networked virtual environment for the United States Army. The effort is organized into concurrent engineering teams responsible for each system component. Prototypical users who are formally called a User Optimization Team are an integral part of the development effort. The system under development is the Close Combat Tactical Trainer (CCTT). It is comprised of a network of simulators and workstations which interface with a virtual environment representing real world terrain. The nature of these systems requires user involvement in all phases of systems engineering, software development, and testing. The development organization and the usability engineering approaches used are mosaics of engineering skills, knowledge and HCI techniques.

KEYWORDS: User-Centered Development, User Evaluations, User Optimization Team, Concurrent Engineering, Integrated Development, Spiral System Development

INTRODUCTION

Virtual environments or virtual reality-based systems are starting to migrate from research laboratories into real world applications [5]. The ultimate goal for virtual reality is to provide a common synthetic world in which two or more users can interact. The Close Combat Tactical Trainer (CCTT) is a full scale development of such a system. It is a precursor to an entire family of applications based on similar technology [1] that will have to deal with a complex, but common, set of human-computer interaction (HCI) issues. Such systems could be a primary communications media of the future, will provide much of our entertainment, and will be useful for training people to

work together to accomplish team objectives. This paper discusses the development approach being used in CCTT; it specifically focuses on the usability design issues and how they are being addressed.

THE CHALLENGE

The purpose of CCTT is to train Army units to operate as a team [4]. The training context for CCTT is a virtual environment -- a realistic model of the terrain that is shared by the training audience. Over 50 different interfaces to that virtual environment are being designed, implemented and tested as part of the CCTT development effort. Some of these interfaces are simulators with visual ports into the virtual environment, others are workstations used to control the interaction of intelligent agents which operate in that virtual environment. Developing these user interfaces presents many unique challenges that require us to apply and adapt HCI methodologies developed over the past decade.

In CCTT the training audience operate as a team interacting within a virtual environment. Designing that environment in itself presents numerous challenges. But designing and testing the interfaces to that environment requires the application of proven HCI techniques and technology. The approach we are taking in CCTT and the lessons learned from our experiences will assist others who will have to design interconnected virtual reality systems in the future. Interfacing even a single user to a synthetic world using currently available technology requires a user-centered approach to design, continuous user testing, and an iterative approach. This challenge is compounded in CCTT because we are also developing a large scale system with a complex HCI environment comprised on many numerous and varied user-computer interfaces [7].

The CCTT system will be installed at 32 sites throughout the world. More than 500 simulators and over 300 workstations will be built and fielded. These sites will be located at Army posts in fixed configurations or placed into trailers to provide a mobile capability to train National Guard units in their home towns. The configuration of a single fixed site will depend upon the units which train there, but in general they will have approximately 40 simulators and upwards of 20 workstations available. The system requirement calls for seven different types of simulators and 12 different types of workstations. Each type of simulator is being designed to meet the specific training requirements of a prototypical crew. Four types of workstations are run by contractor personnel and their purpose is to support the training, the other eight types will be used by soldiers to interact with the virtual environment that is shared with and among the simulators.

IEEE has established a standard for this type of interactive virtual environment, the IEEE Standard 1278 for Distributed Interactive Simulations (DIS) [3]. The key to DIS is that each node on the network maintains its own copy of the virtual environment and the state of each object within its area of concern within that environment. A set of protocols, defined by the standard, pass entity state information to other entities so they can update local state information. This allows any simulator or workstation on the network to continuously display a correct view of the synthetic world. Each node is in itself a view port into that synthetic world.

Designing the interface for each node on the network is our main challenge. Prior to contract award we conducted an analysis of the CCTT system concept to identify all potential user-computer interaction (UCI) components, characterize their target users, and to determine what type of task analysis was required to support design of each interface. That pre-design analysis determined that within CCTT there would be 57 different interfaces. We considered an interface as any medium used to communicate with the system or operate within the virtual environment. Our notion is somewhat unique because, in addition to traditional *screen and input devices*, our notion of interfaces includes those UCI where users must cognitively (and manually) respond as if they were operating actual equipment.

Four different types of interfaces were identified:
- crew stations that are simulator components,
- workstation interfaces,
- visual displays, and
- control devices to move the user through the environment (e.g., a geoball).

Three types of user categories were identified:
- soldier/trainees,
- leader/trainers, and
- contractor operators of the system.

Tasks to be performed at each interface are based upon either an assessment of what the operator has to accomplish or what tasks are targeted for training in the system. Both types of analysis are necessary to the design of the interfaces in CCTT.

DEVELOPMENT APPROACH

Another challenge for this program is managing the HCI design effort within a large engineering organization. The CCTT development team has approximately 300 engineers. Most of them are collocated in an Integrated Development Facility in Orlando, but some hardware engineers work in the manufacturing facility, also in Orlando. The visual system engineering and development is taking place in Salt Lake City. The effort is organized using a concurrent engineering (CE) approach. There are four CE teams, each responsible for one type of system component:
- workstations,
- semi-automated forces (SAF),
- simulator modules, and
- visual system.

Teams are further broken down to build specific products (e.g., there is a module CE team building the M1 Abrams Tank simulator). Human Factors engineers are an integral part of each CE team. Although not formal members of sub teams they must interact with them to support design analysis and evaluations. To assist in this process a three man detachment of Army experts (soldiers) work in our integrated development facility full time. Figure 1 is a conceptual overview of that organization.

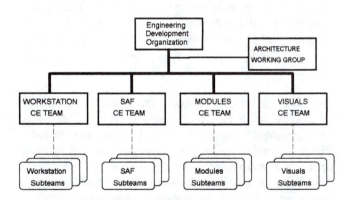

FIGURE 1. The CCTT Development Organization is organized into Concurrent Engineering Teams by system component

An Integrated Development Team (IDT) was formed to build and test CCTT. That team is a mosaic of skills and knowledge; it is comprised of system and software engineers and users organized around a concurrent engineering approach. There are four concurrent engineering (CE) teams developing each of the major types of components. Each team is comprised of members from all engineering disciplines needed for a project this complex: systems, software, hardware, human factors, and test. Figure 2 shows the interrelationship of engineering disciplines and CE teams.

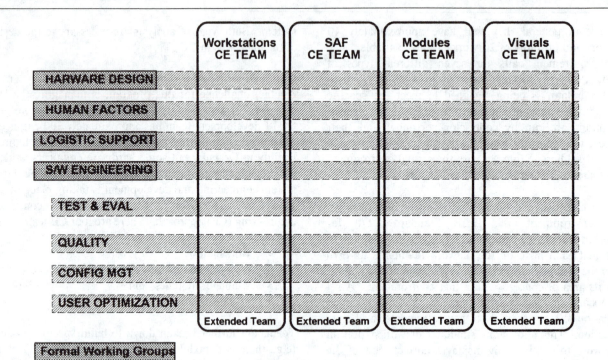

FIGURE 2. Engineering and support disciplines are active participants
in the CCTT development effort as members of each of the CE Teams

We extended the CE concept by including members of the user community on each team [6]. A User Optimization Team comprised of three subject matter experts is part of our overall integrated development approach. These three experts are active members of the CE teams. They are involved on a daily basis with system requirements analysis, high level design, implementation, and evaluation.

The system is being developed using a spiral methodology to produce seven incremental system builds. The builds integrate those components of CCTT developed to date. Figure 3 graphically depicts the general content of each spiral build and completion dates. An exit condition for each build is a user evaluation. These evaluations are conducted as scenario-based training exercises prototypical of situations that will be used on the objective system. The results of the user exercise are analyzed by human factors engineers for technical shortcomings in the design or implementation. Those deficiencies are reported back to engineering to be addressed prior to government acceptance of the system for independent formal testing.

MAJOR DESIGN ISSUES

The CCTT development effort will take three years from contract award until delivery of a complete system for user acceptance testing. A number of issues were identified early in the program which would impact the success of

that test and appropriate HCI approaches selected to address those issues.

Early User Involvement
It was imperative to have early user involvement in the design process [9]. To achieve that we implemented several initiatives.

•   First, the contractor asked the government to assign prototypical users to be part of the development team. We called this concept a *user optimization team*. Three soldiers are assigned to work full time with our engineering staff in Orlando, Florida. They are experts in collective training and combat operations, armored tactics and vehicles, and infantry tactics and equipment. They will be part of the CCTT development effort through government acceptance testing.

•   Second, we recognized the need for expertise in other areas of Army operations. Engineers doing systems engineering or software design and development needed easy access to this expertise, but full time staffing in these areas could not be justified. An electronic network of subject matter experts was established to connect the IDT with 9 different Army Schools and Centers. This network is known as the SME-Net. Engineers can access that network to post questions using an email bulletin board. Anyone on the network can

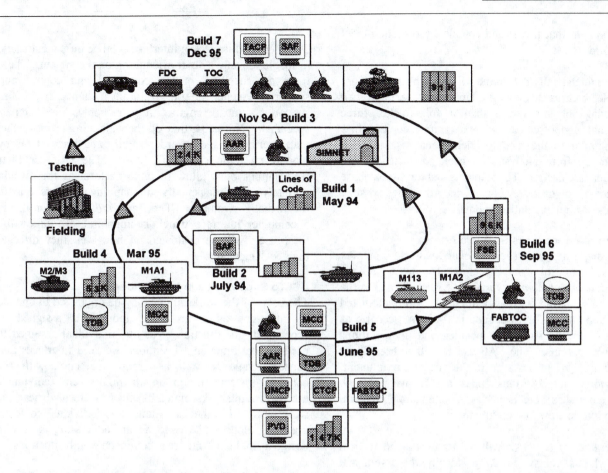

FIGURE 3. CCTT is being developed as a series of seven incremental spiral builds spread over a two year period. Dates shown in the figure represent the completion of each build to include a user evaluation.

respond or participate in the discussion, similar to an INTERNET news group. There is a single office responsible for managing the system, that office monitors the discussion and has the authority to determine a final response if there is any uncertainty or disagreement.

• Finally, periodic formal reviews of design concepts are conducted for panels of user community representatives. Some of these conform to standard government required reviews (e.g., preliminary design review or PDR), but others are special reviews established because of the uniqueness of CCTT (e.g., reviews of the visual models that will appear in the virtual environment).

Simulator Fidelity
Simulator module fidelity is a major HCI issue for CCTT. At first glance it appears to be a fairly simple problem, just make a simulator that looks and operates like a real vehicle. The challenge is to keep the cost of that simulator reasonable. Flight simulators have been developed following a full fidelity approach for last decade, but they cost upward of tens of millions of dollars per copy. CCTT

cannot afford such costly modules in the required quantities. A selective fidelity approach is used instead.

An up front task analysis determined the task steps which could and should be performed to accomplished effective training in CCTT. Based on that analysis, a determination is made for each control, display or indicator in the actual vehicle to determine if it has to be functional, pictorially depicted (e.g., using a decal), or can be left out all together. The results of that analysis were reviewed by the user community and then used to build prototype engineering development models which are subject to further user review.

Virtual Environment Design
The virtual environment in which soldiers train using CCTT must be a high fidelity replication of real world terrain. The sparse and often artificial worlds used in most laboratory systems and entertainment-based virtual reality systems are inadequate. There is a danger that training in an unrealistic synthetic world will result in soldiers acquiring incorrect skills, ones which may cause them to make incorrect and lethal decisions in actual combat. On the other hand, we cannot afford the computational cost for

a visual system that provides a completely realistic model of the world.

We had to analytically determine which simplifications are acceptable, such as displaying a building as a simple two dimensional polygon rather than a full fidelity three dimensional model that can be entered and occupied. In a process similar to the module fidelity analysis discussed above, an up front analysis has been conducted of the terrain database design. The implemented terrain database is subjected to user review to identify any possible oversights or implementation anomalies.

Controlling Autonomous Software Agents

Another major design issues associated with this complex system is the control of intelligent agents. In CCTT, they are called *semi-autonomous forces* or SAF. To add to the realism of a large virtual battlefield, semi-automated friendly and enemy forces operate in the same scenario as do trainees in the simulator modules and at operations center workstations. The SAF can be either friendly or enemy units. When observed in the virtual environment, manned forces and SAF must appear and behave similarly. The design issues here is a classic automation problem, where to interject the human into the loop.

The operators of the SAF will be highly trained tactical experts who use the system daily. Ideally, the design will allow operators to control a large number of individual vehicles. These forces cannot operate autonomously, some human intervention is required. The IDT analyzed the real world execution of military tactics and determined which decision points can be automated and which must be controlled by a human operator [2]. The problem is complicated by the design issue of how the UCI can present all of the required information, while providing easy control and minimizing the operator workload. The SAF workstation is a MOTIF interface running on a high fidelity X-Station 19 inch monitor. The interface includes a graphic depiction of unit organizations, task execution matrices, and report windows. The screen also shows a topographical display of the virtual battlefield, the units in the exercise, follows their movement, their status and their firing activities.

Training Effectiveness for Workstation Operators

The challenge of insuring effective transfer of learning from the virtual to the real environment is related to the above issue of controlling intelligent agents. Some workstations, those located in the Operations Center component of CCTT, must control these semi-automated forces while at the same time allowing the operator to perform tasks required of him in a battlefield environment. In other words, we need an interface that allows control of agents by the operator, yet is not so highly automated that it impedes transfer of training.

The users of this class of interfaces will be infrequent users, available for only a short orientation on the system. These users are members of the command and control and support elements of the training audience, e.g., fire support, combat engineering, resupply, and repair specialists. Normally, they do their planning using acetate map overlays, dispense orders orally and track status by monitoring radio communications. The UCI for their workstations must allow them to perform as they would normally, but while actually commanding synthetic units in the virtual environment. They must *talk* to their units via a computer interface which is unnatural. Additionally, these users will be evaluated on how well they directed their SAF elements.

UCI To Support Group Interactions

The users of the workstations may be organized to operate in situations as a group of individuals assigned to accomplish one function or as one individual assigned to perform that function. Therefore, we need interfaces that support single as well as group interaction with the computer system. In a group situation, several individuals are performing tasks manually, they are located near the workstation. In other situations, one individual could be operating alone. The issue we are addressing is how to design a single interface to accommodate both situations.

Balancing Commonality and Usability

A UCI style guide was written early in the project to influence consistency and commonality among the different concurrent engineering teams. Trading off commonality and usability becomes an issue in a system this complex because there is a range of users -- expert to novice. Different users will use the same information for different reasons. Should the same information be presented to different users in the manner most useful to them? For example, a contractor operating the system for a training exercise is interested in summary information related to fuel levels for vehicles. A logistics officer, however, is interested in fuel levels in terms of precise gallons. For consistency sake, we have to consider how this type of information be should presented.

User Acceptability

Systems based on synthetic environments whose purpose is to train specific skills or knowledge must be demonstrated to have user acceptability for their intended purpose. Their success or failure cannot simply be determined by the quality of the virtual environment, the usability of interfaces, or the efficiency of the software. These types of systems must, as a whole, achieve acceptability with the target user community and they must effectively train or teach their users the target skills. In the latter case transfer of learning to situations calling for the skills to be performed in a real environment is the only true measure of success.

To achieve user acceptance we have established close and continuous communications with the user community, some techniques used for doing that were discussed earlier. We also established a series of user evaluations of the system. These evaluations are conducted in the form of exercises in which users are placed into training scenarios and subjected to data collection methods designed to assess the effectiveness of the system. To review, CCTT components are being developed and integrated into an operational system using seven incremental builds each subject to a user evaluation. Each incremental build delivers a partially functional system, and, of course, the overall functionality increases with each build.

User subjects who are not familiar with the system are provided by the Army. The members of the User Optimization Team members are *not* used because of their familiarity with the CCTT design. The subjects are brought to the Integrated Development Facility where, after receiving an orientation and completing a biographical questionnaire, they are put on the system and required to perform prototypical training scenarios designed to teach specific collective or team tasks. During these scenarios data is collected by human factors engineers who observe the training. After each scenario is completed, questionnaires are administered to solicit specific target information about the system. At the conclusion of all scenarios a structured interview at the team level is conducted to establish a consensus opinion when necessary.

PROCESS FOLLOWED AND RESULTS

The CCTT development effort has delivered and evaluated four of the seven incremental builds. A series of *Design and Control Conferences* allow users to review the design approach earlier in the development process. User Exercises at the conclusion of each build are a development team controlled approach to evaluating progress toward achieving the project goals.

As with many typical large-scale system development programs, most CCTT hardware and software engineers have very little understanding of the military training environment for which they are building the system. Conversely, the typical Army user has very little understanding of the complex development environment in which his system is being developed. To help bridge this gap, CE teams conduct Display and Control Conferences early in the design process. These conferences require the participation of a prototypical user for a particular component of the CCTT system. For instance, an M1A1 Tank Commander attended the M1A1 Conference and experienced SAF operators supported the SAF workstation Conference.

At these conferences, the operator tasks are compared to the CCTT preliminary design using early mockups and prototypes. The notable aspect of these conferences is the user involvement at an early phase in the design process. With very simple prototyping tools and materials, designers are able to convey the hardware and software design concepts to users and conduct beneficial discussions. The results of these conferences are then incorporated into the preliminary hardware and software design. Both users and designers came away with a better understanding of how the other group does business

Our data collection approach during the user exercises has been to analyze all feedback to determine specific user comments. User comments are then subjected to a second level analysis to determine which of them address specific technical issues. The set of technical issues are further categorized as to those which are software or hardware deficiencies in the design or implementation versus those which either identify system capabilities scheduled for a later delivery or user requirements that are outside of the scope of the current design. User evaluations for the first three builds of CCTT resulted in 244 user comments, 156 technical issues, and 96 development problems that need to be addressed prior to delivery for formal testing.

Developing any complex computational system is a challenge, one which is based on a virtual environment and has the complexity CCTT is especially challenging. We have instituted an integrated approach that uses the best HCI practices. In making that effort, the following observations are relevant to both CCTT and similar virtual training environments as well as any large-scale complex system with a significant HCI requirement.

- It is critical to get both the customer and ultimate user for a contracted development to understand the importance of their involvement so that they will support initiatives, such as the User Optimization Team. For government programs this is especially difficult when dealing with procurement agencies that are accustomed to traditional waterfall development models.
- Both these personnel and many senior engineers and managers have to be convinced early (and then continually reinforced) that they cannot simply define or accept a requirement specification then go off and build a system to that specification and expect it to be successful. Younger engineers appear more willing to accept new paradigms, such as a spiral development approach or delivering and formally evaluating prototypes in a context where users are part of the team.
- It is also important for the entire concurrent engineering organization to agree on the expected

results as well as the approaches for conducting analyses, design reviews, and user evaluations. We failed to clearly define these early and, as a result, some early efforts were less than satisfactory.

- It is virtually impossible to demonstrate the value of user involvement. No one disputes its inherently value. But, managers understand programmer productivity in terms of lines of code produced and schedule deadlines. For them, the value of identifying problems in the near term rather than fixing them latter is not easily quantified and the concept is not always eagerly supported. One argument we used successfully is the relative cost to fix software deficiencies during spiral development compared to waiting for them to be discovered during acceptance testing, or worse yet, after system delivery.

CONCLUSIONS

CCTT is halfway though a 5 year development effort. It is at an appropriate juncture to report on our efforts to date to apply the appropriate usability engineering [8] and user-centered development techniques. We have learned many lessons of value to the HCI community. A User Optimization Team approach is important for a complex system development. It is crucial that users be brought into the project early, their value is just as critical during systems analysis as during prototype evaluations or testing.

The specific reviews of UCI analysis, design documents and prototypes need to be clearly defined in terms of their objectives, what will be provided, and who needs to review it. It is important to get the user community to support the efforts so they understand what it is they will see and do, and to insure errors and omissions are caught early.

Using operational scenarios is key to successful user evaluations of an incrementally developed system. Scheduled user evaluations of not just the interfaces, but the entire system to insure it works as advertised are critical. The requirement on developers to be prepared for these evaluations motivates them to complete promised functionality on time. It is important that a perspective be maintained that user exercises are for the engineers doing the development; their primary purpose is not for management to audit progress nor for customers to prematurely evaluate the system. Lastly, with a large complex system, the number of user subjects can grow quickly. It is important to determine if feedback from a user is guidance an engineer might consider or rather a user approved position that must be accommodated.

CCTT is the first of an entire class of networked systems in which users interact within a virtual environment. The military envisions using these environments to train their organizations and perform analysis of future concepts and systems. The attraction of virtual reality is not just the realistic immersion of individual users but the promise of interconnected users sharing computer-generated worlds for a host of functions ranging from education to entertainment. This class of system will challenge the HCI community with new and unique design issues. This paper has presented one approach to addressing some of those issues, hopefully it can serve as a starting point for others confronting these challenges in the future.

REFERENCES

1. Alluisi, Earl. A. The Development Of Technology For Collective Training: SIMNET, A Case History, in *Human Factors* 33(3), pp. 343-362, 1991.

2. Bimson, K., Marsden C., McKenzie, F., and Paez Namoi, Knowledge-Based Tactical Decision Making in the CCTT SAF Prototype in *Proceedings of the Fourth Conference on Computer Generated Forces and Behavioral Representation,* Institute for Simulation and Training, Orlando, Florida, pp. 293-301, 1994.

3. Institute for Simulation and Training, *IST-CR-93-15, Standards for Information Technology, Protocols for Distributed Interactive Simulations Applications.* University of Central Florida, Orlando, Florida, 1993.

4. Johnson, W. R., Mastaglio, T. W. and Peterson, P. R. The Close Combat Tactical Trainer Program, in *Proceedings 1993 Winter Simulation Conference, pp.* 1021-1029. ACM, New York, 1993.

5. Macedonia, M. R., Zyda, M. J., Pratt D. R., Barnham, P. T. and Zeswitz, S. NPSNET: A Network Software Architecture for Large Scale Virtual Environments, in *Presence.* Vol. 13, No 4, 1994.

6. Mastaglio, T. W., and Thompson D. R. The CCTT Development Approach: Integrating Concurrent Engineering And User-Centered Development, in *Proceedings of the 15th Industry/Interservice Training Systems and Education Conference,* American Defense Preparedness Association, Washington, DC, pp. 354-360, 1993.

7. Mastaglio, T. W. Developing A Large-Scale Distributed Interactive Simulation System, in *Proceedings 1994 Winter Simulation Conference,* ACM, New York, NY, 1994.

8. Nielsen, J. 1993. *Usability Engineering.* Academic Press, Cambridge, MA, 1993.

9. Schuler, D. and Mamioka A. *Participatory Design: Principles and Practices,* Erlbaum Associates, Hillsdale, NJ, 1993.

Neither Rain, Nor Sleet, Nor Gloom of Night: Adventures in Electronic Mail

Maria Capucciati, Patrick Curran, Kimberly Donner O'Brien, and Annette Wagner

SunSoft

2550 Garcia Ave., MTV 21-225

Mountain View, CA 94043-1100, USA

415 336-5358 or 415 336-7255

Email: maria.capucciati@eng.sun.com, patrick.curran@eng.sun.com

ABSTRACT

This Design Briefing tells the story of the design and implementation of Mailer, an electronic mail application being built as part of the Common Desktop Environment, a UNIX-based desktop. The design is notable in that it incorporates past usability data, new toolkit widgets, and compliance with a user interface style that was being written at the time the interface was being designed. In addition, Mailer is the product of a collaborative effort within and across companies, where the design is orchestrated among software developers, human interface engineers, and technical writers across the hall and across the country.

KEYWORDS: user interface design, electronic mail, design collaboration, Common Desktop Environment.

INTRODUCTION

The interface for this electronic mail application is notable in two respects. First, this e-mail application (Mailer) was the product of a collaborative effort within and across companies, where the design was orchestrated among software developers, human interface engineers, and technical writers across the hall and across the country. In addition, it is a design that incorporates past usability data, new toolkit widgets, and compliance with a user interface style guide that was being written at the time the interface was being designed. The story of the design, and the story of the design process, are described in this Design Briefing.

The design of the Mailer application took place against a background of intra- and inter-company communication and tight deadlines. The Mailer project was a direct result of the Common Open Software Environment agreement, which was joined by SunSoft and several partners (Hewlett-Packard, IBM, and Novell, Inc.) in the Spring of 1993. These partners combined forces to design a unified UNIX-based desktop, referred to as the Common Desktop Environment. The Mailer team was responsible for designing and implementing the e-mail application for this

desktop. Up to this point, the Mailer team had focused on learning about and accommodating the needs of SunSoft's Mail Tool users and was relatively unaccustomed to design interaction with other companies. With this agreement, the priorities and plans of the team changed overnight. One crucial difference was that inter-company teams were formed to manage and mediate decisions which previously had been the domain of the individual development teams.

In terms of the design, the most obvious challenge faced by the Mailer team was the change from SunSoft's OPEN LOOK graphical user interface to Common Desktop Environment Motif (based on OSF Motif 1.2). The two toolkits have different widget sets and behaviors. The design also had to be compliant with the newly-emerging Common Desktop Environment Style Guide, which defined the look and feel of the new environment. The team took advantage of existing usability data in redesigning aspects of the interface which were known to be problematic for end users. The e-mail software developed for the Common Desktop Environment had to satisfy the needs of a variety of end users, from UNIX hackers to novice computer users and users migrating from other e-mail applications, including SunSoft's Mail Tool.

POLITICS AND PARTNERS

Following the Common Open Software Environment agreement, processes were put in place to handle the increasingly complex communications between and within the companies. Inter-company teams were formed to manage and mediate decisions which previously had been the domain of the individual development teams. One of these teams was composed of a human interface engineer from each company. These individuals were responsible for coordinating interface work (design and evaluation) and ensuring a consistent look and feel, a shared user model, and sound design principles among the desktop applications. This coordination took place during a weekly conference call. Designs for the Mailer, as well as all Common Desktop Environment applications were critiqued, approved, and in some cases, debated and altered by representatives of these inter-company teams.

The SunSoft Mailer team also organized a weekly meeting in order to coordinate the range of activities. Software developers, human interface engineers, and technical writers converged to discuss design issues, implementation problems, and schedules that were too complex to be worked out in hallway conversations or e-mail exchanges. In retrospect, although the meeting was useful for resolving issues that dragged on too long via e-mail, it never had a clear charter. Leadership of the meeting fluctuated: it was not clear whether it was meant to resolve design or development concerns.

True design collaboration was clouded by an ongoing debate regarding ownership of the interface, an issue never fully resolved. Some team members thought that all design decisions, no matter how trivial, had to be reached by team consensus. These discussions were aggravated by the fact that the human interface specifications had to describe what was being implemented to a great degree of detail. This frequent discussion of interface details produced extra work for the team and hesitation on the part of the interface engineers.

Although the debates often concentrated on specifics of Mailer (such as interface components or terminology), we also faced issues common to many interface design efforts: designing for user groups with disparate characteristics, designing a new interface that remains usable by users of the old interface, and improving relationships between software developers and human interface engineers. In terms of designing for user groups with disparate characteristics, we found that members of the team seemed to be biased towards designing for particular user groups (such as end users or developers) closer to their own characteristics. Although we were interested in designing a new interface that remained usable by users of the old interface, we ran up against an interesting twist in this attempt: there were people who were comfortable with the functionality of Mail Tool and who attempted to propagate the old user model.

All these interactions were heightened by a sense of immediacy: the team had roughly a year to produce an e-mail application for the Common Desktop Environment.

MAJOR DESIGN ISSUES

Though nearly every aspect of the electronic mail interface was reconsidered, three areas emerged as major design issues: the selection model, the ability to edit received messages, and message filing. Video clips will be shown during the Design Briefing presentation to illustrate some of the more interactive issues (such as the selection model) and some of the more eclectic and illuminating user behaviors.

Selection Model

Deciding upon the means to select and display messages in Mailer turned out to be the most complex design task we faced as a team. We realized that we could not carry over the selection model from SunSoft Mail Tool for several reasons. Some of these reasons seemed to have less to do with selection specifics than with larger design issues about how messages were displayed and the overall format of the main Mailer window. We found that only after these larger issues were decided, could we progress to deciding the selection style.

• There are stylistic and widget differences between OPEN LOOK and Common Desktop Environment Motif. Mail Tool made use of the OPEN LOOK widget known as the "pushpin". This widget, located in the upper left hand corner of the message window, would allow the user to "pin" the window in place. The pin position determined whether the window would be reused or not: a pinned window stayed on the desktop; an unpinned window would be a reusable window to display subsequently selected messages. No comparable widget was available in the Common Desktop Environment Motif toolkit.

• We were aiming for a more object-based user interface style in the Common Desktop Environment. Mail Tool re-used one window for most operations, including message display. For example, opening a new message would cause the contents of the currently displayed message to be replaced by the new message contents. In Mailer, each object, such as a mailbox or a mail message, would have its own independent window, in keeping with the object-based model.

• In the object-based model, each mailbox would open into its own window. This multiple mailbox model introduced the possibility of several mailboxes (and any number of mail messages) being open simultaneously. We needed the ability to associate messages with the parent mailbox from which they were launched to prevent user error and confusion. This led to the design of a combination window, containing both the message headers and the message text.

The first browsing model we considered for Mailer used double-click to display mail messages in the lower section of the combination window. This lower section was referred to as the "message pane". However, in most graphical user interfaces, double-click means "open another window". In addition, the classic browser model uses a single mouse click to select and display (a la the Xerox Smalltalk browser).

Furthermore, comparative analysis data convinced us that we would have to accommodate two predominant reading styles: those users who preferred to browse through messages quickly, re-using one window to do so; and those who preferred opening messages into separate windows.

As part of the design process, the team decided to prototype the two approaches and test them on users. We called our prototypes "Fat Boy" (each message opened into a separate window) and "Little Man" (each message was displayed in the lower message pane). Screen shots of Fat Boy and Little Man are shown in Figures 1 and 2. In Fat Boy, single-clicking only selected a message and double-clicking opened each message into a separate window. In Little Man, however, single-clicking selected **and** displayed the message in the lower message pane. Double-clicking opened each message into a separate window. In addition, we experimented with using a push-button to "hide" and "show" the lower message pane.

Our usability testing showed that users of both camps were satisfied with the dual interaction mode that Little Man gave them. Those who preferred a "reusable window" style easily learned and consistently used the single-click model. We found that test participants who were accustomed to double-clicking to view messages preferred the "separate window-per-message" model. These participants used the Hide button to eliminate the message pane and continued to use the double-click model they were familiar with.

As a result of the user testing data and many other considerations, the team ultimately decided to use a modified version of the Little Man prototype for our final version of Mailer. Single-clicking displayed a message in the lower pane and double-clicking created a new window in which the message was displayed. A screen shot of Mailer is shown in Figure 3.

The decision was not arrived at easily, primarily because some team members were reluctant to endorse a model with which they were not familiar. This unease concerning the decision persisted throughout the design year, forcing some post-decision brainstorming sessions to discuss the validity of our design choice and other issues which arose from the new selection model.

Editable Messages

SunSoft's Mail Tool allows users to freely edit the contents of received mail messages. This quickly became an issue for the Mailer team, especially when it became apparent that the new selection model would allow multiple views of one message to be open simultaneously. Multiple views of one

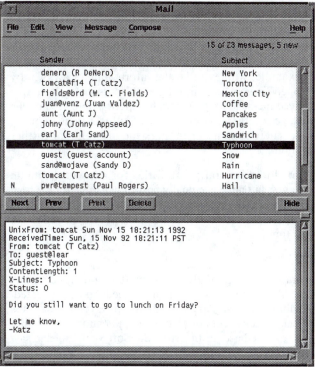

Figure 2. The "Little Man" prototype — text of a message appears in the lower pane of the window.

message introduced compelling questions: What would happen to one message when the other was altered? Would the changes in one message be reflected in the other?

We wrote up four potential scenarios (read-only single view; editable single view; read-only multiple views; and editable multiple views) to assess the benefits and disadvantages of each. Users were asked how they would approach the editing task given multiple views of the same message. The results were inconsistent: users were split on where the edits should occur (in the separate window or in the message pane) and whether changes would be reflected in all views of the message.

The team also performed a comparative analysis of other e-mail applications from various platforms. Most e-mail packages did not allow editing of received mail, a few did, and one allowed it but did not advertise it to the user in the interface or in the documentation.

More interestingly, the idea of editing received mail brought up a host of social, ethical and even legal issues. Many users objected strongly to the ethical implications of editing. Users unfamiliar with editable mail messages reacted with shock that someone could change the contents of a received e-mail. This was seen as a violation of privacy and of original intent. From the legal perspective, some users were

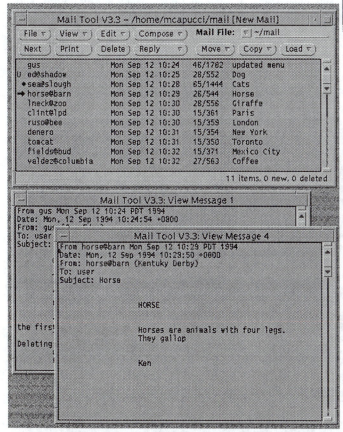

Figure 1. The "Fat Boy" prototype — text of each message appears in a separate window.

concerned with editability in light of companies using electronic mail messages as courtroom evidence or as grounds for dismissal. Several team members who talked to customers on an informal basis also found that editing was viewed as a security concern. To them, being able to alter a received message was the equivalent of being given access to someone else's computer.

However, users who were accustomed to editing were horrified at the thought of giving up the capability. These users had valid reasons for editing received messages. Often messages arrive at their destination in unreadable format: for example, with extra header information or awkwardly wrapped lines of text. Editing enables users to clean up a received message for their own benefit or before sending the message to others. Users also edit subject lines to better reflect their content, or to mark the message for further attention.

We explored several possibilities for allowing editable messages, but none eventually proved workable for the final version of Mailer (mostly due to schedule constraints). We anticipate, however, that this issue will remain active for some time.

Mail Filing

Virtually all users of electronic mail would agree on one thing: there's too much of it and it needs to be organized. The ability to easily and quickly file received messages is critical to the success of an e-mail application. The Mailer team reviewed user testing from SunSoft Mail Tool and saw an opportunity to improve the mail filing operation, which was confusing to many users.

Mail filing confused many users in its previous form in Mail Tool because of its complexity. Mail Tool provided three different methods for mail filing: drag and drop to the File

Manager from the message header list, a special Mail Files dialog, and a customizable set of menus. The customizable menus, once set up, are the fastest method of mail filing and are used primarily by advanced e-mail users. However, many users could not figure out how to set up and use these menus.

Furthermore, the Mail Tool mail filing interface made use of several OPEN LOOK widgets which were not available to us in Common Desktop Environment Motif. There was a pop-up menu and type-in field in which users could choose mail folders that were not in the custom menus. The behavior of this widget did not translate to Common Desktop Environment Motif.

Our goal for Mailer was to simplify the mail filing task so that it was more accessible to all users. Mailer needed to accommodate two filing styles. Frequent filing of messages needed to be accomplished in as few steps as possible. A convenient method was also needed to file messages to mailboxes in infrequently accessed parts of the file system.

To resolve both of these design issues, we created a menu called "Move" (Figure 4). To allow for frequent filing of messages, the user may customize a portion of the Move menu on a property sheet. The user can build a custom list of mailboxes, and/or choose to build a list automatically from the most recently used mailboxes. The custom list and/ or the automatically generated list are displayed in the Move menu. The user selects message(s), pulls down the Move menu, selects a mailbox and releases the mouse button. The messages are moved.

We also gave users a way to move messages to mailboxes not on their Move menu through a choice on that menu called "Other Mailboxes". This brought up the standard Common Desktop Environment Motif file selection box. The file selection box allowed users to move or copy their

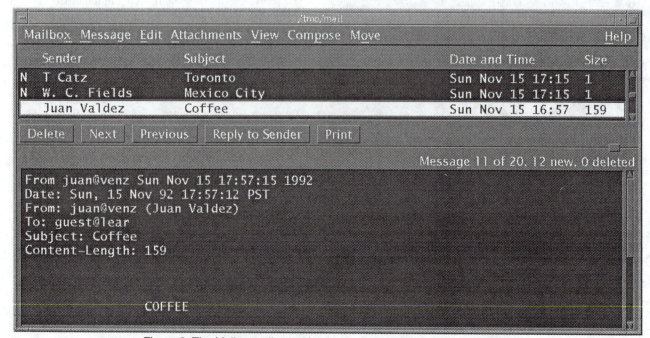

Figure 3. The Mailer application for the Common Desktop Environment.

messages to any mailbox in the file system.

The file selection box represented different functionality from Mail Tool's Mail Files dialog, which not only allowed moving and copying but other actions such as Open, Edit, etc. This was an interesting point in user testing because subjects would consistently go to the Move file selection box for other operations, such as opening a mailbox. This suggested to us that users were perhaps looking for a "universal" type of dialog to perform mail housekeeping

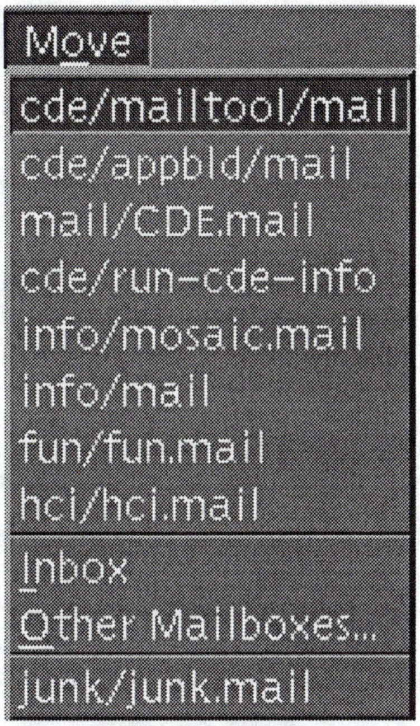

Figure 4. The Move menu is used to file messages into other mailboxes: user-defined mailboxes appear at the top and recently visited ones are appended to the bottom.

functions. Unfortunately, since the file selection box was an inherited widget and would be found in other Common Desktop Environment applications, there were limits to the changes that could be made to it.

The file selection box was not without its benefits. Mail Tool's message filing function assumed that the user would keep all mailboxes in one physical directory. In this event, the mail filing pop-up menu would display all mailboxes. However, many users prefer to keep mailboxes in folders associated with their projects, for example. This led to navigational difficulties. Without a dedicated mailbox directory, the user would have to navigate through layers of pull-right menus to reach a filing destination. The file selection box allows users to get to other mailboxes via a simpler navigation model.

SUMMARY OF USER TESTING

Mailer for the Common Desktop Environment was the subject of many usability evaluations: lab studies, comparative studies, and heuristic evaluations.

Within the first few weeks of working on the Mailer project, we uncovered information about user preferences, and the functionality of various e-mail applications. We conducted two user surveys: one to gauge users' attitudes toward editable messages and a second on users' acceptance of the SunSoft Mail Tool user model concerning the In-Box and deleting messages. We also performed a comparative analysis to monitor how other well accepted e-mail user interfaces supported users' tasks. A human interface engineer performed key tasks on Mail Tool and three other popular e-mail packages and issued a comparative report.

We performed five usability tests in a lab setting. These evaluations fit into an iterative design process: create a design or alternate designs; evaluate them with user tests or heuristic evaluations; and redesign. These served the general and specific goals of periodically checking the usability of Mailer and continuing to evaluate newly implemented features.

To address users' needs who would be migrating from other desktops, a usability study was conducted which focused on the whole Common Desktop Environment. Pertinent results were given to the Mailer team, including information about interactions between Mailer and other desktop applications.

Late in the development cycle, several technical writers and human interface engineers participated in an heuristic evaluation of error messages and dialogs: as in the migration study, the Mailer team was given information pertinent to the Mailer error messages.

A Comments button was added to the main Mailer window in an alpha version and was distributed to internal and external customers. Alpha program participants used this button to send their feedback and suggestions (via Mailer itself) to the Mailer team.

REFERENCES

1. Hewlett-Packard Company, International Business Machines Corp., Sun Microsystems, Inc., Novell, Inc. *Common Desktop Environment Style Guide*. Unpublished manuscript (publishing date Spring, 1995).

2. Open Software Foundation, Inc. *OSF/Motif Style Guide* (revision 1.2). Prentice Hall, Englewood Cliffs, New Jersey, 1993.

3. Sun Microsystems, Inc. *OPEN LOOK Graphical User Interface Functional Specification*. Addison-Wesley, Menlo Park, California, 1989.

ACKNOWLEDGMENTS

We would like to thank the other members of the Mailer team (Bart Calder, Jenny Chang, Maria Cherem, Joe Dipol, Bob Gahl, Neil Katin, Wendy Mui, Dan Pritchett, Satish Ramachandran, Doug Royer, John Togasaki, Ajay Vohra), Leif Samuelsson for his work on developing the prototypes, and Karen Bedard for valuable comments on this paper.

The Interchange™ Online Network:
Simplifying Information Access

Ron Perkins
Senior Interactive Designer
AT&T Interchange Online Network
25 First St.
Cambridge MA 02141, USA
Tel: 617-252-5231
E-mail: ronperkins@ichange.com

ABSTRACT

The AT&T Interchange Online Network is an online service designed to foster a sense of community while making it easy for customers to find information. This briefing describes how numerous design iterations aided by usability testing led to progressive refinement of the interface, specifically the information space layout for navigation. By combining context and content, Interchange allows orientation in a large information space. It becomes possible to understand all that is contained in a specific area at a glance. One design goal was to leverage editorial expertise while simultaneously taking advantage of publishing models extended to a very large online information space. Our overriding objective was to create an elegant, modern, and professional information service that values the time of busy people. Testing showed that even people who had never used an online service successfully navigated the large information space and enjoyed using Interchange. At the time of this writing, Interchange is at a Beta test stage and the design may be modified by the time the briefing is presented.

KEYWORDS: On-line service, information design, information space, electronic publishing, hypertext, hypermedia, interface design, usability testing, information retrieval.

INTRODUCTION

This design briefing describes how Interchange combines logical hierarchical navigation with flexible searching and hypermedia links in an electronic publishing platform that takes advantage of editorial intelligence. Editors, as experts in a specific field know what interests people. The combination of editorial expertise with an easy to navigate information space should provide the best of both worlds.

One of the major problems confronting users of large hypermedia spaces is getting lost in hyperspace. An overview diagram of the information space can help [1]. Work at Bellcore on visualizing information with fisheye views demonstrates the usefulness of the technique but requires a highly structured hypertext [2]. There is some evidence that a hypertext organized *both* as a hierarchy and a network is more efficient than strictly one or the other [3].

The structure of Interchange, which simulates a hierarchy but is actually a network, incorporates both means of navigation using hyper links. Below is a directory page for the Washington Post™ showing a standard navigational structure. The special interest publishers are listed down the left side. Photos and direct links to highlighted stories are in the center. In the Post, under the word 'Extra' is a index listing with all the major sections of this publisher.

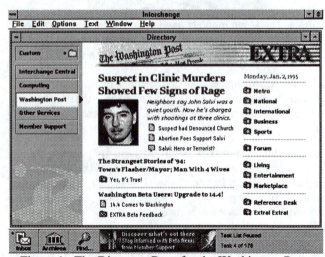

Figure 1. The Directory Page for the Washington Post

INTERCHANGE OVERVIEW

Interchange is an electronic publishing platform developed by Ziff Communications and now owned by AT&T. Following the publishing model of providing very deep special-interest information that made Ziff-Davis successful in the print medium, the challenge was to build an online service that provided customers with both powerful functionality and ease of use. Also, because it offers publishing partners modern editing tools, hypertext linking, and control of membership revenues, it is the first online service built specifically for a network of special interest publishers.

Running under Microsoft Windows™, and accessible via modem over telephone lines (soon over the Internet), Interchange allows compound documents that support a wide range of media types including graphics, styled text, and links to any other item in the service. The document architecture is easily extended to support video and sound. Editors can highlight new content daily, hourly, or by the minute. Users can customize information by organizing their own view of the entire network of services. They can create saved searches that function like simple agents to monitor information constantly. Users can save information off-line and connect to update the information on their PC as well as set up automatic dial-up sessions. Finally, they can participate in online discussions about the topics in each special interest area and talk to editors: this is where the community is formed around shared information.

As a next generation online service designed specifically to take advantage of the Windows environment, the interface is notable for it's simplicity. As a result of numerous design iterations the complexity of this full featured online service is presented with an interface that has a browsable navigational structure, minimal menus, and simple search tools. In fact, computer professionals may find the interface unremarkable for the apparent lack of never-before-seen interface conventions. We consider it a success if customers find the information they need and do not consciously notice the interface itself in the process.

Major feature innovations include:

- True compound documents with graphics, styled text, and links to any object on the service
- Downloading software in the background, while browsing the service
- Powerful searching and saved searches that monitor for new information on the service
- An information space designed for easy access with an emphasis on publishing and community
- Customizable information space

Directory View

The top level directory view of Interchange showing the Central Directory Page open is shown below in Figure 2. This directory view serves as a map or overview of the entire network. Multiple services are available down the left side of the window, listed on buttons below Custom. The index to all of the content in Interchange Central is listed to the right of the buttons. Some links to stories are shown on the right, highlighting some important news, in this case fighting in Russia and the sale of our company.

Figure 2. The Beta design for Interchange's Central Directory

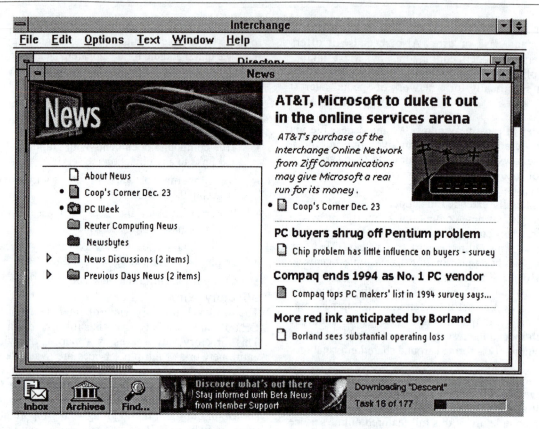

Figure 3. The Beta design for News in Computing.

Opening a News folder maintains the same structure showing the contents list at this new level and more highlighted stories, shown in Figure 3 above.

Competing Design Goals
Customers, editors, development, and management had competing and seemingly incompatible goals.

Customers vs. Editors. For Interchange, efficient access to information for service members was an important goal but at the same time editors needed a way to call attention to timely and important information. These goals were in direct competition. In early usability testing potential members of the service wanted a large table of contents to browse, and mistook the graphics and some figures for advertising, leading designers to think people want lists of data, data, and more data (see 'using customer profiles' below). But editors needed to present articles, software, and news stories as highlights at many levels in the hierarchy of contents, and many customers value this added benefit of information filtering; it simply saves time to have experts highlight important information.

Development's view: a hierarchical structure with links, powerful searching.

Management's view: a use model that affords rapid viewing of many special interest areas by alternately scanning either directories of information or editorially controlled highlights pages.

The current design reflects the above goals. A clearer picture should emerge when the stages of design are described later on, but first it's helpful to know what the present interface is like.

CURRENT DESIGN
The Interchange interface was designed to allow simple access to a vast body of information using a logical structure that is complemented by highlighted or featured items at each level. This standard structure serves to give a sense of place and allow the serendipitous discovery of information of interest. Three aids to information access make this possible:

1. Logical Hierarchy including Highlights. The directory window shown in Figures 1 and 2 serves as an interactive map of the entire network of services. Selecting another special interest directory button on the left changes the directory page on the right. This illusion of a hierarchy seems to work well together with the featured stories, like in a magazine or newspaper, and the featured stories are simply links to items deep in the service.

This model of a comprehensive list of contents combined with highlighted items on a directory page is consistently used throughout Interchange. It gives an overview of the structure at each succeeding level while showing important highlights at the same time. The combination of *context* and *content* allows orientation in a large information space. It is possible to understand all that is in an area at a glance, literally in seconds. At the same time the added value of editorial judgment to bring specific information to the attention of members at a top level has a number of benefits:

- Readers get to see the importance of information without having to search for it.

- There is a sense of place at each level, giving the member a clear and concise sense of "what's going on here" as well as landmarks for navigation

- Members come across information placed by editors that they wouldn't see in a pure hierarchy.

- Members have a common base of information adding to a sense of community.

This sense of a shared common experience and the opportunity to communicate about it is one of the most important aspects of Interchange.

2. Unique Hypertext Links. Members of the service can send each other links, precluding a need to *describe* how to get to a particular point of interest. Links to selected items can take you anywhere deep in the information space. They are the means for editors to give prominence to important information, bringing it to the top of an area. Because links can occur at any level throughout the hierarchy, not only as highlighted items on directories, this network structure allows direct access to related information for any given document or item.

3. Searching and Saved Searches. An online service can be a vast sea of changing data. Searching was carefully designed to promote easy iterative searching to narrow the member's focus to the desired items without requiring mastery of the complexity of Boolean operations and language syntax. Interchange allows searching at any time; it is not modal. You can navigate to an area of interest, search for a specific topic, and then continue navigating with links to a narrowed list of topics. In addition, editors as well as members can save searches to run automatically, collecting any new information matching the search criteria.

Giving Edit Credit

Information organized by editors using the means described above creates an information space that is more than a mere warehouse of data. When experts bring important information to the forefront, it adds value, creates a shared experience throughout the service, and for some reduces the fear of missing information. In addition to highlighted stories on directories, individual stories can have sidebar links to related information. For example, a new software product announcement could have links to the company's history, past products, financial information, and a demo of the product. A news story about O.J. Simpson could link to photos, constitutional law, and even a news clipping of archival football footage. Without editorial support, navigation would be about as exciting as a library card catalog. Observing people using Interchange in the lab, it became increasingly evident that the mix of editorial effort combined with a simple and attractive interface appealed to both experienced online users and novices.

EVALUATION SUMMARY: ITERATIVE TESTING AND DESIGN

Working with a prototype that was about a year ahead of the development schedule, designers successfully tested the interface with both experienced and inexperienced online users well ahead of any code writing. Simple usability procedures mixed with basic consumer testing acceptance criteria were used to obtain qualitative feedback on the design at very early stages. We were trying to answer 3 basic questions for the service:

Do they get it?

Can they use it?

Do they like it?

Testing the appeal and performance of a service is different than testing a software product or tool. Each time a customer uses a service, they can choose not to come back the next time. Instead of paying once for your product, they decide to pay each time they use your service.

Performance was evaluated from task completion data and observation. Acceptability and appeal were measured with consumer testing instruments such as surveys and a line scale. The acceptability scores were measured with line scales to compare the overall appeal of the service with its competitors, a concept statement describing the service, and actual experience performing directed tasks. Our line scale has intervals from Terrible to Excellent and the test respondent marks each rating for their present experience with a service (online users), the concept of Interchange, and experience with the prototype on the same line. Thus each rating is relative, accounting for 'hard graders'.

Ratings Definitions (for Figure 4)
Top Two Categories -- This is an *absolute* measure of how the service was rated. The respondent rates the prototype in the very good to excellent categories on the rating scale.

Two Way Wins -- This is a *relative* measure of how the service compares with a written concept statement describing the service, and with a respondent's prior experience. The respondent rates the service higher than the concept and rates the concept higher than their previous experience with online services, or their perception of online services.

Over 130 people, taken from Ziff subscriber lists, were tested with design prototypes over time as shown below. Informal testing of specific design and content issues also contributed to the evaluation of the design.

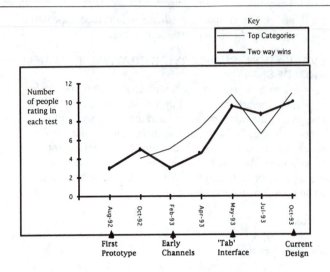

Figure 4. Ratings during iterative testing sequence

Here's how our ratings have looked over time in Figure 4 above. Note that the top-two categories scores bounce around, while the line-scale ratings for two-way-wins represent a steady trend as we refined the design. These scores have been normalized to a sample size of 12, which is the typical number of respondents, though individual studies have had as few as 10 and as many as 15 respondents.

Two groups always tested.
Both experienced and inexperienced online users were in every group tested. We were looking for differences in their performance and overall ratings to make sure we didn't build a service that only either experts or novices would like. Twelve people were tested in each formal lab that we ran, acting sort of like a jury: when we got near unanimous opinions or performance on a design issue, the changes to make were very clear.

Using Customer Profiles
When looking at diverse users, we found it helpful to categorize different types of customers according to their motivations, traits, and skills. This helps when observing their behavior during testing. Without them it would be easy to get pulled in many directions by seemingly contradictory opinions and behavior. For example, if you know someone is a 'database diver', they generally will not value editorially formatted information spaces but will look for powerful searching tools.

Some Anecdotes.

Mac Users vs. PC Users.
Early designs were tested with both Mac and PC users. A curious difference in attitude appeared in each group that probably speaks more to the architecture of each machine. Mac users were confident and told us what was wrong with our design. PC users felt beaten down and said "I can learn this, just give me more time or let me read the manual!".

What kind of an animal is this?
Another way we got a feel for people's perceptions of the service was by asking projective questions like "if this service were a person, what kind of person would it be?" Later, we substituted 'animal' for person and got amazingly consistent answers, once people got over the "this is a stupid question" feeling. The question seems to reveal the most salient aspect of the service to each respondent and whether it is a positive aspect or not. The service was consistently compared to fast cats: "it's a panther, sleek, powerful, and fast". Another frequent comparison was made to animals that "go and get things for you, like a retriever".

Figure 5. Panther, fast, sleek, and powerful.

Figure 6. Retriever, gets things.

EARLY DESIGNS: NO, IT'S THE INFORMATION..
People come to an information service for the content, not the interface. Often the interaction is what needs design more than the interface. The rest of this briefing shows evolutions of the design and will focus on information design issues, specifically the way the highest levels of the service evolved to a simple hierarchy that balances context with content. To get the right mix of content organization and the contextual cues needed for orientation in a hypermedia space, the design moved through various stages including a matrix, a dual card directory (Early Channel Changers), and hybrid directories of information (Tab Design) up to present. Many issues around the use of a table of contents for each successive section of the service to maintain context (directory) will be explored.

Organizing Information in a Matrix
Early attempts at organization used a matrix with publishing products on one axis and different types of information on another. Picking any two points on the axes would show the intersection, in this case, Computing and News, in the center screen as shown in Figure 7.

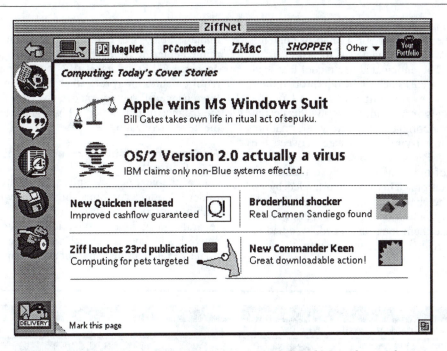

Figure 7. Matrix organization with types of information on the left and Magazine products across the top.

For a large information space this approach turned out to be impractical. If you analyze the matrix, it's very similar to a spatial metaphor, like a cityscape; you have to memorize the locations along each axis. You have to present a continuum to impose some ordinal discipline, like a city street grid to make it useful. Without a continuum, the axes are little more than a new list that requires memorization. We could not identify a rational order for information to make the matrix navigable. Another problem was that some of the axes could not be filled in for all combinations of choices, so this idea was dropped even before lab testing.

Fighting for Space on the Front Page

Satisfying the competing design goals previously mentioned would take more than a matrix. Focusing on the use model that management suggested, the next design addressed scanning special interests and deliberately *separating* navigation and content.

A design using buttons on the left to represent sections of

Figure 8. Early Channel Changers: Highlights page with a separate Directory button (lower right)

the service (like channel changers on a television) and a card that showed either highlights or a table of contents controlled by a Directory button toggle (on the lower right) is shown in Figure 8. The anticipated use model that lead to this design was that a customer could flip through their special interest areas by scanning *either* all of the highlights *or* the contents very rapidly. Testing this design in the lab showed the importance of directory information; nearly everyone got lost trying to navigate between the content and highlights. People had to sidestep highlighted stories to get a view of the content in each section, never developing any orientation. This was confounded by an affordance problem; the channel changers on the left were related to the Directory button on the lower right, but the spatial separation made them look un-related. We had to take the blame for rushing this design into the lab.

Fatigued by arguments from an influential member of management, we tried his idea and may not have implemented it well. Here is an unsolicited warning to designers--listen carefully to management design requirements, interpret rather than implement. Back to Interchange, all subsequent designs were focused on varying the balance of content on a highlights page as this seemed the key to navigating.

Finding the Right Mix: a Balance of Highlights and Context
One of the next designs split special interest areas among folder tabs, and combined highlights with directory lists within each area, as shown in Figure 9. There are two separate issues here: Navigation *between* services on the network and navigation *within* a service. Folder tabs were used as a familiar metaphor for separating services.

Figure 9. The Tab interface that was most successful, combining highlights and context.

The other issue was that of navigating within a special interest area or service. The hypothesis was that the first page (highlights page) could serve as mix of hierarchical order and show some story highlights at the same time, to emerge as a hybrid structure. The contents of a section were listed on the left with highlighted stories on the right. This mix began to work very well in the lab, where people

began to find information without getting lost. The overall ratings went up at this point (see May 93 in Figure 4).

Usability Testing : "How do you know when you're done?" or "It tested well but we are still not happy"
We all know there are limits to usability testing--you cannot test good design into a product or get users to solve your problems. You can only try to reduce the number of problems customers have. Designers are very good at

taking extremely diverse requirements and putting together solutions. One scheme used throughout the design effort for was to remind ourselves of a set of high level objectives, or a mantra. For Interchange, it was 'EMP', standing for Elegant, Modern, and Professional. The tab folder interface was not.

It was generally felt among members of the design group that there was too much graphical complexity at the top of the window in the design previously described. A tool bar, tabs for categories, window borders, and dividing lines for stories added numerous horizontal lines. Lab respondents remarked that it was 'busy'.

Figure 10. Fox, clever.

Moving to the Current Design

The cleaner look of the current design (Figure 2.), with Directory buttons down the side and feature buttons at the bottom, drew excitement from a majority of lab respondents. There was no statistically significant measure for this, in terms of ratings, but everyone observing the labs could 'feel' the difference.

Moving the features and status lines to the bottom gives prominence to the content, which is more important than the interface. There appears to be less interface. A layout grid is used to locate all of the content in the window, and the fonts are quieter. Screens are clean and screen transitions are smooth due to a layout grid. The list of directory buttons showing services on the left are more scannable in vertical list format. Also, there is more room for directory buttons.

SOME OTHER LESSONS LEARNED

These are a mix of other things not really detailed here that we believe we did right and some we did wrong but learned from them just the same:

Start testing before any code is written.

Use a prototype with the same architecture as that planned for the product

Rapidly change the design in early stages. Don't spend time on details to fine tune before testing. It's better to be fast and 80% right than slow and try to be dazzling.

Document your design criteria, changes, and the reasons you changed (you may find the circle completes someday)

Use profiles of customers showing their diversity of values to keep conflicting reports and opinions straight.

Test often with real prospective customers--once a month, if possible, during rapid change periods.

Don't be afraid to make small changes to the interface right in the middle of a test--it's qualitative data anyway.

Get important and influential people to come to watch the testing.

Make testing a fun part of everyone's job; provide good food.

Lastly, and very important; address management proposals for design carefully and present the ideas with an analysis and a better (if possible) design before going into the lab for testing--it will save both time and face.

CONCLUSION

Thanks to constant refinement and iterative testing, the design of Interchange offers efficient access to a large information space. As a result of our taking advantage of editorial intelligence and integrating the mapping of contents with highlights, users found that orientation was less of a problem in hypermedia.

Figure 11. Octopus, has it's tentacles everywhere.

REFERENCES

1. Neilson, Jakob, Hypertext and Hypermedia, Academic Press, 1993.

2. Furnas, G.W., Generalized Fisheye Views, Proc. ACM CHI 1986, 16-23.

3. Mohageg, Michael F, The Influence of Hypertext Linking Structures on the Efficiency of Information Retrieval, Human Factors, 1992, 34(3), 351-367.

ACKNOWLEDGMENTS

The work presented in this briefing reflects the collaborative efforts of many people including Dave Rollert, Marty Gardner, glenn mcdonald, Cynthia Shanahan, Maethee Ratnarathorn, Karen Tichy, Cindy Augat, Andrew Kleppner, Kevin Wells, Andrew Knight, Charles Dao, and Judy Marlowe. Many members of the development team (responsible for architecting and coding the software) participated actively as well. Thanks also to Matt Belge for review comments.

Articulating a Metaphor Through User-Centered Design

H. J. Moll-Carrillo, Gitta Salomon, Matthew Marsh, Jane Fulton Suri, Peter Spreenberg
IDEO Product Development
1527 Stockton Street San Francisco, CA 94133
415.397.1236 vox 415.397.0823 fax
hector@IDEO.com • gitta@IDEO.com • mMarsh@IDEO.com • jane@IDEO.com • spreenberg@IDEO.com

ABSTRACT

TabWorks™ book metaphor enhances the standard Windows™ user interface, providing an alternative way to organize applications and documents in a familiar, easy to use environment. The TabWorks interface was designed collaboratively by IDEO and XSoft and was based on a concept developed at Xerox PARC. This briefing describes how a user-centered approach affected the design of the TabWorks user interface: how the metaphor's visualization evolved and how interaction mechanisms were selected and designed.

KEYWORDS: User-Centered Design, Design Process, Product Design, User Observation, Metaphor, Book, Tab, Application, Document, Container.

uct would use the standard Windows 16-color VGA palette and have a maximum size, including window title, menu bar and borders, of 640 by 480 pixels. *Our task was to create a design that implemented this metaphor in an elegant, usable and economical way within the constraints of the delivery platform.* In other words: the design had to be engaging, easy to use and useful, and require as little of the computer's resources as possible.

The design also needed to accommodate a wide range of users; it would be bundled with all Compaq computers. The main target group, however, were novice and intermediate users. Novices were defined as first-time computer users who expected some "out-of-the-box" functionality. Interme-

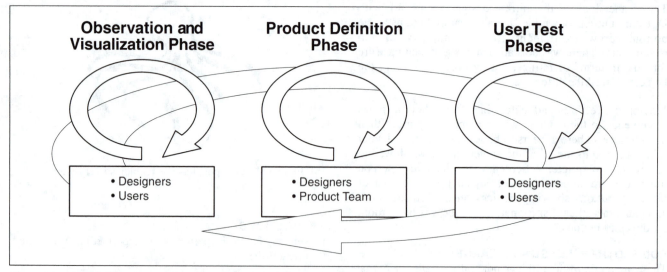

Figure 1. The Idealized Process. Development of the interface followed an iterative process consisting of three phases.

INTRODUCTION

XSoft (a division of Xerox Corporation specialized in document management software) approached IDEO (a product design consultancy) in December of 1992 to assist in development of PC Catalog, a Windows application based on a book metaphor. The design brief stated that the prod-

diate users were those already familiar with computers at home or the office but not proficient enough to significantly customize or troubleshoot their setups.

A User-Centered Design Process

We approached the problem with a user-centered, iterative method, similar to those discussed by Gould and Lewis and by Moggridge [1, 4]. An interdisciplinary team of IDEO interaction and human factors designers worked with XSoft's developers and marketers over a period of nine months. Our goal was to understand the needs of potential end users and, within the scope of our design brief, create interface mechanisms that would best serve them.

Figure 1 shows the three main phases of our iterative design process for this project. Each one is depicted as a cycle because iterations occurred within it. In the Observation/Visualization Phase we applied previous experience, knowledge of related materials and insights gained from informal observation of users' tasks, environment and tools to generate and visualize ideas. During the Product Definition Phase we worked with the XSoft team —using these preliminary ideas— to further refine the feature set and the proposed interface. Features and interface ideas were approved, rejected or found in need of further refinement based on perceived value, implementation constraints and product release dates. In the User Test Phase key product elements were evaluated with users. These three phases were part of a larger iterative process that lead to the final product implementation.

The remainder of this briefing describes our design process and provides examples from each of the three phases to illustrate how and why specific design decisions were made and how we collaborated with XSoft.

OBSERVING AND UNDERSTANDING: GATHERING MATERIAL FOR VISUALIZATION

In order to visualize the metaphor, we wanted a better understanding of what functionality should be provided by the product. We began the process by informally observing users and reviewing as many products that offered similar capabilities as our schedule and budget constraints allowed.

Observing Users

Our design brief was to "visualize and articulate a book metaphor as a mechanism to store and organize documents and applications." However, we prefer to approach design problems from the user's point of view, which requires an understanding of the user's context. This often has the effect of widening the problem's scope but helps us design for the environment in which the final product will exist. Our first step was therefore to observe how users "store and organize documents and applications," independently of the proposed metaphor.

The Physical Environment

Informal observations were conducted to gain insight into what methods and mechanisms were used by naive, intermediate and advanced users to deal with documents and applications in their computer and physical environments. Over 20 Macintosh and Windows users were observed at IDEO and XSoft. Our findings were similar to those noted in other studies [2, 3]; people working with numerous documents always made some attempt at containment. Drawers, piles, binders, folders, boxes, envelopes, rubber bands, clips and other devices were used to keep groups of documents together. We noted three main reasons for containment. These were: *organization/access*, *transportation* and *safekeeping*.

Organization/access refers to making current and relevant documents easy to find and retrieve in order to work with them. Obvious and accessible spatial groupings, such as piles, were used most often as a strategy for containing these items. Users also needed to move documents from one place to another or transfer ownership to others. Folders and binders were often the mechanisms used to complete these transportation tasks. Safekeeping involved the use of drawers and file cabinets to archive or secure items.

The character, style, functionality and interchangeability of real-world containers varied greatly. We observed that most users kept tools and documents close at hand but in distinct groups and containers. For example, pencils and rulers were stored separately from documents. Three-ring binders were one notable exception; some had penholders, rulers, pockets and floppy-disk holders built into them.

The Computing Environment

In contrast, computer interfaces offered few containers. We also observed that users often failed to discover or use their full functionality. Windows 3.1 offers two container strategies: Program Groups within Program Manager and folders within File Manager. Only advanced users customized the containers in Program Manager extensively. They also used File Manager and found little difficulty moving from one to the other, even though the look and behavior of containers was quite different in each of these applications. Intermediate users tended to make use of File Manager only, creating directories and sub-directories represented by folder icons in an outline-style hierarchy. Naive users did little customization, relying solely on the Program Manager Groups provided. These users seldom created containers on their computers, though they did so easily in the real world. All Macintosh users observed created containers. They constructed flat hierarchies that quickly filled their desktops with numerous folders and files, or a complex hierarchy of nested folders requiring significant navigation. We observed that some Windows and Macintosh users never moved beyond default conditions, relying on specific applications (e.g., a word processor) to find their files wherever they happened to reside in the system. These observations suggested that users could benefit from additional container strategies.

In addition, the observations helped us validate the applicability of a book metaphor —specifically a binder metaphor— as a container to organize and access applications and documents within a computer system. We found that collecting diverse but related documents into ringed-binders was *an experience familiar to most users*. It was common among the users we observed to have assembled these collections themselves or to have used them.

Related Products

We also looked at a variety of Windows and Macintosh products with similar or related functionality. Our goal was not simply to compare product features but to understand how their intended markets were related to their chosen metaphors and functionality. Products including Norton Desktop for Windows, Hewlett Packard's Dashboard and Apple's At Ease and At Ease for Workgroups were reviewed.

Norton Desktop for Windows replaces Program Manager with a Macintosh-like "desktop" metaphor that makes use of the entire screen. It does so at the expense of requiring significant amounts of the system's resources. Dashboard uses a more abstract solution. The "dashboard" is a control panel that reduces the amount of screen space required to perform file and program management functions by using buttons that parallel Program Manager Group icons. The metaphor is not expressed much further than the product name and the inclusion of a single gauge to show system resource use. Dashboard speeds up access to File Manager Groups and other system resources. At Ease —developed for the Macintosh Performa line of home computers— substitutes the Macintosh Finder with a system of folders and single-click buttons, trying to hide everything from users except their documents and productivity applications.

A 3-Ring Binder with Tabs

Insights gained from these comparisons were useful in determining the extent of functionality for PC Catalog and how realistic the representation of that functionality should be. We wanted a distinctive product look that expressed all the available functionality while leaving sufficient system resources for users to run their applications. Based on our observations and because it was in keeping with the main goal of PC Catalog —allow users to organize and access documents and applications— we arrived at a book metaphor consisting of a *three-ring binder with labeled tabs*. This container did appear well suited to the functionality goals stated in our brief; at a glance, it would be likely to suggest a containment strategy and what functionality to expect.

VISUALIZING THE METAPHOR

Inspired by the experiences described above we set out to visualize how the binder metaphor might be expressed on-screen. During the observations we noted common binder elements easily recognized by users; these suggested possible implementation styles and functionality for the computer interface. Figures 2 and 3 depict some early iterations of the design during these phase.

Using a Physical Mock-Up

We began by using a real binder to explore how it worked and how we could represent its elements and functions on-screen. We created different tab and page arrangements using a variety of tab and page styles. Using this mock-up we were able to try out many possible layouts and functions more realistically than with paper sketches and much more expediently than computer simulations would have allowed. We played out functions such as opening the book cover, selecting a tab to open the book to that section as well as adding and removing tabs and pages and moving items from one place to another, while asking ourselves how they could be best translated to the flat medium of the computer screen —within the constraints stated in our design brief.

Using the mock-up helped us to determine which representations were the most economical and advantageous. We quickly noticed that the representation of a realistic binder would require extensive use of perspective and foreshortening, which would be difficult given the small color palette (the standard Windows 16 colors), low resolution and the fact that we wanted to devote as much screen space as possible to functional aspects. Instead, we decided to explore simpler representations of the binder's elements, relying on small details (e.g., the depiction of rings, hints of the cover texture framing the inside of the book, dog-eared page corners, etc.) and subtle modeling to suggest a more three-dimensional look.

We knew that performance issues curtailed our use of animation in the interface but we tried to identify elements that

would be useful and implementable, such as page or tab turning effects. Experience with the mock-up (and later computer simulations) suggested that some of the easy and transparent interactions of the real-world object would translate into repetitive, and potentially annoying, animations on-screen and would require abstract and un-intuitive control mechanisms to manipulate a fully three-dimensional book layout. Using animation also meant punishing users of less capable machines —those with limited storage capacity and slower processor speeds— by either substantially increasing the size of the installed application to include pre-rendered sequences or accepting the inadequate performance of real-time animation.

The mock-up also revealed alternatives for binder/tab behavior and layout. When laid flat and opened, the left hand side of the binder seldom contained useful information, aside from the (possibly hidden) text of previous tabs. Furthermore, on the right hand side of the open binder, tabs were only visible if they were in the topmost row; looking at additional tabs required rotating the binder or peering along its side. In an on-screen interface, the tabs would have to be spread in order to view them all simultaneously.

Constructing Software Prototypes

These experiences with the mock-up helped us to understand elements and behavior for the product's user interface. We then built design prototypes using Macromedia Director to demonstrate how these elements —covers, pages, tabs, bookmarks, paper clips, pockets, sticky notes— would be expressed in a computer interface. Some prototypes were not interactive, they explored alternative looks for these elements (e.g., Two rows of tabs or ten? Horizontal or vertical tabs? Color as a highlighting strategy? More or less realistic representation of the binder rings?). In other prototypes we used animation to simulate specific interaction sequences, such as clicking on tabs, moving from page to page, adding elements by dragging icons from other windows into the book or launching applications from a button strip feature.

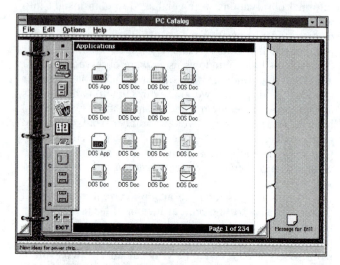

Figure 2. In an early design left and right page turning controls were separate. The tab title and page number appeared at the top and bottom, respectively. The "ButtonStrip" on the left included pull-out panels to access disk icons and system controls, in addition to application launching buttons.

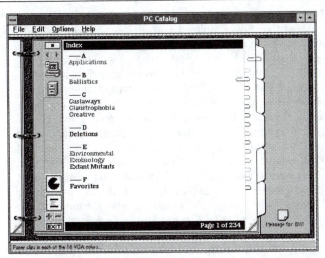

Figure 3. In this early design, application launching and system information buttons were both attached to the "ButtonStrip" on the left. Color paper clips were explored as a mechanism to mark specific pages within tabs. In this image the paper clips are shown in different locations to facilitate discussion about how they might be used.

DEFINING THE PRODUCT

We used the design prototypes in brainstorm sessions and presentations with the XSoft development team to identify key features thought essential to the metaphor and to eliminate un-implementable functionality. We modified but retained some features. For example, functionality associated with the binder rings (clicking on them to add or delete tabs and pages) was dropped due to implementation issues. The rings themselves, however, remained to be tested as identifying elements of the metaphor.

Some functionality did not seem to work well within the book metaphor. For example, including extensive system information and controls (as in Dashboard) cluttered the book and reduced the amount of space available for documents and applications.

We also considered how the binder-metaphor *shell* functionality could be integrated in upcoming versions of Windows. What longevity and perceived value would a shell have? How should the product be positioned in marketing and development terms? Instead of a shell, the product could be positioned as a *container/organizer*.

The XSoft team built working prototypes of the book containing different features that would be tested with users. We built additional design prototypes to explore the evolving visual representations and behaviors of the features being tested. Working prototypes were updated accordingly.

CONDUCTING USER TRIALS

We performed user trials as early as possible to identify usability problems, assess early implementations and obtain feedback from potential users. More extensive user trials were carried out at a later date by XSoft and Compaq using more advanced versions of the product. Their aim was to quantitatively compare TabWorks to Program Manager in terms of performance, preference and user satisfaction. In contrast,

our aim was broadly to find out what characteristics of our design were working well, which features were easy and intuitive to use, what was difficult to use, what caused confusion and how people reacted to the concept as whole. Our intent was to obtain information which could be fed back directly into the design loop, whereas their goal was to evaluate the product.

We initially proposed a two phase evaluation program. First, to perform semi-structured user trials with representative users, and second to carry out field trials where specially selected individuals would take the product away and use it for a period of time. It was anticipated that the semi-structured user trial would provide information which related directly to the product's functionality, whereas the field trial would better explore how the product would fit into a person's existing work pattern, especially with regard to the advantages offered by a metaphor with multiple containers.

Unfortunately, due to time constraints, only the semi-structured user trials could be performed. As a result, their scope was broadened to include a 'free exploration' section where the user was encouraged to 'play' with the product. In addition, another structured part was introduced so that navigation between various containers could be better explored. We chose not to conduct the trials in a usability lab where the moderator and participant are physically separated from each other, and instead set up a dedicated space at IDEO's studio in San Francisco. Being 'face to face' with the participant tended to facilitate observation; especially with regard to being sensitive to gestures, understanding context of behavior and seeing exactly what they were looking at.

XSoft's marketing group wanted to aim the product at wide range of users, from prospective users to experts. We therefore recruited people representing this range. In conjunction with XSoft's marketing group we identified five types of users: prospectives, novices, intermediates, advanced and experts. Prospectives were classified as people who had never used a PC but were about to start. Novices were just learning to use a PC, whereas intermediates were people who were able to do most of what they wanted. Advanced users considered themselves competent using the computer and experts were those who were able to do everything they wanted.

Finding participants was achieved by designing a screener which was then used by a recruitment consultancy. Using a recruitment consultancy enabled IDEO to retain control over participants while reducing cost and time. In addition to the initial screener, a participant profile was developed which was administered at the beginning of the trial. This further investigated their general computing experience, explored how they managed their computer work, determined where they used the computer and for what tasks they used it.

Pilot Trials

We developed the test protocol by performing an initial 'walk through' of product functionality. This led to the immediate identification of potential usability problems. These were recorded and added to a list of other usability issues provided by the rest of the team. As issues for investigation arose, tasks were designed to explore them. Finally, questions were formulated which introduced and then encouraged participants to try the tasks. It was important to phrase the questions in such a way as to be completely 'neutral', i.e., that they did not imply a correct methodology or approach.

The test protocol was tested by performing a series of pilot trials. This helped us rectify any ambiguity that may have existed in the questions, allowed us to approximate how long the trial would last and provided an opportunity to fine tune the instructions which were given to participants. Performing the pilot trials enabled us to also include, and occasionally exclude, certain questions which either needed further exploration or tended to confuse the data being obtained. We also found that prospective and novice users required a tutorial on some of the fundamental aspects of computing, how to use a graphical user interface and how to operate a mouse. It was important to ensure that it was the product participants were reacting to, rather than to computing.

Test Protocol

First, participants completed a profile questionnaire and read the test instructions. Every participant was given an overview of the products' intended use, was reminded to talk aloud, and told that it was the product that was being tested, not them. Participants were shown the product and first impressions were elicited to its overall appearance and functional elements. They were then encouraged to explore the system as they saw fit. As they did this, the moderator asked questions about what they were thinking, what they expected to happen next and what their impressions were. Once subjects began to exhibit a level of comfort and familiarity with the system they were encouraged to begin the structured tasks. The structured tasks explored utilization of the table of contents; opening, using and saving files and applications; making, moving and deleting new sections and pages in the book; and navigating between TabWorks and Windows. Finally, everyday operation of the product was explored in a semi-struc-

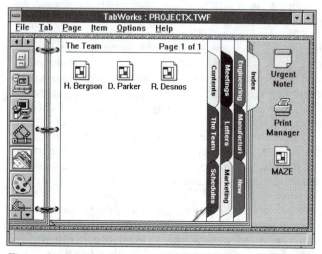

Figure 4a. An Early Design. The "Team" tab is selected. It is in the first row and contiguous with its pages, therefore no tab row rearrangement is necessary.

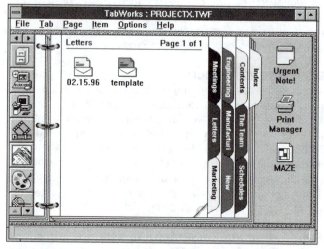

Figure 4b. Selecting the Letters *tab requires tab row rearrangement to move it from the second to the first row so that it will be contiguous with its pages.*

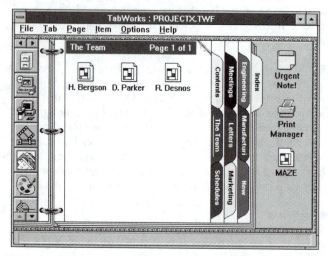

Figure 5a. The Final Design. The "Team" tab is selected. It is in the first row and contiguous with its pages, therefore no tab row rearrangement is necessary.

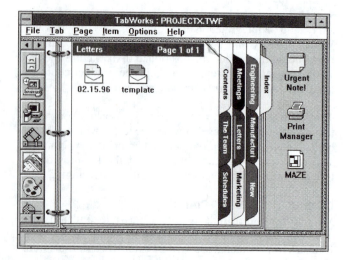

Figure 5b. In this case, when the Letters *tab is selected it retains its position within the tab cluster. A title bar of matching color reinforces the fact that its pages are now shown.*

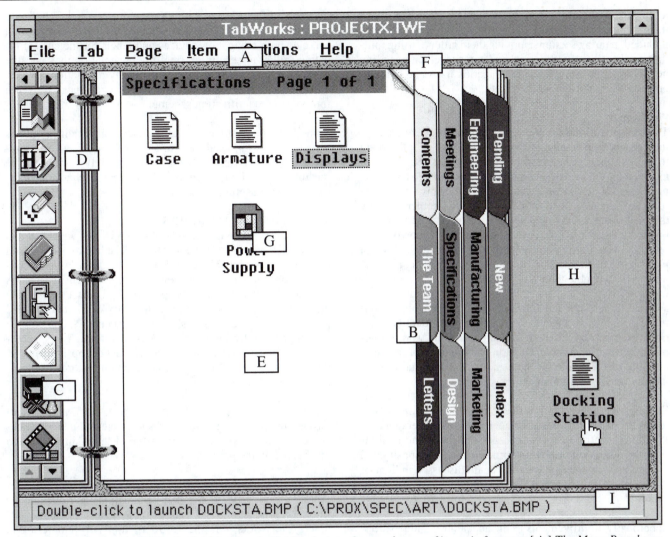

Double-click to launch DOCKSTA.BMP (C:\PROX\SPEC\ART\DOCKSTA.BMP)

Figure 6. The Open Book. The final design for the TabWorks interface and some of its main features: [A] The Menu Bar shows TabWorks' tab-page-item structure. At one point we wanted to use a "Book" instead of a "File" menu to reflect the container hierarchy (Book / Tab / Page / Item). Instead, we decided to follow standard Windows usage. [B] The Tab Cluster includes Contents and Index Tabs —provided by default— that allow the user to see and access the contents of the Book using outline and alphabetical views. The user can create, name, arrange and delete tabs. [C] The scrolling Button Strip provides single-click launch buttons for applications and documents and is available no matter which tab or page is displayed. The user adds items to it using drag-and-drop and can click-drag the buttons to rearrange them. [D] The Rings Area allows the user to move the pointer off the single-click buttons in the Button Strip without launching an item. [E] Pages show the tab titles by default and page titles as an option. The user can add, name and delete Pages. [F] The user moves from page to page using the Page Turning Corners or through menu and keyboard commands. [G] Items are added to Pages using drag-and-drop or from a dialog box. Long filenames are available to the user as well as custom icons. [H] The Holding Area can be used to store temporary item groupings. [I] The Status Bar displays information about the item currently under the hand cursor.

Figure 7. The Book Cover appears while the book is loading, then opens automatically to the display in Figure 6.

tured part of the trial where participants were asked to use the product as they would in their typical work. These tasks included: naming sections, moving them around, using multiple applications, grouping related items together, installing new books, and using special navigation tools and containers such as bookmarks and pockets.

Findings

The trials identified usability issues at three levels: conceptual, general and detailed. Conceptually, some people tended to become confused as to whether the product was a shell or a container. This was especially true for the naive and inexperienced users who would suddenly find themselves on the 'desktop', having 'lost' TabWorks behind another window that had opened. On a general level, experienced users expressed concern that they were working more slowly; they felt they were doing double work with both TabWorks and File Manager. In addition, users expected more contextual help to be available than was provided. A number of usability issues were found which related to general operation, such as page turner size and location, and whether tabs should move or not when selected. It was possible to identify functional elements which were difficult to use and whose functionality was not intuitive. Of particular interest was the functionality of the delete mechanism and the potential to mistakenly remove a number of items at once.

Findings were presented to the combined IDEO and XSoft teams for discussion in a succinct report. We generated new iterations of attributes which needed improvement. In addition, the definition of how the product should be positioned —shell vs. container— was once again discussed in light of the findings. It was agreed that the container strategy would be adopted.

Specifically, improvements to the appearance and positioning of page turners, table of contents, tabs and the book itself were made. Furthermore, changes to the grouping and positioning of pull down menus were made, as well as to interface terminology (e.g., "placements" became "icons" became "items").

THE LIMITS OF THE METAPHOR

User testing and rapid prototyping allowed us to discover where and how to enforce, break and sometimes contradict the metaphor in ways that enhanced its usefulness and usability. We knew, for example, that double-clicking a folder icon on the Macintosh opens a window bearing no resemblance, visually or functionally, to a folder. Users were not bothered by this. Similarly, we wanted to explore the boundaries of the binder metaphor representation.

To our surprise, users' instincts often contradicted our common sense as designers. In an early design, when a user clicked a tab, it moved to the front and the others were rearranged accordingly (see Figures 4a and 4b). Tests showed users were confused by this behavior and found navigation difficult. In subsequent designs (see Figures 5a and 5b) the selected tab didn't change its position. Testing validated this implementation; a stable configuration of tabs made navigation easier. To support this behavior we devised visual and interaction cues to identify the selected tab (using the color of the selected tab in a page title bar and underlining the title on the tab). We had thought the tabs themselves would serve as both navigation devices and labels for the current location in the binder. In the final implementation they became —as in our physical mock-up— primarily navigation devices. The inclusion of a color tab title bar on every page proved a more viable identification scheme.

Some elements that were initially functional remained only as visual cues in the final design. For example, user tests indicated the binder rings and the cover strongly reinforced the metaphor; they didn't need to be functional mechanisms in order to serve a useful purpose.

ARTICULATING A METAPHOR

As a result of these design-test-redesign cycles we chose and defined key elements of the metaphor. The book *cover* opened to display three *rings* binding a set of divider *tabs*, each containing one or more *pages*. Pages contained *items* —icons representing documents or applications. Users could create multiple books. Users could name their books, divider tabs and pages and add or delete tabs, pages and items which could be rearranged in different ways. An area next to the rings, accessible at all times, kept frequently used applications or documents handy. Figure 6 shows the main features of the final user interface.

The initial working name for the product, PC Catalog, went through many revisions. TabWorks was selected late in the development process and was indicative of the main interaction mechanism in the product.

TabWorks shipped in November of 1993, giving users of Windows 3.1 new ways to organize applications and documents. Based on product reviews to date, it met a need in the marketplace. TabWorks was the result of applying an iterative, user-centered methodology. Collaboration between the XSoft and IDEO teams ensured design decisions were informed by both technological and human interaction concerns. Involving users throughout the process allowed continuous improvement that resulted in an easy to use and useful product.

REFERENCES

[1] Gould, J.D. and Lewis, C. Designing for Usability: Key Principles and What Designers Think. Comm. of the ACM. (March 1985), 300-311

[2] Malone, T. W. How do People Organize Their Desks? Implications for the Design of Office Information Systems. ACM Transactions on Office Information Systems, 1,1, (January 1983), pp. 99-112.

[3] R. Mander, G. Salomon, and Y. Y. Wong. A "Pile" Metaphor for Supporting Casual Organization of Information. Proc. of CHI, 1992 (Monterey, California, May 3-7, 1992), ACM, New York, 1992, pp. 627-634.

[4] B. Moggridge. Design for the Information Revolution. Design DK 4:1992 (Copenhagen, Denmark, 1993), The Danish Design Centre.

Designing a "Front Panel" for Unix: The Evolution of a Metaphor

Jay Lundell
Corvallis User Interface Lab
Hewlett Packard MS 4UE3
1000 N.E. Circle Blvd.
Corvallis, OR 97330
Phone: (503) 715-4119
jay_l@cv.hp.com

Steve Anderson
User Interaction Design Group
Hewlett Packard
1266 Kifer Road
Sunnyvale, CA 94086
Phone: (408) 746-5415
steve@ptp.hp.com

ABSTRACT

The Front Panel component of the Common Desktop Environment is a culmination of several year's effort in designing a "dashboard-like" element for graphical Unix desktop systems. This design was a cooperative effort between graphic design artists, human factors professionals, and software designers, and eventually became a cross-company effort as it was adopted for the Common Desktop Environment. We describe the processes that emerged to support this design, and make observations about how metaphors may evolve over time.

KEYWORDS: Metaphor, Front Panel, Software Design, Visual Design, Workspaces, Dashboard.

INTRODUCTION

It is generally recognized that a metaphor provides a method by which people can quickly learn to use a system. Through metaphors, users infer the appropriate attributes of the target metaphor by mapping the aspects of the real-world source onto the software objects. Thus, users might infer that a trashcan-like object that appears on a computer screen is a container for objects that are to be discarded, and may infer that objects can be dropped into the trash without having to learn explicit instructions on how to throw away an object -- the knowledge for throwing away objects is contained in the user's knowledge about trash cans in the real world. Many critical aspects of metaphors have been identified [3], and processes for designing metaphors have been suggested by several authors [1][2][5]. In this design review, we describe the development of a front panel metaphor for a Unix desktop environment, and we identify some lessons we learned in its development.

DESCRIPTION OF THE FRONT PANEL

The Front Panel is an ubiquitous interface element designed to perform an eclectic range of services for an end user in a "desktop" environment. These services include:

- Quick access to commonly used applications.
- A method for creating multiple virtual workspaces and navigating between those workspaces.
- Drop regions for commonly used system functions, e.g., printing, mailing a file.
- Single location for viewing status information, e.g., new mail has arrived, or the time and date.

This interface is notable for several reasons:

- There have been many design challenges in providing such a wide range of services in a single User Interface component.
- The front panel has had a long history of use in commercially available products, and has seen several interesting developments across versions.
- The front panel has evolved from a rather simple metaphor to a more complex composite metaphor, and this development tells us something about the ways that a metaphor can evolve.
- The front panel is highly visual in appearance, and there are several interesting issues in the interaction between visual design and usability.
- The front panel has had an extensive amount of prototyping, user testing, human factors evaluation, and customer participation in its design, and we shall describe some of the techniques we have used in developing the user interface, as well as show some of the early prototypes, which took the form of animated end user scenarios created with Macromedia Director, a multimedia scripting application.

DEVELOPMENT OF THE FRONT PANEL

The front panel was initially conceived by Steve Anderson as part an attempt to show a collection of varied ideas for extending the "desktop" metaphor. This particular component, which he dubbed the "dashboard", attempted to solve the following user problems:

- quick access to system utilities and services
- single place to monitor system activities
- optimization of screen real estate

In addition, he wanted a component that had a strong visual signature and could be easily understood by users. He used the term "dashboard" because it was an ever-present element full of user controls, found at the bottom foreground of an ever-changing scene.

Its initial form was as a horizontal bar that appeared at the bottom of the screen. On the left side were several iconic buttons that had several of the "core" functions of electronic office workers: phone, e-mail, a rolodex, calculator, clock, etc. On the right side were labeled buttons, where the user could organize his/her own applications. These were labeled because their content was so likely to change over time --- it being much easier to type a new label than design or find a suitable icon. In the prototype, but not shown in the illustra-

tion, these buttons behaved like the fronts of small "drawers." When opened, each button slid upwards, revealing its contents, like small jewel-like objects nested in a shallow tray. At the time, (and even currently) an application that cost hundreds if not thousands of dollars seemed to merit a little extra panache in its presentation to its purchaser/user. This was seen as the place where applications were kept, not the files they would create. Those were to go in the File Space, seen in the background.

At the time, this prototype was not part of a particular project, but was instead an exercise of Steve's in attempting to learn how to use the animation application, then called VideoWorks. If it could serve as a spur to discussions with engineers, so much the better.

Figure 1. from June, 1988: a minimal ever-present "dashboard" that provided quick access to the user's most frequently used items There was also an attempt to look at how window overhead in inactive windows might be minimized, how the active window could be more efectively expressed, where iconized windows might go and what they should look like, and how different types of windows might take on more specialized and distinctive

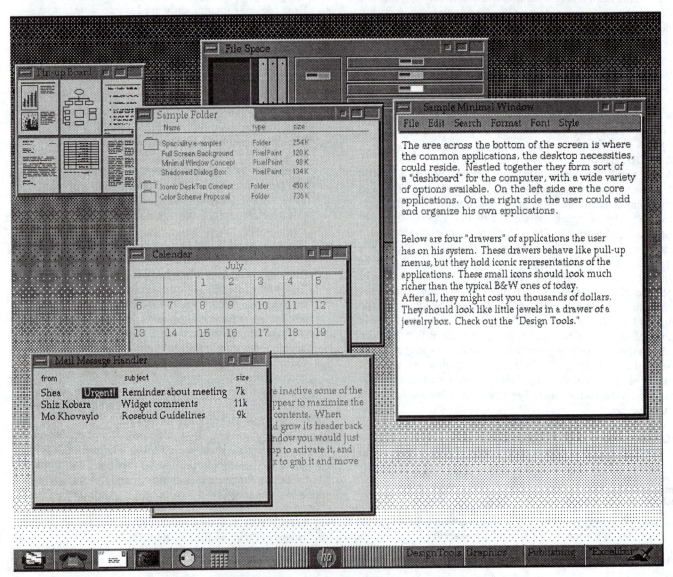

VUE 2.0

The Hewlett Packard Visual User Environment 2.0 project got underway in late 1989 as an attempt to create a "desktop" for Unix workstations. One of the technologies incorporated into this product was to be a window manager that supported multiple workspaces. With multiple workspaces, users could create and maintain a number of different virtual screens of information, and quickly move between these different "desktops". Once it was seen that there also needed to be a means to navigate between workspaces, the dashboard metaphor, which had since been exposed in various parts of the company, became an immediate candidate as a HP VUE component, and the basic premise of a horizontal "control panel" at the bottom of the screen was adopted by the development team.

Design Environment

A wide range of different disciplines were involved in the front panel development, including visual design, human factors, R&D, documentation, and marketing. Because the dashboard concept was such a new idea, direct customer input was difficult to obtain early in the project. The visual design group produced Macromedia Director animations that played out user scenarios. These high-resolution animations were extremely useful in the development and evaluation of the software.

Design Issues

As the team began to design the dashboard, they ran into some problems with the dashboard metaphor. The major problem was that real-world dashboards have many components that

critical. The survey results helped in defining which items should go into the front panel. We also discovered that some services required frequent access, while other items might not require high level access, but still needed to be easily accessible. An example of this is a "logout" capability. Although users may not need to log out more than once a day, it was seen as a critical capability. Thus, we developed a large top row for primary, frequent access and a smaller bottom row for secondary, critical access to tools.

As we explored the capability of the front panel with paper mock-ups and Macromind prototypes, a few fundamental types of "controls" emerged. Some front panel controls needed to be drop zones: the trash can and the printer, in particular. Some controls were simply indicators, i.e., the clock, date, and load controls. Controls had to support a push action: clicking on the control would start up a desktop application or run a command. In addition, we needed a type of control that would change its appearance based on system status, e.g., to show the user that new e-mail had arrived. These controls were called monitor controls.

We developed design rules for the location of controls based on human factors principles of motor control and functional grouping. These principles were:

- Drop zones should be located at "anchor" points --- the end controls or adjacent to the center workspace switches.
- Avoid placing drop zones next to each other to reduce

Figure 2: the VUE 2.0 Front Panel

have no real corollary in the computer world: steering wheels, speedometers, turn signals, etc. These would have little meaning or use in the VUE environment. As an exercise, the design team came up with a number of alternate metaphors: the desk organizer, a briefcase, a stereo panel, a remote control, etc. These all had interesting attributes, but all had too many undesirable properties. Finally, the team changed the term "dashboard" to "front panel" --- a metaphor that lacked some of the concrete and distinctive attributes of the dashboard, but had fewer areas in which the mapping failed. Thus, in this case, a less concrete metaphor seemed to be more useful.

As the front panel began to take shape, it became possible to ask more specific questions: What goes into the front panel, and what are the behaviors of the front panel buttons and controls? Some sort of component had to be provided to allow users to navigate between workspaces. A set of switches were developed for moving quickly between workspaces (see figure 2, the VUE 2.0 front panel). A clock and date seemed useful. Other items were researched in a customer survey which asked users to state which services might be most frequently used, and which services were considered most

errors in dropping on the wrong controls.
- Indicators should be located toward the ends of the front panel, where they would be less likely to be obscured by windows.
- Services that are less frequently accessed should be in the bottom row.
- Personal services such as the Home directory and the Toolbox should be on the right; system services such as the clock and load monitor should be on the left.

Some of these principles are contradictory, e.g., locating drop controls and indicator controls towards the ends. Because there are so many possible combinations of control locations, we have had to adopt the strategy of point testing a few possible designs with users, and have identified designs that appear to be sufficient, if not ideal. We have observed that each time a new version of the front panel is developed, new ways to organize the front panel controls emerge.

An additional feature that came directly out of user testing was the busy light. In early prototype testing, we noticed that users would often watch the disk drive light after clicking a front panel button to see if the application was actually starting up.

Although this worked well for users in our test, we realized that many Unix users do not have the disk drive located at their desk. Therefore, we incorporated a "busy light" graphic directly into the front panel that looked just like the HP disk drive light to provide user feedback.

Visual Features

We had constraints of having only a few colors available, yet we wanted a visual design that used shading effects and had the 3D visual style of Motif. Because icons and graphics in Motif could only be drawn in two colors, the entire front panel was dithered in order to achieve the light/medium/dark coloration that provided the 3D look of the front panel..

The VUE 2.0 front panel had a very "hardware" look to it with large, blocky buttons and the 3D effect. The front panel metaphor carried with it some characteristics that we had not necessarily intended --- it looked heavy, and difficult to move, and its hardware-like appearance gave the impression that it

was not very flexible. Although at a technical level it was simply a window, it did not look like a window, and did not lend itself to the actions available to windows --- minimization, resizing, moving around on the screen.

Reaction

In spite of some of the above limitations, the visual designers entered the product in the Industrial Designers Society of America 1990 annual design awards program, and it won one of 16 Gold awards, the first time a software product had ever been given any design award. As one of the jurors said, "it appears deceptively simple, offering a representation of functionality which is easy to grasp and quick to learn... it seems obvious now that we can see it."

Figure 3. Summer of 1990: a screen capture of the VUE 2.0 product, showing the 3D character of the abutted butoons in the Front Panel. Also featured is the Style Manager, with which the user exerts control of numerous customization features of the environment.

VUE 3.0

In designing the next generation of the product, VUE 3.0, we had an opportunity to solicit customer feedback from the field, and were able to see firsthand how our front panel metaphor was being used.

We were surprised to find the extent to which users had modified the front panel to suit their own tastes. One of the design concerns for VUE 3.0 was the extent to which our customers had ignored our carefully crafted design practices in designing their own front panels. Some of these designs were really bad, and we had concerns not only about the poor visual design of these panels, but about the human factors implications of these panels that had been designed by our users. We decided that the VUE 3.0 front panel had to be more flexible to allow for a greater range of customization, and that the front panel had to manage the customized controls and visuals more gracefully. Thus, the VUE 3.0 front panel had a flexible geometry that would better accommodate icons of different sizes. We also provided multicolor icons for greater visual appeal and so that people who customized their front panels would not have to dither their icons to get the proper colors and shading, which had proven challenging enough for professional graphic designers.

The design team also developed a prioritized list of customer needs for VUE 3.0 as a result of a systematic effort to gather customer information known as QFD (see [6] for more details on this process). One requirement was the desire to put more items into the front panel --- some customers wanted as many as 120 buttons. Another need was for greater visual "pizzazz" --- apparently the award-winning visual appeal of VUE 2.0 had sparked even greater interest in visual design. An answer to both of these requests took the form of sub panels --- animated extensions of the front panel controls that would "slide up" out of the top of the panel. The "slide up" animation conveyed the impression that sub panels were extensions of the front panel, not separate "menus".

User testing and customer feedback provided us not only with a new set of features to add, but allowed us to look critically at whether the features we had provided in VUE 2.0 were actually providing the value we had intended. For example, we had developed a visual semantic for front panel drop zones that we discovered were not being noticed by our users. We replaced this persistent cue with an interactive cue that occurs only as the user drags an icon over a drop zone. This cue was easier for users to notice, and was built in for any drop zone control instead of requiring a special semantic that had to be drawn into the control's visual representation.

Figure 4. Summer of 1992: showing the new features of colored icons and slide up sub panels. The lack of distinct buttons gave us greater design flexibility in terms of arrangements and spacings.

COMMON DESKTOP ENVIRONMENT

The advent of the COSE (Common Operating System Environment) effort in late 1992 saw the front panel concept adopted as part its the Common Desktop Environment, and additional changes to the metaphor were made as a result of VUE 3.0 customer feedback, user testing, and extensive human factors evaluation from the four different companies involved in the joint effort.

Ease of configurability was a critical issue, and we developed "drop zones" by which the end user could install new controls. Users could also "promote" controls on the sub panels into the front panel, and could delete controls as well.

Screen real estate was a critical issue, and since sub panels had become pervasively used, we removed the bottom row, as its

"secondary access" function was largely met by the sub panel design. Additionally, we changed the default number of workspaces from six to four, since research had shown that four was acceptable to most users, and the CDE front panel allowed easy creation of additional workspaces.

A new feature was added to the "monitor" capabilities --- the indicator light, which takes the form of a lighted up or down arrow at the top of the Front Panel. This allowed users to see an indication of a change of status for controls that are "hidden" on sub panels. Thus, a user might have a sub panel containing several in-boxes --- Fax, e-mail, voice-mail, etc. The status indicator allows the user to see from the main front panel if one of the in boxes has new messages, and to see via an LED-like graphic which of the in-boxes has changed when the sub panel is posted.

Figure 5. Current: the Common Desktop Environment version of the Front Panel, with the bottom row of buttons removed. Other changes include smaller icon sizes, drop zones at the top of slide up sub panels for drag and drop configuration, and 4 workspace

LOOKING AHEAD

In any software product, the actual implementation falls short of some of the design goals, for a myriad of reasons. But with the visual prototyping techniques we have employed, we can create very realistic visions of what we think the solution calls for. One example is the area of Front Panel and sub panel configurability. The addition of the drop zone in the sub panel gives the user a simple and nearly intuitive way of installing items into the sub panel. Steps we couldn't take would allow greater ease of further configuration--- rearranging their order, intuitive deletion, allowing and positioning separators, setting default behavior, etc. While we made significant progress in enabling users to do the most basic configuration, the prototypes show that we still have potential for going further.

SUMMARY

We have learned a great deal about the interaction between visual design, usability, and metaphor development throughout the development of the front panel. Some of the lessons we have learned are:

Let the Tasks Drive the Metaphor

Our initial visual approach was very hardware-like, with large modular buttons. We soon realized that this was an inhibitor to easy configurability. Users needed easy methods of altering the front panel to suit their needs, including changing the orientation from horizontal to vertical, and adding and removing controls. Thus, we modified the metaphor, making it more abstract, and softened the visual design to give it a more flexible appearance. We also added drop zones to allow users to install new buttons via drag and drop. In this case, the changes to the metaphor were accepted by users because the new metaphor accommodated the tasks users wanted to perform.

Don't Be a Slave to Self-Consistency

Grudin [1] has pointed out that consistency just for the sake of consistency can be misguided. We observed that there was a tendency to rigidly define the behavior of the front panel metaphor at the expense of efficacy. For example, pressing a front panel button initially would always bring up a new instance of a window. However, we realized that in some cases, users did not want a new instance of a window, but instead wanted a button to bring a current window to the top of the window stack if it had already been opened previously.

Thus, we allowed the apparent inconsistency that some buttons always created new windows, while others did not. Users do not seem bothered by this, and often do not even notice the difference, as long as the "inconsistent" aspect performs the behavior suitable to the task

Subtle Visual Cues Can Make a Big Difference

When we "softened up" the visual design of the front panel and relaxed the 3D button-like appearance of the front panel icons, we were concerned that the icons would be seen as having the same dragability characteristics as the icons in the File Manager, i.e., users would think that the front panel icons were could be moved around on the desktop. In an attempt to create a visual distinction to reflect the behavioral one, we gave the front panel icons an "etched in" appearance which successfully conveys the impression that the icons are imprinted into the front panel and cannot be moved around. The subtle visual cue appeared to work in this case, and has been adopted in the Common Desktop Environment as a visual cue for all non-draggable icons.

Subtle Visual Cues Can Make Little Difference

As we previously mentioned, our initial version had a visual semantic for indicating buttons that were drop regions vs. no-drop regions. In user testing, we noticed that these cues were not being picked up by users. Since the cues added visual complexity to the design with no apparent benefit, they were dropped from the design, with no measurable decrease in user performance or learning. User testing is the only reliable way to determine what works and what does not work as subtle visual cues, but our feeling is that subtle cues are effective when they reinforce the basic metaphor, but are ineffective when they are used to suggest additional functions that are outside the metaphor or are inconsistent with the metaphor.

Don't Design Metaphors in a Vacuum

The front panel exists in an environment that contains many other metaphors. As each design issue came up we had to anticipate problems in the interaction between the front panel design and aspects of other metaphors in the environment. For example, the initial design called for users to open applications with a single click on the front panel button. This caused problems when the Front Panel is used in the context of the File Manager, as users open files with a double-click. Thus, users were always getting two instances of their applications because they would double click on front panel buttons. We changed the design such that either a single or double click would produce the same results.

Let the Visual Designers use their tools to do their thing. First, and not after the fact. Having the interactive prototypes was crucial to exposing and selling the concepts embodied here. It is extremely rare for software developers to be able to produce conceptual things like this in the short time spans and with the visual appeal that visual design professional can bring to such an effort. Admittedly, having the right people available at the right time, with their objectives and responsibilities properly aligned, is a considerable challenge. But all of the ideas adopted here existed in visual prototpye form at least one or two years before they were ever incorporated into the actual products. And many more have not yet made it into products. None of them would have ever made it if the visual designer role was to execute nice icons at the end of the development cycle.

REFERENCES

1. Carroll, J.M., Mack, R.L., & Kellogg, W.A. Interface metaphors and user interface design. In M. Helander (Ed.), Handbook of Human-Computer Interaction, Elsevier Science Publishers B.V., 1988.

2. Erickson, T.D. Working with interface metaphors. In B. Laurel (Ed.), The Art of Human-Computer Interface Design, Addison-Wesley Publishing Co., Inc. 1990.

3. Gentner, D. Structure mapping: A theoretical framework for analogy. Cognitive Science, 7, 155-170.

4. Grudin, J. The case against user interface consistency. Communications of the ACM, vol. 32, number 10, 1164-1173, 1989.

5. Mountford, S.J. Tools and techniques for creative design. In B. Laurel (Ed.), The Art of Human-Computer Interface Design, Addison-Wesley Publishing Co., Inc. 1990.

6. Rideout, T., & Lundell, J. Hewlett Packard's Usability Engineering Program. In M. E. Wiklund (Ed.), Usability in Practice. AP Professional, 1994.

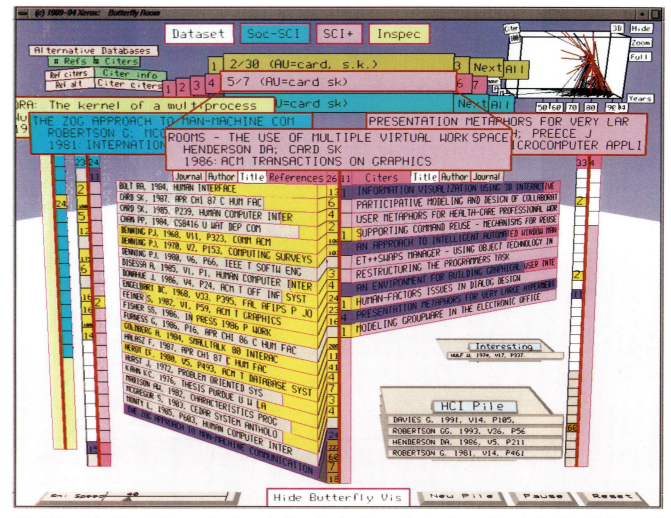

Mackinlay, Color Plate 1: This snapshot shows the Butterfly visualizer application for searching citation links. Link-generating queries support automatic creation of asynchronous query processes that grow a visualization of the search space. Metaphorically, the Butterfly visualization is like an information landscape that can be watched and pruned by the user to grow the search in fruitful directions.

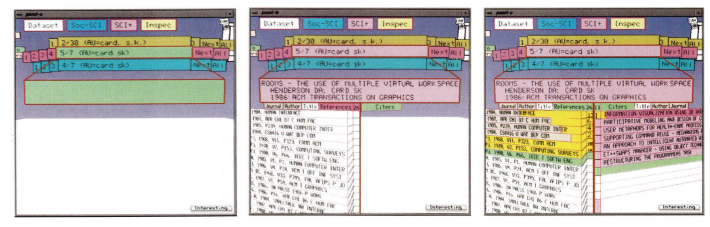

Mackinlay, Color Plate 2: These three panels shows how asynchronous query processes automatically grow a butterfly. Query processes are shown in pale green. The left panel shows a query process retrieving the database record associated with the butterfly's article. The middle panel shows a query process retrieving the citers for the article. The right panel shows two query processes retrieving information about the references and citers.

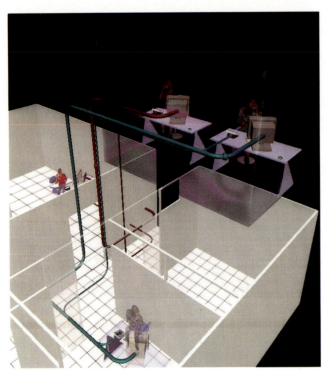

Plate 1.
Rebecca's view of the environment.

Plate 2.
John's view of the Virtual Meeting Room.

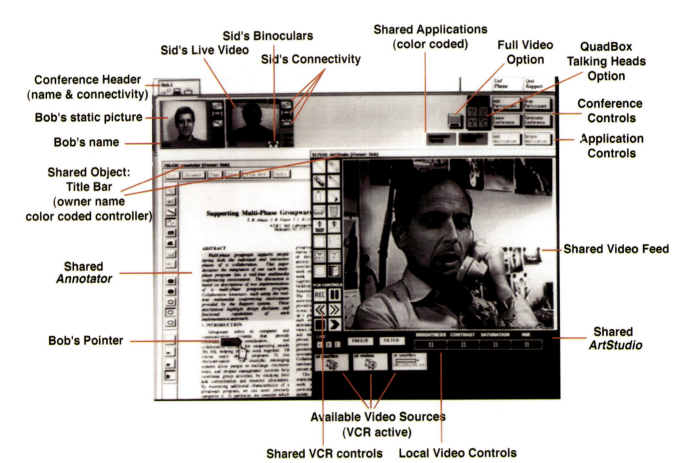

Plate 3.
Bob's view of the Virtual Meeting Room.

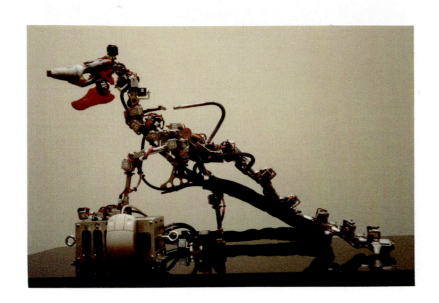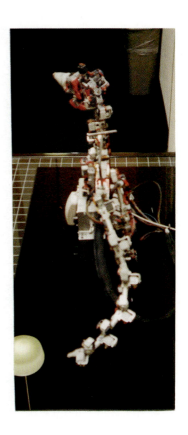

Color Plate 1: Close-ups of one of the *Velociraptor* armatures. The mouse in the pictures gives a sense of scale.

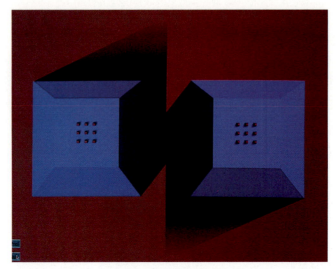

Ware, Color Plate 1

Ware, Color Plate 2

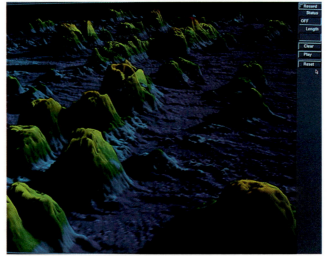

Ware, Color Plate 3

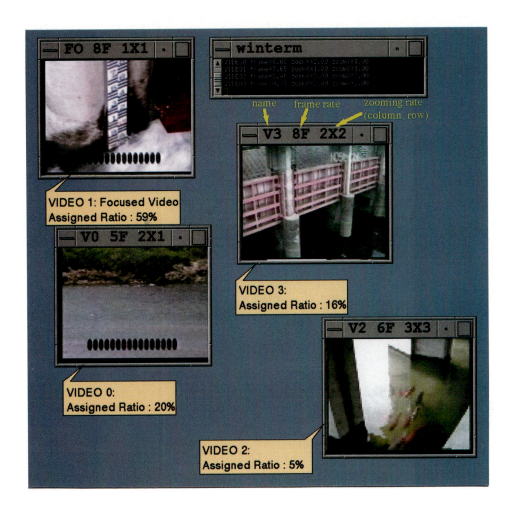

Yamaashi_PLATE 1 : A simulated view of multiple video images with the User-Centered Video through ISDN-1500

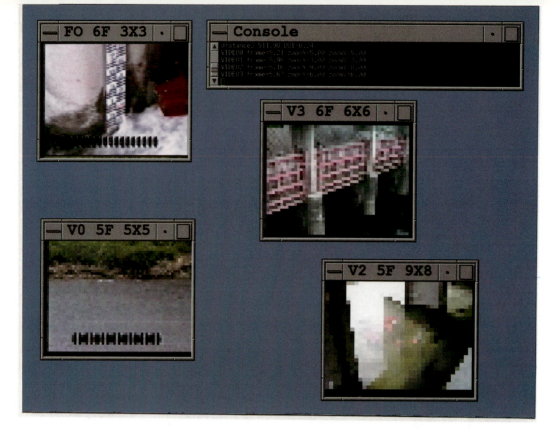

Yamaashi_PLATE 2 : A simulated view of multiple video images with the User-Centered Video through ISDN-64

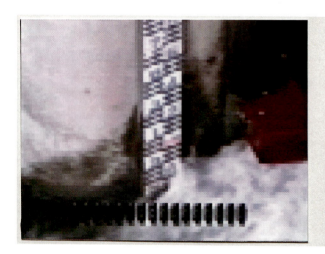

Yamaashi_PLATE 3 : A magnified view with no region of interest

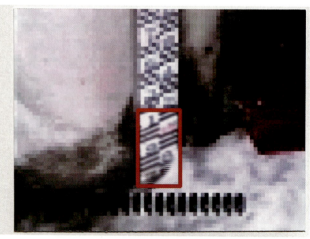

Yamaashi_PLATE 4 : A magnified view with the region of interest shown with the red rectangle

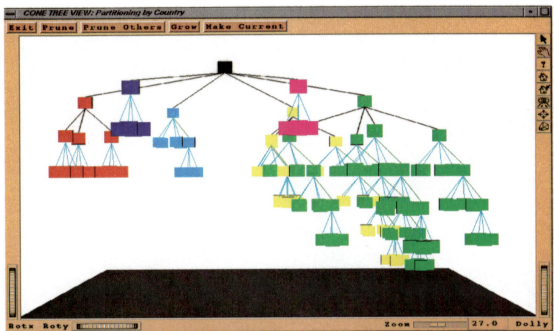

Color Plate 1: A 3D tree view of a hierarchy of the automobile database. Initial partitioning by the attribute *Country*. Node colors represent different countries and the link colors different link types.

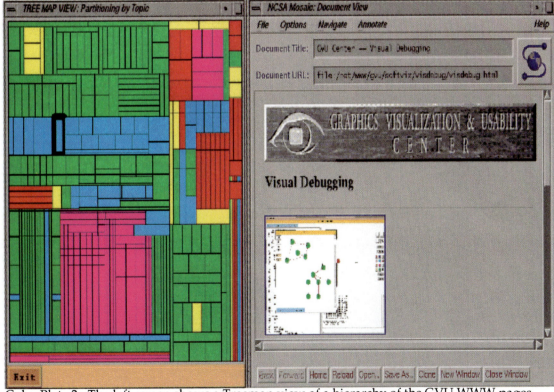

Color Plate 2: The left screen shows a Treemap view of a hierarchy of the GVU WWW pages. Initial partitioning is by the attribute *Topic*. Colors represent different types of authors. The selected node is *visdebug.html*. The corresponding WWW page is shown on the right.

Plate 1. The map display after the user requested to see Cambridge.

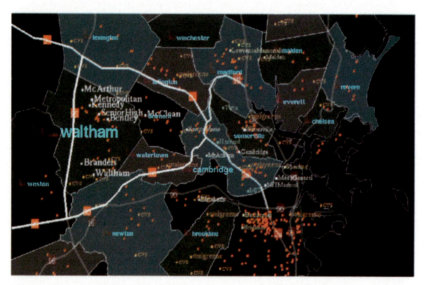

Plate 2. The map display after the user requested to see Waltham.

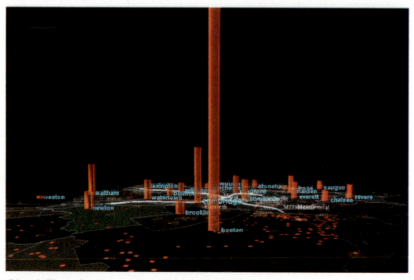

Plate 3. Three dimensional view of the map display after the user requested to see crime data.

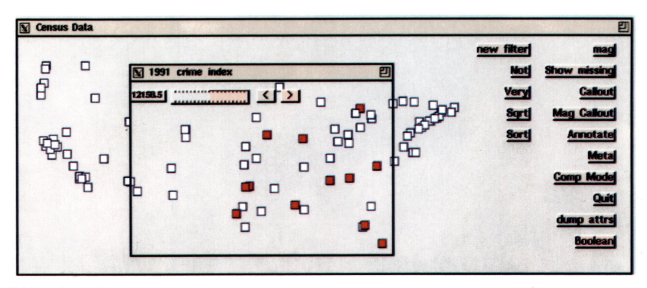

Fishkin & Stone, Color plate 1: Displaying a database of US census data. Cities are displayed as boxes at a point proportional to their longitude and latitude. A filter screening for cities with a high crime rate is positioned over the central United States.

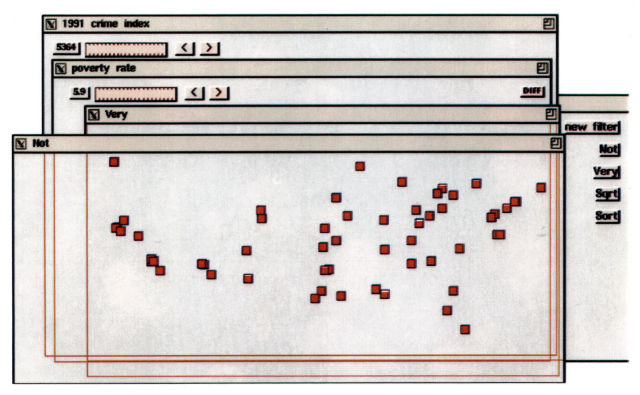

Fishkin & Stone, Color plate 2: Real-valued filters support more powerful queries. The redder the city, the greater the extent to which poverty and crime rates are NOT VERY DIFFerent.

Figure 1. Prototype Help System (Information & Access Methods)

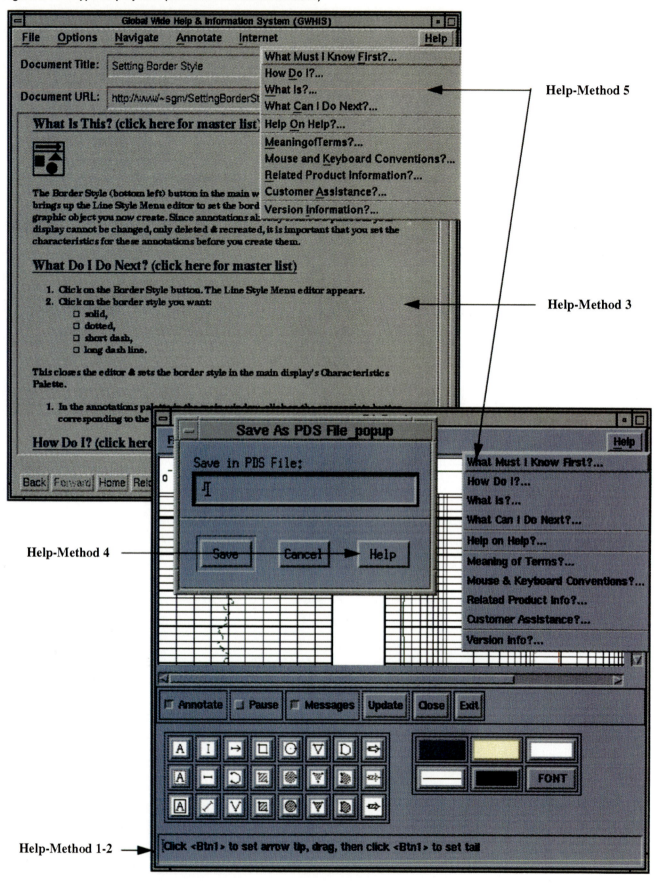

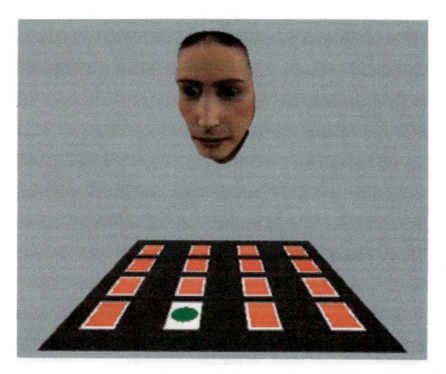

Color Plate 1. A scene image of the card game experiment with a synthesized face.

Color Plate 2. A scene image of the card game experiment with a 3D arrow.

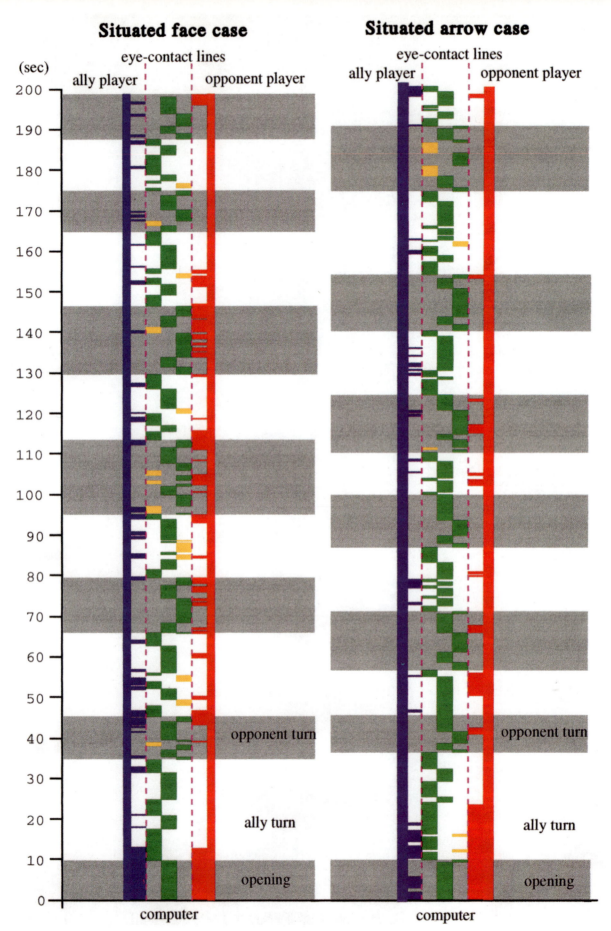

Color Plate 3. Comparison of time charts of the two games. Left: a situated face. Right: a situated arrow.

The First Society in Computing

Join ACM/SIGCHI Today!

The ACM **Special Interest Group on Computer-Human Interaction (SIGCHI)** focuses on how people communicate and interact with computers. SIGCHI serves as a forum for the exchange of ideas among computer scientists, human factors engineers, psychologists, social scientists, UI designers, and end users. SIGCHI offers its members the Member Plus package and co-sponsors a number of conferences and workshops each year, including the annual CHI conference. Membership includes the *SIGCHI Bulletin*, published quarterly.

ACM Associate and Student Member Dues include a subscription to the monthly *Communications of the ACM* You may join as an Associate Member and convert to Voting Member status by requesting a "Self-Certification" form from ACM's Member Services Department.

1

✔ **Check the option which applies to you...**

❑ I'm an ACM member. I want to become a SIGCHI member and receive the *SIGCHI Bulletin* (Amount due: **$30**)

❑ I'm an ACM Student member. I want to become a SIGCHI member and receive the *SIGCHI Bulletin* (Amount due:**$10**)

❑ I want to become a SIGCHI member only and receive the *SIGCHI Bulletin* (Amount due: **$57**)

❑ I want to become a member of both ACM (as an Associate Member) and SIGCHI, and receive the *SIGCHI Bulletin* (ACM dues: $82; SIGCHI dues: $30—Total amount due: **$112**)

❑ I want to become a member of both ACM (as a Student Member) and SIGCHI, and receive the *SIGCHI Bulletin* (ACM Student dues: $25; SIGCHI dues: $10—Total amount due: **$35**)

❑ I would like a subscription to the *SIGCHI Bulletin* only (Amount due: **$57**)

❑ I'm a SIGCHI member and would like to add the SIGCHI **Member Plus Package**. This package consists of the following proceedings: *UIST: User Interface Software and Technology (Nov.); Bonus Conference Proceedings (Nov.); CHI: ACM Conference on Human Factors in Computing Systems (May)* (Only available to SIGCHI members—Amount due:**$30**)

2

Name _____ ACM Member# _____

Mailing Address _____

City/State/Province _____

Country/Zip/Postal Code _____

Email _____

Phone _____ Fax _____

3

Payment Information

❑ Check *made payable to ACM, Inc (In addition to U.S. Dollars, ACM accepts bank check and Eurocheque payments in several European currencies. For currencies accepted and conversion rates, see contact information below.)*

❑ Credit Card: ❑ AMEX   ❑ VISA   ❑ MasterCard

Credit Card# _____ Exp. Date _____

Signature _____

Please send me:
❑ Information about ACM and SIG membership
❑ An ACM Publications and Services Catalog (204940)
❑ Local Activities Guide

Remit to: *ACM*, PO Box 12115, Church St. Station New York, NY 10257 USA

Prices include surface delivery charge. Expedited Air Service, which is a partial air freight delivery service, is available outside North America. Contact ACM for further information.

Contact Information

To order any ACM product, have your credit card handy and call the ACM Order Department at: **1-800-342-6626** *from the USA & Canada* **+1-212-626-0500** *from the NY metro area & outside the USA Fax:* **+1-212-944-1318** *Email:* **acmhelp@acm.org**
ACM Online Products and Services Publications Catalog:
The WWW URL is: *http://info.acm.org/catalog/*
The Homepage URL is: *http://info.acm.org/*
The SIGCHI Homepage URL is: *http://info.sigchi.acm.org/sigchi/*

For inquiries regarding membership, contact ACM:
ACM Member Serv. Dept.
1515 Broadway
New York, NY 10036 USA
Phone: +1-212-626-0500
Fax: +1-212-944-1318
Email: acmhelp@acm.org

ACM European Service Center
Avenue Marcel Thiry 204
1200 Brussels, Belgium
Phone: +32 2 774 9602
Fax: +32 2 774 9690
Email: acm_europe@acm.org

AAPCHI95

JOURNALS

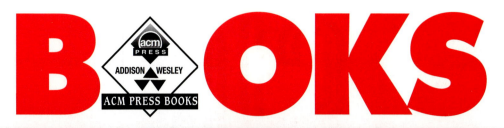

BOOKS